Introduction to System-on-Package (SOP)

Introduction to System-on-Package (SOP)
Miniaturization of the Entire System

Rao R. Tummala

Madhavan Swaminathan

New York Chicago San Francisco
Lisbon London Madrid Mexico City
Milan New Delhi San Juan
Seoul Singapore Sydney Toronto

The McGraw·Hill Companies

Cataloging-in-Publication Data is on file with the Library of Congress.

McGraw-Hill books are available at special quantity discounts to use as premiums and sales promotions, or for use in corporate training programs. To contact a special sales representative, please visit the Contact Us page at www.mhprofessional.com.

Introduction to System-on-Package (SOP)

Copyright ©2008 by The McGraw-Hill Companies, Inc. All rights reserved. Printed in the United States of America. Except as permitted under the United States Copyright Act of 1976, no part of this publication may be reproduced or distributed in any form or by any means, or stored in a data base or retrieval system, without the prior written permission of the publisher.

2 3 4 5 6 7 8 9 0   DOC/DOC   0 1 4 3 2 1 0 9 8

ISBN 978-0-07-145906-8
MHID 0-07-145906-5

Sponsoring Editor
Stephen S. Chapman

Acquisitions Coordinator
Alexis Richard

Editorial Supervisor
David E. Fogarty

Project Manager
Aparna Shukla

Copy Editor
Lucy Mullins

Proofreader
Deepa Pathak

Indexer
Kevin Broccoli

Production Supervisor
Richard C. Ruzycka

Composition
International Typesetting and Composition

Art Director, Cover
Jeff Weeks

Information contained in this work has been obtained by The McGraw-Hill Companies, Inc. ("McGraw-Hill") from sources believed to be reliable. However, neither McGraw-Hill nor its authors guarantee the accuracy or completeness of any information published herein, and neither McGraw-Hill nor its authors shall be responsible for any errors, omissions, or damages arising out of use of this information. This work is published with the understanding that McGraw-Hill and its authors are supplying information but are not attempting to render engineering or other professional services. If such services are required, the assistance of an appropriate professional should be sought.

About the Authors

Rao R. Tummala is the Distinguished and Endowed Pettit Chair Professor and founding Director of the Micro and Nanosystems Packaging Research Center at Georgia Tech. He was a former IBM Fellow, a former President of IEEE CPMT and IMAPS Societies, an IEEE Fellow, and a member of the National Academy of Engineering in the United States and India. Dr. Tummala received many industry, academic, and professional society awards, including *IndustryWeek's* award as one of the 50 Stars in the United States. He authored 5 books, 425 technical papers and 72 patents and inventions.

Madhavan Swaminathan is Joseph M. Pettit Professor in Electronics in the School of Electrical and Computer Engineering, and Deputy Director of the Microsystems Packaging Research Center, Georgia Tech. He is the co-founder of Jacket Micro Devices, a leader in integrated RF modules and substrates for wireless applications and SoPWorX, a company specializing in EDA software for SOP applications. Prior to joining Georgia Tech, he was with IBM working on the packaging of supercomputers. He has written more than three hundred publications, holds fifteen patents and has been honored as an IEEE Fellow.

Contents at a Glance

1	Introduction to the System-on-Package (SOP) Technology	3
2	Introduction to System-on-Chip (SOC)	39
3	Stacked ICs and Packages (SIP)	81
4	Mixed-Signal (SOP) Design	151
5	Radio Frequency System-on-Package (RF SOP)	261
6	Integrated Chip-to-Chip Optoelectronic SOP	321
7	SOP Substrate with Multilayer Wiring and Thin-Film Embedded Components	377
8	Mixed-Signal Reliability	443
9	MEMS Packaging	495
10	Wafer-Level SOP	535
11	Thermal SOP	605
12	Electrical Test of SOP Modules and Systems	659
13	Biosensor SOP	717
	Index	749

Contents

	Foreword	xvii
	Preface	xix
1	**Introduction to the System-on-Package (SOP) Technology**	**3**
	1.1 Introduction	4
	1.2 Electronic System Trend to Digital Convergence	5
	1.3 Building Blocks of an Electronic System	7
	1.4 System Technologies Evolution	8
	1.5 Five Major System Technologies	11
	1.5.1 System-on-Board (SOB) Technology with Discrete Components	11
	1.5.2 System-on-Chip (SOC) with Two or More System Functions on a Single Chip	11
	1.5.3 Multichip Module (MCM): Package-Enabled Integration of Two or More Chips Interconnected Horizontally	13
	1.5.4 Stacked ICs and Packages (SIP): Package-Enabled IC Integration with Two or More Chip Stacking (Moore's Law in the Third Dimension)	13
	1.6 System-on-Package Technology (Module with the Best of IC and System Integration)	18
	1.6.1 Miniaturization Trend	22
	1.7 Comparison of the Five System Technologies	23
	1.8 Status of SOP around the Globe	26
	1.8.1 Opto SOP	26
	1.8.2 RF SOP	28
	1.8.3 Embedded Passives SOP	29
	1.8.4 MEMS SOP	29
	1.9 SOP Technology Implementations	29
	1.10 SOP Technologies	33
	1.11 Summary	34
	Acknowledgment	34
	References	34
2	**Introduction to System-on-Chip (SOC)**	**39**
	2.1 Introduction	40
	2.2 Key Customer Requirements	42
	2.3 SOC Architecture	44
	2.4 SOC Design Challenge	50
	2.4.1 SOC Design Phase 1—SOC Definition and Challenges	50
	2.4.2 SOC Design Phase II—SOC Create Process and Challenges	57

	2.5	Summary	76
		References	76
3	**Stacked ICs and Packages (SIP)**	81
	3.1	SIP Definition	82
		3.1.1 Definition	82
		3.1.2 Applications	82
		3.1.3 CEO Figure and SIP Categories	82
	3.2	SIP Challenges	85
		3.2.1 Materials and Process Challenges	85
		3.2.2 Mechanical Challenges	87
		3.2.3 Electrical Challenges	88
		3.2.4 Thermal Challenges	89
	3.3	Non-TSV SIP	93
		3.3.1 Historical Evolution of Non-TSV SIP	93
		3.3.2 Chip Stacking	96
		3.3.3 Package Stacking	113
		3.3.4 Chip Stacking versus Package Stacking	120
	3.4	TSV SIP	121
		3.4.1 Introduction	121
		3.4.2 Historical Evolution of 3D TSV Technology	124
		3.4.3 Basic TSV Technologies	126
		3.4.4 Different 3D Integration Technologies using TSV	134
		3.4.5 Si Carrier Technology	141
	3.5	Future Trends	143
		Acknowledgments	144
		References	144
4	**Mixed-Signal (SOP) Design**	151
	4.1	Introduction	152
		4.1.1 Mixed-Signal Devices and Systems	153
		4.1.2 Importance of Integration in Mobile Applications	155
		4.1.3 Mixed-Signal Architecture	156
		4.1.4 Mixed-Signal Design Challenges	157
		4.1.5 Fabrication Technologies	159
	4.2	Design of Embedded Passives in RF Front End	160
		4.2.1 Embedded Inductors	161
		4.2.2 Embedded Capacitors	166
		4.2.3 Embedded Filters	167
		4.2.4 Embedded Baluns	171
		4.2.5 Filter-Balun Networks	175
		4.2.6 Tunable Filters	178
	4.3	Chip-Package Codesign	180
		4.3.1 Low Noise Amplifier Design	181
		4.3.2 Concurrent Oscillator Design	184
	4.4	Design of WLAN Front-End Module	191

Contents

4.5	Design Tools		194
	4.5.1	Synthesis of Embedded RF Circuits	195
	4.5.2	Modeling of Signal and Power Delivery Networks	198
	4.5.3	Rational Functions, Network Synthesis, and Transient Simulation	204
	4.5.4	Design for Manufacturing	208
4.6	Coupling		214
	4.6.1	Analog-to-Analog Coupling	214
	4.6.2	Digital-to-Analog Coupling	222
4.7	Decoupling		227
	4.7.1	Need for Decoupling in Digital Applications	228
	4.7.2	Issues with SMD Capacitors	229
	4.7.3	Embedded Decoupling	230
	4.7.4	Characterization of Embedded Capacitors	235
4.8	Electromagnetic Bandgap (EBG) Structures		239
	4.8.1	Analysis and Design of EBG Structures	242
	4.8.2	Application of EBGs in Power Supply Noise Suppression	246
	4.8.3	Radiation Analysis of EBGs	248
4.9	Summary		250
	Acknowledgments		251
	References		251

5 Radio Frequency System-on-Package (RF SOP) — **261**

5.1	Introduction		262
5.2	RF SOP Concept		262
5.3	Historical Evolution of RF Packaging Technologies		265
5.4	RF SOP Technologies		267
	5.4.1	Modeling and Optimization	267
	5.4.2	RF Substrate Materials Technologies	268
	5.4.3	Antennas	269
	5.4.4	Inductors	278
	5.4.5	RF Capacitors	282
	5.4.6	Resistors	288
	5.4.7	Filters	295
	5.4.8	Baluns	297
	5.4.9	Combiners	298
	5.4.10	RF MEMS Switches	300
	5.4.11	RFIDs	305
5.5	Integrated RF Modules		308
	5.5.1	WLAN	308
	5.5.2	Intelligent Network Communicator (INC)	310
5.6	Future Trends		312
	Acknowledgments		313
	References		314

6 Integrated Chip-to-Chip Optoelectronic SOP 321

- 6.1 Introduction 322
- 6.2 Applications of Optoelectronic SOP 323
  - 6.2.1 High-Speed Digital Systems and High-Performance Computing 323
  - 6.2.2 RF-Optical Communication Systems 324
- 6.3 Integration Challenges in Thin-Film Optoelectronic SOP 325
  - 6.3.1 Optical Alignment 326
  - 6.3.2 Key Physical and Optical Properties of Thin-Film Optical Waveguide Materials 326
- 6.4 Advantages of Optoelectronic SOP 331
  - 6.4.1 Comparison of High-Speed Electrical and Optical Wiring Performance 331
  - 6.4.2 Wiring Density 332
  - 6.4.3 Power Dissipation 334
  - 6.4.4 Reliability 335
- 6.5 Evolution of Optoelectronic SOP Technology 336
  - 6.5.1 Board-to-Board Optical Wiring 336
  - 6.5.2 Chip-to-Chip Optical Interconnects 339
- 6.6 Optoelectronic SOP Thin-Film Components 341
  - 6.6.1 Passive Thin-Film Lightwave Circuits 342
  - 6.6.2 Active Optoelectronic SOP Thin-Film Components 354
  - 6.6.3 Opportunities for 3D Lightwave Circuits 355
- 6.7 SOP Integration: Interface Optical Coupling 357
- 6.8 On-Chip Optical Circuits 363
- 6.9 Future Trends in Optoelectronic SOP 365
- 6.10 Summary 365
- References 366
- Table 6.1 References 374

7 SOP Substrate with Multilayer Wiring and Thin-Film Embedded Components 377

- 7.1 Introduction 378
- 7.2 Historical Evolution of Substrate Integration Technologies 380
- 7.3 SOP Substrate 381
  - 7.3.1 Drivers and Challenges 381
  - 7.3.2 Ultrathin-Film Wiring with Embedded Low-K Dielectrics, Cores, and Conductors 384
  - 7.3.3 Embedded Passives 415
  - 7.3.4 Embedded Actives 430
  - 7.3.5 Miniaturized Thermal Materials and Structures 434
- 7.4 Future SOP Substrate Integration 435
- Acknowledgments 437
- References 437

8 Mixed-Signal (SOP) Reliability — 443
- 8.1 System-Level Reliability Considerations — 445
  - 8.1.1 Failure Mechanisms — 446
  - 8.1.2 Design-for-Reliability — 447
  - 8.1.3 Reliability Verification — 449
- 8.2 Reliability of Multifunction SOP Substrate — 450
  - 8.2.1 Materials and Process Reliability — 450
  - 8.2.2 Digital Function Reliability and Verification — 458
  - 8.2.3 RF Function Reliability and Verification — 461
  - 8.2.4 Optical Function Reliability and Verification — 463
  - 8.2.5 Multifunction System Reliability — 467
- 8.3 Substrate-to-IC Interconnection Reliability — 468
  - 8.3.1 Factors Affecting the Substrate-to-IC Interconnection Reliability — 469
  - 8.3.2 100-μm Flip-Chip Assembly Reliability — 471
  - 8.3.3 Reliability against Die Cracking — 476
  - 8.3.4 Solder Joint Reliability — 476
  - 8.3.5 Interfacial Adhesion and Effect of Moisture on Underfill Reliability — 478
- 8.4 Future Trends and Directions — 482
  - 8.4.1 Extending Solder — 483
  - 8.4.2 Complaint Interconnects — 484
  - 8.4.3 Alternative to Solder and Nano Interconnects — 484
- 8.5 Summary — 486
- References — 487

9 MEMS Packaging — 495
- 9.1 Introduction — 496
- 9.2 Challenges in MEMS Packaging — 496
- 9.3 Chip-Scale versus Wafer-Scale Packaging — 497
- 9.4 Wafer Bonding Techniques — 499
  - 9.4.1 Direct Bonding — 500
  - 9.4.2 Bonding Using Intermediate Layers — 500
- 9.5 Sacrificial Film-Based Sealing Techniques — 505
  - 9.5.1 Etching the Sacrificial Material — 505
  - 9.5.2 Decomposition of Sacrificial Polymers — 509
- 9.6 Low-Loss Polymer Encapsulation Techniques — 514
- 9.7 Techniques Utilizing Getters — 516
  - 9.7.1 Nonevaporable Getters — 516
  - 9.7.2 Thin-Film Getters — 517
  - 9.7.3 Improving MEMS Reliability through Getters — 520
- 9.8 Interconnections — 522
- 9.9 Assembly — 524
- 9.10 Summary and Future Trends — 527
- References — 528

10 Wafer-Level SOP 535
10.1 Introduction 536
10.1.1 Definition 536
10.1.2 Wafer-Level Packaging—Historical Evolution 537
10.2 Buildup Wiring and Redistribution 540
10.2.1 IC-Package Pitch Gap 540
10.2.2 Redistribution Layers on Si to Close the Pitch Gap 543
10.3 Wafer-Level Thin-Film Embedded Components 544
10.3.1 Embedded Thin-Film Components in the ReDistribution Layer (RDL) 544
10.4 Wafer-Level Packaging and Interconnections (WLPI) 548
10.4.1 Classes of Wafer-Level Packaging and Interconnections (WLPI) 552
10.4.2 Rigid Interconnections 560
10.4.3 WLSOP Assembly 585
10.4.3 WLSOP 590
10.5 Wafer-Level Probing and Burn-In 591
10.6 Summary 595
Acknowledgments 595
References 595

11 Thermal SOP 605
11.1 Fundamentals of Thermal SOP 606
11.1.1 Thermal Implications of SOP 607
11.1.2 System-Level Thermal Constraints in SOP-Based Portables 609
11.2 Thermal Sources in SOP Modules 610
11.2.1 Digital SOP 611
11.2.2 RF SOP 613
11.2.3 Optoelectronic SOP 615
11.2.4 MEMS SOP 617
11.3 Fundamental Heat Transfer Modes 618
11.3.1 Conduction 618
11.3.2 Convection 623
11.3.3 Radiation 626
11.4 Fundamentals of Thermal Characterization 629
11.4.1 Numerical Methods for Thermal Characterization 629
11.4.2 Experimental Methods for Thermal Characterization 637
11.5 Thermal Management Technologies 637
11.5.1 Thermal Design Methodologies 638
11.6 Power Minimization Methodologies 648
11.6.1 Parallel Processing 649
11.6.2 Dynamic Voltage and Frequency Scaling (DVFS) 649

		11.6.3 Application-Specific Processors (ASP)	650
		11.6.4 Cache Power Minimization	650
		11.6.5 Power Harnessing	651
	11.7	Summary	651
		Acknowledgment	651
		References	652
12	**Electrical Test of SOP Modules and Systems**		**659**
	12.1	SOP Electrical Test Challenges	660
		12.1.1 Objectives of the HVM Test Process and Challenges for SOPs	662
		12.1.2 HVM Test Flow for SOPs	663
	12.2	Known Good Embedded Substrate Test	664
		12.2.1 Substrate Interconnect Tests	664
		12.2.2 Testing Embedded Passives	671
	12.3	Known Good Embedded Module Test of Digital Subsystems	677
		12.3.1 Boundary Scan—IEEE 1149.1	677
		12.3.2 Multi-gigahertz Digital Test: Recent Developments	681
	12.4	KGEM Test of Mixed-Signal and RF Subsystems	685
		12.4.1 Test Strategies	685
		12.4.2 Fault Models and Test Quality	688
		12.4.3 Direct Measurement of Specifications Using Dedicated Circuitry	689
		12.4.4 Alternate Testing Methods for Mixed-Signal and RF Circuits	690
	12.5	Summary	707
		Acknowledgments	707
		References	707
13	**Biosensor SOP**		**717**
	13.1	Introduction to Biosensor SOP	717
		13.1.1 SOP: A Highly Miniaturized Electronic System Technology	717
		13.1.2 Biosensor SOP for Miniaturized Biomedical Implants and Sensor Systems	718
		13.1.3 Building Blocks of Biosensor SOP	723
	13.2	Biosensing	723
		13.2.1 Microchannels for Biofluid Transport	723
		13.2.2 Biosensing Element (Probe) Design and Preparation	724
		13.2.3 Probe-Target Molecular Hybridization	727
	13.3	Signal Conversion	730
		13.3.1 Nanomaterials and Nanostructures for Signal Conversion Components	730
		13.3.2 Surface Modification and Biofunctionalization of Signal Conversion Component	734
		13.3.3 Signal Conversion Methods	735

13.4	Signal Detection and Electronic Processing	741
	13.4.1 Low-Power Application-Specific Integrated Circuits (ASICs) and Mixed-Signal Design for Biosensor SOP	741
	13.4.2 Bio- SOP Substrate Integration Technologies	744
13.5	Summary and Future Trends	745
	13.5.1 Nano Bio-SOP Integration Challenges	745
	References	746
Index		749

Foreword

In 1994, it was my pleasure as the incoming president of Georgia Tech to participate in the inauguration of the Institute's first National Science Foundation center of excellence, the Microsystems Packaging Research Center. Directed by Professor Rao Tummala, the Center was designed to take an innovative new approach to packaging chips into ultraminiaturized single component systems that would serve multiple purposes. Through the auspices of the Center, and for the first time at Georgia Tech, experts from different disciplines on our campus were linked with like minded experts from other universities and industries around our nation and the world to create a powerful team whose combined efforts magnified the impact of the individuals themselves. It was a bold undertaking, requiring not only the best technical ideas, but also a new approach to system testbed research involving a large diverse team of faculty and students and managing information. It was a high risk, but high payoff strategy.

Not all of the faculty experts Georgia Tech needed were in place at the outset and this led to recruiting new talented people who were glad to join us because of the exciting prospects that lay ahead. We also recognized the need to develop cutting edge facilities, including extensive special-purpose clean room infrastructure. These investments, along with the leadership of Professor Tummala, led to the elevation of Georgia Tech to a leadership and pioneering position among universities in the important field ultraminiaturized systems based on the System-on-Package platform concept.

The SOP technology described in this book was funded by NSF, industry, and the Georgia Research Alliance over a 12-year period from 1995–2007. During this period, the Center involved 55 faculty and senior researchers from electrical, mechanical, materials science, and chemical engineering departments at Georgia Tech and 160 global companies. A remarkable 600 Ph.D and MS students took part in the pioneering SOP R&D, and they are now acting as the industry leaders in driving the next generation developments using the SOP technologies that they developed.at Georgia Tech.

The SOP technology introduced in this book promises over the next two decades to miniaturize electronic and bio-electronic systems by a factor of a thousand to a million. It introduces the idea of a System Integration Law, known as the Second Law of Electronics for miniaturization of the entire system. It complements the well known Moore's Law for integrated circuits which applies principally to a small part of the system. Because of the miniaturization that SOP enables, along with the increased number of system functions enabled, it is expected to lead to new generation products in consumer electronics, healthcare, energy, and automotive industries.

Faculty on our campus and throughout the world will find this book a resource in educating future generations of students in the exiting world of SOP technology. I congratulate Professor Tummala and his team for bringing the Georgia Tech's first National Science Foundation center of excellence, and for taking it to the next level by producing this outstanding book.

Wayne Clough
President
Georgia Instutute of Technology

Preface

The System-on-Package (SOP) is an emerging system miniaturization technology in contrast to System-on-Chip (SOC) at IC level and System-in-Package (SIP) at module level. The SOC accomplishes miniaturization primarily by shrinking lithographic dimensions from microscale in 1980s to nanoscale currently. The miniaturization in SIP is accomplished by thinning ICs from their original 800 micron thick wafer dimensions to 50 microns and stacking as many as 10 of these, one on top of the other, in 3D form. These are then interconnected by either wire bond or flipchip technology. The SIP thus miniaturizes more than SOC. The recent Through Silicon Via (TSV) developments further miniaturizes SIP by replacing flipchip with pad to pad bonding. However, both SOC and SIP miniaturize a tiny part of the system since the number of ICs or its size in a typical system such as a cellphone is a small fraction—10–20% of the entire system. Therefore, both SOC and stacked SIP technologies, address a small part of the total system.

This book introduces the SOP concept for miniaturizing the entire system. It argues for the benefits of system miniaturization which include lower cost, higher electrical performance and better thermo-mechanical reliability, than the current approach of discrete component packaging. The SOP concept described in this book has two fundamental drivers for miniaturization—1) reduction from three level hierarchy of IC, package and board to a two- level hierarchy of IC and system package, which integrates IC package and system board into one and 2) miniaturization of system components from their current milliscale size discrete components to ultrathin-film embedded components at micro to nano scale. The SOP concept proposes two locations for these thin-film system components, one within the CMOS IC and the other in the system package. In addition, the SOP makes a compelling case for what should be integrated in CMOS and what components should be integrated in the system package, based on optimized cost, performance, and functionality.

The *Introduction to SOP* book is based on 12 years of research at Georgia Tech Packaging Research Center (PRC), funded by the National Science Foundation (NSF) under its Engineering Research Center (ERC) program, the State of Georgia, and more than 100 companies from throughout the world. This research was performed by an interdisciplinary team of 25 faculty and 500 graduate students from electrical, mechanical, materials, and chemical engineering departments.

The SOP book is the first fundamental and technological book written to meet both the industry and the academic needs. It tries to define, compare, and contrast the three primary electronic system technologies namely, SOC, SIP, and SOP for their impact on miniaturization, cost, electrical performance, and reliability. It has a total of 13 chapters to bring about the total system perspective of SOP technology. It starts with defining the SOP concept in the first chapter, comparing and contrasting it with SOC and SIP in the next two chapters. It then systematically describes each of the SOP technologies starting with SOP design, and integration of digital, RF and optical functions in the system package. The next set of chapters describe the SOP materials and process fabrication technologies, assembly, test and reliability by using a variety of interconnection technologies. The last chapter, Biosensor SOP, introduces the unique application of SOP concept for a variety of biomedical applications.

We are grateful to NSF ERC, Georgia Tech, and Georgia Research Alliance for their financial support and to our 40 academic and research faculty and to industry colleagues for their technical contributions that are included in this book. We are thankful to the PRC staff, particularly to Reed Crouch who coordinated the book project from start to finish. We are also grateful to Aparna Shukla for her professional editing and to Steve Chapman, the Publisher at McGraw-Hill, who ensured the high quality of the book. We thank our wives, Anne and Shailaja, for their patience and full support during the course of the book.

PROF. RAO R. TUMMALA
PROF. MADHAVAN SWAMINATHAN
Georgia Institute of Technology
Atlanta, Georgia

Introduction to System-on-Package (SOP)

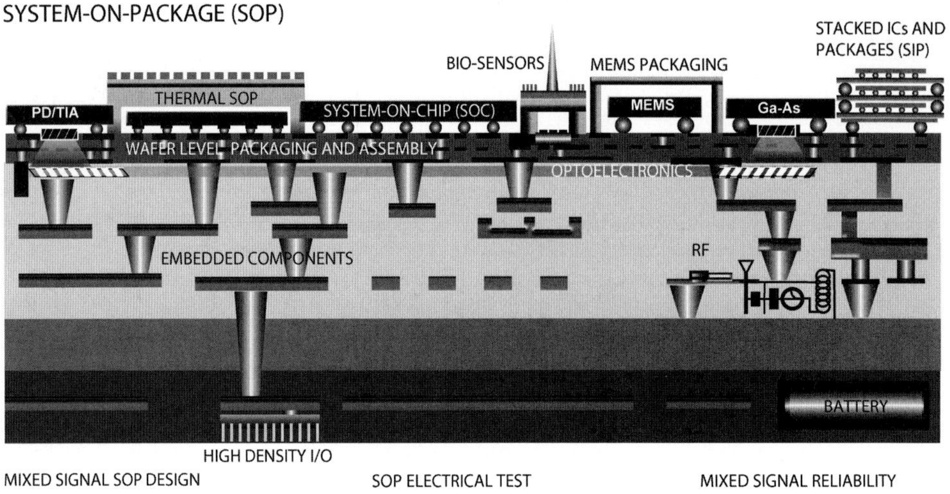

SYSTEM-ON-PACKAGE (SOP)

STACKED ICs AND PACKAGES (SIP)

BIO-SENSORS

MEMS PACKAGING

PD/TIA

THERMAL SOP

SYSTEM-ON-CHIP (SOC)

MEMS

Ga-As

WAFER LEVEL PACKAGING AND ASSEMBLY

OPTOELECTRONICS

RF

EMBEDDED COMPONENTS

BATTERY

HIGH DENSITY I/O

MIXED SIGNAL SOP DESIGN   SOP ELECTRICAL TEST   MIXED SIGNAL RELIABILITY

CHAPTER 1

Introduction to the System-on-Package (SOP) Technology

Prof. Rao R. Tummala and Tapobrata Bandyopadhyay
Georgia Institute of Technology

1.1	Introduction 4	1.7	Comparison of the Five System Technologies 23
1.2	Electronic System Trend to Digital Convergence 5	1.8	Status of SOP around the Globe 26
1.3	Building Blocks of an Electronic System 7	1.9	SOP Technology Implementations 29
1.4	System Technologies Evolution 8	1.10	SOP Technologies 33
1.5	Five Major System Technologies 11	1.11	Summary 34
1.6	System-on-Package Technology (Module with the Best of IC and System Integration) 18		References 34

The primary drivers of the information age are microsystems technologies and market economics. Gigascale integration of microelectronics, gigabit wireless devices, terabit optoelectronics, micro- to nano-sized motors, actuators, sensors, and medical implants and integration of all these by the system-on-package concept leading to ultraminiaturized, multi-to-mega function are expected to be the basis of the new information age.

This book is about system-on-package (SOP) technology in contrast to system-on-chip (SOC) technology at the integrated circuit (IC) level and stacked ICs and packages (SIP) at the module level. In this book, SIP is defined as the stacking of ICs and packages. Thus SOP is considered as an inclusive system technology of which SOC, SIP, thermal structures and batteries are considered as subset technologies. System-on-package is a new, emerging system concept in which the device, package, and system board are miniaturized into a single-system package with all the needed system functions. The SOP technology can be thought of as the second law of electronics for system integration in contrast to Moore's law for ICs.

This chapter introduces the basic concept of SOP. It reviews the characteristic features of a system-on-package and compares it with traditional and other major system technologies. It provides insight into the status of global research and development efforts in this area. Finally, it outlines the different technologies involved in making SOP-based products. The chapter concludes with an overview of all these basic SOP technologies, which form the chapter titles of this book.

1.1 Introduction

The concept of SOP originated in the mid-1990s in the Packaging Research Center at the Georgia Institute of Technology. The SOP is a new and emerging system technology concept in which the device, package, and system board are miniaturized into a single-system package with all the needed system functions. The SOP is described in this book as the basis for the second law of electronics for system integration in contrast to Moore's law for IC integration. The focus of SOP is to miniaturize the entire system, such as shown in Figure 1.1, which includes

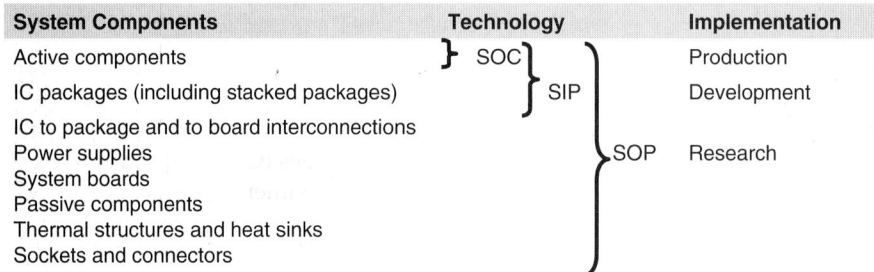

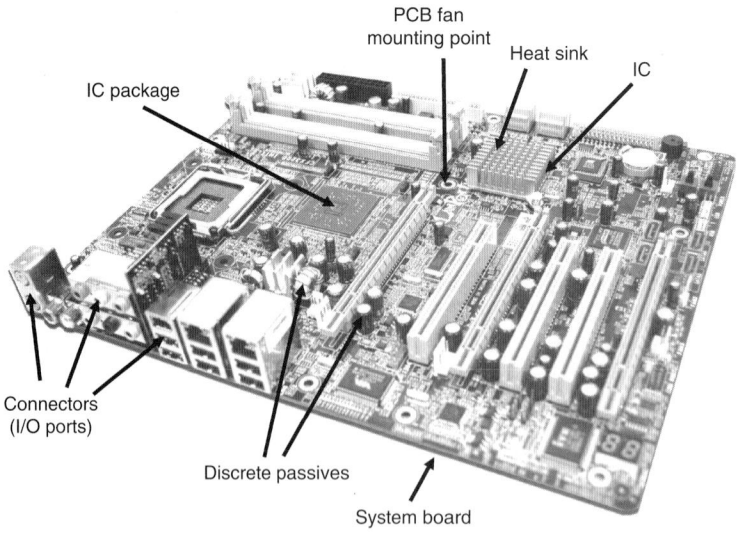

FIGURE 1.1 A typical example of a system with all its system components—DFI LanParty UT RD600. (*Courtesy:* dailytech.com)

Introduction to the System-on-Package (SOP) Technology

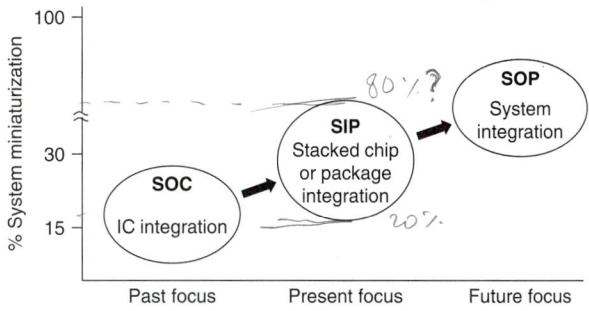

FIGURE 1.2 The miniaturization trend in ICs since the 1960s to systems around 2020.

The initial focus of SOP is on miniaturization and convergence of the package and system board into a system package, hence the name system-on-package. Such a single-system package with multiple ICs provides all the system functions by codesign and fabrication of digital, radiofrequency (RF), optical, micro-electro-mechanical systems (MEMS), and microsensor functions in either the IC or the system package. The SOP thus harnesses the advantages of the best on-chip and off-chip integration technologies to develop ultraminiaturized, high-performance, multifunctional products. Figure 1.2 depicts the miniaturization trend that started at the IC level in the 1960s at the microscale level and continued on to reach the expected level below 40 nanometers (nm). This is referred to as "SOC." The single-chip package miniaturization took place in a similar manner but at a slower rate until chip-scale packages (CSP) and two-dimensional (2D) multichip modules (MCMs) in the 1990s and three-dimensional (3D) SIPs a decade later were introduced. This is referred to as module-level miniaturization. The system-level miniaturization began subsequently.

1.2 Electronic System Trend to Digital Convergence

The combination of microelectronics and information technology (IT), which includes hardware, software, services, and applications, has been a trillion-dollar industry. It has been acting as the driving engine for science, technology, engineering, advanced manufacturing, and the overall economy of the United States, Japan, Europe, Korea, and other participating countries for several decades. Of this trillion-dollar worldwide market, hardware still accounts for more than $700 billion. Of this $700 billion, the semiconductors constitute about $250 billion and microsystems packaging (MSP), defined as both packaging of *devices* and systems but excluding semiconductors, accounts for about $200 billion. The simplistic way to define MSP is as the bridge between devices and end-product systems as depicted in Figure 1.3.

The MSP market of $200 billion, accounting for more than 10 percent of the entire IT market, is a strategic and critical technology, unlike in the past. It controls the size, performance, cost, and reliability of all end-product systems. It is, therefore, the major limiting factor and a major barrier to all future digital-convergent electronic systems. The MSP, in the future, involves not just microelectronics but also photonics, RF, MEMS, sensors, mechanical, thermal, chemical, and biological functions.

From cell phones to biomedical systems, the modern life is inexorably dependent on the complex convergence of technologies into stand-alone portable products designed to provide complete and personal solutions. Such systems are expected to have two

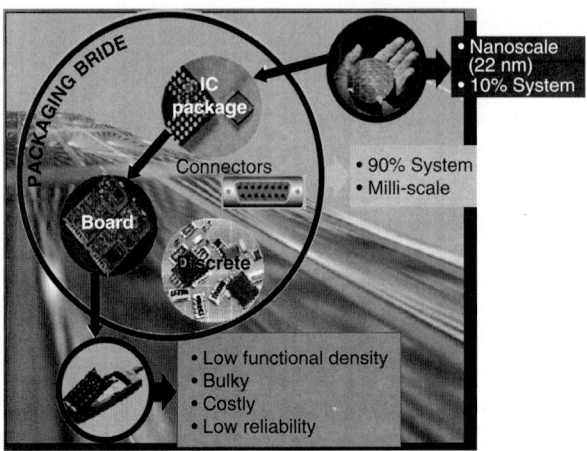

Figure 1.3 Packaging is the bridge and the barrier between ICs and systems.

criteria—size of the system and functionality of the system as shown in Figure 1.2. Computers in the 1970s were bulky, providing computing power measured in millions of instructions per second (MIPS). The subsequent IC and package integration technologies in the 1980s paved the way for systems with billions of instructions per second (BIPS), which further led the way for smaller and personal systems called PCs. The technical focus of these small computing systems by IC integration to single-chip processors, and package integration to multilayer thin-film organic buildup technologies, together with other miniaturization technologies such as flip-chip interconnection technology led to a new paradigm in personal and portable systems—cell phones. This trend, as shown in Figure 1.4, is expected to continue and to lead to highly miniaturized, multifunction-to-megafunction portable systems with computing, communication, biomedical, and

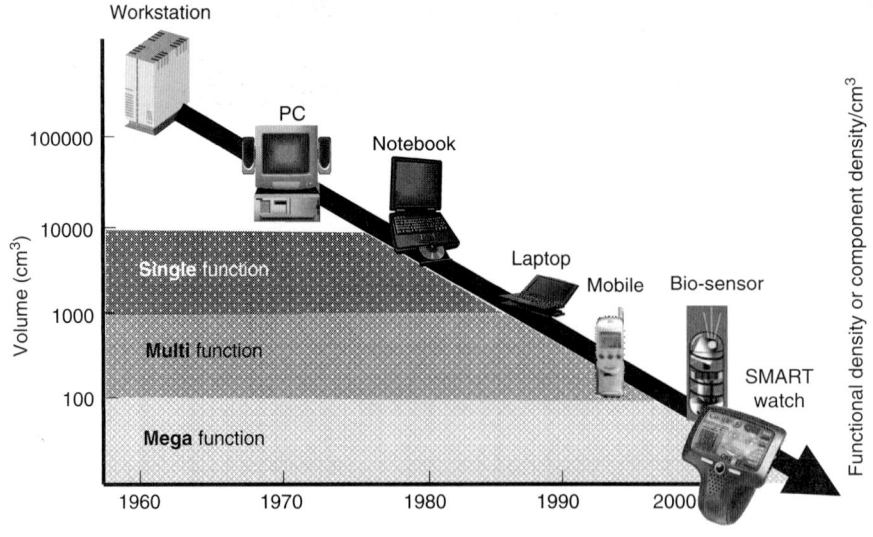

Figure 1.4 Electronic system trend toward highly miniaturized digital convergence.

consumer functions. Figure 1.4 shows some examples of electronic systems in the past and others projected in the future. This trend is expected to continue to megafunction systems that are about a cubic centimeter in size with not only computing and communication capabilities but also with sensors to sense, digitize, monitor, control, and transmit through the Internet to anyone anywhere.

1.3 Building Blocks of an Electronic System

The basic building blocks of an electronic system are listed in Table 1.1. The table also outlines and contrasts the traditional elements of an electronic system with SOP-based components of these building blocks.

Building Blocks	Traditional Technology	SOP-based Technology
Power sources	DC adapter, power cables, power socket	Embedded thin-film batteries, microfluidic batteries
Integrated circuits	Logic, memory, graphics, control, and other ICs, SOCs	Embedded and thinned ICs in substrate
Stacked ICs in 3D/Packaged ICs in 3D	SIPs with wire bond and flip chip	Wire-bonded and flip-chip SIPs. Through silicon via (TSV) SIPS and substrates
Packages or substrates	Multilayer organic substrates	Multilayer organic and silicon substrates with TSVs
Passive components	Discrete passive components on printed circuit board (PCB)	Thin-film embedded passives in organics, silicon wafer and Si substrate
Heat removal elements	Bulky heat sinks and heat spreaders. Bulky fans for convection cooling	Advanced nano thermal interface materials, nano heat sinks and heat spreaders, thin-film thermoelectric coolers, microfluidic channel based heat exchangers
System board	PCB-based motherboard	Package and PCB are merged into the SOP substrate
Connectors/sockets	USB port, serial port, parallel port, slots [for dual in-line memory modules (DIMM) and expansion cards]	Ultrahigh density I/O interfaces
Sensors	Discrete sensors on PCB	Integrated nanosensors in IC and SOP substrate
IC-to-package interconnections	Flip chip, wire bond	Ultraminiaturized nanoscale interconnections

TABLE 1.1 Building Blocks of a Traditional Electronic System versus an SOP-based System

Chapter One

Building Blocks	Traditional Technology	SOP-based Technology
Package wiring	Coarse wiring Line width: 25 µm Pitch: 75 µm	Ultrafine pitch, wiring in low-loss dielectrics Line width: 2–5 µm Pitch: 10–20 µm
Package-to-board interconnects	Ball grid array (BGA) bumps, tape automated bonding (TAB)	None!
Board wiring	Very coarse-pitched wiring (line width/spacing: 100–200 µm)	No PCB wiring. Package and PCB are merged into the SOP substrate with ultrafine pitch wiring.

TABLE 1.1 *(Continued)*

1.4 System Technologies Evolution

The barriers to achieving the required miniaturization are circled in Figure 1.1. These are the bulky IC packages, discrete components, connectors, cables, batteries, I/Os, massive thermal structures, and the printed wiring boards on which all these are assembled. This approach to system integration is called system-on-board (SOB). It constitutes 80 to 90 percent of traditional electronic system size and more than 70 percent of the system manufacturing cost. In general, as shown in Figure 1.5, all the system barriers can be addressed by three main approaches:

1. IC integration toward system-on-chip (SOC)
2. Package-enabled module-level integration by 3D stacked ICs and packages (SIP) and 2D multichip modules (MCMs)
3. System integration by system-on-package (SOP) as presented in this book

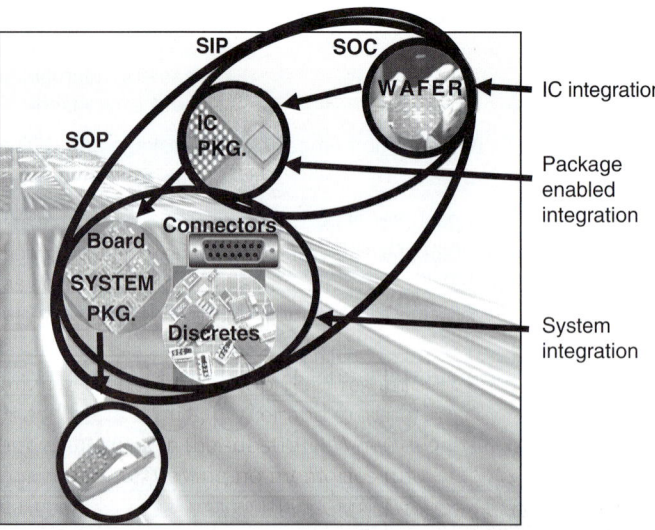

FIGURE 1.5 Three main integration approaches to address the system barriers.

Introduction to the System-on-Package (SOP) Technology

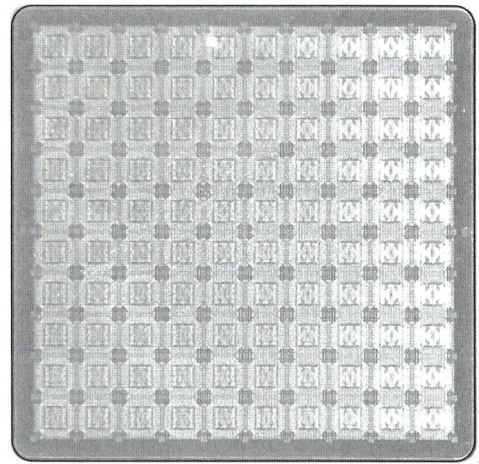

(a) Industry's first MCM (IBM), 1982   **(b)** 61 Layer LTCC/Cu-MCM (IBM), 1992

FIGURE 1.6 Examples of multichip modules. (*Courtesy:* IBM)

The on-chip integration is referred to as SOC, and it is expected to continue as long as it is economical. In the 1980s and 1990s, companies like IBM, Hitachi, Fujitsu, and NEC developed highly sophisticated subsystems called MCMs [1], as illustrated in Figure 1.6. The MCMs are three-dimensional structures during their fabrication; with as many as 60 to 100 layers of metallized ceramic prefired sheets stacked one on top of the other, interconnected by highly conductive metals such as molybdenum, tungsten, or copper. The finished MCMs, however, look like 2D structures, ultrathin in the Z dimension as compared to the X and Y dimensions.

Before MCMs were put into production, so-called wafer scale integration (WSI) was attempted in the 1980s by Gene Amdahl to bring the package and IC onto a single large silicon carrier. This subsequently led to the so-called silicon-on-silicon technologies using complementary metal-oxide semiconductor (CMOS) tools and processes both by IBM and Bell Labs. Both were abandoned at that time for a variety of reasons but began to reemerge recently for a different set of applications. The emergence of the cell phone in the 1980s and its need for miniaturization, since then, required a different concept than the two-dimensional SOCs or MCMs.

The concept of stacking thinned chips in the third dimension has been called stacked ICs and packages (SIP) [2] wherein ICs are thinned and stacked one on top of the other. Such an interconnected module is then surface-mount bonded onto a system board. Most of the early versions of SIP were interconnected by wire bonding. More recent versions of this technology began to use flip chip as well as silicon-through-via connections to further miniaturize the module. The latest versions of SIP are often referred to as 3D packaging, which includes

- Stacked ICs with silicon-through vias (with flip chip or copper-to-copper bonding)
- Silicon ICs on silicon wafer board
- Wafer-to-wafer stacking

The ultraminiaturized systems such as "Dick Tracy's watch" in Figure 1.3 with dozens of functions requires yet another major paradigm in system technology. This paradigm is based on the concept of system-on-package, which originated in the mid-1990s at the NSF-funded Packaging Research Center at the Georgia Institute of Technology [3].

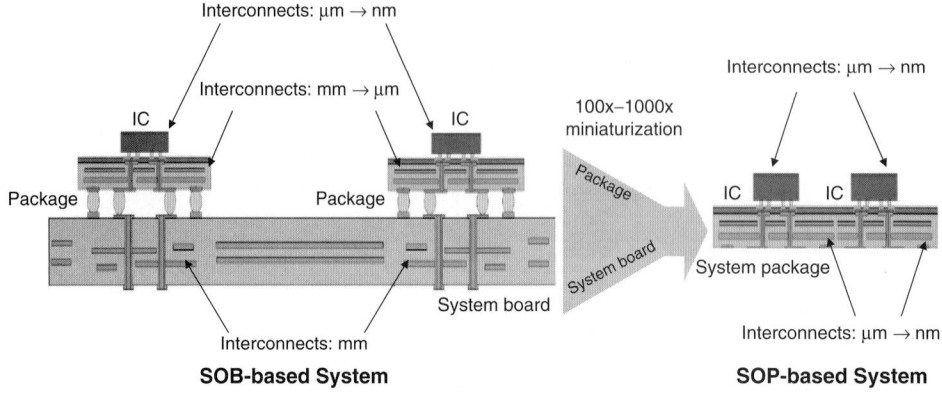

Figure 1.7 A comparison between three-tier SOB-based and two-tier SOP-based systems.

The SOP technology concept has two characteristics. First, it combines the IC, package, and system board into a system package (as shown in Figure 1.7), hence its name system-on-package. The second key attribute of SOP is its integration and miniaturization at the system level just like IC integration at the device level. Unlike SIP, which enables IC stacking without real package integration, SOP integrates all the system components either in ICs or packages as ultrathin films or structures that include the following [4]:

- Passive components
- Interconnections

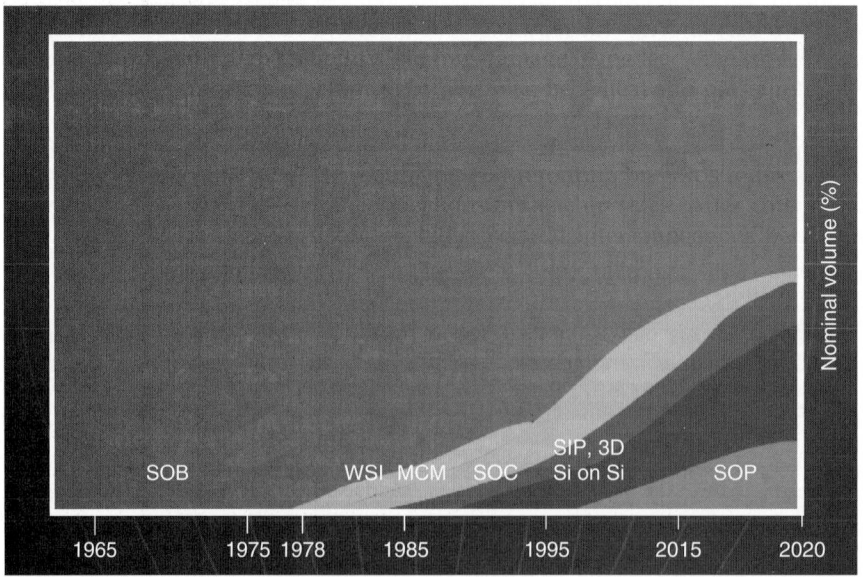

Figure 1.8 Historical evolution of the five system technologies over the past 50 years.

- Connectors
- Thermal structures such as heat sinks and thermal interface materials
- Power sources
- System board

Such a single-system package provides all the system functions such as computing, wireless and network communications, and consumer and biomedical functions in one single module. Figure 1.8 depicts the historical evolution of the five system technologies during the last 50 years as well as the expected projection during the next 15 years.

1.5 Five Major System Technologies

The five major system technologies for electronic digital convergence are schematically illustrated in Figure 1.9a and b:

1. *System-on-board (SOB)*. Discrete components interconnected on system boards.
2. *System-on-chip (SOC)*. Partial system on a single IC with two or more functions.
3. *Multichip module (MCM)*. Package-enabled horizontal or 2D integration of two or more ICs for high electrical system performance.
4. *Stacked ICs and packages (SIP)*. Package-enabled 3D stacking of two or more thinned ICs for system miniaturization.
5. *System-on-package (SOP)*. Best IC and system integration for ultraminiaturization, multiple to mega functions, ultrahigh performance, low cost, and high reliability.

1.5.1 System-on-Board (SOB) Technology with Discrete Components

The current approach to manufacturing systems involves fabricating the components separately and assembling them onto system boards, as illustrated previously in Figure 1.3. The strategy to miniaturize the systems in this traditional approach has been to reduce the size of each component by reducing the input-output (I/O) pitch, wiring, and insulation dimensions in each of the layers. But this approach presents major limitations to achieving digital convergence, as explained earlier. The IC packaging that is used to provide I/O connections from the chip to the rest of the system is typically bulky and costly, limiting both the performance and the reliability of the IC it packages. Systems packaging, involving the interconnection of components on a system-level board, is similarly bulky and costly with poor electrical and mechanical performance.

1.5.2 System-on-Chip (SOC) with Two or More System Functions on a Single Chip

Semiconductors have been the backbone of the IT industry, typically governed by Moore's law. Since the invention of the transistor, microelectronics technology has impacted every aspect of human life by electronic products in the automotive, consumer, computer, telecommunication, aerospace, military, and medical industries by ever-higher integration of transistors as indicated in Figure 1.9, and at an ever-lower cost per transistor. This integration and cost path has led the microelectronics industry to believe

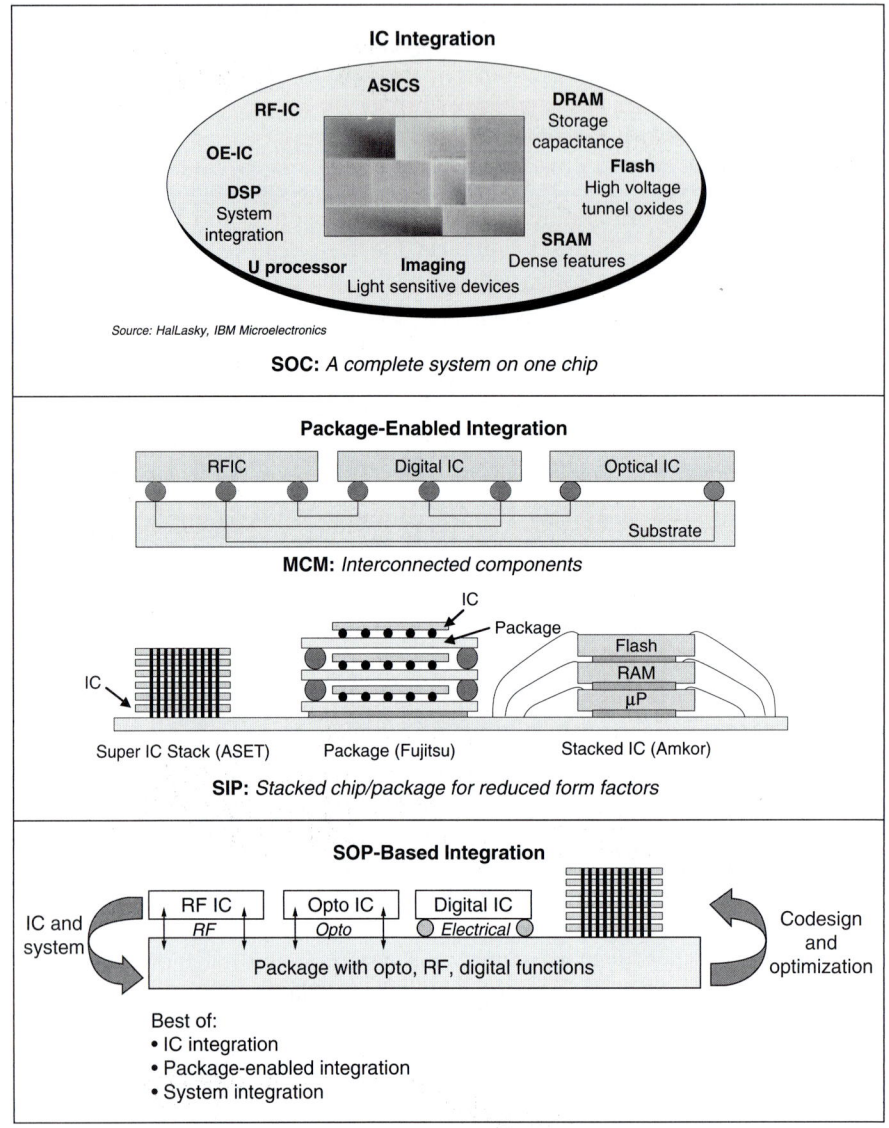

FIGURE 1.9 (*a*) IC and package-enabled integration interconnecting two or more ICs. (*b*) SOP: True package and IC integration.

that this kind of progress can go on forever, leading to a "system-on-a-chip" for all applications to form complete end-product systems.

The SOC schematic shown in Figure 1.9a, for example, seeks to integrate numerous system functions on one silicon device horizontally, namely the chip. If this chip can be designed and fabricated cost effectively with computing, communication, and consumer functions (such as processor, memory, wireless, and graphics) by integrating the required components (such as antennas, filters, switches, transmitting waveguides, and other

Introduction to the System-on-Package (SOP) Technology

components required to form a complete end-product system), then all that is necessary to package such a system is to provide protection, external connections, power, and cooling. If this can be realized, SOC offers the promise for the highest performance and the most compact, lightweight system that can be mass-produced. This has been and continues to be the road map [8] of IC companies.

So the key question is whether SOC can lead to cost-effective, complete end-product systems such as tomorrow's leading-edge cell phones with digital, wireless, and sensing capabilities or biomedical implants. Researchers around the world, while making great progress, are realizing that SOC, in the long run, presents fundamental limits for computing and integration limits for wireless communications and additional nonincremental costs to both. Among SOC challenges are the long design times due to integration complexities, high wafer fabrication costs and test costs, and mixed-signal processing complexities requiring dozens of mask steps and intellectual property issues. The high costs are due to the need to integrate active but disparate devices such as bipolar, CMOS, silicon germanium (SiGe), and optoelectronic ICs—all in one chip with multiple voltage levels and dozens of mask steps to provide digital, RF, optical, and MEMS-based components.

It is becoming clear that SOC presents major technical, financial, business, and legal challenges that are forcing industry and academic researchers to consider other options for semiconductors and systems. For the first time, industry may not invest in extending Moore's law beyond 2015. This is leading the industry to explore alternative ways to achieve systems integration wherein semiconductor integration is pursued, not only horizontally by SOC, but also vertically by SIP via 3D stacking of bare or packaged ICs and by SOP. More than 50 companies are pursuing SIP as indicated in Chapter 4.

Hence, a new paradigm that overcomes the shortcomings of both SOC and traditional systems packaging is necessary. The SOP technology described in this book makes a compelling case for the synergy between the IC and the package integration by means of the SOP concept, which can also be applied to SOCs and SIPs, as well as to silicon wafer, ceramic, or organic carrier platforms or boards.

1.5.3 Multichip Module (MCM): Package-Enabled Integration of Two or More Chips Interconnected Horizontally

The MCM (Figure 1.6) was invented back in the 1980s at IBM, Fujitsu, NEC, and Hitachi for the sole purpose of interconnecting dozens of good bare ICs to produce a substrate wafer that looked like the original wafer, since larger chips could not be produced with any acceptable yields on the original silicon wafer. These original MCMs were horizontal or two-dimensional. They started with so-called high-temperature cofired ceramics (HTCCs)—multilayer ceramics, such as alumina, metallized and interconnected with dozens of layers of either cofired molybdenum or tungsten. These then were replaced with higher-performance ceramic MCMs called low-temperature cofired ceramics (LTCCs)—made of lower-dielectric-constant ceramics such as glass-ceramics, metallized with better electrical conductors such as copper, gold, or silver-palladium. The third generation of MCMs improved further with add-on multilayer organic dielectrics and conductors of much lower dielectric constant and sputtered or electroplated copper with better electrical conductivity.

1.5.4 Stacked ICs and Packages (SIP): Package-Enabled IC Integration with Two or More Chip Stacking (Moore's Law in the Third Dimension)

Here, SIP is defined as a vertical stacking of similar or dissimilar ICs, in contrast to the horizontal nature of SOC, which overcomes some of the above SOC limitations, such as latency, if the size of the chips and their thicknesses used in stacking are small. SIP is also defined often as the entire system-in-a-package. If all the system components (for example, passive components, interconnections, connectors, and thermal structures such as heat sinks and thermal interface materials), power sources, and system board are miniaturized and integrated into a complete system as described in this book as SOP, then there is no difference between SIP and SOP. The intellectual property issues as well as yield losses associated with dozens of sequential mask steps and large-area IC fabrication are also minimal. Clearly, this is the semiconductor companies' dream in the short term.

But there is one major issue with this approach. The SIP, defined above as stacking of ICs, includes only the IC integration and hence addresses only about 10 to 20 percent of the system by extending Moore's law in the third dimension. If all the ICs in the stack are limited to CMOS IC processing, the end-product system is limited by what it can achieve only with CMOS processing at or below nanoscale. The above fundamental and integration barriers of SOC, therefore, remain. There are clear major benefits, however, to SIP: simpler design and design verification, a process with minimal mask steps, minimal time-to-market, and minimal Intellectual Property (IP) issues. Because of the above-mentioned SIP benefits, however limited, about 50 IC and packaging companies alike have geared up in a big way to produce SIP-based modules (Figure 1.10).

SIP Categories

The SIP technology can be broadly classified, as shown in Figures 1.10 and 1.11, into two categories: (1) stacking of bare or packaged ICs [9–12] by traditional wire-bond, TAB, or flip-chip technologies, and (2) stacking by through-silicon vias (TSVs), without using wire bond or flip chip. SIP and 3D packaging are often meant to be the same and are loosely referred to as the vertical stacking of either bare or packaged dies. In this book, however, 3D package integration refers to stacking of ICs by means of TSV technology.

SIP by Wire Bonding Three-dimensional integration of bare dies can be done using wire bonding as shown in Figure 1.12. In this approach, the different stacked dies are interconnected using a common interposer (or package). The individual dies are connected to this interposer by wire bonds. Wire bonding is economical for interconnect densities of up to 300 I/Os. However, it suffers from the high parasitic inductance of the wire bonds. There is a lot of inductive coupling between the densely placed wire bonds which results in poor signal integrity.

SIP by Flip Chip and Wire Bonding In this 3D integration technique, as shown in Figure 1.13, the bottom die of the stack is connected to the package by flip-chip bonds. All other dies on the top of it are connected to the package using wire bonds. This eliminates the wire bonds required for the bottom die, but still suffers from the high parasitics of the wire bonds for the upper dies.

SIP by Flip Chip-on-Chip The bare dies are flip-chip bonded with each other in this approach of 3D integration as shown in Figure 1.14a and b. The dies are arranged face-to-face with the Back End of Line (BEOL) areas of the dies facing each other. The bottom die is usually bigger than the top die. The bottom die is connected to the package by wire bonds.

3D Integration by Through-Silicon-via Technology Three-dimensional integration enables the integration of highly complex systems more cost-efficiently. A high degree of

$$\Phi_s = \chi_s + (E_c - E_i) - (E_F - E_i)$$

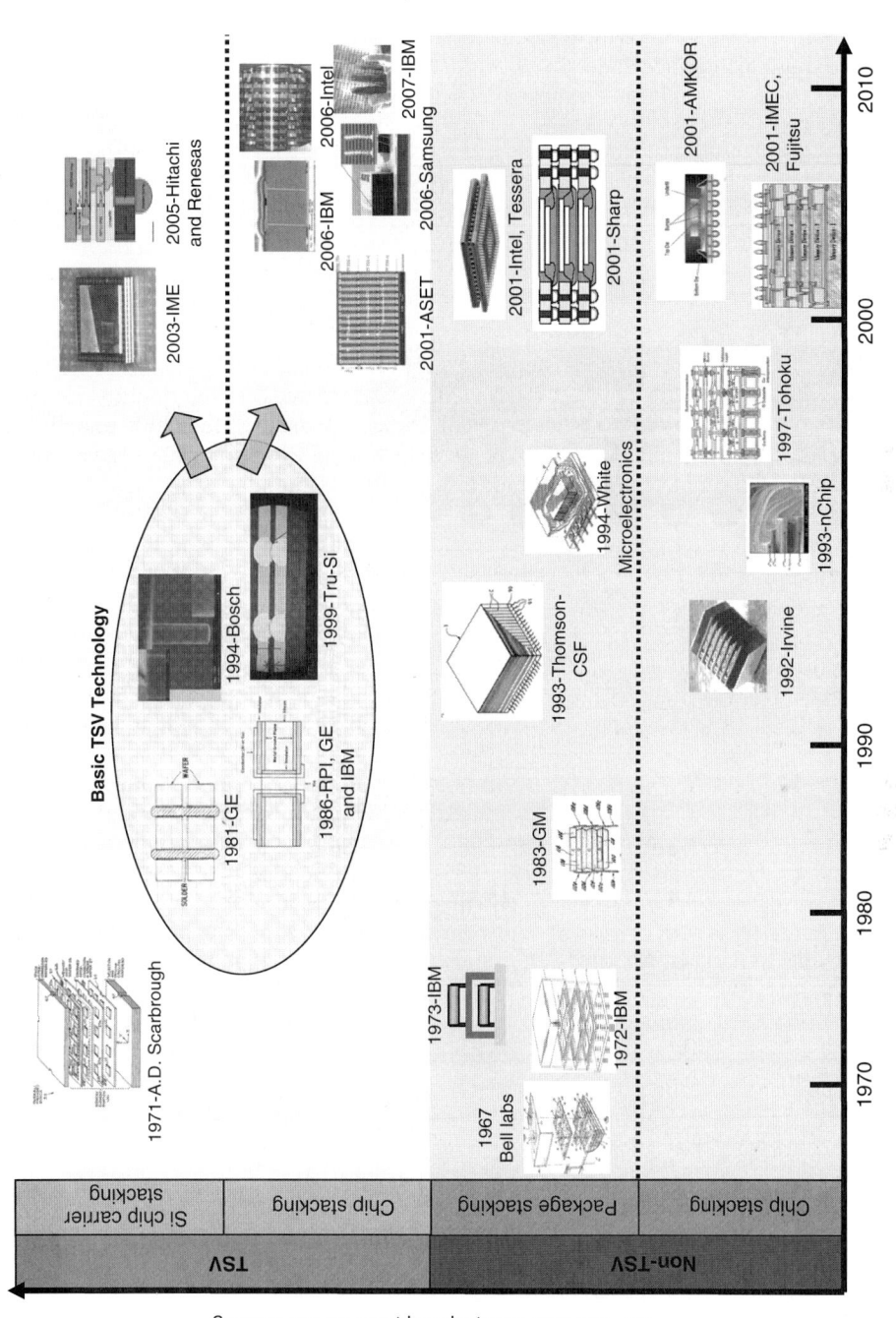

FIGURE 1.10 Emerging stacked IC and packaging technologies.

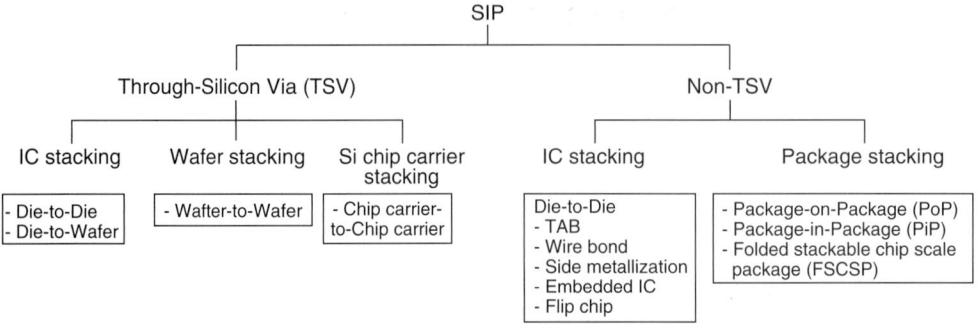

FIGURE 1.11 Different integration approaches in SIP.

FIGURE 1.12 Three-dimensional integration using wire bonding.

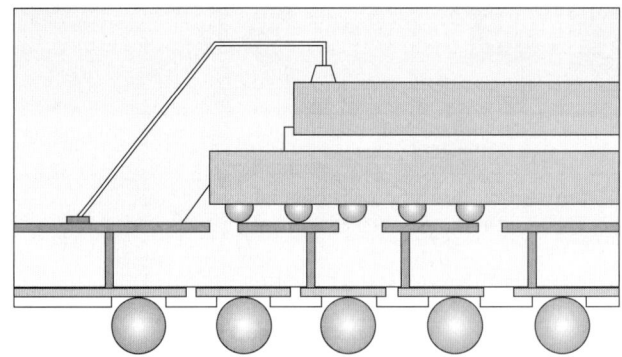

FIGURE 1.13 Three-dimensional integration using a combination of flip-chip and wire bonding.

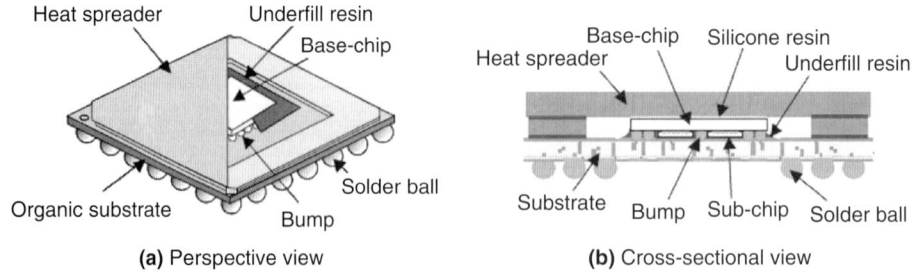

FIGURE 1.14 Three-dimensional integration by the flip chip-on-chip approach. (*a*) Perspective view. (*b*) Cross-sectional view. [13]

miniaturization and flexibility for the adaptation to different applications can be achieved by using the 3D integration technologies. It also enables the combination of different optimized technologies with the potential of low-cost fabrication through high yield, smaller footprints, and multifunctionality. Three-dimensional technologies also reduce the wiring lengths for interchip and intrachip communication. It thus provides a possible solution to the increasingly critical "wiring crisis" caused by signal propagation delays at both the board and the chip level.

It is possible to stack multiple bare dies using die-to-die vias and TSVs as shown in Figure 1.15. The latter run through the silicon die [Front End of Line (FEOL) and BEOL] and are used to connect stacked dies. There are various technologies for via drilling, via lining, via filling, die (or wafer) bonding, and integration of the 3D stacked dies (or wafers). TSV technology can potentially achieve much higher vertical interconnect density as compared to the other approaches for 3D integration discussed above.

The dies can be bonded in a face-to-face or in a face-to-back. In the face-to-face die stacking, two dies are stacked with their BEOL areas facing each other. In the face-to-back die stacking, two dies are stacked with the BEOL areas of one die facing the active area of the other die. Face-to-face bonding enables a higher via density than face-to-back bonding because the two chips are connected by die-to-die vias which have sizes and electrical characteristics similar to conventional vias that connect on chip metal routing layers. On the other hand, in face-to-back bonding, the two chips are connected by TSVs which are much bigger than the BEOL vias. However, if more than two chips are to be stacked, then TSVs are necessary even for face-to-face bonding.

Three-dimensional integration was initially introduced by stacking Flash (NOR/NAND) memory and SDRAM for cell phones in one thin CSP. This was later extended to Memory/Logic integration for high performance processors. Stacking of an ASIC digital signal processors (DSPs) and RF/analog chips or MEMS are the next logical developments in 3D packaging.

Si Substrate or Carrier

The concept of the silicon chip carrier was developed in 1972 [14] at IBM where a Si substrate was used as a chip carrier instead of insulating organic or ceramic substrates. Initially, the chips were connected to the chip carrier by perimeter connections such as wire bonding. Later, the connections were replaced by flip-chip connections. Lately, TSVs have been used in the chip and the carrier. The TSVs help to develop a high-density

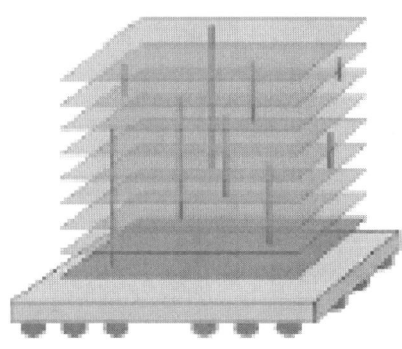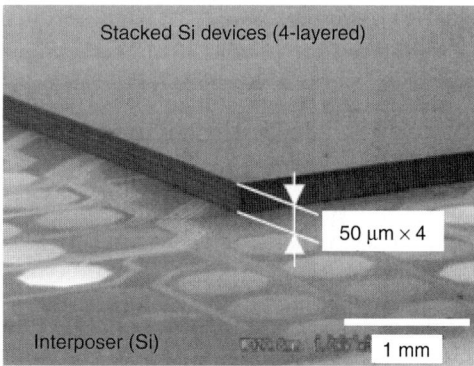

FIGURE 1.15 Three-dimensional integration with through-silicon-via technology.

Figure 1.16 Package-in-package (PiP) structure. *Left:* PiP package stack of two packages (four chips). *Right:* PiP with a package and a die stack (four dies). [16]

interconnection from the chip to the carrier and from the carrier to the board. Presently, silicon chip carrier technology involves through-silicon vias (TSVs), high-density wiring, fine pitch chip-to-carrier interconnection, and integrated actives and passives. The TSVs can also be used to stack the Si chip carriers on top of one another [15].

SIP by Package Stacking Three-dimensional integration is also possible by a vertical stacking of individually tested IC packages. There are two topologies: package-in-package (PiP) and package-on-package (PoP). PiP, as shown in Figure 1.16, connects the stacked packages by wire bonds on a common substrate. In PoP, as shown in Figure 1.17, the stacked packages are connected by flip-chip bumps.

1.6 System-on-Package Technology (Module with the Best of IC and System Integration)

If, in fact, the system components such as batteries, packages, boards, thermal structures, and interconnections are miniaturized as described above with nanoscale materials and structures, this should lead to the second law of electronics [17]. The SOP described in this book is exactly that, and it (Figure 1.18) achieves true system integration, not just with the best IC integration as in the past but also with the best system integration. As such, it addresses then the 80 to 90 percent of the system problems that had not been addressed, as described earlier. In contrast to IC integration by Moore's law, measured in transistors per cubic centimeter, the SOP-based second law addresses the system integration challenges as measured in functions or components per cubic centimeter.

Figure 1.18 illustrates the evolution of these two laws during the last 40 years. As can be seen, the slope of the first law of electronics is very steep, driven by the unparalleled growth in the IC integration from one transistor in the 1950s to as many as a billion by 2010. The growth in the system integration, however, is very shallow as measured in components per square centimeter (cm^2) on system-level boards to less than 100/cm^2 in today's manufacturing. This slow growth, however shallow, required

Figure 1.17 Package-on-package (PoP) structure with two packages (four chips). [16]

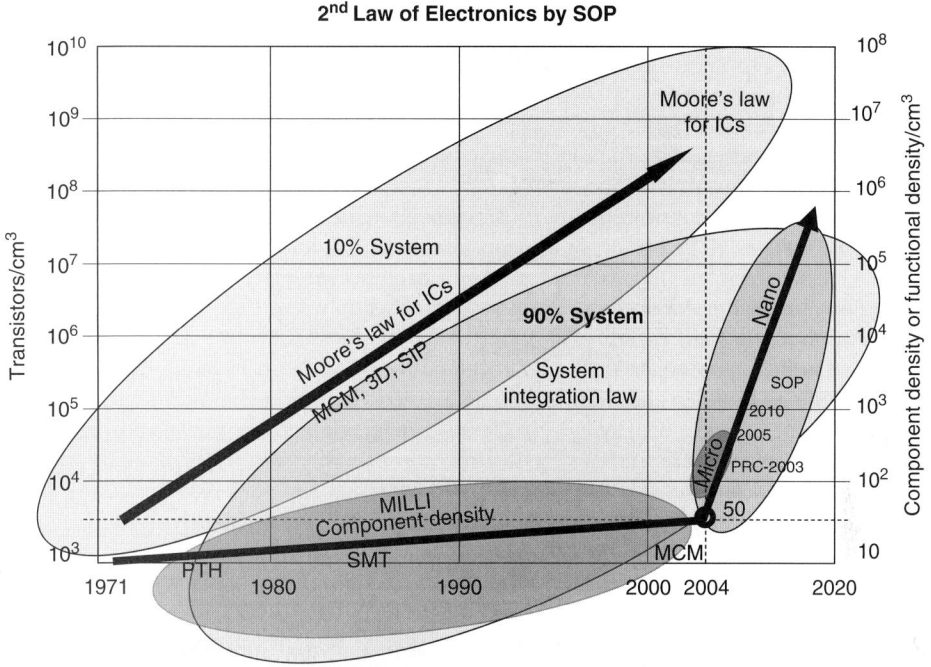

FIGURE 1.18 Second law of electronics achieves true package integration combined with the best of IC integration.

a number of global package developments, as illustrated in Figure 1.19. These developments can be summarized as package size reductions enabled by I/O pitch reductions enabled by wiring and via dimensions.

In the SOP concept, "the system package, not the bulky board, is the system." While "systems" of the past consisted of bulky boxes housing hundreds of components that performed one task, the SOP concept consists of small and highly integrated and microminiaturized components in a single system package or module with system

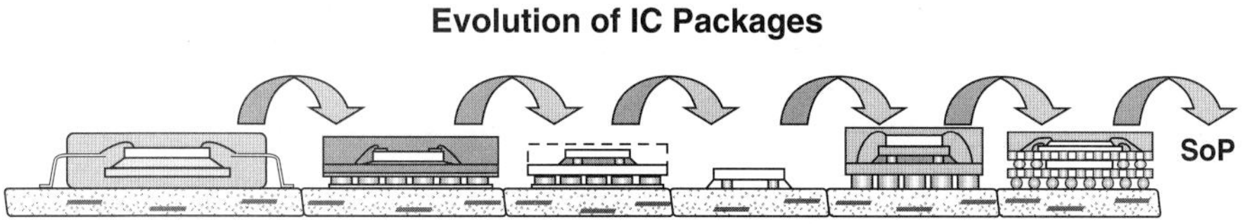

Evolution of IC Packages

| | | | | | SIP | | |
| | | | | |-----------|-----------| |
Family	QFP	BGA	FC-BGA	DCA	Non-TSV	TSV	SoP
IC	Wire bond	Wire bond	Solder ball	----	WB, FC	TSV	Flip chip
Package	Leadframe	Substrate	Substrate	Thin-film	Substrate	Substrate	Substrate
Si efficiency	30	50	75	100	>100	>100	>100

FIGURE 1.19 IC packaging evolution. (*Courtesy:* Infineon)

functions for computing, communication, consumer, and biomedical applications—in a small system package no greater than the size of Intel's Pentium processor package (Figure 1.19). Thus, SOP can be thought of as the "package is the system."

As such it combines the package and system board (as shown in Figure 1.2) into a system package. The fundamental basis of SOP is illustrated in Figure 1.20, which consists of two parts—the digital CMOS or IC part with its components and the system package part with its components. What is new and different about SOP is the system package part that miniaturizes the current milliscale components in this part to microscale in the short term and nanoscale in the long term (Figure 1.18). Thus SOP reduces the size of the 80 to 90 percent of the non-IC part of the system by a factor of 1000 in the short term (from milli to microscale) and in the long term by a factor of a million (from milli to nanoscale).

The SOP paradigm brings synergy between CMOS and system integration, and this synergy overcomes both the fundamental and integration shortcomings of SOC and SIP, which are limited by CMOS. While silicon technology is great for transistor density improvements from year to year, it is not an optimal platform for the integration system components such as power sources, thermal structures, packages, boards, and passives. These are highlighted in Figure 1.20. Two good examples for which CMOS is not good are front-end RF electronics and optoelectronics. This system-package driven size reduction has benefits of higher performance, lower cost, and higher reliability, just like with ICs. The cost advantages of system integration over digital CMOS integration for the same components are exemplified in Figure 1.21. In general, costs of any manufacturing technology can be simply viewed as throughput-driven cost and investment-driven cost.

In theory, there should be other factors such as yield and materials and labor. Most major thin-film technologies including liquid-crystal displays (LCDs), plasma panels,

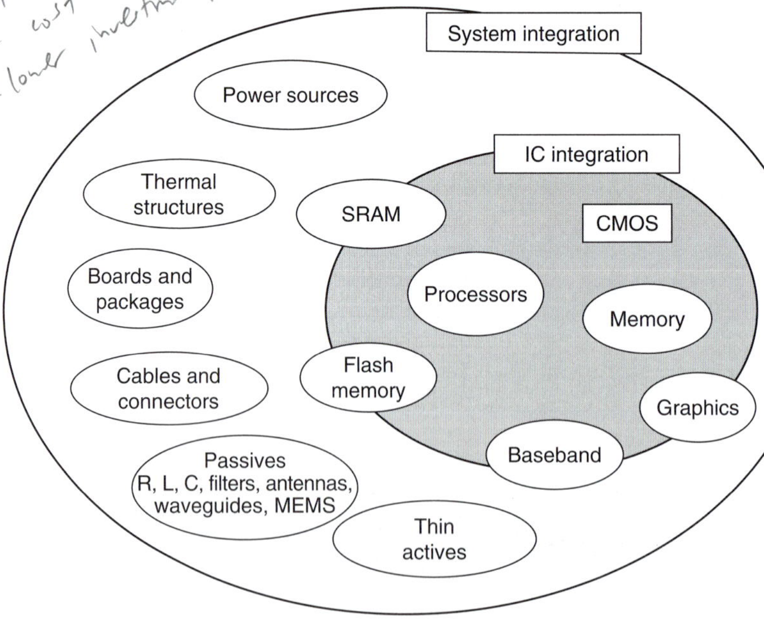

FIGURE 1.20 Fundamental basis of SOP with two parts: the digital CMOS IC regime and system regime.

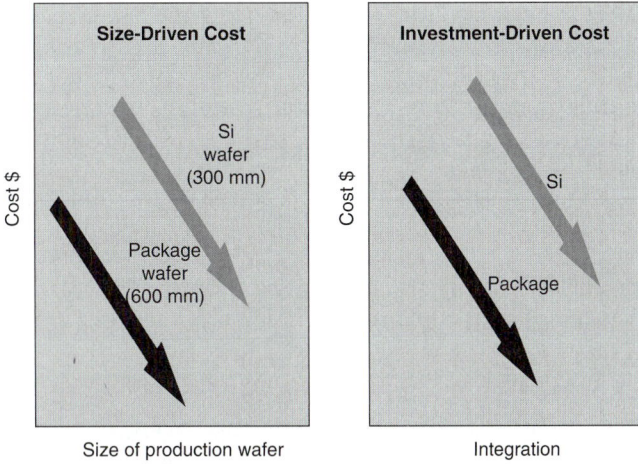

FIGURE 1.21 Cost advantages of package integration over digital CMOS integration, for the same components.

as well as front-end and back-end IC technologies yield better than 90 percent by manufacturing companies that practice rigorous manufacturing. The raw material costs are a small fraction, often less than 5 percent of the final cost of the product. While the labor costs can be high, most advanced factories are automated, thus minimizing these costs. The throughput-driven cost has two elements—the size of the panel and the number of panels per unit time. The SOP-based system package integration has one unique advantage in this respect in that the typical size of the panel is about 450 × 550 millimeters (mm) in size compared to 300 mm for CMOS. This translates into a factor of almost 3× advantages over on-chip manufacturing. The package integration cycle time, however, is longer than the CMOS cycle time because the speed at which the SOP wafers are produced in relation to CMOS wafers is slower. Package integration, however, more than makes up for this deficiency by lower cost investment for the SOP package integration factory (by a factor of 5 to 10) in relation to the CMOS factory.

In addition to financial advantages, SOP offers technical advantages in digital, wireless, and optoelectronic-based network systems. In the computing world, the SOP concept overcomes the fundamental limits of SOC. As IC integration moves to the nanoscale and wiring resistance increases, global wiring delay times become too high for computing applications [18]. This leads to what is referred to as "latency," which can be avoided by moving global wiring from the nanoscale on ICs to the microscale on the package.

The wireless integration limits of SOC are also handled well by SOP [19–20]. The RF components, such as capacitors, filters, antennas, switches, and high-frequency and high-Q inductors, are best fabricated on the package with micron-thick package dimensions rather than on silicon with nanoscale dimensions. To meet the need for the amount of decoupling capacitance necessary to suppress the expected power noise associated with very high performance ICs that use more than 100 watts (W) per chip, a major portion of the chip area would have to be dedicated to the decoupling capacitance alone. Semiconductor companies are not in the capacitor business; they are in the transistor business. The highest Q factors reported on silicon are about 25 to 60, in contrast to 250 to 400 achieved in the package.

Optoelectronics, which today finds use primarily in the back plane and is used for high-speed board interconnects, is expected to move onto the SOP package as chip-to-chip high-speed interconnections replacing copper, thereby, addressing both the resistance and crosstalk issues of electronic ICs. Optoelectronics, as it moves into silicon as silicon photonics by Intel, is viewed, not as CMOS technology, but as an SOP-like heterogeneous technology.

The SOP is about system integration enabled by thin-film integration of all system components at microscale in the short term and nanoscale in the long term. As such, the system package integration that SOP enables can be applied to CMOS ICs as overlays; applied as thin films on top of silicon wafers (TFOS), silicon carriers, ceramic, and glass substrates; or embedded into multilayer ceramics, packages, or board laminates.

1.6.1 Miniaturization Trend

The single most important parameter for digital convergence is system miniaturization. It is now generally accepted that miniaturization leads to

- Higher performance
- Lower cost
- Higher reliability
- Higher functionality
- Smaller size

Figure 1.22 depicts the historical evolution to miniaturization technologies as a function of the fraction of the system miniaturized using that technology. The miniaturization originated at the device level soon after the discovery of the transistor,

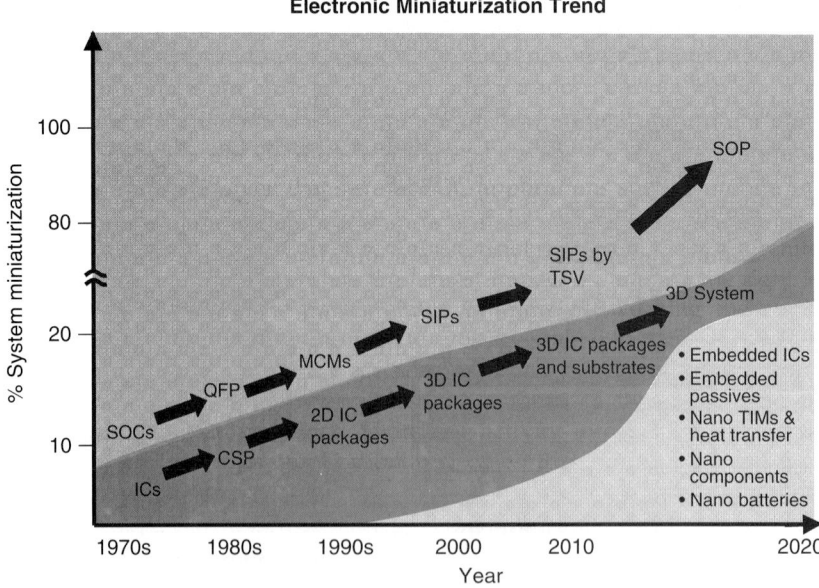

FIGURE 1.22 Historical evolution of miniaturization technologies during the last four decades.

leading to nanometer nodes currently from micrometer nodes in the 1970s. This miniaturization is expected to continue through at least 32 nm and perhaps beyond. The miniaturization in IC packages, however, was not so dramatic. As can be seen from Figure 1.18, the dual in-line packages with only I/Os in centimeter size in the 1970s migrated to Quad-Flat Pack (QFP) with I/Os on all four sides of the package in the 1980s. Both are lead frame based, making them bulky. The next wave in miniaturization led to solder ball attach and surface mount assembly to the board and was typically achieved with ball grid arrays. The IC assembly miniaturization followed a similar path starting with coarse-pitch peripheral wire bond, then finer pitch, and then to area array wire bond by some companies. Further miniaturization at the IC level was brought about by a major breakthrough by IBM, commonly referred to as "flip chip." The flip-chip miniaturization that started in the 1970s at the millimeter pitch, is paving the way to 10- to 20-micron pitch by 2015. The so-called chip scale package that was no more than 20 percent larger in size than the packaged ICs was the next miniaturization technology currently implemented at the wafer level. Further miniaturization has been accomplished with bare chips by so-called chip-on-board or flip-chip MCM technologies.

The next wave in miniaturization has been achieved by 2D MCMs for ultrahigh computing performance, as shown previously in Figure 1.5. Two factors contributed to this miniaturization: (1) the highly integrated substrate and its multilayer fine line and via wiring dimensions, and (2) 2D dimensions with as many as 144 bare chips interconnected in 100- to 144-mm size substrate. The market need for cell phones changed this 2D approach to 3D, achieved by stacking as many as 9 thinned chips to date with the potential to stack 20 or more by 2015. Two major factors contributed to this miniaturization: (1) thinned chips to 70 microns in thickness and (2) shorter and finer-pitch flip-chip assemblies. The next paradigm in miniaturization is being achieved by so-called through-silicon-via technology, as described above, and pad-to-pad bonding, replacing the flip-chip assembly.

The fraction of the system miniaturized by the above IC-based and Moore's law driven technologies, as shown in Figure 1.22, is typically about 10 to 20 percent of the system, leaving the remaining 80 percent in a bulky state. This 80 percent consists of such system components as passives, power supplies, thermal structures, sealants, intersystem interconnections, and sockets. This is what SOP is all about, miniaturizing these components from their milliscale to microscale in the short term and nanoscale in the long term.

1.7 Comparison of the Five System Technologies

Figure 1.23a lists the system drivers as being miniaturization, electrical performance, power usage, thermal performance, reliability, development and manufacturing cost, time-to-market, and flexibility. Figure 1.23b compares each of the above system technologies against the same parameters, showing the strengths and weaknesses of each of the technologies.

The SOC is a clear technology leader in electrical performance and power usage, and while it is a miniaturization leader at the IC level, it is not a leader at the system level, as can be seen in Figure 1.23b. This is due to the fact that the system technologies such as power supplies and thermal structures are not miniaturized. The high development cost, longer time-to-market, and limited flexibility are its major weaknesses. In addition, complete integration of RF, digital, and optical technologies on a single chip poses numerous challenges. RF circuit performance, for example, is a tradeoff between the quality factor (Q) of passive components (inductors and capacitors) and

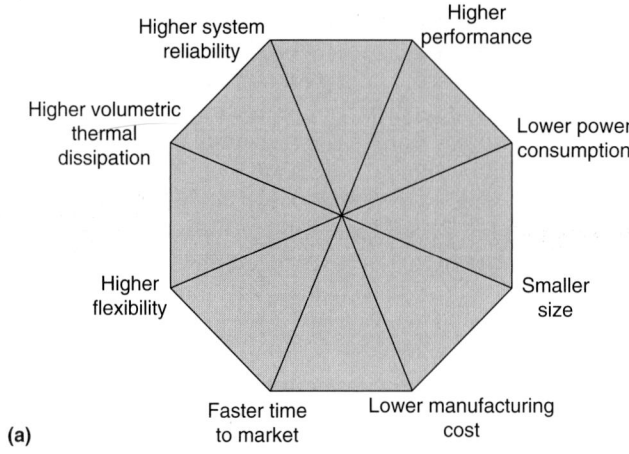

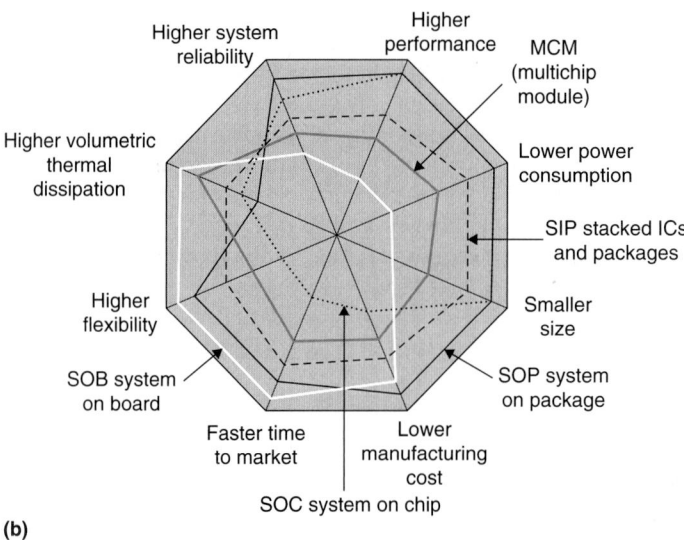

FIGURE 1.23 (*a*) System drivers: miniaturization, electrical performance, power usage, thermal performance, reliability, development and manufacturing cost, time-to-market, and flexibility. (*b*) System technologies compared against system driver parameters showing the strengths and weaknesses of each.

power. Low-power circuit implementations for mobile applications require high-Q passive components. In standard silicon technologies, the Q factor is limited to about 25 due to the inherent losses of silicon [21] and large area usage beyond traditional digital CMOS dimensions. This can be improved by using esoteric technologies such as thick oxides, high-resistivity Si, SiGe, or gallium arsenide (GaAs), which increase the cost substantially. In addition, these passive components consume valuable real estate and occupy more than 50 percent of the silicon area.

Antennas are another example that cannot be integrated on silicon due to size restrictions [20, 22–25]. Another example involves RF circuits that function in the microvolt range. Integration of dissimilar signals requires large isolation between them. On standard silicon, a major concern is substrate coupling caused by the finite resistivity of the silicon substrate. Though solutions have been proposed using high-resistivity silicon or N-well trenches, the isolation levels achieved are insufficient. For multiple voltage levels, distributing power to the digital and RF circuits while simultaneously maintaining isolation and low electromagnetic interference (EMI) can be a major challenge [26].

These issues can be addressed quite easily with SOP using embedded filtering and decoupling technologies [27–31]. The SOP has already been demonstrated with Q values in the range of 100 to 400 using low-loss dielectrics and copper metallization structures that enable low-power solutions. With advances in digital processing speeds, embedded optical waveguides in the package have the potential of bringing photonics directly into the processor. This integration in the package can eliminate the serialization and deserialization of data and therefore provide a compact platform for integration with higher data bandwidth. In synchronous systems that support large ICs, a major problem is the clock skew between various logic circuits on silicon. A potential solution for such problems is the use of embedded optical clock distribution in the package, which is immune to most noise sources [32–41].

The SOC, MCM, and SIP described above have one major shortcoming. They extend Moore's law in two or three dimensions. They address only 10 to 20 percent of system needs and depend on CMOS only for system functions and on packaging for interconnection only. This leads to bulky systems, not because of ICs but because of the lack of system miniaturization. This single-chip CMOS focus at the system level, over the long run, presents fundamental limits to digital systems and integration limits to RF and wireless systems. Thus, while CMOS is good for transistors and bits and certain other components, such as Power Amplifier (PA) and Low Noise Amplifier (LNA), it is not an optimal technology platform for certain other components such as antennas, MEMS, inductors, capacitors, filters, and waveguides.

The SOB, on the other hand, shows its strengths in those areas where SOC is weak but suffers in those areas such as electrical performance and power usage where SOC shines. The SIP is a good tradeoff between these two technologies, and at the same time it is at the heart of semiconductor companies and their need to manufacture as much silicon as possible to justify their wafer fabrication investments. In addition, the SIP addresses the wireless cell phone "sweet spot" application. Therefore, it is not surprising that almost all major IC companies are manufacturing these modules. The major weakness of SIP is that it addresses the system drivers at the module level only and not at a system level. The 80 to 90 percent of the system problems remains unanswered.

The SOP is an even better and more optimized system solution than SIP, as can be seen from Figure 1.24. It addresses at the IC level without compromise by means of both on-chip SOC integration and package-enabled SIP and 3D integration and at the system level by system miniaturization technologies such as power supplies, thermal structures, and passive components, as indicated previously in Figures 1.5 and 1.9b for digital, RF, optical, and sensor components. Unlike SOC, however, no performance compromises have to be made in order to integrate these disparate technologies since each technology is separately fabricated either in the IC or the package and subsequently integrated into the SOP system package. System design times are expected to be much shorter in the SOP concept, as it allows for greater flexibility with which to take advantage of emerging

System technology	Device	Package	System board	End system
SOB	IC	Bulky	Bulky	Bulky
SOC	SOC	Bulky	Bulky	Less bulky
MCM	IC	MCM	Bulky	Miniaturized
SIP	SOC	SIP	Bulky	Miniaturized
SOP	SOC	SIP	System integration →	Highly miniaturized nano-micro system

FIGURE 1.24 Size comparisons of the five system technologies.

technologies. Nevertheless, SOP must successfully overcome a different set of challenges, namely infrastructure and investment challenges.

1.8 Status of SOP around the Globe

SOP is the ability to integrate disparate technologies to achieve diverse functions into a single package, while maintaining a low profile and a small form factor supporting mixed IC technologies. The SOP accomplishes this by ultrahigh wiring densities with less than 5-μm lines and spaces, in multiple layers, and a variety of embedded ultrathin-film component integrations, achieving greater than 2500 components per square centimeter. In the SOP concept, this is accomplished by codesign and fabrication of digital, optical, RF, and sensor functions in both the IC and the system package, thus distinguishing between what function is accomplished best at the IC level and at the system package level. In this paradigm, ICs are viewed as being best for transistor density, while the system package is viewed as being best for system technologies that include certain front-end RF, optical, and digital-function integration.

Apart from Georgia Tech, SOP research is going on in various universities, research institutes, national labs, and in the research and development (R&D) divisions of various companies across the world. IBM; Sandia National Labs; Motorola; NCSU; and IMEC, Belgium, are actively involved in the embedded passives research. The Royal Institute of Technology (KTH) Sweden, KAIST, the University of Arkansas, and Alcatel are also working on SOP. IME Singapore has worked on optoelectronics mixed-signal SOP. The R&D in SOP is now global as indicated in Figure 1.25.

1.8.1 Opto SOP

The Institute of Microelectronics in Singapore has built an optoelectronic SOP intended for high-speed communications between a network and a home or office [42]. The approach involves optical circuits made of silicon. The system transmitted data at 1 gigahertz (GHz).

Intel has reported developments in silicon photonics, the technique of fabricating high-volume optical components in silicon using standard high-volume, low-cost silicon manufacturing techniques. In 2005, researchers at Intel demonstrated data transmission at 10 gigabits per second (Gbps) using a silicon modulator. Intel and the University of California Santa Barbara (UCSB) demonstrated an electrically driven hybrid silicon laser (Figure 1.26). This device successfully integrates the light-emitting capabilities of indium phosphide with the light-routing and low-cost advantages of silicon.

Introduction to the System-on-Package (SOP) Technology

MEMS
Raytheon, TI,
University of
Arkansas

Thermal
High heat flux-GE, HP, IBM,
SUN, Intel and Fujitsu
Portable-Nokia, Sony,
Motorola, Ericsson and Intel

WLP and assembly
Lead free solders
Amkor, IBM, National
Semiconductor, Fujitsu, K&S,
Unitive, Cookson

**Integrated high density wiring-EIT,
Asahi, IBM**, Ibiden, Matsushita, **Shipley**,
Northrup Grumman, Hitachi, Samsung,
 Materials-Promerus, Shipley, DuPont,
 DOW
 Coating-SCS (Meniscus)
 Curing-Lambda (Microwave)
 Metallization-**ATOTECH**, Technics,
 Shipley

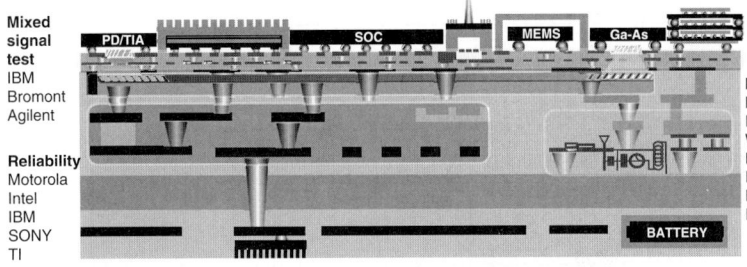

Mixed signal test
IBM
Bromont
Agilent

Reliability
Motorola
Intel
IBM
SONY
TI

Embedded opto
Back Plane-NTT, Siemens,
Infineon
Waveguides-Toray,
Kyocera, DuPont, Shipley,
DOW, GE
MSM Detectors-Fraunhoffer,
Kyocera

Signal and power integrity
EMI-NEC Corp., Toshiba
Power distribution-Sun, Cadence, Intel, and AMD
Signal integrity-IBM and Ansoft
Embedded decoupling-Sanmina, EIT, DuPont
Design tools-Cadence, Sun, Motorola, HRL, Rambus

Embedded RF SOP
Embedded passives-IBM, EKC-DuPont, Sanmina, 3M,
Boeing, Shipley, Motorola, Nokia, Intel, Amkar, Lucent, IMEC,
Kyocera
Antennaes-Asahi, DoD, NASA
Design tools-Cadence

FIGURE 1.25 R&D in SOP is now global.

Recently, IBM researchers have built an optical transceiver (Figure 1.27) in current CMOS technology and coupled it with other optical components, made with materials such as indium phosphide (InP) and GaAs, into a single integrated package only 3.25 by 5.25 mm in size. This compact design provides both a high number of communications channels as well as very high speeds per channel. This transceiver chipset is designed to enable low-cost optics by attaching to an optical board employing densely spaced polymer waveguide channels using mass assembly processes. According to IBM, this

FIGURE 1.26 Hybrid silicon laser. (*Courtesy:* Intel)

FIGURE 1.27 Optical transceiver developed by IBM. (*Courtesy:* IBM)

prototype optical transceiver chipset is capable of reaching speeds at least eight times faster than traditional discrete optical components available today.

1.8.2 RF SOP

At the Interuniversity Microelectronics Center (IMEC), in Leuven, Belgium, Robert Mertens and colleagues are studying the best type of RF antenna to build in an SOP for a range of wireless communications products yet to be introduced. IBM has developed a small, low-cost chipset that could allow wireless electronic devices to transmit and receive 10 times faster than today's advanced WiFi networks. The embedding of the antennas directly within the package helps reduce the system cost since fewer components are needed. A prototype chipset module, including the receiver, transmitter, and two antennas, would occupy the area of a dime. By integrating the chipset and antennas in commercial IC packages, companies can use existing skills and infrastructure to build this technology into their commercial products.

FIGURE 1.28 Wireless chipset module developed by IBM. (*Courtesy:* IBM)

1.8.3 Embedded Passives SOP

The University of Arkansas, in Fayetteville, has developed techniques for burying capacitors, resistors, and inductors in the layers of its SOP board. The university determined that almost all the resistance and much of the capacitance needed for a system can be embedded in the board using vacuum-deposition processes typical of the IC industry.

An example of volume production with embedded passives is Motorola's C650 Triband GSM/GPRS and V220 handsets. Motorola, working with AT&S, WUS, and Ibiden, introduced these handsets with embedded components to the market in June 2004. Motorola's embedded capacitor is fabricated by ceramic-polymer thick-film composite technology [ceramic-filled polymer (CFP) composite] with laser via connection (Motorola has IPs on this structure) with 20- to 450-picofarad (pF) capacitance, 15 percent tolerance, Breakdown Voltage (BDV) > 100 volts (V), Q factor of 30 to 50, and tested up to 3 GHz. Motorola also developed embedded inductor technology with 22-nanohenries (nH) inductance with 10 percent tolerance and resistor technology with 10 megaohms, 15 percent tolerance, trimmed to 5 percent.

A survey done in Japan in May 2005 of all printed wiring board (PWB) and package companies indicated that nondiscrete embedded capacitors were in production by a number of companies since 2004 and were expected to expand rapidly. The other embedded resistors and inductors, either as embedded discrete or as thin film, are already in near production in 2006. The same study shows that the embedded actives are also aimed for production starting from 2006. The survey also indicates that in a 5-year time frame, the embedded actives and passives (EMAP) market is expected to expand tremendously. We believe this growth will be in organic-based buildup of board or package substrate technologies.

There are several basic patents in embedded components technology. They range from thin-film embedding of capacitors, resistors, and inductors to embedding of discrete components. Patents on thin- and thick-film type embedded capacitors have been a hot issue recently. Sanmina-SCI owns U.S. Patent No. 5,079,069 filed in January 1992 and claims the technology for embedded capacitors.

1.8.4 MEMS SOP

A parts synthesis approach (PSA) for 3D integration of MEMS and microsystems leading to system-on-package has been developed at Malaviya National Institute of Technology, in Jaipur, India. This eliminates the interconnection-related problems that arise when MEMS and its associated circuitry are packaged separately.

Amkor has developed solutions that combine multiple chips, MEMS devices, and passives into one package. These solutions are aimed at reducing the cost of MEMS packaging and increasing functionality through greater levels of integration.

1.9 SOP Technology Implementations

SOP is an emerging concept and has been demonstrated so far for limited applications including the mezzanine capacitor in Motorola's cell phone (Figure 1.31), in a conceptual broadband system called an intelligent network communicator (INC) at Georgia Tech (Figure 1.29a and b), and at Intel (Figure 1.32).

The INC testbed acted as both a leading-edge research and teaching platform in which students, faculty, research scientists, and industry evaluate the validity of SOP

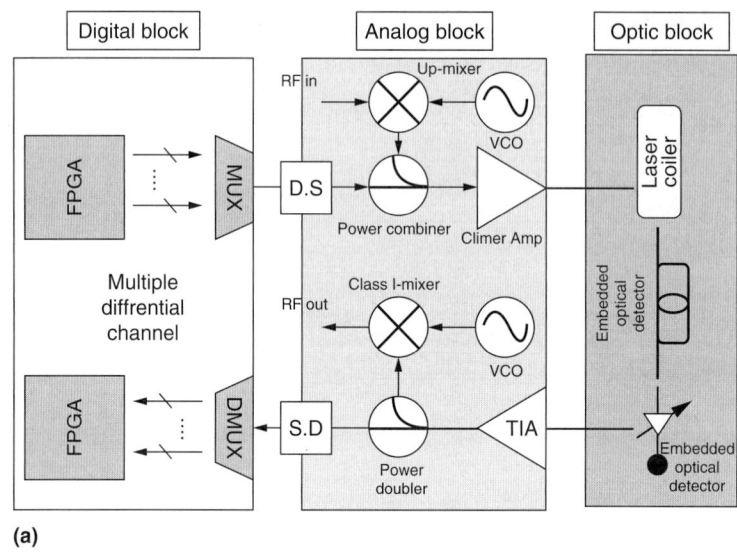

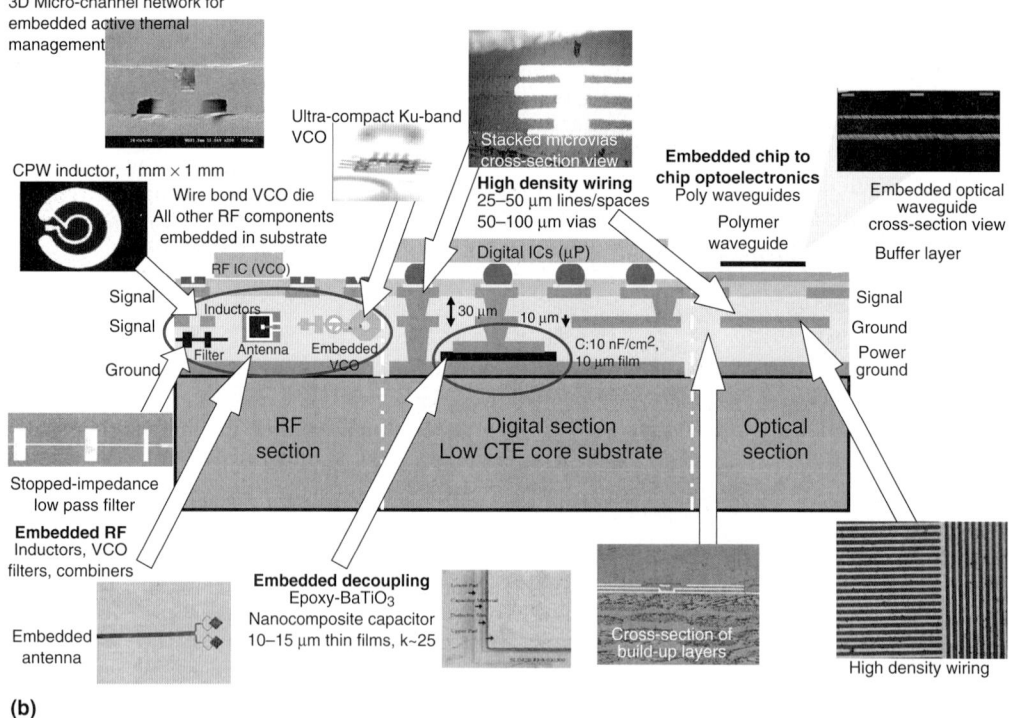

Figure 1.29 (a) A conceptual broadband system called the intelligent network communicator (INC), developed at Georgia Tech. (b) A cross-sectional view of the INC.

technology from design to fabrication to integration, test, cost, and reliability. The testbed explored optical bit stream switching up to 100 GHz; digital signals up to 5 to 20 GHz; decoupling capacitor integration concepts to reduce simultaneous switching noise of power beyond 100 W/chip; design, modeling, and fabrication of embedded components for RF, microwave, and millimeter wave applications up to 60 GHz.

So far, at least 50 companies have taken parts of the SOP technology developed at the Georgia Institute of Technology's Packaging Research Center (PRC) and applied them to their automotive, computer, consumer, military, and wireless applications. A number of test vehicles have also been built over the years for different companies focused on integrating different combinations of analog, digital, RF, optical, and sensor components in a single package.

Japanese companies, such as Ibiden, Shinko, Matsushita, Casio, and NEC, have been active in R&D in EMAP technology for more than 5 years. Casio and Matsushita have already demonstrated embedded passives and IC components in laminate layers. They started this research around 1998–2000. One example of Matsushita's SIMPACT technology developed in 2001 is shown in Figure 1.30 where discrete passives and actives are embedded in dielectric layers. Matsushita indicated that its embedding program uses discretes but will migrate to thin films as the company perfects manufacturing.

In the United States, Intel has been active in EMAP for its RF modules and digital applications and is expected to appear with EMAP products in the market in 2 to 3 years. Companies like 3M and Oak-Mitsui have thin-film capacitor technologies ready for production. GE has been a big player in embedded actives technology for a long time and is now focusing on embedded passives to go with existing embedded active technology. TI is beginning to be a big contender in this research and business. Even in the automotive industry, companies like Delphi are interested in EMAP technology. There is a big interest in Europe too, such as by Nokia.

Motorola uses parts of SOP technology in two models of its GSM/General Packet Radio Service quad-band cell phones to gain about a 40 percent reduction in board area. The module contains all the critical cell phone functions: RF processing, base-band signal processing, power management, and audio and memory sections. Not only does the module free up space for new features, it is also the base around which new cell phones with different shapes and features (camera or Bluetooth, for instance) can be rapidly designed. Motorola calls it a system-on-module (SOM), for which it developed its own custom embedded-capacitor technology. It reports it has shipped more than 20 million SOM-based phones.

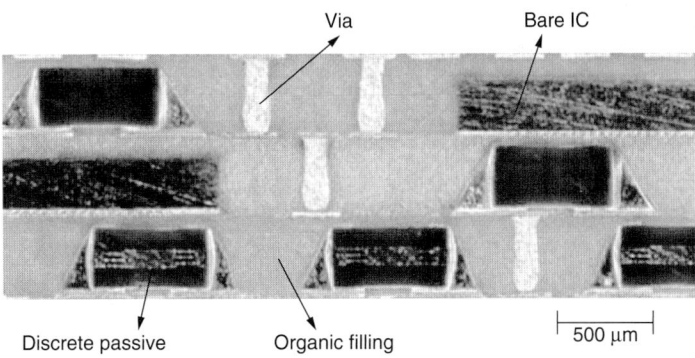

FIGURE 1.30 Matsushita SIMPACT with embedded discrete passives and actives developed in 2001.

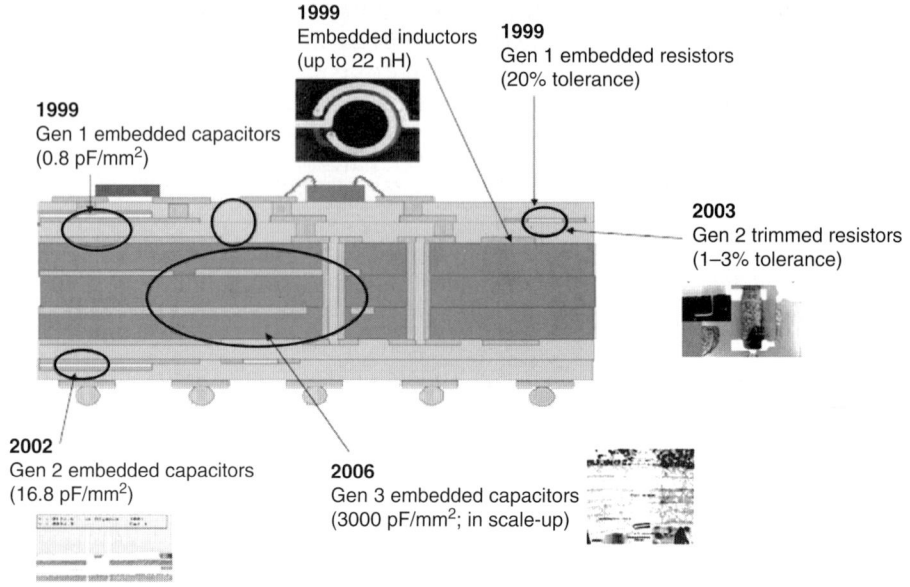

Figure 1.31 SOP technology in production at Motorola. (*Courtesy:* Motorola)

Motorola has been a global leader in both the R&D and manufacturing implementation of RF passives (Figure 1.31). Its first generation of RF capacitor passives was used in its cell phones in the 1999 time frame. The second generation of passives was improved for not only capacitance density but also for process tolerances around 2002. Ferroelectric thin film capacitors are under development in Motorola.

Intel has also reported a 43 percent reduction in the form factor along with increased functionality in its wireless local area network (WLAN) solution (Figure 1.32)

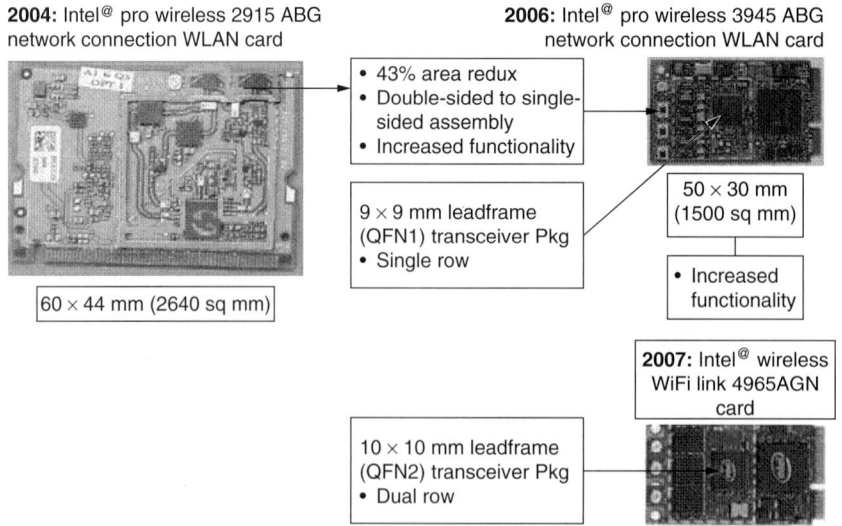

Figure 1.32 SOP implementation in Intel's WLAN and wireless WiFi link cards. [43]

implementation by adopting a top-down approach to the system design, application of self-calibration schemes, modularity approach in RFIC design, and the use of custom board and front-end (FE) elements to reduce the part count [44].

1.10 SOP Technologies

The SOP concept seeks to integrate multiple system functions into one compact, lightweight, low-cost, and high-performance package or module system. Such a system design may call for high-performance digital, RF, optical, and sensor functions as indicated in Figure 1.33 in the SOP concept.

The technologies involved in the SOP concept have been outlined in the different chapters of this book:

- Introduction to the System-on-Package (SOP) Technology (Chapter 1)
- System-on-Chip (Chapter 2)
- Stacked ICs and Packages (SIP) (Chapter 3)
- Mixed-Signal (SOP) Design (Chapter 4)
- RF SOP (Chapter 5)
- Optoelectronics SOP (Chapter 6)
- SOP Substrate (Chapter 7)
- Mixed-Signal Reliability (Chapter 8)
- MEMS (Chapter 9)
- Wafer-Level SOP (Chapter 10)
- Thermal SOP (Chapter 11)
- SOP Electrical Test (Chapter 12)
- Biosensor SOP (Chapter 13)

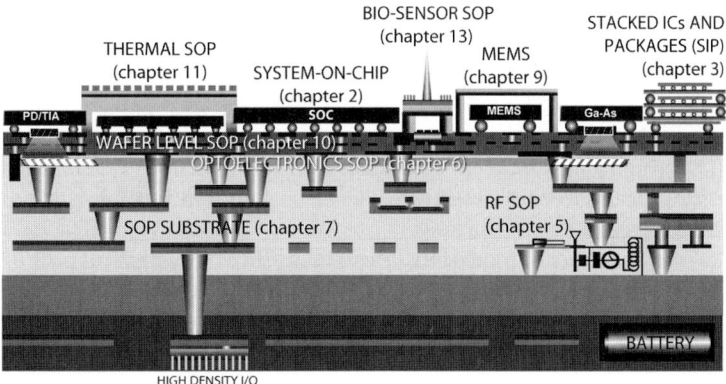

- Introduction to the System-on-Package (SOP) Technology (chapter 1)
- Mixed Signal (SOP) Design (chapter 4)
- Mixed Signal Reliability (chapter 8)
- Electrical Test of SOP Modules and Systems (chapter 12)

Figure 1.33 SOP includes all system building blocks: SOCs, SIPs, MEMS, embedded components in ICs and substrates, thermal structures, batteries, and system interconnections.

1.11 Summary

SOP is about system miniaturization enabled by IC and system integration by ultrathin-film components at microscale in the short term and nanoscale in the long term for all system components. Some of these thin-film system technologies that SOP enables can be used in CMOS ICs as overlays, as thin films on top of silicon wafers (TFOS) and silicon carriers, or on ceramic and glass substrates or embedded into multilayer ceramic or organic laminate packages and boards.

SIP is defined in this book as the stacking of ICs and packages. But since SIP is also often referred to as a total system technology that miniaturizes and integrates all system components such as passives, actives, thermal structures, power sources, and I/Os, if this happens, then SOP and SIP are identical. But so far, this has not been demonstrated.

Acknowledgments

The authors gratefully thank the Georgia Tech PRC team of faculty, engineers, students, and industry advisors for their contributions in the development of the SOP technology. The authors also thank both the Georgia Research Alliance and the National Science Foundation Engineering Research Centers for their funding of SOP technology for more than a decade.

References

1. R. R. Tummala et al., *Ceramic Packaging Technology, Microelectronics Packaging Handbook*. New York: Van Nostrand, 1988.
2. Y. Yano, T. Sugiyama, S. Ishihara, Y. Fukui, H. Juso, K. Miyata, Y. Sota, and K. Fhjita, "Three dimensional very thin stacked packaging technology for SiP," in *Proc. 52nd Electronic Components and Technology Conference*, 2002.
3. K. Lim, M. F. Davis, M. Maeng, S. Pinel, L. Wan, L. Laskar, V. Sundaram, G. White, M. Swaminathan, and R. Tummala, "Intelligent network communicator: Highly integrated system-on-package (SOP) testbed for RF/digital/opto applications," in *Proc. 2003 ElectronicComponents and Technology Conference*, pp. 27–30.
4. R. Tummala, "SOP: Microelectronic systems packaging technology for the 21st century," *Adv. Microelectron.*, vol. 26, no. 3, May–June 1999, pp. 29–37.
5. R. Tummala, G. White, V. Sundaram, and S. Bhattacharya, "SOP: The microelectronics for the 21st century with integral passive integration," *Adv. Microelectron.*, vol. 27, 2000, pp. 13–19.
6. R. Tummala and V. Madisetti, "System on chip or system on package," *IEEE Design Test Comput.*, vol. 16, no. 2, Apr. –June 1999, pp. 48–56.
7. R. Tummala and J. Laskar, "Gigabit wireless: System-on-a-package technology," *Proc. IEEE*, vol. 92, Feb. 2004, pp. 376–387.
8. ITRS 2006 Update
9. H. K. Kwon et al., "SIP solution for high-end multimedia cellular phone," in *IMAPS Conf. Proc.*, 2003, pp. 165–169.
10. S. S. Stoukatch et al., "Miniaturization using 3-D stack structure for sip application," in *SMTA Proc.*, 2003, pp. 613–620.

11. T. Sugiyama et al., "Board level reliability of three-dimensional systems in packaging," in *ECTC Proceedings*, 2003, pp. 1106–1111.
12. K. Tamida et al., "Ultra-high-density 3D chip stacking technology," in *ECTC Proc.*, 2003, pp. 1084–1089.
13. Toshihiro Iwasaki, Masaki Watanabe, Shinji Baba, Yasumichi Hatanaka, Shiori Idaka, Yoshinori Yokoyama, and Michitaka Kimura, "Development of 30 micron Pitch Bump Interconnections for COC-FCBGA," *Proceedings IEEE 56th Electronic Components and Technology Conference*, 2006, pp. 1216–1222.
14. D. J. Bodendorf, K. T. Olson, J. P. Trinko, and J. R. Winnard, "Active Silicon Chip Carrier," *IBM Tech. Disclosure Bull.* vol. 7, 1972, p. 656.
15. Vaidyanathan Kripesh et al., "Three-Dimensional System-in-Package Using Stacked Silicon Platform Technology," *IEEE Transactions on Advanced Packaging*, vol. 28, no. 3, August 2005, pp. 377–386.
16. Marcos Karnezos and Rajendra Pendse, "3D Packaging Promises Performance, Reliability Gains with Small Footprints and Lower Profiles," *Chip Scale Review*, January/February 2005.
17. R. R. Tummala, "Moore's law meets its match (system-on-package)," *Spectrum*, IEEE, vol. 43, issue 6, June 2006, pp. 44–49.
18. R. Tummala and J. Laskar, "Gigabit wireless: System-on-a-package technology," *Proc. IEEE*, vol. 92, Feb. 2004, pp. 376–387.
19. M. F. Davis, A. Sutono, A. Obatoyinbo, S. Chakraborty, K. Lim, S. Pinel, J. Laskar, and R. Tummala, "Integrated RF architectures in fully-organic SOP technology," in *Proc. 2001 IEEE EPEP Topical Meeting*, Boston, MA, Oct. 2001, pp. 93–96.
20. K. Lim, A. Obatoyinbo, M. F. Davis, J. Laskar, and R. Tummala, "Development of planar antennas in multi-layer package for RF-system on-a-package applications," in *Proc. 2001 IEEE EPEP Topical Meeting*, Boston, MA, Oct. 2001, pp. 101–104.
21. R. L. Li, G. DeJean, M. M. Tentzeris, and J. Laskar, "Integration of miniaturized patch antennas with high dielectric constant multilayer packages and soft-and-hard surfaces (SHS)," in *Conf. Proc. 2003 IEEE-ECTC Symp.*, New Orleans, LA, May 2003, pp. 474–477.
22. R. L. Li, K. Lim, M. Maeng, E. Tsai, G. DeJean, M. Tentzeris, and J. Laskar, "Design of compact stacked-patch antennas on LTCC technology for wireless communication applications," in *Conf. Proc. 2002 IEEE AP-S Symp.*, San Antonio, TX, June 2002, pp. II. 500–503.
23. M. F. Davis, A. Sutono, K. Lim, J. Laskar, V. Sundaram, J. Hobbs, G. E. White, and R. Tummala, "RF-microwave multi-layer integrated passives using fully organic system on package (SOP) technology," in *IEEE Int. Microwave Symp.*, vol. 3, Phoenix, AZ, May 2001, pp. 1731–1734.
24. K. Lim, M. F. Davis, M. Maeng, S.-W. Yoon, S. Pinel, L. Wan, D. Guidotti, D. Ravi, J. Laskar, M. Tentzeris, V. Sundaram, G. White, M. Swaminathan, M. Brook, N. Jokerst, and R. Tummala, "Development of intelligent network communicator for mixed signal communications using the system-on-a-package (SOP) technology," in *Proc. 2003 IEEE Asian Pacific Microwave Conf.*, Seoul, Korea, Nov. 2003.
25. M. F. Davis, A. Sutono, K. Lim, J. Laskar, and R. Tummala, "Multi-layer fully organic-based system-on-package (SOP) technology for rf applications," in *2000 IEEE EPEP Topical Meeting*, Scottsdale, AZ, Oct. 2000, pp. 103–106.
26. M. Alexander, "Power distribution system (PDS) design: Using bypass/decoupling capacitors," in *XAPP623 (v. 1. 0)*, Aug. 2002.

27. J. M. Hobbs, S. Dalmia, V. Sundaram, V. L. Wan, W. Kim, G. White, M. Swaminathan, and R. Tummala, "Development and characterization of embedded thin-film capacitors for mixed signal applications on fully organic system-on-package technology," in *Radio and Wireless Conf. Proc., 2002. RAWCON*, Aug. 11–14, 2002, pp. 201–204.
28. M. F. Davis, A. Sutono, S.-W. Yoon, S. Mandal, N. Bushyager, C. H. Lee, L. Lim, S. Pinel, M. Maeng, A. Obatoyinbo, S. Chakraborty, J. Laskar, M. Tentzeris, T. Nonaka, and R. R. Tummala, "Integrated RF architectures in fully-organic SOP technology," *IEEE Trans. Adv. Packag.*, vol. 25, May 2002, pp. 136–142.
29. R. Ulrich and L. Schaper, "Decoupling with embedded capacitors," *CircuiTree*, vol. 16, no. 7, July 2003, p. 26.
30. A. Murphy and F. Young, "High frequency performance of multilayer capacitors," *IEEE Trans. Microwave Theory Tech.*, vol. 43, Sept. 1995, pp. 2007–2015.
31. R. Ulrich and L. Schaper, eds., *Integrated Passive Component Technology*. New York: IEEE/Wiley, 2003.
32. D. A. B. Miller, "Rationale and challenges for optical interconnects to electronic chips," *IEEE Proc.*, vol. 88, 2000, pp. 728–749.
33. S. -Y. Cho and M. A. Brooke, "Optical interconnections on electrical boards using embedded active optoelectronic components," *IEEE J. Select. Top. Quantum Electron.*, vol. 9, 2003, p. 465.
34. Z. Huang, Y. Ueno, K. Kaneko, N. M. Jokerst, and S. Tanahashi, "Embedded optical interconnections using thin film InGaAs MSM photodetectors," *Electron. Lett.*, vol. 38, 2002, p. 1708.
35. R. T. Chen, L. L. C. Choi, Y. J. Liu, B. Bihari, L. Wu, S. Tang, R. Wickman, B. Picor, M. K. Hibbs-Brenner, J. Bristow, and Y. S. Liu, "Fully embedded board-level guided-wave optoelectronic interconnects," *IEEE Proc.*, vol. 88, 2000, p. 780.
36. J. J. Liu, Z. Kalayjian, B. Riely, W. Chang, G. J. Simonis, A. Apsel, and A. Andreou, "Multichannel ultrathin silicon-on-sapphire optical interconnects," *IEEE J. Select. Top. Quantum Electron.*, vol. 9, 2003, pp. 380–386.
37. H. Takahara, "Optoelectronic multichip module packaging technologies and optical input/output interface chip-level packages for the next generation of hardware systems," *IEEE J. Select. Top. Quantum Electron.*, vol. 9, 2003, pp. 443–451.
38. X. Han, G. Kim, G. J. Lipovaski, and R. T. Chen, "An optical centralized shared-bus architecture demonstrator for microprocessor-to-memory interconnects," *IEEE J. Select. Top. Quantum Electron.*, vol. 9, 2003, pp. 512–517.
39. H. Schroeder, J. Bauer, F. Ebling, and W. Scheel, "Polymer optical interconnects for PCB," in *First Int. IEEE Conf. Polymers and Adhesives in Microelectronics and Photonics. Incorporating POLY, PEP and Adhesives in Electronics*, Potsdam, Germany, Oct. 21–24, 2002, p. 3337.
40. M. Koyanagi, T. Matsumoto, T. Shimatani, K. Hirano, H. Kurino, R. Aibara, Y. Kuwana, N. Kuroishi, T. Kawata, and N. Miyakawa, "Multi-chip module with optical interconnection for parallel processor system," in *IEEE Int. Solid-State Circuits Conf. Proc.*, San Francisco, CA, Feb. 5–7, 1998, pp. 92–93.
41. T. Suzuki, T. Nonaka, S. Y. Cho, and N. M. Jokerst, "Embedded optical interconnects on printed wiring boards," in *Conf. Proc. 53th ECEC*, 2003, pp. 1153–1155.
42. Mahadevan K. Iyer et al., "Design and development of optoelectronic mixed signal system-on-package (SOP)," in *IEEE Transactions on Advanced Packaging*, vol. 27, no. 2, May 2004, pp. 278–285.

43. Lesley A. Polka, Rockwell Hsu, Todd B. Myers, Jing H. Chen, Andy Bao, Cheng-Chieh Hsieh, Emile Davies-Venn, and Eric Palmer, "Technology options for next-generation high pin count RF packaging," *2007 Electronic Components and Technology Conference*, pp. 1000–1006.
44. M. Ruberto, R. Sover, J. Myszne, A. Sloutsky, Y. Shemesh, "WLAN system, HW, and RFIC architecture for the Intel pro/wireless 3945ABG network connection," *Intel Technology Journal*, vol. 10, issue 2, 2006, pp. 147–156.

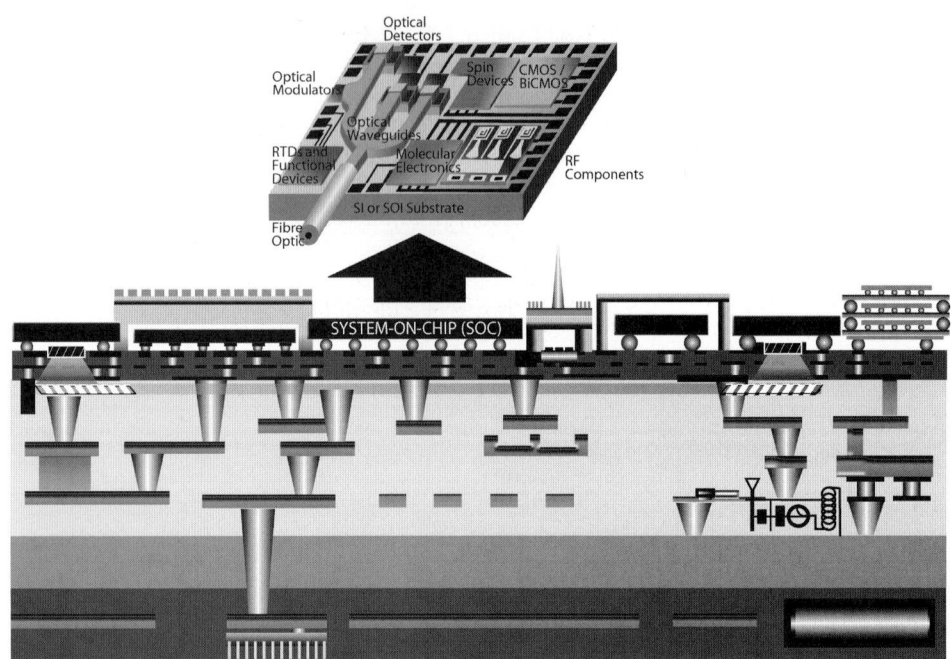

CHAPTER 2
Introduction to System-on-Chip (SOC)

Mahesh Mehendale and Jagdish Rao
Texas Instruments

2.1	Introduction 40	2.4	SOC Design Challenge 50
2.2	Key Customer Requirements 42	2.5	Summary 76
2.3	SOC Architecture 44		References 76

The semiconductor industry has been fueled by Moore's law where the number of transistors in a microprocessor has been doubling every 18 to 24 months. With the possibility of integrating a billion transistors within a single chip, various methodologies are being developed for system-on-chip (SOC) integration. Unlike pure digital systems, the need for heterogeneous integration in the system is becoming important due to the need for mobility-enabled devices. This is creating new challenges for SOC implementation. In this chapter, the customer requirements for a new class of application-specific devices that support mobility are discussed. Issues such as electr-omagnetic interference (EMI), soft errors, environmental concerns, and fault tolerance that affect such systems from an SOC standpoint are discussed. The customer require-ments lead to SOC architectures containing embedded processor cores and multiple cores within the processor. The role of leakage power and the use of multiple threshold voltage libraries along with hardware and software codesign concepts are discussed. This is followed by a discussion on SOC design challenges that include the need for chip-package codesign and hierarchical design flow. The challenges posed by hetero-geneous integration are also discussed warranting an SOP approach presented in this book.

2.1 Introduction

With the advances in semiconductor technology, the number of transistors that can be integrated on a single chip continues to grow. This trend, represented by Moore's law (the number of devices on a chip will double every 18 to 24 months), is projected to hold true through 2010 and beyond per the International Technology Roadmap on Semiconductors (ITRS) [1]. The increasing level of integration enables the implementation of electronic systems, which were earlier implemented using multiple chips on a board, on a single chip—called "system-on-a-chip."

The definition of SOC is thus evolving, where with each generation more and more system components are integrated on a single device. As an example, consider the digital subscriber line (DSL) modem system evolution through three generations, as shown in Figure 2.1.

From an initial system consisting of five chips, memory, and other discrete components, the next-generation DSL solution integrated the analog codec, line driver, and line receiver into a single analog front end (AFE), and in the following generation, the integration was taken even more forward with the communications processor, digital PHYsical layer, and AFE all integrated onto a single-chip digital signal processor (DSP) modem SOC. This journey continues even further, as the system itself evolves along with the SOC evolution. The DSL system needs to provide voice and video

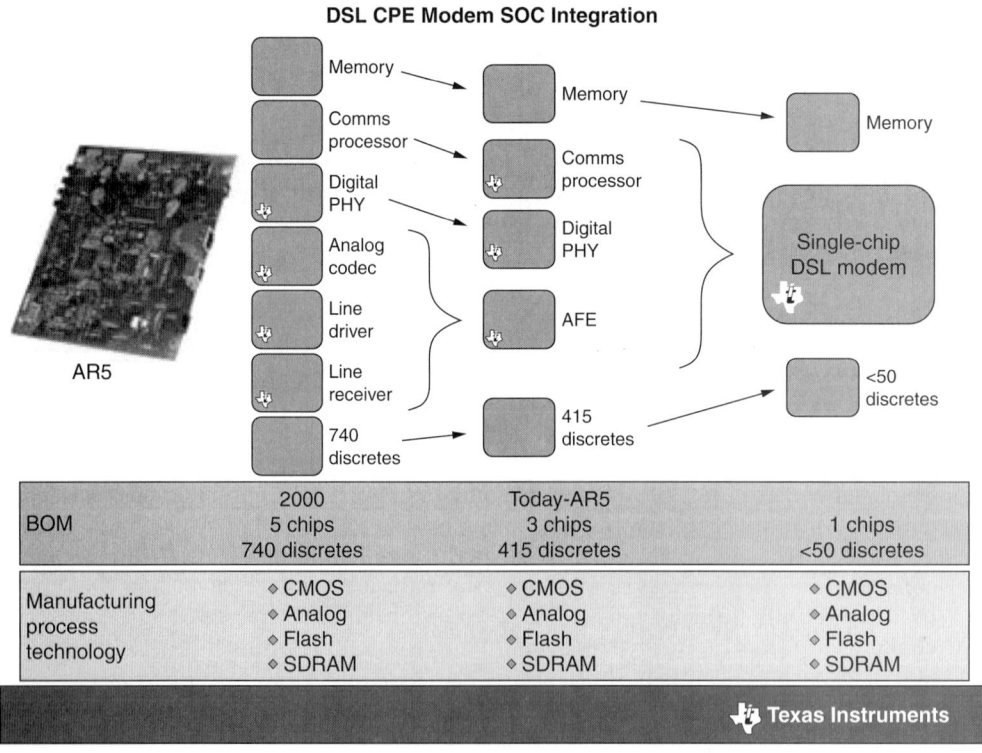

Figure 2.1 Single-chip DSL modem system-on-a-chip.

capabilities as well, and that is driving the next level of SOC integration. Moving forward, the system will evolve into a "triple-play (data, voice, and video) residential gateway" and that will drive SOCs that will integrate wireless LAN (IEEE 802.11) components along with the DSL modem and voice and video processing engines.

Among the important targets for SOCs are the applications fueled by and fueling the Internet era. These applications can be characterized as a convergence of communication (both wireless and broadband wire-line) and consumer (digital multimedia content). These applications also include domains such as telematics that are driving the convergence in automotive space. Figure 2.2 shows a spectrum of these applications.

Across these applications, signal processing is a key common function, and DSP and analog form the key building blocks of these SOCs. In this chapter we will focus on such SOCs, which are built using CMOS technology.

The rest of the chapter is organized as follows. We will start by discussing customer requirements (cost, low power, performance, form factor, etc.) and highlight how SOCs address them by integrating application-specific intellectual properties (IPs) and embedded processor cores. We will present examples of such SOCs to illustrate this. We will then present SOC design as a multidimensional optimization problem and discuss how it can be addressed using concurrent engineering (hardware-software codesign, chip-package codesign, etc.). While CMOS technology scaling enables higher levels of integration, it poses unique challenges for SOC implementation. We will highlight these implications and conclude by discussing trends that will look at optimal system partitioning and hence link to SIPs and SOPs.

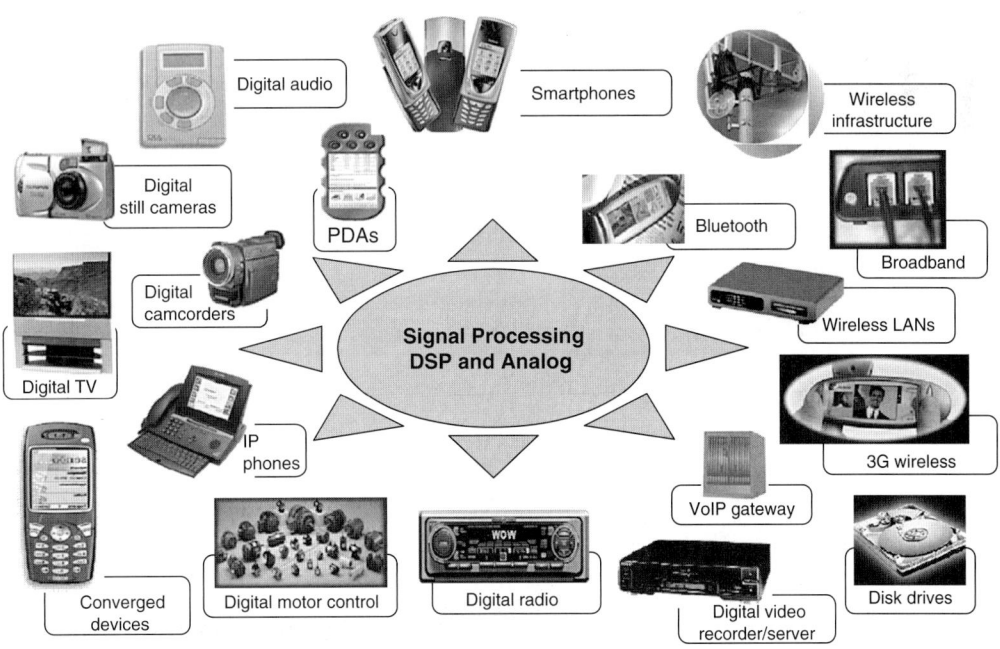

FIGURE 2.2 SOC applications of the Internet era.

2.2 Key Customer Requirements

Before we get into the specifics of SOC architecture and SOC development process, it is important to understand how SOCs address the key customer requirements (Figure 2.3) across these applications. These include:

1. **Cost** While it is obvious that a customer cares for lower cost, it is important to note that the cost applies to the bill of materials (BOM) of the entire system, as opposed to the cost of the SOC chip alone. For example, consider two scenarios for a system that performs a data-intensive application such as video and image processing and hence is built with an SOC and a large amount of off-chip memory. In one case, the external memory interface of the SOC needs to operate at 100 MHz, as opposed to 133 MHz in the other case, to be able to achieve the desired system throughput. The SOC that operates with a 100-MHz interface will need to employ microarchitectural options such as a wider interface (64 bit versus 32 bit) or a higher on-chip memory, which can result in a higher SOC cost. However, at a system level, a 100-MHz interface allows the use of memories with a lower speed grade, which are significantly cheaper than the memories required to interface with a 133-MHz interface. Thus at a system level, the solution that uses a marginally expensive SOC can turn out to be more cost efficient. Later in this chapter we will discuss how such system- and board-level considerations can be comprehended during the SOC definition phase.

2. **Power dissipation** Power dissipation is increasingly becoming a key concern for portable devices such as mobile phones, personal digital assistants (PDAs), digital still cameras, and MP3 players because lower power translates to longer battery life. As mobile phones move from second generation (2G) to 2.5G to 3G, the computing requirements are increasing at a rapid pace, and with that the dynamic and switching power dissipation is also increasing. Battery technology is progressing in terms of energy per dollar, energy per weight, energy per volume, and so forth, but at a relatively slower pace. This is making low power an increasingly important requirement. While deep submicron CMOS technology

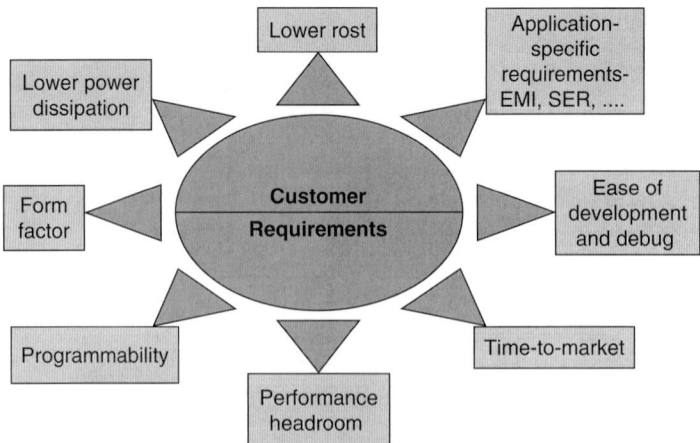

FIGURE 2.3 Customer requirements.

enables the performance and level of integration required for the 3G application, with each new process technology node, the leakage power is also increasing significantly. Since this impacts the standby time, an important consideration for these mobile applications, SOC designers need to employ aggressive power management techniques to reduce leakage power dissipation.

Power dissipation in the "standby mode" is an important requirement for automotive applications as well, where a small component of the system needs to be running continuously even when the car is switched off.

In case of infrastructure devices such as wireless base stations, DSL central offices, and Cable Modem Termination Systems (CMTS), the system employs arrays of SOCs to be able to support thousands of communication channels. The power per channel is hence an important metric for these applications. While performance is the key optimization vector for these infrastructure devices, the performance needs to be pushed while taking the power constraints into consideration.

3. **Form factor** The form factor is an important consideration for handheld portable devices such as mobile handsets, MP3 players, and PDAs. These applications require the system electronics to take up as little board area as possible. This not only drives SOC integration leading to a reduced number of devices on the board but also drives aggressive packaging technologies (such as wafer-scale packaging) to minimize the SOC chip area itself. These applications and other applications, such as those that require a PCMCIA (Personal Computer Memory Card International Association) form factor, demand constraints on the thickness as well. In infrastructure applications, where the system employs arrays of SOCs on a board and multiple boards are built into a rack, the form factor is again an important consideration as it drives the number of channels supported per square inch of board area.

4. **Programmability and performance headroom** In applications where the same devices perform different functions (for example, multifunction devices that operate as a printer-scanner-copier-fax), programmability enables the same hardware to be used to efficiently implement these different functions. Programmability is also required for applications that need to support multiple standards such as, for example, in the video domain where in addition to MPEG2, MPEG4, H.263, there are applications that use proprietary standards as well. For applications where the standards are evolving, programmability is again very valuable as the standards can be supported primarily through software upgrades. The programmability also allows customization, differentiation, and value-added capabilities over the baseline functionality of the system. This customization hence demands the appropriate performance headroom to be able to provide additional capabilities while still meeting the performance requirements of the base functionality. While the programmability is primarily supported by embedding programmable processor cores into the system, hardware programmability is also feasible using field programmable gate array (FPGA) technology.

5. **Time-to-market, ease of development, and debug** In most markets, being the first to introduce a system enables a higher market share and higher margins. This implies that the hardware should be robust to be able to ramp to volume

production quickly and also should provide the necessary hooks to be able to do a quick debug. Most customers also expect that the hardware be bundled with a reference design, lower-level software drivers, and algorithm kernels so that it significantly reduces the development cycle time.

6. **Application-specific requirements** In addition to the above-mentioned requirements, which generically apply to most applications, there are certain requirements that are unique to the specific domain.

   6.1. Electromagnetic interference (EMI) is a key issue in the automotive market and also for mobile handsets. These applications require that the radiation from the device should be under a specified limit and typically specify frequency bands in which the radiation limits are stringent.

   6.2. Soft errors are the transient defects caused during the operation of the device. The most common form of a soft error is flipping of a memory bit. The severity of this error can vary depending on whether the impacted memory contains a program or data. In applications that use a large amount of memory and have stringent robustness requirements, the soft error rate (SER) needs to be managed by providing on-line bit error detection and correction mechanisms.

   6.3. The requirement of building lead-free devices is driven primarily by environmental considerations and is increasingly becoming a mandatory requirement across most markets—especially in Europe and Japan. This primarily drives changes in the packaging technology where the bumps with lead content have been used as they enable processing at a lower temperature than the lead-free bumps.

   6.4. Automotive applications demand systems to have near-zero defective parts per million (DPPM). With the complexity of systems going up both in terms of number of transistors and performance, achieving 0 DPPM is getting increasingly challenging.

   6.5. For mission-critical applications and also infrastructure-type applications with stringent downtime requirements, fault tolerance is an important requirement. Fault tolerance is the ability to detect faults (either transient or permanent) occurring while the device is in operation and the ability to continue to function correctly in the presence of the fault. The fault tolerance requirements are typically addressed by providing redundancy at the SOC and/or at the system level.

2.3 SOC Architecture

Figure 2.4 shows a generic SOC architecture in terms of the key building blocks. These include embedded processor core(s) with associated data and program memory, application-specific hardware accelerators and coprocessors, customer-specific IP, industry-standard interfaces, external memory controllers, and analog or RF IPs.

SOCs address the key customer requirements through the following:

1. **Level of integration** Because SOC involves integrating multiple chips on a single device, the cost of the single device is typically less than the cost of multiple chips. Since it reduces the number of devices on the board, the resultant

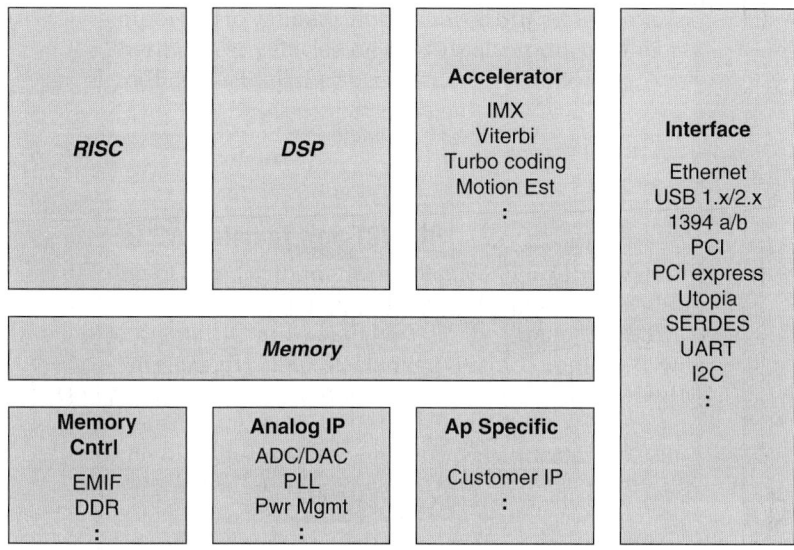

FIGURE 2.4 SOC architecture.

system implementation is simpler (reduced time-to-market) and also enables a smaller form factor. The off-chip interconnects in a "multiple chips on a board" system are replaced by on-chip interconnects within an SOC. This results in a significantly reduced switched capacitance and, hence, lesser power dissipation. It also helps improve the performance, as the interconnect delays across the chip boundaries are significantly higher than the on-chip interconnect delays.

The single-chip DSL modem shown in Figure 2.1 is a good example of how the increasing level of integration has helped reduce cost and the system development cycle time.

Wireless handset electronics has over the years gone through the SOC evolution, where with each generation more and more of the chips on a board are getting integrated into a single device. From the previous-generation four-chip solution, the current-generation single-chip solution has (a) a digital baseband and application processing, (b) a digital RF, (c) an analog baseband and power management, and (d) a static random-access memory (SRAM) integrated with (e) a nonvolatile memory either embedded or stacked. This level of integration is key to reducing the cost, power dissipation, and form factor—all critical requirements in this market.

This SOC evolution in the wireless handset will continue as more and more functionality is becoming integrated, including functions such as digital cameras, wireless local area network (WLAN) and global positioning system (GPS) connectivity, and digital TV functionality. Figure 2.5 shows the convergence of communications, connectivity, and applications on a handheld device. The increasing level of integration will enable a single-chip solution for such devices.

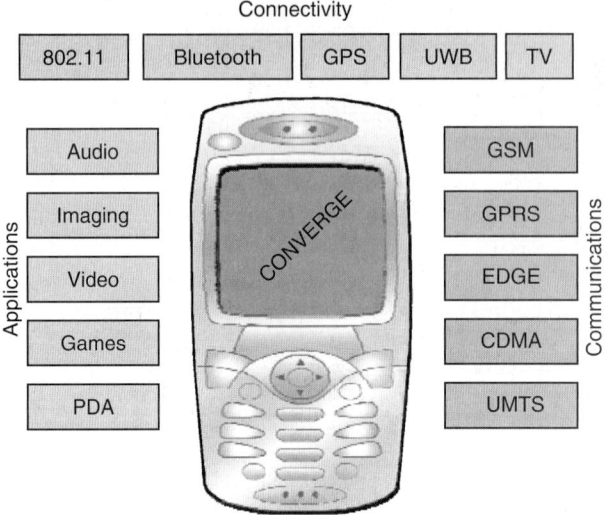

FIGURE 2.5 Convergence in a mobile handset.

2. **Application-specific IP** Because an SOC is targeted to a specific application domain, it enables building and integrating IP modules that perform the specific domain functions efficiently—in terms of performance, power, and die size. These application-specific IP blocks include hardware accelerators and coprocessors that perform some of the performance-critical but standardized functions. Some examples of such IPs include Viterbi and Turbo coprocessors that help significantly improve the number of channels per device in wireless infrastructure space. In the video space, a motion-estimation accelerator can help meet the frames-per-second performance requirement with optimum power and die size. In the digital still camera space, an image-processing pipeline is implemented as a dedicated hardware accelerator—to enable lower power while improving performance parameters such as shot-to-shot delay and picture resolution. The application-specific IPs also include application-specific interfaces such as video ports that conform to BT656 standards and hence can seamlessly interface with video encoder-decoders and multichannel audio serial ports that can directly talk to audio digital-to-analog converters (DACs).

Consider a high-performance audio system. Figure 2.6 shows an implementation based on Texas Instruments' TMS320C6711 general-purpose, 150-MHz floating-point processor. The next-generation system based on DA610 SOC has seven less devices—resulting in a lower cost (due to both a lower bill of material and also a lower cost of manufacturing). DA610 achieves this by on-chip integration of random-access memory (RAM) and read-only memory (ROM), a higher-performance floating-point processor (225 MHz) that eliminates the need for the microcontroller, and by providing multichannel audio serial ports (McASP)—application-specific peripherals with seamless interface with

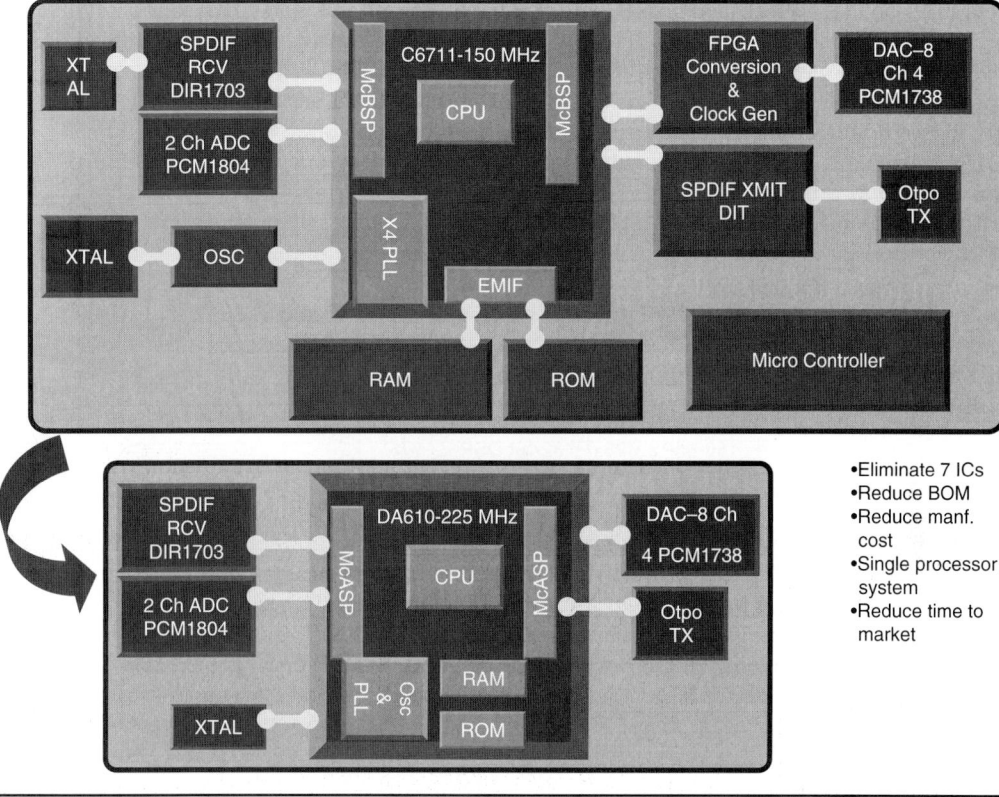

Figure 2.6 SOC for high-performance audio.

audio DACs. The single-processor system also makes the software development and debug simpler, thus enabling a faster time-to-market.

Figure 2.7 shows TMS320F2812—an SOC targeted for the embedded control market. It integrates a high-performance 32-bit digital signal processor (DSP) core customized for a control-type application, 128-kbytes of flash memory, a 12-bit analog-to-digital converter (ADC), and control-specific peripherals. This is another example of an SOC addressing key customer requirements of system cost, programmability, and time-to-market through level of integration and application-specific IPs.

3. **Embedded programmable processor cores** As mentioned earlier, embedded processor cores address the software programmability requirements. Since programmability comes at the expense of area, power, and performance, processor cores are customized and optimized for target application requirements. The customization is done in terms of instruction set architecture, functional units, pipelining, and memory management architectures. For control-dominated code, code size and interrupt latency requirements drive the customization, while for DSP applications, the performance for compute intensive kernels drives the optimization. Depending on the application requirements, SOCs embed one or more processor cores. The cycle time requirements typically drive

48 Chapter Two

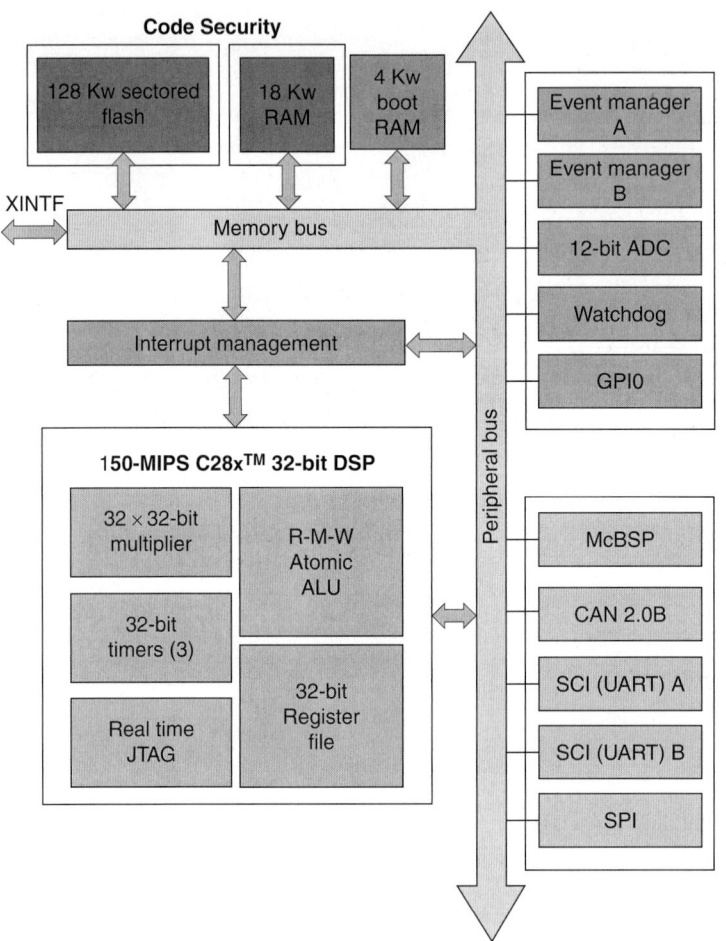

Figure 2.7 SOC for digital control.

the use of prebuilt processor cores. These cores also come bundled with development systems (assembler, compiler, debugger) as well as a preverified software library—of drivers and application code. In cases where the available processor cores do not fully meet the area, power, and performance needs, application-specific instruction-set processors (ASIPs) are used. Such ASIPs typically allow customization of a baseline architecture in terms of application-specific instructions, functional units, and register files (number, width, etc.).

As an example consider the digital media processor TMS320DM642 shown in Figure 2.8. This SOC is based on a programmable DSP that employs the second-generation high-performance, advanced VelociTI Very Long Instruction Word (VLIW) architecture (VelociTI.2), with a performance of up to 4800 million instructions per second (MIPS) at a clock rate of 600 MHz. It has 64 general-purpose registers of 32-bit word length and eight highly independent functional

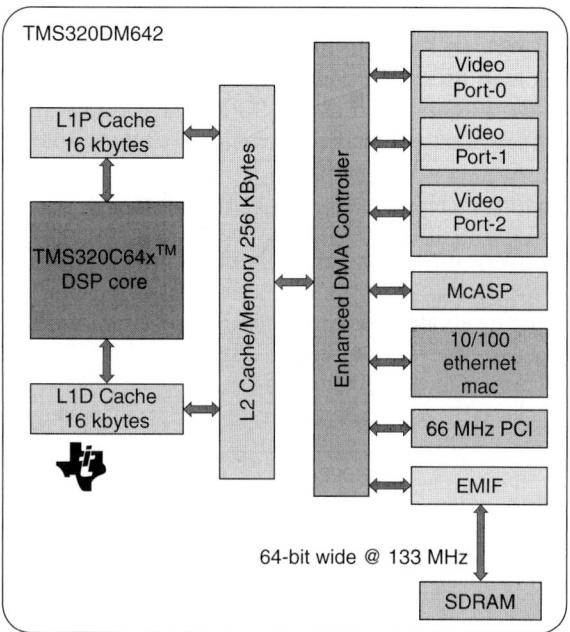

Figure 2.8 Digital multimedia processor.

units—two multipliers for a 32-bit result and six arithmetic logic units (ALUs)—with VelociTI.2 extensions, including new instructions to accelerate the performance in video and imaging. The DM642 can produce four 16-bit multiply-accumulates (MACs) per cycle for a total of 2400 million MACs per second (MMACS), or eight 8-bit MACs per cycle for a total of 4800 MMACS.

The memory subsystem (on-chip storage) consists of a two-level cache-based architecture. The Level 1 program cache (L1P) is a 128-kbit direct-mapped cache, and the Level 1 data cache (L1D) is a 128-kbit two-way set-associative cache. The Level 2 memory/cache (L2) consists of a 2-Mbit memory space that is shared between the program and data space. L2 memory can be configured as mapped memory, cache, or combinations of the two.

The interface engine consists of peripherals including three configurable video ports capable of video input-output or transport stream input, providing a glueless interface to common video decoder and encoder devices. The video port supports multiple resolutions and video standards (e.g., CCIR601, ITU-BT.656, BT.1120, SMPTE 125M, 260M, 274M, and 296M).

The digital media processor's high-performance programmable DSP core optimized for video and imaging applications, a memory subsystem tuned to meet the real-time constraints of various video processing algorithms, and application-specific peripherals such as video ports, make it an industry-leading SOC for high-performance digital media applications including video IP phones, surveillance digital video recorders, and video-on-demand set top boxes.

2.4 SOC Design Challenge

Since an SOC integrates multiple chips of a system onto a single chip, it is targeted to a specific application domain. While the SOC addresses the application requirements in a better way (in terms of cost, power, form factor, and other considerations), building an SOC involves a significant investment, so it's important to understand business considerations that challenge the SOC design process.

Building a complex SOC in an advanced CMOS process typically requires a development cost of more than US$10 million and a cycle time from design start to ready for production of 18+ months. Assuming a 40 percent gross profit margin (GPM), the SOC revenue needs to cross US$25 million to reach break-even, which means that the target available market needs to be in excess of US$75 million to US$100 million. Given that not many such applications exist, the SOC design needs to address the problem of accelerating and maximizing the return on investment, and also being able to address markets with smaller revenue potential. This implies a focus on

- Reducing cycle time
- Reducing development cost (reduced effort)
- Providing differentiation to command a higher GPM
- Reducing the cost of build (COB)

The SOC design challenge is thus an optimization problem along the following vectors:

- Cost (die cost, test cost, package cost)
- Power dissipation (leakage, dynamic)
- Performance (must meet real-time constraints)
- Testability
- DPPM, reliability, yield
- Application-specific requirements—EMI, SER, etc.
- Design effort and cycle time

The conflicting nature of these requirements implies the need to drive appropriate trade-offs. The decisions taken at the SOC definition phase have the highest impact on the optimization parameters. The design effort and cycle time are driven primarily by the chip create phase. The SOC design challenge is hence addressed via a two-phase approach, where in phase I—the SOC definition phase—the microarchitecture-level decisions are taken to meet the key product parameter goals such as die size, power dissipation, and performance. In phase II—the SOC create phase—a platform-based design approach is adopted to reduce the design effort and cycle time. In the following sections we highlight the SOC design challenges in both these phases.

2.4.1 SOC Design Phase 1—SOC Definition and Challenges

As discussed earlier, the customer requirements of cost, power, performance, and form factor apply to the entire system as against the chip alone. The SOC definition phase

Introduction to System-on-Chip (SOC)

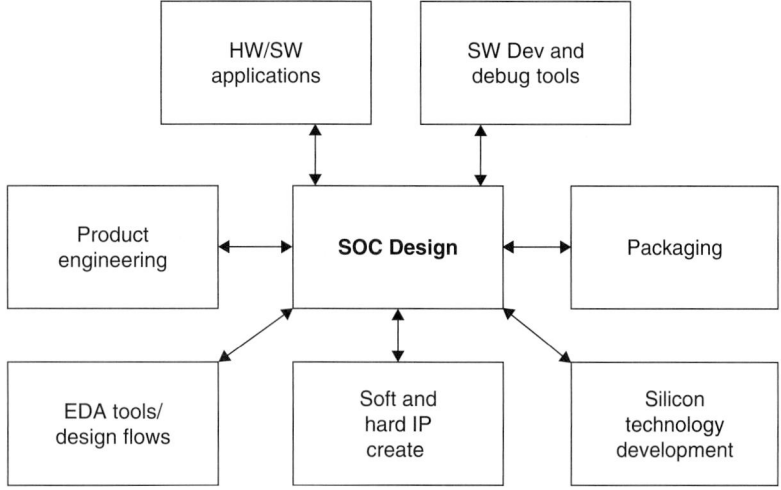

FIGURE 2.9 Concurrent engineering.

hence needs to comprehend system-level implications of the SOC microarchitecture-level decisions. In most cases the SOC and the system are developed in parallel, thus posing concurrent engineering challenges. These challenges if addressed can provide opportunities to drive optimal system definition. Figure 2.9 shows multiple concurrent engineering challenges.

For most DSP applications, real-time performance is a critical system requirement. The system performance is typically determined by the SOC microarchitecture along with the software running on the embedded processor. The SOC definition phase hence involves working closely with the software applications team to profile the code, identify performance bottlenecks, and drive appropriate hardware-software partitioning decisions.

The amount of software running on the embedded processor(s) of an SOC has been increasing over the years. The criticality of a user-friendly application development environment has consequently gone up. An SOC hence needs to provide appropriate hooks in the hardware to enable the software debug. Debug architecture is an important component of an SOC microarchitecture, and it is best defined jointly with the team developing the application development environment for the SOC.

Time-to-market is a big concern for most customers, especially in the consumer electronics space. It's not adequate to build functional silicon; it needs to be followed by product engineering functions that get it ready for volume production. The SOC design team works closely with the product engineering team starting from the SOC definition phase to build appropriate hooks in the SOC microarchitecture and provide necessary information to be able to get the test programs ready just in time for the silicon, thus enabling rapid ramp to volume production.

Electronic design automation (EDA) is a critical enabler to meet the aggressive cycle time goals. Since with each generation the design complexity keeps going up significantly, in many cases the design flow automation and the design methodology gets built concurrently with the chip create process.

While it's desirable, from a design cycle time perspective, to have all the IPs available before the start of the SOC create process, in many cases, the IPs are developed concurrently with the chip. This helps reduce the overall cycle time but makes it critically important for the chip create team to work closely with the IP team, to ensure that the IP is developed to meet the chip requirements.

For SOCs that aggressively adopt the new process technologies, the chip design starts even before the manufacturing process is completely qualified and the transistor and interconnect characteristics are stabilized. The design team hence needs to work closely with the silicon technology development team to be able to adapt quickly to process changes. This concurrent engineering can also be leveraged to tune the manufacturing process so as to meet a critical SOC requirement such as leakage power and/or performance.

The package is a key contributor to the cost, performance, power dissipation, and form factor of an SOC. It is hence becoming increasingly important to do package design concurrently with SOC design.

In the following sections we discuss two examples of these concurrent engineering challenges in further detail.

HW-SW Codesign—Memory Subsystem Definition

The memory subsystem is an important component of an SOC, as it significantly impacts the performance, die size, and power dissipation. Figure 2.10 shows a generic memory subsystem that has two levels of hierarchy.

For an SOC targeting a set of applications, the key objective of memory subsystem design is to meet the performance requirements while minimizing die size and power dissipation. This is a nontrivial task considering the large number of options available

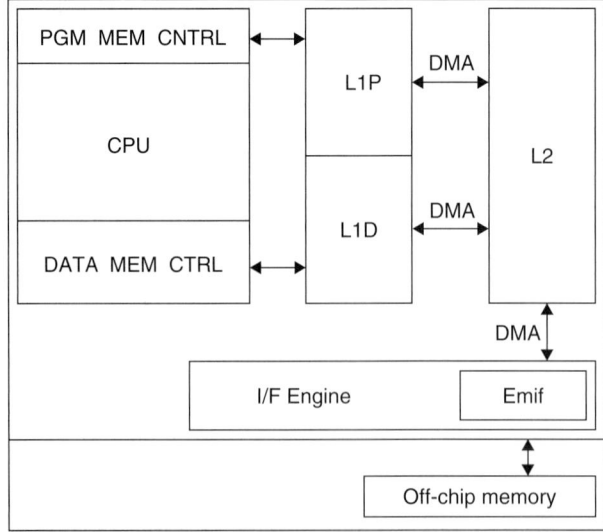

Figure 2.10 Memory subsystem.

related to the logical and physical architecture of the memory subsystem. We list some of these options here based on Figure 2-10:

- Type of memory: SRAM, ROM, flash, embedded DRAM (eDRAM), and Ferroelectric DRAM (FEDRAM) at both the L1 and L2 levels. The decision depends on specific application requirements, availability in a given technology node, performance, and cost.
- For L1 and L2
  - Size (kbits)
  - Unified (program and data) or program-only or data-only or combination
  - Number of physical blocks, size of each block
  - For each block choice between a denser (but slower) or a faster (but bigger) memory
  - Single-port versus dual-port or multiport memory
  - For each physical block—MUX-factor, which decides performance and aspect ratio
  - Cache or mapped or combination
  - In case of cache—type of cache, line size, etc.
  - Clock rate relative to the central processing unit (CPU) and number of wait states which may vary for each physical block
- For external memory interface (EMIF)
  - Type(s) of memory to be interfaced
  - Size and number of physical block of the off-chip memory
  - Width of the EMIF interface (16, 32, or 64 bits)
  - Clock rate

Since the performance and throughput need to be met in the context of an application, the memory subsystem design involves working closely with the applications team. While there can be multiple feasible solutions, an optimal solution is one in which the CPU, memory, and I/O bandwidths are balanced such that none of them becomes a bottleneck. This requires building a model (software simulator) of the instruction set architecture, the memory subsystem, the direct memory access (DMA), the external memory interface, and the off-chip memory. While it is desirable for the model to be cycle accurate, it conflicts with the requirement of faster software simulation to enable performance analysis over a reasonably large number of cycles. The design, application, and software development tools teams have to work closely to make the right trade-offs and adopt appropriate levels of abstraction for different system components.

The challenges in arriving at an optimal memory subsystem increase further if the SOC is targeting applications that are based on different core algorithms. As an example, Table 2.1 shows different applications targeted by the DM642 digital media processor and the key algorithms for each of the applications. The CPU, memory, and I/O bandwidth requirements vary across these applications. The memory subsystem is decided by the application with the most stringent performance requirement, and for other applications the CPU can be run slower (e.g., 500 MHz instead of 600 MHz) at a lower supply voltage thus reducing the power dissipation.

Just as the memory subsystem can be optimized for a given software implementation of an application, the software implementation can also be optimized for a given memory subsystem. The memory subsystem hence needs to be designed concurrently with the application development.

Application	Algorithm
Security systems	4*CIF MPEG4 encode/decode
IP Videophone	H.263 encode
Video servers	Multichannel MPEG2 encode/stream media encode
PVR/home server	MPEG2 encode/decode
IP set-top box/streaming media decode	Streaming media decode

TABLE 2.1 Target Applications and Algorithms for DM642

It can be noted that for an application, there can be multiple solutions possible that meet performance requirements and balance the CPU, memory, and I/O bandwidth. The most optimal solution is then decided based on the cost and power dissipation goal. For example, a smaller L1P can reduce the die size; however, because of an increased number of cache misses, it may require that the CPU be run at a higher clock rate, which in turn would require the chip to operate at a higher voltage resulting in increased power dissipation. In case of video processing applications, which are data intensive, the size of L2 can impact the EMIF bandwidth requirement. While a smaller L2 implies a lower chip cost, increasing the EMIF clock rate, for example, from 100 to 133 MHz may result in a cost increase at the system level due to an increase in the cost of the off-chip memory that needs to run at a higher data rate. In general for the same application throughput different L2 sizes can result in an EMIF bandwidth ranging from say 32 bits at 100 MHz to 64 bits at 133 MHz. Since 64-bit I/O switching results in increased noise (package implications) and higher power dissipation, the decision on the memory subsystem needs to be driven by the desired cost-power tradeoff.

Chip-Package Codesign

Since the package is an important contributor to the cost, power dissipation, and performance of an SOC, in this section we discuss chip-package codesign to optimize system-level objectives. The performance (megahertz) of a chip is dependent on the resistive drop, which in turn is dependent on the package. A flip-chip package provides a lower IR drop than a wire-bond package, and hence supports higher performance.

A flip-chip package, however, is significantly more expensive than a wire-bond package (Figure 2.11). It is, however, possible to limit the cost overhead by using a low-cost flip-chip package.

The flip-chip package cost is driven by the number of layers, standard substrates versus built-up substrates that enable micro-vias (Figure 2.12), substrate size, bump pitch, and other factors. The package selection from a pin-out point of view is a tradeoff between the package cost (substrate size), form factor, and board-level cost. While a smaller ball pitch translates to a smaller package size, the board-level manufacturability

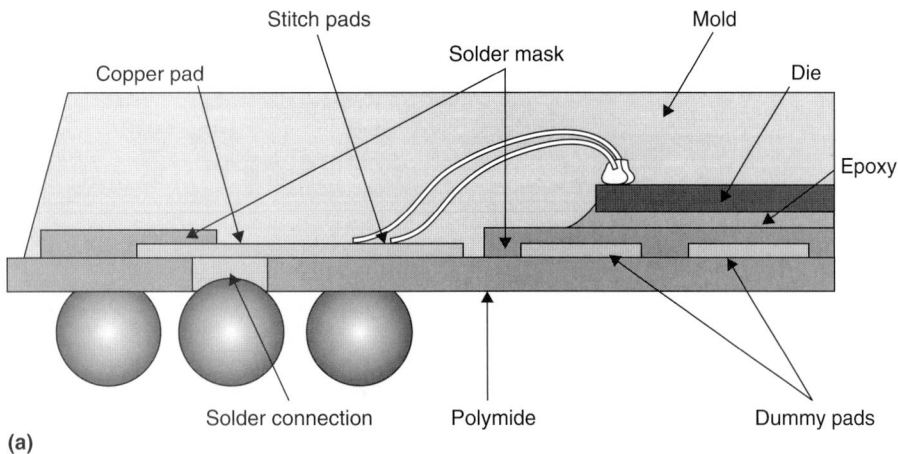

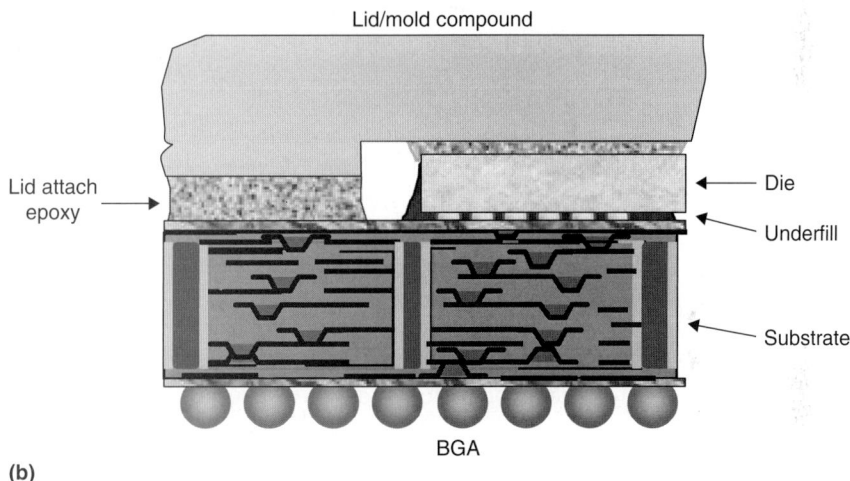

FIGURE 2.11 (a) Wirebond ball grid array (BGA). (b) Flip-chip BGA.

and reliability requirements typically put a lower limit on the ball pitch. While a smaller ball pitch and increased number of ball rows can help reduce the package size, they can make board-level routing difficult, in some cases forcing the number of layers of the board to increase, which may not be the right tradeoff from a system cost point of view.

One of the key steps in package design is the pinout definition—for which the following need to be taken into consideration:

- Board layout considerations drive pin assignment (location, ordering), which eases routing and results in a smaller route length and a board with fewer layers (a low system-level cost).
- Pin assignment also need to comprehend compatibility requirements with respect to the existing devices.

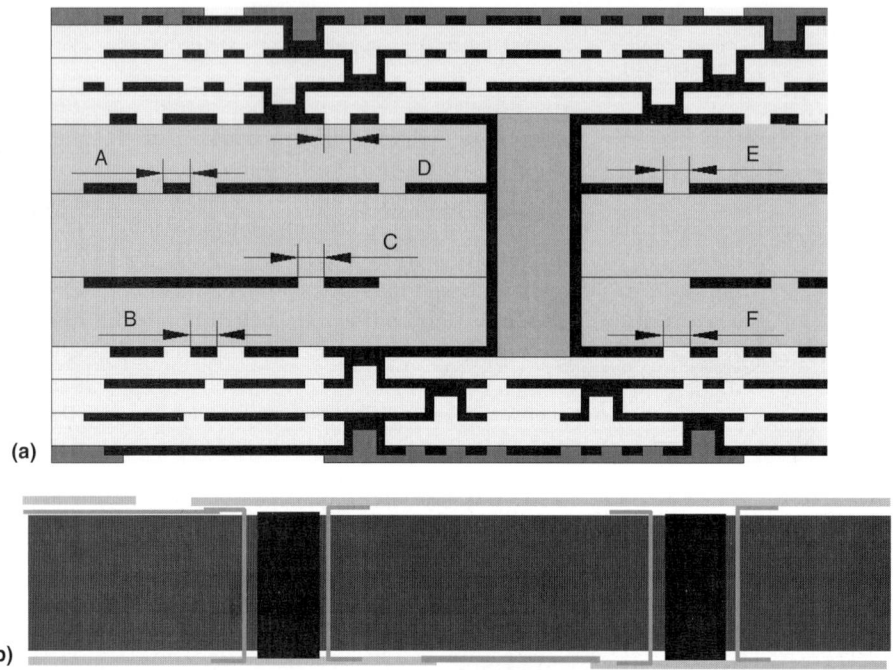

Figure 2.12 (a) Build-up multilayer (6 to 12 layers) substrate. (b) Low-cost PCB-based (2 to 4 layers) substrate.

- Pin assignment drives the bump assignment at the chip level. For a low-cost substrate, the signal bumps are restricted to the outer two rows and the signals are distributed evenly on all four sides so as to ease substrate routing.
- Chip floor plan considerations drive the I/O and hence the pin assignment. These include the location of clock and PLL inputs relative to fast switching I/Os and also assignment to ease chip-level routing congestion.
- The bump pitch and the number of signals on each side translate to a lower bound on the die size. In case the chip is bump limited, the bottleneck can be removed by either reducing the number of I/Os (aggressive pin muxing, or dropping and reducing interfaces) or increasing the core size (adding functionality, increasing L2 size, etc.)
- If the chip power dissipation exceeds the package thermal capacity, techniques such as adding thermal balls at the center can be adopted. These balls, however, take up space on the board where decaps are placed to reduce power supply noise.
- Minimizing the resistive drop requires an adequate number of the following: power and ground pins, power and ground area bumps, and connections (vias) to the ground plane in the substrate. When determining the area bump locations, the requirement of having no memories under the bump and the need to reuse a mega-module (hard macro) that comes with its own bump pattern should be taken into consideration.

The preceding considerations in many cases result in conflicting requirements. The key challenge in chip-package codesign is driving the right tradeoffs. The codesign methodology requires the following capabilities:

- Die size estimation
- Power estimation
- Physical design methodology—unifying chip, substrate, and potentially board-level design
- Chip-package-board electrical modeling
- Package thermal modeling
- Cost model to drive appropriate tradeoffs

While the SOC definition phase provides the opportunity to impact area, power, and performance parameters the most, the SOC design phase aims at achieving maximum technology entitlement for a given SOC microarchitecture. The design phase not only has the most impact on cycle time and effort goals, but it's the phase where reliability and testability aspects are addressed in support of robustness and DPPM goals.

2.4.2 SOC Design Phase II—SOC Create Process and Challenges

In this section we discuss challenges faced during the chip create phase of SOC design and also present approaches to address these challenges. We start by first describing the overall SOC design methodology—specifically the HW-SW codesign aspects. We then focus on various components of the SOC create process starting with chip integration, verification, and design for test considerations. While technology scaling enables an increasing level of integration, it brings with each generation unique design challenges. We hence present implications of technology scaling and discuss abstraction as a mechanism to manage increasing chip create complexities. We then present challenges with the physical design phase of the SOC create process—covering design planning and design closure. We finally discuss challenges (specifically noise) in monolithic integration of analog modules with complex digital logic.

SOC HW-SW Codesign and Architectural-level Partitioning

Since an SOC is built using one or more embedded processors, the HW-SW codesign and verification is an important component of developing an optimal overall solution. Figure 2.13 shows the flow that starts with HW-SW partitioning, the creation of HW and SW components, and finally their system-level integration and verification.

Several modern-day applications, particularly embedded applications like automotive, telecommunications, and consumer electronics, involve both software and hardware components. The software generally influences the features and flexibility, whereas the hardware provides the performance. Traditionally, the software and hardware definitions and descriptions were developed sequentially and very often in isolation, thereby leading to overall system incompatibilities and in many cases suboptimal system architecture. This directly impacts time-to-market due to iterations and rework. To address the growing problems associated with the design of complex SOC and the need for extensibility and configurability in processor-based design, electronic system-level (ESL) methods are becoming very common. A typical system-level design and the components and data flow involved are described in Figure 2.13.

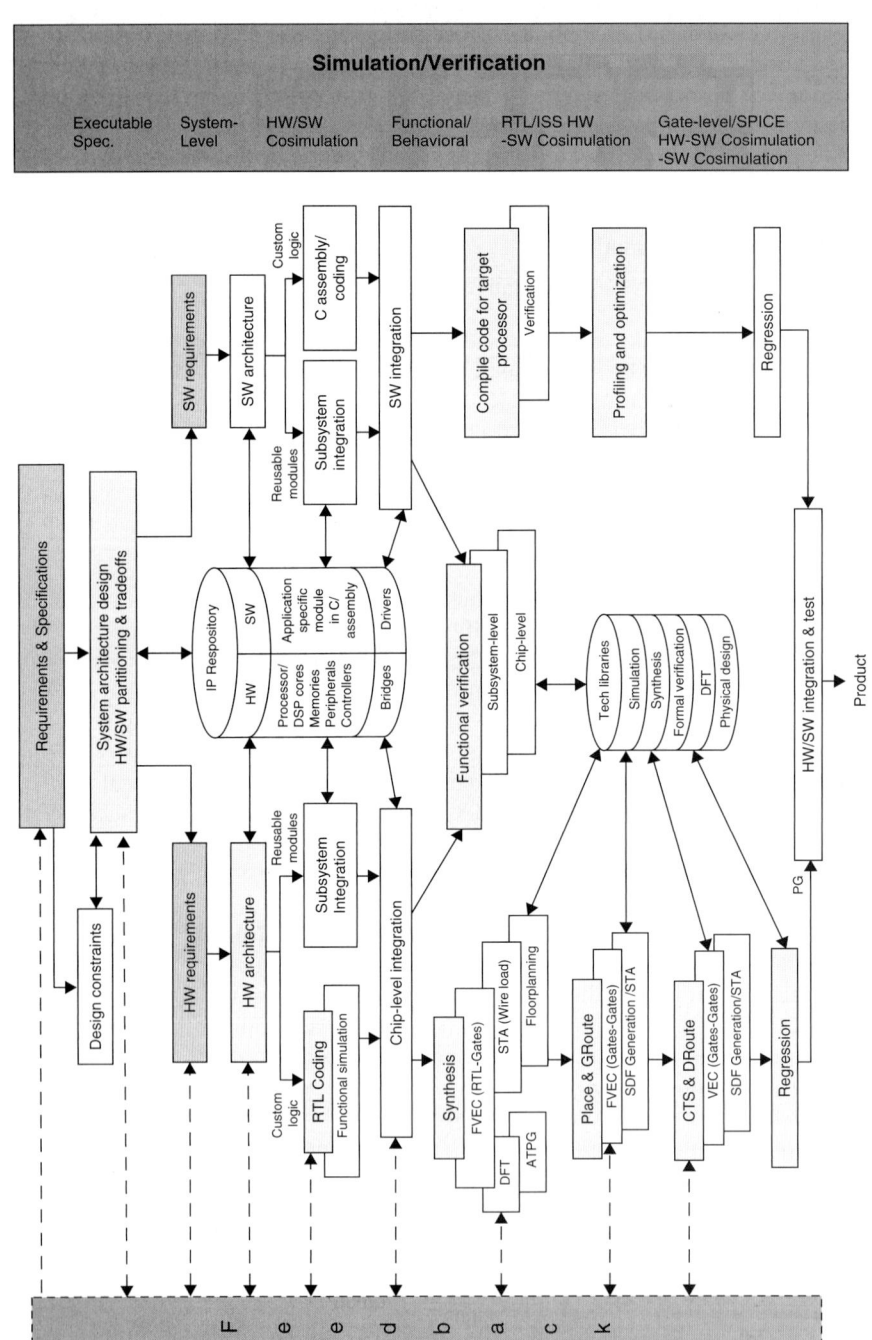

FIGURE 2.13 SOC HW-SW codesign methodology.

The system-level design process typically starts with a description of the system to be built in terms of requirements mainly driven by the applications' "use" context and then translates to capturing the functional aspects of the system in terms of the behavior. The next step is to define the system architecture that primarily involves coming up with the required hardware platform and the partitioning of the system into hardware and software components. This is a very critical phase of system design and has to be accomplished in the context of the constraints involved such as performance, cost, and power. Correct tradeoffs at this stage are absolutely essential to address the overall SOC requirements to meet the customer's product specification. The HW requirements are then translated into a SOC design architecture specification that is typically then described in hardware description languages (HDL) like VHDL or Verilog. The correctness of the specification is verified using simulation tools that verify the functionality of the implemented HW. The HDL description of the HW is then synthesized to a target technology library followed by the physical implementation of the design into mask layers. At every stage of the HW design flow, care is taken using functional equivalence checker tools that the functionality is not changed by downstream logic and physical synthesis and implementation tools. The SW components are typically coded in high-level programming languages such as C/C++ or even assembly level and then simulated for correctness. The key aspect in getting both the HW and SW components, and hence the overall system right, is cosimulation of the HW and SW. This is typically performed using either HW acceleration systems or in-circuit emulation that can co-verify the HW along with the embedded SW components.

SOC Integration

As indicated earlier, just as CMOS scaling drove the PC era for the last two decades, large-scale integration of digital and analog/RF functionality on a SOC is going to fuel the Internet era going forward. This is going to pose a significant challenge if we combine the technology scaling issues outlined in the previous section along with the sheer complexity of the SOC integration challenge.

One of the most critical components of SOC design is the integration of predeveloped pieces of functionality called intellectual property (IP). These IP blocks can offer a huge differentiation to designers building SOC designs for various applications and helps reduce development cycle-time significantly. However, while attempting to integrate such multiple IPs, the SOC designer is faced with tremendous challenges in understanding these predefined functional IPs as well as coping with the issues that need to be dealt with in getting these IPs to talk to the rest of their SOC and in verifying the whole system thereafter. Complicating the problem is the reality that IP developers and SOC designers are geographically distributed across the world. This can, very often, offset the advantage of reducing development cycle-time that the reuse itself brings in. Several initiatives have been started in the industry today to tackle this major IP reuse issue. One such industry consortium called the Virtual Socket Interface Alliance (VSIA) was founded in September 1996 with an attempt to bring together IP [also called virtual components (VC) by this consortium] developers and SOC houses to work together to define standards for design and integration of reusable IP. Several IP-centric bus definitions and interconnect strategies have been suggested to make IP reuse as seamless as possible for the SOC designer. A VC quality checklist was also developed by this consortium to quantify the "readiness" of these components for reuse. This focused on qualifying an IP from both a developer's perspective as well as from an integrator perspective and incorporating as many best practices as possible. Another consortium that was launched at the Design Automation Conference 2004 was the Structure for Packaging, Integrating, and Re-using IP with Tool-flows (SPIRIT) to cover a Register Transfer Level (RTL) encapsulation for automated IP integration and interoperability of IP with multiple toolsets. This included tools for system-level design, verification, and simulation as well as synthesis.

IP cores or blocks can be integrated in three variants on an SOC:

- *Hard.* These blocks are physical design completed and optimized at a particular process node. As a result, while these blocks can differentiate in terms of speed, power, and area, they are the least flexible and portable across technology nodes and SOC designs, given that their physical attributes such as size and aspect-ratio cannot change.

- *Soft.* These blocks are reused as a register transfer–level representation of the IP along with the necessary synthesizable constraints and test benches to implement and verify these blocks in the SOC context. In contrast to hard blocks, these IP components have the most flexibility and portability across SOC designs and are amenable to in-context physical optimization for best SOC power-speed-area parameters.

- *Firm.* This type of IP reuse combines the best advantages of both the above two reuse scenarios where the IP is optimized for power-speed-area careabouts across process nodes. However, given that the physical layout is uncommitted, the IP is configurable to various "use" scenarios.

As evident from these IP reuse variants, the right SOC-level tradeoffs in terms of time-to-market, performance requirements, and portability need to be made in deciding the reuse strategy.

An important development in the area of IP reuse for SOC designs is the concept of platform-based design. This has evinced significant debates and analysis on the advantages and challenges it brings. A platform-based SOC design is based on the fundamental premise that if you have an architecture, a bunch of predefined building blocks, including a processor or a DSP, a standard bus, memory controller, and SW tools, it would be very easy to quickly generate several derivative chips targeted at the application segment. Needless to say, this enables rapid SOC development and time-to-market that several market segments demand. For example, cell phone and automobile manufacturers can very easily deploy a platform-based development model to spin incremental variants of their models into the market without having to design each chip from scratch. A direct benefit of this approach of working at the system level is that it promotes a lot more architectural-level analysis and tradeoffs during SOC design and brings into focus a system-level view to SOC design and verification. A critical benefit of a platform-based SOC development is reduced risk. Given the huge mask costs outlined above in the 90-nm or 45-nm process nodes, any design mistake will result in a costly respin as well as a delay in the product getting to the market by more than 6 months. This is where the SW codevelopment model of the SOC brings in the benefit of end customers being able to simulate their application code on the system even before the SOC design can be sent for fabrication. A platform-based approach therefore allows for such HW-SW concurrent design with SW being ready before silicon, thereby also reducing SW development cycle times. Table 2.2 provides an example of how a platform-based design at 90 nm helps reduce several SOC development cost and time-to-market parameters. In some sense, the "realm" of reuse in a platform-based approach

	Time to Working Silicon	Development Cost	BOM Cost Reduction	Volume Breakover Point	Time-to-market
Non-platform-based 90-nm SOC	2–3 years	$10 million	$40	250,000	3 years
Cell-based, platform-based SOC	6–9 months	$4 million	$40	100,000	1 year

Source: Toshiba America Electronic Components, Inc.

TABLE 2.2 Benefits of Platform-based SOC Design

moves up from just an IP level, which was described earlier, to a HW-SW components level reuse. On the other hand, one of the key drawbacks of a platform approach is the need for a larger up-front investment in terms of both the man-month effort required to deliver a platform as well as the complexity involved. This means that there needs to be a lot of careful early planning and analysis, particularly at the architectural level. All hardware and software platform components need to be preverified all the way to silicon, and this adds to the up-front cost as well. We will describe more of these challenges and careabouts while discussing design abstraction levels in the next section.

The platform-based design development working group of the VSIA Consortium has been working for a while to determine a clear set of platform attributes in order to provide a definition for the platform taxonomy. The two major kinds that have emerged are the "application-driven" and "technology-driven" platforms as shown in Figure 2.14. In the application-driven scenario, different architectural-level families are defined from certain application domains and a product line family typically gets created. A top-down process instantiates the required preverified hardware and software IP modules, and application-specific derivative products are integrated and built on a single chip. This process is independent of the process technology that is used to build such systems. In contrast, a technology-driven platform is built bottom-up with the need to either extend the functionality or performance, or to migrate the design to later technologies, irrespective of the application requirements.

SOC Verification

The functional verification of an SOC has two components, firstly to verify whether the implementation meets the specification and secondly to verify that the specification meets the true intent of the system functionality. Since the system functionality is realized by the software running on the processors embedded in an SOC, HW-SW coverification is an important component of SOC verification. In addition to the software delivered as part of the system solution, the software development environment also needs to be provided so that customers and users can develop and integrate differentiated, value-added software. It's thus important to verify that the appropriate hardware hooks (for debug for example) required for the software development are functioning as desired. Figure 2.15 captures the software environment around an SOC with embedded processor(s). This includes both the software components that run on the SOC and also

62 Chapter Two

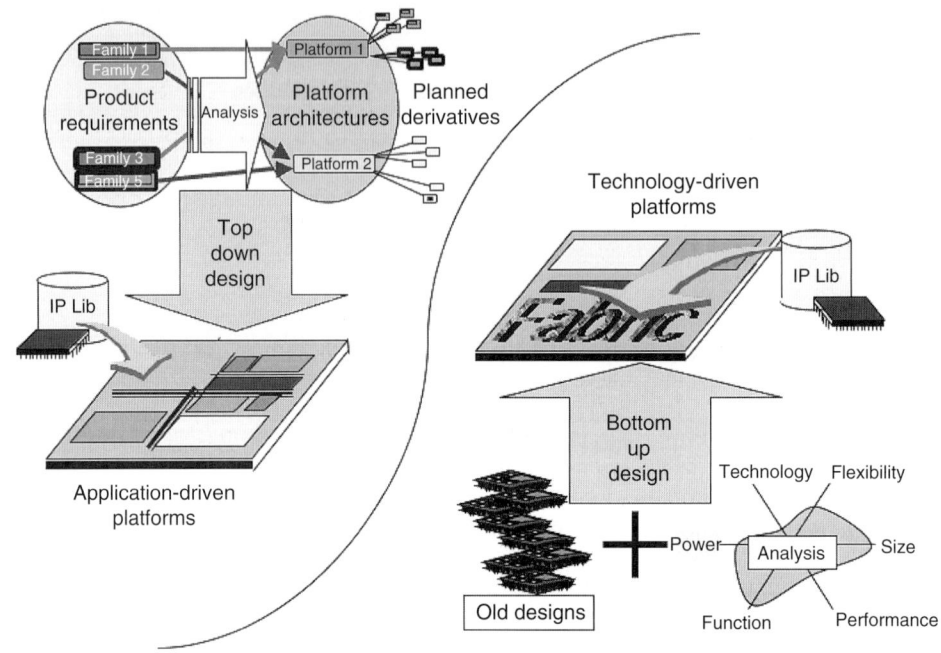

FIGURE 2.14 Major platform types.

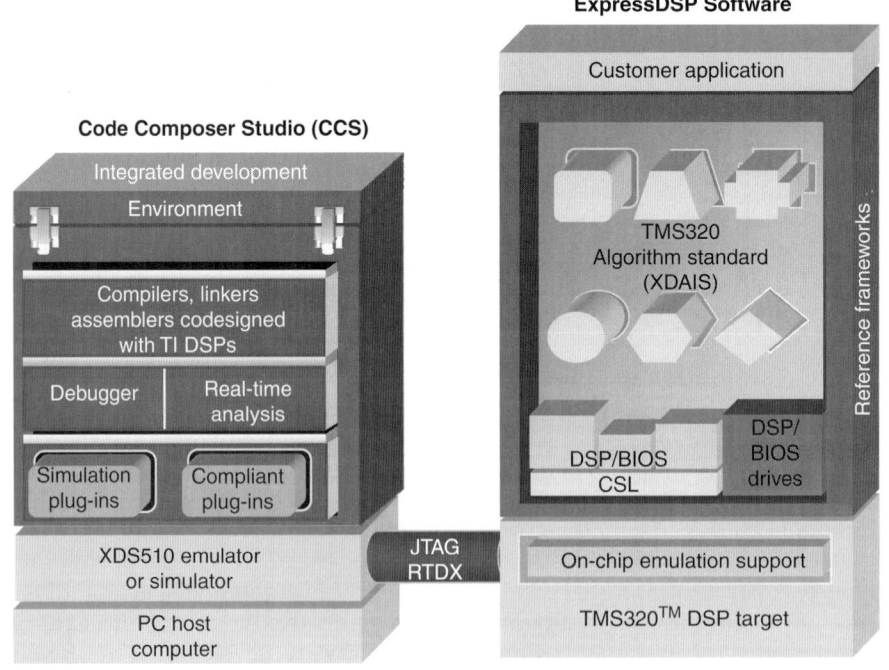

FIGURE 2.15 HW-SW coverification.

the host-side PC-based software used for application development. The SOC verification needs to ensure that these two software components interface and interact with the SOC hardware to provide the desired system functionality.

Design for Test

Given that the SOC technology roadmap is moving feverishly toward smaller silicon feature sizes, the use of newer physical processing methods involving interconnect materials like copper and low-K dielectric, and the integration and reuse of complex IP and memory from various sources, it is becoming increasingly important to ensure the quality and reliability of the silicon used. At the same time the cost involved in measuring these quality levels needs to also come down to reduce the overall SOC cost. It is important that the right set of vectors are generated and applied to not just ensure the ease of detecting manufacturing defects but also to ensure a reduction in the overall test time. The process of integrating features or logic to enable this is called "design for test." While the use of built-in self-test (BIST) techniques for memories have been in use, increasingly designers are adopting BIST techniques to test logic as well, so as to achieve higher quality at a lower cost.

Technology Scaling

Process technology is linearly shrinking at approximately 70 percent per generation. This enables the implementation of a logic function in half the die area compared to the previous technology node, hence lowering the cost.

While every advanced process technology node provides the 70 percent linear shrink, the bond pad pitch (for wire-bond packaging) and bump pitch (for flip-chip packaging) have not scaled accordingly. In addition, I/Os and analog components do not shrink as much as the standard logic. These factors need to be taken into consideration when assessing the cost benefit of moving to a new technology node.

Every new process node comes with an increased reticle cost and increased fabrication cycle time. The wafer manufacturing cost depends on several factors such as the cost of capital involved in the procurement of steppers and scanners and the cost of the process material and fabrication facilities. Wafer throughput also impacts the manufacturing cost, and this throughput is directly dependent on the number and size of the "steps" printed on the wafer. Typically, a 130-nm mask set can cost around US$750,000, while the 90-nm mask set costs over 1 million U.S. dollars.

As indicated in Figure 2.16, technology scaling has resulted in finer geometry sizes, which in turn has caused an increased resistance in both wires and vias. The number of metal layers that are supported in current technology nodes has also increased the cross-coupling capacitance to ground capacitance ratio. Lower device thresholds have

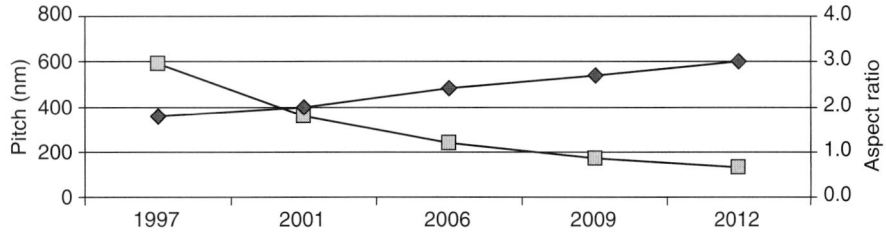

FIGURE 2.16 Interconnect geometry trends. (*Source:* ICCAD 2000 Tutorial)

caused lower noise margins. In addition, due to huge SOC die sizes, the high speed of operation running into gigahertz, and the shrinking of metal layers has caused severe on-chip IR drop issues. IR drops on the power and ground distribution network can severely impact chip performance, including the clock signals. Excessive IR drop on the power grid can cause timing failures in the circuits that designers have to analyze and comprehend. It has been found that a 10 percent voltage drop in a 180-nm design can increase propagation gate delays by up to 8 percent [38].

Hunger for MIPS to fuel the ever-growing demand from applications discussed earlier has pushed up clock rates higher and higher, resulting in several new applications requiring design and circuit-level innovations. This is also the result of the fact that the move to a new process node does not necessarily offer significant performance lift. As the clock frequency increases, the timing margins required for the circuits to operate dependably across process variations decrease. Increased speed and faster transition rates on-chip require a more comprehensive handling of issues such as crosstalk and ground bounce than needed for prior process nodes. Simulation and analysis tools, for example, need to handle timing, signal integrity, and issues such as electromagnetic interference (EMI).

Leakage power continues to dominate newer process nodes such as 90, 65, and 45 nm, as shown in Figure 2.17. This is primarily due to the source-to-drain leakage current that increases with a lowering of the threshold voltage (V_t), increasing temperature, and shorter transistor channel lengths. Also, with gate oxide thicknesses decreasing at such newer process nodes, the voltages across the gate must be reduced to keep the electric fields from becoming too high for the insulating material. Both a lower V_t and gate oxide thickness exponentially increase the transistor leakage current. New design techniques have come on the horizon to tackle the leakage power issue. Several power management techniques are being integrated on the SOC to handle leakage power. One of the most common approaches to address leakage power is the use of multi-V_t libraries that most Application Specific Integrated Circuit (ASIC) vendors provide today at 130 nm and below. In addition to the libraries that support multi-V_t

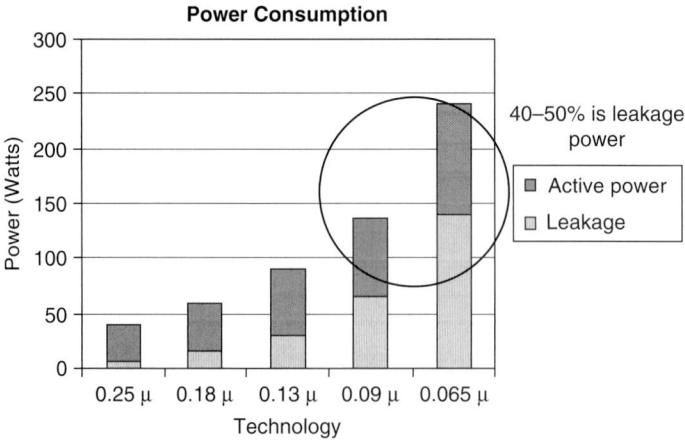

Figure 2.17 Power dissipation trends. (*Source:* Intel.)

cells, design tools need to handle optimization techniques to minimize leakage power by the appropriate usage of cells that have a high V_t. Typically, the cell libraries that support the multi-V_t optimization have both fast cells with lower V_t thresholds (but higher leakage) and slow cells with higher V_t thresholds (and hence a smaller leakage component). One approach commonly adopted during design to reduce leakage power is to deploy the fast cells during the initial logic synthesis, and then swap in the low-leakage cells during subsequent timing optimization on circuit paths that can use these low-leakage (but slower in performance) cells without impacting the timing. While such power optimizations are done, it is very important to also understand the implications to the chip die area (cost) and yield in terms of sensitivity to process variation. As silicon technology scales down, the gate oxide thickness also goes down, and as a result, the oxide layer fails to act as a perfect insulator and leakage current flows through it. This can be overcome by the use of high-dielectric-constant oxides, and as a result of this thick oxide, the leakage current is minimized. In addition to leakage power, the on-chip power density trend is on an exponential rise, as can be seen with the Pentium example: the P-4 had a power density of 46 W/cm^2, which is seven times that of the Intel 486. Several techniques for reducing switching power have been considered in the recent past. Reducing activity, capacitance, and supply voltage are some of the commonly known methods. Design methodologies such as clock gating, power gating, power-aware physical design, and voltage scaling have been deployed to tackle the SOC power efficiency issues for sensitive handheld applications such as wireless mobile phones, PDAs, and personal multimedia players.

As more and more transistors continue to be packed on a single die with demands for higher and higher performance rates due to the technology scaling trends, controlling power is becoming a very critical issue. Several power management techniques have been used to contain this problem utilizing the inherent characteristics of being able to turn off functional modules that are not always needed during chip operation, as shown in Figure 2.18. However, another approach that is being heavily adopted today,

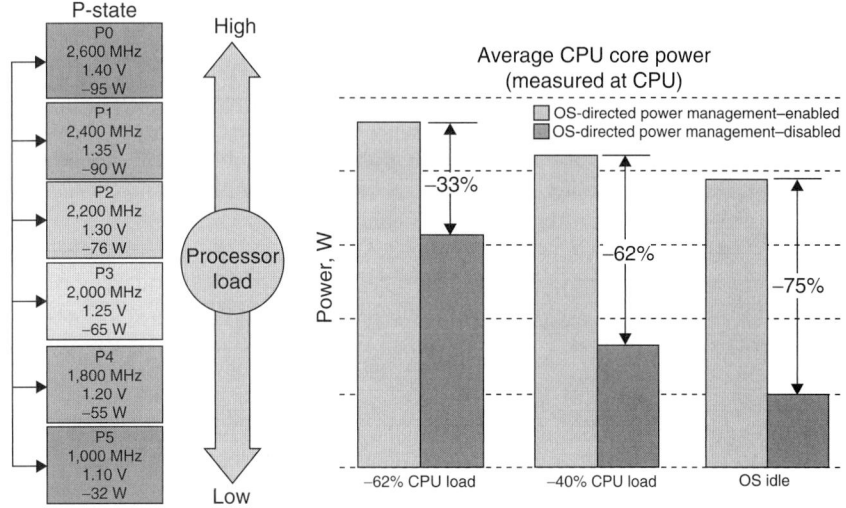

FIGURE 2.18 Processor "performance states."

particularly in the microprocessor space, is to go for multiple CPU engines, popularly called the "dual-core" architectures, where multiple CPU cores are integrated in silicon on the same die. This provides additional flexibility to manage and distribute power, particularly as options to reduce operating voltages and performance become an option. Dual-core processor architectures allow devices to run multiple "threads" at a time and are therefore amenable to what is referred to as thread-level parallelism. Additionally, integrating multiple CPU cores on a single die improves the performance of the circuits since the signals do not have to travel off-chip, and also utilize the board space in the system more efficiently compared to two discrete chips. Even in the digital signal processor SOC world, integrating a RISC digital signal processor with a microcontroller such as ARM or MIPS is very common, and the DSP functions are available as hardware accelerators.

With process nodes at and below 90 nm, in-die variations are becoming a huge issue to tackle, placing SOC manufacturability at risk. If in-die variations and their effects are not modeled correctly, there is a large probability of silicon failure causing mask respin and hence an increased cost. The width variation of a critical wire in layout depends on the width of the wire segment and the spacing to its neighbor. This variation is referred to as selective process bias (SPB). Resistance and Capacitance (RC) extraction engines use a two-dimensional table of width and spacing to model these wire width variations. Design experts now talk about hold failures found on silicon attributed back to metal RC variation between adjacent planes. Given that recent SOC design can support seven to eight metal layers, it becomes computationally prohibitive to use traditional analysis tools to comprehend these effects. At the same time, the traditional single (typically worst-case) corner timing analysis approach is no longer sufficient to handle such in-die or intrachip variations. Hence statistical timing analysis methods and variation-aware timing closure flows are being investigated for 90-nm and below SOC designs. Another increasingly threatening device reliability or, as it is commonly referred to, "chip aging" issue, is the negative bias temperature instability (NBTI), specifically occurring at lower operating voltages. NBTI is known to cause significant V_t shifts, and hence an accurate method to model this effect is required.

Traditionally, yield has been considered a fabrication-only issue. The manufacturability issues of an SOC were limited to the adherence of design rules that the FAB would drive for a particular process. With design feature sizes and spacing rules getting lower than the wavelength of light, process material and lithography effects can considerably alter what gets created by the layout designer versus what gets actually printed on silicon. This, in effect, changes the electrical characteristics of the circuit causing reliability or speed problems. As a result, physical designers need to understand these manufacturing effects and up-front handling of the impact during layout and analysis. This trend is similar to what happened 5 to 10 years back with logic and physical design merging to attack the timing closure problem. Chemical and mechanical polishing (CMP) has been a well-known step during manufacturing to ensure planarization of the silicon surface and hence improve yield. However, this can cause changes in the thickness of the dielectric between metal layers and interconnect resistance, and therefore impact die yield. This problem can be circumvented by postprocessing the layout with strips of dummy metal to even out the metal density on the chip. Insertion of dummy metal can, in turn, impact the timing of critical signals on the chip and can even cause additional parasitic coupling to existing signals that could result in functional problems. Layout designers should, therefore, comprehend the effect of such dummy fill insertion during placement and routing and ensure timing analysis considering the impact of these additional parasitics. This is beginning to be called "yield-driven-layout" in the industry today. Another major cause of yield concern on the chip are the "via" structures added to connect adjacent metal interconnects. Thermally induced stress can impact both the copper interconnects and the low-K materials used in today's process technology given that the dielectric has a large thermal expansion coefficient and poor adhesion. This can cause voids below the via structures and result in poor reliability of the circuit. Layout designers need to take care of this yield issue during physical design. This is typically done by optimizing the interconnect routing to minimize vias as much as possible by ensuring straighter routes, and where vias are added, to insert redundant vias in the layout so as to improve the reliability of the design.

In contrast to the exciting growth opportunities and enablers toward integrating and building complex SOC devices with advanced process technology, and the several huge challenges outlined above, the availability of design engineering talent to support the creation of such complex SOCs has not increased. This has resulted in a huge design productivity gap that needs to be addressed.

SOC design methodology has evolved over the last three decades by trying to keep pace with these advances and the complexity in process technology at submicron nodes. However, the design productivity gap continues to increase, as indicated in Figure 2.19, given that this pace is found not to be sufficient to cope with the complexity growth.

Addressing SOC Design Create Complexity

One of the significant enablers to cope with the various SOC design complexities that were described in the previous sections is design abstraction. Over the last two decades, design engineers have moved up one abstraction layer to another in a bid to comprehend the ever-increasing integration of complex components and functionality on a single chip.

As described in Figure 2.20, each abstraction level has a critical influence on the final SOC behavior, and while clearly the implementation effort and complexity is reduced at higher levels of design abstractions, it is very important to ensure the functional correctness of each level before the next level up can be created and verified. This is one of the fundamental aspects of how SOC design complexity is handled via hierarchical design approaches that will be described later.

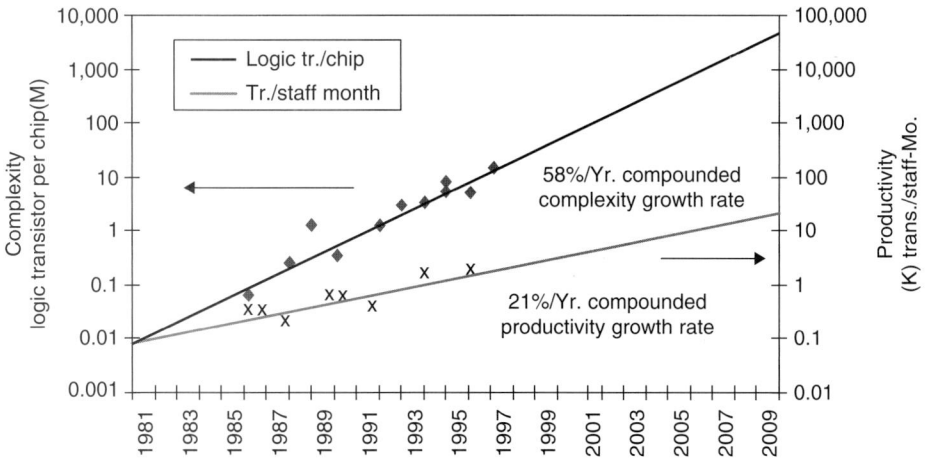

Figure 2.19 Design productivity gap.

68 Chapter Two

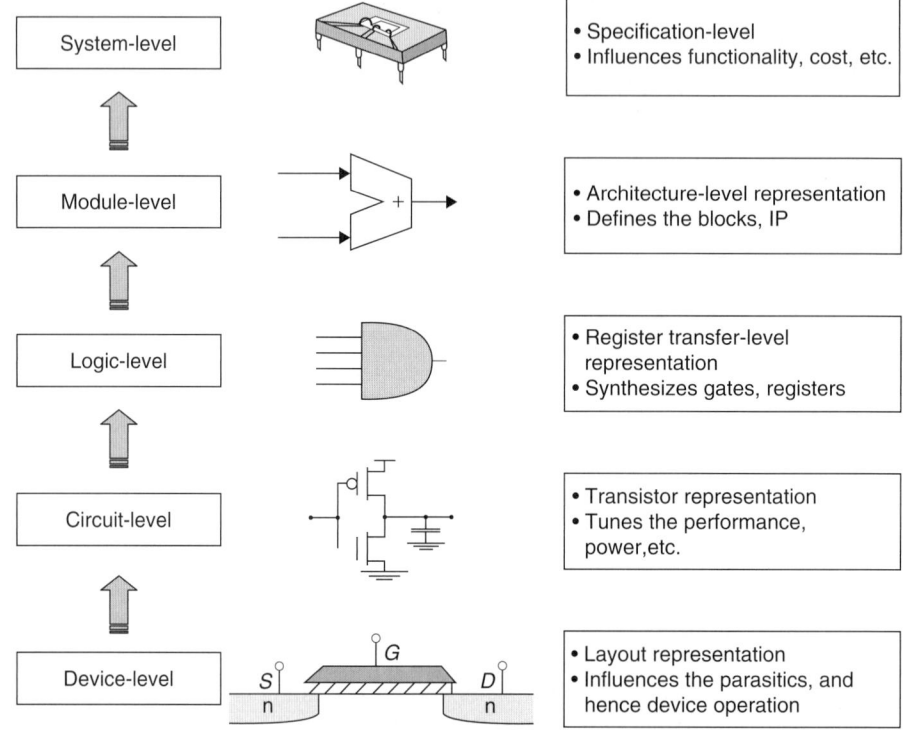

FIGURE 2.20 Design abstraction levels.

Dilemma of SOC Challenges versus Deep Submicron (DSM) Technology Challenges

While we have seen why the complexity involved in enabling SOC integration requires a good level of hierarchical abstractions, the submicron process issues at 90 nm and beyond require an extremely thorough understanding of the underlying issues involved and an accurate handling of these issues. In Figure 2.21, the left-hand box summarizes a whole set of challenges that are typically referred to as the "macroscopic" challenges in SOC design where design abstractions are mandatory to manage and address these issues. On the other hand, the right-hand box lists a whole set of DSM issues that are referred to as "microscopic" SOC design challenges where detailed analyses of these effects are essential to meet the design goals. This, as one would expect, is a contradiction of sorts and has resulted in the development of two key SOC design methodologies to address this dilemma:

1. *Design planning.* Early SOC planning techniques and procedures so that the right level of tradeoffs can be understood and appropriate decisions are taken
2. *Design closure.* Implementation and optimization techniques to meet the performance, area, and power requirements of an SOC in the presence of the DSM effects

Introduction to System-on-Chip (SOC)

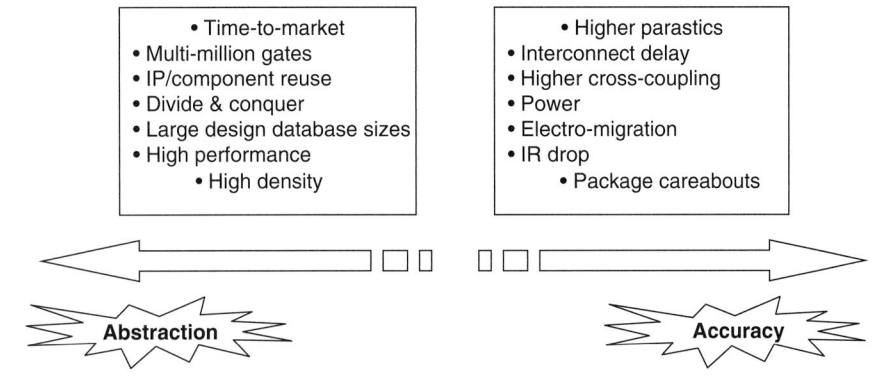

FIGURE 2.21 SOC versus technology challenges.

Design Planning

Design planning is one of the most critical phases of SOC development where upfront tradeoffs and decisions are made so that downstream integration and implementation become seamless toward achieving overall design closure, as illustrated in Figure 2.22. So, in some sense, careful design planning can help address the "macroscopic" challenges described above and avoid costly time-consuming iterations to meet SOC goals. One of the most critical aspects in design planning is the process of estimation. Design planning is all about providing quick, but fairly decent estimates of several of the "microscopic" factors described above, so that downstream silicon implementation can be achieved with minimal surprises and an avoidance of issues requiring costly iterations. This is achieved by a prototyping methodology where initial decisions are forced, physical and timing effects are estimated, and these estimates are checked for

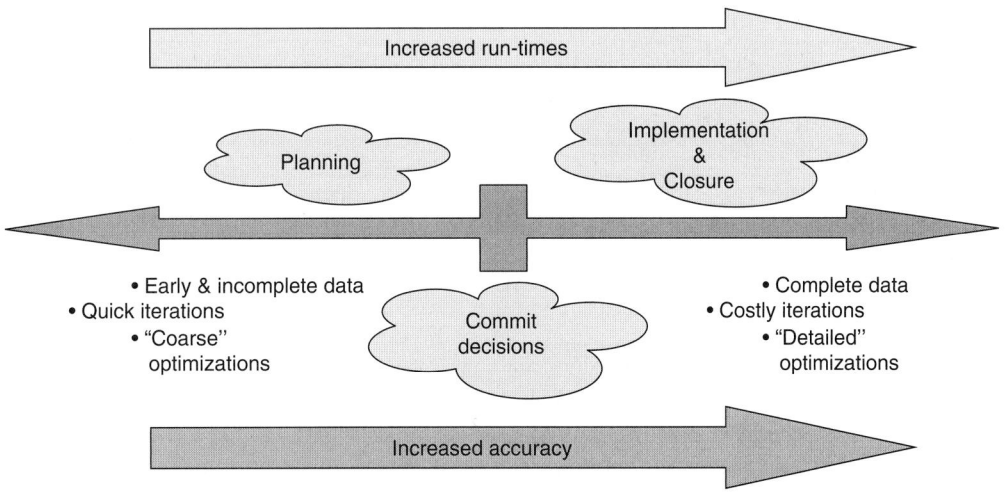

FIGURE 2.22 SOC prototyping and tradeoffs.

feasibility, thereby validating the original decisions. This iterative process continues until the "implementation prototype" is refined sufficiently such that the confidence of committing to the original decisions is extremely high. The obvious challenge in the above methodology is the "quick" versus "accurate" contradiction, and the success of design planning wholly depends on the quality of the involved tradeoffs. Following are some of the SOC design planning activities that fall into this input-estimate-refine process:

- *Floor planning.* The first step in design planning is to physically plan the components on the SOC. This involves defining the areas on the die that will be occupied by the incoming hard IP, memory blocks, mixed-signal or other custom blocks, and external I/O cells. An initial placement of these components is arrived at based on basic connectivity information between them and forms the original "forced" baseline to estimate the overall size, timing, power, and other factors to check for physical feasibility. Based on the estimates obtained, this original floor plan can be refined or tweaked until there is good confidence in achieving critical goals such as the die size (or area), timing, and power.

- *Size estimation.* This process involves calculating the minimum silicon area that would be required to accommodate all the components (logic, memory, I/O cells, IP blocks, interface logic, other special macros, etc.) and the amount of interconnect wire required to connect up all these components. The SOC specification drives the selection of appropriate memory configurations, I/O choices, and the physically ready IP blocks required. Given that this reuse is occurring in the "hard" form, estimating the size of these components is fairly trivial. Estimating the logic area is not as straightforward and requires a decent estimate of the amount of logic to be integrated along with a targeted logic density achievable for that particular logic library and process technology. Estimating the interconnect wiring area requires using a routing efficiency factor that represents the overhead associated in connecting up all the above components meeting the design rules. Other contributions that are typically considered overheads arise from the physical power grid distribution required on-chip to meet the SOC power and performance goals as well as any other special spacing careabouts during physical integration to address issues such as noise, crosstalk, and other such effects.

- *Power estimation.* Given that several SOC applications such as mobile handsets and portable appliances require a very tight control over the power that is dissipated, early estimation of power is crucial right from the SOC architectural or system level. This level has the largest impact on making the right level of tradeoffs to reduce the power either in terms of tweaking the application algorithms or in deciding the need for voltage scaling to address power. Once a technology node and library are selected for the SOC implementation, power estimation is done by determining the amount of switching logic and the switching activity per block. Switching activity information can come from vectors generated by application test cases that represent worst-case SOC working operation.

- *Timing estimation.* Given the fact that metal resistance per unit length is increasing with scaling, while gate output resistance is decreasing, and compounded by the fact that average wire lengths are not coming down, the delay contribution from the interconnect is continuously increasing. This makes timing estimation a very critical component of design planning. Traditional approaches to estimating the interconnect delay were based on the concept of wire-load models (WLM). These models are statistically generated and provide an estimate of the parasitics of the wire in relation to fanout, and therefore can be used to estimate the delay of the interconnect wire on the chip. However, this approach has long been replaced by more realistic interconnect delay calculation models and methods. Given the several process and scaling effects discussed earlier, the RC parasitics of the wire depend on a lot more factors than just fanout; hence, a WLM was considered grossly inaccurate to model these effects. Inaccuracies in such estimates caused timing surprises later during physical implementation, and therefore poor convergence in the design flow. The criticality of the physical aspects impacting the circuit and interconnect delays such as the wire length, coupling capacitance between neighboring wires, and clock signal skews require access to a lot more physical implementation information to reasonably estimate interconnect delays. This pioneered the advent of the physical synthesis technology where the underlying logic synthesis and placement of the SOC components occurred concurrently, thereby providing a lot more accuracy in delay estimation and hence a lot more confidence in the timing feasibility of the design. In addition to the delay estimation, the timing feasibility of the SOC implementation also depends on the design and IP constraints that drive the timing optimization. Design planning also involves verifying the timing specification of the SOC by validating these constraints and budgeting the top-level constraints among the several soft and firm IPs being integrated on the SOC. Timing abstractions of hard IPs are used during this process.

- *Routability estimation.* The process of determining whether the original SOC physical size estimation is sufficient to achieve design closure is done by an interconnect wire routing resource availability versus demand analysis, also called congestion analysis. Given the original floor plan, timing, and other physical constraints, budgets for various soft IP and physical and timing abstractions for the hard IP, a quick power grid distribution, and global placement are done. This global placement is then used as a starting point to virtually route all interconnects between the components keeping in mind the timing or other constraints fed in. This virtual route is a pretty good estimate of the availability and demand for routing resources to ultimately connect up the SOC components, and hence is a good measure of routability. Congestion "hot spots," if any, are tweaked by placement changes and another design planning iteration done to verify the SOC design goals.

Hierarchical SOC Design

As the thirst for modern electronics continues to require larger levels of logic integration on a single chip, it is not uncommon to see SOC designs with over 10 to 20 million gates required to be integrated. This poses significant complexity and challenges through the SOC "create" flow, requiring a divide-and-conquer approach to attacking the problem. Design planning and implementation of such large SOCs is enabled using a hierarchical design methodology as shown in Figure 2.23. The basic principle of hierarchical design is to use the design planning framework described above to break the SOC implementation

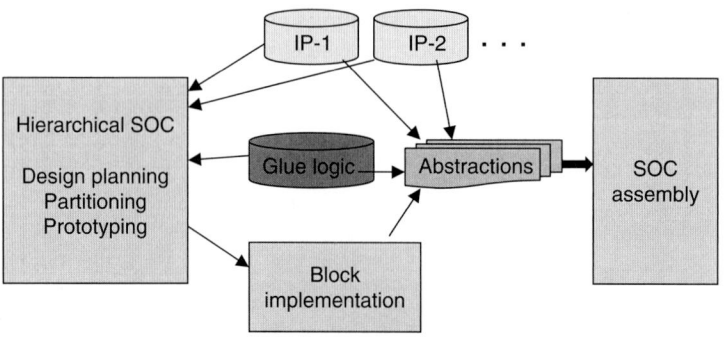

FIGURE 2.23 Hierarchical SOC design.

into independent blocks that can be concurrently taken through design closure with the top-level implementation. As is evident, the incoming hard IP, custom macros, and memories do not get broken up during this process and the design planning process would treat them as "black boxes" with their timing, physical, and other aspects abstracted to enable estimation and feasibility analysis of the SOC floor plan and design closure. Once the SOC logic is broken up into smaller-sized blocks, also called "soft blocks," they are treated exactly like the soft IP being integrated in the design. These soft blocks are referred to as the SOC partitions. Simple guidelines can be used to decide the partitions such as,

- Logic gate count capacity limitations of design tools.
- Timing criticality of the logic requiring more local and controlled design closure.
- Minimizing the impact of design changes on overall design cycle-time by localizing the implementation of the change to those partitions that are affected
- Potential or known reuse of the soft blocks on future SOC in physical form (as hard IP)
- Multiply instantiated soft blocks enabling design closure and physical implementation on such blocks only once

Once the partitioning process is complete, all partitioned "soft blocks" are treated similarly to the rest of the hard IP and macros on the chip, and the design planning techniques discussed above are used to determine the feasibility of the overall SOC floor plan, placement, and partitioning decisions so as to enable seamless assembly of these components (soft blocks, hard IP, memory, I/O, etc.) toward design closure. Given that these components are abstracted using timing and physical models, the focus in hierarchical design is primarily on interconnect optimization and the timing between these components, thereby reducing the complexity.

Design Closure

As indicated in Figure 2.24, the definition of SOC design closure is the simultaneous process of meeting the speed, power dissipation, area, signal integrity, and reliability

Introduction to System-on-Chip (SOC)

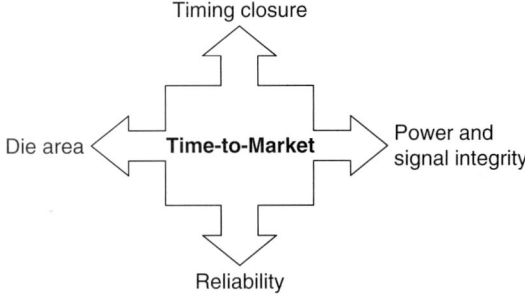

Figure 2.24 What is design closure?

requirements of the device, while at the same time ensuring that the critical time-to-market goals are met. The complexity to enable this has been very well indicated by a study that Collett International did back in 1999, shown in Figure 2.25, when it polled several SOC design teams on the effort in terms of number of iterations they took to solve this concurrent optimization problem and how this problem became worse as DSM effects became more predominant below 180 nm.

As indicated earlier, technology scaling is causing feature sizes to become tinier and tinier, as a result of which the electrical behavior of interconnect wires is becoming more critical. As shown in Figure 2.16, while the wires are getting closer to each other, their current carrying requirements have resulted in increased aspect ratios and thereby much higher coupling capacitance between neighboring signals. When the signals in the neighboring wires switch, the coupling capacitance causes a transfer of charge between them. Depending upon the switching transition, significant crosstalk noise can be generated that can cause both delays in the signal propagation as well as functional problems due to glitches. Considering these physical aspects of the wires during

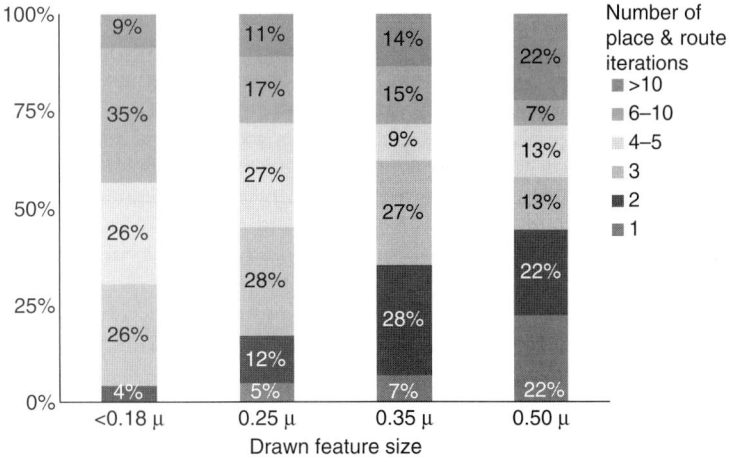

Figure 2.25 Design closure complexity.

placement and routing is therefore critical to avoid such signal integrity issues while timing optimizations are being done. Note the concurrent nature of the solution that this demands. Aggressive scaling of interconnect wires is also increasing the resistance per unit length of these wires and the average current densities. With logic switching at high speeds as well, depending on the magnitude of the current flowing in the power grid and the length, width, and sheet resistance of the power grid, the actual voltage as seen by the switching logic can be much less than the true supply voltage. This slows down the transistor performance characteristics and hence can cause a timing violation in the circuit. This problem can be overcome by designing a robust power grid on the SOC in such a way that a minimum voltage level is guaranteed across the chip, and then ensuring that the performance of the device is met at that voltage. However, as evident, since not all switching logic in an SOC are equally timing critical, this can cause overdesign of the power grid, and hence over-constrain the routing resources required to ensure the SOC routability and area goals are met. Again, note the concurrent nature of the optimizations needed to achieve overall design closure, as defined earlier. Another critical phenomenon in the presence of high interconnect current densities and high-speed SOC components is the signal or power electromigration problem. The migration of metal ions due to the electron wind caused by the current causes voids (opens) or hillocks (shorts) between neighboring interconnect wires, thereby potentially causing functional failures. A common solution to the electromigration problem is to increase the width of the interconnect wires or add more vias to the power grid, again impacting the total routing resources available. Electromigration is not an initial time phenomenon, in the sense that while the device will function at the time it is manufactured, the longer-term life of the device operation is at risk due to these reliability issues. Design closure for SOC designs is therefore a multi-optimization problem, and an integrated approach is required to address all the careabouts. Also critical from a time-to-market perspective is a methodology where the above signal integrity and reliability issues can be avoided during the physical design process as opposed to addressing or fixing them as an afterthought and thereby incurring painful iterations and increased cycle times.

Mixed-Signal Integration

One of the recent challenges in SOC design has been the integration of complex digital circuitry and analog or RF components on the same chip. This has been necessitated by the ever-increasing demand for applications such as wireless handsets, WLAN products, single-chip satellite TV setup boxes, and Bluetooth-enabled products. Integrated mixed-signal components could be high-performance phase locked loop (PLL) blocks, high-speed I/O interfaces, RF modules, or high-speed and high-resolution analog-to-digital converters (ADCs) and digital-to-analog converters (DACs). The substrate is the connecting layer between all circuits in a single piece of silicon. Thus when high-speed digital switching components inject noise or produce spiky signals, they get injected into the common substrate. This noise can impact the sensitive analog circuitry on the same chip. Further complicating the issue is the technology scaling trends that allow higher frequency and reduced voltage of operation as shown in Figure 2.26. All these issues have resulted in silicon failures or reduced yields for such mixed-signal, RF-integrated SOC designs.

The key contributor of the digital noise that gets injected into the substrate in the SOC context is the power supply, given that the CMOS core and I/O logic cause spikes on the supply lines which in turn are connected to the substrate. The other significant contributor is the package bond wire inductance (L) that can increase the $L\,di/dt$ noise

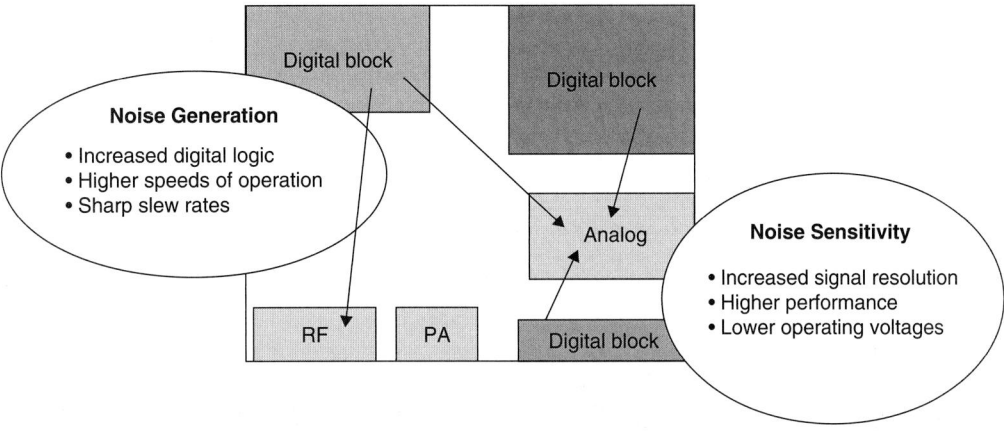

FIGURE 2.26 SOC mixed-signal scaling challenge.

generated, where dI/dt is the current slew rate. Improper power grid structure, high clock speeds and clock skew, and very sharp transition times on the signals can all contribute to the noise that gets generated on the die. Given the severity of the substrate noise issue, several techniques and approaches have been discussed and attempted to minimize, if not eliminate, this issue. The challenges in addressing the issue lie in the fact that in order to analyze the performance degradation of noise on sensitive analog circuits, a good measure of the noise generated is needed, and for the huge, complex SOC designs being talked about here, this process is computationally prohibitive, hence the need to accurately model the noise sources from digital blocks. It is important that the SOC design planning phase captures the process of enabling noise management by careful planning and modeling of various noise sensitivities and ensuring that guidelines are followed to minimize substrate noise injection, thus enabling the successful integration of analog and RF components. Noise sensitivity analysis can involve the identification of the sensitive circuits on the die and a specification of the maximum amount of substrate noise these circuits can tolerate. Common guidelines and techniques that are followed include

- Physically separating the power domains for noisy and sensitive circuitry
- Reducing the impedance on the power/ground network
- Ensuring a good distribution of power and ground pads to minimize the effective inductance
- Minimizing the inductance on the package
- Adding on-chip decoupling capacitors wherever possible
- Placing guard rings that are tied to a quiet supply around sensitive circuits
- Using a low-impedance backside contacting to obtain good noise rejection

One of the technological developments in the area of mixed-signal integration has been in the development of digital RF techniques to overcome the challenges of integrating RF components in advanced CMOS technologies. RF components today take up more than 40 percent of a mobile phone printed circuit board. This will only increase as integration of functions such as Bluetooth, wireless LAN, and GPS are included. While integrating such components on a bipolar CMOS (BiCMOS) process is possible, it cannot match up with the aggressive yield and test cost goals of high-volume applications such as mobile phones. Integration in advanced BiCMOS processes such as SiGe is possible, but typically this technology is one or two process nodes behind the digital process technology.

While CMOS will continue to be the technology of choice for mobile applications, work continues to be needed to address the challenges of keeping up with other alternatives like bipolar technologies (SiGe) in terms of power efficiency and high performance for mixed-signal and RF components, or GaAs processes for power amplifier circuits that all go into a wireless system. Packaging advances will also influence this roadmap moving forward since system-in-package (SIP) and system-on-package (SOP) approaches are fast becoming viable solutions for integrating RF components on a single package rather than on the same die. Breaking the functionality into separate digital and analog components provides the flexibility of rapidly shrinking the SOC devices without getting locked into the shrinking constraints imposed by the analog circuits.

2.5 Summary

In this chapter we presented system-on-chip (SOC) as a way to provide a customized optimal solution for various electronics systems of the Internet era. We discussed key customer requirements and showed how by moving from multiple chips on a board to a single-chip solution, SOCs address the application requirements. Such levels of integration are made feasible by advances in CMOS manufacturing technology. However, with these advances come design challenges too—in terms of verification and testing of these complex systems, which include software running on embedded processors, and also in terms of chip create and design implementation, which maximizes technology entitlement in the presence of deep submicron silicon effects. We highlighted these design challenges and presented approaches to address them.

While CMOS scaling enables increasing levels of integration, single-chip integration may not always be the most optimal solution. This is true for heterogeneous systems that require analog, RF, flash memory, and power management type of components along with digital blocks. Analog design, for example, is not able to leverage CMOS scaling as aggressively as the digital blocks, and in terms of system cost for a given power and performance requirement, an appropriate partition of the system across multiple chips using the SOP concept may provide a better solution.

References

1. "International Technology Roadmap for Semiconductors (ITRS)—2004 Update," http://www.itrs.net/Links/2004Update/2004Update.htm.
2. Vijay K. Madisetti and Chonlameth Arpikanondt, *A Platform-Centric Approach to System-on-Chip (SOC) Design*, Springer, 2005.
3. Henry Chang et al., *Surviving the SOC Revolution—A Guide to Platform-Based Design*, Kluwer Academic Publishers, 1999.
4. Rochit Rajsuman, *System-on-a-Chip: Design and Test*, Artech House, 2000.

5. Wayne Wolf, *Modern VLSI Design: System-on-Chip Design*, 3rd ed., Prentice Hall, 2002.
6. Ricardo Reis and Jochen Jess, *Design of System on a Chip: Devices & Components*, Springer, 2004.
7. Farzad Nekoogar and Faranak Nekoogar, *From ASICs to SOCs: A Practical Approach*, Prentice Hall Modern Semiconductor Design Series, 2003.
8. Michael Keating and Pierre Bricaud, *Reuse Methodology Manual for System-on-a-Chip Designs*, Springer, 2007.
9. Andreas Meyer, "Principles of Functional Verification," Newnes, 2003.
10. Sadiq M. Sait and Habib Youssef, *VLSI Physical Design Automation: Theory and Practice*, McGraw-Hill, 1995.
11. Naveed A. Sherwani, *Algorithms for VLSI Physical Design Automation*, 3rd ed., Springer, 1998.
12. Giovanni De Micheli, *Synthesis and Optimization of Digital Circuits*, McGraw-Hill, 1994.
13. Jan M. Rabaey, Anantha Chandrakasan, and Borivoje Nikolic, *Digital Integrated Circuits*, 2nd ed., Prentice hall, 2002.
14. Neil H. E. Weste and Kamran Eshraghian, *Principles of CMOS VLSI Design*, Addison-Wesley, 1994.
15. John L. Hennessy and David A. Patterson, *Computer Architecture: A Quantitative Approach*, Morgan Kaufmann Pub, 1996.
16. H. B. Bakoglu, *Circuits, Interconnections, and Packaging for Vlsi*, Addison-Wesley VLSI Systems Series. 1990.
17. Jari Nurmi et al., *Interconnect-Centric Design for Advanced SOC and NOC*, Springer Publisher, 2004.
18. Nozard Karim and Tania Van Bever, "System-in-package (SIP) design for higher integration," *IMAP*, 2002.
19. Jun-Dong Cho and Paul D. Franzon, *High Performance Design Automation for Multi-Chip Modules and Packages (Current Topics in Electronics and Systems, Vol. 5)* World Scientific Pub, 1996.
20. John H. Lau et al., *Chip Scale Package: Design, Materials, Process, Reliability, and Applications*, McGraw-Hill, 1999.
21. "System-in-package or system-on-chip," *EETimes*, http://www.eetimes.com/design_library/da/soc/OEG20030919S0049.
22. Giovanni De Micheli and Mariagiovanni Sami, *Hardware/Software Codesign*, Springer, 1996.
23. M. Abramovici, M. A. Breuer, and A. D. Friedman, *Digital Systems Testing and Testable Design*, New York: IEEE Press, 1990.
24. Eric Bogatin, *Signal Integrity—Simplified*, Prentice Hall PTR, September 2003.
25. "System-in-package (SIP): Challenges and opportunities," *Proceedings on the 2000 Conference on Asia and South Pacific Design Automation*, January 2000.
26. Alfred Crouch, *Design-for-Test for Digital ICs and Embedded Core Systems*, Prentice Hall PTR, 1999.
27. M. A. Norwell, *Electronic Testing: Theory and Applications*, Kluwer Academic Press, 1995.
28. Abromovici Miron, Melvin A. Breuer, and Arthur D. Friedman, *Digital Systems Testing and Testable Design*, New York Computer Press, 1990.

29. K. Keutzer, S. Malik, A. R. Newton, J. Rabaey, and A. Sangiovanni-Vincentelli, "System-level design: orthogonalization of concerns and platform-based designs," *IEEE Transactions on Computer-Aided Design,* vol. 19, no. 12, December 2000.
30. Alberto Sangiovanni-Vincentelli et al., "Platform-based design," http://www.gigascale.org/pubs/141/platformv7eetimes.pdf.
31. Alberto Sangiovanni-Vincentelli et al., "Benefits and Challenges for Platform-Based Designs," *Proceedings of the 2004 Design Automation Conference.*
32. G. Carpenter, "Low Power SOC for IBM's PowerPC Information Appliance Platform," http://www.research.ibm.com/arl.
33. Wayne Wolf, "The future of multiprocessor systems-on-chips," *Proceedings of the 2004 Design Automation Conference,* pp. 681–685.
34. Gary Smith, "Platform-based design: does it answer the entire SOC challenge?" *Design Automation Conference, 2004.*
35. A. Sangiovanni-Vincentelli and G. Martin, "A vision for embedded systems: Platform-based design and software methodology," *IEEE Design and Test of Computers,* vol. 18, no. 6, 2001, pp. 23–33.
36. F. Vahid and T. Givargis, "Platform tuning for embedded systems design," *IEEE Computer,* vol. 34, no. 3, March 2001, pp. 112–114.
37. Jan Crols and Michiel Steyaert, *CMOS Wireless Transceiver Design,* Springer, 1997.
38. R. Saleh, S. Z. Hussain, S. Rochel and D. Overhauser, *Clock skew verification in the presence of IR-drop in the power distribution network,* IEEE Transactions on Computer Aided Design of Integrated Circuits and Systems, vol. 19, issue 6, June 2000, pp. 635–644.

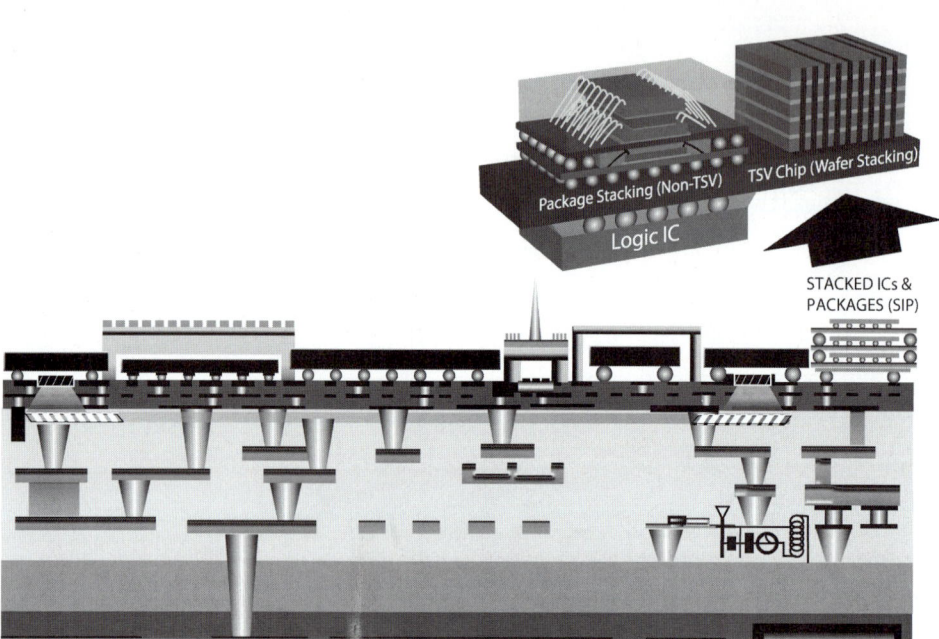

CHAPTER 3

Stacked ICs and Packages (SIP)

Baik-Woo Lee, Tapobrata Bandyopadhyay, Chong K. Yoon, Prof. Rao R. Tummala
Georgia Institute of Technology

Kenneth M Brown
Intel

3.1	SIP Definition 82		3.4	TSV SIP 121
3.2	SIP Challenges 85		3.5	Future Trends 143
3.3	Non-TSV SIP 93			References 144

The ever-increasing demands for miniaturization and higher functionality at lower cost processes have driven the development of stacked ICs and packages (SIP) technologies. The SIP is a single miniaturized functional module realized by the vertical stacking of two or more similar or dissimilar bare or packaged chips. Bringing the chips closer together enables the highest level of silicon integration and area efficiency at the lowest cost, compared to mounting them separately in traditional ways. In doing so, the electrical path length between chips is reduced, leading to higher performance. In addition, this technology allows the integration of heterogeneous IC technologies like analog, digital, RF, and memory into one package, resulting in the integration of more functionality in a given volume. Because of these attributes, SIP technology is emerging as a strong contender in a variety of applications that include cell phones, digital cameras, PDAs, audio players, laptops, and mobile games to be delivered in an innovative form factor with superior functionality and performance.

SIP is being accomplished currently at the bare chip, package, or wafer level by employing either traditional interconnection technologies including wire bonding or flip chip (referred as non-TSV SIP) or advanced assembly technologies such as through-silicon-via (TSV) and wafer-to-wafer bonding (referred to as TSV SIP). This chapter provides a broad overview of a variety of SIP architectures being pursued in the industry. It reviews the SIP challenges up front regarding electrical, materials, processes, mechanical, and thermal issues. It then follows up with a review of the status of each of these in two main areas—SIP by non-through-silicon vias and SIP by through-silicon vias.

3.1 SIP Definition

3.1.1 Definition

The SIP is often referred to and defined as "system-in-package," implying that it is a complete system in a package or module. It is also described often as a multichip module (MCM). But the MCM has been a huge, multibillion dollar market going back to the 1980s and 1990s when IBM, Hitachi, Fujitsu, and NEC poured billions of dollars into developing the 2D MCM technology with as many as 144 ICs on a single substrate to meet the ultrahigh computing needs. This technology is still used and is expected to continue to be used since the 3D technology, described in this chapter, will not solve the thermal problems at 150 to 200 W per chip in a multichip processor system. On the other hand, for any package to be a system, it must fulfill all system functions of a system board. These include not only actives and passives but also multilayer wiring, thermal structures, system I/Os or sockets, and power supplies. But this has not been demonstrated with SIP to date. Most SIP technologies often describe stacking of either the bare chips or packaged chips in three dimensions. This chapter views SIP in this latter context. SIP is defined, therefore, as a 3D module with two or more similar or dissimilar stacked chips. The SIP can be divided into two major categories: (1) interconnection of stacked chips as achieved by traditional chip assembly technologies such as wire bonding, tape automated bonding (TAB), or flip chip and (2) interconnection of stacked chips as achieved by more advanced chip assembly technologies such as through-silicon-via (TSV) and direct bonding of one chip to the other without the traditional wire bonding or flip chip technology. The former stacking is referred to in this chapter as non-TSV and the latter as TSV. The non-TSV technologies can be further classified into chip stacking and package stacking, as described later in this chapter. The TSV technologies, as described in this chapter, can be used to bond not only bare ICs but also wafers and Si chip carriers, thereby ending up with more functional subsystems or complete systems.

3.1.2 Applications

Since SIP includes both similar ICs such as dynamic random access memory (DRAM) and dissimilar ICs such as logic and memory, the applications for SIP are as broad as ICs themselves. These applications, therefore, include high-volume manufacturing for mobile consumer products such as multifunction handsets, MP3 players, video-audio gadgets, portable game consoles, and digital cameras, to name a few.

3.1.3 CEO Figure and SIP Categories

Figure 3.1 shows a summary of how SIP technology has evolved during the last 40 years. SIP technology is divided into two major categories: non-TSV and TSV technologies, as defined earlier. As shown in Figure 3.1, the concept of SIP or 3D integration of ICs was first

Stacked ICs and Packages (SIP)

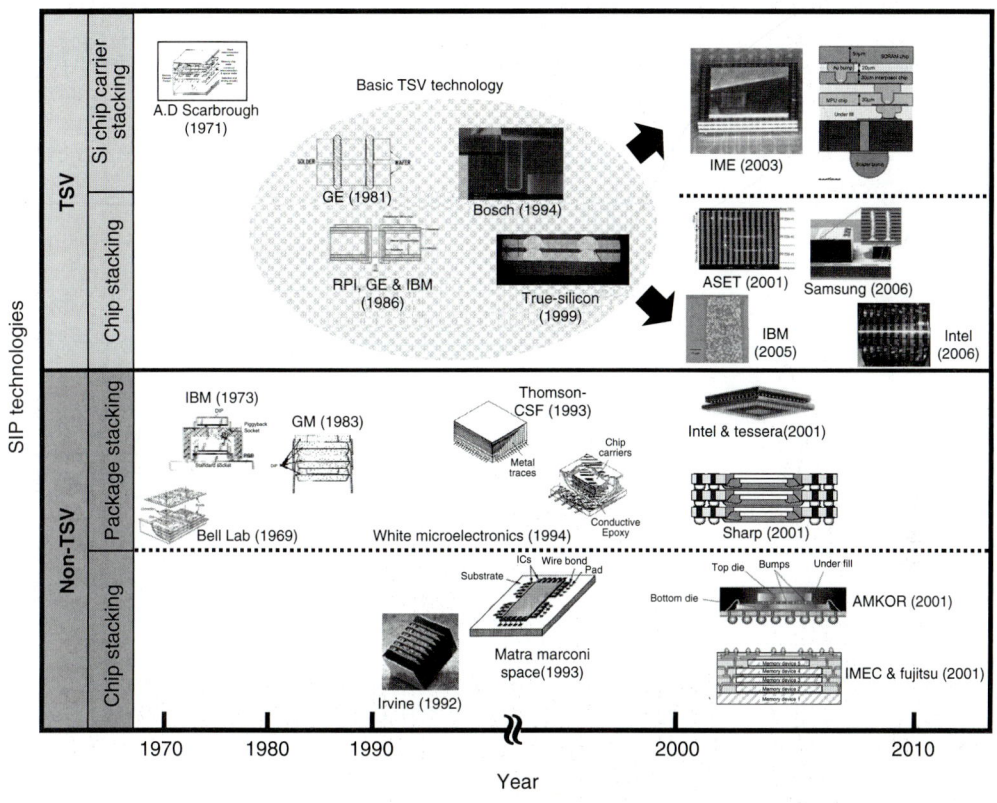

FIGURE 3.1 Evolution of SIP technologies during the last 40 years. (Courtesy of PRC, Georgia Tech.)

introduced about 40 years ago by Bell Laboratories and IBM. A modern 3D chip stacking by non-TSV was, however, successfully introduced by Irvine Sensors in 1992, wherein chips are stacked and interconnected by side metallization. Subsequently, chip stacking by wire bonding technology was widely adopted, since abundant infrastructure was readily available. This led to more advanced types of chip stacking of more than 20 chips. As expected, the wire bonded stacking gave rise to flip chip stacking for higher performance or miniaturization.

Si chip carrier stacking by the TSV technology, as shown in Figure 3.1, was first introduced by A. C. Scarbrough in 1971. About a decade later, GE, IBM, and RPI realized its importance and introduced the TSV technology for chip stacking. The earliest through-silicon vias in this era were fabricated using anisotropic chemical etching on both sides of silicon. As its value in the miniaturization of modules became more evident, a number of companies and research organizations, including Bosch, ASET, Samsung, and TruSilicon, began to explore other more advanced ways to form vias as well as to bond chips with TSVs, as included in Figure 3.1. More recently, companies began to look at TSV as the solution not only for 3D stacking of memories but also as a more complete solution to high-performance systems replacing the traditional ceramic or organic substrates with ultrahigh-density wiring, dielectrics, vias, I/Os, and thin-film components. One such view by IBM is illustrated in Figure 3.2.

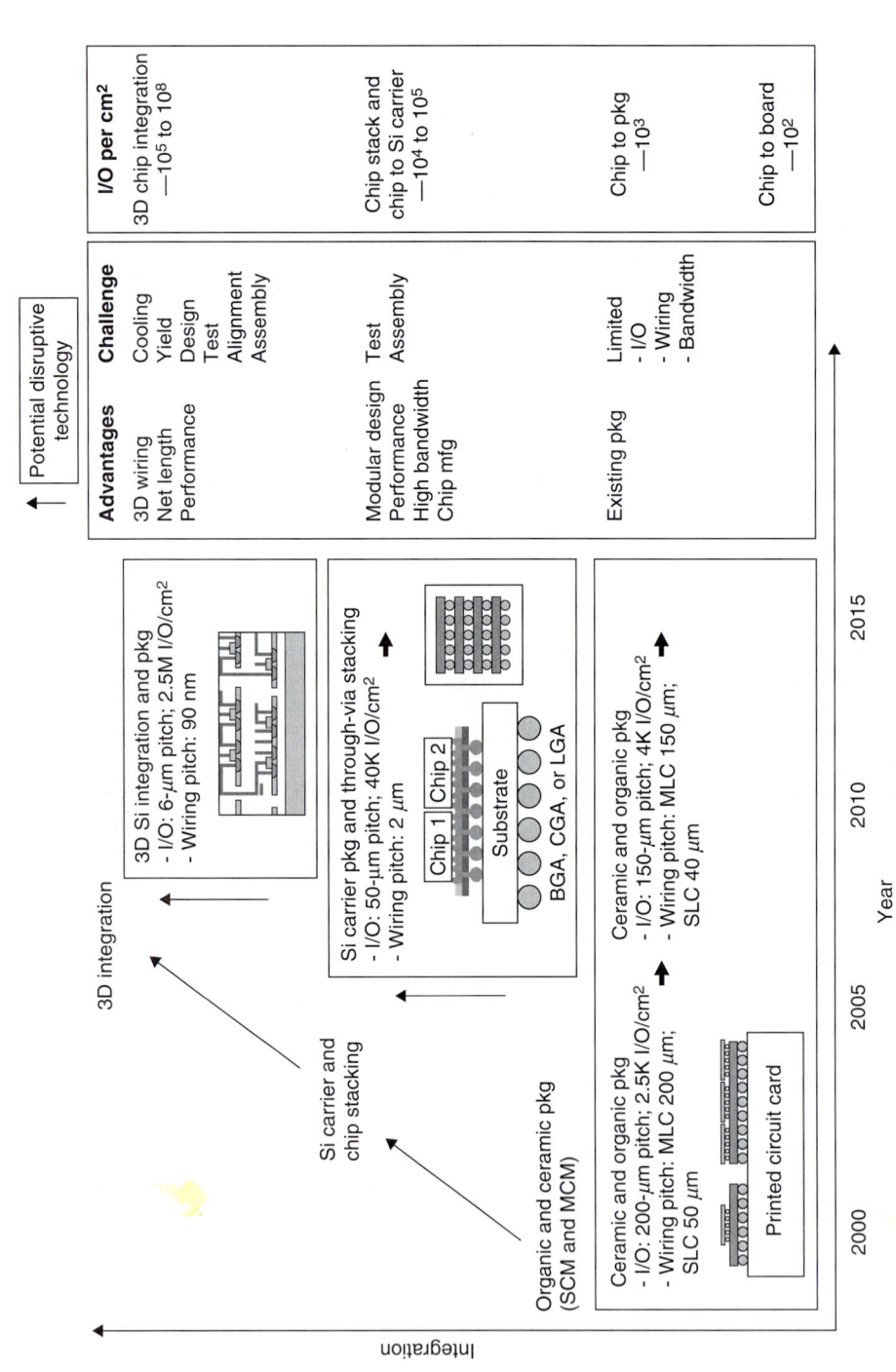

Figure 3.2 A silicon integration comparison for ceramic and organic packaging, silicon carrier and chip stack, and 3D silicon circuits and wiring. (Courtesy of IBM) [1]

Stacked ICs and Packages (SIP)

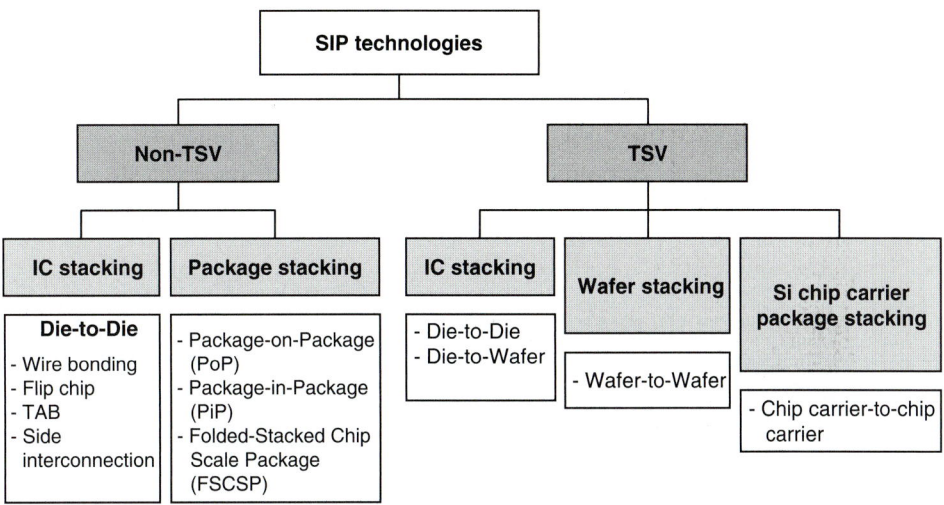

FIGURE 3.3 Classification of SIP technologies into non-TSV and TSV technologies for stacking ICs, packages, wafers, and Si chip carriers.

Figure 3.3 shows the classification of SIP technologies into non-TSV and TSV technologies. The non-TSV technologies include the traditional chip assembly technologies—wire bonding, flip chip, TAB, and side interconnection. The non-TSV technologies also include stacking of package-on-package (PoP), package-in-package (PiP), and folded-stacked chip-scale package (FSCSP) such as by Intel and Tessera.

The TSV technologies, on the other hand, leverage silicon-through-via interconnections for forming 3D structures. Such structures include die-to-die, die-to-wafer, wafer-to-wafer, chip carrier–to–chip carrier, and ultimately silicon circuit board with silicon devices, packages, or interposers.

3.2 SIP Challenges

Figure 3.4 shows major SIP challenges including materials and process, mechanical, electrical, and thermal issues.

3.2.1 Materials and Process Challenges

Materials and processes involved in the fabrication and assembly of SIP are numerous and complex. And, in addition, some of their electrical, thermal, and mechanical behaviors are not well understood either. Therefore, it is a great challenge to understand the needs up front and develop or select materials to fabricate, assemble, and characterize SIP modules with the right combination of materials and processes. For decades, many attempts have been made to characterize important materials parameters that are necessary for producing successful modules with the right combination of electrical, thermal, mechanical, and thermomechanical properties.

Electrical parameters such as the dielectric constant, insulation resistance, electrical conductivity, loss factor, temperature coefficient of capacitance (TCC), and temperature coefficient of resistance (TCR) are very important material properties that affect insulators, resistors, capacitors, inductors, and filters, to name a few. Thermomechanical

86   Chapter Three

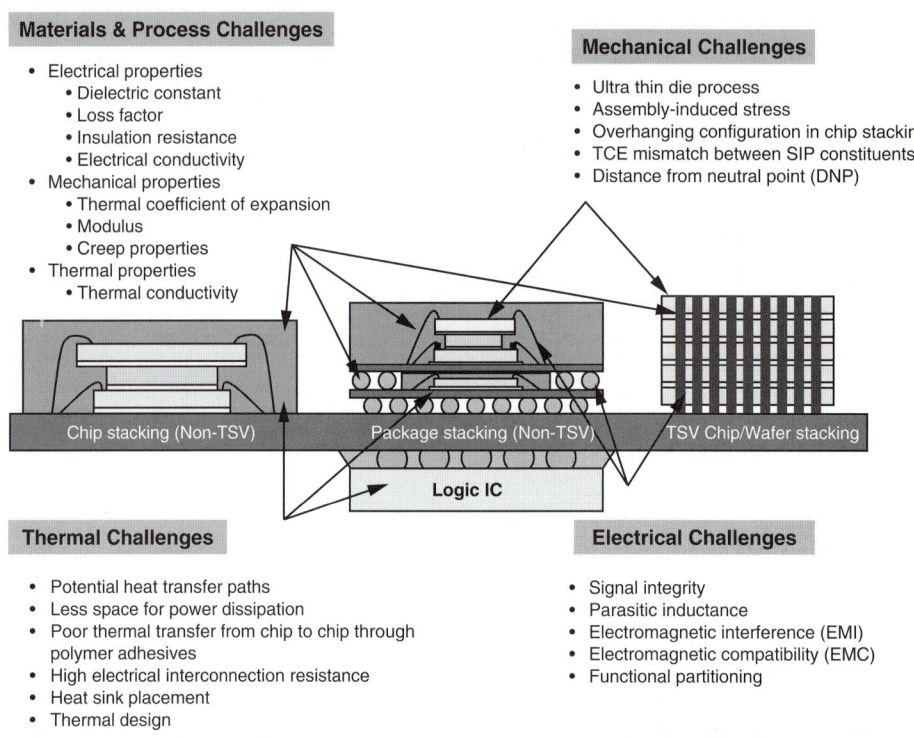

FIGURE 3.4   Major challenges in SIP technologies.

reliability, on the other hand, depends on such thermal and mechanical parameters as the thermal coefficient of expansion (TCE), modulus, and temperature and on time-dependent mechanical properties such as the creep property, fracture toughness, and temperature- and humidity-dependent fatigue properties. Thermal parameters such as thermal conductivity are also a very important property for effective conductive heat dissipation from chips to substrates to modules and systems. In addition, one should also consider all the intrinsic and extrinsic parameters of materials such as the microstructure, porosity, grain size, alloying effects, and physics of failure.

Thermomechanical reliability of interconnection technology has been a major source of reliability problems. Interconnect materials, such as Cu and Al, and bonding and assembly materials, such as lead-free solder and anisotropic conductive film (ACF), have been successfully used with and without underfill encapsulations. In addition, a variety of compliant interconnections that can withstand a TCE mismatch between chips and substrate during thermal cycling have also been developed. While all these and others that are described in Chapter 10 (wafer-level SOP and interconnections) have been successfully used in traditional IC packaging, the challenge remains how to solve the interconnection and assembly reliability of SIPs with stacked ICs with minimal interconnections standoff. The through-silicon via is perhaps the ultimate challenge with little or no interconnection height. Another challenge has to do with costacking of Si and GaAs chips with their different TCEs. Since most SIPs are stacks of Si ICs with TCE around 3 parts per million per degree Celsius (ppm/°C), their assembly to organic substrates with TCE around 16 ppm/°C is another challenge.

3.2.2 Mechanical Challenges

Stacked die packages pose several mechanical challenges that can affect product performance and reliability. First is the relationship between die thickness and die size. Since there are several chip designs for stacking various types of random access memory (RAM), the die size and pad ring must be selected that best match the product requirements. Chip thickness in advanced stacked die configurations are currently at 75 micrometers (μm) for stacks up to seven dies or more. When stacking thin dies in a wire bonded package, particular attention has to be paid to the bonded and nonbonded sides of the overhang, since the wire bonding process imposes significant force on the die during manufacturing. A stacked die at 75 μm usually has little to no overhang to avoid the die cracking during the wire bonding process, whereas an overhang up to 2 mm or more can be achieved when the thickness is allowed to increase to 150 μm or greater.

Silicon functionality and transistor performance can also be adversely affected in thin-die situations in stacked packages if the stack, overhang, and material selection are not chosen carefully. Because of the piezoresistive effects of silicon, assembly-induced stress can adversely affect device performance. TCE mismatch between silicon, substrates, mold compound, and die attach adhesive produce additional thermomechanical stresses. In particular, spacer and adhesive materials play a large role in the total stress applied to the silicon. In addition, these packaging materials are all polymers with widely different mechanical properties (modulus and TCE) below and above their glass transition temperatures. Reducing packaging-induced stress involves, therefore, a proper selection of material properties and processing steps. Evaluating the stress is typically done through device performance after packaging or through up-front finite element models. Finite element models are capable of evaluating the residual stresses generated due to the complex assembly process but need to be validated in each case. Validation is performed through package warpage and in-plane measurements such as Moire interferometry techniques.

Solder joint reliability (SJR) is also an area of great concern in die and package stacking applications. Material selection for solder joints, solder joint design, intermetallic compounds formation, overmold materials, and the substrate core material all play a role in joint fatigue life. A TCE mismatch between the IC and package as well as between the package and board drives the fatigue shear strain in the solder joints. Two competing factors that determine the worst joint reliability are the global TCE mismatch driven by the distance from neutral point (DNP) effect and the local TCE mismatch between the package and the substrate. The ballout pattern and the die sizes are very critical for identifying the worst joints for failure under temperature cycling. Die size and local TCE mismatch are the primary drivers in perimeter array logic packages. For memory packages, the ballout patterns are typically smaller than logic packages since the DNP as the main driver for solder joint fatigue is small. In addition to this, the mold cap height is also an important parameter for thin flexible substrates. The move from eutectic Sn-Pb solder to lead-free Sn-Ag-Cu solder will enhance the temperature cycle performance of the package due to the lead-free solder's better creep properties. An example of a typical shear-driven package-level failure of solder during the temperature cycle is shown in Figure 3.5.

In cyclic bend and drop conditions, the package stiffness plays an important role. A stiffer package results in more forces being transmitted to the solder ball, resulting in faster failure. At high strain rates typically experienced in drop conditions, a much stiffer lead-free solder results in earlier failures than leaded. Compliance of the solder

Chapter Three

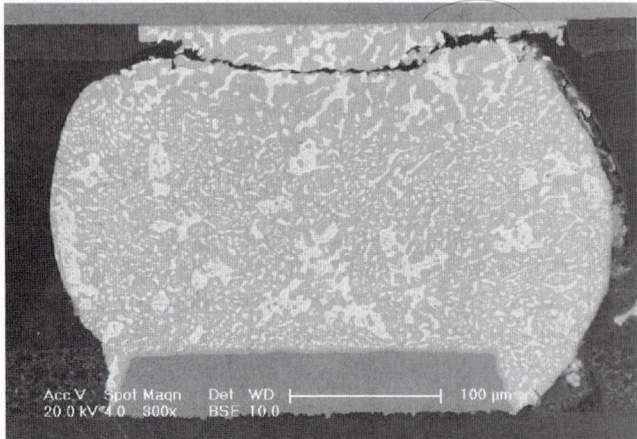

FIGURE 3.5 Shear driven temperature cycle failure at a solder joint.

and the brittle intermetallic interfaces also govern the failure mode in drop conditions. Figure 3.6 shows a typical brittle intermetallic failure.

3.2.3 Electrical Challenges

With the increased die or chip stacking, the density of I/O interconnections is increased dramatically. Moreover, for high-performance requirements, the interconnect speed through

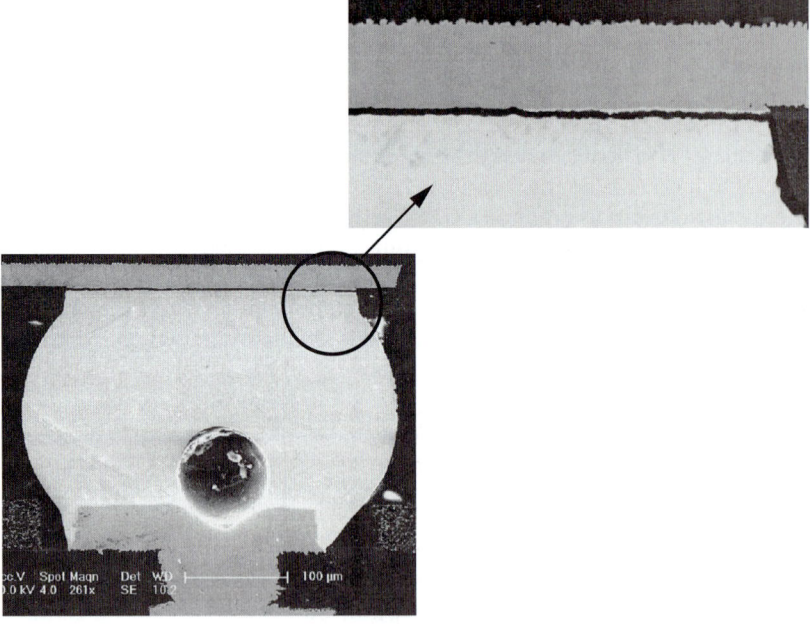

FIGURE 3.6 Failure of a solder joint during drop testing.

each interconnection such as a wire bond, transmission line, via, or solder ball needs to increase. From a signal integrity point of view, a higher interconnect speed means a more difficult package design due to the constraints of package size, layer number, and cost. For example, with the data rate under 50 MHz, a chip-scale package can be treated as a circuit block with R (resistor), L (inductor), and C (capacitor) elements and the impact on signal integrity is limited. However, when frequency goes to 500 MHz or greater, a package is no longer a "small" portion of the signal propagation path, and "full wave" theory behavior must be considered. As a result, the package design and associated technology need to pay specific attention to electrical performance. First, the signal path on the package level needs a well-designed reference and return path. For example, each signal line needs its nearby power or ground path for reference as well as for crosstalk shielding. An excellent reference design means more power or ground connections per signal and that can prove costly. Therefore, accurate predictions for the right ratio of signal to power not only provide good performance, but also the lowest cost. Second, the package design needs to provide a path for higher IC power delivery. For cost and form factor reasons, it would not be desirable to put decoupling capacitors on the package. Parasitic inductance, therefore, from the package needs to be extremely slow in order to minimize voltage fluctuations during circuit switching. The way to keep a clean power supply for stacked die packages is to mainly focus on package V_{ss} (source voltage) and V_{cc} (collector voltage) design for the lowest loop inductance. Although on-die decoupling capacitors help to reduce power noise, it is usually not the first choice due to the added cost factor. Third, electrical package design requires consideration of electromagnetic inference (EMI) and electromagnetic compatibility (EMC). With higher-density wire bonds in place, coupling between wire bonds becomes more significant. The problem becomes more severe when high-power circuits are close to lower-power circuits. For example, when RF circuits and digital circuits are within one package, the electrical design needs to make special considerations for the isolation between digital and RF in order to minimize the EMI and EMC impacts.

3.2.4 Thermal Challenges

As chips and passive components are closely stacked and mounted, thermal management challenges become major bottlenecks. Figure 3.7 shows the trend of stack-die packages

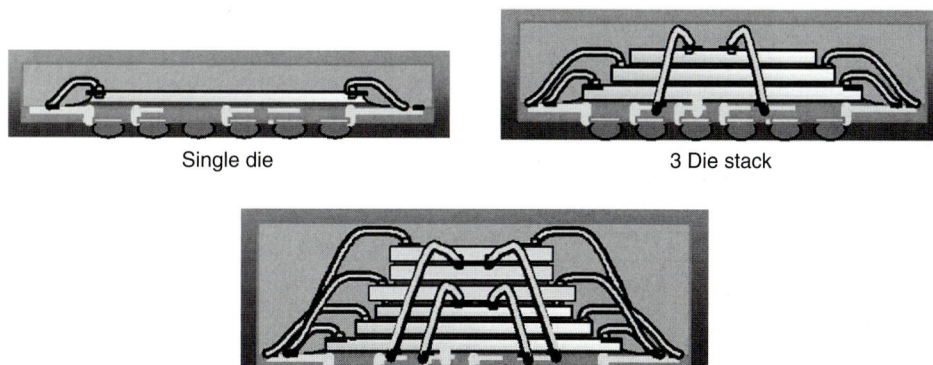

Figure 3.7 Typical trend of stack-die packages.

90 Chapter Three

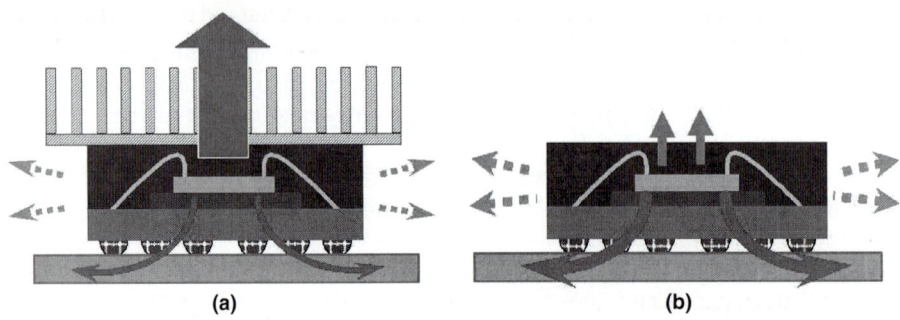

FIGURE 3.8 Different heat transfer paths (a) with and (b) without a heat sink.

that will impose several thermal challenges that include high electrical interconnection resistance, poor thermal transfer from chip to chip through polymeric adhesives, and less space for power dissipation.

The first step in the thermal design of SIP is to understand the potential heat transfer paths. Figure 3.8a shows an example with a heat sink mounted on top of a SIP. In this configuration, the majority of heat generated by the SIP will be conducted to the heat sink, and then to the external ambient by either natural convection or forced-air convection. In addition to that, a small portion of heat is dissipated through the package substrate, vias, solder balls, and then the printed circuit board. Only a very little portion of heat is dissipated through radiation. Figure 3.8b shows an example without a heat sink on top of a SIP. Under this configuration, the majority of heat generated by the SIP is dissipated through the printed circuit board. Natural convection as well as radiation can account for some dissipation through the package surfaces. In this particular configuration, radiation usually plays an important role to help dissipate heat. Neglecting radiation effects under this configuration may result in significant errors. Thus, heat dissipation paths strongly depend on thermal designs. Understanding the potential heat transfer paths and fully utilizing them in SIP designs leads to thermo-mechanical reliability of SIPs.

The second step in the thermal design of SIP is to place hot components close to the main heat transfer paths. Figure 3.9 shows examples of hot component placement under different system designs. If the majority of heat is dissipated through the board or by natural convection, the hot component should be placed close to the package substrate. On the other hand, if the major heat transfer path is from the top surface such as through radiation, the hot component should be placed near the package top.

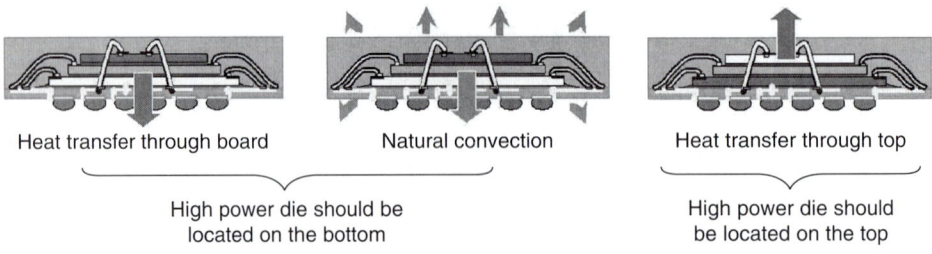

FIGURE 3.9 Examples of the hot component placement under different system designs.

The third step in the thermal design of SIP is to understand the thermal characteristics of SIP. There are two levels of thermal characterization. One is package-level thermal characterization, and the other is system-level thermal performance. The package-level thermal characterization can provide a better understanding of the package thermal behavior due to different packaging architectures, thermal interface materials, and operating environments. The JEDEC JC15 committee has defined several package-level testing standards as described here:

- *JESD51-2*. Integrated Circuits Thermal Test Method Environment Conditions—Natural Convection (Still Air) [2]. The purpose of this document is to outline the environmental conditions necessary to ensure accuracy and repeatability for a standard junction-to-ambient (θ_A) thermal resistance measurement in natural convection.
- *JESD51-6*. Integrated Circuit Thermal Test Method Environmental Conditions—Forced Convection (Moving Air) [3]. This standard specifies the environmental conditions for determining thermal performance of an integrated circuit device in a forced convection environment when mounted on a standard test board.
- *JESD51-8*. Integrated Circuit Thermal Test Method Environmental Conditions—Junction-to-Board [4]. This standard specifies the environmental conditions necessary for determining the junction-to-board thermal resistance, $R_{\theta JB}$, and defines this term. The $R_{\theta JB}$ thermal resistance is a figure of merit for comparing the thermal performance of surface-mount packages mounted on a standard board.

All these testing standards are solely for the thermal performance comparison of one package against another in a standardized environment. This methodology is not meant to predict the exact performance of a package in an application-specific environment. However, the data generated under these standard environments is very useful for numerical model validation, for exchanging package thermal performance between companies, and for quantification of the degradation in thermal performance post reliability tests.

The fourth step in the thermal design of SIP is to utilize thermal simulations to expedite SIP design optimization. Based on the thermal characterization mentioned above, a numerical model can be generated using the commercial computational fluid dynamics (CFD) and finite-element method (FEM) codes. Figure 3.10 shows a typical

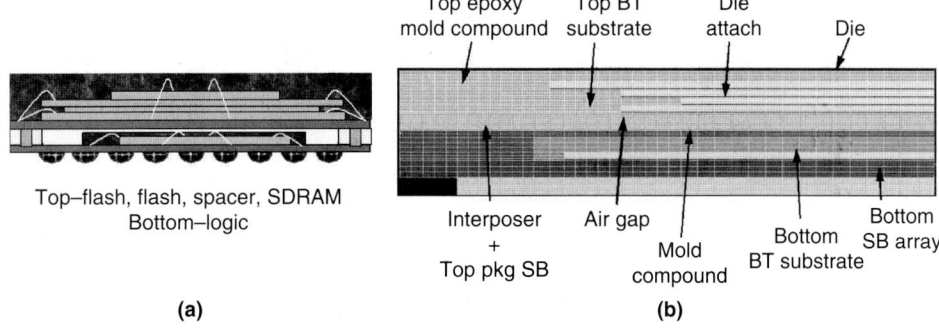

Figure 3.10 A typical example of a SIP thermal model. (*a*) Package cross-sectional view. (*b*) Cross-sectional view of "quarter" thermal model.

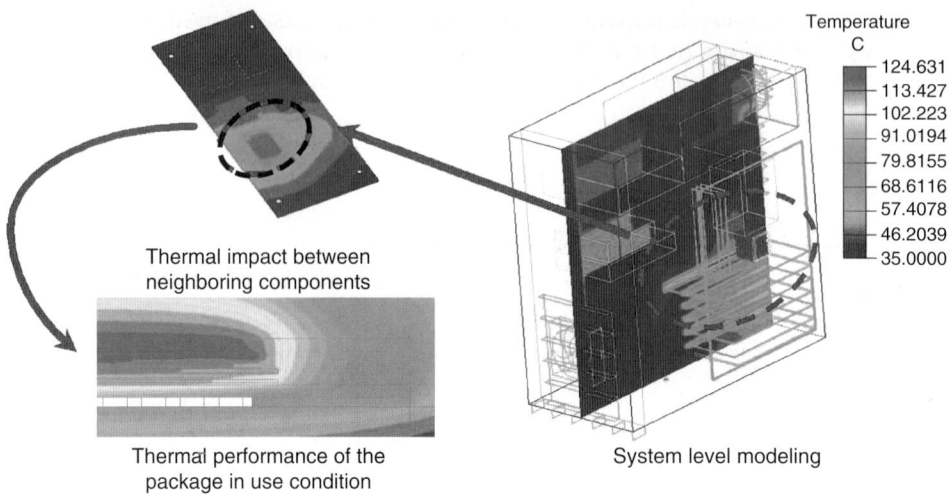

Figure 3.11 An example of a temperature contour predicted from thermal simulation.

example of thermal modeling for a SIP. The model should capture as many details as needed so that any factor that will significantly impact on the SIP thermal performance won't be skipped. Figure 3.11 shows an example of a temperature contour predicted by the thermal simulation. Based on the hot spots predicted by the numerical model, SIP design can be efficiently optimized. Figure 3.12 shows a typical procedure to optimize SIP designs.

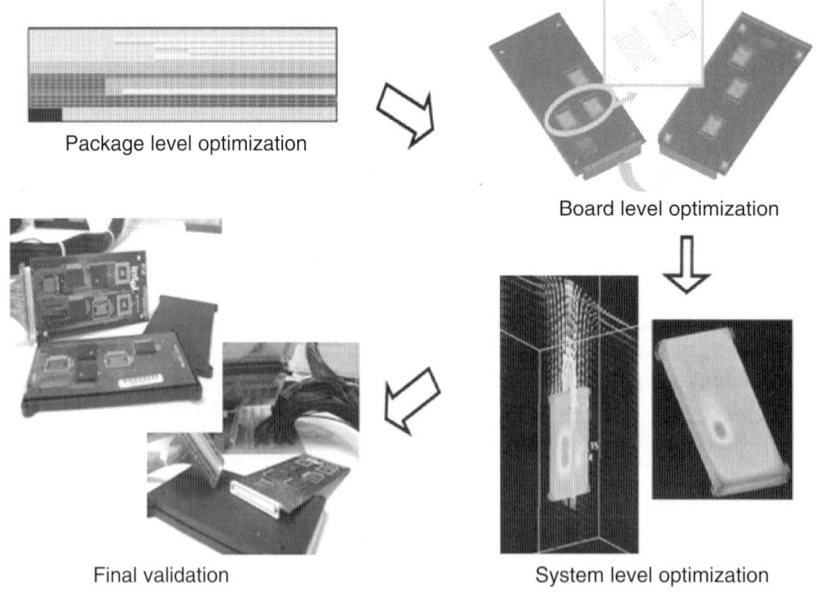

Figure 3.12 A typical procedure for SIP thermal design optimization.

Finally, it is very important to validate the optimized SIP design from thermal simulation by using either thermal test vehicles or the actual products. This is the only way to ensure that the product meets all the specifications.

3.3 Non-TSV SIP

3.3.1 Historical Evolution of Non-TSV SIP

Traditional stacking of chips without through-silicon via (TSV), referred to here as non-TSV SIP, has been developed in close relationship with the evolution of packaging technologies. The overall IC and systems assembly trend in Figure 3.13 reflects this coupling. For example, early chip stackings were accomplished with wire bonding, which has more recently moved to flip chip and which is moving to finer pitch and bumpless interconnections.

In the 1960s, the dual-in-line package (DIP) was developed at the IC level and the pin-through hole (PTH) was developed at the system level to mount the DIPs onto the printed circuit board. The earliest SIP involved the PTH interconnection as shown in Figure 3.14 [6]. Each board is connected by inserting the pins on the board into the holes of connectors, thus yielding a board stacking.

As the DIP and PTH interconnections became more commonly used in the 1970s, the applications of this PTH interconnect in stacking also increased. Figure 3.15a shows chip carrier stacking by utilizing the PTH [7]. The chip carriers are electrically connected through interposers, in which plated notch pins in the interposer periphery are inserted into the holes of the chip carrier. With more common use of DIP, the stacking structure of the DIP was also demonstrated (Figure 3.15b) [8]. The DIP plugs into a so-called piggyback socket, which plugs into a receptacle on the PCB. Beneath the piggyback socket, another DIP plugs directly into the PCB or into a conventional socket.

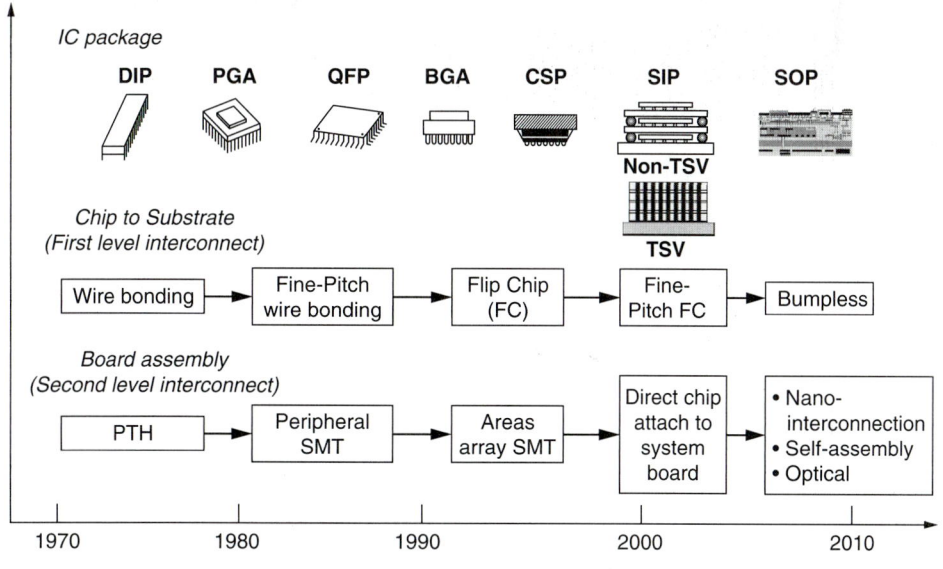

Figure 3.13 Packaging evolution. (Modified from [5])

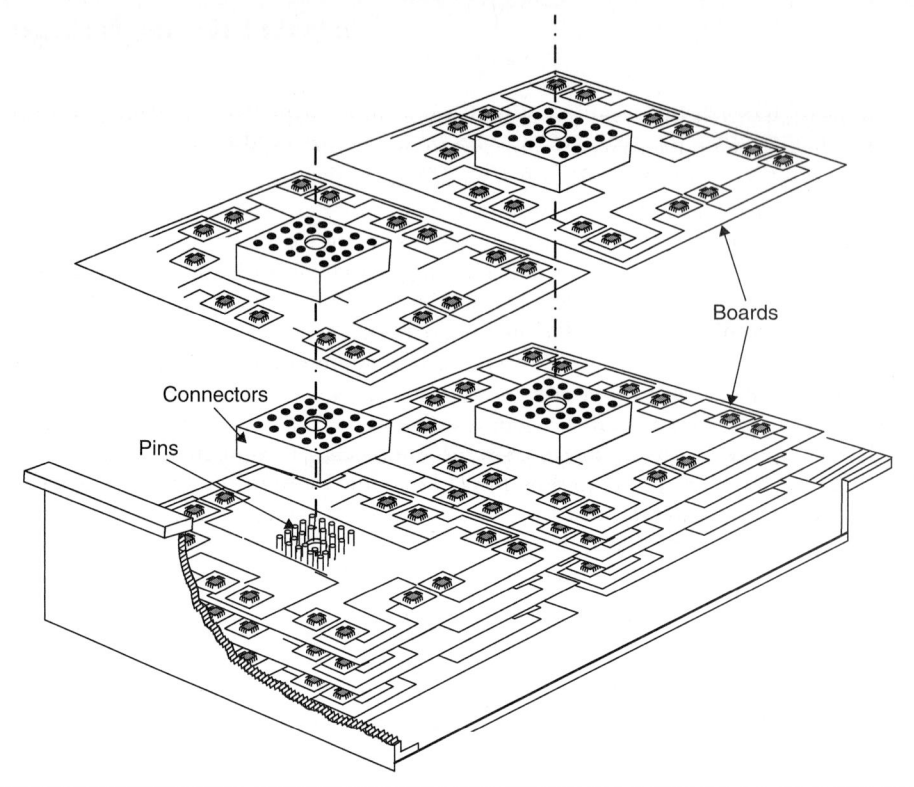

FIGURE 3.14 Early board (patent filed in 1967) stacking by PTH interconnections. [6]

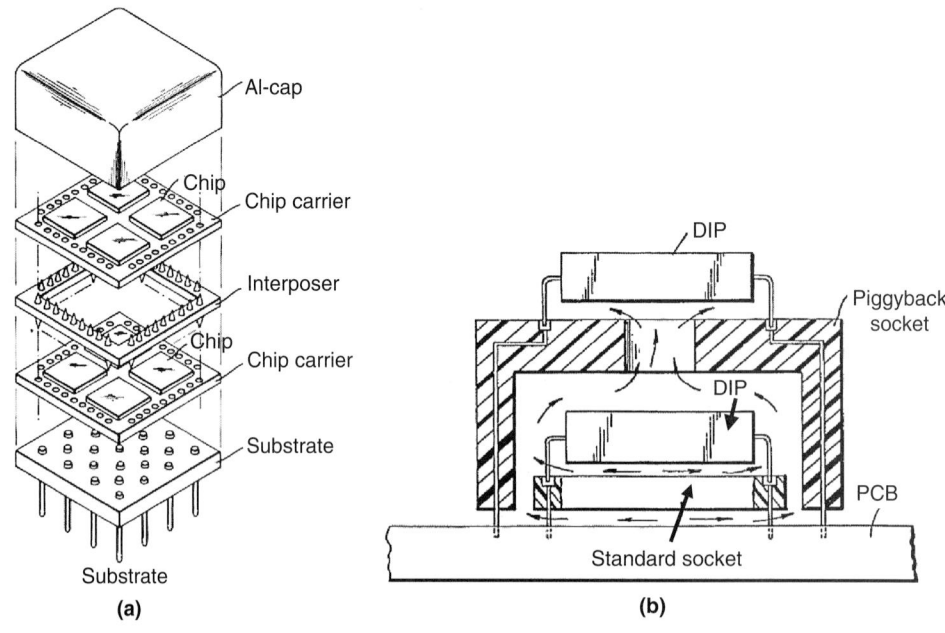

FIGURE 3.15 (*a*) Chip carrier stacking [7]. (*b*) DIP stacking with piggyback socket [8].

In response to the need for higher-density printed wiring boards (PWBs), the 1980s saw the development of surface-mount technology (SMT) and the quad-flat package (QFP). Since packages such as QFP used in SMT have leads and not pins, they can be mounted on both surfaces of the PWB, leading to higher-density packaging. The QFP allows a lead frame to run around on all four sides of a square package, thus enabling higher pin counts. Figure 3.16a shows the DIP stacking with their leads soldered, not needing PTH interconnections or interposers like in Figure 3.15 [9]. Figure 3.16b shows the stacking of J-leaded chip carriers (JLCC) having leads on four sides of the package like for the QFP, in which the leads of the top JLCC are mounted on the pads of the bottom JLCC by solder reflow [10].

Until the 1980s, most stacking technologies involved stacking of boards or completed IC packages such as DIP or JLCC. In this era, packages were, in fact, simply placed one on top of the other in z direction, instead of being mounted on the xy plane of the PCB. There had not been many efforts to reduce either the stack height or the interconnection length between stacked packages as seen in today's true SIP technology. The new generations of SIP technologies with this focus on stack height have started to evolve since 1990.

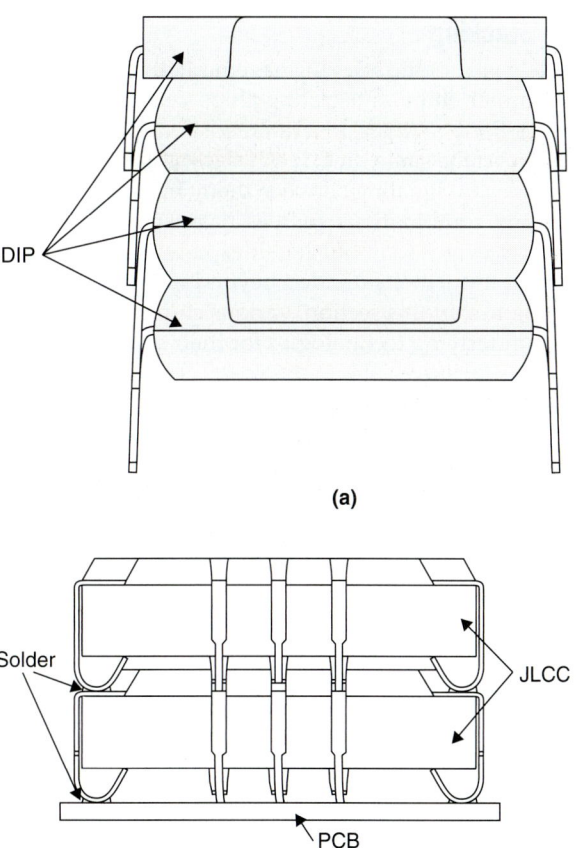

FIGURE 3.16 (a) DIP stacking with soldered leads [9]. (b) JLCC stacking [10].

A new generation of chip stacking began to evolve in the 1990s. It is stacking of bare chips leading to higher stacking density. The electrical performance was also greatly enhanced by employing short interconnections that include wire bonding, tape automated bonding (TAB), or newly introduced side termination methods. Even though the wire bonding technology had been used since the 1970s, it was only in the 1990s when the technology began to be applied to SIP on a commercial scale. A few package stackings were also demonstrated by employing the same configuration as chip stacking with side termination interconnects.

The wire bonded chip stacking interconnection led to the introduction of flip chip interconnection for SIP around 2000 when flip chip became a high-volume assembly technology. Application of embedded IC technology to chip stacking enabled chip-scale package (CSP) stacking. At about this time, some of the limitations of chip stacking technologies were also realized. This led to alternatives to chip stacking, which include package-on-package (PoP), package-in-package (PiP) and folded-stacked chip scale package (FSCSP).

The following two sections describe widely used non-TSV chip and package stacking technologies currently in use.

3.3.2 Chip Stacking

Over the past few years, chip stacking has emerged as an effective solution for integrating similar or dissimilar chips. Integrating chips vertically in a single package multiplies the amount of silicon that can be crammed in a given package footprint, conserving board real estate. At the same time, it enables shorter routing of interconnects from chip to chip, which speeds signaling between them. Initial applications of chip stacking were two-chip memory combinations such as flash and SRAM. Memory stacking remains the most popular even today but includes new variations like flash plus flash. More recently, it has been further extended beyond memories to include the combination of logic and analog ICs. In this section, various chip stacking architectures are introduced with the basic underlying technologies for their chip stacking.

Wafer Thinning, Handling, and Dicing
Advances in chip stacking technologies are enabling the stacking of more dies in a given height. To accomplish this basic goal, a variety of underlying technologies have been developed, which include wafer thinning, thin wafer handling, and dicing as described here.

Wafer Thinning Wafer thinning is an essential process step before chip or package stacking of SIP modules because it reduces stack height and enables the addition of more chips without increasing the overall stack height. Stacking multiple chips helps to minimize the xy package dimensions, while thinning minimizes the total height of the z dimension. Figure 3.17 shows the evolution of wafer production in diameter, thickness, and wafer-thinned dimensions during the last 50 years [11]. Larger wafers require thicker silicon to withstand wafer manufacturing, while new packaging trends such as SIP continue to require thinner final chips. The industry has decreased chip thickness by about 5 percent a year. This trend is expected to continue, leading to wafers as thin as 20 µm by 2015. Figure 3.18 shows the number of chips to be expected in SIP and the Si chip thickness requirement for the stacking [12]. Future SIP definitely requires stacking a larger number of thinner chips.

Stacked ICs and Packages (SIP)

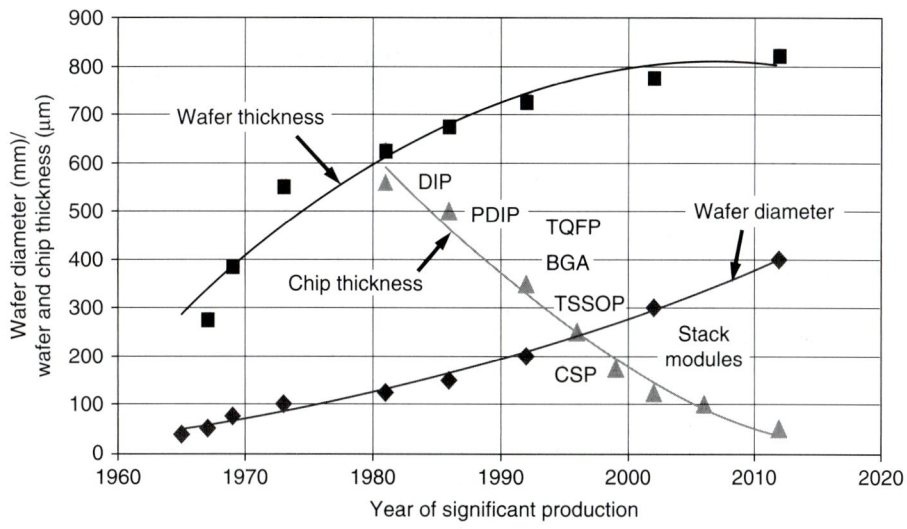

FIGURE 3.17 The evolution of wafer size, its original and final ground thickness during the last 50 years. [11]

Back-grinding was the most efficient way of thinning wafers until recently when it met its limits for acceptable wafer warpage and fragility, at approximately 100 μm. It includes two process steps: (1) coarse grinding and (2) fine grinding. Coarse grinding uses larger diamond particles so as to remove silicon faster for greater throughput, but this process induces substantial wafer damage. Fine grinding removes coarse grinding damage with better surface finish and die strength. Fine grinding uses smaller diamond particles and thus removes silicon at a slower rate than coarse grinding, resulting in a

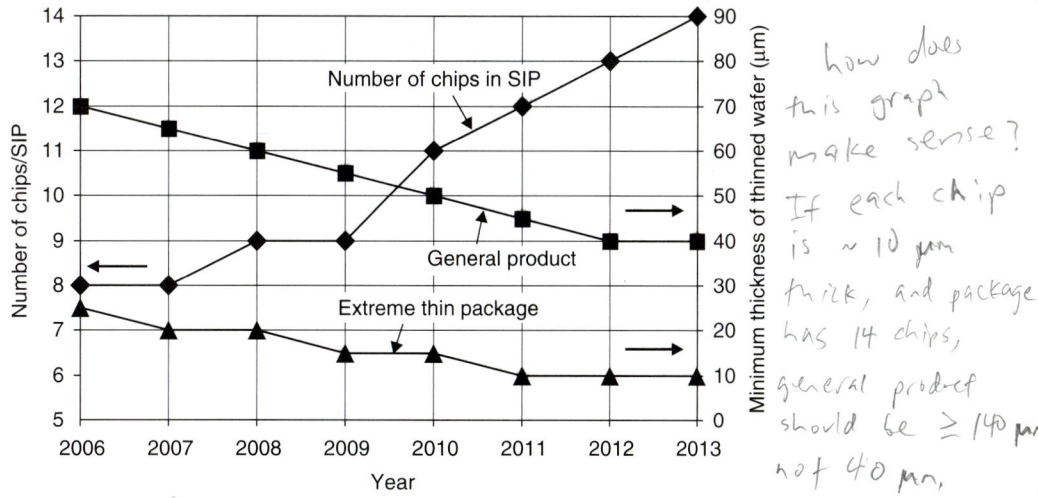

FIGURE 3.18 The SIP technology trend with the number of chips per SIP and Si wafer thickness requirement. [12]

smoother surface. It should be noted that both coarser and fine background wafers produce wafer warpage because of the damaged layer created during the back-grinding process. This damage also becomes a source of cracks that propagate subsequently when the chips are stressed.

The next step after grinding is polishing, which is required to remove or reduce the damage produced by fine grinding, leading to enhancement of wafer strength and wafer warpage. Several polishing methods have been employed, which include chemical and mechanical polishing (CMP), dry and wet polishing, and dry (downstream plasma) and wet etching.

Chemical and mechanical polishing (CMP) [13]. CMP uses a special pad with slurry containing ammonium hydroxide to simultaneously chemically etch and mechanically remove silicon. This synergy between chemical and mechanical processes in the CMP process reduces the mechanical forces required for polishing. The CMP removes most of the damage caused during coarse and fine grinding, restoring both the mechanical strength and wafer bow to their original status, while giving wafers a mirror finish.

Dry polish. Dry polishing uses an abrasive pad without any chemicals, as the name suggests. Wafers that undergo stress relief through dry polish are expected to have higher die strength, lower surface roughness, and lower wafer warpage compared to those undergoing the grinding process only.

Wet polish [14]. Wet polish is a well-developed process used to remove silicon wafer. It uses slurry, composed of SiO_2 and water (around 40 to 50 wt% of SiO_2). Silica particle sizes range from 5 to 100 nm, with an average size of about 30 to 50 nm. This process mechanism is similar to CMP with a lower capital investment and a lower processing cost but with a lower throughput, however.

Dry etching (downstream plasma) [15]. Microwave-excited plasma reacts with the silicon wafer surface to chemically remove silicon. The tool uses a mixture of SF_6 and oxygen gases. Silicon is removed by the chemical reaction: $Si + 4F \rightarrow SiF_4$. The oxygen in the gas mix removes the reaction products. Such a downstream plasma process reduces wafer heating to below the 90°C, preventing face tape burn. It also produces a significantly higher silicon removal rate, as compared to wet etching. This process is contactless and does not require protective tape to be applied to the wafer topside, which reduces the associated processing costs and allows for processing of bumped wafers.

Wet etching. Wet etching can be applied, provided that wafers remain relatively thick to withstand physical handling. Such etchants as HF and HNO_3 are typically included in spite of process control difficulties that limit the wet-etch processes.

Thin-Wafer Handling Secure handling and processing of very thin wafers are generally accomplished by temporarily bonding a rigid carrier onto the wafer front side before thinning. Well-known techniques use polymeric bonding agents like wax, thermally releasable adhesive tapes, or dissolvable glues [16]. Although wax bonding is commonly used, it is a time-consuming process and needs specific cleaning procedures to remove residuals of the bonding layer. Application of thermal release tapes has become a widespread method in order to support wafers with low topographies during thinning processes. A carrier is attached to the wafer to be processed by means of a double-sided adhesive tape with one side thermally releasable. The carrier is removed by heating it to between 90 and 150°C. Dissolvable glue is another widely used bonding material for thin-wafer handling. Its spin-coating allows for very thin and uniform adhesive layers, which is of the utmost importance when wafers of a final thickness are in the range of 10 μm. This method has also been shown to embed surface topography. Thin dies can be released from the carrier without mechanical force by immersing the wafer into a solvent bath that dissolves the glue. Ultraviolet (UV) sensitive bonding materials have also been applied that can be released after UV laser irradiation through a transparent glass carrier such as quartz [17]. However, these polymer-based bonding techniques are limited to temperatures below 200°C. Further increased thermal stability is required to allow process steps like sintering of back-side metal or plasma etching of dielectric layers. A more advanced thin wafer handling concept is based on electrostatic forces, which do not need any polymeric bonding materials [18]. Bonding and debonding of thin wafers onto electrostatic carriers is achieved within a very short time, in a repeatable manner, and without any constraints regarding surface contaminants from bonding agents. It was also shown that this electrostatic attraction state remains active at temperatures even above 400°C.

Thin-Wafer Dicing Singulation of thin wafers is another major process before stacking. The current conventional wafer dicing process uses a diamond-bonded wheel to cut through the full depth of the wafer and into the mounting tape. This mechanical dicing method induces such problems as unacceptably high rates of chipping on the front and back surfaces of the die, delamination of mechanically brittle layers such as low-k inter layer dielectric (ILD), and the formation of microcracks. These cracks and chipping are especially detrimental to thin chips. Several alternative dicing methods for thin dies are being explored including dice before grind (DBG) [19] and laser singulation [20]. DBG has been developed to reduce the breakage of ultrathin chips by chipping. The front side of a wafer is partially diced before grinding and then the back side of the wafer is ground, leading to separation into single dies. As the die is partially diced initially, stress is relieved at the free edges of the die. However, DBG requires a special dicing tool, increasing ownership costs and adding complexity to the process. Laser-based dicing presents a simple dry process that minimizes the handling and processing of thin wafers. No special tapes are necessary as the wafers are diced on standard polyolefin tape, using standard wafer carriers. Laser dicing offers such benefits as minimal chipping, high-yield, small kerf width, and high die strength. However, laser dicing creates large heat-affected zones (HAZs), causing the low-k layer delamination and cracking. To avoid this problem, water jet-guided laser dicing technology is being developed, in which water can reduce the effect of HAZs [21].

Wire Bonded Stacking

The wire bonded stacking method uses the traditional wire bonding technique for the vertical interconnection of stacked chips. Wire bonding is the most popular chip interconnection method in chip stacking because of its existing low-cost infrastructure and flexibility. This stacking technology has been used by a number of companies, including Hitachi, Sharp, Amkor, Intel, and Hynix. Applications of this stacking technology are not only for memory chip stacking such as DRAM, SRAM, and flash EPROMs for mobile applications, but also for the heterogeneous chip stacking of logic and memory chips.

Wire bonded stacking is typically configured in a pyramid or overhang stacking fashion with the same size dies or larger dies than the bottom one, as shown in

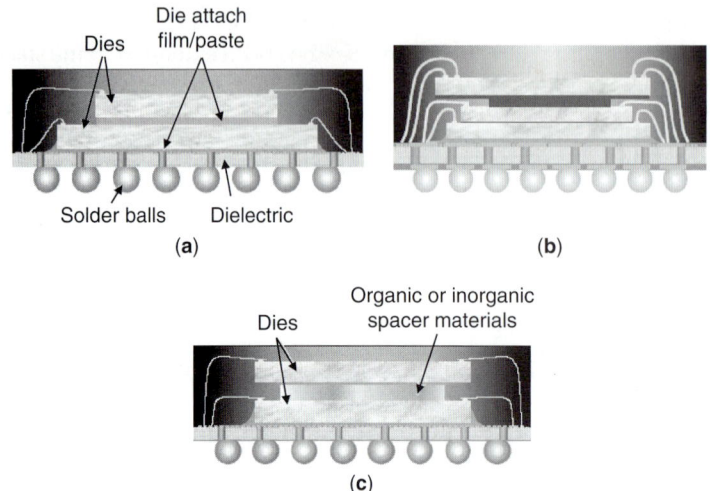

Figure 3.19 Configurations of wire bonded stacking. (a) A pyramid stacking. (b,c) Overhang stacking. (c) is with the same size die but (a) and (b) are not. [22]

Figure 3.19 [22]. The pyramid stacking is the most common die arrangement because wire bonding of a smaller chip over a larger chip can be done very simply with that arrangement. In the overhang stacking, a spacer or rotated die is needed to provide the clearance to enable wire bonding. These configurations are chosen appropriately for the application with the consideration of die thickness, spacer thickness, die overhang, die stack order, wire length, wire profile, and pad placement on the chip. Figure 3.20 shows some examples of the stacked chips by wire bonding. In Figure 3.20a, four chips are stacked with one Si-spacer in between [23], showing a mixed configuration of pyramid and overhang stacking. Figure 3.20b shows an advanced wire bonding capability to stack 20 memory chips, each 25 μm in thickness, maintaining the total height of the stack to 1.4 mm [24].

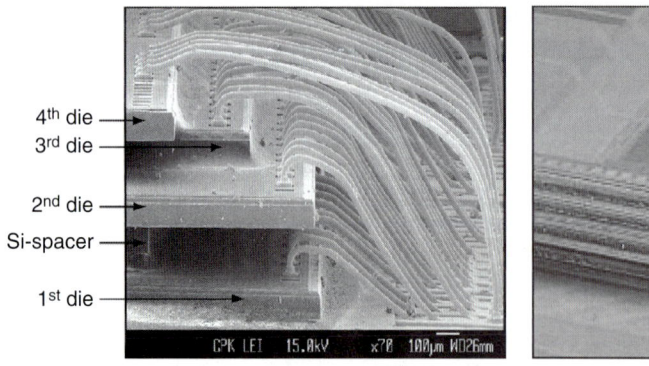

Figure 3.20 Wire bonded chip stacking. (a) ChipPAC's 4+1 stacked chip. (b) Hynix's 20 chip stacking with 25-μm-thick chips.

Stacked ICs and Packages (SIP)

The initial development of wire bonded chip stacking was for low-cost memory chips. However, this stacking technology has been extended to the stacking of logic and memory chips. Figure 3.21 shows the wire bonded chip stacking with one logic IC and two memory chips [25]. This uses assembly technology similar to that used for stacking memory chips. There are, however, some significant differences and challenges, including

- Assembly complexity of logic dies due to increased interconnection layers on it, which introduces new processes for sawing and stacking.
- Higher-density substrates needed to route all the traces due to the higher I/O of the logic processor.
- Integrated silicon and package stresses due to stacking of significantly different silicon chips.

This mixed-chip wire bonded stacking requires much more consideration in wire bonding materials and processes such as adhesives, spacers, and molding materials as well as electrical rerouting than those used for single-die-type wire bonding packages.

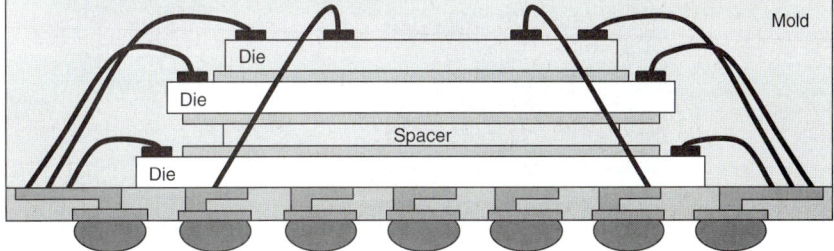

FIGURE 3.21 Intel's logic-memory stacked SIP comprised of one logic (top die) and two memory dies. [25]

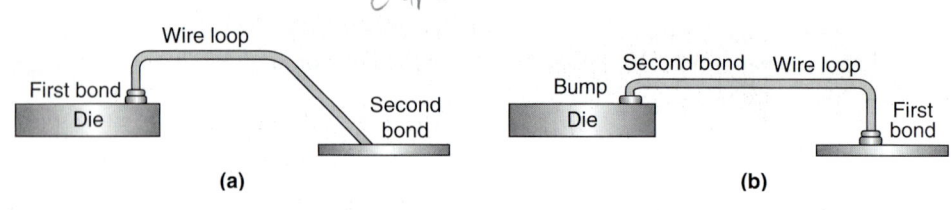

FIGURE 3.22 (a) Forward and (b) reverse wire bonding for chip stacking. [26]

Wire Bonding for Chip Stacking The application of wire bonding for chip stacking poses a few unique challenges due to height restrictions and the increased complexity of stacking configurations. As die thickness decreases, the space between the different wire looping tiers decreases accordingly. The wire bond loop height of the lower tiers needs to decrease to avoid wire shorts between the different layers of the loop. The top layer of the loop also needs to stay low to avoid exposed wire outside the molding compound. The maximum loop height of the device should not be higher than the die thickness to maintain optimal gaps between the loop tiers.

Wires can be bonded with forward or reverse bonding, as shown in Figure 3.22, both of which have separate length limitations. Forward bonding is the traditional approach that can handle long wire lengths and allows for higher-speed assembly. The bond starts from the die and ends at the substrate (Figure 3.22a). One disadvantage of this bonding approach is that the loop height over the silicon can increase the overall thickness of the package. Reverse bonding or standoff stitch bonding, starts at the substrate and ends at the die, creating a low loop height over the silicon and a higher loop height at the substrate side (Figure 3.22b). This allows multiple bond shelves by creating more wire-to-wire space and thus thinner packages. The disadvantage of the reverse wire bonding process is that its manufacturing process takes significantly longer time. Figure 3.23 shows a four-die stack, with the second die containing a forward wire bonding and the top die with a reverse bonding. Reverse bonding on the topmost stacked die can lead to overall packages that are thinner.

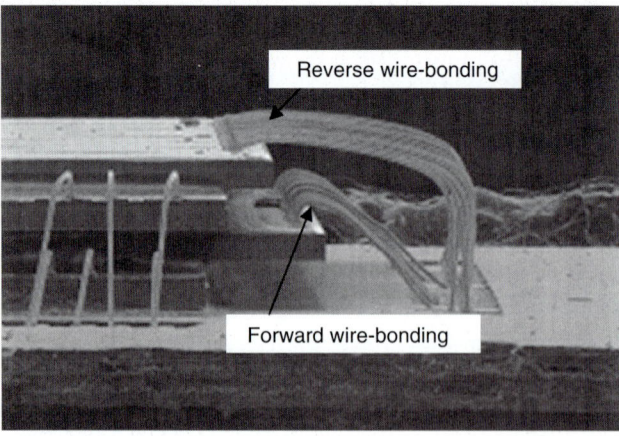

FIGURE 3.23 Wire bond loop height profiles.

The wire bonding process is particularly challenging when an upper die in a stack is larger or overhangs a lower die. Bonding to an overhanging die can cause many problems, including die cracks, loop damage, and inconsistent bump formation, due to die edge bouncing. The maximal overhang length for a package depends on the application and is determined by the die thickness, back-side die defect sizes, properties of the die attachment layers, and the impact and bonding forces in the wire bonding process.

Die Adhesive Two types of die adhesives are used in stacking chips: nonconductive epoxy (NCE) and film adhesive (FA) [27]. NCE is generally lower cost and involves minimal capital investment because it is used with existing die bonders. The weaknesses of NCE processing, however, include control of voids, fillet coverage, epoxy bond-line thickness control, and die tilt—all critical issues for successful die stacking. In addition, resin bleed can contaminate die-bond pads and make wire bonding difficult. The FA technology, on the other hand, can address the above process concerns associated with using NCE in die-stacking applications. Because resin bleed is a major concern when stacked dies are of the same size, the FA is the only workable option. In addition, FA provides a uniform bond-line thickness that is void-free, with 100 percent edge coverage. The FA also acts as a stress absorber between dies. The FA technology, however, requires an initial capital investment in wafer-back lamination and die-bonder modifications, and involves higher materials cost. The increasingly higher demand for higher quality of die-stacking applications can offset these additional expenses.

Spacer Technology Stacking of chips with varying die sizes requires a spacer between the dies when the top die is either the same size or larger than the bottom die, to avoid damage to its wires. Numerous spacer materials have been used, including silicon, adhesive paste, and thick tape. Each presents advantages and shortcomings. Silicon is widely used because of its acceptance, its infrastructure, and its cost-effectiveness. But it has more processing steps. Epoxy with spacer spheres requires fewer process steps, but has more epoxy bleed. Tape has no bleeding, but it is more costly. Epoxy with spacer spheres is preferred for a die with a thickness < 100 μm, because it minimizes the overhanging span of the top of the die and enables its wire bonding [28]. The use of spacers affects mold-cap thickness and total package height. The process capability for controlling wire loop height and mold flow dictates the spacer gap. A larger mold gap works against the trend to thinner package height. Choosing a reasonable spacing gap is important for mold compound flow, since turbulent mold compounds flow inside a mold cavity.

Molding Increased wire density and wire length in wire bonded chip stacking makes molding the stack more difficult than conventional single-die packages. Different layers of wire bond loops that are subjected to varying amounts of drag force can result in differences in wire sweep. This increases the possibility of wire shorts. Further, the variable gaps between various die components make it more difficult in the molding process to achieve a balanced flow without voids free. Molding compound development and selection, as well as gate design and wire layout optimization, are required to achieve a better yield in molding. Low-viscosity compounds and compounds with smaller filler sizes and slower molding transfer speeds show improved wire sweeps. A lateral loop trajectory is known to be able to reduce the mold sweep by predeforming the wire in anticipation of the sweep direction [26]. The change in gate design from conventional bottom-gate to top-center mold gate can also reduce the wire sweep, especially for long wire applications [29].

Electrical Routing Considerations In general, DRAM and flash memory have interconnections on only two sides of the die, splitting the address and data bus. In addition, the memory chips can have different bond options, as 16 or 32 bit, and can even have one- or two-sided bonding options. These various options change the order of signal sequencing on the die and must be accounted for in the wire bonded chip-stack design. Stacking of logic ICs with memory chips also brings electrical routing issues. It is very common to have a logic die that has a flash bus on one side and a double data rate (DDR) bus on the other, stacked with an external DDR and flash that are two-sided. These widely different pad placements make the substrate routing and integration even more difficult.

In order to effectively stack chips, the pad ring sequencing of different die in the stack should be such that it allows the bond wires to land on the bond fingers with minimal overlap or cross. This ensures stackability, routability, the highest electrical performance, and the lowest cost by simplifying the interconnect methodology. This methodology could enable multiple wires to be bonded to the same bond pad. Since only one bond finger is needed for two or more signals, the decrease in bond fingers allows much more substrate routing flexibility.

Flip Chip Stacking

An alternative to the wire bonding interconnection in chip stacking is flip chip. Flip chip has been used for more than three decades to increase the electrical performance by decreasing the electrical length of the interconnection between the chip and the rest of the system and by allowing a higher number of connections by utilizing the entire area of the chip. The flip chip interconnection has been used for chip stacking, either on its own or as a complement to wire bonding. The possible applications of this stacking technology are for high-performance workstations, servers, data communication products, internet routers, and other high-frequency and RF systems.

Flip Chip and Wire Bonding Stacking Flip chip interconnection can be adopted for chip stacking in conjunction with wire bonding. The flip chip configuration may be applied either to the upper die or the lower ones (Figure 3.24), depending on the intent of the design. Flip chipping a top die eliminates the use of long wires for connection to the substrate (Figure 3.24a), while flip chipping a bottom die directly onto the substrate enables that die to operate at a high speed (Figure 3.24b).

The chip stacking with flip chipping of the top die is for chip-to-chip communication. As shown in Figure 3.25a, flip chip interconnections between chips provide the traditional and inherent benefits of flip chip technology such as high-frequency operation, low parasitics, and high input-output (I/O) density in a reduced package footprint. In addition,

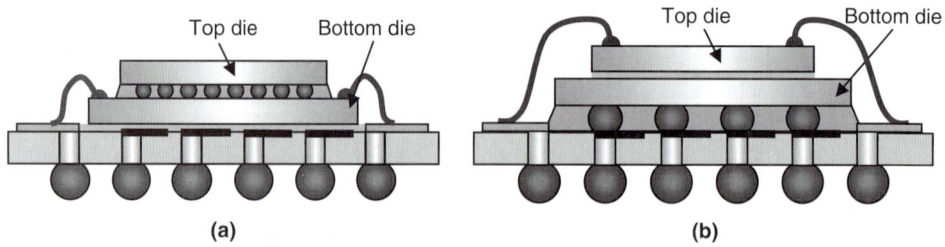

FIGURE 3.24 Two different types of hybrid chip stackings. (*a*) Flip chipping of a top die. (*b*) Flip chipping of a bottom die.

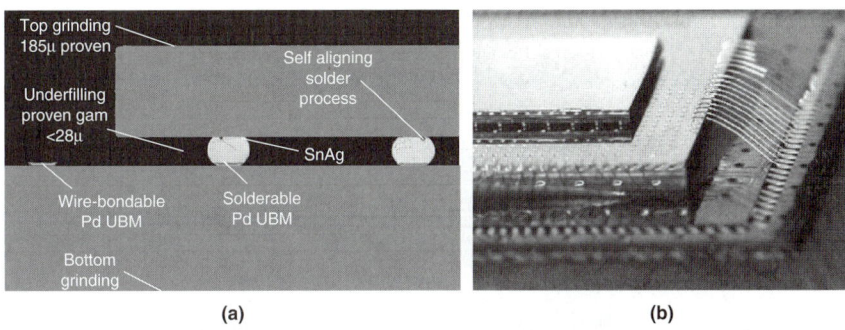

FIGURE 3.25 (a) Two face-to-face chips connected by microbumps in a flip chip architecture. (b) Bottom chip of the stack connected to the substrate through wire bonding. [30]

this short interconnection enables miniaturization by eliminating long wire spans that would otherwise be needed to bond the top chip. For this type of chip stacking, bottom chips need both wire bonding and flip chip pads, as shown in Figure 3.26. In this stacking, the bottom dies are first attached and connected to the substrate by wire bonding. Then, the top dies are attached face down on the front surface of the bottom die. Figure 3.25b shows such stacked chips with the flip chip interconnections between two dies and the wire bonding for interconnection of the bottom dies to the substrate.

Figure 3.27 shows the flip chip bonded bottom die in the stack [32]. In this stacking, the bottom die operates at a higher speed with its high number of I/Os. This stacking method also relieves the bond-finger crowding in one concentrated region of the substrate by redistributing the substrate density to two different regions: the region under the die for the flip chip and further out for the wire bonded chips. This stacking method has been developed for next-generation handsets and is extendable to other products in the future.

Chip-on-Chip (COC) Stacking In COC stacking, flip chip interconnections connect all stacked chips without wire bonding. Figure 3.28 shows the COC stacking structure, in which subchips are flip chip bonded on a base chip with ultrafine-pitch bump interconnection and the base chip is also flip chip mounted on a package substrate. This method allows the connection of a very large number of I/Os in the stack and thus a significant increase in the data transfer speed between the chips.

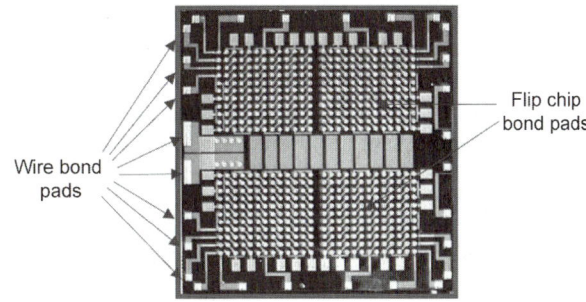

FIGURE 3.26 Bond pads of bottom dies for flip chip and wire bonding stacking with the bottom dies flipped. [31]

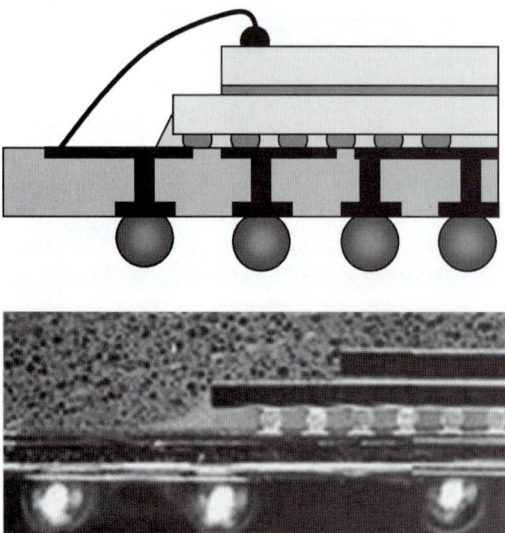

Figure 3.27 Hybrid chip stacking with a DSP flip chip attached to the substrate and an analog or memory chip stacked directly on top and interconnected with wire bonding. [32]

Figure 3.29 shows the base chips embracing subchips, in which very fine pitch (~30 μm) solder bumps were employed for the chip-to-chip interconnections. In this COC stacking, a shorter transmission length with a high number of bumps is critical for high-speed and wideband data transmissions between stacked large scale integration (LSI) chips. Very fine pitch interconnection methods in COC are under development including the following two approaches. A micro solder bump formation method has been developed that uses a molten-solder-ejection technique to produce small solder balls [33]. Fused junction technology is also being developed that can achieve fine-pitch bump connections with high reliability and low damage using lead-free solder without flux, in which the connection is achieved by applying both heat and pressure [34].

Side Termination Stacking

Side termination stacking typically requires metal rerouting on chips to provide edge bonding pads for external electrical connections in the chip stack. After chip stacking, connecting these edge bonding pads provides vertical interconnections between the

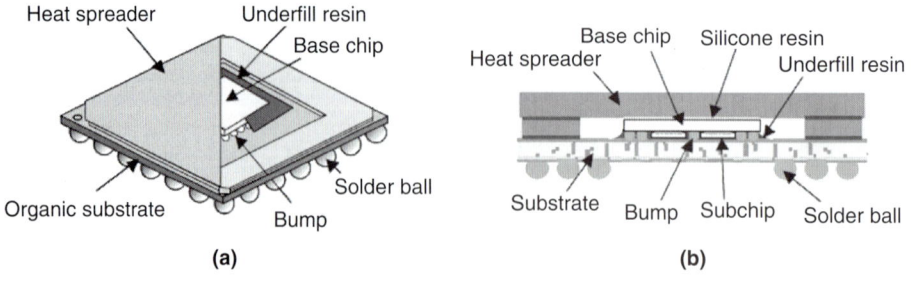

Figure 3.28 Schematic diagram of COC stacking. (a) Perspective view. (b) Cross-sectional view. [33]

Stacked ICs and Packages (SIP) 107

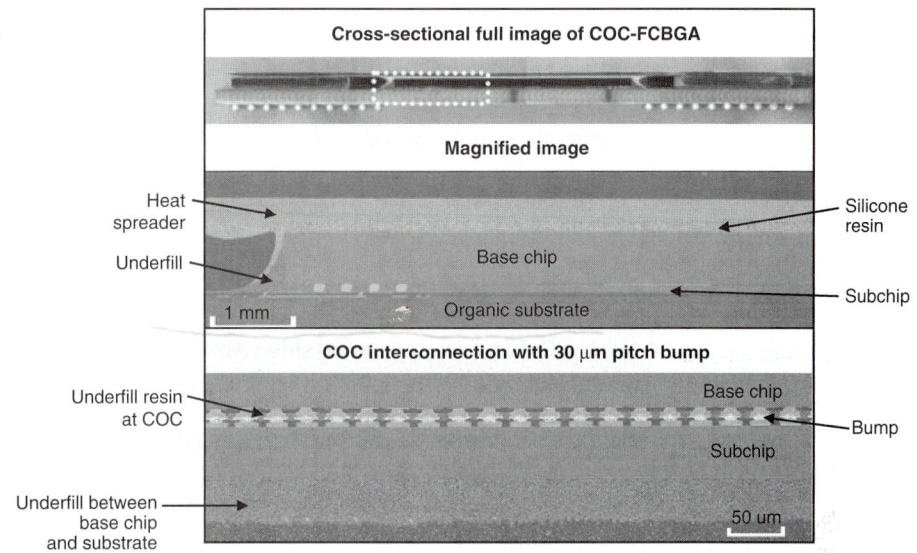

FIGURE 3.29 Cross section of COC stacking with 30-μm-pitch solder bump interconnections. [33]

stacked chips. There are three variations of this side termination stacking method depending on the side interconnection: metallization, conductive polymer, or solder.

Metallization Stacking Stacked bare chips can be electrically connected by metal traces deposited on the side of the stack. Figure 3.30 shows a 19-layer flash memory chip stacking by side-metallization developed by Irvine Sensor [35]. Figure 3.31 shows the process flow of this chip stacking method. Chip pads are first rerouted at the wafer level. Then, the wafer is thinned, and a passivation layer is deposited on the back surface of the thinned wafer. Each chip is singulated from the wafer, and the bare chips are then

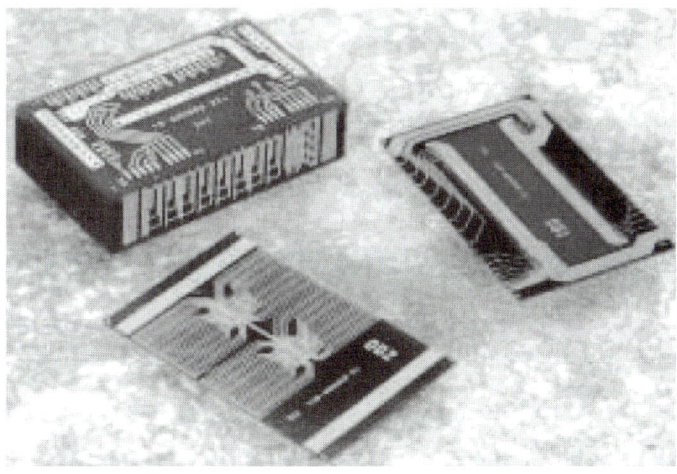

FIGURE 3.30 Flash memory chip stacking by side-metallization. [35]

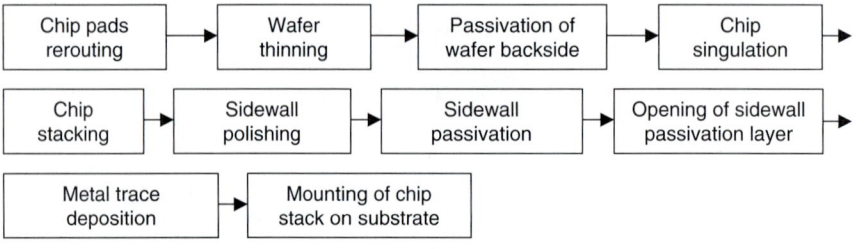

FIGURE 3.31 Process flow for side-metallization chip stacking.

placed one on top of another to form a stack. In the side termination interconnection of the stacked chips, polishing of the sidewall of stacked chips is needed. The passivation layer is again deposited on all the polished sidewalls of the stacked chips, and openings are made in the passivation layer above the desired electrical connection pads. Finally, vertically adjacent chips are electrically interconnected by depositing metal traces on the sidewall of the stack. This stack is then mounted on substrate.

Initially, this metallization stacking method was developed for the stacking of same-sized bare Si chips. But, it was later applied to stack different sized chips [36]. In this case, a compound matrix with the size of a standard wafer is generated, into which thinned chips are molded. This so-called neo-wafer can then be processed by the same processes as used for stacking bare chips on regular Si wafers. Figure 3.32 shows the schematic cross section of the chip stacking by side-metallization with different sized dies. The side-metallization brings all input-output signals to the cap chip at the top of the stack. This neo-wafer stacking allows heterogeneous chip stacking and easily adapts to the change in chip size without substantial retooling.

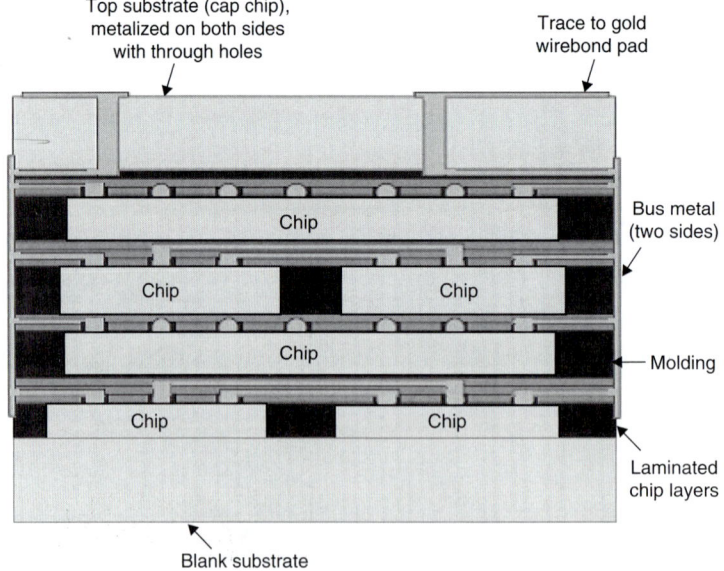

FIGURE 3.32 Schematic cross section of stacking of different sizes and types of chips by side-metallization. [36]

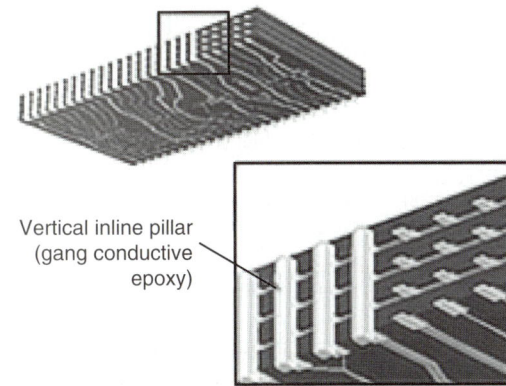

FIGURE 3.33 The vertical interconnection process with conductive polymer for chip stacking. [37]

Stacking with Conductive Polymer Side-metallization for vertical interconnection of stacked chips can also be achieved with conductive polymers. Metallic conducting elements extending from rerouted chip pads, such as a bond wire or a bond ribbon, are embedded within the conductive polymer, thus providing an electrical connection between the stacked chips, as shown in Figure 3.33. Typically, the side-metallization process requires a lithography process to obtain the desired metallization patterns at the small area of the sidewall of stacked chips, which includes application of photoresist (PR) materials, exposure, development, metal etching, and PR strip processes. The use of conductive polymers for side interconnections can eliminate the lithography step in the chip stacking process. The typical conductive polymer can be a conductive epoxy filled with metallic particles such as silver and gold [37].

Stacking with Solder Edge Interconnect Solder balls or bumps have been commonly used for mechanical and electrical interconnections between electronic components including chips, functional modules, and substrates. Figure 3.34 shows the solder

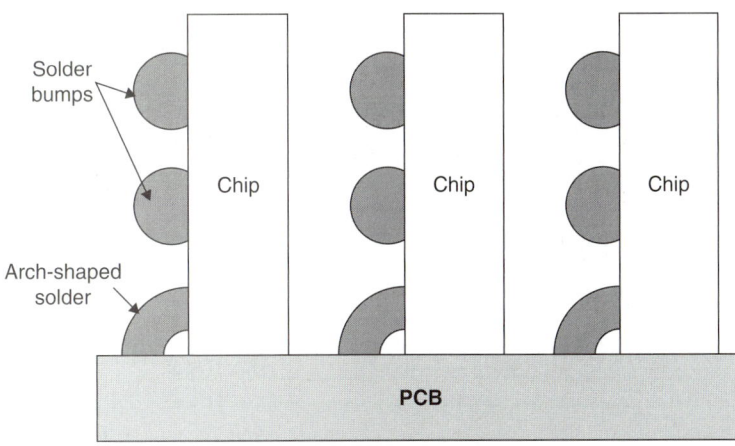

FIGURE 3.34 Chip stacking by arched solder column interconnections. [38]

interconnection for edge mounting of a chip onto a bottom base chip [38]. In this solder interconnection, solder bumps are formed along the edge of a vertically placed chip to contact pads on the bottom base chip during solder reflow, yielding arch-shaped solder column interconnections. This solder interconnection shows the capability to stack chips on a single base chip. The arched solder columns offer several electrical and mechanical advantages. The circular cross section is an excellent geometry for electric signal propagation as it provides a controlled transition for microwave and millimeter wave signals. The arched columns may also provide structural support, while allowing some compliance for improved reliability, in contrast to traditional rigid fillet shapes.

Figure 3.35 shows another different type of chip stacking utilizing solder edge interconnection [39]. In this chip stacking method, two thin chips are first back-to-back bonded with nonconductive adhesive and then these are stacked with solder balls or bumps on the peripheral pads. These peripheral solder bumps also provide bonding sites for external electrical connection. There are two variations of this chip stacking structure: wire-on-bumps (WOB) and bump-on-flex (BOF). In the WOB technology, stacked chips are electrically connected through solder bonding with metal wires including Au and Cu. In the BOF technology, on the other hand, the vertical connection is realized by a flex circuit with Cu lines and pads. These new 3D chip stacking technologies have such benefits as shorter signal path in the vertical direction and 3D stackability of an unlimited number of chips compared to wire bonded chip stacking. In addition, the flex circuit of BOF can achieve more component integration by embedding of thin-film passive components together with chip stacking.

Embedded IC Stacking

Embedded IC technology, in which ICs are embedded into substrates or buildup layers, has been gaining a lot of interest for more miniaturization, higher performance, and more functionality of microsystems. The details of embedded IC technology are described in Chapter 7. These embedded IC technology approaches have also been used in chip stacking. Figures 3.36 and 3.37 show two embedded IC stackings for combining logic and memory functions and stacking memory chips, respectively. The

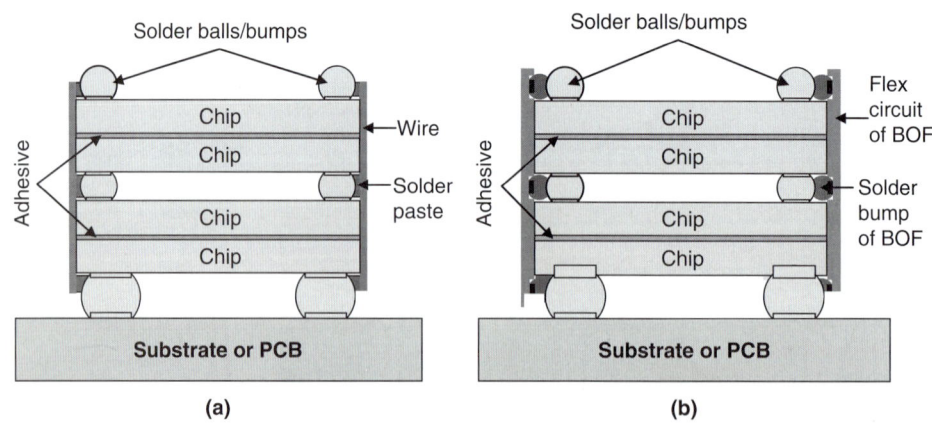

Figure 3.35 Schematic cross sections of 3D chip stackings by (a) WOB and (b) BOF. [39]

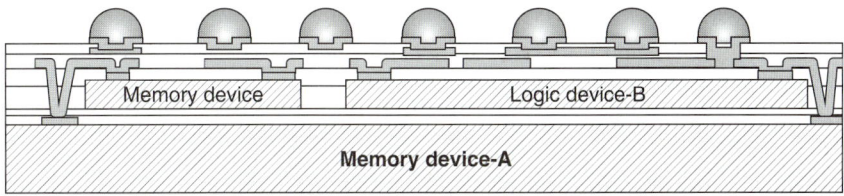

FIGURE 3.36 Embedded IC stacking for high-end applications, in which two logic devices are placed on top of each other, along with a memory chip. [40]

embedded IC stacking concept uses a silicon wafer carrying large base chips as substrate. On this base wafer, completely processed thin ICs are mounted by applying adhesive to its back side. Then, buildup layer interconnections are processed on top of the chips and the wafer, which includes dielectric polymer coating to planarize the mounted thin chips, via formation and metallization to fill the vias, and thin-film metal wiring layer. Both the thin IC mounting and buildup processes are repeated until the necessary numbers of chips are stacked. Once the chip stack process is completed on the wafer, the wafer is diced into single stack modules. Finally, the stack module is mounted onto substrates with solder bumps.

A number of benefits are expected from this embedded IC stacking. The size of the chip stack is equal to or slightly larger than the base chip housed in it, which is almost a chip-scale package (CSP). The chip stacking cost is reduced because all the stacking processes are completed at the wafer level. A short interconnection length between stacked chips improves the electrical performance in the stack. Thin-film passive components such as the capacitor and inductor can be integrated into the chip stack by embedding them into the buildup layers, contributing to the increased functionality of the SIP. However, there is a concern of a lower process yield, since several sequential buildup processes above the embedded chips can accumulate a yield loss associated with each process step.

Great benefits! But high risk, for yield purposes.

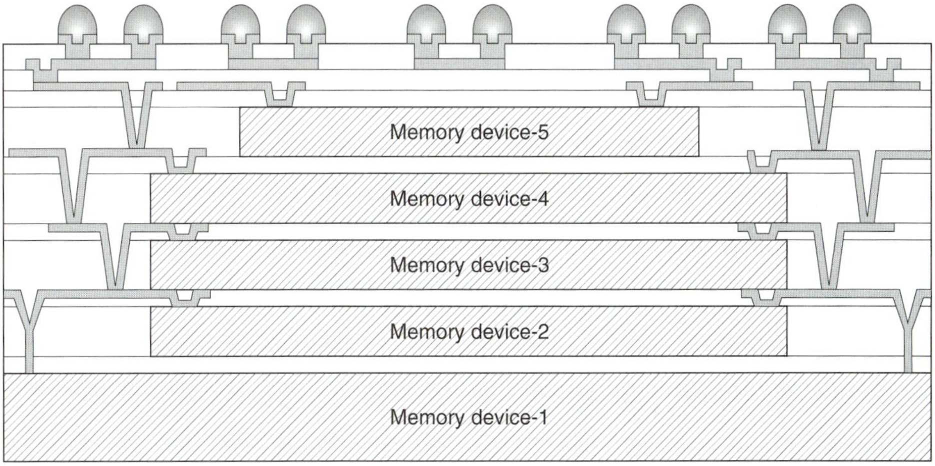

FIGURE 3.37 Embedded IC stacking for a five-layer, high-capacity memory with four chips stacked on top of the base-level memory chip. [40]

TAB Stacking

Tape automated bonding (TAB) has been one of the most common interconnection technologies for chip-to-substrate interconnection (first-level interconnection), along with wire bonding and flip chip. It is based on the use of metallized flexible polymer tapes, in which one end of an etched metal lead is bonded to the chip and the other end of the lead is bonded to the substrate. TAB technology has also been employed in chip stacking due to its several advantages that include the ability to handle small bond pads and finer pitches on the chip, elimination of large wire loops, low-profile interconnection structures for thin packages, improved heat conduction, and ability to burn-in on tape before device commitment. The TAB chip stacking method can be divided further into stacked TAB on PCB [41] and stacked TAB on lead frame [42], as shown in Figure 3.38a and b, respectively. Figure 3.38a shows chips with inner TAB lead bonding that are first stacked and then all outer TAB leads of these stacked chips are

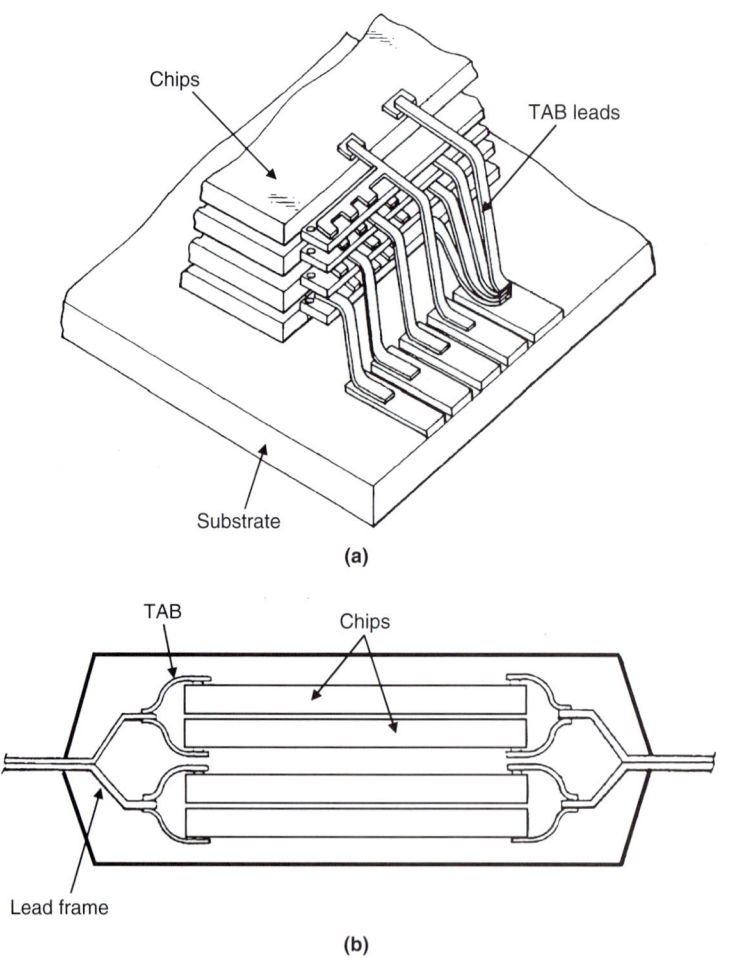

FIGURE 3.38 Two variations of chip stacking by TAB [41]. (a) Stacked TAB on PCB. (b) Stacked TAB on a lead frame [42].

bonded onto PCB, providing the electrical interconnection of stacked chips on PCB pads. Figure 3.38b shows, on the other hand, that chips are first mounted on both surfaces of a lead frame by using TAB and then these lead frames are bonded together, so that the chip stacking structure is finally established.

However, the use of TAB technology for chip stacking has been limited due to a variety of concerns that include increased package size with large I/Os, long interconnection lengths, relatively little TAB production infrastructure, and additional wafer processing steps required for bumping to accommodate TAB.

3.3.3 Package Stacking

While chip stacking provides many advantages such as a small form factor, high performance, and low cost, it has several challenges including lack of chip testability before stacking, lower stacking process yield, and difficulty in integrating dissimilar chips. By employing package stacking technologies, many of these issues can be addressed, as the individual chips are prepackaged, sourced, tested, and yielded separately and then combined once they are known to be good. Package stacking can be realized in many different ways such as package-on-package (PoP), package-in-package (PiP), and folded-stacked chip-scale package (FSCSP).

Package-on-Package (PoP)

PoP consists of individual packaged dies, in which a top package is stacked directly over an existing package. Figure 3.39 shows one of the earliest versions of PoP stacking. In this example, the PoP interconnections are realized by side terminations, similar to chip stacking. At the sidewall of stacked packages, conductive epoxy is applied in Figure 3.39a [43] and metal traces are formed in Figure 3.39b [44].

However, more recent types of PoP stacking structures are more like that shown in Figure 3.40. Stacked packages are connected typically with solder balls providing both clearance and electrical connection. This PoP stacking has been considered a major breakthrough in the package design of mobile applications (Figure 3.41). In a typical PoP, the top package is a multichip package that stacks flash memories and xRAM, while the bottom package is a single-chip package, typically with a logic chip. On the front side of the bottom package, there are land-pads, which are used for electrical communication between the top and bottom packages by mounting the top package on them. The height of the solder balls is adjusted to effectively encompass the logic die and its wire bond loop height.

Figure 3.42 shows the variations of PoP stacking. Figure 3.42a demonstrates PoP stacking realized by very short interconnects. Solder balls for package-to-package interconnections are embedded into the substrate, and chip pads are directly connected to substrate traces by electroplating, often referred to as bumpless interconnects [46]. In Figure 3.42b, chips are molded by polymer materials and electric signals are routed through holes in the molding from the front side of the chips to the back side of the molding. These molded packages are then stacked one on top of the other with solder balls, which allow area array interconnection in PoP [47]. Figure 3.42c shows the modernized PoP with side-metallized interconnections. This packaging uses thin flex film as interposers, and Ni-Au metal traces are patterned by laser etching at the sidewall of the stacked package [48].

A beneficial feature of the PoP stacking is that each individual package can be tested as a ball grid array (BGA) package before it is stacked. In other words, the known good package is ready for the final assembly, leading to yield improvement. This stacking

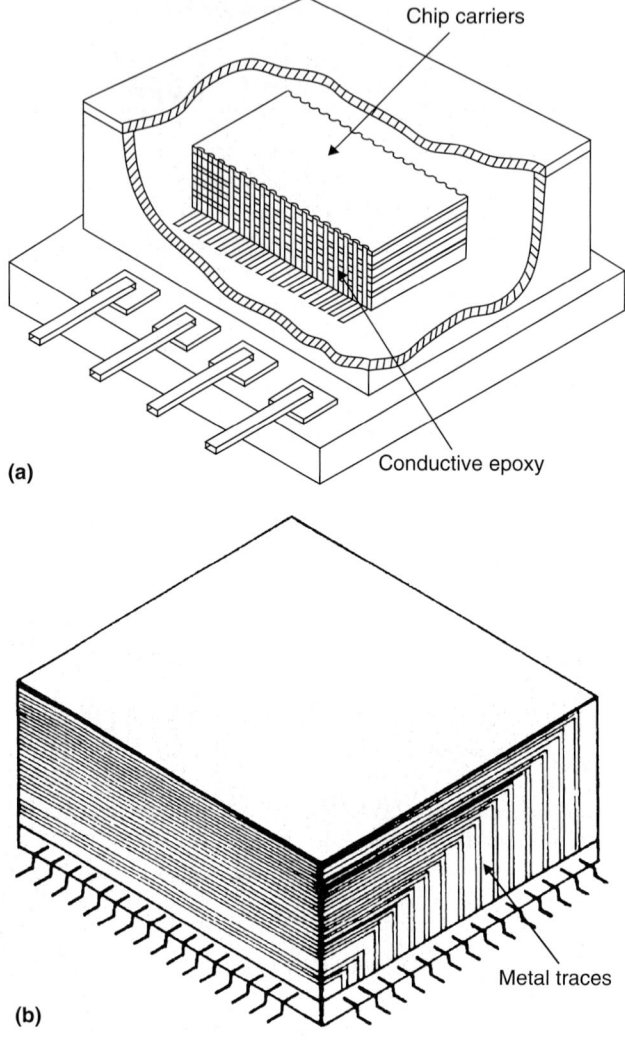

Figure 3.39 The earliest PoP stacking by employing side termination interconnection methods with (a) conductive polymer [43], and (b) metal traces. [44]

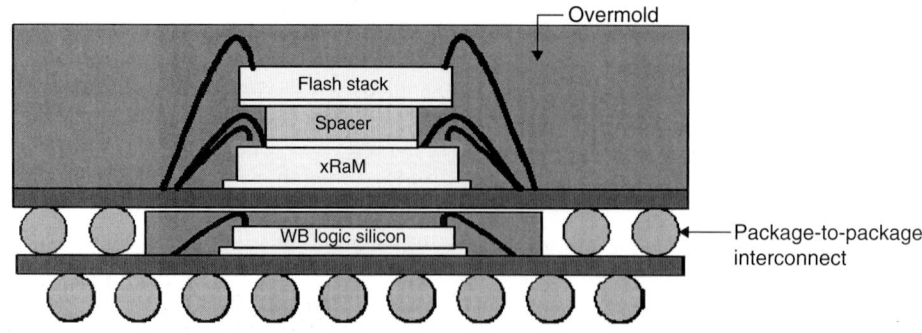

Figure 3.40 Stacked package with ball grid array interconnects, leading to PoP stacking.

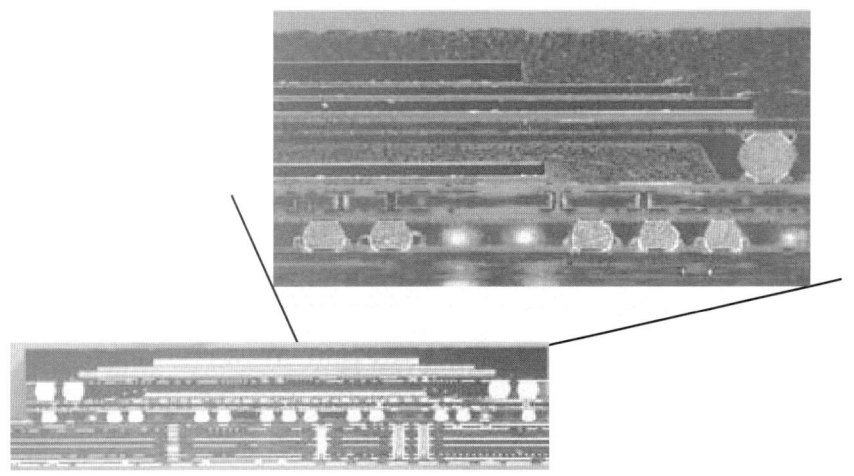

FIGURE 3.41 A cross-sectional view of a four-chip PoP on a mobile handset. [45]

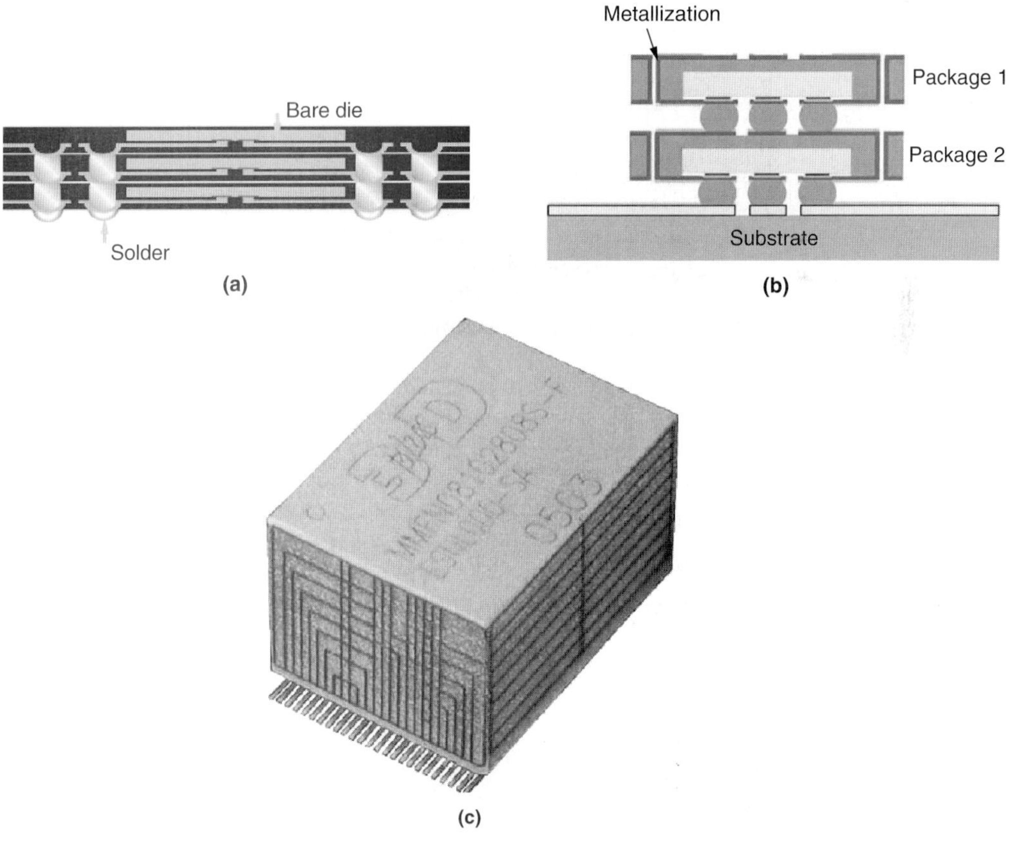

FIGURE 3.42 Variations of PoP stacking by (*a*) embedded solder interconnection [46], (*b*) area array interconnection [47], and (*c*) side-metallization [48].

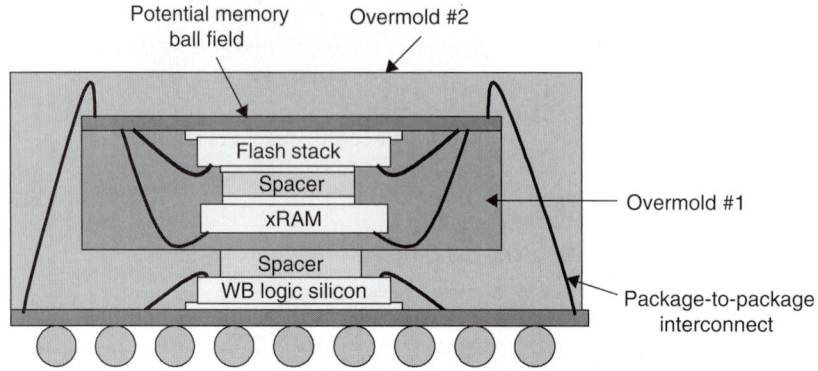

FIGURE 3.43 Schematic cross section of PiP stacking with a wire bond interconnect.

solution is scalable. Stacks can go beyond two packages, as long as the total package height meets the product requirements. The PoP technology provides a high number of connections that allows for stacking of chips with different sizes and functions. However, there are some disadvantages to this technology. As compared to chip stacking, it basically has an additional package substrate, increasing the total height of the PoP package, and is much larger in size due to the interconnect methodology.

Package-in-Package (PiP)

While PiP is very similar to PoP, PiP involves flipping and stacking a tested package onto a base package with subsequent interconnection via wire bonding, as shown in Figure 3.43. In PiP stacking, the top package is an industry-standard memory package without solder balls. This package is flipped over and stacked onto the bottom package, in which the logic die is already bonded. The top package has exposed wire bond pads on the back side of its substrate allowing for a wire bond connection to the bottom package. The entire package is then overmolded. Figure 3.44 shows PiP stacking with an ASIC chip and memory chip stack [49].

This approach allows each package to be tested for a better final test yield as PoP, but has other key benefits as well. First, the top package can be an industry-standard package with the addition of the exposed wire bond pads as the only difference. PiP is slightly thicker than a competing stacked package due to wire bonding interconnections. An overmolded wire bond is much safer than a solder ball interconnect since the solder ball may crack under stress. The connection is also much smaller in the xy plane. Solder balls have a diameter of 300 to 400 µm, while wire bonds are closer to 25 µm. This reduction ratio allows wire bond interconnects to achieve a density of 10 times more than solder ball interconnects. This ratio allows the connections from top to bottom packages to

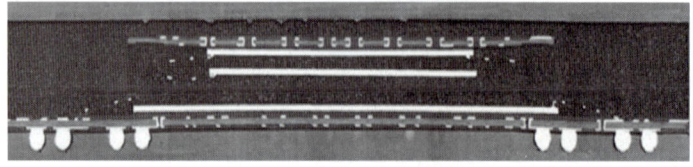

FIGURE 3.44 A cross section of PiP stacking with an ASIC chip (bottom package) and stacked memory (top package). [49]

increase dramatically over other stacked solutions. In addition, this methodology is also not sensitive to bus technology. One-, two-, or four-sided buses can be connected.

However, wires are much thinner and have much higher resistance and much lower current carrying capability than solder balls. Long wires can severely affect high-frequency performance. In addition, since this package uses wire bonds as the primary interconnect, most customers will have to buy this unit as a complete system. It is also doubtful that an industry-standard interface could be obtained given the complex nature of how the interconnect is formed.

Folded-Stacked Chip-Scale Package (FSCSP)

FSCSP uses a flexible, thin-film tape substrate, as shown in Figure 3.45 [50]. A chip is mounted on one-half of the flex substrate with wire bonding or flip chip interconnections. Then an adhesive film is applied onto the top surface of the chip and the remainder of the flex is folded over the chip to provide open land pads on top of the package. Another package is finally stacked on the FSCSP package. Figure 3.46 shows the package stacking with the FSCSP as the bottom package. Instead of a single chip in the FSCSP, multiple chips can also be mounted on the flex substrate, as shown in Figure 3.47 [51]. Folding the substrate creates a stacked package structure.

The FSCSP stacking provides the benefit of testability, flexibility, and a higher process yield similar to PoP or PiP. The folded stack package has only a slightly larger planar dimension than the largest die in the stack, since it does not need the extra package area for solder balls in PoP and wire bonding in PiP for the interconnection between packages. Another advantage is an increased routing density by using the flex tape, since the flex tape substrate process allows finer lines and spaces than PCB.

However, there are still some issues to be resolved before its wide application. One concern is the availability and cost of the double-sided tape substrate, which adds an

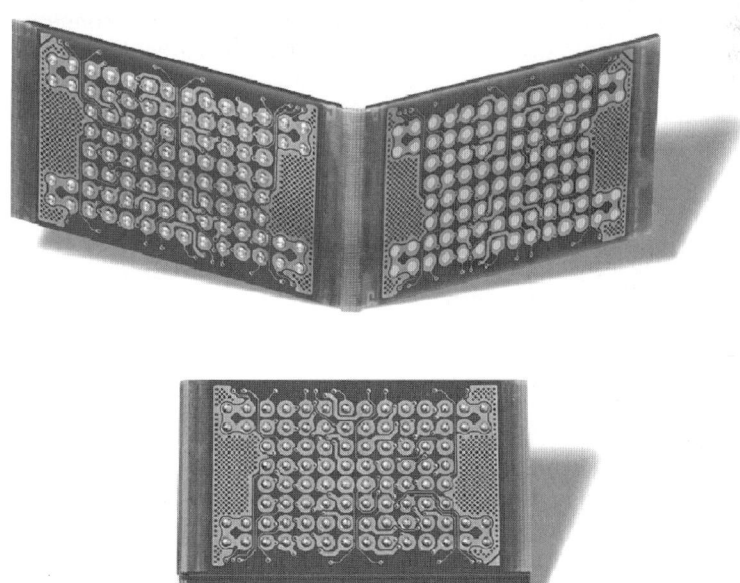

FIGURE 3.45 Unfolded and folded flex substrate for FSCSP stacking.

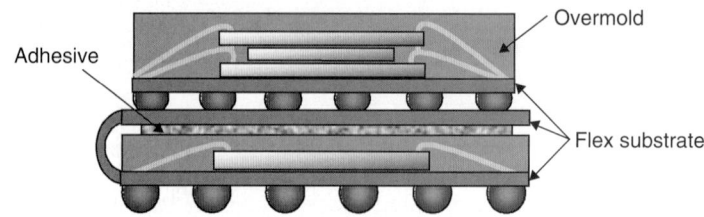

Figure 3.46 Schematic cross section of a package stacking with FSCSP.

extra cost due to the additional process steps and different manufacturing lines it involves. Another concern is related to the routing of bus signals on chips to be packaged. The FSCSP is typically designed to contain logic ICs. This requires that the logic die should have a one-sided bus to facilitate the routing and interconnect to the top package by routing around the fold side. Logic IC designs that have a two-sided or even a four-sided bus are not suitable or adaptable to this package type, consistent with the package designer's desire to take advantage of all four sides to provide the most amount of interconnects in the smallest area. The last concern is associated with the electrical routing of the folded substrate. The substrate is of fixed width, almost equivalent to that of the chip, which restricts the total number of signals capable of being routed to the top package. This may create more of an issue depending on the substrate line and space design rules as well as electrical shielding and power delivery requirements. Figure 3.48 shows one of the variations of the FSCSP stacking. Folding the flex substrate on both sides of the chip can improve the electrical routing of chips and substrates in FSCSP stacking [52].

Table 3.1 compares three package stacking technologies—PoP, PiP, and FSCSP.

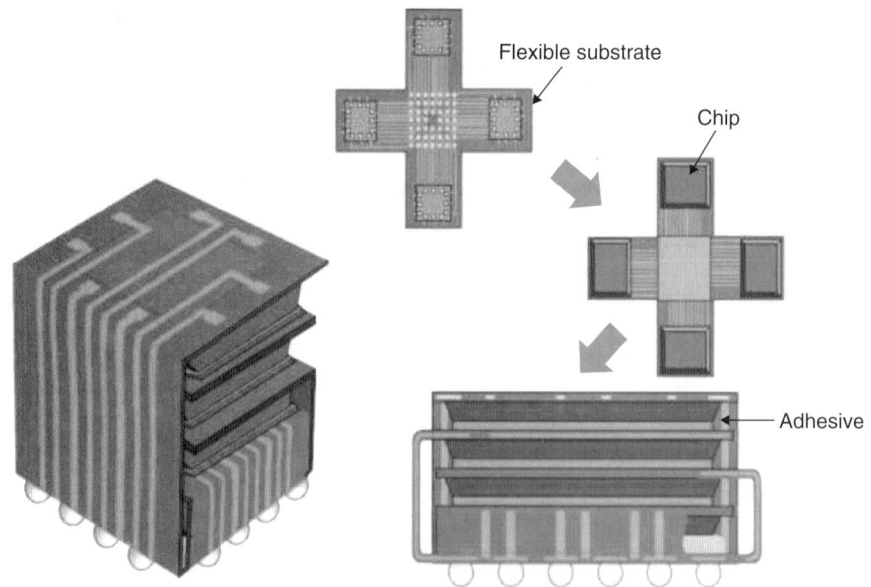

Figure 3.47 FSCSP with multiple chips mounted on the flex substrate. [51]

Stacked ICs and Packages (SIP)

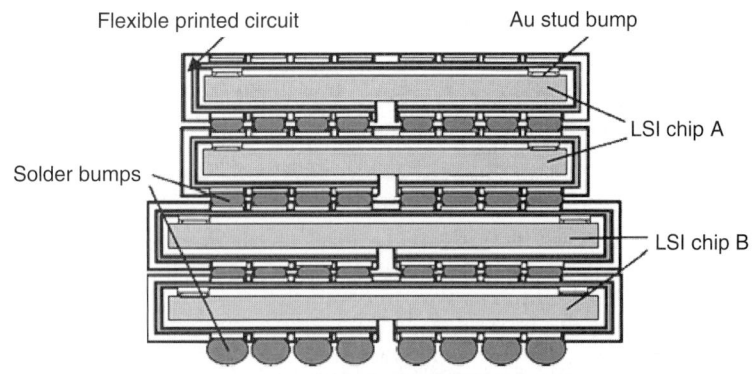

FIGURE 3.48 Variation of FSCSP developed by NEC. [52]

	PoP	PiP	FSCSP
Test and yield	Capable of memory screening test	Capable of memory screening test	Capable of memory screening test
Size and thickness	Thicker package, small *xy* size	Thicker package, small *xy* size	Thicker package, small *xy* size
Silicon bus architecture	Four-sided required for interconnect	One- or four-sided possible	One side only (fold side)
Package-to-package connects	High number of connects but adds package size	High number, limited by only wire bond pitch	150, limited by power and ground ratio and fold size
Design complexity	Most simple, two-bond shells, nice BGA interconnect, large substrate	Less complex than chip stacking; special design tools needed	More simple than chip stacking but one-sided fold makes design complex
Expandability	More interconnects by increasing *xy* size, BGA pitch reduction, and *z* height stacks	Not expandable with additional packages but much more capable for package-to-package connects	Stackable in *z* direction with 2+ packages on packages
Mechanical reliability	More compliant package due to multi-interconnects	Stiffer due to silicon die stack with single interconnect, additional thickness	More compliant package due to multi-interconnects
Electrical performance	Longer trace lengths than chip stacking due to larger *xy* package size	Shortest trace lengths besides chip stacking, long package-to-package wire	Long trace lengths around fold requiring shield

TABLE 3.1 Comparison of Three Package Stacking Technologies: PoP, PiP, and FSCSP

3.3.4 Chip Stacking versus Package Stacking

Table 3.2 compares chip and package stacking for a number of packaging parameters. The ability to house a chip stack in the package that is essentially the same size as the chip itself provides many advantages such as system integration, performance, and cost. Even if the stacking itself carries a cost premium, it typically results in a system-level savings because of smaller boards and other related cost reductions. Another advantage is the use of the existing infrastructure for the chip stacking processes. The chip stacking process yield greatly depends on the availability of the known good die (KGD). One of critical issues in the chip stacking is whether the KGDs can be obtained in wafer form. Thus, chip stacking has been an effective solution for stacking of high-yielding memory chips that don't have the KGD issue. However, there are still concerns about chip stacking including poor testability; low process yield, when stacking large number of chips; low flexibility in heterogeneous chip stacking (logic ICs and memory); and long time-to-market. Package stacking addresses some of these concerns.

When packages are stacked instead of chips, it is possible to test chips before stacking, thus eliminating bad chips in the stack, resulting in a higher stacking process yield. Furthermore, electrical testing of each device enables LSI chips to come from different sources. It also allows flexibility for product upgrades by accommodating the change in die size and design easily. This enables the memory and logic devices to be obtained separately from various, or even competing, vendors while solving the KGD issue. With chip stacking, a new die size and set of pad locations might require an extensive redesign of the package, assembly process, and even the system board to accommodate the changes. In addition, system designers acknowledge that package stacking provides a platform they can reuse for new applications and future generations of products. Thus, it can offer better time-to-market than chip stacking. However, package stacking is typically thicker than chip stacking due to the use of the interposer and solder balls. Lack of infrastructures for package stacking is another concern.

	Chip Stacking	**Package Stacking**
Prospects	• Low package profile available with advanced wafer thinning technology • Existing SMT line infrastructure available • Cost reduction by minimum substrate consumption	• Testability at individual package level for KGD • Greatly increased package stacking yield • Flexible selection of chips to be stacked
Concerns	• KGD required for high product yield • Single sources product • New development needed to change stacked device	• Higher package profile • Lack in infrastructures for package stacking

TABLE 3.2 Comparison of Chip Stacking versus Package Stacking

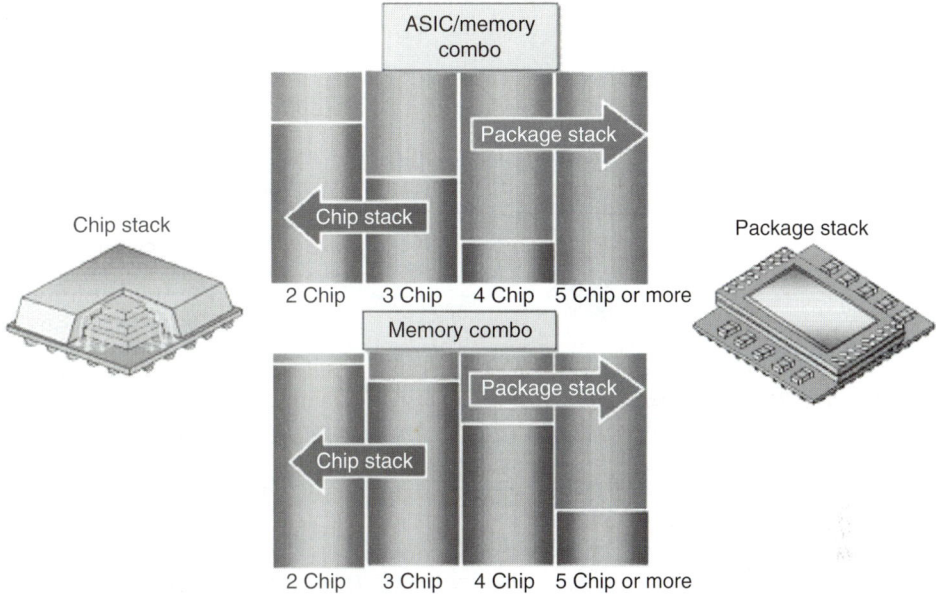

FIGURE 3.49 Selection of non-TSV SIP solutions. [53]

Figure 3.49 shows some guidelines to select chip versus package stackings. For a small number of IC stacking (two or three chips), chip stacking is competitive for both memory stacking and combined stacking of logic IC and memory. For stacking a higher number of chips, chip stacking may still be competitive with a low cost for high-yielding memory chips, but with expensive logic ICs combined with memory chips, package stacking is definitely more preferable. In conclusion, balancing and optimizing cost, flexibility, performance, form factor, and time-to-market will result in the optimal application solution between chip and package stacking technologies.

3.4 TSV SIP

3.4.1 Introduction

The main drivers for 3D interconnections in packaging are (1) size reduction, (2) solving the "interconnect bottleneck," (3) heterogeneous integration of different technologies, and (4) higher electrical performance. International Technology Roadmap for Semiconductors (ITRS) has identified 3D IC stacking as a way to achieve better electrical performance without further shrinking of transistor dimensions. Through-silicon via (TSV) has been identified as one of the major technologies to achieve the above goals by 3D integration. A TSV is a through-via hole drilled in silicon (die, wafer, or Si chip carrier) and filled with conductor material to form vertical electrical interconnections in modules or subsystems. TSVs run through the silicon die and are used to connect vertically stacked dies, wafers, or Si chip carriers.

Figure 3.50 shows the historical trend in system integration leading to TSV-enabled 3D integration. The first 2D interconnect is an example of horizontal integration

Chapter Three

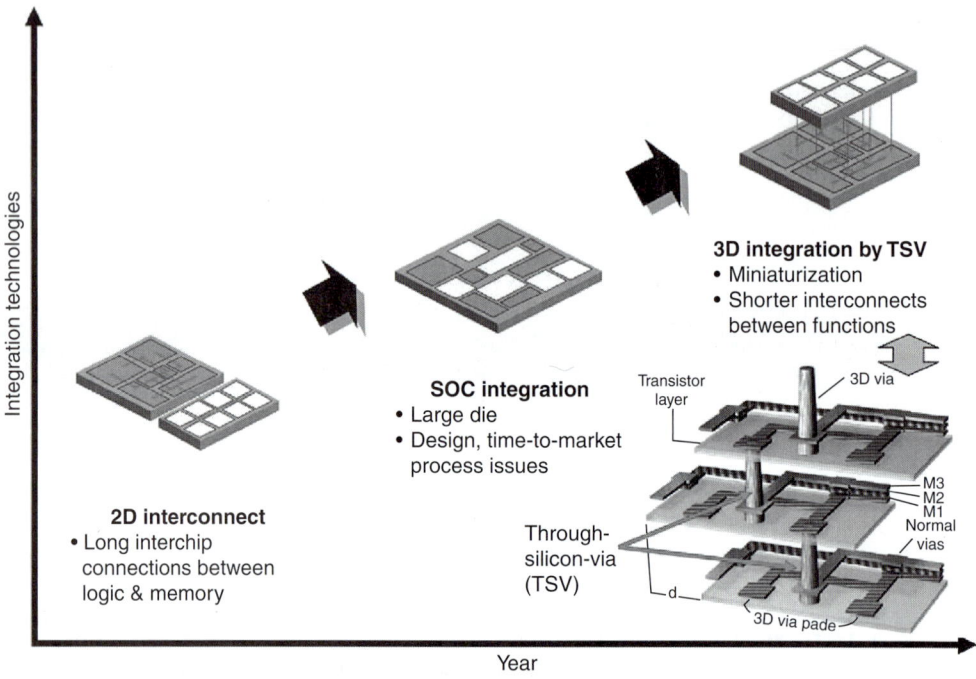

FIGURE 3.50 Three-dimensional integration benefits using through-silicon vias (TSVs) in contrast to MCM and SOC. (Courtesy of IMEC and Dr. P. Garrou.)

achieved by multichip modules (MCMs) in the 1980s. While this served the need at that time, this 2D MCM approach had long interconnections between the chips such as logic and memory in Figure 3.50 with at least a 10-mm interconnection length between these passing through the MCM substrate. Electrical losses were incurred at the chip-to-package interfaces, in addition to the delays in the long substrate wiring. MCMs gave way to system-on-chip (SOC) in the next generation. SOC integrated MCM functions in the same die thus eliminating the long interconnections in the package. However, the SOC still has long global wiring, a few millimeters in length, connecting the blocks in the die.

Through-silicon vias (TSVs) enabled the chips (or wafers) to be stacked vertically, thus reducing the wiring length to the thickness of the die, which is currently at 70 μm. Memory dies can be stacked right on top of the processor die to provide high-speed and low-loss memory-processor interfaces due to the lower parasitics of the TSV vertical interconnections. TSVs can be developed in an area array format thereby increasing the vertical interconnection density. They can also be used for heterogeneous integration of different IC technologies, as shown in Figure 3.51.

Table 3.3 compares TSVs with the traditional wire bonds for a variety of package characteristics. It can readily be seen that TSV technology has several important advantages over wire bonding.

TSVs can be used to stack dies on dies, dies on wafers, or dies on Si chip carriers. It can also be used to stack a wafer or Si chip carrier on top of another wafer or Si chip carrier. Table 3.4 compares die-to-die and wafer-to-wafer stacking characteristics.

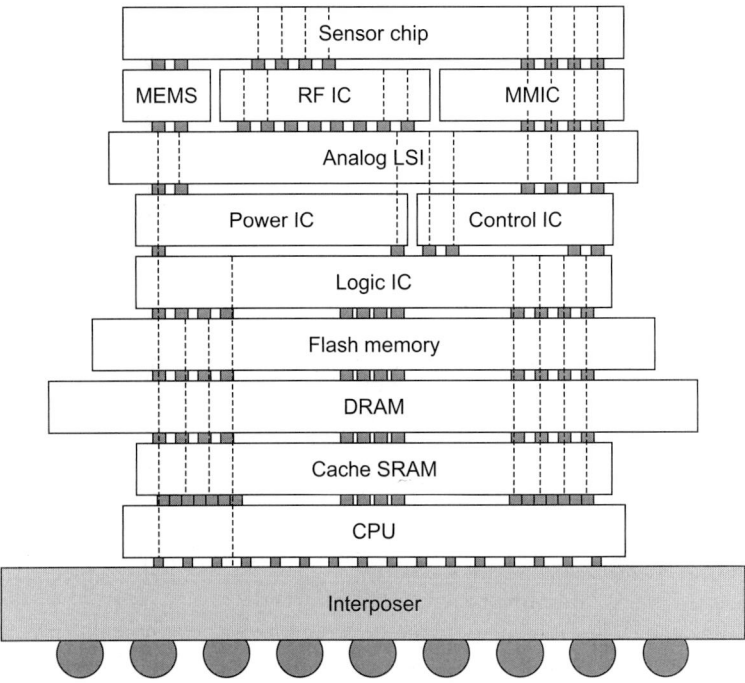

FIGURE 3.51 Heterogeneous integration by 3D TSV technology. (Courtesy: Zycube.)

Characteristics	TSV	Wire Bonding
Interconnection arrangement	Interconnections can be area-array or peripheral	Only peripheral interconnect
Interconnection length	Shorter interconnections	Much longer interconnect length
Electrical parasitics	Much lower electrical parasitics	Higher parasitics
I/O density	Potentially high density achievable	Lower I/O density
Reliability	Higher reliability	Less reliable
Processing	IC fabrication process	Packaging process

TABLE 3.3 Comparison between TSV and Wire Bonding

Die-to-Die Stacking	Wafer-to-Wafer Stacking
1. Different sized dies can be stacked.	1. Individual die sizes must match.
2. Alignment is easier.	2. Alignment is more difficult.
3. It uses known good die (KGD) for stacking; hence, there is a much higher yield.	3. There is a lower yield because of KGD issues.
4. Throughput is lower.	4. Throughput is higher.

TABLE 3.4 Die-to-Die Stacking versus Wafer-to-Wafer Stacking

Figure 3.52 compares the throughput of stacking chip-to-wafer and wafer-to-wafer [54]. It can be seen that for stacking 1000 or more chips per wafer, the wafer-to-wafer stacking process has a much higher throughput, as compared to chip-to-wafer stacking. However, usually this comes at the cost of a much lower yield in wafer-to-wafer stacking.

3.4.2 Historical Evolution of 3D TSV Technology

The earliest development on 3D TSV technology can be traced back to a U.S. patent [55] filed in February 1971 (and accepted in November 1972) by Alfred D. Scarbrough, as shown in Figure 3.53. This patent introduced the concept of wafer stacking with through-wafer interconnects. This shows a 3D wafer stacking arrangement with alternating layers of wafers carrying memory chips and wafers with only interconnection layers.

Figure 3.54 shows a cross-sectional schematic of the wafer stackup. The memory chips are bonded to the chip-carrying wafer. The through-wafer vias are filled with conductors. There are malleable contacts that connect the through-wafer vias on the two wafers (the chip-carrying wafer and the combined interconnection and spacer wafer). The bonding is performed by pressure and temperature to achieve wafer stacking together.

In 1980, T. R. Anthony of GE [56–57] demonstrated through-wafer vias drilled in silicon-on-sapphire (SoS) wafers by a laser drilling technique. In 1981, he subsequently

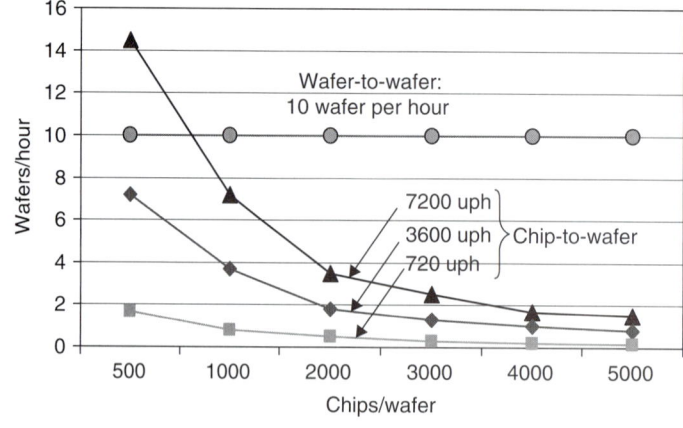

FIGURE 3.52 A throughput comparison between chip-to-wafer and wafer-to-wafer stacking. [54]

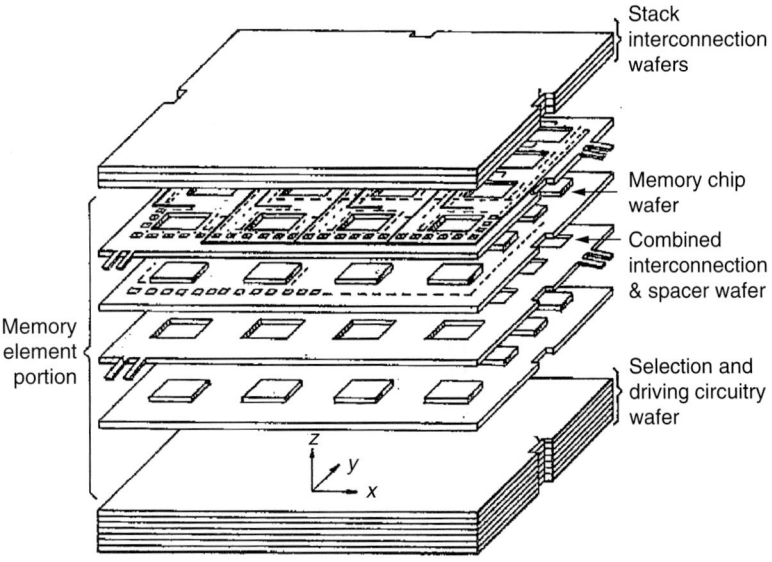

FIGURE 3.53 Perspective view of a multiwafer stacked semiconductor memory. [55]

studied six different techniques of forming the conductors in the drilled via holes, and stacking SoS wafers by wire insertion, electroless plating, capillary wetting, wedge extrusion, electroforming, and double-sided sputtering followed by through-hole electroplating [58].

In 1986, McDonald et al. (RPI, GE, and IBM) [59] used laser drilling to form 1- to 3-mil tapered through-wafer vias and then deposited metal by laser sputtering. These vias were then filled by electroplating. They considered the application of these through-wafer vias for 3D wafer stacking. In 1994, Robert Bosch Gmbh of Germany patented an inductively coupled plasma (ICP) etching process which came to be known as the "Bosch process" [60]. It was later used for drilling nearly straight-walled vias in the wafer. TSV formation by using anisotropic wet etching of silicon with solutions of KOH or ethylenediamine-pyrazine combinations was used as early as 1985 by German Manufacturing Labs (GMTC) of IBM and by others [61–62] in 1995–96. In 1997, Gobet et al. [63] used fast anisotropic plasma etching to form the through-wafer vias. This technique utilized standard photolithography steps.

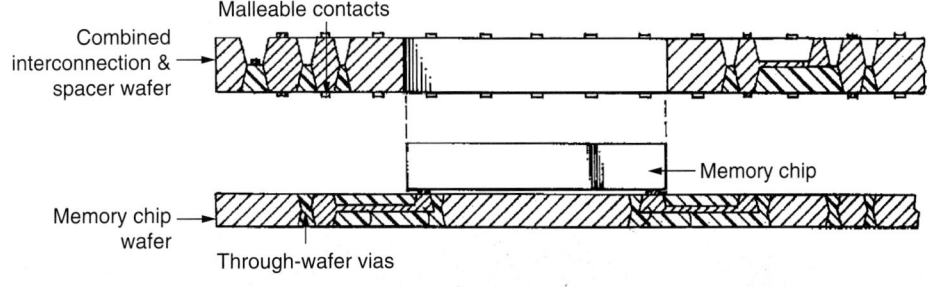

FIGURE 3.54 Cross-sectional schematic view of a multiwafer stacked memory. [55]

There have been quite a few products introduced in the market with 3D TSV technology. Tru-Si Technologies began marketing its Thru-Silicon vias in late 1999 [64]. The Association of Super-Advanced Electronics Technologies (ASET) developed a 3D die-stacked module in which four ultrathin chips (50 μm thick) are vertically stacked and have electrically interconnected Cu-filled through-hole vias [65]. IME developed a 3D silicon chip carrier stacking technology (using TSVs) in 2003 [66]. In 2005, Hitachi and Renesas developed another 3D stacking technology with TSVs with gold stud bumps [67]. In this approach, a compressive force is applied at room temperature to electrically connect the gold stud bumps on upper chips to through-hole-via electrodes in the lower chips. In April 2006, Samsung Electronics announced that it had developed a wafer-level processed stack package (WSP) of high-density memory chips using TSV-based 3D interconnection technology [68]. Samsung's WSP is a 16-Gbit memory solution that stacked eight 2-Gb NAND chips. In September 2006, Intel developed a prototype processor with 80 cores [69]. It used 3D TSV technology to stack 256 kbytes of SRAM directly on top of each of the chip's 80 cores. In June 2007, IBM announced SiGe BiCMOS 5PAe technology, which uses through-silicon vias for 3D stacking [70].

In addition to the above, there are several others who are also actively working in the area of 3D TSV integration technology. Some of them are Micron, Tezzaron, Ziptronics, Lincoln Labs, and RTI in the United States; NEC, Oki, Elpidia, Toshiba, and Zycube in Japan; and IMEC, Fraunhofer IZM, and LETI in Europe.

3.4.3 Basic TSV Technologies

There are several basic technologies for 3D integration by TSVs. The four main TSV processes are (1) via formation, (2) via filling with conductor material, (3) bonding chips with TSVs, and (4) thinning. Figure 3.55 outlines these four different technologies in more detail.

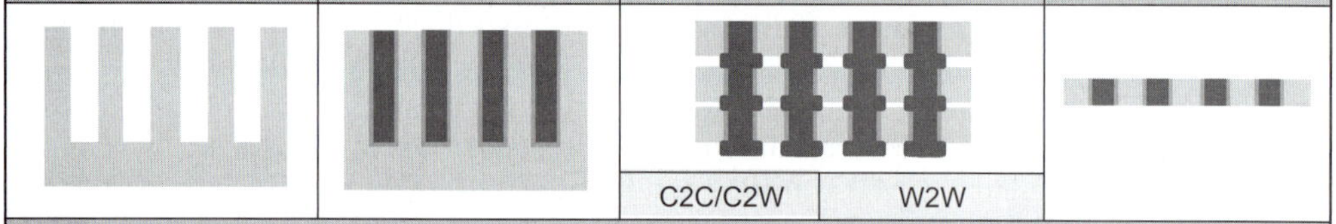

Via drilling	Via filling	Chip/Wafer bonding		Thinning
		C2C/C2W	W2W	
Technologies				
• Laser drilling • Bosch DRIE • Cryogenic DRIE • Wet etching	• Electroplating • CVD • Photolithography	• Adhesive bonding • Metal-metal bonding • Chip alignment	• Adhesive bonding • Metal-metal bonding • Wafer alignment	• Grinding • CMP • Wet etching • Plasma etching
Equipments				
• Laser or DRIE DRIE • Coater • Maskaligner or stepper	• Mask deposition system • Coater • Mask aligner or stepper	• Device bonder • Device aligne	• Waver bonder • Temporary bonder • Wafer aligner	• Thinning equipment • Temporary bonder

FIGURE 3.55 Different TSV technologies. (Courtesy of Yole Developpement.)

Via Drilling

The TSVs can be formed by Bosch-type deep reactive ion etching (DRIE) [60], cryogenic DRIE, laser drilling, or by a variety of wet etching (isotropic and anisotropic) processes. Laser drilling was initially explored in the mid-1980s, as described earlier. Figure 3.56 shows the SEM picture of some TSVs formed using the laser drilling process. The laser drilling creates some silicon "splashes" due to "melting." The laser-drilled vias should be at least 2 µm away from the active devices in order to ensure that the device characteristics are unaffected. It is very difficult to develop vias with diameters less than 25 µm using laser drilling. The natural slope of the via sidewalls varies from 1.3° to 1.6°.

The Bosch process forms TSVs with smooth and straight sidewalls. The alternating passivation and etching steps ensure almost smooth straight sidewalls. Figure 3.57 shows the process steps involved in a typical Bosch process along with an SEM diagram of a TSV developed using this process.

Cryogenic DRIE is very similar to ordinary DRIE. The main difference is that the wafer is cooled to cryogenic temperatures (−110°C), which drastically lowers the mobility of incoming ions, after they have hit the surface. By preventing the ions from migrating, very little etching of the sides is realized. In addition, the anisotropy is dependent on the temperature. This demands the implementation of a powerful cooling system, often with several stages of cooling, which is capable of dissipating the heat generated by the etching process.

Via Filling

Once the TSVs are drilled, insulating films are deposited in order to provide insulation between the silicon and the conductor. These films can be deposited in a variety of ways including thermal, plasma-enhanced chemical vapor deposition (PECVD) using silane, and tetra-ethoxysilane (TEOS)-type oxides, as well as low-pressure chemical vapor

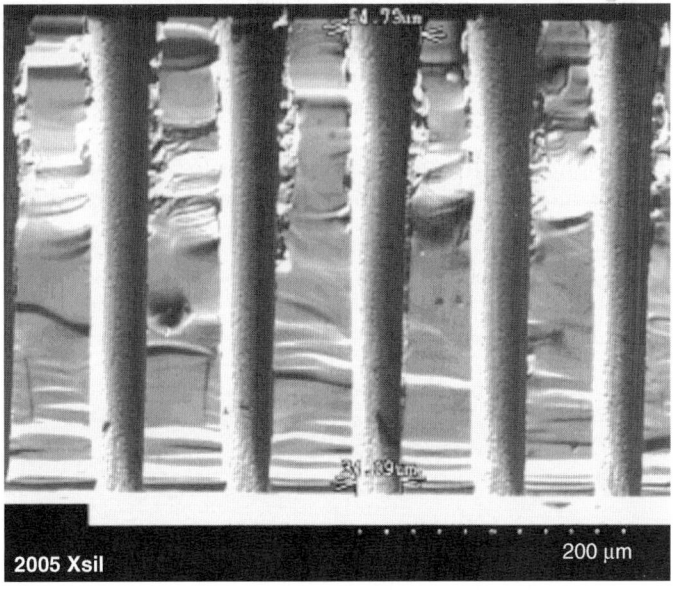

FIGURE 3.56 SEM image of some laser-drilled TSVs developed by XSil.

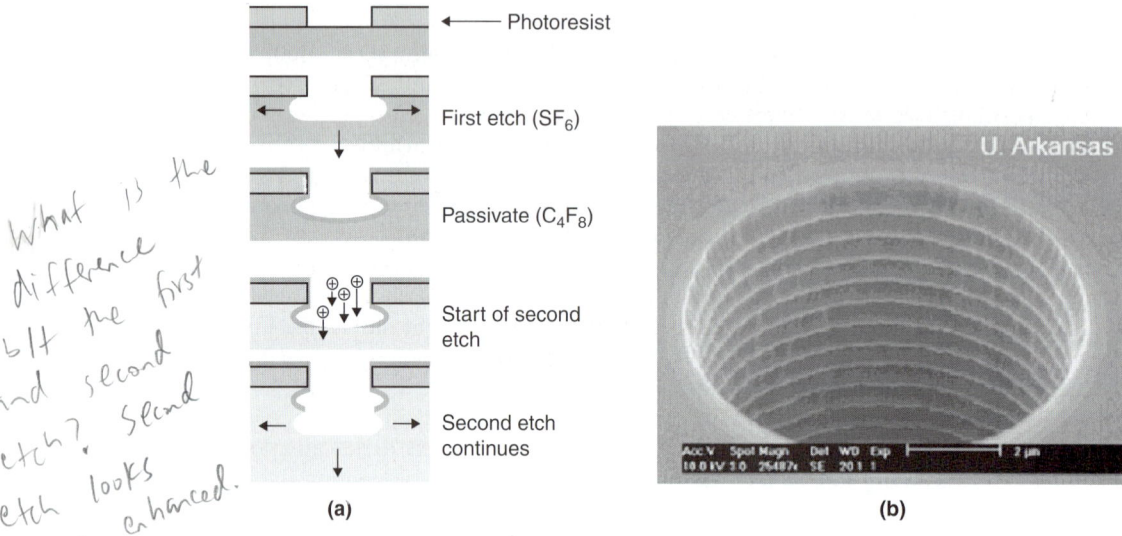

Figure 3.57 (a) Steps involved in the Bosch process. (b) An SEM image of a silicon via drilled by the Bosch process at the University of Arkansas.

deposition (LPCVD) nitrides. The TSVs are ready for metallization after the insulation layer formation.

There are different competing materials that can be used as conductor material in the TSV such as Cu, W, and polysilicon. Cu has excellent electrical conductivity. The deep TSVs can be filled by copper plating or copper paste filling. TSVs, which have relatively small depths, can be fully filled with copper. However, for deep TSVs, the difference in the CTE of Si (3 ppm/°C) and copper (16 ppm/°C) becomes significant. Thermomechanical stresses developed due to this mismatch can result in interlayer dielectric (ILD) and silicon cracking. The thin insulation layer deposited on the TSV sidewalls results in high electrical capacitance, thus degrading the electrical performance of the TSV interconnections. The electroplating process for completely filling large vias is also quite slow.

IMEC (Belgium) uses an approach with a 2- to 5-μm-thick polymer isolation layer. Thereafter the via hole is partially filled by electroplated copper before using polymer to fill the remaining via hole. Figure 3.58 shows the cross sections of some TSVs developed

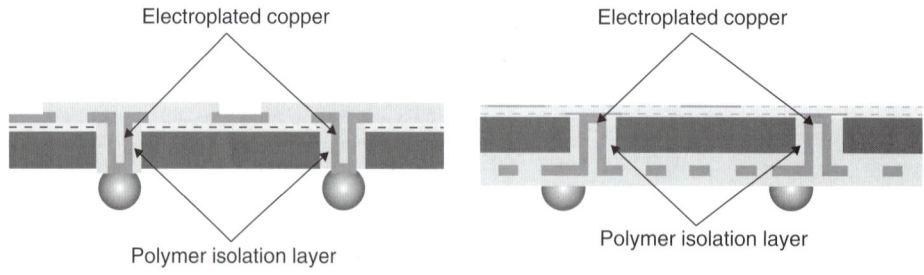

Figure 3.58 Schematic cross-sectional views of IMEC's through-silicon vias with partial copper filling. [71]

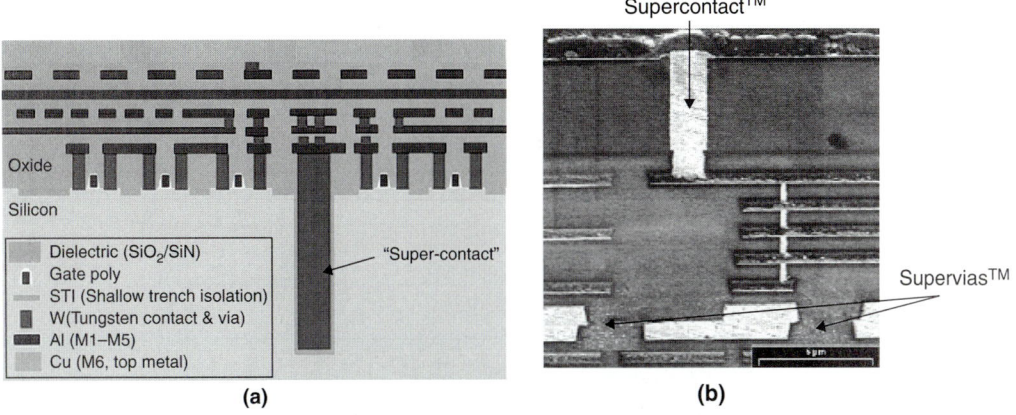

FIGURE 3.59 (a) Schematic cross-sectional view of 3D integration using tungsten-filled TSVs. (b) Cross-sectional SEM of interwafer interconnects in the Tezzaron 3D platform showing tungsten Supercontact and Cu-to-Cu Supervia. [72]

in this approach. The TSVs developed in this method have lower capacitance due to the use of thicker low-k isolation layers. The thermomechanical stresses in the TSV region are reduced due to the relatively smaller percentage of copper in the through-hole structure. This approach is also compatible with wafer-level packaging technologies.

Alternatively, tungsten (W) or molybdenum (Mo) has been used for filling the vias. Although they have lower electrical conductivity than copper, they have lower CTEs than copper (CTE for W = 4.5 ppm/°C; CTE for Mo = 4.8 ppm/°C), which are better matched to the CTE of Si. Thus, the TSVs filled with these metals suffer from much lower thermomechanical stresses than those filled with copper. Figure 3.59 shows the cross-sectional views of 3D integration using TSVs filled with W, developed by Tezzaron [72].

There are different methods of filling the vias with these metals, as shown in Figure 3.60 [73]. Physical vapor deposition (PVD) or sputtering are used for small vias,

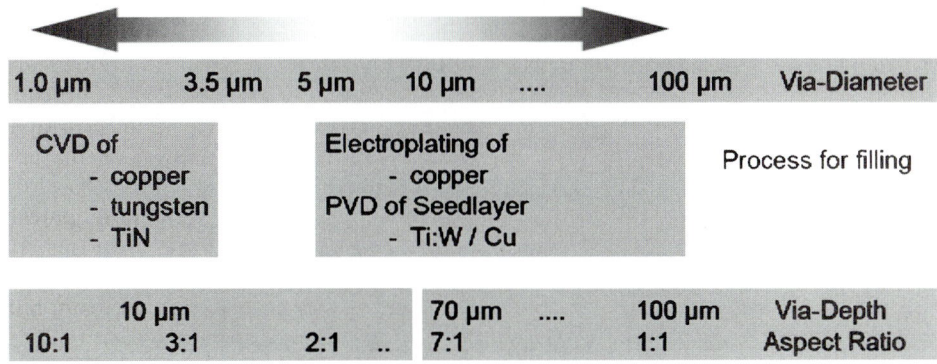

FIGURE 3.60 Various via filling technologies depending on via diameter and aspect ratio. (Courtesy of Fraunhofer, 73.)

but the process is very slow and may not produce a perfectly conformal coating. Laser-assisted chemical vapor deposition (CVD) of Mo or W is considerably faster and is used for filling deep vias. There are also different metal-ceramic composites with lower CTEs (<<16 ppm/°C). It is difficult to fill deep, blind vias with an aspect ratio > 5:1. Specialized processes are required for filling such deep vias. Figure 3.61 shows the SEM pictures of different TSVs [74] fabricated this way.

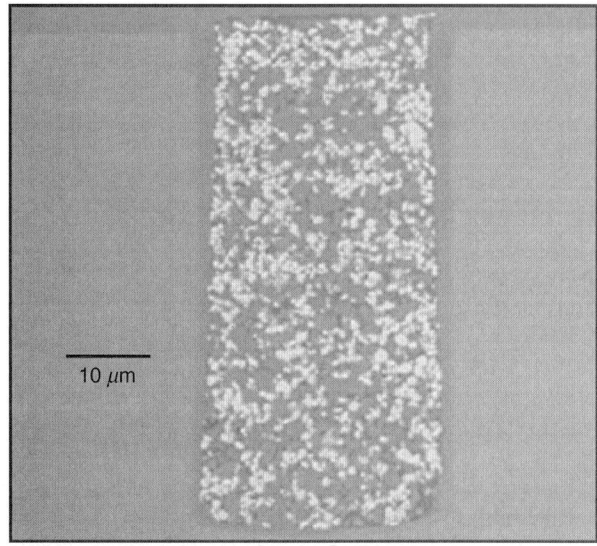

(a)

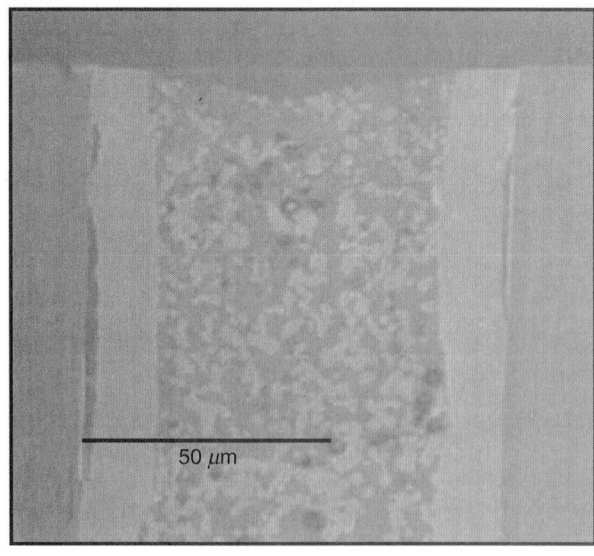

(b)

Figure 3.61 (a) Small copper-ceramic–filled via. (b) A via with partial copper plating and a composite filling. [74]

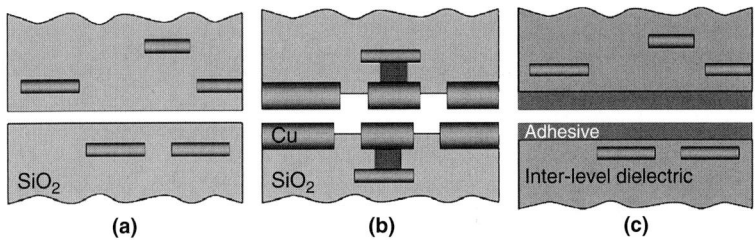

FIGURE 3.62 Different bonding approaches. (*a*) Oxide fusion bonding. (*b*) Metal-metal bonding. (*c*) Polymer adhesive bonding. (Courtesy of P. Garrou.)

Wafer Bonding

There are different ways of bonding die-to-wafer, die-to-die, or wafer-to-wafer. Three generic types are reported in the literature. They include silicon dioxide (SiO_2) fusion bonding, metal-metal bonding, and polymer adhesive bonding. Metal-metal bonding can be of two types: metal (Cu) fusion bonding and metal eutectic bonding such as with Cu-Sn. Figure 3.62 shows examples of these different bonding approaches.

Oxide Bonding Oxide bonding techniques have been developed, for example, by Lincoln Laboratory [75]. The preprocessed wafers with active devices and first-level or multilevel on-chip interconnects are aligned and bonded using silicon dioxide layers. The wafers to be bonded are coated with a low-temperature oxide (LTO) layer deposited by a low-pressure chemical vapor deposition (LPCVD) technique. The surfaces are polished to smooth them to a roughness of < 0.4 nm root mean square (rms). In order to form the bonds, both the surfaces should have a high density of hydroxyl groups (OH) present. The wafers are immersed in H_2O_2 to remove any contaminant and to coat the surfaces with the OH groups. After this, the wafers are rinsed and spin dried in nitrogen. The wafers are aligned and bonded by initiating contact at the center of the top wafer. The bond strength can be increased by a higher-temperature process that creates covalent bonds at the interfaces. Atomic-scale smooth interfaces are needed for adequate bond strength after the wafer bonding process. Figure 3.63 shows a schematic representation of this approach.

IBM used oxide bonding in its 3D integration platform [76]. The wafer bonding process is compatible with back-end-of-the-line (BEOL) wafer processing. Figure 3.64 shows a schematic process flow of the IBM process.

Metal-Metal Bonding

Cu-Sn Eutectic Bonding Bonding with low-melting-point metals such as tin either through diffusion or solder fusion is commonly adapted for 3D Si integration. Vertical interconnections with Cu bump bonding utilizing Cu-Sn diffusion for connecting

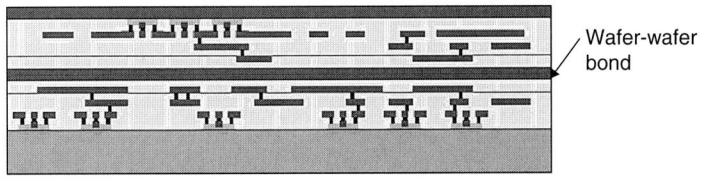

FIGURE 3.63 A cross-sectional view of wafer-to-wafer bonding using the oxide bonding technique. [75]

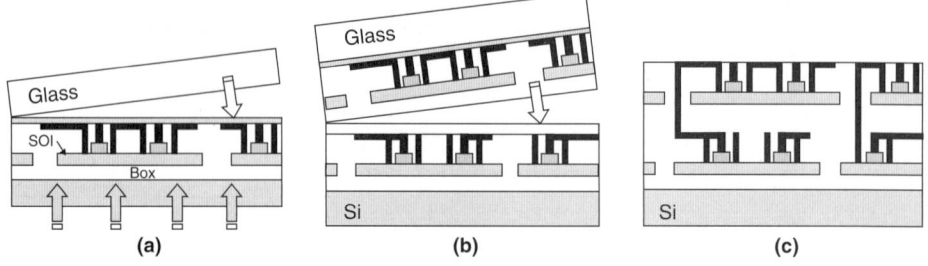

FIGURE 3.64 IBM's Via Last approach of 3D integration using oxide bonding. [77]

Cu-through vias will eliminate the formation of bumps on the chip back surface [78]. High-aspect-ratio copper vias at pitches less than 50 μm, bonded with tin-based contacts have been demonstrated by ASET as shown in Figure 3.65. The bonding reliability was enhanced further with resin encapsulation. By combining Cu pads with lead-free solder plating, IBM has demonstrated reliability at a 50-μm pitch [74].

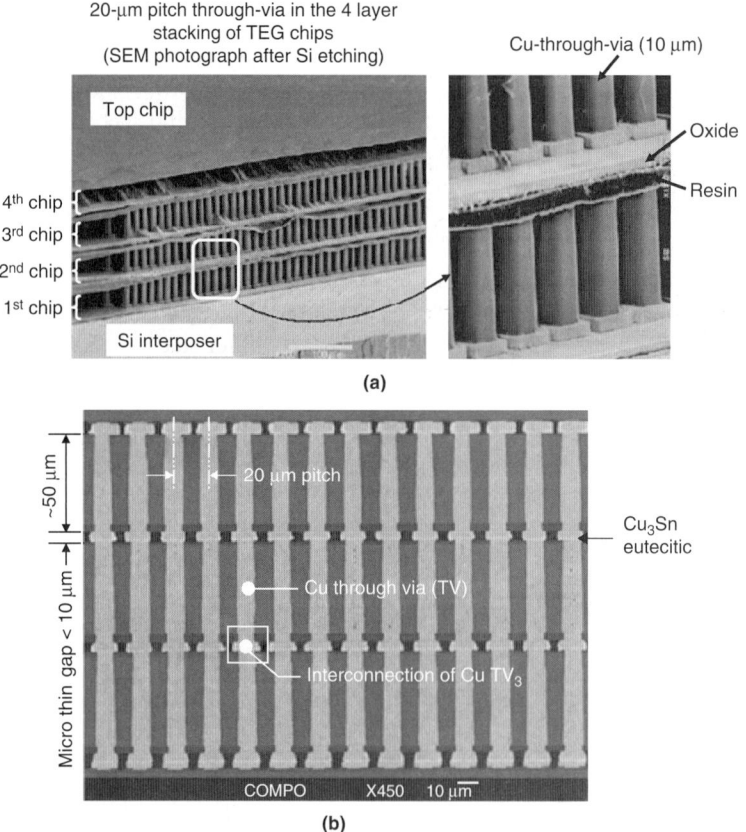

FIGURE 3.65 Demonstration of interchip through vias by ASET, with copper plating and Sn bonding. [78]

Direct Cu-Cu Bonding Direct Cu-Cu bonding eliminates the tin or gold bumping steps as well as several electrical and mechanical reliability issues associated with solders and intermetallics. This approach makes 3D technologies more compatible with standard wafer fabrication processes. Earlier fundamental studies on thermocompression bonding of copper were reported by Reif et al. [79]. The TEM micrographs in Figure 3.66 show the evolution of interface morphologies at different stages of wafer bonding and annealing. They show strong grain growth during bonding and annealing. Initial bonding causes some interdiffusion but does not complete the fusion and grain growth. A postbonding annealing to induce diffusion across the Cu-Cu interface, grain growth, and recrystallization is essential to complete the crystallization.

A recent work by Chen et al. at IBM [80] reported that wafers bonded with a slow temperature ramp rate (6°C/min) have a better bonding quality than those with the fast rate (32°C/min). Their studies also showed that application of a small force prior to temperature ramping and high bonding down-force during bonding enhanced the bonding strength. The quality of the bonded interface improves with the increasing interconnect pattern density, but does not strongly depend on the size of the Cu interconnect. Minute amounts of copper oxides are generally known to impact the bonding of copper to copper. Highest shear strengths were obtained when the surfaces were pretreated with dilute citric acid. IMEC has also extended this process to extremely thin Si containing 10-μm pitch through-silicon vias.

Polymer Bonding Polymer adhesive wafer bonding does not require special surface treatments such as planarization and excessive cleaning. Contaminant particles at the wafer surfaces can be compensated to some extent by the polymer adhesive. Two types of polymer adhesives are mainly used for wafer bonding applications: thermoplastic polymers and thermosetting polymers. The adhesive polymer is applied to both wafer surfaces to be bonded together by spin coating a liquid polymer precursor on the wafer surfaces. The polymer coatings are subsequently heated to remove the solvents and to form the cross-linking in the polymer. The wafers are then carefully aligned together, and bonded under pressure in a vacuum. The wafer stack is then cured in a vacuum to form a strong and reliable bond.

Various polymers have been proposed for adhesive polymer wafer bonding, including negative photoresists [81–82], benzocyclobutene (BCB) [64, 83–85], parylene [76], and polyimides [77, 86]. BCB has outstanding wafer bonding capabilities, chemical resistance, and bond strength. BCB reflow can be minimized by partially curing it prior

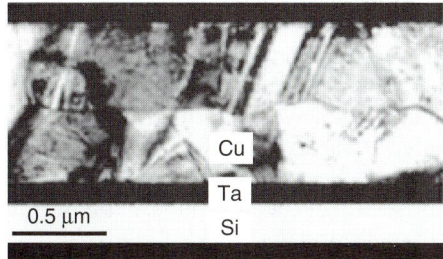 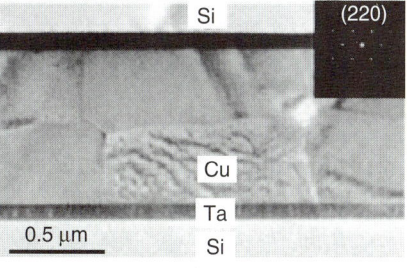

FIGURE 3.66 XTEM studies of Cu-Cu bonding by Reif et al. [79].

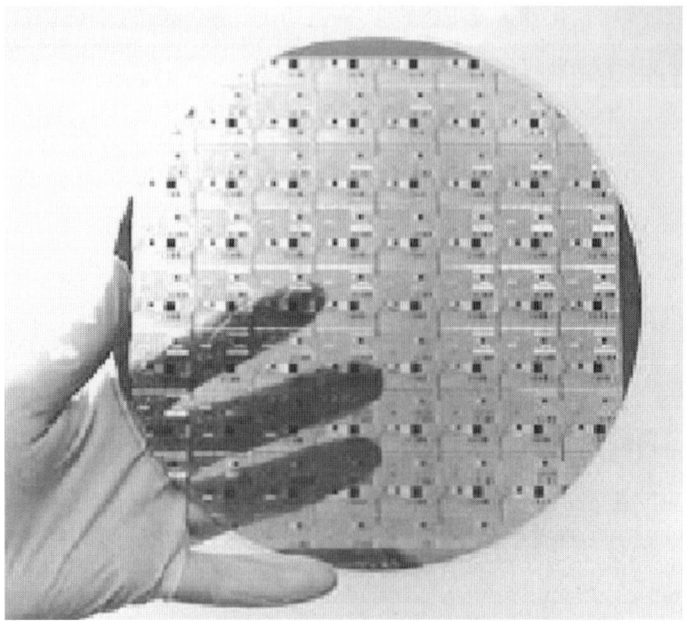

Figure 3.67 A wafer after polymer adhesive bonding. [64]

to wafer bonding. Wafer bonding with partially cured BCB coatings results in very uniform BCB layers and prevents bonding-induced misalignment [84]. Negative photoresists and polyimides can be etched in oxygen plasma. Therefore, they are suitable as sacrificial bonding layers or as adhesives for temporary bonds in 3D integration platforms such as for MEMS applications. Figure 3.67 shows the picture of a wafer with Cu-oxide interconnect structures after bonding to a glass wafer using BCB and removing the Si substrate by grinding, polishing, and wet etching [64].

The advantages of adhesive bonding are several and include compatibility with integrated circuit wafers, relatively low bonding temperatures, the ability to join practically any kind of wafer material, and a lower sensitivity of the bond strength to the presence of interlayer particles. However, the downside is that the wafers are prone to being misaligned during the bonding or curing process.

3.4.4 Different 3D Integration Technologies Using TSV

There are several different ways of developing a 3D integrated system using TSV technology. These processes can mainly be classified in two broad categories: via-first and via-last. In the via-first scheme, the TSVs are formed before the BEOL developed on the carrier. In contrast, in the via-last scheme, the TSVs are formed after the development of the BEOL interconnection layers. Table 3.5 shows the main process steps of these two processing schemes.

There are variations of these two processing schemes that are followed by several companies, organizations, and universities around the world. Table 3.6 provides a brief overview of these different processes along with examples of some of the organizations that are following them.

Step No.	Via-First	Via-Last
1	Drilling of TSV	Development BEOL wiring on wafer
2	Deposition of dielectric	Mounting of wafer on carrier and wafer thinning
3	Formation of passivation layer on sidewalls and via filling with conductor material	Formation of TSV from back of wafer
4	Development BEOL wiring on wafer	Deposition of dielectric
5	Wafer thinning to make contact with TSV	Formation of passivation layer on sidewalls and via filling with conductor material
6	Development of interconnects on back of wafer	Development of interconnects on back of wafer

TABLE 3.5 Comparison of the Via-First and Via-Last Processing Approaches

Via-First Process 1

In this via-first approach for TSV development, the first two wafers are stacked face-to-face. Tezzaron, for example, uses a copper-to-copper vertical interconnect (Super-Via). The process flow is described in Figure 3.68. Most of the processing is done by traditional equipment except the EVG aligner and bonder, which is more common in MEMS fabrication.

The wafer is fabricated with devices (step 1). The intermetal layer dielectric (IMD) is deposited on the wafer. It is planarized by oxide chemical mechanical polishing (step 2). The Super-Via is then etched in the dielectric stack (step 3). In step 4, the Si base is etched to a depth of 4 to 9 μm. In the next step, oxide and the SiN layer are deposited as barrier and passivation layers. Trenches and vias are then drilled for wafer-to-wafer bonding in steps 6 and 7. In step 8, a Ta or TaN layer is deposited as a copper seed layer before filling the vias with copper by electroplating. The excess Ta and Cu are then removed by the CMP process. At this time, the wafer is finished with its BEOL process, which can include a combination of aluminum and copper wiring layers. The Cu pad is

Step No.	Via-First Process 1	Via-First Process 2	Via-Last Process 1	Via-Last Process 2	Via-Last Process 3
1	Via drilling	Via drilling	Bonding	Thinning	Thinning
2	Via filling	Via filling	Thinning	Bonding	Via drilling
3	Bonding	Thinning	Via Drilling	Via Drilling	Via Filling
4	Thinning	Bonding	Via Filling	Via Filling	Bonding
Examples	Tezzaron	IMEC, ASET, Fraunhofer	RPI	RTI	Infineon

TABLE 3.6 Different Processing Approaches for 3D Integration by TSV

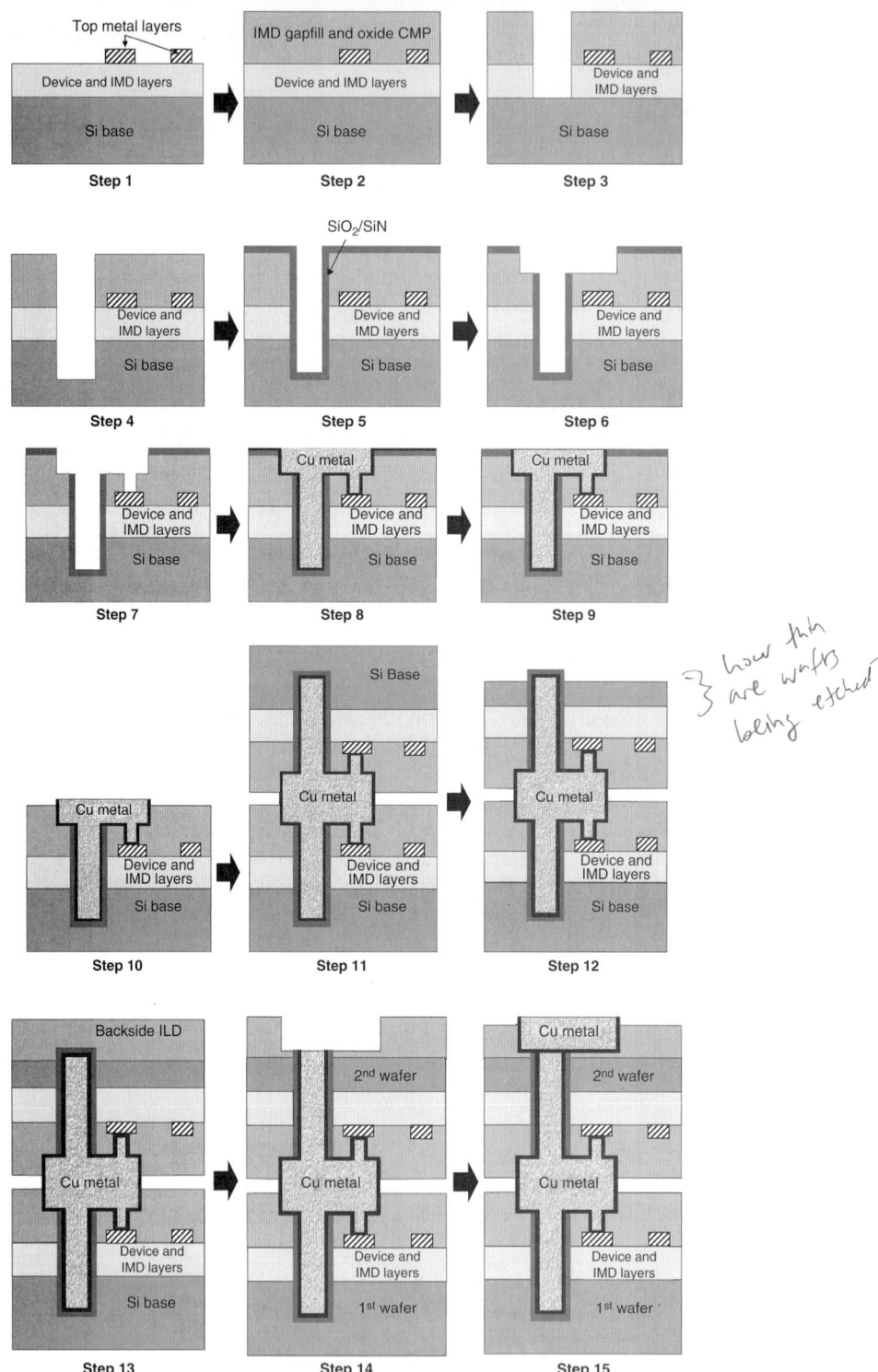

FIGURE 3.68 Processing steps in Tezzaron's 3D integration technology using TSVs. (Modified from [87].)

then grown with electroless deposition to form the wafer-to-wafer contact pads in step 10. Alternatively, the dielectric can also be removed to form these wafer-to-wafer contacts. In the next step, the wafers are aligned and bonded by Cu-Cu thermal diffusion bonding. The base Si layer on the top wafer is thinned by CMP and grinding. Chemical etching is used to remove another 1 µm of the Si layer (step 12). A layer of oxide is deposited by plasma-enhanced chemical vapor deposition (PECVD) on the back side of the thinned wafer to avoid Si contamination while integrating the next wafer on top of this stack (step 13). In step 14, the oxide layer is again etched (like in step 2) to form the trenches for depositing the copper connection. In the last step, the copper pad is formed for connecting the next wafer on top of the stack.

Via-First Process 2
This process used by ASET, Fraunhofer, IMEC, and others is also a via-first approach like the previous one. The difference is that the wafer bonding is carried out after thinning the wafer, unlike in the previous approach in which these steps are carried out in the reverse manner. The ASET process flow [88] is described in Figure 3.69, which also shows the cross-sectional schematic of the stacked structure developed by this process.

The vias are filled with copper by electroplating. CMP is used for polishing. The wafers are bonded either by using gold bumps or by Cu-Sn eutectic bonds. A similar method used by IMEC is shown in Figure 3.70.

The TSVs are drilled using an ICP-RIE Bosch process. The vias are filled with Cu using a modified Cu/ILD damascene process. Standard front-end-of-the-line (FEOL) and back-end-of-the-line (BEOL) processes are used. The wafers are handled with a wafer carrier. The wafers are bonded by Cu-Cu direct bonding.

Via-Last Process 1
In this example of a via-last approach of 3D integration using TSVs, RPI and IBM in collaboration with the University at Albany have developed this SOI-based process [83], as shown Figure 3.71 with a simplified process flow.

The wafers (with active devices and BEOL interconnects) are first precisely aligned and bonded using polymeric adhesives such as BCB. This face-to-face initial wafer bonding step eliminates the need for a "handle" wafer. The polymeric bonding is tolerant to the presence of contaminants at the wafer surface. One of the wafers is thinned down to the etch-stop region by back-side grinding, CMP, and wet etching. Thereafter, high-aspect-ratio vias are drilled in the wafer stack. A barrier layer is deposited before filling the vias with Cu. Another wafer can be added on top of this stack by following similar alignment, bonding, thinning, and via formation steps.

Via-Last Process 2
This is another variation of the via-last process. The difference from the previous approach is that the wafers are first thinned and then bonded, unlike in the previous case where it is done the other way around. The RTI process using this approach is used for stacking thin IC layers [89]. All the 3D processing steps are performed under 250°C. This approach can be used for either die-to-die or die-to-wafer stacking. Figure 3.72 shows the key process steps in the 3D integration process by the via-last process.

The wafer (marked as "IC2") is bonded on a "handle" carrier substrate. It is thinned using back-side grinding and CMP. The individual dies of the IC2 wafer are diced (while still bonded to the carrier substrate). The individual dies of IC2 are carefully aligned and bonded with the IC1 wafer. High-aspect-ratio vias are etched through IC2.

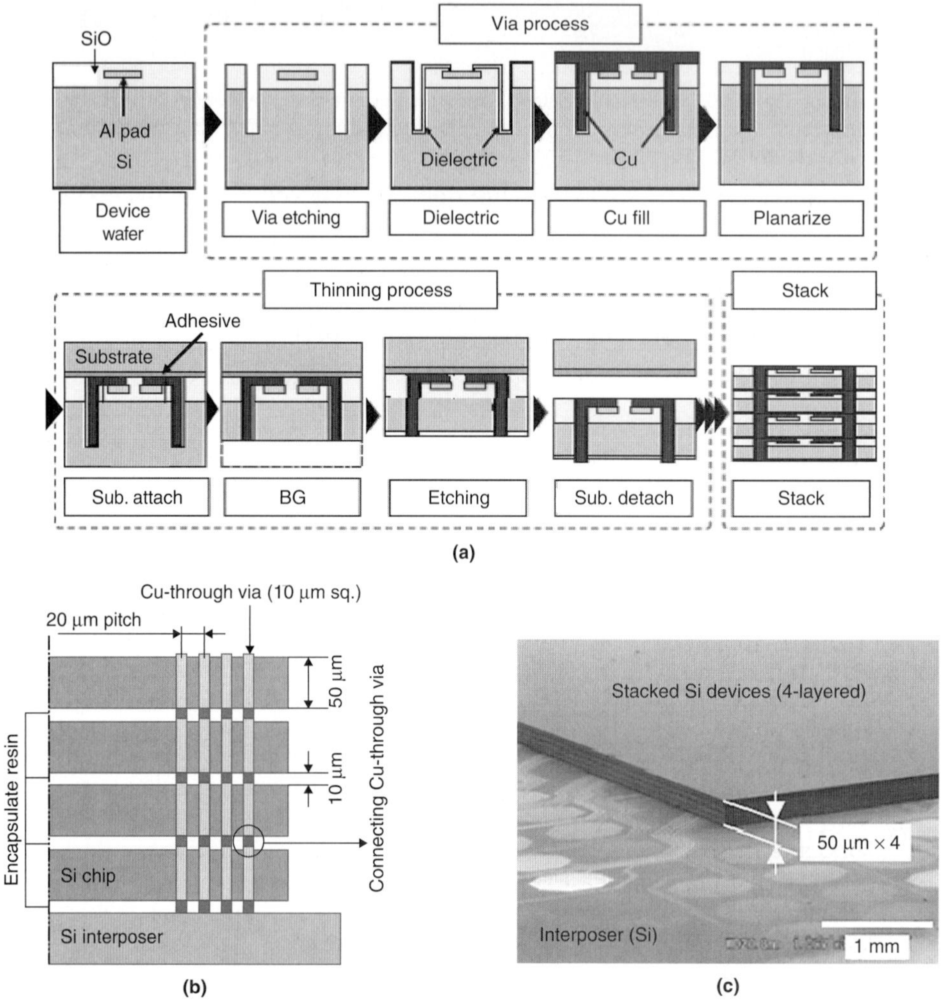

FIGURE 3.69 (a) Processing steps in 3D integration by ASET [88]. (b) Cross-sectional schematic. (c) An SEM picture of the 3D stacked structure [78].

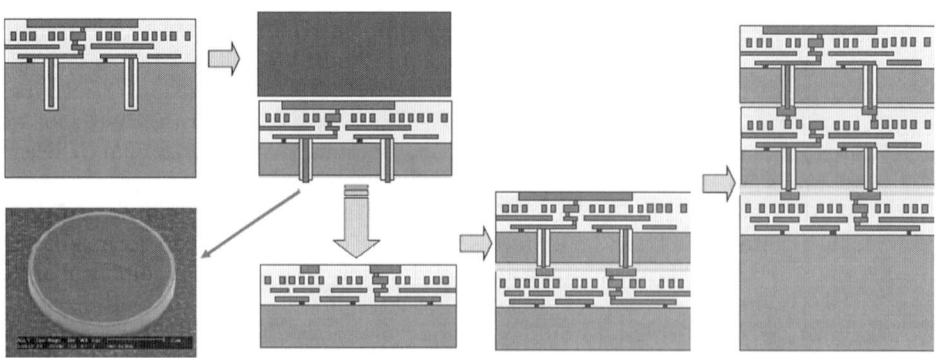

FIGURE 3.70 Via-first approach of 3D integration by IMEC. [71]

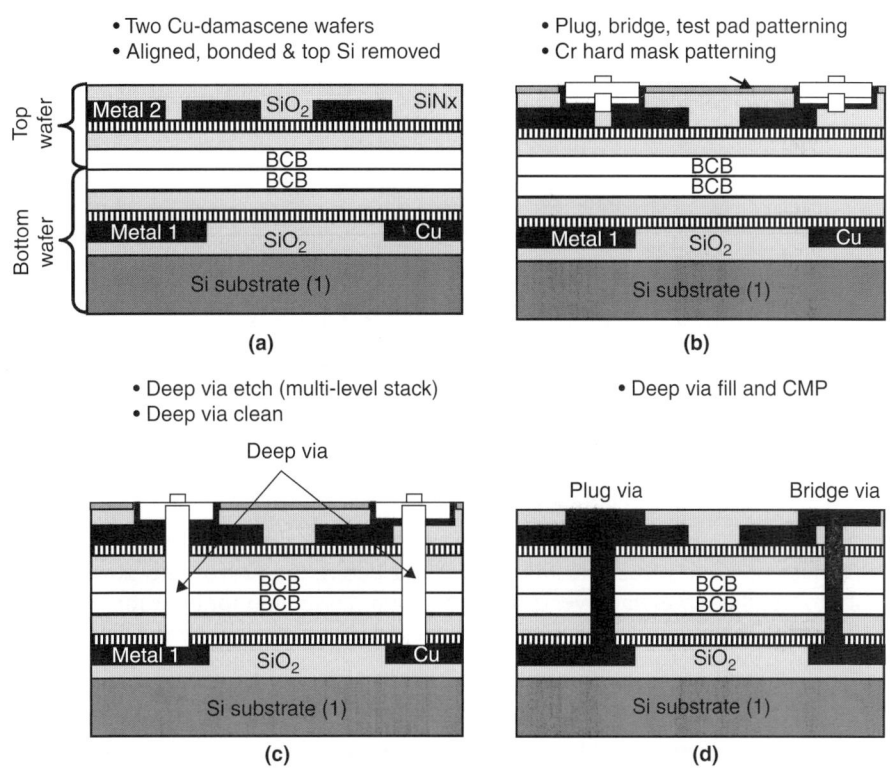

FIGURE 3.71 The process flow diagram for 3D integration with TSVs at RPI, in collaboration with the University at Albany. [83]

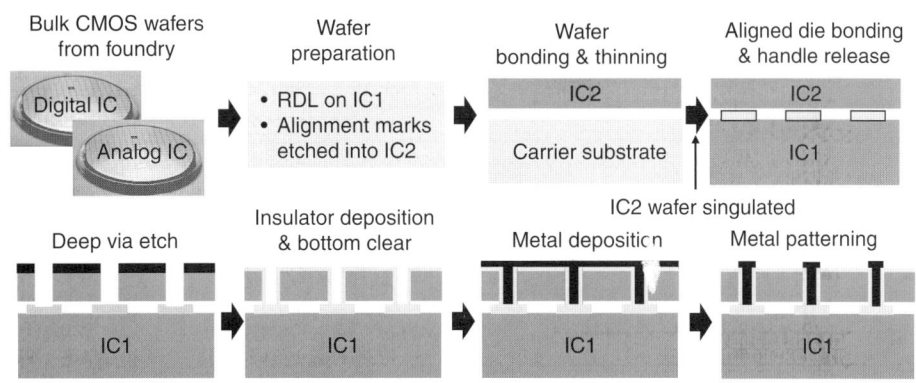

FIGURE 3.72 Schematic process flow for RTI's 3D integration approach. [89]

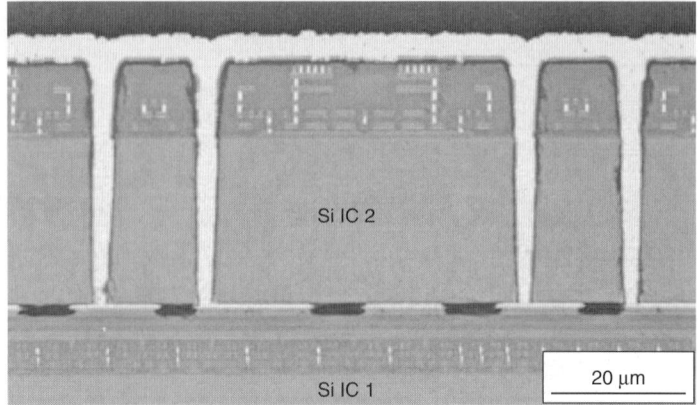

FIGURE 3.73 An SEM cross-sectional view of vertically stacked ICs. [89]

Via surfaces are coated with a conformal coating of insulator material. The bottom surface of this layer is selectively removed before the vias are filled with metal. In the last step, the top metal layer is patterned and passivated for bonding with the next die. Figure 3.73 shows an SEM picture of stacked ICs integrated by this method.

Via-Last Process 3

This approach of 3D integration is followed by Infineon [90]. The wafers are thinned, the vias are formed, and then the wafers are bonded in this process. Figure 3.74 shows the important steps in this process flow.

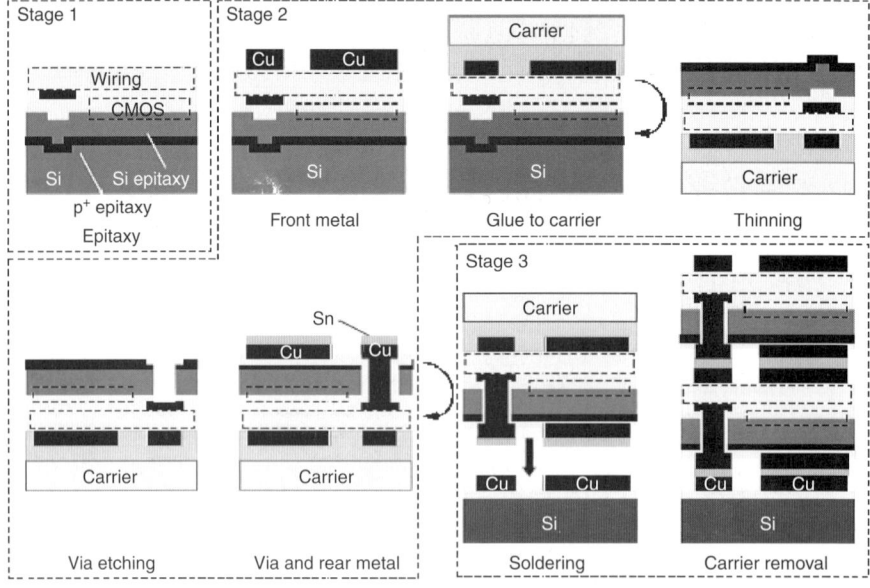

FIGURE 3.74 Process flow for Infineon's 3D chip stacking technology. [90]

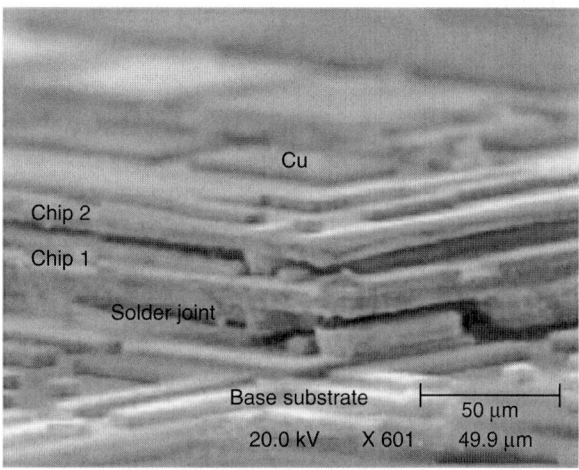

Figure 3.75 SEM picture of a chip stack with two thinned chips soldered in back-to-face technology. [90]

The wafer is first patterned with alignment markers, and a silicon layer is epitaxially grown on it. The wafer is diced and attached to a carrier. The thinning is done by fast mechanical thinning followed by wet chemical etching. After the thinning process, the alignment marks become visible. Anisotropic etching is used to etch the vias in the silicon substrate. An oxide insulation layer is formed on the exposed chip and via surfaces. A Ti-W barrier layer is deposited followed by a seed layer of copper. The vias are filled with copper by electroplating. The wafers are precisely aligned (using the alignment marks), and the bonding is performed by Cu-Sn-Cu eutectic bonding. Finally, the carrier is removed. Another chip can be bonded on top of this stack by using similar process steps as described above. Figure 3.75 shows the SEM micrograph of a stacked IC structure formed by this process.

3.4.5 Si Carrier Technology

The concept of a silicon chip carrier was developed at IBM in 1972 [91], wherein a Si substrate was used as a chip carrier instead of organic or ceramic substrate on which to deposit multilayer polymer-copper wiring. Initially, the chips were connected to the chip carrier by perimeter connections such as wire bonding. Later, the connections were replaced by flip chip connections. Lately, TSVs have replaced both. The TSVs help to develop a much higher density interconnection from the chip to the carrier and from the carrier to the board. Presently silicon chip carrier technology involves through-silicon vias (TSVs), ultrahigh-density wiring, fine-pitch chip-to-carrier interconnections, and integrated actives and passives.

Figure 3.76 shows the cross-sectional view of an Si chip carrier and the process steps involved in developing such an Si chip carrier with TSVs [92]. The chips are flip chip bonded to the chip carrier either by Cu-Cu bonding or solder bumping. The TSV development is carried out by the via-first approach. The TSVs supply the signal and power from the board to the top side of the chip carrier. High-speed and high-density wiring on the chip carrier distributes the signal and power to the chip.

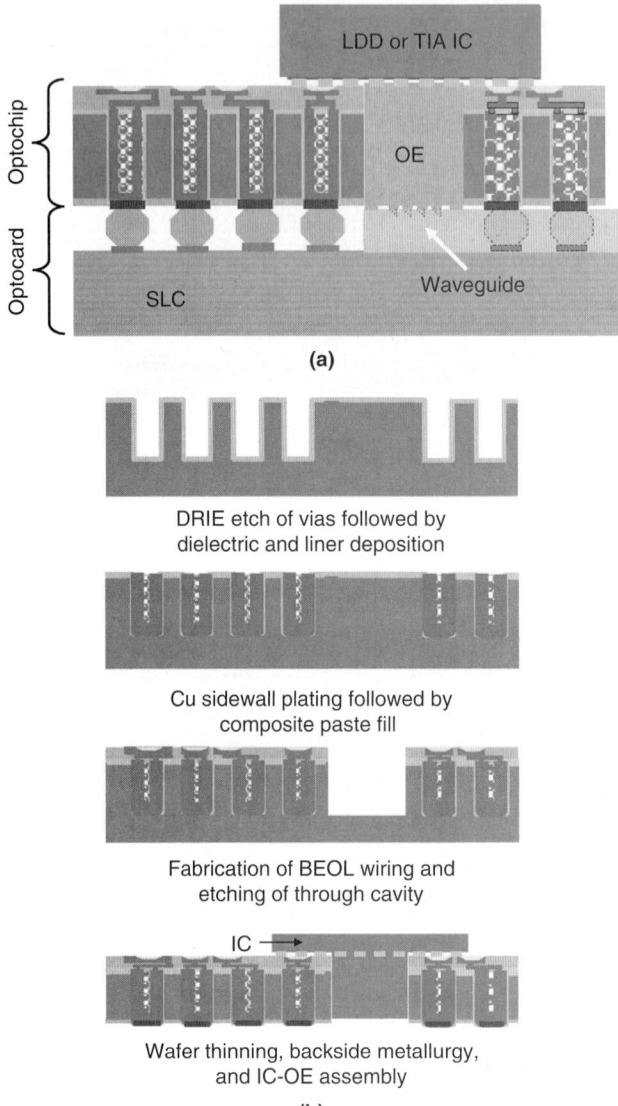

FIGURE 3.76 (*a*) Cross-sectional view of an Si chip carrier (with an optical chip) mounted on a board. (*b*) Process flow for the Si chip carrier. [92]

TSVs in the Si chip carrier can also be used for stacking the individual chip carriers. Figure 3.77 shows the schematic view of such a stacked Si chip carrier module [93]. ICs are flip chip bonded to the individual Si chip carriers that, in turn, are stacked on top of each other. The entire 3D module is finally mounted on a PCB. Solder-via-fill technology is used to develop the vertical interconnects through the TSVs in the chip carriers. The process flow adopted is similar to the via-first process 2, which was described in the previous section.

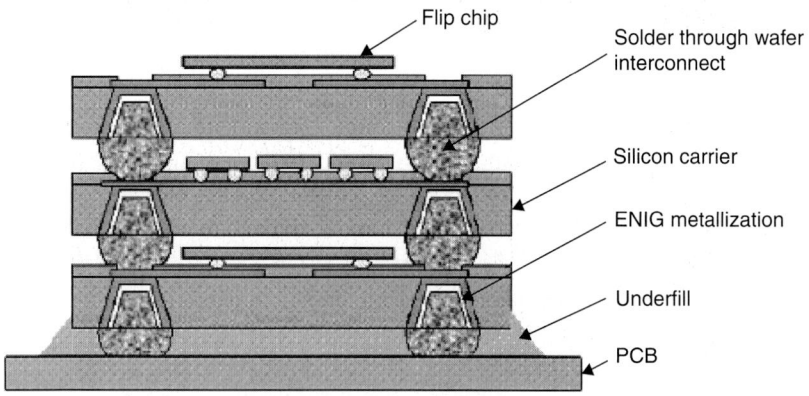

FIGURE 3.77 A perspective view (left) and a cross-sectional view of a 3D stacked Si chip carrier module (ENIG = electroless nickel immersion gold). [93]

There are several attractive features for using an Si chip carrier. Standard Si BEOL fabrication processes can be used to develop high-density wiring on the carrier at lower cost while achieving a higher yield. The thermal coefficient of expansion (TCE) of the chip carrier matches that of the IC. This helps in forming a very reliable connection between the chip and the carrier, even while using highly miniaturized microbumps. Active devices can also be fabricated in the carrier, thus providing a highly integrated, multifunctional system.

3.5 Future Trends

SIP is about the stacking of two or more similar or dissimilar chips to achieve module miniaturization, higher performance, and lower cost than other module options. In essence, the SIP allows Moore's law to continue, not in two dimensions as in the past, but in three dimensions. As seen in this chapter, the stacking began at board level in late 1960 and subsequently moved to module level. SIP began with chip stacking by wire bonding, then by flip chip or side termination in the 1990s. Simultaneously, thinning of chips is introduced to miniaturize the stacking even more. And then came the breakthrough technology, through-silicon via (TSV), driving the miniaturization to the next step.

With this evolution of SIP technology during the last four decades, the future trend for SIP technology in the next decade or two will be clearer if we look at the future electronics market sectors such as mobile products, high-end computers, automobiles, flat-panel high-definition TVs (HDTVs), and sensors for security, health care, and environment. The consumer mobile products will advance to more multifunctionality, thus realizing the digital convergent dream with flexible displays with LED as a light source, thin film, or nano batteries. At the same time, wireless signal speed would go even higher to the 10-GHz range in order to meet ever-increasing transmission speeds for data, voice, and video. The high-end computing would go beyond terabit speed with optical chip-to-chip interconnections. The automotive sector is expected to move to robust and reliable on-board systems with seamless wireless connection to the Internet, satellite, and cellular phones, while driving at high speed. The future flat-panel

HDTVs would be a computing hub for a family that manages and controls all wireless and wired digital data transmissions inside and outside of the home. The future biosensors based on nano materials would find roles for personal security, health care, and environment sensing.

All of these can be referred to as the next generation of miniaturization technologies. In order to meet the above ever-increasing demands for miniaturization, a component density of tens of thousands per cubic centimeter or a functional density that is about 100 to 1000 times more is required. The only way to achieve such parameters is by the SOP concept at the system level as a follow-up to the SIP concept of stacked ICs at the module level. One can project the SOP to consist of

- Ultrathin ICs of 3 to 10 μm in thickness
- Three-dimensional chip stacking of more than 100 memory and logic chips
- System wiring lines and spaces in the submicron range
- Through-silicon-via diameters and their pitch in the submicron range

One such SOP module was proposed by IBM [1], as illustrated in Figure 3.1. The technology challenges to achieve the above goals include mixed-signal electrical design, multifunction materials and processes, ultra high-density wiring, novel thermal management of miniaturized systems, thermomechanical reliability, bumpless interconnections, mixed-signal testing and characterization, and low-cost manufacturing technologies [94]. This book with its 13 chapters reviews the state of the art in all these areas and predicts the next generation of technology evolutions in each.

Acknowledgments

The authors gratefully acknowledge the contribution of Robert M Nickerson, Nasser Grayeli, and Johanna M Swann of Intel Corporation for their constructive and valuable comments.

References

1. J. Knickerbocker, C. Patel, P. Andry, C. Tsang, L. Buchwalter, E. Sprogis, H. Gan, R. Horton, R. Polastre, S. Wright, C. Baks, F. Doany, J. Rosner, and S. Cordes, "Three dimensional silicon integration using fine pitch interconnection, silicon processing and silicon carrier packaging technology," *IEEE 2005 Custom Integrated Circuit Conference*, pp. 659–662.
2. EIA/JEDEC Standard, JESD51-2: Integrated Circuits Thermal Test Method Environment Conditions—Natural Convection (Still Air), December 1995.
3. EIA/JEDEC Standard, JESD51-6: Integrated Circuit Thermal Test Method Environmental Conditions—Forced Convection (Moving Air), March 1999.
4. EIA/JEDEC Standard, JESD51-8: Integrated Circuit Thermal Test Method Environmental Conditions—Junction-to-Board, October 1999.
5. R. R. Tummala, E. J. Rymaszewski, and A. G. Klopfenstein (eds.), *Microelectronics Packaging Handbook*, 3 vols.: *Technology Drivers, Semiconductor Packaging*, and *Subsystem Packaging*, New York: Chapman and Hall, 1997.
6. J. P. Focarie, "Modular Circuit Assembly," US Patent 3,459,998, 1969.

7. R. A. Jarvela, G. E. Lee, and J. W. Schmieg, "Stacked high-density multichip module," *IBM Technical Disclosure Bulletin*, vol. 14, no. 10, 1972, pp. 2896–2897.
8. D. J. McAtee, "Dual-in-line package socket piggyback structure," *IBM Technical Disclosure Bulletin*, vol. 16, no. 4, 1973, p. 1315.
9. P. A. Lutz, P. R. Motz, and E. H. Sayers, "Vertical integrated circuit package integration," US Patent 4,398,235, 1983.
10. IBM Corporation, "Stackable J leaded chip carrier," *IBM Technical Disclosure Bulletin*, vol. 28, no. 12, 1986, pp. 5174–5175.
11. S. Savastiouk, O. Siniaguine, and E. Korczynski, "3-D stacked wafer-level packaging," *Advanced Packaging*, March 2000, pp. 28–34.
12. International Technology Roadmap for Semiconductors (ITRS), 2006 Update.
13. N. R Draney, J. Liu, and T. Jiang, "Experimental investigation of bare silicon wafer warp," *IEEE Workshop on Microelectronics and Electron Devices*, April 2005.
14. Larry Wu, Jacky Chan, and C. S. Hsiao, "Cost-performance wafer thinning technology," *Proc. 53rd Electronic Components and Technology Conference*, May 27–30, 2003, pp. 1463–1467.
15. S. Sandireddy and T. Jiang, "Advanced wafer thinning technologies to enable multichip packages," *Microelectronics and Electron Devices*, 2005. WMED 2005 IEEE Workshop on April 15, 2005, pp. 24–27.
16. C. Landesberger, S. Scherbaum, and K. Bock, "Carrier techniques for thin wafer processing," *International Conference on Compound Semiconductor Manufacturing Technology "GaAs Mantech,"* May 14–17, 2007, pp. 33–36.
17. M. Yan, M. Bartlett, and B. Harnish, "UV Induced Attachment of Ultrathin Polymer Films on Silicon Wafers," *Proc. 8th International Symposium on Advanced Packaging Materials*, March 3–6, 2002, pp. 311–316.
18. K. Bock, C. Landesberger, M. Bleier, D. Bollmann, and D. Hemmetzberger, "Characterization of electrostatic carrier substrates to be used as a support for thin semiconductor wafers," *International Conference on Compound Semiconductor Manufacturing Technology "GaAs Mantech,"* April 2005.
19. Disco Corporation, "Dicing Before Grinding (DBG) Process" available on *http://www.disco.co.jp/eg/solution/library/dbg.htm* (Access date: Dec. 4, 2007).
20. J. Sillanp, J. Kangastupa, A. Salokatve, and H. Asonen, "Ultra short pulse laser meeting the requirements for high speed and high quality dicing of Low-k wafers," *Advanced Semiconductor Manufacturing Conference and Workshop*, 2005 IEEE/SEMI, April 11–12, 2005, pp. 194–196.
21. D. Perrottet, S. Green, and B. Richerzhagen, "Clean dicing of compound semiconductors using the water-jet guided laser technology," *17th Annual SEMI/IEEE Advanced Semiconductor Manufacturing Conference, 2006*, ASMC, May 22–24, 2006, pp. 233–236.
22. Amkor Technology, "Stacked CSP (LFBGA/TFBGA / SCSP) Data Sheet" available on *http://www.amkor.com/products/all_datasheets/SCSp.pdf* (Access date: Dec. 4, 2007)
23. E. J. Vardaman, "Trends in 3-D packaging" *available on http://www.napakgd.com/previous/kgd2004/pdf/vardaman.pdf* (Access date: Dec. 4, 2007).
24. J. Demmin, "Packaging beat: Industry leaders vie for memory-stacking bragging rights," *Solid State Technology, available on http://sst.pennnet.com/articles/article_display.cfm?Section=ARCHI&C=TETAK&ARTICLE_ID=295133* (Access date: Dec. 4, 2007).
25. K. M. Brown, "System in package the rebirth of SIP," *Proc. IEEE Custom Integrated Circuits Conference, 2004*, Oct. 3–6, 2004, pp. 681–686.

26. B. Chylak and I. W. Qin, "Packaging for multi-stack die applications," *Semiconductor International*, June 2004, available on http://www.semiconductor.net/article/CA420735.html (Access date: Dec. 4, 2007).
27. B. Miles, V. Perelman, Y. W. Heo, A. Yoshida, and R. Groover, "3-D packaging for wireless applications," *Semiconductor International*, February 2004, pp. 11–14.
28. M. Karnezos, "Stacked-die packaging: Technology toolbox, step 8," *Advanced Packaging*, vol. 13, no. 8, 2004, pp. 41–44.
29. B. Chylak, S. Tang, L. Smith, and F. Keller, "Overcoming the key barriers in 35 µm pitch wire bond packaging: Probe, mold, and substrate solutions and trade-offs," *27th Annual IEEE/SEMI International Electronics Manufacturing Technology Symposium, 2002*. July 17–18, 2002, pp. 177–182.
30. W. Weber, "Three-dimensional integration of silicon chips for automotive applications," *Mater. Res. Soc. Symp. Proc.*, vol. 970, 2007, p. 0970-Y03-01.
31. D. Zoba, "Stacked flip chip CSP development," *8th Annual KGD Workshop*, September 10, 2001, available on http://www.napakgd.com/previous/kgd2001/pdf/5-3_Zoba.pdf (Access date: Dec. 4, 2007).
32. M. Karnezos, F. Carson, and R. Pendse, "3D packaging promises performance, reliability gains with small footprints and lower profiles," *Chip Scale Review*, January/February 2005, available on http://www.chipscalereview.com/archives/0105/article.php?type=feature&article=f6 (Access date: Dec. 4, 2007).
33. T. Iwasaki, M. Watanabe, S. Baba, Y. Hatanaka, S. Idaka, Y. Yokoyama, and M. Kimura, "Development of 30 micron pitch bump interconnections for COC-FCBGA," *Proc. 56th Electronic Components and Technology Conference*, May 30–June 2, 2006, pp. 1216–1222.
34. Renesas Technology, "Focus on: Packaging new SiP structure: chip-on-chip technology achieves world-leading fine-pitch connections," Renesas edge, vol. 13, 2006, p. 20.
35. K. D. Gann, "Neo-Stacking Technology," available on http://www.irvine-sensors.com/pdf/Neo-Stacking%20Technology%20HDI-3.pdf (Access date: Dec. 04, 2007)
36. K. D. Gann, "High density packaging of flash memory," *Proc. Seventh Biennial IEEE Nonvolatile Memory Technology Conference*, June 22–24, 1998, pp. 96–98.
37. A. Vindasius, M. Robinson, L. Jacobsen, and D. Almen, "Stacked die BGA or LGA component assembly," US Patent 7,215,018 B2, 2007.
38. G. A. Rinne and P. A. Deane, "Microelectronic packaging using arched solder columns," US Patent 5,963,793, 1999.
39. B.-W. Lee, J.-Y. Tsai, H. Jin, C. K. Yoon, and R. R. Tummala, "New 3D chip stacking SIP technology by wire-on-bump (WOB) and bump-on-flex (BOF)," *Proc. 56th Electronic Components and Technology Conference*, May 30–June 2, 2006, pp. 819–824.
40. P. Garrou, "Future ICs Go Vertical," *Semiconductor International*, February 2005, available on http://www.semiconductor.net/article/CA499680.html (Access date: Dec. 4, 2007).
41. K. Hatada, "Stack type semiconductor package," US Patent 4,996,583, 1991.
42. M. Waki, J. Kasai, T. Aoki, T. Honda, and H. Sato, "Semiconductor device having a plurality of chips," US Patent 5,530,292, 1996.
43. H. Shokrgozar, L. Reeves, and B. Heggli, "Stacked silicon die carrier assembly," US Patent 5,434,745, 1995.
44. C. Val, "3D interconnection process for electronic component package and resulting 3D components," US Patent 5,526,230, 1996.
45. L. J. Smith, "Package-on-package: The story behind this industry hit," *Semiconductor International, June 2007, available on http://www.semiconductor.net/article/CA6445430.html* (Access date: Dec. 04, 2007).

46. C. W. C. Lin, S. C. L. Chiang, and T. K. A. Yang, "3D stacked high density packages with bumpless interconnect technology," *IEEE Nuclear Science Symposium Conference Record*, vol. 1, Oct. 19–25, 2003, pp. 73–77.
47. K.-F. Becker, T. Braun, A. Neumann, A. Ostmann, M. Koch, V. Bader, R. Aschenbrenner, H. Reichl, and E. Jung, "Duromer MID technology for system-in-package generation," *IEEE Transactions on Electronics Packaging Manufacturing*, vol. 28, no. 4, 2005, pp. 291–296.
48. C. Val, "Three dimensional interconnection method and electronic device obtained by same," US Patent 6,716,672 B2, 2004.
49. M. Karnezos, "Package-in-package: A 3-D stacked package module," *11th Annual International KGD Packaging and Test Workshop*, September 12–15, 2004, available on http://www.napakgd.com/previous/kgd2004/pdf/karnezos.pdf (Access date: Dec. 04, 2007).
50. Intel Corporation, "Packaging Overview" *available on http://download.intel.com/design/flcomp/packdata/wccp/download/chpt1.pdf* (Access date: Dec. 04, 2007).
51. Tessera Technologies, "Folded die stack" *available on http://www.tessera.com/technologies/products/z_mcp/folded_stacked.htm* (Access date: Dec. 04, 2007).
52. T. Yamazaki, Y. Sogawa, R. Yoshino, K. Kata, I. Hazeyama, and S. Kitajo, "Real chip size three-dimensional stacked package," *IEEE Transactions on Advanced Packaging*, vol. 28, no. 3, 2005, pp. 397–403.
53. Y. Yano, T. Sugiyama, S. Ishihara, Y. Fukui, H. Juso, K. Miyata, Y. Sota, and K. Fujita, "Three-dimensional very thin stacked packaging technology for SiP," *Proc. 52nd Electronic Components and Technology Conference*, May 28–31, 2002, pp. 1329–1334.
54. T. Matthias et al., "3D process integration—wafer-to-wafer and chip-to-wafer bonding," *Mater. Res. Soc. Symp. Proc.*, vol. 970, 2007, p. 0970-Y04-08.
55. A. D. Scarbrough, "3D-coaxial memory construction and method of making," US Patent 3704455, 1972.
56. T. R. Anthony, "The random walk of a drilling laser beam", *Journal of Applied Physics*, vol. 51, 1980, p. 1170.
57. T. R. Anthony and P. A. Lindner, "The reverse laser drilling of transparent materials," *Journal of Applied Physics*, vol. 51, 1980, p. 5970.
58. T. R. Anthony, "Forming electrical interconnections through semiconductor wafers," *Journal of Applied Physics*, vol. 52, no. 8, 1981, pp. 5340–5349.
59. J. F. McDonald et al., "Multilevel interconnections for wafer scale integration," *Journal of Vacuum Science & Technology A (Vacuum, Surfaces, and Films)*, vol. 4, no. 6, 1986, pp. 3127–3138.
60. F. Laermer and P. Schilp, "Method of anisotropically etching silicon," U.S. Patent 5501893, 1994.
61. C. Christensen, P. Kersten, S. Henke, and S. Bouwstra, "Wafer through-hole interconnections with high vertical wiring densities," *IEEE Trans. Components, Packaging and Manufacturing Technol. A*, vol. 19, 1996, pp. 516–522.
62. P. Kersten, S. Bouwstra, and J. W. Petersen, "Photolithography on micromachined 3D surfaces using electrodeposited photoresists," *Sensors and Actuators A*, vol. 51, 1995, p. 51–54.
63. J. Gobet et al., "IC compatible fabrication of through-wafer conductive vias," *Proceedings of the SPIE—The International Society for Optical Engineering*, vol. 3223, 1997, pp. 17–25.

64. Tru-Si Technologies, "Through-silicon vias" available on *http://www.trusi.com/frames.asp?5* (Access date: Dec. 4, 2007)
65. K. Takahashi et al., "Development of advanced 3D chip stacking technology with ultra-fine interconnection," *Proc. 51st Electronic Components and Technology Conference*, 2001, pp. 541–546.
66. V. Kripesh et al., "Three dimensional stacked modules using silicon carrier," *Proc. 2003 Electronics Packaging Technology Conference*, 2003, pp. 24–29.
67. N. Tanaka et al., "Ultra-thin 3D-stacked sip formed using room-temperature bonding between stacked chips," *Proc. 2005 Electronic Components and Technology Conference*, 2005, pp. 788–794.
68. PhysOrg, "Samsung Develops 3D Memory Package that Greatly Improves Performance Using Less Space" available on *http://www.physorg.com/news64161294.html* (Access date: Dec. 4, 2007)
69. S. Vangal et al., "An 80-tile 1.28 TFLOPS network-on-chip in 65nm CMOS," *Solid-State Circuits Conference, 2007. ISSCC 2007. Digest of Technical Papers. IEEE International*, 2007, p. 98.
70. IBM Corporation, "SiGe BiCMOS 5PAe: advanced through-silicon via technology for RF power applications", available on *http://www-01.ibm.com/chips/techlib/techlib.nsf/techdocs/6B994C8F42D91314002572E900707987/$file/5PAe_Aug2207_final.pdf* (Access date: Dec. 4, 2007)
71. B. Swinnen and E. Beyne, "Introduction to IMEC's research programs on 3D-technology," available on *www.emc3d.org/documents/library/technical/IMEC%20Technical%20Review_3D_introduction.pdf* (Access date: Dec. 4, 2007)
72. F. Niklaus, J.-Q. Lu, J. J. McMahon, J. Yu, S. H. Lee, T. S. Cale, R. J. Gutmann, "Wafer-level 3D integration technology platforms for ICs and MEMs," *Proceedings of the Twenty Second International VLSI Multilevel Interconnect Conference (VMIC), T. Wade (ed.), IMIC* 2005, pp. 486–493.
73. A. Klumpp, P. Ramm, R. Wieland, and R. Merkel, "Integration Technologies for 3D Systems" FEE 2006, May 17–20, 2006, Perugia, Italy. Available on *www.mppmu.mpg.de/~sct/welcomeaux/activities/pixel/3DSystemIntegration_FEE2006.pdf* (Access date: Dec. 4, 2007)
74. J. U. Knickerbocker et al., "Development of next-generation system-on-package (SOP) technology based on silicon carriers with fine-pitch chip interconnection," *IBM J. Research and Development*, vol. 49, no. 4/5, 2005, pp. 725–753.
75. J. A. Burns et al., "A wafer-scale 3-D circuit integration technology," *IEEE Transactions on Electron Devices*, vol. 53, no. 10, 2006, pp. 2507–2516.
76. H. Noh, Kyoung-sik Moon, A. Cannon, P. J. Hesketh, and C. P. Wong, *Proc. IEEE Electronic Components and Technology Conference*, vol. 1, 2004, pp. 924–930.
77. K. W. Guarini, A. W. Topol, M. Ieong, R. Yu, L. Shi, M. R. Newport, D. J. Frank, D. V. Singh, G. M. Cohen, S. V. Nitta, D. C. Boyd, P. A. O'Neil, S. L. Tempest, H. B. Pogge, S. Purushothaman, and W. E. Haensch, *Proc. IEDM*, 2002, pp. 943–945.
78. M. Umemoto, K. Tanida, Y. Nemoto, and M. Hoshino, "High performance vertical interconnection for high-density 3D chip stacking package," *Proc. Electronic Components and Technology Conference ECTC*, 2004, pp. 616–623.
79. K. N. Chen, A. Fan, and R. Reif, "Microstructure examination of copper wafer bonding," *Journal of Electronics Materials*, vol. 30, 2001, pp. 331–335.
80. K.-N. Chen, S. H. Lee, P. S Andry, C. K. Tsang, A. W. Topol, Y.-M. Lin, J.-Q. Lu, A. M. Young, M. Ieong, and W. Haensch, "Structure, design and process control for

Cu bonded interconnects in 3D integrated circuits," *IEEE IEDM*, 2007, pp. 13.5.1–13.5.3.

81. F. Niklaus, S. Haasl, and G. Stemme, "Arrays of monocrystalline silicon micromirrors fabricated using CMOS compatible transfer bonding," *IEEE Journal of Microelectromechanical Systems*, vol. 12, no. 4, 2003, pp. 465–469.
82. F. Niklaus, J. Pejnefors, M. Dainese, M. Häggblad, P.-E. Hellström, U. Wållgren, and G. Stemme, "Characterization of transfer-bonded silicon bolometer arrays," *Proc. SPIE*, vol. 5406, 2004, pp. 521–530.
83. J.-Q. Lu, A. Jindal, P.D. Persans, T.S. Cale, and R.J. Gutmann, "Wafer-level assembly of heterogeneous technologies," *The International Conference on Compound Semiconductor Manufacturing Technology, 2003, available on http://www.gaasmantech. org/Digests/2003/index.htm* (Access date: Dec. 4, 2007)
84. C. Christensen, P. Kersten, S. Henke, and S. Bouwstra, "Wafer through-hole interconnections with high vertical wiring densities," *IEEE Trans. Components, Packaging and Manufacturing Technol. A*, vol. 19, 1996, p. 516.
85. J. Gobet et al., "IC compatible fabrication of through-wafer conductive vias," *Proc. SPIE—The International Society for Optical Engineering*, vol. 3223, 1997, pp. 17–25.
86. M. Despont, U. Drechsler, R. Yu, H. B. Pogge, and P. Vettiger, *Journal of Microelectromechanical Systems*, vol. 13, no. 6, 2004, pp. 895–901.
87. S. Gupta, M. Hilbert, S. Hong, and R. Patti, "Techniques for producing 3D ICs with high-density interconnect," *Proc. 21st International VLSI Multilevel Interconnection Conference*, Waikoloa Beach, HI, 2004. pp. 93–97.
88. K. Takahashi et al., "Process integration of 3D chip stack with vertical interconnection," *Proc. 54th Electronic Components and Technology Conference*, 2004, vol. 1, pt. 1, pp. 601–609.
89. C. A. Bower et al., "High density vertical interconnects for 3-D integration of silicon integrated circuits," *Proc. 56th Electronic Components and Technology Conference*, 2006, pp. 399–403.
90. P. Benkart et al., "3D chip stack technology using through-chip interconnects," *IEEE Design & Test of Computers*, vol. 22, no. 6, 2005, pp. 512–518.
91. D. J. Bodendorf, K. T. Olson, J. P. Trinko, and J. R. Winnard, "Active silicon chip carrier," *IBM Tech. Disclosure Bull.*, vol. 7, 1972, p. 656.
92. J. U. Knickerbocker et al., "Three dimensional silicon integration using fine pitch interconnection, silicon processing and silicon carrier packaging technology," *Proc. IEEE Custom Integrated Circuits Conference*, 2005, pp. 659–662.
93. V. Kripesh et al., "Three-dimensional system-in-package using stacked silicon platform technology," *IEEE Transactions on Advanced Packaging*, vol. 28, no. 3, 2005, pp. 377–386.
94. P. Garrou, private communication, 2007.

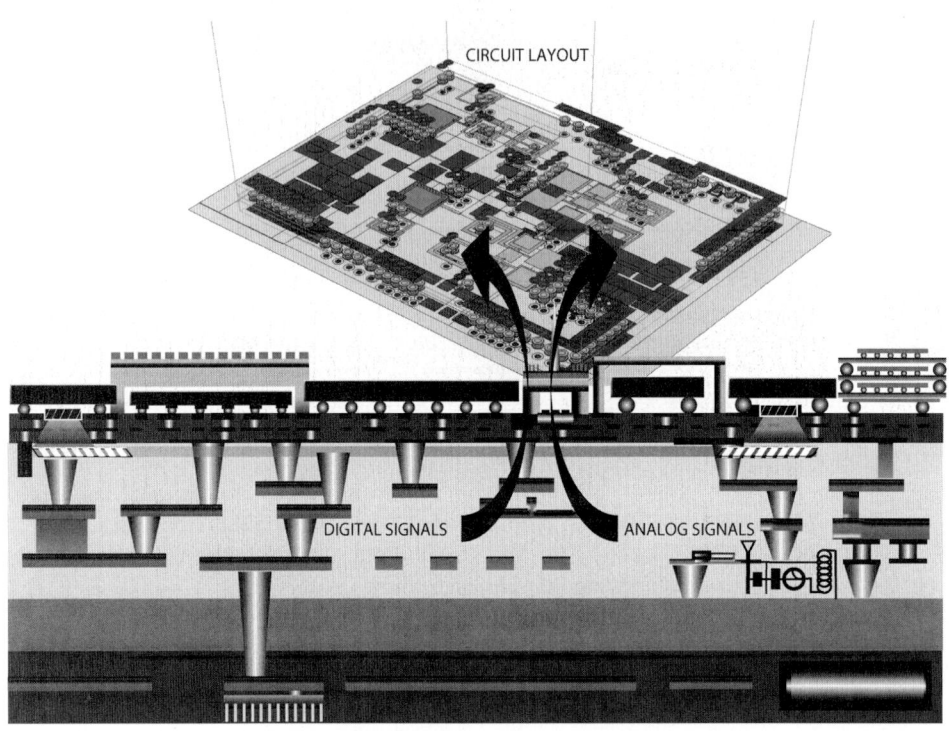

CHAPTER 4
Mixed-Signal (SOP) Design

Madhavan Swaminathan and A. Ege Engin
Georgia Institute of Technology

Vinu Govind
Jacket Micro Devices

Amit Bavisi
Freescale Semiconductor

4.1	Introduction 152		4.6	Coupling 214
4.2	Design of Embedded Passives in RF Front End 160		4.7	Decoupling 227
			4.8	Electromagnetic Bandgap (EBG) Structures 239
4.3	Chip-Package Codesign 180		4.9	Summary 250
4.4	Design of WLAN Front-End Module 191			References 251
4.5	Design Tools 194			

Today's electronic products include a lot of functions in a small area. One enabling technology for this has been the miniaturization of transistors such as in microprocessors with an increase in their performance. At the same time, components that require different signaling domains need to be integrated in the system to achieve the multifunctionality of products. The design of such a system poses a lot of challenges, since the close proximity of components with different specifications, such as a microprocessor and an RF front end, can result in interference between the components. This necessitates a mixed-signal design methodology especially for systems based on the system-on-package (SOP) technology that miniaturizes the system immensely.

The basis of the SOP concept is embedded thin-film components such as capacitors and inductors that are essential for the design of RF front-end modules. The transition from discrete surface-mount components to such embedded components can be quite useful for the system designer. The components can be realized at any desired value or specification (within the boundaries of the technology), and parasitic effects due to interconnections are significantly reduced, resulting in an improvement in performance. For RF integration, the filters provide a lower insertion loss with better rejection while active devices such as oscillators have better phase noise using the SOP technology rather than the SOC. Similarly, for digital integration, embedded decoupling capacitors, which can be connected to switching digital circuits with a low-inductance interconnection, can provide improved signal and power integrity. Above all, the maximum benefit achievable using SOP is the ability to manage electromagnetic interference, a problem of paramount importance in mixed-signal integration. However, the design of embedded components in mixed-signal environments can involve many challenges such as longer design time, requiring extensive electromagnetic analysis. Hence, new technologies need to be supported by new design tools that become necessary to help the designer with the design of such elements and shorten the design time.

A mixed-signal system has to be designed to reduce the interference between components having different signal levels. For example, switching currents from digital ICs need to be isolated from sensitive RF signals. One such technology presented in this chapter is the electromagnetic bandgap (EBG) structure, which can create isolated islands on a power delivery network in a specified frequency range.

This chapter presents advanced design concepts including embedded components that are enabled by the SOP technology. A chip-package codesign methodology is introduced for maximizing performance and area. This chapter also presents algorithms and ideas for new design tools that are necessary for the successful and efficient design of mixed-signal systems. These tools automate the design of embedded passives, help with the analysis of EBG structures, and provide a fast electromagnetic analysis of package interconnections and power delivery networks.

4.1 Introduction

In recent years, the marriage of high-speed computing and wireless communication has emerged as a formidable driving force in the global electronics industry. This has resulted in products with both computing and communication capabilities and has engineered a tremendous surge of interest in the mixed-signal market [1]. Figure 4.1 shows the global mixed-signal market and its composition [2], with communication products being the major driver, constituting $8.9 billion of the total market.

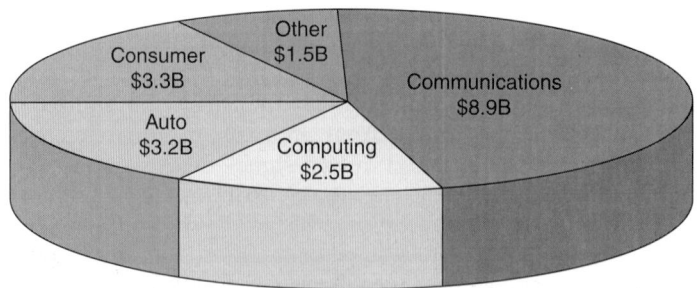

FIGURE 4.1   Mixed-signal market.

With the convergence in communication, computing, and biomedical applications, future devices are expected to become more and more mixed-signal in nature. For example, Intel Corp., the largest microprocessor manufacturer in the world, announced a "Radio Free Intel" initiative, adding communication capability to high-performance computing by the merger of CMOS wireless radios and microprocessor chipsets [3]. Intel envisages a future where a user with a mobile computer can seamlessly move between different wireless networks (long distance as well as short distance), achieving "ubiquitous wireless connectivity" for a computing device. Similarly, Nokia Corp., the world's largest manufacturer of cell phones, has announced the "N-Gage" gaming device platform where customers all over the world can compete with each other wirelessly [4]. With the processing power in these gaming consoles expected to be equal to or more than a consumer laptop, this would represent an unprecedented increase in the computing power of a commercial wireless communication device.

4.1.1 Mixed-Signal Devices and Systems

The term "mixed signal" represents the integration of multiple signal domains. For example, the cellular phone or wireless handset represents a mixed-signal system that supports RF and digital signals. In a handset, the RF section receives the analog signal, which is then down-converted and digitized for processing the data. Another example of a mixed-signal device is an analog-to-digital converter that supports both analog and digital signals. In a nutshell, the processing of data is best done digitally. However, since we live in an analog world, the signals that are transmitted and received are analog in nature. Hence, integration of multiple signal domains is required to enable the convergence of communication and computing, which is the driving force behind the emergence of mixed-signal devices.

In the context of handsets, as the convergence trend continues, next-generation mixed-signal devices and systems will be expected to provide high-performance computing and wireless connectivity to a mobile user. With a proliferation of communication standards geared toward different applications, these computing-communicating hybrids will need to support multiple communication protocols at multiple frequency bands [5] to achieve this goal of ubiquitous connectivity. For example, a mobile user engaging in a video-conference via cell phone can expect the call to be routed over a Wideband Code Division Multiple Access (WCDMA) network as the person walks across the parking lot, with a seamless handoff to a wireless LAN (WLAN) based network as the person enters the office. At the same time, GPS signals from satellites continuously communicate location based information to the phone, while the Bluetooth protocol is used to synchronise the contents of the phone-based calendar with the one on the office computer.

In parallel, the computing industry has been following Moore's law where the number of transistors on the IC has continued to double every 18 months for the last 15 years. However, the scaling beyond the 90-nm node is causing significant challenges associated with leakage and latency causing engineers to develop new architectures. With the trend toward multicore processors and the need for significant memory content with reduced latency in systems, the package is becoming more critical for the functioning of the system. The package now has to support high-speed I/Os containing serial and parallel links with speeds in excess of 3 Gbytes per second (Gbytes/s). With the convergence of computing and communication capabilities, the need for integrating the high-speed microprocessor, memory, and wireless ICs in a single package, with the antenna and RF front-end passives integrated in the package, is becoming very critical.

Various embodiments of such an integrated package are shown in Figure 4.2. Figure 4.2a represents an integrated wireless module containing the RF integrated circuit (IC) and digital baseband. The passive components of the RF front end are integrated in the package. Figure 4.2b is a microprocessor package with the stacked memory assembled on the same package. High-speed interconnections in the package along with embedded decoupling capacitors reduce the latency between the processor and the memory. Figure 4.2c is a package-on-package with high-speed logic and memory in one package, the wireless module in another package, and the two are stacked on each other.

In this chapter, the design of embedded passives for low-power mobile communication systems, such as handsets and laptop computers, is covered. This is followed by the design of packages for microprocessor-type applications with an emphasis on embedded decoupling. The design of any high-frequency package requires design tools, which are covered in detail with an emphasis on signal integrity, power distribution, and RF design. Coupling issues in microsystems with integrated RF and mixed-signal circuits are covered next followed by methods for reducing interference.

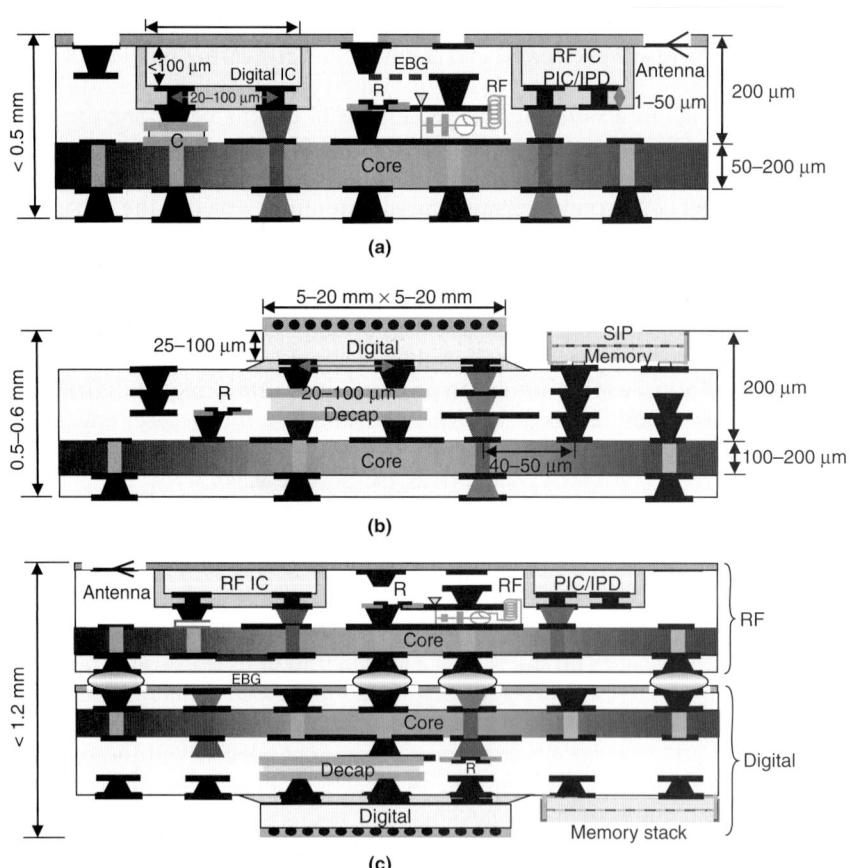

Figure 4.2 Integrated module or subsystem: (*a*) wireless, (*b*) microprocessor, and (*c*) package-on-package.

4.1.2 Importance of Integration in Mobile Applications

In this chapter, mixed-signal design issues in the context of handsets and other mobile applications will be discussed. The market drivers for handsets are (1) size, (2) performance, (3) cost, and (4) reliability, not necessarily in that order. With the trend toward a "world phone" that supports a multiband architecture in handsets, there is a clear need for supporting multiple frequency bands. These include (1) the cellular bands such as 850, 900, 1800, and 1900 MHz to service the Global System for Mobile Communication (GSM)-USA, Extended GSM (EGSM), Digital Cellular System (DCS), and Personal Communication Services (PCS) protocols, (2) Wideband CDMA (WCDMA) band operating at 2.1 GHz, (3) the Wireless Local Area Network (WLAN) bands such as 802.11 a/b/g operating at 2.4 and 5.2 GHz, (4) the Global Positioning System (GPS) band at 1575 MHz, (5) ultra-wideband (UWB) supporting frequencies from 3.1 to 10.6 GHz, as shown in Figure 4.3, and World Interoperability for Microwave Access (WiMAX).

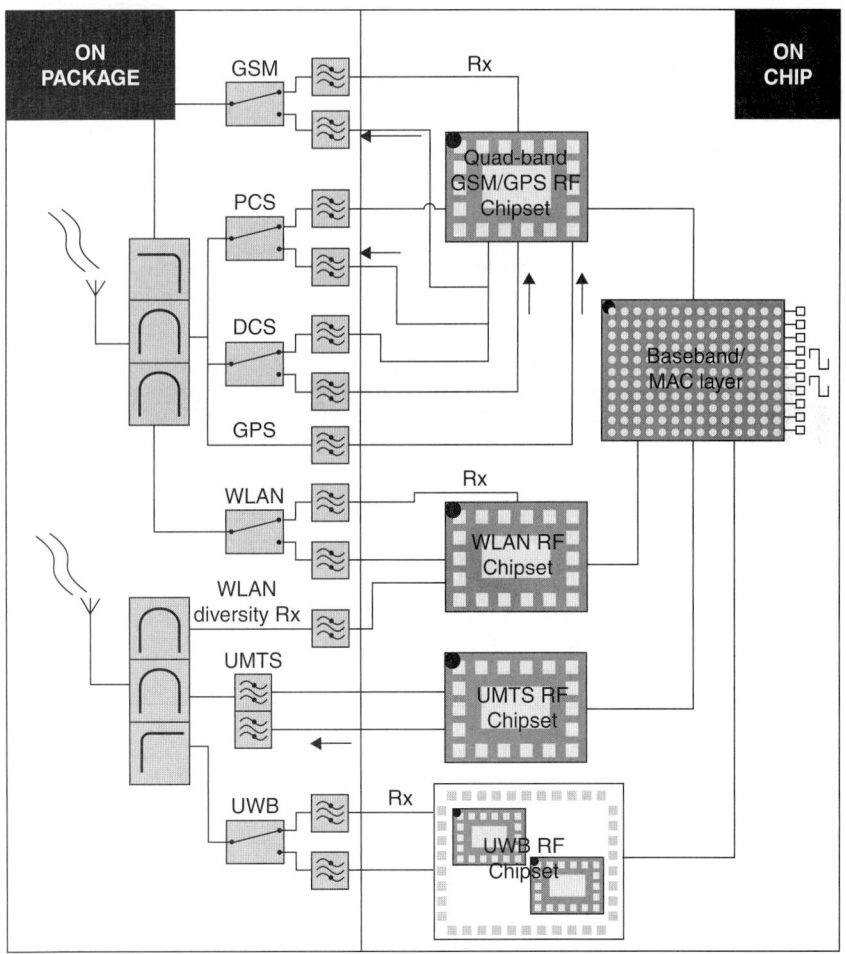

Figure 4.3 Multiband system.

As the technologies for handsets evolve, a major requirement is the height, which cannot exceed 1.2 mm today and is rapidly shrinking. A similar trend can be seen in laptop computers where WLAN and WiMAX with multiple transmit and receive chains are being integrated along with the PCI mini express chipsets.

The implementation of the handset requires two basic devices, namely, (1) actives such as transistors and (2) passive networks, such as inductors, capacitors, resistors, and transmission lines. Though the transistor density has been increasing from one generation to the next, it enables the miniaturization of the IC and not necessarily the system. This is especially true in mixed-signal systems such as handsets, where the size of the system is determined by the passive components in the RF and analog front end. For size and performance reasons, which will be discussed later, it is difficult to integrate every passive component on silicon (or gallium arsenide, silicon germanium, and other IC technologies) and hence, micro-miniaturization of mixed-signal systems requires new packaging technologies. Hence, a system can be partitioned in such a way that the package is used as a platform for passive integration while the transistors are integrated on silicon as in Figure 4.3. Often referred to as system-on-package (SOP), this implementation method offers the best of both IC and package integration. Therefore, it enables system miniaturization. In such implementations, certain circuits contain both package and IC elements, as will be discussed later in this chapter.

4.1.3 Mixed-Signal Architecture

The architecture for a handset is shown in Figure 4.4, which consists of a transmit and receive chain followed by the baseband processor. A single antenna is used to both transmit and receive signals. The channels can be isolated using an RF switch after the antenna for nonconcurrent architectures and through appropriate filtering for concurrent

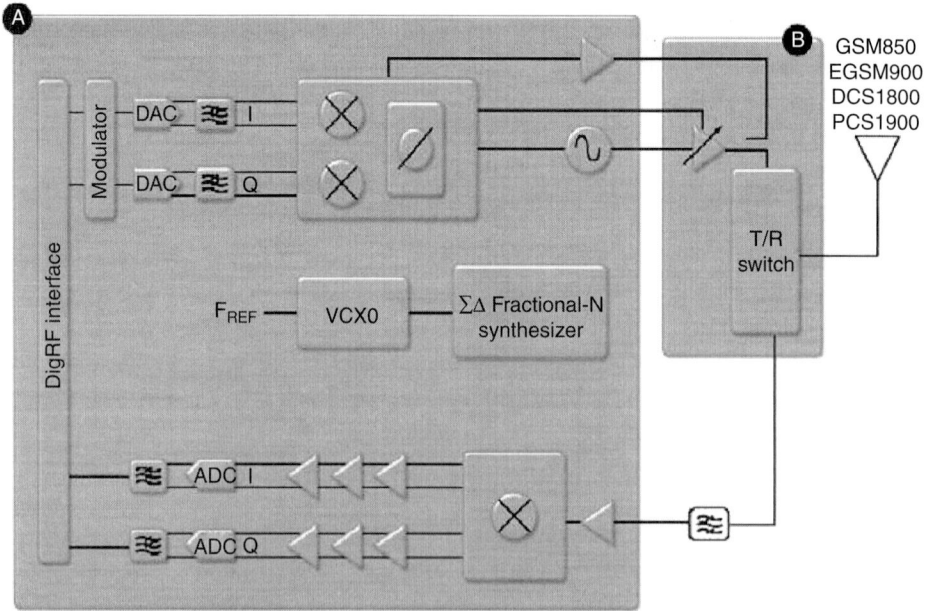

Figure 4.4 Mixed-signal architecture.

architectures (e.g., WCDMA), which is more difficult to implement. The RF front end is critical for the functioning of the radio module and therefore requires diplexers and a high quality factor (Q) filters to manage interference. With the trend toward integration, the transceiver section, which consists of low-noise amplifiers (LNAs), mixers, voltage-controlled oscillators (VCOs), and modulators, is being integrated into a single die using silicon-based CMOS processes. Because of power requirements, the power amplifier (PA) is a separate die that is often implemented using gallium arsenide (GaAs). For managing interference between the digital and RF ICs, the baseband processor and memory that are implemented using digital CMOS are implemented as a separate die. There is however a trend toward the integration of the transceiver with the baseband on a single die, with the RF front end consisting of the antenna, switches, filters, and matching networks outside of the die. Because of the high performance (quality factor, insertion loss, sharp roll-off, and noise containment) required in the RF front-end module, the components in this part of the architecture are often implemented using discrete devices that limit the size of the handset. All the passive components required in a handset that are difficult to integrate within the IC can now be embedded into the package, thereby reducing the form factor and resulting in a SOP solution. The passive devices that are required in a handset that can be embedded in the package include the following: single-band and multiband antennas, diplexers consisting of low-pass and high-pass filters, filter banks with bandpass characteristics, baluns that convert single-ended signals into differential form and vice versa, matching networks, couplers, inductors, and capacitors, to name a few. In addition, the package serves as a platform that supports the ICs and, hence, performs the important role of protecting the die and managing the thermal problems.

4.1.4 Mixed-Signal Design Challenges

Major challenges for the design of handsets with an SOP-based implementation include:

1. The design of highly integrated embedded components for the RF front end. An example are diplexers that combine low-pass and high-pass filters on the receive chain that minimize interference from adjacent frequency bands.

2. Chip-package codesign of active components with embedded components. These include PAs, LNAs, and VCOs that have matching circuits, inductors, and resonators embedded in the package.

3. Managing the analog-to-analog and digital-to-analog coupling. An example is coupling through substrate, which causes interference between the digital and RF ICs.

The parasitic inductance and capacitance of chip-package interconnects become significant when they result in signal degradation at high frequencies. This becomes significant for circuits like LNAs, VCOs, and PAs, which reside at the interface between the package and chip domains. The back-and-forth transfer of signals between the package and chip domain necessitates a design partitioning methodology to decide the distribution of circuit components within the package or within the chip. The use of multiple embedded passives in the substrate can also generate undesirable resonance and feedback, jeopardizing the functionality of the system [6].

Integration at the package level leads to changes in priority at the design phase. With the availability of high-Q passives embedded in the package, the *number* of lumped components becomes less important than the *value* of each of these components. In contrast, designs utilizing discrete passives are more concerned with the total number of passives and not the individual values of each; the cost of assembly depends only on the number of discrete elements to be soldered on board, and the packaged size for different values of capacitance or inductance usually remains the same for commercially available discrete devices. This requires novel designs for RF components using these embedded passives, with priorities shifting from reducing the component count to keeping the value of the passives low.

Noise coupling between digital and analog circuits remains a problematic issue even in SOP-based implementations. Although coupling through the silicon substrate of an SOC chip has been minimized, new noise sources, like electromagnetic interference (EMI) from high-speed signal lines and fluctuations in the power plane, arise and have to be dealt with through careful modeling and analysis.

Wireless radios have incorporated multiple standards and hence have become multiband devices. Apart from noise sources, next-generation mixed-signal systems present many architectural challenges as well. Multiband functionality in components can be achieved in numerous ways. For example,

1. Implementation of devices with wide bandwidths capable of operating at multiple frequency bands.

2. Implementation of multiple single-band devices with matching networks at the input and output producing a single-input–single-output (SISO) component. Each single-band device has a narrow operating bandwidth, and multiband operation is achieved for the component by switching between the different single-band circuits. In these implementations, the component outputs only a single frequency at any given time [7].

3. Implementation of concurrent devices that achieve simultaneous multiband functionality and require sampling one or more frequencies at a time depending on the application [8].

Option 1 is difficult to implement due to technological concerns and the presence of large blocker signals close to the frequency bands of interest. Option 2 is a component-scale replica of the multiband architecture, and as such exhibits problems such as large size and high power consumption. In contrast, option 3 (the use of true multiband devices) leads to lower power consumption and a vastly reduced footprint.

The passive devices like antennas, filters, and baluns can be completely integrated in the package. For the circuit components lying at the interface of the chip and package domains (LNA, VCO, and PA), design partitioning and optimization is required, to take into account the parasitics involved in a chip-package signal transition. Examples of PAs with the output matching networks implemented using embedded passives and VCOs with high-Q embedded inductors for improvement in phase noise has been reported in the literature [9–11].

It is possible to achieve complete integration by integrating passives in the package. However, this approach neglects the fact that compared to on-chip inductors with low Q values and discrete passives with fixed Q values, the use of embedded passives leads to the development of the passive Q as a new variable in circuit design. With Q values ranging from 20 to 200 [12], designers now have a choice in the Q value they need for a particular component. Higher Q values result in new tradeoffs, particularly with respect to device size, and a design partitioning and optimization strategy is thus required to ensure efficient use of the packaging substrate. This has to be incorporated into the design methodology of each circuit, for optimal system performance. As an example, the design of completely integrated CMOS LNAs is possible by using embedded passives in the package. However, the common CMOS LNA design methodology has to be updated to take into account the tradeoff of a higher inductor size for a higher Q.

With the high-sensitivity requirements for radio circuits necessitating the handling of microwatts of input signal power, noise coupling from digital to analog domains has become a major impediment to mixed-signal integration. Noise is generated in digital circuits when many static gates change state simultaneously, causing a spike in current flow through parasitic resistances and inductances in the circuit. This results in a spike in the power supply, which can couple into the analog circuits through a common power distribution system. The noise can also appear in analog circuits through capacitive coupling to and from the highly doped silicon substrates used in SOC systems [13–22].

However, with the use of SOP-based schemes, new noise coupling and propagation mechanisms come into play. With the use of power planes in the package for power distribution, ground bounce and simultaneous switching noise (SSN) become important factors in mixed-signal design. High-speed signal lines also end up radiating energy due to common mode currents, resulting in electromagnetic interference (EMI). The main digital-analog noise coupling mechanisms in mixed-signal SOP-based systems can thus be summarized as follows: (1) through a common power supply, (2) through EMI from high-speed signal lines, and (3) through capacitive or inductive coupling. The low-loss power-distribution networks used in SOP-based systems produce sharp resonances that do not exist in a higher-loss SOC-based power system [23]. In addition, coupling within the analog signal domain through closely spaced multiple embedded passives also results in performance degradation of the RF circuitry.

A major challenge in mixed-signal SOP design is the lack of design tools that can accurately model, analyze, and evaluate complex effects such as signal integrity, power delivery, crosstalk, radiation, quality factor, process variations, and yield. Along with spanning the architectural, transistor, and layout level hierarchies, the tools have to provide rapid turnaround time and therefore enable shorter design cycle time.

4.1.5 Fabrication Technologies

The design methodology used is a function of the technology chosen for fabrication. This is because the chosen technology defines the process ground rules such as line width, line thickness, dielectric thickness, via diameters, and stackups, which are required for physical implementation. For mobile applications, cost-effective solutions are required that provide a significant improvement in size, performance, and reliability. Currently, five different technology platforms are available for integration in mobile applications, namely, (1) use of discrete passive components on printed wiring boards (PWBs), (2) system-on-chip (SOC) where all the passive and active devices are integrated into silicon, (3) low-temperature cofired ceramic (LTCC), (4) thin film on silicon (TFOS), and (5) use of PWB-based organic processes. Since, the package and the PWB can be combined using similar material sets using the last option, this technology has gained in importance lately. Table 4.1 provides a qualitative comparison of the five approaches without providing much detail, purely from a design standpoint. In this chapter, design implementations using the PWB-based organic processes with liquid crystalline polymer (LCP) dielectric materials have been discussed.

Technology	Advantages	Disadvantages
Discretes on PWB	• Low-to-mid Q passives • Readily available	• Low density, large size • Component variation
SOC	• Compact • High integration in digital circuits	• Low-Q passive • No single technology platform covers all mixed-signal system requirements
LTCC	• High Q passives • High integration • High density	• Coefficient of Thermal Expansion (CTE) mismatch (The CTE mismatch for LTCC is with the PCB) • Shrinkage • Future scalability • Lack of metal planes
TFOS	• High density	• Low Q passives • Low integration
PCB-based organic processes	• High Q passives • High integration • High density • Availability of large metal planes • Large area manufacturing	• Hermiticity

TABLE 4.1 Comparison of Technologies Based on Design Flexibility

In Table 4.1, the main advantage of using a printed circuit board (PCB) based organic process (such as LCP) is that the substrate can ultimately become the PCB with bare die ICs assembled directly to it. It also lends itself to the use of the PCB infrastructure for fabricating such high-density boards.

4.2 Design of Embedded Passives in RF Front End

The RF front end consists of antennas, diplexers, filters, filter banks, baluns, and matching circuits. These components can use considerable real estate if implemented using discrete surface-mount components. Hence, integration of these components into the IC or package is necessary for reducing the size of the system. Besides the antenna, the performance of the remaining components is dictated by the basic building blocks used to implement them, namely, inductors and capacitors. Because of size restrictions, these components cannot be implemented using transmission line elements in the frequency range from 1 to 10 GHz. The performance of the inductors and capacitors are measured by their unloaded quality factor (Q), which is a measure of the loss in the device. Silicon IC technologies limit the Q of inductors to 5 to 15 due to the semiconducting substrate that generates eddy currents, which decreases inductance and increases losses. In addition, inductors can occupy considerable real estate on silicon, and hence can increase the cost of the process. Therefore, integration of these components into the substrate is necessary. Standard BT (Bismaleimide-Triazine) laminate (such as FR4) based processes cannot be used for integration since conductor losses (such as line profiles, surface roughness, tolerance) and dielectric losses (such as loss tangent) degrade the Q of the inductors and capacitors. Hence, high-frequency dielectric materials with low loss tangents, metals with high conductivity, and processes that provide good surface finish and rectangular line profiles are required. Since, the ICs are mounted on PCBs, a packaging technology that is organic based is preferable since this can ultimately replace the PCB. Called system-on-package (SOP), this enables the assembly of bare die ICs directly on the integrated substrate containing embedded components in thin-film form.

In this chapter, LCP material has been used for integration. This material has a relative permittivity of 2.95 and a loss tangent of 0.002, which are invariant from 1 to 100 GHz; a moisture absorption less than 0.04 percent; and a CTE that is matched to the PCB [24]. The dielectric thickness is in the range of 1 to 8 mils. A parallel process can be used with copper metallization, which results in rectangular line profiles with minimum surface roughness. As an example, a balanced LCP cross section is shown in Figure 4.5. Two balanced LCP layers are circuitized separately, along with micro-vias, followed by lamination of the LCP layers using organic prepreg materials (the core in Figure 4.5). The prepreg has a loss tangent of 0.0035 and dielectric constant of 3.38. Through holes are mechanically drilled and plated to form interconnections. A liquid photoimageable solder mask can be used, and an electroless nickel immersion gold finish can be used to plate on the bond pads and terminals. All processes including lamination (<200°C), electroless and electrolytic copper plating, and dry film photoresists are compatible with standard FR4/PWB manufacturing. The panels can be fabricated on 12 in × 18 in and 9 in × 12 in format using large-area PWB tooling resulting in a low-cost implementation that can be easily scaled to an 18 in × 24 in panel size for further cost reduction.

Since the stackup combines LCP and prepreg layers, the resulting cross section is inhomogeneous. Hence, design methodologies are required that enable the use of LCP layers for embedding inductors and capacitors that require a high Q. The metal on the prepreg layers can be used for routing. A homogeneous layer stack-up has also become possible where LCP with different melt temperatures can be used to bond the various layers together.

4.2.1 Embedded Inductors

An inductor is an essential passive component in the design of RF front-end modules. Inductors are typically implemented as planar spiral coils, although coaxial [25] and

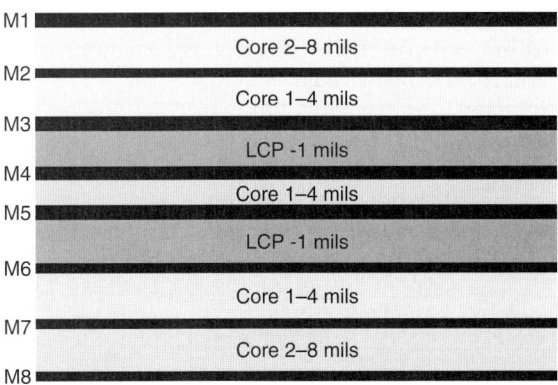

FIGURE 4.5 Cross section of LCP substrate.

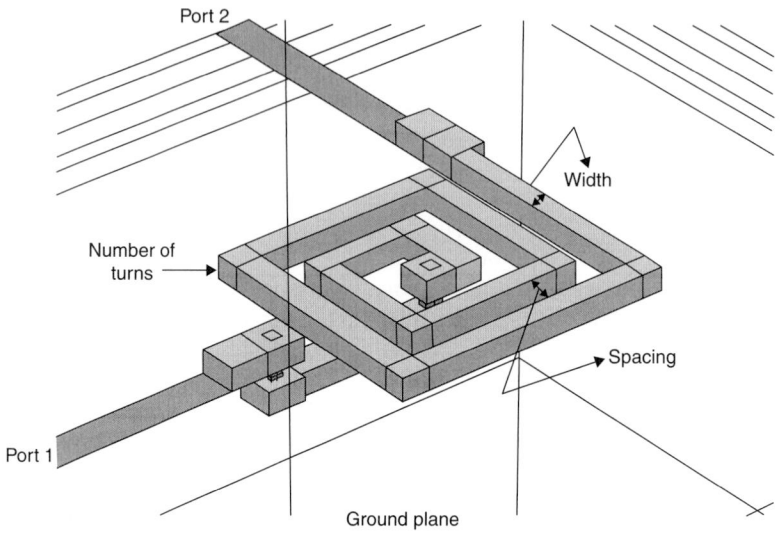

FIGURE 4.6 Single-layer spiral inductor.

wire-wound [26] implementations of inductors are also possible, on one or more metal layers of the substrate. The performance of the inductor is dictated by its inductance at a specified frequency, the unloaded Q at that frequency, and the self-resonant frequency of the inductor. It is preferable to operate the inductor at its maximum Q, which is roughly between 30 to 50 percent of the self-resonant frequency. The design of the inductor requires careful optimization of the parasitics, which occur due to the parasitic capacitance, conductor losses, and dielectric losses.

As an example, consider the inductor shown in Figure 4.6, which is implemented as a spiral on an LCP layer. The broadband equivalent circuit for the two-port inductor is shown in Figure 4.7, which represents a physical model of the inductor geometry. In Figure 4.7, L_s is the inductance, R_s is the series resistance, C_s is the coupling capacitance between the input and output ports, C_{p1} and C_{p2} are capacitors to the ground plane, and R_{p1} and R_{p2} represent the dielectric loss. Resistor R_{sa} is a fudge factor that enables the optimization of the circuit model based on simulated or measured data. For a fixed LCP thickness of 1 mil and distance to the ground plane on either side of roughly 8 mils, typical parasitic values for width = 1 mil, spacing = 1 mil, and turns = 1.5 are C_s = 2.7 femtofarads (fF), C_{p1} = 33.7 fF, C_{p2} = 492.1 fF, L_s = 2.5 nanohenries (nH), R_s = 1.39 ohms (Ω), R_{sa} = 6.6 kΩ, and R_{p1} = R_{p2} = 10 MΩ.

For the two-port inductor model in Figure 4.7, the effective inductance L_{eff} and Q can be calculated as:

$$L_{\text{eff}} = \text{Imag}\left(\frac{1}{2\pi f Y_{nn}}\right) \quad n = 1, 2 \qquad (4.1)$$

$$Q = \left|\frac{\text{Imag}(Y_{nn})}{\text{Real}(Y_{nn})}\right| \quad n = 1, 2 \qquad (4.2)$$

where Y represents the admittance parameters either at port 1 or 2 and f is the frequency.

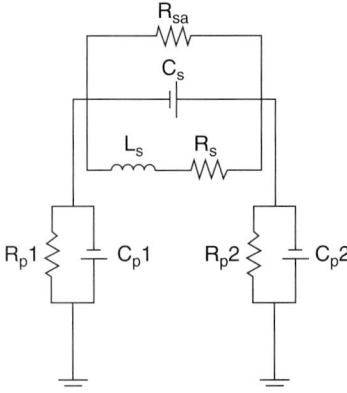

FIGURE 4.7 Equivalent circuit of inductor.

Because of the parasitic capacitance, the effective inductance changes as a function of frequency. For a one-port inductor, either port 1 or 2 is grounded. For the circuit model in Figure 4.7, the equations for L_{eff} and Q can be derived analytically, or the response can be simulated using any circuit simulator. For the inductor parameters defined earlier, the maximum Q attained is 36 in the frequency range between 4 to 5 GHz with an effective inductance of 2.7 nH. Clearly, the performance of the inductor is being limited by the series resistance and the capacitance to ground. These parameters can be minimized by using novel topologies (such as multilayer spirals) and maximizing the distance to the ground plane.

It is possible to optimize the inductor layout using electromagnetic simulators, by investigating new topologies and separating the ground plane from the inductor. The results for a one-layer spiral inductor are shown in Table 4.2 [27]. Inductors A and B are the same size inductors using different layers: inductor A is on the topmost layer M1 in Figure 4.5 for achieving a higher Q factor, and inductor B is embedded on the top LCP layer M3 which is 12 mils below. As shown in Table 4.2, the Q can be increased to 126 from 75. This result shows clearly the scalability of inductor Q using 3D integration.

In Table 4.2, various size inductors have been shown to achieve Q factors in the range of 58 to 126. Sets 1 and 2 are different coupons that were fabricated with the same inductor geometries, which show repeatability in the measurement. As the inductor Q increases, calibration becomes important, since the accuracy of the Q measurements depend on it. In Table 4.2, SOLT (short, open, load, and through) calibration was used to calibrate the Vector Network Analyzer (VNA). Inductor Qs greater than 100 are difficult to measure even with good calibration. Hence, good electromagnetic modeling tools are necessary to confirm the measured values. Oftentimes, the response of a circuit containing the inductor is required to back-calculate the unloaded Q of the inductor.

To further enhance the inductor Q beyond 126, two or more layers are required. Figure 4.8 shows a two-layer spiral inductor where the layers are interconnected in such a way that the inductance is enhanced, the series resistance is reduced, and the ground plane is removed from the inductor. The frequency response of the inductor is shown in Figure 4.9 where a Q value of 165 can be attained at 3.7 GHz. The model-to-hardware correlation is reasonably good. The simulated results were obtained using Sonnet, an electromagnetic solver [28].

Inductor	No. of Turns	Size (mils)	Metal Layer	Inductance (nH)		SRF (GHz)		Q @ GHz	
				Set 1	Set 2	Set 1	Set 2	Set 1	Set 2
A	1	54 × 30	M1	2.74	2.74	9.57	9.52	126 @ 3.68	122 @ 3.6
B	1	54 × 30	M3	2.97	2.92	6.99	6.91	75 @ 2.52	74 @ 2.33
C	1	75 × 51	M1	4.32	4.32	6.57	6.57	122 @ 2.19	119 @ 2.18
D	1	75 × 51	M3	4.74	4.72	4.93	4.88	69 @ 2.12	67 @ 2.01
E	2	45 × 45	M1	9.05	9.33	3.82	3.81	58 @ 1.7	65 @ 1.5
F	3.5	60 × 61	M1	17.7	17.8	2.56	2.55	65 @ 1.38	65 @ 1.05

TABLE 4.2 Summary of Embedded Inductor Results

Mixed-Signal (SOP) Design 165

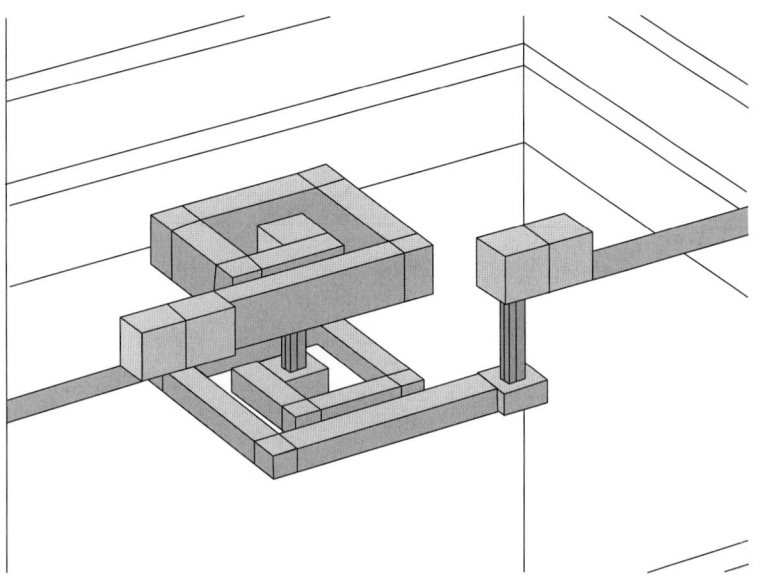

FIGURE 4.8 Two-layer spiral inductor.

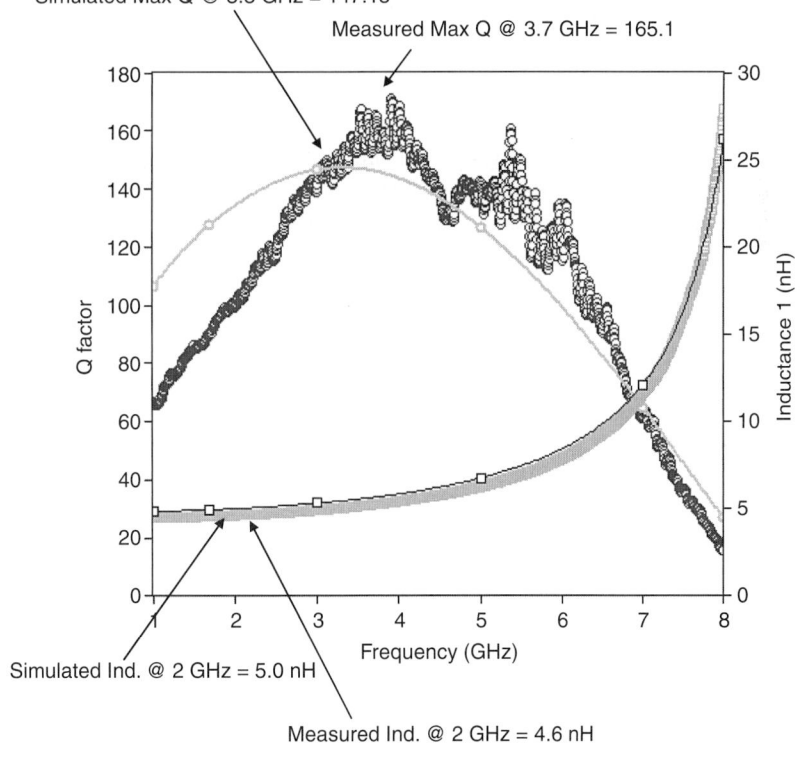

FIGURE 4.9 Frequency response of two-layer spiral inductor.

166 Chapter Four

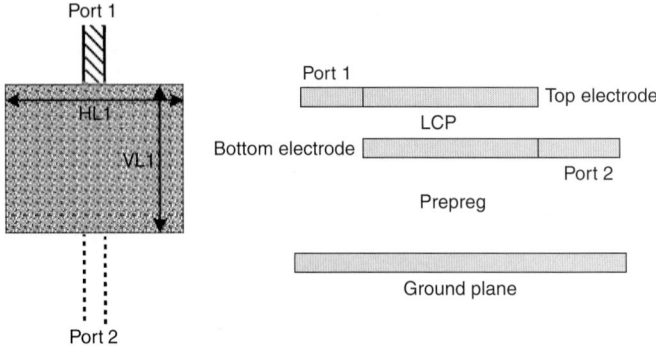

FIGURE 4.10 Capacitor layout.

4.2.2 Embedded Capacitors

Like inductors, the parameters of interest for embedded capacitors are its capacitance at a given frequency, the unloaded Q at that frequency, and the self-resonant frequency of the capacitor. It is preferable to operate the capacitor at a frequency corresponding to its maximum Q. Both the conductor and dielectric loss affect the performance of the capacitor. For a capacitor with perfect conducting plates, the maximum unloaded Q achievable for a capacitor is $1/\tan \delta$, where $\tan \delta$ is the loss tangent of the dielectric material. Since $\tan \delta = 0.002$ for LCP over a broad range of frequencies, the maximum unloaded Q is 500.

Though single metal layers can be used to construct interdigitated capacitors, the parallel plate geometry is preferable due to its smaller size and minimization of conductor loss. The physical structure of the parallel plate capacitor is shown in Figure 4.10 with two electrodes. Additional electrodes can be introduced to increase the capacitance. For example, by adding an additional electrode between the top and bottom electrodes in Figure 4.10 and shorting the top and bottom electrodes, capacitors can be constructed in parallel without increasing area (but by increasing layers). The RF ground plane is also shown in Figure 4.10, which is used as the reference to simulate and measure the frequency response of the capacitor.

The equivalent circuit of the capacitor is shown in Figure 4.11, which includes all of the parasitics. The capacitor C_S is the capacitance between the capacitor electrodes, C_{p1}

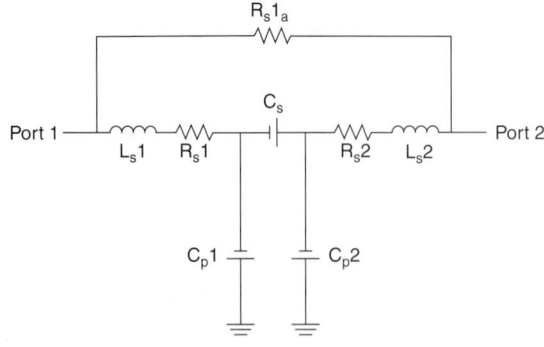

FIGURE 4.11 Equivalent circuit of capacitor.

and C_{p2} are the parasitic capacitances to the ground plane, L_{s1} and L_{s2} are the spreading inductance of the electrode plates, R_{s1} and R_{s2} are the series resistance of the electrode plates due to conductor loss, and R_{s1a} is the dielectric loss. The dielectric loss to the ground plane has not been included in the equivalent circuit since its effect on the capacitor response is negligible. As an example, for a square capacitor with HL1 = VL1 = 23 mils, LCP thickness = 0.92 mils and the distance to the ground plane of ~8 mils, the parameters in Figure 4.11 are C_s = 2.4 pF, C_{p1} = 497 fF, C_{p2} = 46 fF, L_{s1} = 55.9 pH, L_{s2} = 56.6 pH, R_{s1} = 140 mΩ, R_{s2} = 1 mΩ, and R_{s1a} = 66.7 kΩ. These parameters can be derived by optimizing the equivalent circuit parameters to fit the frequency response obtained from Sonnet [28] or other electromagnetic simulators.

Using the equivalent circuit or the frequency response obtained from an electromagnetic simulator, the effective capacitance and Q of the capacitor can be obtained as

$$C_{\text{eff}} = \left(\frac{1}{2\pi f \operatorname{Imag}\left(\frac{1}{Y_{nn}}\right)} \right) \quad n = 1, 2 \tag{4.3}$$

$$Q = \left| \frac{\operatorname{Imag}(Y_{nn})}{\operatorname{Real}(Y_{nn})} \right| \quad n = 1, 2 \tag{4.4}$$

The frequency response of these equations using the parameters extracted is shown in Figure 4.12 from 1 to 9 GHz. As can be seen, the capacitance increases as a function of frequency and resonates around 9 GHz after which it becomes inductive. The maximum Q of 320 occurs at 1 GHz.

4.2.3 Embedded Filters

In addition to several performance improvements, multimode radio architecture provides a method to efficiently conserve both power and real estate in the portable multiband radio design. The multimode radio requires simultaneous multiband operation of the key RF components such as the low-noise amplifier (LNA) [5], oscillator [29], and filters [30–33]. A multimode filter is one that has more than one controllable

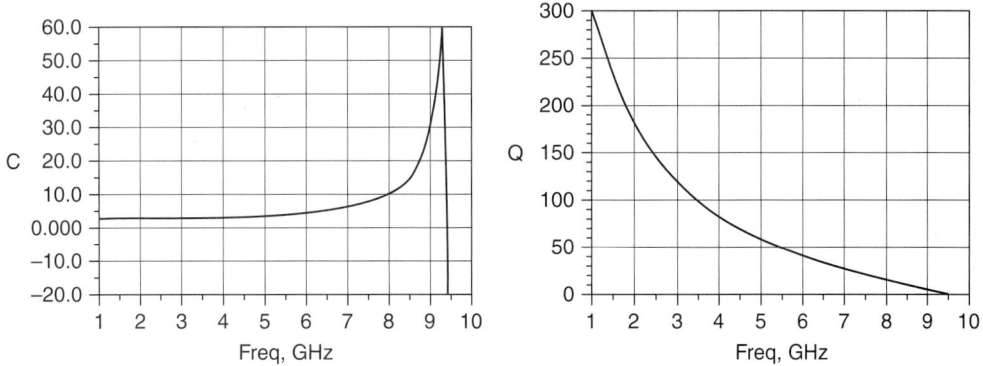

FIGURE 4.12 Frequency response of capacitor. (a) Capacitance. (b) Quality factor.

passband while having only one input and one output port. Such filters are very feasible in the design of diversity-type WLAN (IEEE 802.11a/b/g) where low economics and small form factor designs are of prime importance. Several dual-band filters can be found in the literature [30–33] where dual behavior transmission line resonators (DBRs) are used to synthesize different passbands. The same theory can be used to implement filters with three passbands [33]. Current solutions for a single-band front-end filter use high-quality-factor (Q) packaging technologies based on ceramics and polymers with filter sizes in the range of 6 to 14 mm². For a dual-band filter to be used in commercial wireless products it should, in addition to small insertion losses, occupy a smaller area compared to using two single-band filters. One of the major requirements in the design of dual-band filters, because of the large number of resonators, is the need for high Q technologies. Transmission line based dual-band filters [30–33] are often too large in size to be employed in commercial multimode systems since the physical length of the transmission line has to be at least comparable to half a wavelength, which is large in the frequency band from 1 to 10 GHz, a frequency band where most of the mobile consumer applications are expected to function.

A reduction in the size of the filter is possible by replacing the transmission lines by lumped-element dual-band resonator sections. Figure 4.13 shows the schematic of a dual-band filter that is an efficient method to obtain dual-passband characteristics. It consists of two, fourth-order resonators that are capacitively coupled to each other and to the input-output terminals. The filter is matched using series input capacitors (C_1 and C_2) that also control the two center frequencies. The bandwidth and the passband ripple of the filter at the lower passband are adjusted by controlling the capacitive coupling (C_c) between the two shunt fourth-order resonators. Since the filter uses the same set of passives to operate over a wide frequency range (~3 to 5 GHz), passives with high quality factors (Qs) and stable electrical characteristics over a broad frequency range are required. The basic idea is to synthesize two different inductances from the fourth-order

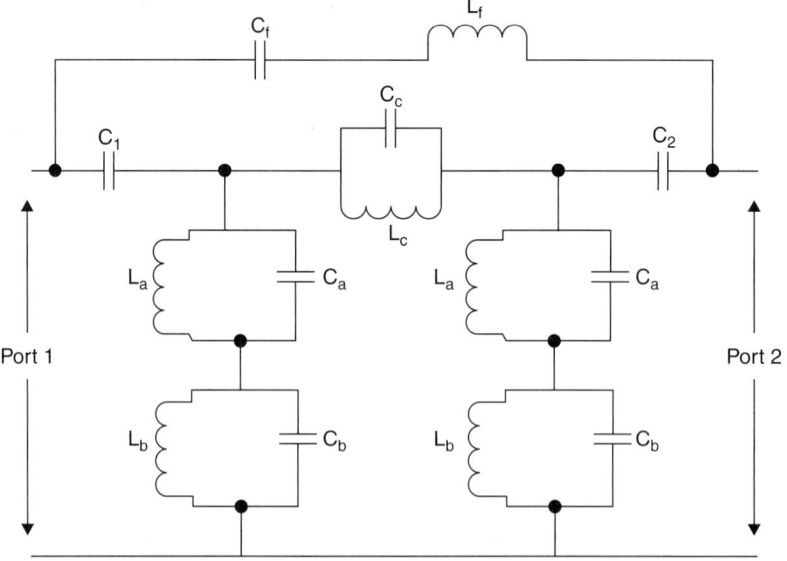

FIGURE 4.13 Dual-band filter.

resonators that will resonate with the matching capacitors. Since the same set of passive elements is used to simultaneously produce two passbands, this method translates to a size and, hence, a cost-efficient solution for the future multiband, multifunctional systems. The cross-coupling components between the input and output terminals provide another transmission zero at the higher passband and thus control the bandwidth at the higher frequency.

A single-input–single-output (SISO) prototype has been designed to pass 2.4 and 5.2 GHz simultaneously. Figure 4.14 shows the layout of the filter in Sonnet [28], a commercially available electromagnetic simulator that is accurate for the kinds of structures discussed in this section. The LCP-based stackup used enables 3D integration of the passives over two LCP sheets. The process combines two, diclad 25-µm-thick LCP layers with multiple low-loss tangent glass-reinforced organic prepreg (core) layers resulting in an eight metal layer stackup. The stackup is balanced (mechanically and electrically) and allows the filters to be designed with ground planes on both the top and bottom sides (stripline configuration), thereby minimizing any losses associated with radiation and at the same time minimizing electromagnetic interference issues. The entire layout is shown in Figure 4.14, which consists of a 25-µm-thick dielectric layer (LCP) with metal (copper) on both sides (diclad) and prepreg (dielectric) on either side. The inductors have been implemented as spirals on a single metal layer with no ground on the opposite layer. The one-port inductors and capacitors are connected to the microstrip ground using the plated through-holes. The fourth-order resonators are laid out to enhance the series mutual coupling between the inductors to minimize the

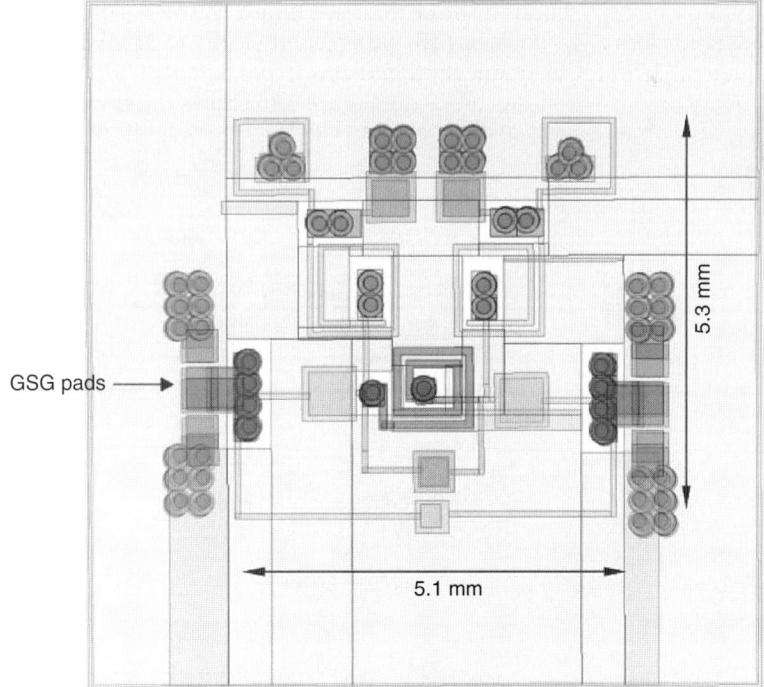

Figure 4.14 Filter layout.

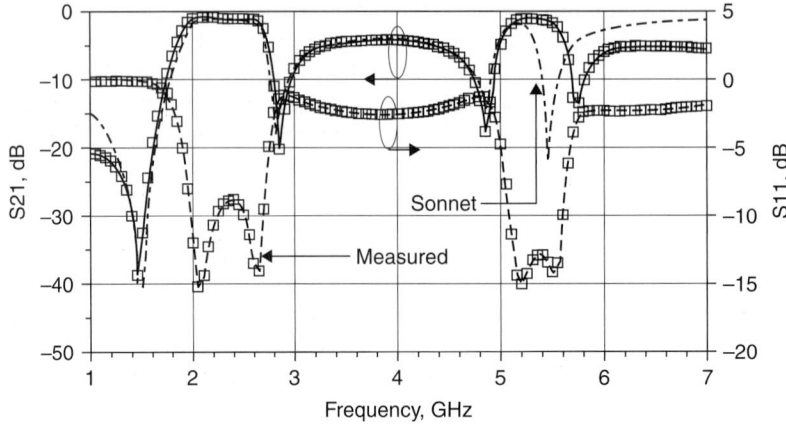

Figure 4.15 Model-to-measurement correlation for filter.

resonator area and maximize the resonator Q. The resonators are symmetrically placed on either side of the capacitors to minimize the inductive coupling.

Figure 4.15 shows the measured and modeled results of the dual-band filter. In simulating the filter performance, modeling of the entire structure including the parasitics becomes critical. The parasitics include via transitions, effect of metal losses, and ground returns affected by the parasitic capacitances. The filter has two controllable passbands at 2.4 and 5.3 GHz, as shown in Figure 4.15. The design of the filter allows the separation of the bands by more than two times. The higher-frequency band exhibits an insertion loss of 1.1 decibels (dB) with a bandwidth of 530 MHz (10 percent of center frequency). The lower-frequency band has a passband of 525 MHz and exhibits a relatively low insertion loss of 1.3 dB. Figure 4.16 shows the current distribution of the

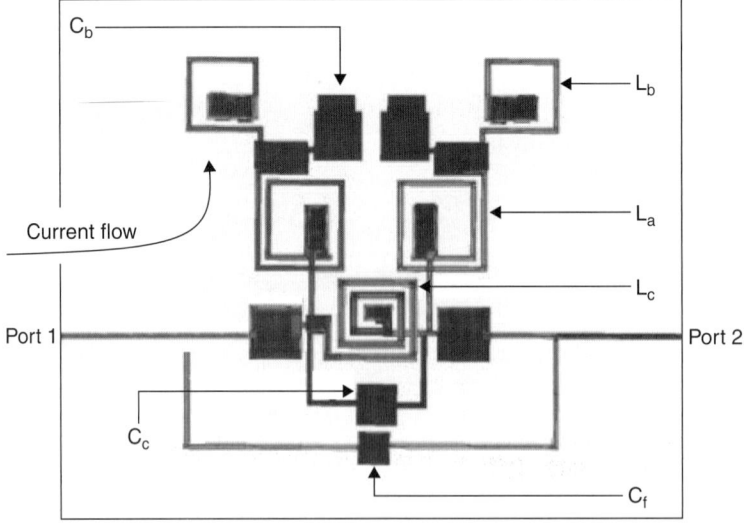

Figure 4.16 Current density in stopband at 0.5 GHz.

filter in the stopbands at 0.5 GHz. From the plots it can be seen that in the stopbands the circuit functions as a short circuit wherein the majority of current is directed through the resonator inductors to ground. Hence, high-Q inductors are required to improve insertion loss and minimize excessive heating, especially in high-power applications. As shown in Figure 4.15, careful modeling of the structure can provide an excellent model-to-hardware correlation. However, not all structures can be modeled using electromagnetic simulators, especially when many more components are integrated, due to enormously large simulation time. Hence, an intermediate step using circuit-level simulation becomes necessary during the design phase that enables rapid changes to the layout, based on performance evaluations. This can be followed by the modeling of an entire layout, which may take a day to complete.

The filter can be scaled in frequency by scaling the resonator network. Addition of transmission zeros is possible by adding a feedback capacitor between input and output or by adding additional resonant networks with a marginal increase in size. Table 4.3 summarizes the results of the frequency scaled dual-band filter by modifying the resonant networks for 1/2.4 and 2.4/5 GHz operation, each supporting different frequency bandwidths. For the 2.4/5-GHz filter, a size of 5.1 mm × 5.4 mm has been achieved, with an insertion loss between 1.1 and 1.5 dB.

4.2.4 Embedded Baluns

Baluns are three-port devices that provide balanced outputs from unbalanced inputs, as shown in Figure 4.17. Electrically, this means that the input signal power is split into two channels that are equal in magnitude but opposite in phase, by 180°. They are thus required in almost all RF architectures, and their design for multiband radio architectures becomes a key challenge in SOP-based integration.

Traditionally, baluns have been implemented using distributed components. A functional balun can be implemented by tapping the differential output across the signal and ground of a transmission line. However, difficulty in controlling the return current path can cause poor amplitude and phase imbalance in the practical implementation of such a device. As a solution, N. Marchand has described a compensated balun, utilizing

References	Center Frequencies (GHz)	3-dB Bandwidth (MHz)	Insertion Loss (dB)	Area (mm²)
Multilayer LCP	1 2.4	80 625	0.8 1.2	6 × 6
	2.4 5	965 1250	1.5 1.2	5.1 × 5.4
	2.4 5	525 500	1.3 1.1	5.1 × 5.4
Alumina	1.5 2	Not applicable	1 4	48 × 24
Organic laminate	2.4 5	Not applicable	2.4 1.8	15 × 8

TABLE 4.3 Dual-Band Filter Performance and Comparison between Technologies

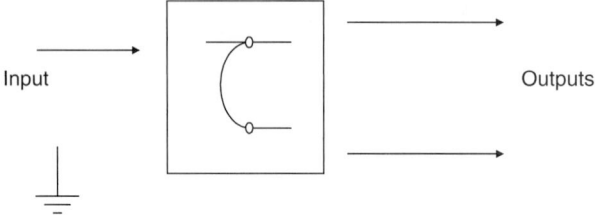

FIGURE 4.17 Functional representation of a balun.

a second transmission line segment with the same electrical properties as the first one to compensate for the effect of the reference ground and thus match the output signals [34]. As shown in Figure 4.18, the Marchand balun uses two $\lambda/4$ coupled-line pairs. With the compensated architecture ensuring good phase and amplitude balance, the design equations primarily deal with ensuring good return loss in the band of interest. The input impedance seen at point d can be derived as [35]

$$Z_d = \frac{Z_{load}}{\frac{Z_{load}^2}{Z_{ab}^2 \tan^2\theta} + 1} + \frac{jZ_{load}^2 Z_{ab} \tan\theta}{Z_{load}^2 + Z_{ab}^2 \tan^2\theta} - jZ_b \cot\theta \qquad (4.5)$$

where θ is the electrical length of the transmission line segments, and the rest of the variables are as shown in Figure 4.18. The impedance seen by the source (Z_{in}) can then be calculated by transforming Z_d with a length of transmission line with characteristic impedance Z_a (assuming lossless operation), as

$$Z_{in} = Z_a \frac{Z_d + jZ_a \tan\theta}{Z_a + jZ_d \tan\theta} \qquad (4.6)$$

The input return loss S_{11} is thus a function of Z_a, Z_b, Z_{ab}, Z_{load}, Z_{source}, and θ. Perfect matching [Im(Z_{in}) = 0 and Re(Z_{in}) = Z_{source}, leading to S_{11} = 0] occurs when Z_a is set equal to ($Z_{source} \cdot Z_{load})^{1/2}$ and the lengths of the transmission line segments are chosen such that $\theta = 90°$.

Owing to the use of distributed elements, the size of distributed baluns becomes prohibitively large at low frequencies. Several methods (impedance variation, capacitive loading, etc.) have been suggested [36–37] for reducing the balun size; however, they all

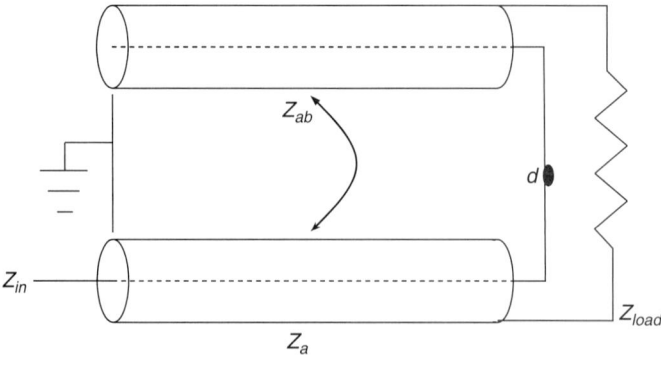

FIGURE 4.18 Marchand balun.

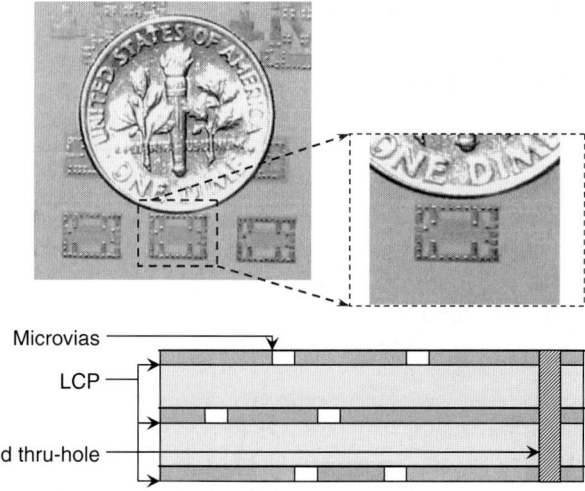

FIGURE 4.19 Wideband LCP balun and stackup.

result in a reduction of the percentage bandwidth also. An architecture that allows for both wider bandwidth and smaller size has been described in [38].

Using an organic LCP-based SOP process, impedance scaling techniques can be used to achieve a WLAN balun for use in the 4.9- to 5.9-GHz WLAN frequency band, achieving a 64 percent reduction in size while maintaining a percentage bandwidth of 53 percent. Figure 4.19 shows the layout of the balun in planar form and the multilayer LCP stackup used for its implementation. The cross section of the balun along the A-A' axis and the implementation of the coupled-line segments in stripline topology is shown in Figure 4.20. In the figure, h represents the total height of the substrate, s the spacing between the coupled lines, and d the spacing between the coupled-line segments. Electromagnetic solvers were used to determine the value of the impedances Z_a, Z_b, and Z_{ab}. Table 4.4 shows the comparison of the fabricated balun with a commercially available Marchand balun, also built on a organic substrate. As can be seen from the table, the present balun

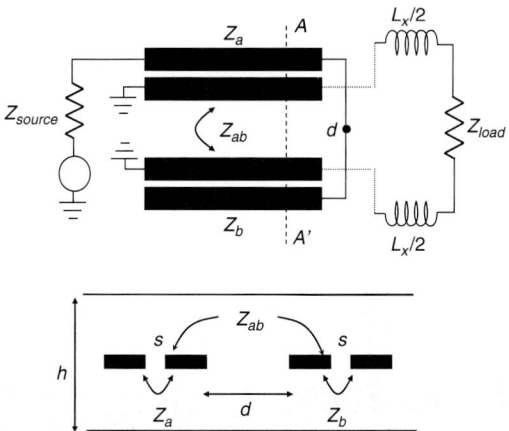

FIGURE 4.20 Layout of the planar balun and cross section along the A-A' axis.

Device	Commercial Balun	LCP Balun
Frequency	4.8–5.9 GHz	4.8–5.9 GHz
Return loss (S_{11}) (min)	16 dB	−15.5 dB
Insertion loss (max)	0.6 dB	0.57 dB
Amplitude imbalance (max)	0.6 dB	0.33 dB
Phase imbalance (max)	5°	6°
Area	9 mm²	5.16 mm²

TABLE 4.4 Performance Comparisons of the Fabricated Balun

implementation compares to the commercially available balun with a 42 percent reduction in size. In addition, the present balun can be embedded into the layers of LCP. Lumped-element implementations using impedance matching networks have been proposed as a means for size reduction [39]. As with any lumped-element approximation, this also results in degradation in performance, particularly in amplitude imbalance across the frequency bands. The lattice topology is a commonly used lumped-element solution for implementing small-size narrow-band baluns, as shown in Figure 4.21. A combination of low-pass and high-pass networks allows the splitting of the input signal into two output signals that are equal in power but with a 180° phase difference. The low-pass and high-pass networks can be implemented with as low a number as four passives (two inductors and two capacitors), leading to small sizes (with the tradeoff of narrow-band operation). The circuit topology also lends itself well to impedance transformation.

The design equations for the device are as follows:

$$Z_0 = \sqrt{R_{source1} \cdot R_{load}} \tag{4.7}$$

$$L = \frac{Z_0}{\omega_0} \quad C = \frac{1}{Z_0 \omega_0} \tag{4.8}$$

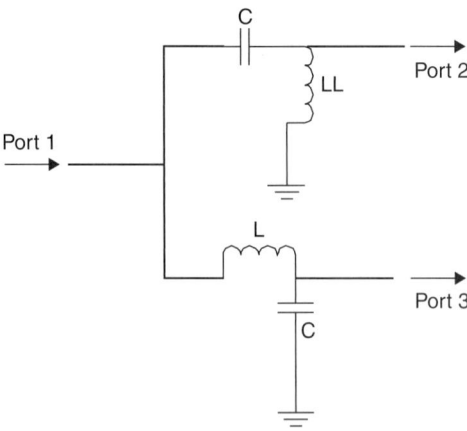

FIGURE 4.21 Narrow-band balun topology with inductors and capacitors.

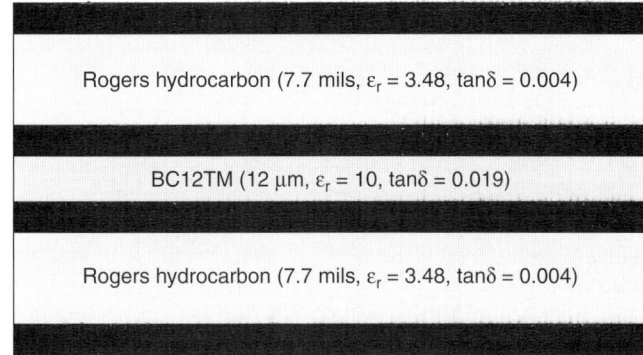

FIGURE 4.22 Stackup for balun implementation using high-K material.

where R_{source} and R_{load} are the source and load impedances, and $\omega_0 = 2\pi f_0$ is the frequency of operation.

To demonstrate the operation of the lumped element balun, a balun operating at 2.44 GHz with a 100-MHz bandwidth has been designed. For an R_{source} and R_{load} of 50 and 100 Ω, respectively, this yields values of 0.92 pF and 4.6 nH for the capacitance and inductance. With the thickness limitation restricting the use of multiple dielectric layers and size limitations restricting the use of low-K materials, it is difficult to realize such baluns using homogenous dielectrics. Figure 4.22 shows a 0.5-mm-thick stackup incorporating a high dielectric constant material (Oak-Mitsui's FaradFlex BC-12TM). With a tan δ of 0.019 and $\varepsilon_r = 10$ (at 1 MHz) and a thickness of 12 μm, the material has been developed for embedded digital decoupling applications. However, the high capacitance density (11 nF/in² at 1 MHz) makes this a suitable candidate for small size low-profile baluns. The lattice topology is particularly suitable for design using this material, as it uses low-pass and high-pass structures that are more tolerant to dielectric losses compared to bandpass structures. The shielded device measures 1.25 mm × 2 mm in area with a thickness of 0.507 mm, 1 dB of insertion loss, an amplitude imbalance of 2 dB, and a phase imbalance of ±10°. Table 4.4 shows the comparison of the fabricated balun with a commercially available Marchand balun, also built on an organic substrate. As can be seen from Table 4.3 the present balun implementation compares to the commercially available balun with a 42 percent reduction in size. In addition, the present balun can be embedded into the layers of LCP.

A third alternative in balun design is the use of transformers. Although compact designs are possible, the performance of the balun in this case is very much dependent on the coupling between primary and secondary coils. An SOP technology with high coupling coefficients, achieved through tight metal-to-metal spacing or low dielectric thickness or a combination of both, is required for the implementation of these baluns.

4.2.5 Filter-Balun Networks

In a receiver, the signal coming in from the antenna is single-ended in nature, but the active circuitry (beginning with the LNA) is usually differential, as shown in Figure 4.23. The single-ended signal is filtered using a bandpass filter and then converted to differential mode using a balun. With SOP-based implementation, a circuit embedded

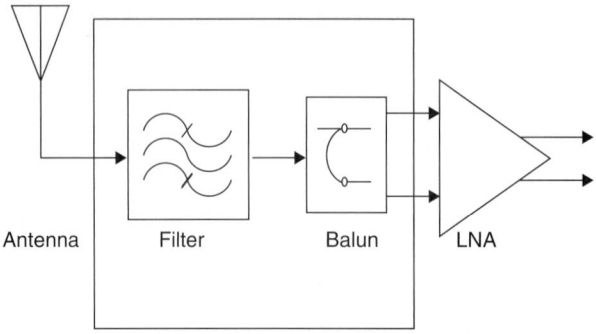

FIGURE 4.23 Filter-balun in a receiver front end.

in the substrate can be used that combines the functionality of both a balun and a filter. Any single-ended circuit can be made into a balanced network (with differential inputs and differential outputs) using network theory [40]. Balanced bandpass filters can also be designed in this fashion. However, this technique also results in an increase in the number of components. It leads to doubling of the capacitance values required in the series path [40], which can lead to large device sizes in embedded circuits where the device size is directly proportional to the capacitance or inductance value required. Lattice filters have also been used in the past to achieve balanced filter topologies [41–42]. Although they provide both frequency selectivity and differential outputs, both these approaches require additional matching circuits for single-ended to differential conversion at the input port. Two alternate approaches are adding frequency selectivity to existing balun circuits and cascading a balun with a bandpass filter.

The Marchand balun by its very nature has a bandpass behavior. The coupled line segments prevent the transmission of signals at low frequencies, while the transmission line behavior causes the signal transmission to fall off after the resonant frequency of the coupled lines. Implementation of the lumped elements in the modified Marchand balun using resonators allows transmission zeroes in the transfer function of the balun, leading to sharper roll-offs for the frequency response. Figure 4.24 shows a modified Marchand balun designed for operation in the 5- to 6-GHz frequency band. To increase

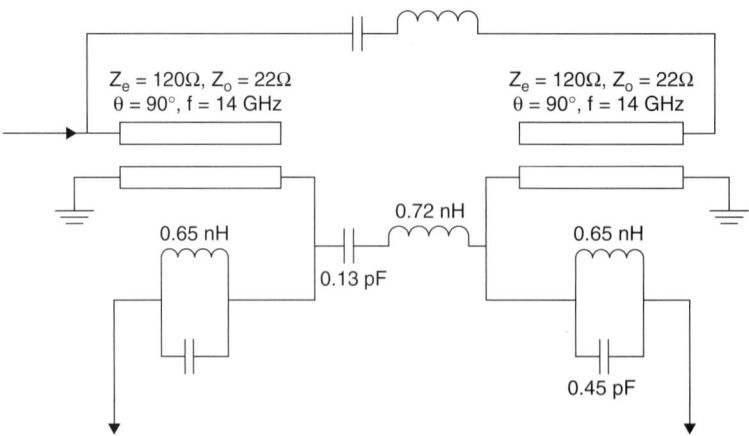

FIGURE 4.24 Integrated Marchand balun and filter.

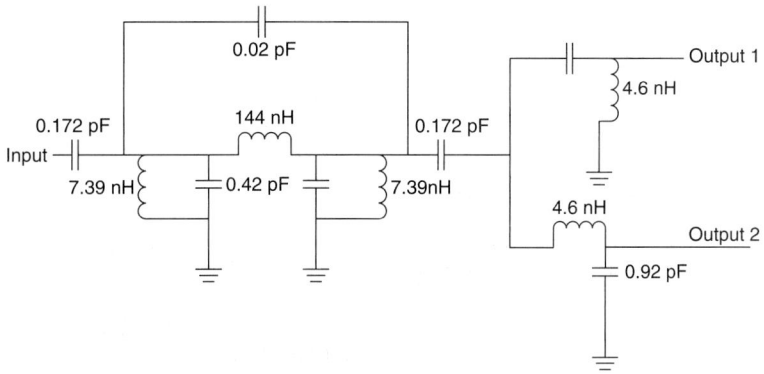

FIGURE 4.25 Cascaded filter balun circuit.

high-frequency rejection, the lumped elements have been implemented using resonators. The capacitors have been replaced with two resonators (consisting of 0.72 nH in series with 0.13 pF and 2.3 nH in series with 0.1 pF) to provide transmission zeros at ~16.5 GHz and ~10.5 GHz. The series inductors at the output have been replaced with 0.6 nH and 0.45 pF in parallel, to provide a block at 9.5 GHz.

The resonators provide enough rejection at the second and third harmonics (10 to 12 GHz and 15 to 18 GHz) of the operational frequency. However, the rejection at frequencies lower than 5 GHz (and especially in the 2.44-GHz band) remains low. It is thus clear that the use of resonators in existing balun designs to provide frequency rejection has limitations. Another option for implementing the filter-baluns is to cascade a bandpass filter with a balun.

Figure 4.25 shows an implementation of a cascaded filter-balun. The filter (described in [43]) consists of capacitively coupled resonators, with a smaller capacitor connected between the input and output terminals connected to introduce two transmission zeros (at 1.8 and 3 GHz). The balun uses the lattice topology and performs 50- to 100-Ω single-differential transformation. Both the filter and the balun have been designed for operation in the 2.4- to 2.5-GHz band. The filter provides 20 dB of rejection at the cell-phone frequency bands, while the balun has been optimized for low loss in conjunction with the filter.

An LCP-based stackup was used to implement the device, primarily because of the high Qs required for the resonators in the filter, as shown in Figure 4.26. To maintain the

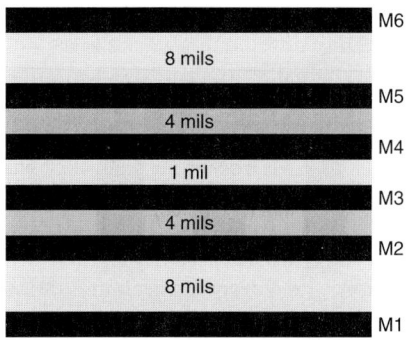

FIGURE 4.26 Stackup.

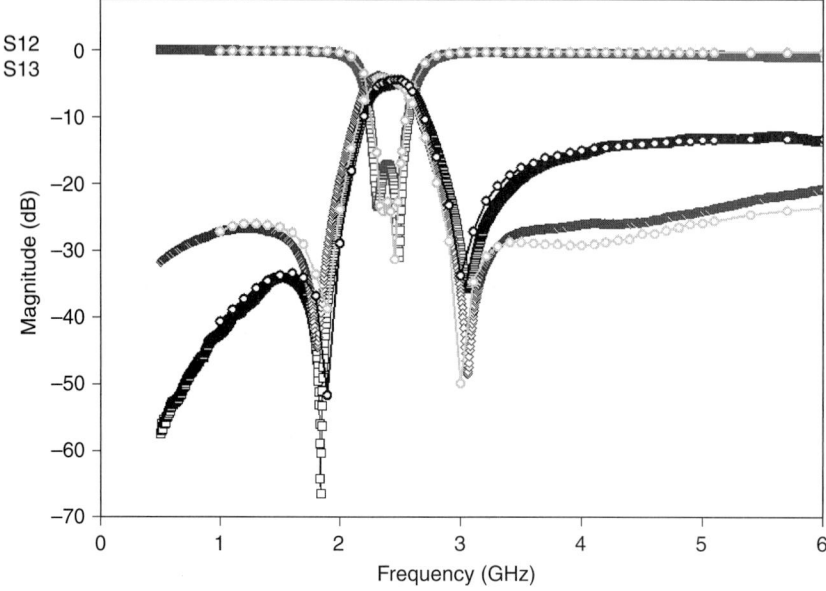

Figure 4.27 Model-to-hardware correlation. Dotted lines are Sonnet simulations, and continuous lines are measured data.

device size under control, the balun has been implemented using all six-metal layers offered by the technology—the capacitors have been implemented on the LCP layer, and the inductors have been implemented as meandering lines on metal layers 2 and 5.

The fabricated devices measured 4 mm × 1.5 mm, with a total thickness of 0.75 mm. Figure 4.27 shows the measured S_{21} and S_{31} of the device. The filter provides a minimum rejection of 20 dB across the cell-phone bands (GSM, EGSM, PCS, and DCS), while maintaining an insertion loss below 2 dB. It is to be noted here that the filter itself has a loss close to 1.7 dB in [43]. By careful design of the balun to minimize return losses, it is possible to keep the total losses in the device to less than 2 dB.

4.2.6 Tunable Filters

Tunable filters are required for in-band tuning and for providing better matching to the circuits at the input or the output of filters. Tunable filters can also be used to improve the yield for correcting parametric defects during manufacturing (such as frequency shifts). Tuning can be achieved by using variable capacitors such as varactor diodes or by using electronic switches. Varactor diodes are reverse-biased diodes whose capacitance can be varied by changing the voltage. These have a low Q (10 to 15 range) and can load the filter, thereby increasing the insertion loss and in-band ripple. Electronic switches such as GaAs and CMOS switches can be used to switch between a capacitor array. These switches also have high insertion loss. The problems associated with these implementations can be solved using MEMS variable capacitors or MEMS switches, respectively. Using polymer MEMS processes, these devices can be integrated into the substrate. In this section, the varactor implementation on embedded filters in LCP substrates has been described. Even though the varactor diodes have a low Q, the performance of the tunable filter can be enhanced by using filters embedded in the

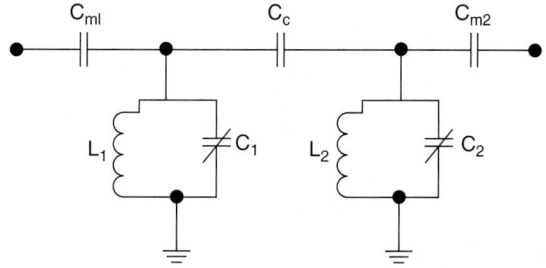

FIGURE 4.28 Tunable filter.

layers of the substrate. Figure 4.28 shows the circuit schematic of the varactor-tuned filter. It is a second-order capacitively coupled Chebychev filter [44]. The circuit functionality is the same as the dual-band filter except in this case the shunt capacitors enable the designer to decouple the matching and passband frequency from the series matching capacitors.

Figure 4.29 compares the measured results (square marker) of the fixed frequency filter (i.e., instead of varactors, embedded capacitors were used) with the results obtained from the electromagnetic simulator Sonnet (solid line). The data line with the × marker is the measured result of a similar filter with the exception that the capacitors C_1 (1.1 pF) and C_2 (1.1 pF) are replaced with silicon varactor diodes from Skyworks Inc. (model SMV 1405). Each abrupt-junction-type diode can provide a capacitance of 2.7 to 0.6 pF over a reverse junction voltage of 0 to 30 V (tuning of 4.2:1). The data of the tunable filter in Figure 4.29 are at a tuning voltage of 6 V where the varactor provides 1.05 pF of capacitance. From Figure 4.29 it can be seen that the addition of the varactor reduces the loaded Q of the filter from 5.5 to 2.6 and, therefore, the bandwidth increases.

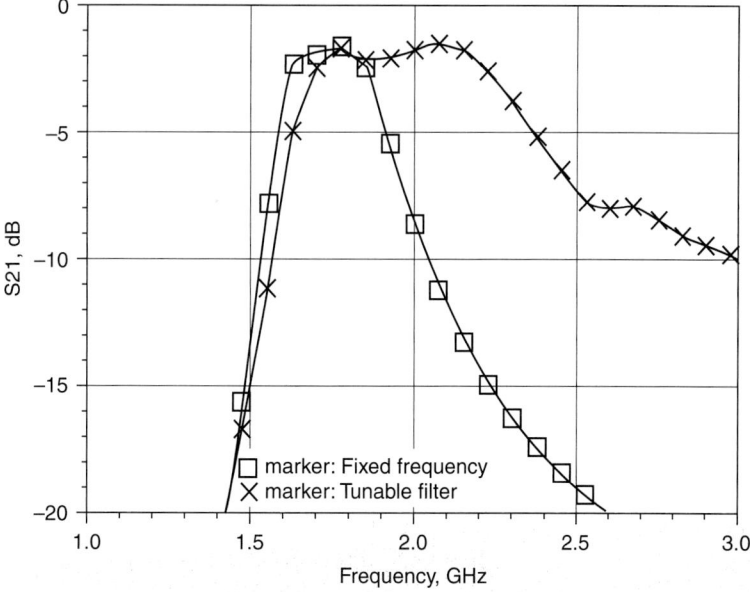

FIGURE 4.29 Fixed-frequency and tunable filter measurements and EM simulations.

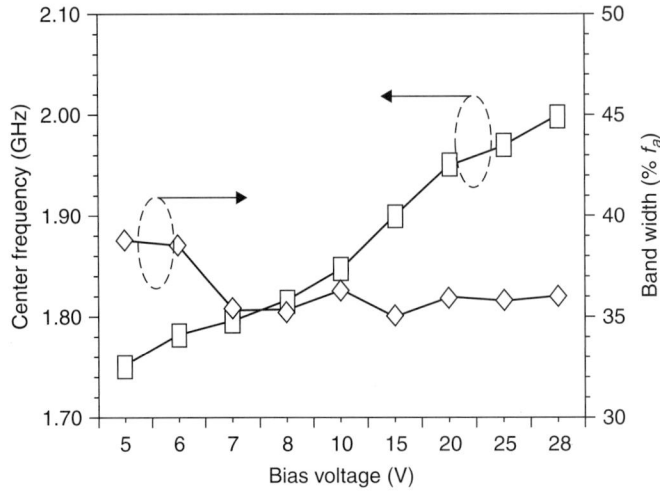

Figure 4.30 Measured tuning performance.

The measured center frequency and filter bandwidth as a function of varactor bias voltage are shown in Figure 4.30. The tuning voltage is supplied through a surface-mount inductive choke to minimize any phase distortion due to bias modulation. A 2.2-pF embedded capacitor (C_g) has been used to provide the RF grounding at the diode-choke junction. Figure 4.30 shows that the filter can be tuned from 1.75 to 2.03 GHz with a tunability of 12 MHz/V. Additionally, Figure 4.30 shows that the bandwidth (and hence the Q) of the filter is almost constant over the entire tuning range even though the Q of the lossy varactor changes with frequency. This characteristic can be attributed to the broadband nature of the Q of the passives embedded in LCP substrates. The widening of bandwidth is also a function of C_g. A high loaded Q is observed with a higher C_g since, the alternating currecnt (ac) resistance between the tuning ports to ground is reduced. The tunable filter occupies a volume of $5 \times 5 \times 0.76$ mm³, whereas the fixed frequency filter occupied a volume of $3.9 \times 5 \times 0.76$ mm³. The response of the filter can be further improved by incorporating transmission zeros by using both capacitive and inductive coupling [44].

4.3 Chip-Package Codesign

Chip-package codesign represents the concurrent design of the IC and package such that together, the IC package can support the performance specifications. For digital ICs, this could be the I/O speed or bit error rate (BER), while for an RF IC this could be the noise figure, phase noise, power, etc. Chip-package codesign represents the partitioning of a circuit, subsystem, or system such that some of the functionality can be embedded in the package while the rest remains on the IC. Oftentimes, chip-package codesign is mistaken to represent the matching of the footprint between the chip and package. Though this is important from a physical design standpoint, it does not represent true chip-package codesign. For digital systems, sizing the I/O driver concurrently with the package interconnect parameters is an example of chip-package codesign since this enables faster I/O speeds and optimized power. Similarly, embedding decoupling capacitance in the package is another example of chip-package codesign since it enables lower noise, thereby leading to higher performance as measured by the BER. With the trend toward integration, chip-package codesign is becoming a requirement for RF circuits since the RF performance can degrade due to the semiconducting properties of silicon ICs. For example, the integration of inductors using standard silicon processes can generate eddy currents in the silicon substrate, thereby reducing its Q. High Q inductor integration, however, is achievable in the package, and hence placement of these inductors in the package can increase performance. In general, active RF circuits contain a large number of passive components, consisting of inductors, capacitors, and resistors. Resistors are mainly used for biasing, while the inductors and capacitors are necessary for the circuit's high-frequency operation. Though in general, low-loss passive components are required for RF circuits, not every passive component requires a high Q. Hence, during system partitioning, only the devices that require a high Q can be integrated in the package. The passive components embedded in the package can be concurrently designed with the circuit components on silicon, to meet the performance specifications. Another exercise in chip-package codesign is the real estate required for implementing the passive components. If the area occupied by the passives is a large fraction of the transistor area, then it is not economical to integrate these components in silicon and hence can be moved into the package.

In this section, two classes of RF circuits are designed using the chip-package codesign methodology. The first example is a low-noise amplifier, which is the first active stage in any receiver architecture. Here, the gate inductor is embedded in the package for achieving a lower noise figure. This implementation is also more economical since 70 percent of the real estate on silicon can be occupied just by the gate inductor. The second example is a voltage-controlled oscillator (VCO). The phase noise of a VCO is related to the power and Q of the passive components as [45]:

$$L\{\Delta\omega\} \propto \frac{F}{Q^2 \cdot P_o} \qquad (4.9)$$

For a VCO, low phase noise is required. This can be achieved either by increasing P_0, which is related to the direct current (dc) power or by increasing the Q of the passive components. For a mobile application, achieving low power is necessary. Because of the square law dependence, it is therefore necessary to use passive components with a high Q in the VCO circuit. Hence, chip-package codesign of the VCO with passive components integrated in the package is necessary.

4.3.1 Low Noise Amplifier Design

The low-noise amplifier (LNA) is the first active device of any RF front-end architecture. Essential requirements of this amplifier circuit are reasonable gain, a good input impedance match, linearity, and the lowest possible noise figure (NF). If the device is to be used in a portable device, the need for low power consumption also becomes important. The noise factor (F) of an LNA is a measure of the amount of noise added by the circuit to the incoming signal and is defined as the ratio of signal-to-noise ratio (SNR) at the input of the device to the SNR at the output.

The cascode LNA architecture of Figure 4.31 has also been used widely for its low NF and high input-output isolation, particularly for single-chip solutions where the transistor parameters can be strictly controlled. The design process for the inductively degenerated LNA consists of sweeping the NF with respect to the transistor (M1) gate

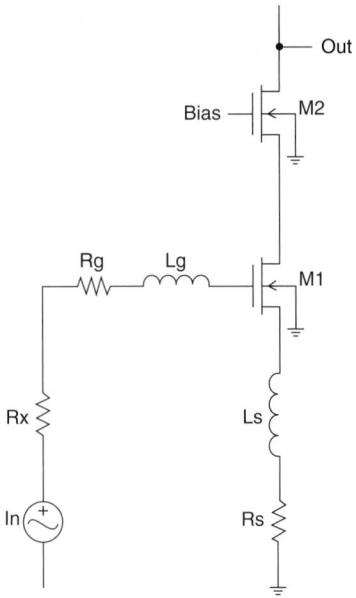

Figure 4.31 CMOS cascode LNA.

width. If the parasitic resistances can be ignored, the real part of the input impedance can be controlled by choosing appropriate values for L_s and can be set to equal the source resistance for impedance match. The gate inductance is then chosen such that L_T (which is the sum of the inductances L_s and L_g) resonates with C_{GS} (gate-source capacitance) at the operating frequency, thus canceling out all the imaginary terms and making the input impedance purely real at the frequency of operation.

Optimization strategies for CMOS LNAs are well known [46–47]. All of these design methodologies have assumed fixed Qs for the inductors. An SOP approach that provides embedded inductors in the package substrate allows the designer an extra design variable, namely, the Q of the inductors. Depending on their contribution to performance specifications like NF and gain, any or all of the three inductors in the LNA circuit can be implemented on-chip or embedded in the package. However, attaining a particular Q also comes with tradeoffs in size and layout. In order to incorporate these into the optimization methodology, it is necessary to derive F as a function of R_g and R_s.

Including all the noise contributions of the field-effect transistor (FET) and that of the parasitic resistances of the inductors, the noise factor (F) can then be derived as [48]:

$$F = 1 + \frac{R_g}{R_x} + \frac{R_{gate}}{R_x} + \frac{R_s}{R_x} + \frac{\beta\omega_0^2 C_{GS}^2(\omega_0^2 L_T^2 + (R_x + R_g + R_{gate} + R_s)^2)}{5R_x g_{do}}$$
$$+ \frac{2c\omega_0^2 C_{GS}^2(R_x + R_g + R_{gate} + R_{ch} + R_s)(R_x + R_g + R_{gate} + R_s)}{g_m R_x}\sqrt{\frac{\beta\gamma}{5}}$$
$$+ \frac{\omega_0^2 C_{GS}^2(R_x + R_g + R_{gate} + R_{ch} + R_s)^2}{g_m^2} \frac{\gamma g_{do}}{R_x} \quad (4.10)$$

where R_x is the source resistance, which is typically 50 Ω; β and γ are bias-dependent noise parameters of the MOSFET; and g_{do} is defined as the drain output conductance evaluated at $V_{ds} = 0$ V. In Equation (4.10), C is the correlation coefficient between the drain and gate noise currents of the FET.

Equation (4.10) shows that F is equally dependent on the parasitic resistances of both gate and source inductors (R_g and R_s). However, in practice, L_s is much smaller than L_g and can be implemented as an on-chip or bond wire. However, depending on the frequency of operation, L_g can be as high as 35 nH. The parasitic resistance of L_g (R_g) is hence a very important contributor to the F of the LNA. As it is impossible to implement this inductor on-chip, an optimum solution is to embed it in the package. The Q of an inductor is a function of the signal loss within the device. The losses in an inductor consist of two components, namely, losses in the metal and losses in the dielectric. It has been shown that the inductor can be optimized for maximum Q. Under these conditions, conductor losses dominate the total loss (and hence the Q). The conductor losses can be reduced by increasing the conductor width (which reduces the series resistance), leading to an increase in the size of the inductor, thus allowing for a tradeoff of larger size for higher Q. By using embedded inductors in the package in place of chip inductors for L_g, the designer has control over the required unloaded Q for this inductor. However, because of the tradeoff with respect to size, using inductors with the maximum Q possible is not a good strategy and could lead to unnecessarily large sizes for the packaged LNA.

Equation (4.10) can be used to find the optimum Q_g, required for a particular NF. Figure 4.32 shows the variation of NF for the optimum transistor gate width, for a 1.9-GHz CMOS LNA designed for the 0.5 μm process. The NF decreases rapidly for increasing Q_g at low values of Q_g, but the rate of change decreases at higher values of Q_g. Hence, there is very little reduction in NF beyond a certain inductor Q. Equation (4.10) and Figure 4.32 provide the minimum tolerable inductor Q required for satisfying the sensitivity requirements of a particular circuit. Figure 4.33 shows the variation in area for different values of Q, in a six-metal layer organic SOP process.

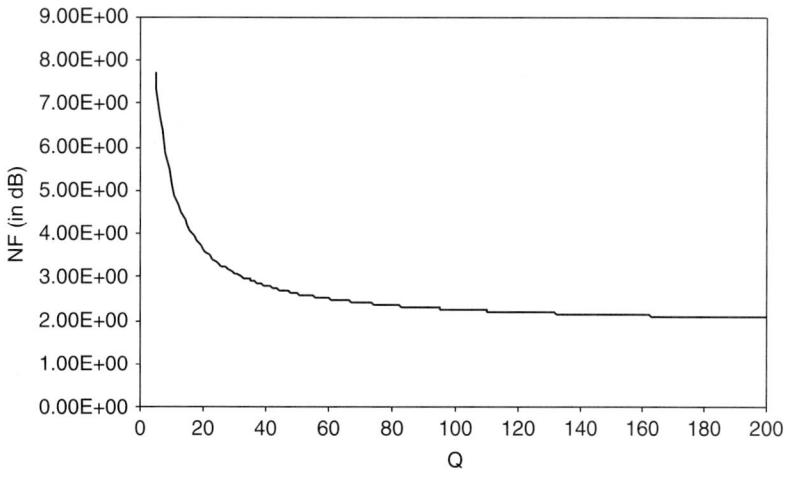

FIGURE 4.32 NF versus Q.

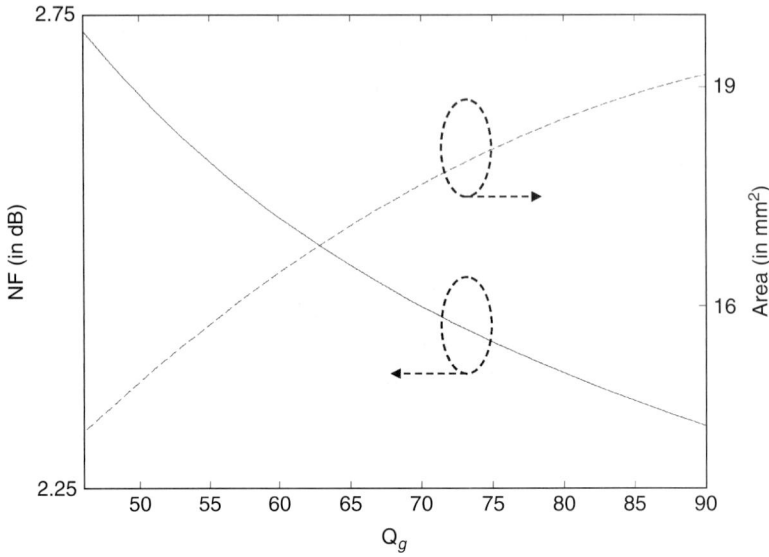

Figure 4.33 NF versus Q versus area.

As an example of the chip-package codesign methodology discussed so far, an LNA for GSM applications is designed for the 0.5 μm CMOS technology, with a standard source resistance of 50 Ω and an operating frequency of 1.9 GHz, leading to inductance values of 9 and 1.2 nH for L_g and L_s, respectively. The parameter L_s is small enough to be implemented on-chip; however, L_g is too high to be implemented on-chip without a drastic increase in the NF of the circuit.

Plotting the NF of the LNA versus its gate inductor Q, the NF decreases from about 5.2 to 2.1 dB as the Q of the gate inductor is increased from 10 to 200. However, on designing, fabricating, and measuring different topologies for the gate inductance (on the organic substrate mentioned previously), it is found that its size increased from 9 mm² for a Q of 110 to 28 mm² for a Q of 170. Since the NF of the LNA is not affected by an increase in Q_g beyond 70 to 90 and since size constraints limit the packaged device to an area of 3.5 mm × 3.5 mm, the inductor that provides the optimum Q for a minimum size is chosen.

Figure 4.34 shows the layout of the LNA, along with the gain and NF numbers. The embedded inductor has a two-loop Coplanar Waveguide (CPW) topology (ground on same layer as inductor), occupies ~9 mm² of area, and has a Q of 110.

4.3.2 Concurrent Oscillator Design

Many embodiments of LCP-based single-band oscillators with low phase noise and low power consumption have been demonstrated in [45, 49–50]. However, next-generation wireless communication radios are required to be frequency agile and need to seamlessly span multiple bands of frequencies to cover different standards globally [51]. Such a concurrent multiband (or multimode) system provides convergence of many standards and enables users to communicate around the globe using only one portable device. Current solutions to the multiband systems are not truly concurrent, in the sense that an electronic switch is used to alternate between many frequencies. Such radio

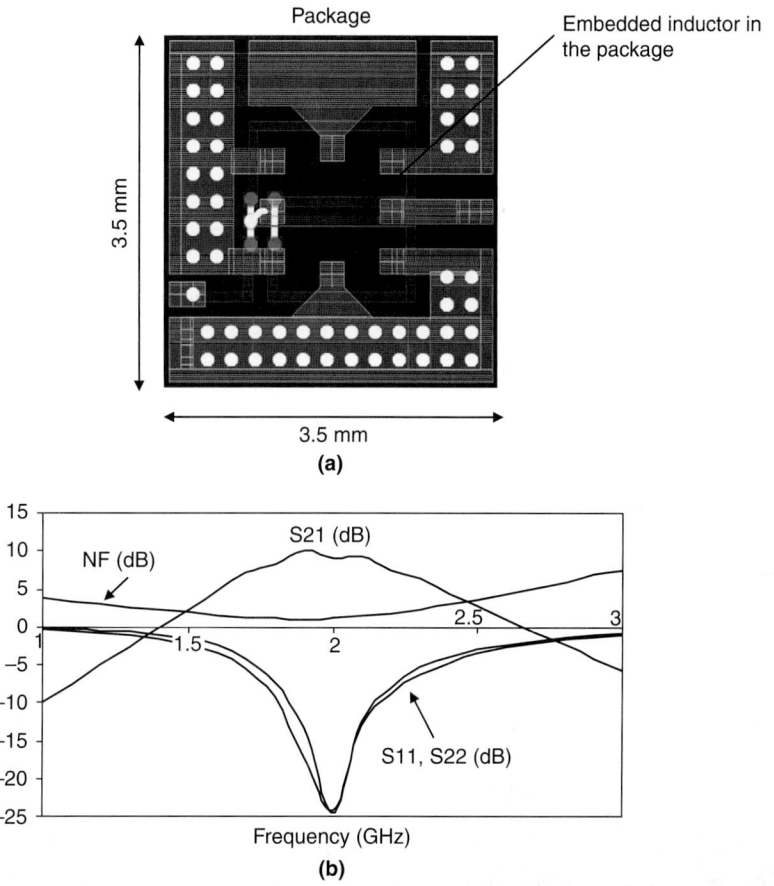

Figure 4.34 (*a*) LNA with an embedded inductor in package. (*b*) Response for GSM application.

architectures require multiple processing blocks that in addition to requiring more silicon real estate place an enormous pressure on the system power budget. Radio architectures of the future, in particular those that are cellular based, would involve more diverse communication standards such as full-duplex communication standards (UMTS, GSM) with Bluetooth/UWB, mixed with half-duplex standards such as WLAN and GPS. Hence, future radio architectures would be required to be multimode with the ability to transmit and receive information over at least two radio channels simultaneously [51–52]. References [5], [52] and [53] discuss the architecture and advantages of concurrent multiband radios. The basic principle behind both the architectures suggested in [5], [52] and [53] is the concurrent multiband operation of each RF front-end block. Hence, multimode transceivers would require innovation at both the circuit level (to minimize power consumption and size) and at the package level (to reduce cost without compromising performance).

One of the key building blocks of a multimode transceiver is a multimode oscillator that is required for both up- and down-conversion. Such a dual-frequency oscillator has its resonant circuit synchronously tuned to two different frequencies. Hence, instead of mechanically switching or discretely tuning the tank circuit, the dual-frequency resonance

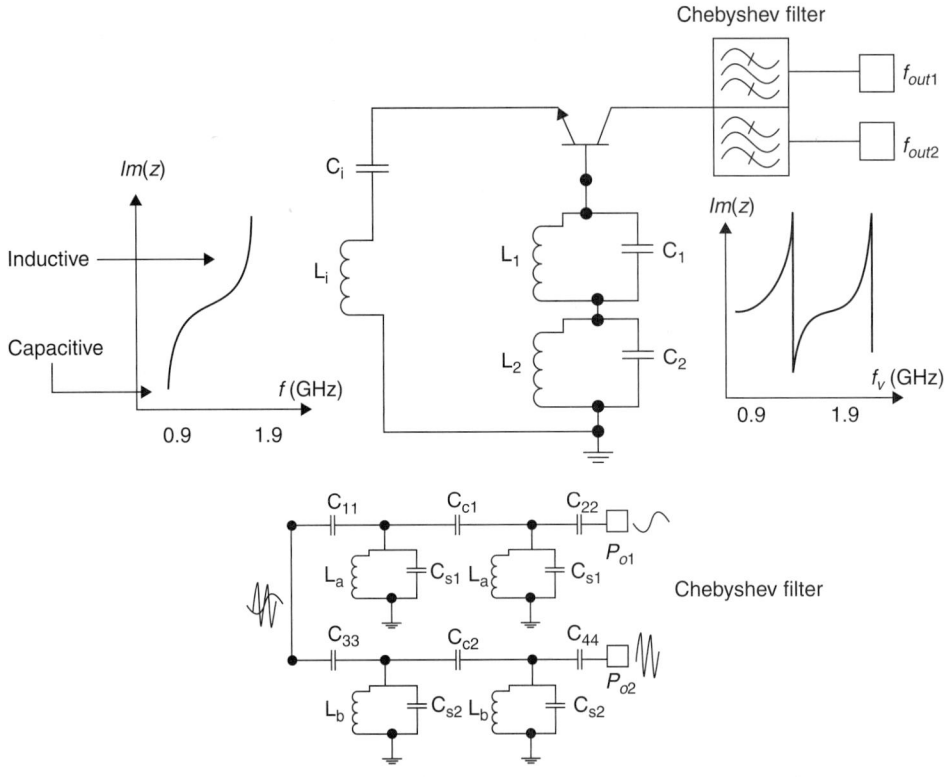

FIGURE 4.35 Concurrent oscillator circuit that generates 0.9- and 1.8-GHz signals.

enables the oscillator to function simultaneously at 1.79 GHz and 900 MHz. The oscillator prototype utilizes high quality factor (Q) lumped-element passive components that are embedded in an organic packaging technology, namely, liquid crystalline polymer.

Figure 4.35 shows the circuit schematic of a possible dual-frequency oscillator. This section presents a method of generating two frequencies simultaneously, similar to the theory presented in [54]. The circuit is a common-base type negative resistance oscillator. The oscillator essentially consists of four components, (1) the fourth-order resonator, consisting of L_1C_1 and L_2C_2, connected at the base terminal of the transistor (base resonator), (2) the second-order series LC resonator (input resonator), consisting of L_i and C_i, at the emitter terminal of the transistor, (3) the output filtering network, and (4) the transistor. Biasing circuitry and other parasitic components are not shown for the sake of circuit schematic clarity.

The concept of negative resistance single-frequency oscillator design, as discussed in [53], is extended to two frequencies to design the dual-frequency oscillator. Depending on the value of the inductance at the base terminal and the load impedance at the collector of the transistor, an effective negative resistance (Z_{in}) is observed at the emitter terminal of the transistor [53]. For sustained oscillations the $S_{in} \cdot \Gamma_r$ product should be greater than 1,

$$S_{in} \cdot \Gamma_r \geq 1 \angle 0° \tag{4.11}$$

Component	Value	Q
L_1	12.5 nH	60 @ 0.9 GHz
L_2	8 nH	57 @ 0.9 GHz
L_i	7 nH	62 @ 0.9 GHz
C_1	0.22 pF	260 @ 1.8 GHz
C_2	2 pF	247 @ 1.8 GHz
C_i	2 pF	253 @ 1.8 GHz

TABLE 4.5 Component Parameters for Oscillator

In Equation (4.11) S_{in} is the reflection coefficient corresponding to the input impedance Z_{in} and Γ_r is the reflection coefficient of the input resonator. The basic idea is to generate instability at two frequencies so that the oscillator circuitry can be adjusted to satisfy Equation (4.11) at the two frequencies. Circuit simulations show that an inductance of 29 nH at 900 MHz and an inductance of 19 nH at 1.8 GHz is required at the base terminal to generate instability at 900 MHz and 1.8 GHz, respectively. For sustained oscillations the load at the collector should also be frequency dependent [46]. Hence, a dual-band Chebychev bandpass filter is used as the matching network at the collector, as shown in Figure 4.35. The filters are designed to match the core of the oscillator at their designed center frequency while providing very low insertion loss (~1.5 dB). The input resonator is designed to provide capacitive reactance at 900 MHz and inductive reactance at 1.8 GHz. The reactance of the input resonator cancels the reactance observed at the emitter terminal thereby satisfying the oscillation conditions in Equation (4.11).

Prior to design, it is important to estimate the component values required and their corresponding unloaded quality factors that provide the necessary performance. This can be evaluated using any RF circuit simulator such as Advanced Design System (ADS) from Agilent Technologies or Spectre from Cadence. The results are tabulated in Table 4.5, for the VCO, and Table 4.6 shows the component values used in the design of the Chebychev filter.

Component	Value	Q
$C_{11}, C_{22}, C_{33}, C_{44}$	0.6 pF	260 @ 0.9 GHz
C_{c1}	0.18 pF	269 @ 0.9 GHz
C_{c2}	0.23 pF	262 @ 0.9 GHz
C_{s1}	1.5 pF	245 @ 0.9 GHz
C_{s2}	1 pF	247 @ 1.8 GHz
L_a	13 nH	59 @ 0.9 GHz
L_b	4.2 nH	79 @ 1.8 GHz

TABLE 4.6 Component Parameters for Chebyshev Filters

188 Chapter Four

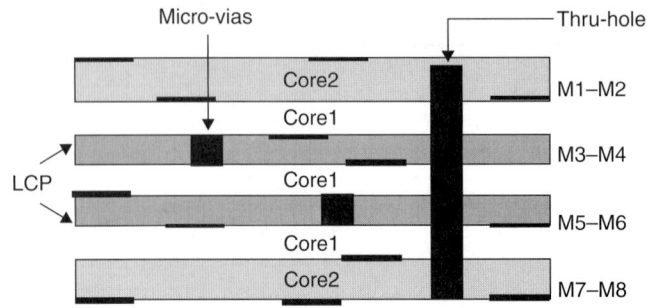

FIGURE 4.36 Eight-metal package cross section containing inductors and capacitors.

In Table 4.5, the unloaded Q values at the 0.9- and 1.8-GHz frequencies are provided, which leads to the overall base resonator Q of 48 at 0.9 GHz and 36 at 1.8 GHz, respectively.

Figure 4.36 shows the cross section of the multiple LCP layer packaging technology used to implement the VCO. This process combines two diclad LCP layers with multiple, low-loss tangent glass-reinforced organic prepreg (core) layers resulting in a multilayer stackup. In total there are eight metal layers with the bottom-most metal layer used as a microstrip-type ground reference. Each diclad LCP layer is 25 µm thick with ½ ounce copper. The LCP dielectric layer has a dielectric constant of 2.95 and a loss tangent of 0.002. The low-loss tangent and thick metal results in high-Q inductors ($Q > 100$) and capacitors ($Q > 200$). Additionally, microvias and buried vias with diameters <100 µm have been used in the design. The layout of the design is shown in Figure 4.37, where each metal layer is shown in a different color. With more components being embedded

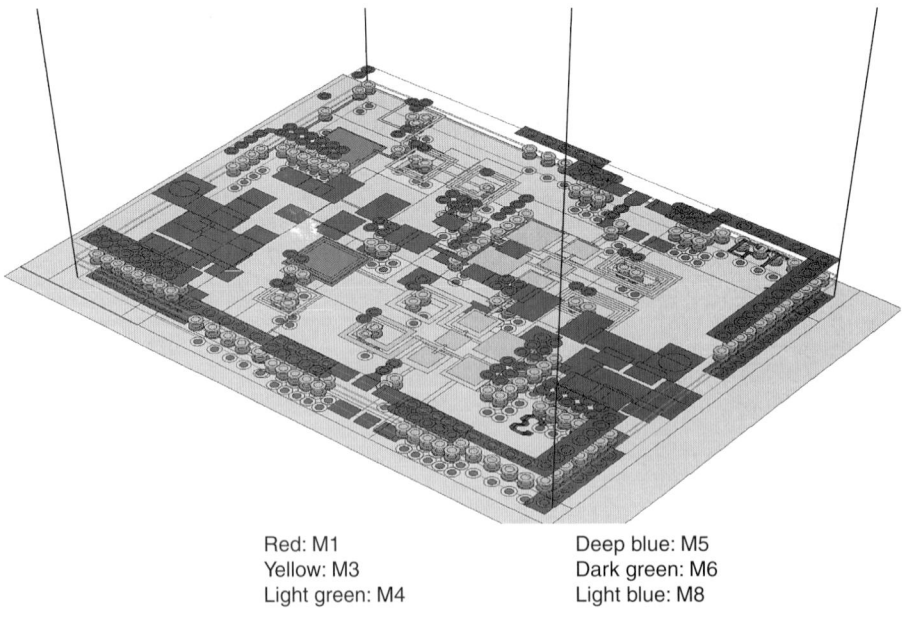

Red: M1
Yellow: M3
Light green: M4

Deep blue: M5
Dark green: M6
Light blue: M8

FIGURE 4.37 Layout of VCO.

as in Figure 4.37, detailed electromagnetic simulations become very time consuming and hence circuit simulators become necessary.

The circuit is powered using two bench-top dc power supplies providing an emitter voltage of –1.5 V and a collector supply of +1 V. Additional filtering of the power supply noise is provided through microwave bias-tees and surface-mount decoupling capacitors. The dual-frequency oscillator has been designed using a silicon bipolar transistor from Agilent Technologies; a SOT-343 packaged bipolar transistor (HBFP 0420).

Under steady-state, the oscillator was biased at 10 mA from the 2.5-V supply. Figure 4.38b shows the measured results at the 900-MHz port. The 900-MHz signal has

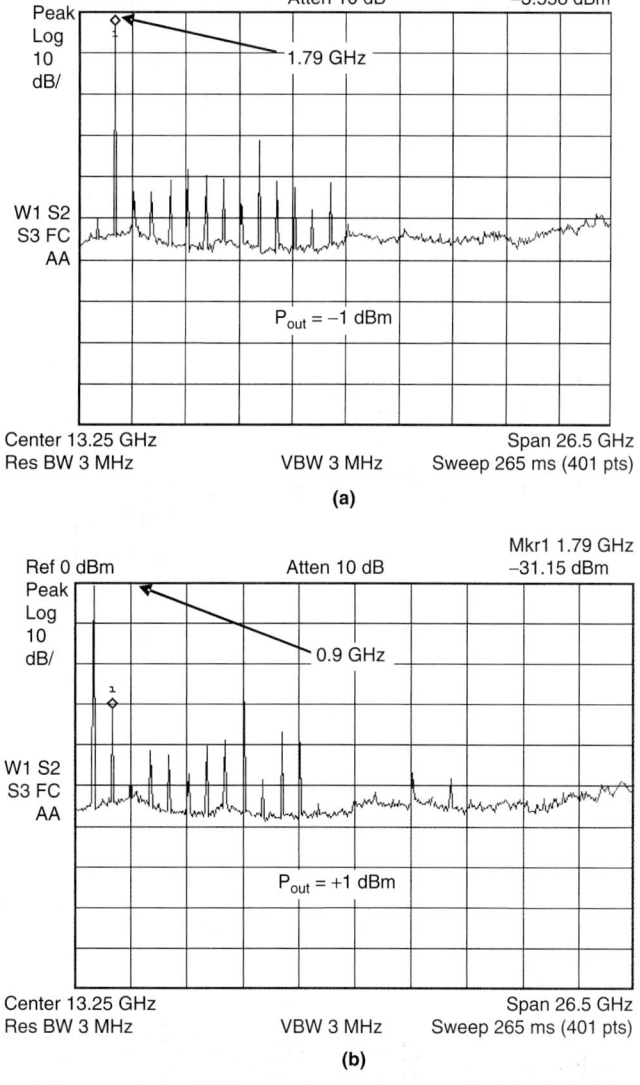

FIGURE 4.38 (a) A 1.8-GHz channel response. (b) A 0.9-GHz channel response.

an output power of +1 dBm after de-embedding the 2-dB loss of the cable. It can be observed that all the higher-order harmonics have been attenuated by more than 30 dB. The 1.8-GHz signal has been measured simultaneously with the 900-MHz signal as shown in Figure 4.38a. This signal has an output power of −1 dBm. At the 1.8-GHz port, the 900-MHz signal is attenuated by at least 50 dB, as shown in Figure 4.38a. The phase noise measurements are performed at only one port at a time to minimize the frequency pulling of the oscillator during phase noise measurements. Both the signals measure a phase noise of ∼−120 dBc/Hz at a 1-MHz offset. All the measurements here have been made using a E4407B spectrum analyzer and a 8594E spectrum analyzer. Each of the two second-order Chebychev filters have adjustable bandwidths and are designed to provide a rejection of at least 30 dB at the other center frequency. Bandpass filters have been used as matching networks to provide harmonic rejection (frequency domain) or to provide a clean time-domain response. Thus the filters, in addition to harmonic rejection, clean up the spectrum, which helps in obtaining a clean and accurate phase noise measurement.

Figure 4.39 shows the photograph of the fabricated prototype of the dual-frequency oscillator with inductors and capacitors embedded in the package and three surface-mount components on the package (transistor and two bypass capacitors). The size of the fabricated VCO measures 10 mm × 14 mm. A similar design with inductors and capacitors integrated in the chip would suffer heavy metal and dielectric losses, leading to reduced performance and higher power.

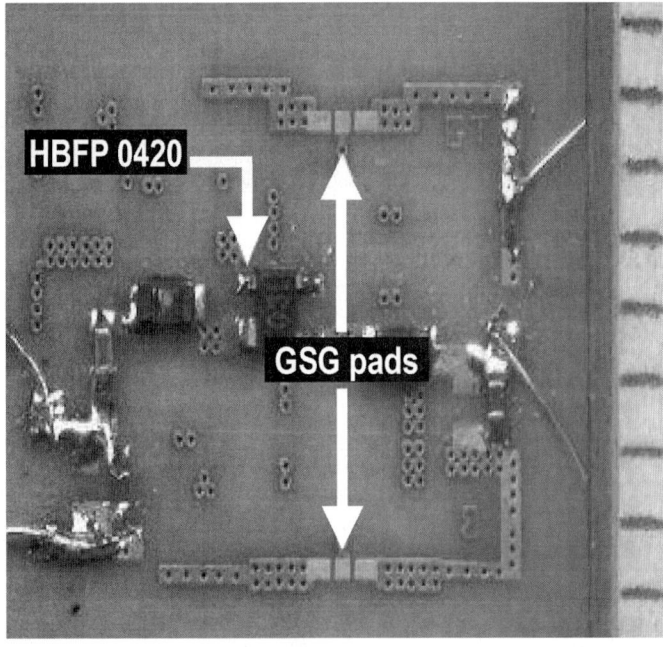

Figure 4.39 Fabricated concurrent oscillator with L and C in the substrate. Only the transistor and two decoupling capacitors are surface mounted.

4.4 Design of WLAN Front-End Module

Based on the aforementioned discussion on individual RF modules on LCP substrate, complete modules for use in RF front-end can be economically and systematically designed. Figure 4.40 shows a 1 × 1 (one transmit and one receive) building block for a dual-band WLAN Multple Input and Multiple Output (MIMO) front end. The integration of this functionality into a front-end module is a critical enabler for designing MIMO in the new, very small PCI Express Mini card form factor. The challenge is to integrate all this functionality into a single module. The Front End Module (FEM) incorporates two power amplifiers in the transmit paths and two low-noise amplifiers in the receive paths, two additional Transmit (Tx) filters baluns, two Receive (Rx) filter-baluns, two diplexers, and one Double Pole Double Throw (DPDT) in 64 square millimeters.

The front-end module approach allows the design to be optimized in several ways. Matching network losses are minimized because in the module interior the designer is not restricted to the 50-Ω impedances used for the I/O terminals. Further performance enhancements can be achieved because the designer has complete control of the precise location of the passives and actives and therefore can take into account all coupling and parasitics in the model.

This level of integration presents two conflicting design challenges. On the one hand the desire for compact size requires unprecedented passive component density.

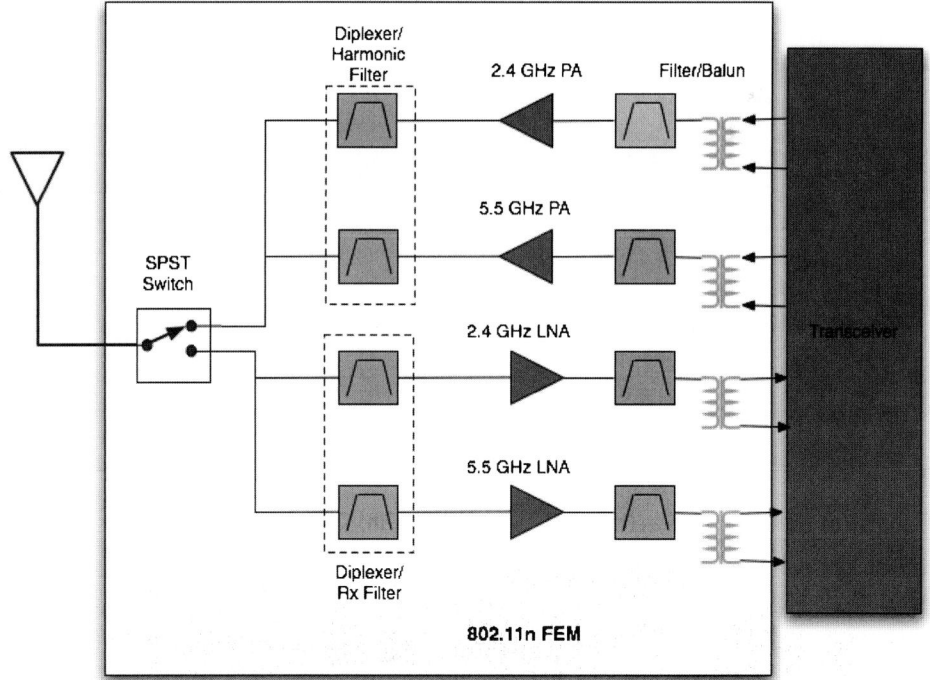

FIGURE 4.40 WLAN MIMO front-end building block.

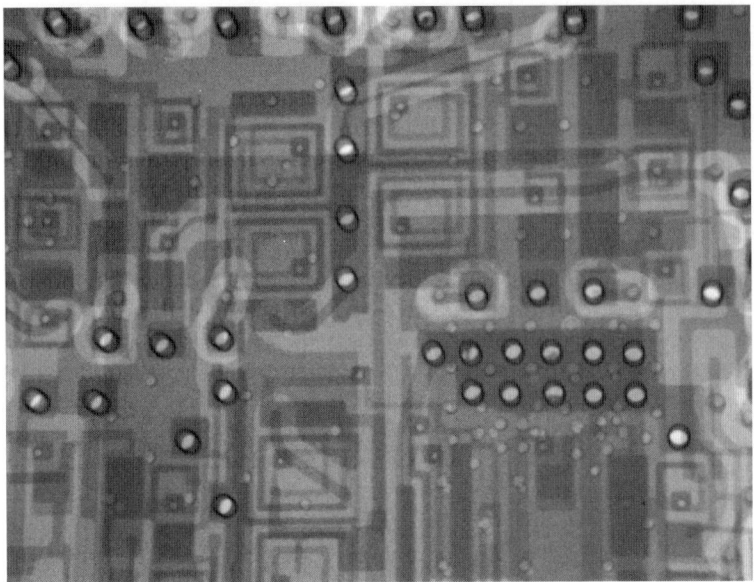

FIGURE 4.41 X-ray of the substrate.

On the other hand the close proximity of these components requires careful design to control crosstalk and coupling. The x-ray image in Figure 4.41 shows some of the 97 embedded inductors and capacitors used to implement the diplexers, filters, and baluns in Figure 4.40. Copper-plated vias have been used to provide a thermal path for the power amplifier heat.

The diplexers, filters, and baluns are embedded in the inner layers of the multilayer organic substrate. The PA, LNA, and switch die are attached on the substrate along with a handful of discretes for decoupling and biasing. This module is less than one-quarter of the size of the discrete design it replaced.

Figure 4.42 shows the assembled module before the overmold is applied. The active devices (dual-band LNA, switch, and PA) are developed by Anadigics, Inc. The module has been designed to achieve an optimum overall performance. In the 2.4- and 5-GHz receive modes the module has a total gain of 15 and 12 dB, respectively. The Rx filtering is split into a preselection stage in front of the LNAs and a high-rejection stage located after the LNAs. The preselection, which is an integral part of the Rx diplexer, is optimized for low loss, such that low receiver NF and high sensitivity is ensured. For the same reasons two low-noise pHEMT LNAs have been used, so that the overall noise figure is better than 3 dB for 2.4-GHz band and better than 3.5 dB for the 5-GHz band. The Tx filters, located in front of the dual-band PA, reject Local Oscillator (LO) spurious signals of the RFIC, while after the PA output, low-loss bandpass filters are located that will reject the harmonics of the Tx signal and the spurs. This approach of minimizing insertion loss and rejection after the amplifiers provides the highest possible output power and current consumption, which is especially important for portable devices.

Mixed-Signal (SOP) Design

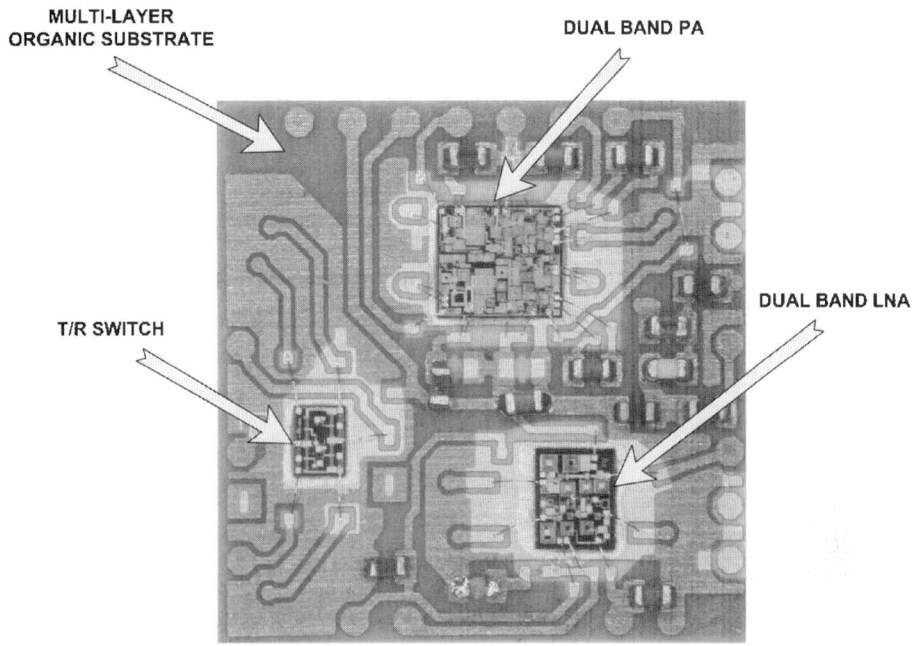

FIGURE 4.42 Assembled MIMO 1 × 1 FEM.

Figure 4.43 shows a cross section of the completed module. The lower portion is the substrate and the upper portion is an epoxy overmold. The substrate has a total of six metal layers. The inductors and capacitors are implemented on a diclad LCP sheet and form the two inner metal layers. Two additional layers are used for grounds that simplify application by shielding critical passives from coupling. The remaining two layers are used for I/O, routing, and die connection on the top layer.

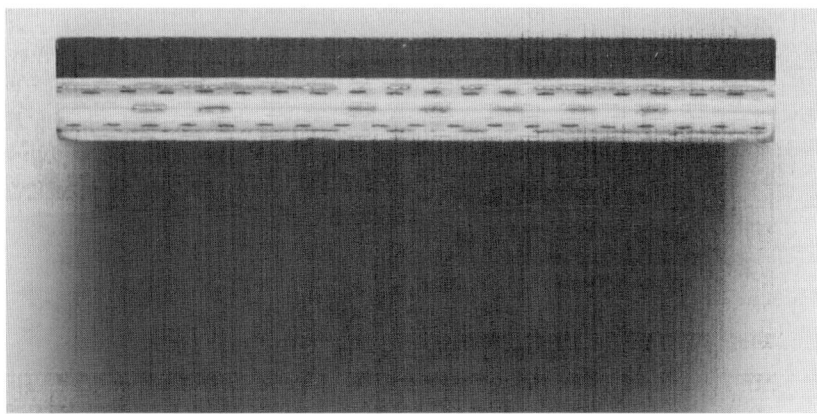

FIGURE 4.43 Module cross section.

4.5 Design Tools

Chip-package codesign requires the simultaneous design of the chip and package that meets specifications with minimum size. This includes system partitioning followed by simulations at three levels of abstraction, namely, the behavioral level, circuit level, and layout level, as shown in Figure 4.44. The electromagnetic analysis at the layout level enables parasitic extraction, which can be used to construct models for simulation at the circuit level. The results of the circuit simulation can be used to construct high-level models for behavioral simulation of the system. Levels 1, 2, and 3 need to exchange information with each other for influencing decisions at each level. For example, the interconnect coupling extracted in Level 3 may result in excessive noise in Level 2, which needs to be used to change the layout in Level 3. Using the three levels of design in Figure 4.44, either a top-down or bottom-up approach can be used for designing

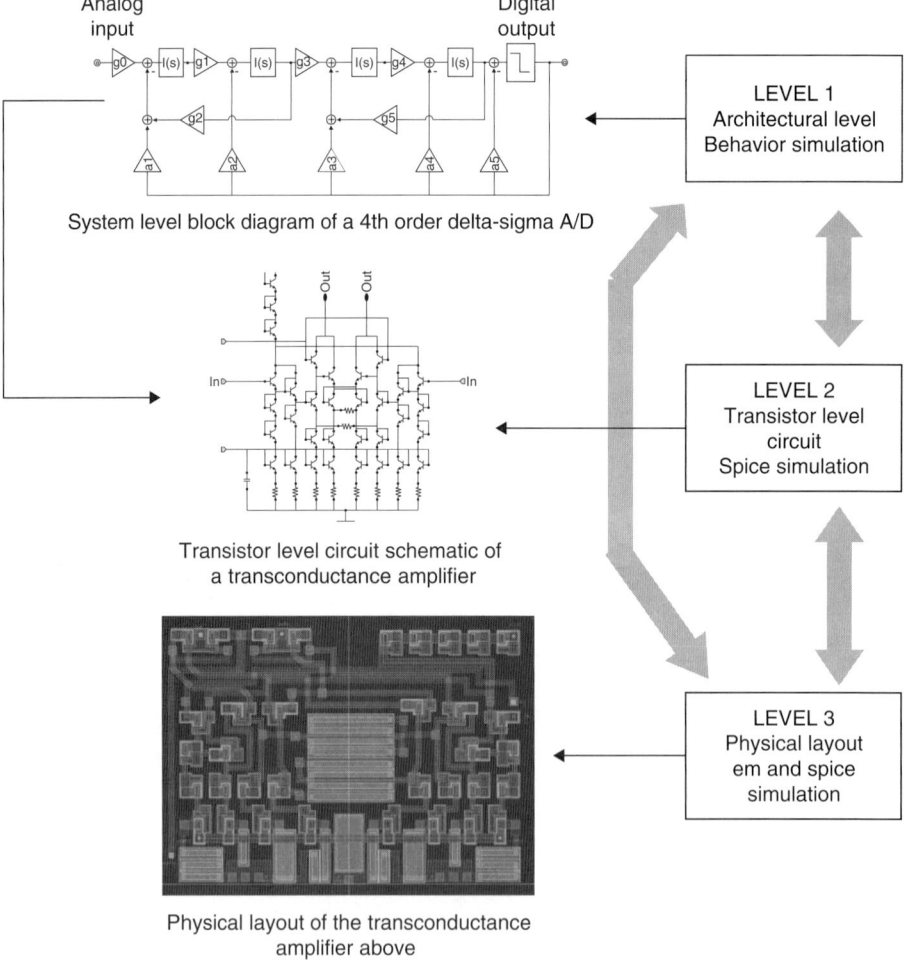

Figure 4.44 Hierarchical design flow.

complex circuits. For design cycle time reduction, an automated top-down approach is preferred where electrical specifications can generate layouts automatically. However, with RF and mixed-signal circuits where a change of 0.1 dB or a frequency shift of 10 MHz can make the circuits unusable, the use of only a top-down approach will not suffice and is also not practical. Therefore, the layouts generated in Level 3 need to be modified and the information passed back to Levels 2 and 1 to ensure that the design meets specifications. Hence, a continuous feedback loop is required between the three levels. This translates into design cycle time, and oftentimes fast simulators are required to ensure that the time required for the iterative process can be minimized. Electronic design automation (EDA) tools that enable both a top-down and bottoms-up approach for the design of SOP-based systems are described in the following sections.

4.5.1 Circuit Sizing of Embedded RF Circuits

Circuit sizing is the process of extracting network- and layout-level parameters for a component or circuit from electrical specifications. It is common in digital designs and is being increasingly used in low-frequency analog circuits. The main reason for this is the scalability of design cells that allows an automated hierarchical design flow. RF designs, however, lack this scalability due to the effects of layout-level parasitics and coupling on circuit performance. A conventional design flow tries to optimize circuit performance at the layout level at the premium of time-consuming electromagnetic iterations for entire circuit layouts. In contrast, a sizing approach can be used to extract physical dimensions of the layout from the electrical specifications by using intermediate steps consisting of circuit-level models, optimization methods, and mapping. In the following section, we consider the design of inductors and filters using such an approach.

Sizing of Inductor Layouts

A. Use of Artificial Neural Networks—Forward Mapping Artificial neural networks (ANNs) have emerged as a powerful alternative to numerical and analytical modeling techniques. A typical neural network consists of weighting functions that are adjusted during training to enable it to map highly nonlinear input-output relationships through a combination of activation states of the neural layers [55]. Multilayer feed-forward perceptrons can be used to implement the neuromodels for RF circuits. In forward mapping, the geometrical parameters of an inductor are mapped to the electrical parameters. This requires a coarse library of inductors with stepwise variations in layout parameters, which can be generated using any electromagnetic simulator. The generated data are used for training and testing the neural network. An initial sensitivity analysis can be performed to identify the dominant geometrical parameters affecting the inductance parameters (L, Q, and self resonance frequency (SRF) at any frequency), which are the typical electrical parameters for an inductor. The majority of the data (up to 80 percent) can be used for training the neural network, while the remaining data can be used for model validation. The number of hidden layers in the neural network structure is adjusted in order to have moderate training as well as limit validation errors. Based on the accuracy of component values required by the design specifications, the ANN-based model can be sampled to further populate the coarse data using interpolation methods. This is possible due to the piecewise monotonicity of the data obtained for inductors for parameters such as L, Q, and SRF as a function of the dominant geometrical parameters [55].

B. Reverse Mapping   Reverse mapping is the process of mapping the electrical parameters to the geometrical parameters. Trained and validated neuromodels can be employed to extract the geometrical parameters that meet the area constraints and maximize the inductor parameters such as quality factor. During reverse mapping, the geometrical parameters are extracted from electrical specifications while satisfying the design constraints. This solution is nonunique as multiple combinations of layout parameters of the inductor can correspond to the same inductance value (with a different Q factor, and SRF). A knowledge base of equations or constraints for the inductor parameters can be included in the model to allow for fast convergence and generate unique solutions in such a multivariable optimization problem. The geometries synthesized by reverse mapping may not be feasible for fabrication. In such a case, the design values are rounded off to the closest processable physical dimensions in accordance with process ground rules. This can be enabled by enforcing constraints on the physical dimensions during synthesis.

As an example consider the inductor in Figure 4.6. The physical dimensions that dictate the electrical response of the inductor are line width, line spacing, and number of turns. The distance to the ground plane is fixed since it depends on the stackup that in general has a fixed thickness. The equivalent circuit for the inductor with parasitic elements is shown in Figure 4.7. The circuit parameters can be extracted using an optimizer that is available in most RF circuit simulators using the frequency response as the input. The frequency response can be generated using an electromagnetic solver. Since, the equivalent circuit parameters in Figure 4.7 are frequency independent, a mapping can be created both in the forward and reverse directions between the variable physical dimensions in Figure 4.6 and circuit parameters in Figure 4.7. The forward mapping generates a functional relationship between the physical parameters (independent variable) as input and the circuit parameters (dependent variable) as output. This is reversed during the reverse mapping. If the data are monotonic, artificial neural networks are a good way to establish this relationship. A functional relationship can now be established between the circuit parameters in Figure 4.7 and the inductor specifications such as L and Q (SRF is rarely used as a specification as long as it is two times the frequency at which the inductor is used). This approach can also be used for capacitors. The synthesis method described can be used to generate inductor and capacitor libraries. This captures the three-level scheme described in Figure 4.44, where Level 3 has the inductor layout. The parasitics of this layout get mapped to the circuit components in Level 2 from which the Level 1 parameters inductance L and Q can be derived.

Scaling of Filter Layouts

A. Forward Mapping Based on Lumped-Circuit Models and Polynomial Fitting   For the layout scaling of RF circuits with reduced design cycle time, it is important to extract accurate lumped-element models that capture the physical effects of layout. For example, in the scaling of an RF bandpass filter, an extensive lumped-circuit model can be generated based on segmentation of a physical layout, as shown in Figure 4.45. The layout is decomposed into circuit sections that are isolated from each other, under the assumption that the coupling is weak. The dotted lines represent the segmented sections. For example, the L resonators are segmented into the coupled inductor section L_cp and uncoupled inductor sections $L1$ and $L2$. This technique allows separate scaling and mapping of geometrical sections that have little electromagnetic interaction between

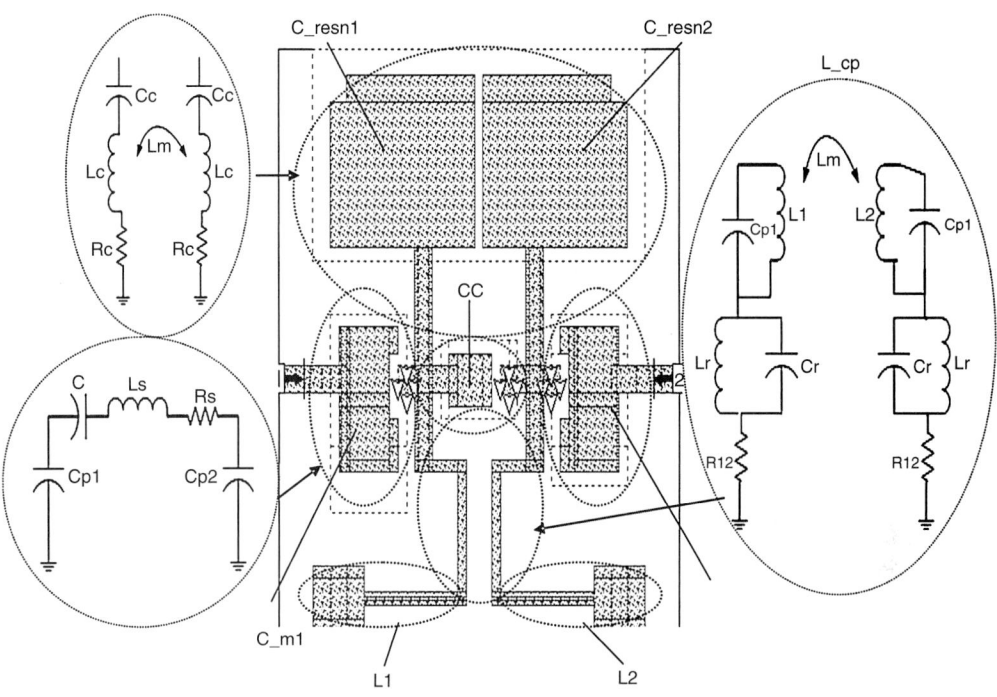

FIGURE 4.45 Segmentation of filter layout into lumped-element models.

them. Based on the two-port and one-port modeling of the sections using an electromagnetic solver, lumped-circuit models that include the effect of parasitics and coupling are developed. Because of the use of segmented models, fast optimization at the circuit level is possible to meet design specifications without losing the effects of physical layout on circuit performance.

The forward mapping is best explained with the help of Figure 4.45. In this figure, the uncoupled inductor section in the lower half of the right lumped inductor model (parallel L_r and C_r in series with R_{12}) can be mapped to the inductor geometry as

$$L_r = -0.0024(\Delta L)^3 + 0.0273(\Delta L)^2 + 0.0674(\Delta L) + 0.8104 \quad (4.12)$$

$$C_r = -0.0009(\Delta L)^3 + 0.0051(\Delta L)^2 - 0.0009(\Delta L) + 0.023 \quad (4.13)$$

$$R_{12} = 0.0007(\Delta L)^3 + 0.111(\Delta L)^2 + 0.1082(\Delta L) + 0.0942 \quad (4.14)$$

where ΔL is the increase in the inductor length of L_1 and L_2. For $\Delta L = 0$, the inductance is ~0.8 nH.

B. Reverse-Mapping Using Polynomial Fitting The scalable lumped-element component models can be combined to perform filter circuit optimization using Agilent's Advanced Design System. At each stage of the optimization process, the desired components are tuned and the corresponding polynomial-mapped geometries and parasitics are updated. At the end of the optimization process, the variable geometries of the components are extracted from the component values of the models using the reverse-mapping functions. As an example the length (ΔL) and spacing (ΔS) of the inductors as well as the width (ΔW) of the capacitors, illustrated in the previous numerical example, is reverse-mapped from the component parameters as shown below

$$\Delta L = 0.039(L_r)^3 + 0.982(L_r)^2 - 0.0674(L_r) + 0.6104 \qquad (4.15)$$

$$\Delta S = 0.0231(C_r)^3 + 0.051(C_r)^2 - 0.0012(C_r) + 0.032 \qquad (4.16)$$

$$\Delta W = -0.0009(k)^3 + 0.351(k)^2 - 0.013(k) + 0.0123 \qquad (4.17)$$

An example of design scaling for a 5.5-GHz bandpass filter is shown in Figure 4.46. In the figure, a reference design is used to develop accurate models. The physical dimensions are then scaled using Equations (4.12) to (4.17) to generate the forward- and reverse-mapping functions. Two examples of scaled designs are shown in Figure 4.46. The scaled designs have been compared to the results obtained from an electromagnetic simulator by modeling the layout. The agreement in the results indicates that the mapping method described can be used for certain topologies to scale layouts from electrical parameters.

4.5.2 Modeling of Signal and Power Delivery Networks

In wireless communication modules, all of the signal processing is done digitally. This is accomplished by digitizing the data through an analog-to-digital converter in the receiver and a digital-to-analog converter in the transmitter. Oftentimes, a common power distribution network (PDN) or a split power distribution network is used to simultaneously power the digital and RF circuits. When the digital circuits switch simultaneously, noise generated on the power distribution can propagate from the

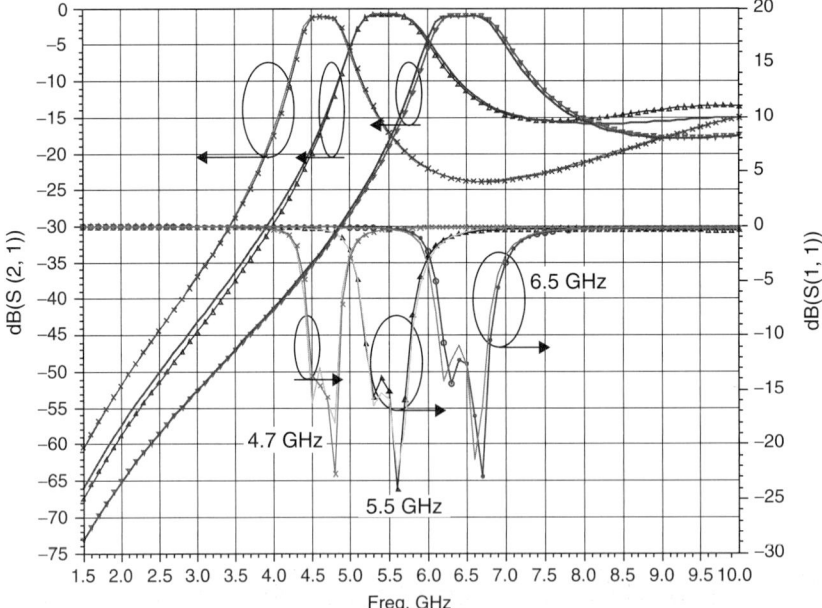

FIGURE 4.46 Correlation between full-wave simulator and scaling. Scaling results were generated from a reference design for a 5.5-GHz filter.

digital to the RF part of the system, either directly through the PDN or through the gap between the split islands. This noise can also couple onto the digital signal lines, causing interference in addition to crosstalk and reflections. It is therefore necessary to simulate the digital signal lines in the presence of power and ground planes. In this section, the modeling of signal and power delivery networks is discussed in the context of mixed-signal modules (such as a mobile wireless communication unit), microprocessors, and other high-speed communication links. A detailed description regarding the design and analysis of power delivery networks can be found in [55a].

Figure 4.47 shows the flowchart of a simulation methodology for a combined signal integrity (SI) – power integrity (PI) simulation. The signal distribution network (SDN) and the power delivery network (PDN) can be modeled separately and combined using the modal decomposition method.

Based on the stackup and the proximity of the power and ground planes, the signal lines can be classified into coplanar, microstrip, and stripline geometries. In all three cases, the coupling between the signal lines and power-ground planes can be analyzed by separating these structures into individual parts, analyzing them separately, and then recombining them to obtain the overall response. This is possible through the modal decomposition technique [55b]. This approach enables the computation of the interaction between SDN and PDN and their effect on signal propagation and noise. The modal decomposition method is described in this section for a stripline interconnection between a power-ground plane, as shown in Figure 4.48. This approach can be generalized for all interconnection structures.

A stripline interconnection at a distance h_2 from V_{dd} and h_1 from Ground can be represented using the equivalent circuit shown in Figure 4.48b. In the figure, the parallel plate waveguide mode propagates between the power-ground plane while the stripline

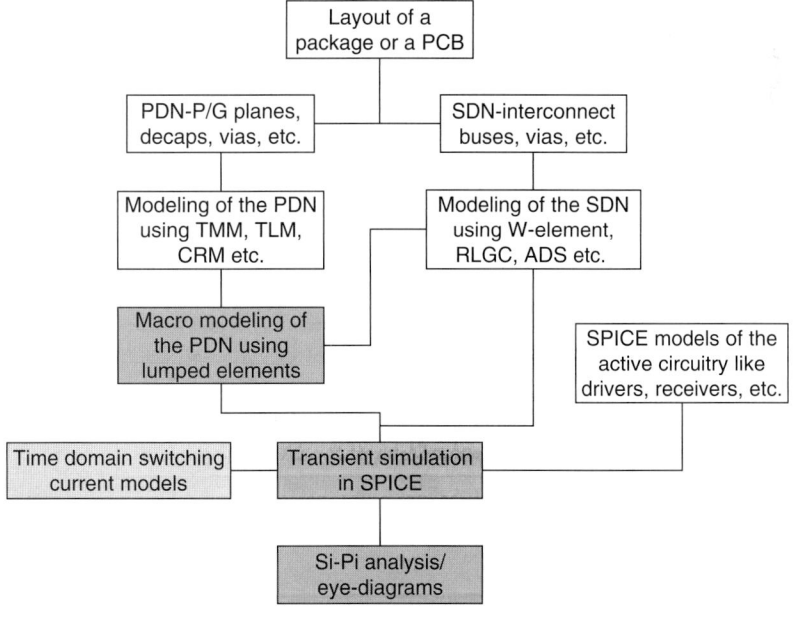

FIGURE 4.47 Flowchart for combined SI-PI simulation.

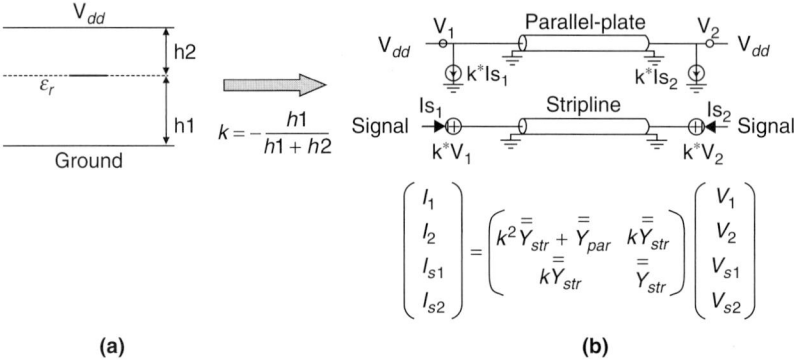

FIGURE 4.48 Modal decomposition of stripline. (a) Cross section. (b) Equivalent circuit.

mode propagates in the direction of the interconnection (assuming ideal planes). The two modes can be combined to obtain the total voltages and currents through voltage and current sources using the coefficient k. One way of representing this structure is by using the admittance parameters, as shown in the figure. The voltages and currents can also be computed in SPICE (or any circuit simulator) by using the equivalent circuit in Figure 4.48b.

The Y parameters of the stripline mode (Y_{str} in Figure 4.48) can be extracted by using a 2D or 3D electromagnetic solver and represented as a transmission line element in SPICE. However, extraction of the frequency response of the power distribution is more difficult (Y_{par} in Figure 4.48). The unit cell (smallest repeatable pattern) for a two-plane pair (three planes) system is shown in Figure 4.49. The equivalent circuit model for the whole system can be constructed by interconnecting such unit cells. The circuit

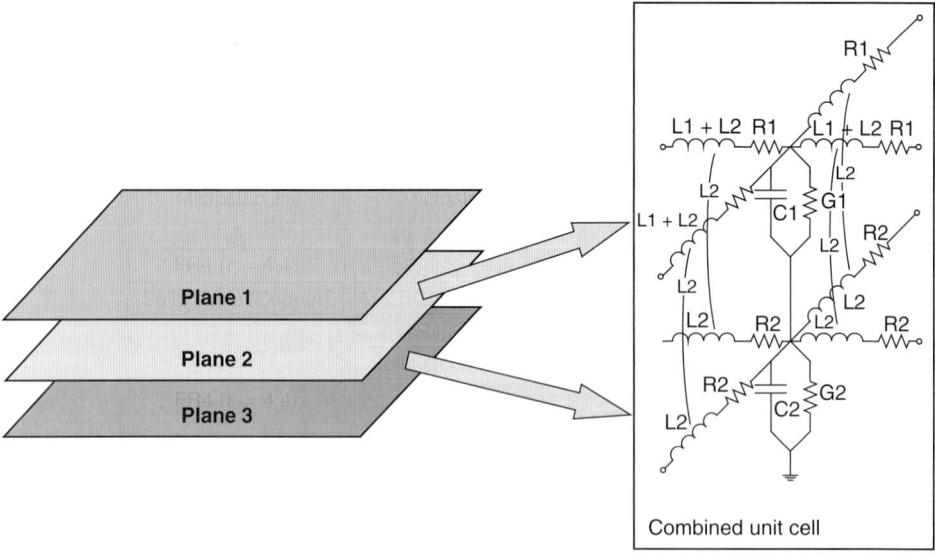

FIGURE 4.49 Circuit model for planes.

parameters can be extracted analytically from the physical dimensions. The nodal admittance matrix for a single plane pair in Figure 4.48 can be written as

$$\overline{\overline{Y}} = \begin{bmatrix} \overline{\overline{A}} & \overline{\overline{B}} & & & & \\ \overline{\overline{B}} & \overline{\overline{A}}+1/Z & \overline{\overline{B}} & & & \\ & \overline{\overline{B}} & \overline{\overline{O}} & \overline{\overline{O}} & & \\ & & \overline{\overline{O}} & \overline{\overline{O}} & \overline{\overline{B}} & \\ & & & \overline{\overline{B}} & \overline{\overline{A}}+1/Z & \overline{\overline{B}} \\ & & & & \overline{\overline{B}} & \overline{\overline{A}} \end{bmatrix} \quad \overline{\overline{A}} = \begin{bmatrix} Y+2/Z & -1/Z & & & & \\ -1/Z & Y+3/Z & -1/Z & & & \\ & -1/Z & Y+3/Z & O & & \\ & & O & O & -1/Z & \\ & & & -1/Z & Y+3/Z & -1/Z \\ & & & & -1/Z & Y+2/Z \end{bmatrix}$$

$$\overline{\overline{B}} = 1/Z \tag{4.18}$$

where Y and Z are the per-unit cell admittance and impedance, which can be obtained from the complex permittivity (ε), permeability (μ), distance between the planes (d), thickness of each plane (t), mesh length (h), and conductivity (σ) as [55c]:

$$Y = j\omega\varepsilon\frac{h^2}{d}$$

$$Z = j\omega\mu d + \frac{2}{\sigma t} + 2\sqrt{\frac{j\omega\mu}{\sigma}} \tag{4.19}$$

In case of multiple plane pairs as in Figure 4.49, Y and Z themselves are also matrices that reflect the multilayered structure.

This formulation is based on the multilayered finite-difference method (MFDM) [55c] and is very useful when analyzing planes with complex shapes and apertures. When a stripline is introduced into the three planes, as shown in Figure 4.49, modal decomposition can still be applied based on superposition. Now, the following matrix has to be added to the admittance matrix of the planes based on modal decomposition:

$$\begin{pmatrix} I_1 \\ I_2 \\ I_3 \\ I_4 \\ I_5 \\ I_6 \end{pmatrix} = \begin{pmatrix} k^2\overline{\overline{Y}}_{str} & (-k^2-k)\overline{\overline{Y}}_{str} & k\overline{\overline{Y}}_{str} \\ (-k^2-k)\overline{\overline{Y}}_{str} & (k^2+2k+1) & (-k-1)\overline{\overline{Y}}_{str} \\ k\overline{\overline{Y}}_{str} & (-1-k)\overline{\overline{Y}}_{str} & \overline{\overline{Y}}_{str} \end{pmatrix} \begin{pmatrix} V_1 \\ V_2 \\ V_3 \\ V_4 \\ V_5 \\ V_6 \end{pmatrix} \tag{4.20}$$

where the currents and voltages are defined in Figure 4.50.

As an example, consider the structure in Figure 4.51, which consists of a stripline between a power-ground plane. The bottom ground plane contains an aperture. Ports are defined at the input (port 1) and output (port 2) of the stripline with an additional

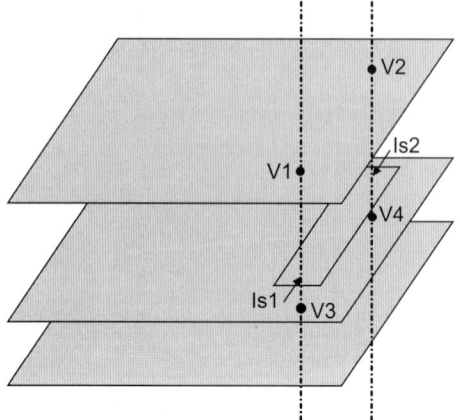

FIGURE 4.50 Stripline in an arbitrary multilayered structure.

port (port 3) defined on the bottom plane with respect to the ideal ground plane. The S parameters are computed in Figure 4.52. The results have been compared between MFDM and Sonnet (an electromagnetic solver).

The good correlation demonstrates the validity of modal decomposition. The coupling between port 1 and port 3 (coupling between the signal line and power-ground plane) is clearly captured in Figure 4.52. For complex structures, modal decomposition along with the finite-difference method provides around two to three orders of magnitude speedup in computation time compared with full-wave solvers. This concept can be extended to analyze a multitude of signal lines in the presence of multiple planes [56].

For the modeling of power-ground planes, Table 4.7 gives a qualitative comparison of some of the available methods that are based on the 2D Helmholtz equation. For planes having complicated structures due to via holes and slits, finite difference method is very efficient and can be used for multilayered structures based on MFDM. The cavity

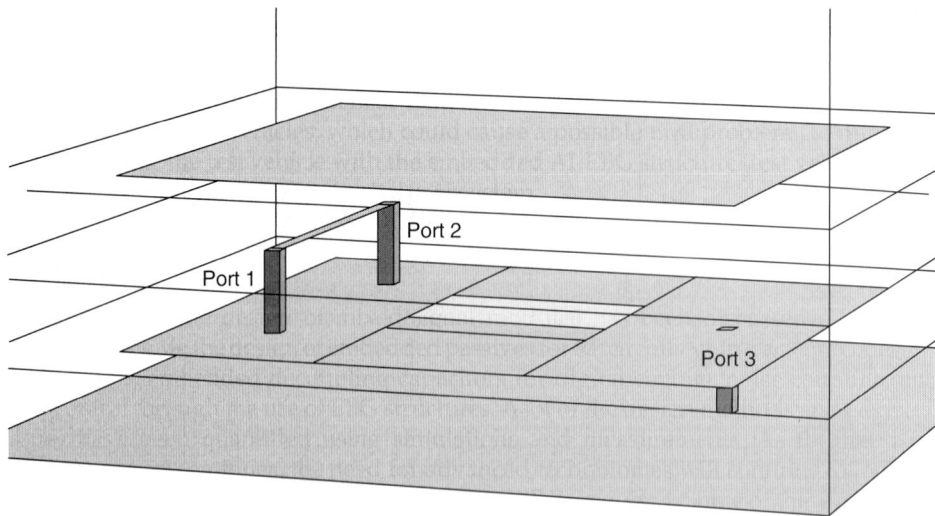

FIGURE 4.51 Test structure.

Mixed-Signal (SOP) Design

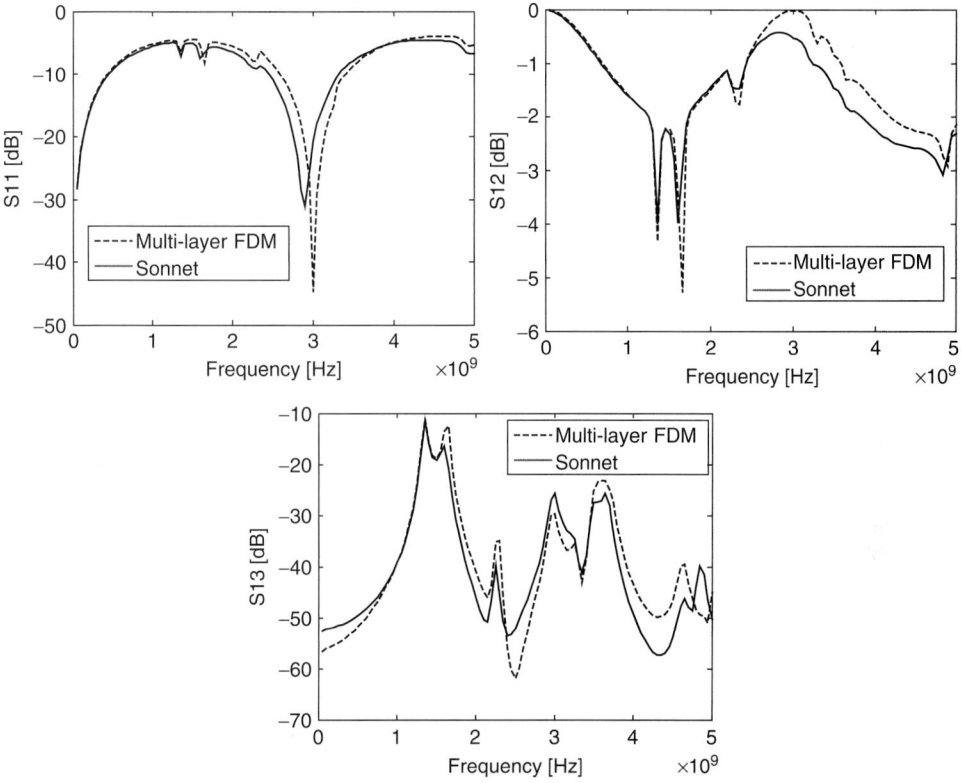

Figure 4.52 Results for test structure.

resonator method, which is based on the Green's function of rectangular planes, provides a circuit representation of the planes; therefore, it is easy to incorporate in circuit simulators. The finite-element method (FEM) allows triangular meshes; however, the fringing capacitance on the open boundaries is still being neglected, so using a triangular mesh may not provide much improvement in accuracy.

	Finite-Difference Method	Cavity Resonator	Finite Elements
Arbitrary plane geometries	Good	Bad	Very good (triangular patches)
Network representation for circuit simulations	Good	Very Good	Bad
Computation of noise voltage distribution on each node	Very good	Bad	Very good
Speed	Good	Depends on geometry	Good

Table 4.7 Comparison of Plane Modeling Methods Based on the 2D Helmholtz Equation

4.5.3 Rational Functions, Network Synthesis, and Transient Simulation

For digital signal lines, transient waveforms need to be generated from the frequency response that can be generated using the methods in the previous section. This is possible by directly simulating the matrix equations developed in the previous section. A more useful approach is to embed the extracted frequency responses in SPICE, as shown in Figure 4.47. This requires the conversion of the frequency response into a SPICE circuit, which is possible using rational functions. For interconnections, the transfer function $H(s)$ (S, Y, or Z parameters) can be represented in the form:

$$H(s) = \frac{\sum_{k=0}^{P} a_k s^k}{\sum_{l=0}^{Q} b_l s^l} \tag{4.21}$$

where $s = j\omega$, ω is the angular frequency, and a_k and b_l are coefficients. Given the frequency response $H(s)$ available at the discrete frequency points, the goal is to compute the coefficients such that the response of the rational function [right side in Equation (4.21)] matches the frequency response at the available discrete points. The coefficients can be computed by solving a matrix equation, the details of which are available in [57–58]. To ensure that the resulting rational function can result in an equivalent circuit that retains all of the properties of the original data, the stability and passivity criteria need to be satisfied [58]. Since interconnections are stable and passive, the poles resulting from Equation (4.21) should reside on the left half-plane and the resulting frequency response from the rational functions should be positive real. These conditions can be satisfied by following a number of methods that have been developed in the EDA community [58]. The resulting rational function, also called as a macromodel or black-box model, can be very useful since these functions can be used to synthesize networks that can be simulated in SPICE or any other circuit simulator.

Network Synthesis

By rewriting Equation (4.21) as a continued fraction, networks can be synthesized. The resulting networks are typically nonunique and nonphysical, meaning that the circuit models do not necessarily match the physical structure. However, the response of the equivalent circuit will match exactly the frequency response modeled.

As an example consider a one-port embedded inductor, whose magnitude and phase of impedance is shown in Figure 4.53. This response can be used to generate a rational function that can be synthesized into an equivalent circuit. Since, the frequency response in Figure 4.52 has only one resonance, the number of poles required is small (in this example, a real pole and a complex conjugate pole pair are required). In general, for embedded inductors and capacitors, the number of poles required is between three to five. Hence, the number of components in the synthesized circuit is small as well. The synthesized circuits for five different one-port embedded inductor designs are shown in Figure 4.54 [57]. In all cases, the inductor is first simulated using an electromagnetic simulator to extract the frequency response, a rational function is then generated from the frequency response and the equivalent circuit is synthesized from the frequency response. Constraints have been imposed to ensure that a relationship could be

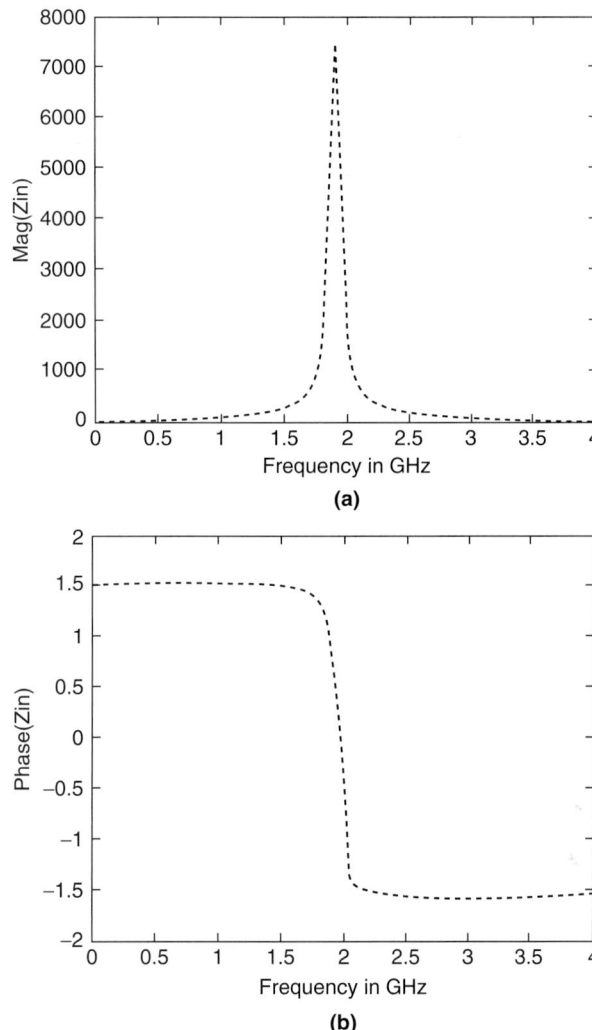

FIGURE 4.53 Frequency response and rational function approximation for one-port inductor. (a) Magnitude. (b) Phase.

established between the physical layout and the equivalent circuit. Interestingly enough, except for inductor L_4, the remaining equivalent circuits have the same circuit topology with different parasitic values. The parasitic values are consistent and scale with the dimensions in the layout.

For arbitrary interconnections containing many resonances in the frequency response, only nonphysical circuit models can be developed [58]. Once the coefficients of Equation (4.21) are computed, the function can be represented in pole-zero form. This can be expanded as low-pass, bandpass, high-pass, and all-pass filters as shown in Table 4.8. Each filter can now be represented as an equivalent circuit. All of these circuits concatenated together results in the overall network.

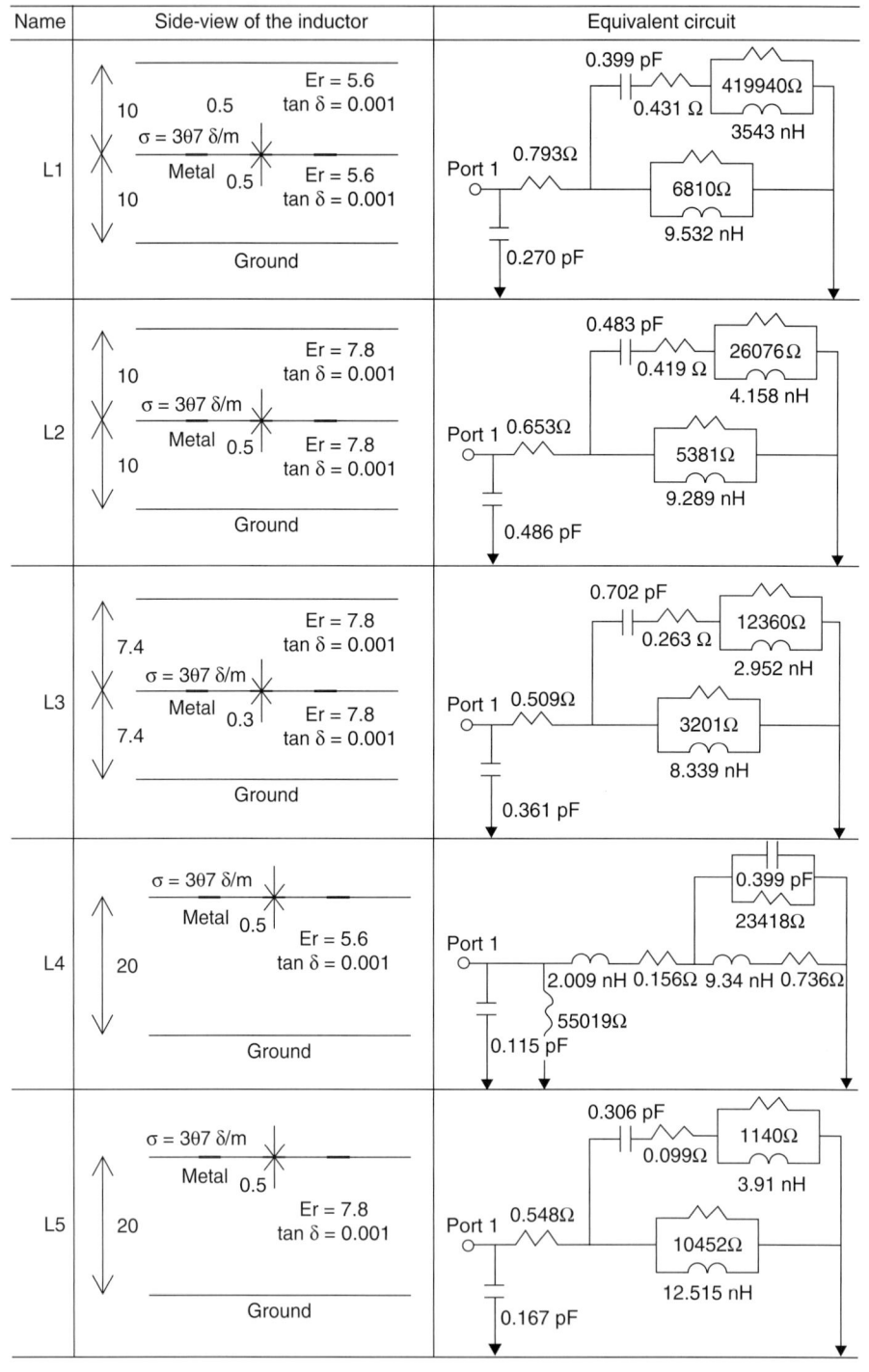

FIGURE 4.54 Networks synthesized for one-port inductors.

TABLE 4.8 Equivalent Circuit Representation of Rational Functions

Low-pass filter	Band-pass filter	High-pass filter	All-pass filter
$Y_m(s) = \dfrac{\gamma_m}{s - p_{mr}}$	$Y_n(s) = \dfrac{2\alpha_n(s - p_{nr}) - 2\beta_n p_{ni}}{(s - p_{nr})^2 + p_{ni}^2}$	$Y_k(s) = \dfrac{s\psi_k}{s - p_{kr}}$	$Y(s) = \delta \quad Y(s) = \eta s$
RD, LD	RS, LS, RP, CP	RH, CH	Rdc, Cac
$RD_m = \dfrac{-p_{mr}}{\gamma_m}$ $LD_m = \dfrac{1}{\omega_o \gamma_m}$	$RS = \dfrac{-\alpha_n p_{nr} + \beta_n p_{ni}}{2\alpha_n^2}$ $LS = \dfrac{1}{2\omega_o \alpha_n}$ $RP = \dfrac{p_{ni}^2(\alpha_n^2 + \beta_n^2)}{2\alpha_n^2(-\alpha_n p_{nr} - \beta_n p_{ni})}$ $CP = \dfrac{2\alpha_n^3}{\omega_o p_{ni}^2(\alpha_n^2 + \beta_n^2)}$	$RH = \dfrac{1}{\psi_k}$ $CH = -\dfrac{\psi_k}{p_{kr}\omega_o}$	$Rdc = \dfrac{1}{\delta} \quad Cac = \dfrac{\eta}{\omega_o}$

Transient Simulation

The networks synthesized can be embedded in a SPICE circuit and simulated in the time domain. As an example, consider differential transmission lines referenced to power and ground planes. The planes can be analyzed using a solver such as Transmission Matrix Method (TMM). The frequency response has been extracted at three ports defined on the planes [59], which represent the positions of the voltage regulator module, the beginning and end points of the transmission line. This results in a 3×3 matrix. The frequency response for two of the admittance parameters is shown in Figure 4.55. The response has been approximated using a rational function, the response of which is also shown in Figure 4.56. A total of 150 complex conjugate pole pairs and 4 real poles were required to approximate the frequency response over a frequency range from DC – 6 GHz.

The synthesized network constructed from the rational function has been used to construct a macromodel (black box) of the power distribution network, as shown in Figure 4.56. The time domain simulation can now be performed in SPICE [60] using the macromodel of power-ground planes, differential drivers, and transmission lines for computing power supply noise and other signal integrity effects. In Figure 4.56, the schematic for simulating power supply noise is shown. The driver model used is a time-dependent resistive switch representing four differential drivers. The drivers are connected to four differential transmission lines. The differential drivers with 0.05-ns rise time and 0.05-ns fall time are powered from port 2. The differential transmission

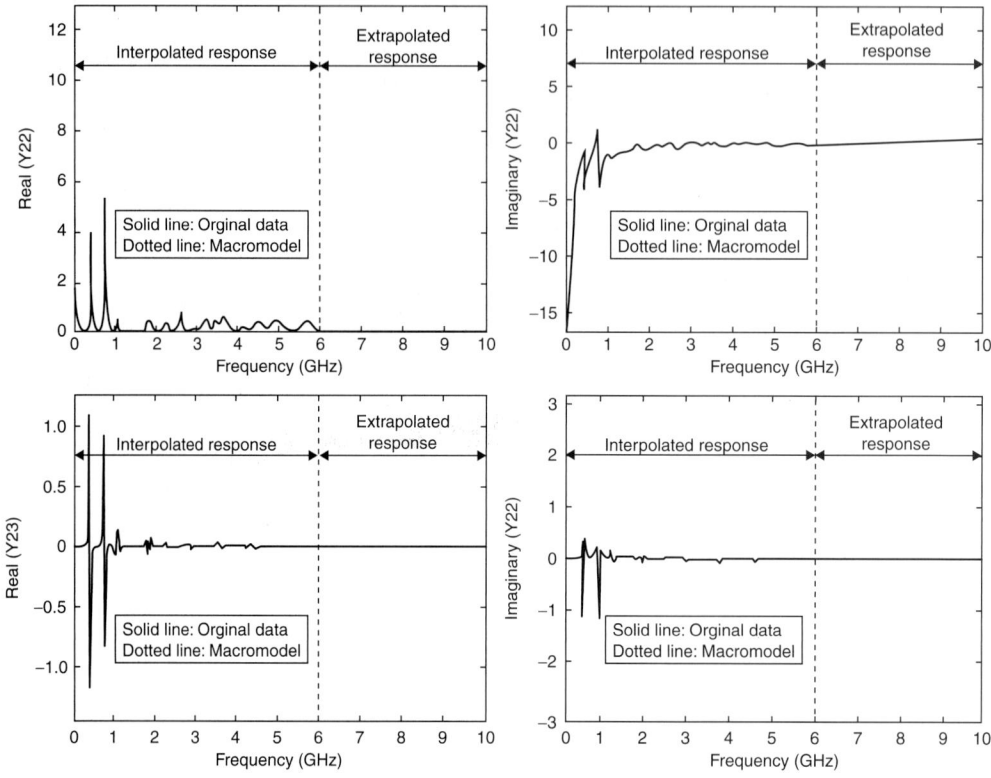

Figure 4.55 Frequency response and interpolated rational function.

lines with 100-Ω characteristic impedance (50-Ω characteristic impedance to ground) and 1-ns delay are connected to the output of the driver. A standard transmission line model available in SPICE is used to represent the transmission lines. The far end of the transmission lines is terminated in 50 Ω for matching and connected to a 0.3-V supply voltage. Port 3 representing a 1.2-V power supply for the slave chip is left unterminated. Hence, the differential transmission lines provide the communication path between the master and slave chips. In Figure 4.56, the voltage regulator module with a 0.6-V supply voltage is connected between port 1 and ground. Using the circuit model in Figure 4.56, the driver output and power supply noise near the driver has been simulated, as shown in Figure 4.57. The spikes in the power supply are caused when the circuits switch simultaneously, and this noise can propagate to the sensitive analog circuits. The primary purpose of this example is to illustrate a methodology whereby the electromagnetic interactions at the layout level can be captured in a circuit simulation.

4.5.4 Design for Manufacturing

The design of wireless circuits for RF frequencies requires precise values of passive components, which is only partially satisfied due to manufacturing variations and therefore results in a yield loss. To alleviate this problem, performance and yield figures for emerging technologies need to be analyzed during the design phase, since fault detection and diagnosis for RF circuits after manufacturing is a time-consuming step in

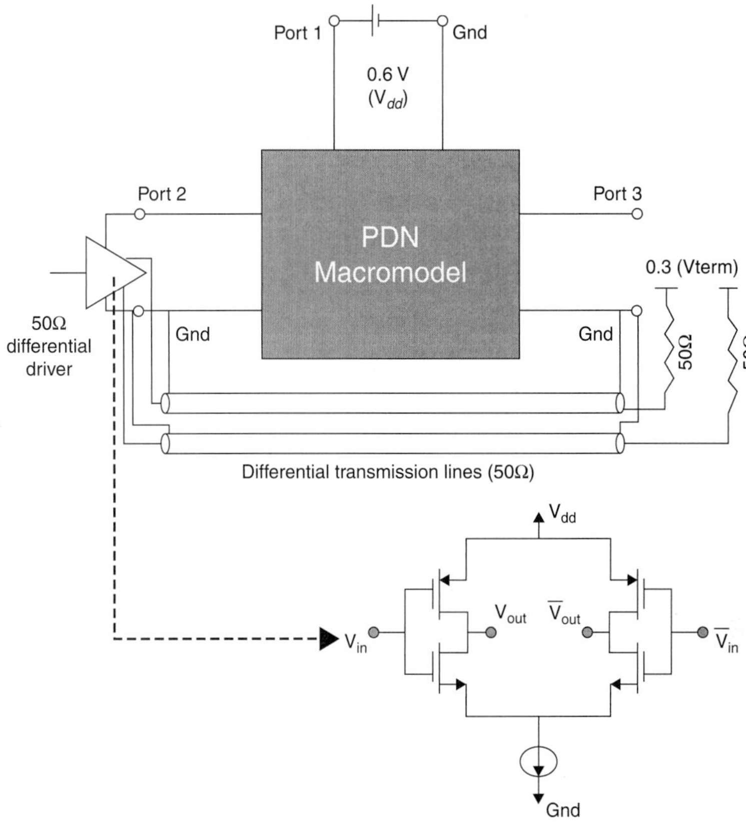

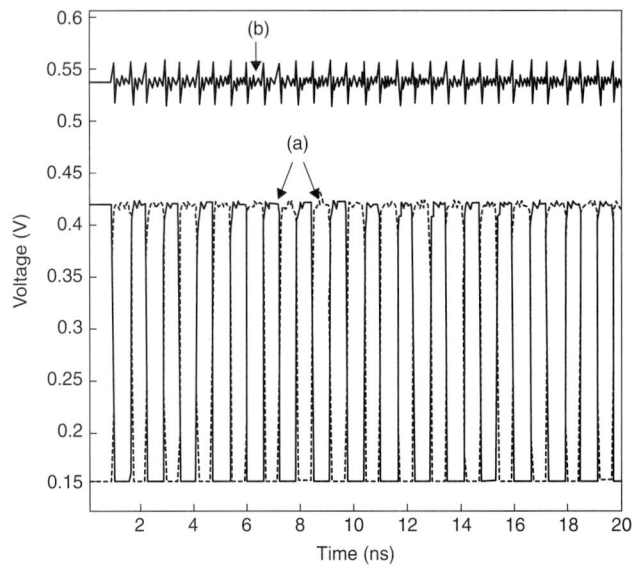

Figure 4.56 Circuit representation.

Figure 4.57 (a) Driver output. (b) Power supply noise.

the design cycle. In this section, the effect of process variations on the electrical performance is discussed along with a framework for including these variations into the design process. The goal is to generate designs that are yieldable under the assumption that catastrophic defects are minimized. Hence, the designs account for parametric variations due to statistical deviations in the process variables.

The statistical analysis and diagnosis methodology are shown in Figure 4.58. The statistical analysis is used to compute the distributions of the specifications given the statistical distributions of the process variables. To enable this computation, design of experiments (DOE) is used to compute sensitivity functions. These functions provide a relationship between the specifications and the process parameters, when the process parameters are varied between $\mu - 3\sigma$ and $\mu + 3\sigma$ values, where μ is the mean and σ is the standard deviation. The diagnosis methodology is used to estimate the process parameter causing a deviation in the specifications, when applied in a manufacturing environment.

Statistical Analysis

In order to map process variations to performance variations, a sequence of electromagnetic analysis can be planned using design of experiment principles. As an

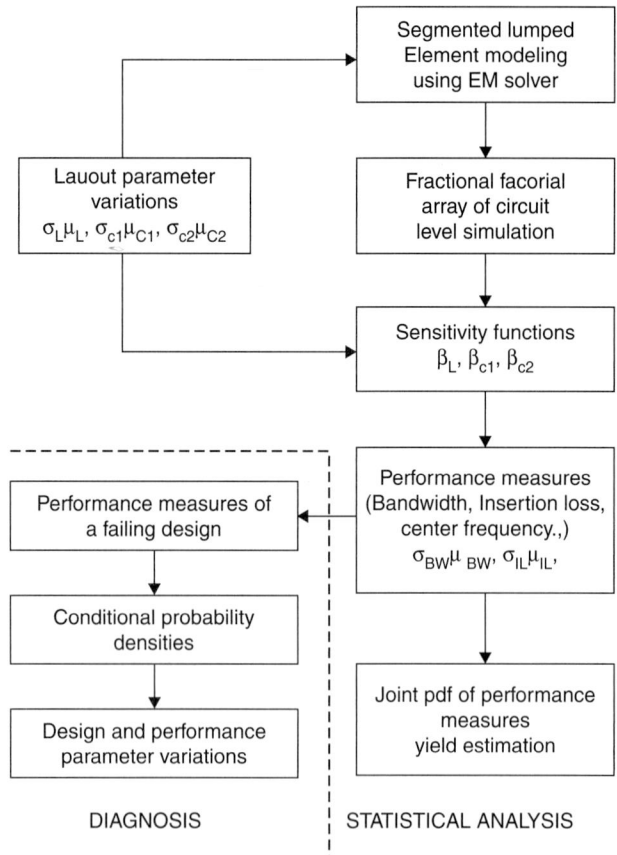

FIGURE 4.58 Statistical analysis and diagnosis methodology.

example, consider the filter in Figure 4.45. Statistical analysis has the following steps to relate the manufacturing variations to the filter performance: (1) Electromagnetic simulations are used to fill the DOE matrix. The DOE matrix contains the filter response when the process variables are varied between the $+3\sigma$, mean, and -3σ values. The filling of the matrix can be done either directly through electromagnetic (EM) analysis or by using an intermediate step containing the circuit models shown in Figure 4.45. Since, most EM simulators work with a grid, having a fine grid that allows the analysis of geometries containing small features (such as small increments in line width) can be difficult and time consuming. Hence, use of circuit models by segmenting the layout as in Figure 4.45 can be more practical. (2) Using regression models that capture the DOE matrix, the filter performance variations can be related to the manufacturing variations using analytical or Monte Carlo methods. (3) Parametric yield can be computed using joint probability density functions and the specifications of the filter. The DOE can be generated using Taguchi array, fractional factorial, or full factorial plans [60a–60d]. These plans relate the process variations to the variations in the electrical specifications and are available in most books on statistical analysis. The DOE can be used to develop sensitivity functions between the process variables and the specifications that provide insight into the process parameters that cause the maximum variation in the filter response [60e–60f].

Traditionally, the parametric yield is estimated by perturbing the independent process variables, which are line width, line thickness, spacing between lines, dielectric thickness, and layer-to-layer alignment. Under the assumption that the process variables have a distribution (gaussian, etc.), random samples from the distribution can be repeatedly chosen to perform circuit simulations to extract the performances as a function of the perturbed process variables. Often called Monte Carlo (MC) analysis, for each process parameter selected at random, the MC method finds a relationship between the process variables and performance parameters, using the sensitivity functions and process distributions. This process is repeated at random many times (e.g., 1000) to obtain a distribution of the performance parameters. This can sometimes be an expensive solution for complex layouts.

As an alternate method, the sensitivity analysis can be used to reduce the amount and time of simulation by extraction of *regression equations* that can be used to compute the distribution of the filter parameters. As an example, consider four process variables each with three levels ($+3\sigma$, μ, $+3\sigma$). A full factorial DOE will need 81 simulations (3^4). Instead, a fractional factorial plan can be used consisting of 27(3^{4-1}) electromagnetic simulations. The elements of the DOE matrix can be coded, where 1's represent their mean and 0 and 2 are $\mu - 3\sigma$ and $\mu + 3\sigma$, respectively, where μ is the mean and σ is the standard deviation. Model parameters of the components can be extracted from an electromagnetic simulator (Sonnet), and the filter response can be generated using the HP-ADS circuit simulator. The statistical distributions of components are highly correlated as they are affected by similar physical parameters, for example, metal linewidth and substrate thickness. Each performance parameter can be approximated using linear and piecewise linear terms forming a regression equation. For example, the following filter performance metric (1-dB bandwidth) can be approximated as shown below:

$$BW\_1dB = 0.1131 - 0.0426(CC) + 0.0023(C\_resn1)$$
$$+ 0.0020(L1)U(L1) - 0.004(\varepsilon_r) \ (R^2 = 0.995) \quad (4.22)$$

where CC, C_resn1, L1resn_L, and C_match are the component dimensions. Here R^2 represents regression coefficients and U is the unit step function. An R^2 value close to 1 indicates good predictive capability. Since the variations of the layout parameters are independent of each other, the probability density functions (PDFs) of the performance (for example, BW_1dB) are computed by convolution of the PDFs of the layout parameters [60g–60h].

Using sensitivity functions and convolution of the probability density functions of the process and material parameters, the statistical parameters of the filter performance in Figure 4.45 can be computed as in Table 4.9.

In Table 4.9, Min_attn is the minimum attenuation in the passband, ripple is the allowed ripple in the passband, F1 and F2 are specific frequencies for the filter, BW_1dB is the 1-dB bandwidth, and BW_3dB is the 3-dB bandwidth.

Parametric Yield

Parametric yield is defined as the percentage of the functional filters satisfying the performance specifications. Here, multiple constraints need to be met, for example, bandwidth, ripple, and center frequency. However, because of the manufacturing variations, certain parameters get shifted in the frequency-amplitude spectrum. In such cases, the joint probability density functions of the performance metrics are approximated using the multivariate normal distribution, which is defined as [60g–60i]:

$$f_Y(Y) = \frac{\text{Exp}\{-1/2([Y]-\mu_Y)^T[\text{Cov}(Y,Y)]^{-1}([Y]-\mu_Y)\}}{(2\pi)^2 |\text{Cov}(Y,Y)|^{1/2}} \qquad (4.23)$$

where Y is the vector of performance measures, μ_Y is the expected value for the vector Y, and Cov is the covariance of the performance metrics. The yield is computed as the integral of $f_Y(Y)$ over the acceptable region of performance. For example, the yield for the filter with a specific set of design constraints is computed as [60j]:

$$\int_{2.35}^{\infty}\int_{-\infty}^{2.8}\int_{2.45}^{\infty}\int_{-\infty}^{\infty}\int_{30}^{\infty} f_Y(Y) d_{f\_1dB\_1} d_{f\_1dB\_2} d_{min\_attn} d_{attn\_2.1GHz} = 55\% \qquad (4.24)$$

where the constraints include a bandwidth of at least 2.35 to 2.45 GHz, maximum attenuation of 2.8 dB, and minimum attenuation of 30 dB at 2.1 GHz. In this example, 55 percent of the filters manufactured on a panel will pass the specifications.

Filter Performance	Mean (μ)	Standard Deviation (σ)
Min_attn (dB)	2.1714	0.0743
Ripple (dB)	0.4894	0.10613
F1 (GHz)	2.3525	0.0437
F2 (GHz)	2.4271	0.0474
BW_1dB (GHz)	0.1139	0.0041
BW_3dB (GHz)	0.135	0.0065

TABLE 4.9 Statistical Parameters of Filter

Probabilistic Diagnosis

As a result of the statistical variations in the process parameters during batch fabrication, some embedded circuits may display unacceptable variations in performance. For parametric defects, the information extracted from the aforementioned statistical analysis can be utilized as a diagnosis tool. Using the diagnosis methodology, the most probable layout parameters causing unacceptable variations in performance can be systematically searched.

For explaining the diagnosis approach, let [X] and [Y] be the random vectors for m layout parameters and n performance metrics, respectively. The functional relation between [X] and [Y] ($R^m \rightarrow R^n$) is obtained by sensitivity simulations explained in the previous section. If n is less than m, then a unique solution of [X] does not exist for a measured set of unacceptable performance [Y]. Hence, the real parameter(s) causing the failure cannot be decided. However since all design parameters are associated with PDFs, the most probable solution can be searched. The conditional PDF of the parameter vector [X] for measured performance y is defined as

$$f(X \mid Y = y) = \frac{f(X,Y)}{f(Y)} \quad (4.25)$$

where $f(X, Y)$ is the joint PDF of the random vector of the design parameters and performance measures $[X^T Y^T]^T$. In Equation (4.25), $f(y)$ is the joint PDF of the performance which is computed in Equation (4.24). Then, the expected value of $f(X \mid Y = y)$ is the most probable parameter set causing the failure.

Let $\tilde{Y} = [P^1 P^2 \ldots P^n]^T$ be the set of unacceptable performance metrics. Equations for the performance metrics can be written by subtracting the intercept term from $\tilde{Y}$ as shown [60g]:

$$Y = \beta X + \varepsilon \quad (4.26)$$

where X is the parameter vector, Y is defined as the performance vector, and β is the sensitivity coefficient matrix without the intercept terms. The error column ε is a gaussian random vector with a zero mean computed from the approximation errors. Since X and Y are gaussian random vectors, a new random vector z can be defined as $Z_{mx1} = [X^T Y^T]^T$. The PDF of Z is equivalent to the joint PDF of X and Y,

The expected value of the conditional PDF can be computed as

$$E[X \mid Y = y] = \mu_X + \text{Cov}(X,Y)[\text{Cov}(Y,Y)]^{-1}(Y - \mu_Y) \quad (4.27)$$

The diagnosis technique does not give the *exact* statistical variation of the layout parameters during batch fabrication, but it captures the dominant variations.

As an example, consider the filter variations in Table 4.9. Let's assume that after manufacturing, the filter parameters are measured as min_attn = 1.9933 dB, ripple = 0.6513 dB, and F2 (higher side of 1-dB cutoff frequency) = 2.53 GHz. For this filter, the center frequency has shifted to a higher frequency and hence does not pass the intended band. The measured results can now be applied as described in this section to capture the significant parameter variations. Table 4.10 shows the estimated (from diagnosis methodology) manufacturing variations based on the measurements. The dominant parameter variations that cause the shift in the filter response have been captured. The diagnosis can be improved through additional measurements on the filter response that have little correlation with each other.

214 Chapter Four

Layout Parameter	Estimated Parameters	Actual Parameters
L1	$\mu + 2.26\sigma$	$\mu + 2.29\sigma$
C_resn1/C_ren2	$\mu - 1.66\sigma$	$\mu - 1.34\sigma$
CC	$\mu - 0.35\sigma$	$\mu - 0.71\sigma$
C_m1/C_m2	$\mu + 2.27\sigma$	$\mu + 1.92\sigma$

TABLE 4.10 Probabilistic Diagnosis for Filter

4.6 Coupling

A mixed-signal system is subject to coupling between the digital and analog domains and also within the analog domain. The modules or signals can couple from adjacent signal routing lines, through the substrate, or electromagnetically between passive components. This section discusses coupling and its undesirable effects on system or module performance.

4.6.1 Analog-to-Analog Coupling

With higher levels of system integration, multiple passives embedded in the package are necessary. An SOP-based receiver could then contain embedded passives for both the LNA and VCO. As mentioned earlier, multiple embedded passives in the package leads to system-level issues like feedback and resonance, many of which are not apparent in an SOC implementation. To better understand these effects, circuits in the 2.1- and 2.4-GHz frequency bands are discussed, as described in this section.

Figure 4.59 shows the schematic of the LNA, using a discrete HBFP-0420 dual emitter transistor in a SOT-343 package and the impedance transformation networks

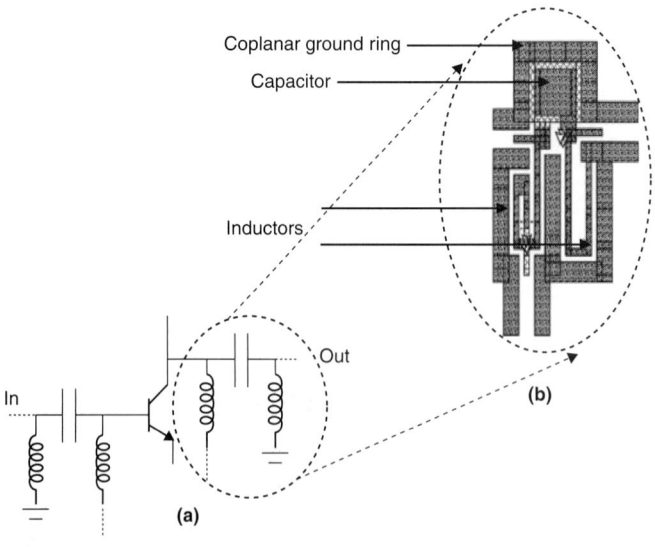

FIGURE 4.59 (*a*) LNA with impedance-matching network. (*b*) Layout of output pi network.

implemented using high-Q embedded inductors and capacitors. The transistor is biased in the common emitter configuration. The input and output of the transistor are matched to 50 Ω by using LC pi networks, which are embedded in the package. The output pi is designed for maximum power transfer and thus performs impedance transformation from the complex conjugate of the collector impedance to 50 Ω. The input pi is designed for a minimum noise figure and presents the Z_{opt} to the gate of the transistor.

To study the effect of ground return current, pi's with different reference ground layouts have been modeled and implemented, and their effect on the LNA performances has been analyzed. Figure 4.60a shows two of the topologies used to implement the output pi, and Figure 4.60b shows the electromagnetic simulations for both the pi layouts. As can be observed (in Figure 4.60b), for the frequency band of interest, there is minimal difference in the S parameters for the two topologies. However, Figure 4.61 shows the measured response of the amplifier circuits for the two pi topologies. The

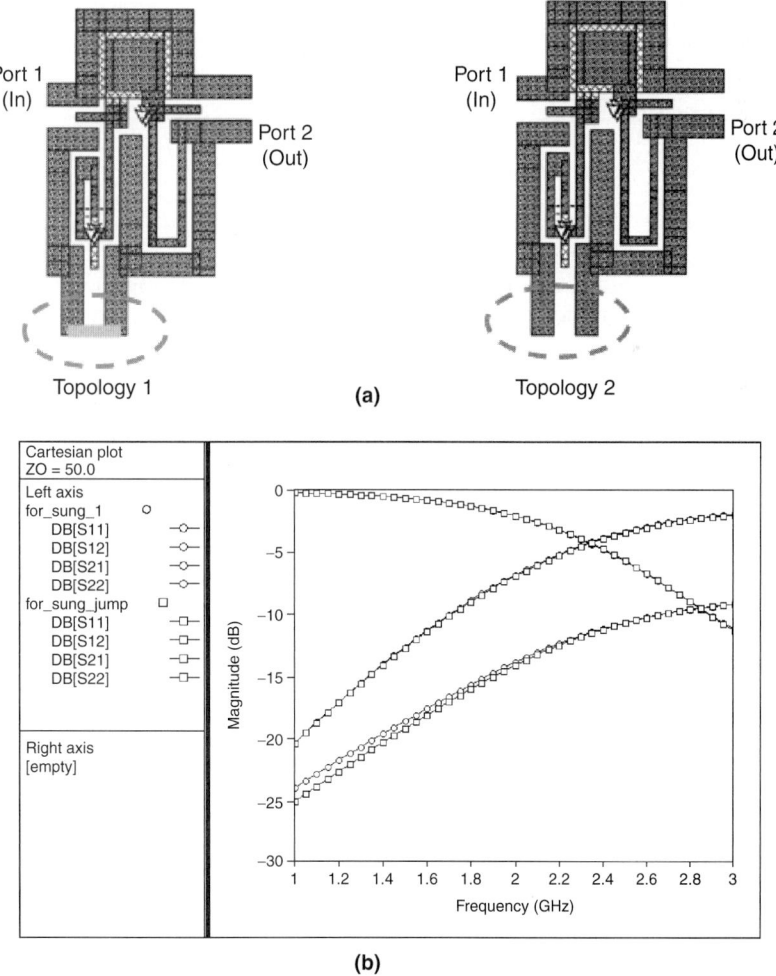

FIGURE 4.60 (a) Two reference ground layouts for the output pi. (b) Electromagnetic simulations.

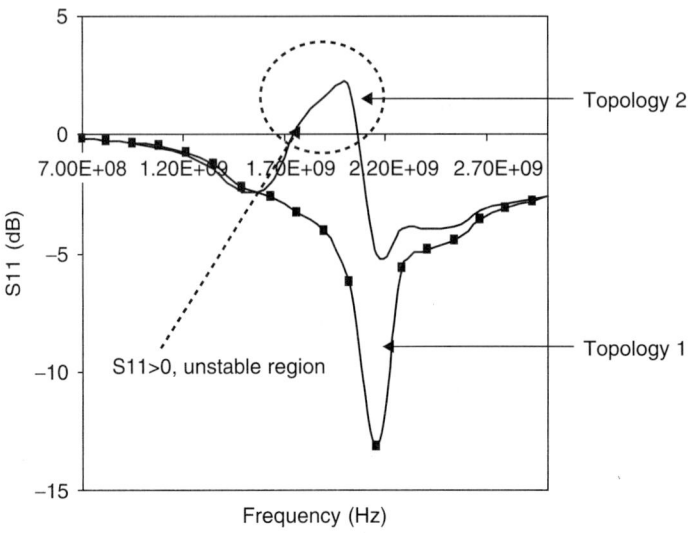

FIGURE 4.61 Measured S_{11} values for the LNA with different output pi.

change in routing for topology 2 causes the amplifier to move into the unstable region of operation, which cannot be predicted by simply simulating the pi's alone using full-wave electromagnetic solvers. The instability is caused by the influence of return currents on the transistor circuit.

With the layout of the reference ground resulting in such drastic changes in system performance, it becomes necessary to model its effect at the design stage, so that any system-level instability problems can be identified and rectified. This involves the incorporation of the reference ground layout into the design and the development of a good simulation methodology.

Field solvers like HFSS [60k] and Sonnet [28] can be used to obtain an *n*-port S-parameter file for the entire layout, which can then be used in a circuit-based simulation tool like Agilent ADS [60l]. However, current modeling tools do have limitations when providing solutions for internal ports, especially for devices configured in a CPW topology. Instead, the effect of the reference ground layout can be modeled as a mutual inductance between the inductors of the input and output pi's, with the coupling coefficient depending on both spatial orientations of the circuit components as well as the return current paths.

Electromagnetic simulations of the complete layout for the unstable LNA shows considerable coupling between the input pi and one of the inductors of the output pi. The reference ground layout (and hence the return current path) for pi topology 2 results in current crowding and signal coupling between the input and output pi's, leading to positive feedback and instability. The ratio of the current densities in the input and output pi's translates to a coupling coefficient of ~0.2, which, when used in ADS, modeled the instability (as shown in Figure 4.62).

With rerouting of the excess current to prevent coupling (through the use of jumpers), it is possible to stabilize the amplifier. It is important to note that the

Mixed-Signal (SOP) Design

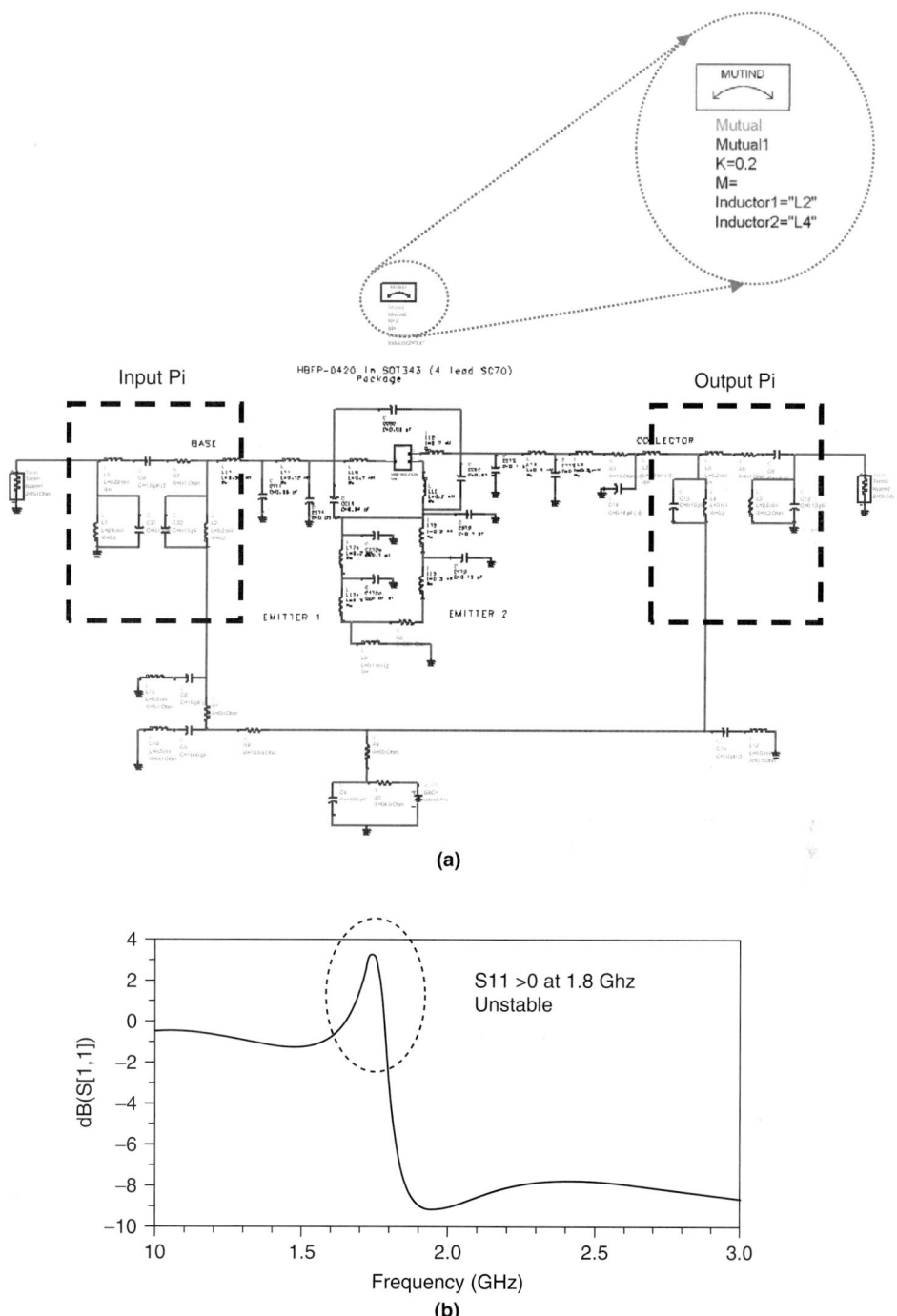

FIGURE 4.62 (a) ADS model of LNA with mutual inductance. (b) Modeled S_{11} result showing instability.

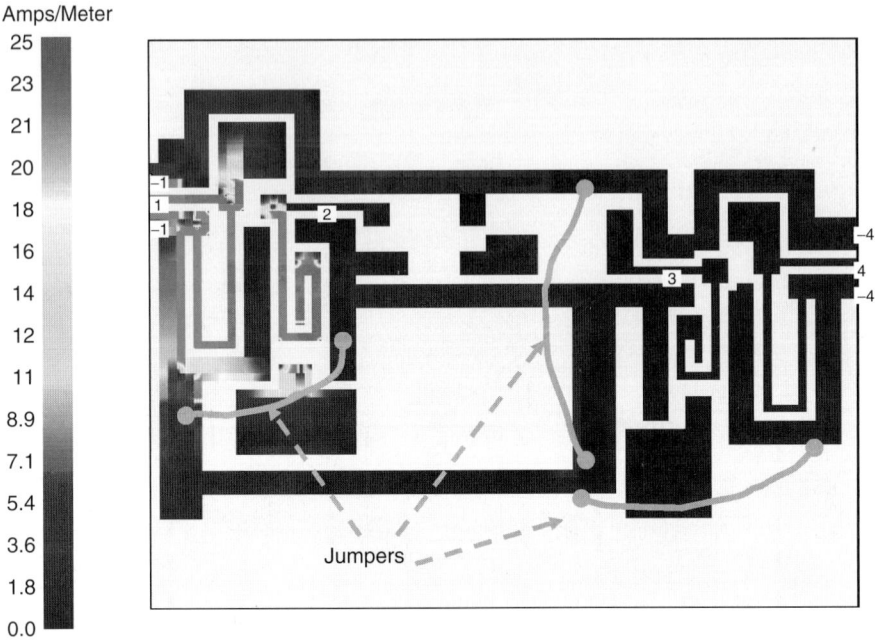

FIGURE 4.63 Electromagnetic simulations of the unstable layout with the use of jumpers for current rerouting.

electromagnetic simulation of the LNA layout in Figure 4.63, now with better ground routing through the use of jumpers, exhibits a coupling coefficient of less than 0.05. Measured results for this LNA showed stable operation and a gain of 12 dB at 2.1 GHz, proving that the instability in the earlier case was indeed because of return current routing.

Another example of analog-to-analog coupling (or coupling between passive components) can be observed in the VCOs. In general, VCOs use multiple passive components for accurate frequency control and higher frequency stability. One such example is considered here of a transformer-feedback VCO (TVCO). Figure 4.64 shows the schematic of the TVCO [50].

The TVCO uses multiple passive components that are subject to both intercomponent magnetic and electrical coupling. As a result, the transformer's frequency response in its out-of-band (stopband) region is of consequence and is important in the oscillation frequency of the VCO. Hence, in the design phase of an oscillator it is important to verify the oscillation condition over a broad frequency range [50].

The transformer in the TVCO has multiple components that are physically placed close together to minimize the area of the oscillator. In the case where the transformer components are distributed on different metal layers and connected through via-holes the parasitic EM coupling between the components is minimized. However, depending on the physical placement of the components, both inductors and capacitors can couple energy. As a result, parasitic passbands at frequencies far away from the desired frequency in the transformer can be generated. The parasitic passbands cause an

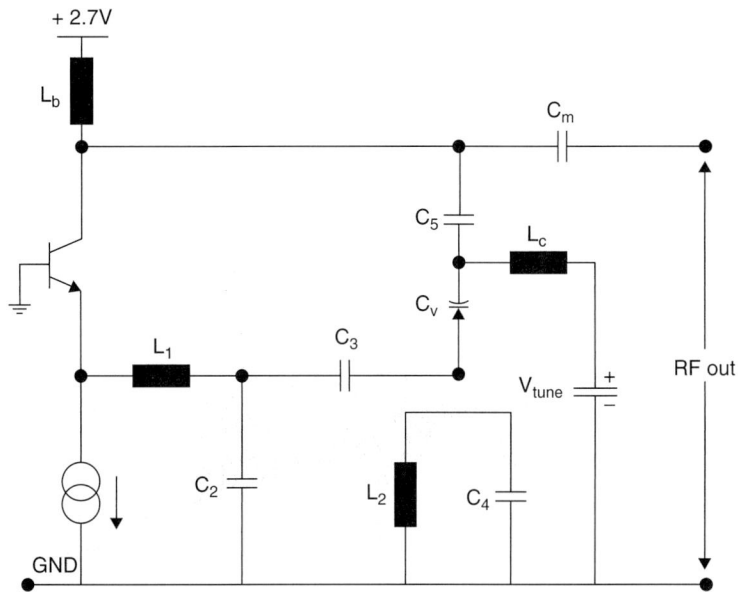

FIGURE 4.64 Schematic of the TVCO [50].

undesired shift in the frequency of oscillation of the TVCO. A few possible scenarios have been investigated based on the physical layout of the transformer as follows:

1. Figure 4.65 shows the photograph of a TVCO (TVCO2) with all the resonator components closely packed on the top two metal layers. *TVCO2 is exactly identical to TVCO1 except that the components are fabricated on only the top two metal layers.* Figure 4.66a and b show the measured response of the transformers used in the TVCO1 and TVCO2. The sampled data in Figure 4.66a and b are the two-port measured S parameters of only the transformer used in the design of TVCO2, and the solid data are for the resonator used in TVCO1. In the frequency

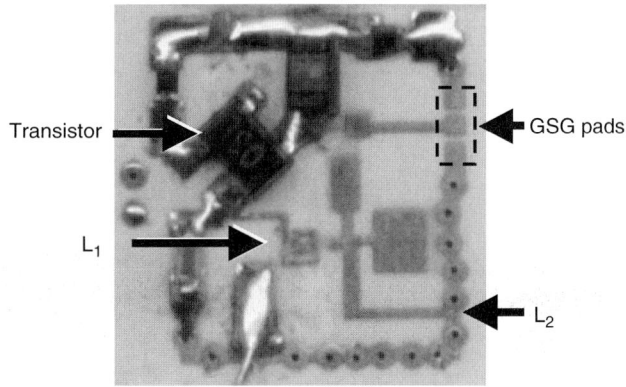

FIGURE 4.65 Photograph of TVCO2 using only two metal layers.

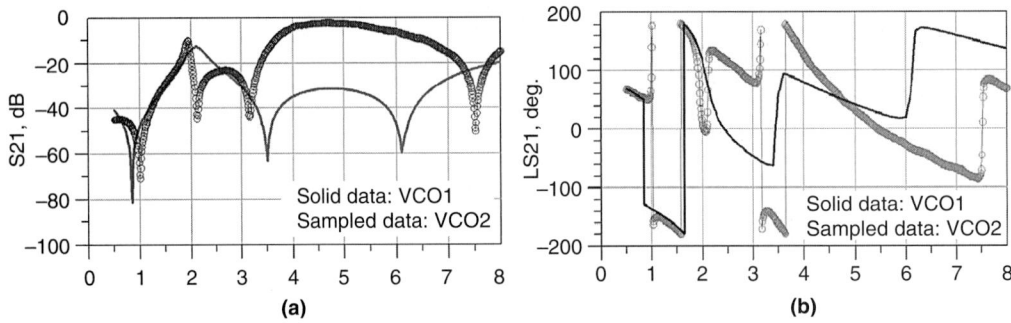

Figure 4.66 Transformer frequency response. (a) Magnitude response of VCO1 and VCO2. (b) Phase response of VCO1 and VCO2. VCO1 shows no out-of-band coupling.

band of interest (around 2 GHz) there is minimum discrepancy between the two responses. However, in the case of TVCO2 the resonator elements couple around 6 GHz. This is evident from the low insertion loss in the transformer magnitude response. Based on modeling, it can be determined that the parasitic inductance of capacitor C_2 (2.5 pF) couples to inductor L_2 (2 nH) [due to the low self-resonant frequency (~5 GHz) of C_2]. Additionally at 6 GHz, it can be observed that the phase response matches well to satisfy the oscillation (Barkhausen) criterion (loop gain = $1\angle 0°$). These effects coupled with the broadband nature of the transistor G_m causes the center frequency of TVCO2 to shift to 6.1 GHz. The measurement results of TVCO2 are shown in Figure 4.67 using an HP 8563E spectrum analyzer. The measured center frequency has shifted from 1.9 to 6.1 GHz because of the coupling and the phase characteristics of the transformer.

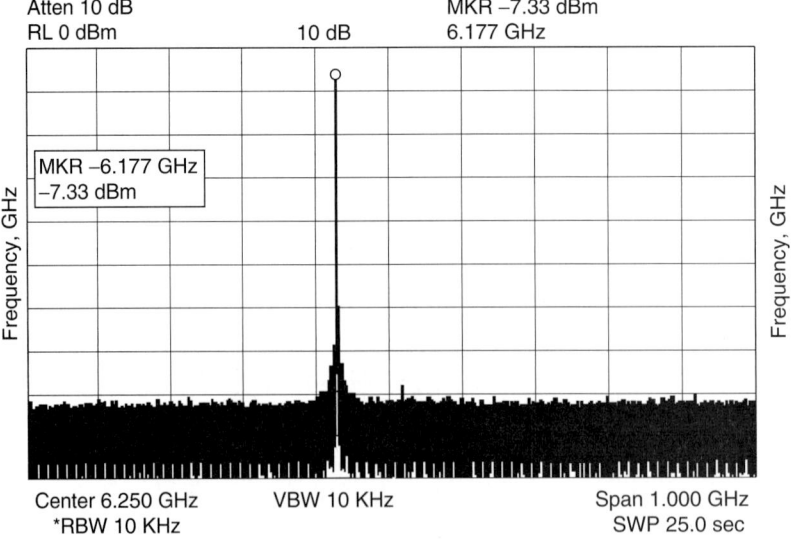

Figure 4.67 Measured spectrum of VCO2. The f_0 shifted to 6.1 GHz due to EM coupling.

The EM coupling between resonator elements of TVCO2 can be reduced by separating the components (especially C_2 and L_2) spatially in the planar direction or by moving from microstrip- to stripline-type designs. Hence, on comparing the results for TVCO1 and TVCO2 it can be concluded that 3D separation of the resonator elements on multiple LCP layers reduces both the undesired EM coupling and VCO size.

2. Figure 4.68 shows the current density (Sonnet simulations) results of two transformers (namely, type 1 and type 2) with different spacing between the

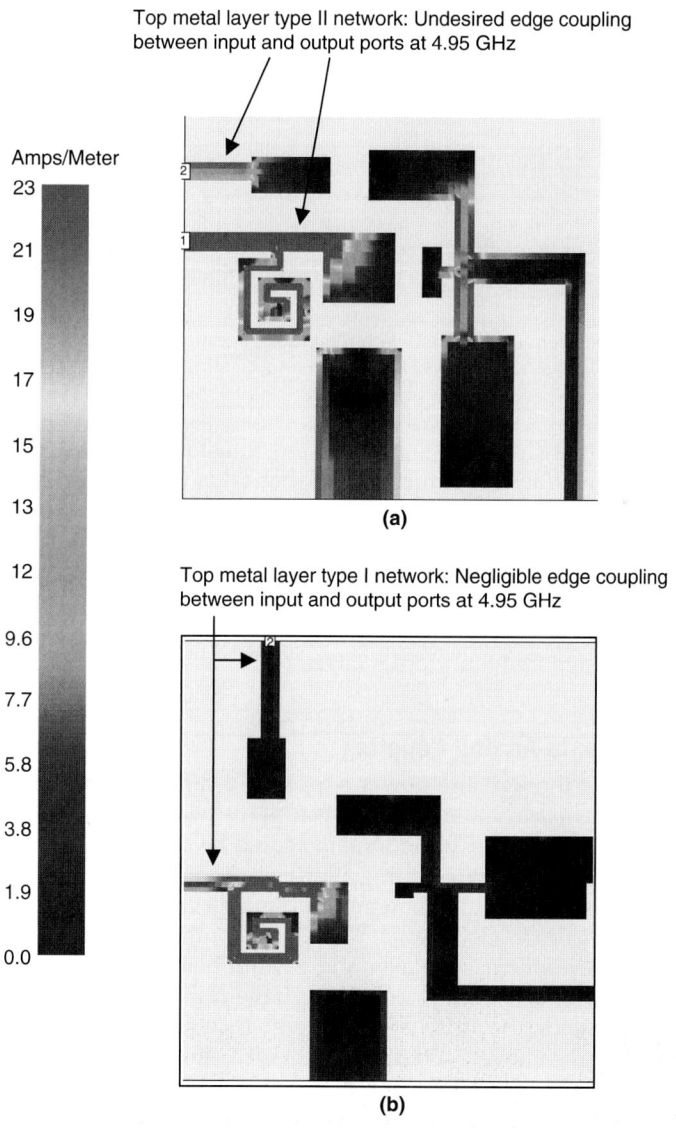

FIGURE 4.68 Sonnet simulation results illustrating coupling between capacitors C_1 and C_m.

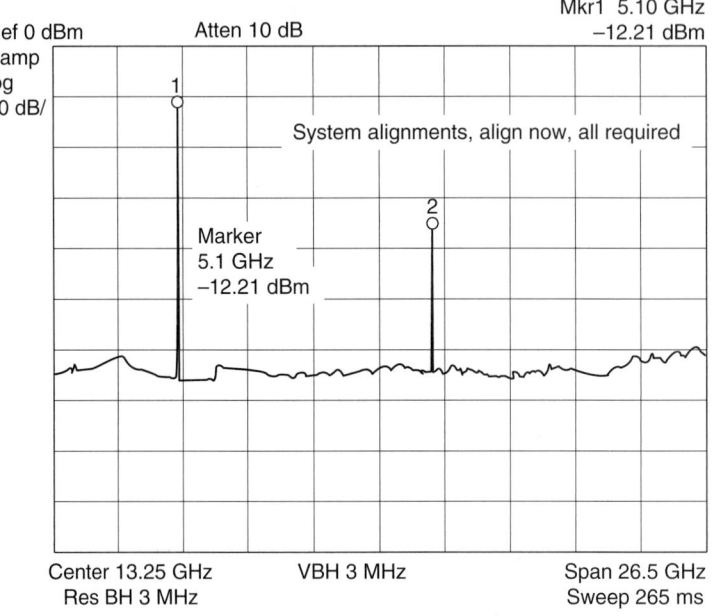

FIGURE 4.69 Spectrum of the TVCO with the type 2 transformer design. The f_o shifted to 5.1 GHz from the design frequency of 1.9 GHz.

capacitors C_1 and C_m. In both the transformer types, C_1 is the capacitance connected across inductance L_1 that adds a transmission zero for harmonic rejection. From the results it can be observed that at nonharmonic frequencies (4.95 GHz) there is considerable current at the output terminal (C_m) that can cause a significant shift in the magnitude of the loop gain and phase shift around the loop. Figure 4.69 shows the measured spectrum of the TVCO with the type 2 transformer. The fundamental frequency has shifted to 5.1 GHz from the design frequency of 1.9 GHz.

4.6.2 Digital-to-Analog Coupling

Noise coupling through the power supply is a difficult problem to solve, primarily because of the *physical* connection that it provides between the RF and digital subdomains. Ideally, the RF subdomain should be separated from the digital circuitry such that there exists *no* coupling between them. As the frequencies of operation increase, however, electromagnetic coupling becomes important and it becomes impossible to *completely* isolate *any* two regions of the system.

As frequencies increase, power and ground planes are required to support a low-inductance power delivery system. Though the planes have low-inductance properties at high frequencies, they also couple energy, especially at their resonant frequencies. Several techniques have been applied to solve this isolation problem. The classical method is to use split planes. If a dc connection is required, the split planes can be connected using a low-pass functional block. This requires additional components and provides only marginal isolation. A very promising method for isolation is based on electromagnetic bandgap (EBG) structures, which are periodic patterns on the power-ground planes. EBGs generally provide better isolation and do not require any additional components. This is described in a later section.

In this section, various coupling mechanisms through the power-ground planes between the digital and analog domains are presented. These include coupling through the splits as well as horizontal and vertical coupling between the power-ground planes.

Split Planes

Split power and/or ground planes (with the use of multiple power supplies) have been applied for isolating the various regions of the power-ground planes [61]. However part of the electromagnetic energy can still couple through the gap, especially at higher frequencies [62]. There is increased coupling at the resonant frequencies of the split planes. Hence, this method only provides marginal isolation (–20 to approximately –60 dB) at frequencies above 1 GHz and becomes ineffective as system operating frequencies increase. With the high sensitivity requirements of long-distance communication protocols (–102 dBm for GSM900, –116 dBm for WCDMA), the system-level isolation requirements are much higher. Further, as systems become more and more compact, multiple power supplies also become a luxury that the designer cannot afford.

With the restriction to use a single power supply for both digital and RF circuits, the need for a low-pass functional block that provides dc connectivity throughout the system but prevents the transfer of high-frequency noise components arises. In such a scenario, the analog/RF and digital subsystems would be powered using separate sections of a common power distribution system (power planes), with the filter blocking transfer of high-frequency signal power between the sections. Several schemes involving split power planes connected using a lumped inductor, a printed inductor, or a ferrite bead have been suggested [63–64]. However, all of them offer maximum isolation in the order of –40 dB, with significantly lower isolation numbers at resonant frequencies of the discrete components.

As an example, Figure 4.70 shows the point-to-point isolation obtained in a system using the Murata BLM18GG471SN1 ferrite bead. As can be observed, the maximum value of isolation is obtained at ~1 GHz and does not go below –25 dB. Note that this is a "high-performance" ferrite bead optimized for operation at 1 GHz.

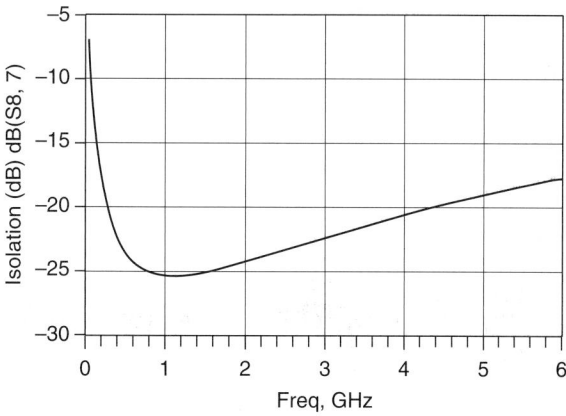

Figure 4.70 Isolation of a ferrite bead.

Continuous Planes

Power-ground planes in electronic packages can be a major factor for noise coupling. Excessive supply voltage fluctuations cause signal integrity problems. In addition, noise voltage that gets coupled to the edge of the board may cause significant electromagnetic interference. Hence, accurate estimation of the performance of power-ground planes is critical in a mixed-signal system.

Figure 4.71 shows the simulated S parameters between two ports on the opposite edges of a 47 mm × 47 mm board. The dielectric between the planes is FR4 with a dielectric constant of 4.6, a loss tangent of 0.02, and a thickness of 50 μm. The magnitude of the transmission coefficient for both a solid plane and an EBG patterned plane are shown (EBGs are discussed in a later section). There is a large amount of coupling between the two ports in case of the solid plane. The EBG patterned plane, on the other hand, suppresses the coupling significantly from about 1 to 4 GHz, which can be designated as the "stopband" of the EBG. Hence if the RF circuits attached close to one of the ports on this board operate within this frequency range, they will be unaffected from the noise generated by switching digital circuits attached close to the other port.

Aperture Coupling

A solid plane made of a perfect conductor of infinite lateral dimensions would completely shield the fields on one side from the other side. Therefore, there would be

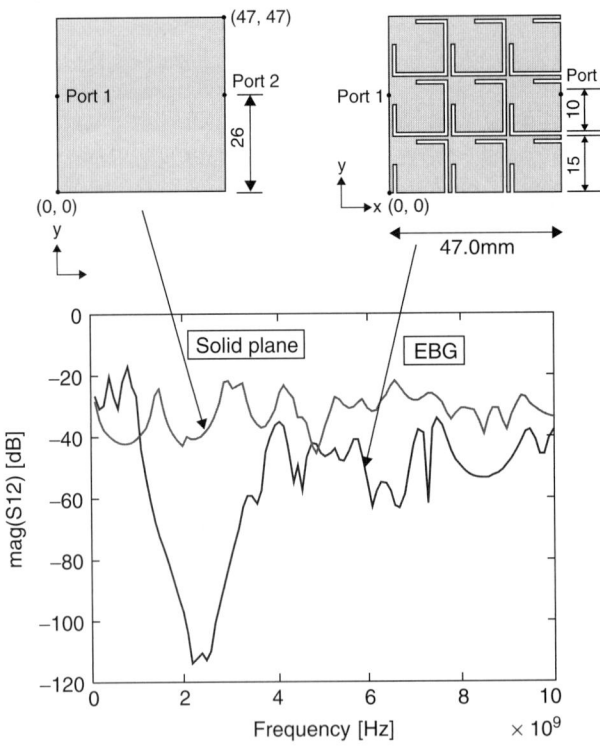

Figure 4.71 Coupling between two points on a solid plane and an EBG patterned plane (units in millimeters).

no need to consider multiple plane pairs. In reality, however, planes at the same dc level have to be connected with vias to each other in order to reduce the effective inductance of the planes. Such a via has to go through a via hole in a plane having a different dc level, in order to avoid a short circuit. Through this via and via hole, fields in different plane pairs get coupled to each other. As a result of this, there can be noise coupling not only in the transversal direction between two planes, but also vertically from one plane pair to another through the apertures and via holes. Coupling of multiple plane pairs through such vias has been analyzed using the cavity resonator model [65], the transmission matrix method [66], and coupled transmission lines [67].

In a multilayered stackup, the field penetration through the conductors can be neglected for frequencies, where the skin depth is much smaller than the plane thickness. At lower frequencies, this field penetration has to be taken into account [68]. Generally, skin effect is pronounced above several megahertz for commonly used copper planes in packages.

In addition, planes generally have irregular geometries. There can be large apertures and splits in planes. Fields in different plane pairs can get coupled through these apertures. This can be regarded as a coupling by means of a wraparound current on the edges of the planes. For narrow slots, a transmission-line based model has been proposed to take into account this interlayer coupling [69]. Electric and magnetic polarization currents have also been considered to compute coupling through electrically small cutouts [70].

Consider the three-plane structure in Figure 4.72 with a hole in the middle plane (plane 2). Port 1 is defined between planes 1 and 2; and port 2 is between planes 2 and 3. In such a three-plane structure, there are three plane pairs. These plane pairs are coupled at their boundaries. Current flowing into the boundary of a plane pair can spread into other plane pairs, which results in a wraparound current.

Figure 4.73 shows the magnitude of current density at 1.5 GHz, which was simulated using Sonnet, for the structure in Figure 4.72. Only port 2 is excited, which is between the middle and bottom planes. The large amount of current on the top plane indicates coupling of the plane pairs through wraparound currents.

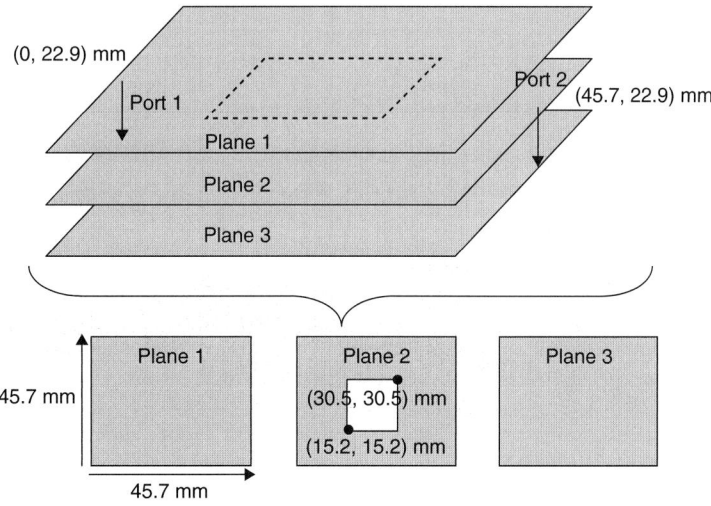

FIGURE 4.72 Three-plane structure with a hole in the middle plane.

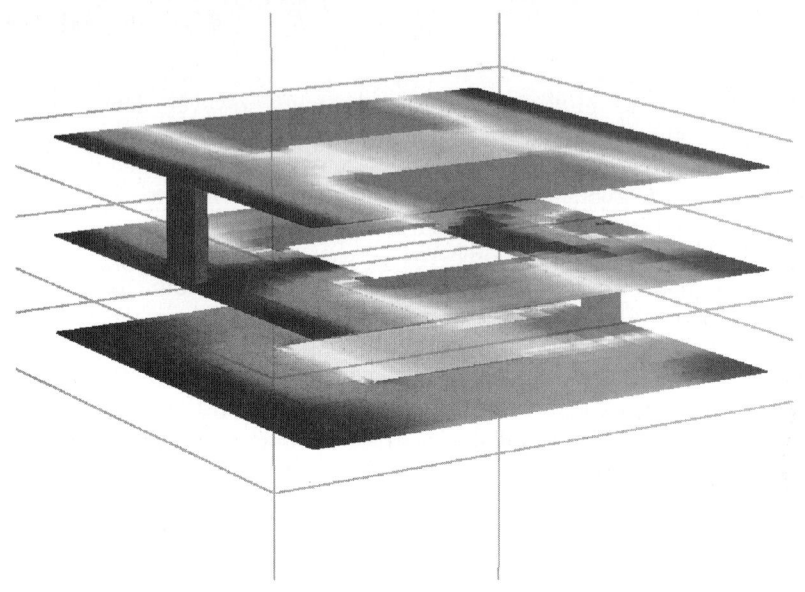

Figure 4.73 Simulated wraparound current when port 2 is excited.

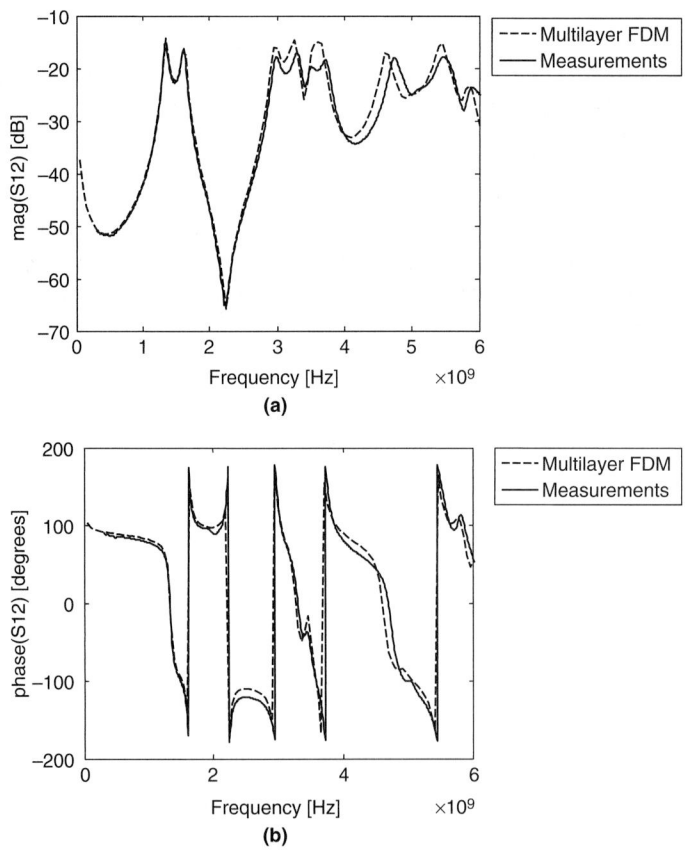

Figure 4.74 Transmission coefficient (S_{12}). (*a*) Magnitude. (*b*) Phase.

The multiple plane structure in Figure 4.72 was fabricated using FR4 dielectric layers with $\varepsilon_r = 4$, $\tan \delta = 0.02$, and thicknesses of 5 mils for each layer. Figure 4.74 shows that there is excellent agreement regarding the transmission coefficient S_{12} obtained from the measurement and simulations using a multilayered finite-difference method (FDM). S_{12} is solely due to the coupling through the aperture and could be very accurately captured. This is a large amount of coupling, which could cause a signal integrity problem if, for example, the top and bottom planes were assigned to different voltage levels.

4.7 Decoupling

The design of power distribution networks (PDN) is critical for the proper functionality of a system. A major challenge in the design of the PDN is to maintain the impedance of the network below the calculated target impedance over a broad frequency range. To reduce the impedance, decoupling capacitors along with low-impedance interconnections and planes are used in a typical PDN. The target impedance is given by [71]

$$Z = \frac{V_{\text{core}} \times 5\%}{I_{\text{avg}} \times 50\%} \quad (4.28)$$

where V_{core} is the core voltage of the active device and I_{avg} is the average current drawn by the device. The noise voltage that can be tolerated on the PDN is assumed to be 5 percent of the core voltage V_{core}. Also 50 percent of the switching current is assumed to flow during the rise and fall time of the clock edge, respectively, to give a 100 percent switching current over the whole clock period [71]. The target impedance must be met at all frequencies where current transients exist. The operations that may cause these current transients involve data transfer to and from the hard disk and memory or on-chip processing. This translates into a frequency range that varies from direct current to multiples of the chip operating frequency. Fast switching speeds of these circuits cause an increase in the current demands. This current is supplied by the PDN, and if improperly designed could lead to excessive power supply fluctuations in the PDN. A methodology to ensure that the PDN is properly designed is to make sure that the target impedance is met over the whole frequency band. Decoupling capacitors aid in the design of PDNs as highlighted in [72]. In this section, decoupling approaches are described in the context of a microprocessor. The microprocessor can be a stand-alone package on the board or may be integrated with a wireless device on the same package. In the second option, the microprocessor could represent the baseband IC that supports signal processing and computing functions.

Power delivery decoupling in today's systems is primarily achieved by using voltage regulator modules (VRMs) and surface-mount discrete capacitors (SMDs). The VRM is a dc-to-dc converter; it senses the voltage near the load and adjusts the output current to regulate the load voltage. VRMs are effective until the lower kilohertz region after which they become highly inductive in their behavior. Surface-mount capacitors provide decoupling from the kilohertz region until several hundred megahertz. SMDs start becoming ineffective above this frequency because of the increased effect of loop inductance associated with the current flow from the capacitors to the switching circuits and back again to the capacitors. A methodology for reducing the loop inductance would be to place the decoupling capacitors as close to the switching circuits as possible.

One possible solution is to provide on-chip decoupling above 100 MHz. Previous work has illustrated the effect of on-chip capacitance for decoupling switching circuits [73–76]. The main disadvantage with this approach is the low capacitance value associated with on-die capacitance. This renders them effective at frequencies much above 100 MHz. The amount of on-die capacitance can be increased, but this would compromise the amount of real estate available for including logic circuits. Another approach investigated in [77–81] is to include planar embedded capacitors within a board or package. In this approach, the capacitor is assumed to be a thin dielectric layer within the package or the board. Another innovative method presented in [82] uses a capacitance interposer between the chip and the board to provide decoupling for the switching circuits.

A very promising method for decoupling is the use of discrete thick- or thin-film capacitors arranged in a capacitive array that can be used within a package to provide decoupling for high-performance circuits in the frequency range from 100 MHz to 2 GHz. Using an array of capacitors allows for control of the resonant behavior of each capacitor for broadband decoupling.

4.7.1 Need for Decoupling in Digital Applications

The power densities of microprocessors have grown over the years due to the increase in the number of transistors and the increase in the processor operating frequency. The major contributors to the power dissipation in the sub-100-nm technology nodes are the active and static power dissipation.

The active power dissipation of a processor is given by [83]

$$P_{active} = \alpha C V_{core}^2 f \tag{4.29}$$

where V_{core} is the core voltage of the processor, α is the activation factor of the processor, C is the capacitance that is switched in each clock cycle, and f is the frequency of operation of the processor. The static power dissipation of the processor is given by

$$P_{static} = V_{core} \times I_{leakage} \tag{4.30}$$

where $I_{leakage}$ is the total leakage current of the processor.

The power dissipation of a processor can be calculated using the product of the average current of a processor and the core voltage. The average power dissipation of a processor is given by

$$P = V_{core} \times I_{avg} \tag{4.31}$$

The estimated power dissipated for the 65-nm node cost performance processor from [85] is 103.6 W and V_{core} is 0.9 V. Using Equation (4.31), the average current I_{avg} drawn by the processor is 115.1 A. The target impedance for this processor is calculated by substituting the value of V_{core} and I_{avg} in Equation (4.28). Table 4.11 lists the different parameter for processors in the 90-, 65-, and 45-nm nodes.

From the table it is evident that the target impedance is decreasing with an increase in the technology nodes. As mentioned before, the methodology for meeting the target impedance is to place decoupling capacitors in the PDN. The number and type of decoupling capacitors required depends on the frequency band to be targeted and the equivalent series resistance of each individual capacitor. The number of capacitors of each type can be decided by the ESR of each type of capacitor and the target impedance to be met given by

$$N_{cap} = \text{target impedance} / ESR_{cap} \tag{4.32}$$

Mixed-Signal (SOP) Design

Year	Feature size (nm)	Power (W)	V_{core} (V)	I_{avg} (A)	Target Impedance (mΩ)
2004	90	84	1.2	70	1.7
2007	65	103.6	0.9	115.11	0.781
2010	45	119	0.6	198.33	0.302

TABLE 4.11 Target Impedance for Different Technology Nodes

4.7.2 Issues with SMD Capacitors

This section briefly describes the decoupling schemes used in today's PDNs and their limitations. To illustrate the decoupling methodologies, a system simulation with different decoupling components is performed. The schematic of the setup is shown in Figure 4.75.

The setup includes a 10 cm × 10 cm board with a dielectric thickness of 0.8 mm and metal thickness of 30 μm. A 4 cm × 4 cm package and a 11.8 mm × 11.8 mm processor mounted on the package are also included in the simulations. The input port in the simulation is the processor looking into the package and the board. SMDs and VRMs are the decoupling components used in today's system. The PDN has been designed to

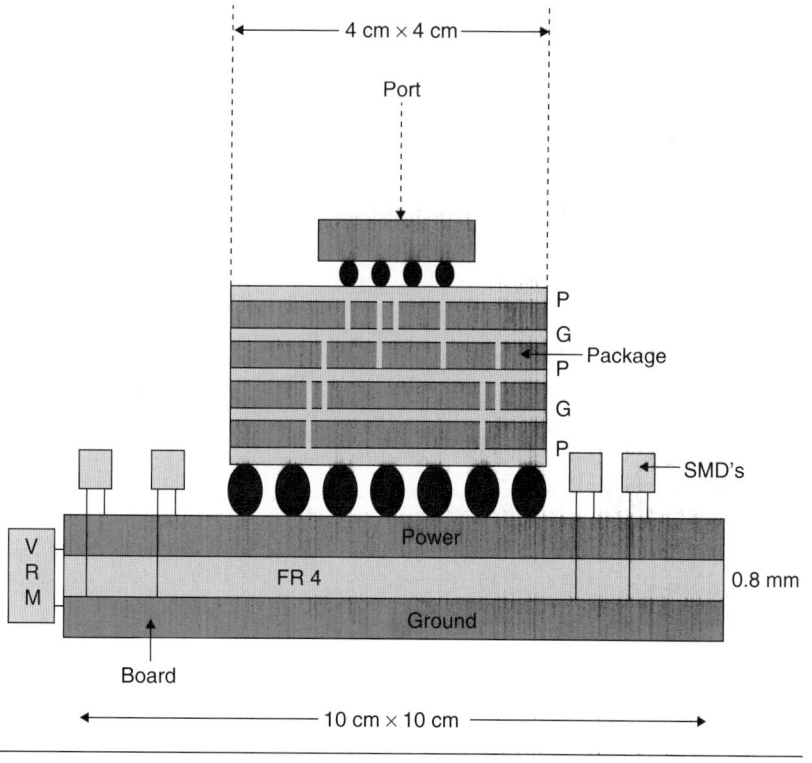

FIGURE 4.75 Decoupling schemes in today's systems.

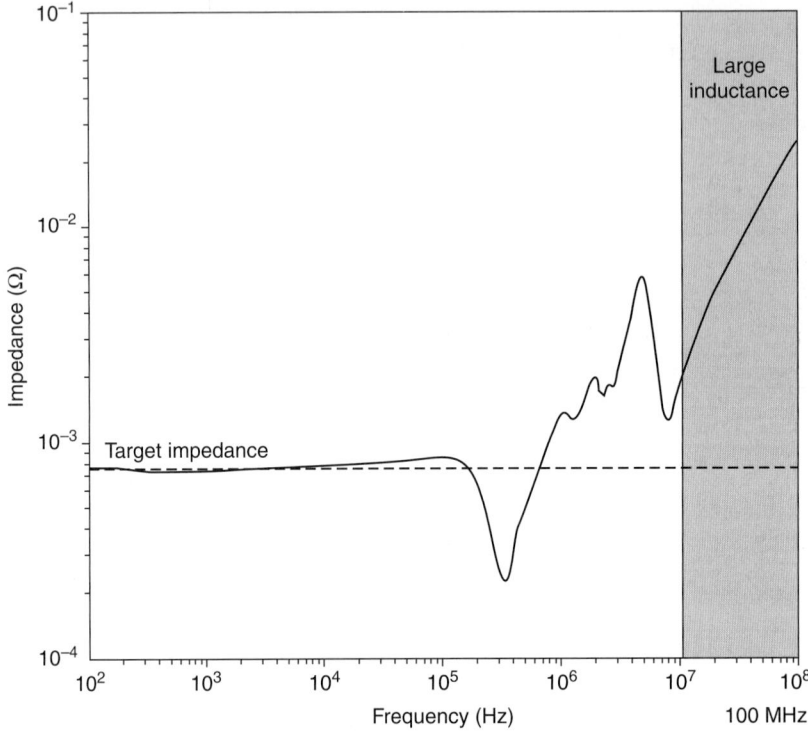

FIGURE 4.76 Decoupling limitations of SMD capacitors.

meet the target impedance of 0.78 mΩ for the 2007 processor using the above-mentioned components. The SMDs are spread out on the board close to the package, with the lower-value series inductance capacitors placed closer to the package. This has been done to reduce the effect of the spreading inductance of the planes on the capacitor performance. As seen from Figure 4.76, at frequencies close to 100 MHz it becomes increasingly difficult to meet the target impedance. This can be attributed to the increased effect of the loop inductance between the active circuits and the capacitors.

In order to provide effective decoupling above 100 MHz, it is essential to reduce the effective inductance associated with the capacitors. One solution is to embed capacitors within the package. Introduction of embedded capacitors reduces the current path enabling decoupling above 100 MHz.

4.7.3 Embedded Decoupling

The focus of this section is on the performance of embedded capacitors. Embedded capacitors can be broadly classified into two categories:

1. *Planar capacitors.* The power-ground plane in the package or the board is used as a thin high-K (or low-K) capacitance layer.
2. *Individual thick- or thin-film capacitors.* Different-sized capacitors are used in a layer of the package for decoupling.

Planar Capacitors with Thin Dielectrics

The planar capacitors are used as power-ground planes, as well as a reference for the transmission lines. Figure 4.77 shows two active chips on a package connected to a PCB. Transmission lines connecting one of these chips to the other one, or to other components on the PCB, have to be connected by signal vias. When such a signal via passes through a power-ground plane, the noise voltage on the power supply at that point gets coupled to the signal voltage, which will degrade the signal waveform. The noise voltage between the planes can be effectively reduced by using power-ground planes separated by thin dielectrics (which will be called here "planar capacitors"), since the power plane impedance is linearly proportional to the dielectric thickness. Even if there is no via transition for a particular signal net, there will be a return current generated by the switching circuit driving that transmission line. A broadband low-impedance power-ground system is necessary to supply enough current during such switching periods. As a result, planar capacitors help to improve the signal integrity, especially for high-speed signaling. Planar capacitors have been implemented in various products including server boards [80–83,87] since the 1990s.

For core decoupling, the capacitance of planar capacitors is generally not sufficient to provide the necessary charge to the switching circuits. However, they provide a low-impedance path for discrete SMD capacitors (or discrete embedded capacitors as will be shown in the next section), improving the effectiveness of the SMD capacitors. According to a study done by several original equipment manufacturers (OEMs), up to 75 percent of the surface-mount discrete decoupling capacitors can be removed by using an ultrathin loaded laminate material between the power-ground planes [97]. According to this study, the ratio of the removed discrete capacitance to the added planar capacitance was on the order of 10.

Measurements in the time domain also show that the power bus noise voltage can be smaller in a board containing a planar capacitor layer (with no SMD capacitors), as compared to a board having a thick FR4 layer between the power-ground planes, which include a number of SMD capacitors. Although the high-K materials introduce more resonances in a given frequency range, these resonances are damped due to the thin dielectrics [98]. Electromagnetic simulations have also shown that the plane resonances decrease by using thin dielectrics with or without high-K materials instead of a thick FR4 layer, even after removing a significant number of SMD capacitors [99].

In general, planar capacitors have to be supplemented by discrete components having a higher capacitance for core decoupling. The next section covers such embedded individual capacitors.

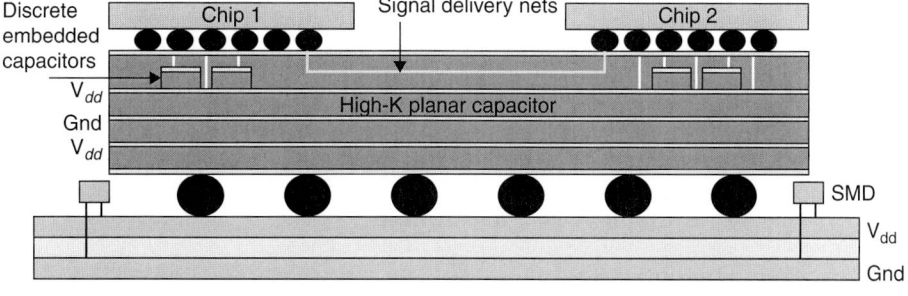

FIGURE 4.77 A package design with planar and discrete embedded capacitors.

Embedded Individual Capacitors

In order to overcome the disadvantages, thick- (or thin-) film capacitor layers in the package can be used for decoupling as shown in Figure 4.78. Embedded individual capacitors can be realized in these layers. These capacitors are of variable sizes, have different capacitances, and therefore resonate at different frequencies. The proximity of these capacitors to the active device reduces the loop inductance as compared to SMDs; therefore, they are effective in targeting frequencies above 100 MHz. Another advantage of using embedded individual thick-film capacitors is the high value of capacitance that can be obtained by using this technology. By proper selection of the sizes of the capacitors, the midfrequency band from 100 MHz to 2 GHz can be targeted for decoupling.

There are various available processes and materials for embedded individual capacitors. Novel polymer-ceramic nanocomposite dielectrics have been used to fabricate thin-film capacitors with a thickness of 10 μm and a dielectric constant of 30 to achieve a capacitance density up to 2.6 nF/cm^2. To support power levels above 200 W, ultrathin-film high-K dielectrics are being developed using low-temperature processes. Synthesis methods such as hydrothermal and sol-gel, with rapid thermal processing, have demonstrated films with a thickness <1 μm and capacitance densities of about 500 nF/cm^2. The low process temperatures are well suited for integration on low-cost organic SOP substrates [100]. As another example of embedded discrete capacitors, Motorola has developed a mezzanine-type capacitor with a density of 16.8 pF/mm^2 (@12-μm dielectric thickness) that has been applied in RF designs [101]. Oak Mitsui has demonstrated discrete embedded capacitors with a density of 17 pF/mm^2 (@16-μm dielectric thickness) [99]. As a case study for the design of a package with embedded decoupling capacitors, we use DuPont embedded discrete capacitors in the following paragraphs.

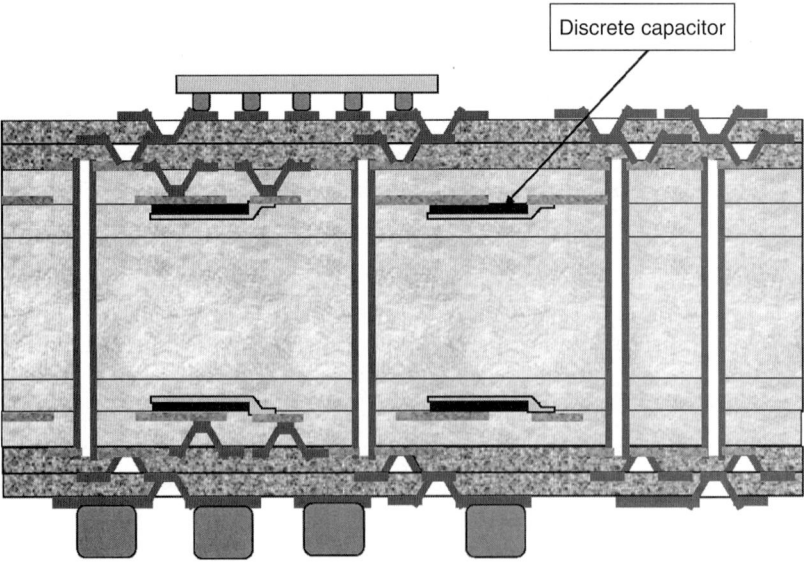

FIGURE 4.78 Discrete capacitor layers in a package.

To meet the target impedance between 100 MHz and 2 GHz, a package capacitive network has been designed with the embedded thick-film capacitors available from DuPont [84]. An important requirement of these capacitors is to be able to place them as close to the switching circuits as possible. The capacitive network can be designed with all the discrete capacitors placed directly under or around the die shadow in the two layers allocated to them in the package. The size of the die here is 11.8 mm × 11.8 mm [85] for a cost performance processor for the 65-nm node. The placement of the capacitors under the die shadow is done to reduce the effect of the spreading inductance of the planes in the package. The network consists of eighteen 1 mm × 1 mm capacitors and eighteen 0.75 mm × 0.75 mm capacitors in each discrete capacitor layer. The discrete capacitors in the lower layer of the package are connected to the power and ground bumps of the processor via blind and through vias. The layout of one of the discrete layers is shown in Figure 4.79. The network is designed such that majority of the vias from the capacitors connect directly to the flip-chip solder bumps of the processor. The vias were modeled using FastHenry, an inductance extraction program [86].

The embedded capacitor network was designed to target the frequency band between 100 MHz and 2 GHz. To provide decoupling over the entire band, VRMs, SMDs, and on-chip capacitance are also required. SMDs and on-chip capacitance in this example are effective below 100 MHz and above 2 GHz, respectively. The complete frequency response is shown in Figure 4.80. The figure shows the frequency band over which each decoupling component is effective.

VRMs are effective until the lower kilohertz region, SMDs provide decoupling from the kilohertz region until around 100 MHz, and on-chip capacitance is used above 2 GHz. It is evident that target impedance on the order of 1 mΩ can be met over a broad

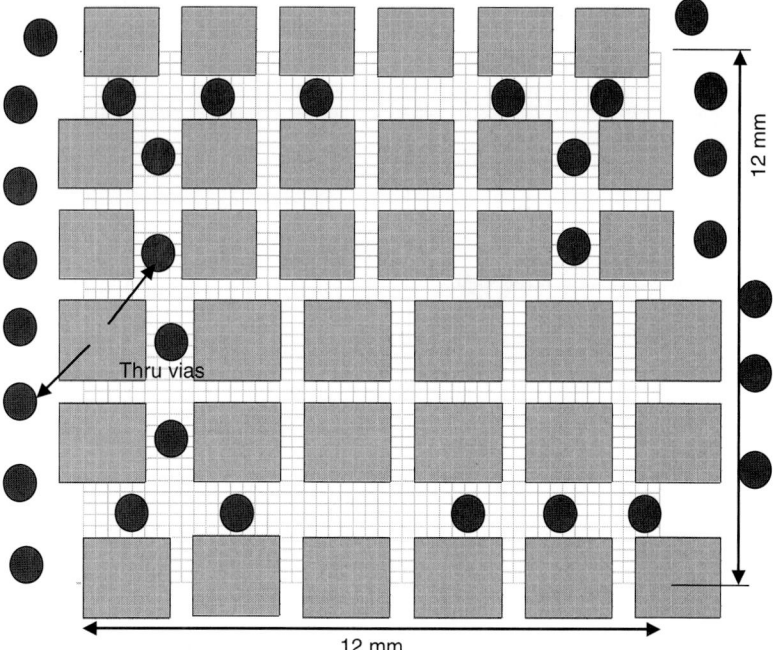

Figure 4.79 Layout of a discrete capacitor layer in the package.

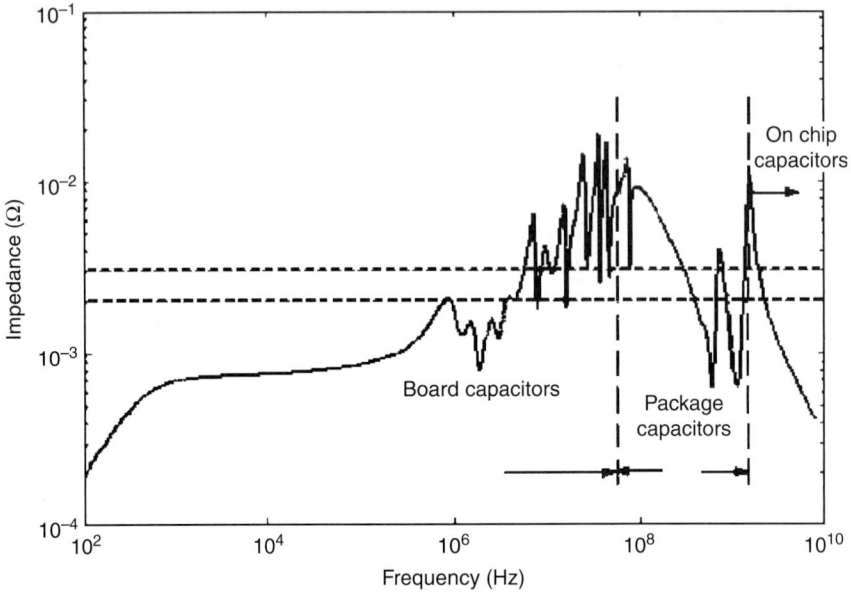

FIGURE 4.80 Impedance profile with VRM, SMD, embedded package, and on-chip capacitor.

frequency range from DC to multiples of the chip operating frequency using the combination of different decoupling components.

Next, the time domain performance of the capacitive network is investigated. The simulations were carried out to capture the performance with a 2 GHz clock as the input to the system. The rise time and fall time of the current pulse is 50 ps, respectively, and the time period of the clock is 500 ps as shown in Figure 4.81. The

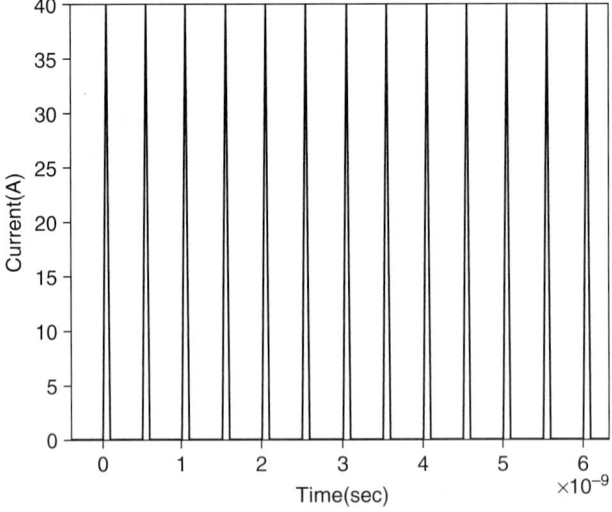

FIGURE 4.81 Current source.

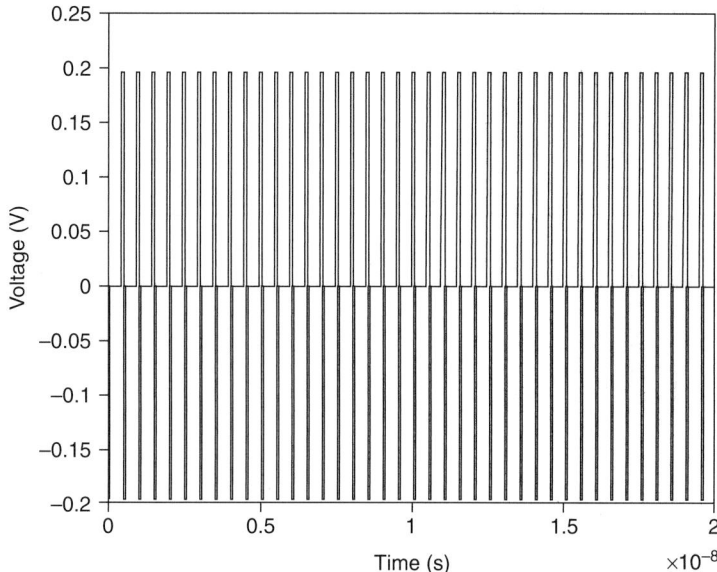

FIGURE 4.82 Noise with VRM, on-chip, and SMD capacitors.

target impedance to be met depends on the magnitude of the current pulse. In this simulation the target impedance chosen was 2.5 mΩ. Therefore, from Equation (4.28), the magnitude of the current pulse calculated assuming a core voltage of 1 V is 40 A. To obtain the time domain response of the system, the Fourier transform of the input current pulse train is multiplied with the frequency domain data of the PDN. The inverse-Fourier transform of the resultant frequency spectrum is then used to obtain the time domain response. The system simulation was initially carried out with a VRM, SMDs, and on-chip decoupling capacitors. The time domain performance of the system is shown in Figure 4.82. To highlight the improvement with embedded package capacitors, the system was simulated with the VRM, SMDs, package, and on-chip capacitors. The response of the system is shown in Figure 4.83. A performance improvement of five times is clearly seen with embedded capacitors included in the simulations.

4.7.4 Characterization of Embedded Capacitors

Accurate measurement methods and development of accurate models of embedded capacitor layers are required to ensure that integration provides good performance. A two-port frequency domain measurement methodology is necessary to accurately measure the impedance of the capacitance structures [87].

This section describes a measurement example using such a two-port measurement methodology. The measurement equipment used here is Agilent's 8720ES vector network analyzer (VNA) with a bandwidth of 50 MHz to 20.5 GHz and 500-μm GS-SG Cascade probes. A standard SOLT (short, open, load, and through) calibration was used

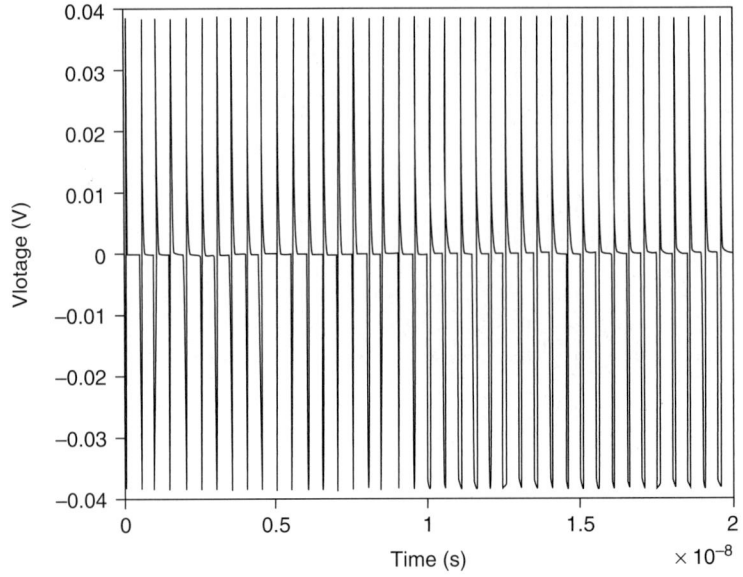

FIGURE 4.83 Noise with VRM, SMD, embedded package, and on-chip capacitors.

by using the Impedance Standard Substrate (ISS). The basic equations used to characterize the capacitor structures can be written as

$$\text{Re}(Z_{11}) = 25 \times \frac{\text{Re}(S_{21}) \times \left[1 - \text{Re}(S_{21})\right] - \left[\text{Im}(S_{21})\right]^2}{\left[1 - \text{Re}(S_{21})\right]^2 + \left[\text{Im}(S_{21})\right]^2} \tag{4.33}$$

$$\text{Im}(Z_{11}) = 25 \times \frac{\text{Im}(S_{21})}{\left[1 - \text{Re}(S_{21})\right]^2 + \left[\text{Im}(S_{21})\right]^2} \tag{4.34}$$

where, $\text{Re}(Z_{11})$ and $\text{Im}(Z_{11})$ are the real and imaginary parts of the device under test. The measurement setup for characterizing the capacitance structure is shown in Figure 4.84. Probe 1 is the transmitter, and probe 2 measures the voltage drop across the device in one measurement cycle. The functions of the ports are reversed in the next measurement

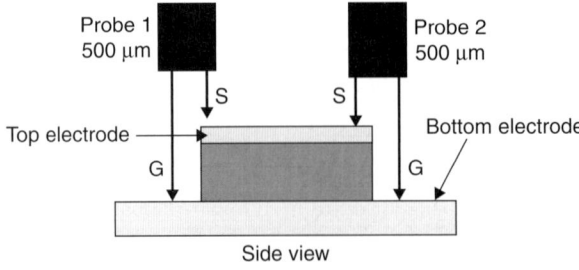

FIGURE 4.84 The measurement setup for characterizing the capacitors.

Mixed-Signal (SOP) Design 237

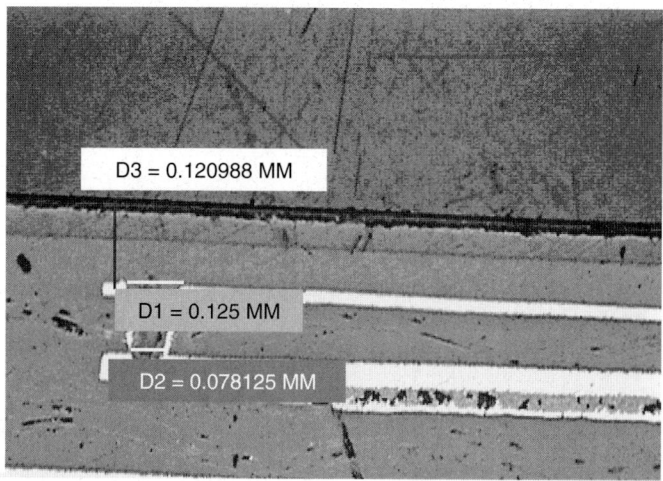

FIGURE 4.85 Embedded capacitor in laminate.

cycle of the VNA. S_{21} is the insertion loss measured across the device at each frequency point. The real and imaginary parts of the impedance over the measured frequency band can be obtained using these equations.

The measured thick-film capacitors are available from DuPont [84]. They are compatible with the standard FR4/BT laminate printed wiring board technology and can be integrated in BT laminate, as shown in Figure 4.85. The cross section of the capacitor is shown in Figure 4.86. The thickness of the dielectric is in the range of 20 to 24 μm, and the dielectric constant is 3000. The loss tangent of the dielectric at 1 MHz is less than 0.05. The top copper foil and the bottom electrodes are 35 μm and 3 to 5 μm thick, respectively. These capacitors are available on a copper foil with discrete patterned dielectrics and electrodes. The process ground rules define the maximum and minimum sizes of the capacitors, which translate to 0.5 and 3 mm a side, respectively.

Figure 4.87 shows the impedance profile of different capacitors that were measured using the two-port methodology. Figure 4.88 shows the extracted dielectric constant value, which remains relatively constant as observed from the figure.

The capacitors were modeled using the transmission matrix method (TMM) [88]. In modeling the capacitor structures the fringing effects have been ignored because of the large aspect ratio of these structures. Figure 4.89 shows the model-to-hardware

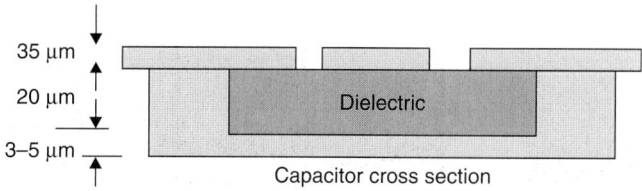

FIGURE 4.86 Cross section of the capacitor.

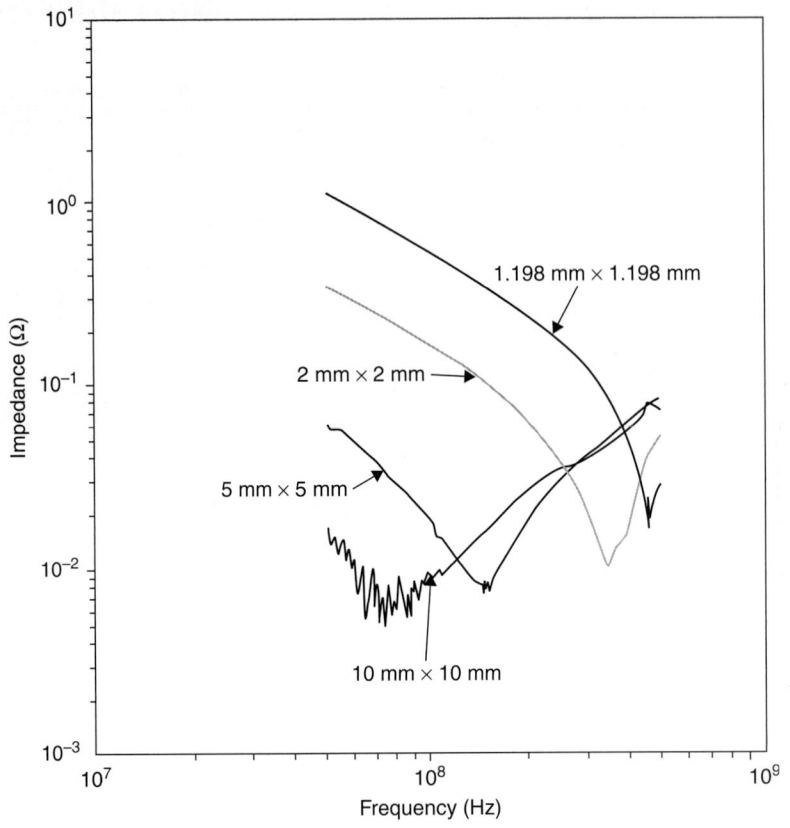

FIGURE 4.87 Frequency response of different size capacitors.

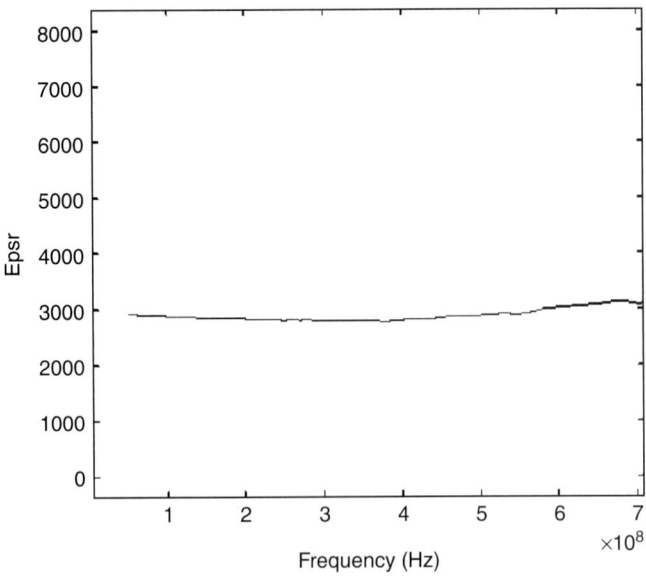

FIGURE 4.88 Extracted dielectric constant versus frequency.

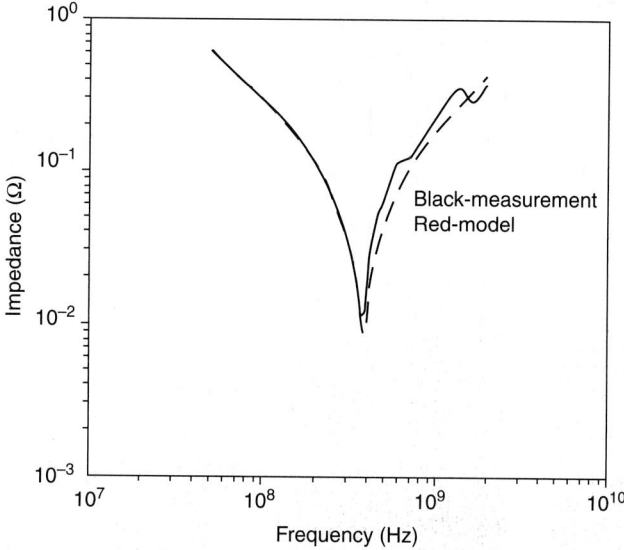

FIGURE 4.89 Model versus measurement correlation for 2 mm × 2 mm capacitor.

correlation of a 2 mm × 2 mm capacitor that was measured and modeled. Ports were appropriately defined in the model to obtain the correct model-to-hardware correlation. The definition of the ports corresponds to the exact coordinates of the probes on the measurement structure. The parasitic inductance of the probe set up has to be included in the model to obtain a good correlation between the model and measurement. The details of the parasitic inductance extraction and model-to-hardware correlation are available in [89].

The capacitance, inductance, and resistance of the various sized capacitors have been extracted from the measured results and are listed in Table 4.12.

4.8 Electromagnetic Bandgap (EBG) Structures

Electromagnetic bandgap (EBG) structures are a distributed means for implementing RF isolation. They are periodic structures in which the propagation of electromagnetic waves is forbidden in certain frequency bands [62]. In the past, EBG-based designs have

Capacitor Size (mm × mm)	ESC (nF)	ESL (pH)	ESR (mΩ)
	2.84	42.6	16
2 × 2	8.772	23.8	10.36
5 × 5	53.93	22.1	7.22
10 × 10	191	24.1	5

TABLE 4.12 Extracted Parameters from Measurements

been used for achieving RF isolation in antenna design [90]. And more recently, Kamgaing and Ramahi [91] have used a three-layer "mushroom-type" EBG to suppress the propagation of SSN in digital applications. With good point-to-point isolations and the ability to use a single power supply, EBGs have the potential to solve the power-supply based noise coupling in SOP-based mixed-signal integration.

An EBG structure proposed in [62] provides excellent isolation of more than 60 dB. This EBG structure consists of a patterned power-ground plane pair and requires no additional vias, which are necessary in the embedded EBG structure [91]. Therefore, standard printed circuit board fabrication techniques are easily applicable, which is a cost-effective solution.

Figure 4.90 shows examples of EBG structures with one-dimensional (1D) lattice and two-dimensional (2D) lattice formed in a power-ground pair. These lattices consist of large metal patches and small metal branches connecting adjacent large patches. The EBG pattern may be applied to either the power plane or the ground plane depending on the design.

One-dimensional and two-dimensional dispersion-diagram analysis can be used to estimate the stopband characteristics. This analysis is available for any EBG structure if a unit cell of the EBG structure is represented as a multiport network. Since the analysis focuses on the unit cell, it considerably reduces the calculation time compared with electromagnetic (EM) calculation of the entire EBG structure.

Figure 4.91 shows various unit-cell structures that have been applied as EBGs in power-ground planes. These represent only a few of the possible shapes that can be applied as EBGs. Any other unit cell that can be applied as a periodic pattern on the power-ground planes would exhibit EBG characteristics such as alternating stopband and passband.

One characteristic that is common to all the four structures in Figure 4.91 is that the unit cells are connected with each other through narrow bridges. Hence, it is convenient to analyze them as multiport networks. The network parameters of such a multiport can be obtained using measurements of unit cells or electromagnetic simulations.

Compared to the "mushroom-type" EBG of [64], the two-layered EBGs in Figure 4.91 use only two metal layers and do not require any microvias. Most importantly, they provide a much higher performance in terms of isolation. Figure 4.92 shows the

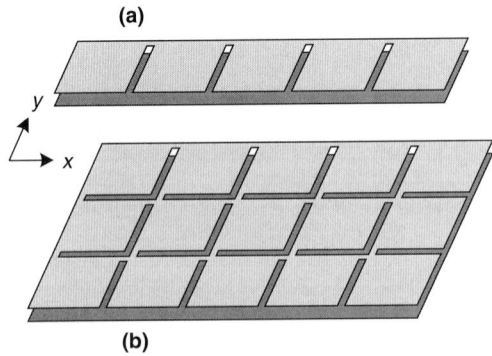

FIGURE 4.90 EBG lattice. (*a*) One dimension. (*b*) Two dimensions.

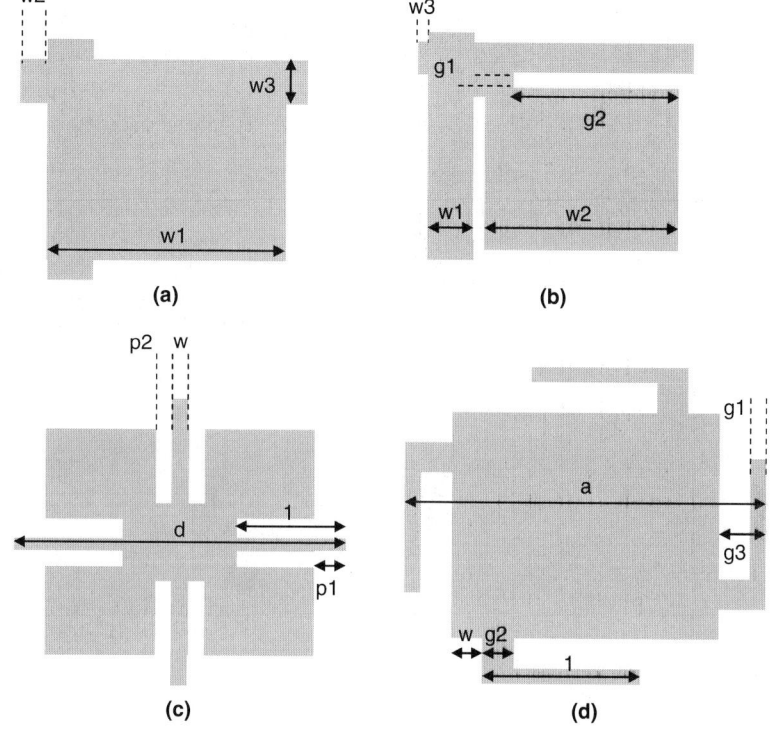

FIGURE 4.91 Examples of two-layered EBGs. (*a*) Alternating impedance EBG. (*b*) Slit EBG. (*c*) Low period coplanar EBG. (*d*) L-bridged EBG.

photograph of an AI-EBG implementation in a two-layer FR4 process (tan $\delta = 0.02$, $\varepsilon_r = 4.4$ at 1 MHz). As can be observed in Figure 4.93, the EBG provides ~80-dB isolation in the 2.4- to 3.5-GHz frequency band.

It is evident that an EBG-based power distribution scheme has the potential for suppressing power supply based noise propagation in mixed-signal systems. It also enables the use of a single power supply for the module, reducing the size and cost of the device.

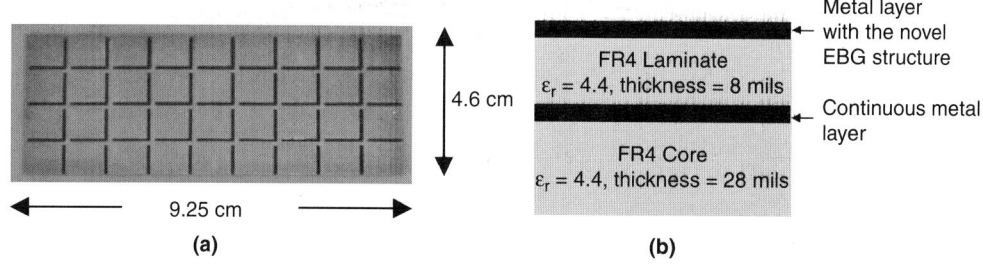

FIGURE 4.92 (*a*) Photograph of an AI-EBG implementation. (*b*) FR4 stackup used.

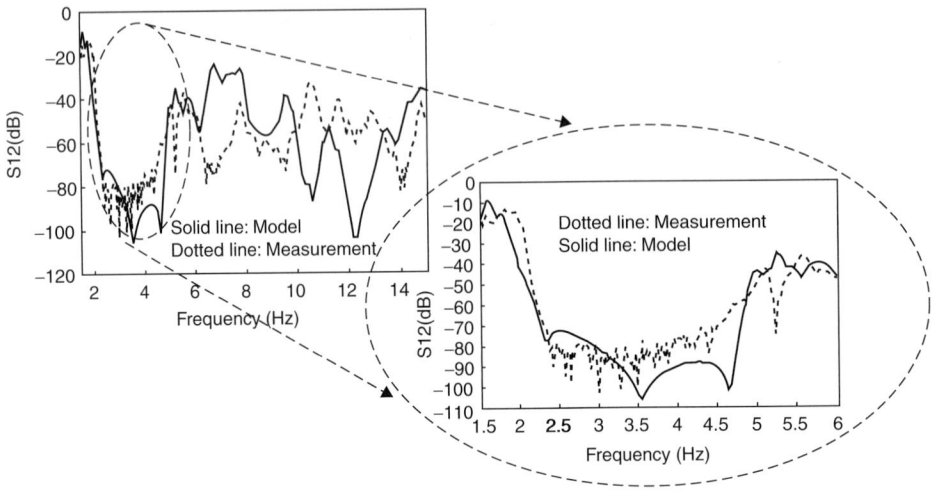

FIGURE 4.93 Model-to-hardware correlation for isolation (S21) in the structure.

4.8.1 Analysis and Design of EBG Structures

Dispersion diagrams can be used to identify the passbands and stopbands of periodic structures [92]; hence, they are very useful in designing EBGs. The dispersion diagram of an EBG is obtained by simulating only one unit cell of the EBG, so it is numerically a very efficient procedure. It enables a systematic methodology for design of EBGs: the unit cell structure with the required stopband frequencies can be identified using the dispersion diagram. Unit cells are then cascaded with each other to form an EBG structure until the required isolation levels are obtained based on the frequency response of the EBG.

One-Dimensional Unit-Cell Analysis

The dispersion diagram for a 1D unit cell is obtained using the relation

$$\cosh(\alpha d + j\beta d) = \frac{Z_{11} + Z_{22}}{2Z_{12}} \qquad (4.35)$$

where α and β are the attenuation and phase constants, d is the distance between the ports, and Z_{11}, Z_{12} are the self and transfer impedances, respectively, between an input port and output port of a unit cell. Perfect conductors and a lossless dielectric are assumed in modeling the unit cell.

This analysis has the advantage of estimating the stopband of a periodic EBG structure by knowing the Z parameters of the unit cell, without any calculations for the entire EBG structure. The Z parameters of a unit cell can be calculated not only by full-wave EM solvers but also by using SPICE models. The latter can be solved more efficiently by applying the transmission matrix method (TMM) [93], or multilayered finite-difference method (MFDM) [94]. Thus, the analysis based on the dispersion diagram enables considerable calculation time reduction.

In order to show that the stopband obtained from the dispersion diagram is accurate, the 1D EBG structure in Figure 4.94 is shown. Figure 4.94b is a photo of a DUT used for

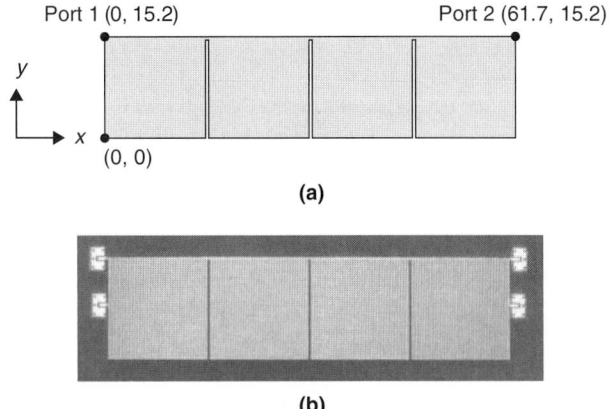

Figure 4.94 (a) One-dimensional EBG lattice of 4 × 1 geometry (unit in millimeters). (b) Photo of the DUT used for measurement.

measurement. The unit cell has the following parameters: $w_1 = 15.2$ mm (600 mils), $w_2 = 0.127$ mm (5 mils), $w_3 = 0.254$ mm (10 mils), and dielectric thickness = 0.127 mm (5 mils). In calculating the Z parameters of the unit cell, material losses were neglected and the dielectric constant ε_r is 4.0. Figure 4.95a shows the dispersion diagram for this example. In the figure, the colored area indicates the stopband since the plots in the dispersion diagram correspond to wave propagation toward the x direction (Γ–X along the x axis represents the propagation constant). Figure 4.95b shows the measured S_{21} (solid curve) together with S_{21} simulated by Sonnet emCluster [28] (dashed curve). In the Sonnet

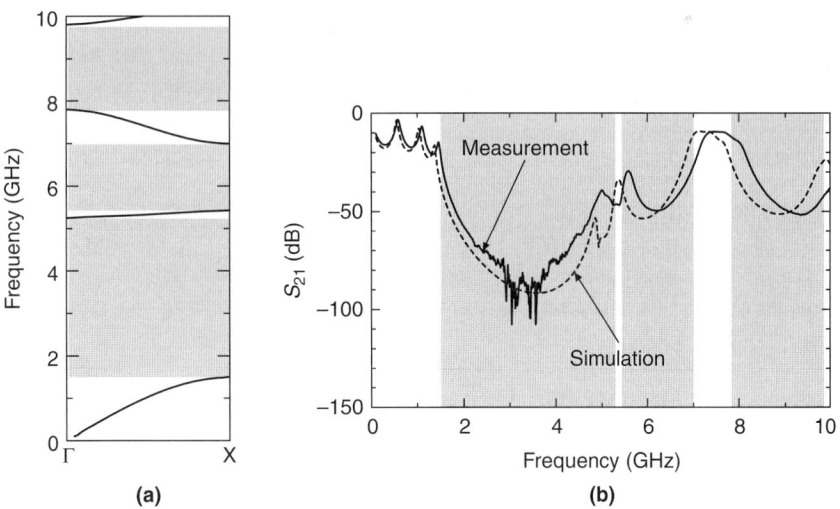

Figure 4.95 (a) Calculated dispersion diagram of the 1D EBG unit cell with patch size = 15.2 mm, branch size = 0.254 mm, dielectric thickness = 127 mm, and $\varepsilon_r = 4.0$. The colored area shows the stopband of the EBG structure. (b) Comparison of transmission coefficient S_{21} between the measurement and simulation with Sonnet. The colored area shows the stopband of the EBG structure predicted with the dispersion diagram.

calculation, a dielectric loss of tan $\delta = 0.02$ and a copper conductivity of $\sigma_c = 5.8 \times 10^7$ siemens per meter (S/m) were used. The stopband predicted with Figure 4.95a is superposed on Figure 4.95b. This figure indicates that the stopband predicted from the dispersion diagram has good agreement with the measured and simulated data.

Two-Dimensional Unit-Cell Analysis

If the unit cells are connected with each other to form a 2D lattice, the 1D dispersion diagram analysis becomes inaccurate, since wave propagation can be possible in any direction on the plane in a 2D EBG, as opposed to the linear wave propagation in a 1D EBG structure. Hence, the 1D analysis needs to be extended for 2D structures. Figure 4.96 shows the port definitions for a 1D and 2D EBG. Now, focusing on the x and y directions, the following network matrix becomes useful:

$$\begin{pmatrix} V_1 \\ I_1 \\ V_2 \\ I_2 \end{pmatrix} = F \begin{pmatrix} V_3 \\ -I_3 \\ V_4 \\ -I_4 \end{pmatrix} \tag{4.36}$$

Based on this matrix, the following eigenvalue equation can be used to generate the 2D dispersion diagram:

$$\left\{ F - \begin{pmatrix} e^{\gamma_x d_x} & & & \\ & e^{\gamma_x d_x} & & \\ & & e^{\gamma_y d_y} & \\ & & & e^{\gamma_y d_y} \end{pmatrix} \right\} \begin{pmatrix} V_3 \\ -I_3 \\ V_4 \\ -I_4 \end{pmatrix} = 0 \tag{4.37}$$

In this equation the propagation factor $e^{-\gamma_x d_x}$ is for the interval d_x in the $+x$ direction and $e^{-\gamma_y d_y}$ is for the interval d_y in the $+y$ direction.

Based on the 2D dispersion diagram analysis, the stopband of the 2D EBG structure with slits (case b) shown in Figure 4.97a and b has been investigated. The unit cell has the following parameters: $w_1 = 0.25$ mm, $w_2 = 14.73$ mm, $w_3 = 0.13$ mm, $g_1 = 0.25$ mm, and $g_2 = 7.62$ mm. The dielectric used was FR4 with a thickness of 127 µm. Figure 4.97c shows the 2D dispersion diagram that is plotted accounting for the Brillouin zone [94a]. The colored area of Figure 4.97c shows a complete stopband indicating no wave

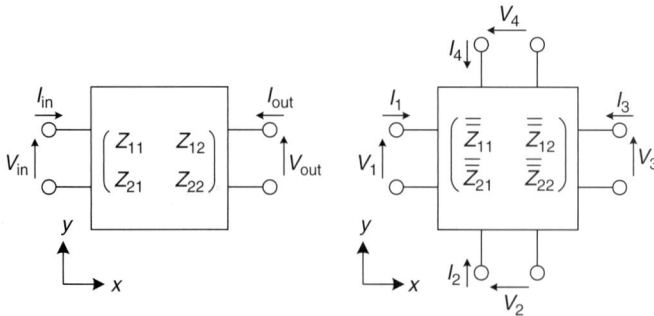

Figure 4.96 Network representation—two port (left) and four port (right).

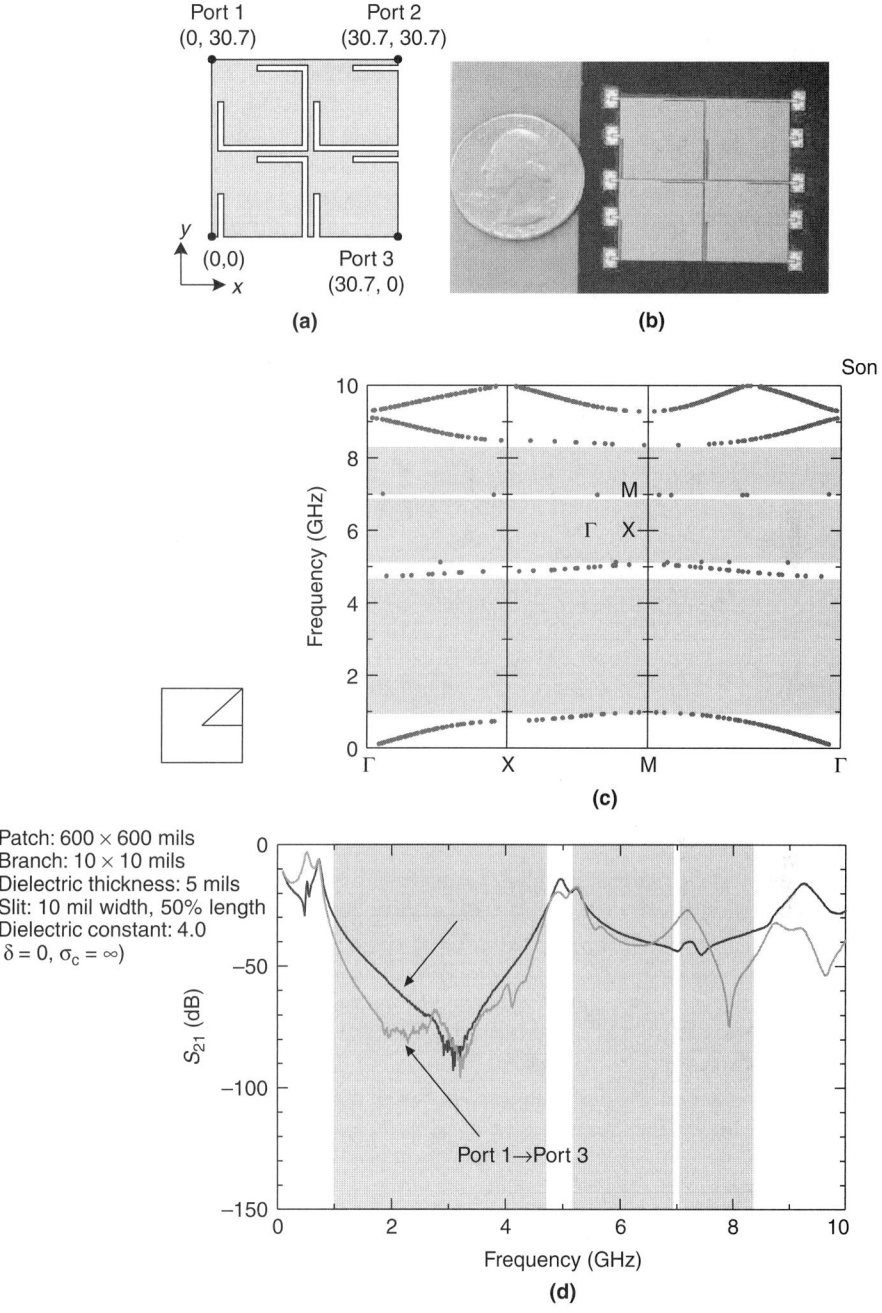

FIGURE 4.97 (a) 2D EBG. (b) Photograph. (c) Dispersion diagram. (d) Measured S_{12}.

propagation toward any directions. This complete stopband is regarded as the stopband in the 2D EBG structure. Figure 4.97d shows the measured data of the transmission coefficient S_{21} from port 1 to port 2 and to port 3 shown in Figure 4.97a together with the colored area, which indicates the stopband predicted with the dispersion-diagram analysis. Figure 4.97d shows that the 2D dispersion-diagram analysis provides a good prediction of the stopband.

4.8.2 Application of EBGs in Power Supply Noise Suppression

The EBG structure with the frequency response in Figure 4.98 is used on a testbed to isolate an LNA from an FPGA. The isolation in the frequency band of interest of the LNA (centered around 2.14 GHz) is more than 90 dB; therefore, this EBG would suppress almost all direct in-band noise coupling.

Figure 4.99 shows the photograph of the fabricated mixed-signal test vehicle. It consists of an LNA and FPGA, where the ground plane has been patterned with the AI-EBG based structure. The FPGA drives four 50-Ω microstrip lines, with terminations implemented using 50-Ω 0603 resistors.

Figure 4.100 shows the comparison of LNA output spectrum for the two test vehicles—the blue line represents noise coupling through ordinary power-ground plane pair, while the red line represents the noise coupled through the Alternating Impedance (AI) EBG-based power distribution system. At low frequencies, where the EBG does not provide much isolation, both measurements remain similar. However at ~2 GHz (where the stopband of the EBG begins), a clear distinction can be seen in the amount of coupled noise power. For the frequencies of 2.1 GHz and above, there is virtually no noise power transferred in the test vehicle with AI-EBG structures, exhibiting superior EMI control.

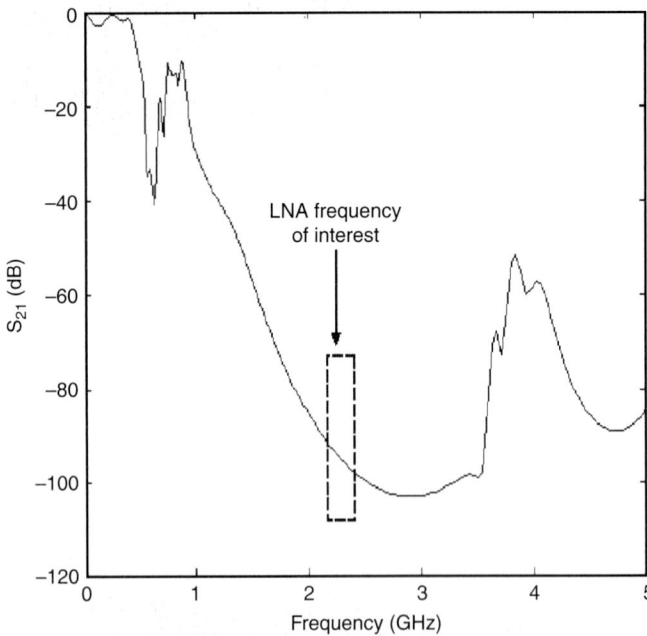

Figure 4.98 Simulated transmission coefficient of the EBG.

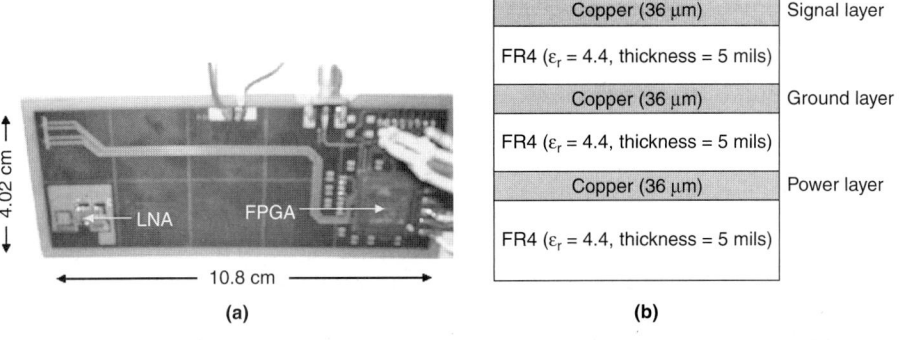

FIGURE 4.99 (a) Photograph of test vehicle. (b) Cross section.

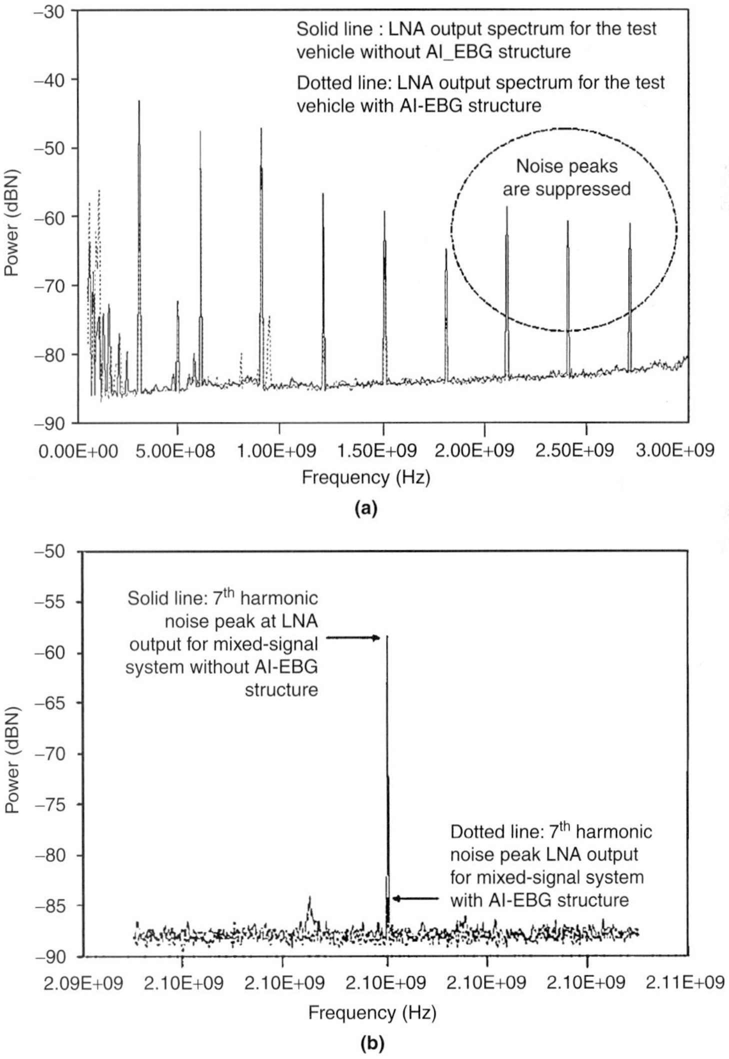

FIGURE 4.100 (a) Comparison of LNA output spectrum with and without Al-EBG. (b) Comparison of seventh harmonic noise at LNA output.

4.8.3 Radiation Analysis of EBGs

The AI-EBG structure has periodic gaps that can radiate energy if the AI-EBG structure is used as a reference plane. This is because return currents are forced to flow around the splits, and charges therefore accumulate on their edges [95], which create a source of far field radiation [96]. Hence, the splits or gaps in a reference plane can be considered as slot antennas that can interfere with other devices. This causes an electromagnetic interference (EMI) problem. In this section, near-field and far-field analysis for three test vehicles are shown through simulations and measurements to better understand the radiation mechanisms of the AI-EBG structure.

Design of Three Test Vehicles

Three test vehicles have been designed and fabricated for radiation analysis. The first test vehicle is a microstrip line on a solid plane, the second test vehicle is a microstrip line on an AI-EBG structure, and the third test vehicle is a microstrip line on an embedded AI-EBG structure. The third test vehicle is designed to suppress noise in mixed-signal systems without any EMI problems. This is possible since the solid plane is used as a reference plane for the microstrip line in this embedded AI-EBG structure. In Figure 4.101, the cross sections of these three test vehicles are shown.

The top view of these three test vehicles is also shown in Figure 4.102. The dielectric material of the test vehicles is FR4 with a relative permittivity $\varepsilon_r = 4.4$, the conductor is copper with conductivity $\sigma_c = 5.8 \times 10^7$ S/m, and the dielectric loss tangent is $\tan \delta = 0.02$.

The copper thickness for the microstrip line, solid plane, and AI-EBG plane in the test vehicles is 35 µm; the dielectric thickness between two conductors is 5 mils; and the dielectric thickness of the most bottom layer is 28 mils. For the AI-EBG structures in the second and third test vehicles, the size of the metal patch is 1.5 cm × 1.5 cm and the size of the metal branch is 0.1 cm × 0.1 cm. It should be noted that the size of the metal patches in the first column near the SMA connector is 1.3 cm × 1.5 cm.

EMI Simulations and Measurements

The full wave solver (Sonnet) has been used for EMI analysis of the three test vehicles. Figure 4.103 shows the current density for the AI-EBG plane in test vehicle 2. The slits due to the EBG structure on the reference plane cause return path discontinuities as can

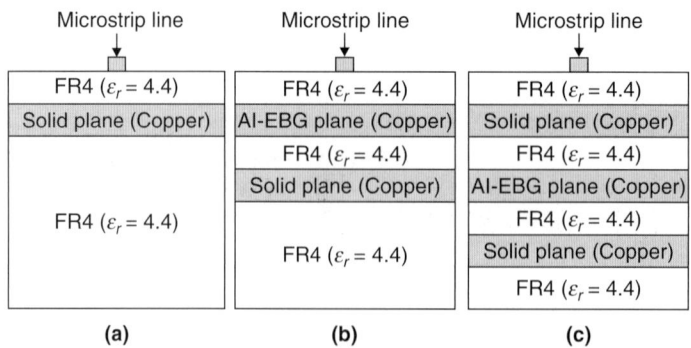

Figure 4.101 Cross section of three test vehicles. (*a*) Microstrip over ground plane. (*b*) Microstrip on AI-EBG. (*c*) Microstrip with embedded AI-EBG.

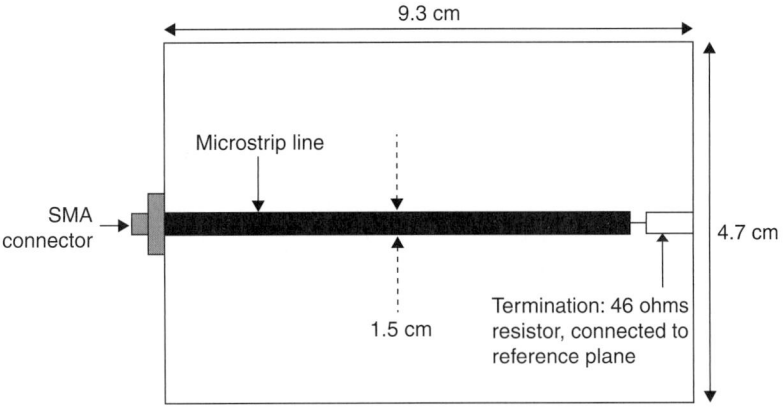

FIGURE 4.102 Top view of test vehicle.

be seen in the figure. Test vehicles 1 and 2 behave as common microstrip lines, due to the solid reference plane underneath.

Far-field measurements have also been done for the test vehicles. The far-field measurements were carried out using the Anritsu MG3642A RF signal generator (bandwidth: 125 kHz to 2080 MHz), Agilent E4440A spectrum analyzer (bandwidth: 3 kHz to 26.5 GHz, and the antenna in anechoic chamber. Figure 4.104a shows the measurement setup for the far-field measurements. Since the RF signal generator works properly up to 2 GHz, the far-field measurements were also done up to 2 GHz. The distance between the equipment under test (EUT) and antenna was 3 m in this case. The RF signal generator was connected to the EUT as a source, and the spectrum analyzer, which was connected to the antenna, recorded the field intensity from the surface of the test vehicles. In this measurement, the radiation intensity from test vehicle 2 is the maximum among three test vehicles, as shown in Figure 4.104b, and test vehicles 1 and 3 showed almost the same radiation intensity because a solid plane was used as a reference plane for these two test vehicles.

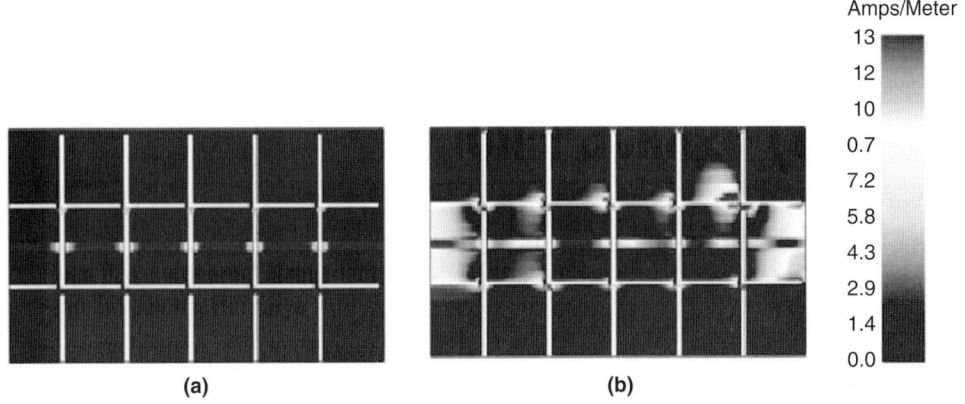

FIGURE 4.103 Simulation result for test vehicle 2: (a) 300 MHz and (b) 2.7 GHz.

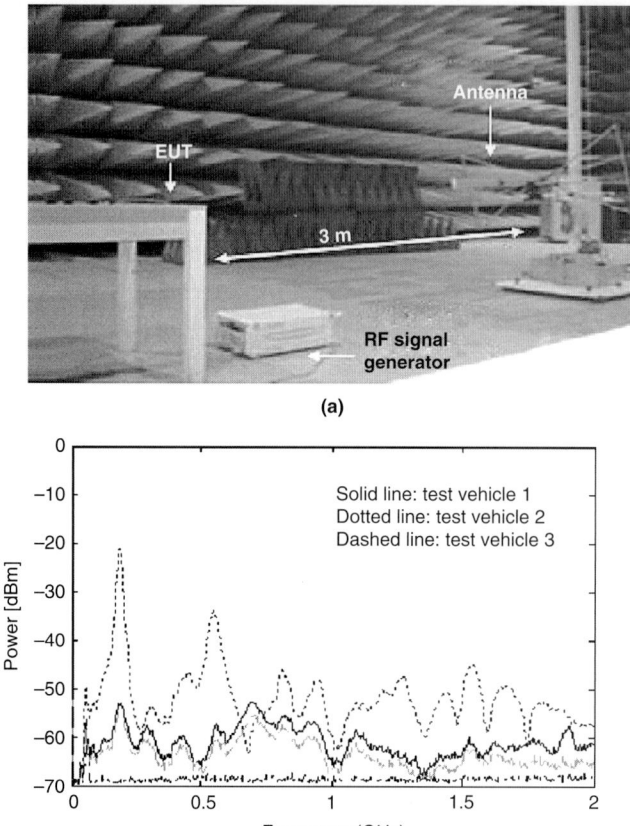

Figure 4.104 (a) Far-field measurement setup. (b) Far-field measurements.

As a conclusion, test vehicle 2 (an AI-EBG plane as a reference plane) showed the maximum radiation intensity in near-field and far-field simulations and measurements among the three test vehicles, which could cause a possible EMI problem. To minimize EMI problems, the test vehicle with the embedded AI-EBG structure (test vehicle 3) can be used to suppress noise in mixed-signal system.

4.9 Summary

In this chapter the design of mixed-signal modules has been discussed. The issues discussed include the design of embedded passives for RF circuits, chip-package codesign, design tools, embedded decoupling capacitors for digital circuits, noise coupling, and EMI control through the use of EBG structures. A lot of the design methods and electrical issues have been quantified using simulations and measurements. As the trend for miniaturization continues, the need for advanced technologies will continue to increase and clever design methods and methodologies will be required. Some of these advanced ideas have been captured in this chapter.

Acknowledgments

The authors would like to thank Sidharth Dalmia for the section on WLAN Front End Modules. The authors would also like to acknowledge all the students, engineers, and visiting scholars of the Epsilon Research Group and Packaging Research Center, Georgia Tech. In particular, the authors would like to acknowledge the contributions of George White, Jinwoo Choi, Venky Sundaram, Raj Pulugurtha, Yoshitaka Toyota, Takayuki Watanabe, Rohan Mandrekar, Krishna Srinivasan, Krishna Bharath, Souvik Mukherjee, and Prathap Muthana.

References

1. The Electronics Industry Report, Prismark Partners LLC, 2004.
2. World Semiconductor Trade Statistics, http://www.wsts.org.
3. Intel Corporation, http://www.intel.com/labs/features/cn09031.htm.
4. N-Gage, http://www.n-gage.com.
5. H. Hashemi and A. Hajimiri, "Concurrent multiband low-noise amplifiers-theory, design, and applications", *IEEE Trans. Microw. Theory Tech.*, vol. 50, no. 1, January 2002, pp. 288–301.
6. V. Govind, S. Dalmia, J. Choi, and M. Swaminathan, "Design and implementation of RF subsystems with multiple embedded passives in multi-layer organic substrates," *Proceedings of the IEEE Radio and Wireless Conference (RAWCON)*, Boston, MA, August 2003, pp. 325–328.
7. S. Wu and B. Razavi, "A 900-MHz/1.8-GHz CMOS receiver for dual-band applications," *IEEE Journal of Solid-State Circuits*, vol. 33, December 1998, pp. 2178–2185.
8. D. M. Pozar and S. M. Duffy, "A dual-band circularly polarized stacked microstrip antenna for global positioning satellite," *IEEE Antennas and Propagation*, vol. 45, November 1997, pp. 1618–1625.
9. A. Raghavan, D. Heo, M. Maeng, A. Sutono, K. Lim, and J. Laskar, "A 2.4 GHz high efficiency SiGe HBT power amplifier with high-Q LTCC harmonic suppression filter," *IEEE MTT-S International Microwave Symposium Digest*, vol. 2, June 2002, pp. 1019–1022.
10. S.–W. Yoon, M. F. Davis, K. Lim, S. Pinel, M. Maeng, C.–H. Lee, S. Chakraborty, and S. Mekela, "C-band oscillator using high-Q inductors embedded in multi-layer organic packaging," *IEEE MTT-S International Microwave Symposium Digest*, vol. 2, June 2002, pp. 703–706.
11. A. , S. Dalmia, and M. Swaminathan, "A 3G/WLAN VCO with high Q embedded passives in high performance organic substrate," *Proceedings of the IEEE Asia Pacific Microwave Conference (APMC)*, Seoul, S. Korea, November 2003, pp. NA.
12. P. Pieters, K. Vaesen, G. Carchon, S. Brebels, W. de Raedt, E. Beyne, M. Engels, and I. Bolsens, "Accurate modeling of high-Q spiral inductors in thin-film multilayer technology for wireless telecommunication applications," *IEEE Transactions on Microwave Theory and Techniques*, vol. 49, April 2001, pp. 589–599.
13. A. L. L. Pun, T. Yeung, J. Lau, F. J. R. Clement, and D. K. Su, "Substrate noise coupling through planar spiral inductor," *IEEE Journal of Solid-State Circuits*, vol. 33, June 1998, pp. 877–884.

14. M. Xu, D. K. Su, D. K. Shaeffer, T. H. Lee, and B. A. Wooley, "Measurement and modeling the effects of substrate noise on the LNA for a CMOS GPS receiver," *IEEE Journal of Solid-State Circuits*, vol. 36, March 2001, pp. 473–485.
15. N. K. Verghese and D. J. Allstot, "Computer-aided design considerations for mixed-signal coupling in RF integrated circuits," *IEEE Journal of Solid-State Circuits*, vol. 33, March 1988, pp. 314–323.
16. D. K. Su, M. J. Loinaz, S. Masui, and B. A. Wooley, "Experimental results and modeling techniques for substrate noise in mixed-signal integrated circuits," *IEEE Journal of Solid-State Circuits*, vol. 28, April 1993, pp. 420–429.
17. T. Liu, J. D. Carothers, and W. T. Holman, "Active substrate coupling noise reduction method for ICs," *IEEE Electronic Letters*, vol. 35, September 1999, pp. 1633–1634.
18. M. Felder and J. Ganger, "Analysis of ground-bounce induced substrate noise coupling in a low resistive bulk epitaxial process: Design strategies to minimize noise effects on a mixed-signal chip," *IEEE Transactions on Circuits and System—II. Analog and Digital Signal Processing*, vol. 46, November 1999, pp. 1427–1436.
19. R. C. Frye, "Integration and electrical isolation in CMOS mixed-signal wireless chips," *Proceedings of the IEEE*, vol. 89, April 2001, pp. 444–455.
20. M. Nagata, J. Nagai, T. Morie, and A. Iwata, "Measurements and analyses of substrate noise waveform in mixed-signal IC environment," *IEEE Transactions on Computer-Aided Design of Integrated Circuits and Systems*, vol. 19, June 2000, pp. 671–678.
21. M Nagata, J. Nagai, K. Hijikata, T. Morie, and A. Iwata, "Physical design guides for substrate noise reduction in CMOS digital circuits," *IEEE Journal of Solid-State Circuits*, vol. 36, March 2001, pp. 539–549.
22. J. Briaire and K. S. Krisch, "Principles of substrate crosstalk generation in CMOS circuits," *IEEE Transactions on Computer-Aided Design of Integrated Circuits and Systems*, vol. 19, June 2000, pp. 645–653.
23. J. Mao, "Modeling of Simultaneous Switching Noise in On-Chip and Package Power Distribution Networks Using Conformal Mapping, Finite Difference Time Domain and Cavity Resonator Methods", PhD Dissertation, School of Electrical and Computer Engineering, Georgia Institute of Technology, 2004.
24. D. C. Thompson, O. Tantot, H. Jallageas, G. E. Ponchak, M. M. Tentezeris, and J. Papapolymerou, "Characterization of liquid crystal polymer (LCP) material and transmission lines on LCP substrates from 30 to 110 GHz," *IEEE Trans. Microwave Theory and Tech.*, vol. 52, April 2004, pp. 1343–1352.
25. http://www.skyworksinc.com/products_detailpop2.asp?pid=8723, March 2006.
26. http://www.coilcraft.com/prod_rf.cfm, March 2006.
27. W. Yun, A.Bavisi, V. Sundaram, M. Swaminathan, and E. Engin, "3D integration and characterization of high Q passives on multilayer LCP substrate," *IEEE Asia Pacific Microw. Conf.*, December 2005, pp. 327–330.
28. http://www.sonnetusa.com.
29. A. Bavisi, V. Sundaram, M. Swaminathan, S. Dalmia, and G. White, "Design of a dual frequency oscillator for simultaneous multi-band radio communication on a multi-layer liquid crystalline polymer substrate," *IEEE Radio Wireless Symp.*, January 2006, pp. 431–434.

30. V. Palazzari, S. Pinela, J. Laskar, L. Roselli, and M. Tentezeris, "Design of an asymmetrical dual-band WLAN filter in liquid crystal polymer (LCP) system-on-package technology", *IEEE Microw. and Wireless Comp. Lett.*, vol. 15, March 2005, pp. 165–167.
31. R. Bairasubramaniam, S. Pinel, J. Papapolymerou, J. Laskar, C. Quendo, E. Rius, A. Manchec, and C. Person, "Dual-band filters for WLAN applications on LCP technology," in *IEEE MTT-S Int. Microwave Symp. Dig.*, June 2005.
32. C. Quendo, E. Rius, and C. Person, "An original topology of dual-band filter with transmission zeros," *IEEE IMS*, June 2003, pp. 1093–1096.
33. C. Quendo, E. Rius, A. Manchec, Y. Clavet, B. Potelon, J.-F. Fanvennec, and C. Person, "Planar tri-band filter based on dual behavior resonator (DBR)" *EUMC*, October 2005, pp. NA.
34. N. Marchand, "Transmission-line conversion transformers," *Electronics*, vol. 17, no. 12, December 1944, pp. 142–145.
35. G. Oltman, "The compensated balun," *IEEE Transactions on Microwave Theory and Techniques*, vol. 44, no. 3, March 1966, pp. 112–119.
36. K. S. Ang, Y. C. Leong, and C. H. Lee, "Analysis and design of miniaturized lumped-distributed impedance-transforming baluns," *IEEE Transactions on Microwave Theory and Techniques*, vol. 51, March 2003, pp. 1009–1017.
37. J.-W. Lee and K. J. Webb, "Analysis and design of low-loss planar microwave baluns having three symmetric coupled lines," *IEEE MTT-S International Microwave Symposium Digest*, June 2002, pp. 117–120.
38. V. Govind, W. S. Yun, S. Dalmia, V. Sundaram, G. E. White, and M. Swaminathan, "Analysis and design of compact wideband baluns on multilayer liquid crystalline polymer (LCP) based substrates," *IEEE MTT-S International Microwave Symposium (IMS) Digest*, Long Beach, CA, June 2005.
39. D.-W. Lew, J.-S. Park, D. Ahn, N.-K. Ahn, C. S. Yoo, and J.-B. Lim, "A design of the ceramic chip balun using the multilayer configuration," *IEEE Transactions on Microwave Theory and Techniques*, vol. 49, January 2001, pp. 220–224.
40. R. K. Feeney, *Private Communication*, Course RF Engineering I, Georgia Institute of Technology.
41. K. Wang, M. Frank, P. Bradley, R. Ruby, W. Mueller, A. Barfknecht, and M. Gat, "FBAR Rx filters for handset front-end modules with wafer-level packaging," *Proceedings of the IEEE Symposium on Ultrasonics*, October 2003, pp. 162–165.
42. J. Kaitila, M. Ylilammi, J. Molarius, J. Ella, and T. Makkonen, "ZnO based thin film bulk acoustic wave filters for EGSM band," *Proceedings of the IEEE symposium on Ultrasonics*, October 2001, pp. 803–806.
43. S. Dalmia, "Design and Implementation of High-Q Passive Devices for Wireless Applications using System-on-Package (SOP) based Organic Methodologies," PhD Dissertation, School of Electrical and Computer Engineering, Georgia Institute of Technology, Atlanta, 2002.
44. S. Dalmia, V. Sundaram, M. Swaminathan, and G. White, "Liquid crystalline polymer (LCP) based lumped-element bandpass filters for multiple wireless applications," *IEEE Int. Microw. Symp.*, June 2004, pp. 1991–1994.
45. A. Bavisi, S. Dalmia, V. Sundaram, M. Swaminathan, and G. White, "Chip-package codesign of integrated voltage-controlled oscillator in LCP substrate," *IEEE Trans. Adv. Packaging*, vol. 29, issue 3, Aug. 2006, pp. 390–402.
46. D. K. Shaeffer and T. H. Lee, "A 1.5-V 1.5-GHz CMOS low noise amplifier," *IEEE Journal of Solid-State Circuits*, vol. 32, May 1997, pp. 745–759.

47. J.-S. Goo, K.-H. Oh, C.-H. Choi, Z. Yu, T. H Lee, and R. W. Dutton, "Guidelines for the power-constrained design of a CMOS tuned LNA," *Proc. Int. Conference on Simulation of Semiconductor Processes and Devices (SISPAD)*, Seattle, WA, September 2000, pp. 269–272.
48. V. Govind, S. Dalmia, and M. Swaminathan, "Design of integrated low noise amplifiers (LNA) using embedded passives in organic substrates," *IEEE Transactions on Advanced Packaging*, vol. 27, February 2004, pp. 79–89.
49. A. Bavisi, V. Sundaram, and M. Swaminathan, "A miniaturized novel feedback LC oscillator for UMTS type applications in a 3-D stacked liquid crystalline polymer technology," *Int. Journal RF and Microw. Comp. Aided Engg.*, Wiley Interscience, published on-line on February 2006., pp. NA.
50. A. Bavisi, V. Sundaram, and M. Swaminathan, "Design of a system-in-package based low phase noise VCO using 3-D integrated passives on a multi-layer LCP substrate," *35th European Microwave Conf.*, October 2005.
51. UMA Technology, [on-line document], available at http://www.umatechnology.org.
52. H. Hashemi, "Integrated Concurrent Multi-band Radios and Multiple Antenna Systems," Ph.D. Thesis, California Institute of Technology, Sept. 2003.
53. A. Bavisi, V. Sundaram, M. Swaminathan, S. Dalmia, and G. White, "Design of a dual frequency oscillator for simultaneous multi-band radio communication on a multi-layer liquid crystalline polymer substrate," *IEEE Radio Wireless Symp.*, January 2006, pp. 431–434.
54. J. S. Schaffner, "Simultaneous Oscillations in Oscillators," *IRE Tran. Circuit Theory*, vol. 1, no. 2, June 1954, pp. 2–8.
55. Simon Haykin, Neural Networks: *A Comprehensive Foundation*, 2nd ed., Prentice Hall, 1998.
55a. Madhavan Swaminathan and A. Ege Engin, *Power Integrity Modeling and Design for Semiconductors and Systems*, Prentice Hall, December 2007.
55b. A. E. Engin, W. John, G. Sommer, W. Mathis, and H. Reichl, "Modeling of striplines between a power and a ground plane," *IEEE Transactions on Advanced Packaging*, vol. 29, no. 3, August 2006, pp. 415–426.
55c. A. E. Engin, K. Bharath, and M. Swaminathan, "Multilayered finite difference method (M-FDM) for modeling of package and printed circuit board planes," *IEEE Transactions on Electromagnetic Compatibility*, vol. 49, no. 2, May 2007, pages 441–447.
56. Rohan Mandrekar, "Modeling and Co-simulation of Signal Distribution and Power Delivery in Package Based Systems," PhD thesis, Georgia Institute of Technology, May 2006.
57. Kwan Choi, Modeling and Simulation of Embedded Passives Using Rational Functions in Multi-layered Substrates, PhD thesis, Georgia Institute of Technology, August 1999.
58. Sung-Hwan Min, "Automated Construction of Macro Models of Distributed Interconnect Networks," PhD Thesis, Georgia Institute of Technology, 2004.
59. Jinwoo Choi, "Noise Suppression and Isolation in Mixed-signal Systems Using Alternating Impedance-electromagnetic Bandgap Structure (AI-EBG)," PhD thesis, Georgia Institute of Technology, December 2005.
60. http://www.synopsys.com/products/mixedsignal/hspice/hspice.html.
60a. G. Taguchi, Introduction to Quality Engineering. Dearborn, *MI: Distributed by American Supplier Institute,* Inc., 1986.

60b. M. S. Phadke, *Quality Engineering Using Robust Design*. Englewood, NJ: Prentice Hall, 1989.

60c. Tomas Berling, Per Runeson: Efficient Evaluation of Multifactor Dependent System Performance Using Fractional Factorial Design. *IEEE Trans. Software Eng.* 29(9): 769–781 (2003)

60d. John Neter, William Wasserman: Applied Linear Statistical Models: *Regression, Analysis of Variance, and Experimental Designs*. Homewood, Ill., R. D. Irwin, 1974.

60e. J. P. C. Kleijnen, "Sensitivity analysis and optimization in simulation: design of experiments and case studies," in IEEE Proc. of Winter Simulation Conference 1995, pp. 133–140, 1995.

60f. R. L. Mason, R. F. Gunst, and J. L. Hess, Statistical design and analysis of experiments: with applications to engineering and science. New York: Wiley Eastern Limited, 1989.

60g. A. Leon-Garcia, *Probability and Random Processes for Electrical Engineering*. Toronto: Addison-Wesley, 1989.

60h. A. Papoulis, *Probability, Random Variables, and Stochastic Processes*. New York: McGraw-Hill, 1984.

60i. S. M. Ross, *Introduction to Probability Models*. San Diego, CA: Harcourt Academic Press, 2000.

60j. S. Mukherjee, M. Swaminathan, E. Matoglu, "Statistical Analysis and Diagnosis Methodology for RF Circuits in LCP Substrates," IEEE Transactions on Microwave *Theory and Techniques*, 2005.

60k. HFSS(tm) v10.1, Ansoft Corporation.

60l. http://eesof.tm.agilent.com/applications/sip-b.html

61. H. Liaw and H. Merkelo, "Signal integrity issues at split ground and power planes," *Proc. IEEE Electronic Components and Technology Conference (ECTC)*, May 1996, pp. 752–755.

62. J. Choi, V. Govind, and M. Swaminathan, "A novel electromagnetic bandgap (EBG) structure for mixed-signal system applications," *Proc. IEEE Radio and Wireless Conference (RAWCON)*, Atlanta, GA, September 2004, pp. 243–246.

63. M. Swaminathan, J. Kim, I. Novak, and J. P. Libous, "Power distribution networks for system-on-package: status and challenges," *IEEE Transactions on Advanced Packaging*, vol. 27, May 2004, pp. 286–300.

64. Y. Jeong, H. Kim, J. Kim, J. Park, and J. Kim, "Analysis of noise isolation methods on split power/ground plane of multi-layer package and PCB for low jitter mixed mode system," *Proc. IEEE Topical Meeting on Electrical Performance of Electronic Packaging (EPEP)*, October 2003, pp. 199–202.

65. S. Chun, M. Swaminathan, L. D. Smith, J. Srinivasan, Z. Jin, and M. K. Iyer, "Modeling of simultaneous switching noise in high speed systems," *IEEE Transactions on Advanced Packaging*, vol. 24, May 2001, pp. 132–142.

66. J. Kim and M. Swaminathan, "Modeling of multilayered power distribution planes using transmission matrix method," *IEEE Transactions on Advanced Packaging*, vol. 25, no. 2, May 2002, pp. 189–199.

67. H. H. Wu, J. W. Meyer, K. Lee, and A. Barber, "Accurate power supply and ground plane pair models," *IEEE Transactions on Advanced Packaging*, vol. 22, August 1999, pp. 259–266.

68. J. Mao, J. Srinivasan, J. Choi, M. Swaminathan, and N. Do, "Modeling of field penetration through planes in multilayered packages," *IEEE Transactions on Advanced Packaging*, vol. 24, no. 3, August 2001, pp. 326–333.
69. R. Ito and R. W. Jackson, "Parallel plate slot coupler modeling using two dimensional frequency domain transmission line matrix method," *Proc. IEEE EPEP*, 2004, pp. 41–44.
70. J. Lee, M. D. Rotaru, M. K. Iyer, H. Kim, and J. Kim, "Analysis and suppression of SSN noise coupling between power/ground plane cavities through cutouts in multilayer packages and PCBs," *IEEE Transactions on Advanced Packaging*, vol. 28, no. 2, May 2005, pp. 298–309.
71. SungJun Chun, "*Methodologies for Modeling Simultaneous Switching Noise in Multilayered Packages and Boards,*" PhD Dissertation, Georgia Institute of Technology, April 2002.
72. Larry Smith, Raymond Anderson, Doug Forehand, Tom Pelc, and Tanmoy Roy, "Power distribution system design methodology and capacitor selection for modern CMOS technology," *IEEE Transactions on Advanced Packaging*, vol. 22, no. 3, August 1999, pp. 284–291.
73. Bernd Garben, George A. Katopis, and Wiren D. Becker, "Package and chip design optimization for mid-frequency power distribution decoupling," *Electrical Performance of Electronic Packaging*, 2002, pp. 245–248.
74. Om P. Mandhana and Jin Zhao, "Comparative study on the effectiveness of on-chip, on package and PCB decoupling for core noise reduction by using broadband power delivery network models," *Electronic Components and Technology Conference*, 2005, pp. 732–739.
75. Nanju Na, Timothy Budell, Charles Chiu, Eric Tremble, and Ivan Wemple, "The effects of on-chip and package decoupling capacitors and an efficient ASIC decoupling methodology," *Electronic Components and Technology Conference*, 2004, pp. 556–567.
76. Tawfik Rahal-Arabi, Greg Taylor, Matthew Ma, Jeff Jones, and Clair Webb, "Design and validation of the core and I/O's decoupling of the Pentium$_R$ 3 and Pentium$_R$ 4 Processors," *Electrical Performance of Electronic Packaging*, 2002, pp. 249–252.
77. Richard Ulrich, "Embedded resistors and capacitors for organic-based SOP," *IEEE Transactions on Advanced Packaging*, vol. 27, no. 2, May 2004, pp. 326–331.
78. Istvan Novak, "Lossy power distribution networks with thin dielectric layers and/or thin conductive layers," *IEEE Transactions on Advanced Packaging*, vol. 23, no. 3, August 2000, pp. 353–360.
79. Hyungsoo Kim, Byung Kook Sun, and Joungho Kim, "Suppresion of GHz range power/ground inductive and simultaneous switching noise using embedded film capacitors in multilayer packages and PCBs," *IEEE Microwave and Wireless Components Letters*, vol. 14, no. 2, February 2004, pp. 71–73.
80. K. Y. Chen, William D. Brown, Leonard W. Schaper, Simon S. Ang, and Hameed A. Naseem, "A study of high frequency performance of thin film capacitors for electronic packaging," *IEEE Transactions on Advanced Packaging*, vol. 23, no. 2, May 2000, pp. 293–302.
81. Joel S. Peiffer, William Balliette, and 3M Company, "Decoupling of high speed digital electronics with embedded capacitance," *38th International Symposium on Microelectronics*, September 2005.

82. Josh G. Nickel, "Decoupling capacitance platform for substrates, sockets and interposers," *DesignCon*, 2005.
83. Prathap Muthana, Madhavan Swaminathan, Rao Tummala, Venkatesh Sundaram, Lixi Wan, S. K. Bhattacharya, and P. M. Raj, "Packaging of multi-core processors: tradeoffs and potential solutions," *Electronic Components and Technology Conference*, 2005, pp. 1895–1903.
84. P. Muthana, A. E. Engin, M. Swaminathan, R. Tummala, V. Sundaram, B. Wiedenman, D. Amey, K. Dietz, and S. Banerji, "Design, modeling and characterization of embedded capacitor networks for core decoupling in the package." *IEEE Transactions on Advanced Packaging*, vol. 30, no. 4, pp. 809–822, Nov. 2007.
85. International Roadmap for Semiconductors (ITRS)—2004 Update. http://public.itrs.net.
86. M. Kamon, M. J. Ttsuk, and J. K. White, "FASTHENRY: a mutipole accelerated 3D-inductance extraction program," *IEEE Transactions on Microwave Theory and Techniques*, vol. 42, issue 9, part 1-2, September 1994, pp. 1750–1758.
87. Istvan Novak and Jason R. Miller, "Frequency dependent characterization of bulk and ceramic bypass capacitors," Poster Material for the *12th Topical Meeting on Electrical Performance of Electronic Packaging*, October 2003, pp. 101–104.
88. Joong Ho Kim and Madhavan Swaminathan, "Modeling of irregular shaped power distribution planes using transmission matrix method," *IEEE Transactions on Advanced Packaging*, vol. 24, no. 3, August 2001.
89. Prathap Muthana, Madhavan Swaminathan, Rao Tummala, P. M. Raj, Ege Engin, Lixi Wan, D. Balaraman, and S. Bhattacharya, "Design, modeling and characterization of embedded capacitors for midfrequency decoupling in semiconductor systems," *Electromagnetic Compatibility*, 2005, pp. 638–643.
90. D. Sievenpiper, R. Broas, and E. Yablonovitch, "Antennas on high-impedance ground planes," *IEEE MTT-S International Microwave Symposium (IMS) Digest*, June 1999, pp. 1245–1248.
91. T. Kamgaing and O. M. Ramahi, "A novel power plane with integrated simultaneous switching noise mitigation capability using high impedance surface," *IEEE Microwave and Wireless Components and Letters*, vol. 13, January 2003, pp. 21–23.
92. R. E. Collin, *Foundations for Microwave Engineering*, IEEE Press, 2001.
93. J. Kim and M. Swaminathan, "Modeling of irregular shaped power distribution planes using transmission matrix method," *IEEE Transactions on Advanced Packaging*, vol. 24, no. 3, 2001, pp. 334–346.
94. A. E. Engin, M. Swaminathan, and Y. Toyota, "Finite difference modeling of multiple planes in packages," *Proc. 17th International Zurich Symposium on Electromagnetic Compatibility*, Singapore, March 2006, pages 549–552.
94a. Yoshitaka Toyota, A. Ege Engin, Tae Hong Kim, Madhavan Swaminathan, and Swapan Bhattacharya, "Size reduction of electromagnetic bandgap (EBG) structures with new geometries and materials," ECTC 2006, pages 1784–1789.
95. T. E. Moran, K. L. Virga, G. Aguirre, and J. L. Prince, "Methods to reduce radiation from split ground planes in RF and mixed-signal packaging structures," *IEEE Transactions on Advanced Packaging*, vol. 25, no. 3, August 2002, pp. 409–416.
96. P. Fornberg, A. Byers, M. Piket-May, and C. Holloway, "FDTD modeling of printed circuit board signal integrity and radiation," *IEEE International Symposium on Electromagnetic Compatibility*, August 2000, pp. 307–312.

97. Joel S. Peiffer, "Ultra-thin, loaded epoxy materials for use as embedded capacitor layers," *Printed Circuit Design & Manufacture*, April 2004, pp. 40–42.
98. M. Xu, T. H. Hubing, J. Chen, T. P. Van Doren, J. L. Drewnial, and R. E. DuBroff, "Power-bus decoupling with embedded capacitance in printed circuit board design," *IEEE Trans. Electromag. Compat.*, vol. 45, no. 1, February 2003, pp. 22–30.
99. John Andresakis, Takuya Yamamoto, Kaz Yamazaki, Yoshi Fukawa, and Glenn Bennik, "Simulation of resonance reduction in PCBs utilizing embedded capacitance," *IPCWorks* 2005.
100. R. R. Tummala, M. Swaminathan, M. M. Tentzeris, J. Laskar, S. Gee-Kung, Chang Sitaraman, D. Keezer, D. Guidotti, Zhaoran Huang Kyutae Lim, Lixi Wan, S. K. Bhattacharya, V. Sundaram, Fuhan Liu, and P. M. Raj, "The SOP for miniaturized, mixed-signal computing, communication, and consumer systems of the next decade," *IEEE Transactions on Advanced Packaging*, vol. 27, issue 2, May 2004, pp. 250–267.
101. John Savic, Robert T. Croswell, Aroon Tungare, Greg Dunn, Tom Tang, Robert Lempkowski, Max Zhang, and Tien Lee, "Embedded passives technology implementation in RF applications," *IPC Printed Circuits Expo*, 2002.

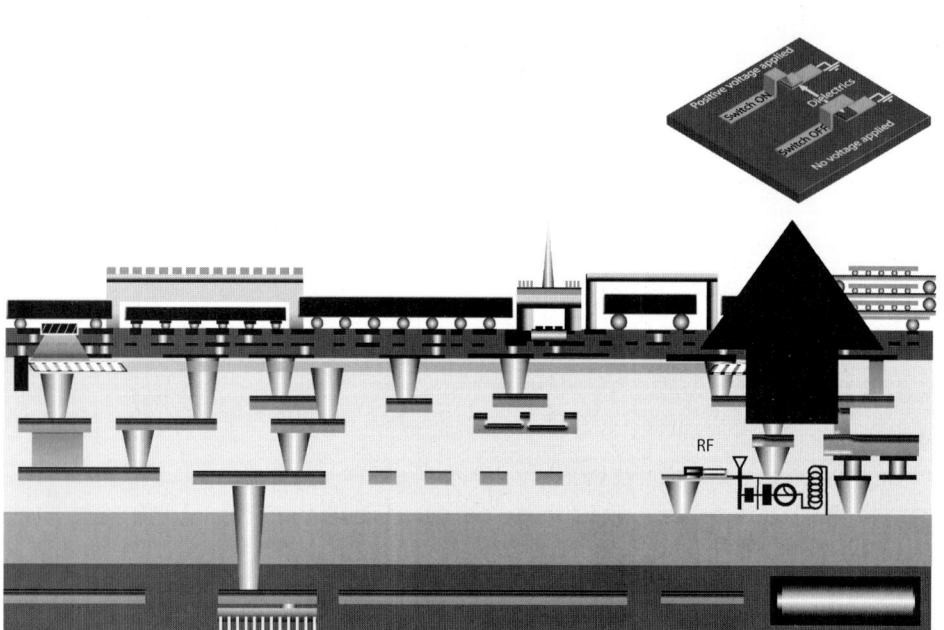

CHAPTER 5

Radio Frequency System-on-Package (RF SOP)

John Papapolymerou, Manos Tentzeris, Joy Laskar,
and Swapan Bhattacharya
Georgia Institute of Technology

5.1	Introduction 262	5.4	RF SOP Technologies 267
5.2	RF SOP Concept 262	5.5	Integrated RF Modules 308
5.3	Historical Evolution of RF Packaging Technologies 265	5.6	Future Trends 312
			References 314

The demand for increasingly higher rates of data, voice, and video together with miniaturization of portable and wireless technologies (digital, analog, RF, and optical) has driven the need for high-performance applications such as personal communication networks, wireless local area networks (WLAN), "last-mile" RF-optical networks, and millimeter-wave sensors. These RF and wireless applications have defined a trend toward more flexible and reconfigurable systems, since they impose very stringent specifications never reached before in terms of low noise, high linearity, low power consumption, small size and weight, and low cost. The electronics packaging industry has proliferated in this area to a point where this technology is at least equally, if not more, important than the semiconductor technology it is supposed to serve.

In this chapter, the RF SOP component technologies, which include antennas, inductors, capacitors, resistors, filters, baluns, power dividers, MEMS switches, and MEMS capacitors, into highly integrated systems are presented. The substrate technologies for these components include low-temperature cofired ceramics (LTCCs) and organic technologies such as liquid crystal polymer (LCP). Two demonstration vehicles are discussed for WLAN and personal communicator applications. This chapter is not limited to just consumer applications but also includes space and defense applications over multiple frequency bands.

5.1 Introduction

RF denotes the radiofrequency spectrum, which ranges from 300 KHz to 300 GHz. The SOP is a system miniaturization technology. RF SOP, therefore, is a miniaturization technology, based on the SOP concept of embedded thin-film components, for wireless systems. The SOP concept has two fundamental bases—miniaturization and optimization of components between ICs and substrates for performance and cost. In today's RF communications, the volume applications are in the wireless systems and the main drivers are cost, functionality, and size. Miniaturization of thin-film components and their integration into RF modules are the key elements to meet these three demands. These costs, functionalities, and size pressures are the primary drivers behind SOP, and SOP addresses these drivers by thin-film embedded components, nanotechnology-based batteries, thermal structures, and interconnections. The SOP concept provides a system technology platform that enables a higher degree of miniaturization than can be achieved by the complementary module-level SIP and SOC technologies. The fundamental basis of SOP is about integration, which leads to higher performance and reliability, lower cost, and reduced size just like in CMOS wafer fabrication.

This chapter describes the progress in SOP-based RF components in ceramics and organics and on silicon wafers. The RF components include inductors, capacitors, resistors, antennas, filters, switches, baluns, combiners, and radiofrequency identification (RFID) implemented in both ceramic and organic technologies. The historical evolution of RF and wireless technologies and the future trend and directions in RF SOP technologies are discussed. Design, modeling, simulation, and component integration challenges pertinent to the RF SOP technology are reviewed.

5.2 RF SOP Concept

There has been an emerging trend to combine computing, communicating, sensing, and biomedical functions into one system package called SOP, as described in this book. Figure 5.1 depicts how the SOP concept can be applied to miniaturize the RF systems

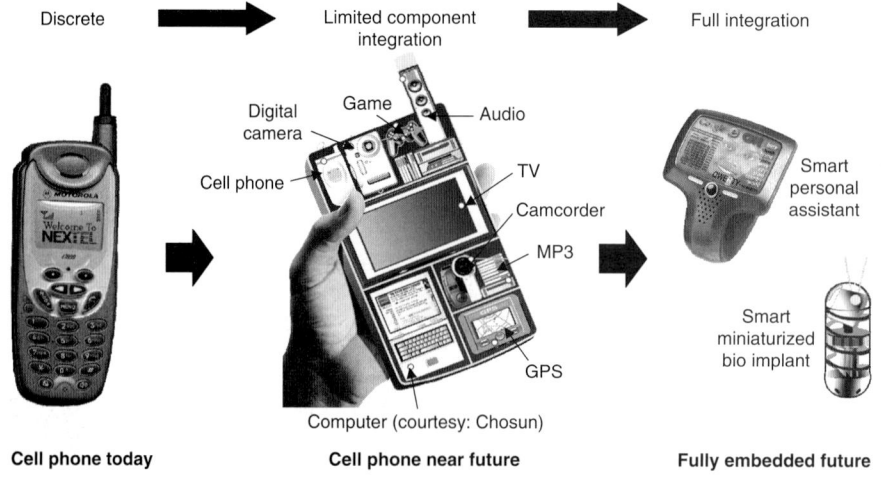

FIGURE 5.1 RF miniaturization and functionality trend.

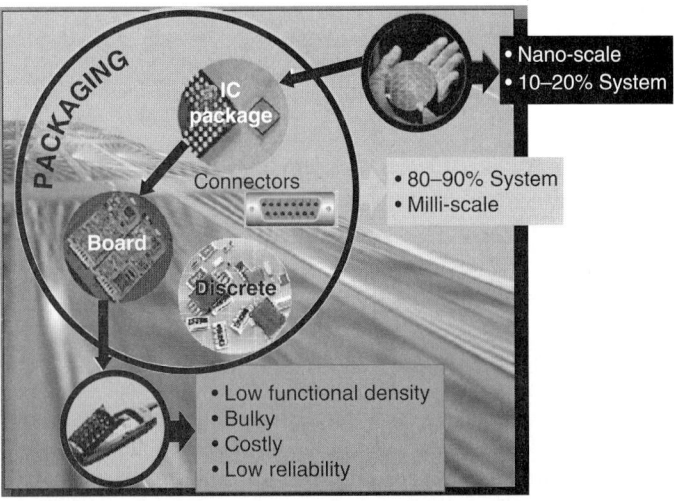

FIGURE 5.2 Packaging is the barrier to future RF systems.

showing the trend from a discrete component-based system such as the cell phone to more convergent and miniaturized systems capable of performing a variety of functions including, but not limited to, wireless phone, wireless networking, navigation systems, and sensor systems [1–3].

The barriers to achieving these ultraminiaturized systems with dozens of functions are not in the digital or CMOS silicon but in the system packaging area, as shown in Figure 5.2. RF systems utilize passive components for matching, tuning, filtering, and biasing. For example, a mobile phone is composed of only about 6 to 10 active components but as many as 400 to 600 passive components, depending upon the level of system integration. These passive components currently are all surface-mount devices (SMDs), which account for more than 90 percent of the system components and occupy more than 80 percent of the system board area [4]. The SOP enables reduction in size this non-CMOS part of the RF system by a factor of a thousand as the RF components are miniaturized from their current thick-film-based milliscale to thin-film-based microscale technologies. Since this non-CMOS part of the system is about 80 percent of the total size of the system and 70 percent of the cost of the system, the SOP dramatically improves both the size and cost. If, in the future, these non-CMOS components are further miniaturized to nanoscale, RF systems can be further miniaturized by another factor of a thousand, leading to the possibility of megafunction systems in the same size scale as today's handsets.

Figure 5.3 shows an RF communication system where the SOP concept can be implemented to enhance performance and reduce size. In the baseband section, the major functions, such as the microprocessor, DSP, static random access memory (SRAM), and flash, are based on silicon technology, and their progress is in line with the advancements in SOC from 65 to 22 nm and 3D integration technologies that use through-silicon vias for chip stacking. However, in the RF front-end section, the situation becomes more challenging. The RF system requires unique components, such as filters, low-loss power amplifiers, and high linearity RF switches [5]. CMOS is excellent for baseband but is not an optimal platform for the RF front end. Here, the SOP offers a solution that cannot be achieved either by SOC or traditional SIP technologies. In simple

264 Chapter Five

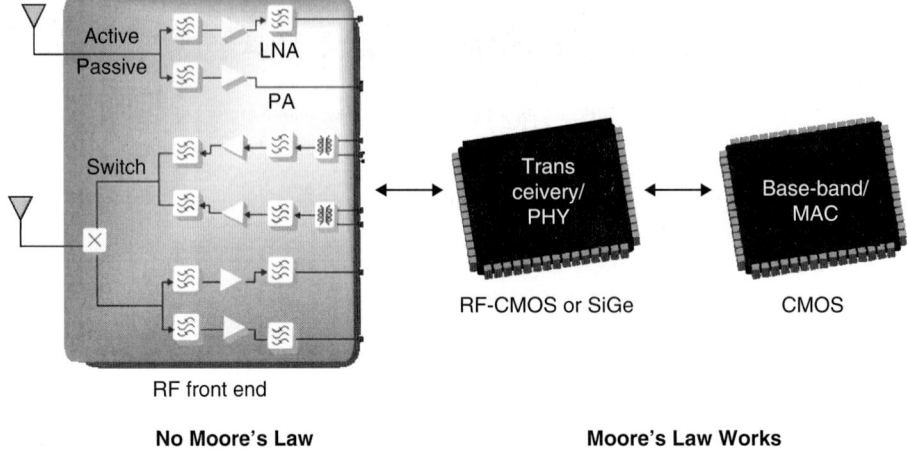

FIGURE 5.3 CMOS limitations in the RF front end and how SOP can address miniaturization.

terms, SOP combines the best of both baseband and RF domain solutions. The baseband section is dominated by CMOS technology for which Moore's law for transistors applies and can be taken advantage of, while the RF front end is driven by more than Moore's law such as with embedded thin-film components for antennas, filters, baluns, oscillators, mixers, and amplifiers, which can be packaged efficiently utilizing the SOP concept.

Figure 5.4 shows an illustration of the distribution of components between the chip and package in an RF front-end module that utilizes the SOP concept [6–10]. The RF

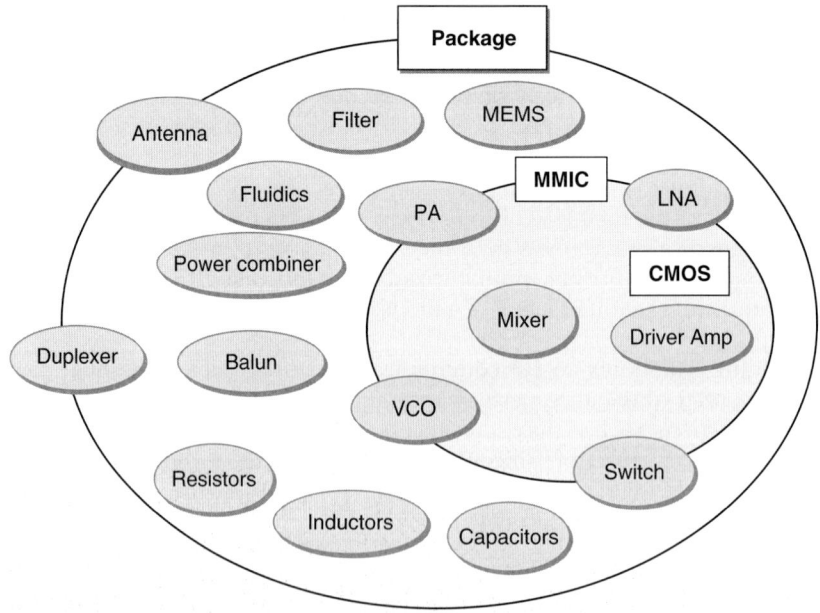

FIGURE 5.4 RF SOP concept–semiconductor versus package partition.

front-end module is the foundation of many applications identified in the beginning of this chapter, and the integration of its many different components poses rigid challenges. The SOC concept tries to achieve the integration of all the system components on a single chip. Such an approach has two major limitations—cost and performance of the RF system. The high cost is due to the high cost of digital and RF mixed-signal ICs. The performance limitation is due to poor component properties such as low Q factor inductors that can be achieved on silicon. The SOP concept addresses both of these. It allows for diverse ICs, not requiring monolithic integration, and it allows for the best component properties to be achieved, optimizing between IC and the package. In addition it allows for miniaturization of the entire RF system. This optimization concept between the two is illustrated in Figure 5.4, showing the best regimes for each. For example, package integration, which does not exist in the current manufacturing of cell phones and PDAs other than through bulky discrete components, can be achieved by microscale thin-film embedded components such as filters, switches, antennas, inductors, capacitors, and resistors as described later in this chapter.

5.3 Historical Evolution of RF Packaging Technologies

Historically, wireless communication systems have been on a path of ever-increasing functionality, lower cost, and smaller size. For example, the first generation (1G) of cellular telephony used analog modulation techniques and was only capable of voice communications. The second generation (2G) used digital modulation technologies, and although limited data communication rates were possible, it was primarily geared to voice communications. The enhancements brought on by digital technologies allowed more users to have simultaneous access, and hence a rapid adaptation followed [5]. The evolution to the third-generation (3G) cellular standards increased the available bandwidth for data communications, but at a substantial increase in the cost of required infrastructure. The sales of cell phones is approaching about a billion units and given a world population of six billion, it is expected to grow at a rate of 20 to 30 percent annually, thus driving a substantial portion of RF component technology. Moreover, handsets are evolving beyond the traditional voice services and are rapidly becoming multimedia terminals with Internet access, and concurrently with "slimmer" size as is evident from the marketing of the Motorola Razor and Samsung's latest 11.9-mm thin ultraslim phone. The latest cell phone that captured the imagination of the consumer and shocked all the other manufacturers is the iPhone from Apple with the most easy-to-use user interface.

To keep up with the above trend in miniaturization and functionality, packaging has gone through dramatic changes in the past three decades from bulky packages to thin-film components and devices. This historical evolution in packaging technology is depicted in Figure 5.5. In the 1970s, the components were bulky and the device packaging was double-sided DIP, which migrated to four-sided quad-flat package (QFP) with interconnects in the board made of drilled holes followed by printed-through-hole connections. The next-generation technologies began to implement surface-mount technologies (SMT) for discrete passive and active components. The next step involved integration of discrete components themselves into individual prepackages leading to integrated passive devices (IPDs) that helped reduce the form factor. With the development of multichip modules (MCMs) in the 1980s with as many as 100 to 144 chips flip-chip bonded onto a single ceramic or thin-film substrate, the package

266 Chapter Five

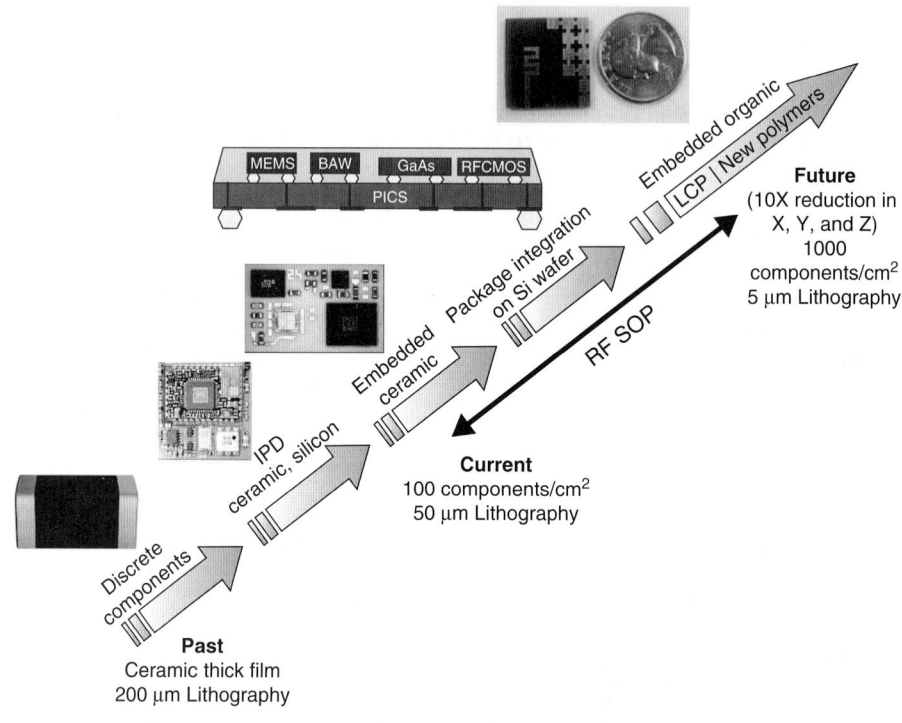

FIGURE 5.5 Historical evolution of RF SOP.

integration improved by a factor of 5 in comparison with discrete packages previously. Next-generation integrations moved on to the wafer itself by so-called wafer-level packaging, which led to chip-scale packaging (CSP), defined as a wafer package that is only 20 percent bigger than the chip it packages. The next set of developments that can further improve package integration involved bare chip packaging on the board and chip stacking of similar or dissimilar ICs, often referred to as SIP. This type of stacked IC packaging began to address the limitations of SOC by minimizing or eliminating the need for heterogeneous integration of functions within a single IC.

The discrete components such as capacitors and filters followed a similar miniaturization trend. The primary technology for discrete components has been the ceramic thick-film technology with a sequential buildup of layers or cosintering of ceramics and metal interconnections. The next step was IPDs as described above integrating several individual components into a single package. The most dramatic improvements, however, began to emerge as embedded components, initially as embedded discretes and more recently as embedded thin-film components. Embedded thin-film components fall into several categories—embedding in ceramics, organic packages, or boards on silicon or glass wafers. The SOP technology described in this book is about total thin-film embedding—embedding of passives, actives, thermal structures, and power sources. With the SOP concept embarking on a relentless increase in component density through thin-film embedded integration (a similar trend as increased transistor density in silicon), the miniaturized RF systems such as the iPhone, are well on their way, as shown in Figure 5.5.

5.4 RF SOP Technologies

SOP offers an ideal platform for miniaturization of RF front-end components better than can be achieved in alternative technologies such as SOB, SIP, or SOC, as described in Chapter 1. However, there are a multitude of challenges that need to be addressed to enable this SOP-based miniaturization. These include (1) design, modeling, and simulation, (2) materials and processes for thin-film components, and (3) reliability of the fabricated substrates, as depicted in Figure 5.6.

5.4.1 Modeling and Optimization

The optimization of RF SOP requires an effective modeling of complex structures that involves mechanical motion and wave propagation. Because of computational constraints, many commercial simulators utilize various approximations to provide fast and relatively accurate results. Popular commercial EM simulation tools such as high-frequency structure simulator (HFSS) [11], Sonnet [12], microstripes [13], IE3D [14], often limit the size or type of circuits that can be modeled. Either the approximations used limit their applicability to specific problems or the simulation time takes too long. To solve complex 3D problems, custom simulators employing full-wave techniques are used. Using a custom code, approximations can be made selectively and the effect on accuracy can be determined. Popular simulation techniques [15] include the method of moments (MoM), finite-element method (FEM) in the frequency domain, finite-difference time-domain (FDTD) method, transmission-line matrix (TLM) method, and multiresolution time-domain (MRTD) [16] method. Frequency domain methods are often used to simulate complex structures and can naturally handle frequency-dependent parameters such as loss. Alternatively, time-domain simulation techniques allow the use of simple grids for complex structures, parallelize well on inexpensive hardware, and through the use of a Fourier transform can give the results for a wide frequency band using a single simulation [15–16]. Both types of simulators can be used on most problems, although not with the same complexity.

Modern RF 3D modules and packages demand a high level of compactness and functionality. Full-wave EM numerical tools require computational complexity, which

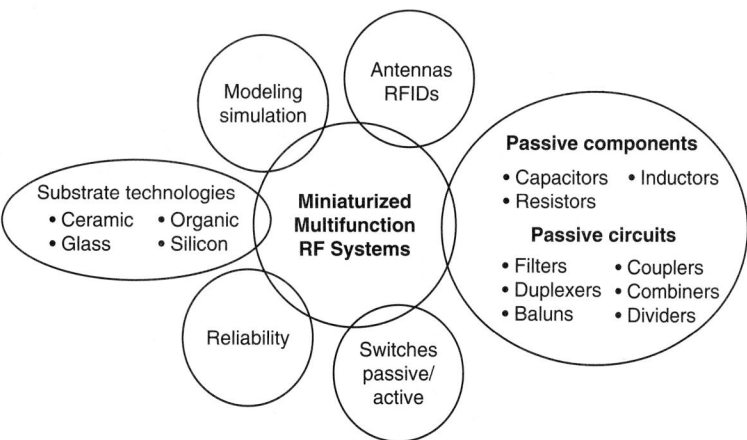

FIGURE 5.6 RF SOP technological challenges.

render this kind of design approach unpractical. Also, the low-frequency RF packaging design process often requires scalable equivalent circuits for the package itself. As the problems associated with the integration involve more and more factors to be considered, the design and optimization of such systems requires more comprehensive and sophisticated tools. The current design and optimization methods, using the commercially available electromagnetic simulators, do not take into account the specific effect of each of the factors involved in the design process, the degree to which these factors interact with each other, and their ranges of values. Only this type of thorough understanding of the entire system can enable the optimization and synthesis of any module under different given conditions. For example, a combination of design of experiments (DOE) and response surface methods (RSM) can be implemented [17]. In such cases, first the factors that affect the performance of the system and the output figures of merit have to be identified. The next step involves the design of a factorial experiment with center points based on a design space for these factors to determine the effect of each of the parameters, identify their interaction, and determine which ones are significant for each of the outputs. The experiment is run using electromagnetic simulations and/or microwave measurements, and the outputs are recorded and input into statistical analysis software. After the statistical analysis of the data, significant factors are identified for all figures of merit, and then RSM statistical methodology is applied for optimization. The result is an explicit set of equations that show how the outputs depend on the input variables, which are used to simultaneously optimize the figures of merit. Within the design space of the experiment, the optimized figures of merit and the required design parameters are identified.

The nonlinearity of the system, combined with the lack of analytical input-output description, suggests the use of soft computing algorithms also. Genetic algorithms can be utilized as an optimization method of this kind. These algorithms search the parameter space stochastically generating solutions that are close to the optimal. They are efficient for problems where small perturbations in the optimal solution lead to an abrupt increase of the error.

These techniques can be applied to any type of design, especially in complex RF microsystems and packages where the number of factors increases and it is extremely difficult to optimize using only electromagnetic simulators. It gives a thorough understanding of the system behavior and integrates geometrical, material, and functional parameters altogether. The approach is generic and independent of the choice of the electromagnetic simulator and statistical analysis software.

5.4.2 RF Substrate Materials Technologies

As mentioned before, central to the theme of the SOP approach is the development of highly miniaturized systems, novel integrating technologies, and the suitable material and component technologies with which to integrate. The substrate material platforms should provide excellent high-frequency electrical properties, mechanical and chemical resistance, and thin-film multilayer capabilities, and be cost competitive. The prominent packaging technologies, which can satisfy all these requirements, fall into two categories—ceramic substrates [18] and organic substrates [19].

The substrate technologies that include ceramics and organics are discussed in Chapter 7 in detail. Ceramics were the primary focus until 2000, but organic technologies began to provide a combination of low cost and high performance to generate both homogenous and heterogeneous multilayer SOP architectures [20]. Ceramic substrates include low-temperature cofired ceramics (LTCC) and high-temperature cofired ceramics (HTCC), while organics include a variety of polymers such as liquid crystal polymer (LCP) [21].

LCP has recently received much attention as a high-frequency substrate material [21–22]. It has impressive electrical characteristics that are environmentally invariant. It provides a nearly stable dielectric constant (~2.97), has a very low dielectric loss [23], and temperature stability up to 125°C [24] across a very wide frequency range up to 110 GHz. Its coefficient of thermal expansion (CTE) can be engineered to match copper, silicon, or GaAs. Being a polymer, it can be processed considerably cheaper than ceramic materials [25]. It is flexible, recyclable, and impervious to most chemicals; has low water absorption [26]; and is physically stable up to 315°C. The availability of two types of LCP substrates with different melting temperatures makes it possible to realize limited multilayer architectures. The core LCP layer, commercially available as R/flex 3850 from Rogers, has a melting point around 315°C, while the bond LCP layer, commercially available as R/flex 3600, has a melting point around 285°C. The bond and the core LCP layers are identical in other characteristics except for their melting points. As a result, LCP is one of the rare organic technologies that allows homogenous multilayer constructions.

While LCP is an emerging RF materials technology, HTCC and LTCC were proven and widely used for RF and microwave systems [27–28] for several decades. LTCC, which cofires with appropriate low-temperature conductors such as AgPd, Cu, or Au around 850°C, allows a multilayer stackup of up to 100 layers. It possesses a combination of electrical, thermal, chemical, and mechanical properties that cannot be found in most other material groups. Some of its characteristics that are applicable are

- Stable dielectric constant over a wide range of RF, microwave, and millimeter-wave frequencies
- Low dielectric loss up to millimeter-wave frequencies
- Engineerable coefficient of thermal expansion (CTE)
- Vertical integration with small vias and lines in a large number of layers
- Very low water and moisture absorption properties

Despite allowing a stackup of a large number of layers, companies like IBM, Kyocera, TDK, and NTK have developed manufacturing technologies with very high yields. HTCC, based on aluminum oxide, also called alumina, was the primary workhorse from the 1960s until LTCC was developed in the 1980s. It is cofired at about 1600°C using cofirable conductors such as molybdenum or tungsten with higher melting points, which confer to this technology a superior stability and reliability in harsh environments. The drawback of these conductors is their higher losses at high frequencies.

5.4.3 Antennas

System-on-package (SOP) can give flexibility to the front-end module by integrating all functional blocks using multilayer processes and novel interconnection methods. One of the major issues, however, is to integrate antennas with a high module efficiency and low cost. Furthermore, the physical sizes of the antennas for low-frequency applications such as cellular communication and WiFi pose serious challenges because of size. Fabricating an antenna directly on the package has the advantages of reduced losses and can result in compact module size. However, integration in the package has other issues that need to be solved, which include narrow bandwidth characteristics due to high dielectric constant substrates (which are preferred for integrating capacitors because of the inherent size advantages) and interference between the antenna and the rest of the RF blocks in highly integrated modules.

With the integration of inductors and capacitors in the substrate, the integration of the antenna now becomes the major size limiter. The gain of an antenna is a function of its electrical size. Typical antennas such as patch antennas require an electrical size of $\lambda/2$, where λ is the electrical wavelength. This requirement makes the antenna have a very large physical size. For example, at 2 GHz, the electrical wavelength in FR4 dielectrics (dielectric constant = 4) is 74 mm, which translates into an antenna physical size of approximately 37 mm. Several technologies are available today where the RF front end from the antenna port to the transceiver can be integrated in a size of 5 mm × 5 mm or less (see Chapter 4) with multiple transmit-receive chains, which is approximately seven times smaller than the antenna itself. Hence, the antenna now becomes the size limiter in the RF front end.

The important attributes of an antenna for consumer applications are (1) small physical size, (2) suitable bandwidth, and (3) good gain.

1. *Physical size.* A printed antenna on a substrate has a size that is a function of the material properties of the substrate. The physical size of the antenna scales as

$$l \propto \frac{\lambda}{\sqrt{\mu_r \varepsilon_r}}$$

where l is the physical size of the antenna, λ is the antenna's electrical size in air, μ_r is the relative permeability, and ε_r is the relative permittivity. With a dielectric material of $\varepsilon_r = 9$, the size of the antenna can be reduced by a factor of $1/3$ (since $\mu_r = 1$) as compared to the antenna in air. As the dielectric constant is increased, electromagnetic fields get trapped in the substrate, leading to substrate modes that diminish the bandwidth and gain of the antenna.

2. *Bandwidth.* The bandwidth (BW) of an antenna is typically described as a percentage of the center frequency, which is calculated as:

$$BW = \frac{F_{High} - F_{Low}}{F_{Center}} \times 100$$

where F_{high} is the highest frequency, F_{low} is the lowest frequency, and F_{center} is the center frequency. The bandwidth determines the range of frequencies where the antenna would operate correctly. In consumer applications, both narrowband (5 percent BW) and broadband (30 percent BW) are used depending on the filtering technologies used.

3. *Gain.* Gain is always measured with reference to a standard antenna such as an isotropic or dipole antenna. When the reference is an isotropic antenna, the units used are dBi. Since an antenna only redistributes power that oftentimes is non-uniform in space, the gain changes as a function of direction and therefore has both positive and negative values. For consumer applications, an omnidirectional radiation pattern is often preferred. The gain of an antenna is a function of many factors, the most important of which is antenna matching. The transmission line that feeds the antenna has to be well matched to the antenna input to ensure a maximum transfer of power. This can be a very challenging task in antenna miniaturization.

To summarize, antennas of small physical size require materials with high permeability and permittivity. However, such materials generate parasitic substrate modes that diminish the gain of the antenna. Similarly, antennas with good gain require matching at the antenna input port for the maximum transfer of energy from the transmission line to the antenna. With high-permittivity and high-permeability materials, this becomes difficult since very narrow traces are required to obtain an impedance of 50 Ω. With such conflicting requirements, the design of antennas can be very tricky. Various approaches have therefore been proposed by several researchers for the miniaturization of these antennas such as combining multiple frequency bands into a single antenna element [29–30], use of conformal antennas, use of magneto-dielectric materials , and inclusion of electromagnetic bandgap structures to improve performance. Another important element that requires special attention while integrating the antenna in the module is the suppression of parasitic backside radiation or crosstalk, which can otherwise couple to sensitive RF blocks in the substrate. This is measured as the front-to-back ratio, which is the ratio of the maximum directivity of the antenna to its directivity in the backward direction.

Three important antenna technologies (multiband antennas, conformal antennas, and antennas on magneto-dielectric substrates) for WiFi applications are illustrated here for miniaturizing the antenna size and integrating it with the rest of the RF front-end module.

Multiband Antennas

As multiple communication standards are integrated into the RF front end, there is a need for antennas that support multiple frequencies. Along with passing the required frequencies, these antenna elements should also minimize interference by suppressing the adjacent frequency bands. To minimize antenna size, a multiband antenna (instead of multiple single-band antennas) can be constructed by controlling the length of the antenna elements. This is illustrated in Figure 5.7, which shows a triband antenna developed on RT-duroid substrate. The antenna resonates at 900 MHz (cellular), 2 GHz (802.11b/g), and 5 GHz (802.11a) and has an almost omnidirectional radiation pattern.

Conformal Antennas

In SOP, technologies that combine rigid and flexible substrates are possible. This concept can be used to embed the RF front end in the rigid part of the module, while the antenna that is patterned on the flexible substrate can be folded or made to conform to the rigid part. This approach reduces the size of the module containing the integrated antenna. An example of a meander monopole antenna is shown in Figure 5.8, which has been used due to its ability to achieve the required length of current path for a specific resonant frequency within a compact size. The substrate of the antenna is composed of two layers of dielectric, as shown in the figure. The top layer is a 25-μm-thick LCP layer with a size of 18 mm × 25 mm and the bottom layer is a 508-μm-thick rigid, glass-reinforced organic prepreg layer (core layer) with a size of 18 mm × 9 mm. The LCP layer has a dielectric constant of 2.95 and a loss tangent of 0.002. The core layer has a loss tangent of 0.0037 with a dielectric constant of 3.48. As can be seen from the figure, the 16-mm-long portion of the LCP layer is not supported by the core layer and can be easily conformed due to the flexibility of the LCP. The antenna is printed on the flexible portion of the LCP, making it possible to bend, fold, and roll the antenna, as shown in Figure 5.9, leading to a compact antenna design and integration with the module.

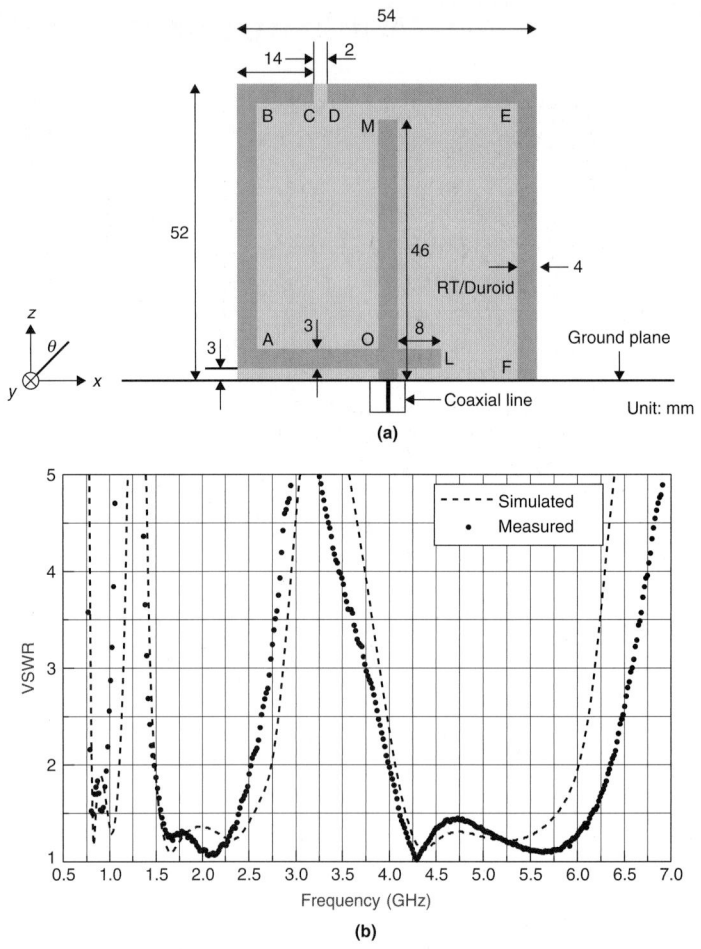

FIGURE 5.7 (a) Schematic of triband antenna on SOP substrate. (b) Return loss results.

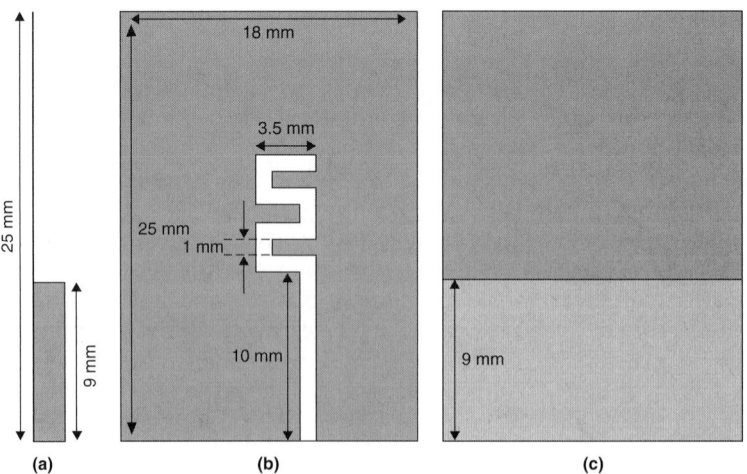

FIGURE 5.8 The details of the designed antenna. (a) Side view. (b) Top view. (c) Bottom view.

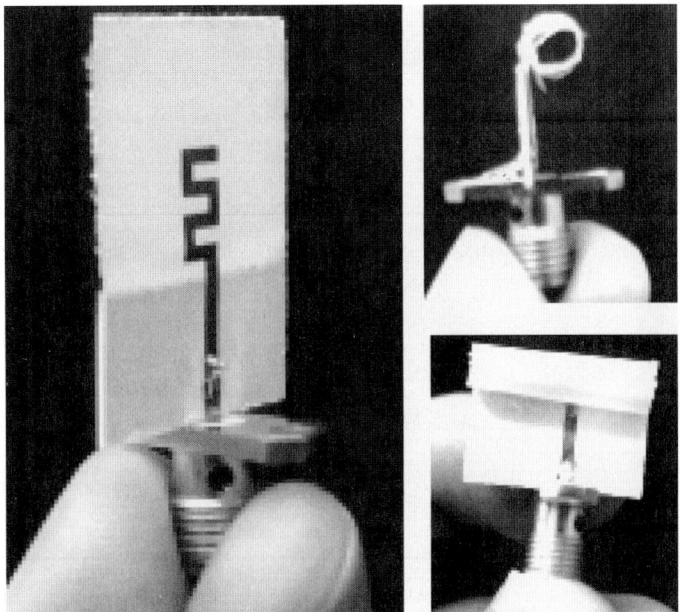

FIGURE 5.9 Photograph of the fabricated prototype in straight and rolled cases.

The total length of the antenna used is 16.5 mm, which is equal to $0.33\lambda_0$ (λ_0 is wavelength in air) at the resonant frequency. The dimensions and the photos of the fabricated prototype are shown in Figures 5.8 and 5.9, respectively. The antenna is excited with a 50-Ω microstrip line that is printed on the rigid part of the substrate, supported by the core layer. The ground plane of the signal line covers the backside of the core layer with a size of 18 mm × 9 mm.

The return loss measurements of the antenna for the folded and rolled cases are shown in Figure 5.10. As seen in the figure, the measured return loss of the antenna does not change much when the antenna is folded or rolled. The simulated far-field patterns of the antenna are presented in Figure 5.11. The antenna has a nearly

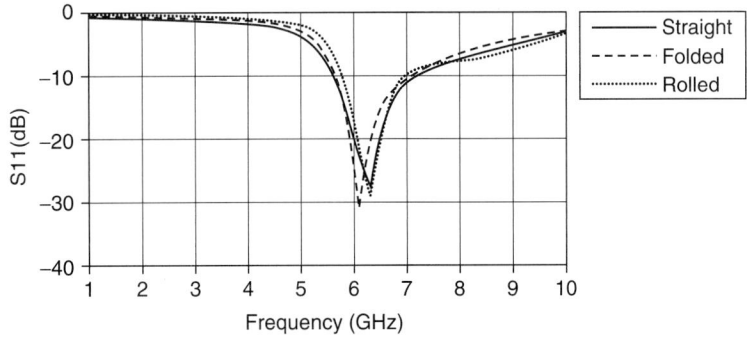

FIGURE 5.10 Measured return loss of the antenna for different configurations.

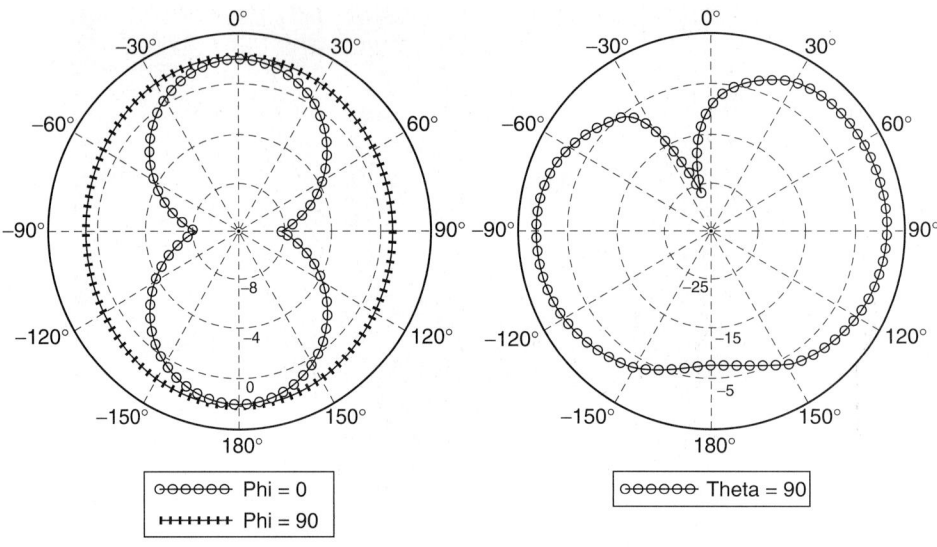

FIGURE 5.11 Simulated far-field pattern of the antenna at 6 GHz.

omnidirectional pattern like that of a dipole, and the simulated peak gain of the antenna is 2.5 dBi at a resonant frequency of 6 GHz.

The integration of the antenna in the module is shown in Figure 5.12a where the flexible substrate containing the antenna is folded over a metallic box. The front-end filters and matching circuits embedded in the substrate are protected by the metallic box from RF interference. The use of the meander line antenna in Figure 5.12b reduces the antenna size and is designed in such a way that the metallic box does not act as a ground plane for the antenna. Without careful design, the presence of the metallic box can make the antenna directional and reduce its bandwidth.

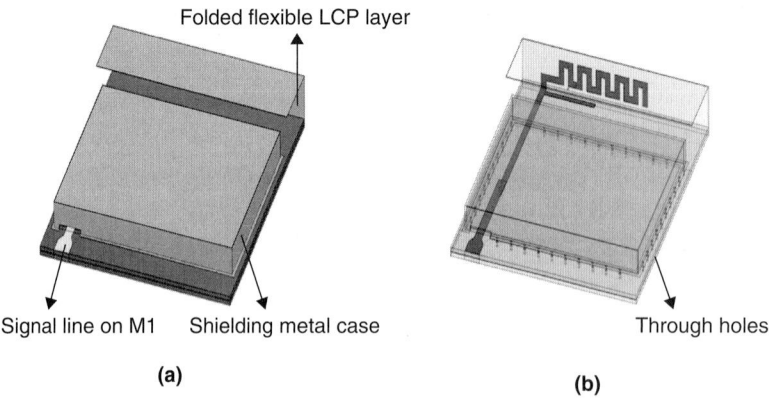

FIGURE 5.12 (a) Integration of a conformal antenna with a module using LCP substrate. (b) Mechanical structure after folding showing the position of the antenna.

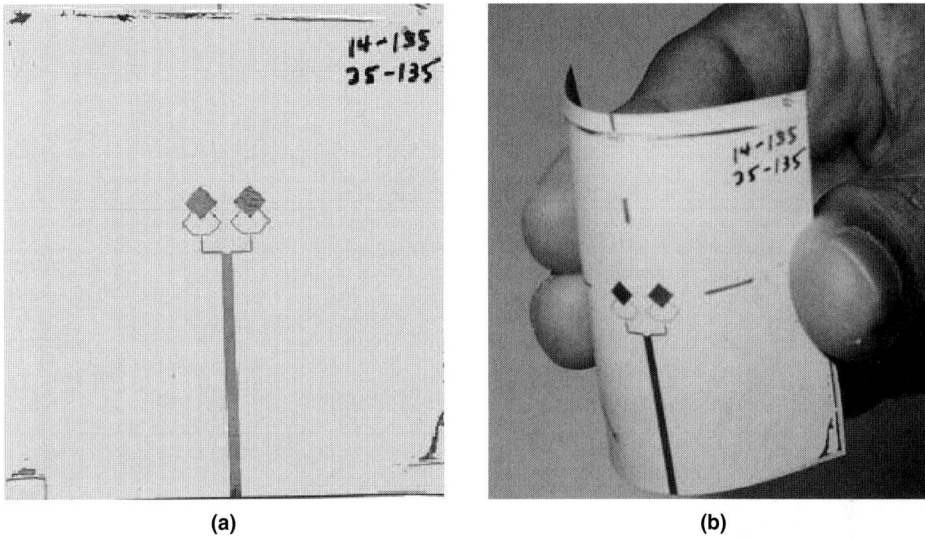

Figure 5.13 Photograph of the (a) fabricated antenna arrays and (b) demonstration of mechanical flexibility.

Conformal antennas can also be used for higher-frequency applications such as the dual-frequency (14 and 35 GHz), dual-polarization microstrip conformal antenna utilizing LCP technology, as shown in Figure 5.13 [31]. This particular antenna array has been developed for space radar applications and is the first demonstration of a multilayer antenna array on an organic technology. The bonding of LCP layers is the most critical step in the fabrication process and has to be understood thoroughly to create multilayer LCP structures reliably. Several experiments were carried out to optimize the temperature, the tool pressure, and the process times to achieve good bonding while preventing shrinkage, formation of bubbles, and melting of core layers. The bubbles can result in air gaps that can affect the array performance at millimeter-wave frequencies. The top layer, visible in the picture, shows the 35-GHz antennas. The 14-GHz ones are similar and embedded in the bottom layer.

The simulated and measured return loss plots for the 14- and 35-GHz arrays are shown in Figure 5.14a and b, respectively. The simulated and measured 2D radiation patterns for the 14-GHz array and the 35-GHz array are shown in Figure 5.15a and b. These results show the excellent radiating characteristics of antennas developed on LCP technology.

Antennas on Magneto-Dielectric Substrates

As mentioned earlier, when only the dielectric constant of the material is increased to reduce the antenna size, the electromagnetic fields get trapped in the substrate. Hence, the antenna behaves more like a capacitor than a radiator of energy. A similar phenomenon can be seen if only the permeability of the substrate material is increased. However, if both the permittivity and permeability of the substrate material is increased simultaneously and matched to each other, then the radiation characteristics of the antenna can be maintained while decreasing size. This material, called the magneto-dielectric material, has large benefits for reducing the size of antennas for consumer

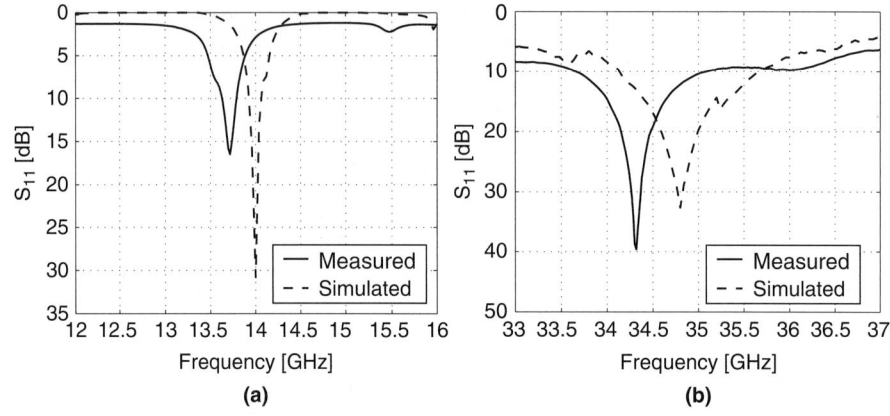

FIGURE 5.14 Return loss plots of the antenna arrays developed on multilayer LCP technology: (a) 14-GHz array and (b) 35-GHz array.

applications. Because of the magneto-dielectric properties, it is easier to match to such an antenna.

Figure 5.16 shows an example of a meander line antenna fabricated on an NiZn ferrite composite (magneto-dielectric) material for VHF (30 to 300 MHz) applications.

The properties of the magneto-dielectric layers are defined by their complex permeability, complex permittivity, and frequency characteristics. Figure 5.17 shows

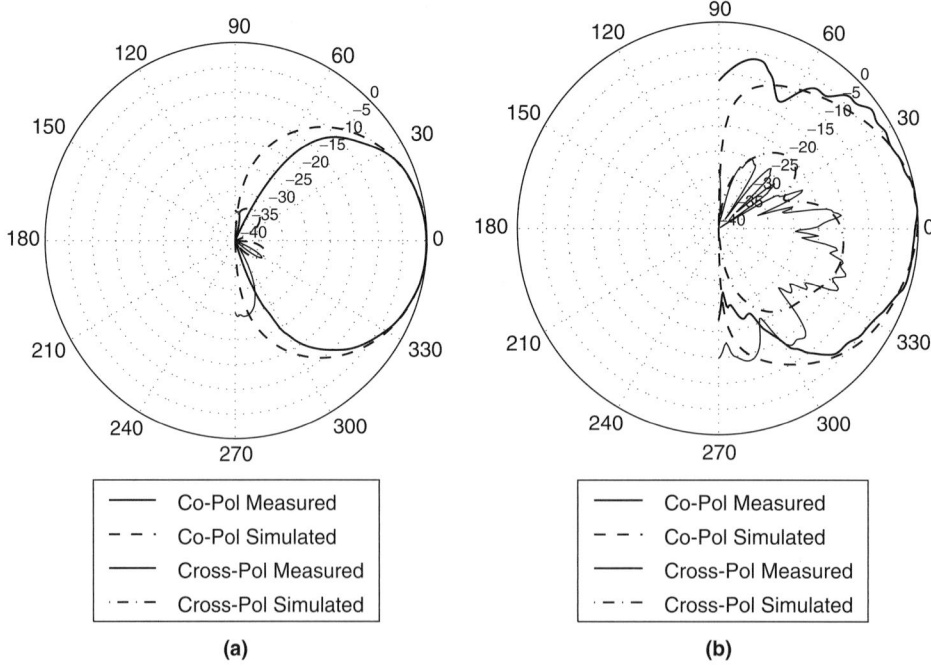

FIGURE 5.15 Radiation pattern plots of the antenna arrays developed on multilayer LCP technology: (a) 14-GHz array and (b) 35-GHz array.

Radio Frequency System-on-Package (RF SOP) 277

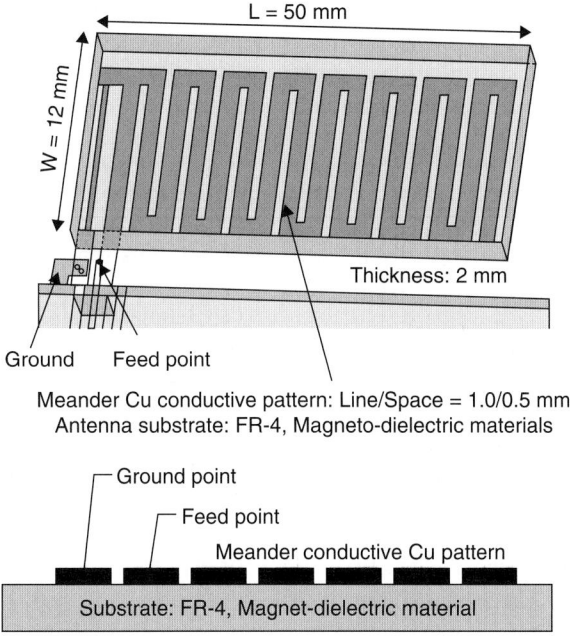

FIGURE 5.16 Meander line antenna on NiZn ferrite composite.

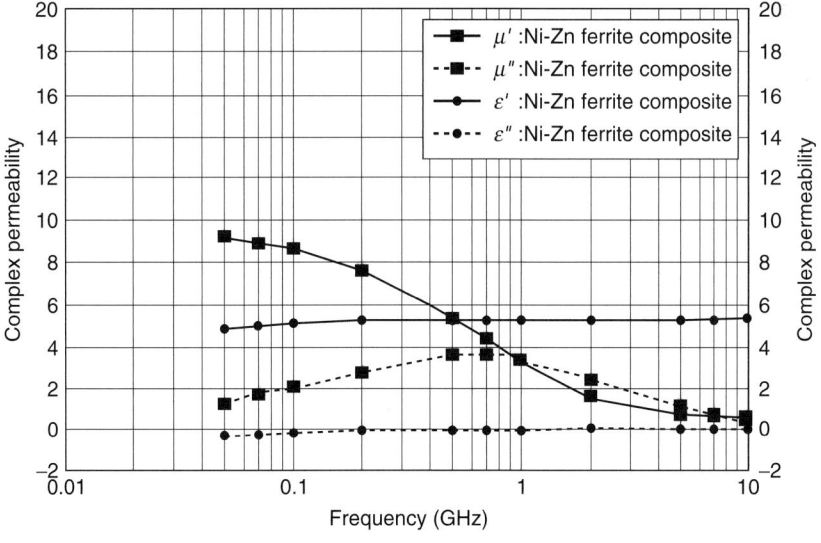

FIGURE 5.17 Frequency characteristics of magneto-dielectric material (NiZn ferrite composite).

the frequency characteristics of the magneto-dielectric materials. The ferrite composite is a composite of NiZn ferrite powder and dielectric resin. The magneto-dielectric substrate has ~5.3 of relative permittivity from 100 MHz to 10 GHz. The dielectric loss is less than 0.1. The relative permeability of this material is ~5 to 8 in the VHF band with a reasonably low loss.

Though the concept of using a magneto-dielectric material to miniaturize antennas is very compelling, the properties of these materials degrade at frequencies beyond 1 GHz, as shown in Figure 5.17, where the material almost becomes nonmagnetic at high frequencies. Hence, the synthesis of new materials that have matched properties is required beyond the 1-GHz range for WiFi applications.

Additionally, novel soft-hard surfaces (SHS) have been researched for the miniaturization and elimination of the substrate-crosstalk modes improving the efficiency of patch antennas and suppressing the backside radiation by 10 to 15 dB [32]. Stacked patch configurations have been explored for the realization of broadband antennas [33]. As in the examples provided here, a combination of novel design and implementation techniques needs to be explored to develop fully integrated, high-performance RF SOP systems.

5.4.4 Inductors

The integration of RF front-end modules has been a key driver for the development of next-generation wireless communication systems. An important component of these high-performance RF modules is the inductor, which forms a critical part in the design of filters, voltage-controlled oscillators, power amplifiers, and low-noise amplifiers. The figures of merit when dealing with inductor performance are

- Inductance (L)
- Self-resonant frequency (SRF)
- Quality factor (Q)

Efficient RF inductors need a very high Q associated with a considerable high SRF. Since Q is affected by the package parasitics, a careful choice of the substrate material such as silicon, LTCC, and organics, used to model and fabricate these inductors, is of utmost concern.

Silicon and ceramics have been the primary materials for inductors until recently. Si is the most mature technology because of its use in wafer fabs, and it offers excellent surface and planarity for a variety of needs including flip-chip bumping and bonding technologies, has good thermal conductivity, and can be multilayered. However, silicon has not been a very successful platform for the design and fabrication of high-Q inductors, primarily due to the losses associated with the semiconducting substrate. Therefore, certain modifications need to be done. Some of these approaches use high-resistive silicon oxides such as silicon-on-insulator (SOI), thick dielectric layers, or thick and multilayer conductor lines. Single- and double-level inductors have been fabricated on silicon by many investigators [34]. A typical silicon substrate used for the single- and double-level inductors is shown in Figure 5.18.

In this approach, a 3D process has been implemented in single layers and multilayers for a spiral inductor. As expected, the inductance observed in the case of a single-layer inductor was higher when compared to that of a double-layer inductor. For a line width of 20 µm, spacing of 5 µm, and substrate thickness of 625 µm, a quality factor of 52.8 was

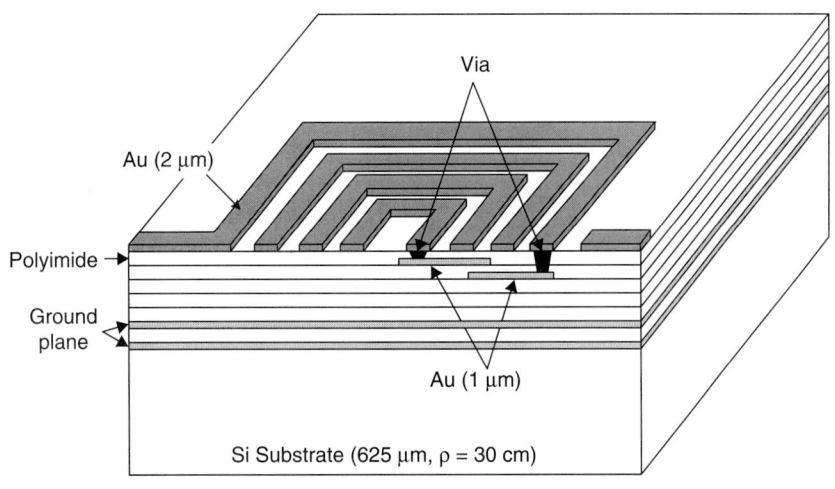

FIGURE 5.18 An example of inductors on Si substrate. [34]

observed at a frequency of 13.6 GHz and an SRF of 24.7 GHz. A high-speed complementary bipolar process has also been used to build inductors with a Q of over 12 for use in wireless applications [35]. In this example, 16 square inductors were built on a test array with outer dimensions of 300 μm. A spacing of 4 μm was used for each inductor. An oxidized porous silicon layer of 25 μm thickness was used on an SiO_2 substrate to obtain a high-performance planar inductor [36]. A Q of 13.3 was obtained for a 6.29-nH inductor with a self-resonant frequency of 13.8 GHz. Instead of direct oxidization of bulk silicon, the oxidation process of porous silicon was used to make the thick oxide layer.

A quality factor of up to 30 and self-resonant frequency higher than 10 GHz was obtained when polysilicon spiral inductors encapsulated with copper were suspended over 30-μm-deep cavities in the silicon substrate beneath [37]. The metallization process simultaneously coated the inner surfaces of the cavities with copper to form a good radiofrequency ground and an electromagnetic shield.

To achieve cost and size reductions, low-cost manufacturing technologies for RF inductors were developed by utilizing a passive integration process using copper metallization with benzocyclobutene (BCB) interlayer dielectric [38]. In this example, a 10-μm-thick copper plating process for low-loss inductor fabrication and interconnections was used. The fabricated inductor library showed a maximum quality factor in the range of 30 to 120 and inductance values in the range of 0.35 to 31.5 nH around 4 GHz.

Most inductors fabricated on organic substrates are designed as onboard components [39–40]. The majority of these structures have been designed as either microstrip or coplanar lines, each with its own set of advantages and limitations. The microstrip configuration provides good power handling capability and low dispersion. However, inductors require vias to provide connection to the ground plane. These vias, not only add process steps, but also introduce process variations and parasitics. Coplanar waveguide structures on the other hand, make it easier to add shunt and series elements as compared to their microstrip counterparts.

As discussed above, LCP is a very attractive material as a high-frequency circuit substrate due to its ultralow loss and low dielectric constant over a high-frequency range, near hermetic sealing as a result of superior moisture barrier properties, flexible interconnections, and microvias for achieving high-density interconnections [39]. A maximum Q in excess of 70 was reported in the C band. The Nelco N4000-6 type is another organic material on which planar CPW loop inductors with a high Q of 85 in the 5-GHz range were demonstrated [40].

Another method to achieve very high Q, a maximum Q of 100 for a 3.6-nH inductor at 1.8 GHz, involved standard FR4 substrate using a buildup layer of Dupont Vialux material [41–42]. The process used was a low-cost process based on large-area multi-chip module-laminated (MCM-L) technology. The self-resonance frequency was 10.6 GHz. About 150 variations of the inductor designs were built on this testbed, and the structures were examined for variations in parameters such as line width, spacing, ground separation, and the number of inductor turns. Dupont Vialux also produced some very good results for cascaded loop inductors. A Q factor of 103 was demonstrated for a, 11-nH inductor at 2.2 GHz with a self-resonant frequency of 3.6 GHz [41]. A unique design using cascaded loops was used as shown in Figure 5.19.

The substrate used for these inductors was FR4 with buildup layers of Dupont Vialux dielectric. These inductors used microstrip designs. A high Q of 180 has been obtained in the frequency range of 1 to 3 GHz for an inductance range of 1 to 20 nH [42]. Microstrip loop, microstrip spiral, and CPW loop inductors with a hollow ground plane were designed. Only one layer of Dupont Vialux material was used. This was done to minimize via registration and alignment problems. A Q of 110 was obtained for the microstrip loop inductor with a width of 6 mils and spacing of 4 mils. A microstrip spiral inductor gave a Q of 170 at 2.4 GHz with an area of 3.2 mm^2 and SRF of 8.5 GHz. The highest Q was for a CPW loop inductor with a Q of 180 at 2.2 GHz and occupying an area of 9 mm^2 with an SRF of 5.5 GHz. The fabricated loop inductors are as shown in Figure 5.19. Using the multi-chip module-deposited (MCM-D) approach, IMEC has developed inductors with a Q factor greater than 100 [43].

LTCC, as discussed above, has been the best choice for RF modules both in mobile phones and base station applications. It offers compact, high performance, and high functionality in microwave packaging applications. Its two main advantages are its ultralow loss and a manufacturing process that allows multilayers to be fabricated in

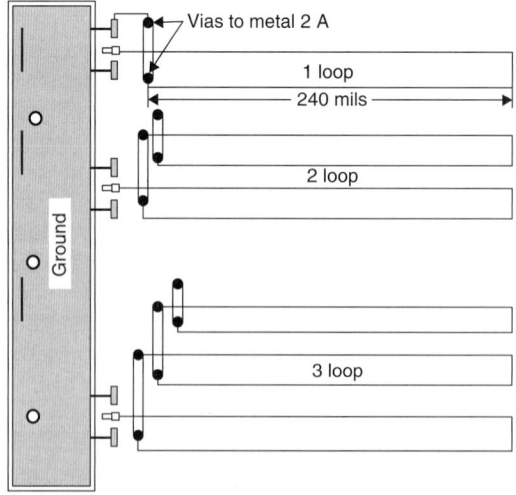

FIGURE 5.19 Schematics of loop inductors. [42]

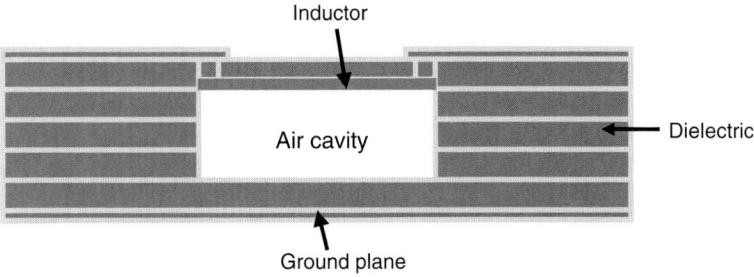

FIGURE 5.20 Inductor on an LTCC platform with an enclosed air cavity. [43]

high volume. A number of LTCC designs have been reported in the literature. One fully embedded LTCC spiral inductor incorporating an air cavity between the spiral and ground plane has been reported to achieve a quality factor of about 51 with an SRF of 9.1 GHz [44]. The air cavity employed under the spiral reduces the shunt parasitic capacitance of the inductor resulting in a high Q factor and a high SRF of embedded inductors. The inductors were designed using a low-loss LTCC dielectric of 114 μm thickness and silver conductor of 12 μm thickness. The spiral inductors, with an air cavity incorporated, were fully embedded in a five-layer LTCC block as well as those without an air cavity. The cross section of such a structure is shown in Figure 5.20.

Another example of LTCC resulted in a quality factor of 93 at 1.1 GHz and an SRF of 3.11 GHz when 3D helical inductors with circular turns were designed on 2a 0-layer LTCC-951-AT ceramic [45]. The 3D helical inductors occupied less space. An inductance of about 9.6 nH was reported. Such a helical inductor fabrication is shown in Figure 5.21. In addition to occupying less space, the helical configuration reduces the coupling capacitance by increasing the distance between the top turns and the underlying turns, thereby preventing a considerable reduction in SRF. For higher Q, MEMS technologies have been implemented by numerous authors [46–53]. The fabricated devices exhibit very high performances such as Q values above 100 and self-resonance frequencies as high as 50 GHz [46].

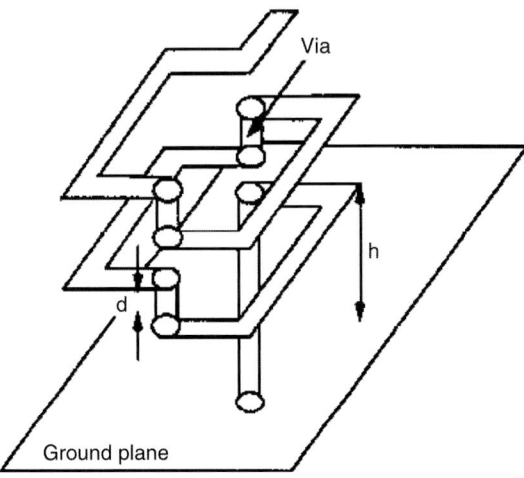

FIGURE 5.21 Diagram of a 3D helical inductor. [44]

	Quality Factor	Inductance (nH)	Frequency (GHz)
Silicon			
Low resistivity	52.8 [94]	1.38	13.6
High resistivity	30 [95]	4	1–2
Micromachined	150 [79]	1	8–23
Wafer-level packaging	38 [80]	1	4.7
LTCC	93 [81]	9.6	1.15
Organic laminate	180 [42]	4.8	2.2

TABLE 5.1 Comparison of Technologies for Inductor Integration

As mentioned earlier, the parameters of interest for the design of inductors are its inductance, Q factor, and self-resonant frequency (SRF). The design of inductors with any substrate can be optimized using a combination of full-wave electromagnetic solvers such as method-of-moment based tools and quasi-TEM approaches. Quasi-TEM approaches are faster compared to full-wave solvers, and at lower frequencies (< 8 GHz) they provide a better approximation for the associated loss in devices with thicker metallization (>10 μm). For passives that use a circular topology, full-wave solvers such as from Sonnet are typically used for purposes of optimization.

In summary, Table 5.1 compares the various technologies for fabricating inductors with the highest quality factor. The corresponding inductance and frequency of operation are also listed in the table. This table does not represent the ultimate limit for that technology but provides an indication of what has been achieved. Silicon-based processing includes inductor fabrication on low- and high-resistivity silicon and micromachining. Wafer-level packaging refers to the realization of inductors above the passivation layer using thin-film postprocessing techniques on the silicon wafer. LTCC and organic laminates are the substrate technologies discussed earlier.

5.4.5 RF Capacitors

RF applications such as filters and resonators need stringent tolerance, a low temperature coefficient of capacitance (TCC), and a high Q (quality factor). This is in sharp contrast to decoupling capacitors that do not have such stringent requirements but require capacitance densities that are much higher. Current RF capacitors are either polymer-based with low capacitance density or high-temperature vacuum-deposited (metal-organic chemical vapor deposition (MOCVD) thin-film, thick-film LTCC-based composites), which causes limitations for RF integration. Among embedded RF passives, emerging applications with embedded RF capacitors require the development of organic compatible dielectric materials with thermally stable high dielectric constant, low loss, and improved electrical performance. Thin-film and thick-film processes should therefore be engineered to be compatible with the low-cost substrate wiring and other embedded RF component technologies.

Electrical and Material Parameters

Capacitors are one of the most basic elements in RF systems. RF capacitors are needed for filtering (<10 pF) and capacitive coupling (<500 pF). These applications need a capacitance density of about 1 nF/cm². RF capacitors require a Q of ≥ 200 to meet the performance requirements. In addition, for many applications, the capacitance value has to be stable within 0.3 percent over a 100°C range of temperature (TCC of <30 ppm/°C). While the high Q and low TCC of capacitors in LTCC RF modules have been demonstrated for decades [82–83], the dielectric mainly consists of ceramics and glass and requires high-temperature crystallization, which is not congruous with low-temperature organic substrate processing. LTCC technology is also limited by its high cost, incompatibility with large-area processing, and low component density integration capability. Nevertheless, LTCC technology for RF modules is still prevalent because of the low loss, good thermal conductivity, and stability for high-frequency applications. The disadvantages of LTCC technology can be overcome with LCP-based RF components [84]. Hence, there is an increasing trend toward LCP-based RF circuits. However, the low dielectric constant of this material makes the RF components and modules larger in size, which may limit the component integration density; increases coupling between the components; and degrades the total system performance. Furthermore, low-loss and low-TCC polymers such as LCP and PTFE are not easily amenable to thin films, without compromising the electrical properties.

Low-loss and high-Q capacitors have been achieved on a silicon platform using a thin-film BCB buildup structure for RF wafer-level SOP functions [85]. High-K and low-loss pyrochlore thin-film in organic substrate has also been explored [86]. This technology enables complete RF integration for various applications such as matching networks, filters, and even tunable components such as phase shifters. On the other hand, new and novel compositions to achieve high Q and low TCC have been pursued using the composite approach with ceramic fillers and low-loss, high-Q polymers. For example, an LCP-based polymer composite has been engineered to replace LTCC components such as capacitors.

MIM and Parallel-Plate Structures

A typical RF capacitor is the metal-insulator-metal (MIM) capacitor, as shown in Figure 5.22a. Electrical connections are made to both the top plate and bottom plate of the capacitor device. The capacitance of the MIM structure can be calculated using the parallel-plate capacitance formula:

$$C = \varepsilon_o A K / t$$

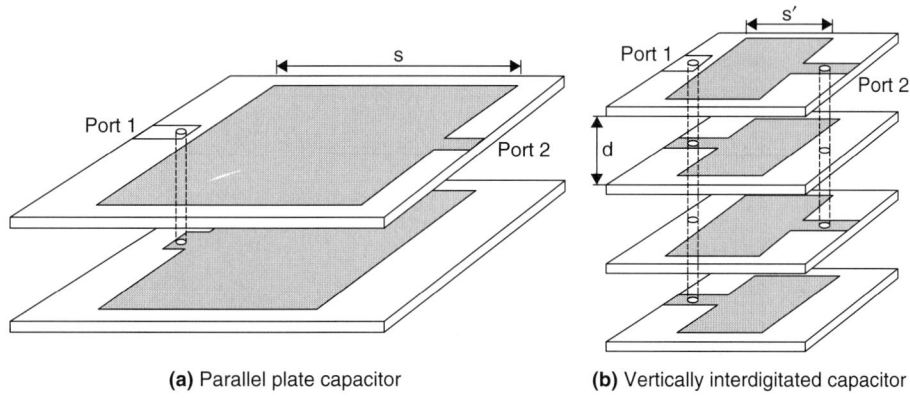

(a) Parallel plate capacitor   (b) Vertically interdigitated capacitor

FIGURE 5.22 Three-dimensional views of a metal-insulator-metal (MIM) and vertically interdigitated capacitor (VIC) configurations. [5]

where C is the capacitance (F), ε_o is the permittivity of free space (8.854×10^{-12} F/m), A is the area (m²), and t is the thickness (m). Insertion of a dielectric between the parallel plates increases the capacitance by an amount proportional to the dielectric constant, K. The dielectric constant is defined by $K = \varepsilon / \varepsilon_o$, where ε is the permittivity of the dielectric. For large capacitors such as the RF ground capacitors, however, the electrode size becomes too large for the MIM configuration to handle. The interdigital topology tends to require a bigger area since the electric flux is generated laterally instead of vertically such as in the MIM, which allows more electrode coverage.

An alternative capacitor implementation to the MIM topology [5] was proposed using the vertically interdigitated configuration (VIC) shown in Figure 5.22b. The MIM structure consisting of a dielectric layer sandwiched between two square plates of widths in Figure 5.22a implements this type of capacitor, neglecting the higher-order excitation mode. This capacitor can also be implemented by a parallel combination of pairs of plates of smaller size. The plate size can be made smaller as more plates are deployed on many dielectric layers. VIC topology occupies nearly an order of magnitude less area than the MIM while maintaining comparable performance.

TCC Properties

The thermal coefficient of capacitance (TCC) is a very important parameter for RF components [88]. Any deviation in component specifications with temperature can adversely affect the frequency selection characteristics of the filter or resonator circuits in RF modules. The TCC is becoming critical for various RF applications because of the tighter design tolerances. The TCC values can be calculated from the measured capacitance data with temperature using the following equation. This definition is used in discrete capacitors and would also be applicable for embedded capacitors:

$$\text{TCC} = \frac{(C_{85°C} - C_{25°C})}{\Delta T \times C_{25°C}} \times 10^6$$

where TCC = temperature coefficient of capacitance (ppm/°C), $C_{85°C}$ = capacitance at 85°C, $C_{25°C}$ = capacitance at 25°C, and ΔT = temperature difference between 85°C and 25°C = 60°C. The TCC can be positive or negative for both polymers and ceramics depending on the material structure. BCB, for example, has a negative TCC behavior over the temperature range of 25 to 125°C, showing its value of about –250 ppm/°C [89]. Ferroelectrics have high-positive TCC, while most paraelectrics have negative TCC. Similarly, polymers such as epoxy and polyimide show a positive TCC unlike certain other polymers. The TCC tolerances for RF components are met by careful selection and engineering of the material compositions. Table 5.2 shows TCC values for typical polymers and paraelectric ceramics.

Material	BCB[88]	PTFE[88]	LCP[90]	SiO$_2$[88]	Al$_2$O$_3$[88, 91]	Ta$_2$O$_5$[92]	TiO$_2$[92]
TCC (ppm/°C)	–250	–100	–42	<100	<390	200–400	–750

TABLE 5.2 TCC Values of Typical Materials

Comparing Materials and TCC Properties

A high dielectric constant (K) invariably comes with either a negative or positive TCC. Fillers can be selected to compensate for each other or compensate with the polymer TCC, leading to a lower TCC. It is important to have a flat TCC over a large temperature range. Hence, paraelectric fillers are preferred to ferroelectric fillers since their capacitance value is much more stable with regard to most processing parameters such as temperature and frequency [93]. Figure 5.23 shows the relation between the TCC and K of various materials and the TCC compensation in the composite system. There is no reliable data on the TCC characteristics of polymers. This is even more complicated because the TCC of capacitors in the buildup layers depends on the substrate thermomechanical properties. Literature data only allude to the properties of bulk materials and thin films but not for powders. This makes the design of low-loss and low-TCC composites even more complicated. Photopolymer composites have been proposed. A tight tolerance can be achieved with thin-film photolithography. Other parameters that control tolerance are thickness control and uniform dispersion of ceramic fillers to prevent material inhomogeneities.

TCC Compensation in Polymer-Based Composites

The filler contribution to the net composite TCC is dependent on its volume fraction and the distribution of filler in the composite material as well. The dielectric constant of composites with fillers perfectly aligned parallel to the electric field strength can be modeled as

$$\varepsilon_c = v_m \varepsilon_m + v_f \varepsilon_f$$

where ε_c, ε_m and ε_f are the dielectric constants of nanocomposite, matrix, and ceramic filler, respectively, and v_m and v_f are the volume fraction of matrix and ceramic filler, respectively. In this case, the TCC of the composite can be obtained by simple differentiation with respect to temperature and by rearranging the equation:

$$\frac{d\varepsilon_c}{dT} = v_m \frac{d\varepsilon_m}{dT} + v_f \frac{d\varepsilon_f}{dT}$$

Dividing the equation by ε_c and rearranging leads to

$$\text{TCC}_C = \frac{1}{\varepsilon_c} \frac{d\varepsilon_c}{dT}$$

$$\text{TCC}_C = v_m \frac{\varepsilon_m}{\varepsilon_c} \text{TCC}_m + v_f \frac{\varepsilon_f}{\varepsilon_c} \text{TCC}_f$$

In this case, the TCC of the filler influences the composite TCC to a large extent because $\varepsilon_f/\varepsilon_c$ is larger than $\varepsilon_m/\varepsilon_c$.

If the composites have fillers aligned perpendicularly to the electric field, the dielectric constant is modeled as

$$\frac{1}{\varepsilon_c} = \frac{v_m}{\varepsilon_m} + \frac{v_f}{\varepsilon_f}$$

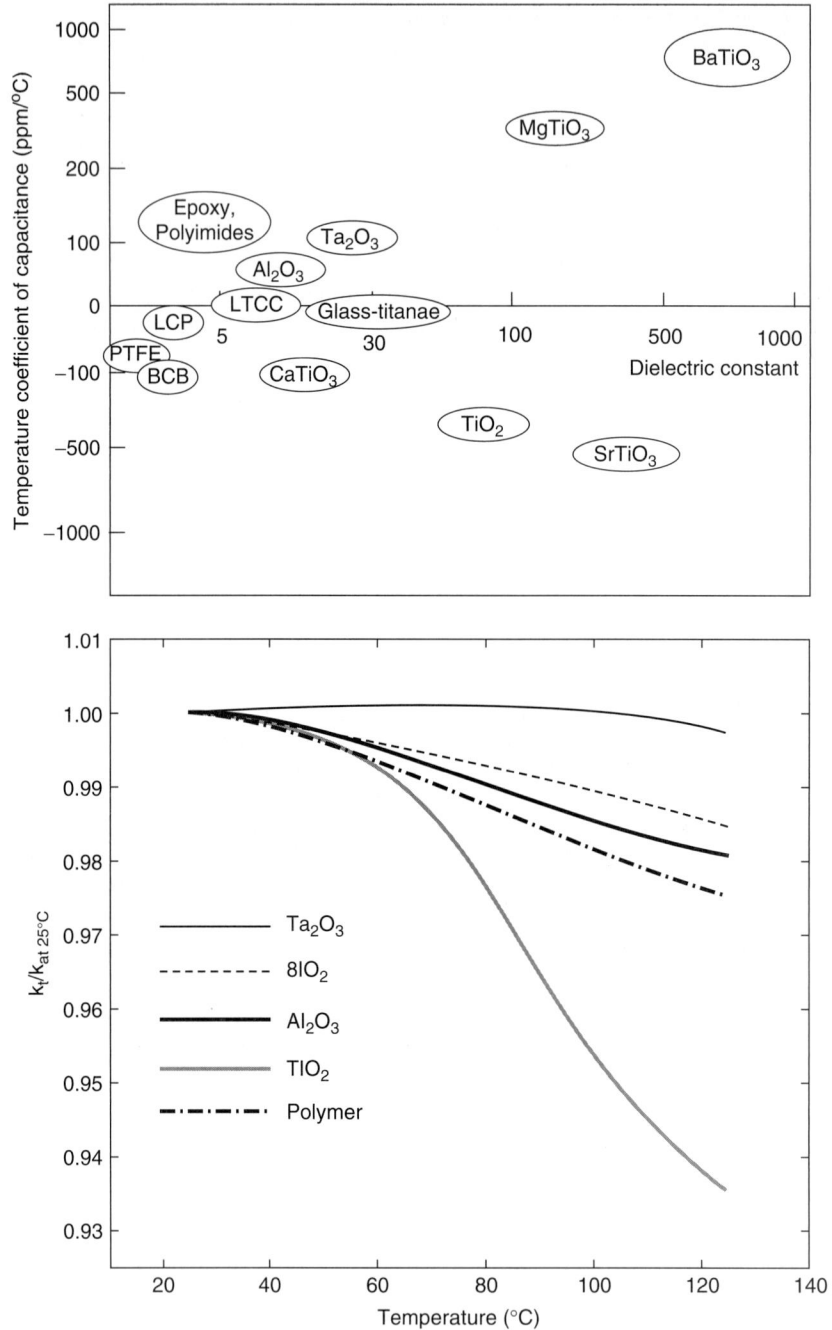

FIGURE 5.23 (a) The TCC versus the dielectric constant of various materials. (b) Temperature stability of the dielectric constant of polymer-based composite films with various fillers.

Differentiating with respect to temperature and rearranging:

$$\text{TCC}_c = v_m \frac{\varepsilon_c}{\varepsilon_m} \text{TCC}_m + v_f \frac{\varepsilon_c}{\varepsilon_f} \text{TCC}_f$$

In this situation, the first term has more effect because $\varepsilon_c/\varepsilon_m$ is much larger than $\varepsilon_c/\varepsilon_f$. Thus, the polymer or filler can significantly affect the composite TCC depending on the filler morphology and arrangement in the composite.

For most particulate ceramic-polymer composites, the behavior is in the middle of these two extreme cases. The modified Lichtenecker's law, commonly used to predict the effective dielectric constant of polymer-ceramic nanocomposites with different volume fractions, is expressed as

$$\log \varepsilon_c = v_m \log \varepsilon_m + k v_f \log \varepsilon_f$$

where k is a fitting constant subject to the composite material and can be related to the distribution of filler in the composite material. Differentiating with temperature, the TCC can be written as

$$\text{TCC}_c = v_m \text{TCC}_m + k v_f \text{TCC}_f$$

The TCC of the composite in this case is linearly dependent on the volume fraction of the filler and the composite. The filler contribution to the net composite TCC is dependent both on k and the volume fraction. For well-dispersed suspensions, there is a strong coupling between the filler particles leading to a higher k. Therefore, the filler has a strong effect on the permittivity and TCC. For aggregated suspensions, the permittivity of the matrix has a stronger influence on the composite permittivity and TCC. In other words, the aggregate composites with widely separate fillers behave similarly to the case where the fillers are perpendicular to the field because of the weak dielectric coupling between one aggregate and the other. The TCC does not scale linearly with the filler content in this case.

The disadvantages of LTCC can be overcome with liquid crystalline polymer based RF circuits. Current organic-compatible embedded capacitor technologies such as epoxy-based composites with ceramic fillers are not suitable for high-performance (higher Q and lower TCC) RF capacitor requirements because they may not achieve a dielectric loss less than 0.02 and a TCC within 300 ppm/°C, even with the best ceramic fillers [54]. Low-loss polymers such as bisbenzocyclobutene (BCB) and polytetrafluoroethylene (PTFE) are easily amenable to thick-film formation and can be filled with ceramic fillers.

Composite-type integrated capacitors for applications demanding precise capacitance control can be based on a simple mixture of polymer and paraelectric ceramic particles. Paraelectric ceramics such as Ta_2O_5, SiO_2, and Al_2O_3 appear to be suitable in these applications. They have a much lower dielectric constant compared to ferroelectric ceramic particles, but their capacitance value is much more stable with regard to most of the processing conditions. Among them, Ta_2O_5 is promising because it has a dielectric constant of 24, relatively higher compared to any other paraelectrics, and exhibits a moderate positive temperature coefficient of capacitance, ranging from +200 to +400 ppm/°C [55]. It is thought that this positive TCC of Ta_2O_5 can compensate the negative TCC of a high-Q polymer itself, giving rise to the nearly zero TCC

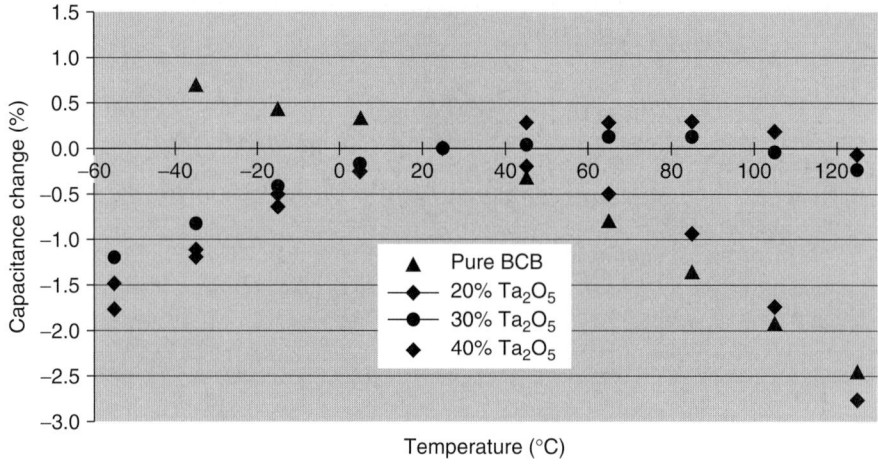

Figure 5.24 Temperature dependence of capacitance of BCB-based composites with various Ta_2O_5 contents. [55]

characteristics. Ta_2O_5 was therefore added to BCB polymer to improve the temperature stability of capacitance, along with improving the capacitance density as shown in Figure 5.24.

5.4.6 Resistors

Resistor Technologies

In general, the RF resistors can be achieved by three major process technologies: screen-printing of polymer thick film (PTF), electroless plating, and direct foil lamination.

The polymer thick films have instability problems in the microwave frequency range due to their moisture absorption. However, these thick films offer much higher resistance that is suitable for the pull-up, pull-down, and circuit isolation. As such, PTF is a mature technology that has been applied into numerous products [56]. For example, Motorola uses carbon-phenolic polymer thick-film (PTF) ink and a screen-printing process to form resistors in the inner layers of high-density interconnect printed wiring boards. The PTF ink is screened on to copper termination pads that are treated with proprietary interface metallurgy to improve reliability and environmental stability. In Motorola product applications, between 8000 to 20,000 resistors have been printed on an 18 in × 24 in panel in a single screening step, resulting in significant economies of scale and associated cost savings. PTF resistors can be trimmed at the inner-layer stage to within 1 percent tolerance.

In contrast to the above thick-film resistors, the thin-film resistors are achieved by electroless plating and direct foil lamination [57]. The electroless plating involves surface preparation of the dielectric medium followed by chemical treatment in order to deposit a thin (usually 0.3 to 1 μm) resistive layer that then can be patterned, and the stubs can be plated to define resistors. Figure 5.25 shows such a resistor comprised of NiWP on an epoxy dielectric [58].

The high-frequency measurements are typically performed with an HP 8510C vector network analyzer and ground-signal-ground (GSG) coplanar waveguide 200-μm-pitch probes. The results of structures with NiP/NiWP fabricated without a ground

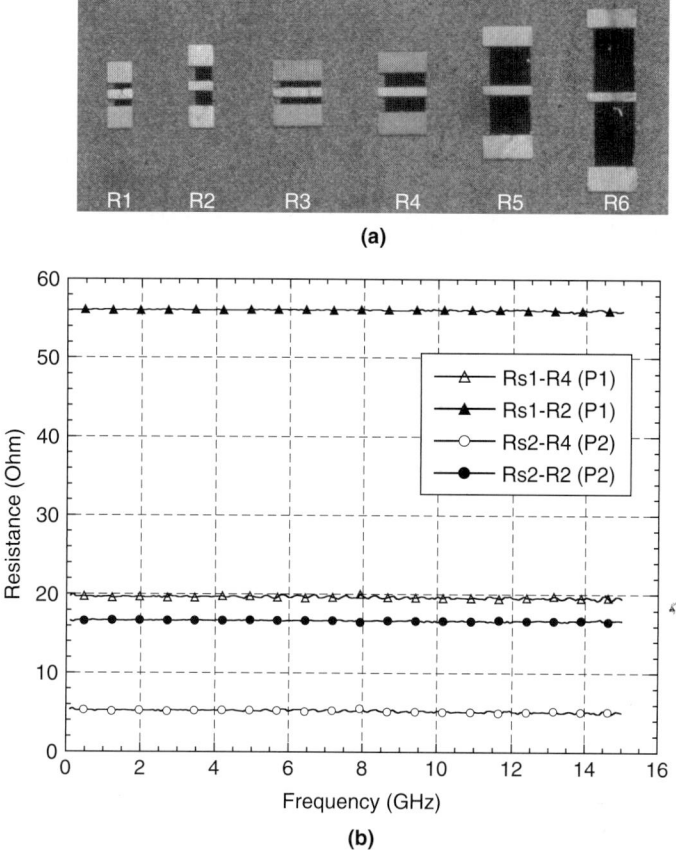

FIGURE 5.25 (a) Photomicrograph of GSG resistor structures, NiP/NiWP resistor film in dark, used in the high-frequency measurements. (b) Measurement results.

plane are presented in Figure 15.24b. The temperature coefficient of resistance in such alloys was near zero—a great advantage to circuit designers. The electroless plating has also been reported on other polymers such as BCB and LCP.

Another approach to forming resistors is the lamination of thin-film resistors predeposited on copper foil. The resistors are printed and patterned after postlamination [60]. Utilizing a lamination process using LCP substrates, various RF resistor structures have been designed, fabricated, and implemented. Measurements up to 40 GHz are presented using the 25 $\Omega/\square$ NiCr sputtered film on copper foil as shown in Figure 5.26.

Applications

Resistors have several applications in high-frequency circuits including in attenuators, terminations, power dividers, and oscillators. The major challenges in the RF resistors are the low profile, substrate smoothness, stability of the properties over a wide frequency range, reproducibility, and availability of low-loss substrate materials.

Termination Resistors Several different termination topologies have been simulated to provide a 50-Ω load with the smallest parasitic response across the broadest range of

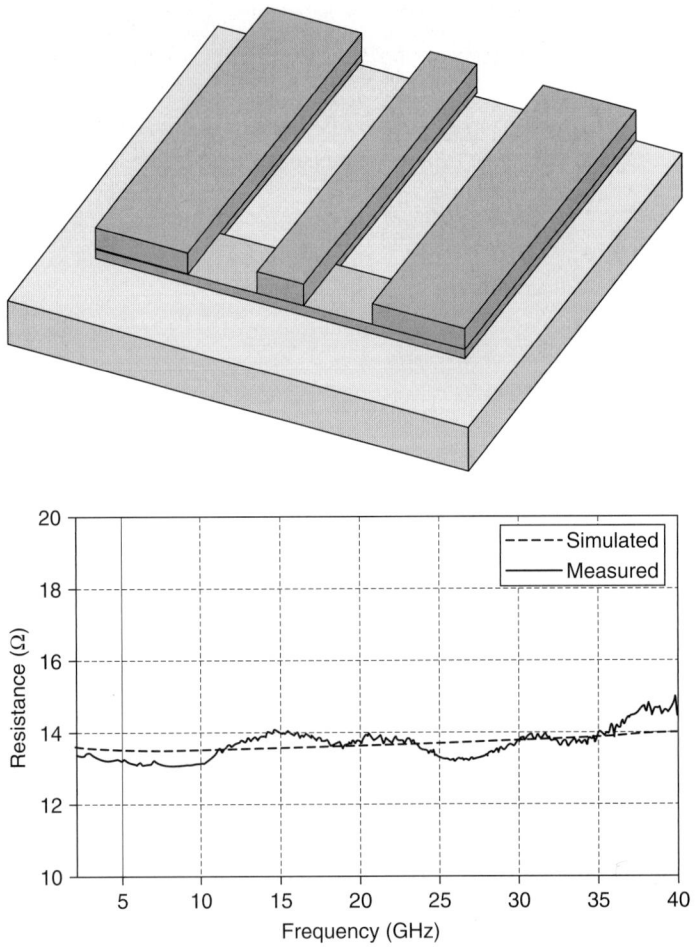

FIGURE 5.26 Designed, simulated, and measured resistance up to 40 GHz. [60]

frequencies. Since the CPW topology is determined to provide the best response, the termination structures are typically fabricated as shown in Figure 5.27 and measured using through, reflect, line (TRL) calibration [60].

Attenuator Attenuators are typically designed and fabricated using the resistor foils. The circuits can be created in two ways, using either a *T*-network or a π-network. Attenuation is expressed in decibels like most other RF system components. To calculate the resistance values needed, the attenuation value should be converted from decibels to magnitude units, and the resulting system of equations should be solved. For a *T*-network, the system of equations is

$$\alpha = 50 R_p (R_s + R_p)(R_s + 50)$$

and

$$50 = [(50 + R_s) \;||\; R_p] + R_s$$

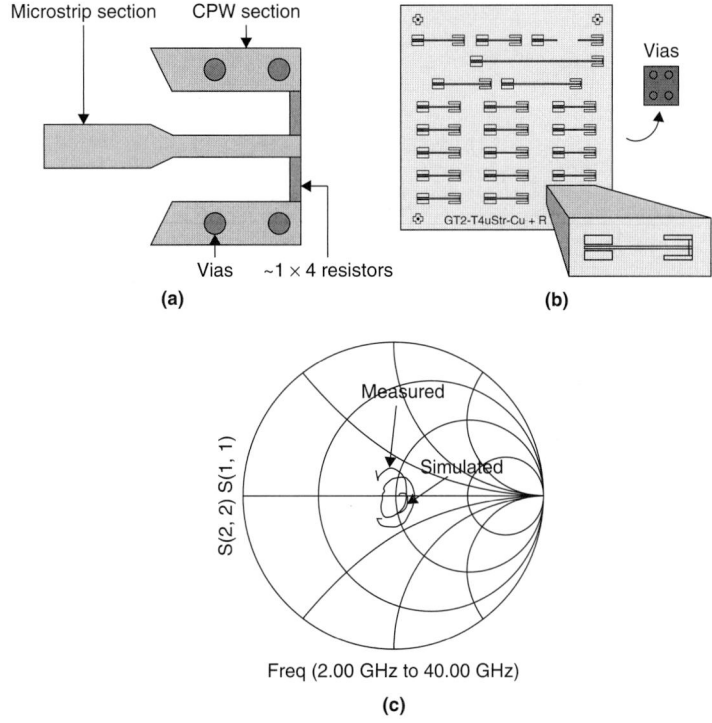

FIGURE 5.27 (a) Designed, (b) fabricated termination structure, and (c) measured and simulated data.

where α denotes the magnitude of the desired attenuation and $||$ is the parallel combination function. A similar set of equations can be made for the π-network. The first equation is the attenuation condition, while the second is the matching condition, ensuring the resistances combine to look like 50 Ω. The layouts of R_s and R_p for both the T-network and π-network topologies are shown in Figure 5.28.

The simulation layout in HFSS and the fabricated circuit can be seen in Figure 5.29. These attenuators are simple and if simulated properly, perform very well even at very high millimeter wave frequencies. The measured data are shown in Figure 5.30.

Wilkinson Power Dividers   Wilkinson power dividers are typically designed across a wide range of frequencies from the X band to the W band. An example of a circuit schematic of a Wilkinson divider can be seen in Figure 5.31. In this example, the resistor

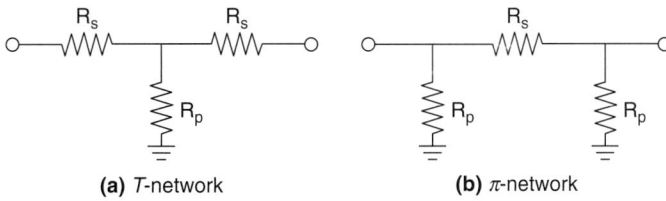

FIGURE 5.28 Attenuator circuit.

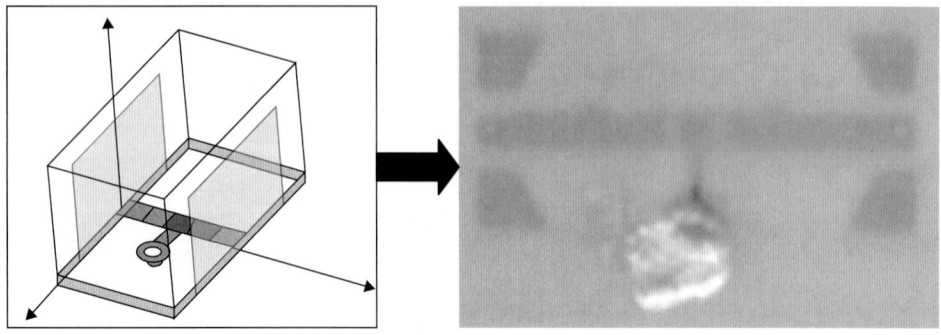

Figure 5.29 Attenuator simulation and fabrication.

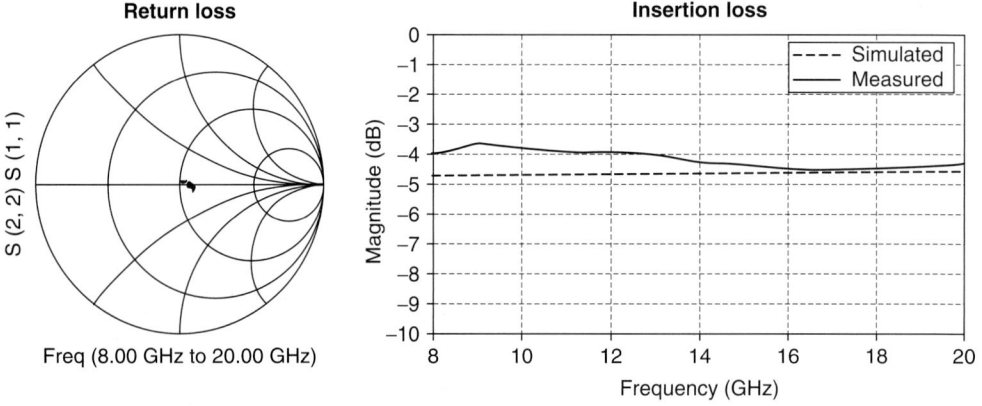

Figure 5.30 Attenuator measurement data. [60]

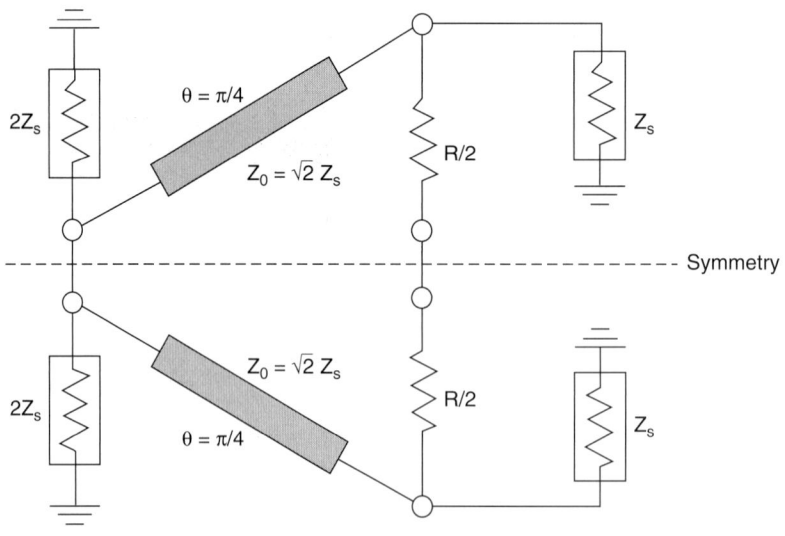

Figure 5.31 Schematic of the Wilkinson power divider.

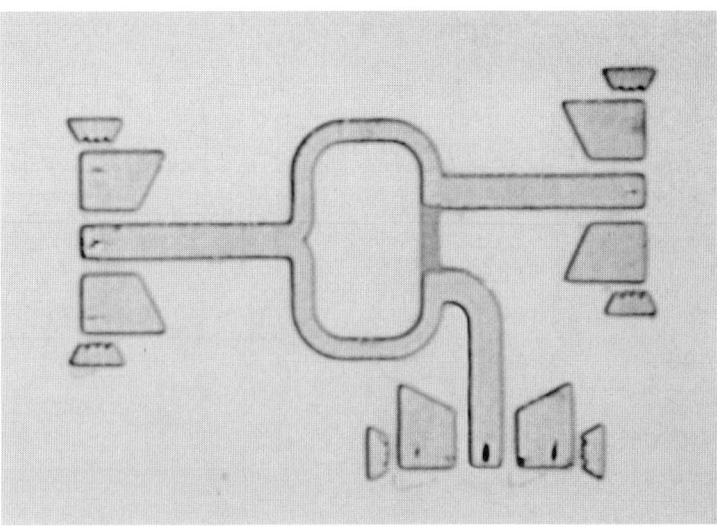
Figure 5.32 Fabricated K_a-band Wilkinson divider.

is placed between the branches of the output ports to isolate any return signals between them. This differential placing makes the Wilkinson divider useful in creating baluns. In the forward case, the divider operates by splitting the input signal with a simple T-style junction. Each path then travels through a $\lambda/4$ transformer with an impedance designed to step the signal up to twice the system impedance. This signal is then placed across a resistance of twice the system impedance through the superposition of odd and even modes. This halves the signal to the original system impedance with half of the original input power on each branch. A photograph of the fabricated K_a band circuit is depicted in Figure 5.32. The measurement data are given in Figure 5.33. An insertion loss of about 0.35 dB is achieved from 28 to 40 GHz.

TCR Properties

One of the most important barriers for embedded resistors is the use of a single materials system to cover the entire range of resistance requirement (near zero to 200 kΩ). In the absence of this single material, several materials are typically used, each with its own advantages and limitations such as the temperature coefficient of resistance (TCR). For example, Polymer Thick Film (PTF) offers low cost and a wide range of resistance values but has problems in temperature and humidity, oxidation at the interface, CTE mismatch, high-frequency stability, and high temperature coefficient of resistance. Resistors predeposited on copper foil, discussed above, are limited to a lower range of resistances. Electroless plated resistors offer cost advantages, but this technology is currently limited to low-value resistors within a small deposited area.

The materials systems used for embedded resistors are generally categorized as metal and alloys, semiconductors, cermets, and polymer thick films [56]. From a processing standpoint, these are classified as thin film, printable, and plated. Among commercial materials, Ohmega-Ply (electroplated NiP), DuPont Interra (screen-printable LaB6), MacDermid M-Pass (electroless plated NiP), Ashai Chemical (polymer thick films), Shipley Insite (doped Pt on Cu by CVD), and Gould Nichrome on copper

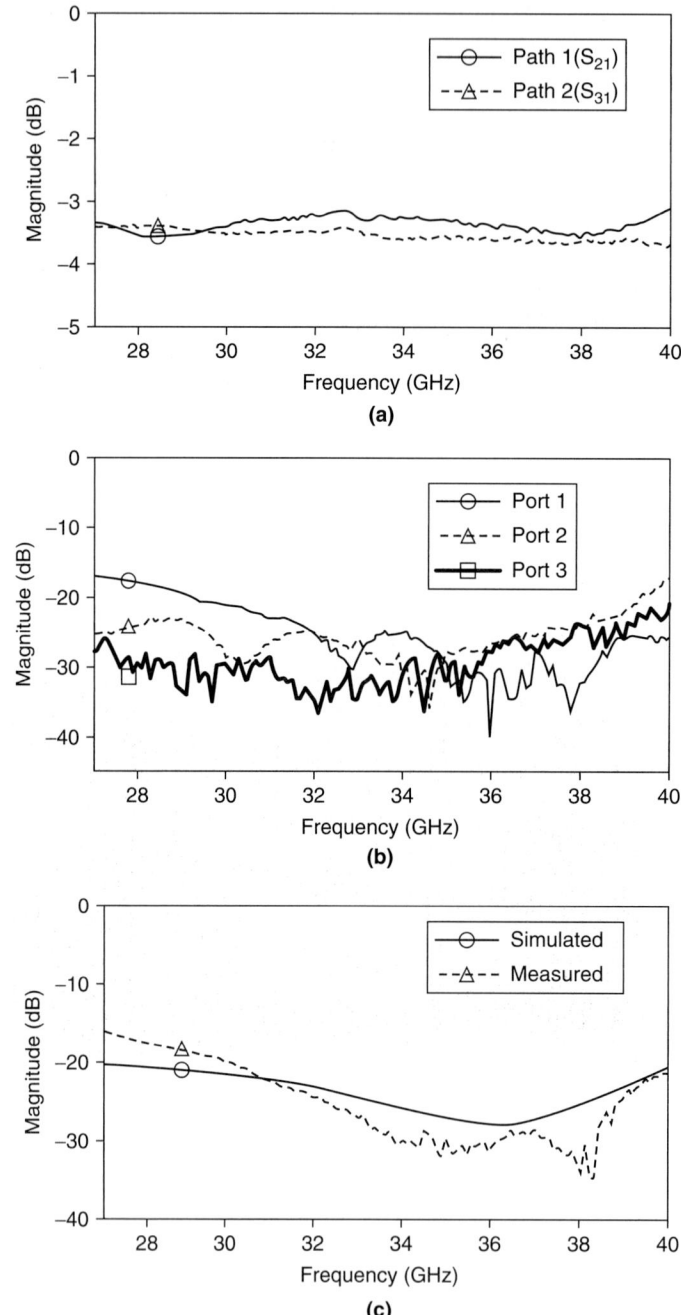

Figure 5.33 (a) Insertion loss, (b) return loss, and (c) isolation measurements [60] of the K_a-band power divider.

foil have received much attention. However, low TCR (<100 ppm/°C) and high tolerance (<5 percent) have not been achieved. Tolerances on the order of 1 to 2 percent are required for analog applications but cannot be achieved without the added cost of laser trimming. Table 5.3 shows the current state-of-the-art in embedded resistors. In most technologies presented in the table, the process tolerances are around 10 to 15 percent without trimming.

5.4.7 Filters

Filters are essential components in many communication systems as they perform the important tasks of channel selection (or rejection) and signal separation. There are three filter technologies available using ceramics, silicon, and organics. Each of these technologies can be used either as discrete components (IPDs) that are surface bonded or as embedded components in the substrate, both as discretes or as thin-film layers. In-package embedded or integrated multilayer filters offer a more attractive implementation than on-chip and discrete filters. The design of filters and their implementation in organic substrates have been explained in detail in Chapter 4. In this section, filters implemented using LTCC and SOP processes are discussed.

An example of an RF image-reject filter uses six layers of LTCC in a stripline configuration as illustrated in Figure 5.34 [61]. Layers 6 and 0 are the top and bottom layers, respectively, serving as the top and bottom ground planes (Figure 5.34a). The two shunt inductors are realized by the U-shaped strips fabricated on layers 4 and 3, which are located two and three layers underneath the top ground plane, respectively. The end strips are connected to both grounds through vias. The VIC topology utilizes two series capacitors. The dumbbell-shaped trace, in this example, is inserted on layer 2 between layers 3 and 1, as the bottom plates of the VIC. The fabricated filter prototype, shown in Figure 5.34b, measures an insertion loss (Figure 5.34c) of 3 dB at 2.4 GHz with a 40-dB rejection at 2 GHz [61].

Georgia Tech has developed several other embedded filters in the SOP process using epoxy materials as the buildup layers. One example shown in Figure 5.35 [62–64] is a bandpass filter for C-band applications consisting of a square patch resonator with inset feed lines. The inset gaps act as small capacitors and cause the filter to have a pseudo-elliptic response with transmission zeros on either side of the passband. This structure also has a tunable bandwidth. The length of the feed lines is determined by the input and output matching requirements. The length of the insets and the distance between them are the main controlling factors, effectively setting the size of the mode-splitting perturbation in the field of the resonator. Measurement results show a bandwidth of 1.5 GHz and a minimum insertion loss of 3 dB at the center frequency of 5.8 GHz [62]. Microstrip pseudo-elliptic bandpass filters, operating in the X band, have been designed and implemented on multilayer LCP technology [64]. Folded open-loop resonators printed on different dielectric surfaces and sharing the same ground plane are coupled through slots etched in the ground plane. Fully canonical filtering and modularity have been achieved through introduction of internal nonresonant nodes. A multilayer configuration is realized through thermocompression bonding of thin sheets of LCP. The designed fourth-order filter exhibits a low insertion loss of 3.2 dB at 9.9 GHz [65]. In addition, a multilayer quasi-elliptic filter using dual-mode resonators has been recently demonstrated on LCP substrate [65]. The filter offers the performance of a four-pole filter by using only two resonators vertically stacked that result in significant space savings. The filter has an insertion loss of about 3.9 dB in the X band. A photograph of the filter along with measured and simulated results are shown in Figure 5.36.

Chapter Five

Name of Company or Organization	Material	Process Approach	Range of Values (Ω/sq)	TCR (ppm/°C)
Intarsia Boeing NTT GE	Ta2N	Sputter	10–100 20 25–125	($\pm$100) (–75 to –100)
Osaka University Metech Acheson Colloids Electra Ashai Chemical W. R. Grace DOW Corning Raychem Corporation Ormet Corporation	Conductive polymer composites	Polymer thick-film process Liquid phase sintering	Insulating to conducting	
Ohmega Ply	NiP alloy	Electroplate	25–500	
Inst. Microelectronics, Singapore	TaSi	DC sputter	10–40 8–20	
University of Arkansas/Sheldahl	CrSi	Sputter		–40
W. L. Gore and Associates	TiW	Sputter	2.4–3.2	
Shipley	Doped Pt on Cu foil	Plasma enhanced chemical vapor deposition (PECVD)	Up to 1000	100
Deutsche Aerospace TICER Technologies	NiCr NiCr NiCrAlSi CrSiO3	Sputter	35–100 25–1000	
Georgia Institute of Technology MacDermid	NiWP NiP	Electroless plate	10–50 25–100	Near zero
DuPont	LaB6	Screen print and foil transfer	Up to 10,000	$\pm$200

TABLE 5.3 Current State-of-the-Art of Embedded Resistors

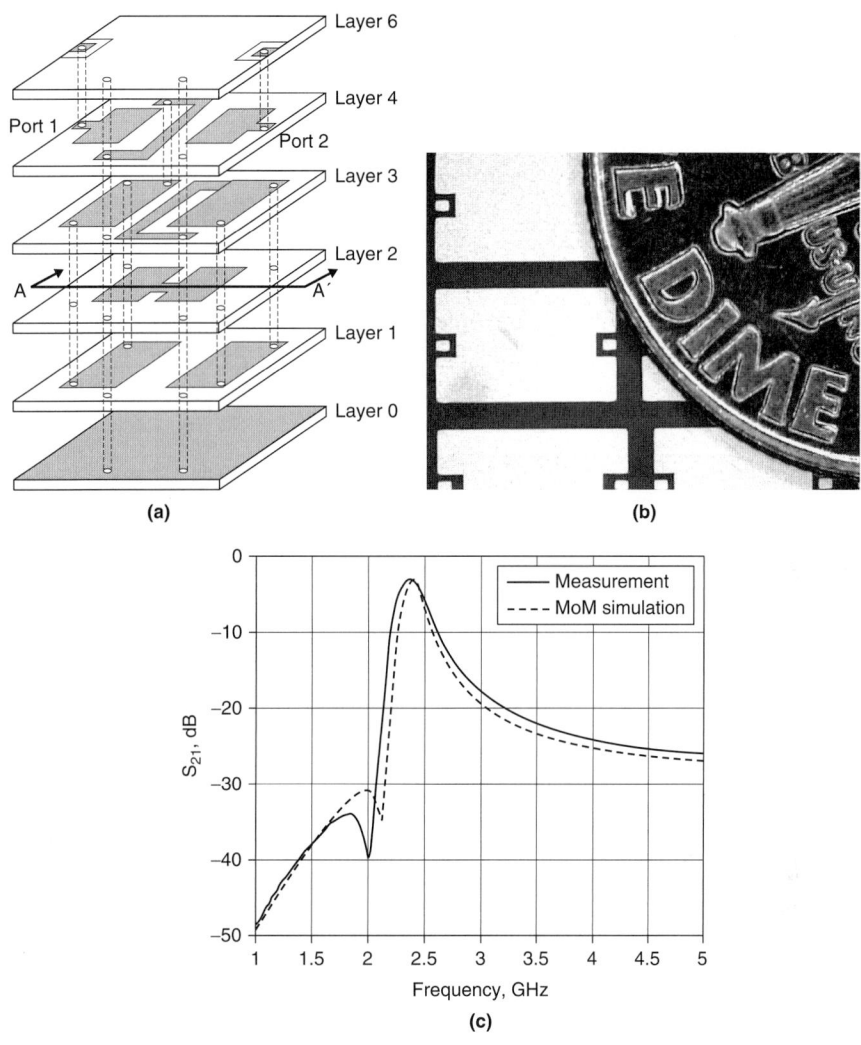

FIGURE 5.34 (a) A three-dimensional layer-by-layer view of multilayer filter structure. (b) Photograph of the fabricated filter. (c) Simulated and measured S_{21} for the filter.

5.4.8 Baluns

Baluns are required in a wide variety of microwave components such as balanced mixers, push-pull amplifiers, multipliers, and phase shifters. The design of baluns and their implementation in organic substrates were discussed in detail in Chapter 4. In this section, the implementation of baluns in LTCC technology is discussed.

Figure 5.37 shows an example of a stripline-type multilayer balun developed on LTCC technology [61]. Two shorted lines are placed next to an open line such that they couple energy from the open line. Each of these lines is shorted to the ground planes (through vias) above and below it, respectively. There is very good agreement between the measurement and simulation as can be seen from the results shown in Figure 5.38.

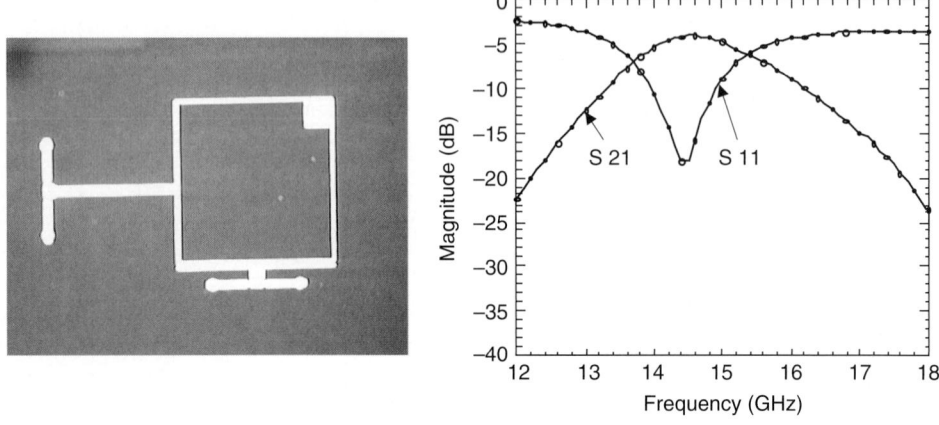

Figure 5.35 Bandpass filter and measured scattering parameters.

An insertion loss of 3.4 dB is measured, while a measured bandwidth of 41 percent is calculated compared with 69 percent for the simulation [62,65]. The measured bandwidth is 75 percent compared with a simulated value of 92 percent.

5.4.9 Combiners

A combiner combines multiple RF signals into a single RF path and is an essential component in any transceiver module. Conventional combiners suffer from limited bandwidth and weak coupling characteristics at the baseband. To overcome these issues, the conventional coupled line coupler has been modified to a vertical coupling structure in the SOP platform, as shown in Figure 5.39a. The output port of the coupled line coupler is used as the input port for the RF signal, and the isolation port is shorted to the ground. Also as designed, a ninth-order Bessel LPF at the input port for the baseband performs as a bandstop filter for the 14-GHz RF signal and reflects the RF signal at the output of the LPF [61,65]. The 14-GHz RF signal has been optimized in-phase

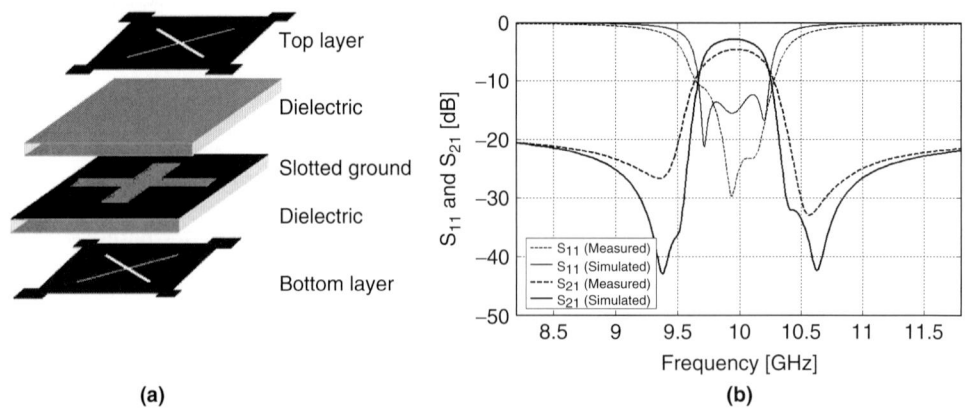

Figure 5.36 (a) Stackup of 3D four-pole filter on LCP. (b) Filter results.

Radio Frequency System-on-Package (RF SOP) 299

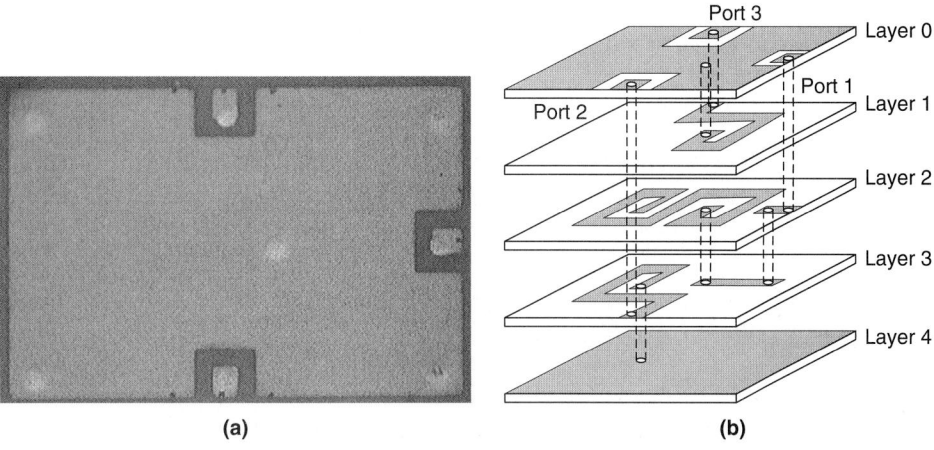

FIGURE 5.37 (a) Photograph and (b) configuration of the stripline multilayer balun.

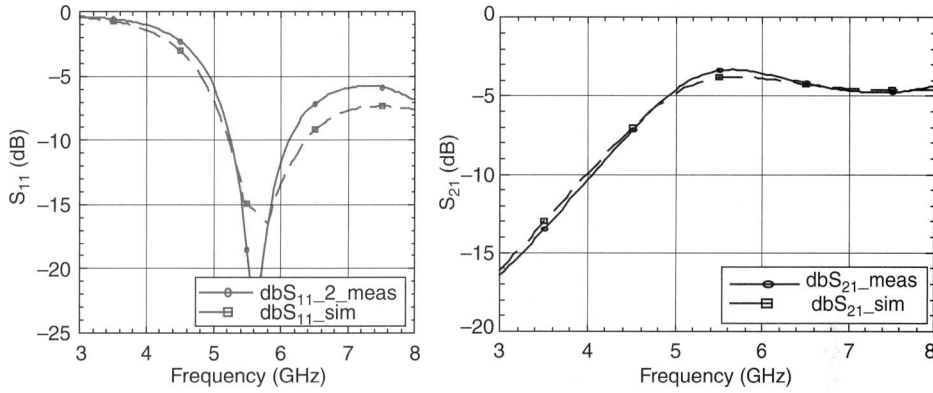

FIGURE 5.38 Measured and simulated results of the stripline balun designed for 5.8 GHz.

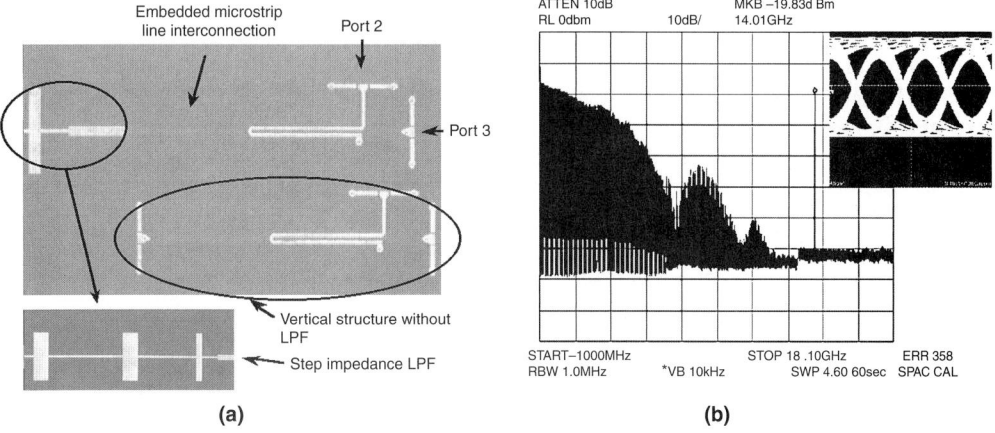

FIGURE 5.39 (a) Picture of fabricated combiner in MLO process, 5 mm in length. (b) Frequency spectrum at the output port.

to obtain a constructive effect at the output port of the combiner by adjusting the embedded microstrip line, which connects the LPF with a vertical coupling structure. The insertion loss of the RF signal at the output port of the combiner was measured as 1.9 dB. The isolation between ports 1 and 2 of the combiner was greater than 10 dB in the baseband and 38 dB at 14 GHz (VNA HP8510). A 3-dB bandwidth of 7 GHz is achieved between ports 1 and 3. Figure 5.39b shows the measured frequency spectrum at the output of the combiner when a 7-Gbyte/s probability random bit sequence (PRBS) with a 14-GHz sinusoidal wave is sent. The figure also shows the measured eye opening of the 7-Gbyte/s PRBS at the output port.

5.4.10 RF MEMS Switches

History and Role of MEMS Switches

MEMS switches as we know them today were conceptualized in the late 1980s and early 1990s. These switches were of great interest to RF engineers for their potential to reduce the total area, power consumption, and cost of their devices. MEMS were initially fabricated exclusively on silicon, since integrated circuit (IC) fabrication at the time was on silicon. RF MEMS enable switching and tuning of front-end circuits for applications of mode switching, antenna tuning, and antenna beam steering with phase shifters.

Early MEMS research was funded and performed by industry leaders looking for applications in a variety of areas, including optics, transportation, aerospace, robotics, chemical analysis systems, biotechnologies, and medical engineering. Devices such as microactuators, microsensors, and microrobots were desired for a plethora of devices, such as for automobile airbags. In time, it was expected that MEMS could be used for flat-panel displays, optical switches, fiber optics, and integrated sensors. It was understood that many reliability issues would have to be solved before these advanced technologies were possible. Early MEMS switches were plagued with electrical and mechanical issues, such as dielectric charging, substrate delamination, creep, and fatigue.

Prior to the new millennium, accurate numerical solvers were not available to MEMS designers. Research was primarily performed by fabricating, testing, and redesigning. This of course is a slow and expensive process. In 1996, the Defense Advanced Research Projects Agency (DARPA) in the United States funded research aimed at improving the computer-aided design technology for MEMS devices. A number of useful software programs were developed from this funding, including MEMCAD (developed by MIT and Microcosm), IntelliCAD (developed by IntelliSense), and CAEMEMS (developed by the University of Michigan). MEMS devices quickly surpassed the RF performance of their solid-state equivalents. Even early MEMS switches had an insertion loss of 0.15 dB at 20 GHz, compared to an on-state insertion loss of approximately 1 dB for a typical GaAs-FET or PIN-diode switch at the same frequency. Today, many of the devices are commercially available. MEMS switches in particular can be purchased with insertion losses as low as 0.1 dB up to 50 GHz with the potential for operating more than 500 billion cycles and handling low watt power levels.

The Holy Grail for many MEMS designers is the cell phone market. Utilizing a bank of RF MEMS switches, capacitors, and inductors, a cell phone could offer unparalleled reconfigurability. The cell phone could conceivably work then at any frequency, on any channel, for any standard, and in any location. Dropped calls would be a thing of the past.

Operating Principle

RF MEMS switches are one type of MEMS device that utilize either a single-supported (cantilever) or double-supported (air-bridge) beam suspended over a metal pad. Since a MEMS switch uses only a single moving part, it is one of the simplest devices in use today. By comparison, a typical sensor can have dozens of moving parts. Switches come in a variety of shapes, sizes, and materials. There are two main types of actuation mechanisms for RF MEMS switches: thermal and electrostatic.

Most materials expand when heated and contract when cooled. This is the basic principle behind thermal switches that use a resistive material on the switch membrane. When electric current is passed through the switch, the resistive material heats up, causing it to expand. This expansion deflects the beam. When the current is reduced (or eliminated), the switch returns to the steady state. This type of switch is not widely used because it is much slower, lossier, has a lower bandwidth, consumes more power, and is more difficult to control than electrostatic actuation. It can have a low actuation voltage, however, which could make it attractive for system-on-chip applications.

Electrostatic actuation relies on the principle that opposite charges attract. With one metal beam suspended above a metal pad, a voltage is applied to the beam while the pad is grounded or vice versa (as shown in Figure 5.40). A static charge will form from the voltage potential, and this creates an electrostatic force between the layers. As the voltage potential is increased, the electrostatic force strengthens. When this force exceeds the beam's ability to resist deflection, the metal layers are pulled together. If the metal layers are allowed to make direct contact, this switch becomes "ohmic" and direct current is able to flow through the switch. If direct contact is prevented by a thin layer of dielectric, usually by silicon nitride, then the switch is "capacitive" and no direct current is allowed to flow. Since a capacitor is the basis for this design, the frequency must be sufficiently high so that the RF energy can pass through. Switches of this type are typically used from 5 to 100 GHz. Varying the thickness of the dielectric layer is one way of tuning the resonant frequency of the switch. Filter designers are particularly fond of capacitive switches since there is no resistance and this gives a higher Q factor.

MEMS switch designers have a number of variables at their disposal that can be optimized for a given application [66]. There is a fundamental tradeoff between isolation, switching speed, and actuation voltage. The best way of improving the isolation of a switch is to increase

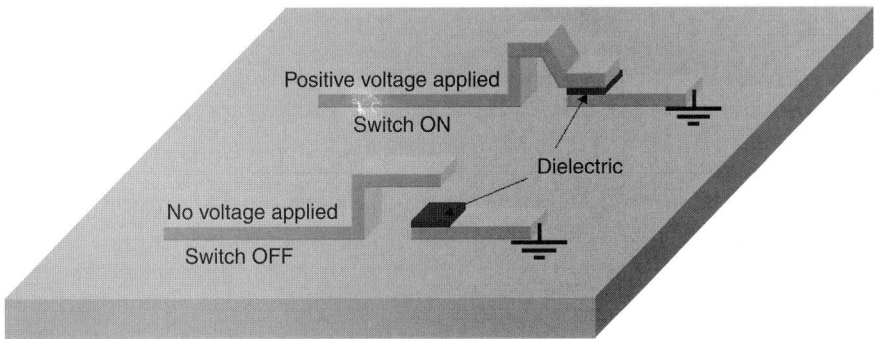

FIGURE 5.40 Basic operation of a single-supported, capacitive MEMS switch. When no voltage is applied to the membrane, no actuation occurs. When a voltage is applied, the voltage potential creates an electrostatic force that pulls the membrane toward the grounded line beneath it. A thin layer of dielectric material prevents direct contact between the layers.

the vertical distance between the beam and the metal pad. This distance is typically 1 to 3 µm. As this distance is increased, the switching time and actuation voltage also increase. Decreasing this distance will likewise decrease the switching time and actuation voltage. The speed and voltage can also be improved by changing the beam material. Very pliable metals, like aluminum, will switch much easier than stiffer metals, like gold. The stiffness of a material is denoted by its Young's modulus, (the higher this value, the stiffer the material).

Equations for predicting the bending of cantilever and double-supported beams have been around for decades. Unfortunately, trying to apply simplistic equations to complex MEMS devices can be cumbersome. One of the most important mechanical parameters of a MEMS switch is the pull-down voltage. This quantity can be estimated by treating the MEMS switch as a mechanical spring. In order to calculate the pull-down voltage, one must equate the electrostatic force pulling down on the beam:

$$f_{down} = \frac{\varepsilon A V^2}{2g^2} \tag{5.1}$$

and the force pushing up from the spring (Hooke's law):

$$f_{up} = -k(g_o - g) \tag{5.2}$$

In these equations, ε is the permittivity, A is the area, V is the voltage, k is the spring constant, g_o is the initial gap, and g is the evaluated gap. We can use these simple, spatially independent equations since we know the charge density (and therefore the force) is uniform across the capacitive region. It is well known that for parallel-plate electrostatic actuation, when the gap reduces to two-thirds of the original gap, the beam becomes unstable and experiences a "pull-in" effect. That is, when the gap reaches a certain threshold, namely two-thirds of the original gap, the switch will snap down. Magnets experience the same effect. As two magnets of opposite polarity are brought closer together, the attractive force is barely noticeable until they reach a certain distance apart. At this point they snap together, and the force between them is great.

Equating Equations (5.1) and (5.2) where the gap (g) is two-thirds of the original gap (g_o) and solving for the pull-down voltage gives

$$V_{PD} = \sqrt{\frac{8kg_o^3}{27\varepsilon A}} \tag{5.3}$$

In order to reduce the pull-down voltage, design engineers can reduce the spring constant, reduce the switch gap, or increase the area of the switch.

Comparison of Technologies: Electromechanical versus Solid State

There are a variety of switching elements available on the market today. In order to choose the right type of switch, one must consider the required performance specifications, such as frequency, bandwidth, linearity, power handling, power consumption, switching speed, signal level, and allowable losses. A comparison of a typical RF MEMS, PIN diode, and FET switching element is summarized in Table 5.4 [67].

PIN diodes are useful switching elements because they are widely available commercially, have fast switching speeds, are low cost, and are rugged. One of the main limitations of PIN diodes is the insertion loss. Above a few gigahertz, PIN diodes can start to have quite a bit of insertion loss. This becomes significant above the X band (8 to 10 GHz)

Parameter	RF MEMS	PIN Diode	FET
Voltage (V)	20–80	±3–5	3–5
Current (mA)	0	0–20	0
Power consumption (mW)	<0.5	5–100	–0.5–0.1
Switching time	1–300 µs	1–100 ns	1–100 ns
Cup (series) (fF)	1–6	40–80	70–140
R_s (series) (Ω)	0.5–2	2–4	4–6
Capacitance ratio	40–500	10	N/A
Cutoff frequency (THz)	20–80	1–4	0.5–2
Isolation (1–10 GHz)	Very high	High	Medium
Isolation (10–40 GHz)	Very high	Medium	Low
Isolation (60–100 GHz)	High	Medium	None
Insertion loss (1–100 GHz) (dB)	0.05–0.2	0.3–1.2	0.4–2.5
Power handing (W)	<1	<10	<10
Third-order intercept (dBm)	+66–80	+27–45	+27-45

TABLE 5.4 Comparison of Electrical Performance for a Typical RF MEMS, PIN Diode, and FET Switch [67]

when the skin effect causes an increase in the resistance of the switch. At any high frequency, diodes tend to have issues with linearity, bandwidth, and power consumption.

Like PIN diodes, FETs are widely used because of their availability, fast speed, low cost, and durability. They consume much less power than PIN diodes, but they are not available for the same frequency range as diodes. That is, they have limited use in the K_a band (26 to 40 GHz) and are practically unusable above the U band (40 to 60 GHz) [67].

MEMS switches are quickly becoming the preferred switching element for RF devices. They offer the lowest insertion loss, highest isolation, extremely high linearity, negligible power consumption, and small size. The switching time, power handling capability, and packaging requirements are the three main limitations with their use. The switching time for a MEMS device usually varies with isolation due to physical constraints (the better the isolation, the slower the switching time). However, for a microwave system that can tolerate a switching time in the micro- to millisecond range, MEMS are suitable. Switching times in the hundreds of nanoseconds have also been demonstrated [68]. Furthermore, if signal amplification in a wireless system can be done right before propagation (thereby eliminating the exposure of high power to the switching elements), then MEMS are also applicable. Packaging MEMS is not as straightforward as packaging solid-state devices, as presented in the MEMS chapter of this book.

Challenges

One of the most difficult challenges for MEMS designers is overcoming dielectric charging. All electrostatic MEMS switches use some sort of dielectric to maintain the voltage potential. Over time, this dielectric will store charge and the switch will remain in the actuated state. This charge naturally dissipates into the substrate, but it can take anywhere from milliseconds to hours for this to occur, depending on the actuation voltage, the substrate material, and the extent of the charging. Lower actuation voltages and materials with good charge mobility tend to dissipate charge faster. Cycle testing is a good metric for determining the lifetime of a switch, but it is not a good indicator of the reliability. Switch-cycle lifetimes in the hundreds of billions of cycles have been demonstrated [69]. A switch that can operate for a hundred billion cycles is sure to have a long lifetime, but if the switch becomes "stuck" after a minute in the actuated state, then it is not a very reliable switch. For a system that is not continuously being reconfigured, this is a big problem. Researchers are currently investigating better ways of removing this charge accumulation [70]. Many techniques being proposed involve the use of novel dielectric materials.

Another solution to the problem of long-duration actuation is to use a hybrid actuation technique, which uses electrostatics and thermal mechanisms. Electrostatic actuation is used first to quickly pull the switch down. The switch is then held down using thermal actuation, and the electrostatics are turned off. This switch combines the fast switching time of an electrostatic switch with the drawback of having a slight increase in the power consumed (from the resistive material). Since the static charge is only present for a very short time, this switch can be much more reliable.

Robustness is another big challenge for engineers. In order for a product to be successful, it needs to be able to survive a certain amount of abuse. Modern cell phones can survive years of abuse from car and home keys, pocket change, multiple falls, and sometimes even brief exposures to water. Could a MEMS switch survive the same? Many experts say that since their size is so small that they are almost unaffected by everyday vibrations. Other experts are more skeptical.

One last limitation worth mentioning is the power handling capability of MEMS. Typically, the geometry of a switch requires the RF signal to travel directly under the metal membrane. As the power level for the RF signal is increased, it can start to have an effect on the membrane. At some power level, the RF signal will become strong enough to "self-actuate" the switch. This is a fundamental limitation, and it is something that is also actively under investigation. At the moment, switches are typically rated in the tens or hundreds of milliwatts [71], although power levels over a watt have been demonstrated [72].

Modeling MEMS switches for optimal electrical, mechanical, and reliability performance can also be a daunting task and is often substituted with a less accurate method. MEMS switches are often designed for optimal electrical properties (such as a low RC time constant) or optimal mechanical properties (such as a low actuation voltage). Often, when multiple physical realms are involved in a problem, the optimal solution method is to use a simulator to solve the problem in the more complicated realm and to combine those results manually with theory from the simpler realm [73–76].

Application

One important application of MEMS switches is in phase shifting for electronically scanned antenna arrays. An example of a 4-bit MEMS phase shifter is shown in Figure 5.41, which has been packaged using an organic flexible low-permittivity substrate (LCP) [73–74]. The microstrip switched-line phase shifter shown in Figure 5.41 has been optimized at 14 GHz for small size and with excellent performance. The improved geometry of the reduced size phase shifter is 2.8 times smaller than a traditional switched-line phase shifter and has a lower loss. For the 4-bit phase shifter, the worst-case return loss is greater than 19.7 dB and the average insertion loss is less than 0.96 dB (0.24 dB/bit or 280/dB). Packaging of MEMS devices can be very tricky since the package can deteriorate its performance. In the example shown in Figure 5.41, the addition of the LCP package has a negligible effect

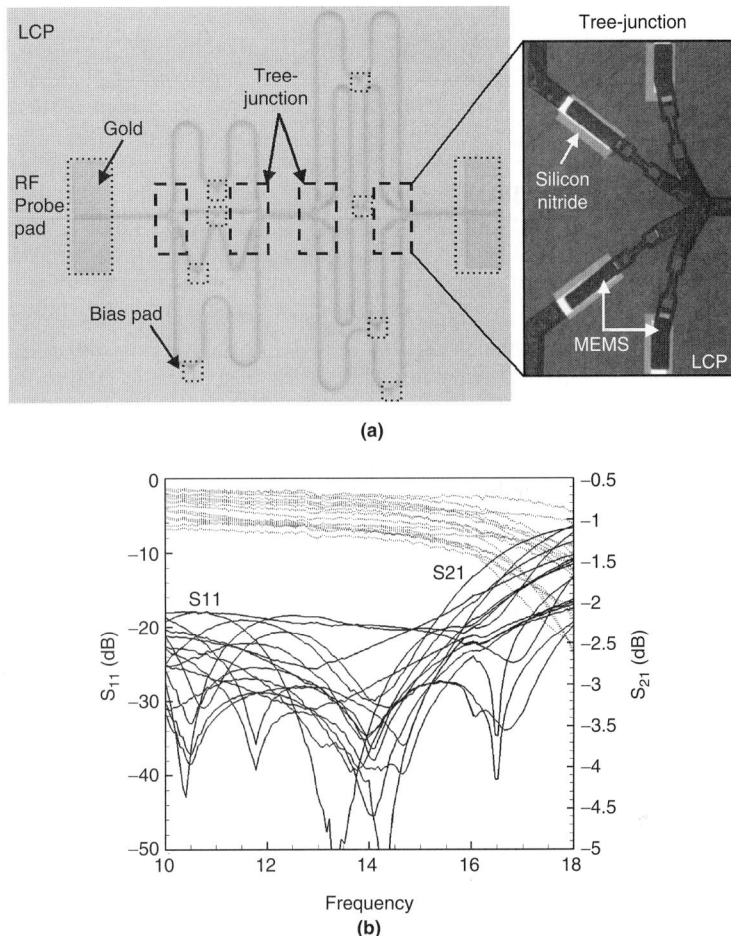

FIGURE 5.41 (a) Fabricated MEMS phase-shifter substrate. The superstrate has been removed, and cutouts represent the location of the cavities and probing windows. (b) Measurement results for MEMS switches bonded using tension or epoxy. The presence of the epoxy adds a minimal amount of insertion loss. [76]

on the phase-shifter performance due to its low-loss characteristics, but enables the device to remain flexible and provides protection against various environmental conditions. A comparison of the measured loss for a tension and epoxy bonded MEMS switch is shown in Figure 5.41b. The average variations in S_{11} and S_{21} are 3.69 dB and 0.087 dB, respectively, at 14 GHz.

5.4.11 RFIDs

The demand for radiofrequency identification tags (RFIDs) has rapidly increased due to the need for automatic identification in various areas, such as item-level tracking, access control, electronic toll collection, and vehicle security. Compared with the use of lower-frequency tags in the LF and HF bands, which suffer from a limited read range (1 to 2 feet), RFID tags in the UHF band see the widest use due to their higher read range (over 10 feet) and higher data transfer rate. Two major challenges exist in today's RFID technology advances toward the practical level. One is the design of tag antennas with higher efficiency and better impedance matching for IC chips with high capacitive reactance. This is essential to maximize the RFID system performance. Another major challenge is the realization of ultra-low-cost RFID tags. Economical applications require the cost of individual tags to drop down to one or two cents. Three types of antennas specifically designed to enhance RFID performance (as compared to the antennas described earlier which were for WiFi applications) are described below [77–78].

Meander Line

An RFID antenna is usually built to achieve half-wavelength resonance at first for efficiency optimization. This is approximately equal to the maximum length, when the dipole antenna is stretched from one end to the other. In the UHF band, a half-wave dipole antenna is almost 16 cm in free space. For the purpose of reducing tag sizes, the meander line structure is an attractive choice. The arc-shaped configuration is a modification of the meander line. It exhibits a smoother transition along the radiating arms. This class of antenna provides the largest size reduction (down to a quarter wavelength) at the desired frequency at the expense of a slight decrease in gain (usually 5 percent lower) due to a less effective radiating length.

Dual Polarization Structure

Antennas in UHF RFID tags are linearly (vertical or horizontal) polarized. In the presence of environmental reflections, which cause the multipath effect, the transmitted or received plane waves undergo polarization direction changes. For instance, a vertically polarized transmitted wave can reach a tag at its blind spot, namely null in the radiation pattern. This causes the RFID tag not to be read. In order to prevent this, polarization diversity has to be utilized, requiring the use of both vertical and linear polarized antennas. These two antennas are identical in dimensions and shape, so the identical signals arriving at these two different branches are in phase and uncorrelated, as shown is Figure 5.42. This is critical for the demodulator in the IC so that no phase difference occurs when the same data are retrieved from the combined reception of the two antennas.

Figure 5.43 shows a 3 in × 3 in dual polarized antenna. The shorting stub that connects the top left legs (RF port) of the design both provides inductive conjugate matching and is used to dc-short the two orthogonal dipole antennas. The bottom right

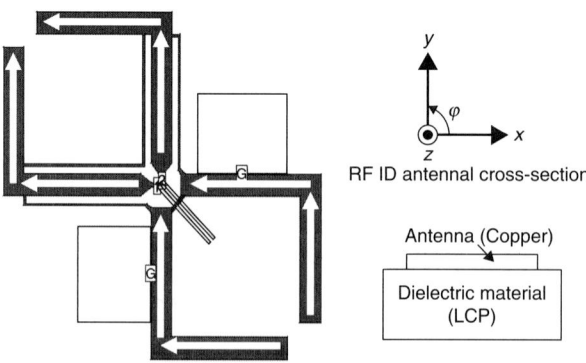

FIGURE 5.42 Dual polarization antenna (arrows indicate current flow directions).

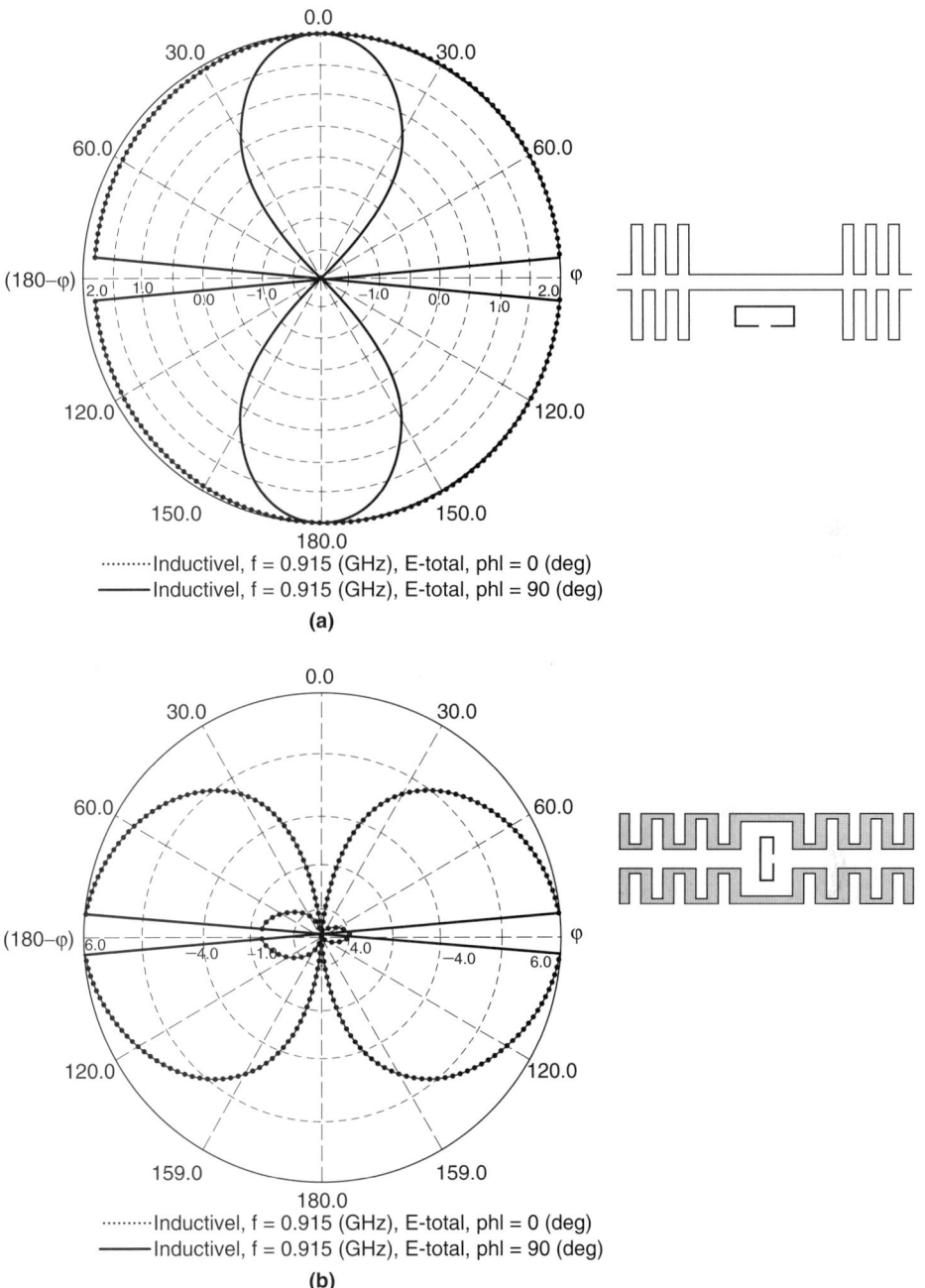

FIGURE 5.43 Two dual radiating bodies RFID antennas with radiation patterns. (*a*) Directivity = 2.69 dBi, efficiency = 86.8 percent. (*b*) Directivity = 5.62 dBi, efficiency = 79.9 percent.

legs are connected to ground to achieve signal-ground excitation. The input impedance of this dual antenna is 20 + j112.5 Ω to match to an IC input impedance of 20 − j113 Ω. The radiation efficiency is 98 percent since the current flow as shown in the figure adds up constructively for the far-field electromagnetic radiation, something that results in an optimized performance in terms of the read rage and environment versatility. The radiation pattern of the dual polarized antenna is omnidirectional in the yz plane with a maximum directivity of 2.25 dBi.

Dual-Body Configuration

The design of a high-directivity RFID antenna is an effective method to increase the read range of a tag. However, the performance of most RFID antennas is constrained by their intrinsic dipole nature, and a high directivity, especially for narrow-beamwidth conveyor belt applications, is very difficult to achieve. Adding another radiating body is a solution to this. A new topology, named dual-body configuration, is presented in Figure 5.43. Two meander line arms are placed in the same side of the feeding loop as shown, and a directivity of 2.69 dBi is achieved. In Figure 5.39b, two meander line arms are placed in each side of the feeding loop. In this case, the current directions are opposite along the arms and the radiation patterns cancel out each other in most of the directions. Thus, in this inductively coupled RFID antenna, the radiating energy is focused directionally in a dumbbell shape, and a high directivity of 5.62 dBi is observed with 79.9 percent radiation efficiency. In general, a highly increased effective range is expected to be achieved with RFID antennas in such a configuration. In addition, a dual-body configuration can also be used to enhance the antenna bandwidth performance. Radiating bodies can resonate in adjacent frequencies by adjusting the length of each arm. In this way, an antenna can resonate at multifrequencies to enlarge the bandwidth and make possible multiband and multistandard RFID tags.

5.5 Integrated RF Modules

5.5.1 WLAN

Wireless LANs are becoming extremely popular since they provide seamless connectivity without the need for wires. With the trend toward including mobility in laptop computers using PCI and mini-PCI express cards, the need for miniaturization of radios becomes very important. With the progress being made in multiple-input and multiple-output (MIMO) architectures, there is a compelling need for including multiple radios in laptop computers without increasing the space available on the cards. SOP technology can be used to miniaturize such systems at low cost, especially for systems operating at 5.8 GHz (IEEE 802.11a) or 2.4 GHz (IEEE 802.11b) without compromising RF performance.

An example of an SOP-based 3D antenna-integrated transceiver module is shown in Figure 5.44. Three different subsystems, that is, the transceiver, filter, and antenna, are vertically stacked and connected through vias (Figure 5.44a). The presented module utilizes 20 LTCC layers. The antenna, filter, and transceiver utilize 8, 10, and 2 layers, respectively. The total size of the module is 14 mm × 19 mm × 2 mm, including all the RF functional blocks [29–30]. The grounds are connected through vias to suppress the unwanted parasitic modes.

Radio Frequency System-on-Package (RF SOP)

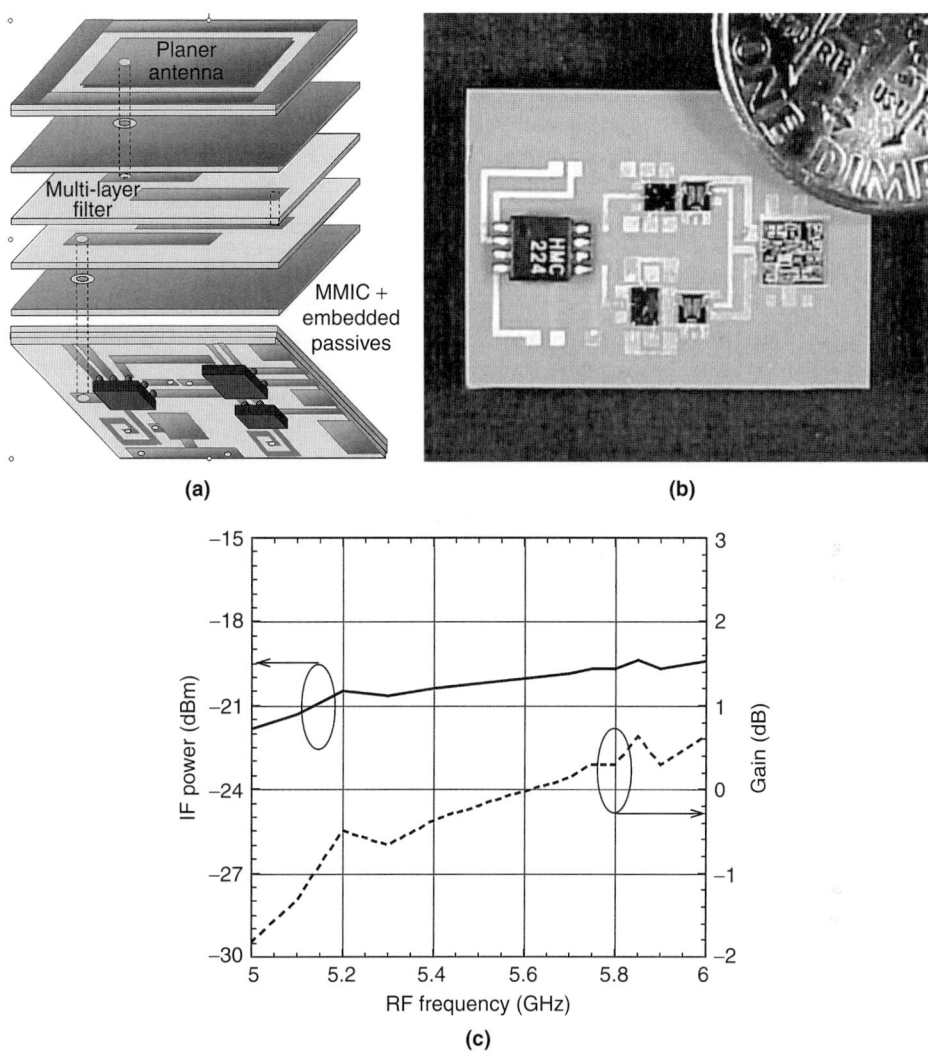

FIGURE 5.44 (a) A schematic view of a multilayer RF front-end architecture. (b) A 3D integrated LTCC RF front-end module for a 802.11a WLAN. (c) Measured gain of the receiver versus frequency.

To utilize the space of the module effectively, the geometries of the passive components have been chosen very carefully. A cavity backed patch antenna (CBPA) has been designed for the module. A three-section coupled stripline filter has been designed to be embedded inside the LTCC package with its input and output ports connected to the antenna and the duplexer switch through vias. The RF functional blocks, including PA, LNA, mixers, and VCO are attached on the bottom of the LTCC board. The specifications of the functional blocks are determined and verified through system simulations based on the IEEE 802.11a standard. A photograph of the integrated module is shown in Figure 5.44b. To evaluate the performance of the module, each of the sections in the module was fabricated and measured with the performance of the receiver shown in Figure 5.44c.

5.5.2 Intelligent Network Communicator (INC)

A highly integrated mixed-signal testbed was developed to demonstrate the concept and realization of the advanced system-on-package concept [6]. This experimental system called Intelligent Network Communicator (INC) deals with three different statuses of the signals (digital, RF, and optical) in a single packaging platform. The INC transmits and receives the high-speed digital signal and wireless signal over the embedded optical waveguide channel. The system has been fabricated by utilizing advanced packaging and assembly processes, and full functionality has been demonstrated successfully. Before the final test, each of the subblocks has been separately developed and tested. The test results clearly show that the developed system performance meets the goal. The digital block generated up to 3.2 Gbytes/s of data stream, and the RF block has less than −1.5 dB of insertion loss up to 6 GHz. The optical block achieved 10 Gbytes/s throughput over the embedded optical waveguide built on the low-cost organic substrate.

The INC system configuration is shown in Figure 5.45. At the digital block, a multi-gigabit pseudo random digital bit sequence is generated by using field-programmable gate array (FPGA) and compared with the received signal after passing through the analog and optical blocks. The multichannel signals from FPGA (Virtex 50E, Xlinx) are converted into a serial data stream by the transceiver IC (TLK 2701, TI), which also includes mux and dmux and is then fed to the analog block. The FPGA at the receiver stage compares the known input data bit stream and recovered data from the receiver to evaluate the system performance. The FPGA has been programmed to generate 16 parallel data (150 Mbytes/s/c/s) channels, which are then fed to a mux that converts the parallel data to a 2.488 Gbytes/s serial signal. This signal is the input to the RF section, which has been integrated into the same board. To reduce the interference from the digital part, a separate ground and power was designed for the RF block. A coplanar waveguide and matching network was used for RF input and for conversion of the differential signal to a single-ended signal.

At the analog block, two narrowband RF signals, namely, the 802.11a/b wireless LAN signal and the voltage-controlled oscillator single tone signal (5 to 6 GHz), are combined with the multi-Gbytes/s digital data stream from the digital block. The high-frequency component of the digital signal was truncated by using an embedded low-pass filter in the board, before combining with the RF signals. A mixed-signal combiner for combining the digital and RF signal was designed and embedded in the multilayered organic board. A voltage-controlled oscillator IC was specially designed using the MESFET process to generate the single tone signal. The VCO utilized an embedded high-Q inductor in the substrate to reduce the phase noise of the IC. The embedded inductor was optimized to obtain the highest performance at 5 GHz. The combined electrical signal was then fed to the input of the optical modulator.

At the optical block, the RF and digital signal were modulated and converted to the optical domain using the Mach-Zehnder modulator and vertical-cavity surface-emitting laser (VCSEL) direct modulation scheme, whose wavelengths were 1550 and 870 nm, respectively. The optical signal was initially transmitted through a multimode optical fiber channel and coupled into the embedded optical waveguide using the butt-coupling method. In order to integrate long (5 to 15 cm) polymer waveguides on flexible FR4 boards that contain two metal layers separated by a low-temperature insulating polymer layer, critical technical issues (which included board flexibility, long-range board nonplanarity, short-range roughness and coefficient of thermal expansion matching) had to be addressed.

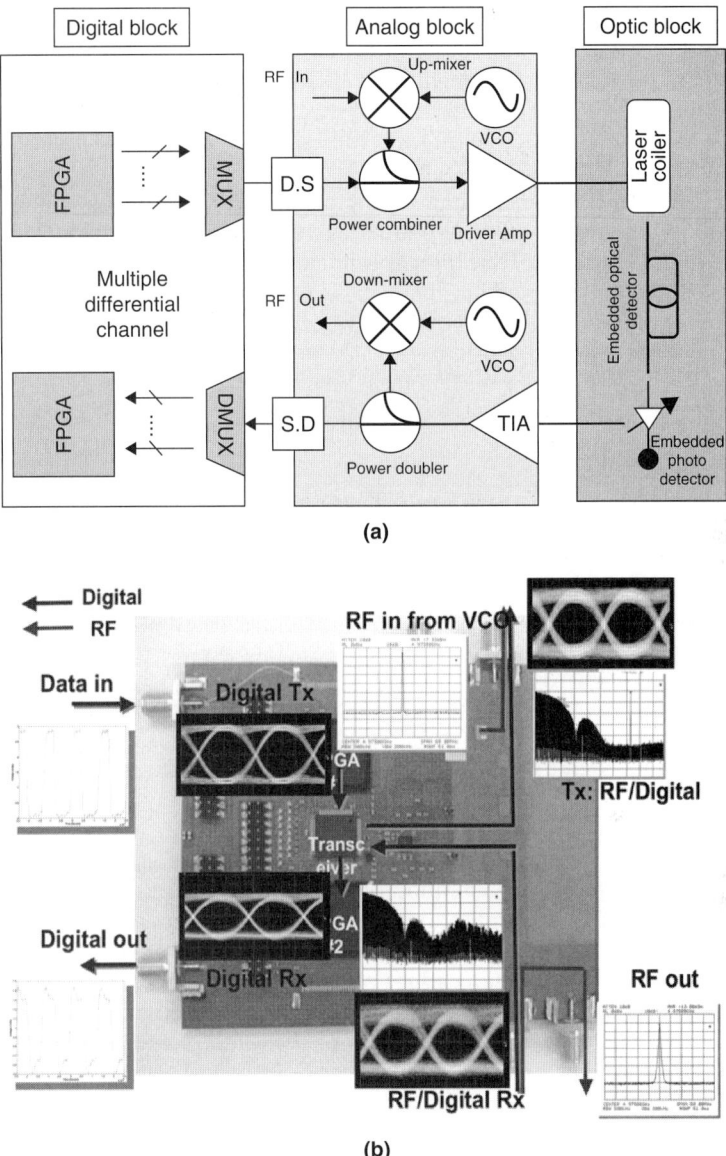

FIGURE 5.45 (a) INC system configuration. (b) Fabricated system-level prototype with measured signal characteristics.

At the receiver end, the electrical signal is obtained using the photodetector. The electrical signal has been amplified using TIA and separated using an embedded mixed-signal splitter that is identical to the combiner. The recovered digital signal is then diverted to separate channels and transmitted to another FPGA for the purpose of comparing it with the transmitted signal.

The RF-analog block of the INC system has been used to focus on the study of mixed-signal architectures, which can provide enhanced bandwidth, frequency, and data rates in real time. The goal of this block was to combine high-speed broadband digital signals up to 10 Gbytes/s and RF carrier frequencies in the range of 5 to 14 GHz. Various passive and active components for the realization of efficient RF-digital interfaces have been implemented in this design, along with design rules that were developed for various frequencies.

A voltage-controlled oscillator module has been developed and successfully demonstrated for 5.8-GHz operation in the INC system. The standard cross-coupled VCO has been fabricated using a commercial MESFET process without inductors. The bare die IC has been wire-bonded to the two embedded inductors that were embedded in the organic substrate. The high-Q inductor helped reduce the phase noise of the VCO significantly, as explained in Chapter 4. The VCO module integrated in the INC testbed showed a phase noise of 110 dBc at a 6-MHz offset frequency and −10 dBm of output power.

5.6 Future Trends

With rapid growth in wireless applications, coupled with the tremendous advances in the silicon process technologies, one can envision the emergence of a fully integrated RF CMOS radio on a tiny piece of silicon. However, this is unlikely because of the cost and technical challenges that include the lossy nature of silicon, low process yields at 45- to 22-nm lithography, and thermomechanical reliability issues. The recent trend has been the development of multiband and multimode wireless solutions enabling effective utilization of bandwidth and tuning to the appropriate data rate whenever necessary and with freedom to use different bands in different countries. Software radios are emerging as platforms for these multiband, multimode personal communication systems, and cognitive radios further enhance their flexibility for personal services through a radio knowledge representation language. Cognitive radio is a paradigm for wireless communication in which either a network or a wireless node changes its transmission or reception parameters to communicate efficiently without interfering with licensed users. This alteration of parameters is based on the active monitoring of several factors in the external and internal radio environment, such as the radiofrequency spectrum, user behavior, and the network state. The cognitive radio language represents increasingly complex knowledge of radio etiquette, devices, software modules, propagation, networks, user needs, and application scenarios in a way that supports automated reasoning about the needs of the user. Nonetheless, cognitive radios are already making headway in the real world. Companies such as Intel for example are exploring reconfigurable chips that will use software to analyze their environments and select the best protocols and frequencies for data transmission in highly congested traffic.

Highly integrated system technologies will be required to reduce the system chip size. These technology developments include (1) system-on-chip (SOC) in Si and SiGe with improved performance at the device and circuit level than in the past, (2) stacked chip approaches separating ICs into individual functional units but realizing the miniaturization by stacking of thinned chips, (3) embedded thin actives in the package or board, (4) embedding of discrete passives, and finally (5) embedding of all passive and active components in thin-film form. The use of embedded thin-film components has already started on silicon or glass wafer by such semiconductor companies as Philips and ST, in ceramics by Murata and TDK in Japan, and in organics by a number of package companies in Japan in partnership with Motorola. The system-on-package is clearly emerging as the strategic direction primarily driven by the need to miniaturize the system and improve functionality. Recent developments in advanced 3D packaging technologies such as TSV will provide additional opportunities for significant miniaturization and power consumption in the SOP concept. The SOP concept further allows heterogeneous integration of different devices with different architectures and differently processed characteristics. Recent work at the Georgia Institute of Technology demonstrated packaged microwave amplifiers within LCP [73]. A 13- to 25-GHz GaAs bare die, low-noise amplifier is embedded inside a multilayer LCP package made from seven layers of thin-film LCP. This new packaging topology has inherently unique properties that could make it an attractive alternative, in some instances, to traditional metal and ceramic hermetic packages. The active device, in this example, is enclosed in a package consisting of several laminated CO_2 laser-machined LCP superstrate layers. Measurements demonstrate that the LCP package and its 285°C packaging process have minimal effects on the monolithic microwave integrated circuit RF performance. These findings show that both active and passive devices can be integrated together in laminated multilayer packages. This active and passive compatibility demonstrates a unique approach to form compact, vertically integrated (3D) RF system-on-package modules [73].

New challenges such as RF-MEMS integration and packaging have been at the leading edge of the packaging research. Various approaches have been developed by universities and leading packaging companies. Nevertheless, there is still a lack of standardization, leading to excessive cost for the development of final products. Eventually, the ideal packaging technology will emerge with the standards for environmental protection, hermetical sealing, accelerated testing, and mechanical stability, among others.

Innovative shielding solutions are also emerging using metallized cavities, or electromagnetic bandgap (EBG) structures as well as advanced simulation platforms to predict electromagnetic interference and crosstalk within complex and miniaturized microsystems. EBG topologies are also emerging for use in confining radiating field of antenna elements, thus leading to more efficient isolation of the rest of the module, while achieving miniaturization of the antenna.

Also emerging are new design approaches based on optimization algorithms such as design of experiments (DOE), feed-forward neural networks, and genetic algorithms to generate comprehensive models taking into account the fabrication and layout parameters and their impact on electrical performance in multilayer configurations.

In addition to smart and miniaturized systems packaging, new materials and chemical processes are the most important technologies that can lead not only to miniaturization by ultrathin films but also can achieve RF component properties that have never been achieved. The thin-film nanomaterials with unparalleled properties in capacitance, inductance, and resistance will lead to a wide variety of new applications that have not been imagined so far.

Acknowledgments

The authors would like to acknowledge the financial support of the Packaging Research Center, NSF, NASA, and DARPA. The authors would also like to thank Ramanan Bairavasubramanian, Nickolas Kingsley, Stephen Horst, Daniela Staiculescu, Stephan Pinel and Kyutae Lim who contributed to the technologies reported in this chapter and Dhanya Athreya for the inductor section write up and Jin Hyun Hwang for the capacitor section write up.

References

1. R. Tummala and V. Madisetti, "SOC vs SOP," *IEEE Design, Test, and Comp.*, vol. 16, June 1999, pp. 48–56.
2. R. Tummala and J. Laskar, "Gigabit wireless SOP technology," *IEEE Proceed.*, vol. 92, February 2004, pp. 376–87.
3. K. Lyne, "Cellular handset integration, SiP vs SOC," *IEEE Custom Integrated Circuit Conference*, 2005, pp. 765–70.
4. iNEMI 2004 Roadmap.
5. J. Laskar, B. Matinpour, and S. Chakraborty, *Modern Receiver Front-Ends*, Wiley-Interscience, 2004.
6. K. Lim, M. F. Davis, M. Maeng, S. Pinel, L. Wan, J. Laskar, V. Sundaram, G. White, M. Swaminathan, and R. Tummala, "Intelligent Network Communicator: Highly integrated system-on-package (SOP) testbed for RF/digital/opto applications," *Proc. 53rd Electronic Components and Technology Conference*, New Orleans, LA, 2003, pp. 1594–98.
7. R. Tummala, V. Sundaram, G. White, P. M. Raj, F. Liu, and S. Bhattacharya, "High density packaging for 2010 and beyond," *4th International Electronics Packaging and Technology Conference*, Singapore, December 2002, pp. 1–10.
8. V. Sundaram, F. Liu, S. Dalmia, G. White, and R. R. Tummala, "Process integration for low-cost system on a package (SOP) substrate," *Proc. 51st Electronic Components and Technology Conference*, Orlando, FL, 2001, pp. 53–40.
9. V. Sundaram, S. Dalmia, J. Hobbs, E. Matoglu, M. Davis, T. Nonaka, J. Laskar, M. Swaminathan, G. E. White, and R. R. Tummala, "Digital and RF integration in system-on-a-package (SOP)," *52nd Electronic Components and Technology Conference*, San Diego, CA, 2002, pp. 646–50.
10. R. Tummala, G. White, and V. Sundaram, "SOP: microelectronics system packaging technology for 21st century; prospects and progress," *Proc. 12th European Microelectronics and Packaging Conference*, Harrogate, England, 1999, pp. 327–35.
11. http://www.ansoft.com/products/hf/hfss/.
12. http://www.sonnetusa.com/.
13. http://www.microstripes.com/.
14. http://www.bay-technology.com/ie3d.htm.
15. T. Itoh, *Numerical Techniques for Microwave and Millimeter-Wave Passive Structures*, John Wiley & Sons, 1989.
16. M. Krumpholz and L. P. B. Katehi, "New time domain schemes based on multiresolution analysis," *IEEE Trans. Microwave Theory Tech.*, vol. 44, April 1996, pp. 555–61.
17. N. Bushyager, L. Martin, S. Khushrushahi, S. Basat, and M. M. Tentzeris, "Design of RF and wireless packages using fast hybrid electromagnetic/statistical methods," *Proc. IEEE-ECTC Symposium*, May 2003, pp. 1546–49.
18. O. Salmela and P. Ikalainen, "Ceramic packaging technologies for microwave applications," *Proc. Wireless Communications Conference*, August 11–13, 1997, pp. 162–64.
19. M. F. Davis, A. Sutono, S.-W. Yoon, S. Mandal, N. Bushyager, C.-H. Lee, K. Lim, S. Pinel, M. Maeng, A. Obatoyinbo, S. Chakraborty, J. Laskar, E. M. Tentzeris, T. Nonaka, and R. R. Tummala, "Integrated RF architectures in fully-organic SOP technology," *IEEE Transactions on Advanced Packaging*, vol. 25, issue 2, May 2002, pp. 136–42.
20. S. Pinel, K. Lim, M. Maeng, M. F. Davis, R. Li, M. Tentzeris, and J. Laskar, "RF system-on-package (SOP) development for compact low cost wireless front-end systems," *European Microwave Conference*, September 23–27, 2002.

21. S. Pinel, M. Davis, V. Sundaram, K. Lim, J. Laskar, G. White, and R. Tummala, "High Q passives on liquid crystal polymer substrates and BGA technology for 3D integrated RF front-end module," *IEICE Transactions on Electronics*.
22. L. M. Higgins-III, "Hermetic and optoelectronic packaging concepts using multilayer and active polymer systems," *Advancing Microelectronics*, vol. 30, July 2003, pp. 6–13.
23. D. C. Thompson, O. Tantot, H. Jallageas, G. E. Ponchak, M. M. Tentzeris, and J. Papapolymerou, "Characterization of liquid crystal polymer (LCP) material and transmission lines on LCP substrates from 30–110 GHz," *IEEE Trans. Microwave Theory Tech.*, vol. 52, April 2004, pp. 1343–52.
24. D. C. Thompson, J. Papapolymerou, and M. M. Tentzeris, "High temperature dielectric stability of liquid crystal polymer at mm-wave frequencies, " *IEEE Microwave Wireless Compon. Lett.*, vol. 15, September 2005, pp. 561–63.
25. C. Murphy, Rogers Corporation, private communication. January 2004.
26. B. Farrell and M. S. Lawrence, "The processing of liquid crystalline polymer printed circuits," *IEEE Electronic Components and Technology Conf.*, May 2002, pp. 667–71.
27. B. Hunt and L. Devlin, "LTCC for RF module," *IEE Seminar on Packaging and Interconnects at Microwave and mm-Wave Frequencies*, June 2000.
28. C. Q. Scrantom and J. C. Lawson, "LTCC technology: where we are and where we are going-II," *Technologies for Wireless Applications, IEEE MTT-S Symp.*, February 1999, pp. 193–200.
29. J. P. Gianvittorio and Y. Rahmat-Samii, "Fractal antennas: a novel antenna miniaturization technique, and applications," *IEEE Antennas and Propagation Magazine*, vol. 44, issue. 1, February 2002, pp. 20–36.
30. R. L. Li, G. DeJean, M. M. Tentzeris, and J. Laskar, "Novel multi-band broadband planar wire antennas for wireless communication handheld terminals," *IEEE Antennas and Propagation Society International Symposium*, vol. 3, June 2003, pp. 44–47.
31. G. DeJean, R. Bairavasubramanian, D. Thompson, G. E. Ponchak, M. M. Tentzeris, and J. Papapolymerou, "Liquid crystal polymer (LCP): A new organic material for the development of multilayer dual frequency/dual polarization flexible antenna arrays," *IEEE Antennas and Wireless Propagation Letters*, vol. 4, May 2005, pp. 22–26.
32. R. Li, G. DeJean, M. Tentzeris, and J. Laskar, "Integration of miniaturized patch antennas with high dielectric constant multilayer packages and soft-and-hard-surfaces (SHS)," *Proc. 2003 IEEE-ECTC Symposium*, May 2003, pp. 474–77.
33. M. Tentzeris, R. Li, K. Lim, M. Maeng, E. Tsai, G. DeJean, and J. Laskar, "Design of compact stacked-patch antennas on LTCC technology for wireless communication applications," *Proc. 2002 IEEE AP-S Symposium*, vol. 2, 2002, pp. 500–503.
34. B. Piernas, K. Nishikawa, K. Kamogawa, T. Nakagawa, and K. Araki, "High-Q factor three-dimensional inductors," *IEEE Transactions on Microwave Theory and Techniques*, vol. 50, issue 8, August 2002, pp. 1942–49.
35. K. B. Ashby, I. A. Koullias, W. C. Finley, J. J. Bastek, and S. Moinian, "High Q inductors for wireless applications in a complementary silicon bipolar process," *IEEE Journal of Solid-State Circuits*, vol. 31, issue 1, January 1996, pp. 4–9.
36. Choong-Mo Nam and Young-Se Kwon "High-performance planar inductor on thick oxidized porous silicon (OPS) substrate," *IEEE Microwave and Guided Wave Letters*, vol. 7, issue. 8, August 1997, pp. 236–238.

37. H. Jiang, Y. Wang, J.-L. A. Yeh, and N. C. Tien, "Fabrication of high-performance on-chip suspended spiral inductors by micromachining and electroless copper plating," *Microwave Symposium Digest., 2000 IEEE MTT-S International,* vol. 1, 2000, pp. 279–82.
38. Dong-Wook Kim, In-Ho Jeong, Ho-Sung Sung, Tong-Ook Kong, Jong-Soo Lee, Choong-Mo Nam, and Young-Se Kwon, "High performance RF passive integration on Si smart substrate," *2002 IEEE MTT-S International Microwave Symposium Digest,* vol. 3, 2002, pp. 1561–1564.
39. M. F. Davis, S.-W. Yoon, S. Pinel, K. Lim, and J. Laskar, "Liquid crystal polymer-based integrated passive development for RF applications," *2003 IEEE MTT-S International Microwave Symposium Digest,* vol. 2, issue 8, June 13, 2003, pp. 1155–58.
40. S. Dalmia, Lee Seock Hee, V. Sundaram, Min Sung Hwan, M. Swaminathan, and R. Tummala, "CPW high Q inductors on organic substrates," *Electrical Performance of Electronic Packaging,* 2001, vol. , issue 2001, pp. 105–108.
41. S. H. Lee, S. Min, D. Kim, S. Dalmia, W. Kim, V. Sundaram, S. Bhattacharya, G. White, F. Ayazi, J. S. Kenney, M. Swaminathan, and R. R. Tummala, "High performance spiral inductors embedded on organic substrates for SOP applications," *2002 IEEE MTT-S International Microwave Symposium Digest,* vol. 3, 2002, pp. 2229–32.
42. S. Dalmia, F. Ayazi, M. Swaminathan, Min Sung Hwan, Lee Seock Hee, Kim Woopoung, Kim Dongsu, S. Bhattacharya, V. Sundaram, G. White, and R. Tummala, "Design of inductors in organic substrates for 1-3 GHz wireless applications," *2002 IEEE MTT-S International Microwave Symposium Digest,* vol. 3, 2002, pp. 1405–08.
43. K. C. Eun, Y. C. Lee, J. W. Lee, M. S. Song, and C. S. Park, "Fully embedded LTCC spiral inductors incorporating air cavity for high Q-factor and SRF," *Proc. 54th Electronic Components and Technology Conference,* vol. 1, June 2004, pp. 1101–03.
44. G. Carchon, S. Brebels, K. Vaesen, W. De Raedt, and E. Beyne, "Spiral inductors in multi-layer thin film MCM-D," presented at IMAPS Europe, Krakow, Poland, September 4–6, 2002.
45. S. Dalmia, Kim Woopoung, Min Sung Hwan, M. Swaminathan, V. Sundaraman, Liu Fuhan, G. White, and R. Tummala, "Design of embedded high Q-inductors in MCM-L technology," *2001 IEEE MTT-S International Microwave Symposium Digest,* vol. 3, 2001, pp. 1735–38.
46. S. Pinel, F. Cros, S. Nuttinck, S.-W. Yoon, M. G. Allen, and J. Laskar, *2003 IEEE MTT-S International Microwave Symposium Digest,* vol. 3, June 8–13, 2003, pp. 1497–1500.
47. Yong-Jun Kim and Mark G. Allen, "Surface micromachined solenoid inductors for high frequency applications," *IEEE Transaction on Component Packaging and Manufacturing Technology,* Part C, vol. 21, issue 1, January 1998.
48. C. H. Ahn and M. G. Allen, *IEEE Transactions on Industrial Electronics,* vol. 45, issue 6, December 1998, pp. 866–76.
49. K. Yanagisawa, A. Tago, T. Ohkubo, and H. Kuwano, "Magnetic micro-actuator," *Proc. 4th IEEE Workshop on Microelectromechanical Systems,* Nara, Japan, 1991, pp. 120–24.
50. H. Guckel, K. J. Skrobis, T. R. Christenson, J. Klein, S. Han, B. Choi, E. G. Novell, and T. W. Chapman, "On the application of deep X-ray lithography with sacrificial layers to sensor and actuator construction," *J. Micromech. Microeng.,* vol. 1, no. 4, 1991, pp. 135–38.
51. B. Wagner, M. Kreutzer, and W. Benecke, "Linear and rotational magnetic micromotors fabricated using silicon technology," *Proc. IEEE Microelectromechanical Systems Workshop,* 1992, pp. 183–89.

52. H. Lakdawala, X. Zhu, S. Santanham, L. Carley, and G. Fedder, "Micromachined high Q inductors," *IEEE J. Solid State Circuits,* vol. 37, 2002, pp. 394–403.
53. C. Chi and G. Rebeiz, "Planar microwave and mm wave lumped elements and coupled line filters using micromachined techniques," *IEEE Transactions on Microwave Theory and Techniques,* vol. 43, 1995, pp. 730–38.
54. Y. Rao, J. Yue, and C. P. Wong, "Materials characterizations of high dielectric constant polymer-ceramic composite for embedded capacitor for RF application," *Active and Passive Elec. Comp.,* vol. 25, 2002, pp. 123–29.
55. J. Hwang, I. Abothu, P. Raj, and R. Tummala, "Organic-based RF capacitors with ceramic-like properties," *57th Electronic Components and Technology Conference,* Reno, May 29–June 1, 2007.
56. R. Ulrich, "Integrated Passive Component Technology," R. Ulrich and L. Schaper, Edited, IEEE Press, 2003.
57. S. K. Bhattacharya, M. Varadarajan, P. Chahal, G. Jha, and R. Tummala, "A novel electroless plating for embedding thin film resistors on BCB," *Journal of Electronic Materials,* vol. 36, no. 3, March 2007, pp. 242–44.
58. S. K. Bhattacharya, M. Varadarajan, P. Chahal, G. Jha, and R. Tummala, "A novel electroless plating for embedding thin film resistors on BCB," *Journal of Electronic Materials,* vol. 36, no. 3, March 2007, pp. 242–44.
59. P. Chahal, R. Tummala, M. Allen, and G. White, "Electroless Ni-P and Ni-W-P thin film resistors for MCM-L based technologies," *ECTC,* 1998, pp. 232–39.
60. S. Horst, S. K. Bhattacharya, J. Papapolymerou, and M. Tentzeris, "Monolithic low cost Ka band Wilkinson power divider on flexible organic substrates," *57th Electronic Components and Technology Conference,* Reno, May 2007.
61. K. Lim, S. Pinel, M. Davis, A. Sutono, C. Lee, Deukhyoun Heo, A. Obatoynbo, J. Laskar, M. Tentzeris, and R. Tummala "RF-system-on-package (SOP) for wireless communications," *IEEE Microwave Magazine,* March 2002, pp. 88–99.
62. R. Tummala, M. Swaminathan, M. Tentzeris, J. Laskar, G. Chang, S. Sitaraman, D. Keezer, D. Giudotti, R. Huang, K. Lim, L. Wan, S. Bhattacharya, Sundaram, F. Liu, and P. M. Raj, "SOP for Miniaturized Mixed-Signal Computing, Communication and Consumer Systems of the Next Decade," *IEEE Transaction on Advanced Packaging,* vol. 27, no. 2, 2004, pp. 250–67.
63. K. Lim, M. F. Davis, M. Maeng, S-W. Yoon, S. Pinel, L. Wan, D. Guidotti, D. Ravi, J. Laskar, E. Tentzeris, V. Sundaram, G. White, M. Swaminathan, M. Brook, N. Jokerst, and R. Tummala, "Development of intelligent network communicator for the mixed signal communications using the system-on-a-packaging technology," *Asia-Pacific Microwave Conference Digest,* 2003, pp. 1003–06.
64. M. Maeng, K. Lim, Y. Hur, M. Davis, N. Lal, S.-W. Yoon, and J. Laskar, "Novel combiner for hybrid digital/RF fiber-optic application," *2002 IEEE Radio and Wireless Conference,* 2002, pp. 193–96.
65. R. Bairavasubramanian and J. Papapolymerou, "Fully canonical pseudo-elliptic bandpass filters on multilayer liquid crystal polymer technology," *IEEE Microwave and Wireless Components Letters,* vol. 17, issue 3, March 2007, pp. 190–92.
66. S. Senturia, *Microsystem Design,* Kluwer Academic Publishers, 2001.
67. NASA 2003 Internal Report.
68. D. Mercier, K. Van Caekenberghe, and G. Rebeiz , "Miniature RF MEMS switch capacitors," *2005 IEEE MTT-S International Microwave Symposium Digest,* June 2005.
69. Radant 2006 MEMS Switch, http://www.radantmems.com.

70. G. Papaioannou, M. Exarchos, V. Theonas, G. Wang, and J. Papapolymerou, "On the dielectric polarization effects on capacitive mems switches," *IEEE MTT-S*, 2005.
71. Radiant MEMS 2006, http://www.radantmems.com.
72. B. Ducarouge, D. Dubuc, F. Flourens, S. Melle, E. Ongareau, K. Grenier, A. Boukabache, V. Conedera, and P. Pons, "Power capabilities of RF MEMS," *International Conference on Microelectronics*, 2004.
73. D. Thompson, M. Tentzeris, and J. Papapolymerou, "Packaging of MMICs in multilayer LCP substrates," *IEEE Microwave and Wireless Components Letters* [see also *IEEE Microwave and Guided Wave Letters*], vol. 16, issue 7, July 2006, pp. 410–12.
74. N. Kingsley, "Development of miniature, multilayer, integrated, reconfigurable RF MEMS communication module on liquid crystal polymer (LCP) substrate," Doctoral dissertation, Georgia Institute of Technology, Atlanta, GA.
75. N. D. Kingsley, G. Wang, and J. Papapolymerou, "Comparative study of analytical and simulated doubly-supported RF MEMS switches for mechanical and electrical performance," *Applied Computational Electromagnetics Society Journal*, vol. 21, no. 1, March 2006, pp. 9–15.
76. N. D. Kingsley and J. Papapolymerou, "Organic 'wafer-scale' packaged miniature 4-bit RF MEMS phase shifter," *IEEE Transactions on Microwave Theory and Techniques*, vol. 54, issue 3, March 2006, pp. 1229–36.
77. S. Basat, S. K. Bhattacharya, A. Rida, T. Vidal, R. Vyas, L. Yang, and M. Tentzeris, "Characterization of paper substrates for ultra-low-cost integrated RFID tags for chemical, pharmaceutical, and bio-sensing applications," *57th Electronic Components and Technology Conference*, Reno, May 2007.
78. S. Basat, S. K. Bhattacharya, L. Yang, A. Rida, M. Tentzeris, and J. Laskar, "Design of a novel high-efficiency UHF RFID antenna on flexible LCP substrate with high read-range capability," *IEEE Antennas and Propagation Society International Symposium*, July 9–14, 2006, Albuquerque, NM, pp. 1031–34.
79. Mina Rais-Zadeh, Paul A. Kohl, and Farrokh Ayazi, "High-Q micromachined silver passives and filters," *International Electron Devices Meeting*, vol. , no. , December 11–13, 2006, pp. 1–4.
80. G. J. Carchon, Walter De Raedt, and E. Beyne, "Wafer-level packaging technology for high-Q on-chip inductors and transmission lines," *IEEE Transactions on Microwave Theory and Techniques*, vol. 52, issue 4, April 2004, pp. 1244–51.
81. A. Sutono, A. Pham, J. Laskar, and W. R. Smith, "Development of three dimensional ceramic-based MCM inductors for hybrid RF/microwave applications," *1999 IEEE Radio Frequency Integrated Circuits (RFIC) Symposium*, vol. , Iss. , 1999, pp. 175–78.
82. J. Harada, Y. Sugimoto, Y. Higuchi, and Y. Sakabe, "Novel LTCC system of cofired low/high materials for wireless communications," *IMAPS/ACerS 2nd International Conference and Exhibition on Ceramic Interconnect and Ceramic Microsystems Technologies*, Denver, CO, April 2006.
83. T. Oda and M. Tomita, "New LTCC technology with integrated components—'developments and future trends'—module miniaturization for wireless network applications," *IMAPS/ACerS 2nd International Conference and Exhibition on Ceramic Interconnect and Ceramic Microsystems Technologies*, Denver, CO, April 2006.
84. L. Jauniskis, B. Farrell, A. Harvey, and S. Kennedy, "LCP PCB-based packaging for high-performance protection," *Advanced Packaging*, October 2006, pp. 40–42.
85. Kai Zoschke, Jürgen Wolf, Michael Töpper, Oswin Ehrmann, Thomas Fritzsch, Katrin Scherpinski, Herbert Reichl, and Franz-Josef Schmückle, "Fabrication of application

specific integrated passive devices using wafer level packaging technologies," 2005, pp. 1594–1601.

86. P. M. Raj, K. Coulter, J. H. Hwang, I. R. Abothu, S. Wellinghoff, M. Iyer, and R. Tummala, "A low temperature process to integrate high K-low loss pyrochlore films in organic substrates for thin film RF capacitors," *IMAPS Advanced Technology Workshop on Integrated/Embedded Passives*, San Jose, CA, November 15–16, 2007.

87. L. Wang, R. M. Xu, and B. Yan, "MIM capacitor simple scalable model determination for MMIC application on GAAS," *Progress in Electromagnetics Research*, PIER 66, 2006, pp. 173–78.

88. A. G. Cockbain and P. J. Harrop, "The temperature coefficient of capacitance," *Brit. J. Appl. Phys.*, ser. 2, vol. 1, 1968, pp. 1109–15.

89. J. H. Hwang, P. M. Raj, I. R. Abothu, C. Yoon, M. Iyer, H. M. Jung, J. K. Hong, and R. Tummala, "Organic-based RF capacitors with ceramic-like properties," *57th Electronic Components and Technology Conference*, Reno, NV, May 29–June 1, 2007.

90. D. C. Thompson, J. Papapolymerou, and M. M. Tentzeris, "High temperature dielectric stability of liquid crystal polymer at mm-wave frequencies," *IEEE Microwave and Wireless Components Letters*, vol. 15, issue 9, 2005, pp. 561–63.

91. R. K. Ulrich and L. W. Schaper, *Integrated Passive Component Technology*, Wiley-Interscience, IEEE Press, NJ, 2003, p. 80.

92. R. Kambe, R. Imai, T. Takada, M. Arakawa, and M. Kuroda, "MCM substrate with high capacitance," *Proc. 1994 International Conference on Multichip Modules*, Denver, CO, April 13–15, 1994, pp. 136–41.

93. R. Ulrich, L. Schaper, D. Nelms, and M. Leftwich, "Comparison of paraelectric and ferroelectric materials for applications as dielectrics in thin film integrated capacitors," *International Journal of Microcircuits and Electronic Packaging*, vol. 23, no. 2, 2000, pp. 172–80.

94. B. Piernas, K. Nishikawa, K. Kamogawa, T. Nakagawa, and K. Araki, "High-Q factor three-dimensional inductors," *IEEE Transactions on Microwave Theory and Techniques*, vol. 50, issue 8, August 2002, pp. 1942–49.

95. L. Zu, Lu Yicheng, R. C. Frye, M. Y. Lau, S.-C. S. Chen, D. P. Kossives, Lin Jenshan, and K. L. Tai, "High Q-factor inductors integrated on MCM Si substrates," *IEEE Transactions on Components, Packaging, and Manufacturing Technology,* Part B: *Advanced Packaging* [see also *IEEE Transactions on Components, Hybrids, and Manufacturing Technology*], vol. 19, issue 3, August 1996, pp. 635–43.

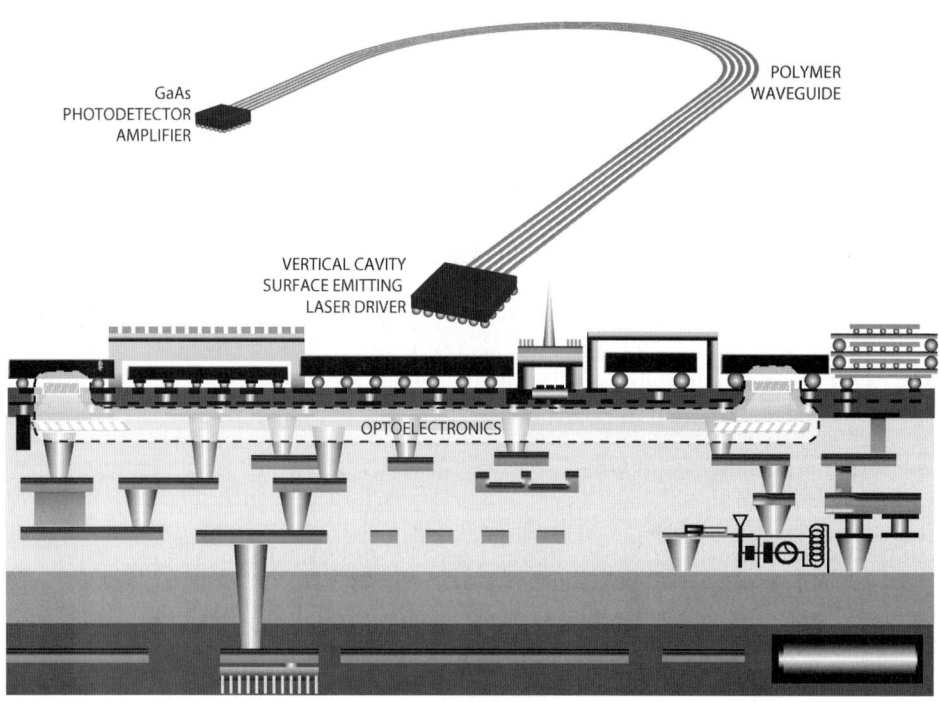

CHAPTER 6

Integrated Chip-to-Chip Optoelectronic SOP

Prof. Gee-Kung Chang, Prof. Thomas Gaylord, Ricardo Vallalaz, and Daniel Guidotti
Georgia Institute of Technology, Atlanta, Georgia

Prof. Ray T. Chen
University of Texas, Austin, Texas

6.1	Introduction 322	6.7	SOP Integration: Interface Optical Coupling 357
6.2	Applications of Optoelectronic SOP 323	6.8	On-Chip Optical Circuits 363
6.3	Integration Challenges in Thin-Film Optoelectronic SOP 325	6.9	Future Trends in Optoelectronic SOP 365
		6.10	Summary 365
6.4	Advantages of Optoelectronic SOP 331		References 366
6.5	Evolution of Optoelectronic SOP Technology 336		Table 6.1 References 374
6.6	Optoelectronic SOP Thin-Film Components 341		

While it is true that modern economies are built on energy, agricultural, and industrial foundations, without any one of which the economy would collapse, it is equally true that timely and detailed information makes it possible for the modern economy to function efficiently, and information has become a foundation in and of itself. Not surprisingly, machines that gather, generate, process, and disseminate information also have to become more efficient at their tasks. Key measures are megawatt for power generation, crop yield for agriculture, production efficiency for industry, and bandwidth for information. In industrialized economies, all but the first have seen large improvements in recent years with information growing the fastest, as measured by percent growth per year. One of our most efficient conduits of information became useless during the last half of the twentieth century and was replaced by a conduit having a much higher bandwidth. Long-distance telephone lines were replaced with optical fibers and so too went local data transmission soon after. To date, even copper interconnects in modern computers are becoming inadequate and ways are being sought to replace them with an optical architecture. The principles of SOP that encompass codesign, cointegration, and ultraminiaturization are timely for the task at hand. In this chapter we offer an up-to-date view of various attempts to introduce optoelectronic architecture in modern information networks in order to stay ahead of the information bandwidth, and we will attempt to make sensible predictions on near-term trends.

6.1 Introduction

Optoelectronic SOP is based on the principles of miniaturization and cointegration to bring about a higher-performance system at a lower cost. Optoelectronic SOP accomplishes this goal by embedding thin-film optoelectronic components in digital and analog circuits to achieve high functional density. Examples of thin-film optoelectronic components may include lasers, detectors, waveguides, gratings, microlenses, micromirrors, and optical amplifiers, all of which can comfortably exist within a thickness of 30 μm. This chapter describes the evolution toward optoelectronic SOP from the board-to-board and chip-to-chip optical interconnects being developed today through the integration of discrete components, into a highly integrated multiprocessor optical network. The advantages of optoelectronic SOP, even in its early stages of evolution, as described in this chapter are higher performance at lower cost for the system as a whole through the use of simplified materials, simplified digital integration, expanded architectural options enabled by the independence of optical bandwidth on distance, and miniaturization.

Optoelectronic SOP systems rely on a strategy of codesign and cointegration of optical, digital, and radiofrequency functions. Each laser, photodetector, amplifier, and passive waveguide interface operates at peak performance, and no compromises are made because of limitations due to incompatibilities with the integration process. At the same time, SOP designers and integrators seek ways to increase functionality by ultraminiaturization. This usually means developing and implementing thin-film optical technology in the form of thin-film lasers and thin-film photodetectors as well as thin-film flexible optical transceivers and full color organic light-emitting diodes (OLEDs), high-resolution displays that can be embedded in low-power cell phones, PDAs, and wireless mobile image processors that can be folded four ways and can fit neatly into a vest pocket. Clearly, conventional flex copper interconnects are subject to severe crosstalk limitations, particularly at higher bandwidths. Shielded copper lines and surface-mount components with large capacitance values, often used for electrical isolation, become less practical because of the high interconnect density and low power requirements of portable digital electronics.

While optoelectronic SOP is most commonly associated with highly miniaturized consumer and sensor products, the SOP concepts can also be applied to larger systems since they promote miniaturization, expanded design options, and greater packing efficiency. For example, in a copper data bus one may replace copper lines with optical waveguides. This can support longer data links with no appreciable signal degradation, narrower buses with increased bit carrying capacity per line, less crosstalk, less shielding, and fewer decoupling capacitors, all on an inexpensive substrate with fewer vias. However, the system is still not fully optimized because this architecture does not take into consideration the advantages and strengths of integrated optics. In order to optimize the system, one can remove, from the processor, the power-hungry electronic serializer-deserializer and the

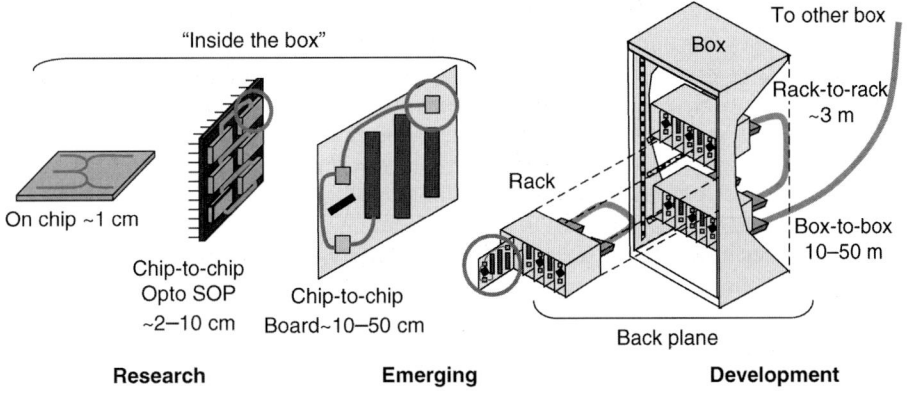

FIGURE 6.1 Optical interconnect hierarchy is depicted in a generic evolution. This may be a stand-alone computer or router or one of a cluster. High-speed local area connections or high-speed communication with other cluster components is already being done optically using 1310-nm edge-emitting lasers. Optoelectronic SOP will bring mixed optical and electronic signaling inside the box, inside the rack, on board, and eventually on the processor itself.

noisy copper bus drivers. These can be replaced with fewer, much faster CMOS laser drivers and photodetector amplifiers. The data bus now can consist of fewer, high-speed optical links that can extend to much longer distances with the elimination of a number of multiple interconnect levels, vias, capacitors, and shielding essential in copper technology. Additionally, flexible optical interconnects can be implemented at full bandwidth in three dimensions for board-to-board or chip-to-chip communication. With existing technology, an optimized system is easily scalable to a channel pitch of 125 µm with a single-channel, single-color bandwidth of 40 Gb/s.

The progression of optical interconnects toward the processor is depicted in Figure 6.1. Because the bandwidth of copper lines is intrinsically and strongly dependent on line length, optical interconnects find increasing applications "inside the box." Modified telecom-like optical transceivers are being introduced in the field for server banks and supercomputer nodes for high-bit-rate digital signaling between boxes. Optical backplanes are being developed for high-speed digital signaling among racks in the box, while processor-to-processor optical communication on boards or between boards within racks inside the box is at the emerging stage. On-chip optical clock distribution is thought to be essential to minimize clock skew and jitter at internal clock speeds above 10 GHz. While replacing copper lines with optical ones can modestly increase performance, the greater impact yet to be realized is in the architectural simplification that is driven by digital-optical codesign. Some of these points will be discussed in the following paragraphs.

6.2 Applications of Optoelectronic SOP

6.2.1 High-Speed Digital Systems and High-Performance Computing

Massively parallel computing and high-end servers will continue to require increasingly higher data transport rates that reach terabit-per-second levels between "boxes" or "processing nodes." This can only be achieved by optical and electronic codesign and integration at the package and chip level in the future. It is envisioned that computers and routers will routinely process and share data at aggregate data rates of several terabits per second using short-reach optical transceivers that employ optical interconnects between chips on a multiprocessor board, and between boards "inside the box." The term "inside the box" generally refers to the inner portion of one of many node enclosures that constitutes a supercomputer. It may also signify the enclosure that defines the physical space of a stand-alone computer, such as a blade center server. Optical interconnections "outside the box" will continue to use parallel, telecom-type optical transceivers modified to accommodate high channel density.

Optoelectronic SOP provides options not available in conductive data transport: (1) Since the optical signal bandwidth is independent of distance over distances of several meters, extended memory may be placed in a separate box. (2) Optical links do not require shielding and separation to reduce crosstalk to the extent that electrical lines do. Less than 10 µm of cladding separation is sufficient at all data bandwidths. (3) Because parallel optical links can be high density, potentially on a 60-µm pitch, and flexible, typically capable of a 1-cm turn radius, they can, in principle, support direct processor-to-processor wiring on a multiprocessor core module. (4) High-density flexible optical interconnects will diminish the dependence of high-speed signal routing through the backplane.

Optical-digital signaling "inside the box" will eventually migrate to optical-digital signaling at the chip level. This paradigm shift will probably occur when the internal processor clock speed increases to ≥10 GHz and clock synchronization and processor temperature becomes difficult to manage. At that time, optimal codesign will integrate on- and off-chip optical signaling at bit rates that are commensurate with that of an internal processor clock. In this scenario, synchronization on- and off-chip will become equally demanding. Alternatively, even if asynchronous processing is adopted, the need for increasing bandwidth capacity will not diminish. One needs to be aware that, in the near future, the total information transfer per year of the Internet is expected to surpass the zettabyte (10^{21} bytes) level. It is currently at the exabyte (10^{18} bytes) level [1]. To meet this jump in information density, the Ethernet electrical and optical infrastructure companies are busy developing 100-Gb/s capabilities worldwide [2].

6.2.2 RF-Optical Communication Systems

Optics is an effective enabling technology in the digital domain. Not only does optics transcend the limitations of copper-based signaling (bandwidth × distance), but it also transcends the limitation of impedance matching in three dimensions. Impedance matching limits the transmission lines in two dimensions. Any long unshielded line along the signal path acts as discontinuities, and impedance matching becomes difficult. Flexible, high-speed, high-density copper interconnects increase the challenge because of radiative signal losses, interference, and pulse distortion [3–4]. Flexible optical interconnects are subject to fewer and less severe limitations: (1) The optical leakage is independent of bit rate. This is not true for bent copper lines whose radiative losses increase with frequency [3–4]. For example, polymer waveguides having a relatively small index contrast of 0.03 between core and cladding can support a bend radius of 1 cm with about –0.12 dB loss [5]. (2) Straight or curved copper lines are notorious for picking up and generating crosstalk primarily through capacitive coupling. Design rules on minimum separation and minimum shielding depending on the application are necessary. On the other hand, crosstalk between two adjacent waveguides occurs when the evanescent field of one overlaps the other, which translates to a distance of the order of a few wavelengths of light in the medium of propagation.

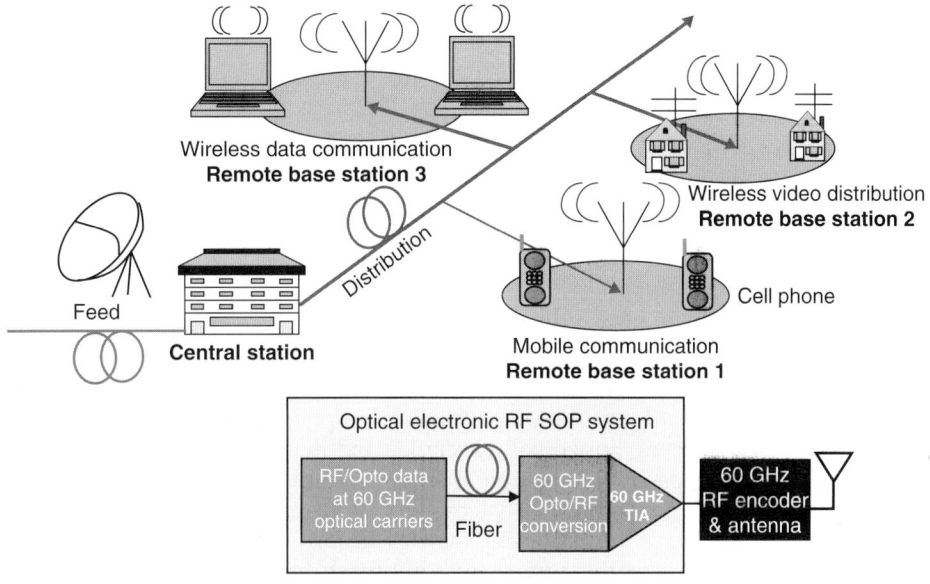

FIGURE 6.2 High-frequency RF information is distributed over long distances by first converting the RF coding to optical coding, transporting the information over optical fibers, and recovering the RF signal for broadcast at the destination. A land signal is returned by recycling the power available in the baseband optical signal.

Hence, parallel waveguides need only be separated by a few micrometers. This flexibility and the density characteristics of optical waveguides will eventually find application in laptops, mobile palm computers, and cell phones primarily due to effective isolation from electromagnetic interference (EMI).

Optical frequency isolation in mobile communication SOP hardware is a leading-edge concept that should find commercialization in the future. To date the concept of opto-RF SOP is embodied in the fast-growing field of "RF over fiber." RF waves are strongly absorbed in the atmosphere above frequencies of a few gigahertz. Yet, high RF frequencies are necessary for transmitting data at bit rates above a few gigabytes per second. A solution to this problem is to convert the high-frequency, amplitude- and phase-modulated RF waves into identical amplitude- and phase-modulated optical waves. At the destination, the optical signal is converted to the original RF signal and broadcast over a local area such as in a conference room, as depicted in Figure 6.2. At the source, the optical signal is encoded by readily available external modulators. At the destination, high-frequency photodetectors and microwave amplifiers can be used to restore and broadcast the original RF message. In some configurations, it is possible to recirculate the optical power and use it for a land-based return signal. This bidirectional RF over fiber technology is summarized in [6].

6.3 Integration Challenges in Thin-Film Optoelectronic SOP

The concept of transporting data optically over short distances of less than 1 m is relatively new when compared to conductive information transport. Thus the processes, materials, and architecture needed for integrating optoelectronic actives into passive lightwave circuits are quite new and tentative. Nevertheless, two major challenges emerge rapidly. The first, optical alignment, has no equivalent in conductive technologies, while the second, the choice of optical waveguide material, can be roughly compared to aluminum versus copper interconnects in that the former was a well-established technology, while the latter is a critically better conductor, but implementation problems such as oxidation barriers, electromigration, electrochemical deposition, and planarization have to be systematically solved. The optical signal distribution architecture, whether interboard or backplane, for example, is yet to be determined. Each approach has its advantages and shortcomings.

6.3.1 Optical Alignment

The difficulty in optical alignment is determined, quite simply, by the choice of optoelectronic components. There are two integration approaches. The first makes use of available components, most notably vertical cavity surface-emitting lasers (VCSELs) and planar photodiodes, and attempts to steer the optical beam from the vertical emission direction into a planar, horizontal waveguide and then out of the waveguide and back into a vertical propagation direction in order to couple into a top-viewing photodiode. Each out-of-plane 90° turn generally requires a 45° mirror and collimating lens and incurs mode coupling losses. Nevertheless, the vast majority of the integration effort around the world is done using this cumbersome approach, in no small part due to the fact that since the mid-1980s all high-speed laser sources operating in the 850- to 980-nm range are of the VCSEL type. Prior to that period, GaAs edge-emitting lasers were available at this wavelength. The 850- to 980-nm range happens to coincide with an absorption minimum for most optical polymers known today, as discussed in section III B.

The guiding principle for the second integration approach, practiced exclusively by Georgia Tech, is that of keeping light propagation strictly planar. Thus edge-emitting lasers and edge-viewing photodetectors (EVPD) are end-coupled to the waveguide. No lenses or mirrors are used in this integration process [5], and a coupling efficiency of laser-to-waveguide of 70 percent has been reported [7]. For the present time, edge-emitting laser sources at 1310 nm and 10 Gb/s offer a good compromise between speed, availability, and optical absorption. These are generally used for local area network applications. As of this date, the only one edge photodetector available commercially is a refractive-type monitor photodetector that operates between 1300 and 1600 nm at 1 Gb/s. Georgia Tech has designed an EVPD with an epitaxial PIN structure grown on a mesa sidewall and operating at 850 nm. For reference it is pointed out that Hitachi has reported an edge-emitting laser coupling to polymer optical waveguides as a means of coupling laser emission into optical fibers for local area network application [8].

6.3.2 Key Physical and Optical Properties of Thin-Film Optical Waveguide Materials

All optical waveguides are composed of organic polymer material with only one exception known to the authors. The exception is PPC Electronic AG, which uses glass laminates both for electric multilayer circuit boards and for optical waveguide circuits [9].

From the perspective of performance and fabrication the desirable properties of organic polymer waveguides can be summarized as follows:

1. From a survey of published data (Table III), a practical lower limit for the absorption coefficient seems to be about 0.01 dB/cm at wavelengths roughly between 800 and 1000 nm, about 0.1 dB/cm between 1000 and 1300 nm, and

about 0.3 dB/cm between 1300 and 1600 nm, limited by the onset of O—H, C—H, and C=O stretch vibration overtones [10].

2. The glass transition temperature of core and cladding materials is desirable to be above 300°C, in which case polymer decomposition should occur by sublimation above a temperature of 350°C in order to maintain the waveguide shape and compositional integrity over time, stress, and temperature.

3. A low bulk modulus is desirable in order to prevent cracking in flexible waveguides; however, the polymer should have a flow rate that is slow compared to time periods of years.

4. The monomer should be a liquid in the absence of solvents, and polymerization should be initiated by a UV-activated catalyst. This is desirable from the point of view of processing efficiency but is not essential.

5. A viscosity between 1000 and 10,000 cP may be desirable from the point of view of a monomer dispensing on a substrate by spin coating or monomer extrusion to obtain a 30- to 50-μm core and top cladding to adequately cover the core. A different, lower viscosity polymer may be used as undercladding prior to core formation.

6. The strain-induced birefringence of the core polymer should be less than $\sim 10^{-4}$, and the thermal optic coefficient should be less than $\sim 10^{-4}/°C$.

7. The coefficient of thermal expansion (CTE) of both the core and cladding polymer should match that of the substrate, if possible, but should be no greater than 150 ppm/°C when the substrate is rigid and has a CTE of about 20 ppm/°C. Flexible substrates such as polyimides may be able to accommodate a larger mismatch depending on the substrate thickness.

8. The difference in the index of refraction between the core and cladding should be tunable over a large range without notable attenuation. A combination of high contrast is desirable for high confinement and lower losses in short radius turns. A practical range in the index of refraction of organic polymer waveguides is between 1.47 and 1.55 at 1310 nm.

9. From the point of view of efficient processing, no adhesion promoter should be necessary.

10. Other desirable attributes of the polymer waveguide are low moisture content to minimize O—H attenuation, inexpensive base chemicals, long shelf life, and environmental friendliness.

Numerous polymer materials have been investigated over the years for application as low-loss optical waveguide material. With the exception of laminated glass waveguides [9], all polymers thus far investigated have a generic near-IR absorption spectrum that can be generically represented by the spectral absorption shown in Figure 6.3 in the wavelength range of interest. While, in general, optical absorption in the best polymers is 10^5 greater (in decibel units) than in the best optical fibers, the best reported organic optical polymers are those with normalized absorption in the following ranges: 0.02- to 0.05-dB/cm loss in the 850- to 980-nm band, 0.2 to 0.3 dB/cm near 1310 nm, and 0.4 to 0.5 dB/cm near 1550 nm.

Broad categories of optically suitable polymers are polyimides, olefins [11], polycarbonates [12], polymethylmethacrylates, polycyanurates, siloxanes, ORMOCERs, benzocyclobutanes (BCB), and fluorinated versions of these. See, for example, [13]. For a recent comprehensive review of polymer properties, see [10,14].

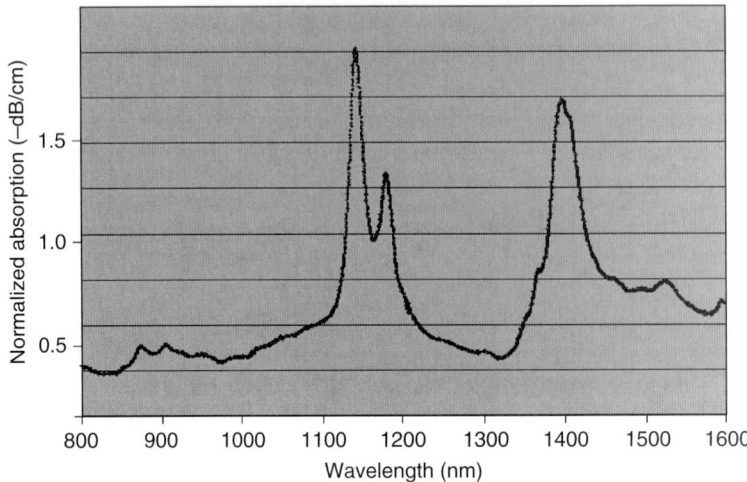

FIGURE 6.3 Representative spectral absorption common to most organic optical quality polymers.

There are four basic power-loss mechanisms associated with organic polymer waveguides.

1. Electronic absorption is mostly due to the absorption of light by hydrogen atoms and is effective in the 2000- to 4000-angstrom (Å) band [14] and contributes mostly to absorption in the visible range.

2. Absorption by overtones of fundamental molecular vibrations (mostly stretch vibrations) happens to fall in the range of 1100 to 1600 nm. This includes O—H, C—H, and C=O stretch vibrations. By introducing fluorine or other atoms into the bond, the frequency of the fundamental stretch absorption is shifted along with a corresponding shift in the overtones.

3. Light scattering from a number of sources, both intrinsic and extrinsic, includes compositional or density fluctuations, mostly due to nonuniform solvent evolution, particulate contamination in the original liquid monomer, processing imperfections such as bubbles, cross-section nonuniformity, and sidewall roughness (generally associated with reactive ion etching processes).

4. Finally, stress-induced birefringence is due primarily to film stresses that arise because of differences in the thermal expansion coefficients of the various polymer layers on the PCB and the board core material itself as well as by actual board bending and warping during thermal cycling, and scales as the elastic modulus of the waveguide material and the coefficients of its isotropic piezo-optic tensor. Generally, birefringence in multimode waveguides contributes to power loss by rotating the polarization of a guided mode into the polarization of an unsupported mode.

Table 6.1 lists commonly used polymer materials, suppliers, optical properties, basic integration processes, thermal properties, and references for further reading and investigation.

Ref.	Supplier	Material	Process	850 nm	1300 nm	1550 nm	n (632 nm)	Tg °C or sublimation	Comments
T1, T2	Allied Signal	Acrylate	UV, RIE	0.02	0.2	0.5	1.3–1.6	25	Hard cladding to contain the core polymer
T1	Dow Corning Telephonics	Halogenated acrylate	laser ablation	0.01	0.03	0.07		−50	
T3	NTT	Cross-linked silicone from methyl and deuterated (d-phenyl) group	RIE			0.23	1.472–1.532		n at 1550 nm
T4		Silicone	RIE		0.16	0.52	1.543–1.395		n at 1300 nm
T5, T6		UV epoxy resin	UV, RIE	0.08	0.5	4.72	1.48–1.6	~200	
T1		Halogenated acrylate		0.02	0.07	1.7			
T1, T7		Deuterated polysiloxane	RIE		0.17	0.43	1.5365 1.5345	400	n at 1310 nm n at 1550 nm
T8, T9, T1, T10		Halogenated polyimide	RIE		0.3	1.0	1.51–1.52	335	n at 1300 nm
T1, T11	Dow	Benzocyclobutanes (BCB), cyclotene resins	RIE laser ablation	0.5	0.8	1.5	1.552–1.561 1.537–1.544 1.535–1.543	350	n at 630 nm n at 1300 nm n at 1550 nm
T1, T12		Perfluorocyclobutane	UV	0.18	0.25	0.25	1.4878	400	n at 1300 nm
T1	DuPont, Optical CrossLinks	Acrylate (polyguide)	UV, RIE laser ablation	0.18	0.2	0.6			
T13, T14	Micro Resist Technology, GmbH, Berlin, Germany	Organically modified ceramics (ORMOCER), Fraunhofer IZM	UV, RIE laser ablation	0.06	0.23	0.55	1.5214–1.538	250	n at 830 nm

TABLE 6.1 Summary of Commonly Used Polymers for Passive Lightwave Circuits

Ref.	Supplier	Material	Process	850 nm	1300 nm	1550 nm	n (632 nm)	Tg °C or sublimation	Comments
T15	MicroChem	Epoxy novolak resin	RIE, UV		0.22	0.48	1.575		n at 1550 nm
T16, T17	RPO Pty. Ltd. Acton, Australia	Inorganic polymer glass (IPG)	UV	0.1	0.24	0.4	1.491–1.543 1.474–1.528	300	n at 630 nm n at 1550 nm
T18	Polyset	Epoxy siloxane oligomer	UV	0.05	1.4		1.45–1.55	400	
T19	NTT	EPOXY	UV	0.1				200	
T20	Dow Corning	Siloxane	UV	0.06				200	
T21	Shipley / Rohm&Haas	Siloxane	UV	0.015	0.110	0.5	1.510	250	n at 850 nm
T22	Kyocera Corporation	Siloxane (TiO$_2$ doped)	UV		0.14	1.9	1.4440–1.5823	400	n at 1300 nm
T23	Fraunhofer IZM	Polycyanurate	UV, RIE		0.2	0.4	1.444–1.4545	250–400	
T24	TeraHertz Exelis	True Mode (acrylate)		0.04	0.4	0.5	1.45–1.58	150	

TABLE 6.1 (Continued)

In the remainder of this chapter we will consider the development of digital-optical SOP technology as it applies to high-performance computing, servers, and routers.

6.4 Advantages of Optoelectronic SOP

6.4.1 Comparison of High-Speed Electrical and Optical Wiring Performance

There are three forces that drive SOP technology for computing, namely, (1) performance, (2) power, and (3) cost. These translate into seven technology improvements that are made possible by introducing optoelectronics architecture into digital systems. These improvements are (1) nearly invariant (bandwidth × distance) product over the entire network, (2) interconnect density that is far greater than copper densities, (3) negligible crosstalk and insensitivity to simultaneous switching noise (SSN), (4) three-dimensional optical wiring, (5) direct, high-speed processor-to-processor optical links that lead to architectural simplicity by reducing the use of much slower copper bus lines, number of printed circuit board (PCB) layers and vias, and capacitors, (6) greatly reduced node crossing delays in multiprocessor networks by direct optical wiring and long-reach off-chip synchronization, and (7) minimum number of components used for noise suppression [15].

It is well known that the bandwidth of copper interconnects is intrinsically limited by numerous factors such as skin effect, inductance, capacitance, and EM radiation, as well as extrinsic factors such as the dielectric susceptibility of the insulator which may cause frequency dispersion and signal attenuation, crosstalk, and power supply noise. The intrinsic limitation of the (bandwidth × distance) product for unequalized copper lines is summarized in Equation (6.1), where B_{max} is the maximum bandwidth capacity, A is the cross-sectional area of the copper line, and l is its length [16].

$$B_{max} \leq 2.4 \times 10^7 A/\rho \quad \text{Gb/s} \qquad (6.1)$$

A graphical representation of Equation (6.1), namely the dependence of the bandwidth of two unequalized copper transmission lines on cross-sectional area and distance, is shown in Figure 6.4 and compared with the bandwidth for a multimode (MM) polymer waveguide carrying a single wavelength (1310 nm) with an attenuation coefficient of −0.2 dB/cm.

In comparison, a single, long-distance optical fiber carrying 100 colors, each having a 10-Gb/s capacity, has an aggregate bandwidth of 1 Tb/s. Under development are 100-Gb/s laser modulators and detectors with similar speed to support the Ethernet infrastructure [2]. The ability to move massive quantities of data at several Tb/s over long distances makes it possible to contemplate real-time distributed, task-specific, parallel computing. Realizing the potential advantages of an optical network architecture that links multiple nodes (processors, memory banks, I/O buffers), a number of authors have simulated a variety of optical bus designs and data passing protocols in symmetric multiprocessing (SMP) and massively parallel processing (MPP) machines. In SMP (servers), all processors share the same memory via one or more buses, and each CPU takes on the next available task. In MPP (supercomputers), problem segments are solved in locked step. Each CPU contains its own copy of the operating system and application and has access to its own memory. Tasks are assigned to each CPU, and communication between MPP subsystems takes place via high-speed interconnects where the optics solution can play a role.

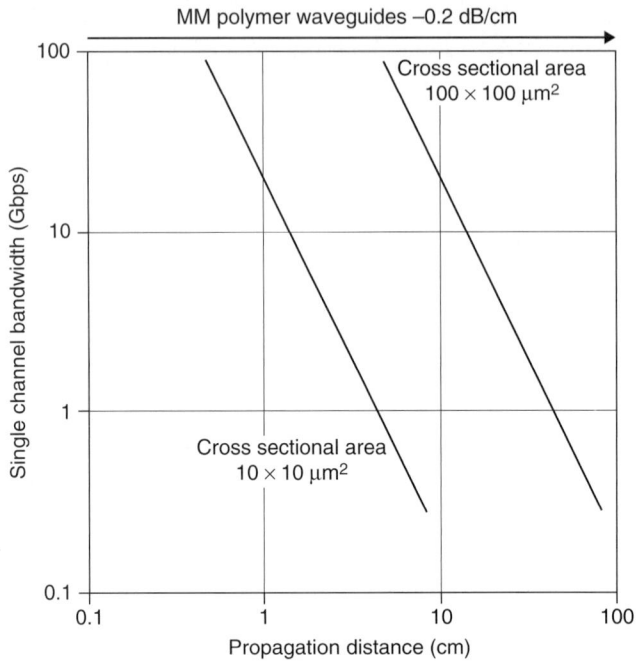

FIGURE 6.4 Plot of Equation (6.1) with two values of cross-sectional area for unequalized copper transmission lines. It also illustrates the lack of dependence of bandwidth on distance for a multimode optical waveguide.

To summarize connectivity, network access latency, bus usage, and scalability [17–20] are important issues in computer system design. In such systems an optical local area network connecting several workstations or computing nodes can provide for scalability (adding more processors), high connectivity (long-distance direct wiring where needed), high bandwidth, reduced network latency (by transcending node hierarchical copper connection by virtue of the independence of optical bandwidth on distance), and high bus utilization, depending on optical network design and choice of protocols.

6.4.2 Wiring Density

In large multiprocessor machines there are thousands of low-speed parallel copper data links compacted in a small footprint that generates a large amount of crosstalk. The number of links is determined by the interconnect architecture; the number of processors [21] (approximately 100 for large server clusters); the bus width, which is determined by the processor I/O speed; board losses; and intrinsic copper losses [16]. In addition to these design issues there are four major challenges for off-chip digital signaling [22], namely,

1. The demand for off-chip data link performance of ~10 Gb/s by 2010.
2. The demand for 1000 high-speed signal I/Os per chip.
3. The increasing gap between processor speed and SDRAM access delay will drive demand toward wider copper buses and an increased number of layers per board.

4. Network node crossing delays due to the hierarchical construct inherent in copper interconnects and the need for data synchronization at each node, and node crossing protocols, will result in increased data transfer delays that affect the performance of the entire system.

These four challenges can become potential bottlenecks and result in both opportunities and challenges for the emergence of a mixed copper and optical interconnect architecture in which copper is used for very short reach and low-speed wiring, while optics is used for high-speed, data broadcast, long-reach synchronization, and long-reach point-to-point wiring.

Limitations for Copper-Wire Interconnects

The design of copper interconnect density on a printed circuit board (PCB) depends on distance, bit rate, and board dielectric properties. This is deduced directly from Equation (6.1), in which it is seen that the maximum bit-carrying capacity of a copper line increases as the cross-sectional area and decreases as the square of the line length. The cross-sectional area has to be large enough to accommodate the design bit rate over the design distance. This leads to many wiring layers with vias, via shielding, and decoupling capacitors. While a polymer optical waveguide density of 500/cm (20-μm pitch) on PCBs is practical, the actual optical channel density is limited by the pitch of the laser and PD arrays, which is generally 250 μm but can decrease to 125 μm. Laser array driver chips and PD array amplifiers are millimeter size. An optimized system will have optical I/O drivers and amplifiers that become part of the processor I/O thereby replacing copper bus drivers, multiplexers, and demultiplexers. The bare die lasers and PDs will then become the principal limiting factor determining the optical wiring density. However, since the bit rate carrying capacity of optical interconnects far surpasses that of copper and is independent of distance, the number of high-speed processor I/Os will actually be lowered, decreasing the pressure on I/O ports and wiring density below the level predicted by ITRS [22]. Hence, optical interconnects offer a paradigm shift as compared to the predictions by Rent's rule [23]. However, as discussed below, the most effective method for interboard or intraboard optical wiring may be through the use of high-density, flexible optical interconnects that are directly and electrically pluggable next to a processor.

Opportunities for Board-to-Board Optical Interconnects

In a typical blade server application [24] the edge connector (with hundreds of pins) connects the system board with the backplane or the centerplane. The insertion force for each pin may be in the range of 0.3 to 0.8 N, and a system board with 1000 pins will require a total insertion force of 73 kg to connect to the backplane. Given the material set, it will be difficult to increase the pin density. At the same time, state-of-the-art low-profile optical transceivers can at best provide 12 channels [25]. These optical transceivers, just as telecom transceivers, are assembled with discrete components such as mirrors, lenses, and possibly optical isolators. The optical assembly is done largely by hand, making these devices expensive and prohibitive to scale up to 100 channels [26]. With the development at Georgia Tech of the flexible optical strap having embedded optical actives, it is possible to foresee an optical interconnect density of 500/cm. This is because in the Georgia Tech flexible optical wiring, lasers, photodetectors, and waveguides are end-coupled and "self-aligned" to the waveguide during the waveguide fabrication process either in arrays or individually and contain no lenses and no mirrors when edge-viewing photodetectors are used. The term "self-aligning" means that the active devices and passive lightwave circuits are all simultaneously aligned during the process of waveguide core path definition, accomplished by selectively cross-linking monomers either by direct UV laser writing, projection lithography, or proximity lithography through a mask. The optical wiring density is limited primarily by the dimensions of laser bare dies and photodetector bare dies. To overcome this potential limitation, the actives can be staggered in a two-dimensional array, and waveguide escape routes can be designed to converge in a parallel array. It should be noted that, because optical interconnects can transcend electrical wiring hierarchy, board-to-board optical wiring can originate and terminate on any board edge or anywhere on the board front or back surfaces, as compared to electrical wiring which is defined at a board's edge. For optical interconnects, it is simpler to refer to the total number of *interboard* optical links, which are separate from the total number of *intraboard* optical links.

6.4.3 Power Dissipation

Simulations of the power dissipation by optical and electrical interconnects, on-chip and on-board, inevitably compare one copper line with one optical link. The one-for-one comparison is not realistic. The task of comparing power dissipation between electrical and optical interconnects has to take the following into account:

1. The electrical bus aggregate bit carrying capacity: width, length, and bit rate per line. This should be compared with the number, usually much lower, of optical lines that are needed for the same aggregate bandwidth.
2. The aggregate CMOS power dissipated in driving the electrical bus, including line equalization, multiplexers, and demultiplexers, versus the CMOS or bipolar power dissipated in driving the (fewer) number of laser drivers and PD amplifiers that replace the electrical bus.

The determining factors are distance and aggregate bandwidth. The type of laser used is a perturbation effect. Two independent authors have shown that when all components in the optical link (laser drivers and photodetector amplifiers) are taken into consideration, there is little difference in the power consumption of optical links based on VCSELs or based on edge-emitting lasers (EELs) [20, 27–29]. The important difference between VCSELs and EELs is the fact that readily available commercial VCSELs roll off rapidly after 3.3 Gb/s [30], while readily available commercial EELs roll off after 10 Gb/s. On the horizon, Agilent and Emcore are developing 20-Gb/s VCSELs, in part with DARPA funding, while NTT has reported operation of a directly modulated 40-Gb/s EEL [31].

Receiver equalization can extend the range of a 10-Gb/s, differential microstrip transmission line to about 0.75 m on an FR4 or PCL-FR3 [32] substrate using non-return-to-zero (NRZ) serial bit encoding. The equalization method is deterministic and compensates mostly for intersymbol interference (ISI) phase shift noise due to frequency-dependent skin effect and dielectric losses. For example, the MAX3804 receiver equalizer is designed to do just that and can significantly decrease deterministic jitter. The Maxim MAX3804 has a footprint of 3 mm × 3 mm and dissipates 115 mW per differential copper line [33].

A 10 Gb/s per channel, 1 × 12 VCSEL driver from Helix AG dissipates 120 mW per channel, and a 10 Gb/s, 1 × 12 photodetector amplifier, also from Helix AG, dissipates

160 mW per channel. Therefore, a single optical channel dissipates 280 mW, or about 2.5 times the power dissipated for a single equalized copper differential channel. However, while the differential copper channel can compensate for transmission line losses up to 0.75 m on FR4 [28], the only distance limitation for the optical channel is that dictated by the optical absorption of the polymer used to form the waveguide channel, and distances of many meters can easily be optically linked and at bit rates that are higher than 10 Gb/s. Depending on the application, the optical alternative may be preferred. In the case of flexible, interboard interconnects, the optical alternative may offer much higher channel density and flexibility over impedance-matched coaxial cable or individual optical fibers or even optical fiber-base transponders [25].

6.4.4 Reliability

1310-nm EELs and 850-nm VCSELs

Failure in a population of lasers or chips is measured as failures in time (FIT) over 10^9 of device service hours. Thus a FIT of 10 signifies 10 devices have failed in 10^9 hours. The rate of failures need not be monotonic. There are numerous reports regarding the reliability of AlGaInAs-based transmitter lasers operating at 1310 nm [34–35]. Generally the failure rate increases monotonically with time of use [36].

In [34], the mean-time-to-failure, or median life, is estimated at 82,000 hours (9.4 years) at 85°C, which is significantly less than Telcordia standards but sufficient for board-level integration of less than roughly 100 lasers. Wearout failure rates at 40°C after 5, 10, and 20 years of service can be calculated using the same lognormal model. A FIT value of 11—that is 11 failures per billion device hours—is calculated for lasers with 5 years of service. Corresponding FIT values after 10 and 20 years are 29 and 60, respectively. These FIT numbers show that the lasers have sufficient long-term reliability for use in the applications described here. The reliability, allied with the attractive price of FP lasers, makes these devices competitive in short-haul 10-Gb/s communications.

The failure rate for 850 nm, 2.5-Gb/s VCSELs is more complicated and displays a rapid increase in FITs after 5 years of service. To date, 10-Gb/s VCSELs have a lower yield, higher failure rates, and a higher price than 10-Gb/s, 1310-nm EELS [37].

CMOS Processors

In comparison, CMOS processors display the typical bathtub failure rate distribution in time. In the infant mortality region where manufacturing defects can cause rapid failure during burn-in, the goal is to reach 200 FITs during the first 3 months. Ideally the failure rate should remain at 200 FITs for up to 7 years of service. Thereafter the onset of the wearout region depends on diffusion processes in older CMOS technology and on gate oxide hot electron wearout in later, thin oxide technologies, starting at the 100-nm node [38].

Optical Polymers

Reliability data on polymer waveguides has not risen to the same level as that of lasers. This is clearly because the polymer materials that have been developed for lightwave circuits (see Section VI), generically polycarbonates, acrylates, polyimides, olefins, polymethylmethacrylates (PMMAs), polycyanurates, siloxanes, organically modified ceramics (ORMOCERs), and BCB, have not had the field exposure and testing necessary to assess their reliability for high-end server application wherein component reliability is measured in terms of 10^9 hours of service. PMMA-based fibers are in service in the consumer market as digital home appliance interfaces, home networks, and automotive networks where reliability requirements are less demanding.

The most comprehensive data comes from Corning on fluorinated acrylates [39]. Exposure to air at 100°C for 270 days shows no measurable change in the index of refraction. Thermal aging at 170°C for 12.6 days decreases propagation losses, and optical aging under 130 mW in a waveguide with a 7 μm × 7 μm cross section (260 kW/cm^2) at 1319 nm for 117 days shows no changes.

6.5 Evolution of Optoelectronic SOP Technology

Because of increasing bandwidth requirements and the limitations of copper interconnects, optoelectronic data communication has migrated from fiber-based long-distance communications in the 1980s to local area optical networks in the late 1990s where it has abruptly stopped. The technical reasons for the abrupt stoppage are fairly easy to understand. The replacement of copper cables outside the box requires relatively few optical channels, and until now one could borrow from long-haul and LAN optical transceiver technology, and with minor modifications adapt that technology to "box-to-box" data sharing in a high-performance computing environment. Standard optical transceiver packs, though bulky and having a bandwidth of only 2.5 Gb/s per channel over 12 channels, can be used over several hundred meters of parallel plastic or glass optical fiber ribbons [40]. An optical transmitter module obtains power and signals from a transmitter board, converts the signals to optical coding and transmits them optically. The signals are then received by a receiver module where they are converted back to digital signals. Optical transceivers have become miniaturized for this application [26] but still offer very low interconnect density. This is where further progress toward "optics to the processor" has slowed; the pool of ready-made solutions has been exhausted. At this point, high-density optical interconnect technology to the processor has become the subject of research and development in many laboratories around the world [15]. The next evolutionary step is to extend optical signaling between cards as in blade center servers, then between processors on a card, and eventually integrated optics on the processor chip itself, as depicted in Figure 6.1.

6.5.1 Board-to-Board Optical Wiring

The ideal optical backplane or midplane consists of optodigital cards that plug in to optical and electrical ports as is done with electrical cards in Nuclear Instrumentation Module (NIM) bins used in photon counting and nuclear counting, a concept that has migrated to rack-mounted computers such as blade servers. The concept of plug-in optical cards has been previously attempted by General Electric [41–43] and a consortium of universities and industry companies and organizations who leveraged MCM technology to package VCSEL arrays and use MT connectors for optical coupling. Reliable optical plug-in cards are difficult to achieve because of the difficulty of achieving and maintaining reliable optical alignment in a mechanically and thermally harsh environment. In addition, the optical channel density is roughly commensurate with electrical pin density. Consequently, the plug-in optical backplane remains an unsolved technical challenge in reliability and edge density. An intermediate solution to high-speed interboard optical interconnects that seems to be gaining momentum for enhancing the performance of high-end machines is the miniaturized optical transceiver module that uses either optical fiber array ribbons or optical polymer WG array ribbons, as shown in Figures 6.5 and 6.6.

FIGURE 6.5 Optical PC bus extension adapter cards and an optical fiber cable. [44]

IBM was the first company to offer a commercially available optical bus for high-speed optical communication between two computers over several meters via card-to-card optoelectronic data transceivers as shown in Figure 6.5. In 1993, the IBM Tokyo Research Laboratory developed a technology for extending the I/O buses of personal computers. A PC bus is extended by means of an optical fiber, preserving full

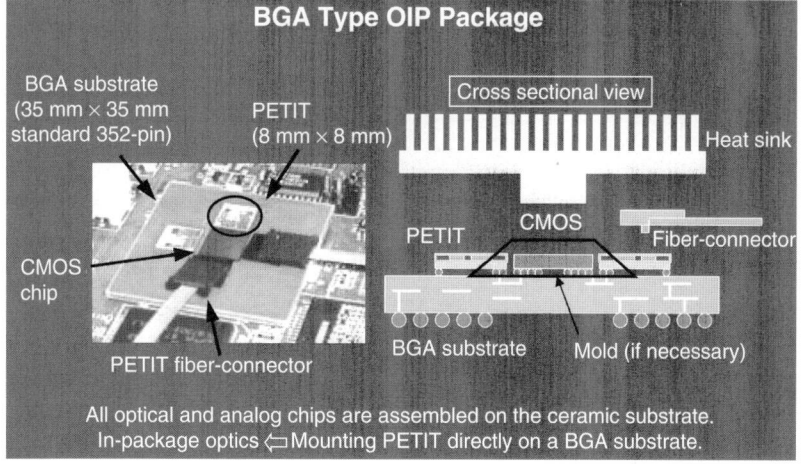

FIGURE 6.6 Multichannel optical transceiver fiber ribbon developed at NEC using optical fibers and miniaturized MT connectors [26].

compatibility with an ISA bus or a Micro Channel bus. Ordinary PC add-in cards can be placed several meters from a PC, to which they are connected via the optical link [44]. The link converts the parallel signals on the PC's I/O bus into serial format, preserving the coherency of the bus protocol, and then transmits them via an optical fiber. Since the optical communication link is perfectly transparent to add-in cards and software, programs can directly access add-in cards placed at a distance from the PC as if they were installed in the PC box. The maximum distance is 100 m for the ISA bus and 10 m for the Micro Channel bus.

Unfortunately the work on card-to-card optical links for enhancing network connectivity and performance stopped at this point. A few years later both NTT and NEC published results of their development work on optically linking printed circuit boards in their mainframes. In this case, either optical fiber arrays or arrays of polymer waveguides are used in conjunction with miniaturized optical transceivers containing arrays of lasers, photodetectors, and associated amplifiers and laser drivers. The transceivers used for card-to-card optical communication are miniaturized versions of bulkier optical transceivers used in the optical communication industry. The substantial decrease in form factor is made possible by running the lasers uncooled. The work from NEC [26,42–43], using optical fibers, is shown in Figure 6.6, while that from Optical CrossLinks [45] and NTT [46–47] using polymer waveguides are shown in Figure 6.7a and b.

In all three cases, the optical interconnects contain 45° beam steering mirrors and, in some cases, microlenses. It is precisely this construction, borrowed from the telecom industry, which makes scaling difficult for interboard optical interconnects even without thermoelectric (TE) coolers and hermetic seals. The problem is largely based on cost. Neither the telecom industry nor the emerging computer optoelectronics industry have been able to find a way to quickly and reliably align optical fibers, lenses, mirrors, and lasers or PDs, except manually or with robotic assistance, in some cases.

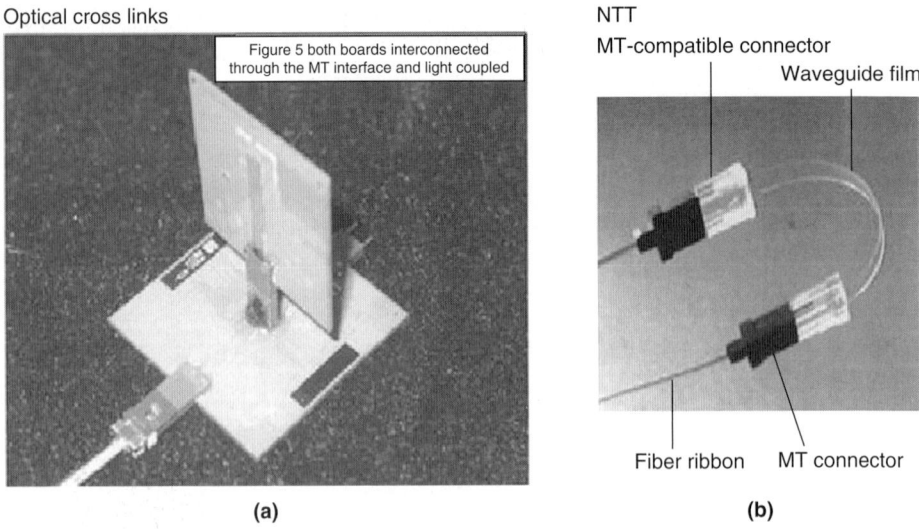

Figure 6.7 (a) Multichannel flexible optical interconnect polymer waveguide ribbon from Optical CrossLinks, using MT connectors. (b) Flexible multichannel polymer waveguide ribbon connected to optical fibers via MT connectors from NTT.

6.5.2 Chip-to-Chip Optical Interconnects

Pioneering work in chip-to-chip optoelectronic integration on real-world, high-performance computer motherboards was carried out by R. Chen in a university-industry consortium [48–50] intended to improve the performance of mainframe computing at the time. A Cray T-90 motherboard, shown in Figure 6.8, was used as the testbed for the development of a wide area optical clock distribution. Numerous innovative technologies were applied or developed for the first time. Long polymer waveguides were formed by direct laser writing, thin-film Si or GaAs MSM photodetectors were embedded in the lightwave circuit for the first time, and surface relief gratings were fabricated for the first time directly on each polymer waveguide for beam steering, and arrays of VCSELs were used as sources.

Since this pioneering work, a number of researchers around the world have made great strides in designing and fabricating chip-to-chip optoelectronics interconnects for high-rate digital data transport over wide areas at the board integration level. The Fraunhofer Institute, IZM, has developed the concept of the "optical pin," which is intended to reliably couple lasers and detectors to the lightwave circuit in the PCB and is compatible with surface-mount technology (SMT) [51]. Later NTT developed a similar hybrid carrier "opto bump" that is also designed to provide compatibility between electrical and optical surface-mounted components (SMT) [52]. The key element of the SMT electrical-optical circuit board (EOCB) concept is the formation of an additional optical layer consisting of multimode waveguide structures. Waveguides are incorporated within the circuit board optical layer by a number of methods from lamination, hot embossing, reactive ion etch (RIE), and photolithography, and standard printed wiring board fabrication technologies. Multimode waveguides are used to meet SMT assembly tolerances in order to interface to common surface-mount packages and

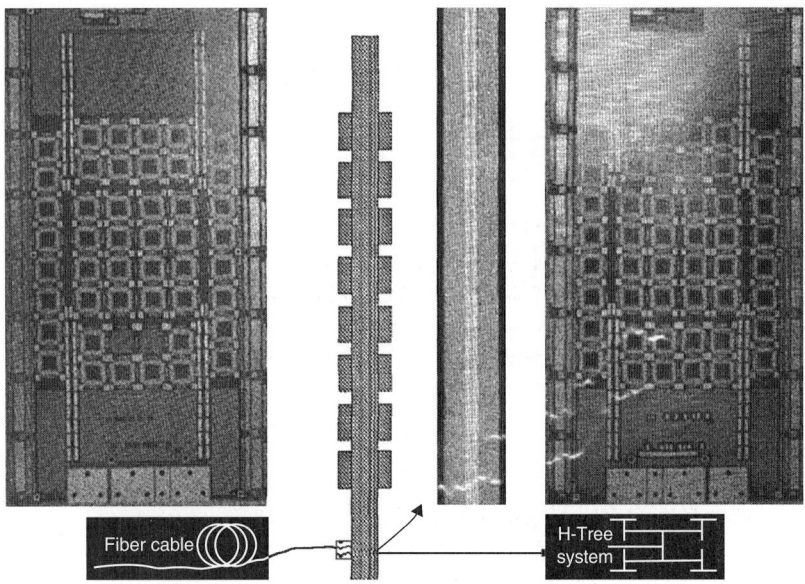

FIGURE 6.8 Photograph of Cray T-90 multiprocessor supercomputer board, 26.7 cm in length and having 52 vertical integration levels and 1 to 48 electrical clock signal distribution at 500 MHz.

FIGURE 6.9 Top view of the ETHZ-IBM demonstrator board.

comply with pick-and-place SMT assembly tolerances. Optoelectronic devices have to fit within these process tolerances.

End-to-end optoelectronic integration on printed circuit boards has been reported by the Swiss Federal Institute of Technology, Zurich (ETHZ), and IBM Rüschlikon, over four channels at 10 Gb/s per channel, using 850-nm VCSEL, PIN PDs, and 45° beam turning mirrors [53]. A photograph of the demonstrator board is shown in Figure 6.9.

Flexible optical interconnects, similar to those of Optical CrossLinks, are just being developed to meet the need for high-density, high-speed, interboard, and intraboard optical interconnections. The latest paper to appear is that from Ray Chen, SCI [54]. The resulting structure is shown in Figure 6.10. As in all previous cases, 45° end mirrors are used here also along with VCSELs and top (or bottom) viewing photodetectors.

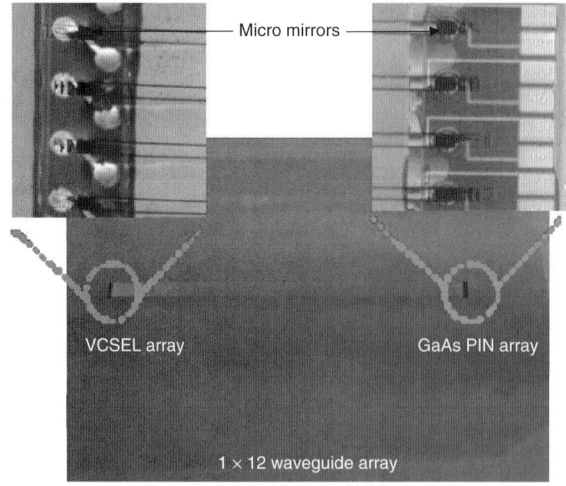

FIGURE 6.10 Integrated VCSEL and PIN detector arrays on a flexible optical waveguide film. [54]

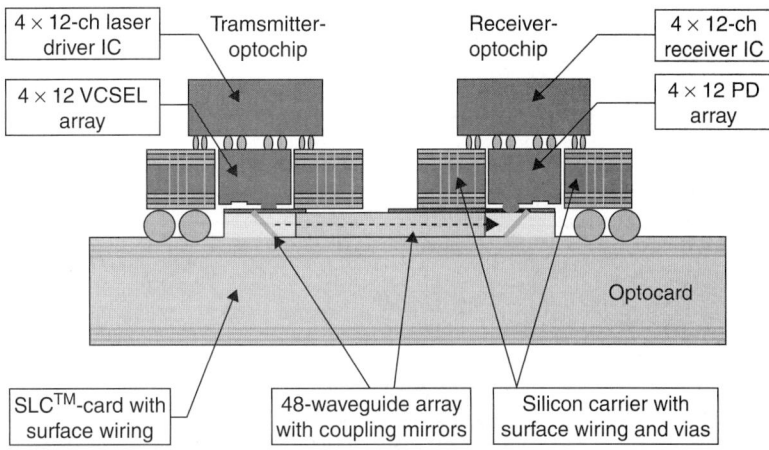

FIGURE 6.11 Schematic cross section of the main components in the IBM Terabus showing the flip-chip hierarchy. The microlens array is not clearly shown. See [55] for additional detail.

Finally, the massively parallel optical interconnect Terabus project developed at IBM has 48 VCSEL channels, each channel having a bit rate of about 20 Gb/s. A sketch of the architecture is shown in Figure 6.11. Consistent with the rest of the industry, designers of the Terabus have taken the discrete component integration approach to the level of arrays of components, but the optical alignment is still accomplished either by hand or is left to the law of surface tension in the flip-chip array reflow process [55].

6.6 Optoelectronic SOP Thin-Film Components

In this section, we summarize the challenges and opportunities that will be encountered over the next 15 years toward the implementation of optical interconnects inside the box as backplanes and as flexible high-speed, high-density chip-to-chip optical interconnects.

Digital signaling over nonequalized microstrip copper transmission lines loses bandwidth carrying capacity both as the inverse square of the line length and as the cross-sectional area shrinks [16]. In addition, microstrip copper lines must maintain impedance matching and shielding, be sufficiently separated to prevent crosstalk, and be isolated from the power plane in order to prevent coupling to simultaneous switching noise or ground bounce [56]. Finally, because of these constraints on electrical digital signaling, high-density and high-speed off-board copper interconnects are difficult to design. High-performance systems will therefore be the first to migrate to high-density optical interconnects for high-speed signaling in two and three dimensions in order to meet performance targets of terabytes per second per unit volume at low power.

The optical solution is sketched in Figure 6.12 in order to point out the necessary components. A laser driver converts a digital input signal into a series of optical pulses by the direct modulation of current source I through the laser. The modulated analog optical signal is then coupled to a waveguide and carried to a photodetector which is coupled to the receiving end of the same waveguide. The photodetector current is converted to a voltage signal by a transimpedance amplifier (TIA) which is maintained at a digital level by a limiting amplifier (LA) with auto gain control. A filter, a clock

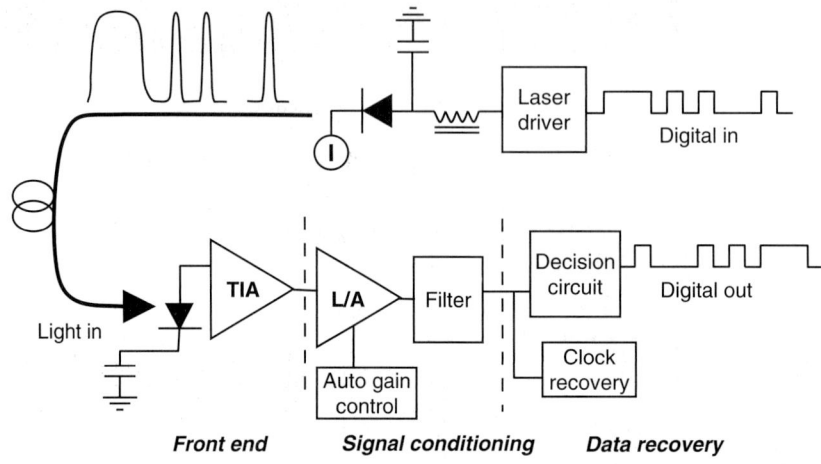

FIGURE 6.12 Key components employed in optical interconnects. A digital signal is converted into an analog signal by the direct modulation of current *I*. Light pulses are emitted in accordance with the digital modulation. The light signal is introduced into and transmitted along a waveguide to a photodetector that absorbs the light and generates a proportional current.

recovery, and decision circuits may form part of the digital data recovery portion of the circuit. The circuitry is virtually identical to that of data optical communications. The requirements on the bit error rate are more stringent for computing applications, being at least 10^{-15}.

6.6.1 Passive Thin-Film Lightwave Circuits

Planar lightwave circuits are used to convey information by manipulating packets of photons. The idea is similar to electric circuits that make use of packets of electric charge toward the same end. Each approach has fundamental and practical strengths and limitations. Planar lightwave circuits consist of light-confining and light-guiding functions (waveguides), power dividing functions (MMI, Y-splitters), power combining functions (Y-combiners), and wavelength combining and spreading functions (waveguide phase array gratings). Ultimately, a useful lightwave circuit also includes interfaces to electrical-optical devices such as sources of light; optical amplifiers; photodetectors; and light amplitude, phase, and wavelength modulators. Thus a planar or nonplanar lightwave circuit consists of two parts, the passive lightwave circuit and the active optoelectronic components for producing, detecting, and encoding information on the light carrier. How well the optical interface between these two parts is designed and constructed determines the optical interconnect reliability, performance, and cost.

In this section we will consider only some aspects of lightwave circuits as they apply directly to optoelectronic SOP technology and will indicate sources of more general treatments along the way.

The prevailing state-of-the-art for implementing optical-digital signaling among processors is exemplified by the optical electric circuit board (OECB) [51] of which the IBM Terabus is an extension [55]. In both cases, active optical components—lasers, laser drivers, photodetectors, and associated amplifiers—are bonded to a peripheral ball grid array (PBGA) package that may eventually also contain a memory controller with a multiplexer-demultiplexer. The entire package is then flip-chip bonded to the transceiver portion of a circuit board. In this approach optical coupling from an array of VCSEL sources to the lightwave circuit is accomplished by microlens relays and waveguides with 45° end mirrors. A similar optical coupling arrangement is used for the photodetector array. The active optical alignment is largely a manual operation of pick-and-place tools, and displacement tolerances have to conform to prevailing pick-and-place and flip-chip assembly tolerances with deviations as large as ±10 µm. This approach, borrowed from Telecom optical packaging, has notable assembly drawbacks, and the scaling costs are prohibitive. A major drawback to scalability is the use of the PBGA package to carry the optical I/Os, as shown in Figure 6.11. The PBGA takes up valuable board space for relatively few optical I/Os, can only be placed a few centimeters away from the nearest processor, and limits the number of I/Os; therefore "optics to the processor" becomes limited to "optics near the processor" with few I/Os. A potentially scalable solution may be to place flexible optical I/Os peripherally around the processor by electrically pluggable means as depicted in Section V.C.

Thin-Film Optical Waveguides

The operation of thin-film waveguides is based on total internal reflection [57]. A high-index thin-film immersed in a medium with a lower index can confine and guide light provided it has certain minimum dimensions with respect to the wavelength of light. Thin-film waveguides can be patterned as channels with the use of standard photolithographic techniques and can be patterned into integrated optics devices such as power splitters and directional couplers to distribute and route signals.

It is a substantial challenge to design and fabricate high-density optical interconnects on system boards. On the other hand, optical interconnects greatly reduce the density of copper bus lines and the need for noise suppression capacitance, thus reducing the design and layout complexity and increasing the available board space for useful components. With an optimal buffer layer design [58], low-multimode or single-mode waveguides can be fabricated on a 20-µm pitch directly over metal levels on PCBs, as shown in Figure 6.13a and b. With the development at Georgia Tech of high-speed EVPDs and the availability of bare die EELs, the necessity for out-of-plane beam turning elements (and associated losses) is greatly diminished. EVPDs, along with presently

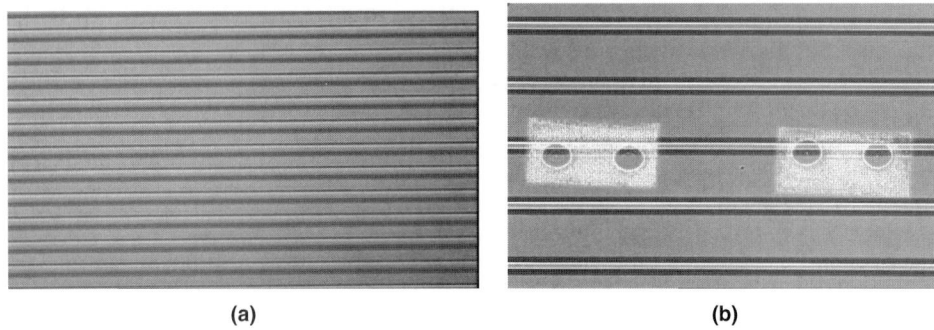

Figure 6.13 (a) Waveguide array with 10 µm wide × 5 µm thick cores on a 20-µm pitch formed on a PCB. (b) Waveguide array (50 µm wide × 25 µm tall cores on a 250-µm pitch) formed on a buffer layer over copper pads and microvias on a PCB.

available EELs, will significantly simplify integration of the optical lightwave circuit by eliminating the need for microlenses, micromirrors, and tedious active alignment and uncertain passive alignment by dead reckoning.

Thin-Film Optical Power Splitters

Optical power splitters can be useful in optical broadcast applications such as in the simultaneous delivery of short pulses to effect synchronization of logical operations either on-chip or on-board. In-plane beam turning single-mode waveguides and Y- or T-splitters can be fabricated having a turn radius of the order of 1 μm in cases for which the difference in refractive index for core and cladding is sufficiently high, as reported by Kimerling for silicon waveguides [59]. Optical broadcasts using multimode interference (MMI) devices [60] and PHASAR or arrayed waveguide grating structures [61] have previously been characterized and reported.

High-speed data transmission at 40 Gb/s through a 1×8, polymer MMI power splitter for a total throughput bandwidth of 320 Gb/s has been demonstrated [62]. The device is designed with multimode waveguides for a better input coupling efficiency for high-speed testing. The MMI splitter was fabricated using a negative tone photosensitive inorganic polymer glass (IPG) made by RPO, Pty. Ltd., Australia (core $n = 1.525$) on a silicon substrate with a 5-μm-thick silicon dioxide (SiO_2) undercladding (lower cladding $n = 1.467$) as the buffer layer. A composite image is shown in Figure 6.14. The insertion loss is 4.12 dB, and the channel uniformity is 0.21 dB. The measured NRZ BER per channel at 40 Gb/s is 10^{-10}.

The eye-opening diagram for one of the channels is shown in Figure 6.15, while the BER test results are shown in Figure 6.16.

Additional passive and dynamic structures that are useful in full lightwave circuit implementation are found in Murphy [63]. These include add/drop switches, thermo-optic switches, and nonlinear devices for optical mixing.

Thin-Film Diffractive Couplers for Optical Interconnects

Diffraction gratings can produce coupling into and out of optical waveguides. They are compact and flat, making them compatible with the planar device technology found in

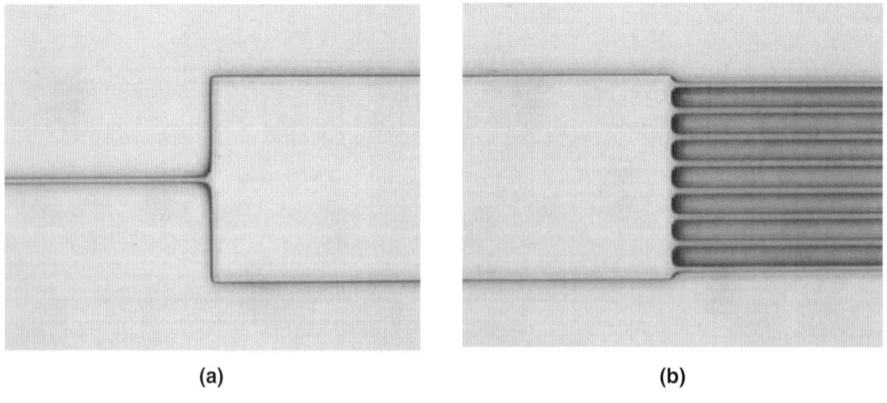

(a)   (b)

FIGURE 6.14 Photomicrograph of (a) the input facet and (b) the output facet of the 1×8 MMI splitter after development. The MMI width is 200 μm. Output waveguides are 10 μm in width with a 15 μm gap in between.

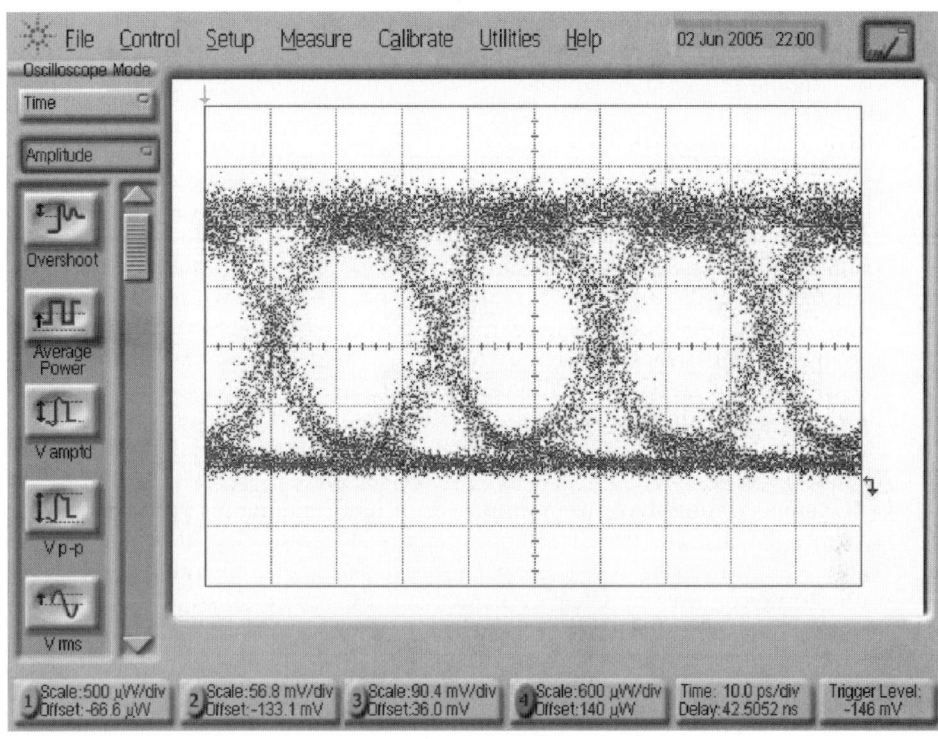

Figure 6.15 The 40-Gb/s optical eye diagram measured at channel 2.

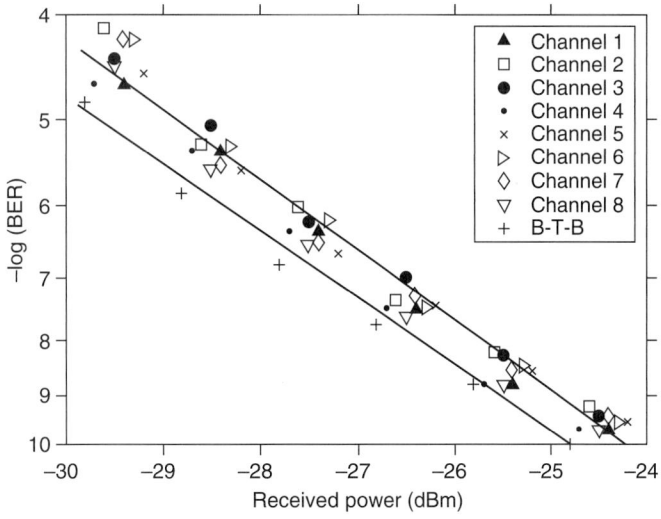

Figure 6.16 The BER testing for a $2^7 - 1$ PRBS NRZ pattern at 40 Gb/s.

microelectronics. They can be designed to focus light that is being coupled out, and in reciprocal fashion to couple in efficiently light from divergent sources. An important consideration for grating couplers is preferential coupling. Since gratings, in general, diffract light into multiple orders, it is important when making an efficient coupler to maximize the power that is diffracted in the desired direction. The ratio of the power coupled in the desired direction to the total out-coupled power is called the preferential coupling ratio.

A volume grating consists of a periodic variation in the index of refraction of a material. Its use as a coupler—volume grating coupler (VGC)—was first demonstrated by Kogelnik and Sosnowski [64] and has been shown to provide high-efficiency coupling and ease of manufacture in a compact device [65–68]. A diagram of a volume grating coupler is shown in Figure 6.17. VGCs with preferential-order coupling and the combination of preferential-order coupling and focusing have been demonstrated [69–71]. VGCs with the slanted grating fringes required for preferential-order coupling have been fabricated through interferometric exposure without the use of complicated chemical processes [72].

Volume gratings have been proposed for use in numerous optical interconnection schemes [73–81]. The use of volume gratings in free-space optical interconnects was explored in the 1980s [73–74]. Their use has also been demonstrated in substrate-mode optical interconnects by Chen et al. [82] and by Yeh et al. [83]. The use of thin-film guided-wave optical interconnects with VGCs for in- and out-coupling has also been proposed [76,81]. Recently, VGCs fabricated with polymer-dispersed liquid crystals have demonstrated electrically switchable out-coupling [84] from polymer waveguides. Focusing, preferential-order VGCs for this application also have been demonstrated [69,71]. More recently, grating-to-grating coupling between two VGCs for board-to-chip interconnects has been demonstrated [85].

The large wavelength bandpass inherent in the VGCs [86] makes them suitable for CWDM applications. The strong effect of the grating thickness on the wavelength bandpass provides some flexibility in the design of the bandpass width. WVGCs are also a robust alternative for use with low-cost sources susceptible to wavelength variations. The FWHM wavelength bandpasses found for high-efficiency couplers range from 173 to 525 nm.

Analysis The analysis of VGCs presented here is based on the rigorous coupled-wave analysis, in conjunction with a leaky-mode approach (RCWA-LM). The structures under analysis and the formulation used to perform the calculations are presented.

Analysis of Volume Grating Couplers for TE and TM Polarized Light This analysis of VGCs begins with the assumption of a multilayer waveguide-coupler structure such as the one shown in Figure 6.17. The m-th layer in the structure (excluding the superstrate and substrate) is treated as a volume grating with grating vector $\vec{K}_m = K_x \hat{x} + K_{m,z} \hat{z}$. This grating vector determines the period and slant angle of the m-th grating. Homogeneous layers can be considered as special cases of gratings with zero modulation. Guided modes supported by the coupler structure are usually leaky modes that radiate power away from the structure. This radiation constitutes the mechanism by which light is coupled out of the waveguide.

Mathematically, the radiation can be represented by a complex propagation constant for each leaky mode. The complex propagation constant is $\tilde{\beta} = \beta - j\alpha$, where β and α are both real; it is calculated using the RCWA-LM [87].

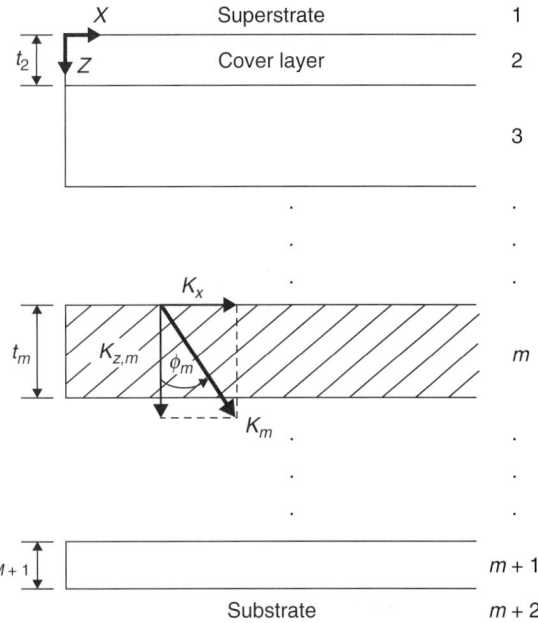

FIGURE 6.17 Schematic diagram for the analysis of coupler structures with an arbitrary number of layers and gratings.

In this analysis, the fields in each region are of the form

$$\vec{U}_m = \hat{y} U_m(z,x) e^{-j\beta x}$$

where $\vec{U}$ represents the $\vec{E}$ field for TE polarized light and the $\vec{H}$ field for TM polarized light.

For the *m-th* waveguide layer (see Figure 6.17)

$$U_m(z,x) = \sum_i S_{m,i}(z) e^{-j\vec{\sigma}_{m,i} \cdot \vec{r}}, \qquad m = 2,\ldots,M+1 \qquad (6.2)$$

where $\vec{\sigma}_{m,i} = \tilde{k}_{x,i}\hat{x} - iK_{m,z}\hat{z}$, with $\tilde{k}_{x,i} = \tilde{\beta} - iK_x$, and $S_{m,i}(z) = \sum_j C_j^m w_{i,j}^m e^{\lambda_j^m z}$,

Where $w_{i,j}^m$ and λ_j^m are the eigenvectors and eigenvalues of the fields in the *m-th* layer, and C_j^m are their respective coefficients, all obtained from RCWA analysis [87]. For the superstrate region

$$\vec{U}_1 = \hat{y} \sum_i R_i e^{-j\vec{k}_{1,i} \cdot \vec{r}} \qquad (6.3)$$

where R_i are the amplitudes of the diffracted orders in the superstrate region and the wavevectors are defined as

$$\vec{k}_{1,i} = \tilde{k}_{x,i}\hat{x} + \tilde{k}_{1,zi}\hat{z}$$

where

$$\tilde{k}_{1,zi} = \sqrt{k_o^2 n_1^2 - \tilde{k}_{x,i}^2}$$

if $Re\{\tilde{k}_{1,zi}\} - Im\{\tilde{k}_{1,zi}\} < 0$, or

$$\tilde{k}_{1,zi} = -\sqrt{k_o^2 n_1^2 - \tilde{k}_{x,i}^2}$$

if $Re\{\tilde{k}_{1,zi}\} - Im\{\tilde{k}_{1,zi}\} > 0$ [88].

For the substrate [$(M + 2)$th layer in Figure 6.17]

$$\vec{U}_{M+2} = \hat{y}\sum_i T_i e^{-j\vec{k}_{M+2,i}\cdot(\vec{r}-d\hat{z})} \tag{6.4}$$

where $d = \sum_m t_m$, with t_m the thickness of the m-th layer; M is the total number of layers in the structure (not including the superstrate and the substrate); and T_i are the amplitudes of the diffracted orders in the substrate region. The wavevectors are

$$\vec{k}_{M+2,i} = \tilde{k}_{x,i}\hat{x} + \tilde{k}_{M+2,zi}\hat{z}$$

where

$$\tilde{k}_{M+2,zi} = \sqrt{k_o^2 n_1^2 - \tilde{k}_{x,i}^2}$$

if $Re\{\tilde{k}_{M+2,zi}\} - Im\{\tilde{k}_{M+2,zi}\} > 0$, or

$$\tilde{k}_{M+2,zi} = -\sqrt{k_o^2 n_1^2 - \tilde{k}_{x,i}^2}$$

if $Re\{\tilde{k}_{M+2,zi}\} - Im\{\tilde{k}_{M+2,zi}\} < 0$ [88].

Using the electric and magnetic field components of the optical waves in the various regions of the structure and the electromagnetic boundary conditions, the problem can be cast as a matrix equation of the form $\overline{\overline{M}}(\beta)V = 0$, where $V = [RC_1...C_M T]^T$ and $\overline{\overline{M}}$ is a $2(M+1)N \times 2(M+1)N$ matrix, where N is the number of diffracted orders retained in the analysis. $R = [R_{-P},...,R_P]$ and $T = [T_{-P},...,T_P]$ are vectors of size N, where $P = (N-1)/2$. These vectors correspond to the complex amplitudes of the diffracted plane waves in the superstrate and substrate, respectively, while the C_m vectors of size $2N$ correspond to the amplitude components of the fields in the inner layers of the structure. For a nontrivial solution of this matrix equation, the determinant of $\overline{\overline{M}}$ must be zero. Thus by solving $\det[\overline{\overline{M}}(\tilde{\beta})] = 0$, where "det" denotes the determinant, $\tilde{\beta}$ can be determined. This is a numerically sensitive problem because there may be multiple $\tilde{\beta}$ solutions corresponding to various modes or nonphysical modes. The Muller method [89] of finding the complex zeros of a mathematical function has been adopted to find an approximate solution. This is feasible provided an adequate initial guess is available. The real propagation constant β of the guided mode in the case of zero modulation in all layers is usually a good starting point, since the real part of $\tilde{\beta}$ is typically close to β (within $10^{-3} \mu m^{-1}$ for index modulations less than 0.05). The Muller method is much more efficient than the sequential quadratic programming [90] techniques previously used, [72,91], thus vastly reducing computation times.

Once $\tilde{\beta}$ has been determined, it is used to calculate the power distribution of the radiation diffracted outside the coupler system. Since det $[\overline{\overline{M}}(\tilde{\beta})] = 0$, the components of $V = [RC_1 ... C_M T]^T$ can be determined only as a function of a common arbitrary constant. For simplicity, this constant can be selected to be unity. In the present case, $R_1 = 1$ (any component of V can be selected without affecting the final result). In this way the R_i and T_i amplitudes are values proportional to the first-order superstrate diffracted amplitude R_1. Then, using $\overline{\overline{M}}(\tilde{\beta})V = 0$ and eliminating this variable and one of the equations from the system, and using $\tilde{\beta}$ as the solution of this reduced system, relative values for all the R_i and T_i coefficients can be obtained. These coefficients are then used to calculate the relative power distribution among the propagating diffracted orders. Relative values for each Poynting vector (PV) are calculated according to

$$PV_{sup,i} = |R_i|^2 \, Re\{-k_{1,zi}\}$$

for reflected orders (into the superstrate), and

$$PV_{sub,i} = |T_i|^2 \, Re\{k_{M+2,zi}\}p$$

for transmitted orders (into the substrate), where $p = 1$ for TE and $p = n_1^2 / n_{M+2}^2$ for TM polarization. Next, the relative values of nonpropagating orders $|Re\{k_{x,i}\}| > k_o n_1$ for superstrate and $|Re\{k_{x,i}\}| > k_o n_{M+2}$ for substrate orders) are set to zero. The relative power calculation is completed by normalizing the relative powers of the propagating orders with respect to their sum. The fraction of the out-coupled power that is directed into the desired order is called the preferential coupling ratio and is designated $\eta_{\ell,i}$, [69] where ℓ can be either the superstrate (sup) or the substrate (sub), and i is the diffracted order. The resulting equation is

$$\eta_{\ell,i} = PV_{\ell,i} / \left(\sum_i PV_{sup,i} + \sum_i PV_{sub,i} \right)$$

The coupling efficiency ($CE_{\ell,i}$), defined as the fraction of the guided power P_o at the beginning of the coupler that is diffracted into the desired order after a certain length L, can be calculated once $\tilde{\beta}$ and the power distribution are known, and it is given by [69]

$$CE_{\ell,i} = \eta_{\ell,i}(1 - e^{-2\alpha L}) \qquad (6.5)$$

Design The design method to obtain grating vectors for a given structure is presented next. Also, a detailed fabrication procedure for VGCs is presented, including the design of the fabrication configuration, its alignment, sample preparation, exposure process, and curing.

Grating in the Waveguide versus Grating in the Cover Layer VGC structures have been reported in the two configurations shown in Figure 6.18a and b. Both structures are based on a glass substrate and have an air superstrate. In both cases, light in the waveguide is incident (from the left) on the grating coupler. In the configuration of Figure 6.18a, the VG is located in a layer adjacent to the high-index waveguide layer ("VG in cover layer" configuration). In the configuration of Figure 6.18b, the VG is embedded in the waveguide layer ("VG in waveguide" configuration). The structure in Figure 6.18a can be designed to support several leaky modes, with only one confined to

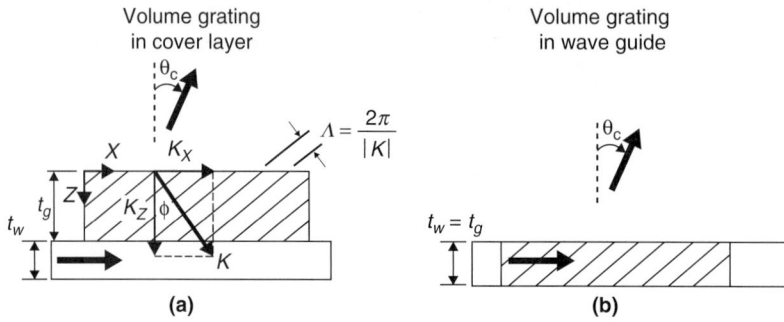

Figure 6.18 Diagrams of the two VGC configurations discussed in this thesis. (a) Volume grating in the cover layer. (b) Volume grating in the waveguide. The grating vector **K**, as well as the period Λ and slant angle ϕ, are shown. The outcoupling angle is θ_c. The thickness of the grating layer is t_g, and the thickness of the waveguide layer is t_w.

the waveguide region, while the structure in Figure 6.18b can be designed to support only the fundamental leaky mode at the design wavelength λ_o.

Grating Vector Design Given the indices of refraction of the substrate, waveguide, and grating layers, and the superstrate, as well as the thicknesses of the waveguide and grating layers, the desired grating vector components can be determined once the out-coupling angle is chosen.

The out-coupling angle θ_c is given by

$$\theta_c = \sin^{-1}\left(\frac{\beta - K_x}{k_o n_{sup}}\right) \qquad (6.6)$$

The value of K_x is thus set for a given β (which in turn depends on the structure parameters) and outcoupling angle. In the case of focusing couplers, K_x must vary along the length of the coupler to produce the various outcoupling angles that result in a focusing output beam. The $K_x(x)$ needed to focus to a line at $(x = x_f, z = z_f)$ is given by [72].

$$K_x(x) = \beta + \frac{k_o n_{sup}(x - x_f)}{\sqrt{(x - x_f)^2 + z_f^2}} \qquad (6.7)$$

Since the K_x component of the grating vector is used to determine the out-coupling angle, only the K_z component is left to optimize the strength of the coupling [72]. Unlike the case of bulk diffraction where the incident wavevector is unambiguous, the case of guided-wave diffraction has three obvious options for the incident wavevector to be used in determining the Bragg condition. The first is the propagation constant β of the mode being coupled, and the second and third options are the upward and downward propagating plane wave components of the mode. It has been found that the Bragg condition using the propagation constant β of the mode produces the most efficient coupler at the design wavelength. This case can be called the guided wave Bragg condition (GWBC). This differs from the well-known Bragg condition for bulk diffraction in that

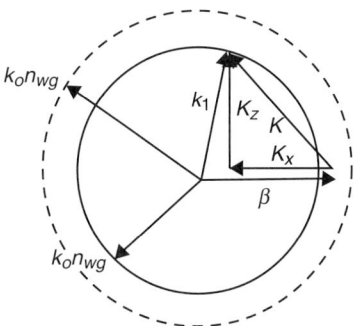

FIGURE 6.19 The guided wave Bragg condition (GWBC).

the incident wave in this case is the mode guided in the waveguide. Figure 6.19 shows a diagram of the GWBC. The equation describing the GWBC is

$$K_z = \sqrt{(k_o n_g)^2 - (\beta - K_x)^2} \tag{6.8}$$

The HRF-600X material used for fabricating the gratings described here incurs a shrinkage of $\delta = 3\%$ during exposure and curing. This does not noticeably affect the K_x component of the grating because of its large dimensions in the plane. However, the K_z component is affected, and after exposure and curing, $K_z = K_{z,orig}/(1-\delta)$, where $K_{z,orig}$ is the originally recorded K_z component. Therefore, precompensation is needed for this shrinkage. This is accomplished by recording $K_{z,orig}$ to obtain K_z as the final result.

This procedure can be used to design couplers with any out-coupling angle. It can be used in the design of focusing couplers and polarization-dependent and polarization-independent couplers.

Volume Grating Coupler Fabrication In this section all the phases of volume grating fabrication are described. First, the design of the interferometric recording configuration used to fabricate the gratings is presented. Next, the alignment procedure for the recording configuration is covered. Then, the sample preparation and recording are described. Finally, the postexposure curing of the VGCs is presented.

Grating Recording Configuration Design In order to fabricate the VGCs described here, the interferometric recording configuration shown in Figure 6.20 is used. Light from a single-line 363.8 *nm* laser is spatially filtered and then collimated to obtain a uniform phase-front. The light is then redirected to a polarizing beam splitter that is preceded by a first half-wave plate and then followed by another half-wave plate. This configuration produces two beams whose power ratio can be accurately controlled. These two beams are then redirected with mirrors toward the recording sample. A prism must be placed in front of the sample in order to be able to achieve the angles between the beams that are required to produce the K_x and K_z values discussed above. Antireflection-coated fused silica prisms are used for this purpose. Finally, another prism is placed behind the sample in order to minimize reflections from the sample-air interface that could affect the pattern being recorded. A photograph of the interferometric recording configuration is shown in Figure 6.20. With fixed recording wavelength and refractive indices of the samples and prisms, the angle between the two recording beams and the angle of rotation

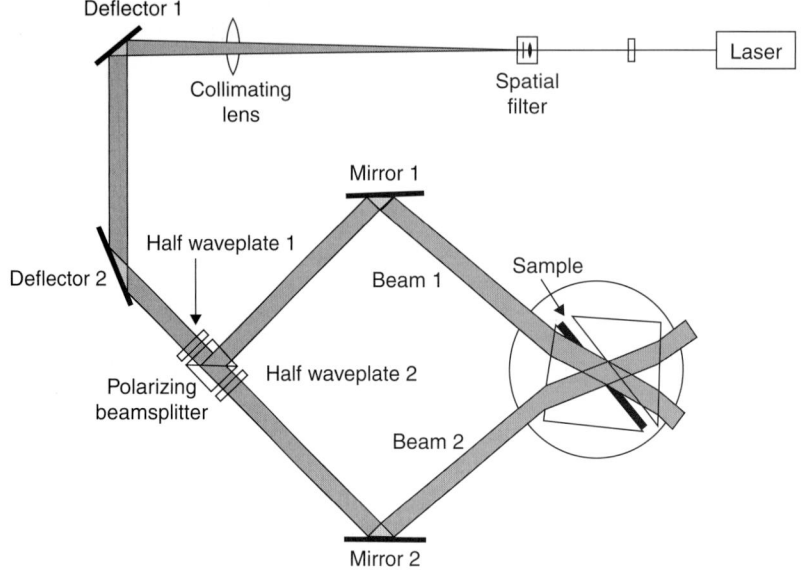

FIGURE 6.20 Diagram of the grating recording configuration.

of the sample (with respect to these two beams) that produces the required K_x and K_z can be determined. First, the wavevector components of the two interfering beams required to produce the grating vector $K = K_x + K_z = k_{1,w} - k_{2,w}$ are calculated, where $k_{1,w}$ and $k_{2,w}$ are the wavevectors of the two writing beams.

From the preceding equation, it is surmised that

$$K_x = k_{x1,w} - k_{x2,w}$$

and

$$K_z = k_{z1,w} - k_{z2,w}$$

It is also known that $|\vec{k}_{1,w}| = |\vec{k}_{2,w}| = k_{o,w} n_{g,w}$, so

$$k_{o,w}^2 n_{g,w}^2 = k_{x1,w}^2 + k_{z1,w}^2$$

and

$$k_{o,w}^2 n_{g,w}^2 = k_{x2,w}^2 + k_{z2,w}^2$$

With these four equations and with the design values of K_x, K_z, for the recording wavelength of $\lambda_w = 363.8$ nm and the index of the grating material at the recording wavelength of $n_{g,w} = 1.535$ [72], the following wavevectors are determined: $k_{x1,w}$, $k_{z1,w}$, $k_{x2,w}$, and $k_{z2,w}$.

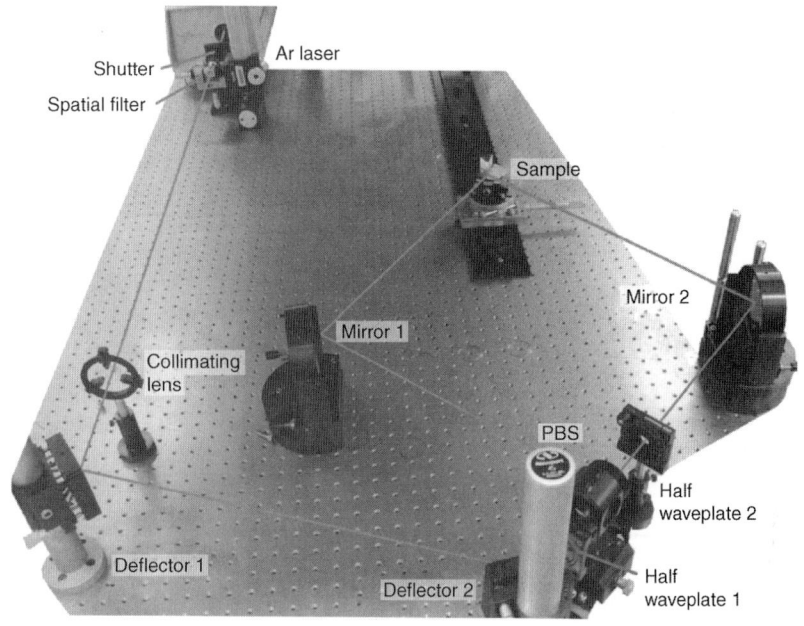

FIGURE 6.21 Photograph of the grating recording configuration. The path of the laser beam has been traced over the image.

From these wavevectors, the incidence angles of the recording beams are

$$\theta_{g,1} = \tan^{-1}\left(\frac{k_{x1,w}}{k_{z1,w}}\right), \tag{6.9}$$

and

$$\theta_{g,2} = \tan^{-1}\left(\frac{k_{x2,w}}{k_{z2,w}}\right). \tag{6.10}$$

By applying Snell's law at the grating-air interface, it is found that sufficiently oblique angles of incidence in the grating material cannot be produced from air. Thus, a prism is used so that the beams with the necessary incidence angles can impinge on the sample. A right angle, fused silica prism, with an antireflection coating, is used. The index of the prism at the recording wavelength $\lambda_w = 363.8\ nm$ is $n_p = 1.47$. Figure 6.22 shows a diagram of the air-prism-grating configuration. Applying Snell's law at the grating-prism interface, the needed beam angles in the prism, $\theta_{p,1} = \sin^{-1}[\sin(\theta_{g,1})n_{g,w}/n_p]$ and $\theta_{p,2} = \sin^{-1}[\sin(\theta_{g,2})n_{g,w}/n_p]$, are obtained. From the geometry of the prism and Snell's law, the following angles for the air-prism interface are obtained: $\theta_{air,1} = \sin^{-1}[n_p \sin(\theta_{p,1} - 45°)]$ and $\theta_{air,2} = \sin^{-1}[n_p \sin(45° - \theta_{p,2})]$. The sum of these two angles is the total angle between the two beams, $\Delta\theta = \theta_{air,1} + \theta_{air,2}$, while the difference between the two is the rotation angle of the sample with respect to the bisector of the two beams, $\psi = (\theta_{air,1} - \theta_{air,2})/2$.

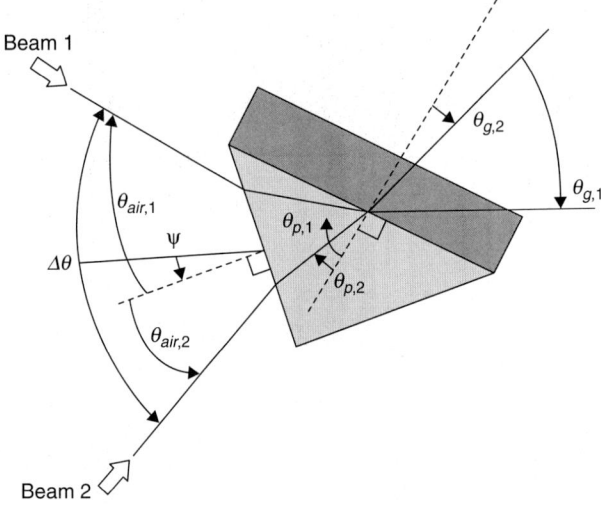

FIGURE 6.22 Diagram of the air-prism-grating configuration. Use of the prism allows the recording of gratings with slant angles that are not possible with an air-grating interface alone.

Once the proper design of the recording configuration has been determined, the alignment of the recording configuration and the recording of the grating can be performed.

6.6.2 Active Optoelectronic SOP Thin-Film Components

The performance of edge-emitting lasers, VCSELs, and photodetectors has greatly advanced over the past 10 years in bandwidth and wavelength selection, while laser reliability has also improved. The geometrical relationship between electrical contacts and the light propagation direction of optoelectronic devices is based on their historical evolution. This geometrical relationship is detrimental to the easy integration of OE components and the passive lightwave circuits. Suitable, off-the-shelf OE devices have electrical contact pads that lie in a plane perpendicular to the light propagation direction. This forces the integrator to design, fabricate, and assemble light beam turning components such as micromirrors (and collimating lenses) so as to cause the light beam to propagate in a plane that is parallel to the plane containing the electrical contact pads.

Would the integration process not be easier if the plane of the electrical contact were to be parallel to the beam propagation direction? In that case, EELs and EVPDs could be simply end-coupled to waveguides, and micromirrors and microlenses would become unnecessary. It is in fact this guiding principle that guides the Georgia Tech Interface Optical Coupling (IOC) approach to optoelectronic SOP integration [91]. In IOC integration, the interface fabrication process is kept as simple as possible at the expense of developing high-performance EELs and EVPDs at wavelengths of optimal transmission, generally between 850 and 980 nm for most optical polymer waveguide materials.

The ultimate goal of SOP is to achieve the highest possible functionality in the smallest physical volume. This requires the development of nanotechnology and nano-integration processes of which thin-film technology is a subset. Aside from the human-machine interface, there is no physical reason why today's laptop capabilities cannot be fitted in a volume as large as a button on one's shirt.

The electrical and physical gate length of CMOS transistors has been in the low nanometer range for at least two decades, and attempts to stack CMOS transistors in the third dimension and interconnect them with nanometer-size electrical wiring has been summarized in [92]. Clearly, the main challenge will be that of thermal dissipation.

In his pioneering work, Mitsumasa Koyanagi has developed parallel optical interconnects in three dimensions at the chip level for rapid access to high-density CMOS memory [93].

Heterogeneous integration of CMOS transistors, thin-film MSM photodetectors, and thin-film edge-emitting lasers has previously been reported [94] and then subsequently in [95–97].

Recently, thin-film VCSEL arrays have been fabricated [98–100] by an etch lift-off process similar to that used for thin-film MSM and thin-film EEL lift-off, [94] demonstrating the feasibility of wafer-level heterogeneous thin-film integration.

The main challenge to the realization of practical 3D optoelectronic SOP, at the chip level by thin-film stacking, remains dissipation of heat that is generated by CMOS transistors, which can be used both for signal processing and as laser drivers and photodetector amplifiers.

Free-space or holographic optical signaling is not discussed here because this technology has not yet gained a foothold in chip-to-chip signaling or board-to-board signaling, whereas it has found on-chip application [101].

Instead we will discuss only guided wave optics and its application to high-speed interchip and interboard optical signaling. We will first review the general optical properties of the most promising and most reported optical polymer materials. Fabrication processes for optical lightwave circuits will be touched upon in general terms along the way.

6.6.3 Opportunities for 3D Lightwave Circuits

In 2D optical interconnects it is generally understood that the optical signal is transported in a plane that contains both the transmitter and receiver modules as exemplified by the Fraunhofer Institute's optical electric circuit board (OECB) [51] and the IBM Terabus [55]. In 3D optical interconnects the transmitter and receiver modules lie in generally orthogonal planes or on widely separated parallel planes as exemplified in [102–103], and in Figures 6.5 to 6.7 and 6.23c. The optical backplane, which resembles the familiar copper backplane, and alternative large-scale signal distributions are well summarized in [15,41,104] and references therein.

We make only the following passing comments regarding the optical backplane approach. (1) If the optical backplane is purely passive, then lasers, drivers, PDs, and PD amplifiers have to be placed on the circuit board along with existing components. (2) If conventional optical interface technology is used in addition (e.g., micromirrors and microlens relays), then on-board alignment challenges still exist as they do for the EOCB and Terabus technologies. At the backplane, the alignment problems can be shifted to passive coupling such as MT connectors in which precision translates into cost and restricted scalability [105].

The Georgia Tech approach shown in Figure 6.25, uses polymer optical waveguides and the GT IOC integration process, but in one critical area the strategy is similar to the NEC chip-to-chip or board-to-board parallel optical interconnect approach that uses optical fibers [26]. The common strategy is in the interface to the chip or board that is electrical, pluggable, and high speed. Unlike NEC, however, the GT IOC method uses no micro-optical components but relies on the IOC process to interface optical waveguides to EELs and PDs. The GT optical transceiver contains prealigned laser and PD arrays and no MT connectors are used anywhere. The pluggability is at each end of the transceiver and is to be accomplished by the use of high-speed pins.

The entry of "optics inside the box" may not be in the form of rigidly embedded lightwave circuits on printed circuit boards that optically transport digital signals from chip-to-chip in a cointegrated optoelectronic circuit layer, as is envisioned by the OECB [51] or Terabus [55] strategy. The best solution for chip-to-chip and board-to-board optical interconnects may be in the form of an electrically pluggable, thin-film flexible, active optical transceiver comprising a flexible array of waveguides with embedded optical actives such as lasers (VCSEL or EELs) and photodetectors (EVPDs or conventional PINs). This approach, first published by Yoshikawa at NEC [26] using optical fibers, 45° mirrors, and VCSELs, is shown in Figure 6.6. An equivalent embodiment is sketched in Figure 6.23a and b for the case of the Georgia Tech IOC integration process. Since bending losses can be negligible [106], it becomes quite feasible to implement 3D optical interconnects by simply using flexible, pluggable optical lightwave circuits (transceivers) to connect stacks or rows of multiprocessor boards, as in blade servers or supercomputer boxes. What is important is to go beyond Yoshikawa's pioneering work by designing scalability, mass manufacturability with high optical interconnect density, and channel bandwidth in such a way that flexible optical interconnects are reliable, scalable, and inexpensive.

The GT IOC integration process promises to be inexpensive because optical alignment, the most troublesome aspect of optical interconnects, is trivial and is accomplished during the waveguide fabrication process on multiple channels simultaneously. In addition, the

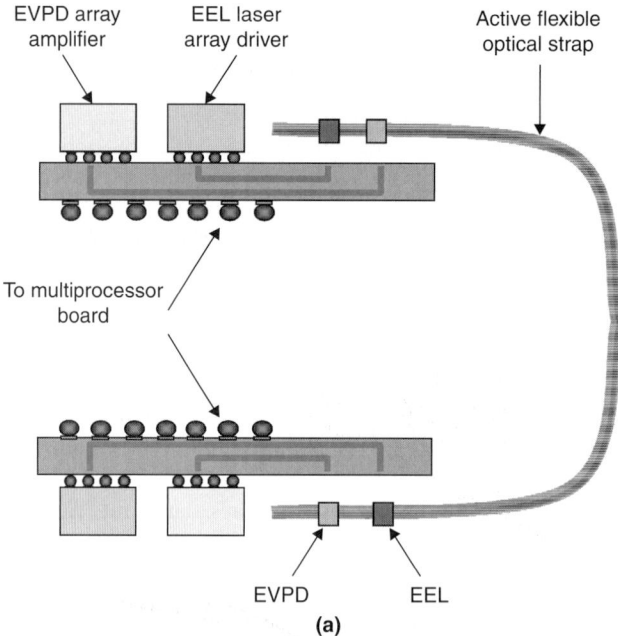

Figure 6.23 (a) Schematics of the interboard (intraboard) transceiver showing the basic components: the laser array driver, the photodetector array receiver means for electrical connections next to the processor, and a detachable optical strap with embedded laser and photodetector arrays for easy field repair and upgrade. (b) Processor peripherally surrounded by flexible optical interconnects. Using conventional polymer waveguide technology, existing EEL or VCSEL technology can be extended to a pitch of 125 μm and easily aligned by the IOC method. (c) An example of the application of a flexible, 3D optical interconnect in high-speed communication between two multiprocessor boards with external memory.

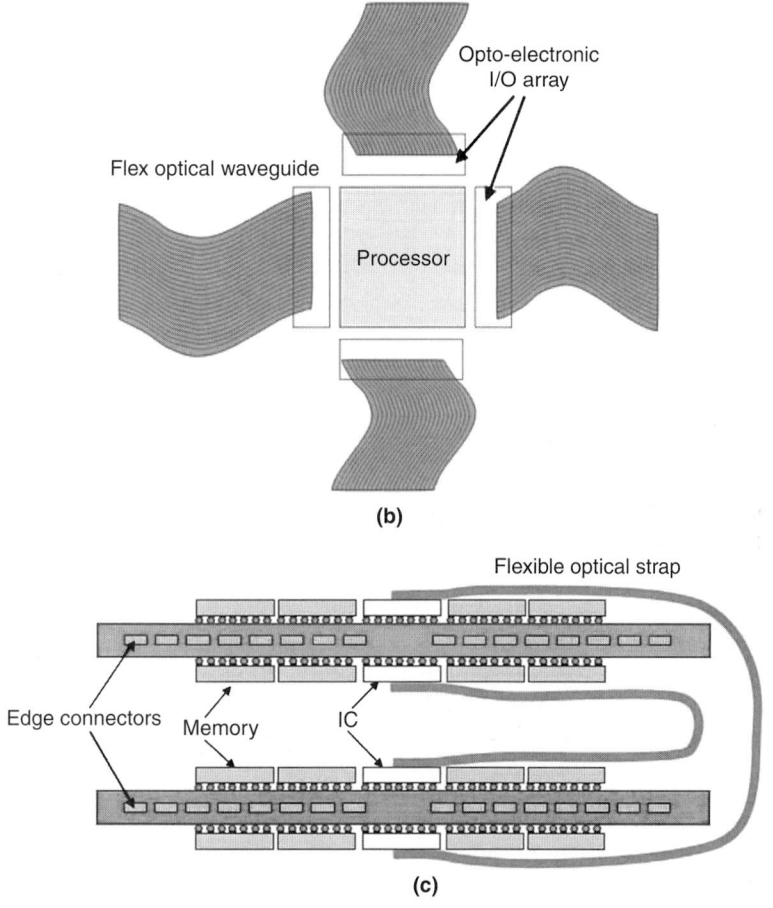

Figure 6.23 *(Continued)*

optical strap that contains embedded waveguides prealigned to lasers and detectors is assembled as a separate part and electrically interfaced to the transceiver module, as shown in Figure 6.23a. Each end of the transceiver is designed to be electrically pluggable to a board location, as shown in Figure 6.23c.

6.7 SOP Integration: Interface Optical Coupling

The interface between optical waveguides and OE active devices and passive optical fibers occupy the most critical position in the integration hierarchy. As can be seen in Figures 6.6, 6.10, and 6.11, fabrication of optical interconnects by the industry and research institutions at large follows the telecom industry practices of discrete component integration for linking the passive lightwave circuit with optical actives by the free-space propagation of light and the use of micromirror and microlens relays. These additions to the lightwave circuit are designed to change the direction of propagation of light and to collimate or focus light. Inevitably, substantial manual intervention is required to affect the optical alignment of the

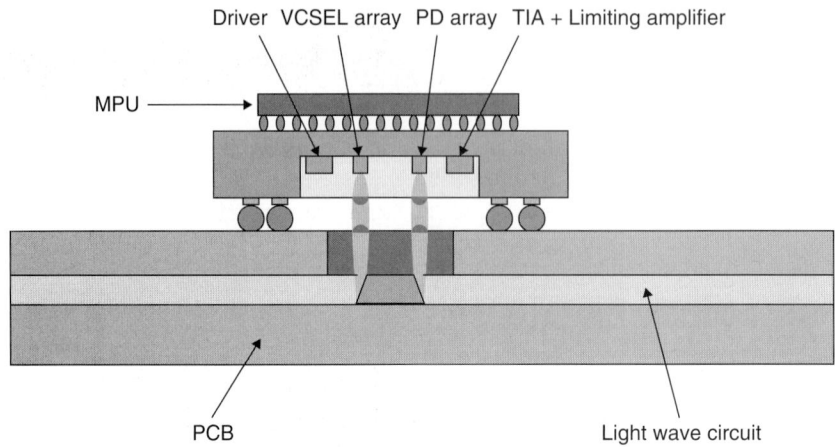

Figure 6.24 Concept sketch of the cross section of an EOCB being developed by a number of computer companies in order to bring "fiber to the processor." The board is depicted to contain an embedded, passive lightwave circuit that carries optical signals to and from a chip that is not a processor. See Figure 6.11. Because of the choice of VCSELs and top- or bottom-viewing PDs, a 90°, out-of-plane beam turn is required and is generally accomplished by using 45° end mirrors and collimating lenses. Optical alignment is usually a main concern.

added components. The results are high insertion loss, uncertain alignment stability, and high production cost. The sketch in Figure 6.24 is representative of the industrywide approach and many publications can be found on this integration strategy, for example, [26,42–43,46–47,49,51,53,107]. One exception in this trend is the use of evanescently coupled thin-film MSM photodetectors reported in [94–97].

In contrast, the guiding principle of the IOC process is to make the optoelectronic integration process simple and elegant as sketched in Figure 6.25. Thus, edge-emitting lasers and edge-viewing PDs are used when available. Alternatively PIN top-viewing photodetectors can be used provided one also provides a 45° end mirror on the waveguide directly over the PIN active area (see Figure 6.28). In either case the optical alignment is accomplished automatically and simultaneously for one or multiple optical channels during the waveguide fabrication process, and light never leaves the waveguide until detected. The IOC process proceeds generally as follows. A spacer or buffer layer is first spun onto a board or laminated thereon, a lower cladding layer is formed, and EELs and EVPDs (or PIN PDs) are placed as needed. A core polymer layer is then spun on. At this point a lithographic mask is used to define the path of the

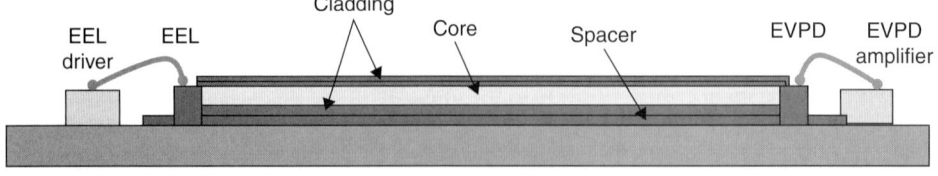

Figure 6.25 In the IOC optoelectronic integration process, the core polymer forms an interface with the EEL waveguide front facet and an interface with the active area of a conventional PD or, preferably, with the active area of an EVPD. The optical alignment occurs during the core formation process by aligning all components through a waveguide mask.

waveguide core. By aligning the EEL emission waveguide with the active area of the EVPD or PIN PD and exposing it to UV light, optical alignment and coupling are automatically achieved between the emitter and detector. This works for one or many channels simultaneously.

At Georgia Tech, we have developed the IOC process in order to keep the integration simple and elegant. Ideally, no micromirrors or microlenses are necessary. The primary consideration is to deliver the optical signal from laser to detector without light ever leaving the optical waveguide using a strictly in-plane lightwave circuit. From the point of view of low-cost manufacturability using standard PCB processes, low insertion loss, and high modulation depth, it is advantageous to lithographically align and end couple edge-emitting lasers and edge-viewing photodetectors directly into the lightwave circuit without having to construct out-of-plane beam steering devices. EELs are widely available, while EVPDs are in development at Georgia Tech.

Consistent with the IOC integration process, GT has demonstrated a single end-to-end optical channel operating at 10 Gb/s using a directly modulated DFB edge-emitting laser transmitter at 1310 nm, end-coupled to a waveguide and communicating, in the absence of available EVPDs, with a top-viewing PIN PD coupled to the same WG by a 45° WG end mirror. The PD, TIA, and limiting amplifier (LA) (Figure 6.26), and the laser and

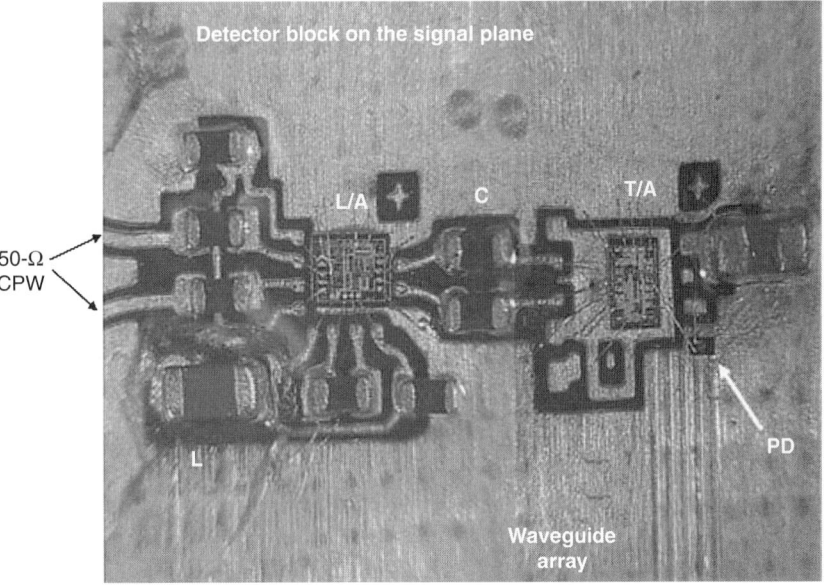

FIGURE 6.26 Top view of the detector block in the digital optical testbed. *C* are surface-mount capacitors of which there are eight. Some are used for signal coupling and some as dc power filters in conjunction with the surface-mount inductor *L*. PD indicates the photodiode (250 µm × 460 µm × 150 µm thick) that is coupled to the transmitting laser by one of the five waveguides in the indicated array. TIA is the transimpedance amplifier (1.3 mm × 0.86 mm) and LA is the limiting amplifier (1.1 mm × 1.3 mm). All components are wire-bonded to pads in the signal plane. The electrical output of the detector block is measured via 50-Ω coplanar waveguides and board-mounted SMA connectors. The top surface of the board, the signal plane, is gold plated, hence the yellowish appearance. The waveguide structure consists of a lower cladding and waveguide core (sometimes with top cladding), fabricated on a polymer buffer layer.

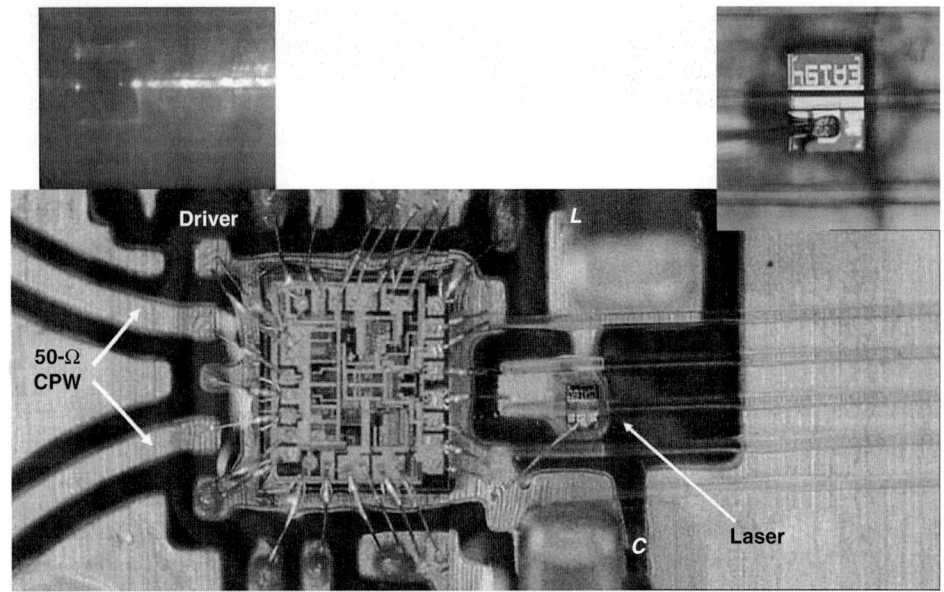

FIGURE 6.27 Top view of the laser driver block in the signal plane showing placement of the laser, laser driver, and surface-mount filter components L (inductor) and C (capacitor). Wire bonding is used to connect components to signal plane pads. Differential 50-Ω coplanar waveguides provide the drive signal to the laser driver from an external source via SMA edge connectors. Laser dimensions are 254 μm × 250 μm × 100 μm. Driver dimensions are 1.6 mm × 1.6 mm. The thickness of the copper metallurgy is 60 μm. Note that the buffer layer planarizes the 60-μm deep insulation gap between the laser pad and the ground plane, allowing for a smooth surface on which to construct the waveguide array. Light coupling from the laser to the waveguide is shown in actual operation as the infrared image in the upper left inset. A visible image is shown in the upper right inset where the wire-bond connection to the input pad is also shown. [108]

laser driver (Figure 6.27) are integrated on the same FR4 substrate as the waveguides. A single-ended random bit stream is input to the laser driver via SMA edge connectors and on-board coplanar waveguides (CPWs). The electrical output from the LA is analyzed by a digital communication analyzer via an output CPW and output SMA edge connector [108]. The GT testbed consists of an FR4 polymer composite material that is laminated with copper planes to form four levels of metal interconnects: signal plane, ground plane, power planes, and backside ground plane. The receiver and transmitter blocks are shown in Figures 6.26 and 6.27, respectively. Details can be found in [108] and references therein. The end mirror coupling the WG to the PD is shown in Figure 6.28.

Figure 6.29 shows the electrical eye-opening diagram at 10 Gb/s, obtained with an HP 83480A digital analyzer, for the single ended output of the detector block. The 1310-nm light is generated by the laser block shown in Figure 6.27.

In order to demonstrate that the Georgia Tech IOC process is particularly well suited for the simultaneous alignment of multiple optical channels, the output waveguides from a 1 × 4 polymer multimode interference (MMI) optical power splitter were simultaneously aligned to the four receiver photodiodes during the fabrication of the MMI itself. The MMI was constructed on a silicon wafer in which the photodetectors were embedded in deep trenches etched in the same silicon wafer [109]. Wire bonding

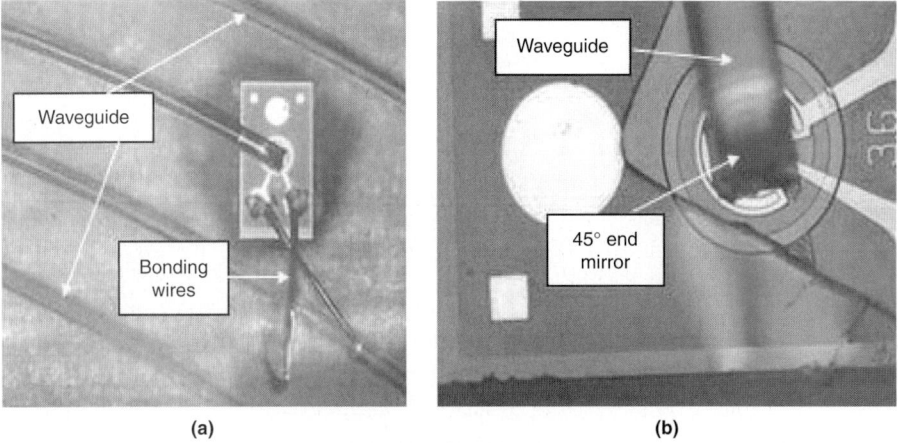

FIGURE 6.28 View of waveguide coupling to a 150-μm-thick PD. (*a*) Panoramic view of the PD placement in one of the waveguides in the array of five. (*b*) Magnified view of the waveguide end mirror on the PD. The waveguides are 50 μm wide × 60 μm tall and have no top cladding. They are formed by spin-on polymer and lithography, and sit on a lower cladding polymer layer, which sits on a buffer layer formed on top of the metallurgy. [108]

then connected each PD output on the silicon wafer to each optical amplifier block on an FR4 board with SMA-type edge connectors used to input the signal into a digital signal analyzer. Acceptable eye opening at 10 Gb/s is reported [109]. Figure 6.30 shows a top view of the interface between the optical channel on silicon and the electrical channel on FR4. Figure 6.31 shows a top view of the four optical waveguides, the four PDs, the four PD amplifier blocks, and the four output coplanar waveguides.

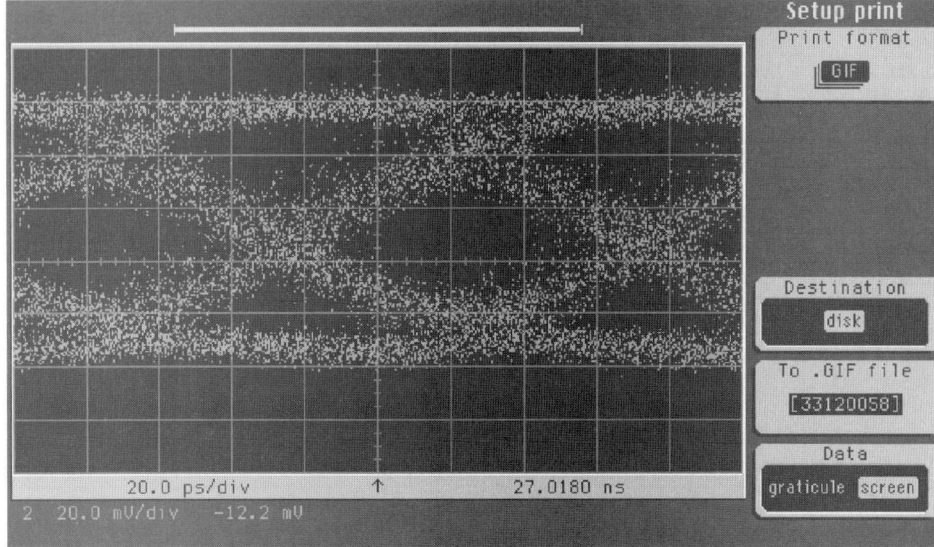

FIGURE 6.29 Single-ended, 10-Gb/s transmission eye measurement.

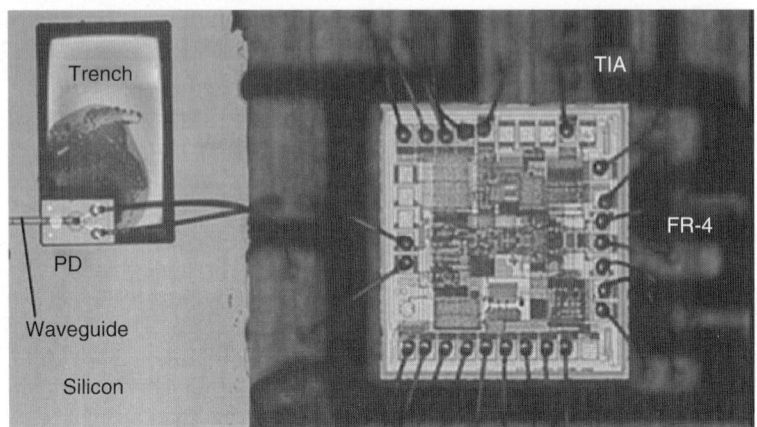

Figure 6.30 Close-in top view of the optical coupling between a waveguide and PIN PD on silicon and between the PD and a first-stage transimpedance (TIA) amplifier on FR4 board.

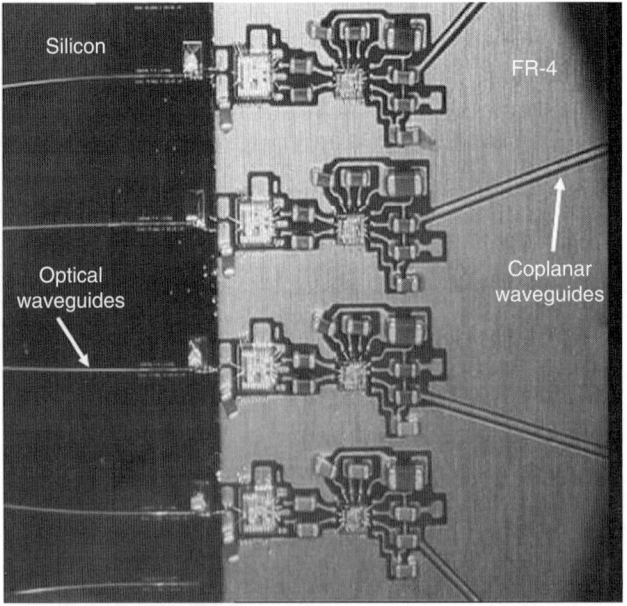

Figure 6.31 Panoramic view of the four optical waveguides simultaneously aligned to four PDs on silicon. Each PD is wire-bonded to a TIA first-stage amplifier and a second-stage limiting amplifier that form each detection block. Each digital output is carried to an SMA connector at the edge of the board by a coplanar waveguide. Each single-ended output is analyzed by a digital communication analyzer. [109]

6.8 On-Chip Optical Circuits

On-chip passive optical circuits and photonic structures are becoming commercialized by a number of start-up companies. For example, Lightsmyth is developing silicon holographic nanostructures to produce spectral comparators, optical multiplexers, and dynamically reconfigurable planar diffractive structures on a chip [110].

Luxtera is working on CMOS-compatible optical modulators and on coupling fibers directly to photonic circuits on a chip [111].

NEC has successfully demonstrated the integration of a nano-photodiode and small amplifier on a silicon chip. In addition, NEC has demonstrated WDM multiplexers and demultiplexers that occupy only 100 μm^2 on-chip and multiwavelength transmission through a 1-μm silicon waveguide. These innovations can lead to unprecedented enhancements in data processing on a chip and between processor cores. These results were reported at the International Solid State Circuit Conference in 2006. Integration of laser sources and laser drivers, however, was not discussed [112].

Intel supports an extensive silicon photonics research program [113]. An example of Intel's pioneering work is the silicon optical modulator based on the MOS capacitor [114]. Even though this particular modulator has a very small aperture and a high capacitance that restricts modulation to the low gigahertz, it nevertheless demonstrates the principle of CMOS compatibility with optical functions. A second example is the all-silicon Mach-Zehnder modulator [115]. Even though this device has a high insertion loss and low depth of modulation, it is reported to have a modulation frequency of 10 Gb/s and represents a significant contribution to CMOS integration of an optical device.

An electrically driven, directly modulated, silicon-based laser is the key optical function that is eluding the realization of a completely CMOS compatible electronic-photonic chip for optical routing and optical encoding. Nevertheless, progress is being made by Intel, having demonstrated the operation of a silicon Raman laser or waveguide amplifier [116]. While not electrically pumped and rather long, the waveguide Raman laser is a prototype example to inspire continuing work. As for completing the optical circuit with CMOS-compatible detection, the concept of receiverless optical detection for synchronizing pulse distribution [117] merits further exploration.

On-chip optical circuits are most probably suitable for optical signal encoding, routing, multiplexing-demultiplexing, and for low-skew, low-jitter optical clock distribution at rates above 10 GHz utilizing CMOS-compatible MMI power splitters. In addition to the customary waveguide material such as silicon, silicon nitride, III-V compounds, the chalcogenide glasses offer the opportunity to define lithographically waveguides and other passive optical elements that have been little explored [118–119].

For packaged processors, the possible use of optics is thought to become necessary when clock synchronization integrity is no longer achievable by digital means. At what clock frequency this will occur (10 or 50 GHz) is not clear since the multicore alternative and architectural efficiency can achieve higher performance at slower clock rates. The question of bringing an optical clock signal onto the processor that is packaged and in thermal flux may be a formidable one. While a laser source can be placed nearby, numerous obstacles have to be overcome, such as

1. Do passively mode-locked, cooled semiconductor lasers have sufficient stability and reliability? Should one use an external modulator?
2. Does the clock signal need to be amplified as it travels through waveguides on the processor? How many times? Does light have to go off-chip to be amplified?

3. How are the SiGe detectors and TIAs going to be integrated on the chip? How many are needed?
4. What is the expected timing skew? Presumably the synchronizing optical pulses are generated somewhere on the motherboard and carried to the processor via an optical fiber or waveguide. Then the task is to efficiently couple the off-chip optics to the on-chip clock signal distribution through the processor package.
5. How is the optical clock signal going to be coupled to what optical transport mechanism on the chip? The authors of the work cited in [120] and [121] demonstrate one approach that uses tapered waveguide–to–phononic crystal coupling. Another method for optical signal distribution on-chip is to use silicon waveguides, as shown by Takeshi Doi at Hiroshima University [122] and later by Kimerling [59]. Free-space, holographic optical signal distribution was demonstrated by Psaltis in conjunction with an optically programmable FPGA system that has applications in database search and optical image processing [123].
6. What is the thermal overhead of introducing active optics, photodetectors, amplifiers, and possibly optical amplifiers on a processor?

Early experimental investigations demonstrate the feasibility of on-chip optical signaling, but also point out great technical difficulties. Goossen has used heterogeneous integration to bond VCSEL arrays on digital chips [124], in addition to heterogeneous integration of multiple quantum well (MQW) modulators on-chip [125], while Koyanagi has used the chip thinning and stacking method to build a 3D digital and optical integrated structure [126]. More recently, Intel has demonstrated the operation of a MOS-based capacitor optical modulator [127].

Because of interconnect scaling [128], global and medium interconnects incur substantial signal propagation delay due to increasing resistance and capacitance [129]. This problem is partially remedied by segmenting long interconnects and inserting repeaters [128], particularly for clock signal distribution. While optimally spaced repeaters do minimize thermal dissipation and the rate of increase of latency with technology node, the 2005 ITRS projects global signaling delays of nearly 1 ns/mm by 2013. A related growing problem is that of processor temperature. Clock signal distribution accounts for approximately 50 to 60 percent of the dynamic power dissipated by a processor [130]. Processor temperatures are sensed internally and are used to slow the clock rate or decrease the supply voltage; both remedies greatly compromise processor performance. Bringing optical signal distribution may not help the thermal situation. On the other hand, if global signaling were to be transferred to an off-chip optical interposer package that sits between the processor and the package wiring redistribution, this would, in one stroke, greatly decrease global signaling delay and cool the chip by removing two main sources of heat: the global clock distribution and global signaling from the chip and onto the optical interposer. The optical interposer can consist of lasers, laser drivers, PDs, and PD amplifiers, and draw its power from the board supply. The net result is optical cooling of the processor and substantially reduced latency.

Finally, a Cornell group stacked SOS chips containing amorphous transistors and demonstrated 3D optical communication through a stack of chips by using homogeneously integrated MSM PDs and heterogeneously integrated VCSELs.

FIGURE 6.32 A stack of silicon-on-sapphire [131] chips communicate optically by "free-space" optical transmission through the transparent sapphire substrate stack.

6.9 Future Trends in Optoelectronic SOP

For the foreseeable future, data processing will continue to be digital and silicon-based CMOS technology will continue to improve by reducing leakage current, increasing gate dielectric constant, and decreasing its thickness to a single atomic layer, and by reducing the resistance of contact metallurgy. Along with the above improvements, advances in the on-chip and off-chip hardware, architectural efficiency, and improvements in the instruction set and programming efficiency, will contribute to the thermal management of the processor.

A significant breakthrough would be the transition to all-optical I/Os. That is, a digital processor will receive and transmit data only by optical means, while dc power and a low-speed signal will continue to be transmitted over copper interconnects. The processor, which may contain more than one core, will have direct optical access to the network over the 100-Gb/s Ethernet being developed worldwide. The readily recognizable roadblocks to direct optical network access are our inability to effectively bring optical signals on and off a chip and our equally formidable inability to merge silicon CMOS processing with III-V compound semiconductor processing, even at the point of wafer bonding at the back end of the line.

"All-optical I/Os" and direct optical access to the 100-Gb/s Ethernet is a natural evolution of modern information technology built on an SOP platform of miniaturization and cointegration of optimized disparate technologies.

6.10 Summary

The role of optics in the transport of information has been discussed in terms of its historical, economic, and technical contents. The transition from copper interconnects to optical interconnects up to the chip level is slow but unavoidable. The role of optoelectronic SOP is to make this transition efficient, economical, and timely. The major technological obstacles are (1) direct optical I/O access from the processor to the network, (2) cointegration of silicon-based CMOS digital and III-V compound semiconductor analog functions on the same substrate, and (3) management of the thermal loads. The SOP concepts of codesign, cointegration, and miniaturization will contribute to the evolution of information technologies at the system, board, and chip levels. The interface optical coupling process, developed at Georgia Tech, can be considered as an example of cointegration of the passive optical lightwave circuits with thin-film optoelectronic active devices such as lasers and photodetectors. This method can be practiced at the board level by using optically defined polymer waveguides and at the chip level by using high-index, optically defined chalcogenide waveguides. It will usher in a new era of system integration technology that harnesses the advantages of optical interconnects for next-generation computing and communications systems.

References

1. http://www.sims.berkeley.edu/research/projects/how-much-info-2003.
2. E. Lach and K. Schuh, "Recent advances in ultrahigh bit rate ETDM transmission systems," *J. Lightwave Technol.*, vol. 24, 2006, p. 4455.
3. S. Lee, M. Hayakawa, and N. Ishibashi, "Radiation from bent transmission lines," *IEICE Transactions on Communications*, vol. E84-B, 2001, p. 2604–09.
4. T. Shiokawa, "FDTD analysis of the transmission/radiation characteristics of 90° bent transmission lines," *Electron. Commun. Jpn. 1, Commun. (USA)*, vol. 87, 2004, p. 11.
5. Daniel Guidotti, Jianjun Yu, Markus Blaser, Vincent Grundlehner, and Gee-Kung Chang, "Edge viewing photodetectors for strictly in-plane lightwave circuit integration and flexible optical interconnects," *56th Electronic Components and Technology Conference*, San Diego, CA, May 2006.
6. Jianjun Yu, Zhensheng Jia, Jianguo Yu, Ting Wang, and Gee Kung Chang, "Novel ROF network architecture for providing both wireless and wired broadband services," *Microwave and Optical Technol. Lett.*, vol. 49, 2007, p. 659.
7. M. Oda, D. Guidotti and G.-K. Chang, "In-plane optical interconnection with high coupling efficiency between optical chips and waveguides," *2006 IEEE LEOS Annual Meeting Conference*, October 29–November 2, 2006, Montreal, Quebec, Canada.
8. Takuma Ban, Reiko Mita, Yasumobu Matsuoka, Hirokazu Ichikawa, and Masato Shishikura, "1.3 µm four channel × 10-Gb/s parallel optical transceiver with polymer PLC platform for very-short-reach applications," *IEEE J. Sel. Topics in Quantum Electron*, vol. 12, 2006, p. 1001.
9. http://www.ppc-electronic.com.
10. Hong Ma, Alex K.-Y. Jen, and Larry Dalton, "Polymer-based optical waveguides: materials, processing and devices," *Adv. Mater.*, vol. 14, 2002, p. 1339.
11. G. Khanarian and H. Celanese, "Optical properties of cyclic olefin copolymers," *Optical Engineering*, vol. 40, 2001, p. 1024.
12. Jang-Joo Kim and Jae-Wook Kang, "Thermally stable optical waveguide using polycarbonate," *Proc. SPIE*, vol. 3799, 1999, p. 333.
13. Louay Eldada and Lawrence W. Shcaklette, "Advances in polymer integrated optics," *IEEE J. Sel. Topics in Quantum Electronics*, vol. 6, 2000, p. 54.
14. J. D. Dow and D. Redfield, "Toward a unified theory of Urbach's rule and experimental absorption edges," *Phys. Rev. B*, vol. 5, 1972, p. 549.

15. Steffen Uhlig and Mats Robertsson, "Limitations to and solutions for optical loss in optical backplanes," *J. Lightwave Technol.,* vol. 24, 2006, p. 1710.
16. D. A. B. Miller and H. M. Ozaktast, "Limit to the bit-rate capacity of electrical interconnects from the aspect ratio of the system architecture," *J. Parallel and Distr. Comp.* vol. 41, 1997, p. 42.
17. Martin H. Davis, Jr., and Umakishore Ramachandran, "Optical bus protocol for a distributed shared memory multiprocessor," *Proc. SPIE,* vol. 1563, 1991, p. 176.
18. Jacques Henri Collet, Daniel Litaize, Jan Van Campenhout, Chris Jesshope, Marc Desmulliez, Hugo Thienpont, James Goodman, and Ahmed Louri, "Architectural approach to the role of optics in monoprocessor and multiprocessor machines," *Appl. Optics,* vol. 39, 2000, p. 671.
19. Constantine Katsinis, "Models of distributed-shared-memory on an interconnection network for broadcast communication," *J. Interconnection Networks,* vol. 4, 2003, p. 77.
20. Avinash Karanth Kodi and Ahmed Louri, "All-photonic interconnect for distributed shared memory multiprocessors," *J. Lightwave Technol.,* vol. 22, 2004, p. 2101.
21. IBM J. of Research and Development, vol. 49, nos. 2/3, 2005.
22. *International Technology Roadmap for Semiconductors,* 2003 Edition, Executive Summary, Tables 4c, 4d, p. 54.
23. E. Diaz-Alavrez and J. P. Krusius, "Probabilistic prediction of wiring demand and routing requirements for high density interconnect substrates," *IEEE Trans. Adv. Packaging,* vol. 22, 1999, p. 642.
24. Alan Charlesworth, "The Sun Fireplane Interconnect," *IEEE Micro,* vol. 22, 2002, p. 36.
25. T. Tamanuki, Z. Shao, N. Huang, C. Keller, and M. Ito, "Multi form pluggable 8-channel Tx/Rx module for OC-3/OC-12," *LEOS 2003. The 16th Annual Meeting of the IEEE,* vol. 2, 2003, pp. 569–70.
26. Takashi Yoshikawa and Hiroshi Matsuoka, "Optical interconnects for parallel and distributed computing," *Proc. IEEE,* vol. 88, 2000, p. 849.
27. Efstathios D. Kyriakis-Bitzaros, Nikos Haralabidis, M. Lagadas, Alexandros Georgakilas, Y. Moisiadis, and George Halkias, "Realistic end-to-end simulation of the optoelectronic links and comparison with the electrical interconnections for system-on-chip applications," *J. Lightwave Tech.,* vol. 19, 2001, p. 1532.
28. Osman Kibar, Daniel A. Van Blerkom, Chi Fan, and Sadik C. Esener, "Power minimization and technology comparisons for digital free-space optoelectronic interconnects," *IEEE, J. Lightwave Technol.,* vol. 17, 1999, p. 546.
29. Hoyeol Cho, Pawan Kapur, and Krishna C. Saraswat, "Power comparison between high speed electrical and optical interconnects for interchip communication," *IEEE J. Lightwave Technol.,* vol. 22, 2004, p. 2021.
30. Christina Carlsson, Hans Martinsson, Richard Schatz, John Halonen, and Anders Larsson, "Analog modulation properties of oxide confined VCSELs at microwave frequencies," *IEEE J. Lightwave Technol.,* vol. 20, 2002, p. 1740.
31. Kenji Sato, Shoichiro Kuwahara, and Yutaka Miyamoto, "Chirp characteristics of a 40 Gb/s directly modulated distributed-feedback laser diode," *J. Lightwave Technol.,* vol. 23, 2005, p. 3790.
32. Composite PCB material made by Polyclad, Inc., http://www.Polyclad.com/laminate.
33. See the documentation for the MAX3804 from http://www.maxim-ic.com.
34. Chung-En Zah, R. Bhat, B. N. Pathak, F. Favire, Wei Lin, M. C. Wang, N. C. Andreadakis, D. M. Hwang, M. A. Koza, Tein-Pei Lee, Zheng Wang, D. Darby, D. Flanders, and J. J. Hsieh, "High-performance uncooled 1.3-μm $Al_xGa_yIn_{1-x-y}AsInP$ strained-layer

quantum-well lasers for subscriber loop applications," *IEEE J. Quantum Electron.*, vol. 30, 1994, p. 511.

35. T. Ishikawa, T. Higashi, T. Uchida, T. Yamamoto, T. Fujii, H. Shoji, M. Kobayashi, and H. Soda, "Well-thickness dependence of high-temperature characteristics in 1.3-μm AlGaInAs-InP strained-multiple-quantum-well lasers," *IEEE Photonics Technology Lett.*, vol. 10, 1998, p. 1703.
36. Pekko Sipilä, "Febry-Perot lasers offer low cost 10 Gb/s ethernet," *Compound Semiconductors*, October 2004, pp. 37–3.
37. Jim A. Tatum et al., "The VCSELs are coming," *Proc. SPIE*, vol. 4994, 2003, pp. 1–6.
38. J. H. Stathis, "Reliability limits for the gate insulator in CMOS technology," *IBM J. Res. and Dev.*, vol. 46, 2002, p. 265.
39. Constantina Poga, McRae Maxfield, Larry Shacklette, Robert Blomquist, and George Boudoughian, "Accelerated aging of tunable thermo-optic polymer planar waveguide devices made of fluorinated acrylates," *SPIE*, vol. 4106, 2000, p. 96.
40. Zarlink, http://products.zarlink.com/product_profiles/ZL60101.htm.
41. Y. S. Liu, R. J. Wojnarowski, W. A. Hennessy, J. P. Bristow, Yue Liu, A. Peczalski, J. Rowlette, A. Plotts, J. Stack, M. Kadar-Kallen, J. Yardley, L. Eldada, R. M. Osgood, R. Scarmozzino, S. H. Lee, V. Ozgus, and S. Patra, "Polymer optical interconnect technology (POINT) optoelectronic packaging and interconnect for board and backplane applications," *Proc. 46th Electronic Components and Technology Conference*, 1996, p. 308.
42. Takashi Yoshikawa, Sohichiro Araki, Kazunri Miyoshi, Yoshihiko Suemura, Naoya Henmi, Takeshi NHagahori, Hiroshi Matsuoka, and Takashi Yokota, "Skewless optical data-link subsystem for massively parallel processors using 8 Gb/s × 1.1 Gb/s MMF array optical module," *IEEE Photonics Technol. Lett.*, vol. 9, 1997, p. 1625.
43. T. Yoshikawa, I. Hatakeyama, K. Miyoshi, K. Kurata, J. Sasaki, N. Kami, T. Sugimoto, M. Fukaishi, K. Nakamure, K. Tanaka, H. Nishi, and T. Kudoh, "Optical interconnection as an IP macro of a CMOS library," *HOT 9 Interconnects. Symposium on High Performance Interconnects*, 2001, pp. 31–35.
44. http://www.research.ibm.com/trl/projects/mobtech/osmcat/index_e.htm.
45. A spin-off of DuPont, Optical CrossLinks, Inc., produces acrylate-based passive, optical waveguide structures and passive flexible optical ribbons (http://www.opticalcrosslinks.com).
46. Makoto Hikita, Satoru Tomaru, Koji Enbutsu, Nakoi Oba, Ryoko Yoshimura, Mitsuo Usui, Takashi Yoshida, and Saburo Imamura, "Polymeric optical waveguide films for short-distance optical interconnects," *IEEE J. Sel. Topics in Q.E.*, vol. 5, 199, p. 1237.
47. Kohsuke Katsura, Mitsuo Usui, Nobuo Sato, Akira Ohki, Nobuyuki Tanaka, Nobuaki Matsuura, Toshiaki Kagawa, Kouta Tateno, Makoto Hikita, Ryoko Yoshimura, and Yasuhiro Ando, "Packaging for a 40-channel parallel optical interconnection module with an over-25-Gbit/s throughput," *IEEE Trans. on Adv. Pack.*, vol. 22, 1999, p. 551.
48. Suning Tanp, Ting Li, Feiming Li, Linghui Wu, M. Dubinovsky, R. Wickman, and R.T. Chen, "1-GHz clock signal distribution for multi-processor super computers," *Proc. Third International Conference on Massively Parallel Processing Using Optical Interconnections*, 1996, p. 186.
49. R. T. Chen, Lei Lin, Chulchae Choi, Y. J. Liu, B. Bihari, L. Wu, S. Tang, R. Wickman, B. Picor, M. K. Hibb-Brenner, J. Bistrow, and Y. S. Liu, "Fully embedded board-level guided-wave optoelectronic interconnects," *Proc. IEEE*, vol. 88, 2000, p. 780.
50. R. T. Chen, L. Wu, F. Li, S. Tang, M. Dubinovsky, J. Qi, C. L. Schow, J. C. Campbell, R. Wickman, B. Picor, M. Hibbs-Brenner, J. Bristow, Y. S. Liu, S. Rattan, and C. Noddings,

"Si CMOS process compatible guided-wave multi-GBit/sec optical clock signal distribution system for Cray T-90 supercomputer," *Proc. Fourth International Conference Massively Parallel Processing Using Optical Interconnections,* 1997, pp. 10–24.

51. D. Krabe, F. Ebling, N. Arndt-Staufenbiel, G. Lang, and W. Scheel, "New technology for electrical/optical systems on module and board level: the EOCB approach," *Proc. 50th Electronic Components and Technology Conference,* Las Vegas, NV, May 21–24, 2000, p. 970.

52. Y. Ishii, S. Koike, Y. Arai and Y. Ando, "SMT-compatible large-tolerance 'OptoBump' interface for interchip optical interconnections," *IEEE Transactions on Advanced Packaging,* vol. 26, 2003, p. 122.

53. C. Berger, U. Bapst, G.-L. Bona, R. Dangel, L. Dellmann, P. Dill, M. A. Kossel, T. Morf, B. Offrein, and M. L. Schmatz, "Design and implementation of an optical interconnect demonstrator with board-integrated waveguides and microlens coupling," *2004 Digest of the LEOS Summer Topical Meetings: Biophotonics/Optical Interconnects & VLSI Photonics/WGM Microcavities,* 2004, pp. 19.

54. C. Choi, Yuije Liu, L. Lin, Li Wang, Jinho Choi, David Haas, Jerry Magera, and Ray T. Chen, "Flexible optical waveguide film with 45-degree micromirror couplers for hybrid E/O integration or parallel optical interconnection," *Proc. SPIE,* vol. 5358, 2003, p. 122.

55. J. A. Kash et al., "Chip-to-chip optical interconnects," *2006 Optical Fiber Communication Conference and National Fiber Optic Engineers Conference,* Anaheim, CA, March 5–10, 2006, p. 3.

56. Digital circuits are subject to simultaneous switching noise (SSN) arising primarily from switching transistors and line and via inductance. This causes unwanted reference plane and power plane bounce and unwanted spurious circuit triggering. A common solution is to add isolation capacitance between the power plane layer, the signal plane layer, and the reference plane layer. SSN amplitude increases with the number of switching transistors, and larger capacitance is sought in order to reduce the unwanted effect.

57. Katsunari Okamoto, *Fundamentals of Optical Waveguides,* San Diego: Academic Press, 2000.

58. Gee-Kung Chang, Daniel Guidotti, Fuhan Liu, Yin-Jung Chang, Zhaoran Huang, Venkatesh Sundaram, Devarajan Balaraman, Shashikant Hegde, and Rao Tummala, "Chip-to-chip optoelectronics SOP on organic boards or packages," *IEEE Trans. Adv. Pkg.,* vol. 27, 2004, p. 386.

59. L. C. Kimerling, "Silicon microphotonics," *Appl. Surf. Sci.,* vol. 159–160, 2000, p. 8.

60. Lucas B. Soldano and Erik C. M. Pennings, "Optical multi-mode interference devices based on self-imaging: Principles and applications," *IEEE, J. Lightwave Technol.,* vol. 13, 1995, p. 615.

61. Meint K. Smit and Cor van Dam, "PHASAR-based WDM-devices: Principles, design and applications," *IEEE, J. Sel. Topics Quantum Electron.,* vol. 2, 1996, p. 236.

62. Y.-J. Chang, G.-K. Chang, T. K. Gaylord, D. Guidotti, and J. Yu, "Ultra-high speed transmission of polymer-based multimode interference devices for board-level high-throughput optical interconnects," *Proc SPIE,* vol. 6126, 2006, p. 112.

63. Edmund J. Murphy (ed.), *Integrated Optical Circuits and Components Design and Applications,* Basel, Switzerland: Marcel Dekker, Inc., 1999.

64. H. Kogelnik and T. P. Sosnowski, "Holographic thin film couplers," *Bell Syst. Tech. J.,* 49, 1970, pp. 1602–08.

65. W. Driemeier, "Bragg-effect grating couplers integrated in multicomponent polymeric wave-guides," *Opt. Lett.* 15, 1990, pp. 725–27.
66. Q. Huang and P. R. Ashley, "Holographic Bragg grating input-output couplers for polymer waveguides at an 850-nm wavelength," *Appl. Opt.*, vol. 36, 1997, pp. 1198–1203.
67. M. L. Jones, R. P. Kenan, and C. M. Verber, "Rectangular characteristic gratings for waveguide input and output coupling," *Appl. Opt.*, vol. 34, 1995, pp. 4149–58.
68. V. Weiss, I. Finkelstein, E. Millul, and S. Ruschin, "Coupling and waveguiding in photopolymers," *Proc. SPIE*, 3135, 1997, pp. 136–43.
69. S. M. Schultz, E. N. Glytsis, and T. K. Gaylord, "Design of a high-efficiency volume grating coupler for line focusing," *Appl. Opt.*, vol. 37, 1998, pp. 2278–87.
70. S. M. Schultz, E. N. Glytsis, and T. K. Gaylord, "Volume grating preferential-order focusing waveguide coupler," *Opt. Lett.*, vol. 24, 1999, pp. 1708–10.
71. S. M. Schultz, E. N. Glytsis, and T. K. Gaylord, "Design, fabrication, and performance of preferential order volume grating waveguide couplers," *Appl. Opt.*, vol. 39, 2000, pp. 1223–31.
72. S. M. Schultz, Ph.D. thesis, Georgia Institute of Technology, 1999.
73. J. W. Goodman, F. I. Leonberger, S. Y. Kung, and R. A. Athale, "Optical interconnections for VLSI systems," *Proc. IEEE*, vol. 72, 1984, pp. 850–66.
74. M. R. Feldman, S. C. Esener, C. C. Guest, and S. H. Lee, "Comparison between optical and electrical interconnects based on power and speed considerations," *Appl. Opt.*, vol. 27, 1988, pp. 1742–51.
75. R. K. Kostuk, M. Kato, and Y. T. Huang, "Polarization properties of substrate-mode holographic interconnects," *Appl. Opt.*, vol. 29, 1990, pp. 3848–54.
76. F. Lin, E. M. Strzelecki, and T. Jannson, "Optical multiplanar VLSI interconnects based on multiplexed waveguide holograms," *Appl. Opt.*, vol. 29, 1990, pp. 1126–33.
77. F. Lin, E. M. Strzelecki, C. Nguyen, and T. Jannson, "Highly parallel single-mode multiplanar holographic interconnects," *Opt. Lett.*, vol. 16, 1991, pp. 183–85.
78. M. R. Wang, G. J. Sonek, R. T. Chen, and T. Jannson, "Large fanout optical interconnects using thick holographic gratings and substrate wave propagation," *Appl. Opt.*, vol. 31, 1992, pp. 236–49.
79. J. H. Yeh and R. K. Kostuk, "Substrate-mode holograms used in optical interconnects: design issues," *Appl. Opt.*, vol. 34, 1995, pp. 3152–64.
80. C. C. Zhou, S. Sutton, R. T. Chen, and B. M. Davies, "Surface-normal 4 × 4 nonblocking wavelength selective optical crossbar interconnect using polymer-based volume holograms and substrate-guided waves," *IEEE Phot. Technol. Lett.*, vol. 10, 1998, pp. 1581–83.
81. E. N. Glytsis, N. M. Jokerst, R. A. Villalaz, S. Y. Cho, S. D. Wu, Z. Huang, M. A. Brooke, and T. K. Gaylord, "Substrate-embedded and flip-chip-bonded photodetector polymer-based optical interconnects: analysis, design, and performance," *J. Lightwave Tech.*, vol. 21, 2003, pp. 2382–94.
82. R. T. Chen, S. Tang, M. M. Li, D. Gerald, and S. Natarajan, "1-to-12 surface normal three-dimensional optical interconnects," *Appl. Phys. Lett.*, vol. 63, 1993, pp. 1883–85.
83. J. H. Yeh and R. K. Kostuk, "Free-space holographic optical interconnects for board-to-board and chip-to-chip interconnections," *Opt. Lett.*, vol. 21, 1996, pp. 1274–76.
84. S. Tang, Y. Tang, J. Colegrove, and D. M. Craig, "Fast electrooptic Bragg grating couplers for on-chip reconfigurable optical waveguide interconnects," *IEEE Phot. Technol. Lett.*, vol. 16, 2004, pp. 1385–87.

85. A. V. Mule, R. A. Villalaz, T. K. Gaylord, and J. D. Meindl, "Quasi-free-space optical coupling between diffraction grating couplers fabricated on independent substrates," *Appl. Opt.,* 43, 2004, pp. 5468–75.
86. R. A. Villalaz, E. N. Glytsis, T. K. Gaylord, and T. N. Nakai, "Wavelength response of waveguide volume grating couplers for optical interconnects," *Appl. Opt.,* vol. 43, 2004, pp. 5162–67.
87. M. G. Moharam, E. B. Grann, D. A. Pommet, and T. K. Gaylord, "Formulation for stable and efficient implementation of the rigorous coupled-wave analysis of binary gratings," *J. Opt. Soc. Amer. A,* vol. 12, 1995, pp. 1068–76.
88. M. Neviere, "The homogeneous problem," Chapter 5 in *Electromagnetic Theory of Gratings,* Berlin: Springer-Verlag, 1980, pp. 123–57.
89. D. E. Muller, "A method for solving algebraic equations using an automatic computer," *Math. Tables and Other Aids to Comp.,* vol. 10, 1956, pp. 208–15.
90. The Mathworks Inc., "Matlab, ver. 6.5.1," 2003.
91. Zhaoran Rena Huang, Daniel Guidotti, Lixi Wan, Yin-Jung Chang, Jianjun Yu, Jin Liu, Hung-Fei Kuo, Gee-Kung Chang, Fuhan Liu, and Rao Tummala, "Hybrid integration of end-to-end optical interconnects on printed circuit boards," to appear in *IEEE CPMT,* 2007.
92. A.W. Topol, D.C. La Tulipe, Jr., L. Shi, D. J. Frank, K. Bernstein, S. E. Steen, A. Kumar, G. U. Singco, A. M. Young, K. W. Guarini, and M. Ieong, "Three-dimensional integrated circuits," *IBM Journal of Research and Development,* vol. 50, no. 4–5, July–Sept. 2006, pp. 494–506.
93. Hirofumi Kuribara, Hiroyuki Hashimoto, Takafumi Fukushima, and Mitsumasa Koyanagi, "Multichip shared memory module with optical interconnection for parallel-processor system," *Jpn. J. Appl. Phys.,* vol. 45, 2006, pp. 3504.
94. C. Schwartz, S. Xin, and W. I. Wang, "Thin film transfer of InAlAs/InGaAs MSM phototetector or InGaAsP lasers onto GaAs or Si substrates," *Proc. SPIE,* vol. 1680, 1992, p. 161.
95. Sang-Yeon Cho, Sang-Woo Seo, Nan Marie Jokerst, and Martin A. Brooke, "Board-level optical interconnection and signal distribution using embedded thin-film optoelectronic devices," *J. Lightwave Tech.,* vol. 22, 2004, pp. 211–18.
96. Z. Huang, Y. Ueno, K. Kaneko, N. M. Jokerst, and S. Tanahashi, "Embedded optical interconnections using thin film InGaAs metal-semiconductor-metal photodetector," *Elec. Lett.,* vol. 38, 2002, pp. 1708–09.
97. Elias N. Glytsis, Nan M. Jokerst, Ricardo A. Villalaz, Sang-Yeon Cho, Shun-Der Wu, Zhaoran Huang, Martin A. Brooke, and Thomas K. Gaylord, "Substrate-embedded and flip-chip-bonded photodetector polymer-based optical interconnects: analysis, design, and performance," *J. Lightwave Tech.,* vol. 21, 2003, pp. 2382–94.
98. Chulchae Choi, Lei Lin, Yuije Liu, and Ray T. Chen, "Performance analysis of 10-μm-thick VCSEL array in fully embedded board level guided-wave optoelectronic interconnects," *J. Lightwave Technol.,* vol. 21, 2003, p. 1531.
99. Kenji Hiruma, Masao Kinoshita, Seiki Hiramatsu, and Takashi Mikawa, "Epitaxial lift-off of GaAs/AlGaAs films with vertical cavity surface emitting laser for high-density packaging of opto electronic interconnections," *Jpn. J. Appl. Phys.,* vol. 43, 2004, p. 7054.
100. Kenji Hiruma, Masao Konoshita, and Takashi Mikawa, "Improved performance of 10-μm-thick GaAs/AlGaAs vertical-cavity surface-emitting lasers," *J. Lightwave Technol.,* vol. 23, 2005, p. 4342.

101. J. Mumbru, G. Panotopoulos, D. Psaltis, Xin An, Gan Zhou, and Fai Mok, "Optically reconfigurable gate array," *Proc. AIPR 2000, 29th Applied Imagery Pattern Recognition Workshop,* 2000, p. 84.
102. Makato Hikita, Satoru Tomaru, Koji Enbutsu, Naoki Olba, Ryoko Yoshimura, Mitsuo Usui, Takashi Yoshida, and Saboro Imamura, "Polymer optical waveguide films for short-distance optical interconnects," *IEEE J. Sel. Topics in Quantum Electron.,* vol. 5, 1999, p. 1237.
103. A. L. Glebov, M. G. Lee, and K. Yokouchi, "Integration technologies for pluggable backplane optical interconnect system," *Opt. Eng.,* vol. 46, 2007.
104. Christoph Berger, Marcel A Kossel, Christian Menolfi, Thomas Morf, Thomas Toifl, and Martin L. Schmatz, "High-density optical interconnects within large-scale systems," *Proc. SPIE,* vol. 4942, 2003, p. 222.
105. B. Bauknecht, J. Kunde, R. Krabenbuhl, S. Grossman, and Ch. Bosshard, "Assembly technology for multi-fiber optical connectivity solutions," *Proc. IEEE/LEOS Workshop on Fibers and Optical Passive Components,* Palermo, Italy, 2005, p. 92.
106. Daniel Guidotti, Jianjun Yu, Markus Blaser, Vincent Grundlehner, and Gee-Kung Chang, "Edge viewing photodetectors for strictly in-plane lightwave circuit integration and flexible optical interconnects," *Proc. 56th Electronic Components & Technology Conference,* San Diego, CA, May 30–June 2, 2006, p. 7.
107. Han Seo Cho, Kun-Mo Chu, Saekyoung Kang, Sung Hwan Hwang, Byung Sup Rho, Weon Hyo Kim, Joon-Sung Kim, Jang-Joo Kim, and Hyuo-Hoon Park, "Compact packaging of optical and electronic components for on-board optical interconnects," *IEEE Trans. Adv. Packaging,* vol. 28, 2005, p. 114.
108. Gee-Kung Chang, Daniel Guidotti, Zhaoran Rena Huang, Lixi Wan, Jianjun Yu, Shashikant Hegde, Hung-Fei Kuo, Yin-Jung Chang, Fuhan Liu, Fentao Wang, and Rao Tummala, "High-density, end-to-end optoelectronic integration and packaging for digital-optical interconnect systems," *Proc. SPIE Conf. on Enabling Photonics Technologies for Defense, Security and Aerospace Applications,* Kissimmee, FL, vol. 5814, March 28–April 1, 2005, pp. 176–90.
109. Yin-Jung Chang, Daniel Guidotti, Lixi Wan, Thomas K. Gaylord, and Gee-Kung Chang, "Board-level optical-to-electrical signal distribution at 10 Gb/s," *IEEE Photonics Technology Letters,* vol. 18, no. 17, 2006, pp. 1828–30.
110. http://www.lightsmyth.com.
111. http://www.luxtera.com.
112. K. Ohashi, J. Fujikata, M. Nakada, T. Ishi, K. Nishi, H. Yamada, M. Fukaishi, M. Mizuno, K. Nose, I. Ogura, Y. Urino, and T. Baba, "Optical interconnect technologies for high-speed VLSI chips using silicon nano-photonics," *International Solid State Circuit Conference,* Session 23.5, 2006.
113. Mike Salib, Ling Liao, Richard Jones, Mike Morse, Ansheng Liu, Dean Samara-Rubio, Drew Alduino, and Mario Paniccia, "Silicon photonics," *Intel Technology Journal,* vol. 8, 2004, p. 143.
114. Ansheng Liu, "Optical amplification and lasing by stimulated Raman scattering in silicon waveguides," *Journal of Lightwave Technology,* vol. 24, 2006, p. 1440.
115. L. Liao, D. Samara-Rubio, M. Morse, A. Liu, D. Hodge, D. Rubin, U. D. Keil, and T. Franck, "High Speed Silicon Mach-Zehnder Modulator," *Optics Express,* vol. 13, 2005, pp. 3129–35.
116. A. Liu, R. Jones, L. Liao, D. Samara-Rubio, D. Rubin, O. Cohen, R. Nicolaescu, and M. Paniccia, "A high-speed silicon optical modulator based on a metal-oxide-semiconductor capacitor," *Nature,* vol. 427, 2004, pp. 615–18.

117. A. Bhatnagar, C. Debaes, H. Thienpont, and D. A. B. Miller, "Receiverless detection schemes for optical clock distribution," *Proc. SPIE—The International Society for Optical Engineering,* vol. 5359, 2004, p. 352.
118. Klaus Finsterbusch, Neil J. Baker, Vahid G. Ta'eed, Benjamin J. Eggleton, Duk-Yong Choi, Steve Madden, and Barry Luther-Davies, "Higher-order mode grating devices in As_2S_3 chalcogenide glass rib waveguides," *J. Opt. Soc. Am. B,* vol. 24, 2007, p. 1283.
119. M. L. Anne, V. Nazabal, V. Moizan, C. Boussard-Pledel, B. Bureau, J. L. Adam, P. Nemec, M. Frumar, A. Moreac, H. Lhermite, P. Camy, J. L. Doualan, J. P. Guin, J. Le Person, F. Colas, C. Compere, M. Lehaitre, F. Henrio, D. Bose, J. Charrier, A.-M. Jurdyc, and B. Jacquier, "Chalcogenide waveguides for IR optical range," *SPIE,* vol. 6475, 2007, p. 277.
120. P. Sanchis, J. Garcia, J. Marti, W. Bogaerts, P. Dumon, D. Taillaert, R. Baets, V. Wiaux, J. Wouters, and S. Beck, "Experimental demonstration of high coupling efficiency between wide ridge waveguides and single-mode photonic crystal waveguides," *IEEE Photonics Technology Letters,* vol. 16, 2004, p. 2272.
121. Sharee J. McNab, Nikolaj Moll, and Yurii A. Vlasov, "Ultra-low loss photonic integrated circuit with membrane-type photonic crystal waveguides," *Optics Express,* vol. 11, 2003, p. 2927.
122. Takeshi Doi, Akihito Uehara, Yoshiyuki Takahashi, Shin Yokoyama, and Atsushi Iwata, "An experimental pattern recognition system using bi-directional optical bus lines," *Jpn. J. Appl. Phys.,* vol. 37, pt. 1, 1998, p. 1116.
123. J. Mumbru, G. Panotopoulos, D. Psaltis, Xin An, F. H. Mok, Suat ay, S. L. Barna, E. R. Fossum, "Optically programmable gate array," *Proc. SPIE,* vol. 4089, 2000, p. 763.
124. A. V. Krishnamoorthy, K. W. Goossen, L. M. F. Chirovsky, R. Z. Rozier, P. Chandramani, W. S. Hobson, S. A. Hui, L. Lopta, J. A. Walker, and L. A. D'Asaro, "16 × 16 VCSEL array flip-chip bonded to CMOS VLSI circuit," *IEEE Photonics Tech. Lett.,* vol. 12, 2000, p. 1073.
125. Keith W. Goossen, "Optoelectronic/VLSI," *IEEE Trans. Advanced Packaging,* vol. 22, 1999, p. 561.
126. Mitsumasa Koyanagi, Takuji Matsumoto, Tamio Shimatani, Keiichi Hirano, Hiroyuki Kurino, Reiji Aibara, Yasuhiro Kuwana, Norihiko Kuroishi, Tetsuro Kawata, and Nobuaki Miyakawa, "Multi-chip module with optical interconnection for parallel processor system," *IEEE Intl. Solid-State Circuits Conf. (ISSCC) Digest of Technical Papers,* 1998, p. 92–3, 421.
127. Ansheng Lin, Richard Jones, Ling Liao, Dean Samara-Rubio, Daron Rubin, Oded Cohen, Remus Nicoleasku, and Mario Paniccia, "A high speed silicon optical modulator based on a metal-oxide-semiconductor capacitor," *Nature,* 2004, p. 615.
128. Jeffrey A. Davis and James D. Meindl (eds.), *Interconnect Technology and Design for Gigascale Integration,* Boston: Kliewer Academic Publishers, 2003.
129. *International Technology Roadmap for Semiconductors,* 2005 Edition, http://www.itrs.net/Common/2005ITRS/Home2005.htm.
130. Sungjun Im, Navin Srivastava, Kaustav Benerjee, and Kenneth E. Goodson, "Scaling analysis of multilevel interconnect temperatures for high-performance ICs," *IEEE Trans. Electron. Devices,* vol. 52, 2005, p. 2710.
131. J. Jiang Liu, Zaven Kalayjian, Brian Riely, Wayne Chang, George J. Simonis, Alyssa Apsel, and Andreas Andreou, "Multichannel ultrathin silicon-on-sapphire optical interconnects," *IEEE J. Sel. Topics Quantum Electron,* vol. 9, 2003, p. 380.

Table 6.1 References

T1. L. Eldada et al., "Advances in polymer integrated optics," *J. Sel. Topics in Quantum Electron.*, vol. 6, 2000, p. 54.

T2. J.-F. Viens, C. L. Callender, J. P. Noad, L. Eldada, and R. A. Norwood, "Polymer-based waveguide devices for WDM applications," *Proc. SPIE,* vol. 3799, 1999, p. 202.

T3. S. Toyoda, N. Ooba, M. Hikita, T. Kurihara, and S. Imamura, "Propagation loss and birefringence properties around 1.55 µm of polymeric optical waveguides fabricated with cross-linked silicone," *Thin Solid Films,* vol. 370, 200, p. 311.

T4. T. Watanabe et al., "Polymeric optical waveguide circuits formed using silicone resin," *J. Lightwave Technol.,* vol. 16, 1998, p. 1049.

T5. K. Enbutsu, M. Hikita, S. Tomaru, M. Usui, S. Imamura, and T. Maruno, "Multimode optical waveguide fabricated by UV cured epoxy resin for optical interconnections," *Proc. Fifth Asia-Pacific Conf. Commun. and Fourth Optoelectronics and Commun. Conf. (APCC/OECC),* vol. 2, pt. 2, 1999, p. 1648.

T6. S. Tomaru, K. Enbutsu, M. Hikita, M. Amano, S. Tohno, and S. Imamura, "Polymeric optical waveguides with high thermal stability and its application to optical interconnections," *Optical Fiber Commun. Conf. and Intl. Conf. Integrated Optics and Optical Fiber Commun. (OFC/IOOC) Technical Digest,* vol. 2, 1999, p. 277.

T7. M. Usui et al., "Low loss passive polymer optical waveguides with high environmental stability," *J. Lightwave Technol.,* vol. 14, 1996, p. 2338.

T8. J. Kobayashi et al., "Single mode optical waveguides fabricated from fluorinated polyimides," *Appl. Optics,* vol. 37, 1998, p. 1032.

T9. T. Matsuura et al., "Heat-resistant flexible film optical waveguides from fluorinated polyimides," *Appl. Optics,* vol. 38, 1999, p. 966.

T10. S. Ishibashi and H. Takahara, "Optical waveguide components using fluorinated polyimides," *Proc. SPIE,* vol. 3799, 1999, p. 254.

T11. http://www.dow.com/cyclotene.

T12. G. Fischbeck, R, Moosburger, C. Kostrzewa, A. Achen, and K. Petermann, "Single mode optical waveguides using a high temperature stable polymer with low losses in the 1.55 µm range," *Electron. Lett.,* vol. 33, 1997, p. 1 518.

T13. R. Buestrich et al., "ORMOCERS for optical interconnection technology," *J. Sol-Gel Sc. and Technol.,* vol. 20, 2001, p. 181.

T14. M. Popall, A. Dabeck, M. E. Robertsson, G. Gustafsson, O-J. Hagel, B. Olsowski, R. Buestrich, L. Cergel, M. Lebby, P. Kiely, J. Joly, D. Lambert, M. Schaub, and H. Reichl, "ORMOCERs—New photo-patternable dielectric and optical materials for MCM-packaging," *Proc. 48th Electronic Components and Technology Conference,* May 25–28, 1998, Seattle, WA, 1998, p. 1018.

T15. Yujie Liu, Lei Lin, Chulchae Choi, Bipin Bihari, and R. T. Chen, "Optoelectronic integration of polymer waveguide array and metal-semiconductor-metal photodetector through micromirror couplers," *IEEE Photonics Technol. Lett.,* vol. 13, 2001, p. 355.

T16. RPO Pty. Ltd., Acton, Australia, http://www.rpo.biz.

T17. Gee-Kung Chang, Daniel Guidotti, Fuhan Liu, Yin-Jung Chang, Zhaoran Huang, Venkatesh Sundaram, Devarajan Balaraman, Shashikant Hegde, and Rao Tummala, "Chip-to-chip optoelectronics SOP on organic boards or packages," *IEEE Trans. Adv. Pkg.,* vol. 27, 2004, p. 386.

T18. Polyset, Inc., Mechanicsville, New York, http://www.polyset.com.

T19. Y. Ishii, S. Koike, Y. Arai, and Y. Ando, "SMT-compatible large-tolerance 'opto-bump' interface for interchip optical interconnects," *IEEE Trans. Adv. Packaging*, vol. 26, 2003, p. 122.

T20. A. W. Norris, J. V. DeGroot, T. Ogawa, T. Watanabe, T. C. Kowalczyk, A. Baugher, and R. Blum, "High reliability of silicone materials for use as polymer waveguides," *Proc. SPIE*, vol. 5212, 2003, p. 76.

T21. M. Moynihan, C. Allen, T. Ho, L. Little, N. Pugliano, J. Shelnut, B. Sicard, H. B. Zheng, and G. Khanarian, "Hybrid inorganic-organic aqueous base compatible waveguide materials for optical interconnect applications," *Proc. SPIE*, vol. 5212, 2003, p. 50.

T22. Yuriko Ueno, Katsuhiro Kaneko, and Shigeo Tanahashi, "A new single-mode optical waveguide on ceramic substrate utilizing siloxane polymer," *Proc. SPIE*, vol. 3289, 1998, p. 134.

T23. H. H. Yao, N. Keil, C. Zawadzki, J. Bauer, M. Bauer, and C. Dreyer, "Polymeric planar waveguide devices for photonic network applications," *Proc. SPIE*, vol. 4439, p. 36.

T24. Exxelis: www.exxelis.com; Block 7, West of Scotland Science Park, Glasgow, G209 0TH. Contact: Navin Suyal, n.suyal@exxelis.com.

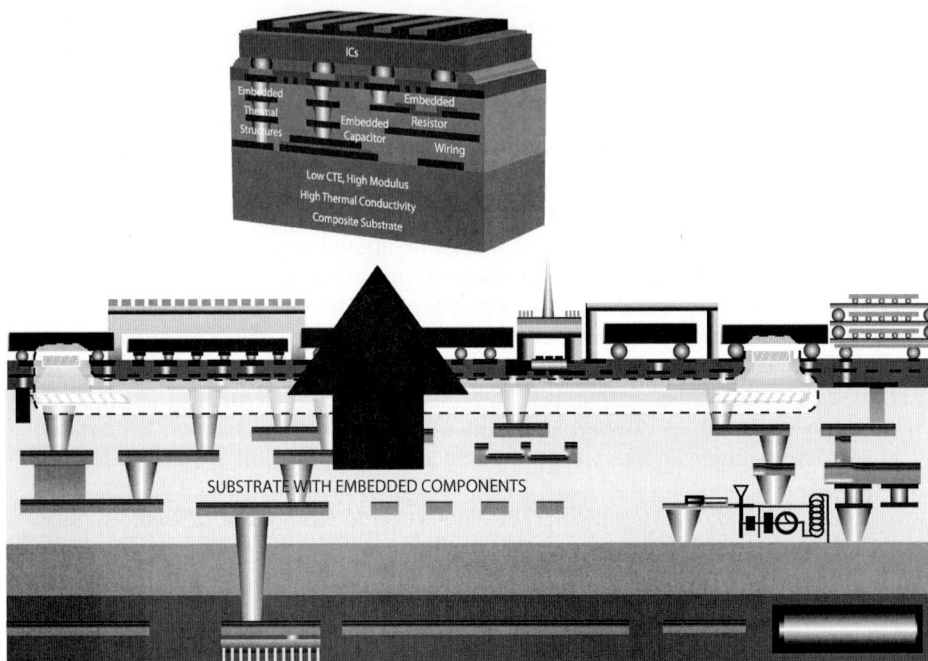

CHAPTER 7

SOP Substrate with Multilayer Wiring and Thin-Film Embedded Components

Venky Sundaram, Fuhan Liu, Ganesh Krishnan, George White,
Rao Tummala, Paul Kohl, P. Markondeya Raj, and Baik-Woo Lee

Georgia Institute of Technology, Atlanta, Georgia

7.1	Introduction 378	7.3	SOP Substrate 381
7.2	Historical Evolution of Substrate Integration Technologies 380	7.4	Future SOP Substrate Integration 435
			References 437

SOP is a system-level technology with ultraminiaturized thin-film and embedded components. Such a system has a wide variety of exciting applications due to its small size and high integration. This miniaturized system is integrated on the SOP substrate, which is the backbone on which the entire SOP concept is based. Thus advances in substrate technology play a crucial role in achieving the aim of microminiaturized multifunctional systems. Traditionally, the substrate has played a passive role. With SOP, the substrate is also the platform for system function integration.

The relentless trend toward miniaturization results in the need for higher wiring, component, and functional density on the substrate. Improvements in process and material technology are key enablers in achieving these goals.

Ultrahigh-density SOP substrates have to simultaneously support digital and RF speeds in the 20- to 50-GHz range, posing additional challenges in electrical and mechanical design. The need for fabrication of 5- to 10-µm structures requires materials of unprecedented electrical and mechanical properties far beyond today's FR4, bismaleimide triazine (BT), and other boards. All of these requirements have to be satisfied by an integrated solution for high density, high speed, low cost, and high reliability, necessitating revolutionary substrate technology combining novel designs, materials, processes, structures, and test methods. Some of the latest developments in materials technology are being leveraged to provide new materials for the core and the buildup dielectric. Passives, which take up vast amounts of surface real estate, are being embedded as thin-film components within the substrate. Different approaches and fabrication methods for embedded passives are discussed. Active ICs are also being embedded to increase the component density. To handle the large heat flux due to higher integration, new thermal technologies are being developed. Novel processes, which allow the efficient integration of the various SOP components, are also briefly discussed. Substrate technology has seen a lot of exciting developments and will provide the platform for increased integration, higher reliability, better energy efficiency, and lower cost.

7.1 Introduction

The historical evolution of package substrates, as shown in Figure 7.1, started with big and bulky, low-density boards with discrete passive components and packaged ICs. The introduction of microvia buildup substrates in the early 1990s started a new paradigm that is beginning to bridge the gap between submicron IC interconnections and milliscale PWB interconnections. Chip-scale packaging (CSP), system-in-package (SIP) technologies through stacked ICs, and development of smaller discretes such as

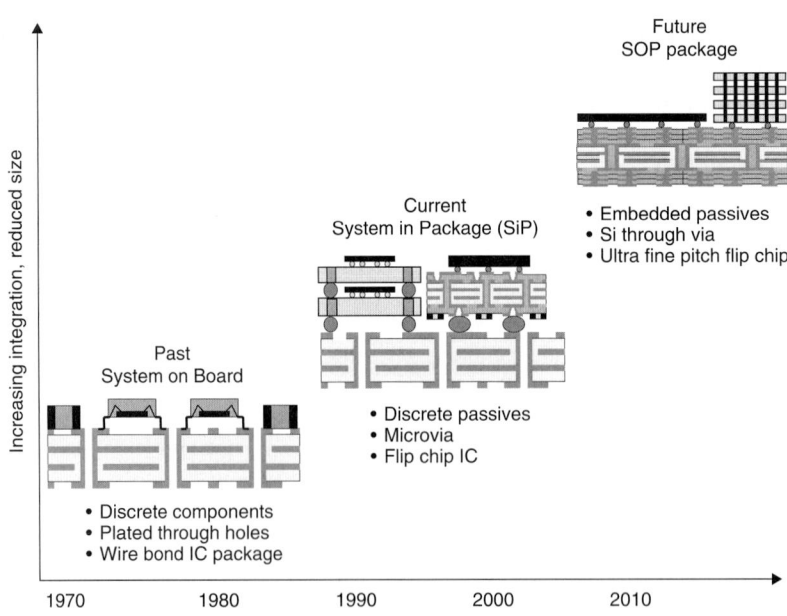

FIGURE 7.1 Trends in package substrates with an increasing level of component integration.

SOP Substrate with Multilayer Wiring and Thin-Film Embedded Components

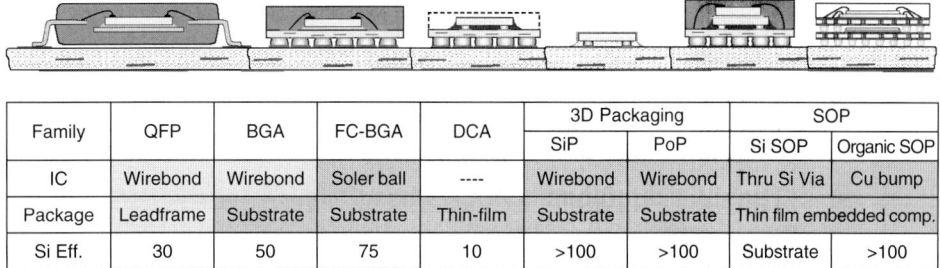

Family	QFP	BGA	FC-BGA	DCA	3D Packaging		SOP	
					SiP	PoP	Si SOP	Organic SOP
IC	Wirebond	Wirebond	Solder ball	----	Wirebond	Wirebond	Thru Si Via	Cu bump
Package	Leadframe	Substrate	Substrate	Thin-film	Substrate	Substrate	Thin film embedded comp.	
Si Eff.	30	50	75	10	>100	>100	Substrate	>100

FIGURE 7.2 Evolution of package form factors and corresponding silicon efficiency.

0201 and 01005 led to further miniaturization of packages and systems. More recently, 3D packaging enabled by through-silicon via (TSV) and package-on-package (POP) stacking is beginning to contribute to the system miniaturization. This evolution is illustrated in more detail in Figure 7.2. But these are all examples of package-enabled integration. There is no true package integration leading to an exponential reduction in system size. The SOP substrate technology described in this chapter is such a technology aimed at system miniaturization. It achieves this system miniaturization in two ways:

1. Converge systems from a three-level (IC, package, and board) hierarchy to two (IC and system package).
2. Miniaturize system components such as conductors, dielectrics, passive and active components, and thermal structures from milliscale to micro- and nanoscales.

This chapter reviews the status of SOP substrate technologies in these areas.

As shown in Figures 7.1 and 7.2, SOP substrate integration consists of embedded passives and active devices interconnected by high-density and fine-pitch multilayer thin-film wiring. In the SOP concept, there is only one level of substrate, which combines the functions of system motherboards and package substrate interposers into a single ultrahigh-density system package with embedded components. Such a SOP substrate is expected to be implemented in two distinct substrate platforms:

1. Extension of current microvia organic substrates with or without traditional cores
2. Bringing this package integration and extending it onto the silicon platform

The latter is referred to as Si SOP. The major difference between the two platforms is the improved dimensional and thermal stability of the silicon carrier substrate leading to a higher wiring and component density.

This chapter describes four SOP substrate integration technologies (Figure 7.3) that are applicable to both the above platforms:

1. High-density wiring including conductors, dielectrics, core substrate materials, and processes
2. Embedded thin-film passive component materials and processes for capacitors and resistors
3. Embedded active ICs in the substrate
4. Integration of thermally conducting materials and structures

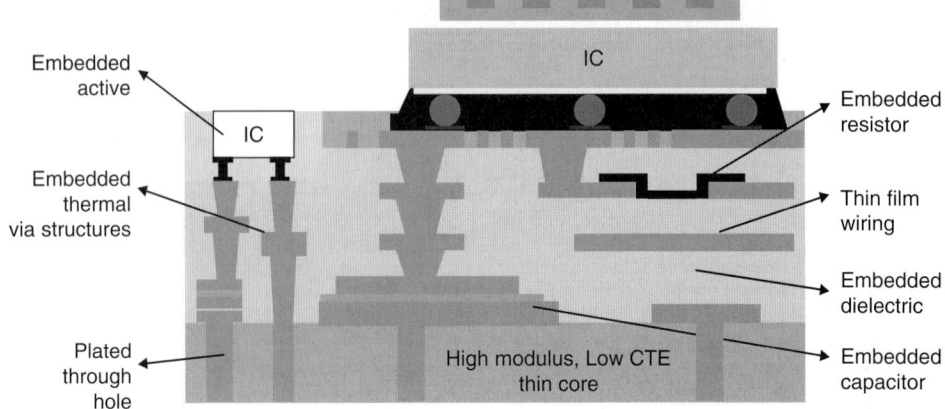

FIGURE 7.3 Key Elements of SOP substrate with embedded components.

7.2 Historical Evolution of Substrate Integration Technologies

Figure 7.4 illustrates a historical perspective of key milestones in multilayer thin-film wiring, embedded passives, and embedded active components.

The advent of high-density thin-film wiring, driven by the wiring demands of advanced ICs and market-driven performance of high-performance computers, led to highly sophisticated multichip modules (MCM) initially in high-temperature ceramics and subsequently in low-temperature ceramics and in the follow-on generations to ultrahigh-density Cu-polyimide multilayer wiring on low-temperature cofired ceramic (LTCC) substrates by IBM, Fujitsu, Hitachi, and NEC. An entirely different approach using spin-on polyimide dielectric on silicon substrate was developed by Bell Labs.

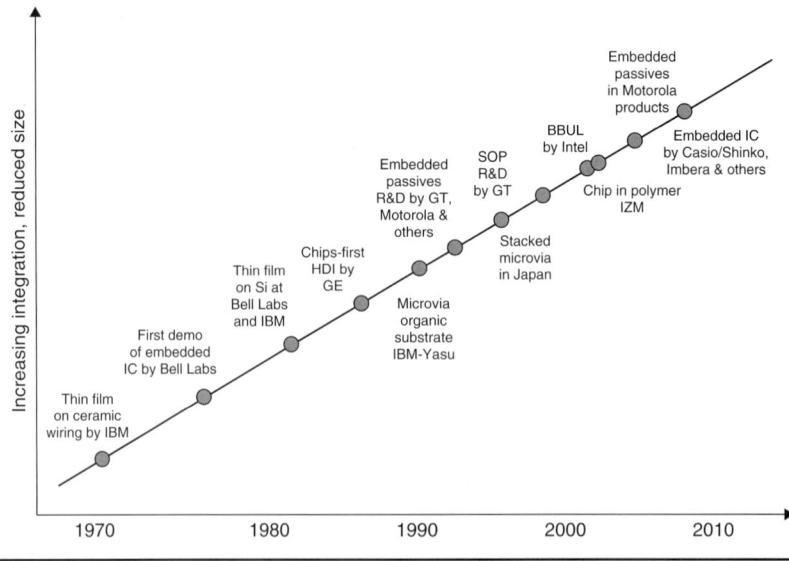

FIGURE 7.4 Key milestones in the evolution of substrate integration technologies.

Both these substrate technologies were derived from semiconductor processing. These technologies were expensive for four reasons:

1. The high fabrication investments
2. Low-volume need
3. Small wafer size
4. Expensive materials and processes

The above experience in MCM technology led to a major innovation in the early 1990s by IBM Japan that overcame the above four major issues. The IBM Japan team adopted the original MCM thin-film technology and applied it to printed wiring board using low-cost materials and processes, a larger area of fabrication, a lower-investment facility, and higher-volume manufacturing. This new technology, referred to as surface laminar circuitry (SLC), is widely considered as the pioneering high-density buildup organic substrate development that enabled the widespread use of flip-chip for mainstream IC packaging. In the late 1990s and early 2000s, the density of microvia organic substrates was further enhanced by advancements in fine-line conductor processes, lasers, and stacked microvias developed by many groups worldwide.

Embedding of active ICs in package substrates was pioneered by GE in the late 1980s with an innovative process to deposit buildup wiring layers on top of ICs embedded in plastic-molded carriers. The IC-to-package interconnections were fabricated directly on the IC pads with microvias. The original embedded IC technology has been further improved in the past two decades with technologies such as the bumpless buildup layer (BBUL) by Intel, chip-in-polymer by Fraunhofer IZM, and the embedded wafer-level package (e-WLP) by Casio and many other groups. Passive components, mainly capacitors, resistors, and inductors, have been embedded as thick and thin films into organic, ceramic, and silicon substrates pioneered by the Georgia Tech Packaging Research Center, Motorola, and the University of Arkansas, among others worldwide.

7.3 SOP Substrate

The SOP substrate has four key elements as referred to in Figure 7.3:

1. Wiring with embedded dielectric, core, and conductors
2. Embedded passives for digital, RF, and optical applications
3. Embedded actives
4. Embedded thermal structures

7.3.1 Drivers and Challenges

The main drivers for SOP substrates and embedded components are listed in Figure 7.5. Primary system driving forces fueling advancements in substrate integration fall into four main categories:

1. Higher electrical performance
2. Cost reduction accelerated by high-volume consumer products
3. Improved reliability
4. Low cost

382 Chapter Seven

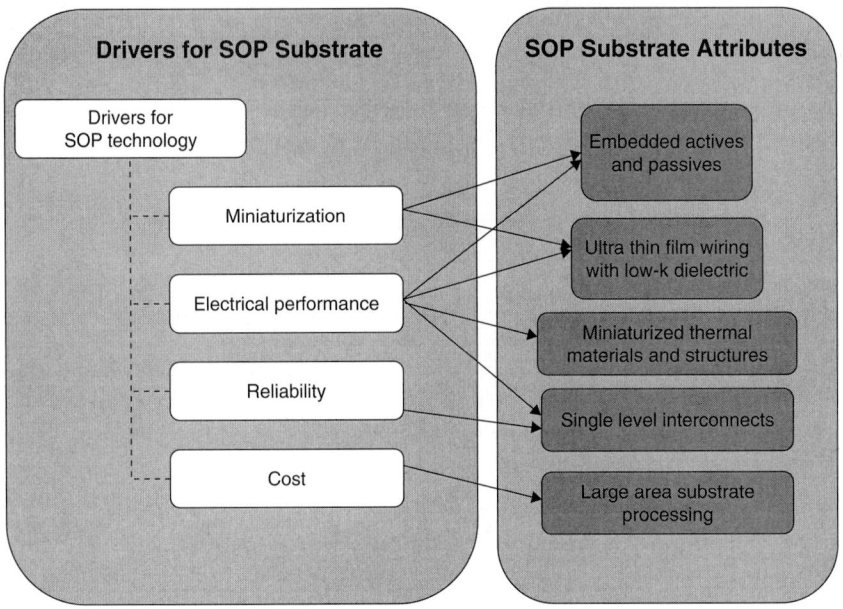

Figure 7.5 SOP substrate and embedded component drivers.

Electrical Performance
The need for higher electronic performance translates into a variety of SOP attributes in signal and power distribution.

Signal Integrity Signal speed is inversely proportional to the square root of the dielectric constant. The dielectric must also have a very low loss (< 0.001) to minimize dielectric signal losses. The use of embedded actives and passives significantly reduces the interconnection length between ICs and passive components. This reduction of interconnections increases the performance of SOP substrates by reducing delays and by minimization of losses. By transferring global wiring from ICs to the substrate, further signal performance increases can be derived.

Power Distribution The need for faster electronics with very high functional density results in a tremendous increase in power requirements, which can be as high as 100 to 200 W per chip. Power integrity to support such high power levels with low ΔI noise translates to embedded decoupling in the package with > 0.1-µF capacitance. The SOP miniaturization leads to increased power density at the system package level unlike with traditional approaches. Innovative heat dissipation solutions including miniaturized thermal structures are required to handle this heat efficiently.

Miniaturization
The need for ultraminiaturization and for extremely high component density of systems has resulted in the need to eliminate bulky passive components by embedding them as thin films. As the material and process technologies are advanced from thick films to micro- and nano-scale thin-film technologies, the current component density of 100 per cubic centimeter should increase 10-fold or higher within the next two decades.

Reliability

Reliability is a very important aspect of any system, including SOP-based systems. The SOP substrates can be expected to be much more reliable than traditional substrates due to the reduction of interconnections and better material properties as the defect-prone thick films are replaced with micro- and nanoscale structures with the same size micro-nano defects.

Cost

Cost reduction in SOP substrates drives the need for large area processing. To produce SOP systems at low cost, the SOP technology should be compatible with low-cost and large-area manufacturing technology. In a manner similar to the present-day production of ICs, SOP substrates are expected to be produced in 300-, 450-, or 600-mm panels. Layers of ultrathin dielectrics and conductors as well as thin-film components, including capacitors, resistors, inductors, filters, switches, and waveguides, are deposited on the wafer. Active ICs are also integrated into the substrate at this stage. Completed wafers are electrically tested and diced. More ICs, stacked as SIPs, can then be connected to the top surface of the substrate if required. In essence, the low cost of SOP comes from large area processing of high thin-film integration, which results in small form factor usage at the system level.

The drivers listed above lead to the following set of challenges in each of the four SOP substrate integration technologies:

High-Density Wiring Challenges

- Substrate core with low thickness for miniaturization, high stiffness, or high modulus to minimize warpage with buildup layers, and low coefficient of thermal expansion (CTE) to minimize stress on the interconnection joint between the IC and the substrate thus improving reliability
- Ultrafine-line and pitch conductors with line size and via reduction for fine-pitch flip-chip routing while maintaining acceptable conductor resistance and required impedance control
- Dielectrics with low dielectric constant, low thickness of films for high signal speed, and low-loss tangent for minimizing signal loss and crosstalk noise

Embedded Passive Challenges

- High-density embedded thin-film decoupling capacitors for power noise control
- High-precision thin-film resistors embedded in the substrate for line termination and signal and power impedance and noise management

Embedded Active Challenges

- High-yielding processes for high-I/O-count ICs embedded in the package substrate for reducing thickness (miniaturization) and interconnect length (high performance)

Thermal Management Challenges

- Integration of small-form-factor thermal structures
- Heavy copper planes and thermal via structures in the core substrate

7.3.2 Ultrathin-Film Wiring with Embedded Low-*K* Dielectrics, Cores, and Conductors

The primary purpose of thin-film wiring is to interconnect I/Os of active and passive components. This is accomplished by four major substrate technologies:

1. High-modulus and dimensionally stable core materials
2. Advanced low-*K* thin-film dielectrics
3. Thin-film conductors
4. Multilayer wiring integration processes involving all the preceding three technologies

Core Substrate Materials

The core provides mechanical support and warpage control during substrate manufacturing. The processing of ultrathin embedded thin-film components is not feasible without an ultrathin but extremely high modulus core. Electrically, the core provides vertical via interconnections by means of plated through holes.

Attributes of Core Materials The main attributes of core materials relate to thermal management, CTE management, rigidity, and flatness. The electrical properties of the core play a key role in the power distribution network and also for signal integrity in case of active ICs embedded in the core. The processability of the core material for fine pitch through hole interconnections is another important attribute.

Thermomechanical Properties

High elastic modulus. The need to process 10 or more layers of thin films onto cores requires core materials that warp very little, thus requiring a high modulus for the core as given by Stoney's equation:

$$\rho \prec E_s \Rightarrow \text{Warpage} \prec \frac{1}{E_s}$$

where ρ is the radius of curvature of the substrate and E_s is the elastic modulus of the substrate.

It can be seen from the preceding equation that a higher modulus core results in a higher radius of curvature and consequently lesser warpage, as illustrated in Figure 7.6. Materials with a higher elastic modulus such as C-SiC have much lower substrate warpage as buildup layers of wiring and embedded components are deposited on them.

Low CTE. As power consumption of ICs increases, the temperature rise in the IC and the junction between the IC and the substrate goes up leading to high enough thermal mismatch stresses to fail the joint. Silicon, the most commonly used IC, expands at a rate of 3 ppm/°C, and traditional FR4 and BT cores have a TCE around 17 ppm/°C. Organic underfills with lower TCE fillers such as silica have typically been used to address this problem. The ideal TCE of core material is around 6 to 10 ppm/°C, about halfway between silicon and the FR4-based organic board.

High thermal conductivity. The core should possess high thermal conductivity to aid in dissipating heat from the chip through multiple layers of organic-based wiring. Metal cores are the best in this respect.

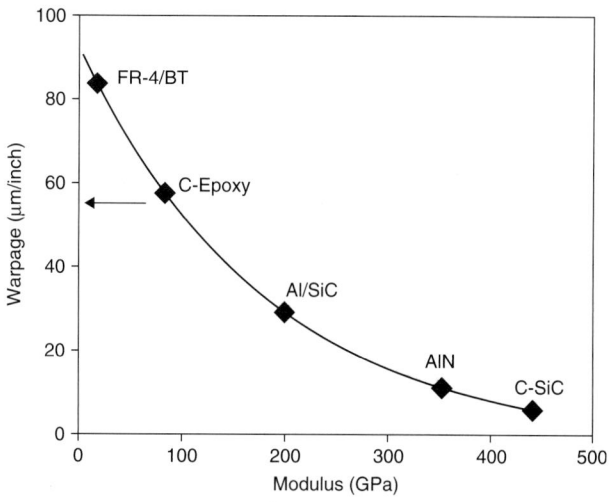

FIGURE 7.6 Warpage of SOP substrate as a function of elastic modulus.

Processability. The core material should not only have a high modulus and low TCE but also should be low cost and be thin-film processable in large areas (600 mm) with through vias.

Electrical Properties

Low dielectric constant. The core should have a low dielectric constant to allow for fast signal propagation. As signal speed is inversely related to the square root of the dielectric constant, decreasing the dielectric constant will result in a higher speed.

Low loss. To prevent large changes in impedance in high-frequency applications, the loss factor should be low.

Classification of Core Materials   Core materials can be classified into conducting and insulating, as shown in Figure 7.7.

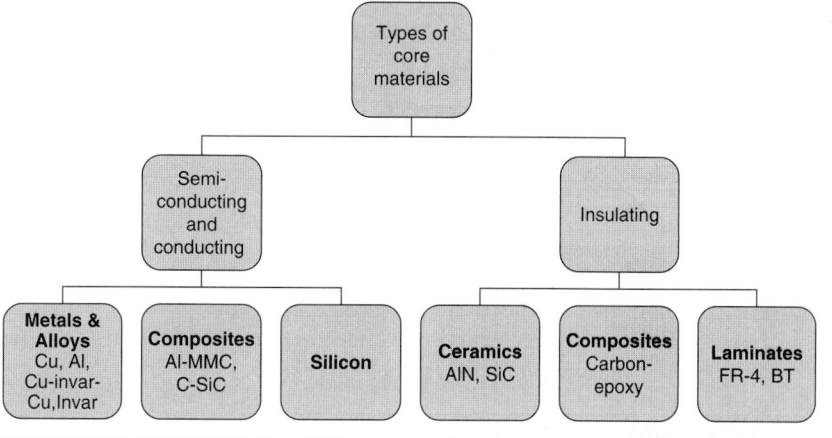

FIGURE 7.7 Types of core materials.

Semiconducting and Conducting Cores

Metal cores. Pure metal cores such as Cu and Al provide high thermal conductivity and good stiffness. They can address thermal and warpage issues. However, they have a large CTE mismatch with the die material. This results in reduced package reliability.

Composites. Metal matrix composites possess many attractive properties such as machinability and high thermal conductivity, but do not easily meet stiffness requirements. Commercially available Al-matrix composites filled with carbon cloth reinforcement are one example of a material that is available and that meets most of the requirements except stiffness. Invar and Cu-invar-Cu meet all the requirements. It has been in volume production for high-performance server applications for more than a decade.

Another material that has been developed more recently is based on carbon-SiC for which a manufacturing process (patented by Starfire Systems, Inc.) has been demonstrated to yield a large area and the required thinness, stiffness, and Si-matched CTE. Composite panels of carbon fibers and silicon carbide matrix are formed from commercially available carbon fiber fabrics and felts and a liquid polymeric ceramic precursor. The polymeric precursor is a highly branched polycarbosilane, which decomposes on firing to 850°C to give amorphous silicon carbide. This preceramic polymer allows for the design and fabrication of advanced ceramic matrix composites at low temperatures, in large-area sheets with a tailorable CTE and modulus.

Si core. Here, Si as a wafer is used as a core with through-silicon vias (TSV) and organic buildup layers. While Si is attractive because of its smoothness, TCE, and modulus, it has two drawbacks—its high electrical loss and brittleness.

Insulating Cores

Ceramics. Ceramic materials have high thermal conductivity, low TCE, and high stiffness. Ceramics such as AlN and SiC are good examples in this category. They suffer from high cost and lack of large-area availability.

Laminates. Laminate materials are those composed of a core clad with conductor layers on both sides. The most commonly used organic laminates are FR4 and BT. However, these materials do not posses a high enough modulus to provide warpage control for highly integrated SOP substrates. In addition, they have a very high CTE mismatch with Si resulting in large stresses in the solder joint.

Low-CTE organic laminate materials have been developed with advanced fillers such as Kevlar-aramide, with negative to low CTE. Nonwoven aramide-reinforced laminate systems have tunable in-plane CTEs that reduce the CTE mismatch between the IC and laminate substrate. These materials also have high laser drillability for vias due to the absence of woven glass fiber reinforcement.

Composites. The composite approach is often used to obtain the desirable properties of two different classes of materials. Carbon composite cores provide many advantages such as high in-plane thermal conductivity, low CTE, and high rigidity. For example, pitch carbon has diamond-like stiffness, and a pitch carbon epoxy can yield a stiffness of 200 GPa when the reinforcement is more than 60 volume percent, but the high filler loading results in a brittle composite material. In addition, these materials present processing difficulties and are not easily drillable.

Through-Via Processing Through vias are typically generated in core substrates by mechanical drilling or laser ablation. Advanced mechanical drilling can produce

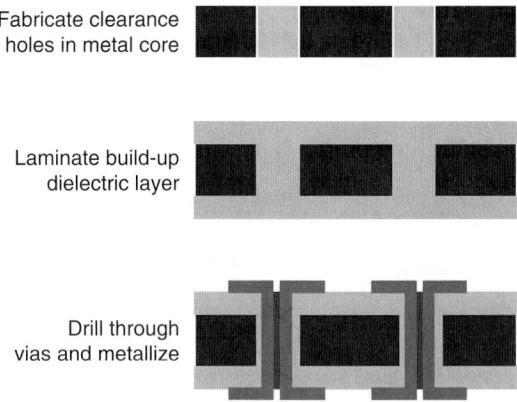

FIGURE 7.8 Process flow for a conducting core.

through holes of up to 100 μm. Laser via processes are explained in more detail in the section on dielectrics.

Conducting Core The process flow for a conducting core is illustrated in Figure 7.8. Since the core is electrically conducting, it is essential to isolate the core from the buildup wiring layers. Clearance holes are fabricated in the metal core by drilling. The buildup dielectric is then laminated, thus filling the vias, and smaller through vias are then formed and metallized within the clearance holes.

Insulating Core The formation of through vias in insulating substrates is illustrated in Figure 7.9. The conductor is first patterned and then etched with the required pattern. Through vias are then formed where required by drilling through the dielectric. The conductor is etched away from the areas where vias are to be formed. The vias are then metallized and plugged.

Thin-Film Buildup Organic Dielectrics

A dielectric serves to isolate conductor layers from each other and also provides vertical a through-via interconnection (Z interconnection). A typical SOP substrate consists of multiple layers of conductors separated by an insulating material of very high insulating resistance. Dielectric materials perform this function. Additionally, two adjacent metal layers are connected to each other through vias in the dielectric. These vias are metallized to make the interconnection.

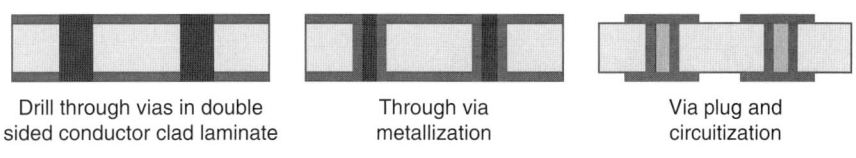

FIGURE 7.9 Process flow for an insulating core.

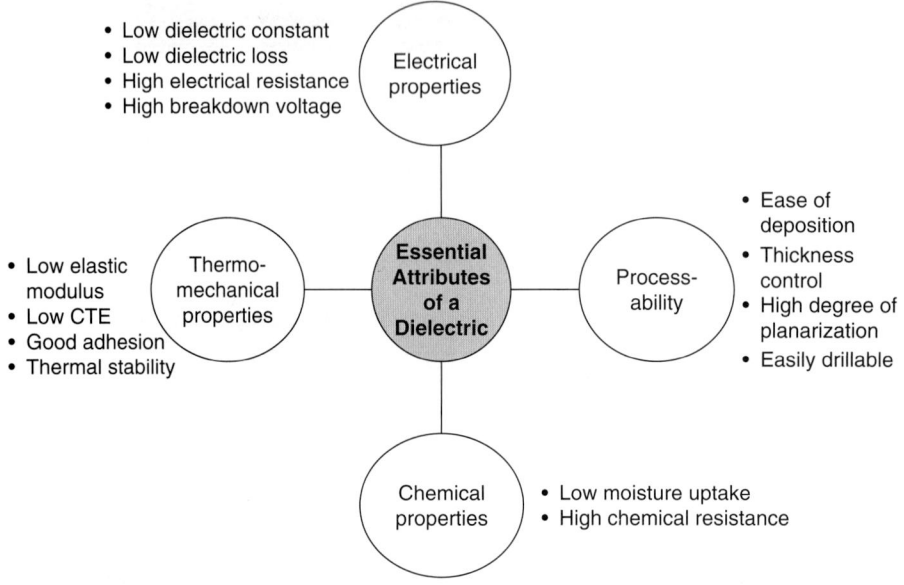

FIGURE 7.10 Essential attributes of a dielectric.

Attributes of Dielectric Materials The important properties that a dielectric material should possess are schematically illustrated in Figure 7.10. Each of these properties is discussed below.

Electrical Properties

Low dielectric constant. Signal speed is inversely proportional to the square root of the dielectric constant as given by

$$v_p = c/\sqrt{\varepsilon}$$

where v_p is the signal propagation velocity, c is the speed of light, and ε is the dielectric constant. Thus decreasing the dielectric constant contributes to a higher signal speed and consequently better electrical performance.

Low dielectric loss. When a sinusoidal voltage is applied at low frequencies, the polarizations inside a dielectric material develop completely before the field reverses. The time-variant polarization is equivalent to an alternating current, which leads the voltage exactly by 90°. Physically, the energy loss can be viewed as resulting from the molecular friction that opposes the molecular motion leading to an energy loss during polarization. Dielectric loss is a measure of the electrical energy dissipated during one polarization cycle. Energy losses are important in high-frequency signal transmission for digital and RF functions, not only because they represent a lack of efficiency, but also because energy losses change the impedance of the circuit.

High electrical resistance. A dielectric is an insulating material between two conducting layers and, as such, must have a minimum value of resistance to be useful. If a dielectric were to have an insufficient value of resistance, electrical performance would be severely affected.

High breakdown voltage. A dielectric must be able to withstand high voltages without breaking down. Breakdown of a dielectric material is a phenomenon where the high magnitude of the electric field breaks bonds within the material.

Thermomechanical Properties
Low elastic modulus. It is desirable to have dielectrics with a low elastic modulus. A low elastic modulus will impart low stress and thus contribute to higher reliability.

Low CTE. A low coefficient of thermal expansion closer to silicon will result in lower stresses on the IC, substrate, and the interconnection between the two.

Good adhesion. Since SOP substrates have multiple layers of wiring, good adhesion between the conductor and dielectric materials is necessary, leading to a more reliable package.

Thermal stability and high glass transition temperature. Dielectrics must have a high decomposition and glass transition temperature in order to withstand assembly processes, which can be as high as 260°C for lead-free solders and higher for AuSn and other alloys.

Chemical Properties
Low moisture uptake. Uptake of moisture can drastically alter the properties of the dielectric. Dielectrics must have low moisture uptake so as to prevent mechanical and electrical degradation both during processing and during product use.

High chemical resistance. Dielectric materials must have resistance to a wide variety of chemicals so that they can easily be processed for a variety of structures.

Processability
Good drillability. Because of the large number of vias used in SOP substrates, drillability is an important parameter. A good dielectric should be easily drillable. In most cases drillability is affected by the fillers used in the dielectric material. Fillers are used in dielectric materials for a wide variety of reasons including CTE control and matching with the IC. Dielectric fillers must be carefully evaluated for their drillability before incorporation.

Ease of deposition and thickness control. The dielectric must be easily deposited for buildup either as a liquid or dry film, with low process time and temperature. It must also be possible to control the thickness precisely, and the dielectric should have good flow behavior to allow for planar layers.

High degree of planarization. The formation of fine lines and spaces is contingent on the presence of a flat dielectric surface. A high degree of planarization (DOP) of the dielectric is required both for photolithography and metallization. Referring to Figure 7.11, the DOP is defined as

$$\text{DOP} = \left(1 - \frac{T_2 - T_1}{T_0}\right) \times 100\%$$

Classification of Dielectric Materials Figure 7.12 shows the evolution of dielectric materials during the last few decades. Initially, ceramic substrates with thick-film technology were utilized by sequential buildup technologies. These gave rise to cofired thick-film structures with metals forming the foundation for multilayer ceramics (MLC). This technology gave way to organic thin-film technology in the 1980s because of the

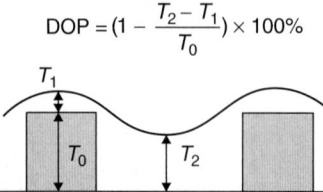

FIGURE 7.11 Degree of planarity.

lower dielectric constant and ease of processing thin films. Today, most high-density packages use low-cost epoxy-based dielectrics and low-cost organic core substrates (e.g., FR4 epoxy fiberglass boards) [1]. Epoxies are thermosets that are widely used in substrates due to their excellent adhesion, good thermal stability, low processing temperature (<150°C), and low cost. However, epoxies also have higher dielectric constants (3.5 to 5.0) than many other polymer dielectrics and have high water uptake (0.3 to 1.0 wt%).

These shortcomings of epoxies led to the development of a new class of low-loss thermosets such as polyimide, BCB, and polynorbornene. Around the same time, PTFE (Teflon) also began to be used. PTFE is a thermoplastic with an extremely low dielectric constant of ~2.1. However, it presents significant processing challenges due to its poor adhesion resulting in high cost. Another thermoplastic, LCP, started gaining acceptance in the late 1990s. However, its high processing temperature has been an issue. To optimize both the processing cost and performance, a new class of thermosets such as polyphenyl ethers (PPE) and hydrocarbon-ceramics were developed.

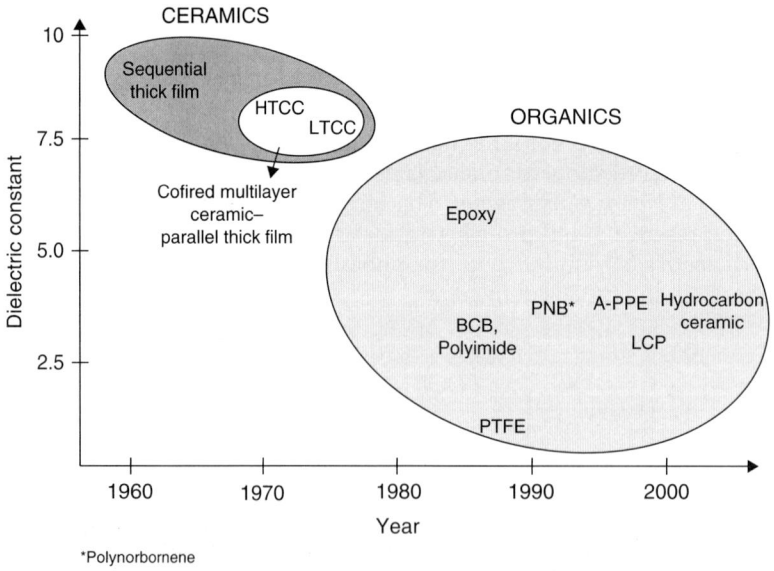

FIGURE 7.12 Historical evolution of dielectric materials.

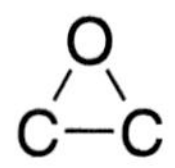

FIGURE 7.13 Structure of epoxy group.

A number of dielectric materials that are typically used in organic substrates are described in more detail below.

Thermosets Thermosets are polymer materials that are cured to form a rigid cross-linked structure. The energy required for curing may be supplied in the form of heat or through a chemical reaction.

Epoxy. The epoxy group is a three-member ring as shown in Figure 7.13. The presence of a highly strained three-member ring gives epoxies high reactivity to a wide range of organic structures, making it a very versatile polymer.

The synthesis of epoxy utilizes bisphenol A and epichlorhydrin as reagents. In the presence of alkali, they generate the structure shown in Figure 7.14. A typical example of an epoxy buildup dielectric widely used in the industry is Ajinomoto Build-up Film (ABF).

Advantages: Excellent adhesion, good solvent resistance, good thermal stability, low cost.

Disadvantages: High dielectric constant (ε_r = 3.5 to 5), high moisture uptake.

Polyimide. Aromatic polyimides have the general structure shown in Figure 7.15. The rigidity due to a large number of aromatic groups provides high T_g and imparts good mechanical properties. The classic synthesis of polyimide via polyamic acid is illustrated in Figure 7.16. Di-anhydride and di-amine are typically reacted to form polyamic acid. The second step involves cyclodehydration either at a high temperature or by utilizing a dehydrating agent.

Advantages: High T_g, excellent solvent resistance, low loss, good adhesion.

Disadvantages: Moisture absorption results in change in dielectric constant.

BCB. Benzocyclobutene is a low-loss polymer used in high-frequency applications. One of the synthesis routes for production of BCB is illustrated in Figure 7.17.

The BCB hydrocarbon is produced by pyrolysis of α-chloro-o-xylene [2]. Subsequently, bromination is carried out to produce 4-bromo-BCB, which is reacted with

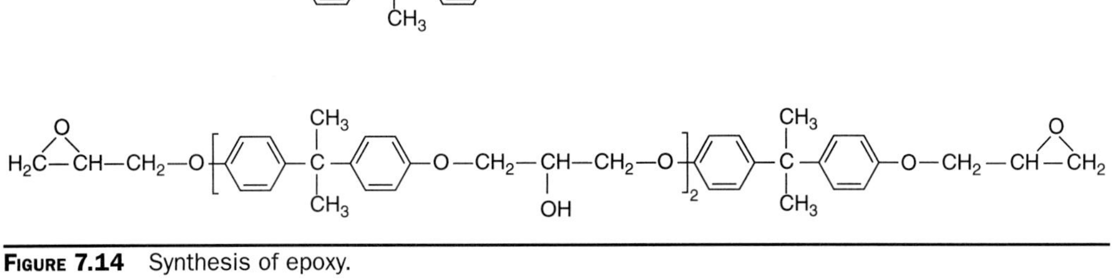

FIGURE 7.14 Synthesis of epoxy.

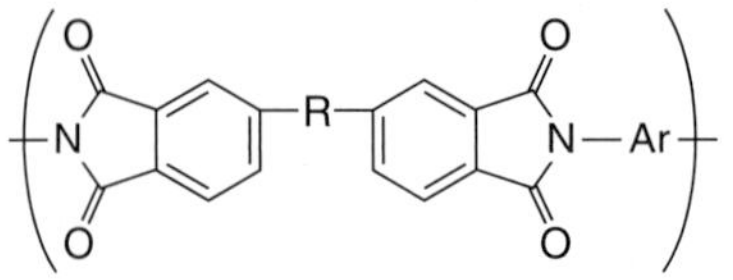

FIGURE 7.15 General structure of aromatic polyimides.

divinyltetramethylsiloxane to produce DVS-bis-BCB. This monomer is then thermally polymerized. B-stage DVS-bis-BCB (incompletely cured) has been marketed by Dow Chemical as Cyclotene.

Advantages: The advantages of BCB include low dielectric constant (ε_r = 2.50 to 2.65), low loss factor (<0.001), high chemical resistance, low curing temperature, high T_g (>350°C), low moisture uptake, and high degree of planarization.

Disadvantages: High CTE (45 to 52 ppm/K), weak adhesion and high cost.

Polynorbornene. Norbornene is a bridged cyclic hydrocarbon. The molecule is illustrated in Figure 7.18. Polynorbornene can be prepared by a variety of processes including ring opening metathesis polymerization and vinyl-addition polymerization of norbornene. The polymer obtained in each case is different. The polymer obtained by vinyl-addition polymerization is marketed as Avatrel. Figure 7.19 shows the production of polynorbornene by vinyl-addition polymerization.

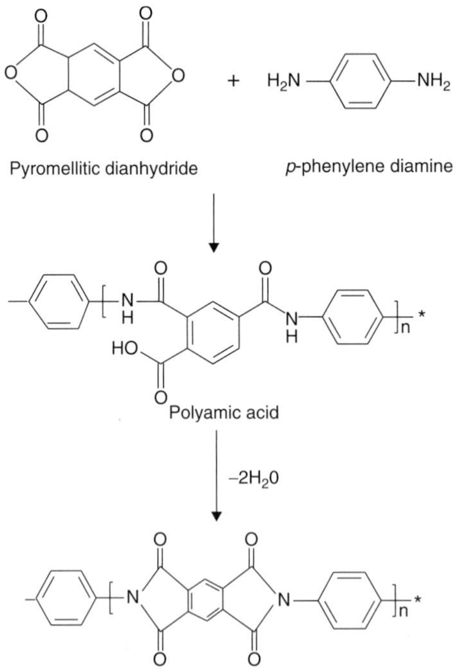

FIGURE 7.16 Synthesis of polyimide.

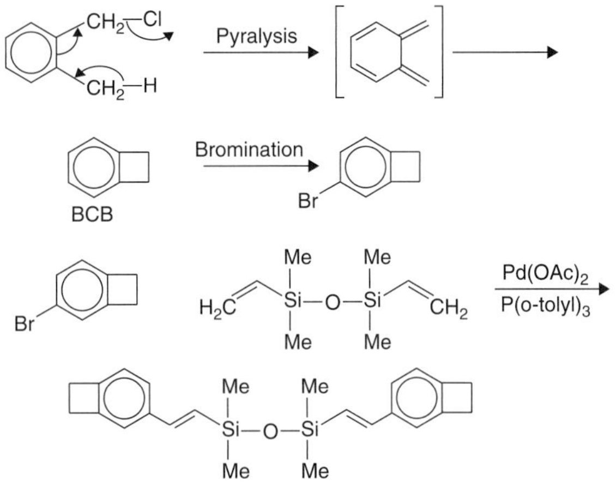

FIGURE 7.17 Synthesis of BCB. [2]

Advantages: Low dielectric constant ($\varepsilon_r = 2.50$), low loss, low moisture uptake.

Disadvantages: Very high CTE (180 ppm/K), weak adhesion and high cost.

Thermoplastics Thermoplastics are polymers that do not undergo a curing process, but soften at high temperatures.

PTFE. Polytetrafluoroethylene is a synthetic fluoropolymer. The synthesis of PTFE is illustrated in Figure 7.20. CF_4 may be subjected to high pressure resulting in emulsion polymerization due to free radical catalysis. Alternatively, the hydrogen atoms on polyethylene can be directly substituted with F atoms using F_2 gas.

PTFE has excellent dielectric properties. Its dielectric constant is ~2.1 and loss is < 0.0002. The advantages and disadvantages of PTFE are summarized below.

Advantages: Excellent dielectric properties (low dielectric constant and low loss), excellent chemical resistance, low CTE (7 ppm/K).

Disadvantages: Extremely poor adhesion, expensive processing, high melt temperature leading to high processing temperatures, melt processing limits wiring density due to poor dimensional stability.

LCP. Liquid crystals are a phase of matter that exhibits properties between those of a conventional liquid and that of a crystalline phase. The material may flow, but it still exhibits ordering of the constituent molecules. Liquid crystalline polymers (LCPs) are a class of wholly aromatic polyester polymers. Because of the orientation of these rigid

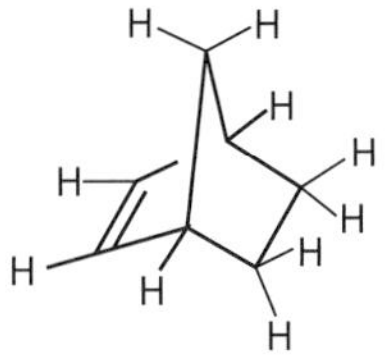

FIGURE 7.18 Norbornene molecule.

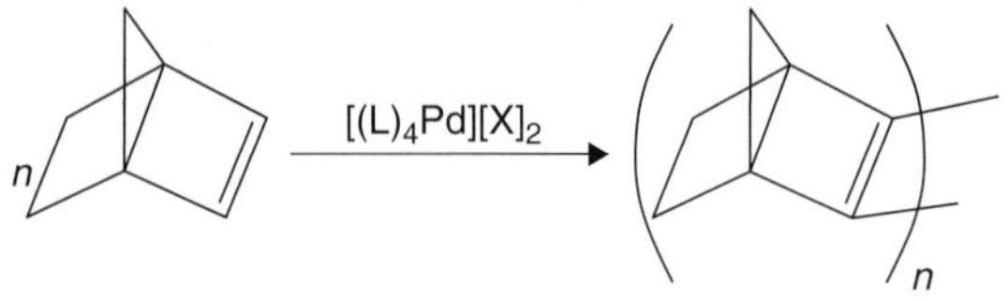

FIGURE 7.19 Polymerization of norbornene.

molecules during processing results, LCPs are termed self-reinforcing. They have many desirable properties and are especially used in high-frequency applications because of their low dielectric loss.

Advantages: Low dielectric constant (~2.9), low loss (<0.005), low moisture uptake, tunable in-plane CTE.

Disadvantages: Expensive processing, high processing temperature due to high melting point.

Dielectric Processes The processing of dielectrics involves the steps shown in Figure 7.21. The dielectric is first deposited on the patterned core. This may be done by spin or meniscus coating if liquid dielectrics are used or by lamination of dry film. The dielectric is then cured and vias are drilled. Each of these processes is detailed below.

Deposition The deposition of the dielectric on the patterned core may be done in a variety of ways depending on the material used and the processes available. Here we consider two such processes.

1. *Spin-coating.* Spin-coating is used if the dielectric is deposited from solution. Examples include BCB and polynorbornene. Spin-coating can typically yield thicknesses from 2 to 20 μm. The thin-film coating is obtained by rotating the substrate at high speed (300 to 10,000 rpm) after dispensing the solution on the core. Empirically it is seen that the thickness is proportional to the square root of the revolutions per minute.

2. *Lamination.* A laminate is constructed by unifying two layers of materials. The layers in question here are the core and the dielectric in the form of a dry film.

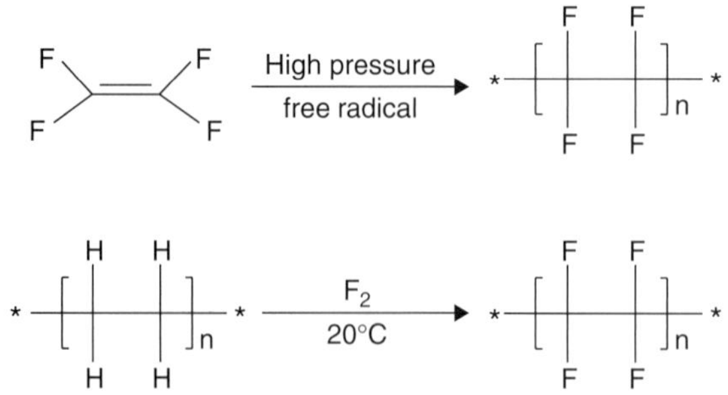

FIGURE 7.20 Synthesis of PTFE.

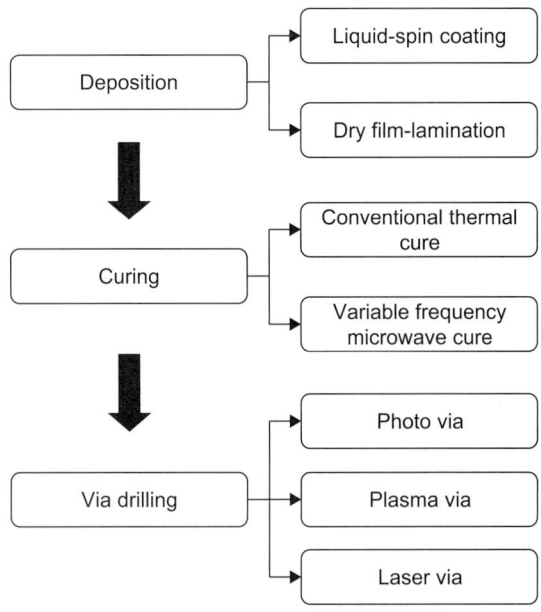

FIGURE 7.21 Processing of dielectrics.

The dielectric is given adhesion treatment so that it will adhere to the core and the conductor that is deposited on it subsequently. Lamination is generally carried out at temperatures > 70°C. At this temperature the dielectric adheres to the underlying core and subsequent processing can be carried out.

Curing Curing is only required for thermoset materials. All thermosets are subjected to a cure cycle, which results in cross-linking.

Traditional curing process. Traditional curing involves subjecting the dielectric to a curing cycle in a conventional oven, which results in cross-linking of the polymer. The material is now "set" due to the cross links. However, many of the existing higher-performance polymers require high thermal processing temperatures that are well above the degradation temperature of traditional substrates. For example, the recommended cure profile for polyimide is shown in Figure 7.22. Not only is the temperature more than the degradation temperature of FR4, the time required is almost 5 hours. To address these issues of throughput and high temperature, new low-temperature processes such as microwave curing have been developed.

Variable frequency microwave (VFM) curing process. Variable frequency microwave (VFM) curing of high-performance polymers has been investigated as a low-temperature curing alternative to conventional heating in a thermal oven [3–7]. The unique feature of VFM heating, as compared to conventional heating, is the ability to quickly and repeatedly step through a range of frequencies. This stepping process provides a time-averaged uniformity in the energy distribution throughout the cavity thereby eliminating the nonuniformities in temperature that occur in single-frequency microwave chambers [8]. The VFM technique also allows metals and conducting materials to be placed in the microwave cavity. By cycling through thousands of frequencies in less than 1 second,

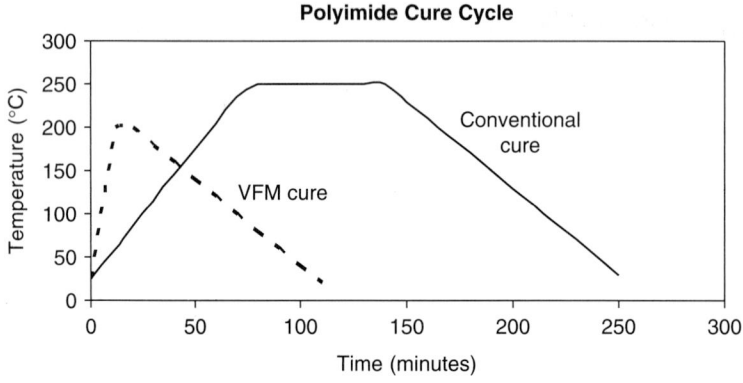

FIGURE 7.22 Conventional and VFM curing profile for polyimide.

the residence time of any established wave pattern is on the order of microseconds and problems with charge buildup and arcing are eliminated [9].

Tanikella [3] demonstrated the feasibility of rapid curing polyimides on organic substrates using VFM processing. Organic boards, such as FR4, are not significantly heated by microwave energy, but the precursor solutions of polyimide couple the microwave energy efficiently. As a result, full curing of the polyimide precursor is achieved without thermal degradation of the temperature-sensitive organic board. For example, Table 7.1 shows the extent of imidization achieved in a particular polyimide film (HD Microsystems PI2611), whose monomeric system consists of biphenyltetracarboxylic acid and phenylenediamine processed on an FR4 board using both conventional heating in a thermal oven and VFM processing.

It can be seen from Table 7.1 that a higher extent of imidization can be achieved by VFM processing for a much shorter cure time as compared to conventional thermal curing. For example, a 4-hour thermal furnace cure at 175°C gives 50 percent imidization, while a 5-minute VFM cure at 200°C gives an extent of imidization of 92 percent. Further, a 5-minute VFM cure at 200°C gives 100 percent imidization without degradation of the epoxy board. Only 73 percent imidization is achieved in a film cured for an hour in a conventional thermal furnace at 200°C, and films cured for an hour at 250°C achieve

Cure Method	Ramp Rate (°C/min)	Temperature (°C)	Hold Time (min)	Imidization (%)
Thermal cure	3	175	60	31
	3	175	240	50
	3	200	60	73
	3	250	60	100
VFM	15	175	5	92
	15	200	5	100

TABLE 7.1 Extent of Imidization Achieved in Polyimide PI2611 Films Cured on Blank FR4 Substrate

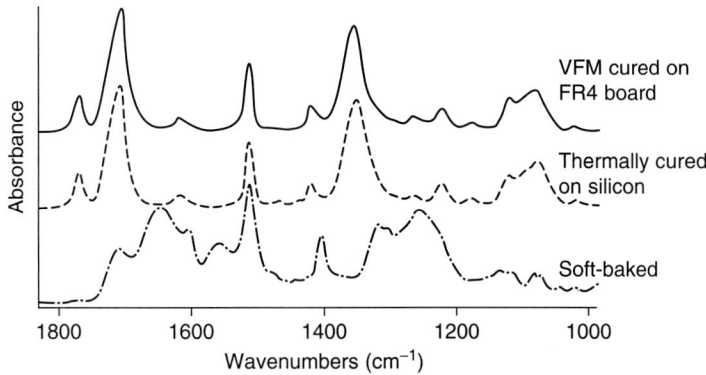

FIGURE 7.23 Infrared spectra of PI2611 films. (*a*) Soft-baked. (*b*) Thermally cured on silicon ramped at 3°C/min to 350°C and held for 1 hour at 350°C. (*c*) VFM cured on FR4 substrate, ramped at 15°C/min to 200°C and held at 200°C for 5 minutes.

100 percent imidization, but the FR4 board is decomposed. Moreover, Fourier transform infrared analysis confirms that there are no differences in chemical structure between a fully imidized system processed using VFM curing at 200°C compared with a film processed in a conventional thermal oven at 350°C (see Figure 7.23). Thus a high-performance polymer dielectric can be fully processed on a temperature-sensitive organic board, without board degradation.

Via formation processes Vias play a key role in achieving higher I/O density in a given area. The ability to consistently produce vias of the required diameter with excellent tolerance has tremendously enhanced the density of substrates with a large number of I/Os.

The blind and buried via technology increases the wiring efficiency dramatically over the traditional plated through hole MLB technology. Staggered vias are most commonly used as shown in Figure 7.24a. However, staggered vias have certain limitations because of the conformal structure of the microvia. The occupied real estate is large, the surface is nonplanar, and the signal path is longer. A stacked conformal via, hole stacking, shown in Figure 7.24b was then developed. The real estate is reduced, and the signal path is shortened, but there is still a hole on the surface. Stacked, filled nonconformal vias, shown in Figure 7.24c, overcome the disadvantages of the staggered and stacked conformal microvias. They provide a planar surface, shortest signal path, and smallest inductance within a small area. The flat surface is critical for ultrafine-line formation and ensures the availability of maximum real estate for routing.

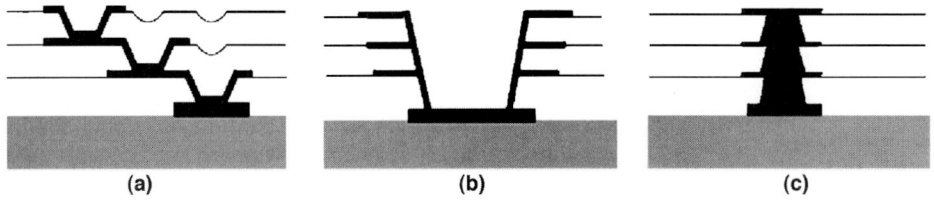

FIGURE 7.24 Various types of via structures. (*a*) Staggered conformal. (*b*) Stacked conformal. (*c*) Stacked nonconformal via structures.

Via formation processes fall into three categories as shown in Figure 7.25:

Photo via process. The photo via process (Figure 7.25a) utilizes photosensitive dielectrics. The vias are defined with a process similar to that used for patterning photoresists. The dielectric is exposed through a mask to form the vias. They are then subsequently developed and cured.

Plasma via process. Plasma-etched via (PEV) (Figure 7.25b) technology applies vacuum processing to remove the dielectric layer. All vias are generated simultaneously. A patterned layer of conductor serves as the mask—that is, the dielectric is etched through the openings of the conductor and etching stops at the inner conductor layer. PEV is a very flexible process: in addition to through and blind vias, it can create slots, windows, stepped windows, slanted vias, and unique structures as well.

Laser via process. The laser via process (Figure 7.25c) has been the most successful among the via formation technologies. Similar to PEV, laser vias can be generated through the polymer films by applying the patterned copper layer for masking by exposing the entire surface with lasers. This technology is economical for mass production due to its high throughput. The different types of lasers and their properties are illustrated in Figure 7.26.

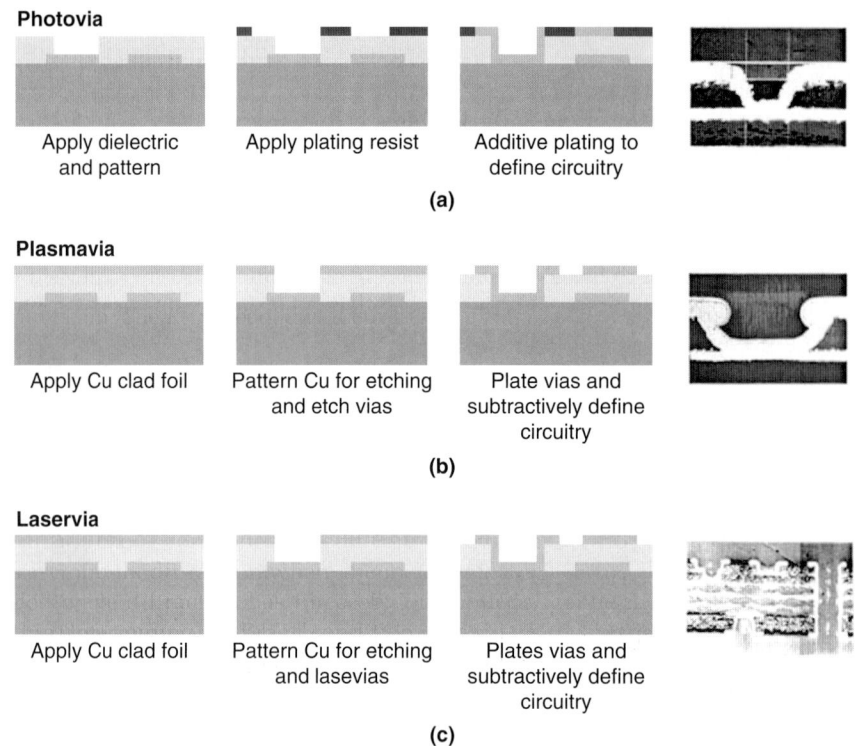

FIGURE 7.25 Via formation processes.

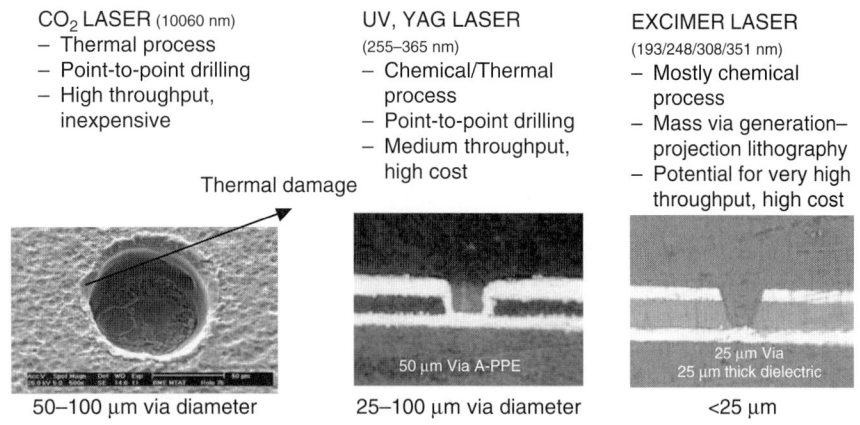

Figure 7.26 Comparison of various laser via processes.

A large number of dielectric materials and processes have been discussed. The major properties of some SOP dielectrics are summarized in Table 7.2.

Embedded Conductors

The function of the conductor is to carry the power and the signals. A judicious choice of conductor material and process leads to ultraminiaturization, better electrical and thermal performance, lower cost, and better reliability.

Attributes of Conductor Materials A conductor is a material that conducts electric current. In the SOP substrates with fine lines and vias, the conductor must have high electrical conductivity. At the same time, it must be easily processable. Among metals, silver has the highest thermal conductivity. But electromigration of silver results in electrical shorts, and hence copper is preferred. Figure 7.27 shows the evolution of conductor materials and processes. Initially, thick-film ceramic and glass buildup technologies used silver-palladium for sintering at 800°C. The cofiring of multilayer ceramics using high-temperature cofired ceramics (HTCC) in the 1980s required cofirable metals at 1600°C with such metals as molybdenum and tungsten that sintered at this temperature after screen-printing onto so-called green sheets. The HTCC led to so-called low-temperature cofired ceramics (LTCC) for two reasons—higher electrical conductivity of the conductor and lower dielectric constant of the ceramic. The conductor was thick-film copper, and the ceramic was either crystallized glass-ceramic or glass added to alumina ceramic. The demand for smaller and higher-performance packages led to the use of lower-dielectric-constant organics that were patterned with copper defined by subtractive etching. The SOP utilizes ultrathin electroplated copper defined by semi-additive plating or additive plating. Semi-additive plating makes use of a seed layer on which electroplating is subsequently performed. Recently, carbon nanotubes have been suggested as alternative conductor materials. Because of their excellent electrical properties based on their ballistic transport processes, controlled carbon nanotube growth may allow unprecedented performance in SOP systems.

Conductor Processing Processing of conductors entails their deposition to electrical and dimensional requirements.

Dielectric Material	Dielectric Constant @ 1 GHz	Loss Tangent @ 1 GHz	Modulus (GPa)	X,Y CTE (ppm/°C)	Availability	Via Formation	Via Metallization
Epoxy	3.5–4.0	0.02–0.03	1–5	40–70	Film, RCC, liquid	UV, CO_2 laser, photo	Electroless copper
Polyimide	2.9–3.5	0.002	9.8	3–20	Film, liquid	Excimer laser, photo	Sputter seed
PPE	2.9	0.005	3.4	16	RCC	UV, CO_2 laser	Electroless copper
BCB	2.9	<0.001	2.9	45–52	Liquid	Photo, RIE	Sputter seed
LCP	2.8	0.002	2.25	17	Laminate	UV laser, mech. drill	Electroless copper
Polynorbornene	2.6	0.001	0.5–1	83	Liquid	Photo, RIE	Sputter seed

TABLE 7.2 Dielectric Properties and Processes

SOP Substrate with Multilayer Wiring and Thin-Film Embedded Components 401

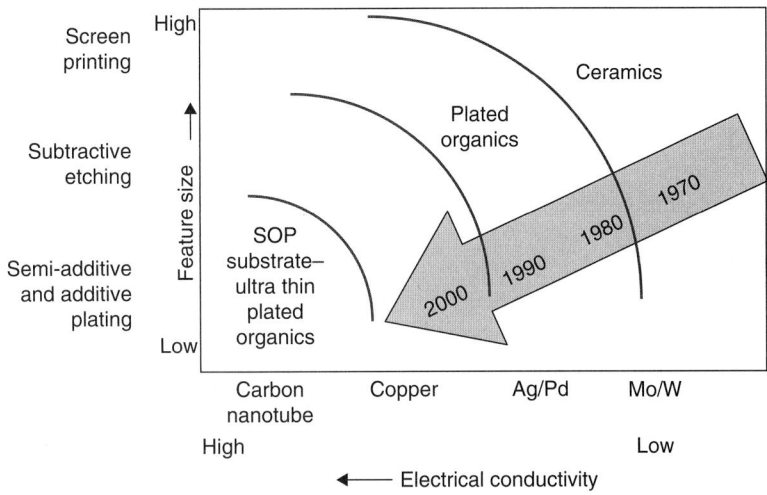

FIGURE 7.27 Historical evolution of conductor materials and processes.

Subtractive etching. The subtractive process (Figure 7.28) uses copper-clad laminates and subtracts the unnecessary pattern of copper from the surface by chemical etching, leaving the copper traces. Photolithography is used to control the areas to be etched. Photoresist is deposited on the entire surface of copper and exposed to make the required pattern. After development, only those areas where copper needs to stay will have photoresist. These areas will be protected during the etching process by the photoresist layer.

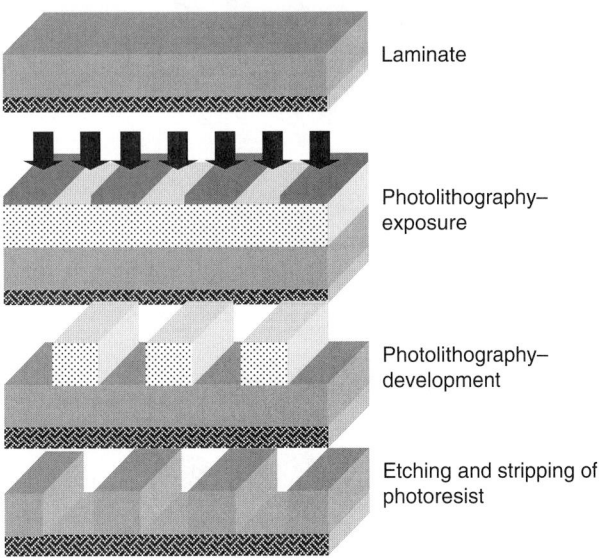

FIGURE 7.28 Subtractive etching process flow.

The main advantage of the subtractive process is the excellent adhesion of the copper foil to the substrate. Limitations of the subtractive etching process stem from the isotropic nature of the etching process. This results in an undercut of the line and reduces the resolution of the line that can be attained. To attain finer resolution, the thickness of the copper-clad laminate should be reduced. However, there is a practical limit to how thin a laminate can be used.

Semi-additive plating. To overcome the disadvantages of subtractive etching, the semi-additive plating process was designed. Figure 7.29 shows a process schematic of semi-additive plating. A very thin seed layer of copper is plated on the dielectric using electroless plating or sputtering. The function of the seed layer is to provide an electrically conductive seed surface for the higher-throughput electroplating process. The seed layer is then patterned using photolithography, and electroplating is then performed to generate the required pattern. The photoresist is then stripped and the seed layer etched away. This will of course result in removal of copper from the patterned areas too. So this should be taken into account while designing the process.

The thickness of the seed layer determines how fine the resolution can be. The thinner the seed layer, the smaller the line and space resolution that can be achieved. This is due to the inherently isotropic nature of wet etching. As the copper gets etched down, etching of the sidewalls also takes place leading to the formation of trapezoidal lines.

Ultrafine lines and spaces. The formation of ultrafine lines and space (<10 μm) requires strict control and optimization of interfacial roughness of dielectrics, seed layer formation, photolithography, and electroplating processes. Advances in all three are discussed below.

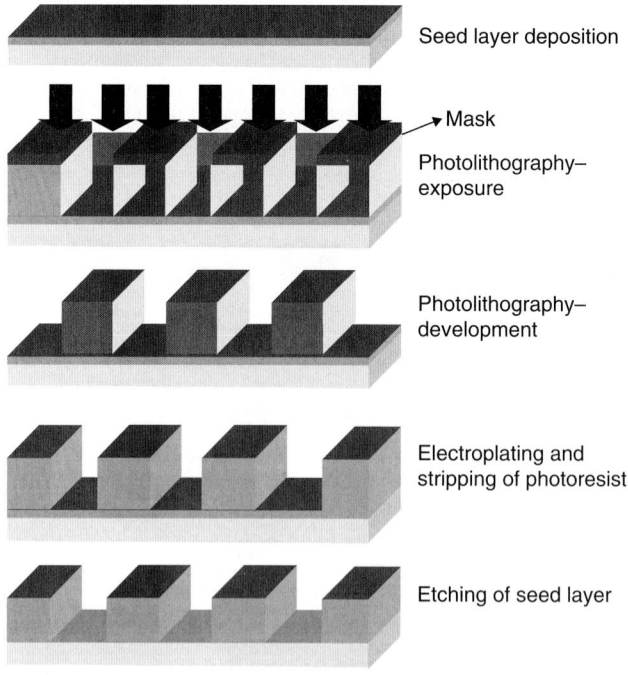

FIGURE 7.29 Semi-additive plating.

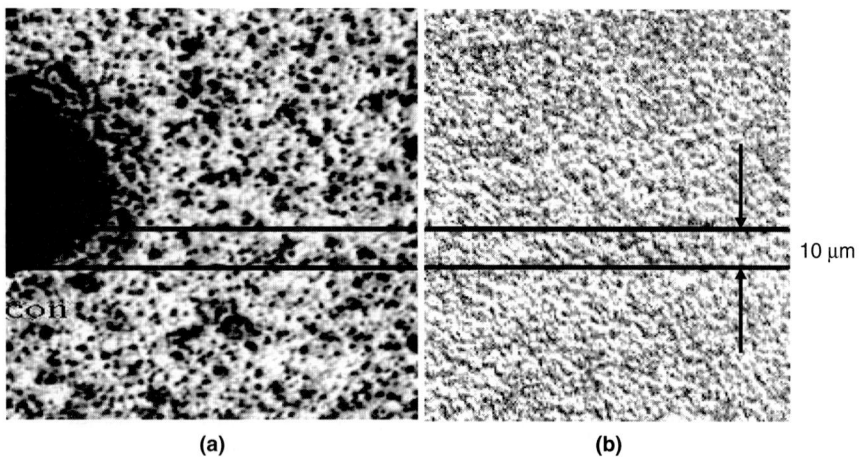

FIGURE 7.30 High-magnification optical micrographs of surface of buildup dielectric after (a) permanganate desmear, and (b) plasma desmear showing an overlay of a 10-μm line width.

Interfacial roughness. Interfacial roughness is an essential aspect of both fine-line lithography and metallization processes. Here, some new process developments for developing interfacial roughness are discussed. Current roughness treatments are based on permanganate desmear or other wet-etch technology. The dielectric surfaces obtained after roughening by these processes make them unsuitable for ultrafine-line formation.

Comparisons of typical epoxy dielectric surfaces obtained by permanganate desmear treatment and CF_4/O_2 plasma roughening is shown in Figure 7.30. Permanganate or other wet-etch processes result in a dielectric surface with a roughness on the order of 2 to 3 μm depth and large pits, as seen in Figure 7.30a.

Multilayer thin-film wiring on such a surface would result in latent defects in the traces and inconsistent dielectric thickness between metal layers as shown in Figure 7.31. The plasma treatment on the other hand produces a fairly uniform roughness on the surface that is typically < 1 μm deep.

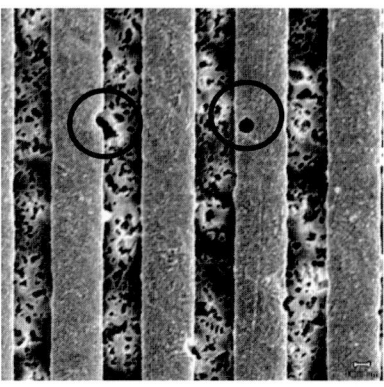

FIGURE 7.31 Scanning electron microscope (SEM) micrograph of defects in 10-μm lines caused by deep pits due to chemical desmear.

Seed layer plating. As has been indicated before, the seed layer is obtained either by sputtering or electroless plating. Electroless plating is a low-cost and batch-processing technique suited for high-volume manufacturing. However, traditional electroless baths have low deposition rates and use formaldehyde, a carcinogen. Also, the high pH of traditional electroless copper baths can degrade some types of photoresists. Furthermore, due to the smoother surface profile of plasma-treated dielectrics, older colloidal-based activation processes for copper plating do not work. Palladium activation is required for the latest generation of SOP dielectrics. The palladium particles are small enough to occupy the tiny pores of plasma-treated dielectrics and provide good activation.

To address the above issues, formaldehyde-free electroless copper plating chemistry with low pH and high deposition rates (3 to 4 µm/h) has been developed to meet the wiring and low-cost needs of SOP packages. Electroless copper plating involves the reduction of Cu ions to copper metal and the surface catalyzed oxidation of a reducing agent [10, 11]. The catalytic oxidation of formaldehyde increases with hydroxide concentration and is only effective at pH above 11. Several electroless copper solutions using nonformaldehyde reducing agents have been reported. The composition of one such bath based on hypophosphite is shown in Table 7.3 [12-17].

However, the inherent drawback of using hypophosphite as the reducing agent is the weak catalytic activity for the oxidation of hypophosphite on copper. While the initial substrate surface is palladium-activated, once it is coated with copper, the reaction slows because copper is not a catalytic material. One way to compensate for the poor catalytic activity of copper is to add nickel ions to the solution. The codeposited nickel in the copper deposit serves to catalyze the oxidation of hypophosphite, thus increasing the overall deposition rate [18]. Thiourea (TU) and diphenylthiourea (DPTU) have been shown to increase the deposition rate of electroless copper plating solutions that use N-(2-hydroxyethyl)ethylenediaminetriacetic acid trisodium salt hydrate (HEDTA) as the complexing agent and sodium hypophosphite as the reducing agent.

Figure 7.32 shows the surface morphologies of the copper deposits from the electroless solutions with and without additives (TU and DPTU). The topography of the copper deposited from the hypophosphite electroless copper plating solution is relatively rough with small growth colonies, which results in higher resistivity. TU and DPTU make the copper deposits more uniform, and the growth colony size is increased thus reducing resistivity.

$CuSO_4 \cdot 5H_2O$	0.04 M
$NaH_2PO_2 \cdot H_2O$	0.12 M
HEDTA	0.08 M
H_3BO_3	0.48 M
$NiSO_4 \cdot 6H_2O$	400 ppm
Polyethylene glycol	200 ppm
pH	9.3
T (°C)	70

TABLE 7.3 Composition and Operating Conditions of the Electroless Copper Plating Solution

FIGURE 7.32 Surface morphologies of the copper deposited from (a) basic electroless copper solution, (b) electroless copper solution containing 0.5 ppm thiourea, (c) electroless copper solution containing 1.0 ppm DPTU.

Low-stress electroless copper plating. Higher wiring density requirements of SOP substrates are forcing the need for smoother conductor surfaces for impedance control and less signal attenuation due to skin effect. But this results in poorer mechanical adhesion between the copper and dielectric. To counter adhesion problems and to obtain reliable SOP substrates, low-stress electroless plating processes have been developed. These processes produce fine-grained deposits with excellent adhesion characteristics. Furthermore, the deposits are blister-free and have excellent coverage of the dielectric surface.

Photolithography processes for sub-10-μm lines and spaces. The formation of ultrafine-line structures requires very controlled photolithography processing. The three primary problems that have to be overcome to generate fine lines and spaces are as follows:

1. *Nonplanar surface effects on photoresist patterning.* This effect is illustrated in Figure 7.33, which shows a 25-μm-wide line in the buildup layer running across a set of 100-μm lines and spaces on the underlying metal layer. It can be observed that the resist is not fully washed away in the "valleys" of the surface created by less than 100 percent degree of planarity of the dielectric layer.

2. *Bridge effect.* This is caused by scattering of the UV light at the edge of the fine line due to local roughness in the dielectric surface as shown in Figure 7.34. These can be caused by filler particles in the dielectric or by overetching during the chemical surface treatment process for electroless copper plating.

3. *Adhesion-related effects.* Surface roughness is a critical factor for both fine-line lithography and metallization. The displacement of the photoresist pattern and the lift-off of photoresist strip from the seed layer are more prevalent when the strip becomes smaller in geometry, as shown in Figure 7.35. This occurs due to the smaller contact area between the photoresist features and the seed layer resulting in weaker adhesion. Because of the same reason, plated copper lines tend to peel off from the dielectric film as they become finer. To counteract this, surface roughness treatments are necessary to improve the adhesion.

406 Chapter Seven

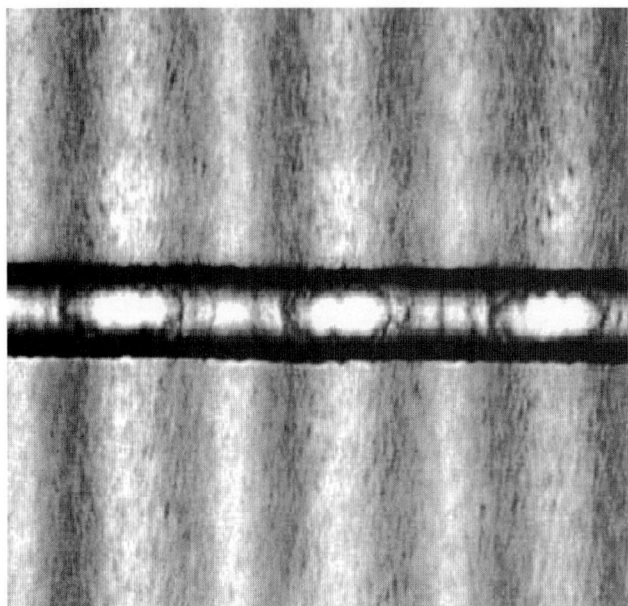

FIGURE 7.33 Fine-line opening problem on photoresist caused by undulating surface.

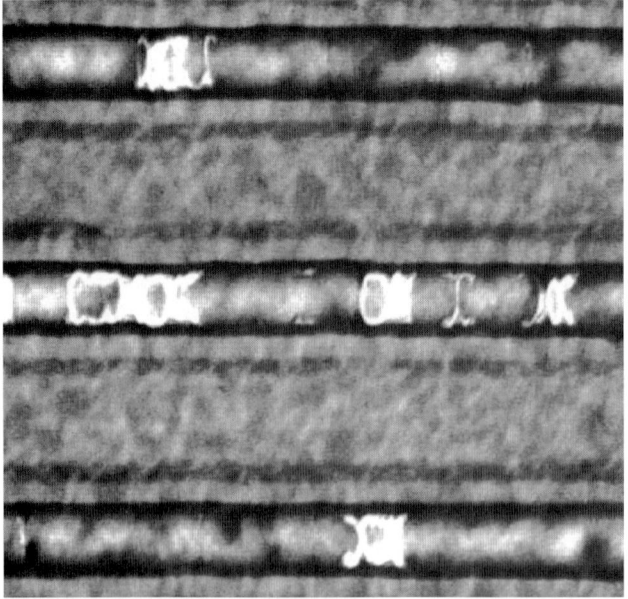

FIGURE 7.34 "Bridge" effect showing a very thin photoresist film remaining in the opening.

SOP Substrate with Multilayer Wiring and Thin-Film Embedded Components

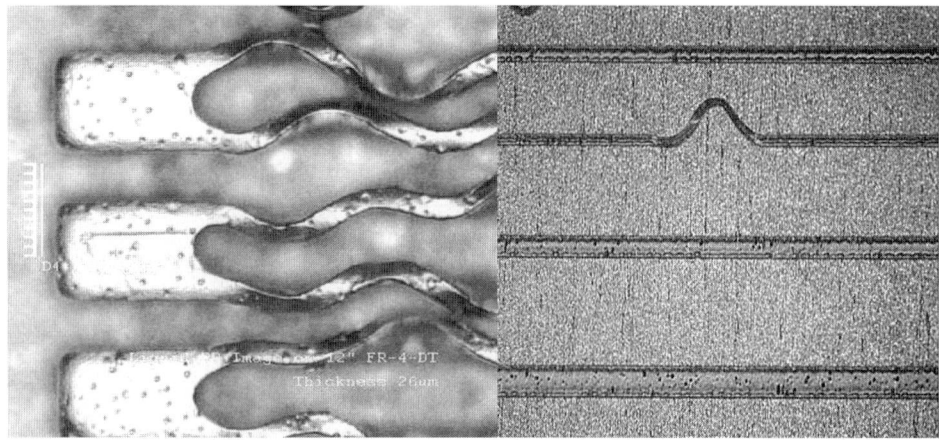

FIGURE 7.35 Fine-line photoresist lifting caused by narrow contact area and insufficient adhesion.

Electrolytic plating process. Electrolytic plating, or electroplating, is the process of using electric current to coat an electrically conductive object with a relatively thin layer of metal. A schematic of electroplating is show in Figure 7.36. In electroplating, the anode dissolves into solution. The dissolved ions then travel through the electrolyte to deposit on the cathode due to the electric field as follows:

Electroplating can be classified into two categories:

1. *DC plating.* DC plating is the most common method of electroplating used in substrate manufacture. To influence the grain structure formation, grain refiners and wetting agents as well as brighteners are added to the electrolytic bath, which is primarily composed of copper sulfate and sulfuric acid. During the

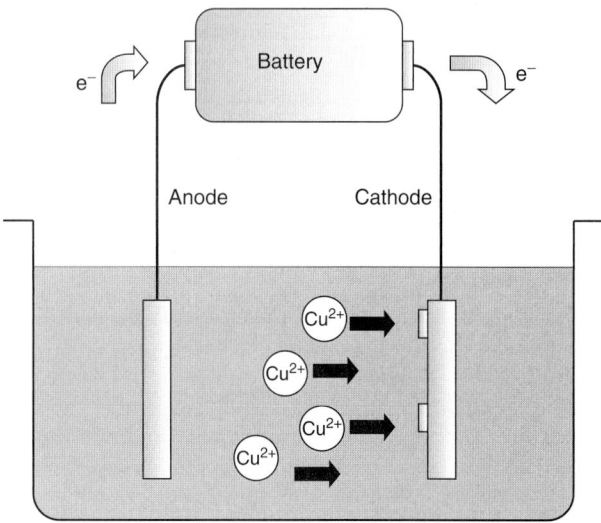

FIGURE 7.36 Schematic of electroplating.

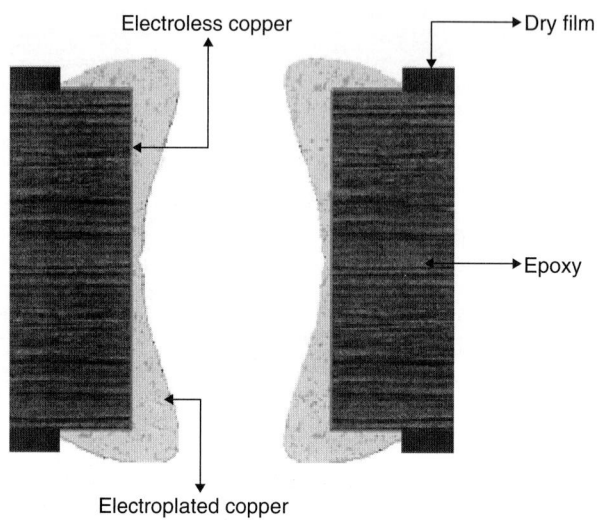

FIGURE 7.37 Dogboning of vias and through holes.

electroplating process, the copper ions are distracted from the solution. To maintain the copper ion concentration at a constant level, copper metal is dissolved in the electrolyte. However, during the anodic reaction some side effects are observed. In some cases, copper anode material is covered with an unknown layer. This layer is slightly soluble in sulfuric acid and is capable of blocking electric current. The anode thus becomes passive or is polarized. This reduces the efficiency of plating.

2. *Pulse reverse plating.* Pulse reverse plating is primarily used for plating blind vias and through holes with a high aspect ratio. Because of the depth of the via or through hole, more copper is deposited on the edges of the hole (high-current-density areas) than at the center (low-current-density areas) resulting in the so-called dogboning of the structure. This is illustrated in Figure 7.37. To counter this problem, pulse reverse plating is used. DC plating can be used to plate such structures, but the current densities that must be used are much lower resulting in higher processing time. In pulse reverse plating, instead of constant current, a waveform is used. A typical waveform is illustrated in Figure 7.38. The time ratio forward/reverse is about 20, but the reverse current density is about three times as high as the forward current density. The frequency is usually about 50 Hz.

Throwing power in a through hole is defined as the copper thickness in the middle of the hole divided by the copper thickness on the surface of the substrate near the hole. Reverse pulse plating is the key to achieving a high throwing power while keeping the processing times short. The high-current-density areas are shielded to reduce the amount of copper being deposited in these areas, so dogboning is avoided.

When the plating starts, copper is deposited in a dogbone fashion during the forward pulse as shown in Figure 7.37. During the reverse pulse, one of the organic components in the electrolyte is adsorbed at the high-current-density areas, shielding these areas (Figure 7.39). When the next forward pulse starts, some copper will be

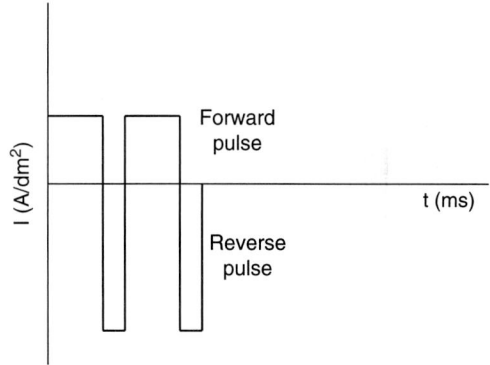

FIGURE 7.38 Typical waveform for pulse reverse plating.

deposited in the shielded areas, but more in the unshielded areas in the hole instead as shown in Figure 7.40. As the forward pulse proceeds, the organic shield is gradually broken down in order to allow some deposition of copper in the high-current-density areas. The result will be a copper deposit of uniform thickness in the hole.

By optimization of the above-mentioned processes such as photolithography, interfacial roughness treatment, and electroless and electrolytic plating, sub-10-µm lines and spaces have been generated. This is shown in Figure 7.41.

Multilayer Wiring Integration Processes

All the components for producing thin-film multilayer wiring are discussed above. This section will describe how all the above components are processed and integrated to form multilayer wiring.

Sequential Process An example of a sequential buildup process is shown in Figure 7.42. The process starts with a double-sided core. Through-hole vias are then drilled and metallized. The vias are then plugged by conductive or nonconductive paste or

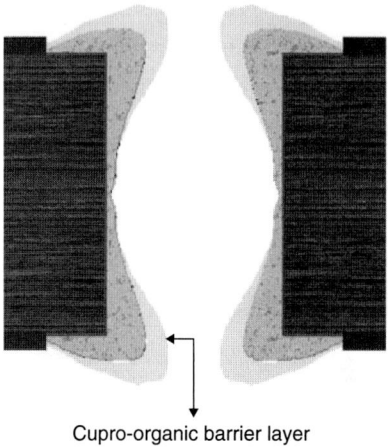

FIGURE 7.39 Deposition of barrier layer during the reverse pulse.

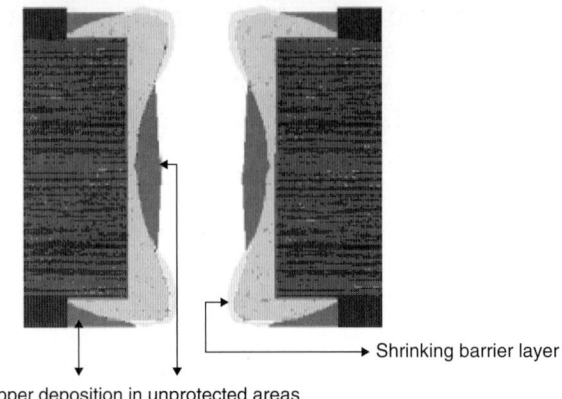

FIGURE 7.40 Copper deposition and barrier layer shrinkage during forward pulse.

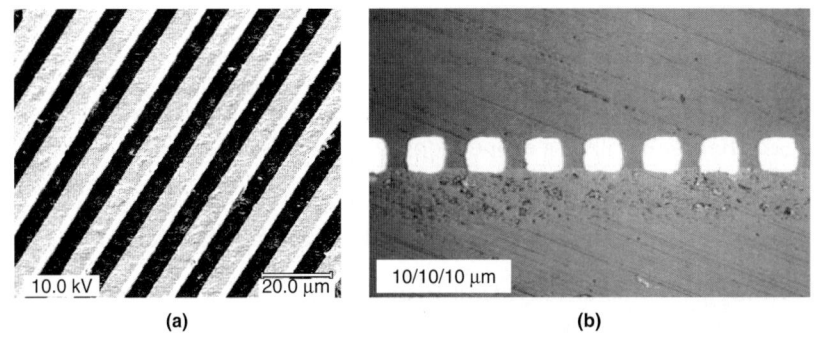

FIGURE 7.41 SEM and microsection view of 10-μm lines and spaces.

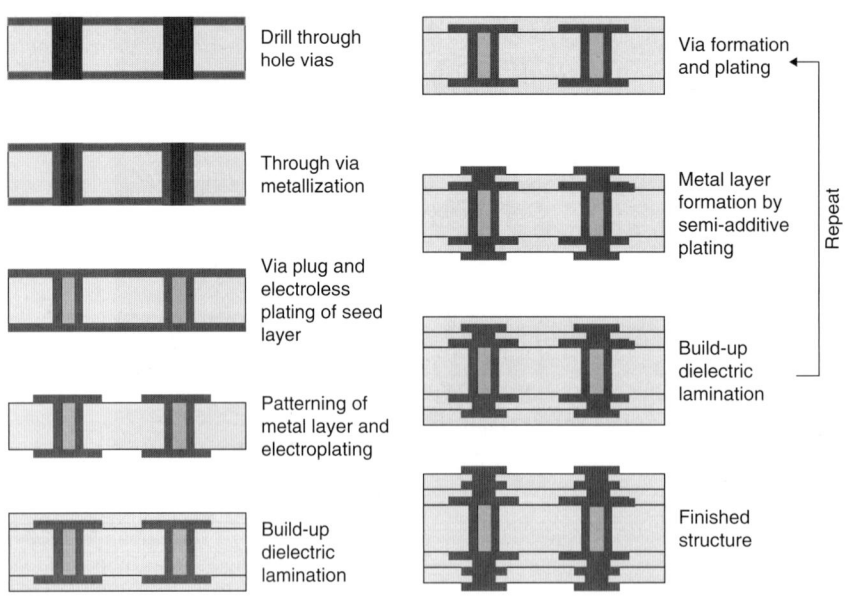

FIGURE 7.42 Sequential buildup process flow.

may be plated by electrolytic plating to form a conductive post. The core is then patterned by semi-additive plating.

A dielectric layer is then deposited on the core. Vias are formed and made electrically conductive. This can be done in three ways:

1. Sidewall metallization
2. Stud plating
3. Copper fill

Of these, stud plating and copper fill processes can be used to generate ultrahigh-density stacked vias.

Subsequently, the thin-film wiring in the buildup layer is generated by semi-additive plating. A seed layer is deposited by electroless plating which is then used for pattern electrolytic plating. The seed layer is then removed. The next layer of dielectric is deposited and the process continued until the entire substrate is finished. The main advantage of sequential processes is the increased wiring density achieved due to the presence of blind and buried vias. The primary disadvantage, of course, is reduced throughput due to the serial fabrication of the layers.

Stud Plating A schematic of the process sequence for stud plating is shown in Figure 7.43 [19]. This process utilizes photoresist to define the via structure. The first metal layer is formed using additive or subtractive processing. This layer is then protected using a barrier layer that is unreactive to the etching chemistry used. A seed layer of conductive material is then developed over this barrier layer. Alternatively, both the seed layer and the barrier layer can be combined together as one metal that is not etched out by the etching solution. Electrolytic plating is then performed to achieve the desired via height. Photoresist is then patterned on the panel-plated layer to define the vias. The panel-plated layer is then etched out with the photoresist protecting the via studs. If the seed layer and barrier layer are different, the seed layer is also etched out. The previous metal layer is protected by the barrier layer. The dielectric is then deposited and cured. The dielectric surface is roughened to expose the studs. This layer can then be patterned by semi-additive plating. The entire process can then be repeated for the next layer. A typical four-metal-layer structure fabricated this way is shown in Figure 7.44. This process is scalable and can be extended to microvias in the 10- to 15-μm range.

Copper-Filled Via Once the vias are drilled in each dielectric layer, the sidewalls are first metallized by electroless plating. The vias are then filled by pulse reverse plating. As has been explained previously, pulse reverse plating is required to generate sufficient throwing power for deep via structures. Stacked via structures can be generated by aligning the vias in the design. Figure 7.45 shows an example of a copper-filled stacked via structure.

Parallel Process In parallel processes, a stiff core is employed for each of the layers to be built and the layers are built in parallel. Finally, they are aligned together and laminated. The main disadvantage with parallel processes is that tolerance requirements for aligning multiple layers result in a lower wiring density. Parallel processes have the advantage of fast production times due to the simultaneous fabrication of all the layers.

412 Chapter Seven

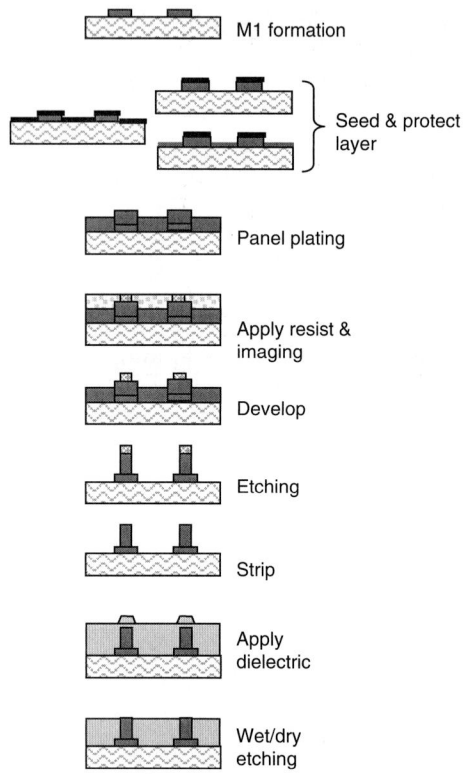

Figure 7.43 Process flow for stud plating—a method for producing stacked vias.

Figure 7.44 Cross section of stacked vias by stud plating.

SOP Substrate with Multilayer Wiring and Thin-Film Embedded Components

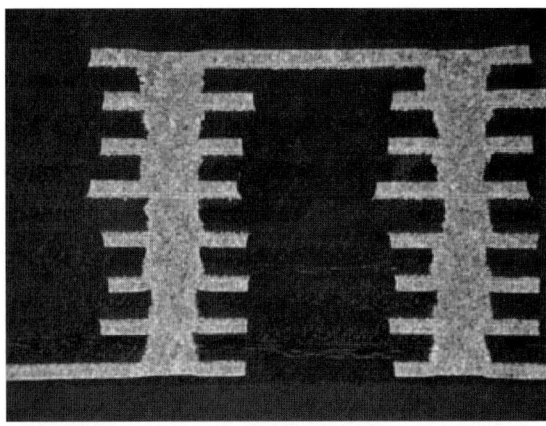

FIGURE 7.45 Cu-filled stacked via. (Courtesy Ibiden.)

A sample process flow for parallel processes is shown in Figure 7.46. Vias are drilled in the core and filled with a conductive paste. Copper is then laminated and the metal layer patterned by subtractive etching. This process is carried out in parallel for all the layers. Once all the layers are finished, they are aligned and laminated.

Paste Via Process After the via holes are generated, these vias are filled with a conductive paste such as a silver paste. Paste vias are very inexpensive, but are less reliable, than wall-metallized or copper-filled vias. For generating stacked-via structures, the layers can be designed so that the vias line up to form a stacked structure.

A number of parallel processes have been used in the industry. Prominent examples include ALIVH (all layer internal via hole) by Matsushita and B²IT (bumped buried interconnection technology) by Toshiba.

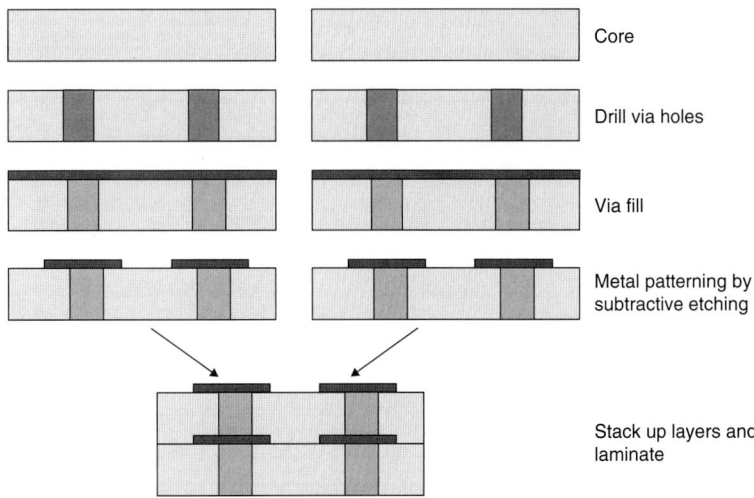

FIGURE 7.46 Process flow for parallel integration.

Parallel and sequential processes are often combined in fabricating a single substrate as required. This allows the attainment of the optimum between cost and production time.

Coreless In a bid to reduce profile thicknesses, coreless substrates were introduced. Coreless substrates present new challenges because of their warpage and processability issues. Many coreless substrates start with a core on which the entire package is built. Subsequently, the core is removed by chemical etching or mechanical grinding processes. Multilayer thin substrate (MLTS) from NEC, Japan, is a typical example [20, 21]. The entire structure is fabricated on a metal carrier that is subsequently etched away. This technology completely eliminates plated through holes, which significantly reduce wiring density. This method also eliminates handling problems arising due to warpage issues. Signal and power integrity were also found to improve using this technology [22].

Figure 7.47 shows a schematic of coreless substrate fabrication. The first metal layer is patterned on a solid copper core by additive processing. Nickel, gold, and another nickel layer are electroplated over the patterned copper layer. These layers aid in the

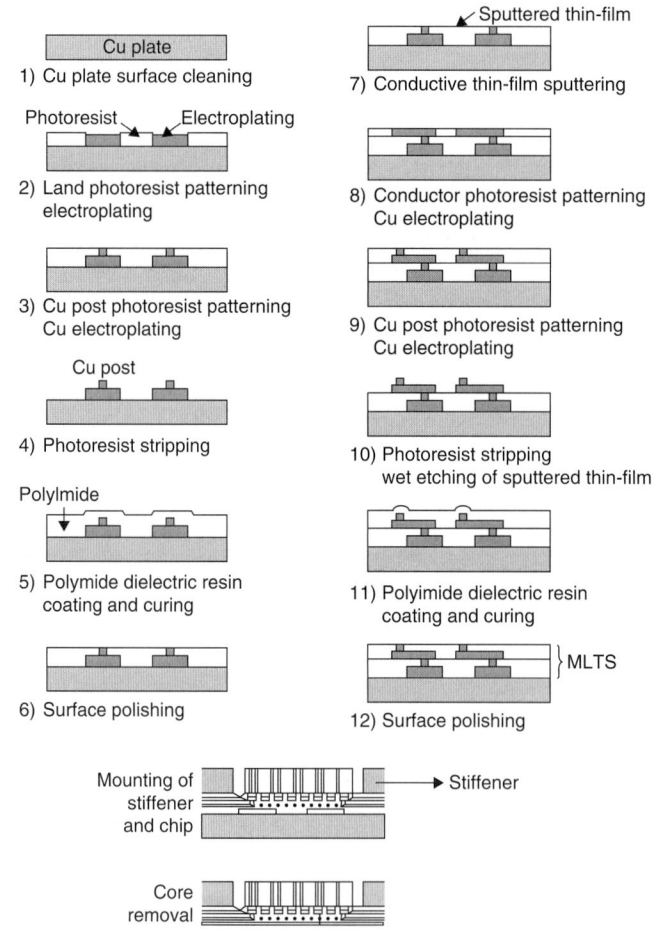

Figure 7.47 Coreless substrate fabrication.

subsequent removal of the core layer. Via structures are then patterned by photolithography, and electroplating is performed to generate the via studs. Photoresist is then stripped and the dielectric deposited. The dielectric is then polished to expose the studs. The next metal layer is then obtained by semi-additive plating. Via structures are again patterned by photolithography and electroplating. After stripping of the photoresist, the dielectric is deposited, cured, and polished. The process can then be repeated for the next buildup layer. When the substrate is completely fabricated, a stiffener is attached and the chip is mounted on the substrate. The stiffener is used so as to prevent warpage once the core is removed. The solid copper core is then etched away. The patterned metal layer is protected due to the presence of barrier layers of Ni and Au, which are insensitive to the etchant used.

7.3.3 Embedded Passives

Embedded passives are driven by miniaturization and better performance, as mentioned before. Decoupling capacitors enable the power integrity by reducing or eliminating the switching noise between different circuits. These are the most critical passive components for high-speed digital systems. For RF circuits, capacitors are also critical for dc-block, matching networks, and oscillator feedback components. Termination resistors are required in high-speed digital systems with low-loss lines. High-speed packages need as many termination resistors as the chip I/Os. The bias or divider resistors are intended to set the operating voltage. Similarly, feedback resistors are critical to improve the stability of an amplifier by reducing the output of an amplifier based on the magnitude of the input signal.

Embedded Capacitors

Attributes of Capacitor Materials The critical properties of any capacitor dielectric are its dielectric constant and breakdown voltage. In thin-film technologies, leakage current becomes critical because of the extremely high field strengths applied to the film. For the RF components, the thermal and frequency stability of the properties along with ultralow loss are the major challenges.

Dielectric Constant When an electric field is applied across a dielectric or an insulator, the positive charges are displaced toward the negative end of the electric field and vice versa. This displacement induces polarization and hence a higher electric flux density inside the material compared to the field created in a vacuum by the same electric field. Electrical flux density is a measure of the electric field strength and charge distribution inside the material. Relative permittivity or dielectric constant is the ratio of the flux density in the material to that in the vacuum. The increased polarization or flux density also manifests as a higher charge storage capacity or capacitance inside the material. Macroscopically, polarization can be estimated by measuring the capacitance of the material.

Electronic polarization is the chief mechanism for the dielectric constant in polymeric materials. Inorganic solids with dielectric constants in the range of 5 to 10 show additional ionic polarization. Certain ionic crystals undergo polarization even in the absence of a field (spontaneous polarization). These are called ferroelectric materials. The polarization results from the asymmetric crystal structure in these ionic compounds, where certain ions are slightly displaced from their electrically neutral positions. Ferroelectric behavior is seen below a certain temperature, referred to as the curie temperature, above which the material becomes paraelectric. The transformation from the paraelectric to ferroelectric phase is associated with a crystallographic change from a symmetric to an asymmetric phase. In case of barium titanate, the transformation occurs from a cubic to tetragonal phase, leading to ferroelectricity. The ionic displacement can generally occur only in certain crystallographic directions. The dipoles exist as spontaneously polarized regions called domains. The domains can have the same possible crystallographic directions as the dipoles that constitute them. The existence of domains and the domain size are guided by the minimization of total free energy. Ferroelectric materials show dielectric constants on the order of 1000 to 20,000 in bulk form. The dielectric constant of ferroelectrics is dependent on the grain size, and hence, most of the nanostructures and submicron ferroelectric thin films may not show such high dielectric constants. The dielectric constant of $BaTiO_3$ is found to be highest when the grain size is about 1 μm. For grain or crystal sizes below 100 nm, the dielectric constant is not expected to be more than a few hundred.

Dielectric Loss Similar to low-K dielectrics used for signal transmission, low dielectric loss is important for high-K dielectrics used for embedded capacitors. The losses are characterized by including a resistive element along with the capacitance in the equivalent circuit of the embedded capacitor. Nonpolar polymers and ceramics have the lowest dielectric loss. Traditional epoxies and ferroelectrics show high loss (Table 7.4).

Breakdown Voltage (BDV) The breakdown voltage of a typical ceramic or polymer is about 300 to 1000 V/μm (3 to 10 MV/cm). However, introduction of defects lowers the BDV to much lower values. Further, higher-dielectric-constant materials show lower BDVs intrinsically. This is summarized in Figure 7.48. BDV is a serious concern for thin high-K ceramic films. For these materials, the value can vary from 30 to 200 V/μm depending on the process-induced defects.

Temperature Coefficient of Capacitance (TCC) The TCC is becoming critical for various capacitor applications because of the tighter design tolerances. The TCC can be positive or negative in both polymers and ceramics depending on the material structure. Ferroelectrics have a high positive TCC, while most paraelectrics have a negative TCC. Similarly, polymers such as epoxy and polyimide show a positive TCC unlike certain other polymers. The TCC tolerances for RF components are met by careful selection and engineering of the material compositions.

Classification of Capacitor Materials Figure 7.49 shows the various classes of capacitor materials and their key properties.

Material	BeO	Titanates	Alumina	AlN	Cordierite	Epoxy
Dielectric constant	6.6	20–10,000	9.2	8.3	4.9	3.5–4.5
Loss Tangent	2.0	1.5–300	2	3–10	10	150–250

TABLE 7.4 High-Frequency Losses of Typical High-K Dielectrics and Epoxy

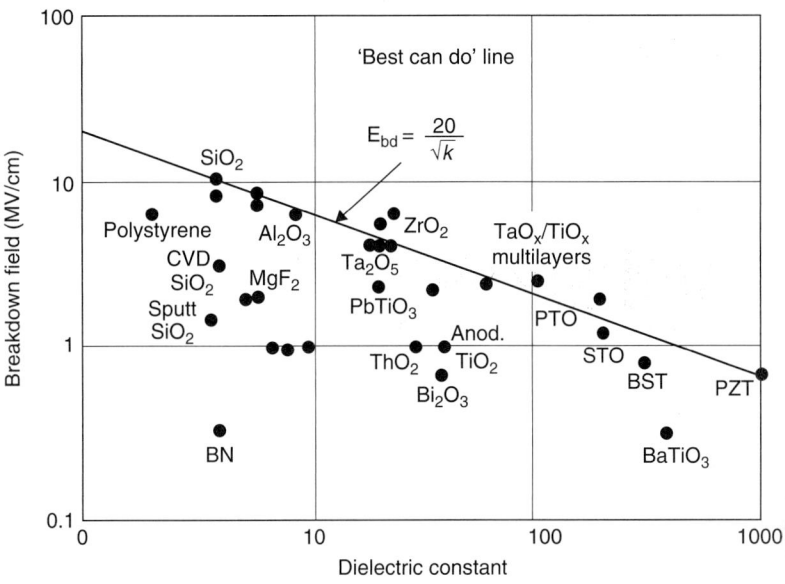

FIGURE 7.48 BDV of various materials. (Courtesy Rich Ulrich, University of Arkansas.)

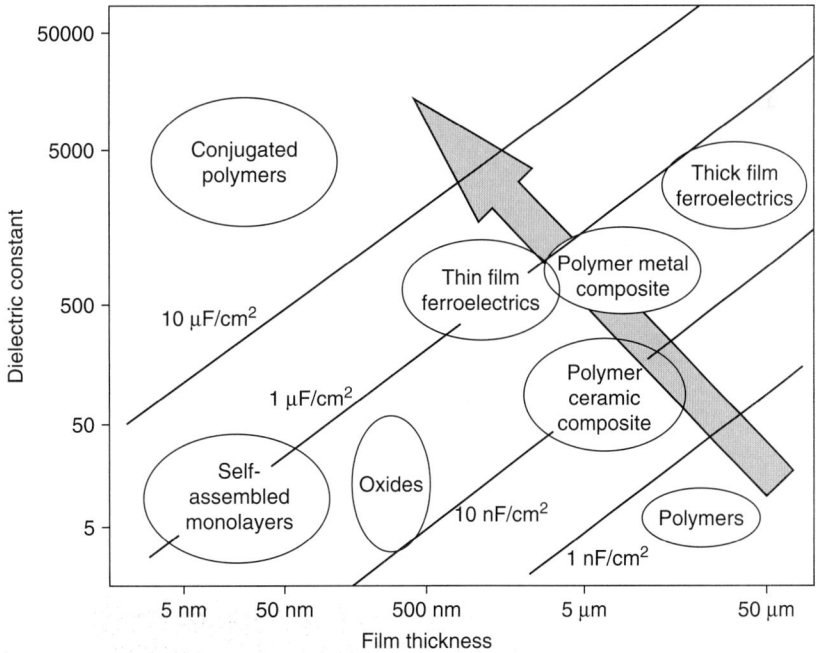

FIGURE 7.49 Capacitor materials.

Polymers The challenge of embedding capacitors in organic packages has been pursued by several academic and industrial groups over the past decade. Closely spaced power and ground planes with thin epoxy dielectrics to dampen the power supply noise and provide decoupling have been commercially used for about two decades. In these applications, the dielectric thickness is typically 16 μm, while thinner power-ground dielectrics are continuously emerging in the market.

Polymer Ceramic Composites Thin polymer films cannot result in capacitance densities of more than nF/cm². This limitation can be overcome with high-dielectric-constant ceramics. Ferroelectrics such as barium titanate have dielectric constants that are 1000 times higher than polymers in the megahertz range and 100 times higher in the gigahertz range. Unfortunately, ceramic crystallization is a high-temperature process that is not compatible with organic packaging. Therefore, most of these materials are confined to discrete or surface-mounted components that are cofired at high temperature before assembling onto a system board or package. High-dielectric-constant materials are ceramic-based, whereas packaging is moving toward low cost based on large-area organic buildup technology. This incompatibility can be partially overcome with the polymer ceramic compositing approach. The rationale for polymer-ceramic composites is illustrated in Figure 7.50. This technology can easily lead to a fivefold improvement in capacitance density, but much further improvements are limited by their defect-free processability and reliability requirements. Embedded capacitor technology is hence driven toward ultrathin-film component integration.

Advanced ceramic thin-film technology can raise capacitance densities into the μF/cm² range for the first time, which can lead to dramatic enhancements in digital package performance, cost, and size. New and innovative ceramic thin-film synthesis routes can crystallize high-dielectric-constant inorganic materials at low temperatures directly onto organic substrates. The high-temperature processing limitation of ceramics can also be overcome by carrying out all the high-temperature steps on a carrier foil that is then subsequently integrated into the organic board.

Ceramic polymer composites do not lend themselves to the formation of thinner film with higher K because the filler particles themselves show a lower dielectric constant when they are in the micro-nano dimensional range. The adhesion strength of an epoxy-barium titanate nanocomposite with copper reduces from 6 to 8 MPa at

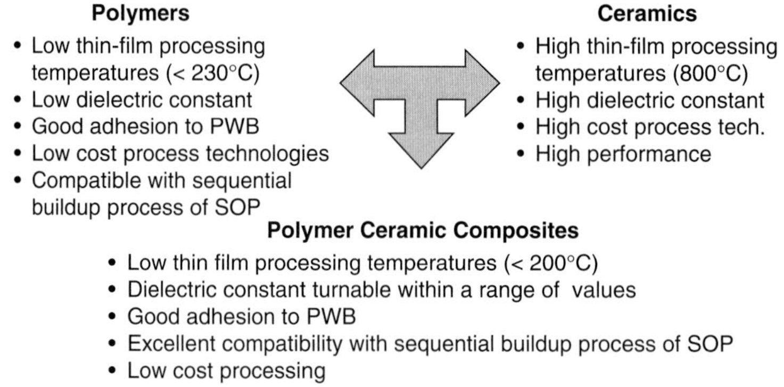

Polymers
- Low thin-film processing temperatures (< 230°C)
- Low dielectric constant
- Good adhesion to PWB
- Low cost process technologies
- Compatible with sequential buildup process of SOP

Ceramics
- High thin-film processing temperatures (800°C)
- High dielectric constant
- High cost process tech.
- High performance

Polymer Ceramic Composites
- Low thin film processing temperatures (< 200°C)
- Dielectric constant turnable within a range of values
- Good adhesion to PWB
- Excellent compatibility with sequential buildup process of SOP
- Low cost processing

FIGURE 7.50 Rationale for polymer ceramic composites.

20 percent filler to less than 2 MPa at the higher filler content [23]. Taking into account the requirements for processability and reliability, the capacitance of polymer-based embedded capacitors is less than a few nF/cm^2. High-K ferroelectric powders also exhibit an intrinsic high-frequency relaxation behavior that gives rise to frequency dependent dielectric constant and loss. This led to the development of polymer composites interspersed with conducting fillers.

Composites with Conducting Fillers Newer capacitor concepts such as supercapacitors and nanocapacitors can overcome the limitations of existing polymer-based capacitors. These concepts rely on nanostructured electrodes for high surface area per unit volume and electrical double layer and interfacial polarization resulting in ultrahigh capacitance densities. Fillers such as carbon black, when dispersed in a polymer, were shown to achieve high capacitance and an effectively higher dielectric constant on the order of thousands, presumably from the giant interfacial polarization and effective increase in the electrode surface area from the nanometallic particles. However, the high dielectric constant resulting from interfacial polarization is not shown to be stable (Figure 7.51) in the high-frequency range, as reported by many independent studies [24-27]. High-frequency measurements of carbon black epoxy composites previously show that the dielectric constant reduces from 10^3 at 10 kHz to less than 10^2 at 100 MHz. Though not expected to be stable at gigahertz frequencies, invoking supercapacitive structures with high-surface-area electrodes and double-layer/interfacial polarization with nanometallic electrodes and thin electrolytes can yield a capacitance higher than 100 µF/cm^2 per single layer. While more complete characterization (high-frequency properties, thin-film processing capability, capacitance density, resistivity, and leakage) is under way to qualify this material for decoupling, it is apparent that current polymer-based capacitors are suitable for mid-low frequency power supply and are not suitable for high-frequency decoupling.

Thin-Film Ceramics Ceramic thin films are ideal to provide ultrahigh capacitance densities. It is well known that ferroelectric films can easily exhibit high capacitance

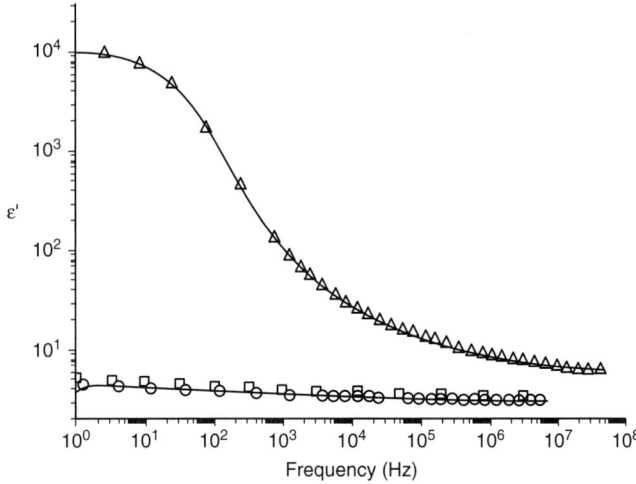

Figure 7.51 Frequency dependence of polymer composites with conducting fillers. [26]

densities in the range of µF/cm² up to gigahertz frequencies and are ideally suited for decoupling applications. Barium Strontium Titanate (BST) with a dielectric constant of 300 to 500 and a film thickness in the range of 100 to 300 nm typically yields capacitance densities of 2 to 3 µF/cm² over a wide frequency range (above 5 GHz). The properties of thin-film ferroelectrics and oxides are shown in Figure 7.49.

Embedded Capacitor Processes Capacitor processes can be broadly classified into thick, thin, and ultrathin films. Conventional ceramic and PWB packaging processes that are typically used to build layers of 20 to 100 µm are classified as thick-film processes. The main processes to process these thick films are doctor blading or screen-printing. For higher capacitor performance, thickness reduction is the main driver. Hence, new classes of thin composite films (10 to 20 µm typically) have emerged. The ultrathin-film processes (< 500 nm) are in turn pursued in a wide range of options that include polymer coating, vapor deposition, and solution deposition followed by crystallization. The capacitor processes are summarized in Figure 7.52.

Polymer-Ceramic Composite Processing Polymer ceramic composite technology consists of incorporating high-dielectric-constant ceramic particles in a polymer matrix and processing the dielectric in a thin-film dielectric form. The organic matrix allows low-temperature processing and large-area dielectric deposition in which the fillers provide an improved dielectric constant. Polymer-ceramic composite technology is a low-cost option for embedded capacitors in organic laminates because of the elimination of expensive processing steps such as CVD and sputter deposition.

Randomly mixed particulate composites do not show linear scaling of dielectric constant with volume fraction. Ceramic polymer composites typically follow log-log behavior in the low filler content region and much worse behavior in the high filler content region. After imposing the processing limitations and reliability constraints, this technology cannot typically achieve more than 50 nF/cm² of capacitance density even with a 2-µm film. Commercial composite dielectrics show a dielectric constant

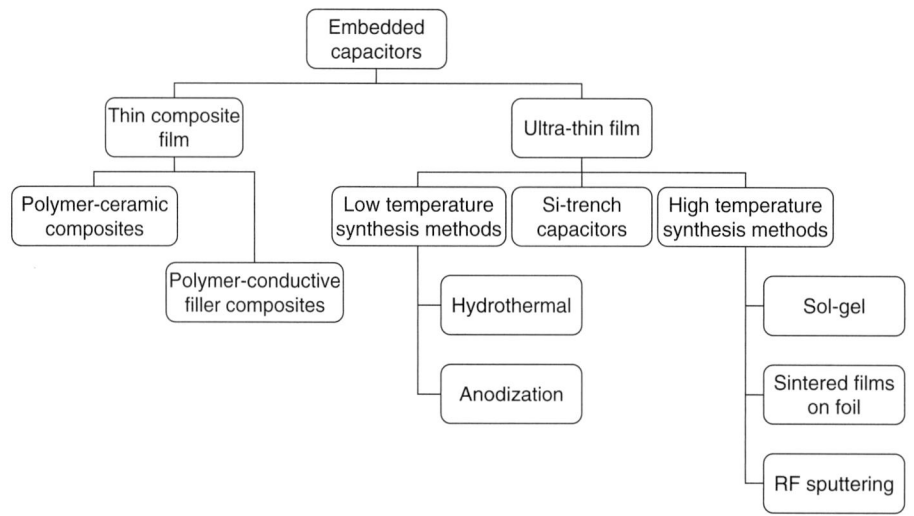

FIGURE 7.52 Classification of capacitor technology based on process.

between 15 and 30. Typical capacitance densities range between 1 to 5 nF/cm². The dielectric constant of composites can be improved with higher filler content, improved filler packing and uniform dispersion of particles in the polymer matrix. This is accomplished using two approaches:

1. *Bimodal/multimodal distribution to increase the packing density.* Most ceramic engineers are aware that the packing density of particles can be improved by mixing powders of different sizes. The finer particles are expected to fill in the vacancies formed by bigger particles. By properly selecting the ratio of different particle sizes with a multimodal distribution, even 90 percent packing density can be theoretically achieved. In this way, the dielectric coupling between different ceramic particles can be further enhanced leading to a higher dielectric constant.

2. *Colloidal chemistry concepts to improve dispersion and prevent aggregates of particles.* The dispersion of particles in a polymer can be controlled by modulating the particle surface with surfactants, invoking various suspension stabilization mechanisms. By creating steric hindrance between particles in combination with electrostatic repulsion (the combined effect is known as electrosteric stabilization), low-viscosity suspensions can be created with high filler loading leading to high packing efficiency of particles in a polymer. The most common example is to use an acidic dispersant (phosphate ester) that can specifically interact with the basic surface of barium titanate. In polar solvents, a certain fraction of the adsorbed polymeric chain desorbs into the solvent forming the counter ion, leading to the electrical double layer causing electrosteric stabilization.

By optimizing the dispersion methodology, filler content, and filler size distribution, dielectric constants as high as 135 have been engineered by a Georgia Tech team [28-32]. A capacitance density of ~35 nF/cm² with a 3.5 µm film was achieved on PWB substrates with high yield. Commercially available composites have a much lower dielectric constant but guarantee reliability and manufacturability. Table 7.5 summarizes the properties of three commercially available polymer-ceramic composites.

Ceramic film deposition has always relied on high-temperature sol-gel processing or high-cost vacuum deposition methods such as RF sputtering and CVD. These cannot

Property	3M (C-Ply)	DuPont (HK)	Oak-Mitsui (Faradflex)
Capacitance density	0.9–1.75 nF/cm² (@ 1 kHz)	0.12–1.75 nF/cm² (@ 1 MHz)	0.15–1.7 nF/cm² (@ 1 MHz)
Loss tangent	0.006 (@1 kHz)	0.003–0.01 (@ 1 MHz)	0.015–0.019 (@ 1 MHz)
Thickness	8–16 µm	8–25 µm	8–24 µm
Dielectric constant	16	3.4–15	4.4–30

TABLE 7.5 Summary of Properties of Commercial Ceramic-Polymer Composite Dielectric Laminates

be adopted by the packaging community because the process is not compatible with the large-area organic substrates. Novel manufacturing techniques are being developed to integrate ceramic thin films in organic substrates. These can be classified as low-temperature and high-temperature synthesis methods.

Thin-Film Ceramic Processing These processes are further categorized into low- and high-temperature synthesis techniques.

Low-Temperature Synthesis Methods
Anodization. Ultrathin paraelectric films of tantalum oxide were developed for decoupling applications by Ulrich et al. [33]. In this technique tantalum is sputtered on a suitable substrate and subsequently anodized in a mixture of ammonium hydroxide and tartaric acid. The resulting film has a dielectric constant of 21 to 28. This translates to a capacitance density of 200 nF/cm^2 for a 100-nm-thick film. The capacitance densities, though not as high as that for ferroelectric films, are stable at gigahertz frequencies due to paraelectric behavior and can effectively address certain decoupling requirements.

Hydrothermal process. Hydrothermal processing involves crystallization in a highly alkaline medium at temperatures generally ranging from 70 to 150°C. It is a versatile technique that has been used for synthesis of a wide range of materials like hematite, quartz, barium, and strontium titanates and more recently for carbon nanotubes. Hydrothermal synthesis of quartz was demonstrated several decades ago and is currently used commercially. Quartz synthesis involves temperatures close to 300°C and pressures of several tens of atmospheres and hence requires steel pressure vessels. However, a number of other materials like barium titanate have been synthesized using this technique under milder conditions by reacting a suitable titanium source with barium ions in an alkaline bath at 95°C. The source of titanium is typically evaporated titanium or an organic precursor coated on a suitable electrode metal. A pH > 13 is required for stability of barium titanate in an aqueous solution. The films synthesized on organic precursors are typically porous and result in poor yield of capacitors and low capacitance densities. Hence the spun-on precursors are pyrolyzed at temperatures as high as 300°C to densify the precursor prior to hydrothermal process [34]. Metallic Ti films give dense barium titanate films but require expensive processes.

The Georgia Tech PRC team did extensive research on synthesizing and characterizing barium titanate hydrothermal films on titanium and titanium-coated copper-clad laminates. The as-synthesized films exhibit a high capacitance of 3.0 µF/cm^2 and a loss tangent of 0.30 to 0.80. The high loss is generally attributed to hydroxyl groups entrapped in the as-synthesized films. Additionally, the hydroxyl groups also contribute to a high dielectric constant in as-synthesized films through an additional polarization mechanism. After post-hydrothermal treatments, a capacitance density on the order of 500 to 1000 nF/cm^2 with a loss of 0.05 has been achieved. The thinnest films (100 nm) have a breakdown voltage (BDV) of 3 to 4 V, while a BDV greater than 10 V can be achieved with multiple coatings. These films can be easily integrated onto organic packages because of their low-temperature processing (Figure 7.53).

High-Temperature Synthesis Methods
High-Temperature Processing on Foils and Film Transfer. Integrating ceramic films on PWB is always hindered by the high processing temperatures of the ceramics. Embedding ceramic films into PWB with low-cost technology would therefore require synthesizing films using high-temperature processes on a suitable high-temperature carrier such as

FIGURE 7.53 Cross section of a hydrothermal film (top grains) on a Ti layer.

a metallic foil that can also serve as an electrode, followed by integration of the foil in PWB using a conventional lamination process [35].

RF sputtering. RF-sputtered barium strontium titanate thin films are integrated in organic substrates using polyimide as a carrier film. Shioga et al. [36] demonstrated $(Ba_{0.7}Sr_{0.3})TiO_3$ thin films with a capacitance density of 0.1 to 2.0 µF/cm² depending on the substrate temperatures that were varied from 260 to 400°C [37]. Breakdown voltages as high as 10 V were achieved. The films are predominantly amorphous due to low deposition temperatures. However, they exhibit excellent dc leakage characteristics. NEC reported a similar approach using RF sputtered pure and manganese-doped strontium films on polyimide-coated silicon wafers. Manganese is reported to improve the dc leakage characteristics of sputtered films [38].

Sintered films on copper foils and foil lamination. The most prominent example is screen-printed high-K barium titanate thick-film technology from DuPont. These thick films are deposited by screen-printing high-K paste on copper foils followed by firing at 900 to 1000°C in reducing atmosphere. Such films have a dielectric constant as high as 3000 at 10 kHz. However, the capacitance is limited because of the high 12-µm thickness of the film. The foil is patterned to electrically separate the plates to form the interconnects (Figure 7.54). Novel dielectric formulations based on glass additives that do not dissolve the ferroelectric material during the firing step were developed and shown to be compatible with board-level etching processes. They exhibit a capacitance density of around 45 nF/cm² with a dielectric constant of around 1000. Their dielectric loss is typically 1.1 to 1.5 percent at 100 kHz, with a breakdown voltage around 900 to 1200 V.

Sol-gel processes. The main advantage of this approach is the ability to make thinner films with precise control of the film composition by ease of introducing dopants to engineer the dielectric properties like the loss tangent and dc leakage characteristics. Other advantages include relatively low cost, compared to vacuum-based processes, and the potential for large-area deposition. However, oxidation of the metal electrode during pyrolysis and high-temperature postsintering remain as major barriers to successful implementation of this technique. Oak-Mitsui recently patented a process of using a nickel barrier to prevent oxidation of copper foil during sintering [39]. A 2-µm nickel coating was electroplated on copper making it a low-cost process. The electrical properties of Ca-doped PZT films on electroless Ni-coated copper films were investigated in another study. The leakage currents were within 6 to 10 A/cm² for field strengths ranging from 100 to 200 kV/cm [40].

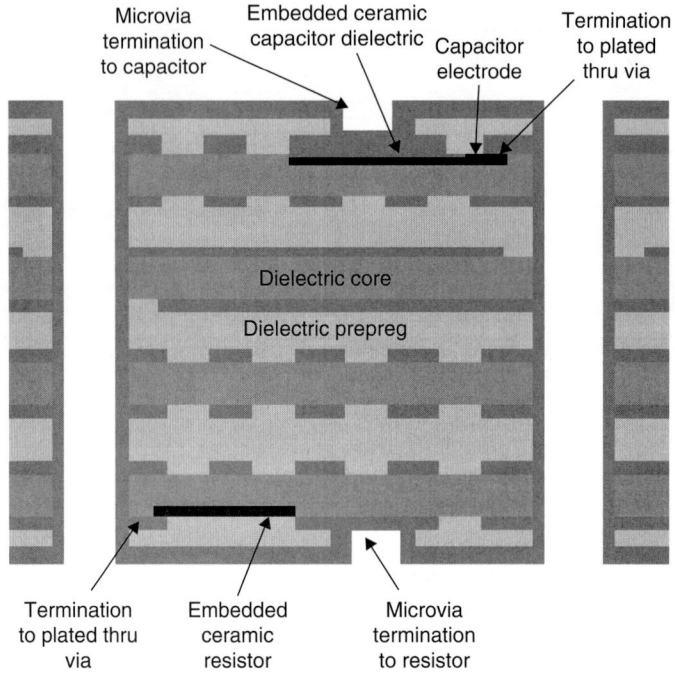

FIGURE 7.54 Cross section for Dupont's Interra process.

Materials based on $BaTiO_3$ and $SrTiO_3$ compositions display the desirable high dielectric constants, low leakage currents, and low dielectric losses, but the high cost of oxidation-resistant electrodes (such as Ni-coated copper and Ni) has been an issue. Replacing these materials with base metals such as Cu is the current interest to reduce fabrication cost. Currently, efforts are under way to synthesize sol-gel Barium Titanate thin films on copper foils. Control of oxygen partial pressure is critical to prevent base metal oxidation while minimizing oxygen vacancies in the ceramic thin film at the same time. Figure 7.55 indicates that a partial pressure between 10^{-9} to 10^{-12} atm is needed to completely pyrolyze the sample without oxidizing the copper.

Si trench based capacitors. Thin-film capacitor technology when combined with high-surface-area electrodes can lead to ultrahigh capacitance density 3D capacitors. There is an increasing trend to introduce Si carriers with micromachined trench capacitors to increase the capacitance density by 10 times. Semiconductor companies are taking advantage of this technique by deep-etching trenches in Si and conformally coating the Si with oxide or oxynitride (Figure 7.56). Though the dielectric constant is low, silicon oxynitrides can be thinned down to less than 30 nm leading to capacitance densities as high as 10 $\mu F/cm^2$ with high aspect ratio Si trench technology.

Embedded Resistors

The need for resistors in digital and RF applications is in the range of 1 Ω to 200 kΩ with TCR < 100 ppm/K and tolerance < 10 percent, which is difficult to achieve with a single source material. Although there are a wide variety of materials available, the choice is limited when the low-cost and low-temperature needs of SOP are taken into account.

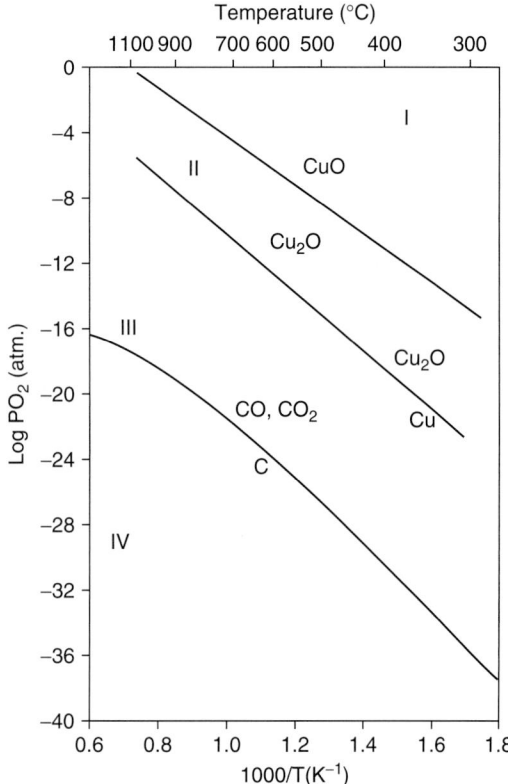

FIGURE 7.55 Phase diagram showing the ideal oxygen pressure for cofiring ceramic film on copper foil. [41]

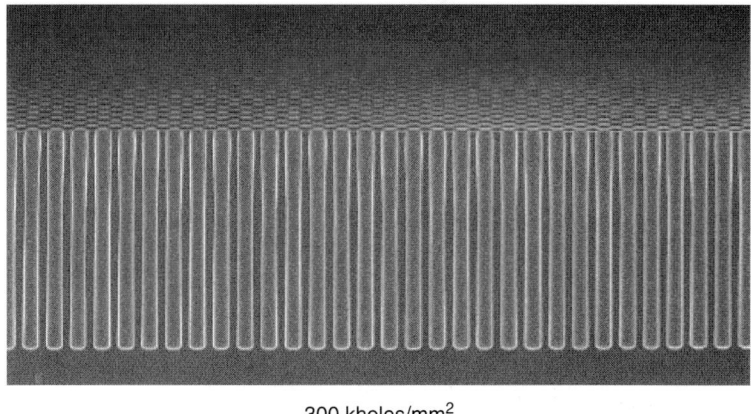

300 kholes/mm²

FIGURE 7.56 Silicon trench capacitors from Philips. (Courtesy S. Bardy.)

In addition to material challenges, there are also challenges in process integration. Some of these are listed in Table 7.6. The materials systems used for embedded resistors are generally categorized as metal and alloys, semiconductors, cermets, and polymer thick films. Obviously, a single material system cannot cover the entire range of resistivity as well as fulfill the other key parameters. For example, polymer thick film (PTF) offers low cost and a wide range of resistance values but has problems in temperature and humidity, oxidation at the interface, CTE mismatch, and high-frequency stability. Resistors predeposited on copper foil are limited to a lower range of resistance. Electroless-plated resistors offer cost advantages, but again this technology is currently limited to low-value resistors within a small deposited area. Laser trim is required for better tolerance control, which is an expensive proposition.

Among commercial materials, Ohmega-Ply (electroplated NiP), DuPont Interra (screen-printable LaB_6), MacDermid M-Pass (electroless plated NiP), Asahi Chemical (polymer thick films), Shipley Insite (doped Pt on Cu by CVD), and Gould Nichrome on copper foil have received much attention. However, low TCR (< 100 ppm/°C) and high tolerance (< 5 percent) have not been achieved. Tolerances on the order of 1 to 2 percent required for analog applications cannot be achieved without the added cost of laser trimming. Table 7.7 shows the current state-of-the-art in embedded resistors. In most technologies presented in this table, the process tolerances are around 10 to 15 percent without trim.

From a processing standpoint, resistors are classified as

- Thin film—sputtered and plated
- Thick film

Polymer Thick-Film (PTF) Resistors PTF resistors are used for low-cost applications such as consumer products. Motorola uses carbon/phenolic polymer thick-film (PTF) ink and a screen-printing process to form resistors in inner layers of high-density interconnect printed wiring boards (HDI PWBs). The PTF ink is screened on to copper termination pads that are treated with proprietary interface metallurgy to improve reliability and environmental stability. The ink is then cured in a conveyorized oven at a peak

Processing Challenges	Material Property Challenges
Large-area fabrication	Low TCR (<100 ppm/C)
Good reproducibility and yield	High-frequency stability up to 10 GHz
Material stability	
Variations in the deposited film	
Contact resistance	
Trimming	
Low-cost fabrication processes	

TABLE 7.6 Challenges in Embedded Resistor Technology

SOP Substrate with Multilayer Wiring and Thin-Film Embedded Components

Organization	Material	Process	Range of Resistance (Ω/sq)	TCR (ppm/°C)
• Intarsia • Boeing • NTT • GE	Ta_2N	Sputter	10–100 20 25–125	($\pm$100) (–75 to –100)
• Osaka University • Metech • Acheson Colloids • Electra • Asahi Chemical • W. R. Grace • Dow Corning • Raychem Corporation • Ormet Corporation	Polymer-metal composites	Polymer thick-film process Liquid phase sintering	Insulating to conducting	
• Ohmega Ply	NiP alloy	Electroplate	25–500	
• IME, Singapore • AT & T Bell Labs	TaSi	DC sputter	10–40 8–20	
• University of Arkansas/Sheldahl	CrSi	Sputter		–40
• W. L. Gore and Associates	TiW	Sputter	2.4–3.2	
• Shipley	Doped Pt on Cu foil	PECVD	Up to 1000	100
• Deutsche Aerospace • Gould Electronics	NiCr NiCr NiCrAlSi	Sputter	35–100 25–100	
• Georgia Institute of Technology • MacDermid	NiWP NiP	Electroless plate	10–50 25–100	
• DuPont	LaB_6	Screen-print and foil transfer	Up to 10K	$\pm$200

TABLE 7.7 Current State-of-the-Art of Embedded Resistors

temperature of approximately 230°C to form the resistors (Figure 7.57). The resistors are subsequently buried in the inner layers of the PWB to achieve an exceptionally compact 3D circuit. Since screen-printing is a "mass forming" process, the process cost is largely independent of the number of resistors printed in one screening step. In Motorola product applications, between 8000 and 20,000 resistors have been printed on an 18 in × 24 in panel in a single screening step, resulting in significant economies of scale and associated cost savings.

The PTF inks are available in different sheet resistances ranging in value from 35 Ω/sq to 1 MΩ/sq. By using different sheet resistance inks, resistors ranging in value from 20 Ω to 1 MΩ can conveniently be printed on the same layer. Furthermore, by blending inks of different sheet resistances any sheet resistance in between can be achieved and the printed resistor size can be minimized. The tolerances of screen-printed resistors range between 15 to 20 percent when environmental variations (temperature, humidity, etc.) are taken into account. While this number may seem high, many resistor applications in handheld products (pull-up and pull-down resistors, etc.) do not require tight tolerances and 20 percent tolerance is acceptable.

PTF resistors manufactured with stability-promoting terminations drifted less than ±10 percent following 500 hours of 85 percent relative humidity/85°C temperature exposure, and this shift was almost entirely reversed after a 125°C, 3-hour bake. Resistors have also been subjected to 5 reflow cycles (225°C peak) followed by 500 cycles of liquid-to-liquid thermal shock and air-to-air thermal shock (–55 to 125°C), resulting in net shifts in resistance of less than ±4 percent.

Lasers are prevalently used in HDI PWB manufacturing to form laser-drilled microvias. Many of the same laser manufacturers (ESI for example) have developed laser systems that can be used for laser trimming of PTF resistors (Figure 7.58). Some recent publications have shown that PTF resistors can be trimmed at the inner-layer stage to within 1 percent tolerance, resulting in a net tolerance of the embedded resistors after environmental variations of 5 percent.

Thin-Film Resistors

Plated Resistors   Electroless plating has been developed, optimized, and implemented on epoxy dielectrics at Georgia Tech PRC. NiP and NiWP compositions deposited by electroless plating cover low resistances in the range of 5 to 1000 Ω with TCR in the range of 50 to 100 ppm/°C. Figure 7.59 shows NiWP thin-film resistor values achieved

Figure 7.57   Screen-printed polymer thick-film resistor.

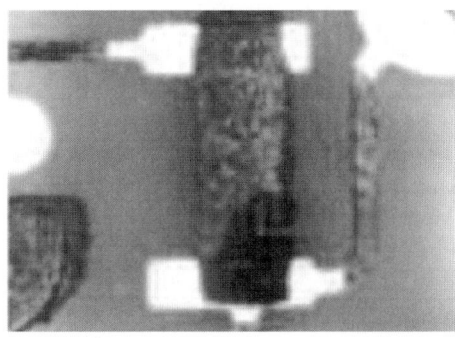

FIGURE 7.58 Laser-trimmed PTF resistor.

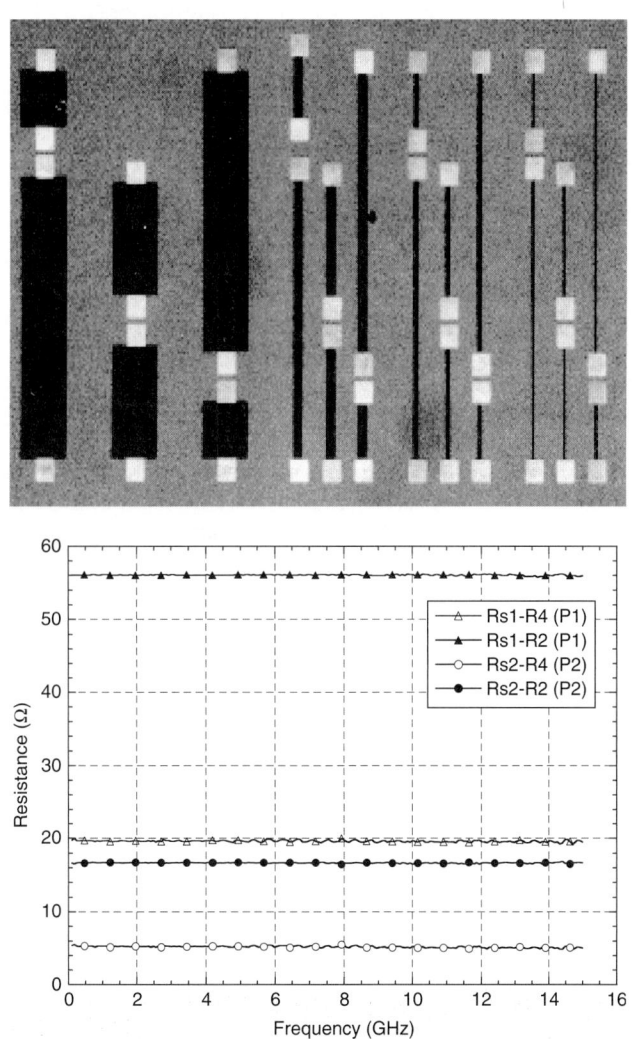

FIGURE 7.59 Photomicrograph of resistor structures, NiP/NiWP resistor film, and the measured resistance at frequencies up to 15 GHz.

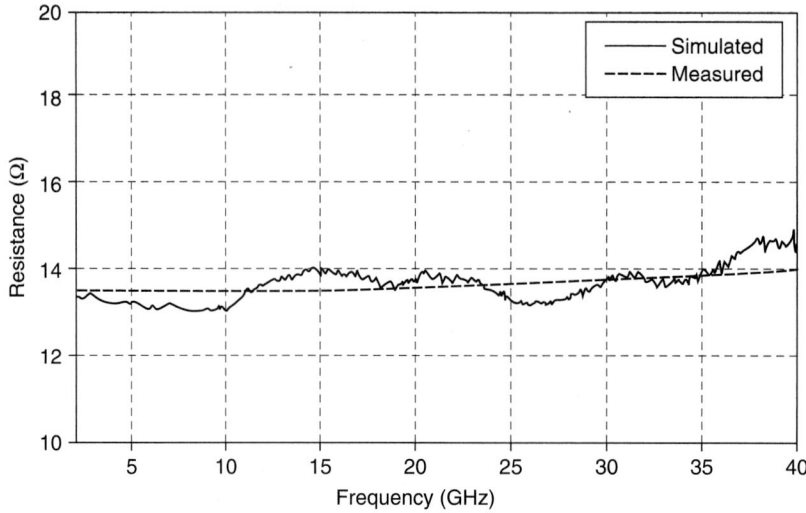

FIGURE 7.60 Measured resistance up to 40 GHz.

on epoxy dielectric. This process has also been optimized for low-loss polymers such as liquid crystalline polymers (LCP) and benzocyclobutene (BCB). The electroless plated resistors are commercially available through MacDermid and Ohmega-Ply.

Sputtered Resistors The alternative approach for thin-film resistors is the direct lamination of predeposited resistors on carrier Cu foil. Upon lamination on selected dielectrics, the resistors are patterned using a two-step etch process. GT PRC has qualified this process on epoxy and BCB dielectrics. The resistance ranges from 25 Ω to 50 kΩ using multilayer process with a tolerance of less than 10 percent. Conventional polymer thick film (PTF) is utilized for resistances greater than 10 kΩ.

Another approach is the lamination of thin-film resistors predeposited on copper foil. The resistors are printed and patterned after postlamination. Using the lamination process on LCP substrate, various RF resistor structures have been designed and fabricated. Measurements up to 40 GHz are presented in Figure 7.60 on an LCP substrate using the 25-Ω/sq NiCrAlSi sputtered film on copper foil to form the resistors.

7.3.4 Embedded Actives

Embedded actives is a package technology to achieve ultraminiaturization of modules and systems with one or more buried chips in organic, ceramic, and other packages. Embedding of active ICs can be classified into three categories as shown in Figure 7.61.

- Chip-first
- Chip-middle
- Chip-last

Chip-First Embedded Actives

In the chip-first approach, embedding starts typically with ICs and the buildup of wiring takes place on top of the ICs. The demonstration of this chip-first embedded

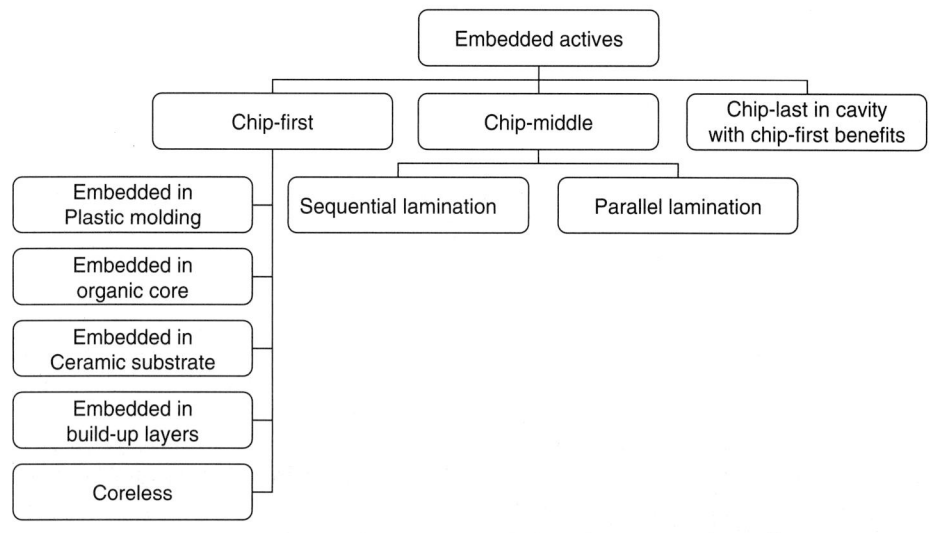

FIGURE 7.61 Approaches to embedded actives.

active dates back to 1975 by Yokogawa [42]. Multiple semiconductor chips are mounted face up on Al metal substrate and forced toward the substrate with a press-jig, which leads to partial embedding of the chips into the substrate. More recent types of chip-first embedded actives have been developed by a number of researchers [43-52] since the early 1990s, the most well known being by GE. In the GE process, plastic is molded around chips and then a multilayer interconnect is built up over the top of the chips using polyimide films. The vias are formed using lasers [50]. The basic structure of the plastic-encapsulated high-density interconnect (HDI) module is shown in Figure 7.62a.

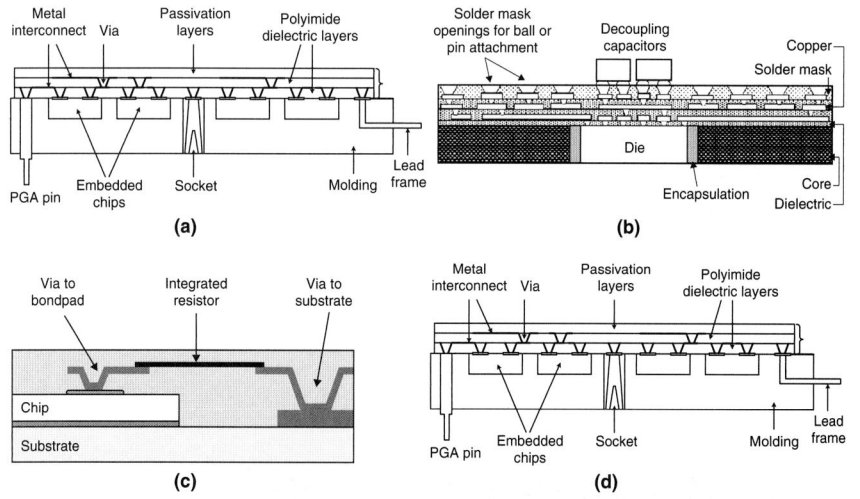

FIGURE 7.62 Examples of chip-first embedded actives. (*a*) Plastic encapsulated embedded active. (*b*) Bumpless buildup layer (BBUL) active. (*c*) Chip-in-polymer active. (*d*) Coreless embedded active.

Compliant materials may be used around the chips to reduce thermal stress due to CTE mismatch between the chip and the molding materials [43].

Organic cores such as BT laminate and FR4 have also been utilized instead of plastic molded substrates [51]. This is based on a cavity-based approach with a microprocessor chip placed inside the cavity in the BT. Figure 7.62b shows the schematic cross section of such an embedded active package. Buildup layers have also been processed on both surfaces of the core layer with this technology [45]. Another cavity-based approach developed by Virginia polytech involves embedding power MOSFET chips into the cavities of ceramic substrates [46].

Fraunhofer IZM and the TU Berlin team has introduced the so-called chip-in-polymer (CIP) technology concept [52]. Chips are mounted on the substrate by die bonding and then are embedded inside a film of dielectric layer (Figure 7.62c). Additionally, resistors can also be integrated into the package by deposition of very thin resistive metal films.

A flexible embedded active structure without a core has also been developed [48]. Chips are placed face up on spin-coated flexible polyimide films and are then embedded into the buildup layers by polyimide coating followed by metallization (Figure 7.62d).

A chip-first embedded active technology has also been developed by embedding wafer-level packages (WLP) into boards [49].

Chip-Middle Embedded Actives

In the chip-middle approach, embedded chips end up in the middle of the buildup substrate, Shinko's approach being one of the representative examples [53]. A chip is placed face down onto a buildup layer like in an SMT process and is fully embedded with subsequent buildup layers (Figure 7.63a).

Another chip-middle active approach involves a laminated structure with active ICs (Figure 7.63b). In this approach, multiple layers with active chips or passive components are fabricated separately and then laminated together. Matsushita is making the multiple layers by pressing conventional discrete passive components into a composite substrate made of ceramic powder and thermosetting resin with inner vias filled by conductive via paste, which is a mix of conductive filler, resin, and a hardening

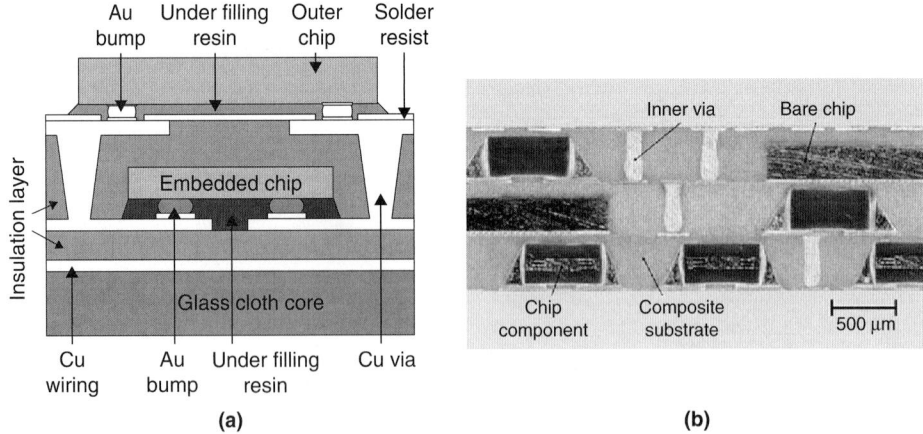

FIGURE 7.63 Examples of chip-middle embedded active. (*a*) Shinko. (*b*) Matsushita's SIMPACT (system-in-module using passive and active components embedding technology).

agent [54]. Nokia uses a patented process in PWB to make the laminated structures [55]. Chips are placed inside the cavity of the PWB, similar to the chip-first embedded active approaches using organic core cavities. Nokia claims with this approach that the electrically conductive layers on a PWB can provide electromagnetic shielding to RF chips embedded inside. A research team in the SMIT center at Chalmers University, Sweden, is also trying to make a laminated structure of embedded active using LCP core [56].

Chip-Last in Cavity with Chip-First Benefits

While current chip-first and chip-middle embedded active approaches can give rise to many advantages such as a small form factor and better electrical performance, they also pose many challenges such as: (1) Serial chip-to-buildup processes accumulate yield losses associated with each process, leading to lower yield and higher cost. (2) Defective chips cannot be easily reworked, therefore requring 100 percent known good dies (KGDs). (3) The metallurgical interconnections in the chip-first approach can fatigue due to thermal stress. In addition, the electrical performance of the chip-middle approach is compromised by long interconnections. (4) Thermal manangement challenges are also dominant, since the chip is totally embedded within polymer materials of substrate or buildup layers.

In order to address the above issues with both chip-first and chip-last EMAP approaches, the Georgia Tech Packaging Research Center (GT-PRC) has proposed an advanced chip-last embedded active approach with chip-first benefits [57]. Here the chips are embedded after all the buildup layer processes are finished. Figure 7.64 shows the schematic structure of PRC's chip-last embedded active approach. Buildup layers with or without cavities are formed on an ultrathin substrate with a core of high elastic modulus and high thermal conductivity. A chip is then embedded directly into the cavity of the wiring and connected to the buildup layer with ultralow profile interconnection technologies, followed by filling with underfill and engineered adhesive materials. If needed, a stiffener, which also acts as a heat sink and EMI shield, is placed on top of the buildup layers. The chip-last interconnections include copper or nickel-based nanostructured interconnections, conductive adhesives, thin solders, or nanobumps with the shortest interconnection and the best electrical and thermo-mechanical properties. It is expected that this chip-last approach for embedded actives

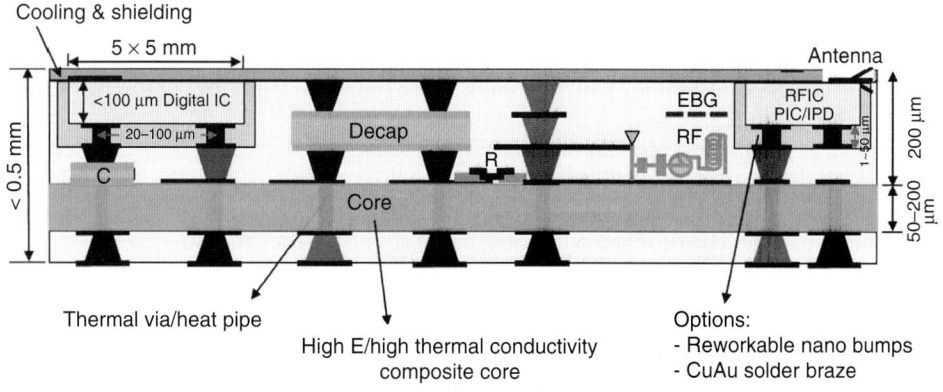

FIGURE 7.64 Chip-last embedded actives and passives.

has many advantages and a few disadvantages, compared to chip-first or chip-middle technologies that include reworkability, higher process yield, and better thermal management.

> *Process and yield.* A lower loss accumulation and higher process yield are expected for multichip applications, since all the processes for substrate, buildup layers, cavity, and embedding the chip are carried out in two parallel operations—substrate and chip. In addition, no complex processes are needed after the chip is embedded, which could otherwise damage the chip.
>
> *Reworkability.* Defective chips can be replaced by the use of reworkable interconnects and appropriate selection and processing of underfill and encapsulation materials after electrical testing.
>
> *Interconnections.* Short interconnections can give rise to an electrical performance as good as direct metallurgical contacts of chip-first embedded active technology. Nanostructured materials with not only high electrical conductivity but also excellent mechanical strength, toughness, and fatigue resistance can provide mechanically reliable interconnects even with a low profile.
>
> *Thermal management.* Since the backside of an embedded chip is directly exposed to air or bonded to a layer of heat sink with highest thermal conductivity, thermal management becomes easier and many possible solutions for cooling can be applied.

7.3.5 Miniaturized Thermal Materials and Structures

A continual increase in performance and a reduction in size is contingent upon solving thermal challenges. In the past, system performance improvements were dictated by silicon technology. But today, thermal challenges are becoming among the most important roadblocks to better performance. Despite the continual increase in power that is forecast, the maximum allowable junction temperature (temperature of the Si die) is almost constant. Thus, the package designer has to carefully balance these requirements.

The main thermal source in the package is the IC itself. Some passives such as resistors also generate heat due to the I^2R effect. With the use of embedded actives and passives, the role of the substrate in heat conduction becomes much more important. The substrate must contribute to providing an efficient cooling mechanism for these two thermal sources.

The general cooling scheme that is in use today for high-power electronics is shown in Figure 7.65.

Thermal solutions that are used in the substrate utilize high-thermal-conductivity materials and thermal vias in order to conduct the heat from the die and embedded components. High-thermal-conductivity materials such as AlN can be used as cores in the substrate. The problem with these materials is that they require expensive processing. Even if these substrates are used, thermal resistance still remains high due to the fact that the thermal path must go through polymeric dielectric with lowest thermal conductivity. To solve this problem, thermal via structures are incorporated in substrates. These vias are filled with high-thermal-conductivity pastes like copper. Thus vias serve a two-pronged purpose in substrates—electrical conductivity and thermal conductivity. By using a cluster of properly designed vias, additional heat sink parts may be eliminated.

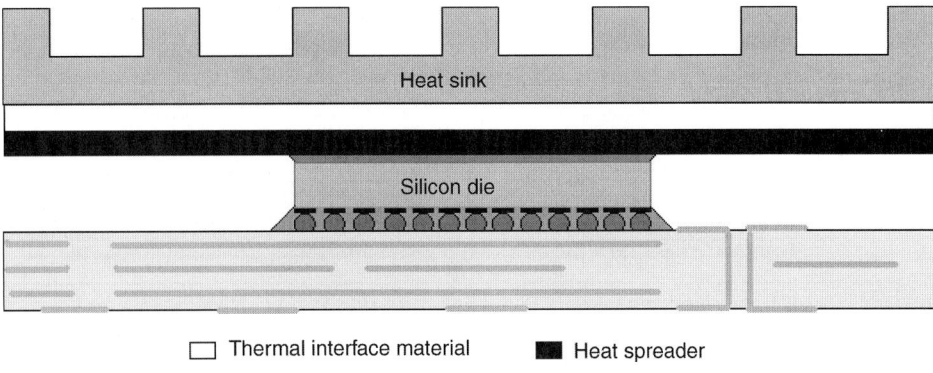

FIGURE 7.65 Typical thermal solution for an electronic package.

The design of such vias requires optimization of various parameters for hole diameter, pitch, and plating thickness. Generally FEA modeling or other analytical models are used to design and validate thermal via structures [58, 59]. The reader is referred to *Fundamentals of Microsystems Packaging* (R. R. Tummala, McGraw-Hill Professional, 2001) for further details of thermal vias.

7.4 Future SOP Substrate Integration

This chapter reviewed leading-edge SOP substrate technologies, namely, high-density wiring, embedded thin-film passive and active components, and thermal structures. The system and IC drivers will continue to add more demands for higher wiring density, higher frequency and performance, thinner and smaller packages, and higher thermal dissipation. For the four substrate elements described in this chapter, key challenges and trends in the future are summarized below.

- Direct chip attach I/O density will increase by 10 times, and corresponding wiring demands on the substrate will drive the ultrafine conductor technology from 10-μm lines and spaces down to 1 μm, matching the back end of the line (BEOL) dimensions on the IC.

- Alternatives to conventional lithography such as imprinting, inkjet printing, and maskless patterning will be necessary to control the cost of ultrafine lines on large-area substrates.

- Core substrates will need to be ultrathin (10 to 50 μm) and extremely flat. Silicon carriers may emerge as a major core material in the future.

- Similar to ultralow-K dielectric integration in on-chip BEOL wiring with a dielectric constant of 2.2 to 2.8, substrate thin-film dielectrics will need to migrate to ultralow-K < 3.0, film thickness of 1 to 5 μm, ultralow loss < 0.001, and microvia interconnections of 1- to 10-μm diameter.

- Embedded capacitor films with a capacitance density on the order of 100 μF/cm^2 will be needed to sustain power noise levels of <10 mV, and technologies such as the Si- trench capacitors must be extended further and leveraged with the nanotechnology.

FIGURE 7.66 Two-chip stacked IC using through-silicon-via (TSV) interconnects.

- New resistor material technology to cover the entire range from a few ohms to a few hundred thousand kilo-ohms, with near-zero temperature coefficient of resistance is desirable.
- Embedding thinned single ICs may not be sufficient, and entire stacks of ICs will have to be buried within the substrate with high interconnect density and yield.

Some of these technology paths have already started to emerge in the global SOP R&D community. Two leading-edge examples are described below. The first is through-silicon-via (TSV) interconnection to achieve the highest density of chip-to-chip interconnects, shown in Figure 7.66. In this approach, through vias are etched in CMOS silicon wafers using deep reactive ion etching or laser ablation, followed by via metallization by sputtering and electroplating. Via diameters down to 2 μm and pitches of 5 to 10 μm have already been achieved, which will allow interconnection directly to the metal levels on the IC without the need for any redistribution layers.

The second approach involves alternate fabrication processes for ultrafine lines and spaces, replacing current lithography and semi-additive plating. These concepts derive from dual damascene techniques, common in IC fabrication, to form vias and traces at the same time by use of chemical or mechanical polishing for planarization of thin wiring layers. Figure 7.67 illustrates a cross section of one such technology wherein 12-μm traces embedded within the underlying dielectric layer can be clearly seen.

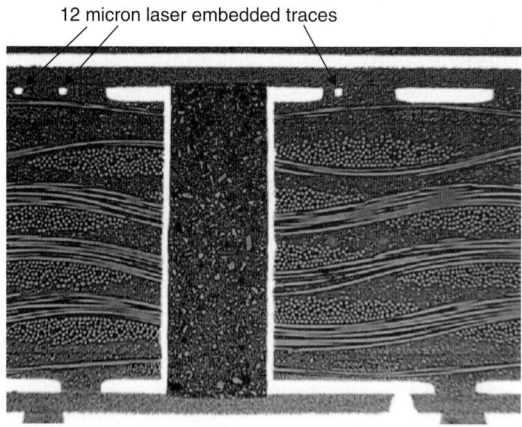

FIGURE 7.67 Embedded traces in dielectric layer by laser ablation and polishing.

In summary, substrate integration with high-density wiring and embedded thin-film components is a critical component of SOP technology, and in spite of the tremendous advances in the building blocks of SOP substrates as described in this chapter, a completely new set of technologies need to be developed in the next decade to keep up with the next generation of convergent systems.

Acknowledgments

The authors would like to thank the many researchers, students and faculty involved in the High Density Substrate and Embedded Passives programs at the Packaging Research Center whose work over the past thirteen years contributed to the breadth and depth of this chapter. Significant contributions were made to the contents of this chapter by Prof. Sue Ann Bidstrup (polymer materials and processes), Prof. Lawrence Bottomley and Dr. Jun Li (plating technologies), Dr. Robin Abothu, Dr. Jin Hwang and Dr. Prem Chahal (embedded capacitors), Dr. Swapan Bhattacharya and Dr. Mahesh Varadarajan (embedded resistors), Prof. Gary May and Boyd Wiedenman (high density substrate processes), and Dr. Chong Yoon (embedded actives). The contributions and support of several company sponsors and partners of the integrated substrate research at PRC are also acknowledged.

References

1. R. R. Tummala, *Fundamentals of Microsystems Packaging*, New York: McGraw-Hill, 2001.
2. Y. H. So et al., "Benzocyclobutene-based polymers for microelectronics," *Chemical Innovation*, vol. 31, no. 12, 2001, pp. 40–47.
3. R. V. Tanikella, "Variable frequency microwave processing of materials for microelectronic applications," *Ph.D. Thesis*, Georgia Tech, 2003.
4. T. Sung, "Variable frequency microwave curing of polymer dielectrics on metallized organic substrates," *M.S. Thesis*, Georgia Tech, 2003.
5. K. Farnsworth et al., "Variable frequency microwave curing of 3, 3', 4, 4-biphenyltetracarboxylic acid dianhydride/P-phenylenediamine (BPDA/PPD)," *Int. J. Microcircuits Electron. Packag.*, vol. 23, 2000, pp. 162–71.
6. K. D. Famsworth et al., "Variable frequency microwave curing of photosensitive polyimides," *IEEE Transactions on Components and Packaging Technologies* [see also *Components, Packaging and Manufacturing Technology, Part A: IEEE Transactions on Packaging Technologies*], vol. 24, no. 3, 2001, pp. 474–81.
7. R. V. Tanikella, S. A. Bidstrup Allen, and P. A. Kohl, "Variable-frequency microwave curing of benzocyclobutene," *Journal of Applied Polymer Science*, vol. 83, no. 14, 2002, pp. 3055–67.
8. R. E. A. Lauf, "2 to 18 GHz broadband microwave heating systems," *Microwave J.*, November 1993, p. 24.
9. B. Panchapakesan et al., "Variable frequency microwave: A new approach to curing," *Adv. Packag.*, September–October 1997, p. 60
10. M. Matsuoka, J. Murai, and C. Iwakura, "Kinetics of electroless copper plating and mechanical properties of deposits," *Journal of the Electrochemical Society*, vol. 139, no. 9, 1992, pp. 2466–70.
11. S. Nakahara, Y. Okinaka, and H. K. Straschil, "Effect of grain size on ductility and impurity content of electroless copper deposits," *Journal of the Electrochemical Society*, vol. 136, no. 4, 1989, pp. 1120–23.

12. L. D. Burke, G. M. Bruton, and J. A. Collins, "Redox properties of active sites and the importance of the latter in electrocatalysis at copper in base," *Electrochimica Acta*, vol. 44, nos. 8–9, 1998, pp. 1467–79.
13. A. Hung and K.-M. Chen, "Mechanism of hypophosphite-reduced electroless copper plating," *Journal of the Electrochemical Society*, vol. 136, no. 1, 1989, pp. 72–75.
14. J. Rangarajan, K. Mahadevaniyer, and W. Gregory, *Electroless Copper Plating Bath*, U.S. Patent 4,818,286, 1989.
15. D. H. Cheng et al., "Electroless copper plating using hypophosphite as reducing agent," *Metal Finishing*, vol. 95, no. 1, 1997, pp. 36–38.
16. A. Hung, "Electroless copper deposition with hypophosphite as reducing agent," *Plating and Surface Finishing*, vol. 75, no. 1, 1988, pp. 62–65.
17. A. Hung, "Kinetics of electroless copper deposition with hypophosphite as a reducing agent," *Plat. Surf. Fin.*, vol. 75, no. 4, 1988, pp. 74–77.
18. P. E. Kukanski et al., "Electroless copper composition solution using a hypophosphite reducing agent," U.S. Patent 4,209,331, 1980.
19. L. Fuhan et al., "A novel technology for stacking microvias on printed wiring board," *Proceedings of the 53rd ECTC Conference*, May 2003, pp. 1134–39.
20. T. Shimoto et al., "High-performance FCBGA based on ultra-thin packaging substrate," *NEC J. of Advanced Technology*, vol. 3, no. 2, 2005, pp. 222–28.
21. T. Shimoto et al., "High-performance FCBGA based on multi-layer thin-substrate packaging technology," *Microelectronics Reliability*, vol. 44, no. 3, 2004, pp. 515–20.
22. S. Jun et al., "Signal integrity and power integrity properties of FCBGA based on ultrathin, highdensity packaging substrate," 2005, pp. 284–90.
23. F. Lianhua et al., "Processability and performance enhancement of high K polymer-ceramic nano-composites," 2002, pp. 120–26.
24. S. Gluzman, A. A. Kornyshev, and A.V. Neimark, "Electrophysical properties of metal-solid-electrolyte composites," *Physical Review B: Condensed Matter*, vol. 52, no. 2, 1995, p. 927.
25. M. S. Ardi, "Ultrahigh dielectric constant carbon black-epoxy composites," *Plastics, Rubber and Composites Processing and Applications*, vol. 24, 1995, p. 3.
26. J. Obrzut et al., "High Frequency Loss Mechanism in Polymers Filled with Dielectric Modifiers," *Proceedings of the MRS Symposium*, vol. 783, Boston, MA, 2003, pp. 179–84.
27. P. M. Raj et al., "High-frequency characteristics of metal-polymer nanocomposite thin-films and their suitability for embedded decoupling capacitors," *6th Electronic Packaging Technology Conference*, Singapore, 2004, pp. 154–61.
28. H. Windlass et al., "Polymer-ceramic nanocomposite capacitors for system-on-package (SOP) applications," *IEEE Transactions on Advanced Packaging* [see also *Components, Packaging and Manufacturing Technology, Part B: IEEE Transactions on Advanced Packaging*], vol. 26, no. 1, 2003, pp. 10–16.
29. J. M. Hobbs et al., "Development and characterization of embedded thin-film capacitors for mixed signal applications on fully organic system-on-package technology," *Proceedings of the IEEE Radio and Wireless Conference (RAWCON)*, 2002, pp. 201–04.
30. T. Ogawa, S. Bhattacharya, and A. Erbil, "Lead-free high-K dielectrics for embedded capacitors using MOCVD," *Proc. International Microelectronics and Packaging Society*, Baltimore, 2001, p. 526.
31. S. K. Bhattacharya and R. R. Tummala, "Next generation integral passives: Materials, processes, and integration of resistors and capacitors on PWB substrates," *Journal of Materials Science: Materials in Electronics*, vol. 11, no. 3, 2000, pp. 253–68.

32. P. Chahal et al., "A novel integrated decoupling capacitor for MCM-L technology", *Proceedings of the 46th ECTC Conference,* 1996, pp. 125–32.
33. R. K. Ulrich and L. W. Schaper, *Integrated Passive Component Technology,* IEEE Press, Piscataway, NJ and Wiley Interscience, Hoboken, NJ, 2003.
34. E. B. Slamovich and I. A. Aksay, "Structure evolution in hydrothermally processed (< 100°C) $BaTiO_3$ films," *Journal of the American Ceramic Society,* vol. 79, no. 1, 1996, pp. 239–47.
35. W. Borland et al., "Ceramic resistors and capacitors embedded in organic printed wiring boards," *Proceedings of the International Electronics Packaging Technical Conference and Exhibition (IPACK 03),* pp. 2003.
36. T. Shioga et al., "Integration of thin-film capacitors on organic laminates for systems in package applications," Presented at *IMAPS Advanced Passives Workshop,* 2005.
37. K. Kurihara, T. Shioga, and J. D. Baniecki, "Electrical properties of low-inductance barium strontium titanate thin-film decoupling capacitor," *Journal of the European Ceramic Society,* vol. 24, no. 6, 2004, pp. 1873–76.
38. S. Yamamichi and A. Shibuya, "Novel flexible and thin capacitors with Mn-doped $SrTiO_3$ thin-films on polyimide films," *Proceedings of the 54th ECTC Conference,* 2004, pp. 271–76.
39. J. A. Andresakis et al., "Nickel coated copper as electrodes for embedded passives devices," U.S. Patent 6,610,417, 2003.
40. T. Kim et al., "Ca-doped lead zirconate titanate thin-film capacitors on base metal nickel on copper foil," *Journal of Materials Research,* vol. 19, no. 10, 2004, pp. 2841–48.
41. Y. Imanaka, *Multilayered Low Temperature Cofired Ceramics (LTCC) Technology,* Springer, New York, NY, 2005.
42. Syunzi Yokogawa, *"Multiple Chip Integrated Circuits and Method of Manufacturing the Same,"* U.S. Patent 3,903,590, 1975.
43. Robert J. Wojnaworski, *"High Density Interconnected Circuit Module with a Compliant Layer as Part of Stress-Reducing Molded Substrate,"* U.S. Patent 5,866,952, 1999.
44. Charles W. Eichelberger et al., *"Electroless Metal Connection Structures and Methods,"* U.S. Patent 6,396,148, 2002.
45. H. T. Rapala-Virtanen et al., "Embedding passive and active components in PCB-solution for miniaturization," *The ECWC 10 Conference at IPC Printed Circuits Expo, SMEMA Council APEX, and Designers Summit 05,* 2005, pp. S16-1 to S16-7.
46. L. Zhenxian et al., "Integrated packaging of a 1 kW switching module using a novel planar integration technology," *IEEE Transactions on Power Electronics,* vol. 19, no. 1, 2004, pp. 242–50.
47. C. Yu-Hua et al., "Chip-in-substrate package, CiSP, technology," *Proceedings of the 6th EPTC Conference,* 2004, pp. 595–99.
48. IMEC, "Smart High-Integration Flex Technologies," http://www.vdivde-it.de/portale/shift/.
49. Casio, "Casio to establish EWLP consortium," http://world.casio.com/corporate/news/2006/ewlp.html.
50. Raymond A. Fillion et al., *"Method for Fabricating Integrated Circuit Module,"* US patent 5,353,498, 1994.
51. R. Mahajan et al., "Emerging directions for packaging technologies," *Intel Technology Journal,* Vol. 6, Issue 2, 2002, pp. 62–75.
52. H. Reichl et al., "The third dimension in microelectronics packaging," *14th European Microelectronics and Packaging Conference & Exhibition,* Friedrichshafen, Germany, 2003, pp. 1–6.

53. M. Sunohara et al., "Development of interconnect technologies for embedded organic packages," *Proceedings of the 53rd ECTC Conference,* 2003, pp. 1484–89.
54. Y. Hara, "Matsushita embeds SoCs, components in substrate," *EETIMES.com,* Sept. 2002, http://www.eetimes.com/news/semi/showArticle.jhtml?articleID=10805530.
55. Lassi Hyvonen, Miikka Hamalainen, "*Shielded Laminated Structure with Embedded Chips,*" U.S. Patent 6,974,724, 2005.
56. *SMIT Center,* http://www.smitcenter.chalmers.se and http://smit.shu.edu.cn.
57. Baik-Woo Lee et al., "*Embedded Actives and Discrete Passives in a Cavity within Build-Up Layer,*" U.S. Patent Application # 20070025092, Filed 2005.
58. R. S. Li, "Optimization of thermal via design parameters based on an analytical thermal resistance model," *Proceedings of Thermal and Thermomechanical Phenomena in Electronic Systems ITHERM,* 1998, pp. 475–80.
59. M. Asai, "New packaging substrate technology, IBSS (interpenetrating polymer network Build up Structure System)," Proceedings of the MRS Symposium, Vol. 445, Boston, MA, 1997, pp. 117–24.

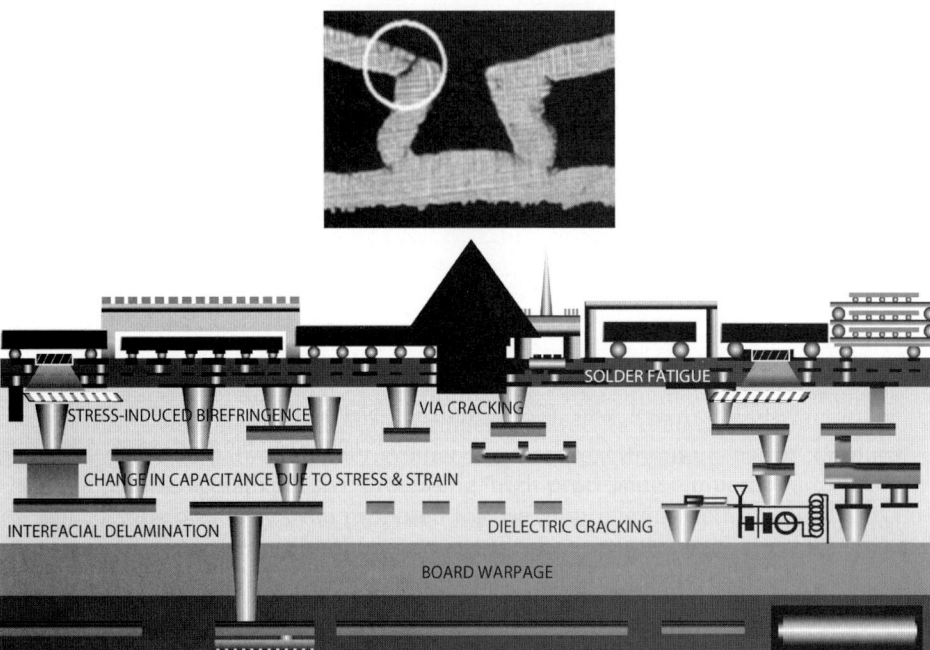

CHAPTER 8
Mixed-Signal (SOP) Reliability

Dr. Raghuram V. Pucha, Prof. Jianmin Qu,
and Prof. Suresh K. Sitaraman
Georgia Institute of Technology

8.1	System-Level Reliability Considerations 445		8.4	Future Trends and Directions 482
8.2	Reliability of Multifunction SOP Substrate 450		8.5	Summary 486
8.3	Substrate-to-IC Interconnection Reliability 468			References 487

System-on-package (SOP) technology, presented in this book, is based on up-front system-level design-for-reliability approaches and appropriate reliability assessment methodologies to guarantee their functional reliability such as for digital, optical, and RF blocks, as well as their interconnections and interfaces. A systems approach to reliability requires systematic design-for-reliability at various levels that include (1) materials and processes, (2) components, (3) subsystems with one or more functions, and (4) multifunction systems as shown in Figure 8.1. Each of these levels needs to be addressed for associated failure modes using a physics-of-failure approach by developing up-front design-for-reliability models and reliability verification strategies. As shown in Figure 8.1, up-front physics-based process mechanics and design-for-reliability models for various individual failure mechanisms are needed to evaluate various design options and materials selection even before the prototypes are made. Experimental material and interface characterization methods are needed to support the developed models and to assess the component reliability. System-level mixed-signal reliability needs to be addressed thorough system-level reliability metrics to relate the component-level failure mechanisms to a system level such as signal and power integrity through physics-based modeling and statistical considerations.

This chapter provides insight into the system-level reliability considerations of highly integrated multifunction systems, discusses the functionality-driven reliability issues, describes the state-of-the art in up-front reliability models and verification methodologies, and predicts the future trends and reliability challenges. The chapter

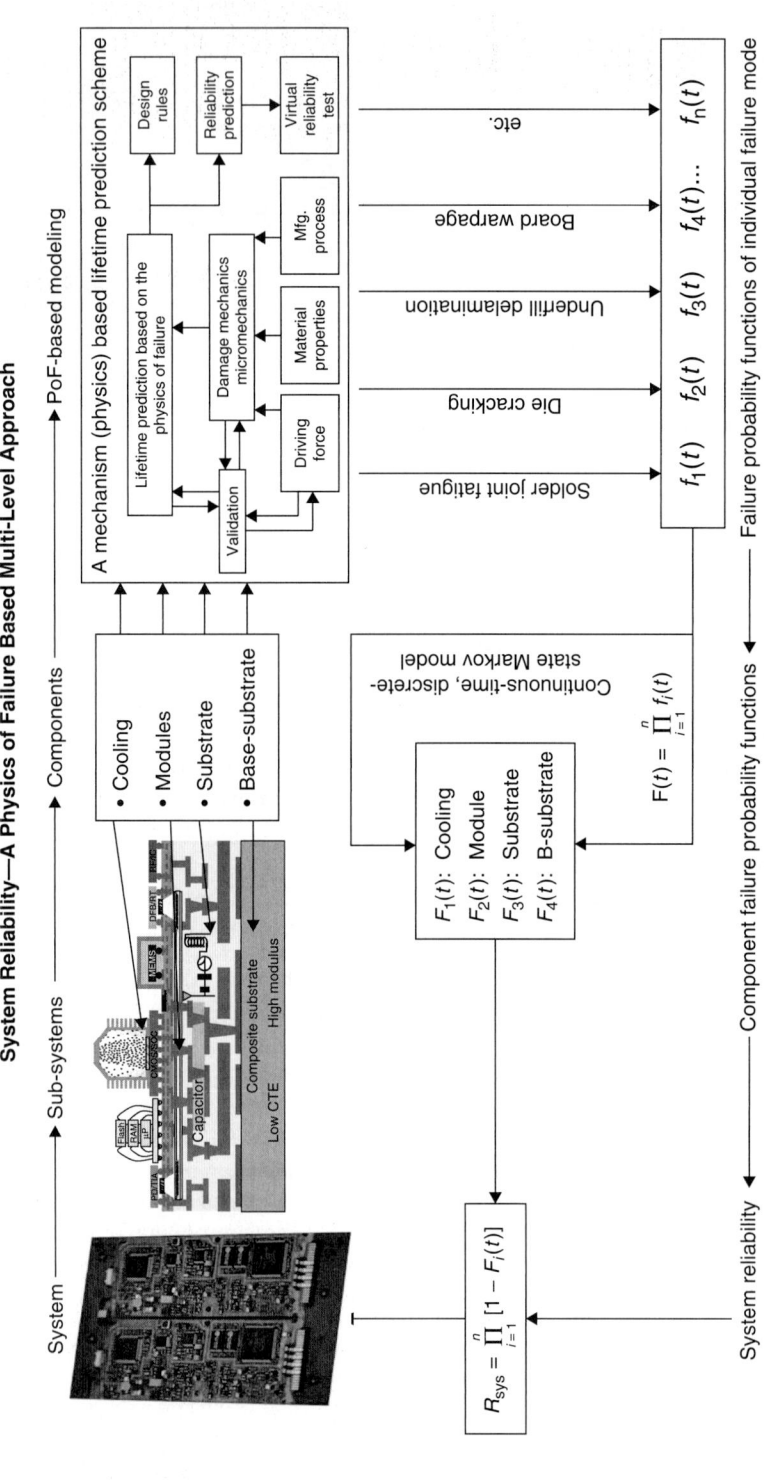

FIGURE 8.1 System reliability—a physics-of-failure based multilevel approach.

presents physics-based reliability models and reliability verification methods for various failure mechanisms associated with digital, optical, and RF subsystem functions of SOP. The chapter also deals with SOP substrate to IC interconnection reliability with advanced core and dielectric materials, solder joint reliability with and without underfill, advanced complaint, and nanostructured interconnections.

8.1 System-Level Reliability Considerations

The SOP technology integrates multiple system functions, such as high-speed digital and high-bandwidth optical, analog, or RF, as well as sensing functions, into one compact, low-weight, low-cost, and high-performance package or module system. The system integration is achieved through microvia and global interconnect (MGI) system boards and wafer-level packaging (WLP) for IC packaging. Alternate materials and processes are used to achieve the system parameters of SOP microsystems. A systems approach to reliability is necessary to understand the material interaction effects on component-level failure mechanisms and their influence on system-level performance. To realize next-generation convergent systems using SOP technology, aggregate system data rates of 40 Gbytes/s, an I/O density of 10,000/cm², a component density of 5000/cm², a wiring density of 6000 cm/cm², a 1000× size reduction, a via size of 5 to 15 μm, an interlayer dielectric with $\varepsilon_r < 3$, a decoupling dielectric constant of 300 to 500, a dielectric loss of 0.0001, and embedded waveguides with an optical energy of around 100 kV/cm² are needed. Although all of these features may not be present in the same microsystem, a highly integrated digital-RF-opto mixed-signal system is likely to have most of the features mentioned. What is required to define and ensure the system-level reliability of such complex multifunctional SOP systems?

The ability to perform required functions under stated conditions and a stated period of time is called "reliability," an inherent characteristic of all systems [1]. Designing complex systems with the satisfaction of target reliability is challenging due to complex assembly structures and logical connections; numerous components and failure modes; limited reliability data and prediction models; a complex system development process; and involvement of multiple design groups. Since the system failure can be caused by the failure of only one of the components, each component in the system must be analyzed, one by one, following assembly structure and logical connections under the usage conditions of the system. The complex system development process, in which various engineering groups work together with different viewpoints of reliability depending on their domain knowledge and the complex failure assessment with multiple failure modes, can also significantly influence the system-level reliability [2].

The traditional approaches of component-level failure data and statistical analysis in predicting the system-level reliability is highly expensive and ineffective due to the system-level functionality-driven changes and challenges in the materials, processes, and packaging configurations. These functionality-driven changes and challenges in the materials, processes, and packaging configurations require new failure data and statistical analyses each time to reassess the system-level reliability. On the other hand the component-level physics-of-failure reliability prediction models for individual functions and associated failure modes alone cannot be used to assess the system-level reliability due to the lack of considering the component-level functional and failure mechanism interactions. Such a systems approach to reliability requires the development of (1) physics-based reliability models for various failure mechanisms associated with digital, optical, and RF functions, and their interfaces in the system, (2) design optimization models for the selection of suitable materials and processing conditions, for reliability as well as functionality, (3) extensive experimental material characterization techniques, and (4) system-level reliability models understanding the component and functional interaction. In the following sections, up-front physics-based process mechanics and design-for-reliability models for various individual failure mechanisms in converged microsystems are presented to evaluate various design options and material selection even before the prototypes are made. Experimental material and interface characterization methods are described to support the developed models. System-level mixed-signal reliability is discussed through system-level reliability metrics to relate the component-level failure mechanisms to system-level signal integrity through physics-based modeling and statistical considerations. Advanced reliability modeling methodologies and algorithms to accommodate material length scale effects, due to enhanced system integration and miniaturization are presented. The future trends in highly converged microsystem packages and associated reliability challenges are introduced.

8.1.1 Failure Mechanisms

The system and package goals of SOP can be achieved through impedance-matched system board and fine-pitch wafer-level packaging (WLP) as shown in Figure 8.2.

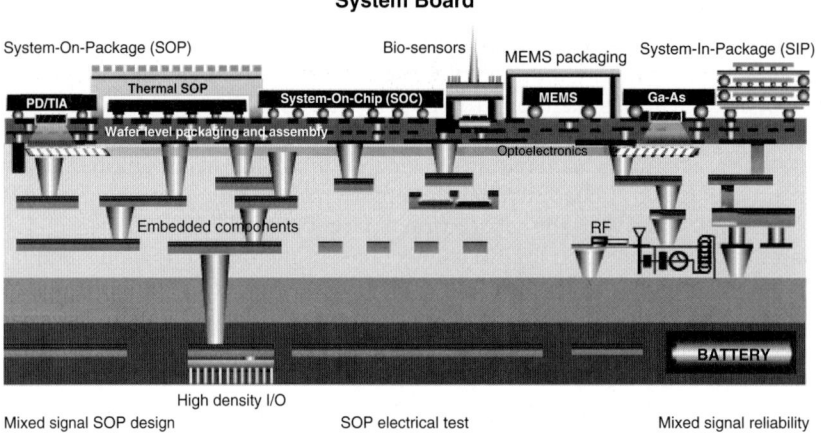

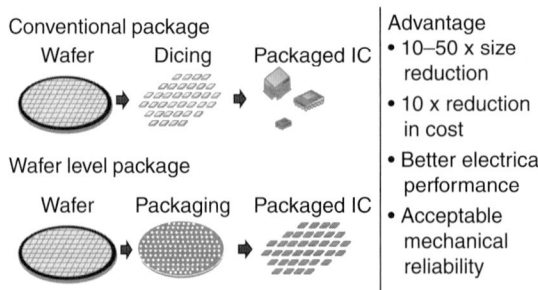

Figure 8.2 Two primary building blocks of SOP: system board and WLP.

As shown in Figure 8.2, many material systems associated with different functional blocks of the SOP system board and the system board–to-IC interconnection pose many material interaction challenges in terms of thermomechanical reliability. The thermomechanical failures are caused by stresses and strains generated within an electronic package due to thermal excursions under environmental and operation temperatures. Because of the mismatch in properties between different materials, thermally induced stresses and strains are generated leading to various failure mechanisms such as solder joint fatigue and cracking, die cracking, via cracking, excessive warpage, and delamination of material layers, as shown in the Figure 8.3. The thermomechanical stress-strain induced failure mechanisms in embedded RF passives and optical components also need to be addressed in highly integrated SOP systems. For example, the temperature excursions during thermal cycling and operating conditions can change the capacitance of the embedded capacitor resulting in undesirable system-level electrical performance. Mechanical or thermomechanical stress causes an anisotropic change in optical properties, known as stress-induced birefringence, leading to distorted optical signals in embedded optical components. Any of these failure mechanisms or their combination occurring at the lowest component level can precipitate a failure mode in the form of an electrical open at the system level. Accurate prediction of the reliability associated with system-level failure modes requires the prior understanding on specific component-level failure mechanisms shown in Figure 8.3. The design-for-reliability of SOP microsystems is more challenging in addressing the material interaction and functional compatibility issues at the component level in developing models for various failure mechanisms and their system-level interaction.

8.1.2 Design-for-Reliability

When a product performs the functions it is designed for, then the product is said to be reliable. When it does not, it is said to be unreliable. To ensure that the electronic systems packaging will be reliable over an extended period of time, two approaches need to be

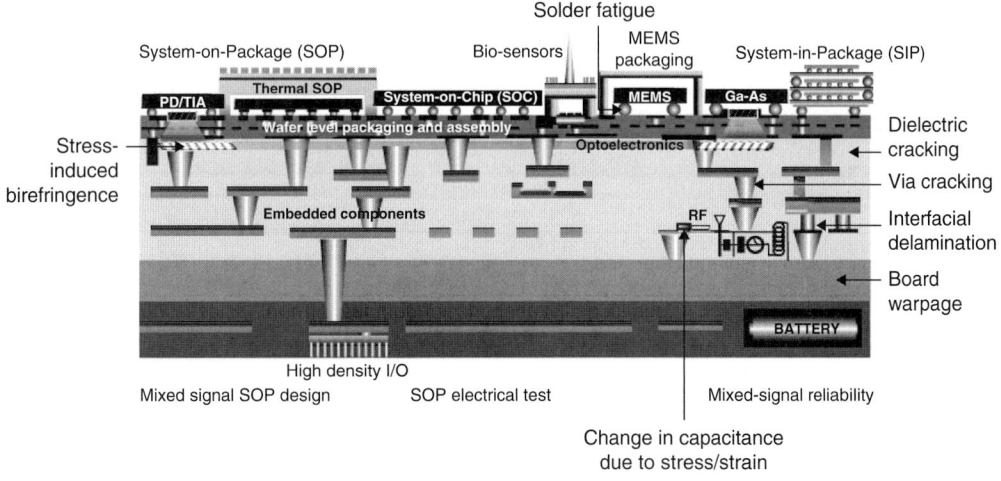

FIGURE 8.3 Potential failure mechanisms in SOP microsystems.

followed: (1) to design the systems packaging up-front for reliability, and (2) to test the systems packaging for reliability, once the system is designed, fabricated, and assembled. In the first approach, one could determine various potential failure mechanisms that could result in product failure, and knowing these fundamental mechanisms, one could create designs and select materials that would minimize or eliminate the chances for the failures. Such an up-front design, even before the system is built and tested, is called design-for-reliability. In the second approach, after a system is built and assembled, the package is subjected to accelerated test conditions for short periods of time by applying higher temperature, higher humidity, higher voltage, higher pressure, and so forth, to accelerate the failure process. This is called testing for reliability or reliability testing, and is discussed in Section 8.1.3. Traditional industrial practice involves testing for reliability, after the IC and the system-level packages are fabricated and assembled. If problems are found in reliability testing, the system is redesigned, refabricated, reassembled, and retested. Such a rebuild and retest process is expensive and time consuming. Therefore, design-for-reliability aims to understand and fix the reliability problems up-front in the design process, even before the IC and the system-level packages are fabricated.

The initiation and evolution of failure mechanisms (Figure 8.3) can severely limit the reliability of advanced material structures like SOP microsystems. The conventional mechanical design based on the ultimate strength of the material approach does not take into consideration the initiation and evolution of failure mechanisms that precipitate different system-level failure modes in SOP structures. The design of such multimaterial structures has to be failure mechanism based, also known as the physics-of-failure approach, as they experience various failure mechanisms during fabrication and end-use operating conditions. The development of design-for-reliability strategy with physics-of-failure based approaches in the early design stages is necessary to enhance reliability and reduce product development cycle time and cost. The design-for-reliability of SOP structures involves understanding the mechanics of materials and the structural response in the presence of various failure mechanisms, extensive up-front reliability modeling, failure analysis, and material characterization.

Several barriers, however, to developing system-level design-for-reliability approaches and reliability assessment methodologies for SOP technology remain. The following are some of the system barriers: (1) Modeling methodology and tools to address system-level reliability do not exist. The lack of such tools is a barrier to be able to assess and enhance system-level reliability through physics-based principles and through optimal up-front selection of materials, processes, and geometry even before prototypes are built. (2) When systems are subjected to thermal excursions and other operating and environmental conditions, the electrical and optical performance of the systems continue to change. However, there is no clear understanding of such changes in the electrical and optical performance metrics. Such a lack of understanding will prohibit the assessment of long-term reliability of the digital-RF-opto integrated systems. (3) The reliability of high-density chip-to-substrate as well as on-substrate microvia and global interconnects, designed to be able to meet the ITRS roadmap requirements for the year 2012 and beyond, is not understood. (4) Reliable base substrate materials, with high modulus (~300 GPa), low CTE (less than 5 ppm), and minimal warpage (1 mil to submil level over an area of 24 in × 24 in) to facilitate high-density thin-film processing as well as no-underfill attach for flip chips, are needed. For multifunction integration, there are no established standards to qualify RF and optical functional blocks along with the digital functional blocks. The barriers include: (1) Physics-based reliability models for various failure mechanisms in optical and RF functional blocks do not exist. (2) Appropriate field-use based qualification methods are not available for RF and optical functional blocks. With increasing functionality and increased integration at the system level, the classical methods of reliability prediction need to be revisited. Fundamental reliability research is also needed to implement design-for-reliability strategies for highly integrated SOP mixed-signal systems as follows: (1) No established techniques exist to measure, understand, and model mechanical behavior of materials at the micron and submicron length scale. (2) There are no appropriate damage metrics defined to quantify the interaction effects of various functional failure modes and system-level signal integrity. (3) In addition to the individual failure mechanisms, there are no established procedures to combine various failure modes at the functional component level to assess system-level reliability. A combination of design-for-reliability strategy with appropriate reliability verification at the component, function, and system level is needed to ensure the reliability of SOP mixed-signal packages.

8.1.3 Reliability Verification

Reliability verification typically involves experimental characterization and accelerated testing. Experimental characterization also plays a significant role in the reliability analysis of electronic packaging and, therefore, is critical in package design and manufacturing. Experimental characterization deals with the material, process, and design issues in electronic packaging to ensure processability and reliability. The reliability of a package should be sufficient to meet the manufacture process requirements and the product life expectancy under application conditions. As electronic technology advances, more novel materials are being used in packages and package sizes are becoming smaller and smaller. The components are typically structures with multimaterials and interfaces in a highly compact and integrated feature. The manufacturing process for the semiconductor devices and the related packages are becoming more and more sophisticated, involving numerous complicated steps. As a result, the process characterizations and reliability analyses are becoming more and more challenging. Together with many analytical and numerical methods used in this area, experimental methods have become more and more important. The experimental methods provide material properties that are needed in the theoretical and numerical modeling and for material selections. They are also critical in the areas of process characterizations and failure analysis for design optimizations. Experimental characterization in electronic packaging is conducted in three major areas: (1) material characterization, (2) process characterization, and (3) stress-strain analysis. On the other hand, to ensure product reliability, extensive reliability tests need to be performed before a new product can be shipped. To perform reliability tests within a reasonable amount of time under well-controlled environment, accelerated tests are commonly carried out in the laboratory environment for collecting reliability data and for product qualification and reliability verification. In accelerated tests, the devices are subjected to much higher "stress" than they would under normal usage conditions. The purpose is to accelerate the failure so reliability data can be collected within a much shorter period of time. This data can be used for reliability verification and reliability model validation. Once the accelerated tests are performed, an acceleration factor is used to convert the time-to-failure under accelerated test conditions to the actual time-to-failure under normal usage conditions. Commonly used accelerated thermomechanical tests are thermal cycle and thermal shock tests. A combination of design-for-reliability strategy with appropriate reliability verification at the component, function, and system levels is needed to ensure the reliability of mixed-signal SOP packages.

8.2 Reliability of Multifunction SOP Substrate

Signal integrity and power distribution are major challenges for the realization of highly integrated SOP microsystems. For maintaining signal integrity at high data rates, novel interconnect topologies using integrated board-level and wafer-level technologies with new materials and processes are necessary. The system boards of SOP technology use (1) low-K dielectrics for better signal integrity, (2) embedded high-K materials for decoupling capacitors for digital function, (3) high-modulus, CTE-matched large-area boards for planarity requirements, (4) staggered and stacked copper microvias for high-speed interconnection, (5) low-K, low-loss materials for embedded RF function, and (6) high-bandwidth, low-loss waveguide materials for embedded optical function. Some of the reliability issues in digital, optical, and RF functions are addressed here through material and characterization, up-front process mechanics, and failure mechanism models and reliability verification. The reliability issues in digital function are discussed through published work and the latest developments in this area. The reliability issues in optical and RF functions are discussed with modeling techniques developed for SOP integrated systems.

8.2.1 Materials and Process Reliability

Materials constitute the heart and soul of converged microsystems. Figure 8.4 shows some of the material systems being studied for SOP microsystems to achieve the desired system-level performance through tailored electrical, mechanical, and thermal properties. These include:

1. Novel inorganic composite materials for FR4 replacement as SOP substrates.
2. Low-loss polymers (tan δ < 0.002) like BCB, polyimide, polyphenyl ether (PPE), Teflon, and novel air-gap polymers for achieving ultralow dielectric constants below 2.0.
3. Novel polymer-ceramic nanocomposite dielectric materials for embedded capacitors, thin-film capacitors, and metal-filled polymers to achieve high capacitance with minimal filler content.
4. High-Q inductor materials in organic substrates.
5. Epoxy-based carbon polymer thick-film resistor materials.
6. Ultem/BCB waveguides and siloxane-based waveguide/cladding materials for optoelectronic applications.
7. No-flow underfill materials and nano-interconnect materials for assembly and wafer-level packaging of SOP.

These material systems associated with several functional blocks of SOP system boards precipitate many failure mechanisms due to material interaction and mechanics challenges during the fabrication process and subsequent assembly [3]. Because of the presence of several new material systems and processing conditions, the traditional build-and-test approach to reliability is not cost effective for developing the multilayered structure fabrication sequence of integrated microsystems. An innovative virtual

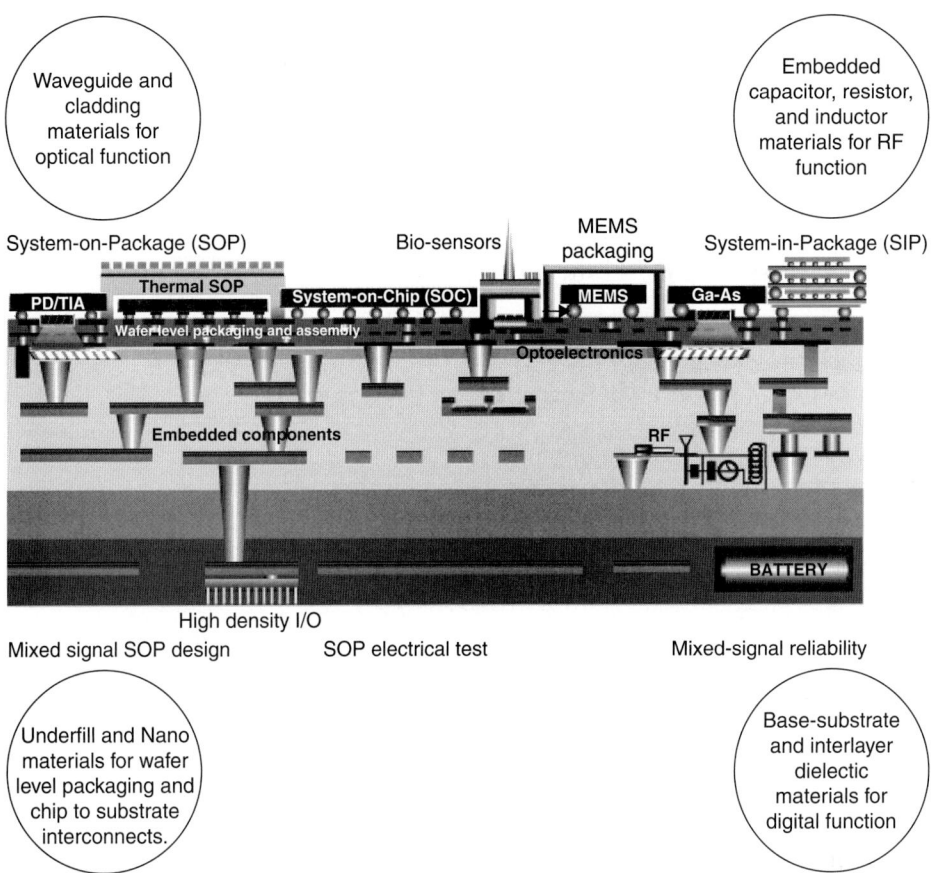

FIGURE 8.4 Material systems for SOP technology.

fabrication methodology is needed for simulating the large-area fabrication of SOP microsystems. The Packaging Research Center (PRC), Georgia Tech, has demonstrated proof of concept testbeds and mixed-signal functional prototype, Intelligent Network Communicator (INC), addressing the functional interface issues inherent in the mixed-signal technologies. The function of the starting convergent system INC is to transmit and receive a high-speed digital signal on an RF carrier wave (wireless signal) concurrently over an optical channel. The INC functional blocks are built on single impedance-matched, high-density interconnect (HDI) microvia board using SOP technology with a set of unified design rules. The process steps in fabricating the INC board are shown in Figure 8.5. In addition to the process steps shown for the multilayer microvia interconnect formation, the buildup high-density interconnect layers for RF include combiners, filters, embedded capacitors, and high-Q inductors. The layered structure for the integration of waveguide in the board is also shown in Figure 8.2. The waveguide fabrication steps include: (1) A layer of polymer is applied on the HDI layer to construct the buffer layer, (2) the optical bottom cladding layer is added, (3) the core layer is then coated and patterned by photolithography, and (4) finally the top cladding layer is applied.

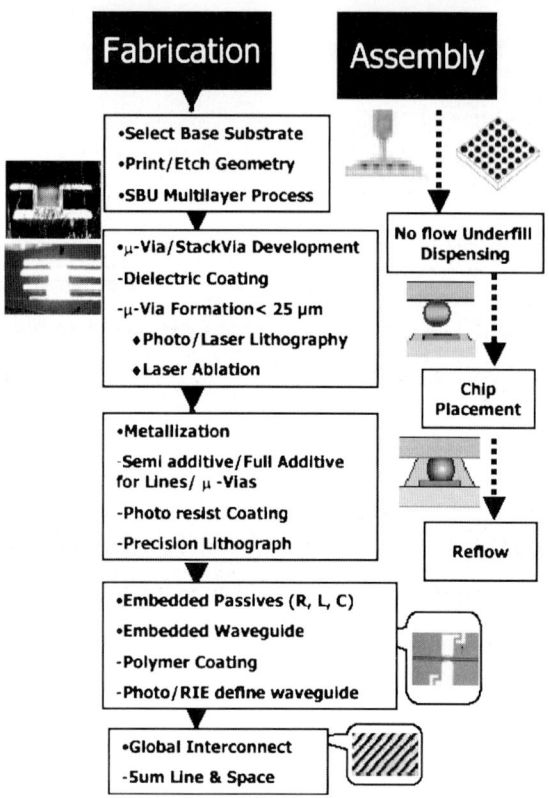

FIGURE 8.5 Typical process flow of SOP system boards.

Process-Induced Stresses and Strains in Digital Functional Boards

The thermal mismatch between the organic substrate and the silicon chip can be alleviated by the use of CTE-matched substrates, which will be discussed later in this chapter. However, the use of organic dielectric materials on such substrates causes large stresses in the dielectric film as well as in the microvias that are fabricated on them. Such stresses are introduced during the sequential buildup process of fabrication. Understanding the evolution of residual stresses and strains during the sequential buildup of a multilayer substrate and the effect of different substrate and dielectric material properties on the residual stresses is necessary in providing up-front guidelines for material selection and process optimization.

The cross sections of the substrate and the plane strain models of staggered microvia test vehicles are shown in Figure 8.6. The sequential process modeling simulates the strains and stresses induced in the structure due to first-level dielectric curing and cooldown, first-level microvia plating at room temperature, second-level dielectric curing and cooldown, second-level microvia plating at room temperature, and solder mask curing. The substrate structure with residual process stresses is then subjected to accelerated thermal cycling (ATC) conditions. The 20-minute ATC consisted of a 5-minute dwell each at high (125°C) and low (−55°C) temperatures and 5-minute transition times.

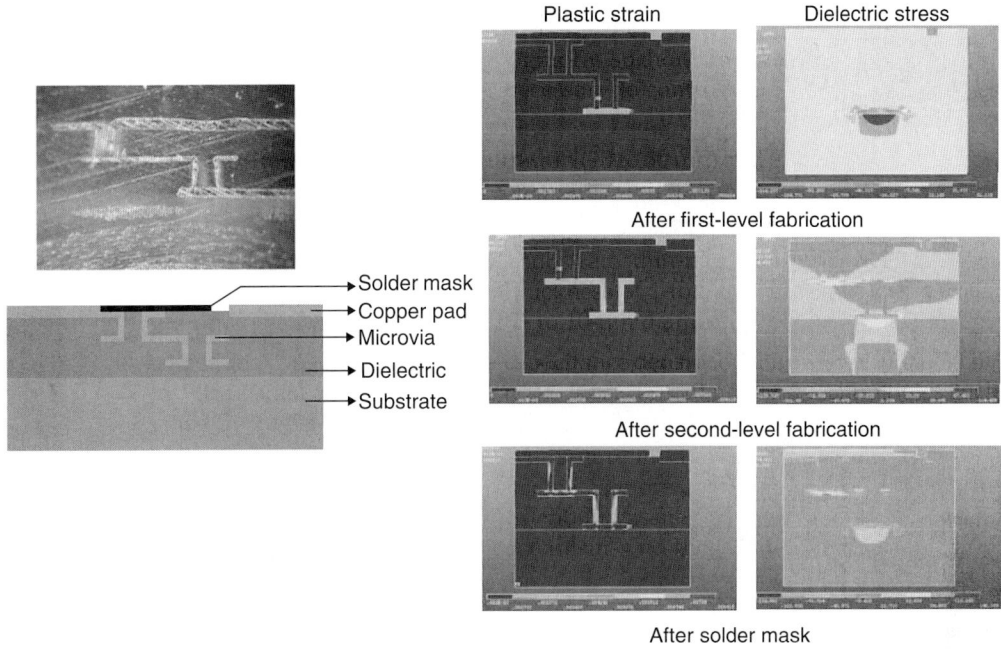

FIGURE 8.6 Microvia board and process-induced microvia strain and dielectric stresses.

These process and ATC simulations are carried out for a number of combinations of base substrate materials and dielectric materials. Figure 8.6 shows the evolution of von Mises plastic strains in the microvias and stresses in the polymer dielectric after the fabrication of the first-level microvia, second-level microvia, and the solder mask.

Figure 8.7 shows the effect of different material properties on the maximum von Mises plastic strain in the microvia during the buildup process. It is observed that the dielectric CTE, the dielectric modulus, the dielectric thickness, and the substrate modulus have a significant influence on the plastic strain experienced by the microvia. It was observed in the models that the average von Mises strain increases as additional layers are being deposited in the buildup process. As can be seen from Figure 8.7, the dielectric CTE and modulus have the predominant effect on the film stress in the dielectric. The average film stresses increase as additional layers are deposited in the buildup process as expected. These results are used in providing up-front guidelines for material selection and fabrication process optimization of mixed-signal SOP systems.

Process-Induced Stresses and Strains in Embedded RF Functional Boards

The SOP-embedded RF components include combiners, filters, embedded capacitors, resistors, and high-Q inductors. The up-front process modeling of a typical embedded capacitor in SOP is described here. The fabrication of the test board begins with the surface preparation of the double-sided copper-clad FR4, and the surface is roughened

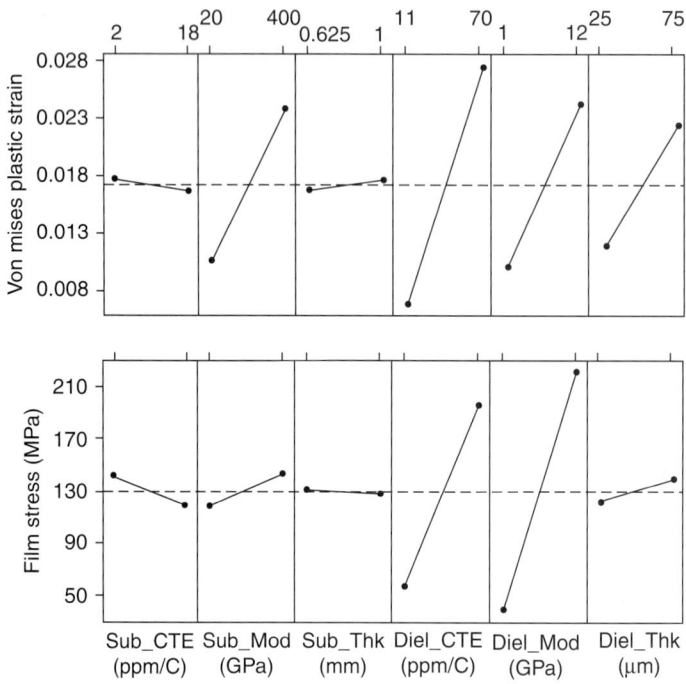

Figure 8.7 Effect of substrate material properties on microvia strain and dielectric stresses.

by microetching as shown in Figure 8.8. Typical process steps between the curing of the dielectric and the end of the fabrication of the top metal layer for capacitors is also presented in Figure 8.8. After baking at a temperature of 110°C, the dry film photoresist is applied by a vacuum laminator and is cured using UV exposure in order to pattern the first metal layer. The photoresist is developed, a subtractive etch is performed to develop the first metal layer, and the remaining photoresist is stripped using a sodium hydroxide solution. The dielectric layer is applied on the patterned metal layer, and the board is baked at 85°C. Vias to connect the bottom metal layer with the top metal layer are created in the dielectric layer by developing openings. These openings are fabricated by applying, exposing, developing, and stripping the photoresist. In order to develop the second metal layer, an electroless seed layer is applied and the substrate is then baked again. The rest of the metal-layer fabrication follows the same process as in the first metal layer. The resistors are formed on the top metal layer by applying the photoresist, developing it, and applying the resistor material in the gaps generated. A dry solder mask is applied, and the board is baked before accelerated cycling. Electrical testing of the passives after fabrication is done in order to ensure that they match their theoretical values.

The fabrication process described above is modeled with the geometric and material models using finite-element analysis. The fabrication process mechanics models follow the deformation induced in the board from the curing of the dielectric to the end of fabrication of the top metal layer. In the process model, it is assumed that the stress-free temperature of the dielectric is the cure temperature of the dielectric, and the stress-free

Mixed-Signal (SOP) Reliability

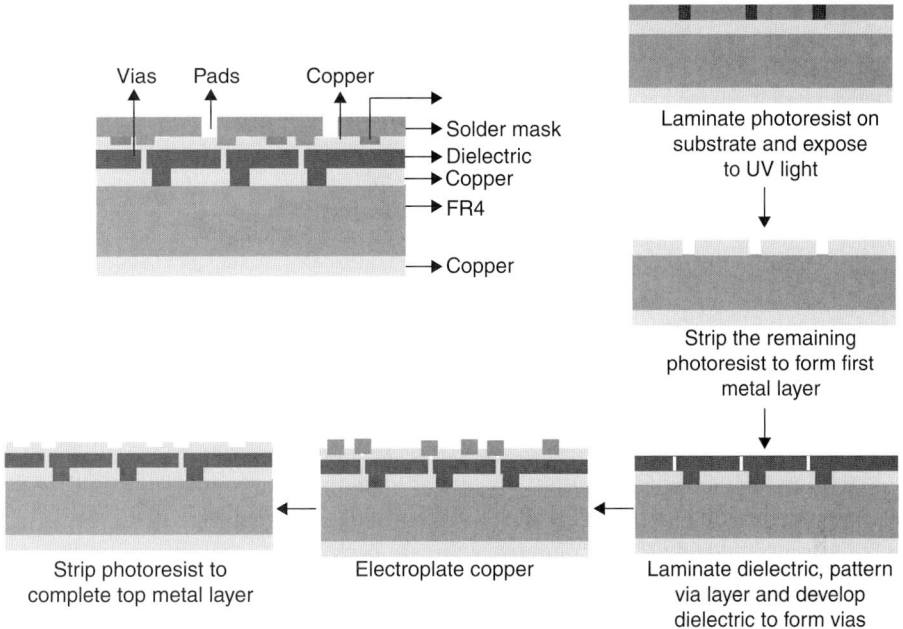

Figure 8.8 Schematic of the fabrication process in embedded capacitors.

temperature of the solder mask is the cure temperature of the solder mask. The copper layers are electroplated at room temperature, and therefore, copper's stress-free temperature is room temperature.

The effect of thermomechanical stresses and deformation induced by the process thermal loading on the electrical characteristics of the embedded capacitor is studied using a global-local modeling methodology. The global model uses layered-shell elements representing the layered geometry with all six capacitors to predict the deformation induced by the process. The local model uses solid elements to represent the critical capacitor of interest for the electromagnetic analysis alone to predict the change in capacitance due to the thermomechanical deformation. The deformation of the layered geometry in the global model due to the thermomechanical loads is transferred appropriately as displacement boundary conditions to the local model to predict the change in capacitance due to the thermomechanical deformation. Electromagnetic analysis is then performed for the capacitors. The local electrostatic model for capacitors consists of the two copper plates with the dielectric sandwiched in the middle. Solid elements are used to represent the geometry of the capacitor structure. Both undeformed and deformed capacitor configurations are considered for the electrostatic analysis. A mesh convergence study is conducted to see the effect of the number of elements and nodes on the predicted capacitance. Figure 8.9 shows the finite-element meshed capacitor and increased capacitance with the number of nodes in the local model. For the thermomechanically deformed capacitor configuration, the deformation of the layered geometry in the global model due to the processing thermal loads is transferred appropriately as displacement boundary conditions to the local model to predict the change in capacitance due to the

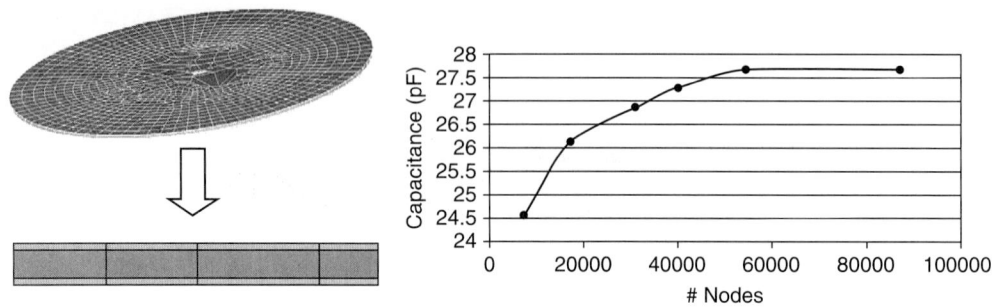

FIGURE 8.9 Convergence study of electrostatic model for embedded capacitor.

thermomechanical deformation. Table 8.1 shows the effect of process-induced thermomechanical deformation on the capacitance. The modeling results of converged capacitance values of the various embedded capacitors at room temperature are also listed in Table 8.1. The capacitance values of both undeformed and thermomechanically deformed configurations after processing are compared to the capacitance values obtained from an analytical solution. The analytical solution refers to the closed-form theoretical value of the parallel-plate capacitance. While the converged capacitance values of the undeformed capacitor matches well with the analytical solution, the capacitance values of the deformed configuration at the end of the processing show as much as 3.12 percent change in capacitance (in capacitor 6) with respect to the undeformed capacitor configuration. The sequentially coupled up-front process modeling of embedded RF components presented here is the first of its kind in the virtual reliability assessment of next-generation convergent microsystems [4].

Process-Induced Misregistration in Embedded Opto Functional Boards

Process-induced residual stresses also need to be quantified for the embedded waveguides. The waveguide core/cladding polymers have a CTE mismatch with the board material, which can cause warpage, misregistration, cracking, and delamination under thermal excursions. Figure 8.10 illustrates the use of finite-element modeling on

Capacitor	Analytical Solution	Undeformed Configuration	Deformed Configuration after Processing
1 (8.25 mm)	29.53	27.72	27.15 (−2.07%)
2 (9.5 mm)	39.16	36.68	35.82 (−2.35%)
3 (10.75 mm)	50.14	46.92	45.76 (−2.48%)
4 (12.0 mm)	62.48	58.68	57.10 (−2.70%)
5 (13.25 mm)	76.18	71.18	69.09 (−2.93%)
6 (14.5 mm)	91.23	85.19	82.53 (−3.12%)

TABLE 8.1 Effect of Process-Induced Deformation on Capacitance (pF)

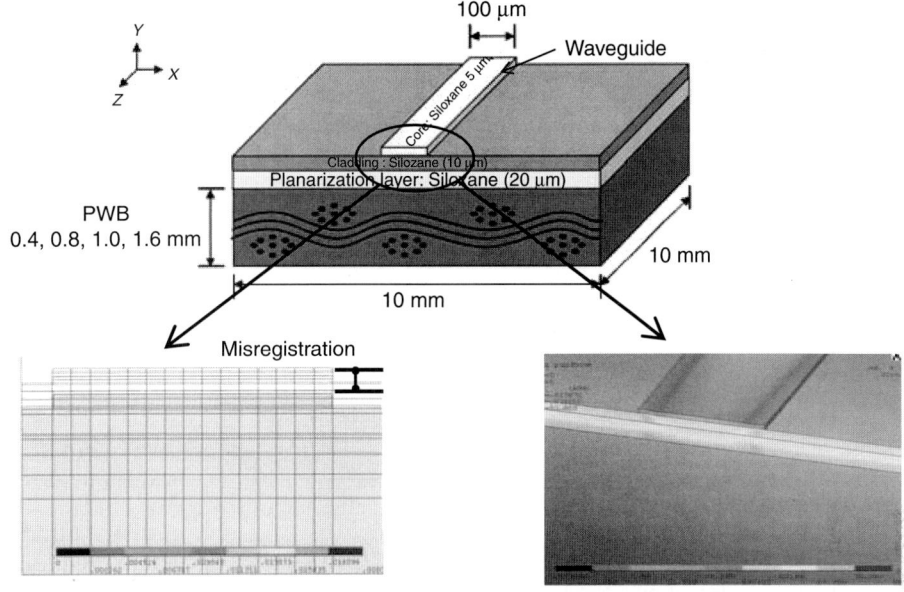

FIGURE 8.10 Process modeling of the polymer waveguides.

the waveguide structure to evaluate the process-induced warpage of the board, vertical misregistration, and stresses in a polymer waveguide material during processing. The model is cooled from a curing temperature of 160°C to room temperature of 25°C.

A symmetry plane is chosen at the midsection along the z axis. Unlike electrical interconnects that have relatively higher tolerance for substrate warpage, for optical interconnects the threshold value is much lower and, therefore, misregistration will lead to optical signal loss. Attenuation of the optical signal can be expected with the bending of the waveguide, which arises as a consequence of the warpage of the entire substrate structure. Figure 8.11 shows the effect of board thickness on the warpage and misregistration during cooldown from the processing temperature to room temperature. The maximum warpage of the board, as expected, decreases with the increase in the board thickness.

Up-front Process Optimization

Efforts to include the process history effects are typically undertaken primarily to predict the residual stress development in thick-section multilayer fabrication [5, 6]. Wu et al. [7] employed the finite-element method without curing and thermal gradient effects to model the fabrication of a multichip module/thin-film interconnect (MCM/TFI) package. A schematic of typical fabrication and assembly process of SOP microsystems is shown in Figure 8.5. A general-purpose integrated process modeling methodology and module has been developed [8, 9] to monitor the evolution of stresses and warpage during the sequential fabrication of the multilayered SOP substrate. This methodology incorporates some fundamental aspects involved in the curing of dielectric polymers such as the curing mechanism, the cure-induced shrinkage, structure-property relationships, and cure process optimization. The ability to monitor

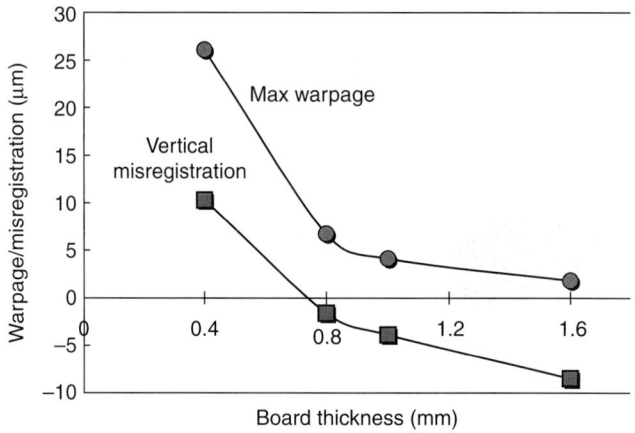

FIGURE 8.11 Effect of board thickness on warpage and misregistration.

the evolution of cure, temperature, and stress and strain distributions during processing allows for (1) the identification of critical regions where premature failure might be induced and the process step at which it occurs, and (2) optimization of the geometric design of the multilayered SOP substrate. The geometric, process, and material optimization thus result in reduced fabrication costs, reduced manufacturing cycle time, improved material properties, and a thermomechanically reliable SOP substrate design.

8.2.2 Digital Function Reliability and Verification

High-Density Wiring Reliability
Reliability of the microvia interconnects in SOP substrates is critical to the realization of high-density package wiring. Ultrafine-line structures on large-area SOP organic substrates using low-cost processes with a line width of 15 to 25 µm and microvia interconnects with diameters of 25 to 100 µm have been achieved in SOP system boards [10]. Physics-based parametric models have been developed to address various microvia failure mechanisms in SOP system boards. These feature-based parametric models take into consideration various process conditions in predicting the residual stresses and material interaction effects in developing predictive models and up-front design guidelines for SOP system board fabrication [11, 12]. These models have been validated using the failure analysis of test boards using accelerated testing conditions as shown in Figure 8.12.

Material Length Scale Effects
Increased functionality and wiring densities and reduced feature sizes in SOP microsystems necessitate advanced modeling and characterization techniques accounting for material length scale effects. For example, as the diameter of the microvia structures reduces to accommodate high-density wiring, the wall thickness of these structures correspondingly reduces to the order of microns, where scale effects in plastic deformation are predominant. Materials display strong scale effects when the characteristic length scale associated with nonuniform plastic deformation is on the order of microns.

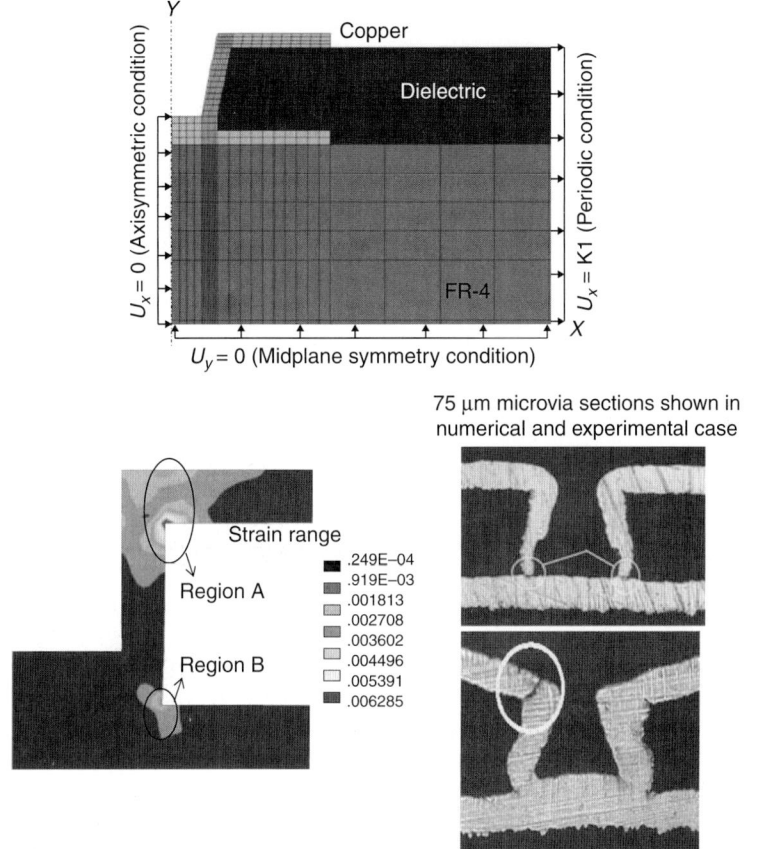

FIGURE 8.12 Microvia reliability modeling and verification.

An algorithm based on mechanism-based strain gradient plasticity (MSG) theory [13] has been formulated and implemented with commercial finite-element software with automated data exchange. The algorithm (Figure 8.13) simulates the mechanical behavior of copper during the cooldown of the substrate from the dielectric cure temperature of 135 to −55°C, which is the lower extreme of the temperature cycle. Figure 8.13 also shows the strain hardening behavior of microvias of different sizes. While there is little hardening in a microvia with a wall thickness of 10 μm (initial), there is substantial hardening in a microvia with a wall thickness of 3 μm. It is significant to note that the length scale when material scale effects are predominant in copper is about 2.8 μm [14]. When the plastic spatial strain gradients are considered in the hardening behavior of copper, it is seen that the strain in microvia could reduce and, therefore, the predicted fatigue life would go up by a factor of 2, especially for microvias with a plating thickness of less than 5 μm [15]. Incorporating material length scale effects in the plastic deformation of microvia structures enhances the up-front design-for-reliability of SOP systems with increased wiring density and reduced feature sizes of vias in the buildup layers.

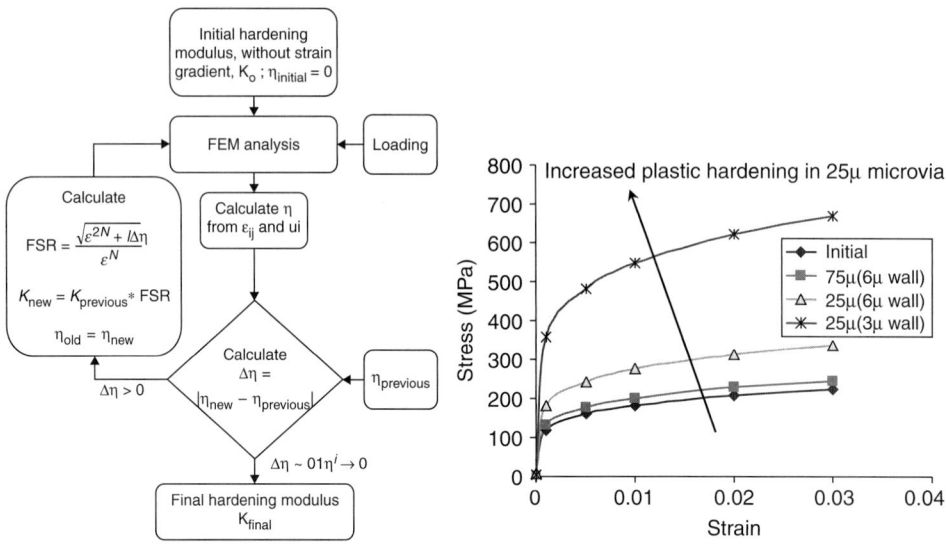

Figure 8.13 The hardening algorithm and increased plastic hardening in copper microvia.

Substrate and Interlayer Dielectric Reliability

Alternate substrate materials are needed to meet the warpage requirements of converged SOP systems. More discussion on the low-CTE and high-modulus alternate base-substrate materials will be presented in Section 8.3, in the context of various factors affecting the substrate to IC interconnect reliability. Although the low-CTE and high-modulus base-substrate materials can result in low warpage and can eliminate the need for underfill, they can potentially cause delamination and cracking in the interlayer dielectric. Experimental results show that dielectrics on low-CTE boards are prone to cracking at the solder mask opening corners, along the edges of the solder mask opening, and across the copper lines [16]. This is due to the high-CTE mismatch between the base substrate and a typical polymer dielectric. Alternate substrate materials are evaluated for SOP system boards, to explore a combination of a base-substrate material and an interlayer dielectric material such that the warpage is minimal, the dielectric will not crack or delaminate, and the flip-chip solder joints, assembled without an underfill, will not crack prematurely during qualification regimes or operating conditions. Nonlinear finite-element models with a design-of-simulations approach [17] are used in arriving at optimized thermomechanical properties for the base substrate and the dielectric materials to enhance the overall reliability of the integrated substrate with flip-chip assembly (see Figure 8.14).

Metal-Polymer Interface Delamination

SOP microsystem packages consist of layers of various materials with dissimilar material properties. Thermally induced stresses in such multilayered structures could reach significant levels near the free edges, possibly leading to interface debonding or delamination, and thus resulting in the failure of the multilayered package. For example, the high-dielectric-constant materials needed for decoupling capacitors in digital function, which typically consist of inorganic-organic nanocomposites or pure inorganic

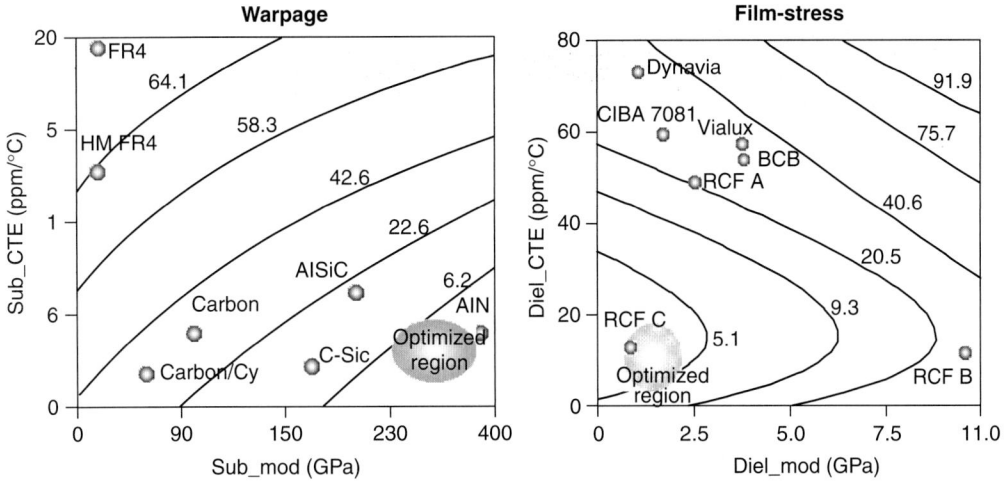

FIGURE 8.14 Response surfaces of warpage and dielectric stress against substrate properties.

films, should have good adhesion with the underlying metal electrodes and compatible thermomechanical properties. The high filler content required with the nanocomposite approach affects the adhesion between the capacitor film and metal and hence limits the reliability. Metal-polymer interface characterization [18, 19], thin-film interface [20], and material characterization [21] with parametric finite-element models have been developed to predict interface integrity in SOP system boards.

8.2.3 RF Function Reliability and Verification

Embedded R, L, C Reliability

Wideband and low-loss interconnects; high-Q multilayer passives including R, L, and C's; board-compatible embedded functions; development and characterization of high-K materials for integral passives; and low-loss, and low-cost boards are some of the features in SOP RF functions for the microwave wireless applications in the 1- to 100-GHz range. At high frequencies, as in the case of SOP systems, the physical dimensions of discrete components become a significant portion of signal wavelength. Thus, the surface-mount components are not well suited for high-frequency applications, as compared to embedded passives where the pad size and overall dimensions can be made small. Multilayer RF packaging on a SOP board is necessary to meet the high-performance requirements of SOP systems at low cost. The materials compatibility and thermomechanical reliability of embedded RF passives is not known for realization of fully integrated R, L, and C in multilayer substrates. Physics-based reliability prediction and testing of embedded passives is needed to realize multilayer RF packaging technology.

Schneider [22] has looked at the reliability of decoupling capacitors and subjected them to a variety of tests, but these capacitors were not embedded and were composed as a polymer ceramic material as the dielectric. Strydom [23] has studied the reliability of integrated inductors from a delamination standpoint. Witwit [24] has looked at the temperature distribution in resistors as dependent on the resistor area and number and size of via holes. However, research on thermomechanical reliability of embedded capacitors and resistors is limited in the published literature.

As integral passives are often composite layers with dissimilar material properties compared to the other layers in the SOP substrate, it is essential to ensure that the integral passive layer will not delaminate, crack, or otherwise fail. Similarly it is essential to ensure that the R, L, and C characteristics of the embedded passives do not deteriorate with thermal cycling. Test vehicle substrates under 1000 thermal cycles between –55 and 125°C and high humidity and temperature conditions at 85°C/85 relative humidity (RH) for 1000 hours showed large changes in the inductance (Figure 8.15). The maximum decrease after thermal cycling in capacitance was 4 percent, in inductance was 20 percent, and in resistance was 12 percent. The change under humidity conditions for capacitance was +10 percent, for inductance was +8 percent, and for resistance was –16 percent.

Figure 8.16 shows cross sections of test vehicles of edge-probed capacitors with microvias after accelerated testing. Although there is no visible sign of cracking or delamination, after accelerated testing, the thermomechanical deformation during

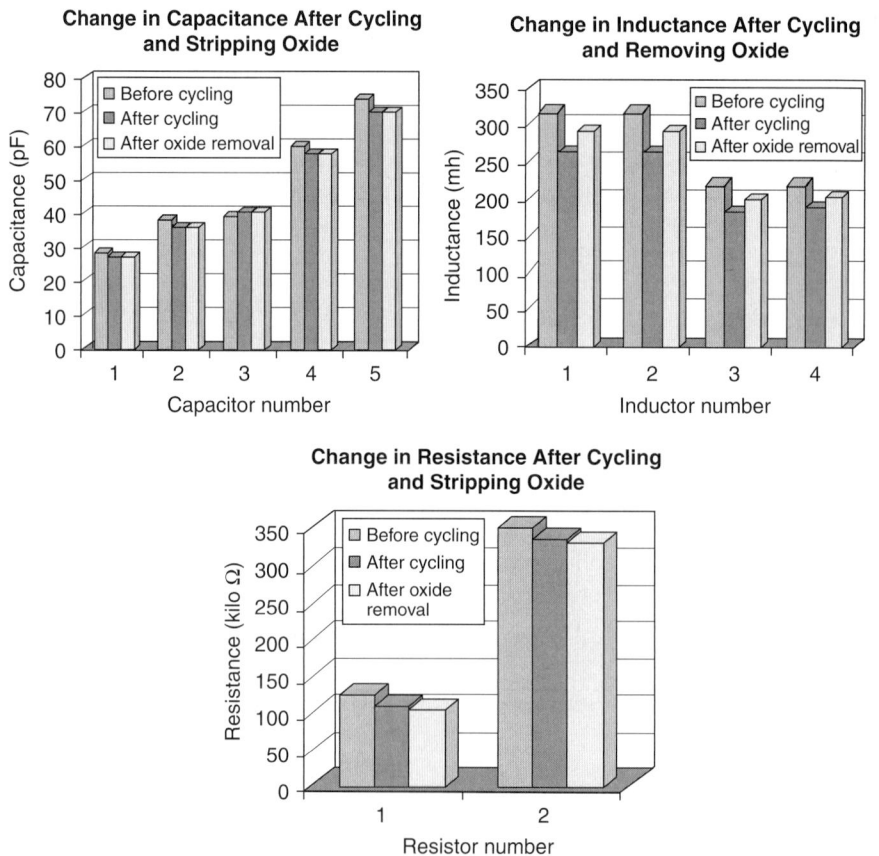

FIGURE 8.15 Changes in R, L, and C electrical parameters after 1000 thermal cycles.

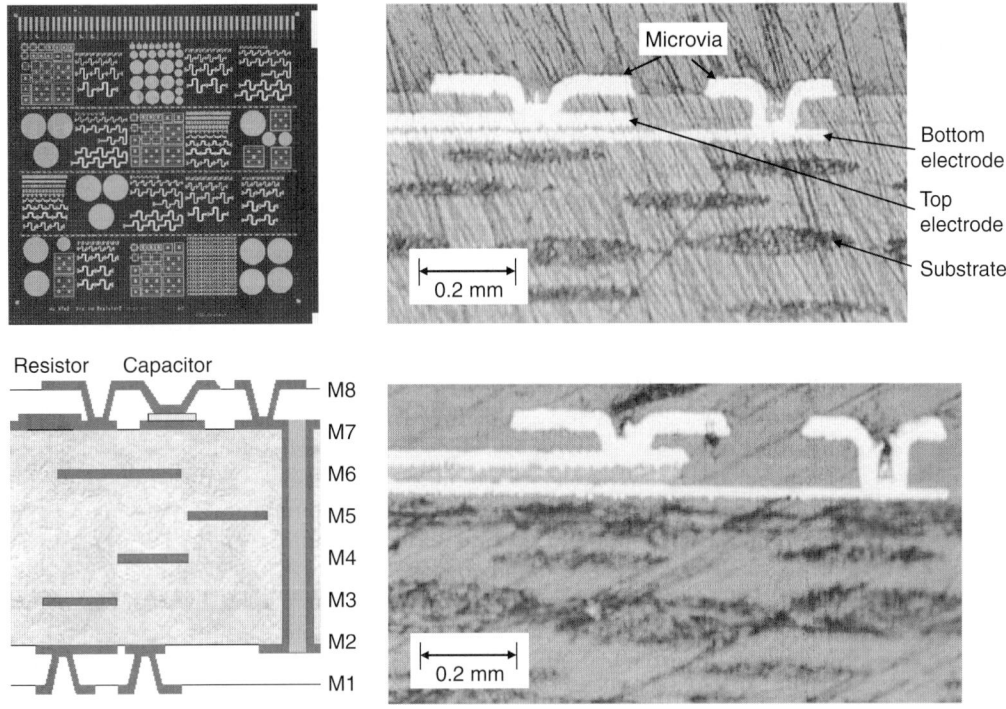

Figure 8.16 Cross sections of capacitors in test vehicles after 1000 thermal shock cycles and 1000 hours of humidity testing.

humidity testing appears to be more severe as indicated by excessive microvia deformation. It should be noted here that both the thermomechanical deformation and change in permeability of the capacitor dielectric contribute to the percentage change in capacitance after accelerated testing [4].

In addition to the build-and-test approaches, the reliability of integral passives under thermal excursions through physics-based models is necessary. The changes in the resistance, inductance, and capacitance of the integral passives due to thermal and mechanical loads need to be quantified to understand their effect on system-level signal integrity. Modeling efforts with coupled thermal and electrical response of embedded RF passive components are briefly discussed here. The physics-based thermomechanical models (Figure 8.17) of the embedded RF test vehicles predict the deformation in the structure and thermally induced stresses during thermal cycling. Using these thermal stresses, the potential for cracking and delamination in each layer can be predicted. Coupled electrostatic models have been developed to predict the changes in electrical parameters with thermal cycling using the thermomechanical deformation induced in the boards during thermal cycling [4].

8.2.4 Optical Function Reliability and Verification

Electrical interconnections at high speeds are limited by several factors, including frequency-dependent signal attenuation, crosstalk, high power consumption, jitter, and

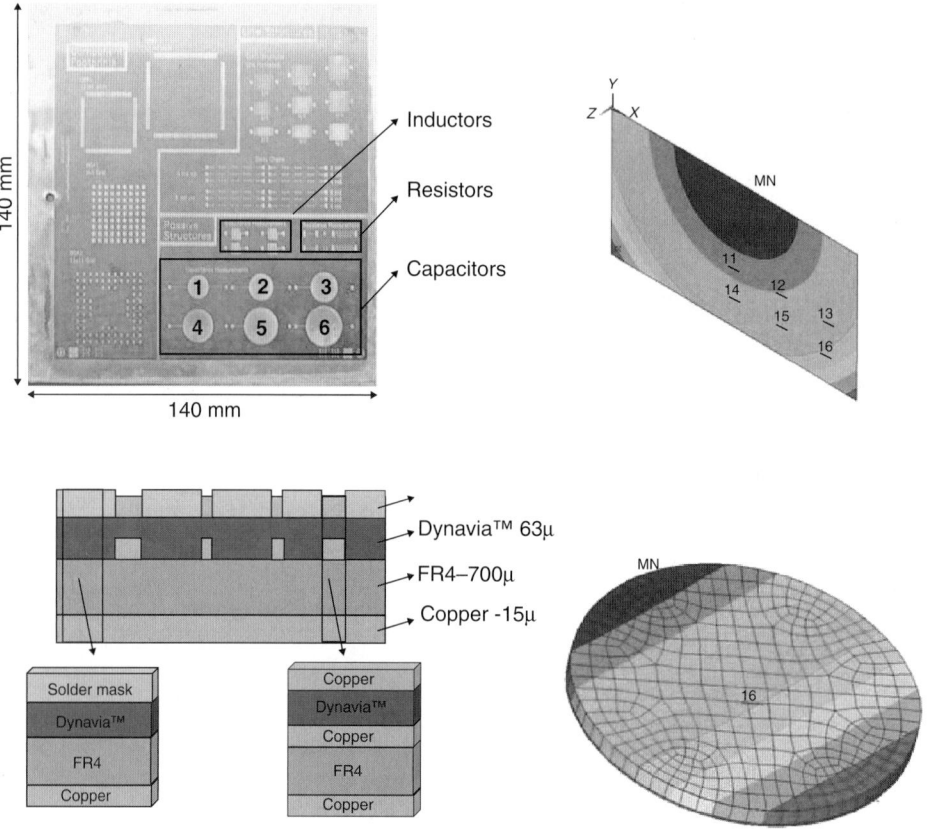

Figure 8.17 Layout of embedded RF test vehicle and finite-element models of embedded resistors and capacitors.

skew [25]. Optical interconnections offer a potential solution to some of the bottlenecks that current electrical interconnection systems face [26]. Key to the success of integrated microsystems like SOP is the ability to integrate optical interconnections into electrical systems, with low power dissipation and a small footprint at high speed. The introduction of new optoelectronic materials and processes, the increasing bandwidth and decreasing loss requirements, the need for reducing cost without compromising reliability, and the continued demand for miniaturization have made the design and development of optical interconnects a challenging and time-consuming process. Models to understand the thermomechanical issues and reliability of optical polymer interconnects and to provide up-front design and material guidelines are needed so that the microsystems integration for chip-to-chip optical interconnections at the backplane and board levels will exhibit maximum performance and be successful in qualification tests. Several design and material parameters influence the stress-strain distribution in the optical systems and thus the overall performance of the systems. Through experiments and finite-element modeling, the waveguide material and structure are studied based on a number of factors, which include thermomechanical stresses, packaging misregistration, stress-optical effects, refractive index stability, and reliability testing. Waveguide optical interconnections are studied that can be embedded with thin-film active optoelectronic devices and integrated onto an electrical interconnection substrate. These optical interconnects on board are for very short-haul backplane and board-level chip-to-chip interconnections. The physics-based models presented here are generic in nature and can be used to perform what-if analysis for a number of integration options by changing the appropriate material properties, processing conditions, and geometry parameters. The models are supported by appropriate material characterization and reliability testing for model validation. Such an up-front reliability analysis is likely to result in reliable optoelectronic structures at a reduced cost and time for next-generation SOP microsystems.

Stress-Induced Birefringence

Apart from thermomechanical concerns, the stress-optical effect causes unwanted birefringence or a split of the fundamental modes in the waveguide. A phenomenon called stress-birefringence arises when a transparent solid material is subjected to a mechanical or thermomechanical load and causes an anisotropic change in optical properties. A plane-strain analysis followed by an optical mode analysis shows the resulting stresses in the waveguide structure and the stress-induced birefringence within the waveguide core. Figure 8.18 shows a plain-strain model of a silica waveguide on a silicon substrate structure and shows that stress-induced birefringence can be made minimum in the waveguide core during processing. The stress-optical coefficient required by the model is calibrated from birefringence measurements on thin-film polymer samples using a prism coupler.

Reliability Testing

Understanding the effect of temperature and humidity on the waveguide material is necessary, as it can lead to a change in material and optical properties. Thermal cycling tests on optical waveguides typically follow the Telcordia standard GR1210, which has been developed for passive optical components and has a maximum temperature of 125°C. A more aggressive thermal cycle, from 0 to 175°C was conducted on a fabricated

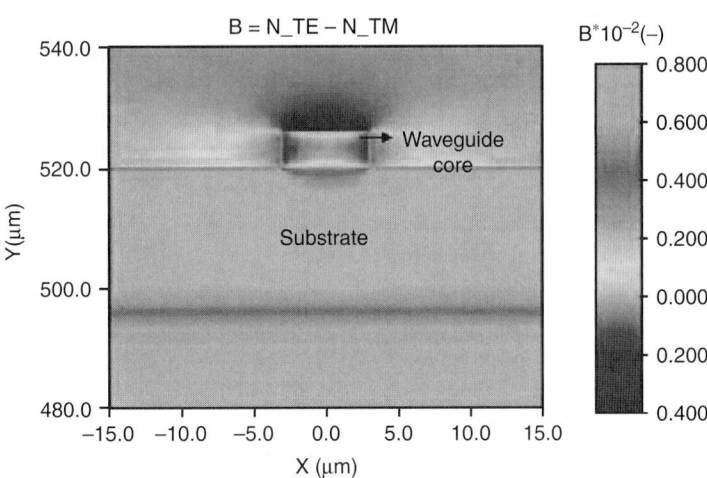

FIGURE 8.18 Modeling of stress-induced birefringence within the waveguide core.

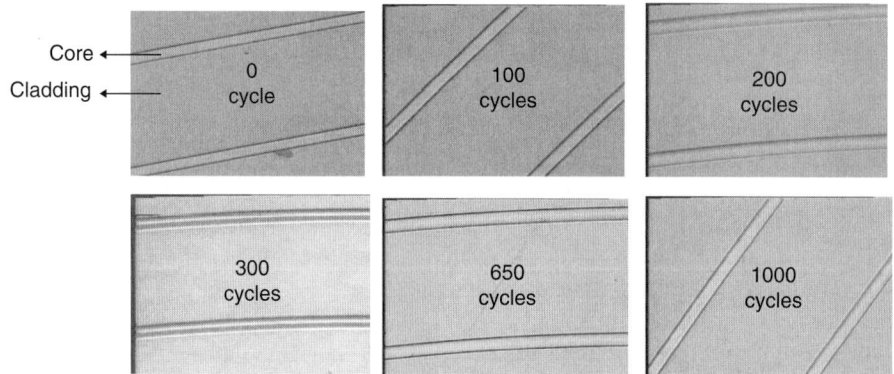

Figure 8.19 Thermal cycling of polymer waveguides on board from 0 to 175°C shows a color change.

waveguide structure. Figure 8.19 shows the effect of the thermal cycling on the polymer materials. There was no cracking or delamination observed in the samples, which were checked using a dark field optical microscope and a Sonoscan, respectively. However, there was a progressive color change in the waveguides during thermal cycling, which is likely due to oxidation of the material. The effect of material changes on optical properties such as optical loss is being studied.

Waveguide Refractive Index Stability

The performance of the optical interconnects depends on the refractive index contrast between the adjoining materials. Tight optical tolerances require an understanding of all the effects that result in a change in that refractive index during operation and reliability testing. A large refractive index difference tends to carry numerous modes simultaneously, leading to undesirable modal dispersion [27]. This refractive index variation should be taken into account when designing the optical interconnects and evaluating their performance. Since a reflective substrate is needed to measure the refractive index, the following experiments were conducted on thin-film siloxane on silicon wafer. Figure 8.20 shows the percentage increase of the refractive index of optical polymer films during thermal aging and thermal cycling experiments. The refractive index of the core/cladding polymers appear to be fairly stable during these experiments.

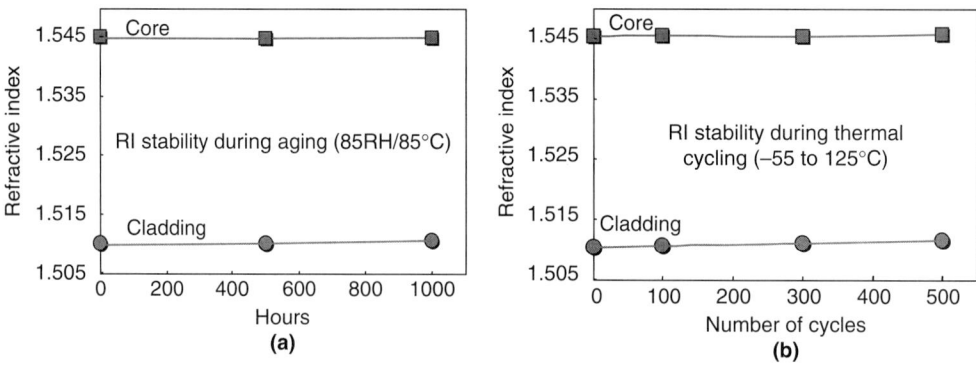

Figure 8.20 Refractive index stability. (*a*) Thermal aging. (*b*) Thermal cycling.

8.2.5 Multifunction System Reliability

As SOP focuses on systems integration and miniaturization, a systems approach to reliability is necessary to relate the component and functional failure mechanisms to system-level signal integrity. System-level metrics need to be identified and defined to understand and predict the reliability of digital, RF, and optical functions and their interfaces taking into consideration physics-based failure prediction methodologies as well as statistical aspects. Figure 8.23 shows a functional SOP system board with digital, embedded opto, and RF components. The function of one component can have interaction in various domains with the other components. In the electrical domain, electromagnetic interference from one component may affect another component in the system. In the thermal domain, the heat generated by one component may increase the operating temperature of other components. In the mechanical domain, the warpage of the substrate due to one package may increase the strain in a nearby package and/or signal integrity in other functions. Therefore, a component must be modeled for reliability in its system environment. Figure 8.21 also shows the system-level reliability assessment modeling strategy with various issues involved [28].

Functional Interaction and Statistical Considerations

While the module-level reliability models primarily focus on one failure mode at a time, in a package of interest, the system-level models address multiple failure modes with several packages on a board with more than one function. The presence of one package on a substrate may affect the reliability of another package or function. Some of the examples of such interactions specific to SOP mixed-signal systems include the following: (1) The presence of a high-performance digital function in proximity to a

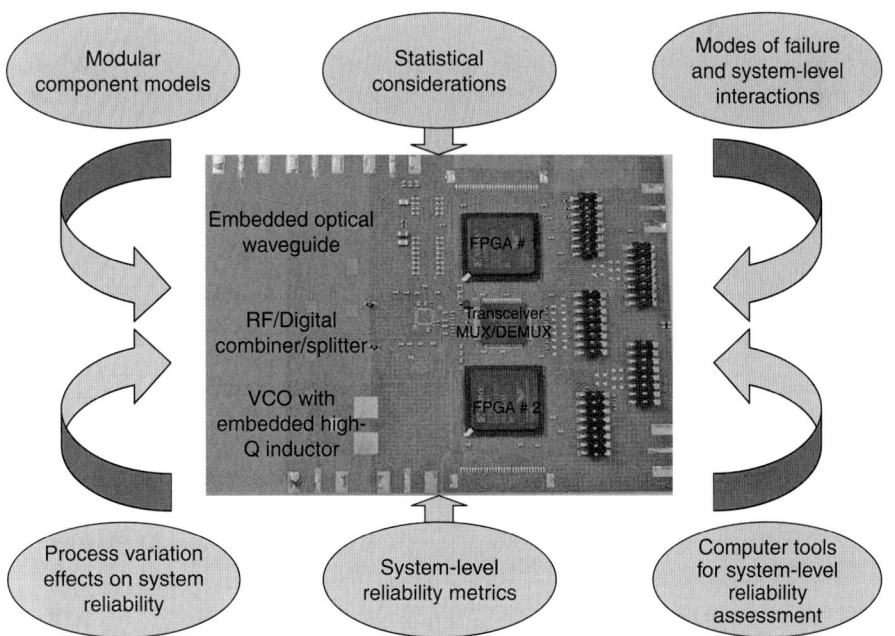

FIGURE 8.21 Functional SOP substrate and system-level reliability strategy.

$$P_s(t) = \prod_{i=1}^{n} P_i(t) \quad \rightarrow \quad P_s(t) = \prod_{i=1}^{n} (\Phi_i(t)) \quad \text{----- (Eq. 1)}$$

Where :

P_i = Percent reliability of component i
P_s = Percent reliability of system
Φ = Cumulative distribution function

$$\Phi(x) = \int_{-\infty}^{x} \frac{e^{-\frac{t^2}{2}}}{\sqrt{2\pi}} \; (\textit{if Normal}) \qquad \Phi(x) = \int_{0}^{x} a\lambda^a x^{a-1} e^{-(\lambda y)^a} \; (\textit{if Weibull}) \quad \text{----- (Eq. 2)}$$

Tabular data exists for Normal and Weibull curves

FIGURE 8.22 Modeling statistical interactions between components in a system.

high-flux optical domain could overheat the optical waveguides and affect the waveguide reliability. (2) The processing of a low-loss interlayer dielectric to achieve digital function target parameters could adversely affect the reliability of the embedded passive layers fabricated. (3) The warpage introduced due to the processing of different layers and due to the operation of different functions would adversely affect dielectric reliability, waveguide misregistation and optical fidelity, and passive layer adhesion. Therefore, failure mode interaction [29] must be considered in assessing the system-level reliability. Some of the fundamental research issues in system-level reliability as shown in Figure 8.21 include (1) developing parametric modular component models, (2) modeling system-level failure mechanisms and their interaction effects, (3) process variation effects on system-level reliability, (4) CAD tools for system-level reliability assessment, and (5) identifying system-level metrics to correlate component and functional failure mechanisms to system parameters. Development of system-level reliability models with multiple functions and failure mode interactions requires tremendous computational resources and time. Innovative modeling techniques augmented with high-performance computing are necessary to ease the computational complexity and reduce the time needed in achieving acceptable results.

Statistical reliability considerations are also necessary when dealing with system-level reliability. Historical component failure data, statistical variations in material properties, fabrication and assembly processes, and other component-to-component variations should be taken into consideration for system-level reliability assessment. In general, the reliability of an entire system is usually less than the reliability of a single nonredundant component of the system (Figure 8.22). Thus by linking the individual component reliability with statistical variations [30] in an integrated system, system-level reliability can be assessed.

8.3 Substrate-to-IC Interconnection Reliability

In this section the various factors affecting the substrate-to-IC interconnection reliability are discussed first, deriving the need for CTE-matched base-substrate materials and compatible dielectrics to improve the interconnect reliability of next-generation packages. The thermomechanical reliability of a 100-μm pitch flip-chip assembly on CTE-matched boards, thermomechanical fatigue models with lead-free solder interconnects, vibration effects in solder interconnects, and interfacial adhesion and underfill reliability are presented next within the framework of wafer-level packaging.

8.3.1 Factors Affecting the Substrate-to-IC Interconnection Reliability

The increasing trend toward miniaturization and higher functionality in microsystems will drive a greater demand for interconnection density at the package and board levels. Two major components of future high-density packaging are the sequential buildup of multiple layers of conducting copper patterns with interlayer dielectrics on a board and multiple ICs flip-chipped on the top layer. A wide range of passives, waveguides, and other RF and optoelectronic components will be buried within the dielectric layers as shown conceptually in Figure 8.23. The interleaved copper and dielectric layers also support the high-density interconnects for power and signal requirements. The board material should therefore meet certain electrical, thermomechanical reliability, HDI processing, and cost requirements. Current substrates impose several fundamental barriers in meeting the reliability and stringent processability requirements of these convergent multifunctional microminiaturized systems. The fundamental barriers of today's substrate materials and the need for novel substrate materials is briefly discussed hereunder in the context of increased functionality, wiring densities, and reduced feature sizes of next-generation microsystems.

Barriers to High-Density Wiring
The routing of future ICs with 10,000+ I/Os requires ultrafine feature sizes of about 10-μm lines and space widths and via-pad diameters ranging from 20 to 35 μm [31, 32].

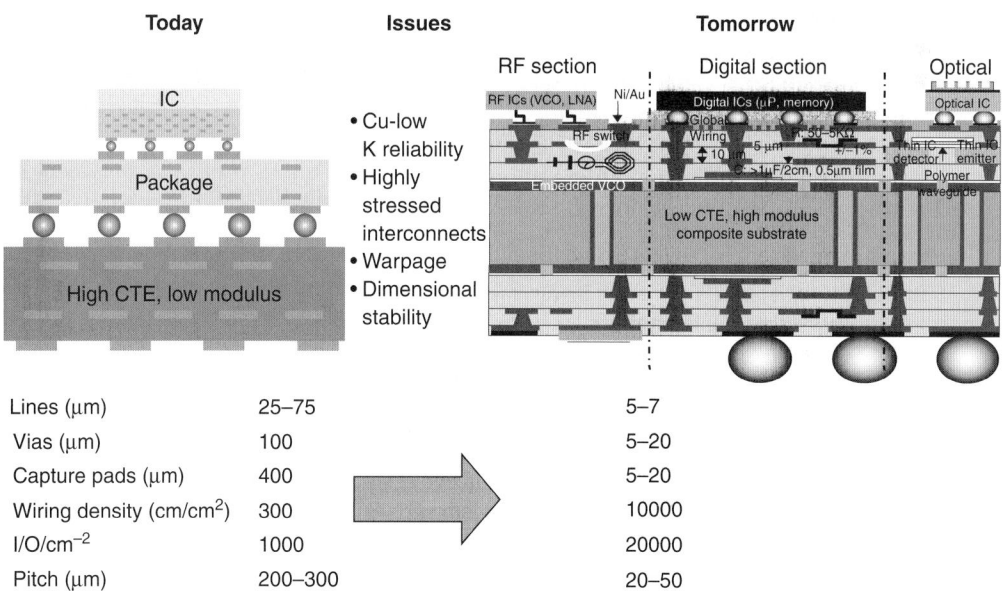

FIGURE 8.23 Comparison of today's asymmetric and tomorrow's convergent high-performance microsystems.

The higher I/O count and wiring density challenges of the Cu low-K devices upstream to the package level drive the need to develop new packaging technology that can support the required number of layers to support high I/O count and I/O density packages as shown in Figure 8.22. There is a need for four to eight layers of 5- to 10-μm wiring for future system boards. Microvia substrate technologies will play a crucial role in the printed wiring board (PWB) industry to accommodate these high-performance requirements. The stringent need to process multiple layers of thin films with via sizes of 20 μm and capture pads less than 35 μm requires layer-to-layer misregistration of less than 10 μm over a 300-mm substrate. This in turn requires warpage control of 5 to 10 μm per 25.4 mm (1 inch) for a 0.65-mm-thick substrate [32]. Substrate warpage is also becoming critical for the emerging 3D packaging technologies such as package-on-package (PoP) technologies. Today's board materials have fundamental limitations in terms of warpage, interconnection stresses, and dimensional stability in order to support the future integrated microsystems.

Barriers to Cu low-K Reliability

Cu low-K structural integrity for on-chip interconnections is another major reliability concern for high-density flip-chip packages. Interfacial delamination is commonly observed in low-K or ultralow-K on-chip interconnects after IC assembly due to the large deformation and stresses generated by the thermal mismatch between the silicon die and the substrate. In the wafer backend process, the crack driving force results from the thin-film residual stresses within each layer and the thermal mismatch stresses within the low-K stacks. During the package and IC assembly process, in addition to the residual stresses and thermal mismatch stresses within the Cu low-K stack, the global thermal mismatch between the package and IC exerts considerable external loads on the on-chip Cu low-K structures. Compared with the oxide, the low-K dielectric is softer, expands more, and adheres weakly to other materials. Cu low-K interfaces are known to have a fracture toughness of 1 J/m^2, compared to oxides and oxinitrides (8 to 16 J/m^2) and Cu epoxy (25 J/m^2). Since the adhesion strength for passivation/low-K dielectric interface is low, it is prone to delamination. Interfacial delamination is known to be a more critical issue after die assembly than for a stand-alone die. The ITRS [33] has identified the UBM integrity and package compatibility as the key areas of challenge in the Cu low-K IC assembly and packaging. Eliminating the CTE mismatch between the package and the IC can minimize or eliminate the Cu low-K reliability problem.

Barriers to Solder Joint Reliability

The 2003 ITRS [33] calls for organic substrates with less than 100-μm area-array pitch in the package or board by year 2010. The reliability of an assembly is affected by the thermomechanical strains and stresses induced in the package due to the differences in thermal expansion coefficients among different components in the assembly under various thermal excursions. These thermomechanical strains result in low-cycle fatigue failure of solder joints, delamination of the solder bumps, and/or cracking of the buildup layers, leading to the failure of the assemblies. A silicon die has a CTE in the range of 2 to 3 ppm/°C, while the conventional FR4 substrate has a CTE in the range of 18 ppm/°C. By employing underfills, thermal stresses at the solder bumps can be effectively reduced to improve the solder bump reliability. Nevertheless, underfill dispensing becomes increasingly more complicated with the narrower and shorter gaps required for future interconnections. Usage of underfill is also known to cause the package to deform, leading to large peeling stresses at the die underfill and die-solder interfaces which significantly impact packaging reliability. Therefore, there is a compelling need to develop cost-effective board materials with a CTE close to that of Si for reliable assembly without the need for an underfill.

Barriers to Microvia and Dielectric Reliability

The reliability of microvia interconnects is critical to the realization of high-density package wiring. Although the new substrate materials with high modulus and low CTE can result in low warpage and eliminate the need for underfill, they can potentially cause delamination and cracking in the interlayer dielectric because of the high CTE mismatch between the dielectric and Si CTE matched substrate. The ideal dielectric in conjunction with the substrate should also lead to minimal interlayer stresses. The ideal combination of electrical (low dielectric constant and low loss) and thermomechanical properties [low CTE, low stiffness, high toughness (elongation for failure)] and thin-film processability is not found in existing dielectric materials. Therefore, appropriate thermomechanical simulations and reliability evaluation are essential to understanding the material compatibility between the substrate materials and buildup dielectric material and associated failure mechanisms.

8.3.2 100-μm Flip-Chip Assembly Reliability

CTE-Matched Core

To address some of the limitations of the existing substrate materials as discussed above, the industry is already migrating to novel laminate materials with advanced fillers. Low-cost epoxy-based laminates with a CTE of 8 to 12 ppm/°C are available from suppliers like Hitachi Chemical (MCL-E-679LDTM), and these laminates also have a 20 to 30 percent higher modulus than FR4 and BT [34]. The current state-of-the-art manufacturing of high-density substrates is for organic chip carriers. EIT's HyperBGATM package has the most leading-edge high speed and high density [35]. HyperBGATM is a 150- to 200-μm pitch package that is flip-chip attachable using 50-μm-thick PTFE-based dielectrics and 50-μm through vias along with 28-μm lines and 33-μm spaces. EIT has also developed Hyper Z technology with the same low-CTE and low-loss organic substrate platform to package 150-μm pitch ICs. DuPont's nonwoven aramid reinforced laminate systems (Thermount laminates and prepregs) have tunable in-plane CTEs that reduce the CTE mismatch between the active devices and the laminate substrate [36]. A similar approach was also reported by AT&S [37]. This results in reduced strains on solder joints during thermal cycling, creating a higher-reliability packaging system. These materials are also shown to have high laser drillability. Organic chip carriers developed in Japan and the Far East use BT resin laminates and have demonstrated 225- to 250-μm pitch flip-chip capability. Leading microvia substrate manufacturers in Japan and Korea such as Ibiden, CMK, Mektron, and Samsung are currently manufacturing substrates with 20- to 25-μm lines and spaces and 40-μm microvias [38]. A hybrid laminate consisting of a high-stiffness, high-thermal conductivity, low-CTE carbon cloth-polymer core (STABLECOR) with outer layers of FR4 layups is being developed by ThermalWorks, Inc.

Recently, thin core or coreless substrate technology is emerging as the choice for the most demanding system applications for thinner profile and superior electrical performance (for example, smaller through-hole-via inductance). The packaging materials industry is responding to this demand with advanced substrate materials such as low-loss epoxy-based laminates (3M [39]), cyanate ester, or epoxy-reinforced with PTFE (Gore's Microlam, Speedboard), the stiffness requirements are still major impediments for this coreless technology unless rigid temporary carriers and stiffeners are used. High-density packages with 15-μm line and spaces were demonstrated by NEC [40] and Fujitsu [41]. The "carriers" used during buildup have to be removed later by etching or grinding making the coreless substrate process expensive. High-stiffness dielectrics need inorganic reinforcement, which makes thinning and laser drilling difficult. Having a rigid core is still the most common way to make high-density buildup substrates. Though organic laminates may meet the CTE requirements, the high-stiffness targets for high-density wiring cannot be addressed with the polymer matrix. Inorganic matrix cores will be examined in the next section.

Low-CTE inorganic substrates with inherently high stiffness are widely used over the past few decades. IBM's glass-ceramic modules can be tailored to have an exact CTE match with Si and hence showed reliability without underfill. Low-CTE boards of metal core (Invar) were employed and found to have better thermomechanical reliability [42]. Metal matrix composites possess many attractive properties such as machinability, high stiffness, and high thermal conductivity. Aluminum matrix composites with CTE less than 6 ppm/°C and high stiffness are also reported [43, 44].

A novel manufacturing process (patented by Starfire Systems Inc.) has been demonstrated to yield large-area thin carbon silicon carbide based composite boards with the required stiffness and Si-matched CTE at low cost. Unlike the conventional ceramic technology based on powder processing, this novel technology uses a polymeric precursor [45] to make the ceramic. This preceramic polymer allows the design and fabrication of advanced ceramic matrix composites at low temperatures by polymer infiltration and pyrolysis (PIP) in carbon fibers and fabrics, in large-area sheets with the required low CTE and high modulus. The in-plane CTE of the boards was measured using a thermo mechanical analysis (TMA) in the temperature range of 0 to 250°C with a ramp of 5°C per minute. TMA data showed that the CTE of the sample lies between 1.5 to 2.5 ppm/°C. The modulus can be made to vary from 80 to 300 GPa depending on the type of reinforcement, fiber content, final hot pressing temperature, and so forth. The properties of fabricated panels are summarized in Table 8.2.

The warpage and the suitable dielectric material that can be used with the high modulus and low-CTE substrate material is briefly discussed in Section 8.2.2. The warpage and dielectric reliability of test vehicles with C-SiC substrates using bumped PB-8 assemblies with and without underfill and microvia reliability in metal-via-metal test vehicles are presented in Banerji et al. [46]. The reliability of such a low-CTE core and associated dielectric material for a 100-μm flip-chip lead-free assembly process with and without underfill [47] is briefly described here in the context of SOP technology.

100-μm Flip-Chip Assembly

The thermomechanical reliability of flip-chip assemblies on C-SiC boards with an interconnect pitch size of 100 μm and a chip dimension of 2 cm × 2 cm is evaluated

In-Plane CTE (TMA)	Out-of-Plane CTE	Modulus	Glass Transition Temperature	Thickness	Projected Cost
2 ppm/°C	4–5.5 ppm/°C	80–300 GPa	850°C	0.5–1.5 mm	11–20 cents/in^2

TABLE 8.2 Properties of C-SiC Substrate

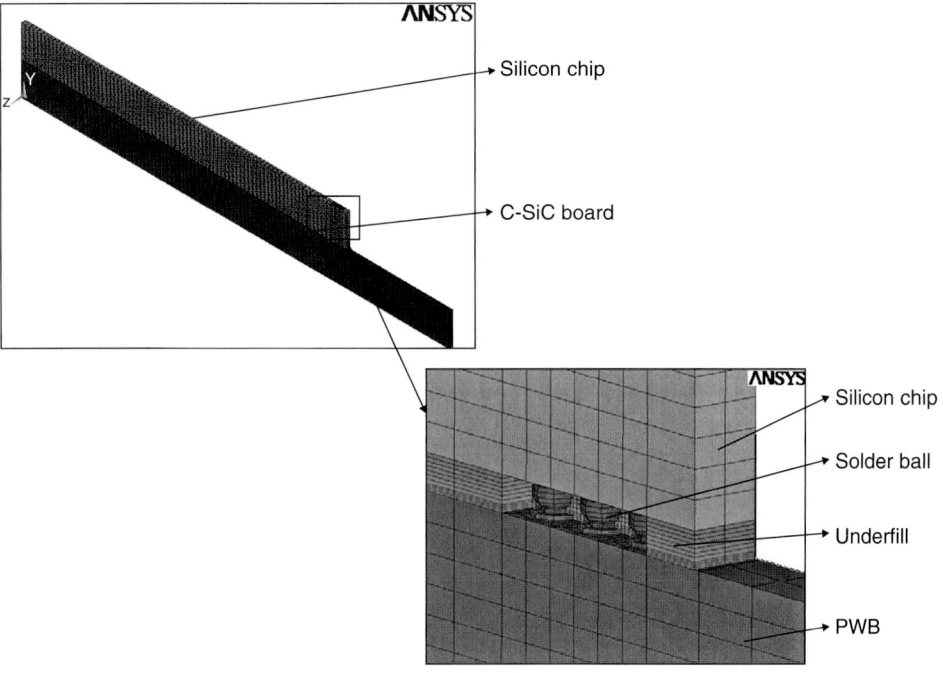

FIGURE 8.24 100-μm pitch assembly models.

using numerical models. The 100-μm flip-chip with a lead-based and an lead-free-based assembly is evaluated with and without underfill on FR4 and low-CTE core C-SiC boards.

Figure 8.24 shows the finite-element model for a 100-μm flip-chip assembly with a 20 mm × 20 mm chip on a C-SiC board. Table 8.3 shows various cases studied to understand the effect of underfilling, board material, interconnect pitch and chip/board ratio on various failure mechanisms such as die cracking and solder fatigue.

PWB	Underfill	Pitch (μm)	Chip/PWB Ratio
FR4	Yes	100	1
FR4	No	100	1
C-SiC	Yes	100	1
C-SiC	No	100	1
C-SiC	No	100	0.75
C-SiC	No	100	0.5
C-SiC	No	200	1
C-SiC	No	200	0.75
C-SiC	No	200	0.5

TABLE 8.3 Underfill Case Studies

		S_{xx} (MPa)	S_{yy} (MPa)
FR4 with underfill	Max	142.50	87.48
	Min	−159.93	−5.43
FR4 without underfill	Max	25.76	28.19
	Min	−36.61	−18.99
C-SiC with underfill	Max	2.138	27.28
	Min	−9.851	−9.23
C-SiC without underfill	Max	0.055	0.740
	Min	−0.587	−0.386

TABLE 8.4 Die Stresses

The axial and peel stresses in the silicon die presented in Table 8.4 indicate that underfilling increases the die stresses and the die stresses are significantly low when low-CTE core substrates are used without underfill. These results demonstrate the significance of high-modulus low-CTE C-SiC as a potential board material for next-generation SOP microsystems packages particularly with inherently weak Cu low-K ICs. Various cases are also studied to understand the effect of underfilling and the effect of board material on stresses near the interconnect regions.

Table 8.5 shows the stresses computed in the interconnect regions for the cases of FR4 with underfill and C-SiC board without underfill. For the FR4 with underfill case, the compressive xx stresses in the interconnect region are higher than in the rest of the die and tensile and compressive stresses in the yy direction are higher than in the rest of the die. For the C-SiC without underfill case, the magnitude of both compressive and tensile stresses in the interconnect region are very small compared to that of the FR4 case. The stress contours shown in Figure 8.25 indicate that the CTE-matched C-SiC board assembly mechanically decouples the die from the package thus reducing the die stresses and interconnect stresses.

To show the improved reliability of CTE-matched substrates without underfill, 100-μm pitch assemblies with FR4 boards underfilled with two types of underfill materials are also analyzed and compared with respect to interconnect stresses and die stresses.

Table 8.6 shows the stresses computed in the interconnect regions and die stresses for the cases of FR4 with two types of underfill materials. For high-modulus low-CTE underfill, interconnect stresses dominate over the stresses in the rest of the die. For

		S_{xx} (MPa)	S_{yy} (MPa)
FR4 with underfill	Max	115.521	249.17
	Min	−462.731	−198.74
C-SiC without underfill	Max	35.26	19.34
	Min	−132.31	−65.77

TABLE 8.5 Stresses Near Interconnect Regions

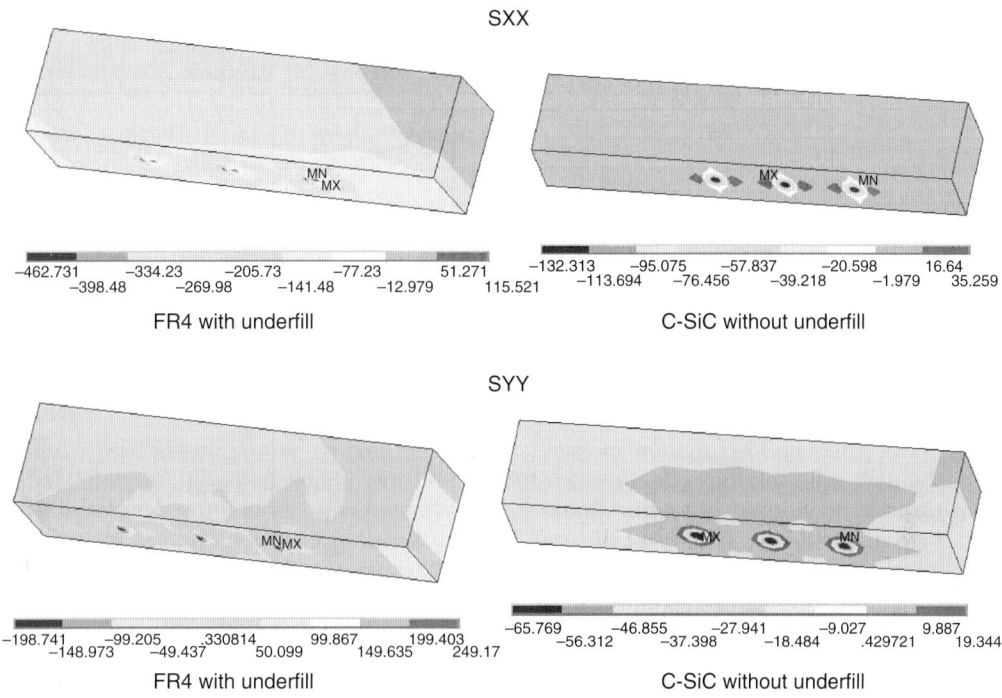

FIGURE 8.25 Stress contours near the interconnect regions for FR4 and C-SiC assemblies.

low-modulus and high-CTE underfill, the interconnect stresses are minimal. Except for the tensile stresses in the y direction, there is a minimal effect of underfill material on stresses in the rest of the die. Table 8.7 shows the effect of the chip/PWB thickness ratio and pitch on the die stresses in the C-SiC assemblies without underfill. Both 100- and 200-μm assemblies and chip/PWB thickness ratios of 0.5, 0.75, and 0.1 are considered. For the CTE-matched C-SiC substrate assemblies without underfill, due to the decoupling of the die from the package, the die stresses are low in magnitude and are independent of the pitch and chip/PWB ratio for all cases considered.

These results demonstrate the need and superiority of low-CTE core and dielectric materials for the 100-μm chip assembly of next-generation SOP technology. The CTE-matched C-SiC board assembly without underfill mechanically decouples the die from

		Die S_{xx} (MPa)	Interconnect S_{xx} (MPa)	Die S_{yy} (MPa)	Interconnect S_{yy} (MPa)
FR4 with underfill (10 GPa, 23 ppm/°C)	Max	142.482	91.151	89.62	235.74
	Min	−159.798	−412.989	−5.61	−193.97
FR4 with underfill (2.5 GPa, 67 ppm/°C)	Max	141.793	2.947	73.56	73.56
	Min	−159.119	−55.647	−4.15	−4.15

TABLE 8.6 Stresses in Underfilled 100-μm Pitch FR4 Assemblies

		S_{xx}		S_{yy}	
		100 μm	200 μm	100 μm	200 μm
C-SiC w/o underfill (ratio = 0.5)	Max	0.055	0.237	0.739	0.170
	Min	−0.588	−0.291	−0.382	−0.473
C-SiC w/o underfill (ratio = 0.75)	Max	0.055	0.237	0.739	0.171
	Min	−0.588	−0.292	−0.383	−0.476
C-SiC w/o underfill (ratio = 1)	Max	0.055	0.237	0.740	0.172
	Min	−0.587	−0.293	−0.386	−0.479

TABLE 8.7 Stresses in 100-μm Pitch CSiC Assemblies without Underfill

the package thus reducing the die stresses and interconnect stresses by an order of magnitude compared to traditional FR4 with underfill technology. Because of the mechanical decoupling of the die and the package in the C-SiC assemblies the stresses are not only less in magnitude but also are not sensitive to the interconnect pitch and the chip/PWB thickness ratio, making it an attractive board technology for the future packaging needs of decreased pitch and die thinning. C-SiC boards with 100-μm lead-free assembly reliability studies establish the superiority of the new board technology that can cater to the next-generation packaging including low-K/Cu IC technology in alleviating the interconnect and die stresses and associated reliability issues.

8.3.3 Reliability against Die Cracking

As described in Section 8.3.2, understanding the die stresses and optimizing the substrate and assembly process parameters for the design-for-reliability against die cracking is very essential in the context of Cu low-K technology. Die cracking during assembly or thermal cycling is a cause of concern in packaging assemblies [48, 49]. An integrated process-reliability modeling methodology has been developed at PRC to determine the stresses at the backside of the die during assembly and subsequent thermal cycling. The modeling methodology is used to understand the effect of material and geometry parameters such as substrate thickness, die thickness, standoff height, interconnect pitch, underfill modulus, CTE, and solder mask CTE on die stresses and thus die cracking. The critical flaw size to induce catastrophic die cracking is calculated using linear-elastic fracture mechanics. Design recommendations, including die thinning and polishing, are made to reduce the tensile stresses on the backside of the die and thus die cracking [50].

8.3.4 Solder Joint Reliability

Low-cycle fatigue of solder joints is a common failure mechanism in electronic packages [51]. Cyclic thermal loading combined with differences in thermal expansion properties in various components of the assembly lead to stress reversals and accumulation of inelastic strain (creep and plastic) in the solder joints. ITRS, 2001, projects that the interconnect pitch will be in the range of 120 μm or less and the number of input-output connections for high-performance microelectronic applications will be 7100 over 310 mm^2 for 2010 and beyond.

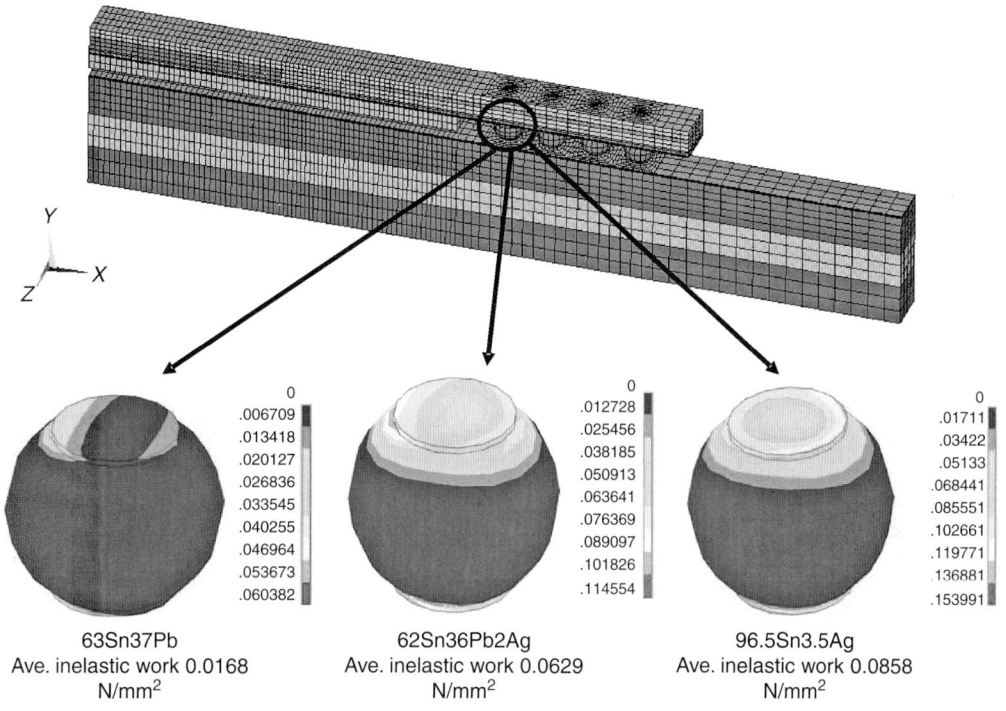

FIGURE 8.26 Accumulated inelastic strain in lead and lead-free solder joints of BGA packages.

With such a fine pitch and with a large number of I/O connections, the reliability of chip-to-substrate interconnects is unknown. The reliability, fatigue life prediction [52], field-use qualification [53, 54], and fatigue life in harsh end-use applications [55] of electronic packages with lead and lead-free solder joints (see Figure 8.26) are studied through physics-based parametric models. In contrast to the build-and-test approach, these models take into consideration the process mechanics of the component assembly, time- and temperature-dependent material behavior, critical geometric features of the assembly, and realistic field-use thermal environments in developing a comprehensive virtual qualification methodology for SOP structures.

Vibration Effects on Solder Joint Reliability

In end-use applications, an electronic component will experience simultaneous thermomechanical and vibration environments [56]. A linear superposition of the low and high cycle damage contributions is not always appropriate. Combined thermal and vibration studies showed that at low vibration frequencies and elevated temperatures inelastic strains can be present and the problem can be considered low cycle [57]. A combined experimental and modeling approach (Figure 8.27) that can accurately determine the interconnect behavior and fatigue life of electronic components under vibration environments is being developed. The experimental work includes vibration testing to determine how the interconnect structure and material properties affect interconnect and system-level reliability. Numerical models are developed and

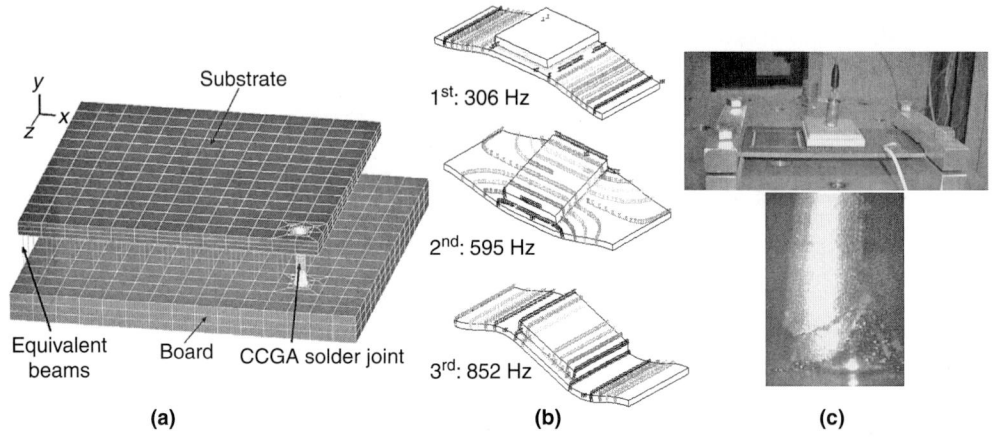

Figure 8.27 (a) Three-dimensional FE model. (b) Mode shapes. (c) Experimental setup and vibration failure in CCGA packages.

calibrated against the experiments and can be used to optimize design parameters and material properties. The effect of solder joint stiffness and package mass on the natural frequency and mode shapes is also investigated [58].

8.3.5 Interfacial Adhesion and Effect of Moisture on Underfill Reliability

Traditionally the addition of a polymer underfill layer to the flip-chip assembly process has provided a solution to many of the thermomechanical concerns in direct chip attachment of electronic packages. However, failure analysis of flip-chip devices subjected to thermal shock testing has revealed that the typical failure mode is delamination at the encapsulant-chip interface, followed shortly by fatigue of the flip-chip solder joints [59]. Once the adhesion between these two surfaces is lost, the flip-chip joints are subjected directly to the strain resulting from the thermal mismatch. As flip-chip technology becomes more prevalent, it becomes necessary to gain a more thorough understanding of the adhesion mechanisms of the interfaces in flip-chip direct chip attachments DCAs [60–62].

Macroscopically speaking, the failure of an interface is observed as an interfacial fracture. Therefore, interfacial adhesion characterizes the resistance of an interface to interfacial crack initiation and growth. The continuum theory of interfacial fracture mechanics is fairly well established for both isotropic and anisotropic elastic bimaterials [63, 64].

Consider an Al-epoxy bimaterial specimen subject to four-point bending as shown in Figure 8.28. The corresponding load versus load-point-displacement curve is shown in Figure 8.29. It is observed that as the load increases, the specimen deforms almost linearly until the load reaches a critical value, P_c, at which point the cracks start to grow causing a sudden drop in the load. Clearly, the total work, U_{total}, done to the specimen up to the point of fracture is given by the area under the curve.

By conservation of energy, the following equation should hold,

$$U_{total} = U_e + U_p + U_a \tag{8.1}$$

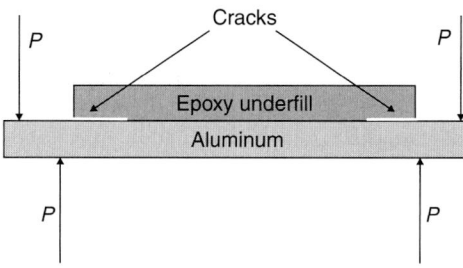

FIGURE 8.28 An Al-epoxy bimaterial specimen under four-point bending.

where U_e is the elastic strain energy stored in the specimen. This part of the energy is reversible in that it relieves itself once the external load is removed. The second term, U_p, represents the energy consumed by certain irreversible processes such as plastic deformation and the heat generated during such inelastic deformation. Both U_e and U_p are dependent on the specimen geometry. The third term, U_a, is used to generate a fracture, or equivalently, to break the interfacial bond to generate a new surface area. Therefore, U_a should be independent of the specimen geometry and should represent the intrinsic interfacial adhesion. In other words, U_a is the amount of energy required to break the interfacial bond.

Rearranging (8.1), one obtains,

$$U_{total} = U_e + U_p + U_a \qquad (8.2)$$

In principle, the left-hand side of Equation (8.2) can be computed for a given bimaterial specimen under a given loading. For example, U_{total} can be evaluated based on the load versus load-point displacement curve. To evaluate U_e and U_p, a transient crack growth calculation is required, which gives the elastic strain energy and the plastic work during an infinitesimal crack growth. Often, an approximation can be made by using the J-integral value obtained for a stationary crack [65]. Thus, once the intrinsic interfacial adhesion on the right-hand side of Equation (8.2) is known for the corresponding interface, Equation (8.2) provides an interfacial failure criterion, since the left-hand side of the equation is computed based on the load, materials properties, and sample geometry.

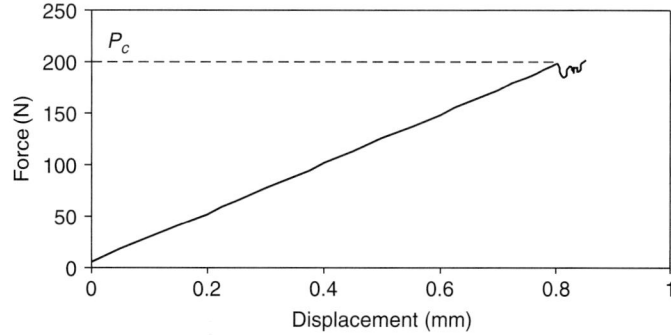

FIGURE 8.29 Typical load versus load point displacement curve for four-point bending specimen.

Generally speaking, there are three intrinsic interfacial adhesion mechanisms for polymers on metallic adherends, namely, physical adsorption, chemical bonding, and mechanical interlocking. Physical adsorption is responsible for the interaction between permanent dipoles in the molecules, the induction effect, and the dispersion effect. The induction effect is the influence of the dipole moments in polarizing neighboring molecules, while the dispersion effect is the result of an interaction between the random motions of the electrons of the materials [66]. Once the adhesive and substrate are selected, the nature and strength of physical adsorption is theoretically predictable. Chemical bonding, on the other hand, is based on the primary covalent bond at the interface. One example of chemical bonding is the chemical reactions created by coupling agents at the interface [67]. Both physical and chemical adhesion mechanisms are on the microscopic scale. At the macroscopic scale, in order to form a strong adhesive joint, mechanical interlocking [68, 69] can be applied, as in the surface treatments applied to metals to provide various topologies. However, strictly speaking, mechanical interlocking is not one of the intrinsic material adhesion mechanisms. It is only a technological means of achieving adhesive bonding as in the case of structural adhesives [70–72].

Based on the above, one may formally express the interfacial adhesion as the sum of the three basic adhesion mechanisms (physical, chemical, and mechanical),

$$U_a = U_{phys} + U_{chem} + U_{mech} \tag{8.3}$$

Effect of Moisture on the Interfacial Adhesion

Environmental factors such as moisture preconditioning can have an adverse effect on adhesion. In what follows, we will use the interface between copper and epoxy-based underfill as an example to develop a simple model for simulating the moisture-induced degradation of interfacial strength/fracture toughness. As discussed earlier, four major mechanisms constitute the primary adhesion forces. They include mechanical interlocking, diffusion theory, electronic theory, and adsorption theory [73]. For the underfill-copper interface, the contributions of interfacial diffusion and electrostatic forces between the adhesive and substrate causing adhesion is far lower than the effects of mechanical interlocking and adsorption. Since the copper substrates in this study were polished to a mirror finish, the effects from mechanical interlocking of the adhesive into irregularities present on the substrate surface will be small compared to the effects from intermolecular secondary forces (i.e., van der Waals) between the atoms and molecules in the surfaces of the adhesive and substrate. Consequently, adsorption theory will dominate the adhesive bonding at the underfill-copper interface.

Provided adsorption theory governs adhesion, and only secondary forces are acting across an interface, the stability of an adhesive-substrate interface in the presence of moisture can be ascertained from thermodynamic arguments. Typically the thermodynamic work of adhesion of an adhesive-substrate interface in an inert medium is positive, which indicates the amount of energy required to separate a unit area of the interface. However, the thermodynamic work of adhesion in the presence of a liquid can be negative, which indicates the interface is unstable and will separate when it comes into contact with the liquid. Thus, the values of thermodynamic work of adhesion of a particular interface with and without moisture provide an indication of the environmental stability of the interface. It follows from Kinloch [73] that the thermodynamic work of adhesion of the epoxy-copper interface is 260.7 mJ/m². If water is present at the epoxy-copper interface, the thermodynamic work of adhesion given becomes –270.4 mJ/m². Therefore, since the work of adhesion is positive before exposure to moisture and negative after exposure, all adhesion of the epoxy-copper interface is lost if water comes into contact with the interface. Based on this theory, a degradation model for the underfill-Cu interface was developed in [62, 74–76]

$$G_{c,wet} = G_{c,dry} \exp\left[\frac{-8C_{sat}r_{debond}^2}{\rho D^2}\right] \qquad (8.4)$$

Equation (8.4) characterizes the loss in interfacial fracture toughness from moisture in terms of key parameters relevant to moisture. Using the value for the density of water at room temperature (0.998 mg/mm³), an average nanopore diameter of 5.5 Å, and the moisture saturation concentration determined from the experimental portion of this study, the number of active nanopores participating, N_N, and the value of r_{debond} can be determined by the intrinsic response of each material system to each level of moisture preconditioning. The results are shown in Table 8.8.

As shown in Table 8.8, the number of nanopores participating increases with the saturation concentration. This is expected since an increase in the saturation concentration would increase the available moisture for transport through the nanopores. In addition, the values for r_{debond} were similar for each moisture preconditioning environment for both respective interfaces, which is also expected since x-ray photoelectron spectroscopy and water contact angle results did not indicate a change in the interfacial hydrophobicity of the copper surface from moisture preconditioning. The slight variation in the values for r_{debond} could in part be attributed to experimental scatter. Since the results were similar, they were averaged to obtain a representative value for r_{debond} in the presence of moisture for each interface.

Using the moisture parameters identified for each interfacial material system, Equation (8.4) was used to predict the interfacial fracture toughness for the underfill-copper interface as a function of increasing saturation concentration. The results are shown in Figure 8.30.

As shown in Figure 8.29, the model accurately predicted the loss in interfacial fracture toughness as a function of increasing moisture concentration. Since Equation (8.4) was based on the physics of adsorption theory, it will yield a loss in interfacial fracture toughness provided there is moisture at the interface, no matter how small the concentration. This contradicts the results of previous studies that have reported a critical concentration of water may exist below which there is no measurable loss in adhesion [77, 78]. Based on the results of adsorption theory, it does not appear possible that a critical concentration of water could exist in theory. It is possible in those studies that other mechanisms for adhesion in addition to adsorption theory governed the adhesion at the interface, which could explain why a critical concentration of water was

Environment	Substrate	Adhesive	C_{sat} (mg H_2O/mm³)	N_N	R_{debond} (mm)
85°C/50% RH	Copper	Underfill	0.0075	1.006E + 13	1.640E − 06
85°C/65% RH	Copper	Underfill	0.0089	1.194E + 13	1.692E − 06
85°C/85% RH	Copper	Underfill	0.0118	1.583E + 13	1.669E − 06

TABLE 8.8 Key Parameters Relevant to Moisture for the Underfill-Copper Interface

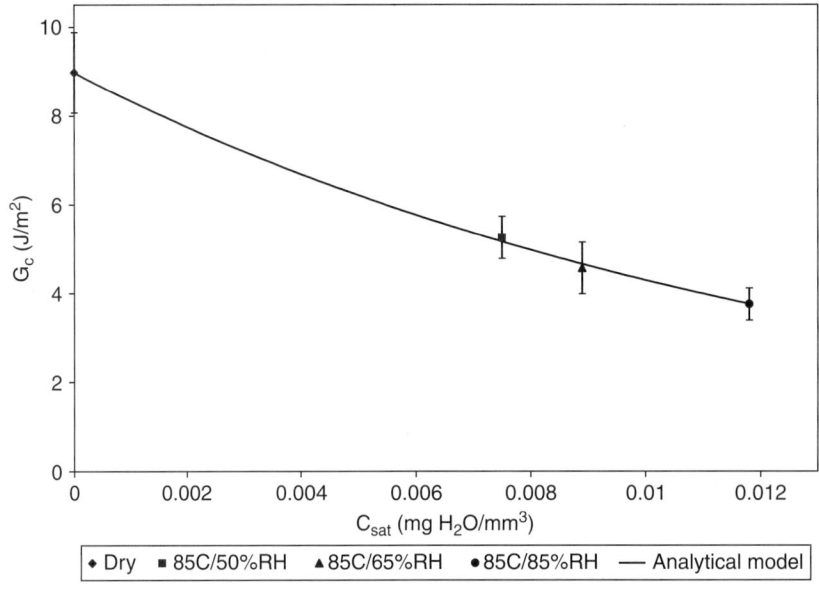

Figure 8.30 Analytical prediction of the loss in interfacial fracture toughness from moisture for the underfill-copper interface.

observed. An additional consideration is the method of testing used to obtain adhesion results. The aforementioned studies used lap shear test specimens to determine the interfacial strength after moisture preconditioning. Because of the lack of a precrack at the interface and the applied load being distributed over the entire bonding area, these test specimens are not as sensitive to interfacial failure; consequently, this possibly also explains why in part a critical concentration of water appeared to exist for low concentrations of moisture. Conversely, interfacial fracture toughness test specimens are designed for interfacial failure through the use of a precrack at the interface, making them more sensitive to environmental attacks at the interface. The work of Wylde and Spelt [79] supports this observation. Using interfacial fracture toughness test specimens with a similar material system previously reported to exhibit a critical concentration of water from lap shear results, they found a decrease in the interfacial toughness from moisture for all concentrations of moisture, including those lower than the previously reported critical concentration of water. Consequently, the provided adsorption theory dominates the adhesive bonding at the adhesive-substrate interface and the assumptions in the development of the model are satisfied. Thus Equation (8.4) should accurately predict the loss in interfacial fracture toughness for a given moisture concentration.

8.4 Future Trends and Directions

The increasing trends in interconnect technology have a profound impact on the reliability of next-generation mixed-signal convergent systems. Some of the recent trends and developments in interconnect technology at PRC and the associated reliability challenges are briefly discussed here.

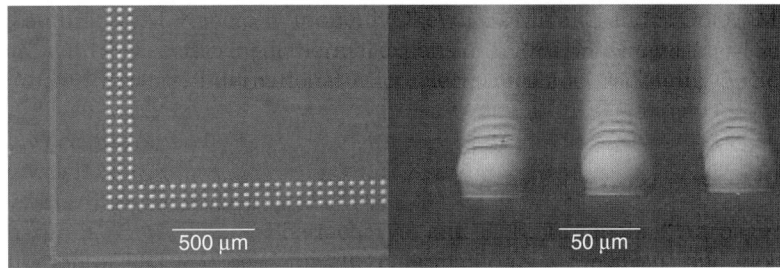

FIGURE 8.31   SEM images of 20 mm × 20 mm ICs with 2256 lead-free solder bumps at a 100-μm pitch.

8.4.1 Extending Solder

The need for highly filled, low-CTE, and high-modulus underfill materials to absorb thermomechanical strains in the ultrafine pitch interconnects places additional demands on underfill processing. Recently PRC has proposed an innovative interconnect and assembly process that overcomes these challenges and also has the potential to solve the yield problems associated with current no-flow underfill processes [80]. Initial process development was performed using lead-free solder interconnects. Based on extensive process parameter optimization, a defect-free interconnect assembly with underfill at 100-μm pitch for a 20 mm × 20 mm IC (see Figure 8.31) has been demonstrated with excellent solder wetting to the substrate pads. This novel approach is also applicable to copper, nickel, gold, or other types of interconnects and enables the use of underfill materials with an optimum combination of thermomechanical properties.

The new interconnect and assembly process (U.S. patent pending) [81] involves depositing the underfill directly on the organic buildup substrate by spin coating into thin underfill film followed by flip-chip placement and reflow to form the interconnections. Figure 8.32 illustrates the process sequence for a ultrafine-pitch flip-chip assembly. An underfill with the best thermomechanical properties with high filler loading can be

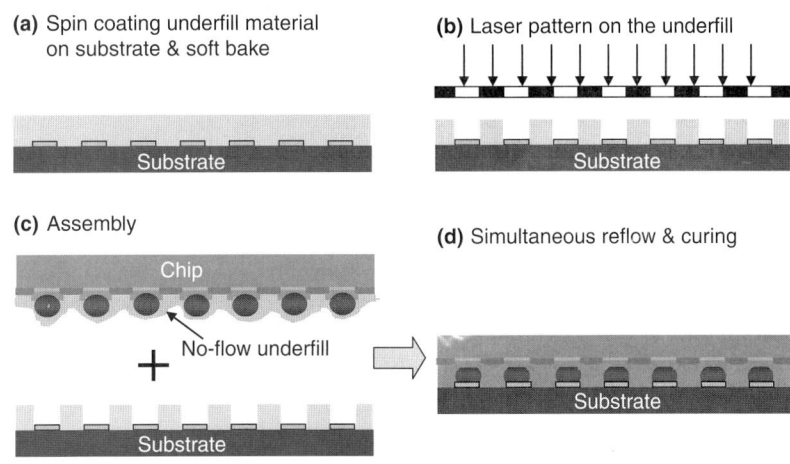

FIGURE 8.32   Process flow for the new underfill and flip-chip assembly process.

selected. Alternate deposition techniques like dry film lamination and localized dispensing may also be used for the thin-film coatings. The underfill is then partially or fully cured, and clean via holes sized to match the flip-chip bump pitch are formed by laser ablation. Photolithography, plasma etching, or other patterning processes are also effective for opening the bump sites in the underfill layer. This process is compatible with most commonly used chip-to-substrate interconnects including standard eutectic solder, lead-free solder, copper-nickel posts and gold stud bumps at a 20- to 50-µm peripheral or area array I/O pitch.

8.4.2 Complaint Interconnects

The reliability, fatigue life prediction [52], and field-use qualification [54] of electronic packages with conventional solder joints has been extensively studied in the past. Wafer-level complaint interconnects called nanolinks are currently studied at PRC for meeting the needs of decreased pitch size for next-generation chip-to-substrate IC packaging. These compliant interconnects [82, 83] have the capability to compensate for thermal expansion in all directions. The fabrication of the compliant interconnects using a LIGA-like technology, optimization of electrical and mechanical properties, and reliability studies (Figure 8.33) for wafer-level applications are in progress.

8.4.3 Alternative to Solder and Nano Interconnects

Next-generation IC-package interconnection requires a technology that is low-cost, reworkable possesses, good electrical properties, good reliability, and is easily testable at the wafer level with good coplanarity. The copper pillar bump offers performance and process advantages to the flip-chip interconnection system. Most semiconductor foundries are now equipped with advanced copper electroplating systems that can be easily integrated with the backend processes. In addition, the flexibility that the bumps can be manufactured in various shapes and sizes further increases the current and heat capacities. As only solder is reflowed to create the joint, a consistent standoff can be maintained through the nonmeltable copper portion, which further enables the downstream processes of flip-chip packaging. It can provide a wider window for

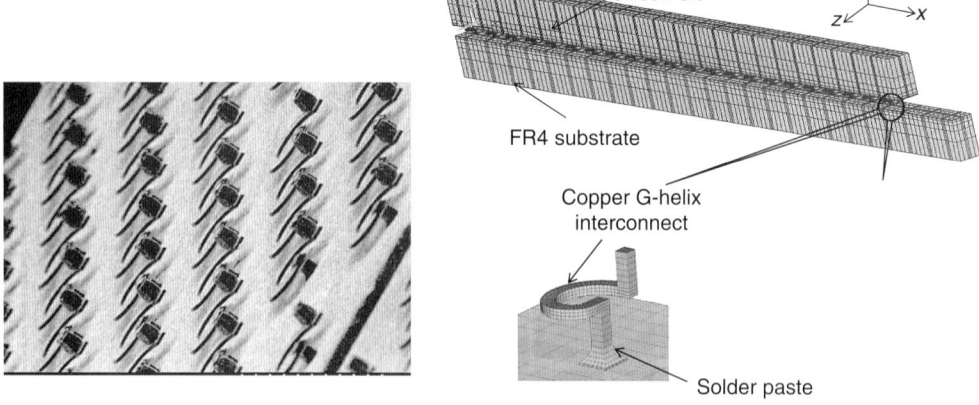

FIGURE 8.33 Lithography-based fabricated complaint interconnects and reliability models.

underfilling process, and also a maskless substrate can be used. The advantages of copper pillar over solder interconnects may also include superior electrical and thermal performance and resistance to electromigration.

Downscaling the traditional solder bump interconnect will not satisfy the thermomechanical reliability requirements at very fine pitches on the order of 50 μm and less because of the poor mechanical properties of solders. Nanostructured interconnects provide a quantum jump in the number of interconnections and yet provide the best electrical and mechanical properties. Several new approaches for nanostructured interconnects through nanomaterials is being pursued for SOP technology. The mechanical properties such as strength and yield stress of nanostructured Ni and Cu were characterized and compared to conventional metallic structures. These structures are being evaluated for interconnect applications. Low-cost chemical solution methods such as electroplating and electroless plating are being used as interconnect fabrication approaches. High-aspect-ratio compliant copper interconnects (1:5, 40-μm pitch) were fabricated by electroplating through SU8 photoresist [84]. In order to lower the interconnect dimensions to nanoscale and also make them reworkable, bumpless interconnects using thin liquid interfaces formed from nanodimensional lead-free solder films was demonstrated [85]. The reworkable solder films (< 200 nm) were fabricated from low-cost chemical solution synthesis approaches such as sol-gel (Figure 8.34).

Polymer matrix nanointerconnects with anisotropic conducting adhesives (ACA) having conducting nanoparticles (< 20 to 30 nm) can be scaled down to pitches close to 1 μm and thinned to submicron dimensions. Though conductive adhesives have several promising characteristics, lower electrical conductivity and poor current carrying capability due to the restricted contact area and poor interfacial bonding of the ACAs and metal bond pads compared to the metallurgical joint of the metal solders are

FIGURE 8.34 High-magnification SEM of the nano lead-free solder interface bonding two silicon wafers.

some of the issues that are preventing its use in high-power devices, such as microprocessor and application-specific integrated circuit (ASIC) applications. Inclusion of nanosized particle fillers has the potential to improve the current density of the ACA joints by distributing current into more conductive paths [86]. In order to enhance the electrical performance of ACA materials, organic monolayers have also been introduced into the interface between metal fillers and the metal-finished bond pad of ACAs [87, 88].

New paradigms in IC packaging and assembly technologies by means of ultrafine-pitch nanostructured and nanoscale interconnect formation and reworkable bonding processes are essential to realizing packaging of future nanoscale devices. This technology can be used for pushing the limits of current interconnect and assembly technologies in terms of pitch, number of I/Os, superior combination of electrical and mechanical properties, as well as reworkability. These nanoscale interconnections are essential to support the advances in nano ICs. Systematic reliability, experimental characterization, and optimization studies are needed to ensure the fatigue life of such alternate interconnects for SOP mixed-signal applications.

8.5 Summary

A potential paradigm shift in electronic packaging is leading to the integration of multiple system functions, such as high-speed digital, high bandwidth optical, analog, or RF, as well as sensing functions, into one compact, low-weight, low-cost, and high-performance package or module system. The system integration in SOP microsystems is achieved through MGI system boards and wafer-level packaging for IC packaging with alternate materials and processes. A systems approach to reliability is presented to provide an understanding of the material interaction effects on component-level failure mechanisms and their influence on system-level performance. Several barriers remain to be able to develop system-level design-for-reliability approaches and reliability assessment methodologies. Design-for-reliability models with process mechanics are presented in this chapter to understand various failure mechanisms in digital, optical, and RF functions to provide up-front design guidelines for alternate materials and fabrication processes. SOP substrate to IC interconnection reliability is presented through 100-µm pitch assembly using CTE-matched advanced core materials. Fundamental characteristics of interface adhesion and the effect of moisture on the underfill adhesion reliability are presented through theoretical models and experiments. The system-level reliability approach is presented to relate the component-level failure mechanisms to system-level signal integrity through physics-based modeling and statistical considerations. The future trends in interconnect technologies and associated reliability challenges are briefly discussed in the context of SOP mixed-signal packages. Developing system-level reliability models with multiple packages and functions and failure mode interactions for future SOP mixed-signal systems requires tremendous computational resources and time. Global-local, submodeling, and domain-decomposition techniques augmented with high-performance computing are necessary to ease the computational complexity and reduce the time needed to achieve acceptable results. The parametric modular reliability models with appropriate material characterization presented in this chapter can be readily extended to study the functional interaction and system-level reliability of next-generation convergent SOP microsystems.

References

1. Bajenescu, T. I., and M. I. Bazu, *Reliability of Electronic Components,* New York: Springer-Verlag Berlin Heidelberg, 1999.
2. Kim, I., S. K. Sitaraman, and R. Peak, "Reliability objects: a knowledge model of system design for reliability," *Proc. IMECE 2005-EPP 79934,* Orlando, FL, November 5–11, 2005.
3. Pucha, R.V., S. K. Sitaraman, S. Hegde, M. Damani, C. P. Wong, J. Qu, Z. Zhang, P. M. Raj, and R. R. Tummala, "Materials and mechanics challenges in SOP-based convergent microsystems," *Micromaterials and Nanomaterials,* a publication series of the Micro Materials Center Berlin at the Fraunhofer Institute IZM, issue no. 3, 2004a, pp. 16–29.
4. Lee, K. J., M. Damani, R. V. Pucha, S. K. Bhattacharya, R. R. Tummala, and S. K. Sitaraman, "Reliability modeling and assessment of embedded capacitors in organic substrates," *IEEE Transactions on Components and Packaging Technologies,* vol. 30(1), 2007, pp. 152–162.
5. Bogetti, T. A., and J. W. Gillespie, "Process-induced stress and deformation in thick-section composite laminates," *Journal of Composite Materials,* vol. 26, no. 5, 1992, pp. 626–660.
6. White, S. R., and H. T. Hahn, "Process modeling of composite materials: residual stress development during cure. Part I. Model formulation, Part II. Experimental validation," *Journal of Composite Materials,* vol. 26, no. 26, 1992, pp. 2402–53.
7. Wu, S. X., C. P. Yeh, K. X. Hu, and K. Wyatt, "Process modeling for multichip module thin films interconnects," *ASME Journal of Cooling and Thermal Design of Electronic Systems,* HTD-vol. 319/EEP-vol. 15, 1995.
8. Dunne, R. C., and S. K. Sitaraman, "An integrated process modeling methodology and module for sequential multilayered substrate fabrication using a coupled cure-thermal-stress analysis approach," *IEEE Transactions—Electronics Packaging Manufacturing,* vol. 25, no. 4, 2002, pp. 326–34.
9. Dunne, R. C., S. K Sitaraman, S. Luo, Y. Rao, C. P. Wong, W. E. Estes, C. G. Gonzalez, J. C. Coburn, and M. Periyasamy, "Investigation of the curing behavior of a novel epoxy photo-dielectric dry film (ViaLux 81) for high density interconnect applications," *Journal of Applied Polymer Science,* vol. 78, 2000, pp. 430–37.
10. Fuhan, L., V. Sundaram, S. Mekala, G. White, D. A. Sutter, and R. R. Tummala, "Fabrication of ultra-fine line circuits on PWB substrates," *Proc. 52nd Electronic Components and Technology Confer ence,* 2002, pp. 1425–31.
11. Ramakrishna, G., F. Liu, and S. K. Sitaraman, "Experimental and numerical investigation of microvia reliability," *Proc. of 8th Intersociety Conference on Thermal and Thermomechanical Phenomena in Electronic Systems,* 2002, pp. 932–39.
12. Mahalingam, S., S. Hegde, R. V. Pucha, and S. K. Sitaraman, "Material interaction effects in the reliability of high density interconnect (HDI) boards," *Proc. ASME International Mechanical Engineering Congress and Exposition,* Washington, D.C., November 16–21, 2003, IMECE 2003 – EPP 41745.
13. Gao, H., Y. Huang, W. D. Nix, and J. W. Hutchinson, "Mechanism-based strain gradient plasticity—I. Theory," *Journal of Mechanics and Physics of Solids,* vol. 47, 1999, pp. 1239–63.
14. Chen, S. H., and T. C. Wang, "A new hardening law for strain gradient plasticity," *Acta Materialia,* vol. 48, 2000, pp. 3997–4005.

15. Pucha, R. V., G. Ramakrishna, S. Mahalingam, and S. K. Sitaraman, "Modeling plastic strain gradient effects in low-cycle fatigue of copper micro-structures," *International Journal of Fatigue*, vol. 26, January 2004b, pp. 947–57.
16. Raj, P. M., K. Shinotani, M. Seo, S. Bhattacharya, V. Sundaram, S. Zama, J. Lu, C. Zweben, G. White, and R. R. Tummala, "Selection and evaluation of materials for future system-on-package (SOP) substrate," *Proc. 51st Electronic Components and Technology Conference*, 2001, pp. 1193–97.
17. Hegde, S., R. V. Pucha, and S. K. Sitaraman, "Alternate dielectric and base substrate materials for enhanced reliability of high density wiring (HDW) substrates," *Journal of Materials Science: Materials in Electronics*, vol. 15, no. 5, 2004, pp. 287–96.
18. Xie, W., and S. K. Sitaraman, "Investigation of interfacial delamination of a copper-epoxy interface under monotonic and cyclic loading: modeling and evaluation," *IEEE Transactions on Advanced Packaging*, vol. 26, no. 4, 2002a, pp. 441–46.
19. Xie, W., and S. K. Sitaraman, "Investigation of interfacial delamination of a copper-epoxy interface under monotonic and cyclic loading: experimental characterization," *IEEE Transactions on Advanced Packaging*, vol. 26, no. 4, 2002b, pp. 447–52.
20. Shan, Z., and S. K. Sitaraman, "Elastic-plastic characterization of thin films using nanoindentation technique," *Thin Solid Films*, vol. 437, 2003, pp. 176–81.
21. Modi, M., and S. K. Sitaraman, "Interfacial fracture toughness measurement for thin film interfaces," *Engineering Fracture Mechanics*, vol. 71, 2004, pp. 1219–34.
22. Schneider, D., "Reliability and characterization of MLC decoupling capacitors with C4 connections," *Proc. Electronic Components and Technology Conf.*, 1996, pp. 365–74.
23. Strydom, J. T., "Investigation of thermally induced failure mechanisms in integrated spiral planar power passives," *37th Industry Applications Conference*, vol. 3, 2002, pp. 1781–1786.
24. Witwit, A. M. R., "Thermal simulation of PCBs with embedded resistors," *International Conference on Simulation*, 1998, pp. 313–16.
25. Sang-Yeon, C., N. M. Jokerst, and M. Brooke, "Comparison of evanescent and directly coupled optical interconnections embedded into electronic interconnection substrates," *Proc. 15th Annual Meeting of the IEEE Lasers and Electro-Optics Society*, vol. 2, 2002, pp. 653–54.
26. Suhir, E., "Microelectronics and photonics—the future," *Proc. 22nd International Conference on Microelectronics*, vol. 1, 2000, pp. 3–17.
27. Suzuki, S., M. Yanagisawa, Y. Hibino, and K. Oda, "High-density integrated planar lightwave circuits using SiO_2-GeO_2 waveguides with a high refractive index difference," *Journal of Lightwave Technology*, vol. 12, no. 5, 1994, pp. 790–96.
28. Pucha, R. V., S. Hegde, M. Damani, K. Tunga, A. Perkins, S. Mahalingam, G. Lo, K. Klein, J. Ahmad, and S. K. Sitaraman, "System-level reliability assessment of mixed-signal convergent microsystems," *IEEE Transactions on Advanced Packaging*, vol. 27, no. 2, May 2004c, pp. 438–52.
29. Ahmad, J., and S. K. Sitaraman, "Modeling methodologies to study PWB assembly reliability," *Proc. 52nd Electronic Components & Technology Conference*, 2002, pp. 1658–64.
30. Yoon, H. J., N. J. Chung, M. H. Choi, I. S. Park, and J. Jeong, "Estimation of system reliability for uncooled optical transmitters using system reliability function," *J. Lightwave Tech.*, vol. 17, no. 6, 1999, pp. 1067–71.
31. Sundaram, V., R. R. Tummala, F. Liu, P. A. Kohl, J. Li, S. A. Bidstrup-Allen, and Y. Fukoka, "Next-generation microvia and global wiring technologies for SOP," *IEEE Transactions on Advanced Packaging*, May 2004, pp. 315–25.

32. Tummala, R. R., M. Swaminathan, M. Tentzeris, J. Laskar, G. K. Chung, S. Sitaraman, D. Keezer, D. Guidotti, R. Huang, K. Lim, L. Wan, S. Bhattacharya, V. Sundaram, F. Liu, and P. M. Raj, "SOP for miniaturized mixed-signal computing, communication and consumer systems of the next decade," *IEEE Component Packaging and Manufacturing Technology (CPMT) Transactions on Advanced Packaging,* May 2004, pp. 250–67.
33. ITRS (International Technology Roadmap for Semiconductors), 2001 and 2003 editions.
34. Takahashi, A., K. Kobayashi, S. Arike, N. Okano, H. Nakayama, A. Wakahayashi, and T. Suzuki, "High density substrate for semiconductor packages using newly developed low CTE build up materials," *International Symposium on Advanced Packaging Materials,* 2000, pp. 216–20.
35. Alcoe, D. J., M. A. Jimarez, G. W. Jones, T. E. Kindl, J. S. Kresge, J. P. Libous, R. J. Stutzman, and C. L., Tytran-Palomaki, "HyperBGA™: a high performance, low stress, laminate ball grid array flip chip carrier," *MicroNews,* vol. 6, no. 2, second quarter 2000, pp. 27–36.
36. Khan, S., C. G. Gonzalez, and M. Weinhold, "Organic, non-woven aramid reinforced substrates with controlled in-plane CTE means more reliable solder joint reliability," *Advances in Electronic Packaging,* vol. 2, 2001, pp. 1345–62.
37. Krziwanek, T. S., "Low CTE materials for printed wiring boards," *Proc. International Symposium & Exhibition on Advanced Packaging Materials,* 2001, pp. 175–80.
38. Microvia Board Technologies, 2002 Electronics Industry Report, Prismark Partners LLC.
39. Qu, S., G. Mao, F. Li, R. Clough, N. O'Bryan, and R. Gorrell, "A new organic composite dielectric material for high performance IC packages," *Proc. Electronic Components and Technology Conference,* 2005, pp. 1373–77.
40. Sakai, J., T. Shimoto, K. Nakase, K. Motonaga, H. Honda, and H. Inoue, "Signal integrity and power integrity properties of FCBGA based on ultra-thin, high-density packaging substrate," *Proc. Electronic Components and Technology Conference,* 2005, pp. 284–90.
41. Koide, M., K. Fukuzono, H. Yoshimura, T. Sato, K. Abe, and H. Fujisaki, "High-performance flip-chip BGA technology based on thin-core and coreless package substrate," *Proc. Electronic Components and Technology Conference,* 2006, pp. 1869–73.
42. Nakamura, K., M. Kaneto, Y. Inoue, T. Okeyui, K. Miyake, and S. Oota, "Multilayer board with low coefficient of thermal expansion," *Proc. 33rd International Symposium on Microelectronics,* International Microelectronics and Packaging Society, 2000, pp. 235–40.
43. www.pcc-aft.com
44. www.alsic.com
45. www.starfiresystems.org.
46. Banerji, S., P. M., Raj, S. Bhattacharya, and R. R. Tummala, "Warpage induced limitation of FR4 and need for alternate materials for microvia and global interconnect needs," *IEEE Components Packaging and Manufacturing Technology (CPMT) Transactions on Advanced Packaging,* vol. 28, issue 1, February 2005, pp. 102–13.
47. Kumbhat, N. P., M. P. Raj, R. V. Pucha, J. Y. Jui-Yun Tsai, S. Steve Atmur, E. Bongio, S. K. Sitaraman, and R. R. Tummala, "Novel Ceramic Composite Substrate Materials for High-Density and High Reliability Packaging," *IEEE Transactions on Advanced Packaging,* vol. 30(4), 2007, pp. 641–653.

48. Popelar, S. F., "An investigation into the fracture of silicon die used in flip chip applications," *Proc. 4th International Symposium and Exhibition on Advanced Packaging Materials, Processes, Properties and Interfaces,* 1998, pp. 41–48.
49. Van Kessel, C. G. M., S. A. Gee, and J. J. Murphy, "The quality of die attachment and its relationship to stresses and vertical die cracking," *Proc. 33rd Electronic Components and Technology Conference,* 1983, pp. 237–44.
50. Michaelides, S., and S. K. Sitaraman, "Die cracking and reliable die design for flip-chip assemblies," *IEEE Transactions on Advanced Packaging,* vol. 22, no. 4, 1999, pp. 602–13.
51. Lau, J. H. and Y. S. Pao, *Solder Joint Reliability of BGA, CSP, Flip Chip, and Fine Pitch SMT Assemblies,* New York: McGraw-Hill, 1996.
52. Pyland, J., R. V. Pucha, and S. K. Sitaraman, "Thermomechanical reliability of underfilled BGA packages," *IEEE Transactions on Electronics Packaging Manufacturing,* vol. 25, no. 2, 2002, pp. 100–106.
53. Sitaraman, S. K., R. Raghunathan, and C. E. Hanna, "Development of virtual reliability methodology for area-array devices used in implantable and automotive applications," *IEEE Transactions on Components and Packaging Technologies,* vol. 23, no. 3, September 2000, pp. 452–61.
54. Pucha, R. V., J. Pyland, and S. K. Sitaraman, "Damage metric-based mapping approaches for developing accelerated thermal cycling guidelines for electronic packages," *International Journal of Damage Mechanics,* vol. 10, no. 3, 2001, pp. 214–34.
55. Pucha, R. V., K. Tunga, J. Pyland, and S. K. Sitaraman, "Accelerated thermal cycling guidelines for electronic packages in military avionics thermal environment," *Transactions of the ASME—Journal of Electronic Packaging,* vol. 126, June 2004d, pp. 256–64.
56. Cole, M. S., E. J. Kastberg, and G. B. Martin, "Shock and vibration limits for CBGA and CCGA," *Proc. Surface Mount International Conference,* 1996, pp. 89–94.
57. Basaran, C., A. Cartwright, and Y. Zhao, "Experimental damage mechanics of microelectronics solder joints under concurrent vibration and thermal loading," *Int. J. Damage Mech.,* vol. 10, 2001, pp. 153–70.
58. Perkins, A., and S. K. Sitaraman, "Vibration-induced solder joint failure of a ceramic column grid array (CCGA) package," *54th Electronic Components and Technology Conference,* IEEE-CPMT and EIA, Las Vegas, NV, June 1–4, 2004.
59. LeGall, C. A., "Thermomechanical stress analysis of flip chip packages," M.S. Thesis, School of Mechanical Engineering, Georgia Institute of Technology, 1996.
60. Olliff, D., J. Qu, M. Gaynes, R. Kodnani, and A. Zubelewicz, "Characterizing the failure envelope of a conductive adhesive," *J. Electronic Packaging,* vol. 121, 1999, pp. 23–30.
61. Kuhl, A. and J. Qu, 2000, "A technique to measure interfacial toughness over a range of phase angles," *J. Electronic Packaging,* vol. 122, 2000, pp. 147–51.
62. Ferguson, T., and J. Qu, "Effect of moisture on the interfacial fracture toughness of underfill/solder mask interfaces," *J. Electronic Packaging,* vol. 124, 2002, pp. 106–110.
63. Qu, J., and J. L. Bassani, 1989, "Cracks on bimaterial and bicrystal interfaces," *J. Mech. Phys. Solids,* vol. 37, 1989, pp. 417–33.
64. Hutchinson, J., and Z. Suo, "Mixed mode cracking in layered materials," *Advances in Applied Mechanics,* vol. 29, 1992.
65. Yao, Q., "Modeling and characterization of interfacial adhesion and fracture," Ph.D. thesis, Georgia Institute of Technology, Atlanta, GA, 2000.

66. Eley, D. D., *Adhesion*, Oxford University Press, London, 1961.
67. Miller, J. D., and H. Ishida, "Adhesive-adherend interface and inter-phases," Chapter 10 in L. H. Lee (ed.), *Fundamentals of Adhesion*, New York: Plenum Press, 1991.
68. Venables, J. D., 1984, "Adhesion and durability of metal-polymer bonds," *J. Mater. Sci.*, vol. 19, 1984, p. 2431.
69. Brockmann, W., O. D. Hennemann, H. Kollek, and C. Matz, "Adhesion in bonded aluminum joints for aircraft construction," *Int. J. Adhes., Adhes.*, vol. 6, no. 3, 1986, p. 115.
70. Yao, Q. and J. Qu, 2002, "Interfacial versus cohesive failure on polymer-metal interface—effects of interface roughness," *J. Electronic Packaging*, vol. 124, 2002, pp. 127–34.
71. Lee, H. Y., and J. Qu., 2003, "Microstructure, adhesion strength and failure path at a polymer/roughened metal interface," *J. Adhesion Science and Technology*, vol. 17, 2003, pp. 195–215.
72. Lee, H. Y., and J. Qu, "Dimple-type failures in a polymer/roughened metal system," *J. Adhesion Science and Technology*, vol. 18, no. 10, 2004, pp. 1153–72.
73. Kinloch, A. J., *Adhesion and Adhesives Science and Technology*, London: Chapman and Hall, 1987.
74. Ferguson, T., and J. Qu, "Moisture absorption analysis of interfacial fracture test specimens composed of no-flow underfill materials," *J. Electronic Packaging*, vol. 125, 2003, pp. 24–30.
75. Ferguson, T., and J. Qu, "Elastic modulus variation due to moisture absorption and permanent changes upon redrying in an epoxy based underfill," *IEEE Component and Manufacturing Tech.*, vol. 29, 2006, pp. 105–111.
76. Ferguson, T., and J. Qu "Effects of moisture on adhesion and interfacial fracture toughness," in E. Suhir and Y. C. Lee (eds), *Micro- and Opto-Electronic Materials and Structures: Physics, Mechanics, Design, Reliability, Packaging*, Springer, Secaucus, NJ, 2006, pp. 431–469.
77. Comyn, J., C. Groves, and R. Saville, 1994, "Durability in high humidity of glass-to-lead alloy joints bonded with an epoxide adhesive," *International Journal of Adhesion and Adhesives*, vol. 14, 1994, pp. 15–20.
78. Gledhill, R., A. Kinloch, and J. Shaw, "A model for predicting joint durability," *Journal of Adhesion*, vol. 11, 1980, pp. 3–15.
79. Wylde, J., and J. Spelt, "Measurement of adhesive joint fracture properties as a function of environmental degradation," *International Journal of Adhesion and Adhesives*, vol. 18, 1998, pp. 237–46.
80. Tsai, J. Y., V. Sundaram, B. Wiedenman, Y. Sun, C. P. Wong, and R. R. Tummala, "A novel 20-100 μm pitch IC-to-package interconnect and assembly process," *Proc. Electronic Components and Technology Conference*, 2006, pp. 263–68.
81. Tummala, R. R., C. P. Wong, V. Sundaram, and J. Y. Tsai, "Novel underfill material and process on package substrate for ultra-fine pitch (10- 30 micron) flip-chip attach," US patent pending, 2007.
82. Zhu, Q., L. Ma, and S. K. Sitaraman, "Design optimization of one-turn helix—a novel compliant off-chip interconnect," *IEEE Transactions on Advanced Packaging*, vol. 26, no. 2, 2003a, pp. 106–112.
83. Zhu, Q., L. Ma, and S. K. Sitaraman "Design and fabrication of β-fly: a chip-to-substrate interconnect," *IEEE Transactions on Components and Packaging Technologies*, vol. 26, no. 3, 2003b, pp. 582–90.

84. Aggarwal, A. O., P. M. Raj, R. J. Pratap, A. Saxena, and R. R. Tummala, "Design and fabrication of high aspect ratio fine pitch interconnects for wafer level packaging," *Proc. of 4th Electronics Packaging Technology Conference,* 2002, pp. 229–234.
85. Aggarwal, A. O., I. R. Abothu, P. M. Raj, M. D. Sacks, and R. R. Tummala, "Novel low-cost sol-gel derived nano-structured and repairable interconnects," *International Microelectronics and Packaging Society,* Boston, MA, November 2003, pp. 943–948.
86. Li, Y., K. Moon, and C. P. Wong, "Enhancement of electrical properties of anisotropically conductive adhesive (ACA) joints via low temperature sintering" *Journal of Applied Polymer Science,* vol. 99(4), 2006, pp. 1665–1673.
87. Li, Y., K. Moon, and C. P. Wong "Adherence of self-assembled monolayers on gold and their effects for high performance anisotropic conductive adhesives," *Journal of Electronic Materials,* vol. 34-3, 2005a, pp. 266–71.
88. Li, Y., K. Moon, and C. P. Wong, "Monolayer protected silver nano-particle based anisotropic conductive adhesives (aca): electrical and thermal properties enhancement," *J. Electronic Materials,* 2005b, p. 34 –12.

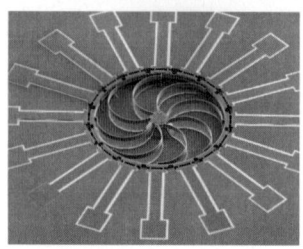
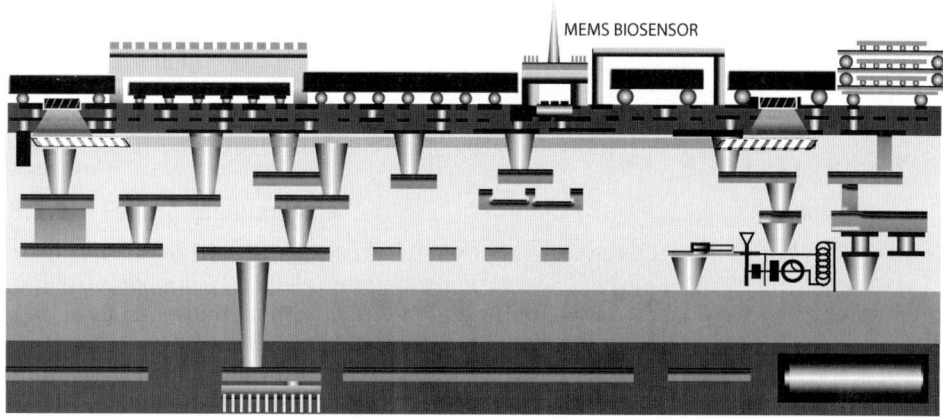

CHAPTER 9

MEMS Packaging

Pejman Monajemi and Farrokh Ayazi
Georgia Institute of Technology

Douglas Sparks
Integrated Sensing Systems, Inc.

9.1	Introduction—496	9.7	Techniques Utilizing Getters—516
9.2	Challenges in MEMS Packaging—496	9.8	Interconnections—522
9.3	Chip-Scale versus Wafer-Scale Packaging—497	9.9	Assembly—524
9.4	Wafer Bonding Techniques—499	9.10	Summary and Future Trends—527
9.5	Sacrificial Film-Based Sealing Techniques—505		References—528
9.6	Low-Loss Polymer Encapsulation Techniques—514		

Micro-electro-mechanical systems (MEMS) are the integration of micromechanical components such as sensors, actuators, and RF elements with their control and read-out electronics on a common silicon or nonsilicon substrate. With efficient and low-cost packaging, MEMS promise to revolutionize nearly every electronic product category by bringing together silicon and non-silicon-based microelectronics, thus making possible the realization of complete systems-on-package (SOP). Sensors include a broad range of, but not limited to, pressure sensors, inertial sensors (accelerometers and gyroscopes), chemical sensors, magnetic sensors, and radiation sensors. RF and microwave MEMS include micromechanical and acoustic resonators, filters, switches, and relays, and RF passives include variable capacitors and inductors. Optical MEMS are the class of MEMS devices used in imaging including micromirrors, light detectors, switches, couplers, and lenses. Bio-MEMS include sensors and actuators implanted in biological environments for medical diagnosis such as DNA sensors, blood pressure sensors, lab-on-a-chip, and microneedles for drug delivery. The MEMS industry is rapidly converging toward NEMS (nano-electro-mechanical systems), which can expand the functionality regime of traditional MEMS to nanoscale.

Wafer-level packaging promises cost-effective and more efficient integration compared to chip-scale packaging, but there remain challenges that need to be overcome. In this chapter, various wafer-level packaging techniques like wafer bonding, thin-film sealing, and polymer encapsulation are discussed. This chapter also reviews key MEMS challenges, which include thermal, mechanical, electrical, chemical, and reliability challenges. Thin-film NanoGetters are described as being capable of maintaining an ultrahigh vacuum in microcavities, which can benefit devices like motion and pressure sensors as well as RF MEMS chips. The chapter concludes with interconnection schemes for MEMS packages to provide proper electrical, optical, fluidic, and chemical interfaces to the outside world and efficient assembly techniques for chip-scale packaged MEMS to system-level circuit boards such as SOP.

9.1 Introduction

Micro-electro-mechanical systems consist of functional components integrated into a silicon or nonsilicon substrate. The mechanical function provides sensing or actuation. MEMS sensors are designed to measure acceleration, rotation, pressure, flow, mass, radiation, temperature, and magnetic field. MEMS actuation mechanisms are used in microfluidic valves and pumps, micromirrors, motors and generators, grippers, ink-jet nozzles, hard-disk heads, mechanical resonators, and optical and electrical switches. MEMS devices can be classified according to their application as sensors, RF and microwave MEMS, optical MEMS (or MOEMS), and bio-MEMS.

There is a broad range of sensors including, but not limited to, pressure sensors, inertial sensors (accelerometers and gyroscopes), chemical sensors (thermal, thermoelectric, electrochemical, and mass sensors), magnetic sensors, and radiation sensors [1]. RF and microwave MEMS include micromechanical and acoustic resonators and filters, switches and relays, and passives including variable capacitors and inductors [2–3]. Optical MEMS are the class of MEMS devices used in imaging including micromirrors, light detectors, switches, couplers, and lenses [4]. Bio-MEMS include sensors and actuators implanted in biological environments for medical diagnosis such as DNA sensors, blood pressure sensors, lab-on-a-chip, and microneedles for drug delivery [5–6].

Interfacing MEMS and electronic circuits is achieved by hybrid and integrated techniques [7]. Hybrid methods include wire bonding and flip-chip assembly. Monolithic integration of MEMS and electronics includes fabrication of an integrated circuit (IC) on the MEMS substrate (pre-CMOS [8]), fabrication of MEMS on top of the IC (post-CMOS [9]), or a combination of the two techniques [10]. Direct integration provides cheaper assembly and packaging and reduced parasitics but requires a higher development cost.

9.2 Challenges in MEMS Packaging

Wafer-level packaging of MEMS represents a challenging and often costly task in microsystem manufacturing. MEMS packaging differs from traditional microelectronics packaging in that the encapsulating cover should not touch the micromachined device. In some applications a vacuum is needed inside a hermetically sealed package (e.g., resonators and gyroscopes), and in other cases the package should protect the sensor

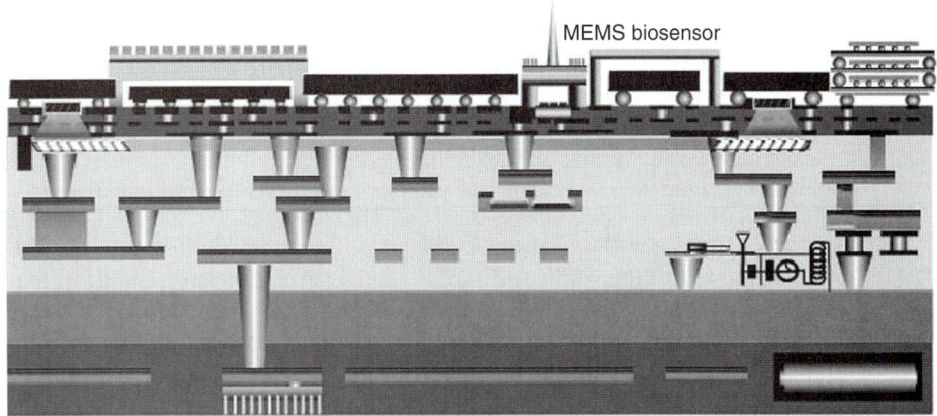

Figure 9.1 Concept of system-on-package (SOP) for integration of thin-film component MEMS packaging [11].

while providing suitable access to the outside environment through a window (e.g., chemical, flow, and optical sensors and actuators). For the system-on-package (SOP) applications shown in Figure 9.1, MEMS should be preferably packaged before integration and assembly onto the substrate [11]. In this chapter, a review of successful chip-scale and wafer-scale MEMS packaging techniques is presented. Wafer bonding is introduced as the mainstream technique for wafer-level packaging, and different bonding methods are reviewed. Alternative packaging techniques to wafer bonding are sacrificial film-based sealing and polymer encapsulation. Various sealing methods including low- and high-temperature deposition of thin films are reviewed. A cost-effective packaging technique using low-temperature decomposition of sacrificial polymers is described. We also present the use of thin-film getters to reduce the pressure inside the package and improve long-term reliability of MEMS devices. Without a getter, the pressure inside the package is limited by adsorption and desorption of surface molecules, which can cause drift and hysteresis and is a potential failure mode in high-performance devices. Various interconnection and assembly schemes are discussed. The chapter ends with a discussion of future trends in MEMS packaging.

9.3 Chip-Scale versus Wafer-Scale Packaging

Most MEMS are fabricated using techniques that leave the mechanical structures exposed after the fabrication process is completed. Open-die MEMS devices are easily destroyed if their unprotected mechanical elements come into contact with a physical object, so physical protection is essential. MEMS are also very susceptible to degradation by small particles, water vapor, stiction, and corrosion and, therefore, need microscopic protection and encapsulation. For instance, a micromechanical switch may collapse because of humidity or may face performance degradation because of absorption of outgassed materials by metal electrodes. The MEMS package creates an air or vacuum cavity over the MEMS active area without impeding its motion or function (deflection, tilt, slide, rotation, or vibration). To ensure long-term reliability, the MEMS package should be hermetic [12]. This is critical for the MEMS operating in biomedical environments or those that require vacuum packaging, such as resonators.

There are two general solutions to MEMS packaging. One approach is using the existing IC infrastructure to package MEMS at the chip level, where release and sealing are performed serially on the individual die after dicing. Figure 9.2a shows ADXL50, a chip-scale packaged accelerometer and interface circuit from Analog Devices, Inc. [13]. The released structure is protected with temporary stand-off housings. After sawing and cleaning the wafer, the temporary covers are removed and the device is housed in a metal can package. Figure 9.2b and c show a diagram of a packaged digital micromirror device (DMD) from Texas Instruments. The DMD is adhesively attached to the package and then wire-bonded to the alumina (Al_2O_3) ceramic header. The optical window is assembled to the ceramic header seal ring [14].

The ceramics have a coefficient of thermal expansion (CTE) in the range of 5 to 9 ppm/°C that is close to the CTE of silicon (2.6 ppm/°C) and can be bonded to silicon

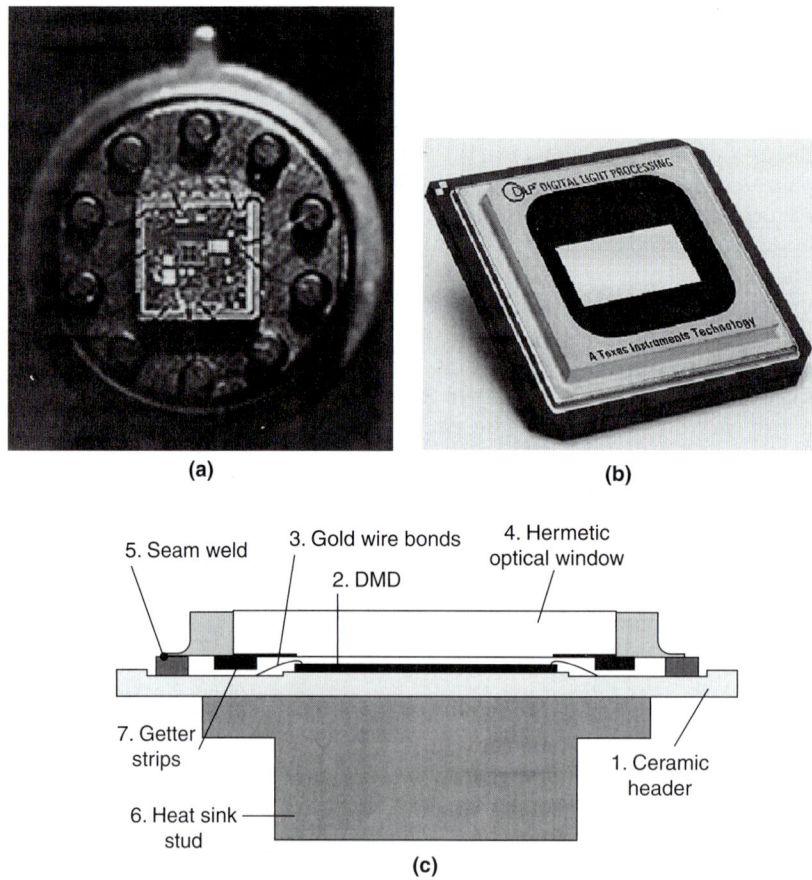

FIGURE 9.2 Chip-scale packaged examples. (*a*) Early accelerometer packages (courtesy of Analog Devices [13]). (*b*) DMD (courtesy of Texas Instruments), (*c*) Diagram of the DMD device [14].

wafers for chip-scale packaging [15]. The package-to-substrate CTE match is required to reduce thermal stress. CTE mismatch is more critical when the die size is larger.

One challenge in chip-scale packaging is selecting the specific equipment required to seal a die [16–17]. Handling a MEMS die prior to packaging is costly and inefficient from a manufacturing standpoint. The alternative approach, known as wafer-scale packaging, includes MEMS release and sealing prior to dicing and assembly. Sealing the die right after release at the wafer level results in a reduction in size, time, and most important, cost.

Wafer-level packaging can be classified into a variety of bonding, thin-film sealing, and polymer encapsulation techniques. Wafer bonding is used extensively for reliable packaging of a wide variety of MEMS and is typically done by bonding a cap with a cavity onto the MEMS wafer [18–42]. Thin-film sealing can be done by surface micromachining to create a thin-film overcoat on top of a sacrificial material. This is followed by sacrificial film removal either by etching [43–51], evaporating [52], or thermal decomposition [53–56]. The sacrificial material can be silicon compounds [43–48], metal [49], water [52], polymer or photoresist [50–51, 53–56]. Other techniques utilize dispensing liquid crystalline polymer (LCP) [57–58]) or polyimide-Kapton encapsulation to seal the MEMS wafer from the topside and backside [59]. Packaging performance can be improved by creation of a cap with absorbent material to maintain the required ambient pressure by absorbing the molecules, as a result of aging and outgassing [60–72]. A successful MEMS packaging should address all the key issues. These include thermal, mechanical, electrical, and chemical management. The cap should have a CTE matched to the substrate to minimize thermally induced stress, be mechanically stable to survive the pressure difference, and if possible, be able to absorb the external shock. For RF applications, the cap should be transparent to RF signals and should add a minimum electrical loss to the MEMS device. Finally the cap should be chemically resistant to the environment especially for devices operating in fluidic and biomedical media.

9.4 Wafer Bonding Techniques

Wafer capping techniques include bonding or transferring a cap from another wafer to the host MEMS wafer [18]. These techniques can be classified according to the cap and bonding material, or the feedthrough type. Wafer bonding includes direct bonding (anodic and fusion bonding) and bonding using intermediate layers (metals or insulators). Metal bonding includes solder bonding [22], eutectic bonding [23], thermocompression bonding (TCB) [24], and rapid thermal processing (RTP) [38]. Metal interface layers have a large CTE mismatch to silicon and do not allow vertical interconnects through the cap. Solder can be easily reflowed to bond a glass or ceramic cap to silicon. TCB can be used to bond glass to silicon using gold or copper as the interlayer. RTP is suitable to bond glass to glass or silicon nitride using aluminum as the intermediate layer for a short duration of time (seconds) at an elevated temperature of around 750°C [38]. Insulator bonding methods include adhesives [25–27] and reflowed glass frit bonding [28]. In the adhesive bonding methods, a near-hermetic package can be obtained due to finite gas permeability of polymers, while reflowed glass frit bonding provides a hermetic package. Table 9.1 lists the wafer-bonding methods used in the MEMS industry and research institutions.

Chapter Nine

Bonding Method		Material	Temperature	Comments
Direct	Anodic	Glass to Si-glass	180–500°C	Surface roughness better than 500 nm; hermetic
	Fusion	Si to Si	>800°C	Surface roughness better than 50 nm; hermetic
Intermediate / Metal	Eutectic	Au to Si	363°C	Does not need smooth surface; hermetic
	Solder	AuSn, PbSn, InSn, or AlSi to Si and glass	183°C, 118°C, 577°C, 800°C	Does not need smooth surface; hermetic
	TCB	Au or Cu to Si	25–250°C	Hermetic, high force required
	RTP	Al to Si_3N_4	750°C (short duration)	Hermetic
Intermediate / Nonmetal	Adhesive	SU-8, BCB, polyimide	<300°C	Does not need smooth surface; nonhermetic
	Melting	Reflowed glass frit	375–410°C	Does not need smooth surface; hermetic

TABLE 9.1 Summary of Wafer Bonding Packaging Techniques

9.4.1 Direct Bonding

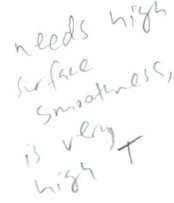

Fusion bonding occurs due to a chemical reaction between OH bonds of the hydrophilic surface [19]. High-temperature annealing is required for increased bond strength that limits its usage for MEMS packaging. Moreover, fusion bonding needs extra smooth surfaces.

Anodic (electrostatic) bonding utilizes sodium-rich glass (e.g., Corning Pyrex 7740 or 8329) to pull the ultraflat silicon and glass into intimate contact by electrostatic force (0.2 to 1 kV) [21]. Glass (generally Pyrex), as a highly hermetic material [11], has a good CTE match to substrate, is electrically insulating, and can maintain a vacuum level for a long time. Glass caps are usually thick and therefore not optimal for thin-film packaging applications. Glass is a transparent and biocompatible material and has a CTE (2.8 ppm/°C for Pyrex 8329) close to that of silicon (2.6 ppm/°C). However, the surface roughness should be better than 1 μm. Moreover, sodium contamination during anodic bonding may cause a significant change to CMOS-based MEMS systems.

9.4.2 Bonding Using Intermediate Layers

Metallic bonding methods are used to seal rough surfaces at low temperatures. These methods include materials and techniques including fluxless solders such as AuSn, eutectic bonding of AuSi, or TCB of thick electroplated gold to Si. Solder can be used to bond rough surfaces like ceramics to silicon. The alloy composition is selected to minimize the melting point. The large CTE mismatch between silicon and metal (14.2 ppm/°C for AuSi and 24.7 ppm/°C for PbSn) is an issue that limits the thermal range of operation. The eutectic bonding reported in [23] is shown in Figure 9.3. Figure 9.3a shows the fully packaged wafer.

MEMS Packaging 501

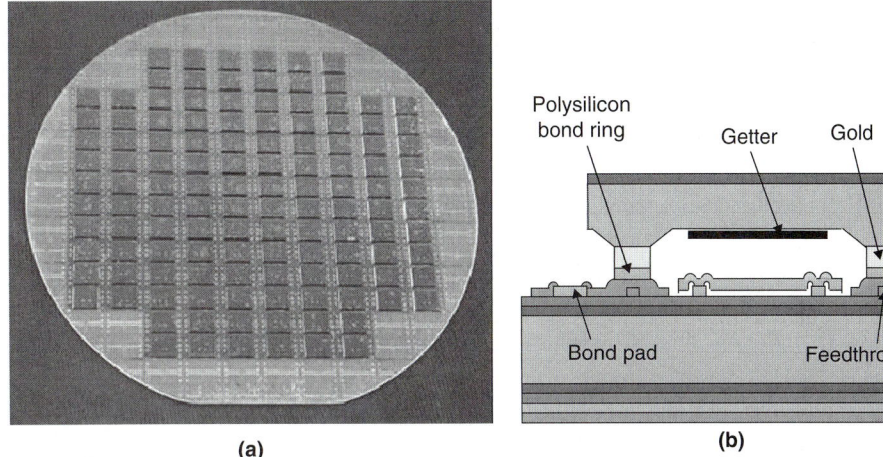

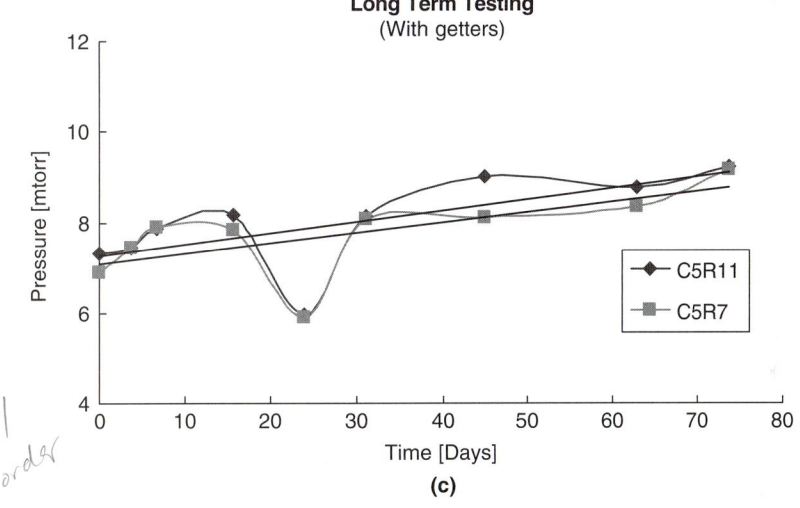

FIGURE 9.3 Gold to silicon eutectic bonding. (*a*) Wafer with 132 vacuum-packaged devices. (*b*) Cross section view after 2.5-μm gold deposition, patterning, bulk etching of the cavity, getter deposition and finally bonding and dicing part of the cap wafer. (*c*) Vacuum package results [23].

The eutectic bonding method starts with deposition and patterning of gold on the cap wafer, followed by etching of the cavity on the wafer. After depositing getters to the cap wafer, the cap is aligned and bonded to the MEMS wafer (Figure 9.3b). Above 363°C and upon pressing the two wafers, polysilicon on the bond rings around the MEMS device diffuses into gold and the eutectic bond is formed. Figure 9.3c shows test results using a Pirani gauge, based on heat transfer from a suspended heater to a heat sink through a gas, where the gas thermal conductivity changes with package pressure. A base pressure of 6.9 millitorr (mtorr) and a leak rate of 8.5 mtorr/year are reported.

Bonding using intermediate layers can be done at much lower temperatures. It can be classified into adhesive bonding, melting the intermediate layer (glass frit), forming a stable intermediate compound (diffusion), or heating up to 50 to 70 percent of the melting point (brazing). The adhesives include polymer, epoxies, or UV photoresists such as polyimides [25], benzocyclobutene (BCB) [26], SU-8 [27], and parylene. The adhesive is generally screen-printed and can be cured at room temperature, and it may require UV exposure (glass wafers) or thermal cure at 80 to 150°C. Reflowed glass frit bonding is another technique that has been used for more than two decades to package microsensors. This technique has been used in pressure sensors, accelerometers, gyroscopes, and switches to join two silicon wafers with a dielectric seal that can conformally cover minor surface steps and even particles [28]. Glass frit bonding has recently been employed to bond silicon to Pyrex to vacuum-seal resonant sensors and Coriolis mass flow sensors [28]. Glass wafers can also be bonded to each other, hence forming an entirely dielectric chip-scale package. The process of forming the glass seal is similar to that used for adhesive sealing. A thixotropic paste is screen-printed onto the bonding wafer. This paste is dried and then fired to burn off the organic compounds. As in the case of the adhesive bond, ceramic particles that do not melt at the sealing temperature can be used to set the wafer-to-wafer gap after bonding. These ceramic particles are often used to adjust the expansion coefficient of the glass-ceramic matrix down to that of silicon. Next in the packaging flow, the device wafer and the cap wafer are aligned and heated in a wafer-bonding system. If the MEMS device is to be vacuum-packaged, the bonder chamber is pumped down during this step. The chamber can also be back-filled with an inert gas to provide damping to devices such as accelerometers or switches. The glass frit is typically reflowed at 375 to 410°C. Pressure on the wafer stack is maintained during a slow cooldown. In the adhesive bonding methods, bonding can be done at low temperatures without causing a large thermal mismatch. However, hermeticity is a consideration, because of outgassing of the polymers or epoxies used in the interface.

Figure 9.4 shows two wafer-level packaged MEMS in production. Figure 9.4a is a hermetic silicon cap bonded on top of the circuit part of the MEMS chip [29]. Figure 9.4b is a packaged switch from Radant MEMS, Inc., by glass frit bonding and horizontal feedthroughs [30].

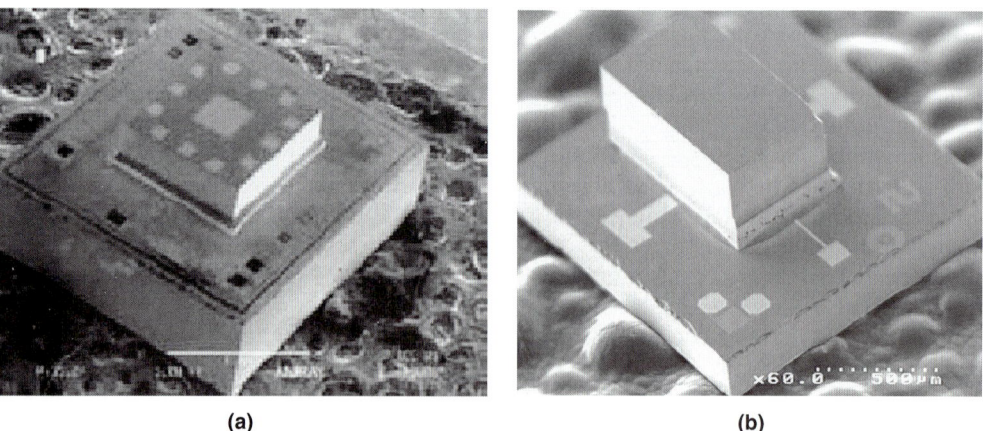

(a)   (b)

FIGURE 9.4 Two recent examples of wafer-level bonding and interconnections. (a) Hermetic packaged accelerometer from Analog Devices, Inc. [29]. (b) Hermetically packaged RF switch from Radant MEMS, Inc. [30].

All the mentioned bonding methods can be performed locally, by concentrating the bond energy to a small ring around the active area to avoid the thermal-sensitive MEMS to be affected [31–36]. Local bonding can be done by heating generated by running current through microheaters, CO_2 laser welding [15], localized chemical vapor deposition (CVD) bonding [32], RF dielectric heating [33], and localized ultrasonic bonding [34]. The local bonding examples include eutectic bonding [23], solder bonding, fusion bonding [35], and plastic bonding [36].

An example of local solder bonding is shown in Figure 9.5 for a vacuum-packaged resonator. Polysilicon deposited by low-pressure chemical vapor deposition (LPCVD) is used to make the resonators and the local microheaters required for local bonding of aluminum deposited on polysilicon to the glass cap. The glass cap with cavity is aligned to the MEMS wafer and brought into contact. Running current through the polysilicon resistors increases the microheater local temperature to around 800°C, where the glass and the aluminum form a solder, and therefore a strong bond is formed. Accessing the MEMS contacts is achieved by horizontal aluminum feedthroughs at the surface. Figure 9.5b shows the package after breaking the cap; the polysilicon bond rings, the glass cap, and polysilicon interconnects can be seen in the figure. Local bonding is fast, reliable, and requires small pressure levels.

Figure 9.6 shows that the Q factor of the packaged resonators does not degrade after 30 weeks.

Metallic bonding can be also used to transfer a microcap from another wafer to the host MEMS wafer [39–42]. Figure 9.7a shows a microcap transferred onto silicon by SiAu eutectic bonding. As shown in Figure 9.7, the cap is created in another wafer by etching the stiffening ribs and the main cavity, followed by depositing the sacrificial PSG and refilling with polysilicon or poly-silicon-germanium (PSG). Then gold is evaporated and patterned to create the bond rings. The cap wafer is bonded to the host MEMS wafer (Figure 9.7a) and the PSG is removed in hydrofluoric acid (HF) (Figure 9.7b) to remove the silicon cap wafer and access the contact pads without the need for dicing the cap [40]. A scanning electron microscope (SEM) view of a bonded microcap is shown in Figure 9.7c [41]. Another method involves transfer of a nickel microcap using sacrificial solder and transient liquid phase bonding [42].

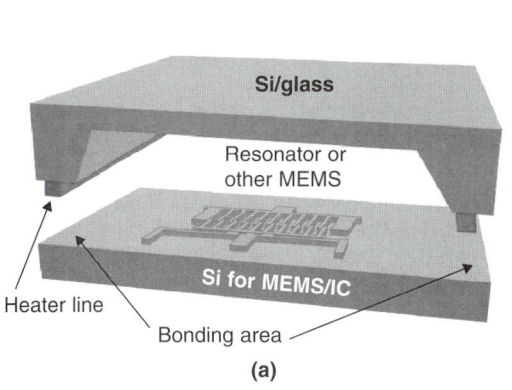
(a)

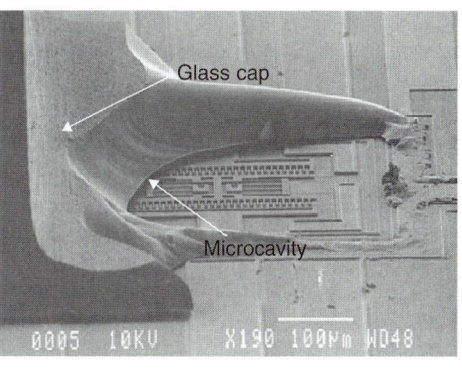
(b)

FIGURE 9.5 Local bonding example. (*a*) Schematic of MEMS resonator, heater line, and cap. (*b*) View of the resonator after breaking the glass cap [37].

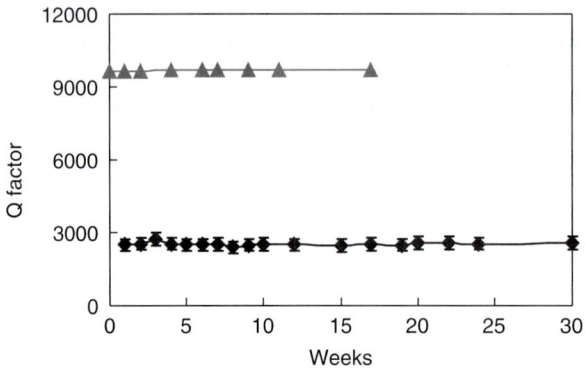

FIGURE 9.6 Hermeticity measurement of two vacuum-packaged resonators by local bonding [37].

Wafer-to-wafer bonding has several drawbacks. First, the anchor region where the cap seals to the MEMS die must be relatively large to ensure a hermetic seal and tolerate the potential misalignments. This translates into a significant increase in die size and die cost. Increasing the package size will result in increasing the parasitic components. Second, the bonded die is thicker than the standard IC die; therefore, the packaged

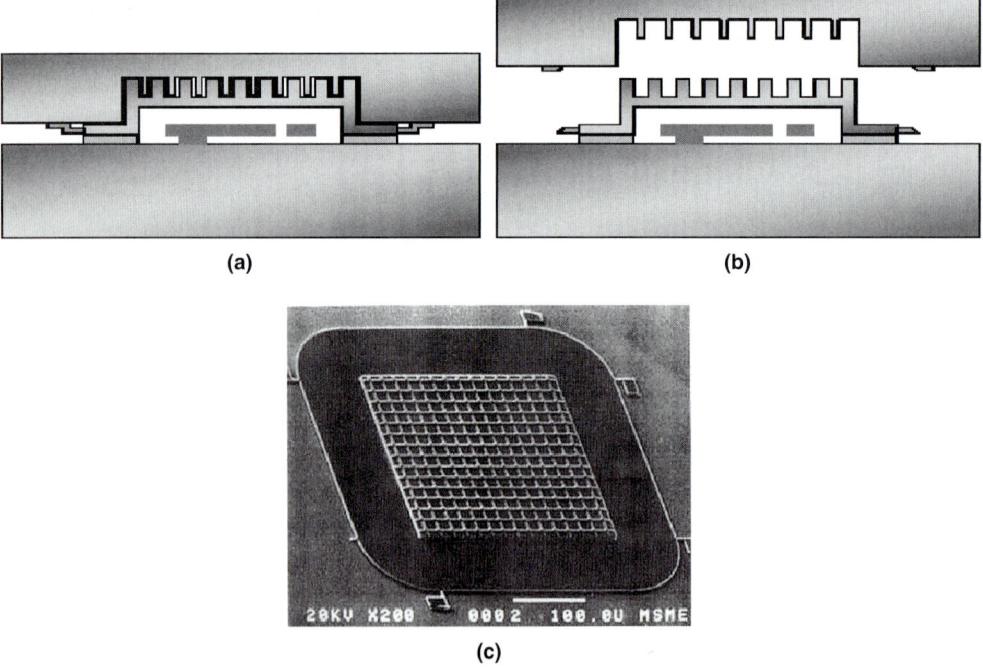

FIGURE 9.7 Microcap transfer. (*a*) The cap is fabricated on a donor wafer, aligned with a MEMS wafer, and bonded to the MEMS wafer by eutectic bonding. (*b*) The wafers are separated, the tethers holding the cap to the donor break, and the cap is transferred to the target [40]. (*c*) SEM view of the microcap-bonded MEMS [41].

MEMS device cannot be housed in a thin profile package. Third, some of the wafer-bonding methods require extra flat and clean wafer surfaces. Consequently, any kind of surface contamination or surface roughness may result in a nonhermetic seal and yield degradation.

9.5 Sacrificial Film-Based Sealing Techniques

A possible alternative for bonding packaging is the formation and sealing of surface micromachined membranes over the MEMS. The advantage of this sealing technique is the reduced thickness and area and therefore a lower-cost batch process. In these methods, a layer of sacrificial material is deposited, followed by deposition of an overcoat. Then the sacrificial material is etched (through the perforations or the porous inorganic overcoat), developed, evaporated, or decomposed (through a permeable polymer overcoat). In the etching methods, another layer of overcoat should be deposited to completely seal the release holes, during which a thin layer of overcoat may penetrate into the MEMS. Therefore, this method is not applicable to MEMS devices that are extrasensitive to contamination (e.g., in silicon beam resonators, the Q factor is very sensitive to extra materials deposited on the beam). The evaporation technique, called the frozen water process, is based on evaporation of frozen water through a photoresist curable at room temperature. In order to pattern the frozen water, hydrophobic and hydrophilic regions should be defined to allow water to be selectively attached to the hydrophilic area under the ambient environment [52].

9.5.1 Etching the Sacrificial Material

In this method, small perforations are made in the overcoat by lithography, or the overcoat is made porous by chemical processing. The sacrificial material is etched by wet etching [43–50] or dry etching [51]. Finally another layer of overcoat is deposited to bridge over the small openings, creating a rigid cap.

Figure 9.8 shows a thin-film vacuum-sealed MEMS pressure sensor reported in [43]. The SEM of a sealed resonator is shown in Figure 9.8a and the process flow is demonstrated in Figure 9.8b to e. The sacrificial layer is boron-doped epitaxial (epi) polysilicon, and the thin-film capsule consists of heavily boron-doped epi-polysilicon. After removing the sacrificial layer, the vacuum sealing is performed in two steps. The first step is sealing the etch holes in the epi-polysilicon reactor. The second step is outgassing of the remaining hydrogen in the capsule by thermal permeation in a high-temperature LPCVD furnace and annealing in a nitrogen-purged furnace at a pressure of about 10 mtorr.

A wafer-level thin-film packaging method has been proposed in [44] that seals the structure under a thin-film cap (30 µm) deposited during wafer manufacturing. Figure 9.9a and b show an accelerometer before and after packaging using this method. The thin-film cap is deposited as one of the last steps of MEMS packaging. The cap hermetically seals the MEMS structure and is sturdy enough to withstand the rigorous conditions of the plastic injection-molding process (pressures up to 1500 psi and temperatures as high as 175°C).

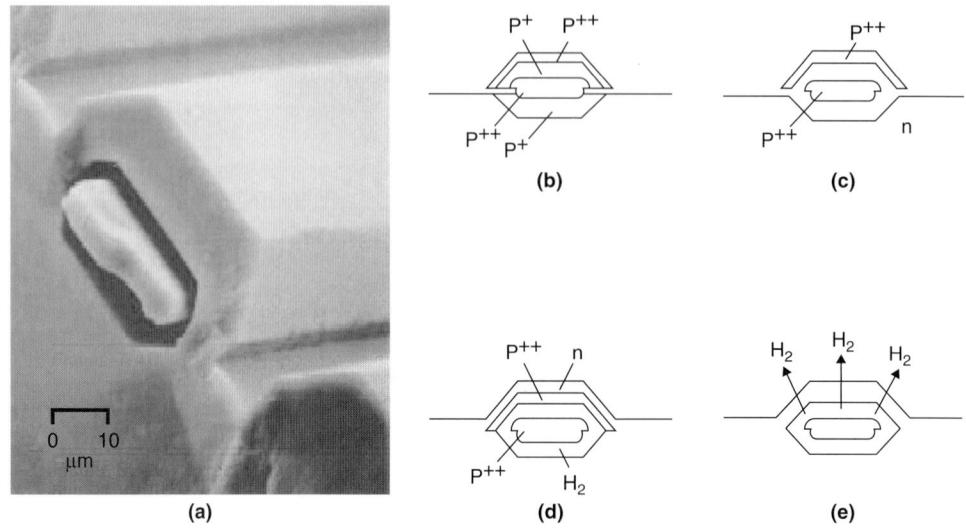

FIGURE 9.8 (a) Cross section of the sealed resonator. (b–e) Process flow for fabrication and packaging of the resonator: (b) after growth of P+ and P++ epi-polysilicon and creating the etch holes, (c) after selective etching of p+ epi-polysilicon, (d) after depositing n-epi-polysilicon to seal the etch holes, and (e) after annealing in N_2 [43].

A similar technique has been developed by epitaxial growth of polysilicon overcoat on top of the sacrificial PSG, as shown in Figure 9.10 [45]. The perforations are created, and the sacrificial PSG is etched using vapor HF to avoid stiction of the overcoat to the accelerometer. These perforations are bridged over by plasma-enhanced chemical vapor depositing (PECVD) silicon dioxide, and the bonding pads are opened. The thin-film overcoats can also be created by reflowing a low-melting-point material such as aluminum, LPCVD and PECVD deposited Ge, or PECVD borophosphosilicate glass (BPSG) to bridge over and seal the perforations in the MEMS capsule.

Sacrificial inorganic films that have been used for packaging include borosilicate glass (BSG), phosphosilicate glass (PSG), silicon dioxide, polysilicon, and metals. An

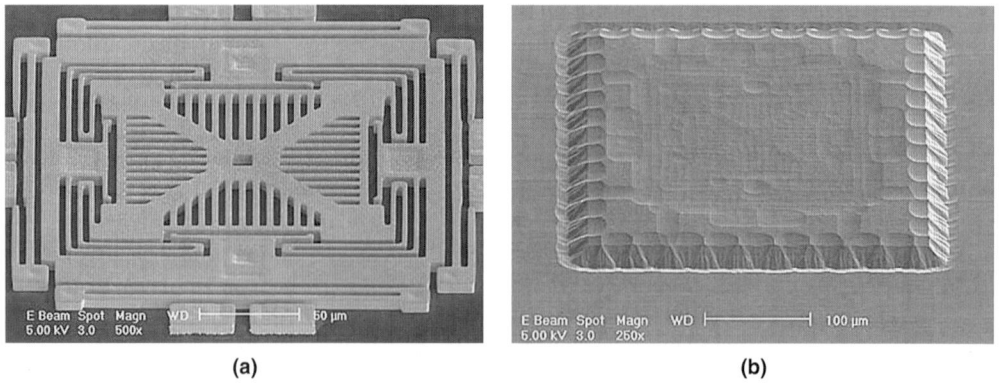

FIGURE 9.9 (a) A three-axis MEMS accelerometer before packaging. (b) The device after thin-film packaging [44].

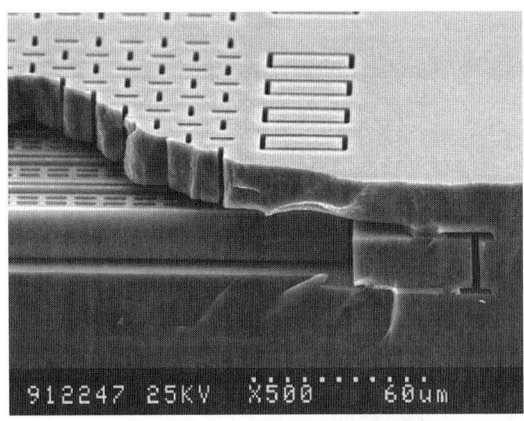

FIGURE 9.10 An accelerometer packaged by epi-polysilicon. [45]

example is shown in Figure 9.11 [46]. PSG defines the sacrificial material (about 7 µm), and 3-µm LPCVD nitride film with perforations is used as an overcoat (microshell). The sacrificial material is removed in HF and dried by using the supercritical dryer to avoid stiction. The holes are covered by depositing another layer of LPCVD nitride at a pressure of 300 mtorr. The final cavity pressure will be on the order of the pressure inside the LPCVD furnace.

Different types of photoresists have been used as organic sacrificial films for MEMS packaging. An example is shown in Figure 9.12 for an electroplated nickel package [50]. The process, as explained in Figure 9.12a, starts with creation of chromium-aluminum access holes around the MEMS device. The sacrificial layer is defined by patterning a thick photoresist to cover the MEMS device and the feedthroughs. Then gold is deposited as the seed layer, followed by electroplating nickel. The access holes are opened by

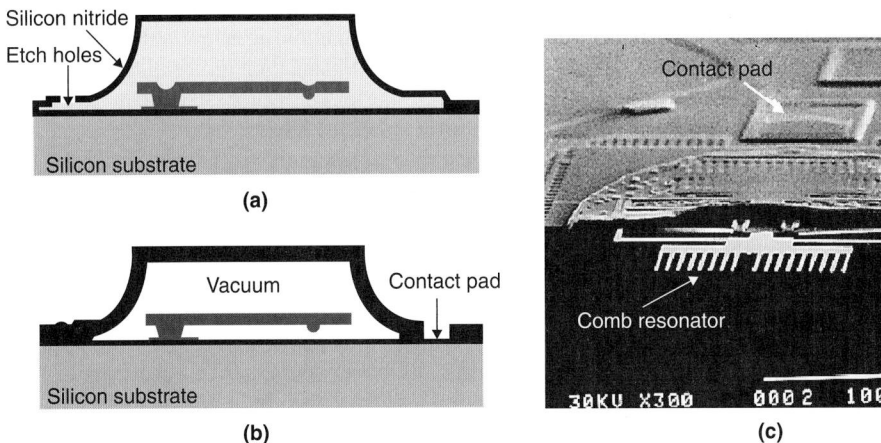

FIGURE 9.11 Vacuum packaging using sacrificial PSG. (*a,b*) Process flow (*a*) after depositing a 1-µm sacrificial PSG layer and a 1-µm low-stress nitride cap and opening the etch holes, (*b*) after removing sacrificial PSG, filling the etch holes with 2-µm nitride and opening the pads. (*c*) View of the packaged resonator [46].

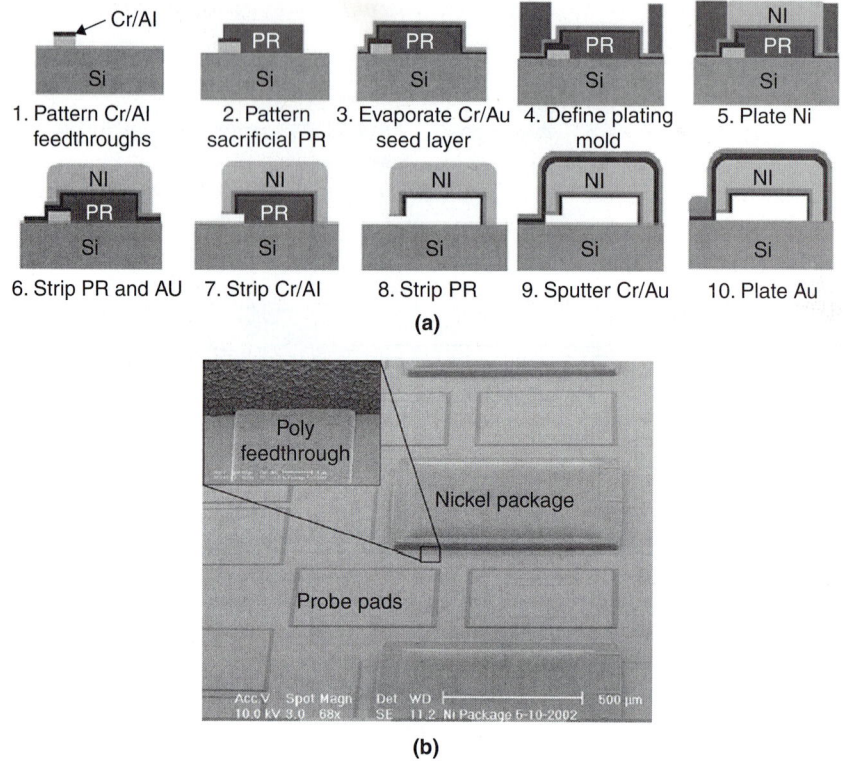

FIGURE 9.12 Vacuum packaging using sacrificial photoresist. (*a*) Process flow. (*b*) View of a packaged Pirani gauge using electroplated nickel. [50]

removing the CrAl, and finally, the sacrificial photoresist is removed in acetone. The access holes are sealed in a vacuum by sputtering chromium and gold, and finally gold is electroplated to complete the packaging. This packaging technique is low-temperature and involves multiple steps of metal processing and sacrificial layer removal. Figure 9.12b shows a packaged Pirani gauge, using 8-μm-thick sacrificial photoresist and 40-μm-thick nickel cap. Polysilicon is used to define the horizontal interconnects.

Another method for sealing includes reactive gas sealing. In this method, the air inside the cavity is consumed by oxidizing the polysilicon overcoat [47]. This packaging method is the highest-temperature technique reported in the literature.

Instead of using perforations in the overcoat, the overcoat can be made porous by special techniques. These methods include electrochemical wet etching of polysilicon, segregation of HF-soluble phosphorus-rich precipitates to the grain boundaries, or forming a sandwich of polysilicon and silicon dioxide layers followed by oxide removal. The electrochemically processed porous polysilicon is shown in Figure 9.13 [48]. A similar method includes deposition of a thin layer of poly-SiGe [41]. All packaging techniques utilizing etching of the sacrificial material through a porous cap have the disadvantage of deposition of extra polysilicon into the MEMS device during final sealing, which can have a detrimental effect on the MEMS performance.

In summary, most of the reported packaging methods based on etching the sacrificial layer are device-specific and have certain limitations. They are either costly (use LPCVD

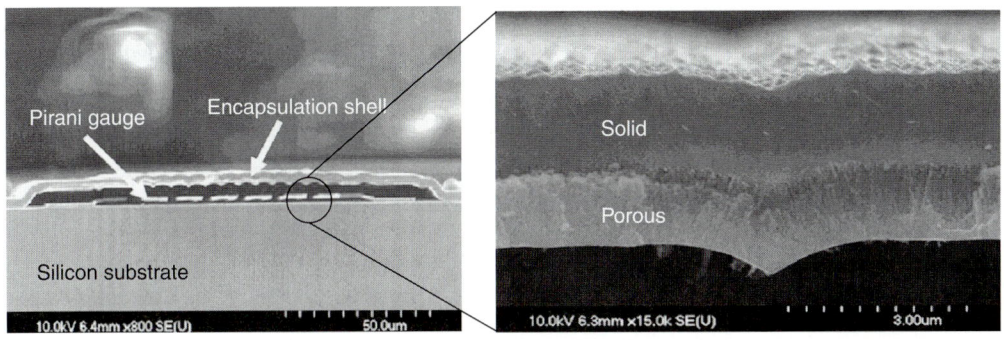

Figure 9.13 Vacuum packaging using porous polysilicon and sacrificial oxide removal. [48]

or complicated chemical processes) or require non-CMOS-compatible high-temperature steps that can introduce stress to the MEMS. In Section 9.5.2, a general-purpose, low-temperature MEMS packaging technique is introduced.

9.5.2 Decomposition of Sacrificial Polymers

This technique involves low-temperature decomposition of a sacrificial polymer through a polymer overcoat. Thermal decomposition is fast, reliable, structurally benign, and CMOS compatible [53–56]. Both the sacrificial polymer and the polymer overcoats can be patterned by photodefinition. This eliminates the need for bond rings and cap-to-wafer aligning and therefore can potentially provide smaller-size packages. Thermal decomposition of sacrificial polymer is performed through a solid perforation-free capsule, which eliminates the steps needed in some other sacrificial-based techniques to seal a perforated protective cover. The packaging sequence is depicted in Figure 9.14.

As shown in Figure 9.14a, a cavity with scalable height is created on top of the movable/resonant part of the device by applying a sacrificial polymer (Unity, a polycarbonate from Promerus LLC). The sacrificial polymer can be applied by spin casting or dispensing. Then

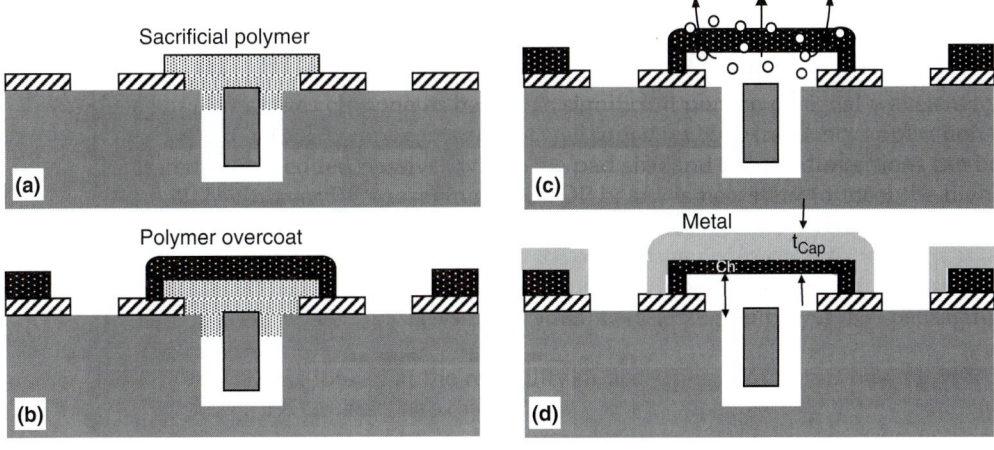

Figure 9.14 Polymer-based packaging sequence flow: (a) after patterning the sacrificial polymer, (b) after overcoat formation, (c) after decomposition, (d) after metal deposition and patterning. [53]

a polymer overcoat with a high glass transition temperature (e.g., Avatrel, a polynorbornene from Promerus LLC) is spin-coated. The polymer overcoat is photodefined to get access to the contact areas (Figure 9.14b). The air cavity is formed by thermolytic degradation of the sacrificial polymer. The by-products are volatile gases (e.g., CO_2) that can permeate through the overcoat polymer at temperatures of 180 to 250°C (Figure 9.14c). To obtain hermeticity and vacuum operation, a thin-film metal layer is formed to cover the polymer overcoat. The metal film can be created by evaporation, or by deposition (sputtering or plating) and etching (Figure 9.14d). The overcoat geometry can be scaled to tailor different-size MEMS. Avatrel has a dielectric constant of 2.5 and a small loss tangent, which is suitable for wideband packaging of RF MEMS components. Moreover, Avatrel is transparent to visible light, which is suitable for optical MEMS encapsulation.

It is possible to create a thick (up to 400 μm) and pinhole-free package over large-area MEMS structures, which is very difficult to achieve by CVD techniques.

Figure 9.15a shows a close-up view of a 15-μm-thick, 8-μm-wide, 150-μm-long silicon-on-insulator (SOI) beam resonator. Figure 9.15b is the same device after

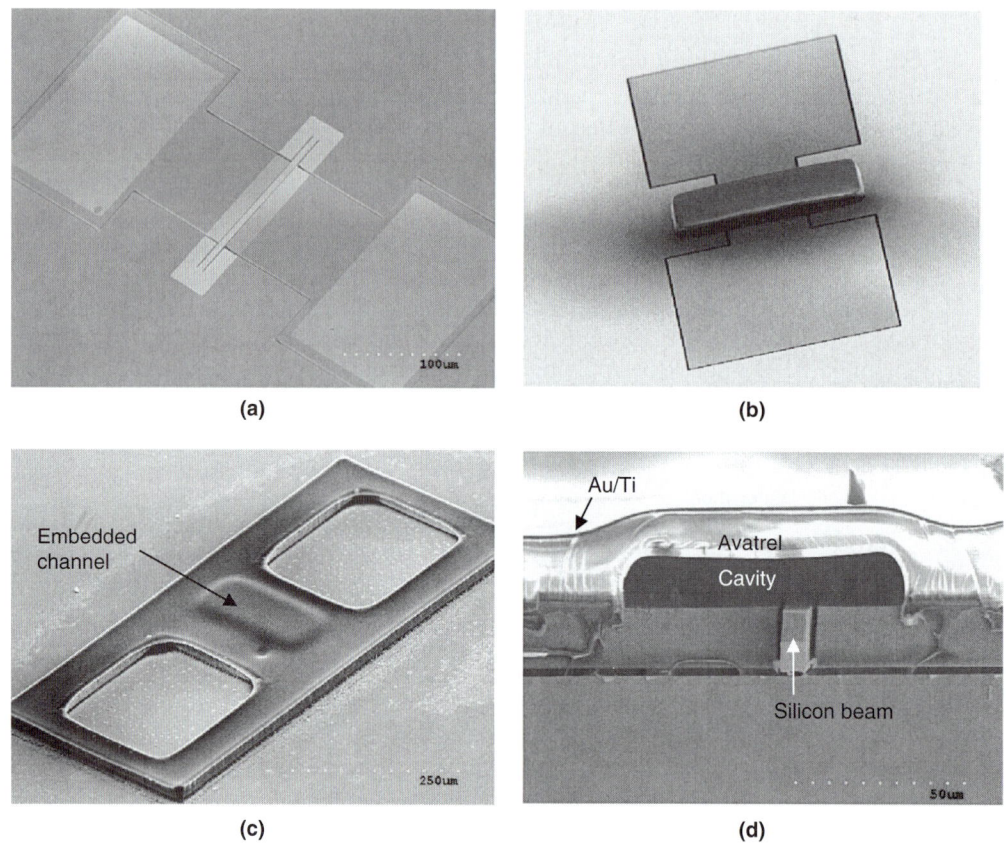

Figure 9.15 View of a vacuum-packaged beam resonator in SOI (a) before packaging, (b) after coating and patterning Unity, (c) after packaging. (d) Cross section of the cavity and 15 μm/1 μm metal-organic membrane. [53–55]

coating and patterning 15-μm thick unity. Figure 9.15c shows the view and cross section of the packaged resonator using 1-μm-thick gold and 15-μm-thick Avatrel [53–55].

Figure 9.15d shows the cross section of the 25-μm-thick silicon beam, the 15-μm-tall embedded vacuum cavity, and the 1 μm/15 μm metal organic cap. The resonators were tested at wafer level before and after packaging. Figure 9.16a and b show the frequency response of the resonator before and after packaging. The Q factor and the resonance frequency (2.6 MHz) of the packaged resonator did not change significantly. This proves that decomposition of the sacrificial polymer does not leave any residues inside the gap.

High-aspect-ratio polysilicon and single-crystal silicon (HARPSS) technology was used to fabricate MEMS sensors including microgravity accelerometers and ring gyroscopes to evaluate the packaging. Unity can be decomposed through a thick polymer overcoat to create a stiff organic cap [54]. A 2-mm-diameter HARPSS gyroscope was encapsulated using the dispensing method to evaluate the method for big and complicated MEMS structures, as shown in Figure 9.17. Figure 9.17b shows the same device after manual dispensing of Unity 200 to cover the rings and the electrodes. Figure 9.17c is the view after forming a thick overcoat cap (120 μm) and decomposing the sacrificial polymer from inside the deep cavity.

Various types of RF tunable capacitors (varactors) were also fabricated and packaged through the mentioned method. Figure 9.18a shows a HARPSS RF varactor made through the self-aligned method. The 0.8-μm capacitive gap is defined by sacrificial silicon dioxide deposition and etching and the 2-μm tuning gap is made by deep reactive ion etching (DRIE). Figure 9.18b shows the same device after encapsulation with 15-μm-thick Avatrel, followed by evaporating 1-μm-thick gold. The evaporated gold creates a hermetic package.

The metal-organic package adds a small loss, as low as 1.4 dB at 1 GHz and 1.5 dB at 5 GHz to the RF MEMS device, as shown in Figure 9.19 [56]. Metal-organic packaging does not require bond rings around the device (as is the case for bonding methods), and the total package is only 10 percent bigger than the varactor.

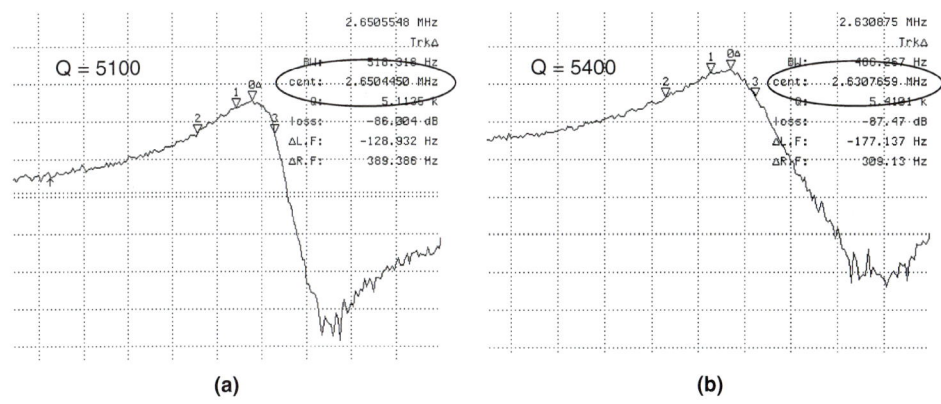

FIGURE 9.16 Frequency response of a 15-μm-thick resonator (a) before packaging and (b) after packaging. [54]

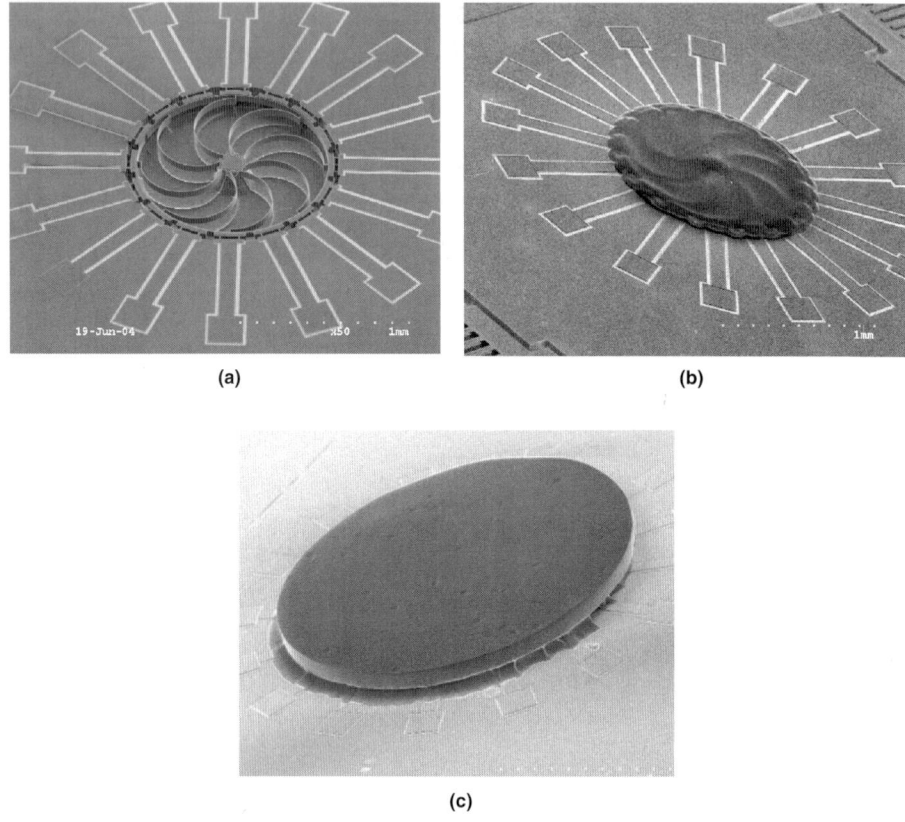

Figure 9.17 (a) Packaged HARPPS gyroscope before packaging, (b) after dispensing Unity, and (c) after Avatrel packaging. [54]

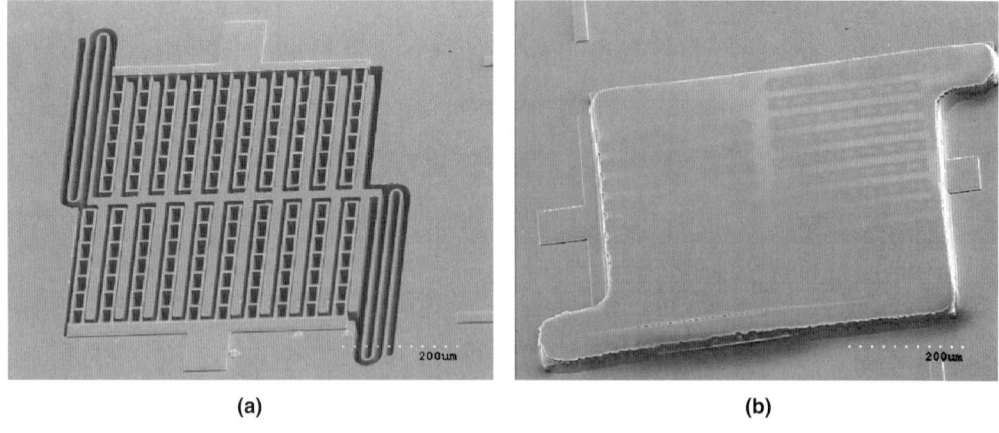

Figure 9.18 HARPSS tunable capacitor (a) before packaging and (b) after packaging using 15-μm Avatrel and 1-μm gold. [56]

MEMS Packaging

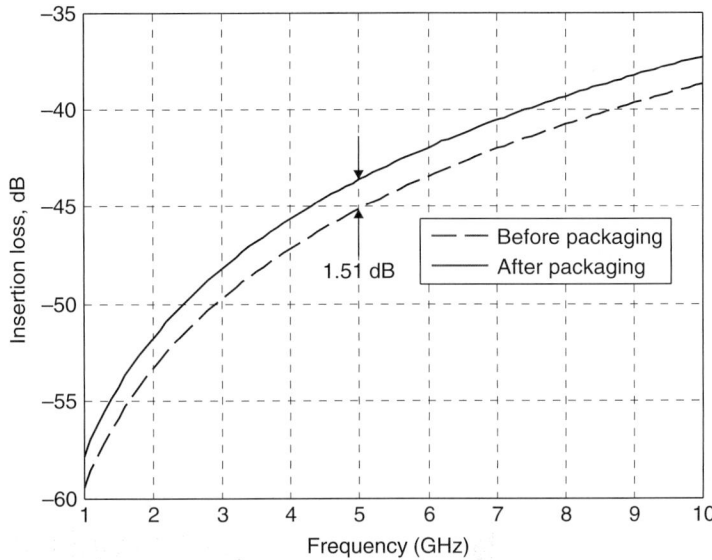

FIGURE 9.19 Measured insertion loss of the HARPSS varactor before and after metal-organic packaging. [56]

Table 9.2 lists five different metal overcoats, as candidates for metal-organic packaging. Chromium has the closest CTE to silicon and the highest modulus and electrical resistance. Gold, platinum, and titanium have very good corrosion resistance and are biocompatible. In order to evaluate the mechanical strength of the metal-organic package, a lateral force can be applied to the package, to break the cap. The thin-film stress of the sputtered metal film deposited at room temperature is always higher than the electroplated film. As mentioned earlier, both metal and Avatrel caps show tensile thin-film stress.

Hermeticity of the polymer package should be characterized to know the minimum time required to create enough vacuum inside the polymer cap before starting the metal deposition. Figure 9.20a shows the permeation rate of the Avatrel overcoat for different thicknesses and different substrate temperatures. Heating the substrate expedites the evacuation and reduces the onset of oscillations (time required to create enough of a vacuum to start resonance with a high Q). This is because of increasing gas diffusivity.

Metal	Chrome	Gold	Aluminum	Copper	Titanium	Platinum
CTE (ppm/°C)	4.9	14.2	23.1	16.5	8.6	8.8
Modulus (GPa)	279	79	70	140	110	168
Resistivity ($\mu\Omega$ cm)	12.5	2.2	2.65	1.68	4.2	10.6
Poisson's ratio	0.21	0.44	0.35	0.34	0.32	0.38
Chemical stability	Good	Excellent	Bad	Bad	Excellent	Excellent

TABLE 9.2 Comparison of Different Metal Overcoats

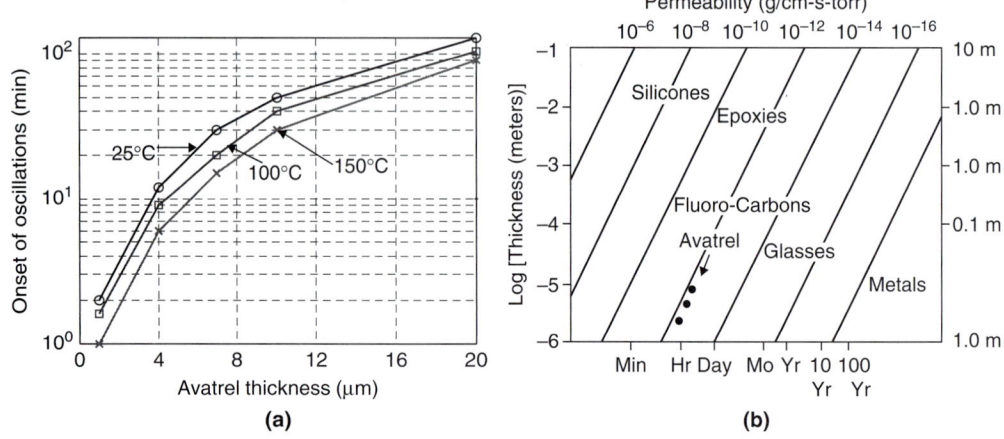

FIGURE 9.20 (a) Plot of time required to create the threshold vacuum for resonators versus Avatrel thickness. (b) Permeation rate of known materials; the dotted line is the Avatrel permeability. [54]

The gas permeability of a material, P_G, is defined as the gas flow rate, F_G, multiplied by the cap thickness, t_{cap}, divided by the cap area, A_{cap}, and the pressure difference, p, as in Equation (9.1) with barrer as the unit [54]:

$$P_G = \frac{F_G t_{cap}}{A_{cap} p} \qquad (9.1)$$

F_G can be approximated in the first order to be inversely proportional to the evacuation time. Using ρ = 1.2506 kg/m³ as the nitrogen volume density, t_{cap} = 10 μm, and A_{cap} = 0.002 mm², Equation (9.1) is solved as P_G = 1.29 × 10⁻¹³ g/(cm · s · torr) (10 barrer). This permeation rate is plotted in the permeability curve of known materials in Figure 9.20b [11]. The Avatrel curve lies within that of the fluorocarbons, with a permeability of about 400 times higher than LCP.

9.6 Low-Loss Polymer Encapsulation Techniques

In these methods, a low-loss polymer is used to seal the MEMS device without wafer bonding a cap or using sacrificial material removal.

The first process is shown in Figure 9.21. The boundaries of a MEMS device are protected by a glass microcap with a cavity and then a semihermetic thermoplastic, such as LCP is dispensed on top of the cap, as depicted in Figure 9.21 [57]. This is similar to the glob-top epoxy sealing for microelectronics. Figure 9.21b and c show an array of varactors before and after LCP encapsulation, respectively. In the second process, LCP is laminated onto silicon to form an enclosure for packaging of RF MEMS switch [58].

LCP has a small dielectric constant (2.49) and loss tangent (0.002) and is attractive for RF MEMS packaging. Also LCP has near-hermetic properties to achieve a low nitrogen permeability of 0.027 barrer (for 2-mil-thick LCP) and small moisture absorption [57–58].

Another process used by Lockheed Martin is shown in Figure 9.22 [59]. The MEMS device is flipped and bonded into a Kapton wafer using epoxy as the intermediate layer, followed by a polyimide coating at the backside. Finally, interconnects are made in the

MEMS Packaging   515

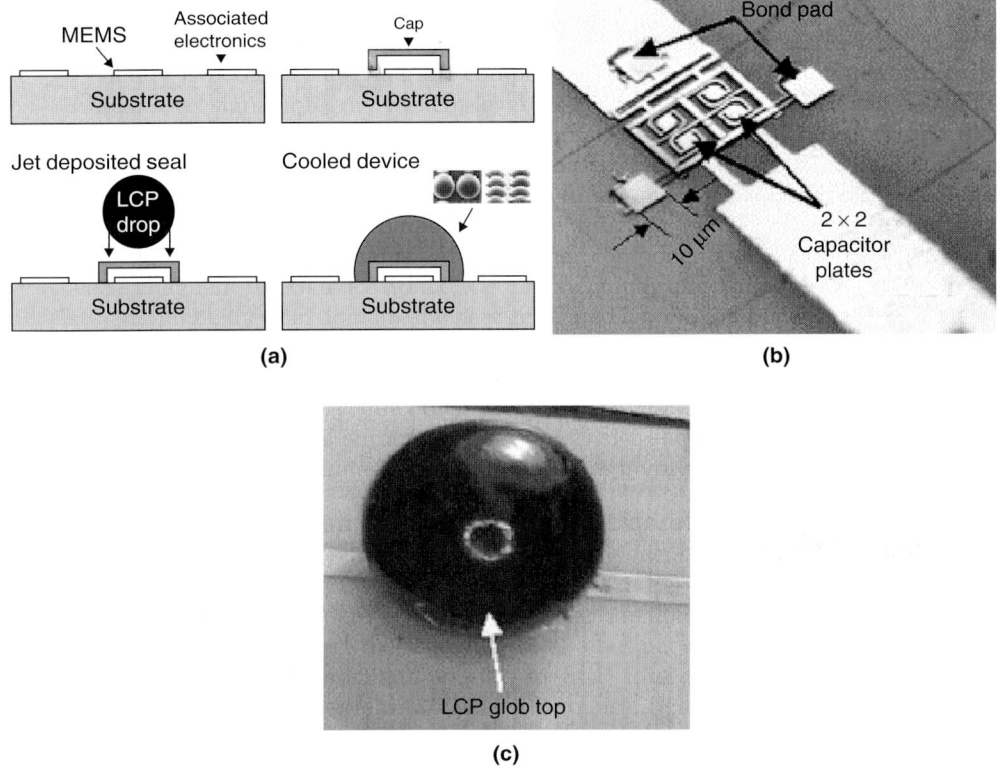

FIGURE 9.21 (a) Process flow for LCP encapsulation. (b) RF varactor before packaging. (c) RF varactor after LCP encapsulation. [57]

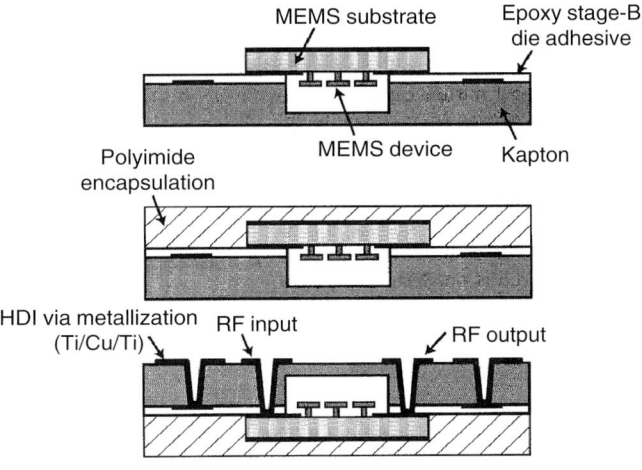

FIGURE 9.22 Polyimide encapsulation method by Lockheed Martin. [59]

Kapton cap by patterning and etching. Kapton has a dielectric constant of 3.4 and loss tangent of 0.0018. This method is not hermetic due to the finite moisture absorption of Kapton.

9.7 Techniques Utilizing Getters

One problem encountered with traditional wafer-to-wafer vacuum bonding is relatively high cavity pressures that change with time and temperature. Anodic bonding is known to generate oxygen and result in cavity pressures of 100 to 400 torr (13 to 53 kPa) [43]. Solder bonding produces cavity pressures around 2 torr due to surface desorption of gases. Baking wafers prior to solder reflow does reduce the amount of adsorbed water but only lowers the microcavity pressure from 2 to 1 torr (133 Pa) [22,60]. While the pressure of the vacuum wafer-bonding system can get down to the microtorr level, ultimately surface desorption after sealing limits the cavity pressure.

While baking can succeed in desorbing gases from surfaces [60,63], temperature and time limit the effectiveness of this method in obtaining ultralow cavity pressures in microsystems. Shallow CMOS junctions, thin-film alloying, and cantilever warpage limit the prebonding bake temperature and time. For solder and eutectic bond approaches, baking can cause metal interdiffusion of the seal layer, resulting in oxidized surfaces as metals like Cr or Ni diffuse through the top nobel metal layer. This can lead to poor sealing in these metal-based bonding processes.

9.7.1 Nonevaporable Getters

To overcome the surface desorption limits on cavity pressure found with traditional wafer bonding, getters have been employed. Metallic getters have been used for decades dating back to vacuum tubes to obtain lower pressures in hermetic packages [64]. Pure metals and alloys of Ba, Al, Ti, Zr, V, Fe, and other reactive metals are used in cathode-ray tubes, flat-panel displays, particle accelerators, semiconductor processing equipment, and other vacuum equipment to lower the pressure [65–67]. These metals trap various gases through oxide and hydride formation and by simple surface adsorption. The capture of oxygen, nitrogen, and hydrocarbons requires elevated temperatures (200 to 550°C), while trapping of hydrogen by the metals occurs at room temperature. Esashi and others first applied getters to MEMS devices in the mid-1990s [43,68]. In these early studies, nonevaporable getters (NEGs) either in tablet or strip form were placed in an extra micromachined cavity or adjacent to the chip in a ceramic package. To maximize the surface area, the NEG is often fabricated using powder metallurgy techniques in which the sintering of the metal particles is just initiated, leaving gaps between metal beads. A high-temperature activation step in a vacuum or a hydrogen-containing reducing ambient is required to remove the surface oxide layer that forms on the metal particles during the sintering process. This activation step is accomplished either through annealing the whole package or by joule heating of the NEG strip. One problem encountered with sintered getters is particle generation. When NEG metal strips are employed, they are typically cut into small segments and hand-placed into a microcavity prior to wafer bonding. The strips often bend during cutting, requiring additional manual handling to straighten the pieces. Particles are generated during handling and the cutting process. The 2- to 3-μm-diameter metal particles can cause electrical shorts, impede motion, and shift resonant frequencies. A frequency shift occurs due to a mass change in the resonator caused by the attached particle. Figure 9.23 shows an example of one way that an NEG can be integrated into a microsystem. In this illustration the

MEMS Packaging 517

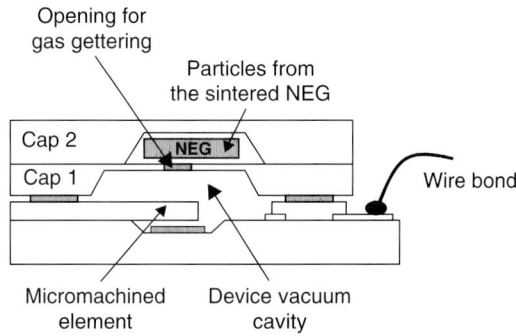

FIGURE 9.23 A cross-sectional illustration of one of the previous methods of integrating NEGs into a vacuum microsystem. [16,70]

NEG is located in a second cavity above the micromachine. An opening in the silicon diaphragm separating the NEG from the resonator provides access between the NEG and resonator chamber. Particles that shed from the sintered NEG can also migrate through this opening to the resonating or tunneling element cavity.

Figure 9.24 shows a photomicrograph of a pressure sensor in which a glass cavity underneath a pressure sensor contains the sintered getter strip [70]. For side-by-side cavity designs the die size area is essentially doubled, while for a vertical integration, such as is illustrated in Figure 9.24, the chip thickness is increased. The size penalty and need for pick-and-place NEG loading also prevents the NEG method from finding use in high-volume MEMS products.

9.7.2 Thin-Film Getters

Because of the particle and handling problems associated with NEGs that were mentioned in the Section 9.7.1, a new vacuum-packaging method and getter were developed. The MEMS integrated, wafer bond packaging process for this new approach is illustrated in Figure 9.25 [70]. A capping wafer, generally made of either silicon or

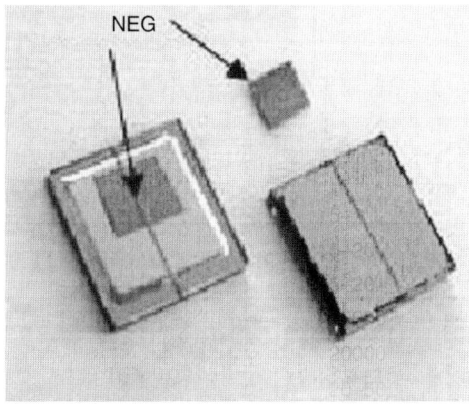

FIGURE 9.24 Anodically bonded, silicon-on-glass pressure sensor, vacuum sealed with an NEG. [16,70]

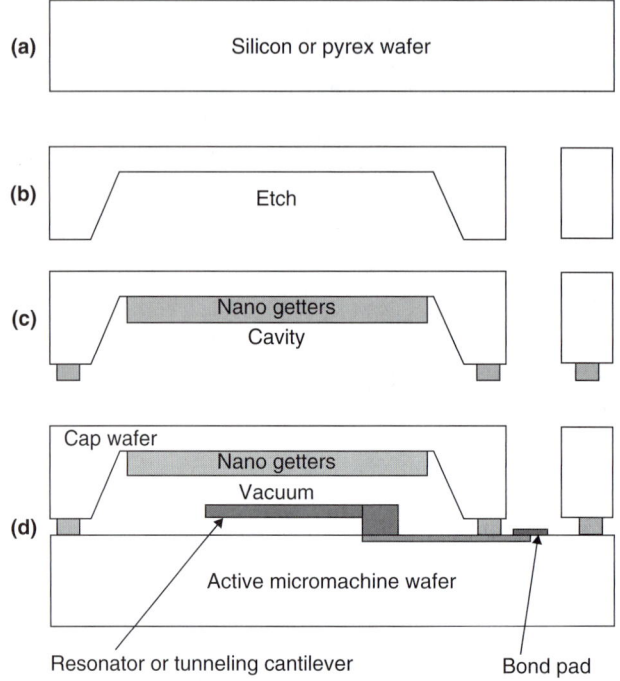

FIGURE 9.25 The process flow for bonding, including the addition of the thin film used to form the chip-level vacuum package. [16,70]

glass, is patterned and etched to form both a cavity that will enclose the active micromachined device and open up access to the electrical bond pads. Next the getter and the sealing material are placed on the cap wafer and activated. Since thin-film deposition techniques are employed in a clean-room environment, the getter called a NanoGetter, is virtually particle-free compared to conventional NEGs formed using powder metallurgy.

A surface texture comparison between the NEG and a NanoGetter film at 200× is shown in Figure 9.26. The thin-film deposition method also enhances the ability to easily integrate the getter into a typical MEMS process flow at the wafer level. For a conventionally wafer-bonded device the process shown in Figure 9.25 would look the same minus the NanoGetter deposition step. Adding the thin-film getter does not impact the chip size. Vacuum wafer-to-wafer bonding is performed next in this process as illustrated at the bottom of Figure 9.25. Other thin-film getter processes are also under development [71].

Through wafer-to-wafer bonding and the use of NanoGetters, a vacuum level under 850 μtorr, resulting in Q values greater than 60,000 for silicon resonators can be obtained. To determine the pressure level corresponding to the measured Q, a resonator was decapped and tested in a vacuum chamber. Figure 9.27 shows how the Q varied with pressure for the relatively wide, U-shaped resonator. With a turbo pump, the chamber pressure could only be pumped down to 790 μtorr (0.10 Pa), which resulted in a Q of 10,350 for the die used to generate the data in Figure 9.27.

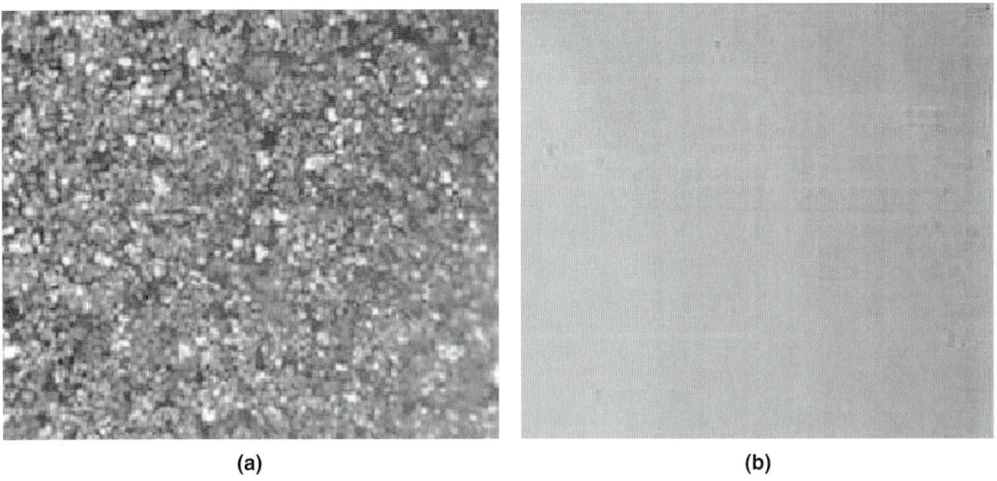

FIGURE 9.26 A pair of 200× optical photomicrographs of an NEG surface (left) and a thin-film NanoGetter surface (right). [70]

A Q value of 40 is obtained for this wide resonator when vacuum-packaged without a getter. The microcavity pressure has been reduced by more than three orders of magnitude through the use of the getter. This getter technology has been applied to chip-scale vacuum packaging of commercially available microfluidic density and chemical concentration meters [72] and micromachined Coriolis mass flow sensors [17] of different flow rate ranges and hence resonator sizes.

To prove the efficacy of thin-film getters a top cap wafer was partially covered with foil during deposition. At the wafer test, it was observed that none of the 18 chips that lacked NanoGetter material had a measurable Q value ($Q < 50$). The chips with the thin-film getter on the same wafer stack had Q values as high as 6760. This controlled experiment with the same wafer proved the effectiveness of the getters in lowering cavity pressure.

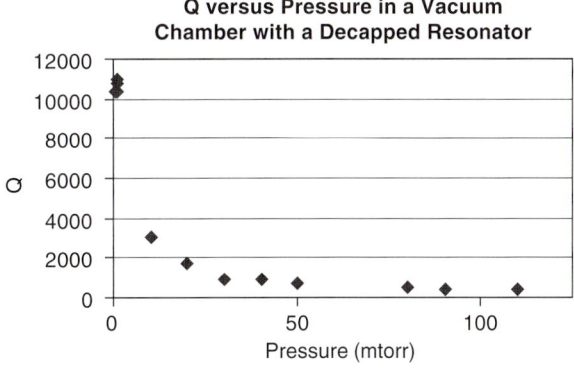

FIGURE 9.27 Plot of Q versus pressure for vertical single-crystal silicon resonators using a vacuum chamber. [16,70]

520 Chapter Nine

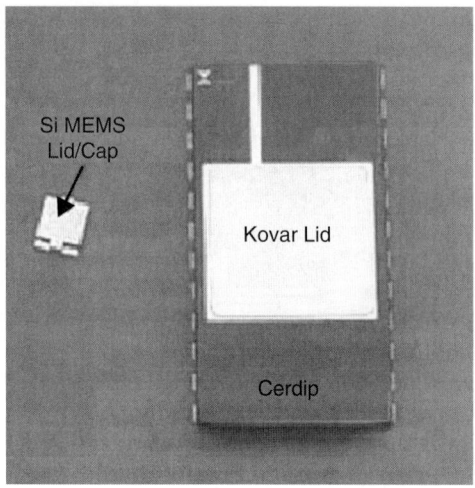

FIGURE 9.28 Chip-level and side-brazed ceramic package using NanoGetters deposited on the underside of the lid. [16,70]

The deposition of NanoGetters is not limited to MEMS chips. Both MEMS die and conventional metal and ceramic vacuum packages, as shown in Figure 9.28, can incorporate a patterned NanoGetter on either the lid or other surfaces.

9.7.3 Improving MEMS Reliability through Getters

This new approach to obtaining an ultrahigh vacuum in a microcavity has the potential to see widespread use in the field of MEMS. The performance of so many resonating devices can be improved through higher Q and gain. High-volume devices like motion and pressure sensors as well as RF MEMS chips can benefit from this technology. In addition to improved device performance, the long-term reliability of resonators, tunneling devices, and pressure sensors can be improved with lower cavity pressures. In automotive applications, sensors and actuators are expected to maintain tight specifications for 5 to 10 years operating at temperatures between −40°C and 85 to 150°C. Long-term Q and sensitivity degradations, which were partially reversible and traceable to adsorbed gases, have been observed with MEMS gyroscopes [22]. Millions of pressure sensors are produced each year for a variety of applications, and they are also prone to temperature hysteresis due to packaging stress and potentially from reversible microcavity gas desorption. Figure 9.29 shows how the NanoGetter could be applied to improve the long-term temperature performance of pressure sensors.

To investigate the reliability of the thin-film getter system a set of long-term tests have been conducted [28,70]. In one test a set of single-crystal silicon resonators using silicon to Pyrex glass sealing were generated. To minimize both the design and packaging sources of Q variation, the same resonator design from the same wafer were used. The Q and frequency values were measured using a probe station at the chip level. No adhesives were used to avoid clamping loss changes due to adhesive aging effects. Since the probe station was employed, the same electronic amplifier was used with

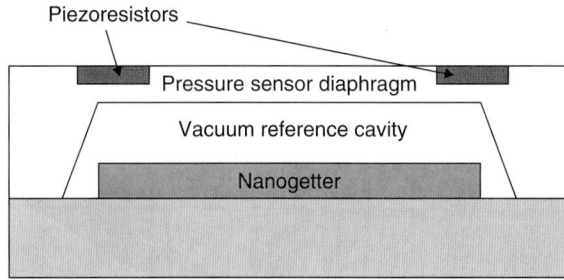

FIGURE 9.29 The application of NanoGetters to a piezoresistive pressure sensor. [16,70]

each MEMS chip reducing another source of variation from the study. The parts were soaked at 95°C, but all measurements were made at room temperature. Figure 9.30 shows how the Q and frequency varied with time in these parts.

The minor difference in frequency can be attributed to variations in room temperature, which would be approximately 3 Hz between the temperatures of 21°C and 25°C. No significant change in either Q or frequency was observed in the parts.

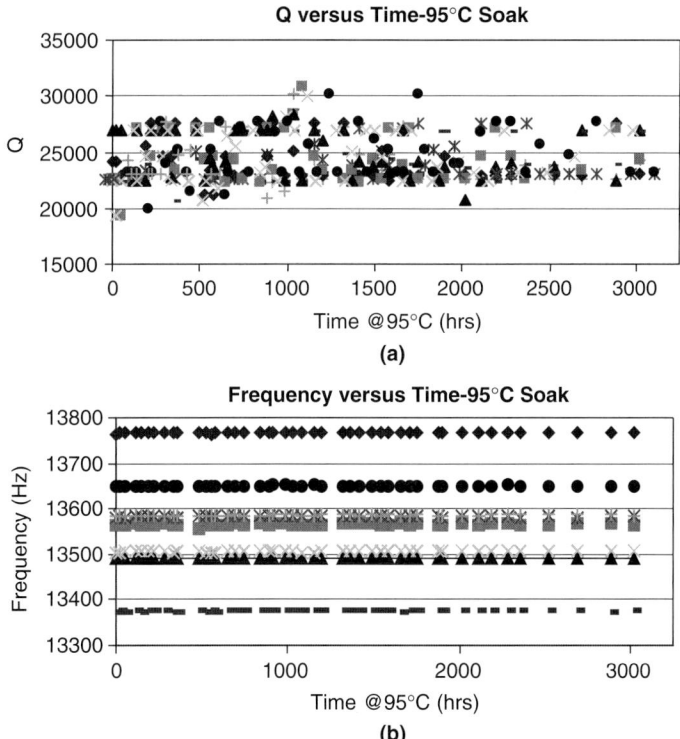

FIGURE 9.30 The room temperature Q and frequency variation observed in the third high-temperature storage test. [28]

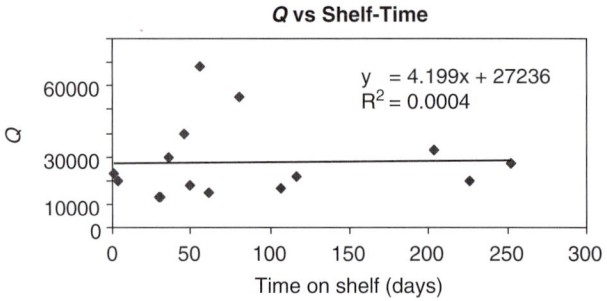

FIGURE 9.31 Getter shelf-life testing using Q values. [28]

The Q variation is due to the resolution of the Q test method using the 3-dB peak width measurement technique. The average Q value at the start was 23,468 with a standard deviation of 1567. At the end of 3000 hours at 95°C the average room temperature Q was 24,711 with a standard deviation of 1923. There is no statistical difference in average Q value after the 3000-hour high-temperature soak.

Shelf life and storage conditions prior to the packaging of any getter are important considerations. Some getters required vacuum canister packaging or nitrogen cabinet storage prior to activation and use. To gauge the shelf life of the thin-film NanoGetter formulation used in this study, historical flow card records and wafer-level test results were reviewed for a number of wafers produced over 2 years. The time between when the thin-film getter was applied to the capping wafer and the time that the wafer stack was vacuum wafer bonded together was compared with the average Q value obtained at test. The wafers were stored in a plastic wafer box in a clean room, no nitrogen purging was employed during storage, and the wafers were stored in air at room temperature. Figure 9.31 shows how the Q value varied as a function of getter wafer storage time. No trend was found versus time.

9.8 Interconnections

MEMS packages are designed to provide a proper electrical, optical, fluidic, and chemical interface to the outside world. The interconnections may include filters to selectively pass desired signals and add new functionality to the MEMS package. For instance, pressure sensors from Infineon Technologies are mounted into a cavity filled with silicone oil or flexible gel [73]. The cavity is hermetically sealed by a membrane. The membrane acts like a filter to filter out moisture, particles, and toxic gases. The media pressure is transferred via silicone oil to the sense element. Another example of filters is the Texas Instruments DMD, in which the lid of the mirror is transparent to desired wavelengths [14]. The electrical interconnect in a MEMS package should have low resistive, capacitive, and inductive parasitics. Low-resistance polysilicon through-wafer interconnects have been used to enable access to the MEMS [74]. Interconnects are one of the main mechanisms for packaging failure. In general, electrical interconnects can be divided into vertical and horizontal feedthroughs.

Horizontal feedthroughs are the easiest and most common interconnections in MEMS packaging, suitable for wire bonding. This kind of feedthrough has been used in many MEMS products including sensors and RF switches [29–30]. All glass frit bonding methods use horizontal interconnects, and examples include accelerometers and gyroscopes from Motorola [75] and Bosch, as shown in Figure 9.32a [76]. In the metallic bonding methods, an insulator is coated on the MEMS lateral feedthrough to provide insulation between the pads. Polysilicon feedthroughs have been used in the glass-bonded packages [77]. The dielectric and feedthrough can be buried under the bond seal ring into the MEMS substrate [78]. Another approach includes making interconnects in a polydiamond panel that can house the MEMS chip [79].

Vertical feedthroughs are proper for flip-chip assembly. This kind of feedthrough can be created inside the cap, for example, through-wafer vias, or in the host MEMS wafer (Figure 9.32b), or in a separate panel that houses the MEMS chip followed by depositing metals over through-wafer vias to get access from the backside. Horizontal feedthrough with increased bonding area adds RF loss (due to parasitic coupling, especially for metal-bonded caps). RF MEMS packaging using vertical RF transitions provide a lower-loss interconnect. Since the pads are at the surface of the packaged MEMS, the die can be surface-mounted using solder balls [80–81]. Also large openings can be created in the cap to form pad holes for direct access to the bonding pads.

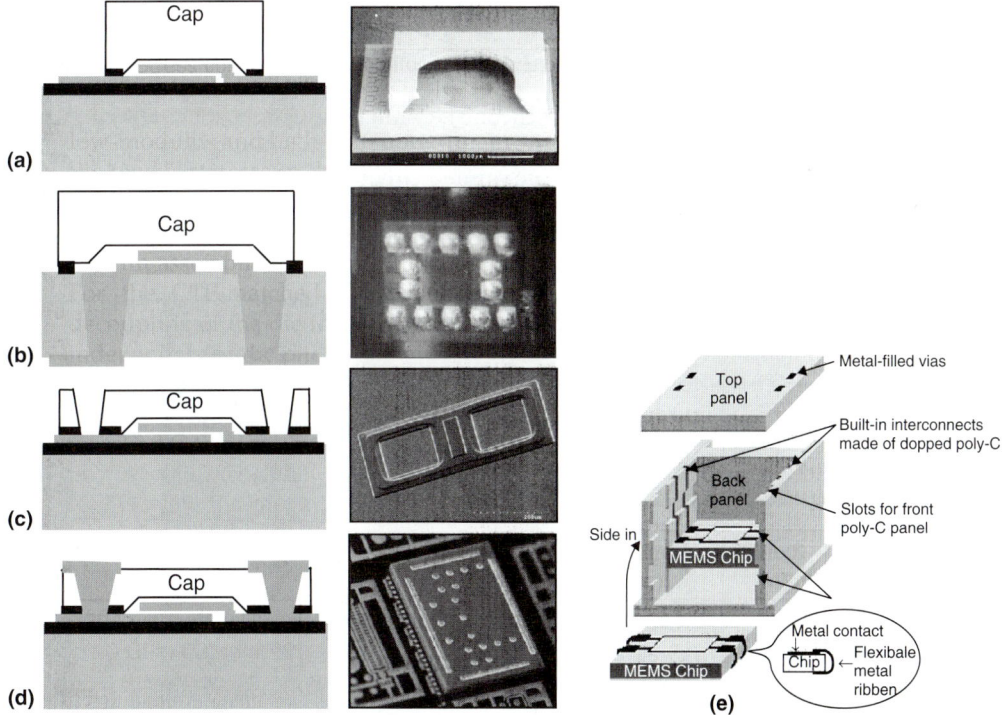

FIGURE 9.32 Classification of MEMS electrical interconnections: (*a*) Horizontal feedthrough in Bosch gyroscope [76]. (*b*) Vertical feedthrough (or transitions) in the substrate [80]. (*c*) Vertical pad-holes in a beam resonator [56]. (*d*) Vertical feedthrough in Intel's ceramic cap [83]. (*e*) Polydiamond panel [79].

Pad holes are efficient for surface micromachined thin-film caps (Figure 9.32c) [53–56]. Another reported MEMS packaging with pad holes is by Agilent [82]. A microcap with through-holes is bonded on top of a MEMS wafer using gaskets surrounding the MEMS bonding pads. Then the cap wafer is thinned until the bonding pads are opened up. Recently, via-hole vertical interconnects made in the ceramic cap have been reported by Intel, as shown in Figure 9.32d [83]. All five categories are schematically shown in Figure 9.32.

9.9 Assembly

Chip-scale packaged MEMS devices are generally mounted to system-level circuit boards or directly attached to other system components. Wire bonding is the most widely used technique to connect packaged MEMS to the printed circuit board (PCB). Some tools have been developed for automated microassembly of MEMS including wire-bonding, die-bonding, and flip-chip bonding machines [84–85]. Figure 9.33 shows an example of an accelerometer, which is mounted to a ceramic board used to control automotive airbag firing signals [61]. This MEMS device is wire-bonded to the circuit board, while other components are soldered to the circuit board. Two wafer-level packaged accelerometers assembled into standard packages from Motorola and Bosch are shown in Figure 9.34. Figure 9.34a is the Bosch accelerometer with silicon cap and the interface circuit in a quad flat package (QFP) [86]. The sensor-ASICs are mounted on a lead frame and interconnections are made by wire bonding. Figure 9.34b is the Motorola accelerometer, packaged by frit glass bonding in a dual-in-line (DIP) plastic package [75]. The sensor and ASIC are mounted to a lead frame using epoxy, wire-bonded, passivated by a thin layer of silicone gel, and followed by final epoxy molding. Another MEMS packaging method by Motorola, called Meso-MEMS, includes direct integration of a switch inside a ceramic printed circuit board, eliminating the need for wire-bonding or flip-chip assembly [87].

FIGURE 9.33 Hybrid ceramic circuit board with a chip-scale packaged MEMS accelerometer [61].

MEMS Packaging

FIGURE 9.34 (*a*) Bosch accelerometer, assembled in a QFP package [86]. (*b*) Motorola accelerometer, assembled in a DIP package [75].

Figure 9.35 shows an example of a microfluidic sensor that is vacuum-packaged at the wafer level. This chip is attached to a stainless-steel fluid component and then wire-bonded to a flex circuit [17,72]. Figure 9.36 gives a general process flow for incorporating sealed MEMS chips into larger system modules.

A variety of assembly options for encapsulated MEMS includes overmolding using epoxies or injection molding using plastic to create a plastic leaded chip carrier package (PLCC), small outline IC (SOIC), system-in-package (SiP), ceramic dual-in-line package (CERDIP), or MicroLead Frame (MLF) [88–89]. The assembly and molding should meet the height requirement of the capped MEMS. The MEMS devices can be arranged in a

FIGURE 9.35 A chip-scale vacuum-packaged microfluidic resonator, directly attached to a stainless-steel fixture and wire-bonded to a flex circuit board. [17,72]

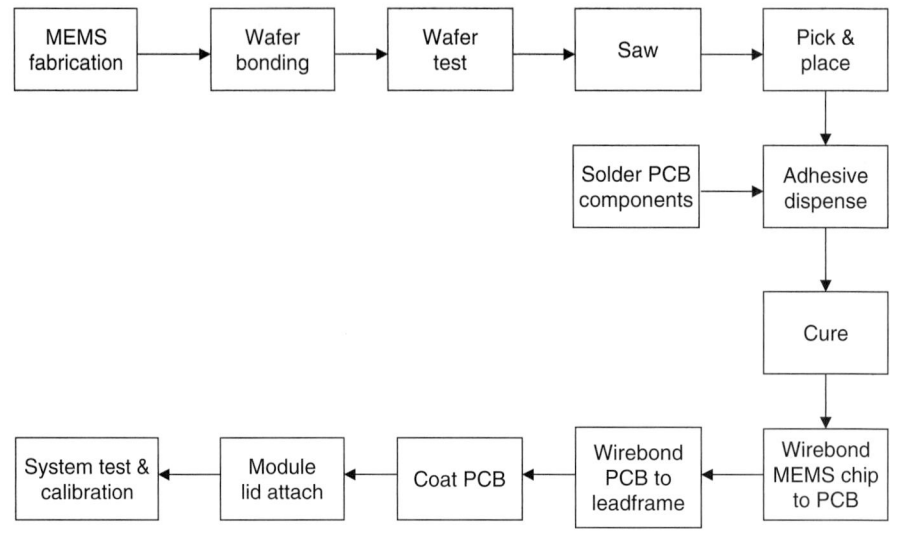

FIGURE 9.36 Process flow for integrating a chip-scale device into a system circuit board.

configuration to be compatible with the lead frame assembly provided by vendors. Shown in Figure 9.37 is a two-dimensional array of polymer encapsulated resonators suitable for plastic molding process. The molding temperature is far below the glass transition temperature of the Avatrel polymer used for packaging.

LCP is a good candidate for near polymer packaging of MEMS intended for lead frame (or leadless) assembly. LCP can be injection-molded to mass-produce packages with fair moisture barrier properties. Plastic, metal, ceramic, or glass molds can be applied on LCP with minimum side effects.

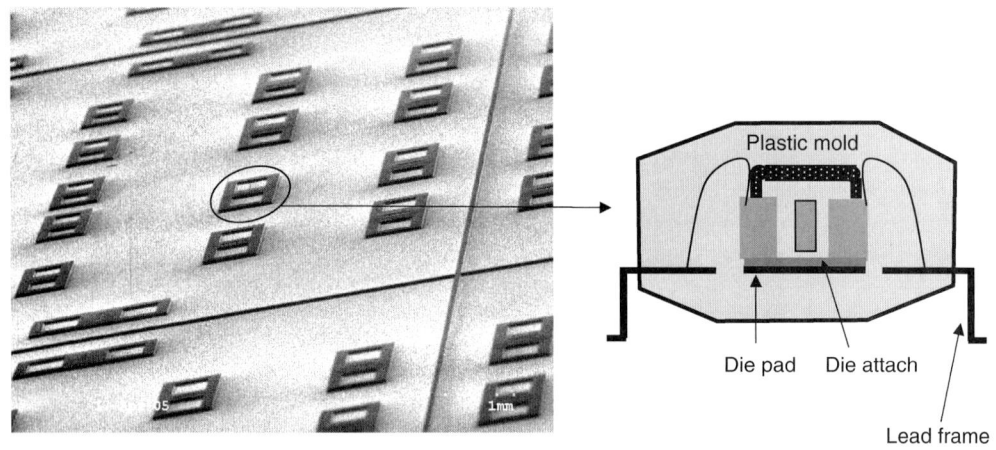

FIGURE 9.37 Array of wafer-level packaged resonators arranged for plastic molding and lead frame assembly [54].

9.10 Summary and Future Trends

Wafer-level packaging of MEMS represents a challenging and costly task in microsystem manufacturing. Various packaging techniques have been proposed and employed by the industry and academia, but more research is needed to address the reliability issues. Hermeticity and stress in the package and their behavior with temperature cycling, the environment and time are important factors in determining reliability. Package reliability depends on a variety of parameters ranging from materials, processes, interconnections, and environment. For example, implantable biomedical devices must survive a harsh environment inside the body with the highest degree of reliability. Some other devices need to survive high levels of shock and vibration.

In order to determine how various factors affect the package, certain parameters (depending on the application) should be monitored for a specific period of time. Accelerated lifetime estimation techniques can be used to predict the lifetime of the package under certain conditions. For implantable MEMS, chemical survivability can be checked by inserting the packaged MEMS in different media to check the performance and verify that no material is transferred through the cap. For vacuum-packaged MEMS, the vacuum level can be monitored over time with temperature cycling. An accurate measurement of vacuum level can be performed using Pirani gauges. Other techniques include the helium leak rate test (MIL883), and monitoring the Q and frequency of high-Q micromechanical resonators.

Besides hermeticity, other issues such as residual stress are also related to the packaging process parameters. Package stress can induce offset and scale factor shifts. Moreover, the die attach material should have a low modulus to reduce stress, with very low outgassing. There are very few reports on the detail characterization of MEMS packages covering all aspects of mechanical, thermal, chemical stability, and aging.

The following issues should be addressed in future developments of MEMS packages:

1. Standardization and cost reduction for reduced time-to-market; elimination of high-temperature processing steps, plasma etching, and deposition tools will have the most effect in cost reduction.
2. Proper codesign of MEMS, interconnects, and package to reduce electrical loss. This is especially critical for packaged RF MEMS devices including switches, filters, and tunable passive circuits.
3. Reduction of the package profile (i.e., area and thickness). With the rapid progress of nanostructures, reducing the package profile will be of interest, for which sacrificial-based thin-film packages have great potential.
4. Long-term reliability evaluation of the package, shear stress characterization, and hermeticity analysis and evaluation. Reliability may depend on the package design, materials, and the residual atmosphere inside the package.
5. For MEMS that need semihermetic packaging, standard IC assembly processes can be used. This requires physical protection of MEMS prior to assembly, which is possible through low-cost polymer encapsulation techniques.
6. Thin-film getters can provide long-term reliability and vacuum stability to hermetic MEMS packages.
7. Mechanical and chemical resistance of the package to environmental excitations including thermal shock, physical impact, vibration, and chemical agents are

important. This requires careful design of the bonded cap and the final overcoat layer in the sacrificial-based methods. For thin-film packages, gold and platinum are good candidates for high survivability. For the bonded capsules, glass is an environment resistant material.

References

1. O. Brand, G. K. Fedder, C. Hierold, J. G. Korvink, and O. Tabata, *Enabling Technologies for MEMS and Nanodevices,* Weinheim: Wiley-VCH, 2004.
2. G. Rebeiz, *RF MEMS Theory, Design, and Technology,* Hoboken: John Wiley & Sons, 2003.
3. H. De Los Santos, *Introduction to Wireless MEMS,* Norwood: Artech House, 2004.
4. M. E. Motamedi, "MOEMS," Bellingham: SPIE Press, 2005.
5. G. A. Urban, *BioMEMS,* New York: Springer, 2006.
6. M. Ferrari, *BioMEMS and Biomedical Nanotechnology,* New York: Springer, 2006.
7. H. Baltes, O. Brand, G. K. Fedder, C. Hierold, J. G. Korvink, and O. Tabata, *CMOS MEMS* Weinheim: Wiley-VCH, 2004.
8. M. A. Lemkin, M. A. Ortiz, N. Wongkomet, B. E. Boser, and J. H. Smith, "A 3-axis surface micromachined $\Sigma\Delta$ accelerometer," *Proc. Int. Solid State Circuits Conference (ISSCC'97),* 1997, pp. 202–203.
9. A. E. Franke, J. M. Heck, T. J. King, and R. T. Howe, "Polycrystalline silicon germanium films for integrated microsystems," *J. Microelectromech. Systems,* vol. 12, no. 2, 2003, pp. 160–171.
10. J. T. Kung, "Methods for planarization and encapsulation of micromechanical devices in semiconductor processes," US Patent No. 5,504,026, issued April 1996.
11. R. Tummala, *Fundamentals of Microsystem Packaging,* New York: McGraw-Hill, 2002.
12. K. Najafi, "Micropackaging technologies for integrated microsystems: applications to MEMS and MOEMS," *Proc. SPIE Micromachining and Microfabrication Process Technology VIII,* vol. 4979, 2003, pp. 1–19.
13. F. Goodenough, "Airbags boom when IC accelerometer sees 50g," *Electronic Design,* 1991, p. 45.
14. J. Faris, and T. Kocian, "DMD™ packaging evolution and strategy," *Proc. Intl. Symposium on Microelectronics (IMAPS '98),* 1998, pp. 108–13.
15. Y. Tao, A. P. Mishe, W. D. Brown, D. R. Dereus, and S. Cunningham, "Laser-assisted sealing and testing for ceramic packaging of MEMS devices," *IEEE Trans. Advanced Packaging,* vol. 26, no. 3, 2003, pp. 283–88.
16. D. Sparks, S. Massoud-Ansari, and N. Najafi, "Chip-level vacuum packaging of micromachines using NanoGetters," *IEEE Trans. Adv. Packaging,* vol. 26, no. 3, 2003, pp. 277–82.
17. D. Sparks, R. Smith, S. Massoud-Ansari, and N. Najafi, "Coriolis mass flow, density and temperature sensing with a single vacuum sealed MEMS chip," *Dig. Solid-State Sensors and Actuator, and Microsystem Workshop,* 2004, p. 75.
18. M. Schmidt, "Wafer-to-wafer bonding for microstructure formation," *Proc. of the IEEE,* vol. 86, no. 8, 1998, pp. 1575.
19. Y. T. Cheng, L. Lin, and K. Najafi, "Localized silicon fusion and eutectic bonding for MEMS fabrication and packaging," *J. Microelectromech. Systems,* 2000, pp. 3–8.

20. B. Lee, S. Seok, and K. Chun, "A study on wafer-level vacuum packaging for MEMS devices," *J. Micromech & Microeng.*, vol. 13, 2003, pp. 663–69.
21. K. Schjolberg, G. U. Jensen, A. Hanneborg, and H. Jakobsen, "Anodic bonding for monolithically integrated MEMS," *Sensors and Actuators A*, vol. 114, 2004, pp. 332–39.
22. D. Sparks, G. Queen, R. Weston, G. Woodward, M. Putty, L. Jordan, S. Zarabadi, and K. Jayakar, "Wafer-to-wafer bonding of nonplanarized MEMS surfaces using solder," *J. Micromech. & Microengr.*, vol. 11, no. 6, 2001, pp. 630–34.
23. J. Mitchel, R. Lahiji, and K. Najafi, "Encapsulation of vacuum sensors in a wafer-level package using a gold-silicon eutectic," *Dig. IEEE Transducers'05*, 2005, p. 86.
24. C. H. Tsau, S. M. Spearing, and M. A. Schmidt, "Fabrication of wafer-level thermo-compression bonds," *J. Electrochem. Soc*, vol. 11, 2002, pp. 641–47.
25. J. Oberhammer, and G. Stemme, "Incrementally etched electrical feedthroughs for wafer-level transfer of glass lid packages," *Dig. IEEE Transducers'03*, 2003, pp. 1832–35.
26. A. Jourdain, P. De Moor, K. Baert, I. DeWolf, and H. A. C. Tilmans, "Mechanical and electrical characterization of BCB as a bond and seal material for cavities housing RF MEMS devices," *J. Micromech. & Microeng.*, vol. 15, 2005, pp. 1560–64.
27. C. T. Pan, H. Yang, S. C. Shen, M. C. Chou, and H. P. Chou, "A low-temperature wafer bonding technique using patternable materials," *J. Micromech. & Microeng.*, vol. 12, 2002, pp. 611–15.
28. D. Sparks, S. Massoud-Ansari, and N. Najafi, "Long-term evaluation of hermetically glass frit sealed silicon to Pyrex wafers with feedthroughs," *J. Micromech. & Microengr.*, vol. 15, 2005, pp. 1560–64.
29. K. P. Harney, "Standard semiconductor packaging for high reliability low cost MEMS applications," *Reliability, Testing and Characterization of MEMS/MOEMS III*, SPIE, vol. 5716, 2005, p. 1.
30. S. Majumdar, J. Lampen, R. Morrison, and J. Maciel, "MEMS switches," *IEEE Instrumentation and Measurement Magazine*, vol. 6, no. 1, 2003, pp. 12–15.
31. L. Lin, "MEMS post-packaging by localized heating and bonding," *IEEE Trans. Advanced Packaging*, vol. 23, 2000, pp. 608–16.
32. G. H. He, L. Lin, and Y. T. Cheng, "Localized CVD bonding for MEMS packaging," *Dig. IEEE Transducers'99*, 1999, pp. 1312–15.
33. A. Bayrashev, and B. Ziaie, "Silicon wafer bonding through RF dielectric heating," *Proc. Sensors and Actuators A*, vol. 103, no. 3, 2003, pp. 16–22.
34. J. B. Kim, M. Chiao, and L. Lin, "Ultrasonic bonding of In/Au and Al/Al for hermetic sealing of MEMS," *Proc. IEEE Microelectromech. Systems Conf. (MEMS'02)*, 2002, pp. 415–18.
35. Y. T. Cheng, L. Lin, and K. Najafi, "Localized silicon fusion and eutectic bonding for MEMS fabrication and packaging," *J. Microelectromech. Systems*, vol. 9, 2000, pp. 3–8.
36. Y. C. Su, and L. Lin, "Localized bonding processes for assembly and packaging of polymeric MEMS," *IEEE Trans. Advanced Packaging*, vol. 11, 2005, pp. 635–42.
37. Y. T. Cheng, W. T. Hsu, and K. Najafi, C. T. C. Nguyen, and L. Lin, "Vacuum packaging technology using localized aluminum/silicon-to-glass bonding," *J. Microelectromech. Systems*, vol. 11, 2002, pp. 556–65.
38. M. Chiao and L. Lin, "Wafer-level vacuum packaging process by RTP aluminum-to-nitride bonding," *Tech Dig. Solid-State Sensors and Actuators Workshop*, 2002, pp. 81–85.

39. Y. M. Johnson Chiang, M. Bachman, and G. P. Li, "A wafer-level microcap array to enable high-yield microsystem packaging," *IEEE Trans. Advanced Packaging*, vol. 27, no. 3, 2004, pp. 490–500.
40. J. Heck and S. Greathouse, "Towards wafer-scale MEMS packaging: a review of recent advances," *Proc. Surface Mount Technology Association International Conference*, 2003, pp. 631–36.
41. M. B. Cohn, Y. Liang, Y., R. T. Howe, and A. P. Pisano, "Wafer-to-wafer transfer of microstructures for vacuum packaging," *Tech. Dig. Solid-State Sensor and Actuator Workshop*, Hilton Head, 1996, pp. 32–35.
42. W. C. Welch and K. Najafi, "Transfer of metal MEMS packages using a wafer-level solder sacrificial layer," *Proc. IEEE Microelectromech. Systems Conf. (MEMS'05)*, 2005, pp. 584–87.
43. M. Esashi, S. Sugiyama, K. Ikeda, Y. Wang, and H. Miyashita, "Vacuum-sealed silicon micromachined pressure sensors," *Proc. of the IEEE*, vol. 86, 1998, pp. 1627–39.
44. L. C. Chomas, Y. N. Hsu, S. Friends, D. Volfson, R. Morrison, H. M. Lakdawala, R. S. Sinha, D. F. Guillou, S. Santhanam, and L. R. Carley, "Low-cost manufacturing/packaging process for MEMS inertial sensors," *Proc. Intl. Symposium on Microelectronics (IMAPS 2003)*, 2003, pp. 398–401.
45. R. N. Candler, W. T. Park, H. Li, and G. Yama, A. Partridge, M. Lutz, and T. W. Kenny, "Single wafer encapsulation of MEMS devices," *IEEE Trans. Advanced Packaging*, vol. 26, no. 3, 2003, pp. 227–32.
46. L. Lin, R. T. Howe, and A. P. Pisano, "Microelectromechanical filters for signal processing," *J. Microelectromech. Systems*, vol. 7, no. 3, 1998, pp. 286–94.
47. H. Guckel, C. Rypstat, M. Nesnidal, J. Zook, D. Burns, and D. Arch, "Polysilicon resonant microbeam technology for high performance sensor applications," *Tech. Dig. IEEE Solid-State Sensor and Actuator Workshop*, 1992, p. 153.
48. R. He, and C. J. Kim, "On-wafer monolithic encapsulation by surface micromachining with porous polysilicon shell," *J. Microelectromech. Systems*, vol. 16, 2007, pp. 462–472.
49. J. Knight, J. McLean, and F. L. Degertekin, "Low-temperature fabrication of immersion capacitive micromachined ultrasonic transducers on silicon and dielectric substrates ultrasonics," *IEEE Trans. Ferroelectrics and Frequency Control*, vol. 51, 2004, pp. 1324–33.
50. B. Stark and K. Najafi, "A low-temperature thin-film electroplated metal vacuum package," *J. Electromech. Systems*, vol. 13, no. 2, 2004, pp. 147–57.
51. D. Forehand and C. L. Goldsmith, "Wafer-level micro-encapsulation," *Proc. ASME Conf. Integration and Packaging of MEMS, NEMS, and Electronic Systems (InterPack'05)*, 2005, pp. 320–24.
52. S. Li, L. W. Pan, and L. Lin, "Frozen water for MEMS fabrication and packaging applications," *Proc. IEEE Microelectromech. Systems Conf. (MEMS'03)*, 2003, pp. 650–53.
53. P. Monajemi, P. Joseph, P. A. Kohl, and F. Ayazi, "A low-cost wafer-level MEMS packaging technology," *Proc. IEEE Microelectromech. Systems Conf. (MEMS'05)*, 2005, pp. 634–37.
54. P. Monajemi, P. Joseph, P. A. Kohl, and F. Ayazi, "Wafer-level MEMS packaging via thermally released metal-organic membranes," *J. Micromech. & Microengineering*, vol. 16, 2006, pp. 742–50.
55. P. Joseph, P. Monajemi, F. Ayazi, and P. A. Kohl, "Wafer-level packaging of micromechanical resonators," *IEEE Trans. Advanced Packaging*, vol. 30, no. 1, 2007, pp. 19–26.

56. P. Monajemi, P. Joseph, P. A. Kohl, and F. Ayazi, "Characterization of a polymer-based MEMS packaging technique," *Proc. IEEE Symposium on Advanced Packaging Materials,* 2006, pp. 139–144.
57. F. Faheem, and Y. Lee, "Flip-chip assembly and liquid crystal polymer encapsulation for variable MEMS capacitors," *IEEE Trans. Microwave Theory and Techniques,* vol. 51, no. 12, 2003, pp. 2562–67.
58. M. J. Chen, A. Pham, N.A. Evers, C. Kapusta, J. Iannotti, W. Kornrumpf, J. Maciel, N. Karabudak, "Design and development of a package using LCP for RF/microwave MEMS switches," *IEEE Trans. Microwave Theory & Techniques,* vol. 54, no. 11, 2006, pp. 4009–4015.
59. G. L. Tan, G. M. Rebeiz, R. Mihailovich, J. DeNatale, B. Taft, N. Karabudak, and B. Kornrumpf, "Low loss RF MEMS phase shifters for satellite communication systems," *Proc. AIAA Int. Communications Satellite System Conf.,* 2002, p. 1.
60. D. Sparks, M. Chia, and G. Q. Jiang, "Cyclic fatigue and creep of electroformed micromachines," *Sensors and Actuators A,* vol. 95, 2001, p. 61.
61. D. Sparks, D. Rich, C. Gerhart, and J. Frazee, "A bi-directional accelerometer and flow sensor made using a piezoresistive cantilever," *Proc. European Automotive Engineers Coop. 6th European Congress,* 1997, p. 1119.
62. Y. Tuzi, T. Tanaka, K. Takeuchi, and Y. Saito, "Effect of surface treatment on the adsorption kinetics of water vapor in a vacuum chamber," *Vacuum,* vol. 47, no. 6, 1996, p. 705.
63. Y. Hirohata, K. Suzuki, and T. Hino, "Gas desorption properties of low-activation ferritic steel as a blanket or a vacuum vessel material," *Fusion Engr. & Design,* vol. 39–40, 1998, pp. 485.
64. T. Giorgi, "An updated review of getters and gettering," *J. Vac. Science Technology A.,* vol. 3, 1985, pp. 417–23.
65. P. Manini, and B. Ferrario, "High-capacity getter pump," US Patent No. 5,320,496, issued June 1994.
66. J. Travis and W. Woodward, "Getter strip," US Patent No. 4,977,035, issued December 1990.
67. F. Ito, "Micro vacuum pump for maintaining high degree of vacuum and apparatus including the same," US Patent No. 6,236,156, issued May 2001.
68. H. Henmi, S. Shoji, K. Yoshini, and M. Esashi, "Vacuum packaging for microsensors by glass-silicon anodic bonding," *Sensors and Actuators A,* vol. 43, 1994, p. 24.
69. Y. Zhang, S. Ansari, G. Meng, W. Kim, and N. Najafi, "An ultra-sensitive high vacuum absolute capacitive pressure sensor," *Dig. IEEE Transducers'01,* 2001, p. 166.
70. D. Sparks, S. Massoud-Ansari, and N. Najafi, "Reliable vacuum packaging using NanoGetters™ and glass frit bonding," *Proc. SPIE Reliability, Testing and Characterization of MEMS/MOEMS III,* vol. 5343, 2004, p. 70.
71. M. Moraja, M. Amiotti, and H. Florence, "Chemical treatment of getter films on wafers prior to vacuum packaging," *Proc. SPIE Reliability, Testing and Characterization of MEMS/MOEMS III,* vol. 5343, 2004, p. 87.
72. D. Sparks, R. Smith, M. Straayer, J. Cripe, R. Schneider, A. Chimbayo, S. Ansari, and N. Najafi, "Measurement of density and chemical concentration using a microfluidic chip," *Lab on a Chip,* vol. 3, 2003, pp. 19–21.
73. A. Gotlieb, M. Schroder, "Pressure sensor and process for producing the pressure sensor," US Patent No. 6,732,590, issued May 2004.

74. V. Chandrasekaran, E. M. Chow, T. W. Kenny, T. Nishida, L. N. Cattafesta, B. V. Sankar, and M. Sheplak, "Thermoelastically actuated acoustic proximity sensor with integrated through-wafer interconnects," *Tech. Dig. Solid-State Sensor, Actuator and Microsystems Workshop*, 2002, pp. 102–105.
75. G. Li and A. A. Tseng, "Low stress packaging of a micromachined accelerometer," *Trans. Electronics Packaging Manufacturing*, vol. 24, no. 1, 2001, pp. 18–25.
76. L. Quellet, "Wafer-level MEMS packaging," US Patent No. 6,635,509, issued October 2003.
77. B. Ziaie, J. A. Von Arx, M. R. Dokmeci, and K. Najafi, "A hermetic glass-silicon micropackage with high-density on-chip feedthroughs for sensors and actuators," *J. Microelectromech. Systems*, vol. 5, no. 3, 1996, pp. 166–79.
78. A. Jourdain, S. Brebels, W. De Raedt, and H. A. C. Tilmans, "The influence of 0-level packaging on the performance of RF-MEMS devices," *Proc. IEEE European Microwave Conf.*, vol. 419, 2001, pp. 403–406.
79. X. Zhu, D. M. Aslam, Y. Tang, B. H. Stark, and K. Najafi, "The fabrication of all-diamond packaging panels with built-in interconnects for wireless integrated microsystems," *J. Microelectromech. Systems*, vol. 13, 2004, pp. 396–405.
80. J. Chae, J. M. Giachino, and K. Najafi, "Wafer-level vacuum package with vertical feedthroughs," *Proc. IEEE Microelectromech. Systems Conf. (MEMS'05)*, 2005, pp. 548–51.
81. A. Badihi, "Ultrathin wafer-level chip size package," *IEEE Trans. Advanced Packaging*, vol. 23, no. 2, 2000, pp. 212–14.
82. R. C. Ruby, T. E. Bell, F. S. Geefay, and Y. M. Deasi, "Microcap wafer-level package," US Patent No. 6,429,511, issued August 2002.
83. J. Heck, L. R. Arana, B. Read, and T. S. Dory, "Ceramic via wafer-level packaging for MEMS," *Proc. ASME Conf. Integration and Packaging of MEMS, NEMS, and Electronic Systems (InterPack'05)*, 2005.
84. K. Boustedt, K. Persson, and D. Stranneby, "Flip Chip as an Enabler for MEMS Packaging," *Proc. IEEE Electronic Components and Technology Conference (ECTC'02)*, 2002.
85. T. T. Hsu, *MEMS Packaging*, London: INSPEC, 2004.
86. J. Marek, "Microsystems for automotive applications," *Proc. Eurosensors XIII*, 1999, pp. 1–8.
87. M. Eliacin, T. Klosowiak, R. Lempkowski, and Ke Lian, "Meso-microelectromechanical system package," US Patent No. 6,859,119B2, issued 2005.
88. K. Gilleo, *MEMS/MOEM Packaging: Concepts, Designs, Materials and Processes*, New York: McGraw-Hill, 2005.
89. L. E. Felton, M. Duffy, N. Hablutzel, P. W. Farrel, and W. A. Webster, "Low-cost packaging of inertial sensors," *Proc Intl. Symposium on Microelectronics (IMAPS'03)*, 2003, pp. 402–406.

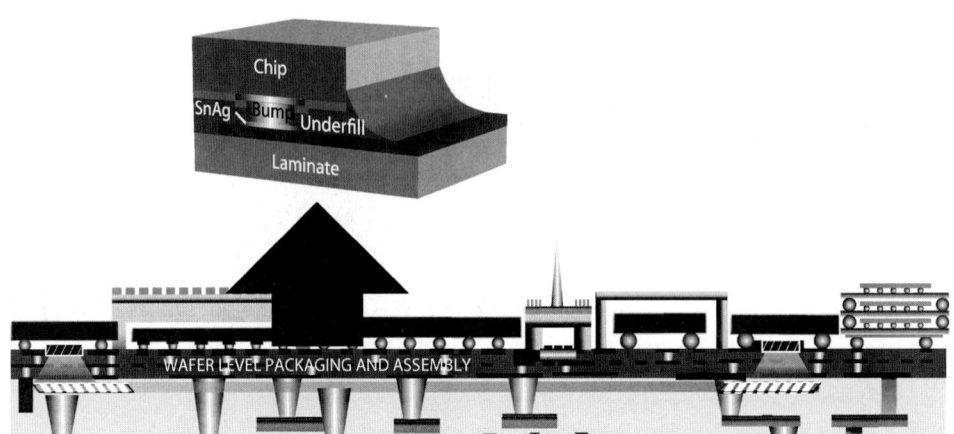

CHAPTER 10

Wafer-Level SOP

P. Markondeya Raj, Zhuqing Zhang, Y. Li, C. P. Wong, and
Rao R. Tummala
Georgia Institute of Technology

10.1	Introduction 536	10.5	3D WLSOP 590	
10.2	Buildup Wiring and Redistribution 540	10.6	Wafer-Level Probing and Burn-In	591
10.3	Wafer-Level Thin-Film Embedded Components 544	10.7	Summary 595	
10.4	Wafer-Level Packaging and Interconnections (WLPI) 548		References 595	

SOP includes two components—ICs and system packages that also act as system boards. In the SOP concept, both are miniaturized and concurrently designed, fabricated, and tested to yield and then interconnected to form system modules with two or more functions of digital, optical, wireless, and sensor functions. Wafer-level SOP (WLSOP) is defined as a silicon-based system package using silicon wafer substrate. SOP technologies such as embedded thin-film components, thin-film wiring, interconnections, and assembly are implemented with both IC and system components on a silicon wafer substrate. The IC part of SOP includes redistribution and interconnections at the wafer level, followed by dicing and assembly onto organic substrates forms the traditional wafer-level packaging (WLP). The system part of WLSOP, which is traditionally achieved with assembly of discrete system components on an organic system board, is achieved by means of embedded thin-film components on silicon wafer. WLSOP therefore integrates and miniaturizes the system by ultrahigh-density multilayer thin-film wiring, thin-film passive and active components, assembly by means of ultrathin interconnections onto the silicon substrate, followed by wafer-level test and burn-in. This chapter reviews all these silicon-based technologies with a special focus on wafer-level interconnections.

Chapter Ten

10.1 Introduction

10.1.1 Definition

WLSOP is defined as a silicon-based system in the SOP concept as illustrated in Figure 10.1. Since SOP is a highly miniaturized multifunction system technology that includes and miniaturizes both devices and system components, WLSOP accomplishes these with advanced CMOS devices in two and three dimensions and goes beyond to include silicon-based system package or board. Advanced devices include 2D integrated circuits (ICs) such as SOCs and 3D stacked ICs by through-silicon vias (TSV) and non-TSVs. System components include silicon substrates with (1) ultrahigh density and fine-line multilayer wiring and input-outputs (I/Os) in organic dielectrics and copper conductors, (2) embedded thin-film passive components, (3) embedded thin active components, and (4) thin thermal structures. In the future, it would include ultrathin power supplies such as nanobatteries and nanoscale external I/Os.

The rationales for WLSOP are several and include the following:

1. To close the IC-to-board I/O gap by means of ultrafine-pitch wiring and an interconnection pitch that cannot be achieved with traditional organic substrate and board technologies.
2. To miniaturize the system by means of ultrathin-film wiring, I/Os, and thermal structures.
3. To enhance system performance because of shorter chip-to chip interconnections enabled by both TSV-based chips and embedded thin-film system components.
4. Minimize the thermomechanical reliability issues by means of an all silicon-based approach by eliminating the global CTE mismatch.

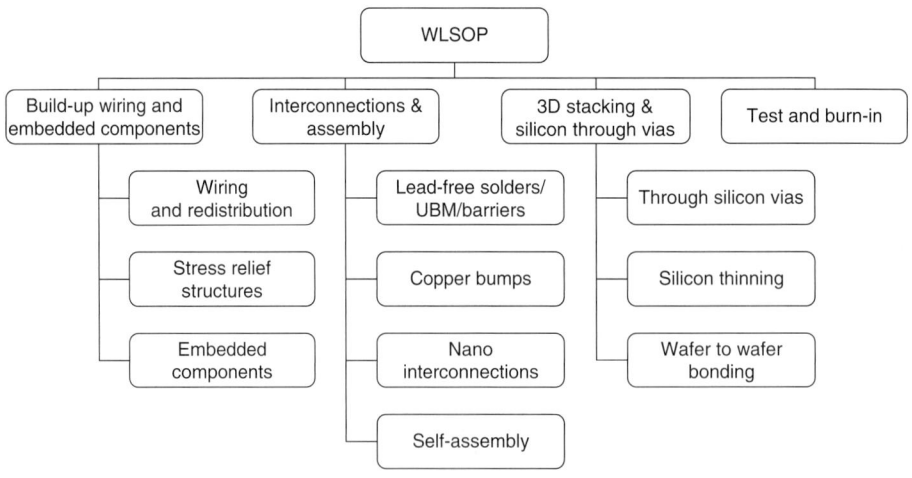

FIGURE 10.1 WLSOP technologies.

As defined above and as illustrated in Figure 10.1, WLSOP includes five major technologies:

1. Silicon substrate with ultrahigh-density multilayer polymer-Cu wiring
2. Wafer-level embedded thin-film components with the best properties at the lowest cost
3. Wafer-level ultrahigh-density interconnections and fine-pitch assembly
4. Active 2D SOCs and 3D SIPS with or without TSV
5. Wafer-level, IC-level, and system-level test and burn-in

This chapter focuses primarily on interconnections and assembly. Buildup wiring and embedded thin-film components, as they relate to WLSOP, are briefly included here but are reviewed extensively in Chapter 7. The electrical test and burn-in, similarly, are reviewed briefly here but are covered extensively in Chapter 12. The SIP technologies are reviewed in Chapter 3.

10.1.2 Wafer-Level Packaging—Historical Evolution

The historical evolution of IC packages is illustrated in Figure 10.2. The traditional IC packages fulfilled two functions: (1) provided I/Os with which to electrically test the ICs and thus guarantee the goodness of those ICs and (2) provided interconnections for assembly to system boards. But this process, as illustrated in Figure 10.3a, involves dicing the ICs and packaging these one at a time making the IC packages typically bulky, costly, and limiting both the electrical performance and the thermomechanical reliability of the

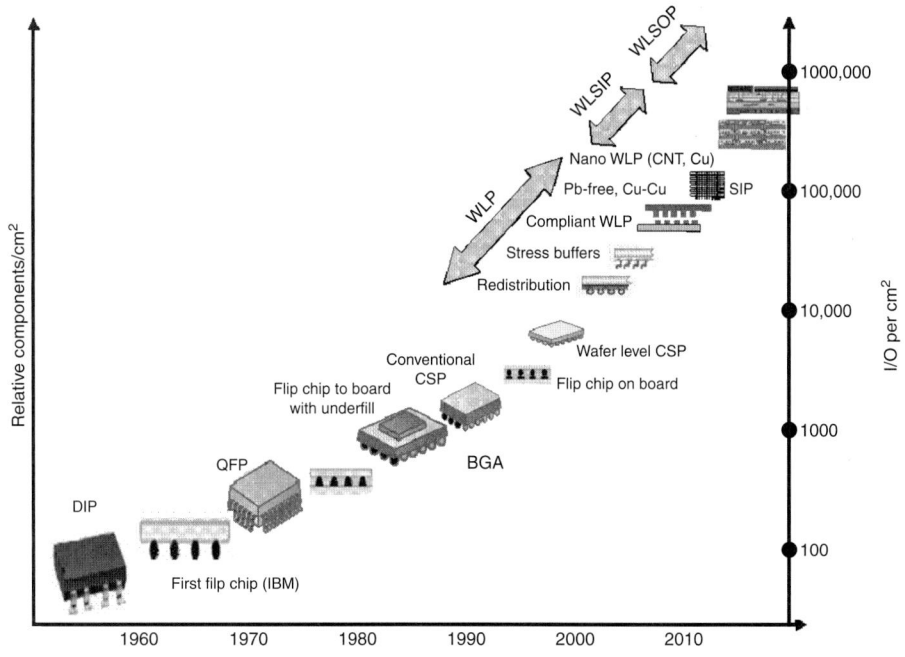

FIGURE 10.2 Historical evolution of IC packaging.

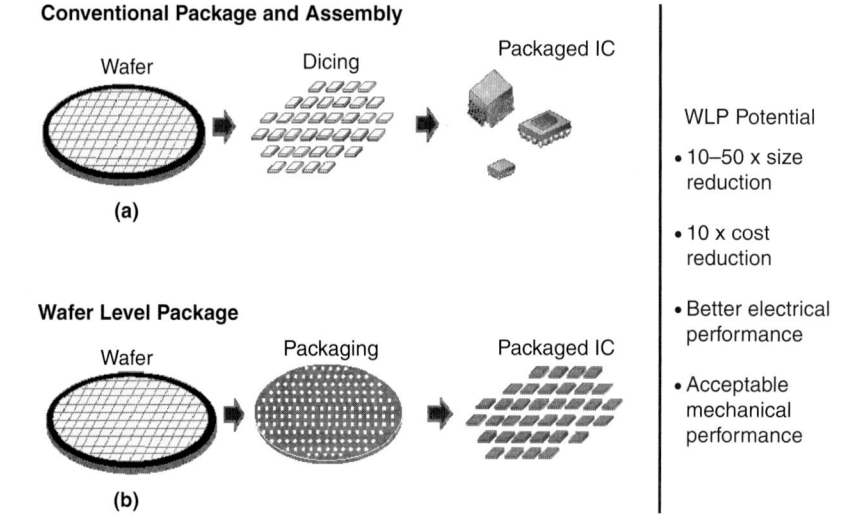

Figure 10.3 A comparison between the conventional packaging and the wafer-level packaging.

system. Wafer-level packaging, on the other hand, as shown in Figure 10.3b, offers several advantages. In this approach, wafer-level interconnections such as bumps, are deposited on the entire 300-mm wafer thereby eliminating packaging of individual ICs. Technologies are being developed to electrically test ICs at the operational frequency at this wafer level. In addition, this approach leads to both chip-scale and bare chip packaging, thus drastically minimizing the size of the IC package.

In the past two decades, there have been rapid advances in IC fabrication and their applications with faster, lighter, smaller, and yet less expensive electronic products. The semiconductor industry crossed the historic transition—nanochips with less than 65-nm nodes. Some of these chips will have more than a billion transistors, which require I/Os in excess of 10,000 and power in excess of 150 W, providing computing speeds at terabits per second. The ITRS roadmap calls for the IC feature size down to 32 nm and the I/O pitch down to 20 μm, as shown in Table 10.1.

As described above, in the conventional discrete IC packaging process, the wafers are diced into individual IC chips first and then the chips are assembled onto a lead frame or an interposer either by wire bonding, tape automated bonding (TAB), or flip chip. The demand for miniaturization and performance improvements has driven the

Year	2007	2009	2010	2015	2020
Wire bond	30	25	25	20	20
Area array flip chip	120	100	90	80	70
Peripheral flip chip	30	25	20	20	20

Table 10.1 2005 ITRS Roadmap for Interconnect Pitch (μm)

size and I/Os. During the last 40 years, IC packages have evolved from dual-in-line packages (DIPs), quad flat packages (QFPs) to area array ball grid array (BGA) packages as illustrated in Figure 10.2. While the traditional wire bonding remains the dominant IC assembly technology, the use of flip-chip assembly became a pervasive technology for high-I/O and high-performance devices. The interposer on which the chip is assembled acted as the redistribution substrate for the I/Os, on one side to meet the chip footprint and on the other side to meet the board footprint. The silicon efficiency, defined as the ratio of the Si area to the package area, has increased to almost 100 percent with the chip-scale packages (CSPs) and to more than 1000 percent with stacked SIP type of packages.

The commonly accepted definition of a chip-scale package is that the package size (length and width of the package) is no more than 1.2 times that of ICs. CSPs, therefore, extended the design concept of BGA packages with a much smaller form factor. A wafer-level package such as CSP, often referred to as wafer-level CSP (WLCSP), is a true chip size package since the package size is the same as the die size. Because of the batch processing nature, WLP has the advantage that, as the wafer size increases or the IC size decreases, the cost per device goes down [1]. The current WLCSPs are mainly designed for small dies with low-I/O devices in the consumer product market. Many of the technologies developed for these applications are based on simple peripheral pad redistribution followed by solder ball attachment. These types of technologies are finding applications in low lead-count devices and integrated passives. In this cost-driven market, the WLP is a technology targeted at lower-cost products where costs per wafer and the packaging of chips per wafer (CPW) are the critical measures for success. The development of 300-mm wafer fabrication technology favors the WLP, which increases the number of chips per wafer by a factor of more than 2 over the previous 200 mm. Some examples of high-volume production of WLSCPs shown in Figure 10.4 are by Texas Instrument's NanoStar [2] and National Semiconductor's MicroSMD [3].

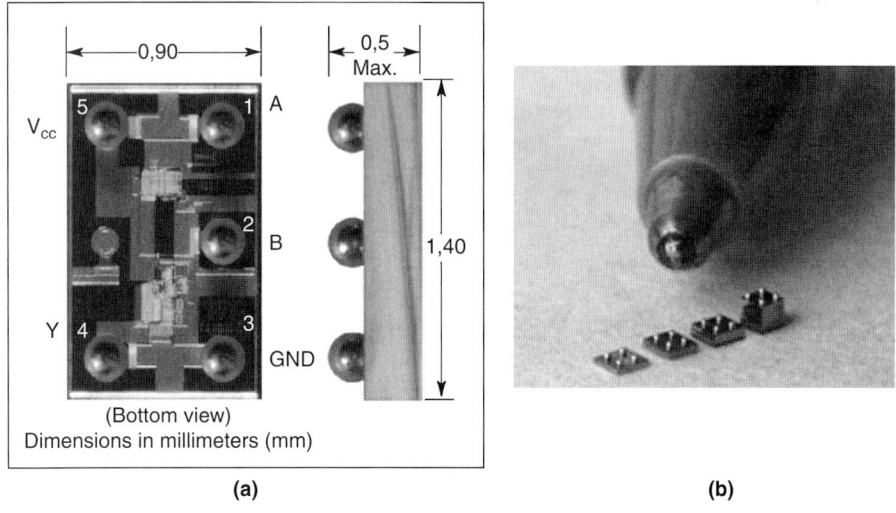

FIGURE 10.4 Examples of WLCSPs. (*a*) NanoStar (Texas Instruments). (*b*) MicroSMD (National Semiconductor).

Figure 10.5 Application of WLCSP in Casio's wristwatch camera. (Source: Casio.)

Another example of the application of WLCSPs is Casio's wristwatch camera as shown in Figure 10.5. One area of rapid adoption of CSP is for memory devices, including flash, static random-access memory (SRAM), and dynamic random-access memory (DRAM). With the advent of universal service bus (USB) ports and memory card readers, the compact flash (CF) memory cards or sticks are increasingly used to transfer data between computers, notebooks, personal digital assistants (PDAs), digital cameras, and even cell phones. The current CF cards are lead-frame packages such as thin small outline packages (TSOPs). However, CSPs are gradually taking over as the product size continues to shrink [4]. The demand for small product size and increasing memory density favors the adoption of WLCSPs.

Even though the current commercial WLPs are for small dies and low-I/O devices, the trend for WLP is pushing toward the applications of bigger die size and higher I/Os. A novel set of WLP interconnection technologies is emerging to address these challenges, as discussed in Section 10.4.

10.2 Buildup Wiring and Redistribution

10.2.1 IC-Package Pitch Gap

The escalating I/O densities of ICs are driven by the need for the highest system performance such as multicore processors, aggregately providing data rates in Tbytes/s. Back-end-of-line (BEOL) wiring refers to the formation of local and global interconnections wiring that interconnects various transistor circuit elements on the chip. In the 65-nm node IC technology, the BEOL interconnect lines vary from 200 (metal layer 1) to 1100 nm (metal layer 8). These interconnections have a thickness-to-width ratio of 1.8 with a low-K interlayer dielectric material. On the other hand, the coarse I/O pitches of even the most advanced package substrate interconnection technologies are still around 100 µm, with lines and spaces around 25 to 30 µm. This is in huge contrast to IC I/O pad pitches, which are as small as 40 µm for state-of-the-art technologies and will reduce to 1 to 20 µm in the future. It is clear that a large interconnect gap exists between the on-chip interconnect technologies and off-chip organic substrate technologies.

Silicon Substrate with Polymer-Copper Wiring

WLSOP offers a solution to the above IC-package gap. The main contributor to this solution is the silicon wafer itself. Unlike organic packages or boards, silicon wafer is flat and smooth and amenable, therefore, for ultrahigh-density wiring with ultrafine pitch. Standard silicon wiring ground rules can be used in the Si system substrate to lower the mismatch between the interconnect pitches on-chip and off-chip to support system integration with active and passive components and submicron BEOL geometries by leveraging and enhancing the previous generations of semiconductor processing tools to achieve dense wiring and meet the future I/O pitch needs.

BEOL Materials and Processes

As mentioned before, for a 65-nm node, the BEOL geometries vary from about 200 nm in layer 1 to 1100 nm wiring in layer 8. Wiring on silicon wafer has used two major process approaches so far. One is a simple extension of BEOL with Al by Bell Labs, and the other is more like polymer-Cu multilayer wiring practiced by IBM and NEC and others for MCMs. Aluminum BEOL has been in production for more than three decades. The Al patterning was initially done with subtractive wet etching but moved to a lift-off process followed by subtractive dry etching to continuously shrink the ground rules. The Al BEOL process consists of

1. Sequential deposition of Ti and Al on a planar surface
2. Subtractive aluminum reactive ion etching
3. Intermetal dielectric deposition to fill and cover the wiring
4. Planarization by chemical mechanical polishing (CMP)
5. CVD tungsten stud via fabrication steps

Subtractive RIE-based aluminum wiring has served the industry from 0.8-µm CMOS through the 0.25-µm generation. Beyond that, aluminum BEOL presents several electrical and processing limits. The process limits are to deal with the defects that arise from the chlorine RIE etch chemistry and shorts between lines from the etch residuals. The electrical limits are generally associated with RC delay. The interconnect RC delay, which determines the clock frequency, can be expressed as

$$RC = \varepsilon \frac{\rho}{M} \frac{L^2}{S}$$

where ρ, M, and L are the resistivity, thickness, and length of the interconnects, respectively, and ε and S are the permittivity and thickness of the InterLayer Dielectric (ILD), respectively. Reduction of the RC delay may be achieved by reducing resistivity or the length of the metal interconnects, increasing metal thickness, increasing ILD thickness, or utilizing a lower permittivity ILD material. The trend toward BEOL wiring for lowering the RC delay resulted in the introduction of copper-based interconnections with low-K interlayer dielectrics. Three characteristics of copper dictate the process technology for the Cu low-K BEOL integration scheme [5]. These are

1. Copper diffuses faster through silicon and silica, which requires diffusion barriers.
2. Copper surface pretreatment is needed for the adhesion of the low-K dielectric materials.
3. The difficulty associated with dry etching prohibits the use of subtractive etching.

To address these constraints, a dual-damascene process was invented by IBM [6]. The process steps involved in this are

1. Etch the dielectric in two steps to form vias and trenches.
2. Use CVD deposit, sputter, or plate copper to fill the vias.
3. Polish and planarize to remove excess copper and barrier metal.

These steps are depicted in Figure 10.6. Another aspect of the Cu low-K BEOL is the insertion of low-K dielectrics. These are materials with lower polarization groups such as C—C, C—H, C—F, and Si—C, or lower density of polarization groups by inducing nanoscale porosity into the materials. The most common materials are based on modified silicon oxycarbides or inorganic-organic hybrid materials. The low-K dielectrics that go with the copper also come with two fundamental problems: (1) inherently weak mechanical strength with decreasing dielectric constant and (2) degradation of

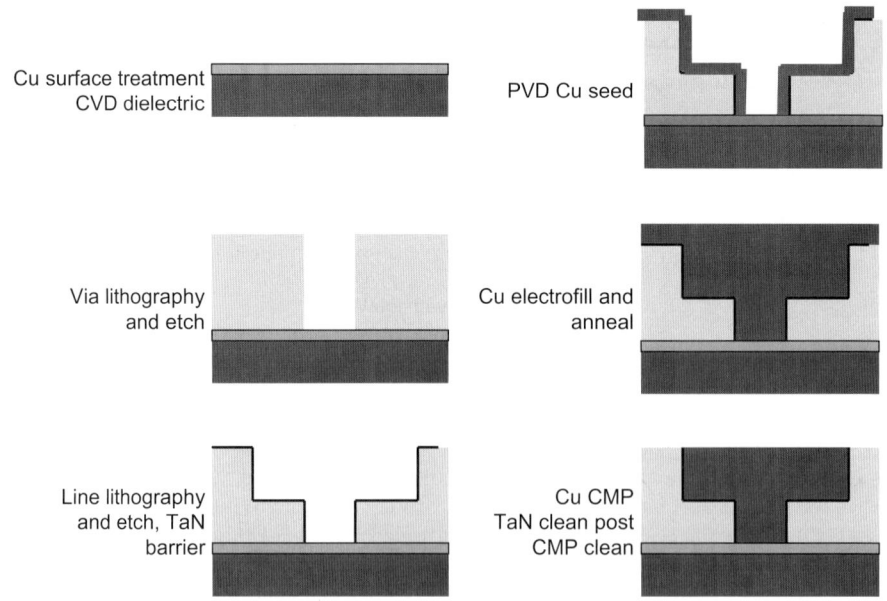

FIGURE 10.6 Schematic of the dual-damascene process flow for submicron copper interconnections.

the low-K dielectric during BEOL processing. While copper, CMP, and dual-damascene processes have been smoothly integrated into the semiconductor industry, the insertion of low-K materials has seen major delays because of the yield and reliability challenges.

Package wiring with about 1-μm lines and spaces can be achieved by utilizing the BEOL processes. However, package design often needs thicker and fatter wires in the 1- to 5-μm range and above. This is due to the fact that the package wiring has to meet the power and current requirements across longer wiring lengths of packages that are often 5 to 10 times bigger than the ICs they package. Furthermore, for high-speed signaling, on-chip wiring typically utilizes *RC* transmission lines that are dispersive and the line delay typically increases quadratically with the interconnection length, necessitating repeaters. Global wiring, on the other hand, needs *LC* transmission lines that essentially behave as waveguides [7]. *LC* transmission lines require larger cross sections (30 μm^2) compared to those of the on-chip global interconnections. Larger cross sections with CMP processes induce defects, and also there are additional reliability concerns because thicker wires need thicker dielectrics to meet the capacitance requirements. In addition, the BEOL processes utilize expensive technologies such as CVD and are generally not cost-effective for packages. Thin-film multilayer polymer-metal buildup layers are typically required to meet this geometry, processing, and cost requirements. Such types of thin-film packaging technologies were originally developed by IBM and NEC in the late 1980s for ceramic MCMs. These technologies were subsequently implemented as surface laminar circuitry onto printed wiring boards by IBM Japan, which paved the way for so-called buildup organic substrates that have been in extensive use for more than a decade.

10.2.2 Redistribution Layers on Si to Close the Pitch Gap

Multilevel thin-film redistribution layers enlarge the wafer pitch to match that of the organic packages with a rerouting process. Such redistribution layers also enhance passivation and partially relieve the WLP stresses leading to better assembly reliability. Redistribution layers have also involved thin-film embedding passive components as described in Section 10.3. In the WLSOP concept, redistribution layers may not be necessary as the package wiring geometries match that of the chip I/Os and the IC packages are codesigned and fabricated for impedance-matched, high-data-rate interconnections with the lowest interconnection distances. However, the traditional wafer-level packaging constituted redistribution of peripheral bonding pads designed for wire-bond to area array interconnections with coarser pitch as well as stress buffering, or stress redistribution features.

Thin-film redistribution has been a cost-effective wafer-level process. One example of this type of package is the Ultra CSP [8]. The manufacturing process of the Ultra CSP as illustrated in Figure 10.7 utilizes two layers of benzocyclobutene (BCB) dielectric and one redistribution layer of Al-NiV-Cu. After the fabrication of the thin-film layers, the solder balls are attached by flux and reflowed. The Ultra CSP concept uses standard IC processing technology thus making it ideal for both insertion at the end of the wafer fab as well as facilitation of wafer-level test and burn-in options. It has also demonstrated that it can be quickly ramped to high-yielding processes as well to integrate easily into the existing semiconductor process flow with minimal incremental capital investment.

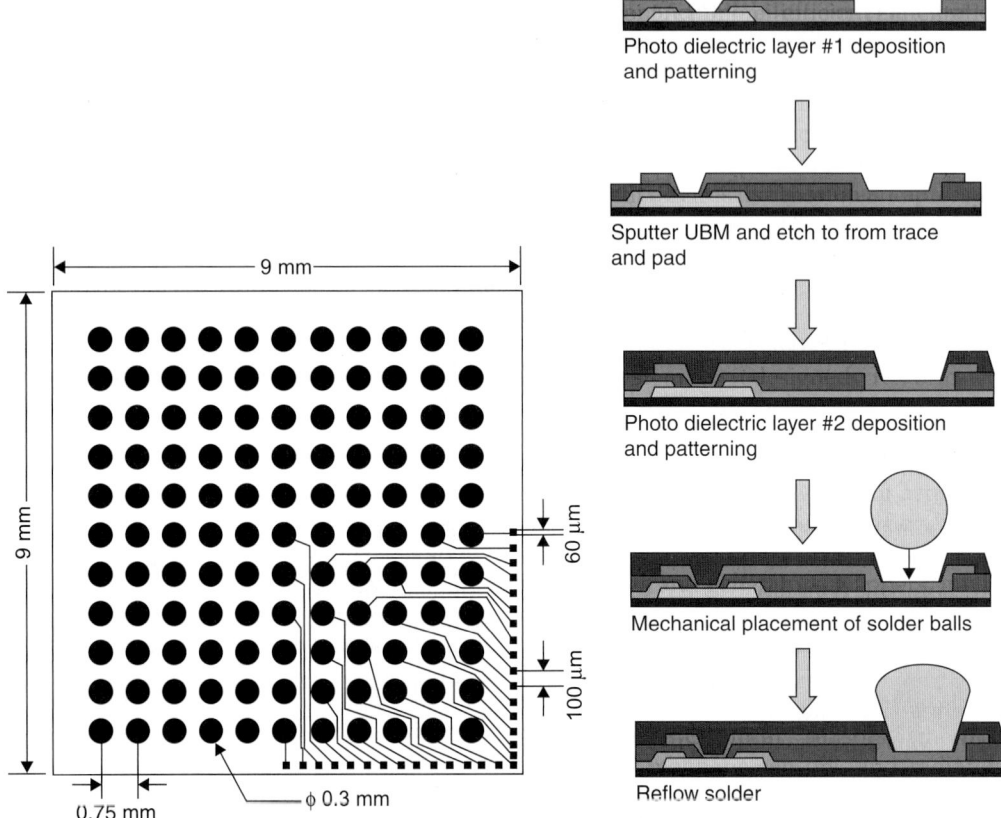

FIGURE 10.7 Processing steps for wafer-level redistribution Ultra CSP.

10.3 Wafer-Level Thin-Film Embedded Components

WLSOP seeks to integrate thin-film components to provide digital functions by means of such thin-film components as decoupling capacitors and termination resistors, and RF functions by means of such thin-film components as resistors, inductors, capacitors, filters, and baluns, as well as embedded power sources such as batteries along with the thin-film wiring and interconnection technologies, discussed above, on a silicon substrate. The silicon wafer substrate provides unique opportunities to integrate all these to achieve miniaturization, which in turn leads to high performance, low cost, high reliability, and functionality.

Approaches to WLSOP Thin-Film Components

The SOP concept of embedded thin-film components can be implemented either in the redistribution layers or in the silicon carrier substrate as presented below.

10.3.1 Embedded Thin-Film Components in the ReDistribution Layer (RDL)

Si-centric thin-film wiring technologies on the active wafers, referred to as redistribution layers above, can include embedded thin-film components. This is schematically shown

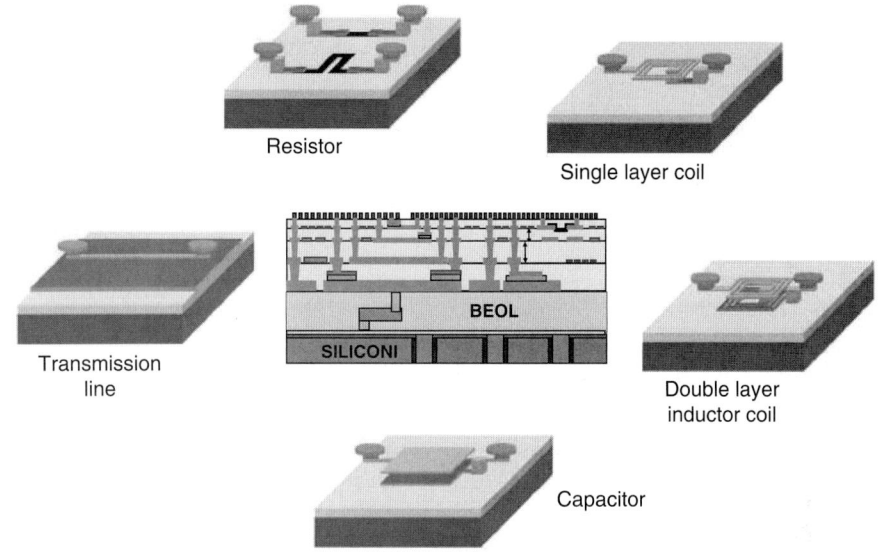

Figure 10.8 Embedded R, L, and C transmission lines in RDL. [8]

in Figure 10.8. Depending on the routing requirements, these thin-film components can be inserted between any two layers providing modularity to the designs. A minimum of two layers is typically needed for embedding inductors and metal-insulator-metal capacitors. For inductors, the first layer typically defines the lateral dimensions with the line width, line spacing, and number of turns, while the second layer provides the underpass that connects the inner terminal of the spiral to the outer terminal.

Inductors

One of the major challenges for thin-film passives on silicon is achieving a high Q. In the RDL layers, it is achieved by thin-film packaging technologies. The silicon substrate losses (Q_{sub}) dominate at high frequencies because of the capacitive coupling through the substrate. The Q of an inductor can be expressed as

$$\frac{1}{Q} = \frac{1}{Q_{sub}} + \frac{1}{Q_{metal}}$$

The Q factor is dependent on the silicon losses by substrate coupling. At low frequencies, the resistive contribution dominates the losses. Therefore, an important avenue to realize high-Q inductors is using thick copper wiring and a sufficient distance away from the silicon substrate, typically achieved by using thick insulating layers. This leads to lower loss because of lower series resistance and higher resonant frequency due to lower parallel capacitance. Si-compatible leading-edge thin-film technologies and 3D microfabrication such as MEMS technologies with high-aspect-ratio copper or silver electrodes (Figure 10.9) with ultralow conductor losses enhance the quality factor of passive components [9].

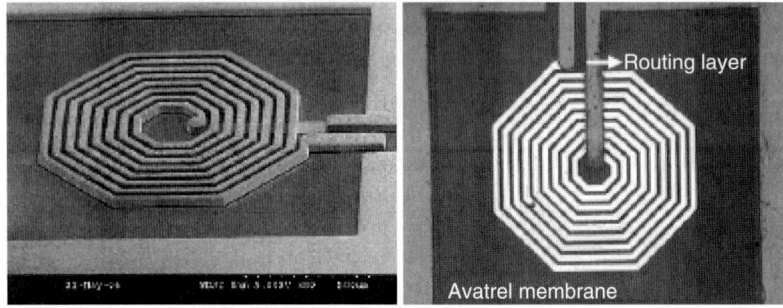

FIGURE 10.9 High-Q MEMS inductors on Si developed at Georgia Tech.

The inherent loss of silicon substrate has been a major impediment to realizing high-performance passive components on a silicon substrate. Sinaga et al. compare the losses on silicon substrate with other candidate substrates as shown in Figure 10.10 [11]. High-resistivity polysilicon (HRPs) with a thick oxide barrier has been used to lower the losses from 0.08 to 0.04 dB/mm, as shown in the figure. These are, therefore, proposed as spacer substrates for RF isolation. Inductors with a spacer that is 100 µm or thicker resulted in a Q of above 40 on silicon. Using these approaches, DIMES-Delft University of Technology has extended the silicon platform for RF integration by building integrated patch antennas, inductors, and isolation trenches, as shown in Figure 10.11 [12].

The Q has also been increased with the use of a polysilicon patterned ground shield below the inductors to lower the substrate loss, as shown by IMEC [13]. MEMS technologies further enhance the quality factor by creating electromagnetic isolation between the circuits. Other approaches include disrupting the path of current by creating trenches for isolation [14], or slicing the substrate with deep high-aspect ratio trenches, thus reducing the substrate's effective permittivity and conductivity, which in turn reduces the electrically and magnetically induced currents as well as the dipole loss [15]. A low-loss substrate on which the inductor is firmly supported can then be realized by subsequently bridging over the open areas through the deposition of a low-loss dielectric

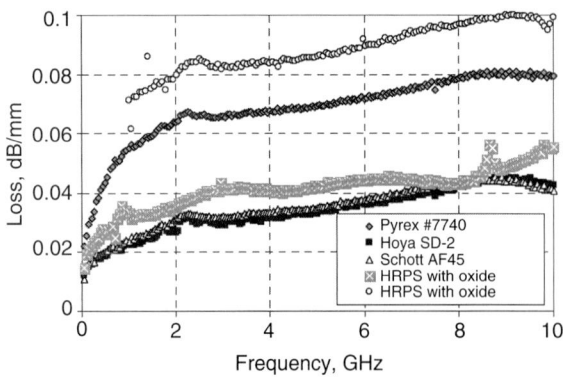
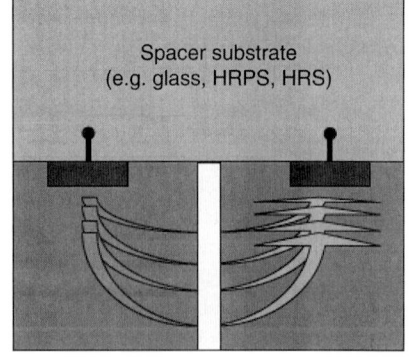

FIGURE 10.10 RF isolation techniques with modified silicon at DIMES.

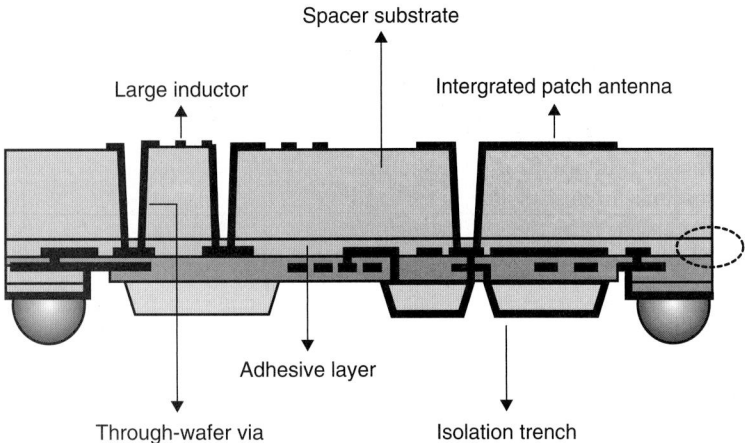

FIGURE 10.11 RF WLSOP on Si at DIMES.

layer such as SiO_2 by PECVD. Selective removal of silicon under the inductor results in Qs above 150 for a 1-nH inductor on CMOS-compatible Si substrates at frequencies ranging from 8 to 23 GHz, as demonstrated by Rais-zadeh et al. [16].

Ultrathin Deposition for Resistors and Capacitors

Resistors Ultrathin-film resistors less than 100 nm thick are easily realized on silicon substrates to achieve sheet resistances above 100 Ω/square by using alloys. NiCr (50 percent Ni) resistors with resistivities varying from 10 to 100 Ω/square is usually dc-sputtered to achieve resistances as high as 100 kΩ. Sputtered TaN resistors (75 Ω/square) are other resistors often used as alternatives. IMEC demonstrated several advances in integrating passive components in the WLSOP concept (Figure 10.12) by sputtering metal-insulator-metal capacitors such as Ta_2O_5 to achieve capacitances of 75 to 200 nF/cm². In this process, under a processing temperature of 250°C, Ta_2O_5 was shown to have a

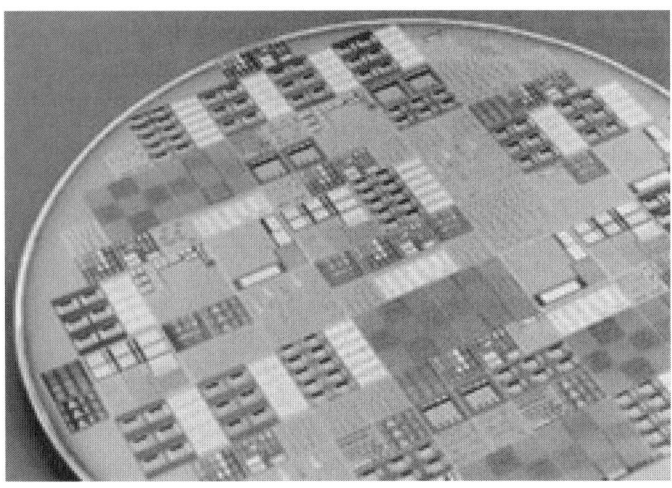

FIGURE 10.12 Demonstration of RF passives with WLP technologies at IMEC.

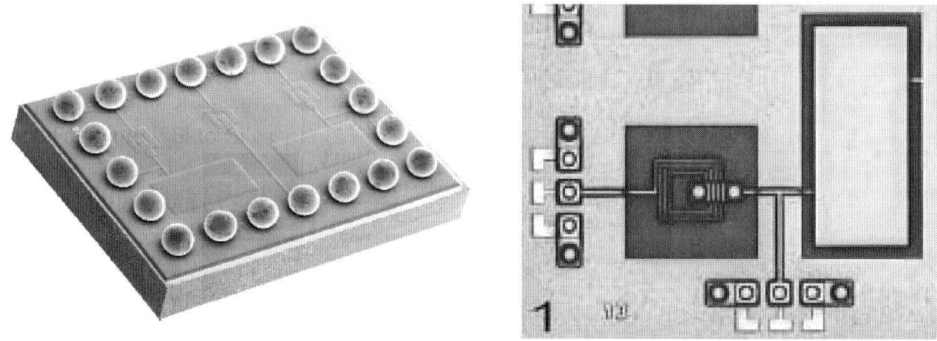

Figure 10.13 RF WLSOP embedded *LC* filter demonstration at Fraunhofer. [9]

breakdown voltage (BDV) of 18 V and a leakage current on the order of nA/cm² at 3 V. With higher processing temperatures of 360°C, a capacitance density of 1 µF/cm² and a BDV of 15 V were achieved with BiTaO using Pt electrodes [17].

Fraunhofer reported RF functions with inductors of 1.5 to 80 nH, resistors of 10 to 150 kΩ, and capacitors of 0.2 to 3 pF and impedance-controlled transmission line structures [9]. In this research, low-pass, bandpass, and bandstop filters were designed with cutoff and center frequencies around the 2.4-GHz band on Si wafers (Figure 10.13).

Capacitors Si carriers with micromachined trench capacitors have been reported to have very high capacitance densities. This is typically achieved by deep-etching the trenches in Si and conformally coating them with oxides or oxynitrides. The pores are then filled with in situ doped polysilicon. Silicon oxynitrides can be deposited to less than 30 nm, but they inherently have low dielectric constants (6 to 8). Novel atomic layer deposition (ALD) processing of such high-K conformal films as hafnium oxide (25) and titanium oxide (~50 to 80) can increase the capacitance densities by 3 to 50 times. However, applying ALD to wafers with deep trenches leads to additional challenges such as roughness on the sidewalls, leading to lower breakdown voltages, and native oxide layers, leading to undesired lower dielectric constants [17]. One approach to overcome these problems is the use of appropriate step coverage with proper microstructure to achieve good insulating layers. As high a capacitance density as 10 µF/cm² with adequate insulation resistance was reported by Philips [18]. A high-resistance silicon wafer substrate with through vias, embedded resistors, capacitors, inductors, and short interconnections for assembly of active dies is introduced by Philips [19].

10.4 Wafer-Level Packaging and Interconnections (WLPI)

WLPI forms interconnections on the entire 300-mm wafer thereby eliminating bumping of individual ICs. The benefits of such an approach are discussed above. To achieve these benefits, however, a variety of challenges have to be addressed. These include electrical and mechanical design; interconnection materials with appropriate electrical, thermal, and mechanical properties; cost-effective interconnection processes; and thermomechanical reliability influenced by heat buildup (ΔT and mismatch in CTE, $\Delta \alpha$). These are shown in Figure 10.14.

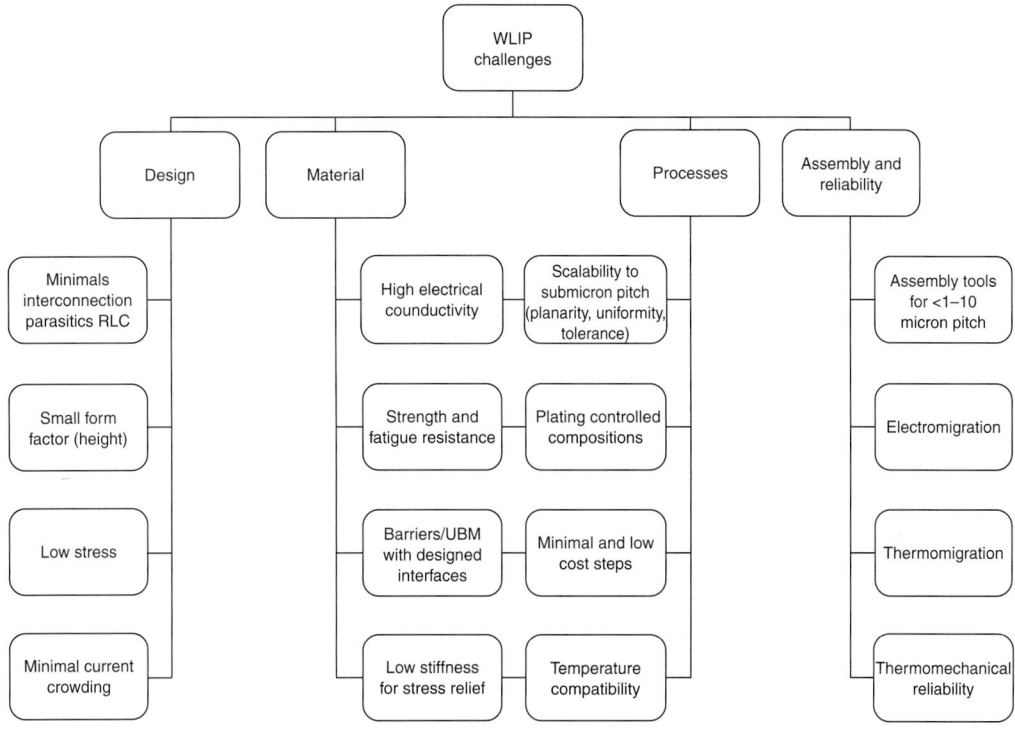

FIGURE 10.14 WLPI Challenges and various technologies to address to address these challenges.

A. Electrical Challenges

The electrical challenges fall into two major categories: (1) interconnection parasitics and (2) electromigration.

The interconnection parasitics are dependent on the interconnection height and the material property. The per-unit capacitance C between two interconnects can be calculated as

$$C = \frac{\pi \varepsilon}{\cosh^{-1}(d/2a)} \left(\frac{F}{m}\right) \quad (10.1)$$

where a is the radius of the interconnect, d is the center-to-center distance (pitch) between two adjacent interconnects, and ε is the permittivity of the material surrounding the interconnects. The total capacitance of the interconnect is $2C$.

The inductance L of the interconnection can be determined as

$$L = \frac{\varepsilon_r}{2C \cdot c_o^2} \left(\frac{H}{m}\right) \quad (10.2)$$

where c_o is the free-space light velocity and ε_r is the relative dielectric constant of the medium surrounding the interconnection such as air or underfill. The inductance of the interconnection is $L \cdot l$ where l is the length of the interconnection. Using $\varepsilon_r = 1$ for air, and $\varepsilon_r = 3.1$ for underfill, the inductance of the interconnections can be calculated.

The resistance R of the interconnect can be calculated as

$$R = \rho \frac{l}{\pi \times a^2} \qquad (10.3)$$

where ρ is the resistivity of the interconnect material.

Figure 10.15 shows the R, L, and Cs of wire bonding or compliant and solder-based flip-chip interconnection technologies as well as a technology described as nano-interconnection with the ultimate interconnection properties.

As the power continues to increase, the electric current through the bump frequently increases leading to a higher bump temperature. The current density scales as $1/d^2$, and this increases the importance of electromigration resistance at fine pitch.

B. Material Challenges

Interconnections require materials with a unique combination of electrical, thermal, mechanical, and chemical properties. In addition, the interconnection materials should be processable in a cost-effective manner in small dimensions to achieve fine pitch and at low temperatures to give strong and reliable metallurgical bonding with the lowest interfacial resistance. To achieve the best electrical interconnections, materials are generally based on high-conductivity metals such as copper or solders that are easy to process and bond at low temperatures. But solders face several challenges that include low mechanical strength, low fatigue resistance, and low electromigration resistance. They also suffer from intermetallic formation with the pad metallizations of the chip and the substrate, further degrading the thermomechanical reliability of the solder joint.

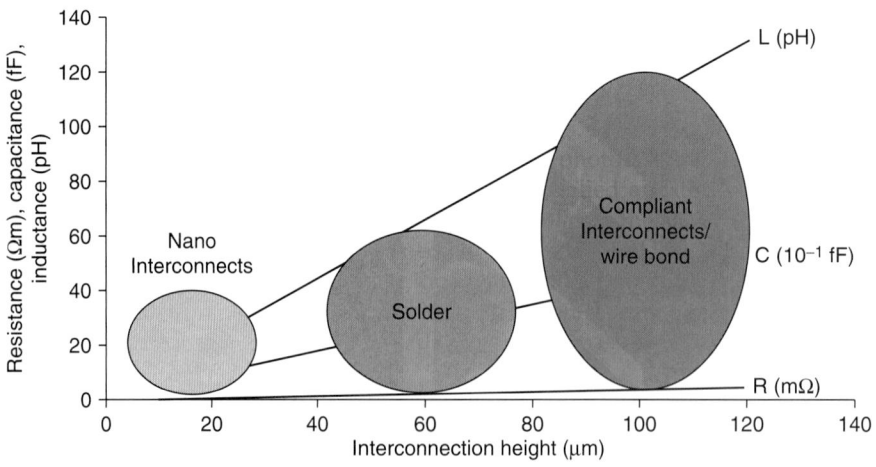

Figure 10.15 Electrical parasitics of nano-interconnects compared with conventional interconnect schemes.

C. Process Challenges

Evaporation, stencil printing, and electroplating are three commonly used techniques for wafer bumping. Evaporation produces highly reliable PbSn solder bumps but is relatively slow especially for high-Sn lead-free solders because of the low vapor pressure of Sn. Stencil printing is comparatively cheaper and is easier to use for lead-free solders. However, because of voiding and volume shrinkage during reflow, stencil printing cannot be used at a finer pitch. Electroplating is a better solder wafer bumping process at finer pitch compared to evaporation and stencil printing to produce copper interconnection structures with dimensions ranging from microns to hundreds of microns, because of the widely characterized processes and standard infrastructure. Moreover, it can easily be extended to larger wafer size. It has a higher yield and better quality and uniformity of solder bumps than stencil printing. However, solder compositions are difficult to control with electroplating frequently compromising the strength and fatigue life.

Stress-relief interconnection provides design flexibility that can be optimized to offer the optimal mechanical and electrical properties. Original compliant metal inter-connection schemes are extended from wire-bonding routes. For fine-pitch compliant metal structures, electroplated 3D MEMS-like processes are frequently used. However, the requirement of the multiple-mask steps for lithography increases the cost of manu-facturing. Simple and cost-effective polymer-based stress-relief schemes are more commercially successful compared to the complex multilayered processes with num-erous steps. The details are discussed in the respective interconnection schemes as required, in later sections.

D. Assembly and Thermomechanical Reliability Challenges

To benefit from the 3D and WLSOP integration, high-precision alignment, cost-effective assembly, and testing must be achieved with a reliability suitable for the targeted applications. The CTE difference between the chip and organic buildup substrate induces stresses in the interconnections, which, if excessive, can result in structural failure and eventually electrical failure of the chip. The estimated fatigue life of an interconnection during thermal cycling is related to the geometry of the interconnection and the load as shown in the following equation:

$$\text{Fatigue Life} \propto \left[\frac{h}{L \Delta \alpha \Delta T} \right]^2 \tag{10.4}$$

where L is the length of the die and h is the height of the interconnection, $\Delta \alpha$ is the difference in CTE between the substrate and the chip, and ΔT is the temperature difference in thermal cycling. While in BGA and CSP packages, a relative thick interposer is used to redistribute I/O and acts as a cushion for stress relief, for WLP, the I/O redistribution is done on the wafer level and the interposer is usually absent. On the other hand, interconnections are moving to thinner geometries at finer pitches, which affects the fatigue life further.

While the electrical requirements demand short interconnects at ultrafine pitch to increase the chip-to-chip and chip-to-package bandwidth at low power, short inter-connects are faced with mechanical challenges. The maximum solder strain at 50-μm pitch is 4 to 8 times higher than that at 200-μm micropitch. Downscaling the traditional solder bump interconnect will not satisfy the thermomechanical reliability requirements at very fine pitches on the order of 50 μm and less because of the poor mechanical properties of solders. Novel low-cost wafer-level packaging technologies are needed,

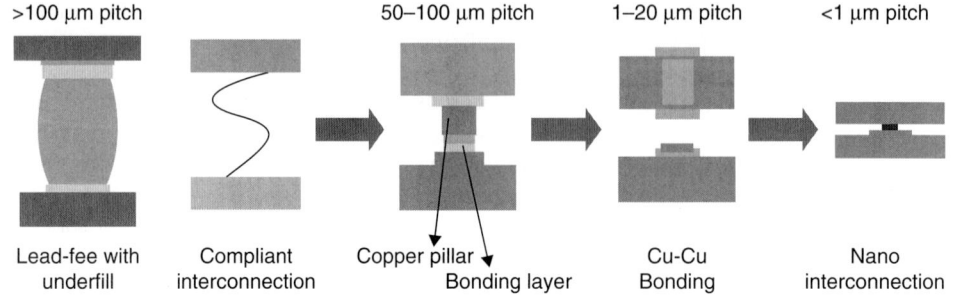

FIGURE 10.16 Wafer-level interconnection trend.

therefore, to address electronic systems with ultrahigh electrical performance while eliminating the reliability problems at fine pitch. The interconnection trend going from lead-free solders to compliant interconnections to copper pillar and thin-film nano-interconnections is shown in Figure 10.16.

10.4.1 Classes of Wafer-Level Packaging and Interconnections (WLPI)

All interconnections can be broadly classified into two types as shown in Figure 10.17: (1) compliant and (2) rigid interconnections. Compliant interconnections are those that provide stress relief by a variety of mechanisms as described below. These can be sub-classified into (1) metal-based and (2) polymer-based. With compliant interconnections,

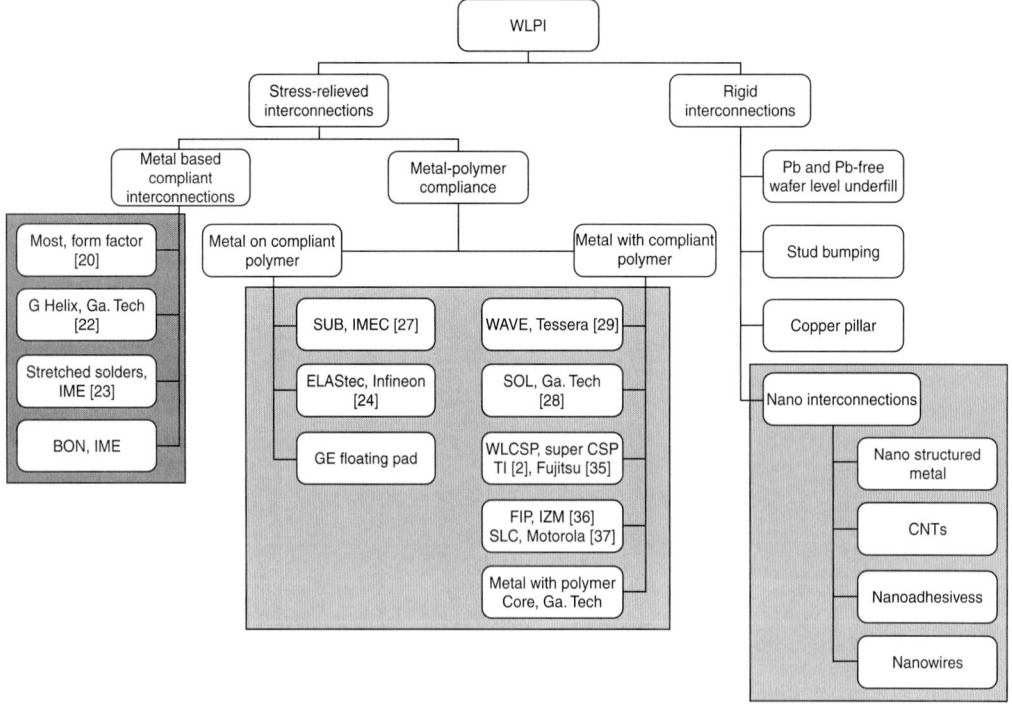

FIGURE 10.17 Classes of WLPI.

the mechanical stress developed in the interconnection due to the mismatch in TCEs between the IC and the substrate are accommodated by the mechanical deformations in the interconnection. A good example is a wire bond. However, compliant interconnections are generally very long and thin and, therefore, have higher values of electrical resistance, inductance, and capacitance. Thus, electrical and mechanical design requirements are often in conflict and the final design is a tradeoff between the two. Rigid interconnections such as solder are typically shorter and cannot accommodate any strain as in the compliant interconnections, often leading to early failure if the strain in the joint is not lowered to appropriate levels such as with underfills. This is considered the single biggest barrier to solder-based fine-pitch interconnections, requiring novel designs, materials, and fabrication technologies. The current approach with lead-free solders using underfill faces major challenges in designing the underfill with appropriate mechanical properties and dispensing such an underfill in fine-pitch and low-standoff-height interconnections with any significant amount of fillers in the underfill polymer.

Metal-Based Stress-Relief Interconnections

Interconnections are divided into two classes: stress-relieved and rigid. Stress-relief technologies modify wafer-level fabrication methods such as RDL, wiring, and interconnection steps in order to mitigate the stresses in the interconnections arising from the CTE mismatch between the IC and the substrate. Two main stress-relief approaches to lower the stresses are reported as presented below. Compliant metallic structures accommodate the CTE mismatch between the die and the substrate during the thermal cycling by easily moving or deforming in the x, y, and z directions to provide stress and strain relief in the interconnection. The deformation keeps stresses in the elastic regime of the interconnection material where the fatigue resistance is much higher, compared to that in the plastic regime. In most cases, z-direction compliancy is also provided to address the substrate coplanarity and wafer testing issues.

Many metal-based compliant interconnect technologies have been developed to improve the interconnection reliability. Examples of this are MicroSpring Contact on Silicon Technology (MOST) by FormFactor [20], and G-Helix Interconnect by Georgia Tech [21]. The MicroSpring technology was first invented for wafer probe cards and land grid array (LGA) sockets. This technology was recently extended to a wafer-level package called MOST, in which the MicroSpring contacts are fabricated directly on silicon at the wafer level as shown in Figure 10.18. The microspring contacts are fabricated by a gold wire–bonding process and are plated with an Ni alloy, called "spring alloy." These contacts are attached to the substrate by soldering, which helps to decouple the CTE mismatch between the silicon die and the board thus resulting in an interconnection reliability better than with solder joints. The microsprings accommodate 1 g of compression force for every 25 µm of displacement and exhibit low contact resistance. So far they have been used at 225-µm pitch. The MOST technology has been integrated into wafer-level test and burn-in as the "wafer on wafer" (WOW) process, as discussed later in the chapter.

Compliant Helical Interconnections   Another example of compliant interconnection is the G-helix (Figure 10.19), which consists of an arcuate beam and two end posts to accommodate the differential displacement in the planar directions (x and z), as shown in [21]. The two end posts connect the arcuate beam to the die and to the substrate. Table 10.2 summarizes its characteristics [22].

Figure 10.18 An SEM picture of microspring interconnections.

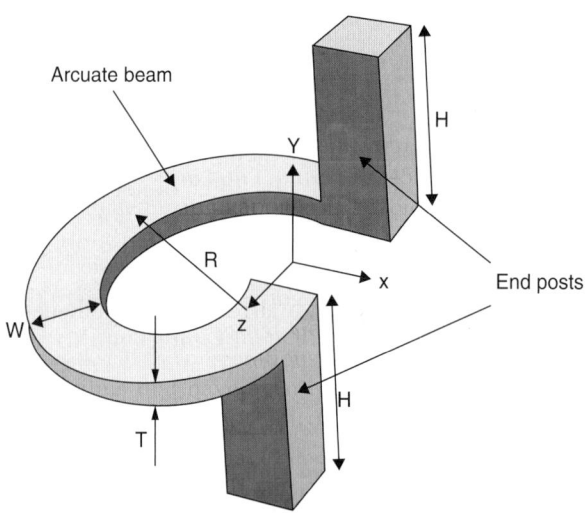

Figure 10.19 Structure of a helical interconnect. [22]

Diagonal mechanical compliance	9.068 mm/N
Vertical mechanical compliance	10.149 mm/N
Maximum von Mises stress (diagonal load)	175.55 MPa
Maximum von Mises stress (vertical load)	172.06 MPa
Electrical resistance	43.63 mΩ
Self-inductance	0.08989 nH

Table 10.2 Mechanical and Electrical Characteristics of Helical Interconnects

The helical compliant interconnections have several advantages:

1. The force exerted by the interconnections on die bonding pads is minimal and thus will not crack or delaminate the low-K dielectric in the die.
2. The interconnections can accommodate the traditional CTE mismatch between the die and the organic substrate as well as the nonplanarity of the organic substrate.
3. The interconnections have been reported to be reworkable and not needing underfill to achieve thermomechanical reliability.
4. The interconnects are wafer-level and are scalable with I/O pitch.
5. The interconnects utilize conventional wafer fab infrastructure for fabrication, and the fabrication process is repeatable with good yield.
6. The interconnects can meet a variety of electrical, mechanical, and thermal requirements by a combination of helix and column interconnects.
7. Lead-free solder can be employed for the interconnect assembly to substrates, and therefore, the technology is environmentally friendly.

Stretched Solder Column Interconnections Compliant interconnections have also been developed by stacking or stretching the solder balls. Stretching the solder column can reduce the structural coupling and increase the in-plane compliance of the interconnection. Using the simple beam theory, the compliance of the interconnection (C) can be related to its length (L) and cross-sectional area (A) as $C \propto L^3/A^2$. In terms of mechanical compliance, the stretched solder column (SSC) interconnection is somewhere between the rigid solder ball and compliant helical interconnects, and may be regarded as a semicompliant interconnection. In this interconnection design, the amount of high-lead solder is first deposited on all the die pads on the wafers. The solder is then melted, stretched, and cooled to form a unique hourglass shape on the entire wafer. With the volume of the solder remaining unchanged by the stretching, the cross section of the stretched solder decreases near linearly with stretching. This translates to more improvement in compliance given by $C \propto L^5$. The stretched hourglass-shaped interconnection alleviates the strain concentration along the critical failure site on the original barrel-shape solder interconnection, leading to further enhancement in thermal cycling reliability (Figure 10.20).

The effect of the height of the SSC was studied with a constant solder volume and a pad diameter at 50 μm [23]. A significant finding was that, with the proposed unique shape of the SSC and for high aspect ratios, the location of maximum strain, and hence the failure site, is shifted away from the usual location at the solder-pad interface toward the center of the interconnection. This is also confirmed by fatigue experiments conducted on the SSC interconnection.

Metal-Polymer–Based Stress Relief

Metal-on-Compliant Polymer Low-stiffness polymers encapsulate the interconnection or support the interconnection to reduce the stresses in the interconnection and interfaces. The stiffness of polymers is much lower than that of metals. Therefore, relatively shorter and smaller metal structures can be designed on compliant polymers with the same stress reduction as long compliant structures, potentially leading to smaller parasitics

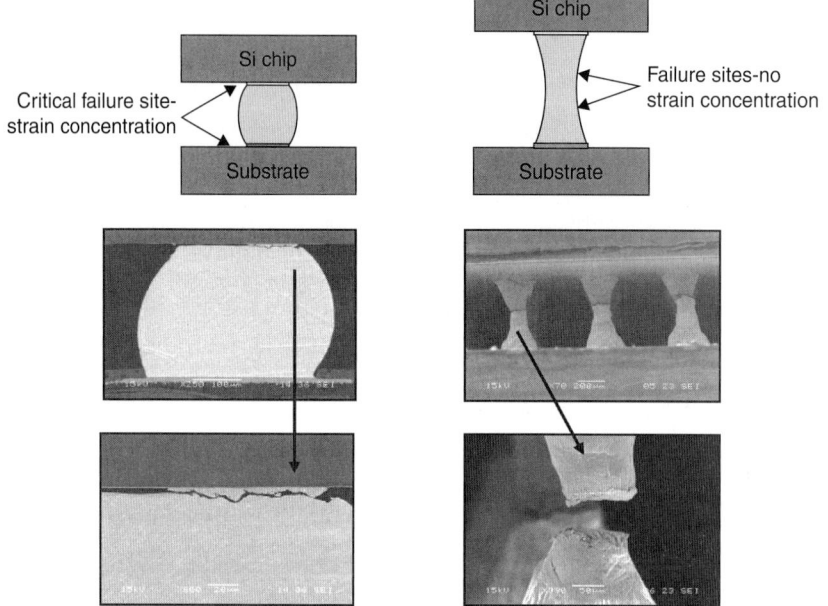

FIGURE 10.20 Schematic of stretched solder column.

and smaller form factors. The Elastic-Bump on Silicon Technology (ELASTec) by Infineon [24], On-Wafer Floating Pad Technology by GE Global Research [25], Silicone Under Bump (SUB) by IMEC and Dow Corning [26] belong to this group. They shared the same characteristic of building the interconnections on an array of polymer islands. A common choice of compliant polymer is silicone due to its low modulus, low stress, excellent thermal stability, low shrinkage, and good moisture resistance. Dow Corning has developed photodefinable and screen-printable silicone material for wafer-level packaging applications. One example by IMEC uses a TiCu metallization partially covering the silicone stress buffer layer. The solder bumps are built on the silicone layer to form an SUB configuration [26]. The ELASTec package also utilizes silicone to provide flexibility. On top of the resilient bumps, a spiral-shaped metallization (CuNiAu) is plated as shown in Figure 10.21 [27].

Compliant Metal and Polymer Interconnections A combination of compliant metal structures and low-stiffness polymers using several elegant combinations of process steps effectively lower the interconnection stresses and meet the reliability requirements. Key approaches are discussed in later sections. Wide Area Vertical Expansion (WAVE) by Tessera and Sea of Leads (SoL) [28] are examples of compliance from both metal-design and low-stress polymeric interlayers.

The basic concept behind the WAVE technology is the placement of a low-modulus encapsulant between a silicon die and its package substrate [29]. The fabrication starts with creating the copper lead on a sacrificial polyimide layer and the solder balls on the die. The polyimide layer is selectively weakened so that the copper lead will easily separate from the polyimide during the injection of the encapsulant material. The die is

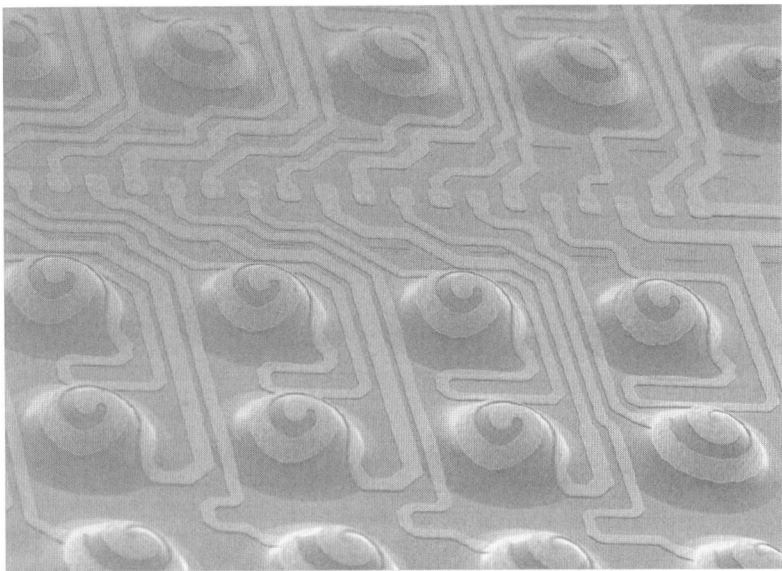

FIGURE 10.21 ELASTec bump with RDL on printed silicone.

then flipped over and soldered to the copper leads. The entire panel is then placed in an injection fixture that injects the encapsulant material between the die and substrate. During the injection, the encapsulant fills and expands the gap between the die and substrate to an extent controlled by the injection fixture. The encapsulant and flexible copper leads enable the relative movement of the die and package terminals in the x, y, and z directions. A cross section of the final device illustrating the flexible link and encapsulant layer is shown in Figure 10.22.

Sea of leads (SoL) extends batch processing of the on-chip interconnect on the wafer to include xyz compliant chip I/Os through the fabrication of "slippery" leads and embedded air gaps in the polymer film. A picture of the package is shown in Figure 10.23. The fabrication process of the SoL is illustrated in Figure 10.24 [30]. There are several methods of allowing the lead movement during the thermal cycling. One method of

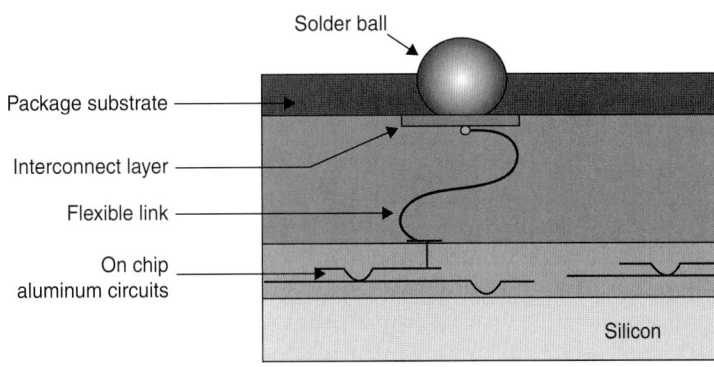

FIGURE 10.22 Tessera's compliant interconnect structure.

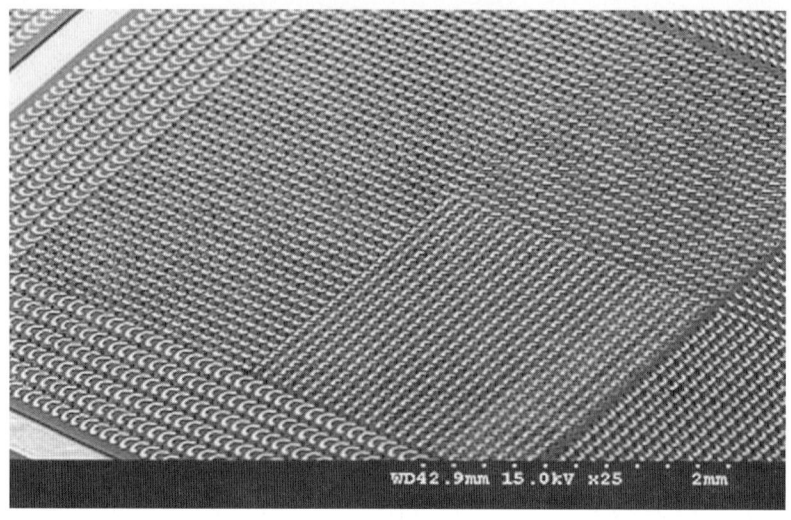

FIGURE 10.23 An SEM picture of an SoL wafer. [30]

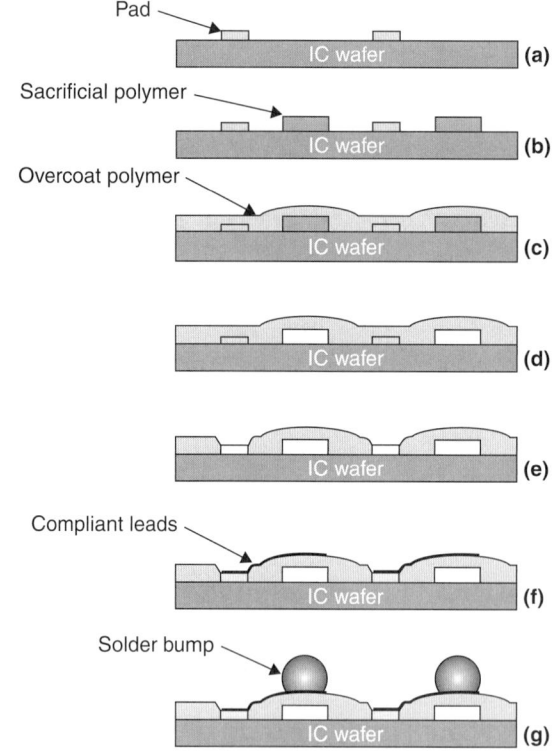

FIGURE 10.24 SoL fabrication process: (*a*) IC wafer with pads exposed, (*b*) sacrificial polymer is applied and patterned, (*c*) overcoat polymer encapsulates the patterned sacrificial polymer, (*d*) sacrificial polymer decomposes to form air gaps, (*e*) vias are etched to expose the pads, (*f*) compliant leads are electroplated after proper seed layer deposition and patterning, and (*g*) solder bumps are fabricated. [30]

fabricating the "slippery leads" uses a seed layer plated onto the leads that is selectively etched when the leads are ready to be released from the surface. The embedded air gaps are created through the decomposition of a patterned sacrificial polymer layer. The density of the SoL package reaches $12 \times 10^3/cm^2$. The package supports high-frequency signals up to 45 GHz.

Metal Interconnections with Polymer Core Current low-stress interconnections for improved mechanical performance at fine pitch use compliant metallic structures with complicated processing steps and compromised electrical performance. So there needs to be an interconnect solution to get good reliability without compromising the electrical properties and mechanical reliability. Polymers with a stiffness 100 to 200 times lower than that of metals are ideal materials for low-stress interconnections. With a polymer core in a conventional solder ball, the stress was estimated to be four times lower at the interface by Movva and Aguirre [31]. Another alternative of this process was investigated by Zhang [32] where the solder balls are dipped into the underfill pool before reflow so that each solder ball is encapsulated in a layer of underfill. The existence of the coating reduces the stress concentration at the solder-pad interface. Therefore, crack initiation and propagation is dramatically delayed and solder ball fatigue life is significantly improved.

In order to downscale to smaller pitches, MEMS-based micromachining of polymers is investigated for interconnection purposes. In a study by Ankur et al., high-aspect-ratio low-CTE polyimide structures with low stress, high toughness, and high strength were fabricated using plasma etching [33]. The etching conditions (oxygen/fluorine ratio, pressure, and power) were optimized to get the required high-aspect-ratio geometry and sidewall definition with a wall angle above 80° leading to an aspect ratio higher than 4. The etching process also leads to roughened sidewalls for selective electroless plating on the sidewalls of the polymer structures. The composite interconnections can then be assembled with thin solders without imparting high solder strains.

Metal Bump with Compliant Polymer Polymer structures are also processed to surround the semirigid or rigid metal bumps to enhance their reliability. One method to accomplish this is to surround the bump with a polymer collar. In a standard WLP bumping process, a flux layer is usually applied before the solder ball placement to facilitate the solder wetting on the bond pads during wafer reflow. In the Polymer Collar WLP, a polymeric material is used instead of the flux and remains after the reflow to build reinforcement around the solder joint neck so as to block the shear deformation of the solder. As such, the reliability can be increased. In the Ultra CSP package with solders of maximum distant to the neutral point (DNP) being 3.18 mm, a 64 percent increase in lifetime was observed with the "polymer collar." Typical products using the Ultra CSP are small packages with a low number of I/Os. In order to increase the solder joint reliability of larger packages, a polymer reinforcement was designed and a technology called Polymer Collar WLP was developed by the former K&S Flip Chip Division [34].

Another example of the extended thin-film redistribution WLP is the Super CSP developed by Fujitsu Ltd. [35]. Figure 10.25 shows the structure of the Super CSP ball grid array (BGA) and land grid array (LGA) type packages. The manufacturing process of the Super CSP involves the formation of the redistribution layer by a polyimide film and electrolytic-plated metal trace. After redistribution, the resist is patterned and the copper posts are formed by electrolytic plating. Then the whole wafer is encapsulated with an epoxy molding compound (EMC), and solder balls or

560 Chapter Ten

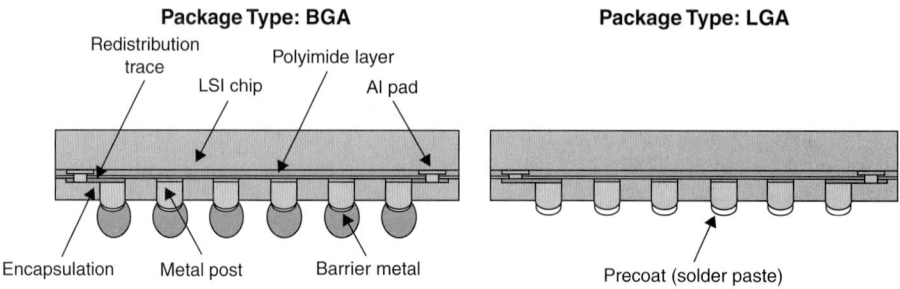

FIGURE 10.25 A schematic of Super CSP structure. (Source: Fujitsu.)

solder pastes are applied on top of the copper posts. The board-level reliability of the Super CSP is good mainly due to high standoffs of the copper post as well as the low CTE of the EMC encapsulation material which effectively reduces the stress occurring in the solder joint interconnect. The Casio wristwatch camera used the Super CSP technology licensed to Shinko.

Similar to the Super CSP, the Fab Integrated Packaging (FIP) developed by Fraunhofer IZM Berlin, uses a stress compensation layer (SCL) that embeds the solder balls before the second solder balls are attached on the top of embedded balls as shown in Figure 10.26. According to the thermal mechanical simulation, the SCL reduces the accumulated equivalent creep strain of the solder balls and also serves as mechanical support for the second solder ball to achieve taller solder heights compared to the standard redistribution technology [36]. The double-bump structure was evaluated by Motorola with different SCL materials [37]. A similar WLCSP was developed by New Japan Radio Co., Ltd., and Saga Electronics Co., Ltd [38]. In addition to "front coating resin," which is similar to the SCL in FIP, a "back coating resin" is applied to the backside of the wafer. Both resins are SiO_2-filled liquid epoxy and are stencil-printed after the creation of the solder post. The board-level reliability results showed that the double bump with the resin coating can enhance the reliability of the package, especially during the low-temperature thermal cycling (–40 to +125°C). The single-bump package (1.23 × 1.19 mm, 400-μm pitch) had an early failure starting at 200 cycles, while the double-bump package did not show any failure for more than 1500 cycles.

Rigid Interconnections

Rigid interconnections are defined as interconnections whose aspect ratio (bump height to bump diameter) is less than 1 and have no additional stress-relieving layers beneath

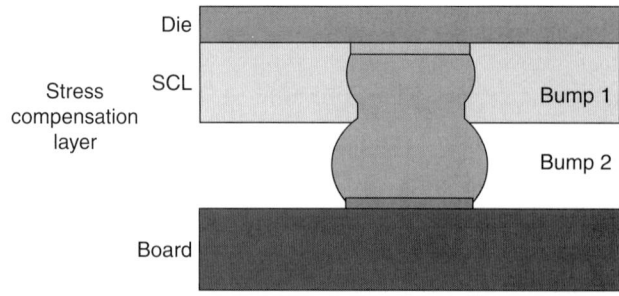

FIGURE 10.26 A schematic of the FIP structure. (Source: Motorola.)

the die and substrate bonding pads. Because of their lowest form factors, best electrical characteristics, and minimal processing steps, these by far provide the leading-edge system performance. The most common rigid interconnections are the controlled collapse chip connections (C4) or flip-chip solder bumps developed by IBM. The global CTE mismatch between the die and the substrate causes high stresses in rigid interconnections during thermal cycling. These were therefore originally implemented only with glass-ceramic substrates that have no CTE mismatch with silicon. The invention of the underfilling process made it feasible to extend these interconnection methods to high-CTE organic substrates. Lead-tin eutectic solders are most widely used for rigid chip-package interconnections. Recent environmental legislations toward toxic materials and consumers' demand for green electronics have pushed the drive toward lead-free solders. A complete discussion on lead-free solders can be found in Shannguan's book [39]. Some of the key lead-free material and processing issues at fine pitch are discussed here. This section also reviews wafer-level underfilling, copper post technology to replace solders, and the stud bumping methods.

PbSn and Pb-Free Solders

Semiconductor packages are considered green when toxic elements such as Pb, Br, Cl, and Sb are not intentionally added during the manufacturing process. They can, however, be present in the finished units as impurities. The well-accepted maximum level for Pb is < 1000 ppm, which was recommended by ITRI, IPC, and DPUG. Many different kinds of "lead-free" solder alloys and soldering processes are being investigated or developed around the world, using multiple combinations of elements like tin, silver, copper, bismuth, indium, and zinc, most of which require increased temperature profiles during the soldering process relative to the well-known tin-lead alloys. Table 10.3 shows some of the common lead-free solders.

Binary Systems	Melting Point (°C)	Ternary and Quaternary Systems
95Sn-5Sb	240	
99.3Sn-0.7Cu	227	
96.5Sn-3.5Ag	221	
	217	96Sn-3.9Ag-0.6Cu (NEMI)
		96.2Sn-2.5Ag-0.6Cu-0.5Sb
		95.5Sn-0.5Ag-1.0Zn
	216	93.6Sn-4.7Ag-1.7Cu
	211	91.8Sn-4.8Bi-3.4Ag
	210	91.0Sn-4.5Bi-3.5Ag-1.0Cu
91Sn-9Zn	199	
	187	77.2Sn-20In-2.8Ag
97In-3Ag	143	
58Bi-42Sn	139	-
52In-48Sn	118	

TABLE 10.3 Lead-Free Solders Being Developed to Replace Traditional Solders

Solder Materials Solders are tin-based materials with low melting temperatures that enable assembly at temperatures less than 250°C. A eutectic solder composition provides the benefit of an alloy with the lowest melting temperature of all alloys comprising the system, and it melts and solidifies at a single temperature. A high Cu content above the eutectic composition of 0.9 wt% Cu in SAC alloys increases the pasty range (the difference between the liquidus and the solidus temperature of the SAC alloy) that complicates the process. Among the several Pb-free candidate solders, the near-ternary eutectic SnAgCu (SAC) alloy compositions, with melting temperatures around 217°C, are becoming consensus candidates. SnAg, and SnCu alloys are also pursued as alternatives. These lead-free solders have higher melting temperatures (208 to 227°C) than eutectic PbSn (183°C). A higher reflow temperature will adversely affect the reliability of packages. Lead-free solders are tin-rich, more brittle, more prone to detrimental intermetallics, and show poor wettability. Flip-chip pitch downscaling will increase the reliability concerns of lead-free solders because of increased stresses and strains, the dominating role of intermetallics, barriers (illustrated in figure 10.27), and surface finish.

Intermetallics in the Solder and Interfaces Intermetallics are metal compounds with a specific chemical composition and a periodic structural arrangement of the constituent elements. Their mechanical properties are usually a compromise between metals and ceramics. Lead-free solders usually constitute a tin-rich phase and interspersed intermetallic Sn-based compounds. The presence of intermetallics in a solder can strongly impact its thermomechanical properties and electromigration behavior. Large pro-eutectic Ag_3Sn plates with or without Cu_6Sn_5 rods that have a strong impact on the mechanical properties may be found for SAC alloys depending on their cooling rate and composition. The excessive growth of large pro-eutectic phases can be attributed to the fact that pure Sn or Sn-rich solders require a large amount of undercooling before the Sn phase nucleates and solidifies, while pure PbSn or Pb-rich solders require a relatively small amount of undercooling. For SnAg, the high Ag content causes the growth of large Ag_3Sn plates that may even physically connect neighboring bumps. In addition, a large volume fraction of Ag_3Sn plates within a single solder bump could significantly alter the mechanical properties and thereby affect the long-term reliability of the solder joint or the reliability of the package to which it is attached.

Intermetallic compounds are also formed at the interfaces during reflow as a result of the interfacial reaction between the UnderBump Metallurgy (UBM) and Sn from the lead-free solders. Intermetallics are essential to form a good metallurgical bond. However, a critical Intermetallic compound (IMC) thickness exists for optimal interfacial reliability. For thinner intermetallics, controlled fracture occurs inside the solder making it more easy to design and predict the failure. As the IMC grows during reflow or aging, the failure shifts to the interface because of the IMC growth. The excessive IMC growth and the brittle nature of the IMC layer are detrimental to the reliability of the solder joint (Figure 10.27). The challenges are multifold for solders at fine pitch. Because of the ratio of the lower solder volume to the bond pad area at finer pitch, the interfacial reaction between UBM and Sn dominates and thus a better UBM is required which can prevent excessive interfacial reaction and promote good wetting as well. CrCu-Cu-Au UBM originally developed at IBM for high-Pb solders may not be equally applicable with eutectic PbSn concentration. Cu-Ni(V)-Al has been shown to be a better UBM for eutectic SnPb [40]. A proper UBM structure with a slow reaction rate with Pb-free solder is extremely important with solders having even higher Sn contents [41–42].

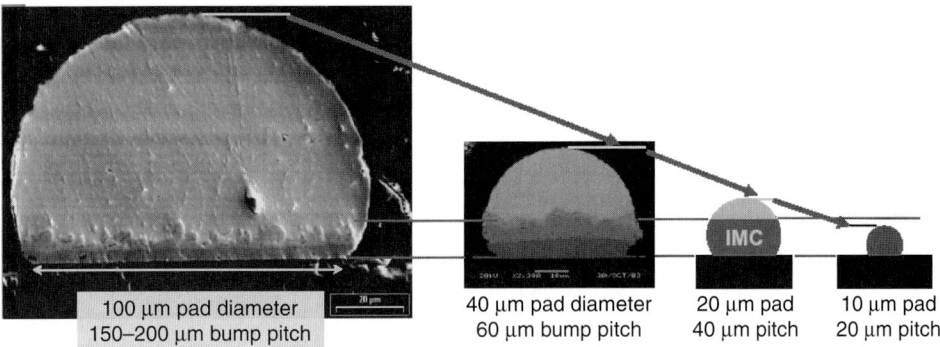

Figure 10.27 Dominating role of intermetallics with scaling of lead-free solders at fine pitch. (Courtesy Eric Beyne, IMEC Belgium.)

The Ni barrier is widely investigated for fine-pitch lead-free solders because of its lower reactivity with tin compared to copper. The reactivity of Ni-based barriers with solders is very sensitive to the alloy composition. Sn-3.5Ag solder has a slower growth rate for Ni_3Sn_4 on both Ni and NiAu substrates, resulting in only about half the IMC thicknesses when compared to Sn on the same substrates. In the cases of the joints with copper-bearing solders, Sn-3Ag-0.5Cu and Sn-3.5Ag-0.8Cu, a single $(Cu,Ni)_6Sn_5$ interface layer grows by fast copper segregation from liquid solder to the interface layer on soldering. The presence of Cu in the solder enhances the formation of the Cu_6Sn_5 intermetallic layer at the interface resulting in the prevention of Ni diffusion to liquid solder. The IMC initially formed can sometimes move away from the surface during multiple reflows. The majority of Ni_6Sn_5 spalls away from the electroless Ni(P) surface finish, while the spalling is absent in the electroplated Ni. Solders containing Cu are known to form $(Ni,Cu)_6Sn_5$ with better adhesion to electroless Ni(P). The sensitivity of lead-free solders to its composition, barriers, and interfaces makes it a relatively complex system compared to lead-tin solders. The surface finish is also critical to improve wetting and pull strength. Electroless nickel gold (ENIG), Ni-Pd/Au, and Ag finish are reported to have better wetting, whereas organic surface protection (OSP) and Hot Air Solder Leveling (HASL) are known to have better pull strength.

Mechanical Properties of Solders Better fatigue properties of SAC when compared to PbSn at low strain ranges and their higher strength favor lead-free solders to PbSn. Lead-free solders are also known to have a higher creep resistance compared to PbSn solders. The higher strength and creep resistance, however, comes at the expense of ductility. At higher strains, SAC cannot readily accommodate deformations and become more susceptible to crack propagation than ductile PbSn. The magnitude of strain energy density for Pb-free solders is less for the same chip-package thermomechanical load. However, the ductility of PbSn solders and its lower susceptibility to intermetallics makes it more reliable in many cases.

Strengthening and toughening agents are investigated to increase the creep and fracture resistance of Pb-free solders. Precipitation of particles along interfaces can significantly reduce grain boundary sliding. Ag_3Sn platelets in tin are known to increase creep resistance by the same mechanism. For stronger solders with more Ag, fatigue cracks favor propagation near to the solder-substrate interface particularly when large nodular Ag$_3$Sn particles are formed near the interface. For weaker solders with low Ag content, the crack propagates within the solder. The alignment of the platelets with respect to the crack determines its resistance to crack propagation. The beneficial effects of a reduced Ag content in SnAgCu alloys are well demonstrated in the literature, in minimizing the formation of large Ag$_3$Sn plates [43]. Aging results in extended interfacial reaction, which leaves an easy path for crack propagation. The strain energy release rate reduces from 25 to 10 J/m² after aging.

Recently, development of new solder materials has been reported by alloying of grain-refining elements into Sn-Ag and Sn-Cu-Ag systems to achieve better mechanical properties. The presence of certain elements can lead to nanoscale precipitation strengthening in the solder [44]. The introduction of additives might also prevent grain boundary sliding during steady-state creep and hence increase its creep strength, although it might be at the expense of creep ductility. Inert, hybrid inorganic-organic, nanostructured chemicals were incorporated into lead-free electronic solders to achieve improved mechanical performance at elevated temperatures and service reliability. Organic derived nano-reinforcements with appropriate functional groups to promote bonding with the metallic matrix efficiently pinned the grain boundary of solder alloys leading to an improvement in properties [45]. More improvements in selecting the barriers, surface finish, and lead-free solder compositions are needed to meet the reliability of the emerging fine-pitch flip-chip requirements.

Electromigration Electromigration is the movement of ions by the transfer of momentum from the moving electrons to the surrounding lattice under an applied field. The rate of electromigration is mainly dependent on the current density, diffusivity of the moving species, and the microstructure. Black's equation provides a semi-empirical guideline for the electromigration failure time. The time to failure t_f is related to the temperature T and the current density J as

$$t_f = \frac{A}{J^n} \exp\left(\frac{E_a}{kT}\right)$$

The current density exponent n and activation energy E_a are determined by experiments. Electromigration can occur both on the on-chip wiring and the chip-package interconnections. This section deals with the later phenomenon. One of the most significant differences between the two is the nonuniform current density that occurs within the chip-package interconnections.

Interconnection scaling impacts not only the intermetallic layers and the fatigue life but also increases the current density, raising electromigration reliability concerns. The electrons enter the solder bump at the edge, creating a localized high-current density called current crowding. The large amount of joule heating and the package junction-to-air factor leads to actual die temperatures above 70 percent of the solder's melting temperature. Since the rate of electromigration damage is roughly proportional to the square of the current density, void nucleation will first occur at the bump edge [46]. The growing void blocks the primary current path, forcing the electrons to flow further along the conductor before entering the solder (Figure 10.28). This continues across the entire solder-silicon interface and leads to failure.

Current crowding, which is the ratio of maximum current density to the average current density in the UBM, is dependent on the UBM thickness. Solder joints with thick

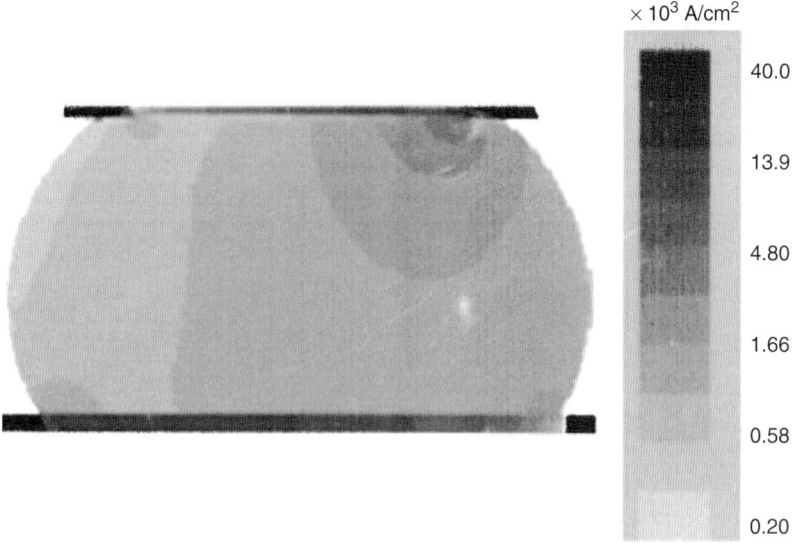

FIGURE 10.28 Schematic of current crowding and void generation [46]. The actual simulation of current density with void is taken from the work of Nah et al. [47]

UBMs were found to have a better ability to relieve the current crowding effect (Figure 10.29). With the formation of voids, the current crowding at the corner is displaced. The ability of the solder bump to dissipate heat is also diminished. This further increases void formation and reduces the contact area for heat transfer, thus producing an increase in the die temperature [47]. Black's equation and the models developed for electromigration in aluminum conductors assume that the current density is fairly uniform and constant. To adapt these models to the solder bump case, additional terms are needed to correct for the nonuniform and changing current path.

Electromigration is dependent on the diffusion under the applied electric fields. In a binary eutectic system, there are four lattice diffusion scenarios to account for the two

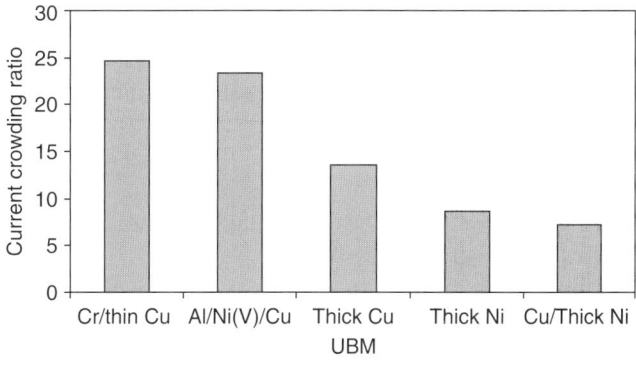

FIGURE 10.29 Current crowding for different UBM interfaces. [48]

Solder	Material UBM	Activation Energy (eV)
High Pb	Thick Cu	1.1–1.3
	Thin Cu	1.8–1.9
Pb-free	Cu UBM	0.64–0.72
	Ni UBM	1.0–1.1

TABLE 10.4 Activation Energies for Diffusion in Lead and Lead-Free Solders [49]

diffusing species and the two crystalline phases (Table 10.4). The relative diffusivities and solubilities affect the electromigration-induced failures. It is documented that tin diffuses faster below 100°C, while lead diffuses faster at above 100°C. In contrast, the Sn-rich Pb-free solders are two-phase solutions with a continuous Sn phase interspersed with intermetallic compounds of the alloying species. The IMCs are fairly resistive, carrying little of the current. Sn diffusing in an Sn-rich phase is the primary diffusion mechanism in this system. Unfortunately, the Sn grains and the IMC grains coarsen over time, often creating platelets normal to the current flow that become divergence planes for ion flux. By interrupting the ion flux, these platelets can nucleate voids by the Kirkendall mechanism on the other side of the platelet. Alternatively, platelets oriented parallel to the current flow create a long uninterrupted grain boundary that forms a continuous high-diffusivity path for migration. Vacancies are thus channeled toward the UBM, facilitating void nucleation and growth [46].

At low temperatures, Sn can migrate and react with the anode. The copper dissolution is faster in SnAgCu compared to PbSn and can accelerate the electromigration. Electromigration cracks are generally located near the UBM-solder interface for both of the high-Pb and Pb-free solder joints. This crack is either a separation between the original UBM and the newly formed IMC, or between the IMC and the bulk of the solder. It can also be caused by a complete disappearance of the UBM material itself. These failure mechanisms are illustrated in Figure 10.30. Interactions between the materials and phases at the UBM-solder interface during EM tests are conducted by Su et al. During the assembly reflow, a thin layer of CuSn IMC is formed in the solder joints and this creates two interfaces: the UBM-IMC interface, and the IMC-solder interface. When a high electric current is applied, the following processes can be viewed as occurring simultaneously [48]:

- Cu or Ni diffuses through the first UBM or IMC interphase and migrates to the second interphase between IMC and solder, reacting with the Sn to form more intermetallics. When there is abundant Cu available, the small amount of Sn in the solder will be consumed rather rapidly by the Cu that diffuses to the IMC-solder interface to form Cu-Sn IMC. Almost all the Sn originally in the solder joint is now in the IMC.
- Cu or Ni will react with the intermetallics to form high NiCu containing phases (Cu_6Sn_5 to Cu_3Sn, Ni_3Sn_4 to Ni_3Sn).
- The IMC, which is generally discontinuous, is driven by the electric current to the bottom of the joint during the test.
- The diffusion fluxes lead to voids and cracking, which further leads to accelerated failures.

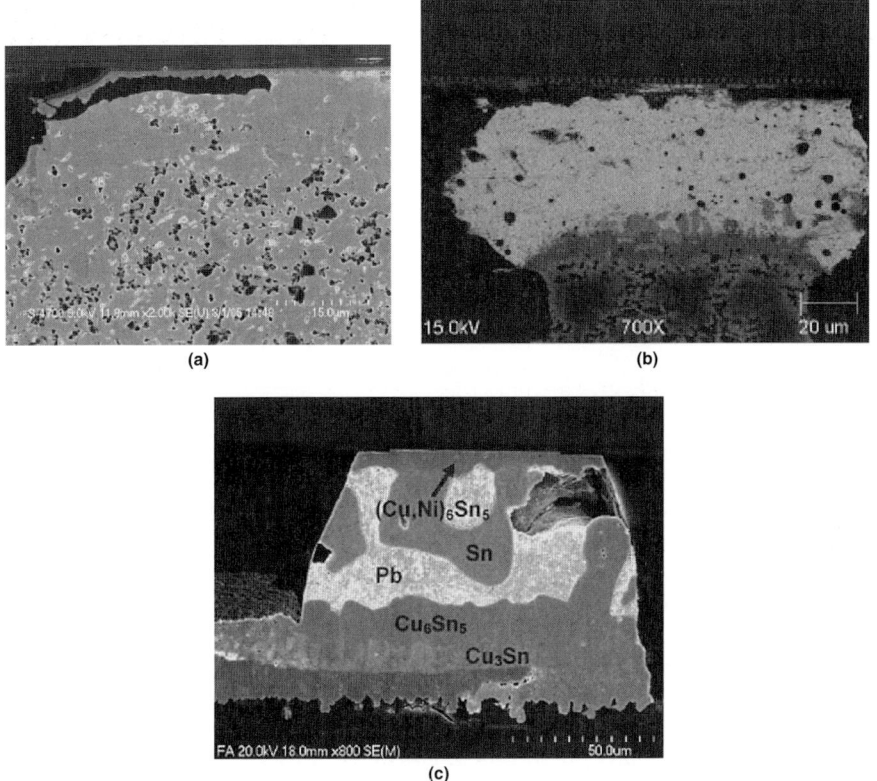

FIGURE 10.30 (*a*) Lead-free bump electromigration testing [50]. (*b*) Failure of evaporated solder joint with thin Cu as the UBM wetting layer. The Cu dissolved leaving a void [49]. (*c*) PbSn structure after high-temperature storage testing [51].

Electromigration will become a significant problem for high-power fine-pitch applications. New UBM, barrier, and interconnection strategies are being developed to improve the electromigration resistance.

Lead-Free Solders at Fine Pitch A number of process challenges for electroplating Pb-free solder bumps at fine pitch are discussed here. A thick photoresist with high resolution is needed to get near-vertical sidewalls to prevent bridging during reflow. A low-resolution photoresist can have a much larger bottom opening than the top opening that can cause solder ball bridging during assembly at fine pitch. An appropriate etchant should be selected that can completely dissolve the seed layer without excessive undercut. At finer pitch, due to the small size of the bumps, the bumps cannot tolerate a lot of undercut or nonuniformity during the etching. The etching nonuniformity also results in an uneven distribution of bumps on the chip. A good stripping chemical is needed to completely remove the photoresist without leaving any residues between the bumps [52]. For fine-pitch interconnection, the surface-to-volume ratio of the bump is high, which makes it difficult to reflow without sufficient flux. On the other hand, flux cleaning becomes difficult at fine pitch because of the small space between the bumps. Hence, a very good fluxless reflow process is required for solders at fine pitch. The

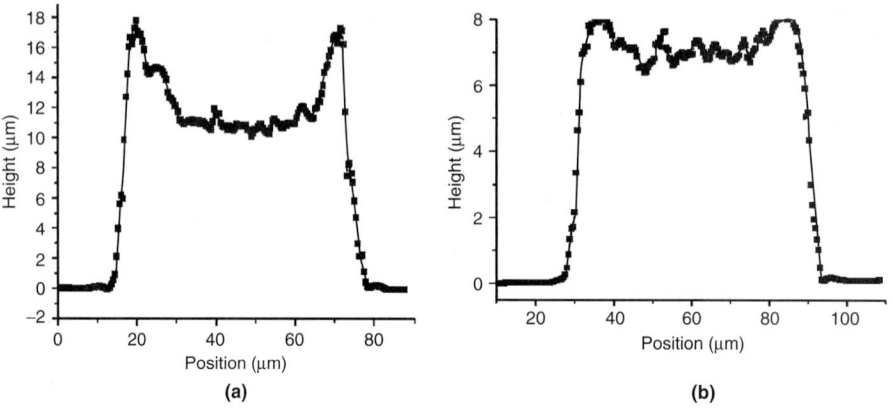

Figure 10.31 Height profile of electroplated lead-free bump. (*a*) Lower current density. (*b*) Higher current density.

uniformity of the bump height/volume ratio within the chip and across the wafer is another issue that can significantly affect the wafer-level testing and flip-chip bonding. In a wafer-level solder plating study at IBM, the uniformity of SnCu bumps was very good in the center region, but not good around the peripheral area. Sn plating is much more sensitive to the plating current change (Figure 10.31). The top morphology is also sensitive to the plating current.

Electroplating of lead-free solders with accurate control in compositions is more difficult than plating pure metals. The large electrochemical potential difference between tin and silver or tin and copper in these alloys requires special handling of the plating bath with appropriate chelating and complexing agents to manage the differences in the reduction potentials. Co-electroplating of ternary and quaternary lead-free alloys will further increase process complexity and cost, since the difficulties of controlling bath chemistry and avoiding contamination multiply with the number of baths required. The role of processing parameters is schematically illustrated in Figure 10.32. Wafer-level plating is now widely established for binary alloys with several companies already plating SnCu (TLMI, Inc.), Aurostan, or AuSn (Technic, Inc.).

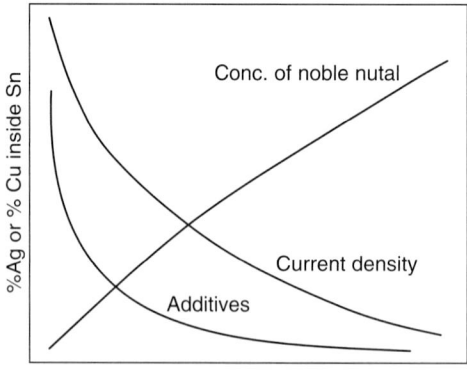

Figure 10.32 Effect of processing parameters on the plating composition. [54]

Several alternate techniques are pursued for wafer bumping. These have been extensively reviewed before [53]. Three of them are briefly mentioned here.

> *Solder injection method.* In IBM's bump transfer process, molten solder is injected into a mold with cavities. In a separate bumping step, the premolded solder bumps are transferred to wafer bond pads.
>
> *Supersolder method.* Solder is chemically generated through a substitution reaction between tin and organolead salt. A paste composed of tin powder, organolead salt, and flux is applied on the wafer and heated to deposit solders on the wafer. No stencils or masks are needed in this technology. Pad oxidation is minimum with this technology because the solder deposition is done in a reducing atmosphere.
>
> *Tacky Dots method.* A photoimageable adhesive is coated on a support and patterned with a photo tool to form a pattern of tacky dots. Solder particles are then coated on the patterned film such that each tacky dot holds only one particle. The patterned solder film is then reflowed and transferred onto a wafer substrate.

Most of these technologies are not reported to be scalable below 50 to 100 μm unlike electroplating technologies. Novel nanotechnology concepts may enable down-scalability of these fabrication routes.

Alternatives to Solder: Copper Interconnections

Copper interconnections utilize copper instead of solder as the primary interconnection material to provide performance and process advantages compared to traditional solders. Copper can be electroplated to various shapes and sizes and therefore has flexible current and heat-handling capabilities. As only solder is reflowed to create the joint, a consistent standoff can be maintained through the nonmeltable copper portion, which further enables the downstream processes of flip-chip packaging. For example, it can provide a wider window for the underfilling process and allows the use of a maskless substrate. Most semiconductor foundries are now equipped with advanced copper electroplating systems that can be easily integrated with the backend processes.

The interconnect signal delay between the IC and the package depends strongly on the interconnect material used. Electrical conductivity of pure copper is $5.96 \times 10^7 \, \Omega^{-1} \cdot m^{-1}$ as compared to a conductivity of $6.9 \times 10^6 \, \Omega^{-1} \cdot m^{-1}$ for PbSn eutectic solders. Solders are weaker materials with much lower yield stress, leading to higher plastic strains and inferior creep resistance compared to copper. Copper, however, has a higher stiffness which induces more stresses in the interfaces.

Thermal and Electromigration Resistance As the power continues to increase, the electric current through the bump increases leading to a higher bump temperature. The bump temperature is reaching a level where electromigration is becoming a severe concern. Joule heating is dependent on the material resistance. As discussed before, rapid dissolution of the UBM and the effect of current crowding cause asymmetrical UBM dissolution. Electromigration-related failures can be controlled with copper pillar bumps. Current crowding occurs at the entrance from die conductor trace into the bump as illustrated in Figure 10.33. In a typical solder bump structure, high current density at the UBM-solder interface causes dissolution of the UBM layer. A copper pillar consists of a UBM layer, a copper segment, and a solder segment as illustrated. After

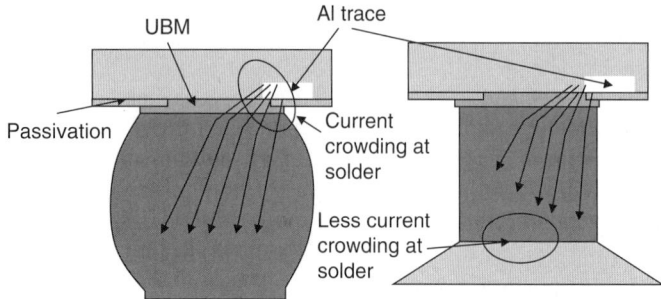

Figure 10.33 Schematic representation of electromigration mechanism in solder and Cu pillar bumps.

electrons pass through the entrance into the bump material, the current density is distributed throughout the bump. The current density at the copper-solder interface is much lower as compared to the metal-solder interface in the solder bump. A study at Intel has shown that solder bumps undergo a sharper increase in temperature compared to copper bumps due to the higher electrical resistivity of solders compared to copper [55]. Thus, copper bumps have better resistance toward electromigration compared to solder bumps.

Copper Post or Pillar Processes Copper interconnections are formed by wafer plating using relatively thick dry film or wet photoresists. The Cu interconnections fabrication process cost using spin-on-photoresist is high since 90 percent of resist material is wasted during spin-coating. Spin-on resists also need multiple coatings to obtain high-aspect-ratio interconnections. Commercial dry film photoresists have good resolution, uniform thickness, ease of processing, and can be easily stripped in diluted alkaline (potassium hydroxide) solution. They are also a low-cost resist material compared to spin-on resists. Dry film thickness layers of 80 and 120 µm can be achieved in a single lamination with very good copper adhesion and excellent chemical resistance to acid electroplating baths such as copper sulphate and solder fluoborate (tin-lead, tin), acid cleaners, ammonium persulfate, and dilute sulfuric acid. The patterned wafer is descummed using O_2 plasma before electroplating to clean the bottom of the mold and to activate the resist molds for copper plating. The wafer is also dipped for 5 minutes into a cleaner solution that contains surfactants to improve the wettability of openings.

Cu pillar fabricated using dry film has a larger contact area at the chip side than the contact area when it is fabricated using spin-on positive photoresist. Because of this larger contact in Cu pillar structure, the maximum von Mises stress in the Cu pillar is less compared to the stress in the Cu pillar fabricated using spin-on photoresist. Hence, Cu pillar interconnections fabricated using such dry film photoresist showed better reliability. The Cu column interconnection fabrication process was demonstrated on 8-in wafers with 50 µm diameter, 120 µm height, and 100 µm pitch by IME, Singapore, on a 20 mm × 20 mm die. Figure 10.34 shows Cu column interconnections at 100-µm pitch. Copper post with solder shows the potential to address the electromigration and reliability issues at fine pitch [56–57].

Stud Bumping
Stud bumping or ball bumping is done by transferring the bump onto a bonding pad from a wire using a modified wire-bonding process. In a typical wire-bonding process,

FIGURE 10.34 Area array of the high-aspect-ratio copper interconnections at 100-μm pitch.

a gold ball is forced down and thermosonically bonded to a die bond pad to form the first connection in a wire bond. With the ball connected, the wire is then fed out and attached to a second surface to complete the connection. The ball bumping process is a variation of this wire-bonding operation. In the ball bumping process, the wire is snapped off after the ball is initially connected to the die. The resulting gold bump (also known as a stud) is firmly connected to the first surface. Because of the maturity of the wire-bonding process, the reliability of these bump connections is well established and documented.

The shape of the bump is an important process characteristic in ball bumping, since it helps define the area of the gold ball that will contact the second surface, and thus, the conduction path. Methods of attaching the gold ball include thermocompression and thermosonic bonding. Gold stud bumping equipment is available to create the desired shape in one step as the ball is being made. The typical bump shape with a tail is the common shape produced by many gold wire bumping machines. For those, a separate coining machine and an additional step to flatten the bump may be required. To achieve the best interconnect across the entire array of contacts on a chip, bump height coplanarity is essential. The term "coplanarity," used in flip-chip bonding, refers to the height consistency that exists across the top of all bumps. Height variations can lead to an uneven distribution of forces, die fractures, and open circuits. Current requirements for coplanarity call for less than 5 μm of variation in bump height across the entire die.

Plated and ball bumps vary primarily in material and geometry. Plated bumps can be solder, nickel, gold, or other materials, while ball bumps are primarily gold. In a comparison of the properties of the two materials, lead (and its alloys) has an electrical resistivity of 22 μΩ · cm while that of gold is 2.19 μΩ · cm. Since conductivity is the reciprocal of resistivity, gold offers an order of magnitude better electrical conductivity than lead. Plating allows for finer pitch and shorter bump height, while ball bumping allows taller bumps of varying top profiles, but at a larger pitch and with a round footprint. Plated bumps require underbump metallization (UBM), while ball bumping requires no additional wafer processing. The cost comparison depends largely on the number of bumps per wafer, because the metal plating cost is based on the number of wafers being bumped, no matter how many bumps are on a wafer [58]. Figure 10.35 compares the relative cost of gold ball bumping versus plating. Based on a 150-mm wafer using electroless nickel and immersion gold, the current cost is between $40 and $75 to plate a wafer. Solder deposition costs between $75 and $120, and electroplated gold between $80 and $120. In gold ball bumping, the cost is based on the number of bumps fabricated. Ball bumping costs include the time to load the wafer, put the bump down, and remove the wafer. This offers a clean, simple, one-step process that can be done in a captive or contract manufacturing facility, while plating requires many complicated process steps. Gold bumping provides flexibility in shape and substrate interconnect methods and conductivity benefits derived from the gold-to-gold connection and strength of the weld. Plating is preferred if a flat bump or a rectangular shape is desired. However,

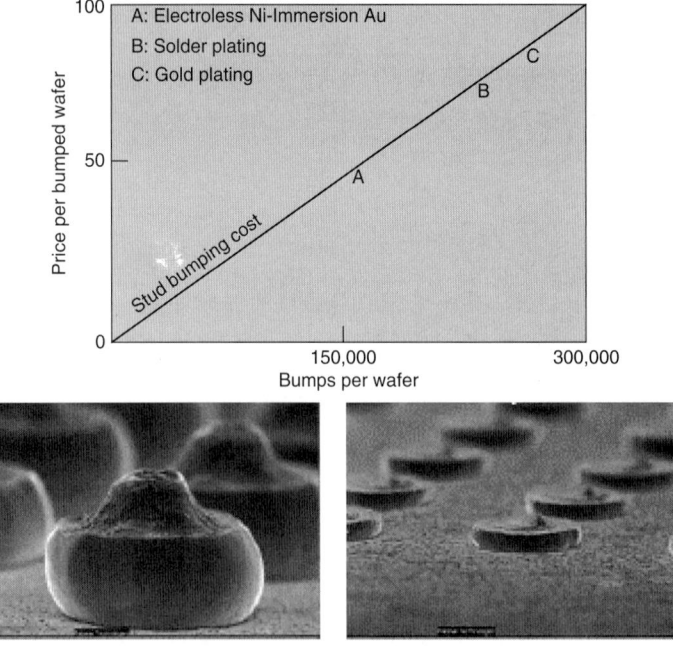

Figure 10.35 Costs of gold ball bumping versus plating. [58]

new flat-top ball bumps have given designers another choice not previously available. When applications call for less than 250,000 bumps a wafer, gold ball bumping is reported to be the more cost-effective choice compared to wafer plating.

Wafer-Level Packaging with Wafer-Applied Underfill

Underfills are polymer-based layers that bond the chip to the package to improve the solder joint reliability by redistributing the global thermomechanical stresses from 100 percent on solder joints to 100 percent on solder joints and underfill polymers. The usage of underfill lowers the solder joint strains typically by a factor of 5 leading to dramatic improvements in reliability. It is noted that in some types of compliant WLPI, a polymeric layer is used on the wafer scale to relieve the stresses and enhance the reliability. However, this polymeric layer usually does not glue with the substrate and cannot be considered as underfill. The wafer-level underfill discussed here is an adhesive to glue the chip and substrate together and functions as a stress-redistribution layer rather than a stress-buffering layer. Underfills are most effective in reducing the solder joint stresses when they have low CTE and high modulus. Typical underfills are loaded with silica to meet the target thermomechanical properties.

In the conventional underfilling process, the underfill is dispensed after the formation of electrical solder joints by a capillary process (Figure 10.36a). Conventional flip chip with

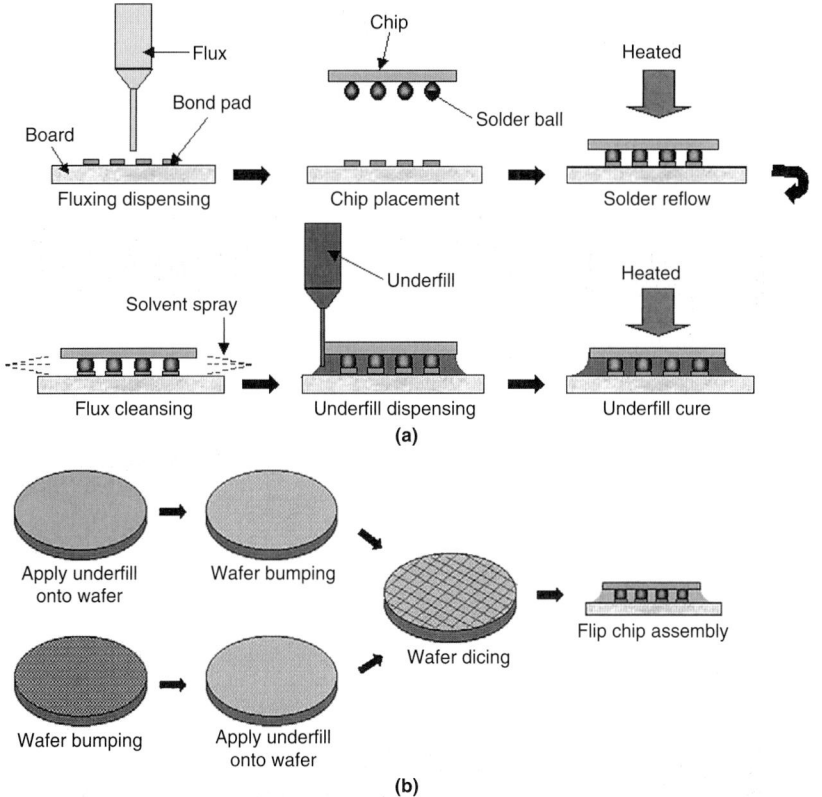

FIGURE 10.36 Capillary underfill versus no-flow wafer-level underfill. (*a*) Traditional capillary underfill process. (*b*) No-flow underfill process steps.

underfill is, however, tedious and requires four separate steps: flux dispensing, solder reflow, underfill dispensing, and underfill cure. For shorter interconnections, the capillary force may not be sufficient to drag the underfill to the center of the chip. To increase the throughput and eliminate some equipment costs required for high-volume assembly, no-flow underfill processes were developed. In the no-flow process, underfill is dispensed on the wafer, eliminating the flux dispensing and cleaning steps. The process combines the solder reflow and underfill curing into a single step (Figure 10.36b).

The majority of current Wafer Level Chip Scale Packages (WLCSPs) do not need underfill because of the smaller package dimensions and coarse pitch. With larger dies and increased I/O count, underfill is becoming necessary as an SMT-compatible flip-chip process to achieve high throughput, low cost, and high reliability [59–62]. It has also been used as WLSCPs as well to enhance their board-level reliability (e.g., MicroFill by National Semiconductor [63–64]). In this process, the underfill is applied either onto a bumped wafer or a wafer without solder bumps, using a proper method, such as printing or coating. Then the underfill is B-staged and the wafer is diced into single chips. In the case of unbumped wafer, the underfill can also be used as a mask for wafer bumping. The individual chips are then placed onto the substrate by standard SMT assembly equipment.

The attraction of the wafer-level underfill lies in the low-cost potential since it does not require a significant change in the wafer backend process and high reliability of the assembly is enhanced with the underfill. However, the wafer-level underfill faces critical material and process challenges including uniform underfill film deposition on the wafer, B-stage process for the underfill, dicing and storage of B-staged underfill, fluxing capability, shelf life, solder wetting in the presence of underfill, desire for no postcure, and reworkability. Since the wafer-level underfill process suggests a convergence of the front end and back end in package manufacturing, close cooperation between chip manufacturers, package companies, and material suppliers is required. Several collaborative research programs have been carried out in this area [65–67]. Innovative ways of addressing the above issues and examples of wafer-level processes are presented in this section.

In most wafer-level underfill processes, the applied underfill must be B-staged before the singulation of the wafer. The B-stage process usually involves partial curing, solvent evaporation, or both, of the underfill. In order to facilitate dicing, storage, and handling, the B-staged underfill must appear solidlike and possess enough mechanical integrity and stability after the B-stage. However, in the final assembly, the underfill is required to possess "reflowability," that is, the ability to melt and flow to allow the solder bumps to wet the contacting pads and form solder joints. Therefore, control of the curing process and the B-stage properties of the underfill is essential for a successful wafer-level underfill process. A study conducted at Georgia Tech utilized the curing kinetics model to calculate the degree of cure (DOC) evolution of different underfills during the solder reflow process [68]. Combined with the gelation behavior of the underfills, the solder wetting capability during reflow was predicted and confirmed experimentally. Based on the B-stage process window and the material properties of the B-staged underfill, a successful wafer-level underfill material and process was developed. A full area array at the 200-µm pitch flip-chip assembly with the developed wafer-level underfill was also demonstrated [69].

The above study shows that the control of the B-stage process of the wafer-level underfill is critical to achieve good dicing and storage properties and the solder interconnect on board-level assembly. One way to avoid dicing in the presence of

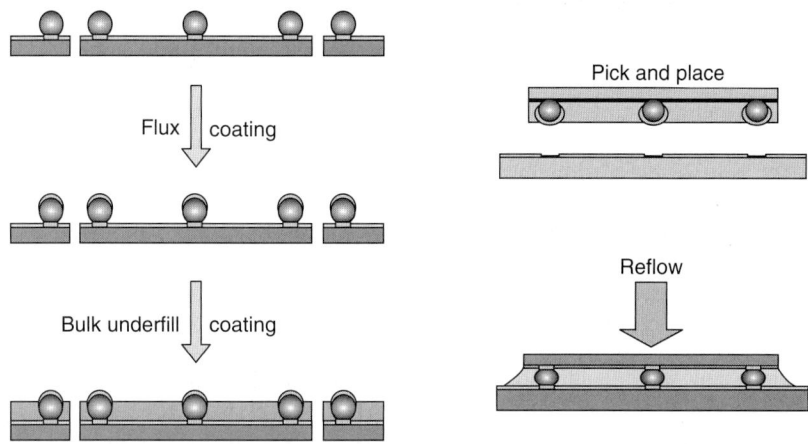

Figure 10.37 A wafer-scale applied reworkable fluxing underfill process.

non–fully cured underfill is presented in Figure 10.37, which is a wafer-scale applied reworkable fluxing underfill process developed by Motorola, Loctite, and Auburn University [65]. In this process, the wafer is diced prior to applying the underfill coating since uncured underfill materials are likely to absorb moisture that leads to potential voiding in the assembly. Two dissimilar materials are applied; the flux layer coating by screen or stencil printing and the bulk underfill coating by a modified screen-printing to keep the saw street clean. The separation of the flux from the bulk underfill material preserves the shelf life of the bulk underfill as well as prevents the deposition of fillers on top of the solder bump so as to ensure the solder joint interconnection in the flip-chip assembly.

Underfill deposition on wafer using liquid material via coating or printing requires subsequent B-staging, which is often tricky and problematic. The process developed by 3M and Delphi-Delco circumvents the B-stage step using film lamination [70]. The process steps are shown in Figure 10.38, in which the solid film comprised of a

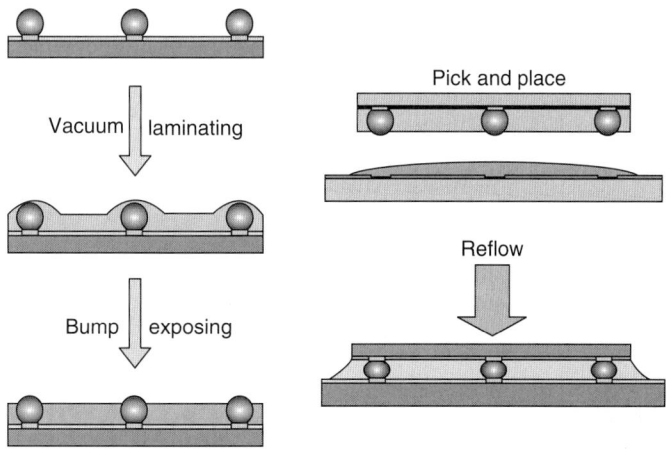

Figure 10.38 A wafer-applied underfill film laminating process.

thermoset-thermoplastic composite is laminated onto the bumped wafer in a vacuum. Heat is applied under a vacuum to ensure the complete wetting of the film over the whole wafer and to exclude any voids. Then a proprietary process is carried out to expose the solder bump without altering the original solder shape. The subsequent SMT assembly is carried out with a curable polymeric flux adhesive preapplied on the board.

Wafer-level underfill can also be applied before the bumping process. A novel photodefinable material that acts both as a photoresist and an underfill layer applied on the wafer-level was reported by Georgia Tech [71]. In the proposed process shown in Figure 10.39, the wafer-level underfill is applied on the unbumped wafer, and then is exposed to the UV light with a mask for crosslinking. After development, the unexposed material is removed and the bump pads on the wafer are exposed for solder bumping. The fully cured film is left on the wafer and acts as the underfill during the subsequent SMT assembly after device singulation. A polymeric flux is needed during the assembly for holding the device in place on the board and providing fluxing capability, a process similar to the dry film laminated wafer-level underfill. In order to enhance the material property, the addition of silica fillers is necessary. In this case, nanosized silica were used to avoid UV light scattering which hinders the photo-crosslinking process. It also resulted in an optical transparent film on the wafer to facilitate the vision recognition during the dicing and assembly process. The photo-definable nanocomposite wafer-level underfill presents a cost-effective way of applying wafer-level underfill and has potentially fine-pitch capability.

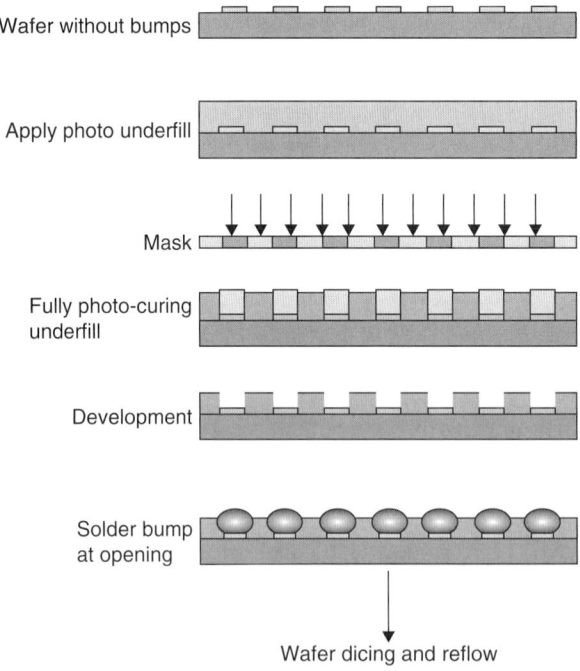

Figure 10.39 A photo-definable wafer-level underfill process.

Nanoscale Interconnections

Nano-interconnections utilize nanostructured or nanoscale materials to improve the interconnection:

- Scalability (submicron to micron pitch)
- Electrical properties (current carrying capability, lower R, L, C)
- Mechanical properties (strength and fatigue resistance)
- Processability (e.g., low-temperature assembly)

compared to their microscale counterparts, and thus offer novel solutions to fine-pitch wafer-level packaging. As an example, metallic single-wall carbon nanotubes (SWCNTs) display ballistic electron transportation, outstanding current carrying capabilities, and thermal conductivities, which minimizes the joule heating and electromigration issues commonly encountered with conventional interconnect materials. The superior electrical and mechanical properties of SWCNTs enable novel chip-to-package interconnections with several orders of magnitude higher interconnect densities and reliability. Nanostructured metals and metal alloys with improved strength and fatigue resistance make interconnections of high electrical performance possible without trading off reliability. Nanometallic particles also show depression of melting point and enhanced fusion at lower temperatures and thus better processability. Combining nanostructured metals with low-stiffness polymer core layers can further lower the stress and enhance reliability. Nanoscale conductive adhesives are also expected to lower the contact resistance, lower the interfacial stresses, improve fine-pitch capability, and provide polymer-compatible metallurgical bonding. Nano-wafer-level packaging, combined with a reworkable interface, wafer-level test, and burn-in is expected to be a cost-effective solution to meet the demands of future systems with aggressive I/O counts, interconnect density, and electrical and mechanical goals. Nanomaterials and processes leading to nano-WLPI are discussed below.

Nanostructured Metal Interconnections

Properties of Nanostructured Metals   Metals such as Ni and Cu when produced in nanocrystalline grain sizes (10 to 50 nm) have been shown to possess dramatically enhanced resistance to deformation and potentially high resistance to fatigue and fracture without significant increases in electrical resistance [72]. Figure 10.40 shows the stress-strain plots for nanocrystalline Cu and Ni. The tensile strength of nanocrystalline Cu was 456 MPa and that of Ni was 897 MPa, with the yield strength being just a bit smaller than the tensile strengths. This represents a five to six times increase in strength of these materials compared to their conventional microstructured forms. This follows the well-known Hall-Petch trend for metals indicating that grain refinement leads to enhanced strengthening. Nanometals are ideally suited for high-density interconnections that are needed for nano-wafer-level packaging. The mechanical properties of nanocrystalline and microcrystalline copper are compared in Table 10.5. On the basis of the stress-strain curve in the plastic deformation range, a depressed strain-hardening effect with respect to that in the conventional polycrystalline Cu and Ni can be noticed.

Nanocrystalline equichannel angular extruded (ECAE) copper and nickel clearly show improved fatigue resistance over their coarse-grained counterparts as shown in Figure 10.41. For electrodeposited nanonickel, however, the crack propagation rate was

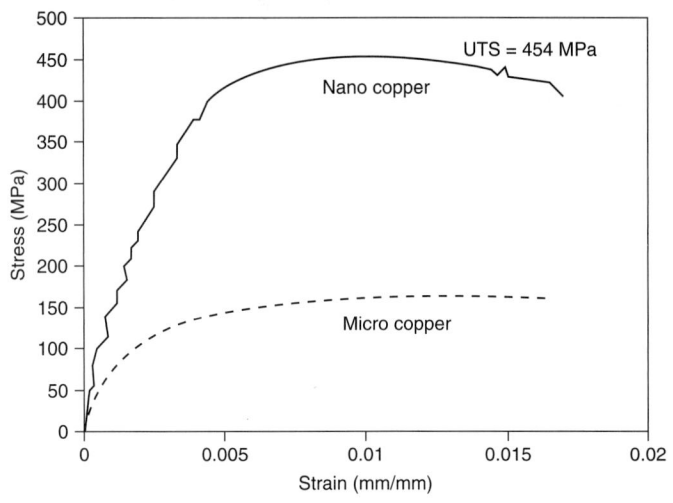

FIGURE 10.40 Stress-strain curves for (a) copper and (b) nickel for tensile tests. The ultimate tensile strength and yield strength have been indicated on the curves.

	Copper	
	Microcrystalline	**Nano**
Average grain size (nm)	4-5 μm	59.13
Activation energy (kJ/mol)	~ 100	33.43
Modulus (GPa)	110	100
Poisson's ratio	~ 0.33	0.259
Ultimate tensile strength (MPa)	~100	454
Yield strength	~100	437
Endurance limit (MPa)	~85	~370

TABLE 10.5 Summary of Results for Nanocopper Compared to Microcrystalline Copper

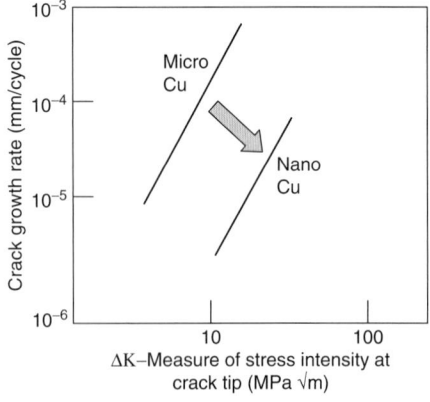

FIGURE 10.41 da/dN versus ΔK behavior for nanocrystalline copper and nickel.

reported to be higher compared to the microstructured nickel. Mechanical properties of nanometals are extensively reviewed by Kumar et al. [73]. They suggest that grain refinement strategies may sometimes also lead to a detrimental effect on subcritical crack growth because it enhances grain boundary–assisted void nucleation and void coalescence at the crack tip.

Figure 10.42 shows the total strain life and high cycle fatigue (HCF, fatigue in the elastic strain regime) results as a function of number of cycles to failure. It can be noticed from this plot that nanocrystalline copper and nickel show enhanced fatigue resistance both in endurance limit and number of cycles to failure as compared to microcrystalline copper and nickel. The low cycle fatigue (LCF, fatigue in the plastic strain regime) performance is, however, reduced because of lower ductility of the nanocrystalline material. This indicates lower resistance to fatigue crack initiation under high strain conditions possibly because of a higher fraction of grain boundaries. Grain boundaries impart improved resistance to fatigue crack nucleation under predominantly high-cycle fatigue loading [72]. The results for grain size measurements, Vickers hardness, tensile tests, fracture toughness tests, and hardness and modulus measurements are compiled in Table 10.5.

It can be clearly noted from the above results that nanocrystalline copper and nickel have superior mechanical and electrical properties and are potential candidate materials for off-chip interconnections. The fracture toughness values are also quite high for these strength levels and are well suited for the present applications. Nanograined copper undergoes grain coarsening over long periods of time even at room temperature. This problem can be solved by doping copper with small amounts of impurity elements below the solid solubility limits to pin the grain growth while not affecting the excellent properties achieved in the nanometer regime.

Metallic nanowires are also considered for nano-interconnections. While nanowires lead to fine-pitch interconnects with better mechanical properties, the benefits in terms of electrical properties are still not clearly established. The electrical resistivity of bulk materials deviates from the bulk properties as the lateral dimension of metals approaches the nanoscale regime. Size effects come into play as the lateral dimension of the wire is in the range of the mean free path of the conduction electrons and below,

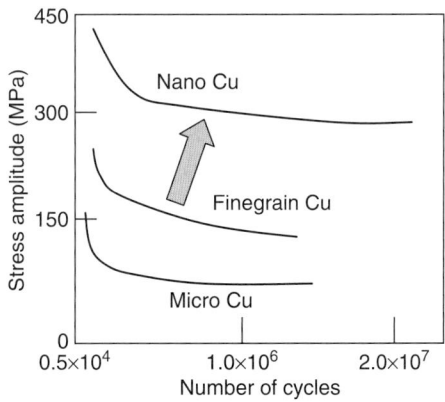

Figure 10.42 Stress versus number of cycles to failure for tension-tension fatigue tests under a load control mode for nanocrystalline and microcrystalline copper.

which is about 40 nm for copper at room temperature. For sizes less than the mean free path, the relaxation time will be lower leading to a higher probability of electron collision and scattering at the external surfaces of the film or wire. Electron scattering at grain boundaries also increases the electrical resistivity of a thin film [74]. Steinhogl et al. recently measured the resistance of Cu nanowires in a silica matrix in the temperature range from 77 to 573 K [75]. A size-dependent increase of the resistivity was found for decreasing wire widths. For the narrow wires, the resistivity is a factor of 2.6 higher than the copper bulk value (1.75×10^{-6} $\Omega \cdot$ cm). Diffusive scattering of the conduction electrons at the surface and the grain boundaries of the wire are attributed to the increase in resistivity. Vertically aligned nanowires can be created by electro-deposition of metal through nanochannels both in inorganic (silicon, alumina) and organic [poly(ethylene tetrapthalate), polycarbonate] templates. The nanochannels can be defined through nanolithography methods such as e-beam, x-ray, and ion-irradiation [76–77].

Processes for Nanostructured Metals Electrochemical plating is the preferred method to produce nanostructures with dimensions ranging from submicron to hundreds of microns, because of the widely characterized processes and standard infrastructure. The grain refinement of electroplated copper can be controlled by incorporation of additives in the aqueous copper sulfate plating bath that regulates and distributes the delivery of copper to the plated surface. Standard additives are a suppressor or carrier (e.g., polypropylene glycol) that controls the plating rate; a brightener, accelerator, or catalyst (compounds with divalent sulfur, such as thioethers, thiocarbamates, dithioethers) that increases the plating reaction; and a leveler (nitrogen-bearing heterocyclic or nonheterocyclic compound) that forces the copper to deposit as small equiaxed crystals by replacing the accelerator and suppressing the high-current density protrusions. In addition to the organic additives, chloride ions are generally added to work in conjunction with the carrier to reduce polarization and help refine the deposit morphology. When consumable anodes are used, chloride ions aid anode corrosion, setting up a uniform and adherent anode film. By proper balancing of these additives and maintaining them within specific parameters, nanograined copper plating can be achieved. Lower current densities and reverse pulse electroplating, as opposed to DC plating, also aid in the structure refinement. As-deposited copper undergoes recrystallization and grain growth, a phenomenon referred to as self-annealing. Sheet resistance of freshly electroplated copper is not stable because of the changes in the grain structure. Though electroplating leads to nanograined copper, incorporation of additives may be essential to retain the nanostructure.

Carbon Nanotubes Carbon nanotubes (CNTs) are a tubular form of graphitic carbon with diameters of 1 nm and above and lengths ranging up to hundreds of micrometers. The tubes can be multiwalled (MWCNTs) or single-walled (SWCNTs). Carbon nanotubes have the capacity to carry very high current densities up to 10^9 A/cm^2, which is orders of magnitude greater than for metals or doped silicon. Nanotubes, like graphite, are excellent heat conductors and will not deteriorate even at elevated temperatures. Carbon nanotubes and their application in micro- and nanoelectronics have been widely researched in recent years. Recent measurements of the properties of carbon nanotube FETs have demonstrated their superior properties compared with silicon-based devices (Table 10.6). These properties have given rise to high hopes for future nanotube-based electronics. However, one of the major problems facing nanotube-based electronics concerns the contacts that are prone to failure.

Materials	Summary of Properties
Single-wall carbon nanotubes [78]	Quantized resistance: 12.6 kΩ Kinetic inductance: 16 nH/μm Quantum capacitance: 100 aF/μm Current carrying capability: 10^9 A/cm^2 Mean free path: 5–10 μm Bending modulus: 600–1200 GPa (diameter: 10 nm) Bending modulus: 200 GPa (diameter > 25 nm) Tensile strength: 1.2 GPa [79]
Cu nanowires	Resistivity: 2.6 times that of bulk copper Mean free path: 40 nm [surface and grain boundary scattering increases resistivity when interconnect and grain size dimensions approach this (e.g., resistivity of 90-nm line is 2.4–3.2 $\mu\Omega \cdot$ cm)] Current carrying capability: 10^8 A/cm^2

TABLE 10.6 Properties of Nanomaterials

PECVD or CVD Carbon Nanotubes and Assembly Most of the research on integrating carbon nanotubes for electronics is focused on horizontal on-chip interconnects. Very few studies are focused on vertically grown nanotubes for chip-package interconnections. Researchers from Oak Ridge National Laboratories (ORNL) produced vertically aligned carbon nanofibers (VACNF) and have shown that their synthesis and assembly is highly controllable in terms of their location, length, tip diameter (20 nm), and shape with prepatterned catalysts [80]. The substrates used were (001) silicon wafers coated with SiO_2 (100 nm) with a thin layer (few nanometers) of metal catalyst, which are then put into a chemical vapor deposition (CVD) tube housed in a furnace at 700°C for VACNF growth. As an alternate low-temperature approach, the xylene-ferrocene based CVD process was also demonstrated for the deposition of aligned carbon nanotubes [81]. The ferrocene can act as the nanotube nucleation initiator, and xylene is the carbon source. The furnace was heated to 600 to 800°C at atmospheric pressure. The reaction time is usually more than 20 minutes. The experimental results showed that the aligned CNTs can be wetted by the molten solder, indicating the feasibility of this novel structure for chip-to-module interconnects. To increase the mechanical robustness of MWCNTs, a slightly modified PECVD process was used by Meyyappan et al. [82]. A silicon wafer with 200-nm Cr or Ta lines is used to deposit a thin Ni catalyst layer. PECVD is used to grow the MWCNT array, and the space between individual MWCNTs is gap-filled with SiO_2, by TEOS CVD. This is followed by CMP to provide a flat-top surface leading to mechanically robust uniform nano-interconnects.

Nanoparticle Conductive Adhesives Conductive adhesives are metal polymer composites with the metal fillers forming a conducting network to enable current flow. There are three types of polymeric adhesives used for interconnections (Figure 10.43):

1. *Anistropic conductive adhesives.* Adhesives with a low metal filler content so as to prevent conduction in the xy plane but enable conduction in the z direction from the pressure-assisted contact between the particles during the assembly process.

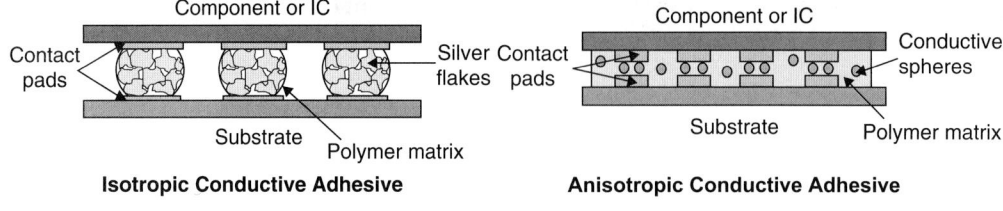

Figure 10.43 Conceptual schematic for ACA and ICA.

2. *Isotropic conductive adhesives.* Adhesives with metal filler content above a threshold limit (known as the percolation limit) forming a conducting network in all directions.

3. *Nonconductive adhesives.* Adhesives with no conducting filler but that enable pad-to-pad contact for electrical connections by compressive stress during assembly and adhesive curing.

Anisotropic conducting adhesives (ACAs) are known to have fine-pitch capability due to the small-size conductive elements in the polymer matrix. Conventional ACAs have conductive particles around 3 to 5 μm, allowing an interconnection pitch down to 40 μm. Using conducting nanoparticles (<20 to 30 nm) enables scaling down of the interconnect pitch close to 1 μm and even to submicron dimensions. Though conductive adhesives have several promising characteristics, lower electrical conductivity and poor current carrying capability are some of the issues that are preventing its use in high-power devices, such as microprocessor and application-specific integrated circuit (ASIC) applications [83]. Instead of a metallurgical joint formed with solder reflow, the electrical conductivity in the case of conductive adhesives relies on the physical contact between the conductive particles and the metal bond pads. Reflowing the conductive elements in conventional conductive adhesives is prohibitive on organic printed circuit boards due to the high melting temperature (T_m) of commonly used fillers, such as Ag, Cu, Ni, and Au.

It has been found that the melting temperatures of materials could be dramatically reduced by decreasing the size of the materials. Nanoparticles exhibit depression of melting point and faster kinetics because of their extremely high grain boundary area/volume ratio. The low-temperature sintering behavior of the nanoparticles is attributed to the extremely high interdiffusivity of the nanoparticle surface atoms, due to the significantly energetically unstable surface states of the nanoparticles. Metals such as Ag and Cu with melting points around 1000°C can be made to melt or sinter below 400°C when the particle size goes down to the nano range. Ag nanoparticles synthesized from $AgNO_3$ precursor with an average particle size of 20 nm showed densification below 200°C in a study by Moon et al. [84]. Nanopowder sintering using copper nanoparticles was also explored by Kwan et al. [85]. To accelerate kinetics, the particles were partially oxidized followed by sintering in a reducing atmosphere. Faster densification because of the oxidation-induced reactivity and electrical properties close to fully dense copper was achieved at processing temperatures less than 400°C.

The use of the nano metal particles in ACAs would be promising for fabricating high-electrical-performance ACA joints through the low-temperature sintering behavior of the nanoparticles since a metallurgical joint can be formed between metal fillers and contact pads. The application of nanosized particles can also increase the number of conductive fillers on each bond pad and result in more contact area between fillers and bond pads. Therefore, the application of nanosized particles has the potential to improve the current density of the ACA joints by distributing current into more conductive paths [86].

In order to enhance the electrical performance of ACA materials, organic monolayers have also been introduced into the interface between metal fillers and metal-finished bond pad of ACAs [87–88]. These organic molecules adhere to the metal surface and form physicochemical bonds, which allow electrons to have a higher current flow with lower electrical resistance. The unique electrical properties are due to their tuning of metal work functions by those organic monolayers. The metal surfaces can be chemically modified by the organic monolayers, and the reduced work functions can be achieved by using suitable organic monolayer coatings. An important consideration when examining the advantages of organic monolayers pertains to the affinity of organic compounds to specific metal surfaces.

Table 10.7 gives examples of molecules preferred for maximum interactions with specific metal finishes; although only molecules with symmetrical functionalities for both head and tail groups are shown, molecules and derivatives with different head and tail functional groups are possible for interfaces concerning different metal surfaces.

The thermal performance of an adhesively assembled chip is also of vast interest as power dissipation in high-performance and high-frequency devices. The current high-performance microprocessor operates in excess of 5 GHz generating enormous heat. Power dissipations have been simulated by Sihlbom et al. for both ICA (isotropic conductive adhesive) and ACA flip-chip joints [89]. They concluded that the ACA flip-chip joint is more effective in transferring heat to the substrate from the powered chip than the ICA joint, because the adhesive thickness is so much thinner than the ICA joints. With the introduction of a suitable organic monolayer treatment, an improved interface property between metal fillers and polymer resins could be achieved, which

Formula*	Compound	Metal Finish
H-S-R-S-H	Dithiol	Au, Ag, Sn, Zn
N≡C-R-C≡N	Dicyanide	Cu, Ni, Au
O=C=N-R-N=C=O	Diisocyanate	Pt, Pd, Rh, Ru
HO-C(=O)-R-C(=O)-OH	Dicarboxylate	Fe, Co, Ni, Al, Ag
(imidazole structure)	Imidazole and derivative	Cu

*R denotes alky and aromatic groups

TABLE 10.7 Potential Organic Monolayer Interfacial Modifiers for Different Metal Finishes

would enhance the thermal conductivity of ACA joints. Studies show that organic monolayer enhanced ACA joints with the best electrical properties also had a higher thermal conductivity. It is also reported that with the addition of high-thermal-conductivity fillers (e.g., SiC, AlN) into the ACA formulation, a higher thermal conductivity could be achieved, which also would render a high current carrying capability [90].

Nonconductive Adhesive (NCA) Bonding The interconnect joints can be formed using nonfilled organic adhesives, that is, without any conductive filler particles. The electrical connection of NCA is achieved by sealing the two contact partners under pressure and heat. Conductive joints with nonconductive adhesives provide a number of advantages compared to other adhesive bonding techniques. NCA joints are not limited in terms of particle size or percolation phenomena. They avoid short-circuiting at fine pitches. Further advantages include cost-effectiveness, ease of processing regarding the possibility of nonstructured adhesive application, good compatibility with a wide range of contact materials, and a low temperature cure. In fact, the pitch of the NCA joint can be limited only by the pitch pattern of the bond pad, rather than the adhesive materials. This NCA has shown a higher current carrying capability than the ACA joints. In addition, the contact resistance was found to be the same order of magnitude as that of the copper foils.

Similar to ACAs, the electrical conductivity for NCA is also achieved through physical contact formed under high pressure, and no metallurgical joints are formed. Therefore, it has the unstable contact resistance problem. The finite-element method (FEM) was applied by researchers to investigate the shear stress distribution induced by the coefficient of thermal expansion (CTE) mismatch. The effect of the temperature variation and the failure mechanism of the adhesive material were also studied by means of FEM [91]. It shows that failures occurring in NCA joints are caused by moisture-induced hygroscopic swelling and stress relaxation. These factors affect the degradation of the compressive force that maintains the mechanical contact in the NCA flip-chip structure. The common failure modes include interfacial delamination, bump-pad opening, as well as cracking. In order to improve the CTE mismatch problem, nonconducting silica fillers were incorporated to enhance the reliability of NCA interconnects by providing a stronger anchoring force while minimizing the shear stress deformation of joining structure [92]. It was reported that the content of nonconducting filler is a key factor that controls the basic properties of NCA materials. The addition of nonconducting fillers could noticeably affect the CTE of NCA and adhesion on flip-chip assemblies. By optimizing the nonconducting filler content, a significantly improved NCA flip-chip assembly reliability was achieved.

Nano-interconnections Summary New paradigms in IC packaging and assembly technologies by means of ultrafine-pitch nanostructured interconnect formation is essential to realize packaging of future nanoscale devices. The critical benefits of nanostructured interconnections are down-scalable bonding technologies, low-temperature bonding, enhanced thermomechanical properties, forgiveness to nonplanarity, and ultra-superior current carrying capability with certain materials. This technology can be used for pushing the limits of current interconnection technologies in terms of pitch, number of I/Os, superior combination of electrical and mechanical properties, as well as reworkability. These nanoscale interconnections are essential to support the advances in nano ICs leading to nanoscale WLSOP.

10.4.2 WLSOP Assembly

WLSOP assembly refers to the integration of SOC and 2D and 3D SIP devices onto the wafer substrate with robust and reliable electrical and mechanical interconnections. For most interconnection approaches such as solder, ACF, nanocopper, and stud gold, the interconnection material itself may be used to bond the device to the substrate. Hence, a flavor of the below discussion is already available in the interconnection sections. For certain other interconnection approaches, a separate bonding layer typically involving solders or adhesives is needed. Assembly approaches can be classified as solder-based and solderless (Figure 10.44). Solder-based approaches are more popular, but solderless approaches are getting more attention recently because of their potential to down-scale to fine pitches with less issues such as fluxing, barriers, UBM, and intermetallics that are common with solders.

Assembly with Solder Reflow

Solder interconnection–based assembly was introduced three decades ago by IBM as the C4 technology. Since then, the success of flip-chip technology (assembly with the active die facing the substrate) resulted in several advances in the solder bump assembly. Solders are well-characterized and well-qualified materials for power-consuming digital applications that use high current densities and large dies (1 to 2 cm^2). They are more forgiving toward nonplanar substrates and are therefore ideal for organic packages. The solder pitch has shrunk to less than 150 μm and is expected to move to even smaller pitches. IBM recently demonstrated semiconductor test chips and silicon carrier test vehicles with 25-μm bump diameters on a 50-μm pitch compared with typical industry standards of 100-μm solder bumps on 200- or 225-μm pitches. Fraunhoffer developed an immersion soldering process for thin solder layers on flip

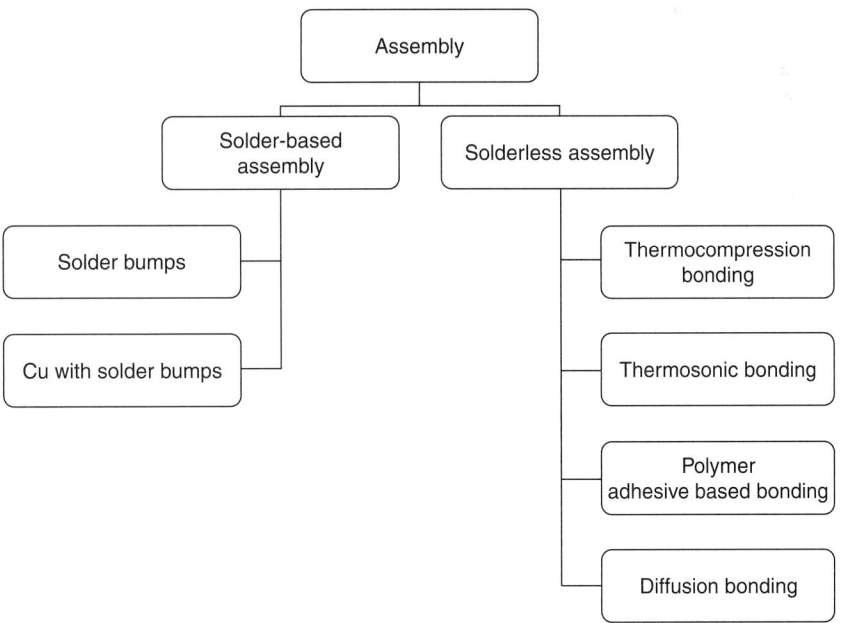

FIGURE 10.44 Assembly and bonding techniques.

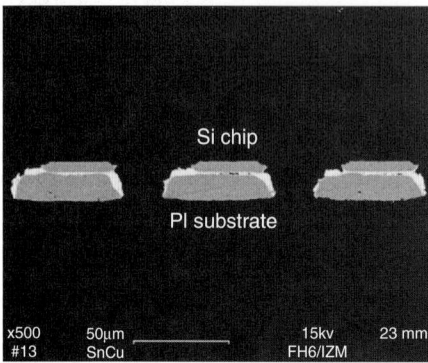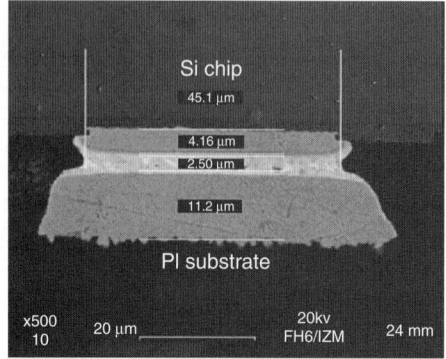

Figure 10.45 SEM picture of an SnCu soldered interconnections 60-μm test vehicle after bonding.

chips with assembly down to 40-μm pitch on thin flexible substrates (Figure 10.45) [93]. They showed that the intermetallic phase formation has a large influence on joint reliability because the intermetallics consume the majority of the solder alloy.

Fine-pitch solder assembly faces several challenges with flux, brittle intermetallics, solder bridging, and additional process steps to deal with UBM and barrier metallurgy and electromigration as discussed in Section 10.4.2. However, solders form the mainstream assembly approach for several high-end processor packaging techniques on organic substrates.

Solder-Based Assembly on High-CTE Rigid Substrates The PRC and IME, Singapore, recently demonstrated reliability with a 20 mm × 20 mm chip with more than 3000 lead-free solder balls with SnCu, Ti-Ni-Cr-Au UBM, and polyimide wafer passivation. The standoff height for solder before reflow was around 35 to 50 μm. A low CTE capillary underfill (CTE < 40 ppm/°C) was used for the testbed with intermediate CTE substrates (7 to 10 ppm/°C). The electrical continuity measurement showed that more than 90 percent of the daisy chains survived 1000 cycles of the air-to-air thermal shock test (−40 to −125°C) [94]. The reliability models clearly establish that underfill with suitably low CTE and sufficiently high modulus is essential to ensure reliability with the intermediate CTE substrates. However, with low-CTE substrates having a CTE of 2 to 3 ppm/°C, 1000 cycle reliability (0 to 100°C) was demonstrated even without underfills with the same 20 mm × 20 mm die at 100-μm pitch.

Solder-Based Assembly with Fine-Pitch Metal Bumps Bonding layers are needed to assemble at low temperatures on organic substrates with copper, gold, or nickel bumps. Solder-based thin bonding interfaces leads to interconnections with the least electrical parasitics. The most prominent example is Intel's copper bump technology [56]. Electroplating and cosputtering processes were used by the PRC to demonstrate 50-μm pitch IC assembly with thin lead-free solder interfaces (5 μm) [95] and nanocopper bumps. Figure 10.46 shows the SEM of a 50-μm pitch die assembled on a substrate. By reducing the interconnect height with the thin bonding interface, an electrical performance similar to chip-first or bumpless interconnects can be achieved. Another example of fine-pitch solder based bonding is the Au bump based AuSn bonding reported by Cubic Wafer and others.

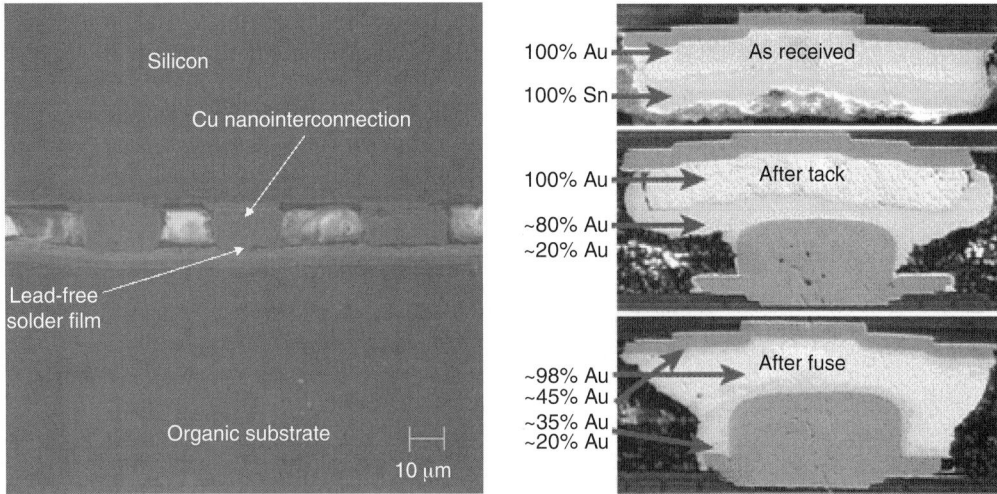

FIGURE 10.46 SEM of nano-interconnect (25-μm Cu, 50 to 60 nm grain size) with barrier metallurgy (NiAu) and an SnCu layer bonded to a substrate (*left*). Au and AuSn based "tack and fuse" bonding developed by Cubic Wafer (*right*).

Thin-film copper-to-copper interconnection with tin interlayers is a major WLSOP assembly technology. This is discussed in detail in Chapter 3.

Solder-Based CNT Assembly   Recently, Georgia Tech achieved breakthroughs in CNT metallization and transfer onto organic substrates. In situ methods for opening CNTs were developed by water-assisted selective etching in order to take advantage of CNT internal walls for electrical and thermal transport [96]. The in situ opening CNTs have obvious advantages over the posttreatment process, since the nanotube walls by posttreatment are inevitably damaged and therefore the electrical and mechanical properties are degraded. With the open-ended CNTs, enhanced solder wetting, CNT metallization with solders, CNT transfer, and assembly at low temperature (<275°C) [97] are feasible. This technique can also be extended to polymer-based CNT transfer and assembly (Figures 10.47 and 10.48).

Solderless Assembly

Solderless assembly is achieved by creating reactive interfaces through applying pressure, temperature, ultrasonic energy, thin reactive bonding layers to create localized melting, deformation, or diffusion to bond the mating surfaces. Gold stud bumping and thermocompression bonding [98–99] is one of the common examples and is shown to be feasible for fine-pitch flip-chip interconnections. Fujitsu Labs announced CNT-based nanobumps for packaging of high-power amplifiers using a CNT-based assembly. Vertically aligned and patterned CNTs were bonded to thin gold pads on AlN substrates using thermocompression bonding. Enhanced thermal conductivity and low inductance compared to gold bumps was realized because of the superior conducting properties of CNTs [100].

Low-Temperature Metal-to-Metal Bonding   Low-temperature metal-to-metal bonding is gaining importance for low-cost applications on organic packages particularly for

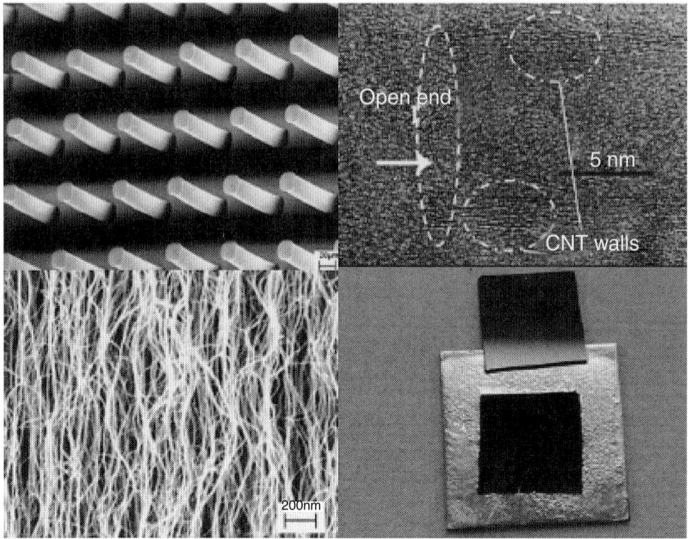

Figure 10.47 Well-aligned open-ended CNT patterns, TEM photo, and the transferred CNT carpet.

mobile products, and also for SIP assembly. Solderless low-temperature bonding is mainly achieved with NCF and ACF approaches. However, the electrical properties are compromised with polymeric adhesives. The PRC has demonstrated 1500 cycle reliability (−40 to 125°C) with nanonickel interconnections using low-stiffness polymer metal composite bonding (ACF) layers [101]. Total system reliability requires stable metallurgicial contacts, stronger adhesion layers, and low-stiffness interlayers to impart

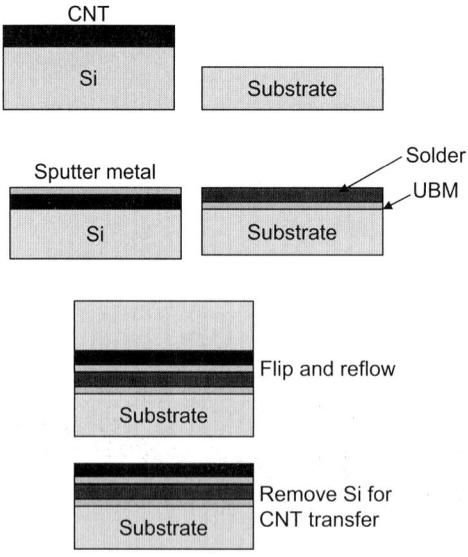

Figure 10.48 Low-temperature CNT transfer technology.

lower stresses on the Cu low-K/UBM and interfaces. Novel nanomaterials can enable low-temperature metallurgical bonding with the best electrical properties without using solders.

High-Temperature Direct Cu-to-Cu Bonding High-temperature direct Cu-to-Cu bonding using thermocompression is a major new paradigm in thin-film copper assembly. This is discussed in Chapter 3.

Placement Techniques

IC assembly at 20- to 50-μm pitch may face several challenges because of the lack of high-accuracy chip-placement systems. The complex assembly systems needed for fine-pitch assembly will compromise speed and cost. These limitations can be overcome by the emerging IC self-assembly technologies. This section briefly describes one example of such assembly technology referred to as fluidic self-assembly.

"Self-assembly" refers to positioning a large number of components simultaneously with microscale precision without alignment using pick-and-place tools. Surface coatings on the interconnects and pads cause high interfacial energy with the surrounding liquid, creating strong capillary forces between the interconnections making it energetically favorable for the chips to self-align. By controlling the surface interactions of interconnects and bonding pads in a liquid medium, the chips can be easily guided to self-assemble on the packages [102]. The liquid media also provides a lubrication to facilitate the movement of the components to the right position. This assembly technique is typically accomplished by preparing the bonding sites with hydrophobic gold and then activating them with a self-assembled monolayer of alkanethiolate precursor. Capillary or surface tension forces are the most dominant forces at microscale and will be the guiding force for self-assembly. Figure 10.49 shows a demonstration of self-assembled ICs. By selectively activating or deactivating specific bonding sites, multibatch assembly is feasible as demonstrated by Xiong et al. [103].

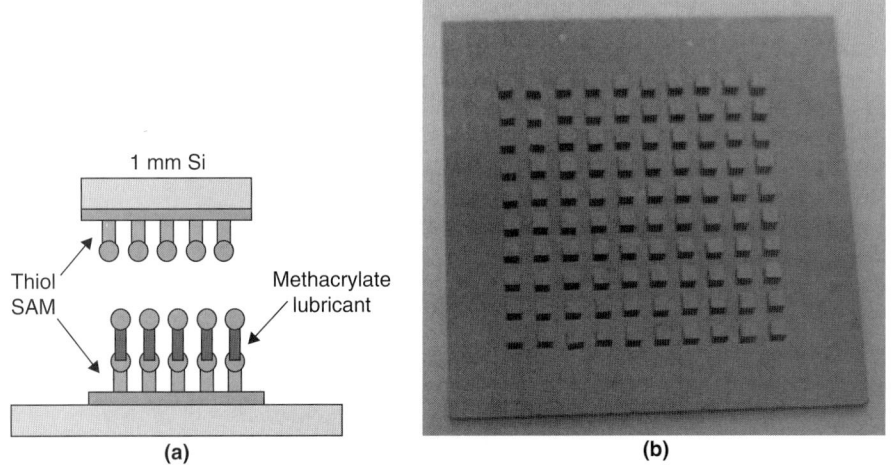

Figure 10.49 Self-assembled Si dies on a patterned Si substrate The dies were modified with thiol SAM and a methacrylate lubricant. (Courtesy IME Singapore).

10.5 3D WLSOP

Horizontal silicon integration faces several challenges because of *RC* delay, cost, fabrication complexity and manufacturability, and several other bottlenecks such as implementation of Cu ultralow-*K* technologies. These are discussed in Chapters 1 and 3. The semiconductor industry is responding to these needs by migrating to 3D Si integration. Heterogeneous device integration (e.g., logic, memory, analog, sensors, microfluidics, and power sources) at wafer scale is pushing the interconnection pitch to 2 to 10 μm and a terabit bandwidth. On the other hand, passive and active functions are continuously migrating to wafer scale to integrate the digital and mixed-signal subsystems in small form factors with leading-edge performance. Three-dimensional integration, with Si through vias and Si-to-Si bonding and embedded wafer-level functions, is evolving as a major paradigm shift to address the system integration demands.

The advantages of wafer-scale 3D Si integration are summarized below [104–105]:

- The vertical interconnection from through vias can realize high-speed data transmission between devices by decreasing the signal delay.
- Short and low-impedance wiring can also lower power dissipation. Moreover, the nature of low impedance achieves stable power supply and stable ground plane with suppressed instabilities.
- Separation of functionality on different wafers allows optimized processing and materials and therefore overcomes fabrication complexity or limitations.
- The die size can be reduced, and more functions (highest density) can be integrated onto a small form factor.
- Three-dimensional Si integration also reduces the number of I/Os, noise, and packaging steps.
- Time-to-market can be reduced due to shortened design verifications and processing development, enabling cross company and cross-industry-sector integration.

The basic operations for 3D WLSOP are (1) TSVs, (2) thinning, (3) wafer-to-wafer bonding, and (4) embedded actives (Figure 10.50). WLSOP with thin-film buildup

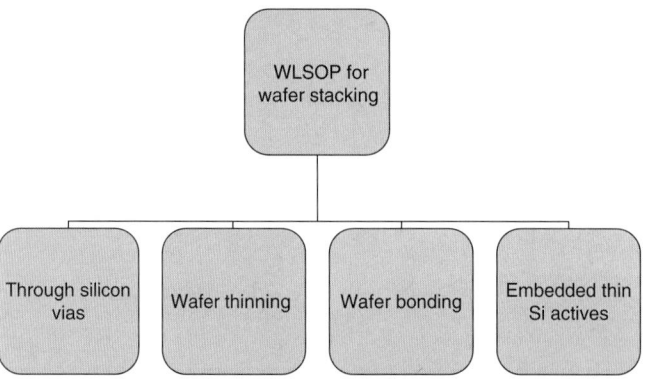

FIGURE 10.50 Emerging trends in 3D WLSOP.

technologies can lead to several material and process compatibility issues for 3D stacking that do not exist in stacking bare wafers. For example, Cu-to-Cu bonding at 400°C is not compatible and may not be feasible with polymer thin-film buildup technologies because of the temperature and planarity issues. The nano and thin-film interconnection strategies discussed in Section 10.5, in addition to the WLSIP technologies discussed in Chapter 3 on SIP, may provide a complete set of interconnection solutions. Chapter 3 provides a complete description of 3D wafer stacking technologies. Chapters 3 and 7 describe embedded dies in the substrate.

10.6 Wafer-Level Probing and Burn-In

The possibility of wafer-level test and burn-in is one of the most attractive attributes of WLSOP. Test at speed and burn-in is an expensive step in the semiconductor business. WLPI, through the design of interconnect on wafer, can enable full wafer test and burn-in to achieve the ultimate low-cost WLSOP. The transfer to full parallel test on the wafer level could dramatically reduce the overall cost. Estimation indicates a savings of up to 50 percent for the transfer from component to wafer test. Figure 10.51 shows the assembly and test process flow for conventional discrete packaging and full wafer-level packaging. The conventional process flow is shown on the left side in which the assembly and tests are done on the component level. On the right side, the integrated assembly and test process on the wafer level is illustrated with test and burn-in performed on the wafer level.

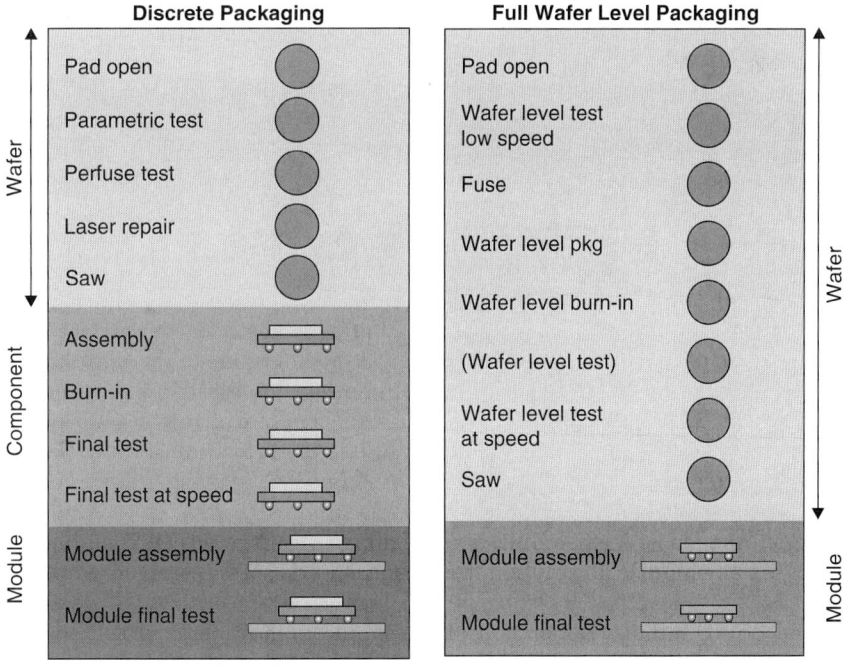

FIGURE 10.51 Assembly and test in discrete packaging and full wafer-level packaging.

Several critical issues need to be addressed before the implementation of wafer-level probing and burn-in. The challenges for the probe card include high-density interconnects onto the wafer, CTE matching of the contactor to silicon, coplanar probe tips, high forces to make electrical connection with low resistance, and uniform load to all the bumps. In addition, precise alignment of the probe to the wafer is needed. Thermal management also becomes critical for wafer level burn-in. All the dies on the wafer should be subjected to a uniform stress. Therefore the voltage, temperature, and the ramping rate need to be carefully controlled. Above all, the main barrier to the success of wafer-level test and burn-in is the cost-performance ratio.

Wafer-level testing is accomplished with either a compliant probe or a compliant bump. Examples of the first approach consisting of a compliant probe are the GoreMate wafer contactor (TEL Probe) and z-axis conductive rubber (TPS probe). Two examples for the wafer-level probing with compliant interconnection approach (ELASTec and WOW) will be discussed later.

TEL Probe

A silicon wafer of completed circuits is placed on a thermal chuck with an extremely flat surface. An electrical contact head, with thousands of contacts, is aligned to the wafer, and contact is made through a sheet of contact material as shown in Figure 10.52. Critical to the process is the unique full-wafer contact material (wafer contact), called GoreMate, that is placed between the contact head and the test wafer. W.L. Gore and Associates, Inc. (GORE) also developed a thermally matched (Inferno) interconnect board, designed to have the same coefficient of expansion as silicon. As estimated from Motorola, through simplification and consolidation of product testing operations, manufacturing

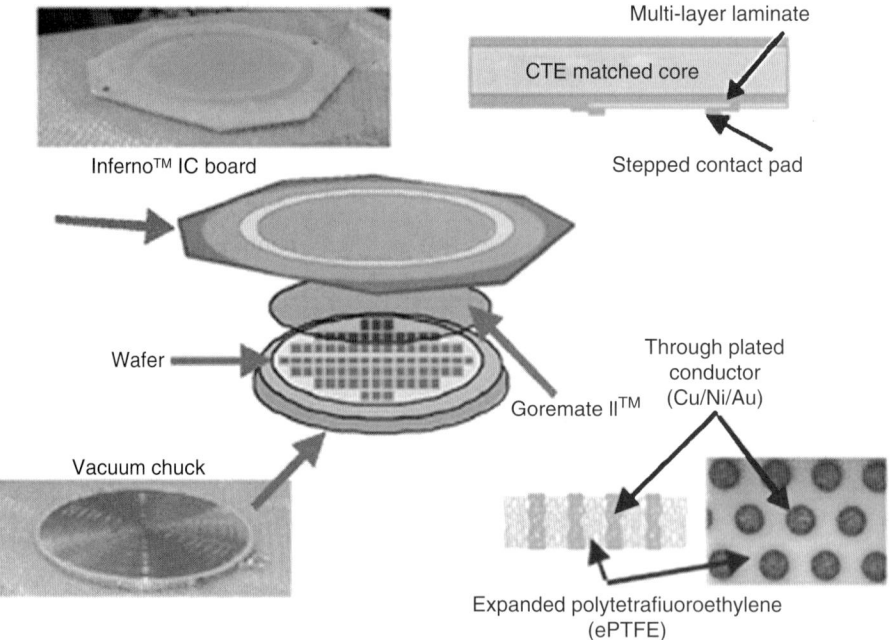

FIGURE 10.52 Motorola wafer-level burn-in strategy.

cost savings are expected to be as high as 15 percent and improvements in the manufacturing cycle time will range up to 25 percent. Motorola announced this wafer-level burn-in technology in 1998 with the partnership of Motorola Semiconductor Products Sector, Tokyo Electron Limited (TEL), and GORE [106]. The developed approach uses TEL wafer-probe technology in a controlled environment and allows each chip on a silicon wafer to be electrically stressed across a range of temperatures from 125 to 150°C.

TPS Probe

A three-part-structure (TPS) probe is used which consists of a glass substrate multilayer wiring board, a compliant z-axis conductor using conductive rubber, and a polyimide membrane with bumps for contacting [107]. The structure of the TPS probe is shown in Figure 10.53. A uniform contact force is provided by the atmosphere when a vacuum is applied between the wafer and the TPS probe through the vacuum valve on the cassette as shown in Figure 10.54. The conductive rubber acts to provide the absorption of the bump height differences. Firm contacts have been achieved on 2756 bumps, which have remained stable up to 125°C. Matsushita Electric Industrial Co. Ltd. has also developed a wafer-level burn-in strategy as shown in Figure 10.53.

Probing on Compliant Interconnections Many compliant interconnection techniques aim at providing flexible bumps that can be pressed down by a low force onto a flat contactor board. ELASTec WLP by Infineon has illustrated the benefit of the resilient bumps. Approximately 2 grams (g) per bump are enough to form a reliable contact, taking into account the height tolerances of bumps and board pads [24]. The compliance of the interconnections also serves to solve the CTE mismatching problem of the wafer and the test board. A good example of WLP enabling wafer-level test and burn-in is illustrated by the WOW (wafer on wafer) technology by FormFactor. WOW is IC industry's first back-end process that provides fully integrated wafer-level package, burn-in, test, and module assembly.

The microsprings of the MOST technology can provide the permanent interconnect onto the final product, as well as temporary connection under pressure during test and burn-in. These microsprings can be located anywhere on the die surface including directly on the bond pads. Figure 10.55 shows the process flow of integrated wafer-level package, burn-in, test, and assembly in WOW. The wafer-level burn-in structure can be seen in Figure 10.56. Silicon wafer is used for building the contactors due to the matched CTE with the wafer under test and also the well-understood interconnect materials and process. However, it is challenging to cost-effectively build a perfectly yielded wafer larger than 200 mm. Therefore, Si tiles with a smaller area are placed on and connected to a backing wafer. The test wafer is clamped against the contactors and tested from 25 to 150°C. The test can also be carried out on a single die level and a multidie (module) level

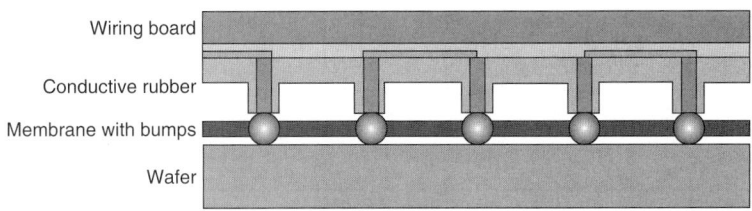

FIGURE 10.53 TPS probe structure of Matsushita wafer-level burn-in.

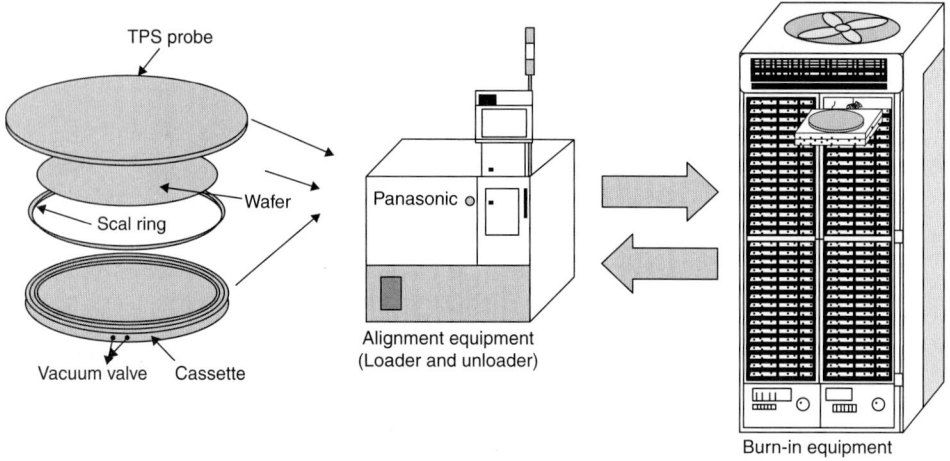

Figure 10.54 Matsushita wafer-level burn-in overview. [107]

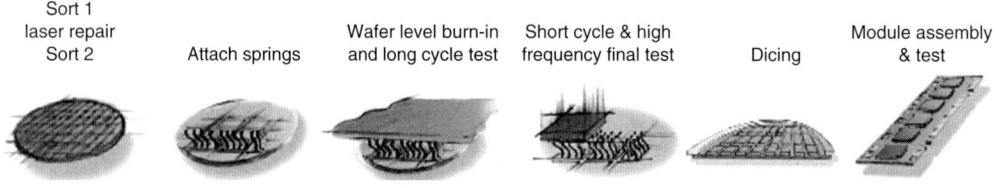

Figure 10.55 Process flow of WOW.

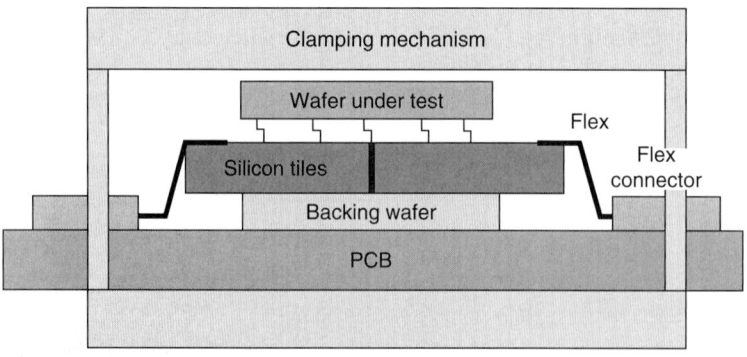

Figure 10.56 WOW wafer-level burn-in structure.

in addition to the wafer level for different testing scenarios. Cost-effective wafer alignment and clamping systems, wafer temperature forcing systems, and wafer-level test and burn-in electronics are to be sought to bring wafer-level test and burn-in solutions.

10.7 Summary

WLSOP was presented as SOP on silicon substrate, implementing such SOP packaging technologies as multilayer thin-film wiring, embedded thin-film components, fine-pitch interconnections and their assembly of stacked and unstacked ICs, and testing on silicon wafers. Such an approach is presented as overcoming the fundamental limitations of organic packages for fine-line wiring and interconnection pitch, thermal management, and warpage. Ensuring signal and power integrity in high-speed ICs requires ultrashort chip-to-substrate interconnections, which in turn lower the fatigue life of the joint and thus do not provide the required mechanical reliability. New paradigms in interconnections that take advantage of the inherent properties of nanomaterials and structures providing the best electrical and mechanical properties have been introduced. These new technologies, combined with wafer-level testing and burn-in, are expected to deliver high-performance and cost-effective WLSOP solutions in the future. Disruptive assembly technologies such as fluidic self-assembly are being developed to provide placement accuracy required for the nano IC assembly at less than 10-μm pitch. WLSOP technologies such as thinning, through vias, and wafer bonding are also gearing up in a big way toward 3D wafer-scale integration for multifunctional systems with the best electrical performance in the smallest form factors.

Acknowledgments

The authors wish to thank their PRC colleagues and co-researchers for their contributions to the technologies included in this chapter.

References

1. P. Garrou, "Wafer-level chip scale packaging: an overview," *IEEE Transactions on Advanced Packaging*, vol. 23, no. 2, May 2000, pp. 198–205.
2. http://www.ti.com/nanostar.
3. www.national.com/packaging/parts/MICROSMD.html.
4. W. H. Koh, "Advanced area array packaging—from CSP to WLP," *Proc. 5th International Conference on Electronic Packaging Technology (ICEPT)*, October 2003, pp. 121–25.
5. Makarem A. Hussein and Jun He, "Materials' impact on interconnect process technology and reliability," *IEEE Transactions on Semiconductor Manufacturing*, vol. 18, no. 1, February 2005, p. 69.
6. Carter W. Kaanta and Susan G. Bombardier, William J. Cote et al., "Dual Damascene—A ULSI Wiring Technology," *Proc. IEEE Eighth International Conference on VLSI Multilevel Interconnection*, 1991, pp. 144–52.
7. Jayaprakash Balachandran, Steven Brebels, Geert Carchon, Maarten Kuijk, Walter De Raedt, Bart K. J. C. Nauwelaers, and Eric Beyne "Wafer-level packaging interconnect options," *IEEE Transactions on VLSI Systems*, vol. 14, no. 6, June 2006, pp. 654–659.

8. P. Elenius, S. Barrett, and T. Goodman, "Ultra CSP™—a wafer-level package," *IEEE Transactions on Advanced Packaging*, vol. 23, no. 2, 2000, p. 220.
9. Kai Zoschke, Jürgen Wolf, Michael Töpper, Oswin Ehrmann, Thomas Fritzsch, Katrin Scherpinski, Herbert Reichl, and Franz-Josef Schmückle1, "Fabrication of application specific integrated passive devices using wafer-level packaging technologies," Proceedings–55th Electronic Components and Technology Conference, 2005, pp. 1594–1601.
10. Yong-Kyu Yoon, Jin-Woo Park, and Mark G. Allen, "RF MEMS based on epoxy-core conductors," *Digest of Solid-State Sensor, Actuator, and Microsystems Workshop*, Hilton Head Island, SC, 2002, pp. 374–75.
11. A. Polyakov, S. Sinaga, P. M. Mendes, M. Bartek, J. H. Correia, and J. N. Burghartz, "High-resistivity polycrystalline silicon as RF substrate in wafer-level packaging", *Electronic Letters*, vol. 41, no. 2, January 20, 2005, pp. 100–101.
12. M. Bartek, G. Zilber, D. Teomin, A. Polyakov, S. M. Sinaga, P. M. Mendes, and J. N. Burghartz, "Wafer-level chip-scale packaging for low-end RF products", in J. D. Cressler and J. Papapolymerou (eds.), Digest of papers, *Topical Meeting on Silicon Monolithic Integrated Circuits in RF Systems*, Atlanta, September 8–10, 2004, pp. 41–44.
13. Snezana Jenei, Stefaan Decoutere, Stefaan Van Huylenbroeck, Guido Vanhorebeek, and Bart Nauwelaers, "High Q Inductors and Capacitors on Si substrate", *IEEE topical meeting on silicon monolithic integrated circuits in RF subsystems*, 2001, pp. 64–70.
14. S. M. Sinaga, A. Polyakov, M. Bartek, and J. N. Burghartz; "Substrate thinning and trenching as crosstalk suppression techniques," *Proc. 3rd European Microelectronic and Packaging Symposium*, Prague, Czech Republic, June 2004, pp. 131–36.
15. Mina Raieszadeh, Pejman Monajemi, Sang-Woong Yoon, Joy Laskar, and Farrokh Ayazi, "High-Q integrated inductors on trenched silicon islands," *18th IEEE International Conference on Micro Electro Mechanical Systems*, Miami Beach, FL, January 30–February 3, 2005, pp. 199–202.
16. Mina Rais-zadeh, Paul A. Kohl, and Farrokh Ayazi, "High-Q micromachined silver passives and filters," *Proc. IEDM*, 2006, pp. 727–730.
17. P. Soussan, L. Goux, M. Dehan, H. Vander Meeren, G. Potoms, D. J. Wouters, and E. Beyne, "Low temperature technology options for high density capacitors," *Proc. Electronic Components and Technology Conference (ECTC)* San Diego, June 1–3, pp. 515–19, 2006.
18. Johan Klootwijk, Anton Kemmeren, Rob Wolters, Fred Roozeboom, Jan Verhoeven, and Eric Van Den Heuvel, "Extremely high-density capacitors with ALD high-K dielectric layers," *Defects in High-K Gate Dielectric Stacks*, Nato Science Series, vol. 220, 2006, pp. 17–28.
19. A. den Dekker, A. van Geelen, P. van der Wel, R. Koster, and E. C. Rodenburg, "The next technology for passive integration on silicon," *Proc. Electronic Components and Technology Conference (ECTC)*, 2007, pp. 968–73.
20. J. Novitsky and D. Pedersen, "FormFactor introduces an integrated process for wafer-level packaging, burn-in test, and module level assembly," *Proc. International Symposium on Advanced Packaging Materials*, 1999, p. 226.
21. Q. Zhu, L. Ma, and S. K. Sitaraman, "Design and optimization of a novel compliant off-chip interconnect—one-turn helix," *Proc. 52nd Electronic Components and Technology Conference*, 2002, p. 910.

22. Karan Kacker, Geoge Lo, and Suresh Sitaraman, "Assembly and reliability assessment of lithography-based wafer-level compliant chip-to-substrate interconnects," *Proc. Electronic Components and Technology Conference,* 2005, pp. 545–50.
23. Andrew A. O. Tay, Mahadevan K. Iyer, Rao R. Tummala, V. Kripesh, E. H. Wong, Madhavan Swaminathan, C. P. Wong, Mihai D. Rotaru, Ravi Doraiswami, Simon S. Ang, and E. T. Kang, "Next generation of 100-μm-pitch wafer-level packaging and assembly for systems on-package," *IEEE Transactions on Advanced Packaging,* vol. 27, no. 2, May 2004, p. 413.
24. H. Hedler, T. Meyer, W. Leiberg, and R. Irsigler, "Bump wafer-level packaging: a new packaging platform (not only) for memory products," *Proc. International Symposium on Microelectronics,* 2003, pp. 681–86.
25. R. Fillion, L. Meyer, K. Durocher, S. Rubinsztajin, D. Shaddock, and J. Wrigth, "New wafer-level structure for stress free area array solder attach," *Proc. International Symposium on Microelectronics,* 2003, pp. 678–92.
26. M. Gonzalez, M. Vanden Bulcke, B. Vandevelde, E. Beyne, Y. Lee, B. Harkness, and H. Meynen, "Finite element analysis of an improved wafer-level package using silicone under bump (SUB) layers," *Proc. 5th International Conference on Thermal and Mechanical Simulation and Experiments in Micro-electronics and Micro-systems,* 2004, p. 163.
27. T. Meyer, H. Hedler, L. Larson, and M. Kunselman, "A new approach to wafer-level packaging employs spin-on and printable silicones," *Chip Scale Review,* July 2004, pp. 65–71.
28. M. S. Bakir, H. A. Reed, P. A. Kohl, K. P. Martin, and J. D. Meindl, "Sea of leads ultra high-density compliant wafer-level packaging technology," *Proc. 52nd Electronic Components and Technology Conference,* 2002, p. 1087.
29. J. Fjelstad, "W.A.V.E.™ technology for wafer-level packaging of ICs," *Proc. Second Electronic Packaging Technology Conference (EPTC),* Singapore, December 1998.
30. M. S. Bakir, H. A. Reed, H. D. Thacker, C. S. Patel, P. A. Kohl, K. P. Martin, and J. D. Meindl, "Sea of leads (SoL) ultrahigh density wafer-level chip input/output interconnections for gigascale integration (GSI)," *IEEE Transactions on Electron Devices,* vol. 50, no. 10, 2003, p. 2039.
31. S. Movva and G. Aguirre, "High reliability second level interconnects using polymer core BGAs," *Proc. Electronic Components and Technology (ECTC '04),* vol. 2, June 1–4, 2004, pp. 1443–48.
32. Q. Zhang, "A novel solder ball coating process with improved reliability," *Proc. Electronic Components and Technology Conference,* 2005, pp. 399–405.
33. Ankur Aggarwal, P. Markondeya Raj, and Rao Tummala, "Metal polymer composite interconnections for ultrafine pitch wafer-level packaging," *IEEE CPMT Transactions on Advanced Packaging,* February 2007, pp. 384–392.
34. D. H. Kim, P. Elenius, M. Johnson, S. Barrett, and M. Tanaka, "Solder joint reliability of a polymer reinforced wafer-level package," *Proc. 52nd Electronic Components and Technology Conference,* 2002, p. 1347.
35. T. Kawahara, "Super CSP™," *IEEE Transactions on Advanced Packaging,* vol. 23, no. 2, 2000, p. 215.
36. M. Topper, J. Auersperg, V. Glaw, K. Kaskoun, E. Prack, B. Beser, P. Coskina, D. Jager, D. Petter, O. Ehrmann, K. Samulewiez, C. Meinherz, S. Fehlberg, C. Karduck, and H. Reichl, "Fab integrated packaging (FIP): a new concept for high reliability wafer-level chip size packaging," *Proc. 50th Electronic Components and Technology Conference,* 2000, pp. 74–80.

37. B. Keser, E. R. Prack, and T. Fang, "Evaluation of commercially available, thick, photosensitive films as a stress compensation layer for wafer-level packaging," *Proc. 51st Electronic Components and Technology Conference,* 2001, pp. 304–309.
38. K. Mitsuka, H. Kurata, J. Furukawa, and M. Takahashi, "Wafer process chip size package consisting of double-bump structure for small-pin-count packages," *Proc. 55th Electronic Components and Technology Conference,* 2005, p. 572.
39. Dongkai Shangguan, *Lead-free Solder Interconnect Reliability,* Materials Park, OH: ASM International, 2005.
40. J. U. Knickerbrocker et al., "Development of next-generation SOP with fine pitch chip interconnection," *IBM Journal of Research and Development,* vol. 49, no. 4/5, July–Sept, 2005, pp. 725–53.
41. Jong-Kai Lin, Ananda De Silva, Darrel Frear, Yifan Guo, Scott Hayes, Jin-Wook Jang, Li Li, and Charles Zhang, "Characterization of lead-free solders and under-bump-metallurgies for lead-free solder," *IEEE Transactions on Electronics Package Manufacturing,* vol. 25, no. 3, 2002, pp. 300–307.
42. Won Kyoung Choi, Sung K. Kang, Yoon Chul Sohn, and Da-Yuan Shih, "Study of IMC morphologies and phase characteristics affected by the reactions between Ni and Cu metallurgies with Pb-free solder joints," *Proc. Electronics Components and Technology Conference,* 2003, pp. 1190–96.
43. S. K. Kang, P. K. Lauro, D.-Y. Shih, D. W. Landerson, and K. J. Puttlitz, "Microstructure and mechanical properties of lead-free solder joints used in microelectronics applications," *IBM Journal of Research and Development,* vol. 49, no. 4/5, 2005, pp. 607–20.
44. K. Mohan Kumar and A. A. O. Tay, "Nano-particle reinforced solders for fine pitch applications," *Proc. 6th Electronic Packaging and Technology Conference,* December 2004, pp. 455–461.
45. Andre Lee, K. N. Subramanian, and Jong-Gi Lee, "Development of nanocomposite lead-free electronic solders," *10th International Symposium on Advanced Packaging Materials: Processes, Properties and Interfaces,* Irvine, CA, March 16–18, 2005.
46. G. A. Rinne, "Electromigration in SnPb and Pb-free solder bumps," *Proc. 54th Electronic Component and Technology Conference (ECTC),* 2004, pp. 974–978.
47. J. W. Nah, J. O. Suh, and K. N. Tu, "Effect of current crowding and joule heating on electromigration-induced failure in flip chip composite solder joints tested at room temperature," *Journal of Applied Physics,* vol. 98, 2005, pp. 13715–13720.
48. T. L. Shao, S. W. Liang, T. C. Lin, and Chih Chen, "Three-dimensional simulation of current density distribution in flip-chip solder joints under electrical stressing," *Journal of Applied Physics,* vol. 98, 2005, no. 4, pp. 44509–44518.
49. Peng Su, Min Ding, Trent Uehling, David Wontor, and Paul S. Ho, "An evaluation of electromigration performance of SnPb and Pb-free flip chip solder joints," *Proc. Electronic Component and Technology Conference (ECTC),* 2005, pp. 1431–36.
50. Stephen Gee, Nikhil Kelkar, Joanne Huang, and King-Ning Tu, "Lead-free and PbSn bump electromigration testing," *Proc. IPACK2005,* ASME InterPACK '05, San Francisco, CA, July 17–22, 2005.
51. Bernd Ebersberger, Robert Bauer, and Lars Alexa, "Reliability of lead-free SnAg solder bumps; influence of electromigration and temperature," *Proc. Electronic Component and Technology Conference (ECTC),* 2005, pp. 1407–15.
52. H. Gan et al., "Pb-free micro joints for the next-generation microsystems: the fabrication, assembly and characterization," *Proc. Electronics Components and Technology Conference,* 2006, p. 1210.

53. John H. Lau, "Low Cost Flip Chip Technologies—for DCA, WLSCP, and PBGA" Assemblies, New York: McGraw-Hill, 1999.
54. J. Y. Kim, J. Yu, J. H. Lee, and T. Y. Lee, "The effect of electroplating parameters on the morphology and composition of Sn-Ag solder," *Journal of Electronic Materials*, vol. 33, no. 12, 2004, pp. 1459–64.
55. David S. Chau, Ashish Gupta, Chia-Pin Chiu, Suzana Prstic, and Seth Reynolds, "Impact of different flip-chip bump materials on bump temperature rise and package reliability," *Proc. International Symposium on Advanced Packaging Materials*, Irvine, CA, March 18–21, 2005.
56. Andrew Yeoh, Margherita Chang, Christopher Pelto, Tzuen-Luh Huang, Sridhar Balakrishnan, Gerald Leatherman, Sairam Agraharam, Guotao Wang, Zhiyong Wang, Daniel Chiang, Patrick Stover, and Peter Brandenburger, "Copper die bumps (first level interconnect) and low-K dielectrics in 65 nm high volume manufacturing," *Proc. Electronic Components and Technology Conference*, 2006, pp. 1611–15.
57. Tie Wang, Francisca Tung, Louis Foo, and Vivek Dutta, "Studies on a novel flip-chip interconnect structure—pillar bump," *Proc. 51st Electronic Components and Technology Conference*, May 2001, pp. 945–49.
58. Daniel D. Evans Jr., "Gold bump technologies—plating versus ball," IMAPS *International Conference on Device Packaging*, March 2005, pp. MP22.
59. S. H. Shi, T. Yamashita, and C. P. Wong, "Development of the wafer-level compressive-flow underfill process and its required materials," *Proc. 49th Electronic Components and Technology Conference*, 1999, p. 961.
60. S. H. Shi, T. Yamashita, and C. P. Wong, "Development of the wafer-level compressive-flow underfill encapsulant," *IEEE Trans. on Components, Packaging, Manuf. Technol.*, Part C, vol. 22, no. 4, 1999, p. 274.
61. K. Gilleo and D. Blumel, "Transforming flip chip into CSP with reworkable wafer-level underfill," *Proc. Pan Pacific Microelectronics Symposium*, 1999, p. 159.
62. S. Shi, and C. P. Wong, "The process and materials for low-cost flip-chip solder interconnect structure for wafer-level no flow process," US Patent 6,746,896 (June 8, 2004), and K. Gilleo, "Flip Chip with integrated flux, mask and underfill," W.O. Patent 99/56312 (November 4, 1999).
63. L. Nguyen, H. Nguyen, A. Negasi, Q. Tong, S. H. Hong, "Wafer level underfill—processing and reliability," *Electronics Manufacturing Technology Symposium*, 2002. IEMT 2002. 27th Annual IEEE/SEMI International 17-18 July 2002 Page(s):53–62.
64. L. Nguyen, H. Nguyen, A. Negasi, Q. Tong, and S. H. Hong, "Wafer-level underfill—processing and reliability," *Proc. 27th Int. Electron. Manuf. Tech. Symp.*, San Jose, CA, July 17–18, 2002.
65. J. Qi, P. Kulkarni, N. Yala, J. Danvir, M. Chason, R. W. Johnson, R. Zhao, L. Crane, M. Konarski, E. Yaeger, A. Torres, R. Tishkoff, and P. Krug, "Assembly of flip chips utilizing wafer applied underfill," presented at IPC SMEMA Council APEX 2002, *Proc. APEX*, San Diego, CA, 2002, pp. S18-3-1–S18-3-7.
66. Q. Tong, B. Ma, E. Zhang, A. Savoca, L. Nguyen, C. Quentin, S. Lou, H. Li, L. Fan, and C. P. Wong, "Recent advances on a wafer-level flip chip packaging process," *Proc. 50th Electronic Components and Technology Conference*, 2000, pp. 101–106.
67. S. Charles, M. Kropp, R. Kinney, S. Hackett, R. Zenner, F. B. Li, R. Mader, P. Hogerton, A. Chaudhuri, F. Stepniak, and M. Walsh, "Pre-applied underfill adhesives for flip chip attachment," *IMAPS Proc. International Symposium on Microelectronics*, Baltimore, MD, 2001, pp. 178–83.

68. Z. Zhang, Y. Sun, L. Fan, and C. P. Wong, "Study on B-stage properties of wafer-level underfill," *Journal of Adhesion Science and Technology,* vol. 18, no. 3, 2004, pp. 361–80.
69. Z. Zhang, Y. Sun, L. Fan, R. Doraiswami, and C. P. Wong, "Development of wafer-level underfill material and process," *Proc. 5th Electronic Packaging Technology Conference,* Singapore, December 2003, pp. 194–98.
70. R. L. D. Zenner and B. S. Carpenter, "Wafer-applied underfill film laminating," *Proc. 8th International Symposium on Advanced Packaging Materials,* 2002, pp. 317–25.
71. Y. Sun, Z. Zhang, and C. P. Wong, "Photo-definable nanocomposite for wafer-level packaging," *Proc. 55th Electronic Components and Technology Conference,* 2005, p. 179.
72. Shubhra Bansal, Ashok Saxena, and Rao Tummala, "Nanocrystalline copper and nickel as ultra high-density chip-to-package interconnections," *Proc. 54th Electronic Components and Technology Conference, IEEE,* Piscataway, NJ, 2004, pp. 1646–51.
73. K. S. Kumar, H. Van Swygenhoven, and S. Suresh, "Mechanical behavior of nanocrystalline metals and alloys," *Acta Materialia,* vol. 51, 2003, pp. 5743–74.
74. A. F. Mayadas and M. Shatzkes, "Electrical-resistivity model for polycrystalline films: the case of arbitrary reflection at external surfaces," *Phys. Rev. B 1,* 1970, pp. 1382–89.
75. W. Steinhögl, G. Schindler, G. Steinlesberger, and M. Engelhardt, "Size dependent resistivity of metallic wires in the mesoscopic range," *Phys. Rev. B, Condens. Matter,* vol. 66, August 2002, pp. 075414/1–075414/4.
76. A. J. Yin, J. Li, W. Jian, A. J. Bennett, and J. M. Xu, "Fabrication of highly ordered metallic nanowire arrays by electrodeposition," *Appl. Phys. Lett.,* vol. 79, 2001, pp. 1039–41.
77. T. Thurn-Albrecht, J. Schotter, G. A. Kastle, N. Emley, T. Shibauchi, L. Krusin-Elbaum, K. Guarini, C. T. Black, M. T. Tuominen, and T. P. Russell, "Ultra-high density nanowire grown in self-assembled di-block copolymer," *Science,* vol. 290, 2000, pp. 2126–29.
78. Azad Naeemi, Reza Sarvari, and James D. Meindl, "Performance comparison between carbon nanotube and copper interconnects for gigascale integration (GSI)," *IEEE Electron Device Letters,* vol. 26, no. 2, February 2005.
79. Yijun Li, Kunlin Wang, Jinquan Wei, Zhiyi Gu, Zhicheng Wang, Jianbin Luo, and Dehai Wu, "Tensile properties of long aligned double-walled carbon nanotube strands," *Carbon,* vol. 43, no. 1, 2005, pp. 31–35.
80. M. A. Guillorn, T. E. McKnight, A. Melechko, V. I. Merkulov, P. F. Britt, D. W. Austin, and D. H. Lowndes, "Individually addressable vertically aligned carbon nanofiber-based electrochemical probes," *Journal of Applied Physics,* vol. 91, no. 6, March 15, 2002, pp. 3824–28.
81. Lingbo Zhu, Yangyang Sun, Jianwen Xu, Zhuqing Zhang, Dennis W. Hess, and C. P. Wong, "Aligned carbon nanotubes for electrical interconnect and thermal management," *Proc. Electronic Components and Technology Conference,* 2004, pp. 44–50.
82. Jun Li, Qi Ye, Alan Cassell, Jessica Koehne, H. T. Ng, Jie Han, and M. Meyyappan, "Carbon nanotube interconnects: a process solution," *Proc. IEEE International Interconnect Technology Conference,* June 2–4, 2003, pp. 271–272.
83. Y. Li, K. Moon, and C. P. Wong, "Electronics without lead," *Science,* vol. 308, 2005, pp. 1419–20.
84. K. Moon, H. Dong, R. Maric, S. Pothukuchi, A. Hunt, Y. Li, and C. P. Wong, "Thermal behavior of silver nanoparticles for low-temperature interconnect application," *J. Electronic Materials,* vol. 34, 2005, pp. 132–39.

85. W. V. Kwan, V. Kripesh, M. K. Gupta, A. A. O. Tay, and R. R. Tummala et al., "Low temperature sintering process for deposition of nanostructured metal for nano IC packaging," *Proc. 5th Electronics Packaging Technology Conference, IEEE*, Piscataway, NJ, 2003, pp. 551–56.
86. Y. Li, K. Moon, and C. P. Wong, "Enhancement of electrical properties of anisotropically conductive adhesive (ACA) joints via low temperature sintering," *Journal of Applied Polymer Science*, vol. 99, Issue 4, 2005, pp. 1665–1673.
87. Y. Li, K. Moon and C. P. Wong, "Adherence of self-assembled monolayers on gold and their effects for high performance anisotropic conductive adhesives," *Journal of Electronic Materials*, vol. 34-3, 2005, pp. 266–71.
88. Y. Li, K. Moon, and C. P. Wong, "Monolayer protected silver nano-particle based anisotropic conductive adhesives (ACA): electrical and thermal properties enhancement," *J. Electronic Materials*, vol. 34, No.12, 2005, pp. 1573–1578.
89. A. Sihlbom and J. Liu, "Thermal characterization of electrically conductive adhesive flip-chip joints," *Proc. IEEE 2nd Electronic Packaging Technology Conference*, Singapore, December 8–10, 1998, pp. 251–57.
90. M. J. Yim, H.-J. Kim, and K.-W. Paik, "Anisotropic conductive adhesives with enhanced thermal conductivity for flip chip applications," *Journal of Electronic Materials*, vol. 34, 2005, pp. 1165–71.
91. L. K. The, E. Anto, C. C. Wong, S. G. Mhaisalkar, E. H. Wong, P. S. Teo, and Z. Chen," Development and reliability of non-conductive adhesive flip-chip packages," *Thin Solid Films*, vol. 462, 2004, pp. 446–53.
92. M.-J. Yim, J.-S. Hwang, W. Kwon, K. W. Jang, and K.-W. Paik, "Highly reliable non-conductive adhesives for flip chip CSP applications," *IEEE Transactions on Electronics Packaging Manufacturing*, vol. 26-2, 2003, pp. 150–55.
93. Barbara Pahl, Thomas Loeher, Christine Kallmayer, Rolf Aschenbrenner, and Herbert Reichl, "Ultrathin soldered flip chip interconnections on flexible substrates," *Electronic Components and Technology Conference*, 2004, pp. 1244–50.
94. Andrew A. O. Tay, Mahadevan K. Iyer, Rao R. Tummala, "Design and development of interconnects for ultra-fine pitch wafer-level packages," *IEEE 6th International Conference on Electronic Packaging Technology*, 2005, pp. 446–453.
95. Ankur Aggarwal, P. Markondeya Raj, Isaac Robin Abothu, Michael Sacks, Rao Tummala, and Andrew Tay, "New paradigm in IC-package interconnections by reworkable nano-interconnects," *Proc. 54th Electronic Components and Technology Conference, IEEE*, Piscataway, NJ, 2004, pp. 451–61.
96. L. Zhu, Y. Xiu, D. W. Hess, and C. P. Wong, "Aligned carbon nanotube stacks by water-assisted selective etching," *Nano Letters*, vol. 5, 2005, p. 2641.
97. L. Zhu, Y. Sun, D. W. Hess, and C. P. Wong, "Well-aligned open-ended carbon nanotube architectures: an approach for device assembly," *Nano Letters*, vol. 6, 2006, pp. 243–247.
98. Takao Yamazaki, Yoshimichi Sogawa, Rieka Yoshino, Keiichiro Kata, Ichiro Hazeyama, and Sakae Kitajo, "Real chip size three-dimensional stacked package," *IEEE Transactions on Advanced Packaging*, vol. 28, no. 3, August 2005, p. 397.
99. Masahiro Sunohara, Kei Murayama, Mitsutoshi Higashi, and Mitsuharn Shimizu, "Development of interconnect technologies for embedded organic packages," *Electronic Components and Technology Conference*, 2003, pp. 1484–89.
100. T. Iwai, H. Shioya, D. Kondo, S. Horose, A. Kawabata, S. Sato, M. Nihei, T. Kikkawa, K. Joshin, Y. Awano, and N. Yokoyama, "Thermal and source bumps utilizing CNTs

for flip-chip-high-power-amplifier," *IEEE International Electronic Devices Meeting*, 2005, pp. 257–260.

101. Ankur Aggarwal, P. Markondeya Raj, Baikwoo Lee, Jack Moon, C. P. Wong, and Rao Tummala, "Reliability studies of nano-structured nickel interconnections on high CTE organic substrates without underfill," *Proc. IEEE Electronics Component and Technology Conference,* Reno, NV, May 2007, pp. 905–913.

102. U. Srinivasan, D. Liepmann, and R. T. Howe, "Microstructure to substrate self-assembly using capillary forces," *Journal of Microelectromechanical Systems,* vol. 10, no. 1, March 2001, pp. 17–24.

103. Xiaorong Xiong, Yael Hanein, Jiandong Fang, Daniel T. Schwartz, and Karl F. Böhringer, "Multibatch self-assembly for multichip integration," *Journal of Microelectromechanical Systems,* vol. 12, issue 2, April 2003, pp. 117–27.

104. J. U. Knickerbocker, P. S. Andry, L. P. Buchwalter, A. Deutsch, R. R. Horton, K. A. Jenkins, Y. H. Kwark, G. McVicker, C. S. Patel, R. J. Polastre, C. Schuster, A. Sharma, S. M. Sri-Jayantha, C. W. Surovic, C. K. Tsang, B. C. Webb, S. L. Wright, S. R. McKnight, E. J. Sprogis, and B. Dang, "Development of next-generation system-on-package (SOP) technology based on silicon carriers with fine-pitch chip interconnection," *IBM J. Research and Development,* vol. 49, no. 4–5, July–September 2005, pp. 725–753.

105. James Jian-Qiang Lu, Ronald Gutmann, Thorsten Matthias, and Paul Lindner, "Aligned wafer bonding for 3-D interconnect," *Semiconductor International,* August 1, 2005, pp. SP.4-8.

106. G. Ganesan and J. Pitts, "Wafer-level burn-in with test", *Advanced Packaging*, May, Vol. 11, 2002, pp. 29–33.

107. Y. Nakata, I. Miyanaga, S. Oki, and H. Fujimoto, "A Wafer-Level Burn-in Technology Using the Contactor Controlled Thermal Expansion," *Proceedings of the 1997 International Conference on Multichip Modules*, Osaka, 1997, pp. 259–264.

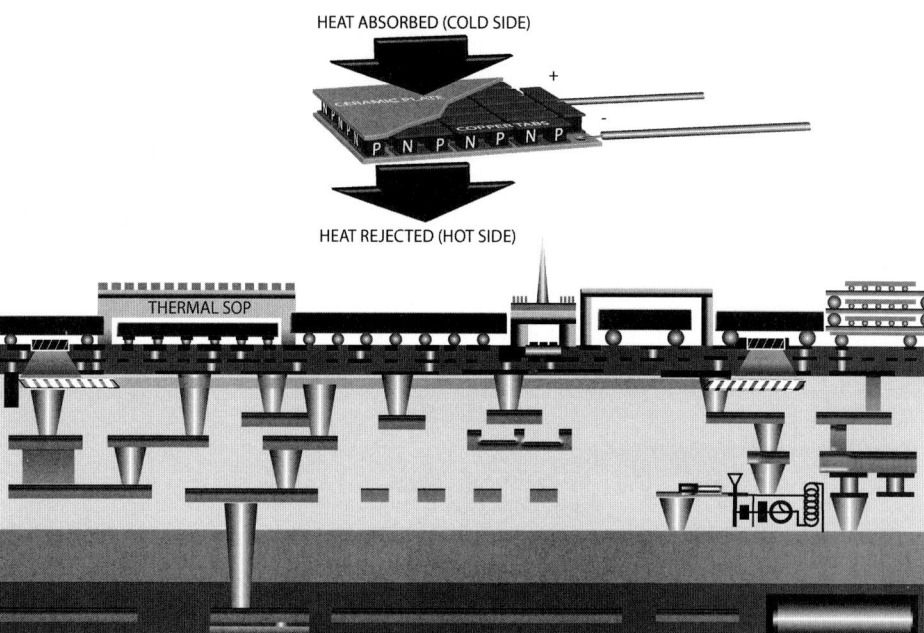

CHAPTER 11
Thermal SOP

Y. Joshi and Gopal C. Jha
Georgia Institute of Technology

M. Patterson
Intel Incorporation

P. Dutta
Indian Institute of Science

11.1	Fundamentals of Thermal SOP 606	11.5	Thermal Management Technologies	637
11.2	Thermal Sources in SOP Modules 610	11.6	Power Minimization Methodologies	648
11.3	Fundamental Heat Transfer Modes 618	11.7	Summary 651	
11.4	Fundamentals of Thermal Characterization 629		References 652	

Thermal SOP deals with effective thermal management of highly miniaturized SOP-based systems. Megafunctionality, microminiaturization, multiple length scale hierarchy, multiple-functional materials, and embedded thin-film active and passive components are the essential features of SOP systems. These result in nonuniform and highly concentrated volumetric heat generation produced by a number of power sources that include not only active ICs such as microprocessors, power amplifiers, and memory devices but also passive components such as resistors. While SOP miniaturization results in an improvement in electrical performance, cost, and some aspects of reliability, it presents unprecedented thermal challenges making the total heat flux densities at the system level very close to the device level, unlike in the current approach of discrete component–based systems. This chapter describes the thermal SOP concept and its thermal implications. Heat sources in SOP modules are identified, and an insight into relevant heat transfer modes is given. An overview of thermal characterization techniques is presented. In addition, this chapter reviews the current state-of-the-art in thermal management technology that is applicable to thermal SOP. Power minimization by efficient power management, as reviewed in this chapter, may mitigate the thermal SOP concerns to a certain degree.

11.1 Fundamentals of Thermal SOP

Ever-growing demands for portability and functionality have always governed electronic technology innovations. Gordon E. Moore foresaw the market demand and set forth the empirical Moore's law [1], or the so-called first law of electronics, which predicts that the number of transistors on a chip will increase as shown in Figure 11.1a, which coupled with shrinking die size contributes to daunting thermal challenges.

The chip power consumption has been increasing despite reductions in the power supply voltage and an increase in the efficiency of devices. According to the International Technology Roadmap for Semiconductors (ITRS-2006), high-performance microprocessors may dissipate 198 W by 2008 (see Figure 11.1b) [2]. This situation can be worse in radiofrequency (RF) and power electronic components. The resulting thermal management challenges can be appreciated by comparing heat fluxes in various electronic components with those in some well-known heat generating systems (see Figure 11.2) [3].

In recent years, the growing fear that Moore's law will slow or collapse with the ever-shrinking dimensions and ever-increasing die size, coupled with the unceasing demand for low-cost microminiaturized devices, such as the iPhone, have lead to the concept of system-on-package (SOP), where the system board and IC package become one and the same. This so-called second law of electronics is expected to enable the electronic devices to achieve unprecedented functionality and miniaturization at reduced cost. The SOP technology, as discussed throughout this book, integrates a number of leading-edge technology waves that include digital, RF, micro-electromechanical systems (MEMS), sensors, and optoelectronics in a highly miniaturized system package. The focus of SOP is thus on miniaturization of system components,

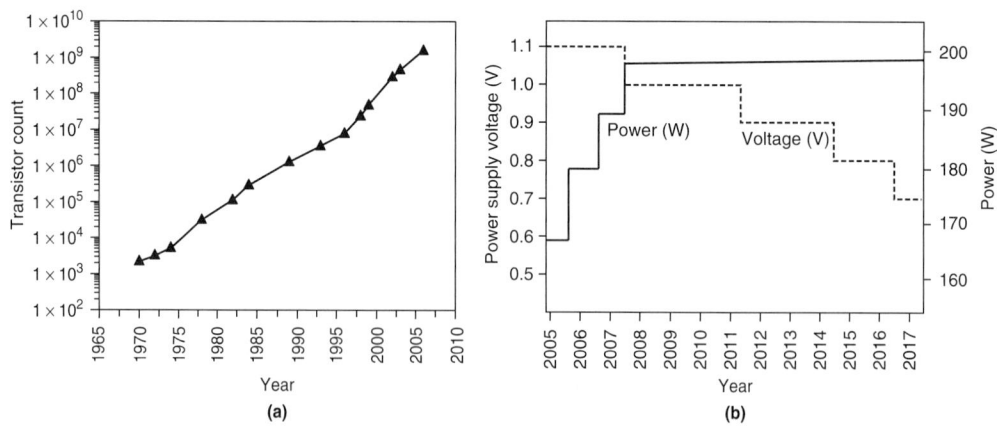

FIGURE 11.1 (a) Transistor count as a function of year [1]. (b) Power supply voltage and maximum power consumption in high-performance processors [2].

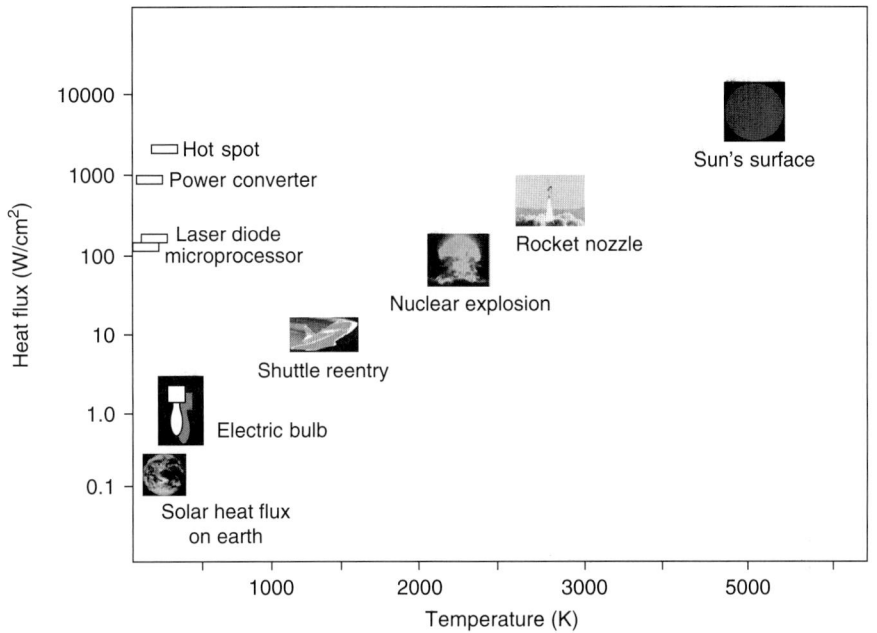

FIGURE 11.2 Comparison of heat fluxes in various phenomena and systems.

including not only actives but also passives, power sources, I/Os, thermal structures, and system I/Os.

11.1.1 Thermal Implications of SOP

The thermal implications of SOP are enormous. Power supply and distribution are another set of challenges that SOP must overcome. Since SOP has two major components—device components and system components, it is convenient to understand the thermal implications in these two parts of SOP technology separately. The thermal aspects of devices in the SOP concept remain the same whether the system is based on SOP or SOB (system-on-board). Perhaps SOP technology may ease the thermal challenges at the device level. This is so because CMOS integration with multiple functional blocks in a large chip with ultrahigh heat flux is not necessary. These can be broken into smaller, easier to design and fabricate higher yielding devices requiring lower power. The system miniaturization of SOP, on the other hand, has a huge impact because of the exponential decrease in the available volumetric system space. An illustration of such an impact is shown in Figure 11.3.

The cooling requirements of SOP systems are dictated by performance and reliability needs. Operating temperatures outside a range can cause deteriorated performance of active semiconductors—the leakage current may increase in DRAM, clock frequency may reduce, and wavelength drift and power drop may occur in optoelectronic modules. The mismatch in the coefficient of thermal expansion (CTE) between ICs and organic substrates (2.8 ppm/°C for Si, ~6 ppm/°C for GaAs, ~25 ppm/°C for eutectic $Sn_{63}Pb_{37}$ solder, and 14 to 20 ppm/°C for organic epoxy fiberglass FR4) generates thermal stress at the solder joints. Repeated thermal cycling due to switching thus can lead to thermal

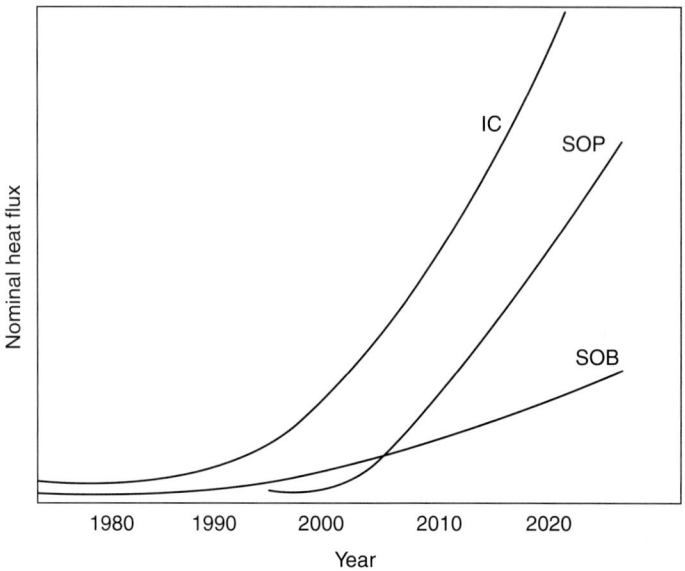

Figure 11.3 Impact of SOP on thermal challenges.

fatigue and ultimate failure. SOP modules and systems may fail due to a variety of causes unless they are designed with the right set of materials and appropriate thermal management solutions. Detailed studies on various failure mechanisms are presented in Chapter 8 [4–5].

The current approach to reliability is based in the root-cause, or physics-of-failure analysis of various failure mechanisms and their temperature dependence [6]. The various operational mechanisms in a given temperature range are ranked to find the dominant one. Outside this range, a shifting of the dominant mechanism may occur. Thus in a broad sense, many failure mechanisms are thermally driven, yet the relationship with the temperature may often be complex. In addition to their impact on reliability, thermal effects also often influence performance parameters. Nonlinear effects in amplifiers and improper operation (wavelength drift and power drop) of lasers in optoelectronic modules are just two examples of such effects.

From the above discussions, it is clear that both operational reliability and functional performance are thermally influenced. An approach often taken by designers of multifunctional microsystems is to limit the maximum "junction temperature," an average measure of the chip temperature during operation. This limit is usually different for commercial and military equipment, as each sees different ranges of ambient temperatures in operation. For handheld and portable devices, the market sometimes demands an even more stringent limit to ensure a greater degree of customer satisfaction. The International Technology Roadmap for Semiconductors (ITRS) provides a projected value of allowable junction and operating temperatures in various microsystems [2]. This is illustrated in Figure 11.4 [2].

As evident from the figure, the junction temperature for single chip-packaged devices can't be allowed to rise above 125°C in low-cost, handheld, and memory devices; above 175°C in devices working in harsh environment; and above 100°C in high-performance and cost-effective devices.

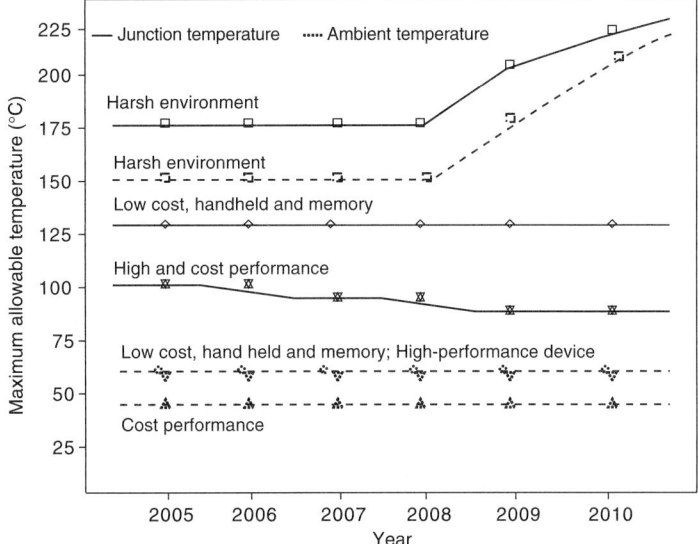

FIGURE 11.4 Maximum allowable junction and ambient temperature as a function of time. [2]

11.1.2 System-Level Thermal Constraints in SOP-Based Portables

SOP has been described in this book as a highly miniaturized system technology for highly portable microsystems of the future. With the continuing drive for size reduction and increased functionality, system-level heat removal is the ultimate limitation in the design of such systems. As an illustration, consider a portable system of dimensions 9.0 cm × 1.85 cm × 4.2 cm exposed to an ambient environment at 25°C. The estimated heat removal capability from such a system by natural convection and radiation for a typical casing surface assumed at a uniform temperature is shown in Figure 11.5. The surface

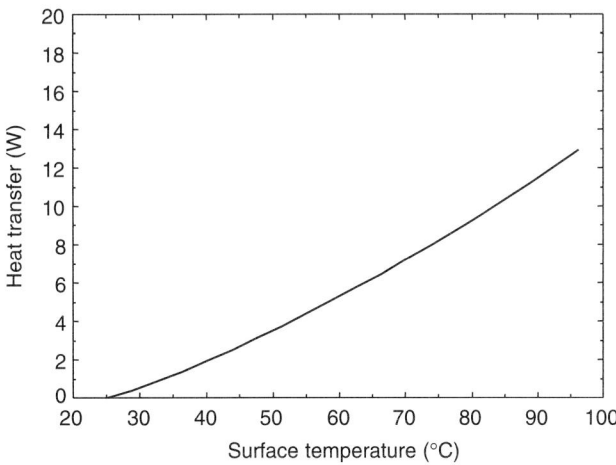

FIGURE 11.5 Heat removal limits for a portable system of size 9.0 cm × 1.85 cm × 4.2 cm by natural convection and radiation to surroundings at 250°C.

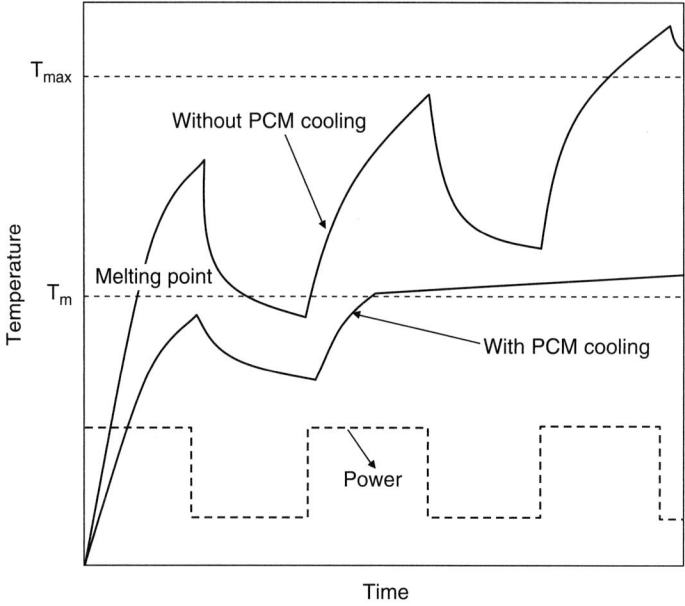

Figure 11.6 Typical response of a microsystem subject to pulsed power dissipation, without and with a phase change material (PCM).

temperature is ~60°C for a heat removal rate of 5 W. A slightly lower surface temperature of ~50°C is obtained for the same heat removal rate if heat conduction from the base surface and natural convection and radiation from other surfaces occur, simulating a handheld device. A related factor to consider is the limited availability of power for cooling. Because of the premium on battery power, the designer would like to use as little of it for cooling as possible. These challenges are significantly more stringent than for high-performance systems, which have plenty of system volume and require highly creative approaches to thermal management.

In general, passive approaches that require no external input power for cooling are preferred. One example is the use of solid-liquid phase change materials (PCM). Examples of PCMs are paraffin and low-melting-point metallic alloys. The typical response of a PCM to pulsed power dissipation is shown in Figure 11.6. In the absence of PCM, the system temperature exceeds a prescribed maximum, prior to achieving a steady state. By incorporating PCM, the temperature increase is arrested, once the melting temperature is achieved due to heat absorption at the near-isothermal phase transition. A number of additional thermal management techniques are discussed in Section 11.5.

11.2 Thermal Sources in SOP Modules

Figure 11.7 illustrates principal thermal sources and thermally sensitive elements in various SOP modules. Detailed discussions are provided in subsequent subsections.

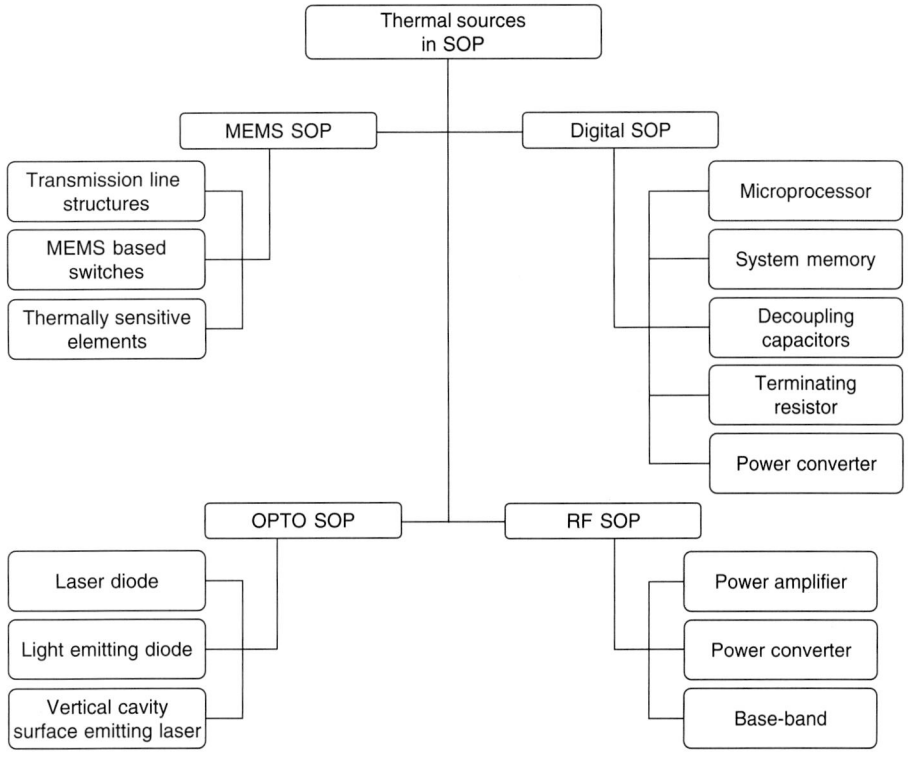

FIGURE 11.7 Thermal sources in SOP modules.

11.2.1 Digital SOP

A digital SOP module consists of various functional units including microprocessors, dynamic random-access memory (DRAM), static random-access memory (SRAM), application-specific integrated circuits (ASIC), and field programmable gate arrays (FPGA). A logic unit in a microprocessor requires quiescent power as well as active power. Quiescent power maintains the logic state and is the source of dissipated power in the receiver and driver. Active power dissipation occurs due to load capacitance and load resistance. A traditional microprocessor runs at a single frequency and voltage. It always remains "on" and consumes full power. For microprocessor chips based on the complementary metal-oxide semiconductor (CMOS) technology, the dynamic heat generation or active power dissipation rate due to switching of the average load capacitance ($C_{average}$) through a voltage (V_{dd}) at the clock frequency (f) is given by,

$$P = N_{transition} V_{dd} I_{leakage} + \frac{1}{2}\alpha N_{transition} C_{average} V_{dd}^2 f \qquad (11.1)$$

where $N_{transition}$ is the number of switching transitions and α is a fitting parameter. In Equation (11.1) the first term represents the power dissipation due to the leakage current through a voltage (V_{dd}) and the second term represents the power associated with the capacitive load.

Shrinking device size and increasing transistor count have been pushing CMOS technology to the nanometer scale, as shown in Figure 11.1. As the transistor gets smaller and faster, the active power dissipation becomes higher. Also, at nanometer length scales, leakage current becomes very critical, as seen in Figure 11.8 [7]. Notably, it already contributes more than 25 percent of the total power dissipation in a microprocessor. This power contribution is speculated to surpass the active power dissipation when the MOS channel length is brought down below 100 nm and the gate dielectric thickness goes below 2 nm [8].

According to ITRS-2006, high-performance microprocessors may dissipate 198 W by 2008 (Figure 11.1b) [2]. A significant amount of dissipated power is contributed by on-chip memory structures. A large area of today's microprocessor chip contains memory elements including a data cache (d-cache), instructional cache (i-cache), branch prediction table (BPT), and translational look-aside buffer (TLB) [9]. For example, on-chip memory structures occupy 30 and 60 percent of the total silicon real estate in Alpha 21264 and Strong ARM processors, respectively [10]. Such incessantly increasing memory capacity, coupled with increasing operating speeds, has made these some of the most energy-intensive components. They have been reported to consume more than 40 percent of the total chip power in many of the microprocessors. The leakage power constitutes a large percentage of total power dissipation in the cache. For example, leakage in the L1 cache constitutes 30 percent of the L1 cache energy, and leakage in the L2 cache constitutes as high as 80 percent of the L2 cache energy [11]. Borker estimates a fivefold increase in the total leakage energy dissipation with every new generation of microprocessors [12].

Spatial and temporal focusing of heat are other characteristics of microprocessors. The performance of a microprocessor (program execution time) is a function of the distance of the logical unit from the memory unit. Therefore, new-generation microprocessors integrate a large area of cache memory with the logic unit on a single chip. Such integration of high-power logic with low-power memory results in spatially localized high-power and high-thermal gradient regions in the chip that are known as hot spots [13]. Heat flux at hot spots is expected to exceed 1000 W/cm^2, making it six times or

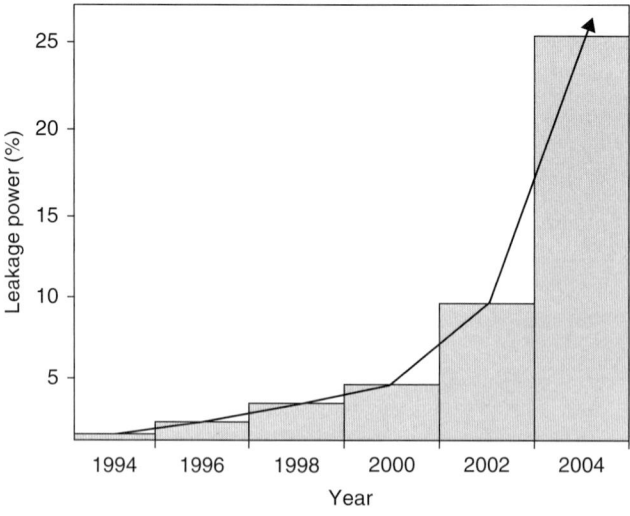

Figure 11.8 Growth in leakage power contribution to the total power dissipation. [7]

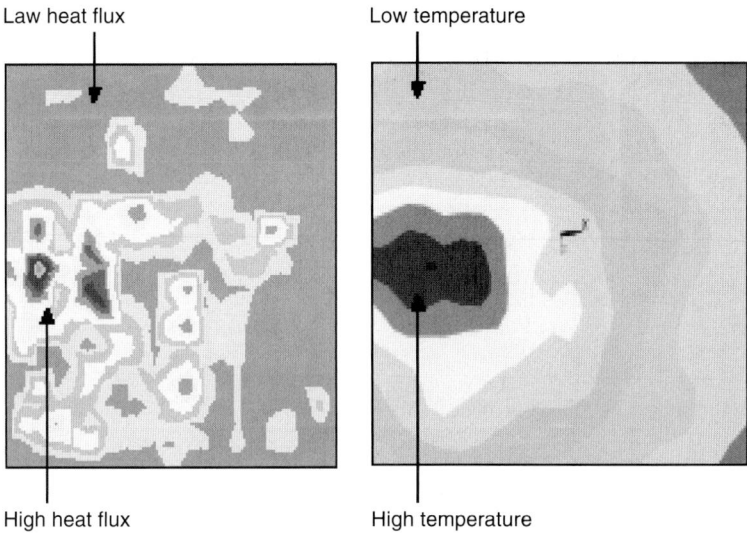

FIGURE 11.9 Distribution of heat flux and temperature on the processor. [14]

more than the average heat flux on the die, and the expected hot-spot temperature to exceed by 30°C or more the average die temperature [13]. A projected heat flux distribution on a microprocessor chip under nonuniform power dissipation is illustrated in Figure 11.9 [14]. Inefficient heat removal from these hot spots may result in a 10 to 15 percent decrease in the operation speed [13]. Temporal localization of hot zones is seen because execution of different instructions uses a different set of transistors on the microprocessor chip, thus concentrating the heat in regions that contain executing transistors at a particular moment. Such spatial and temporal nonuniformity requires on-demand localized thermal management.

11.2.2 RF SOP

A schematic of a typical RF module is shown in Figure 11.10. It consists of transceiver and receiver circuits [4]. In the receiver circuit, the antenna (A) receives the high-frequency signal and routes it to the low-pass filter (LPF) through an active amplifying unit—typically a low-noise amplifier (LNA); the signal is then sent to the digital unit—a microprocessor—via an analog-to-digital converter (ADC). In the transmitter unit, the digital signal is converted to analog, by a digital-to-analog converter (DAC), and is mixed with an intermediate-frequency signal from a local oscillator (LOSC). The output signal is then amplified by an active amplifier, typically a power amplifier (PA) that makes the output signal strong enough to drive the antenna. The passive components include resistors, capacitors, inductors, switches (S), and antennas.

The total dissipated power in a RF module is

$$P_{\text{dissipation}} = P_{\text{input,RF}} + P_{\text{DC}} \tag{11.2}$$

where the first term can be approximated as the summation of the RF input power and RF reflected power and the second term is the product of I_{DD} and the nominal supply voltage (V_{dd}).

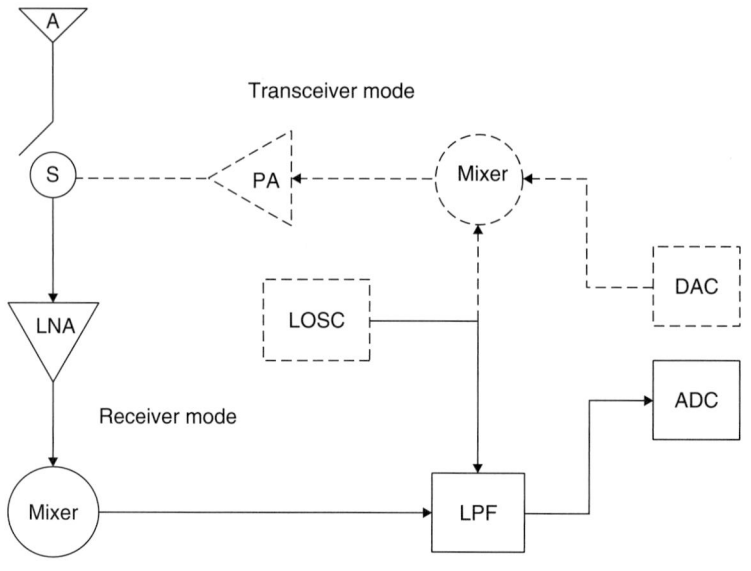

Figure 11.10 Schematic of a typical RF SOP module. [4]

Among the aforementioned components, a significant amount of heat is generated in the microprocessor unit that constitutes the digital segment of the RF module. Characteristics of heat sources in microprocessors have already been discussed in the previous section. Ohmic power dissipation from resistors (used in attenuators, terminations, power dividers, and oscillators, etc.) and interconnections due to the prevalence of skin effect at gigahertz frequencies, and dielectric loss in the RF capacitors are some of the important sources. The most important dissipation source, however, is the PA, which often consumes more than half of the dc power supply in converting the dc power to an RF signal [15]. Power dissipation is directly related to the efficiency of the PAs. Power added efficiency (PAE), defined as the ratio of the difference between the output power and RF drive power to the dc supply power, is

$$\text{PAE} = \frac{P_{out} - P_{RF}}{P_{DC}} \qquad (11.3)$$

The design and biasing of the PA determine the efficiency and thereby the power dissipation. Biasing is important because optimization of the biasing method can significantly decrease the total time when nonzero drain current and drain-to-source voltage coexist and thus can minimize the amount of dissipation. Among various types of PAs, those that are biased as type A and B often produce more heat than those biased as E and F [15]. Another dictating parameter is the type of power transistor used in PAs. These include metal semiconductor field-effect transistors (MESFETs), pseudomorphic high-electron mobility transistors (pHEMTs), enhancement-mode pHEMTs (E-pHEMTs), and heterojunction bipolar transistors (HBTs). The efficiency of a transistor depends on the type of material and the processing technologies used. A silicon carbide MESFET can render 50 percent PAE efficiency for L-band frequencies [16]. A GaAs MESFET on the other hand can give more than 60 percent PAE even for S-band frequencies [17–18]. More than 60 percent PAE can also be achieved in L-band frequencies with an E-pHEMT [19–20]. ITRS-2006 provides a roadmap for efficiency in the linear region (i.e., the region where gain is constant) of various power amplifiers as shown in Figure 11.11 [2]. As seen, the efficiency is projected to be nearly flat. In the face of increasing input power, the thermal issues are of rising concern.

11.2.3 Optoelectronic SOP

Optoelectronic SOP modules are substantially affected by the device and ambient temperatures. Performance, reliability, and efficiency of optoelectronic modules severely depend on the extent of induced strain and displacement due to thermal stress. For example, a lateral displacement as small as 200 nm in the gap between two waveguides can cause a significant loss in the optical coupling efficiency; low-temperature microbending can incur substantial transmission loss; wavelength and output power of laser diodes can be severally affected due to the small increase in the junction/operating temperature; the tiniest change in the MEMS-based optical systems can lead to failure of the device [21]. A good understanding of thermal sources in optoelectronic modules is therefore vital.

A schematic of a typical optoelectronic SOP module is shown in Figure 11.12 [4]. Such modules typically consist of various components, including electrical driver, light sources [light-emitting diode (LED), vertical cavity surface-emitting laser (VCSEL), and laser diode (LD)], coupler, light transmission medium [free space (FS), waveguide (WG), and optical fiber (OF)], light detectors and amplifiers, and electrical receiver [4]. This section is concerned with the high-power light sources that are responsible for almost all power dissipation in optoelectronic modules.

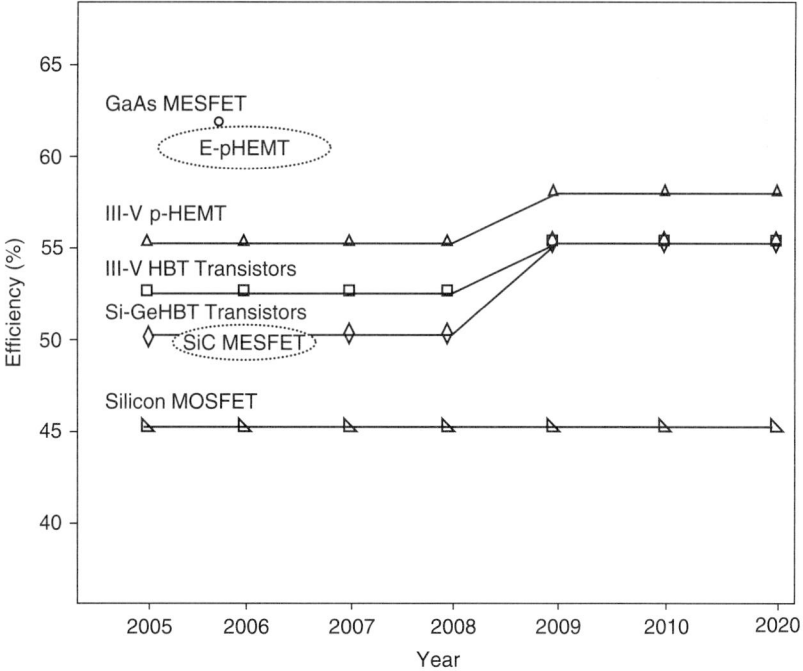

FIGURE 11.11 Comparison of efficiencies of various power amplifiers. [2]

616 Chapter Eleven

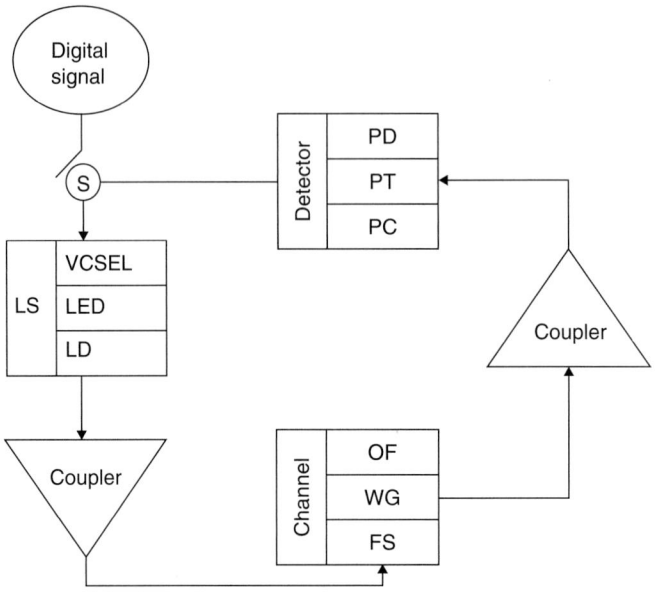

FIGURE 11.12 Schematic of typical optoelectronic SOP module. [4]

LDs are one of the important light sources. Their shrinking size and increasing operating power have resulted in a very high heat flux. A laser diode designed for 100-W optical output power and a typical efficiency of 60 percent generates as high as 80 W of waste heat, which translates to a heat flux of 500 to 600 W/cm² [22]. Dissipation of such a large heat flux is required to maintain a junction temperature below 60°C to ensure a high operative efficiency of the device [22]. The heat source in an LD occupies a very small volume [23], and the thermal resistance includes both spreading and one-dimensional components. Also, different assembly approaches dictate the thermal behavior of the LD. A junction-up assembly of an LD, in which the anode is directly attached to the carrier, poses more severe thermal challenge with more than a 250°C rise of junction temperature, for a power dissipation of 1 W [23].

Distributed feedback (DFB) lasers and VCSELs are other light sources. One of the most important characteristics of DFBs and VCSELs are their lower power consumption and stability over a large temperature range as compared to LDs. However, a small cavity length makes the volumetric heat generation much larger than an edge-emitting laser diode and also increases the thermal spreading resistance in the region close to the active layer [24]. A typical VCSEL generates a heat flux as high as 1 kW/cm² over a limited area of 100 µm² [25].

A high-power light-emitting diode (LED) is another light source used in optoelectronic SOP modules. The power output as well as the radiation spectrum is severely affected by ambient temperature. High-brightness visible light LEDs dissipate approximately 1 W, which corresponds to a heat flux of 100 W/cm². The total heat dissipation in an LED is given by power dissipation due to the LED output current and V_{CE} (output voltage). The residual power dissipation due to the device can be given by

$$P_{\text{dissipation}} = I_{\text{out}} V_{CE} + I_{\text{out}} V_{\text{in}} \tag{11.4}$$

From Equation (11.4), it is clear that the total power dissipation is proportional to the input, as well as the output voltage in the device.

11.2.4 MEMS SOP

MEMS find application in a variety of RF components including transmission line structures, *LC* passives, mixers, power dividers, shunt switches, capacitive switches, couplers, and resonators. High operational speed requires low-loss and low-dispersion transmission line structures in RF circuits. Air-suspended MEMS structures have been identified as a suitable candidate for such applications. These structures pose severe thermal challenges due to self-heating effects at RF frequencies [26]. Such an effect is shown in Figure 11.13 [26]. Evidently, at even 1-W power load, the temperature rise can be as high as 117°C.

Furthermore, increased thermal resistance due to air gaps in these structures limits the thermal dissipation from the active region. The skin effect at the RF signal transmission worsens this situation. Similar concerns are found in shunt capacitive MEMS-based switches. They also produce heat due to the skin effect at a high RF frequency and power.

Other MEMS-based systems in the SOP module could include a micromachined thermal isolation structure, microscale elements for active thermal sensing and control, thermal microactuators, micropumps, and microvalves [27]. Though these structures don't generate significant heat themselves, their performance may be impacted by background temperatures. Integration of these devices with high power dissipating sources in SOP systems becomes very challenging, as operation of these devices demands precise control of background thermal characteristics.

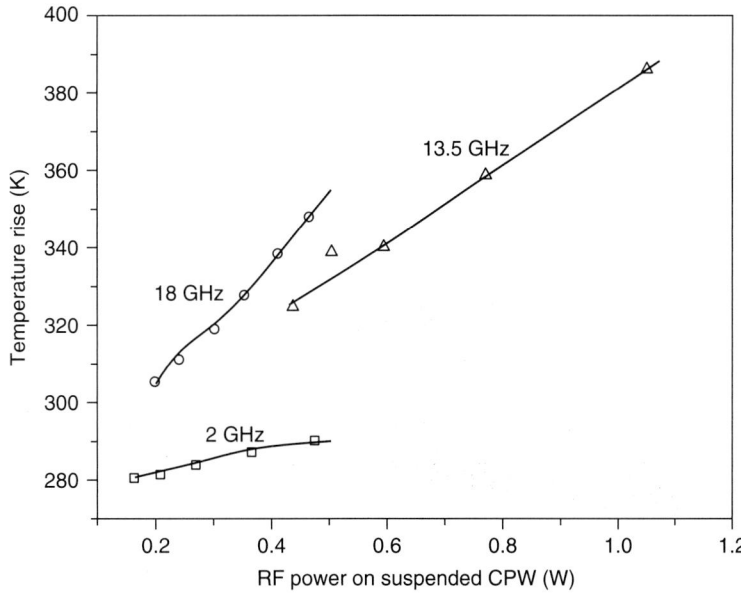

FIGURE 11.13 Skin effect and self-heating effect at RF frequencies in a MEMS SOP module. [26]

11.3 Fundamental Heat Transfer Modes

Effective thermal design and management demand detailed understanding of the heat transfer process involved in SOP systems. Heat transfer occurs via multiple modes in a coupled fashion even for a simple configuration. Figure 11.14 illustrates multiple-mode heat transfer in a substrate, with a single heat source, exposed to an ambient environment.

11.3.1 Conduction

Heat transfer by conduction or diffusion occurs due to random interactions between higher- and lower-energy carriers. Through such interactions, heat is transferred from more energetic molecules to less energetic ones. No bulk movement of the medium occurs. The phenomenological relation for the conduction heat flux vector is Fourier's law:

$$\tilde{q}'' = -k\tilde{\nabla}T \tag{11.5}$$

where k is the thermal conductivity of the medium, which for a nonhomogeneous, anisotropic medium is a tensor with nine components given by $k_{i,j}$. For a homogeneous orthotropic medium, the three components along the orthogonal coordinate directions are k_x, k_y, and k_z. For a homogeneous isotropic medium, only a single value of thermal conductivity k exists.

Fourier's law can be utilized to calculate the heat flux at any location if the temperature distribution $T(x, y, z, t)$ is known. This is obtained by solving the heat diffusion equation, which can be obtained by applying the first law of thermodynamics and Fourier's law to an incremental control volume. In the cartesian coordinate system it is given as

$$\frac{\partial^2 T}{\partial x^2} + \frac{\partial^2 T}{\partial y^2} + \frac{\partial^2 T}{\partial z^2} + \frac{\dot{q}}{k} = \frac{\rho c}{k}\frac{\partial T}{\partial t} \tag{11.6}$$

In general, solution of the heat diffusion equation requires six boundary conditions (two in each direction) and an initial condition. Except for a number of exact solutions under linear condition, this is done numerically as will be discussed later.

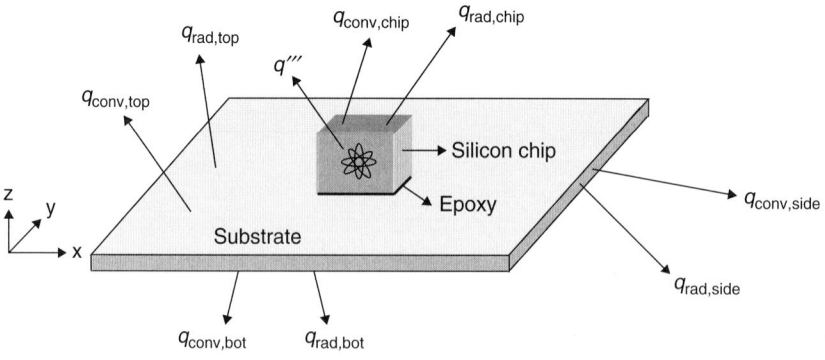

Figure 11.14 Heat transfer in electronic packages via multiple, coupled modes simultaneously.

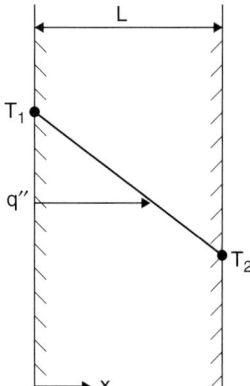

FIGURE 11.15 Temperature profile in 1D steady-state thermal conduction in a homogeneous material in the absence of heat generation.

One-Dimensional Conduction

For steady heat conduction in one dimension in a homogeneous material in the absence of heat generation (see Figure 11.15), Equation (11.6) reduces to

$$\frac{d^2T}{dx^2} = 0 \tag{11.7}$$

With the boundary conditions of prescribed temperature of T_1 at $x = 0$ and T_2 at $x = L$ this can be solved to get

$$T = (T_2 - T_1)\frac{x}{L} + T_1 \tag{11.8}$$

Using Fourier's law, the heat flux is obtained as

$$q'' = (T_1 - T_2)\frac{k}{L} \tag{11.9}$$

The net rate of heat transfer is

$$q = q''A = \left(\frac{kA}{L}\right)\Delta T \tag{11.10}$$

Equation (11.10) forms the basis of the useful resistor network analogy. From Ohm's law in electric circuits, the voltage drop across the conductor V, the current through it I, and its electrical resistance R, are related as

$$V = IR \tag{11.11}$$

In Equation (11.11), if an analogy is drawn between the temperature difference ΔT and V, and the total rate of heat transfer Q and I, the conduction thermal resistance $R_{th,c}$ can be identified as

$$R_{th,c} = \frac{L}{kA} \tag{11.12}$$

In many heat conduction applications, the heat loss at a boundary occurs by convection to a fluid or by radiation to external surroundings. The boundary heat flux under such conditions is given as

$$q'' = h(T_s - T_f) \tag{11.13}$$

where h is the convective or effective radiation heat transfer coefficient and T_f is an appropriate reference temperature in the fluid. More information on the estimation of h will be provided later. Using the resistor network analogy, Equation (11.14) provides the surface convective resistance as

$$R_{th,conv} = 1/Ah \tag{11.14}$$

Heat Flow across Solid Interfaces

Heat conduction across an interface between two contacting solids results in a finite change in temperature ΔT. The thermal interface resistance is defined as

$$R_{t,c} = \frac{\Delta T}{q_x} \tag{11.15}$$

A common approach for specifying the interfacial thermal resistance is in the form of interface conductance per unit contact area:

$$R''_{t,c} = \frac{\Delta T}{q''_x} \tag{11.16}$$

The interface resistance is dependent on the materials involved, the presence of any filler between the interfaces, and the contact pressure at the interface. A compliant filler or thermal interface material (TIM) with bulk resistance R_{Bi} is used to reduce the interface resistance. The overall interface thermal resistance in such cases is

$$R_i = R_{Bi} + R_{Ci} + R_{C2i} \tag{11.17}$$

Common interface materials include organic greases filled with ceramic particles, solid-liquid phase change materials, and solders. In order for the overall resistance to decrease, the sum of these three resistances must be below the bare joint resistance given by Equation (11.15) for $R_{t,c}$. A higher contact pressure can also be used to reduce the interface resistance of a bare joint. One of the most common uses of TIM is in the attachment of microprocessor chips to packages (TIM1) and that of heat sink to microprocessor package (TIM2).

Heat Spreading due to Multidimensional Conduction

Heat flow within electronic chips and packages is multidimensional, particularly near the heat sources, due to their small sizes. This can result in a significant increase in the temperatures within the heat sources and in their close vicinity, compared to one-dimensional estimates. More realistic estimates require the solution of the multidimensional heat conduction equation to determine the heat spreading resistance.

Several analytical solutions of heat flow from a small heat source placed on an extensive substrate have been carried out. To determine the thermal conductivity of a heat generating rectangular parallelpiped on a rectangular substrate of finite thickness requires the solution of the 3D heat conduction equation. A slightly simplified version,

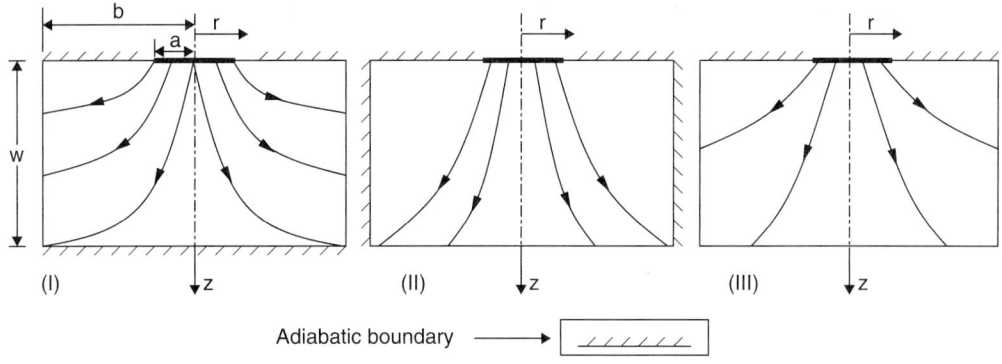

Figure 11.16 Heat spreading analytical solutions in an axisymmetric geometry. [28]

which is often adequate, is the use of an axisymmetric configuration with a heated region on the upper surface seen in Figure 11.16. The heat source simulates the discrete heating provided by a chip. Analytical solutions to this problem, subject to three combinations of isothermal and adiabatic conditions as shown in Figure 11.16, have been given by Kennedy [28].

Example Let us consider a rectangular heated region of 10 mm × 10 mm on a 20 mm × 20 mm heat spreader of thickness 0.3 mm and thermal conductivity of 0.58 W/m · K. If 2D heat spreading is neglected, a 1D thermal resistance can be calculated as $0.3 \times 10^{-3}/[0.58 \times (10 \times 10) \times 10^{-4}] = 0.05$ K/W. Let us compare this with an estimate of the 2D spreading resistance. Based on system thermal considerations, let us assume that it is determined that case II in Figure 11.16 best describes the thermal boundary conditions.

Next we approximate the square heat source and the substrate as equivalent circular regions. The resulting equivalent radii of the heat source and substrate are a and b, respectively, where

$$a = \frac{10}{\sqrt{\pi}} \text{ mm} \qquad b = \frac{20}{\sqrt{\pi}} \text{ mm}$$

With $w = 0.3$ mm, $k = 0.58$ W/m · K, $a/b = 0.5$, and $w/b = 0.027$, we get from Figure 11.17:

$$H = \frac{kT(0,0)}{q''a} = 0.06$$

where q'' is the applied heat flux and $T(0,0)$ is the temperature rise at the center of the heat source. This allows us to calculate the thermal spreading resistance:

$$R_{th} = \frac{H}{k\pi a} = 5.84 \text{ K/W}$$

As can be seen, this value is two orders of magnitude higher than the 1D estimation in this case. The spreading resistance can be included in the overall resistance network to represent the multidimensional heat conduction near the heat source.

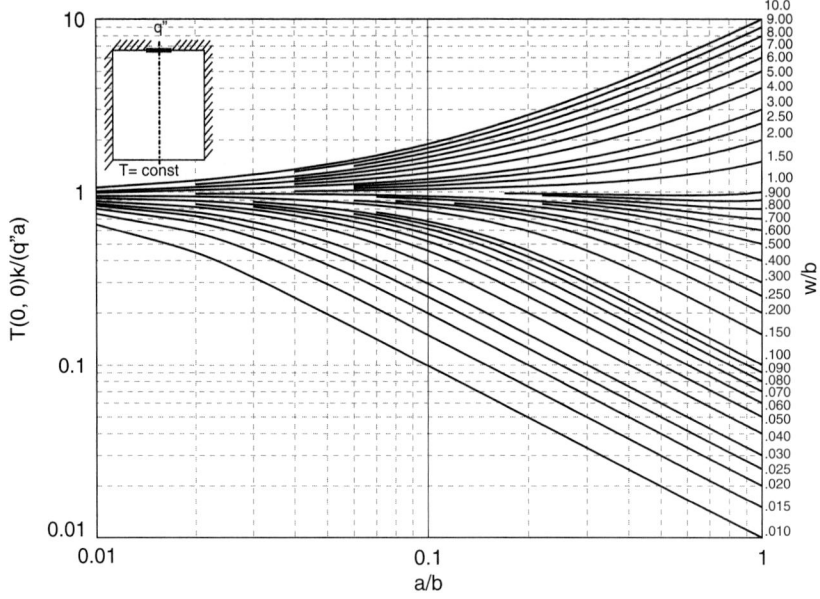

FIGURE 11.17 Spreading resistance factor for case II in Figure 11.16. [28]

Lumped Capacitance Convective Transient The heat conduction equation [Equation (11.6)] may need to be solved for transients arising due to changes in boundary conditions and/or due to heat generation within the components. For example, start-up or shutdown of systems results in transient temperature variations. Another common situation arises when an object at an initial temperature T_i is suddenly heated or cooled by a convection environment at T_f with a heat transfer coefficient h at its boundary.

The general transient thermal response of such a system is determined by the ratio of the internal conduction resistance and the surface convection resistance, termed the Biot number (Bi).

$$\text{Bi} = \frac{hL}{k_s} = \frac{R_{th,cond}}{R_{th,conv}} \qquad (11.18)$$

where L is a characteristic dimension of the object, the ratio of the volume and surface area of the object, and k_s is the thermal conductivity of the solid.

When Bi $\ll 1$, signifying a negligible internal conduction resistance compared to the surface convection resistance, the transient thermal response to a step change in ambient temperature is given by

$$\frac{T - T_i}{T_f - T_i} = 1 - \exp(-t/\tau) \qquad (11.19)$$

where τ is the time constant given by

$$\tau = \frac{hA}{\rho c V} \qquad (11.20)$$

11.3.2 Convection

Convection heat transfer results in the presence of a moving fluid in contact with a surface at a different temperature. As seen in Figure 11.14 an air-cooled printed wiring board in general loses heat by a combination of conduction, convection, and radiation. Convection situations are classified as internal or external depending upon whether the influence of bounding surfaces is important or not. For example, in the case of a circuit card confined within the portable system, the enclosure walls are likely to have a profound impact on heat transfer, whereas in a larger enclosure such as a tower case for a personal computer, the effect of the confining walls may be weaker. A further classification is based on forced or natural modes. Forced convection arises in the presence of a prime mover such as a pump or blower for driving the flow. Natural or buoyancy-induced convection arises in the presence of a body force field such as gravity, working in concert with a density difference produced, for example, by a temperature difference. As such, four circumstances for single-phase convection may be identified: internal forced convection, internal natural convection, external forced convection, and external natural convection.

The rate of heat transfer in convection is given by the phenomenological Newton's law of cooling:

$$q'' = hA(T_s - T_{ref}) \tag{11.21}$$

where T_s is the local surface temperature and T_{ref} is a suitable reference temperature. For external convection the latter is the ambient temperature, while for internal convection in ducts the appropriate reference temperature is the bulk or mean fluid temperature.

The determination of the heat transfer coefficient distribution is the central objective of a convection analysis. In a decoupled conduction analysis, this convection coefficient can be determined by solving the fluid flow problem and subsequently utilized it to determine internal object temperatures. In situations involving combined conduction within an object and convection from its surfaces, the governing equations can be solved for both media simultaneously.

Governing Equations and Nondimensional Parameters

The governing equations in cartesian coordinates (x, y, z) for constant property, incompressible three-dimensional flow ($\vec{V} = u\hat{i} + v\hat{j} + w\hat{k}$) in the absence of mass transfer are

Continuity:
$$\nabla \cdot \vec{V} = 0 \tag{11.22}$$

x momentum:
$$\rho \frac{Du}{Dt} \equiv \rho \left[\frac{\partial u}{\partial t} + \nabla \cdot (u\vec{V}) \right] = -\frac{\partial P}{\partial x} + \mu \nabla^2 u + X \tag{11.23}$$

y momentum:
$$\rho \frac{Dv}{Dt} \equiv \rho \left[\frac{\partial v}{\partial t} + \nabla \cdot (v\vec{V}) \right] = -\frac{\partial P}{\partial y} + \mu \nabla^2 v + Y \tag{11.24}$$

z momentum:
$$\rho \frac{Dw}{Dt} \equiv \rho \left[\frac{\partial w}{\partial t} + \nabla \cdot (w\vec{V}) \right] = -\frac{\partial P}{\partial z} + \mu \nabla^2 w + Z \tag{11.25}$$

Energy:
$$\rho C_p \frac{DT}{Dt} \equiv \rho C_p \left[\frac{\partial T}{\partial t} + \nabla \cdot (T\vec{V}) \right] = k\nabla^2 T + \mu \Phi + \beta T \frac{Dp}{Dt} \tag{11.26}$$

The last two terms on the right-hand side of the energy equation [Equation (11.26)] correspond to the viscous dissipation and pressure stress effects, respectively. Equations (11.22) to (11.26) can be normalized using the following variables:

$$(x*, y*, z*) = \frac{(x, y, z)}{L} \quad (u*, v*, w*) = \frac{(u, v, w)}{U} \quad t* = \frac{tU}{L} \quad \text{and} \quad p* = \frac{p}{\rho U^2}$$

where L and U are appropriate length and velocity scales, respectively. The normalized form of Equation (11.23) in the absence of the body force X is

$$\left(\frac{\partial u*}{\partial t*} + u*\frac{\partial u*}{\partial x*} + v*\frac{\partial u*}{\partial y*} + w*\frac{\partial u*}{\partial z*}\right) = -\frac{\partial P*}{\partial x*} + \frac{\mu}{UL\rho}\left(\frac{\partial^2 u*}{\partial x*^2} + \frac{\partial^2 u*}{\partial y*^2} + \frac{\partial^2 u*}{\partial z*^2}\right) \quad (11.27)$$

Equations (11.24) and (11.25) are similar and not written out for brevity. The Reynolds number emerges as the key nondimensional solution parameter:

$$\mathrm{Re} = \frac{\rho UL}{\mu} = \frac{\rho U^2 L^2}{\mu\left(\frac{U}{L}\right)L^2} = \text{inertia forces/viscous forces} \quad (11.28)$$

In order to nondimensionalize the energy equation, a normalized temperature is defined as

$$T* = \frac{T - T_\infty}{T_s - T_\infty} \quad (11.29)$$

where T_∞ and T_s are the local ambient and surface temperatures, respectively. Both of these, could, in general vary with location and/or time. The resulting normalized energy equation is

$$\frac{DT*}{Dt*} = \frac{1}{\mathrm{Re}\,\mathrm{Pr}}\left(\nabla*^2 T* + 2\mathrm{Ec}\,\mathrm{Pr}\,\Phi* + 2\beta T\,\mathrm{Re}\,\mathrm{Pr}\,\mathrm{Ec}\frac{Dp*}{Dt*}\right) \quad (11.30)$$

where the Prandtl number $\mathrm{Pr} = \nu/\alpha = \mu C_p/k$ is a measure of the ratio of diffusivity of momentum to diffusivity of heat, and the Eckert number $\mathrm{Ec} = V^2/2C_p(T_s - T_\infty)$ defines the relative magnitude of the kinetic energy of the flow and the enthalpy difference. In Equation (11.30), β is the volumetric expansion coefficient of the fluid. For ideal gases $\beta T = 1$, and typically $\beta T \ll 1$ for liquids. The pressure stress term is thus negligible in forced convection whenever viscous dissipation is small.

The wall values of shear stress and heat flux are, respectively,

$$\tau_s \equiv \mu\left.\frac{\partial u}{\partial y}\right|_s = \frac{\mu U}{L}\left.\frac{\partial u*}{\partial y*}\right|_{y*=0} \quad (11.31a)$$

$$q''_s = -k\left.\frac{\partial T}{\partial y}\right|_s = \frac{k(T_s - T_\infty)}{L}\left(-\left.\frac{\partial T*}{\partial y*}\right|_{y*=0}\right) \quad (11.31b)$$

From a practical perspective, the friction drag and the heat transfer rate are the most important quantities. These are determined from the friction coefficient C_f and the Nusselt number Nu defined, respectively, as

$$C_f = \frac{2\tau_s}{\rho U^2} = \frac{2}{\text{Re}} \left(\frac{\partial u*}{\partial y*} \right)_{y*=0} \tag{11.32a}$$

$$Nu = \frac{hL}{k} = \frac{q_s'' L}{k(T_s - T_\infty)} = -\frac{\partial T*}{\partial y*}\bigg|_{y*=0} \tag{11.32b}$$

The solution of the energy equation [Equation (11.30)] allows the determination of the temperature gradient at the wall, which can be utilized to find the heat transfer coefficient from Equation (11.32).

In the presence of buoyancy effects, Equation (11.23) needs to retain the body force term X. This can be combined with the hydrostatic component of the overall static pressure gradient in Equation (11.23). The right-hand side of the resulting momentum equation in the x direction, assuming the gravity acts in the x direction, is

$$-\frac{\partial P_m}{\partial x} + \mu\left(\frac{\partial^2 u}{\partial x^2} + \frac{\partial^2 u}{\partial y^2} + \frac{\partial^2 u}{\partial z^2}\right) + g\beta(T - T_\infty) \tag{11.33}$$

Normalization of the governing equations for flow driven by buoyancy forces near a vertical surface at temperature T_s placed in an environment at temperature T_f requires a new velocity scale, since there is no imposed velocity:

$$U_{nc} \sim (g\beta x \, \Delta T)^{1/2} \tag{11.34}$$

where x is the downstream distance from the leading edge of the plate and $\Delta T = T_s - T_f$.

A resulting new nondimensional parameter is the local Grashof number:

$$\text{Gr}_x = \left(\frac{U_{nc} x}{v}\right)^2 = \frac{g\beta x^3 \Delta T}{v^2} \tag{11.35}$$

The energy equation [Equation (11.26)] in enclosed solid regions within the flow field (e.g., a circuit board or electronic packages) simplifies to the heat conduction equation [Equation (11.6)]. This needs to be solved simultaneously with the energy equation in the flow field. A continuity of temperature and heat flux at various interfaces needs to be imposed.

Phase Change

Liquid-vapor phase change processes are utilized to achieve very high heat transfer coefficients, typically an order of magnitude higher than with single-phase forced convection using liquids. A closed-loop thermal management device utilizing phase change, to be discussed in Section 11.5, is a thermosyphon (see Figure 11.29). The generated vapor travels to the condenser, where it rejects heat to the ambient air. The working fluid now in the form of a liquid is returned to the evaporator. The thermosyphon provides the ability to use a compact evaporator near the spatially constrained heat source, and a larger remote heat exchanger for ambient heat rejection further away. In order to predict the thermal performance of devices such as a thermosyphon it is important to develop convection coefficients for boiling and condensation processes.

Pool Boiling

The process of boiling from a heater into a large, quiescent mass of liquid is termed pool boiling. If the bulk liquid is at the saturation temperature T_{sat}, the resulting process is termed saturated pool boiling. For pool temperatures below saturation levels, the boiling is termed subcooled boiling. Figure 11.18 shows the boiling curve, or logarithmic scale variation, of heat flux with excess temperature:

$$\Delta T_e = T_s - T_{sat} \qquad (11.36)$$

The most desirable region of operation in Figure 11.18 is nucleate boiling, which is characterized by the ability to sharply increase heat flux, with a very modest increase in the excess temperature. This region of high heat transfer coefficients occurs due to the formation and departure of a large number of small bubbles from nucleation sites on the surface. As the bubbles depart, they provide a vigorous stirring of the surrounding liquid. A commonly used expression for the heat transfer coefficient in this regime is the Rohsenow correlation. The temperature difference used in Equation (11.21) in defining the heat flux is now given by Equation (11.36).

Film Condensation

The vapor generated in the evaporator of a closed-loop boiling device, such as a thermosyphon, rejects heat to the ambient and condenses in a condenser. This can occur through several modes, most commonly as the formation of a liquid film over the surface, called film condensation. The temperature difference to be used in Equation (11.21) for the calculation of the heat flux is given by $(T_{sat} - T_s)$.

11.3.3 Radiation

Thermal radiation is part of the electromagnetic spectrum in the range of 0.1 to 10 μm covering part of the ultraviolet region and the entire visible and infrared regions. Energy emission by radiation occurs from any surface at temperatures above 0 K. Once emitted, radiation does not require a medium for propagation, unlike conduction and convection. In addition, surfaces can absorb, reflect, and transmit incident radiation, or irradiation.

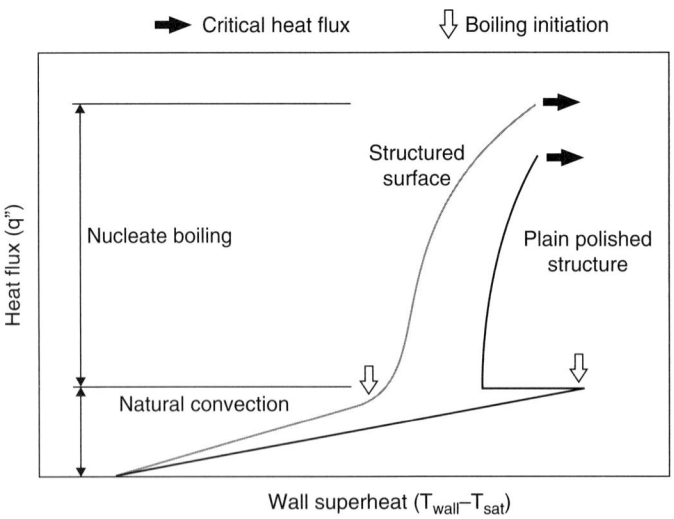

Figure 11.18 Pool boiling curve for a dielectric working fluid.

Energy transfer from surfaces by radiation displays both spectral or wavelength dependent and directional behavior. An ideal surface in radiation is the blackbody, which absorbs all the incident irradiation and emits the most of all surfaces at a given temperature. Real surfaces are characterized by properties that compare them with black surfaces. A common assumption utilized for real engineering surfaces is that they can be approximated as gray, having radiative properties that are wavelength independent.

For opaque surfaces the absorptivity and emissivity are sufficient to characterize radiative transport. In their most general form, these may be defined on a wavelength (spectral) and direction-dependent (directional) basis. Integrated values over wavelength (total) and direction (hemispherical) can also be defined. For the simple analysis presented here, total hemispherical properties defined below are utilized.

$$\varepsilon(T) = \frac{E(T)}{E_b(T)} \qquad (11.37)$$

$$\alpha = \frac{G_{abs}}{G} \qquad (11.38)$$

$$\rho = \frac{G_{ref}}{G} \qquad (11.39)$$

where $E(T)$ and $E_b(T)$ are the respective emissive powers of the object and a black surface at temperature T, and G_{abs} and G_{ref} are the respective absorbed and reflected portions of the irradiation G.

The radiative properties must satisfy the energy balance at an interface:

$$G = G_{ref} + G_{abs} + G_{tr} \qquad (11.40)$$

Since the irradiated energy on an opaque surface must be absorbed or reflected, the total hemispherical absorptivity and reflectivity are related as follows:

$$\alpha + \rho = 1 \qquad (11.41)$$

In addition, Kirchhoff's law provides equality relations between the emissivity and absorptivity, under specific conditions. The equality between the spectral directional emissivity ($\varepsilon_{\lambda,\theta}$) and absorptivity ($\alpha_{\lambda,\theta}$) is unconditional. However, the equality between the total hemispherical emissivity (ε) and absorptivity (α) requires either diffuse irradiation, or a diffuse surface, implying no angular dependence in the incoming radiation intensity and surface properties. Additionally, one of the following conditions needs to apply:

- The irradiation corresponds to a blackbody at the surface temperature.
- The surface emissivity and absorptivity are independent of wavelength, or the surface is gray.

Radiation in Gray Diffuse Enclosures

For a general two-surface enclosure (Figure 11.19):

$$q_1 = \frac{\sigma(T_1^4 - T_2^4)}{\left[\frac{1-\varepsilon_1}{\varepsilon_1 A_1} + \frac{1}{A_1 F_{12}} + \frac{1-\varepsilon_2}{\varepsilon_2 A_2}\right]} \qquad (11.42)$$

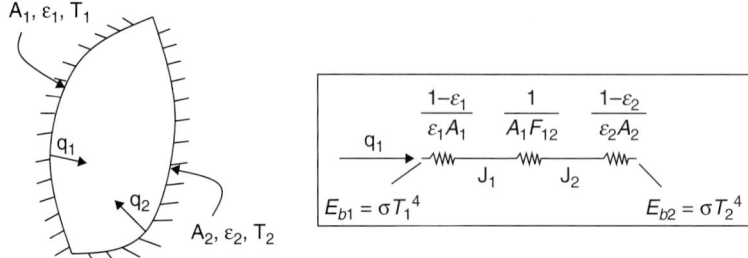

FIGURE 11.19 A two-surface enclosure.

where $\sigma = 5.67 \times 10^{-8}$ W/m²·K⁴ is the Stefan-Boltzmann constant, F_{12} is the fraction of radiation emitted by surface 1 and seen by surface 2, or the view factor from surface 1 to 2, and ε is the hemispherical total emissivity of a surface. Two special cases of this general expression are commonly employed. For a small surface in large surroundings:

$$F_{12} \cong 1 \quad \text{and} \quad \frac{A_1}{A_2} \approx 0$$

resulting in

$$q_1 = \varepsilon_1 A \sigma (T_s^4 - T_{sur}^4) \tag{11.43}$$

The second case corresponds to two infinite parallel plates. In this case,

$$F_{12} = 1 \quad \text{and} \quad A_1 = A_2 = A$$

The net rate of heat transfer is given by

$$q_1 = \frac{\sigma A (T_1^4 - T_2^4)}{\frac{1}{\varepsilon_1} + \frac{1}{\varepsilon_2} - 1} \tag{11.44}$$

Effective Radiation Heat Transfer Coefficient

It is sometimes useful to compare the rates of heat transfer by convection and radiation. The rate of heat transfer by radiation can be expressed in a manner similar to Newton's law of cooling for convection [Equation (11.21)]:

$$q_r = A h_r (T_s - T_{sur}) \tag{11.45}$$

For the small surface in the large-surroundings case, Equation (11.39) can be used to obtain the effective radiation heat transfer coefficient as

$$h_r = \varepsilon \sigma (T_s^2 + T_{sur}^2)(T_s + T_{sur}) \tag{11.46}$$

From Equation (11.46) we can see that for $T_{sur} = 300$ K, $T_s = 330$ K, and $\varepsilon = 0.7$, $h_r \sim 5$ W/m²·K, a value close to the natural convection heat transfer coefficient. This calculation shows that in natural convection cooled systems, radiation may be of the same order of magnitude and cannot be neglected.

11.4 Fundamentals of Thermal Characterization

Effective handling of the heat generation within SOP systems is a key consideration in their design. Thermal design requires understanding of spatial and temporal distribution of temperature fields within the system resulting from multiple-input operating parameters. Materials properties and interface thermal resistances are additional inputs. There are various experimental and numerical techniques for thermal characterization of microsystems.

11.4.1 Numerical Methods for Thermal Characterization

The objective of thermal characterization is to determine the temperature fields within the system, with the appropriate resolution spatially and temporally. As part of the design process of a SOP system it is necessary to characterize the thermal fields through modeling and measurements. Early in the design phase a thermal engineer must construct a model that will evaluate the performance, simulate fail-safe modes, and lead to an optimized design. This ensures a substantial reduction of lead times and costs of new designs by reducing the number of time-consuming and costly testing procedures in many of the steps.

Thermal modeling approaches can be broadly classified into the following three categories. The first is the analytical or closed-form solutions approach, which is suitable for a class of idealized situations involving the chip, package, or module. In this approach, the heat input is usually prescribed as a uniform heat flux or as a uniform volumetric heat generation rate such that exact solutions of linear sets of governing equations can be obtained. The second is the resistor network approach in which 1D and 2D heat conduction resistances in rectangular and radial geometries are utilized to simulate the thermal characteristics of the package. This method is usually used at the package, module, and system levels. The third method is a more general one and involves numerical solutions of the discretized governing differential equations, together with the boundary and initial conditions. This method belongs to the broad subject of computational fluid dynamics (CFD). It can be used for simulations at the package or printed wiring assembly or board (PWA or PWB) level to solve the heat conduction equation in two or three dimensions or for simulations of coupled fluid flow and heat transfer in the entire electronic systems. Since CFD simulations are capable of producing a detailed temperature distribution, they can also be used for evaluating thermomechanical stresses, especially in locations containing packaging materials with unequal coefficients of thermal expansion.

Combined Mode Heat Transfer Modeling

Generally there are multiple modeling levels to be considered in an electronic system. The different levels typically comprise the chip (or device), module (or package), board, and system. These have significantly different length scales, and the cooling–heat transfer considerations (and hence the modeling issues) at each level are unique and are described as follows.

At the chip level, heat transfer is by conduction. The thermal resistance from the junction to the chip is of importance. Although the temperature rise from the junction to the chip may usually not be too large, it cannot be neglected in some very high powered chips. At the module and package level, again, the mechanism of heat transfer is primarily by conduction in solids. The important considerations at this level are the chip and module power dissipation and the module construction—its geometry and material properties. Depending on the complexity of the package and the boundary conditions at the boundary of the package, the solution techniques for calculation of thermal performance could range from analytical closed-form solutions, thermal resistance network method, to CFD (heat conduction only) methods.

At the next level (card and board level), the heat transfer is mainly by convection and radiation. The thermal considerations are component (package) geometry and location on the board, flow type (laminar or turbulent, natural, forced, or mixed convection, etc.), flow rate, flow distribution, and package flow impedance. Radiation heat transfer can become important if cooling is by natural convection. The governing equations (Navier-Stokes and energy conservation equations) are nonlinear, and hence simple analytical techniques cannot be applied. Either CFD techniques or established empirical correlations must be used.

The system-level considerations are the ambient environment (temperature, altitude, and humidity); prime mover (blower/fan/pump) selection considerations such as capacity, physical size, noise level, and location in the machine; and human factors such as acoustic noise, casing temperature, and grill locations. The heat transfer analysis is typically restricted to a simple energy balance at this level.

It is apparent from the above considerations that a complete computational analysis of electronics cooling will require schemes for numerically solving conduction, convection, and radiation heat transfer. Popular methods of computing heat transfer and fluid flow are presented subsequently.

Methodology for Numerical Computation

The fluid flow and heat transfer in any system are modeled by applying conservation laws of mass, momentum, and energy to obtain the required governing differential equations. These equations define the physics of the problem. A complete description of the problem would include

- Governing differential equations
- Geometrical details
- Boundary conditions
- Initial conditions
- Material properties

An analytical solution may be possible only for some restricted cases. However, in most problems of practical interest in electronics cooling, computational techniques are necessary. These computational techniques are generally known as computational fluid dynamics/heat transfer (CFD/HT). Typically, a CFD code has the following three mail components:

1. *Preprocessor.* Deals with the definition of the geometry (computational domain), grid generation (mesh, grids, control volumes, and elements), material properties, physical constants, and boundary and initial conditions. The grid can be classified as a structured grid and unstructured grid. Regular or structured grids consist of families of grid lines with the property that members of a single family do not cross each other and cross each member of the other families only once. This allows the lines of a given set to be numbered consecutively. The disadvantage of structured grids is that they can be used only for geometrically simple solution domains. For very complex geometries, unstructured grids are

more suitable, which can fit an arbitrary solution domain boundary. Unstructured grids are best adapted to the finite-volume and finite-element approaches. The elements or control volumes may have any shape, and there is no restriction on the number of neighbor elements or nodes. In practice, grids made of triangles or quadrilaterals in two dimensions and tetrahedral or hexahedral in three dimensions are most often used.

2. *Solver.* Approximates unknown flow variables by simple functions, discretization, and substitution of the approximate functions into the governing equations, and obtains the solution of the resulting algebraic equations.

3. *Postprocessor.* For graphic visualization of results, such as contour plots, vector plots, 3D plots, and particle tracking.

There are various CFD techniques commonly used for simulation of heat and fluid flow in systems. The techniques differ from each other in the type of discretization method, that is, a method of approximating the differential equations by a system of algebraic equations for the variables at some set of discrete locations in space and time. Some of the most popular techniques include the *finite-difference* method, *finite-element* method, and *finite-volume* method. Fundamentals of CFD analysis using the above techniques can be found in several textbooks [29–31]. Only a brief description of the finite-element and finite-volume methods is presented below. Although the finite-difference method is the oldest method for numerical solution of PDEs, its usage is mainly restricted to simple geometries because of the structured grids used.

Finite-Element Method (FEM) In the finite-element method, the domain is broken into a set of discrete volumes or elements that are generally unstructured; in two dimensions, they are usually triangles or quadrilaterals, while in three dimensions, tetrahedra or hexahedra are most often used. The distinguishing feature of FEMs is that the equations are multiplied by a weight function before they are integrated over the entire domain. In the simplest FEMs, the solution is approximated by a linear shape function within each element in a way that guarantees continuity of the solution across element boundaries. Such a function can be constructed from its values at the corners of the elements. The weight function is usually of the same form. This approximation is then substituted into the weighted integral of the conservation law and the equations to be solved are derived by requiring the derivative of the integral with respect to each nodal value to be zero; this corresponds to selecting the best solution within the set of allowed functions (i.e., the one with minimum residual). The result is a set of nonlinear algebraic equations.

Finite-Volume Method (FVM) The FVM uses the integral form of the conservation equations as its starting point. The solution domain is subdivided into a finite number of contiguous control volumes (CVs), and the conservation equations are applied to each CV. At the centroid of each CV lies a computational node at which the variable values are to be calculated. Interpolation is used to express variable values at the CV surface in terms of the nodal (CV-center) values. As a result, one obtains an algebraic equation for each CV, in which a number of neighbor nodal values appear. The FVM approach is perhaps the simplest to understand and to program. All terms that need be approximated have physical meaning. The FVM, which was originally formulated by Patankar with structured grids [29], is now successfully adapted for unstructured grid systems [29,32], thus making it suitable for complex geometries as well. Since the discretization is performed using an integral form of the governing equations, conservation is always preserved. Moreover, the discretization process in FVM is directly related to the physics of the problem, instead of using variational formulation or functionals (as employed in FEM), which have no easy physical interpretation in problems involving fluid flow and diffusion. It is this particular combination of formulation of a flow problem over control volumes with the geometric flexibility in the choice of grids as well as the flexibility in defining the discrete flow variables that makes FVM suitable for solution of such physical problems involving fluid flow, heat, and mass transfer. Most commercial software such as FLUENT, FLOTHERM, CFX, and STAR-CD are based on this method [33–36].

Consistent with the choice of FVM as the computational technique, it is important to choose a solution scheme accordingly. For the case of incompressible flows, the pressure-density coupling is weak, and hence an explicit equation for pressure in the form of an equation of state is not very useful. The solution is usually obtained using a guessed pressure or velocity field, which is iteratively updated until the velocity field satisfies the continuity equation. This is the basis of the SIMPLE algorithm [29] or its variants such as SIMPLER or SIMPLEC. Since the equations relevant to fluid flow problems are highly nonlinear, direct methods for solution of the algebraic equations (arising out of discretization) would be difficult and would require a large amount of computer storage and time. Hence, for 2D and 3D problems, efficient iterative schemes such as Gauss-Seidel or the line-by-line tridiagonal matrix algorithm (TDMA) are employed. (The latter method, however, is suitable only for structured grid applications.) Also, as an aid for handling nonlinearities, controlled convergence can be achieved by introduction of suitable under-relaxation and over-relaxation parameters in the iterative scheme.

As a demonstration of the finite-volume method for temperature prediction in an electronic system, simulations are performed for the case of an air-cooled laptop enclosure, by considering combined natural convection, conduction, and radiation heat transfer effects (Figure 11.20). The corresponding temperature prediction for various heating loads is shown in Figure 11.21. Details of the simulation can be found in [37].

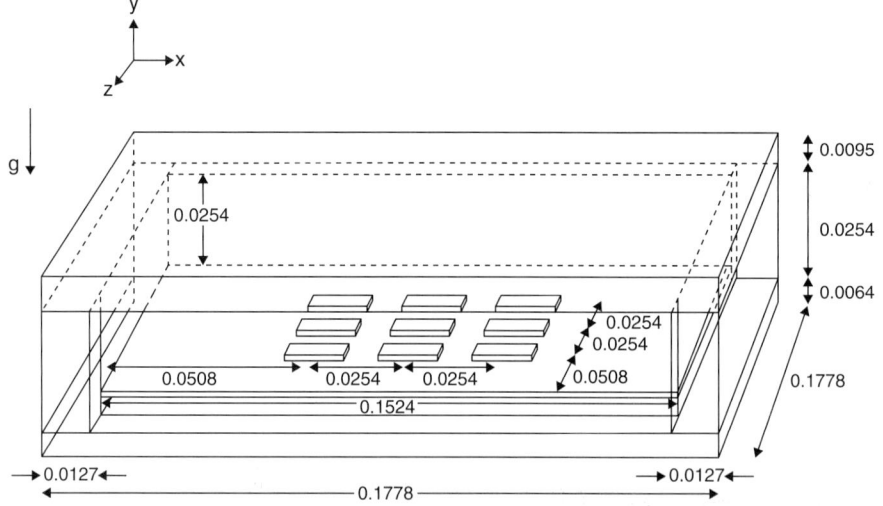

FIGURE 11.20 Three-dimensional finite-volume simulations using a 3 × 3 array of PQFP in a top-cooled enclosure. All dimensions are in meters. [37]

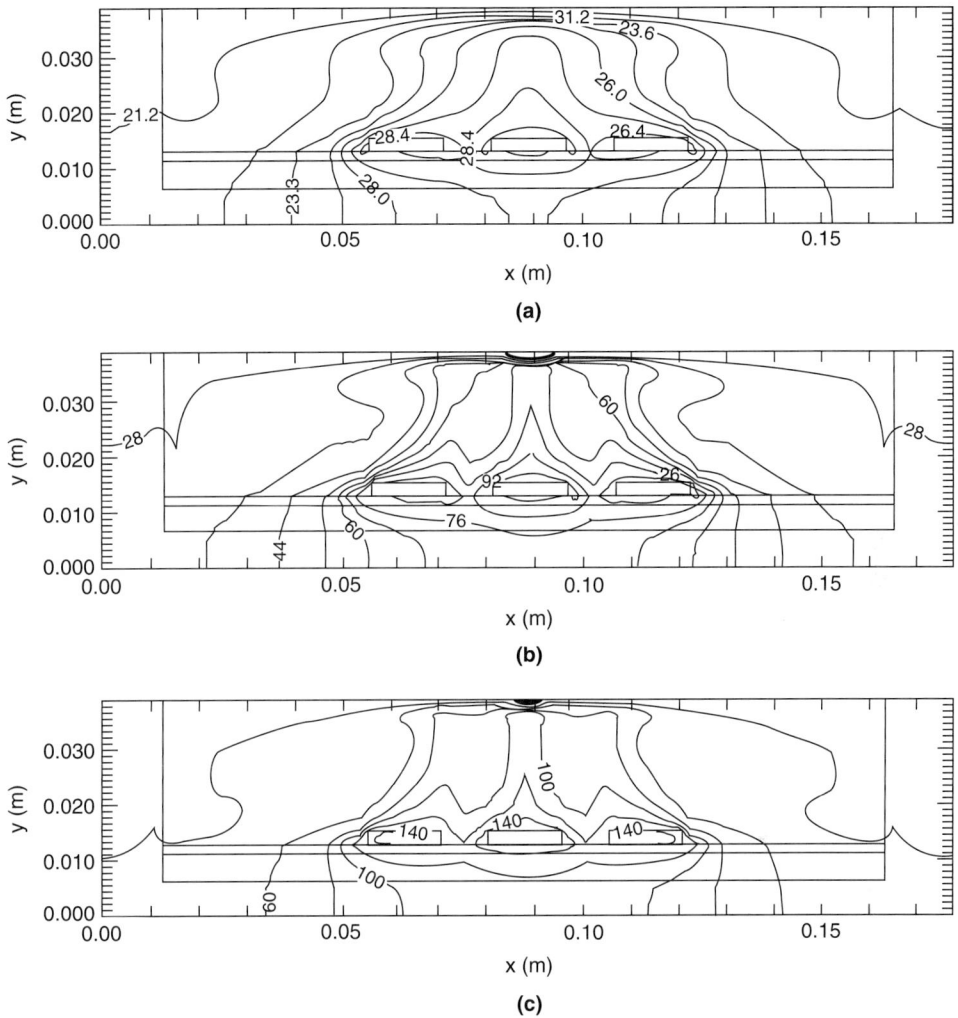

FIGURE 11.21 Isotherms in the central vertical plane. Computations include the combined effects of natural convection, conduction, and radiation: (a) 0.05 W/pkg, (b) 0.5 W/pkg, (c) 1.0 W/pkg. [37]

Limitations of Modeling

In spite of all its advantages, CFD must be used in the design process with the utmost care. In the first place, CFD cannot yet be used to replace experiments completely. Experiments form a critical component of the overall process. It is imperative that any CFD code be validated by experiments for the particular case. There are several reasons for that. The validity of the boundary conditions, the refinement of the grid, or the validity of certain simplifying assumptions can only be checked and confirmed by experiments. There might even be programming bugs that may need to be fixed. For computer codes simulating complicated physics such as turbulence modeling it is sometimes required to check whether the empirically determined constants are valid for the given case.

Available Modeling Tools
Several general-purpose commercial software packages are available. Commercial CFD codes have their own advantages and disadvantages. They are usually so general purpose that they can be used over a wide variety of problems, thus preventing the need for in-house code development to solve a problem. Usually, they come with user-friendly pre- and postprocessing modules. A main disadvantage of commercial CFD software is that it is usually very expensive. Also, since the source code is not available, we cannot enhance the code according to our requirements.

It is beyond the scope of this section to either recommend or evaluate available software packages. Instead, the general methodology followed in some of the most popular commercial codes will be presented, and the key features in selecting a thermal analysis software package will be discussed.

General-Purpose Codes
Available commercial codes are usually based on FEM (e.g., Ansys) [38], and FIDAP [39] or on FVM (e.g. FLUENT [33], FLOTHERM [34], CFX [35], ICEPAK, and STAR-CD [36]). Among the FVM-based codes, a choice of structured as well as unstructured grids is available with FLUENT, CFX, ICEPAK, and STAR-CD. An extensive range of choice of turbulence models is available with most codes. With regard to treatment for the pressure variable, the SIMPLE algorithm or its variants such as SIMPLER, SIMPLEST, SIMPLEC, and PISO are used. With some codes (e.g., FLUENT) the user can choose from a wide range of algorithms.

Dedicated Packages for Electronics Thermal Management
FLOTHERM and ICEPAK belong to a group of packages having dedicated modules for electronics thermal management. They have built-in module databases in the form of parameterized model templates and libraries of a wide range of components used in electronic systems such as heat sinks, fan, vents, and boards. Using these packages, system-level modeling is performed by treating smaller-scale features such as boards, ICs, wire mesh, grills, fans, heat sinks, and transformers in an economical and idealized manner using the "compact models." The packages allow the use of these template models with various levels of details. For example, a typical axial flow fan used in electronics cooling can be represented in increasing levels of accuracy starting from a very simple 2D fan to the very complex 3D fan. The fan curve can be input as a simple linear model or a more accurate nonlinear model. The choice of level of model representation is typically based on the component's significance and computational requirements. It may be noted that these models are designed to produce the same thermofluid effects as their real counterparts. With the general-purpose CFD codes, which may not have these modules, idealized lumped models need to be created separately for each application for the system scale modeling.

Compact Thermal Models
Special considerations are necessary while modeling ICs in board- or system-level models. Though there is considerable interest in recent times to accurately predict the junction temperatures of the ICs, it is still prohibitive to include all the internal details such as die, die attach, and leads for the components. Also, package suppliers own the design of chips and packages and do not want to disclose proprietary details. An efficient and practical way of achieving vendor-neutral data exchange is by using "compact thermal models." A compact thermal model is simply a connected network of

thermal resistances that collectively represent the thermal behavior of the device in any application environment. Systems designers are increasingly requesting that package providers also provide system-independent package thermal models of ICs that can be used within system design tools. Validated compact thermal models of their components will add value to the package supplied to the system customer without disclosing geometric and other proprietary details. Hence the demand for validated boundary-condition independent compact thermal models is increasing.

System-Level Simulation Using Lumped Models While performing system-level modeling, computational resource limitations prohibit the detailed modeling of smaller-scale component- and board-level features. For the purpose of computational economy, it is useful to use lumped or "compact" models along with coarse-grid system-level simulations. Two of the most popular system-level approaches that have been commercialized are described below.

An efficient technique for system-level flow and heat transfer calculations is flow network modeling (FNM). FNM is a generalized methodology for calculating system-level distributions of flow rates, pressures, and temperature in a network representation of the system [15]. The entire system can be represented as a network of flow paths through various passages of system components such as ducts, fans, heat sinks, boards, filters, and small devices. Empirical and known relations provide flow and thermal characteristics of each constituent of the flow network, which are then used to solve for mass, momentum, and energy conservation equations to obtain system-level pressure, velocity, and temperature distributions. In case empirical relations for thermal-fluid characteristics of a constituent are unknown, detailed component-level CFD analysis in a localized manner can be performed in order to establish such relations. A well-known FNM code available commercially is MACROFLOW [40]. Though MACROFLOW is a general-purpose FNM code that can be applied to a variety of engineering problems, special modules for electronics systems such as heat sinks, fans and blowers, grills, and diffusers are available. Several successful applications in electronics thermal management are reported in the literature [41].

Another commercially available package for system-level analysis is SINDA/FLUINT [42]. It is a comprehensive, generalized, lumped-parameter (circuit or network analogy) tool for simulating complex thermal-fluid systems such as those found in the electronics, automotive, petrochemical, and aerospace industries. Originally, this software provided users with the most proven heat transfer and fluid flow design and analysis software in the aerospace industry for decades.

Multiscale Modeling Most modern electronics systems are quite complex from the point of view of thermal and fluid flow analysis, as they possess components of various shapes and sizes, thus leading to a multiscale problem. Under such circumstances, for the purpose of computational economy, it is sometimes useful to use compact or reduced models (such as FNM) with coarse-grid system-level simulations. However, there is a need for detailed component-level thermal modeling for determining local hot spots and for electrical performance and reliability assessments. Following this idea, there has been a recent trend toward multiscale simulations of electronic systems. One approach is to carry out successive levels of computations over increasingly localized computational domains, with just the adequate detail at each level of modeling. In this approach, at the system level, the individual components such as PWBs, heat sinks, and other significant subsystems are simulated with reduced or compact models. In doing so, the model of the entire system is usually performed on a coarse mesh. At this stage, details such as individual copper traces on PWBs, individual leads of packages, and individual fins of heat sinks are ignored or specified as effective thermal and/or flow resistances to avoid excessive computer storage and time. Results from simulations at the system level can serve as boundary conditions for the detailed board-level and component-level thermal modeling. At the component level, a more detailed representation is required. The precise output from the smaller scale modeling (such as pressure drop due to flow over a component) is fed back into the system scale model, and the iteration is continued until convergence.

Evaluation of Thermal Analysis Software for Microsystems
For a thermal design engineer considering an investment in thermal analysis software, there are several questions to ask before selecting a particular product in the market. First, one has to be satisfied with the technical issues regarding a software product for a particular application. If a product is found to be technically suitable, one then has to determine if the technical support offered by the supplier is adequate.

In technical evaluation of thermal analysis software for electronic systems, the user should first look at the modeling methodology used. Since heat transfer in electronics is a fully coupled problem involving all three modes simultaneously, the software should be capable of solving conduction, convection (fluid flow), and radiation. Several problems need turbulence modeling; hence, the software must give the user options to choose from a variety of turbulence models suitable for various applications. Next, one has to evaluate the preprocessor of the software, with regard to the nature of the computational grid and the user interface. Normally, unstructured grid systems permit easy representation of complex geometry. However, if the software offers structured grid as the only option, the user must determine whether the software has the provision to model complex geometry if the application demands so. With regard to the user interface, one has to determine how easy and convenient it is to use the software. Nowadays, geometrical details of components and systems can come from manufacturers and designers in the form of CAD data. The thermal engineer must determine if the preprocessor is able to import CAD data generated using some popular CAD software. Such features can save an enormous amount of preprocessing time. After resolving the preprocessing issues, one needs to know about the nature of the solver. Vendors of general-purpose codes often claim that their products are capable of solving multiphysics problems involving stress, fluid flow, heat transfer, electromagnetics, and so on, but the potential user should verify if the solution methodology is capable of handling nonlinear problems (involving complex fluid flow) with sufficient ease. For such problems, solution convergence is often an issue, and the software should have adequate controlling parameters for numerical stability. For presentation of the results, most CFD software products have built-in postprocessors. It would be useful, however, if an output data can be formatted according to the input requirement of specialized postprocessing packages such as TECPLOT [43].

Before deciding on thermal analysis software for electronic systems, it is important to consider the nature of customer support offered by the vendor. Pertinent questions include accessibility of technical support, level of expertise required by the user, after-sales technical support, training offered by the supplier, and the general reputation of the vendor with regard to technical support. The supplier's familiarity with electronics thermal management problems should also be assessed. Some vendors also provide a full thermal design consultancy service, in which case necessary credentials should be checked. Normally, a supplier who has successfully completed a few thermal design consultancy projects is likely to be familiar with practical design issues associated with a range of electronic hardware.

11.4.2 Experimental Methods for Thermal Characterization

Experimental thermal characterization techniques often involve measurement of a temperature-dependent parameter. Such parameters indirectly estimate temperature at the point of interest. These techniques can be broadly classified into four different categories based on various temperature-dependent parameters, including electrical and optical [44]. Physical contact methods constitute another category. Details of such methods can be found in the review by Blackburn [44].

Electrical Methods Electrical methods are noncontact and usually employ bulky thermal measurement devices. These have low spatial and temporal resolution and are based on averaging of temperature over a large area. A number of temperature-dependent parameters can be used including electrical resistance, PN-junction forward voltage, threshold voltage, leakage current, and current gain [44].

Optical Methods Optical methods are also noncontact, with very high spatial and temporal resolution and can be used to measure very rapid transients. Temperature map construction is also possible with optical methods. A typical optical-based thermal measurement includes focusing of a light beam on the hot region, followed by subsequent measurement of naturally radiated beam, partially reflected beam, or stimulated emitted radiation [44]. Valuable information regarding the local temperature can be extracted using photon and lattice phonon interaction, which governs the relative intensity, energy, and phase of the reflected photons. Various optical methods for temperature measurement include luminescence, Raman scattering, reflectance, infrared, and thermo-optic effect. Infrared radiation among these is the most popular method used for thermal measurement purposes. Despite several advantages, optical methods are not suited for thermal measurement in enclosed space (e.g., in packaged devices).

Physical Contact Methods Unlike electrical and optical methods, physical contact methods measure temperature using actual physical contact of the measuring device with the region of interest. Such methods include thermocouples, scanning thermal probes, liquid crystals, and thermographic phosphors [44]. These can render very high temporal and spatial resolution (< 100 nm). Temperature mapping is also possible.

11.5 Thermal Management Technologies

Thermal management of highly integrated and miniaturized microsystems such as SOP is fundamental to their success. It requires an in-depth understanding of the heat generation patterns and the resulting heat fluxes. In general, the designer must begin with the fundamental step of identifying all sources of heat dissipation in the systems, their total power generation, spatial distribution, as well as time-dependent variability. The second step is to understand the eventual system-level sink for the heat generated. This could range from a person's hand, to room ambient air, to the surrounding sky, and to deep outer space. This source-to-sink consideration thus creates the first set of constraints for the system. For instance, how hot can a device get and still be operable, and must it be in contact with the skin? This goes beyond what's safe to what is desirable to ensure a device's acceptance and preferential use.

Another critical challenge is identifying all of the component thermal limits or constraints. This could include maximum allowed temperatures for individual components, both in the transient state and at steady state. To further complicate the challenge, different parts of the system may be more critical than others based upon the designed performance of that part of the system. The limits also will likely include temperature gradients across components or interfaces. These are driven by both the heat flux and the thermal resistance. The heat transfer path must be understood, and the design must include sufficient heat removal and heat spreading capability to ensure all the defined constraints are met and maximum temperatures or fluxes are not exceeded. The design must consider multiple paths for the heat propagation to properly manage the heat flow out of the electronics. These heat removal components can be as simple as highly conductive thermal vias to pumped two-phase liquid loops.

11.5.1 Thermal Design Methodologies

One or more of the heat transfer modes as discussed in the earlier section may be involved in the thermal management of SOP. Conduction is always present, as the heat propagates through and from the semiconductor, through multiple solid materials out of the system. Convection is sometimes present and depends on the overall system configuration. A mobile phone, for example, will have conduction present all through the system hardware and into the hand that is holding it. Natural or free convection of the air, driven by the buoyancy forces due to the heat released from the component, is also present. If we are considering a desktop computer instead, forced convection is a more typical mode of heat transfer. The systems fans will provide a flow of air over the components. In spacecraft applications, conduction will still be present, as the heat propagates through the different materials, but its final path for heat rejection must be radiation to surroundings. Figure 11.22 illustrates some of the important methods that are useful for the thermal management in system-on-package based devices.

Passive Methods of Thermal Management

High-Conductivity Package Materials As the SOP devices embed various thin-film components at the substrate level, thermal conductivities of the substrate, interlayer dielectric, conductors, underfills, thermal interface materials, and the core become vital parameters. High-thermal-conductivity substrates help in spreading the heat. There are various parameters that govern the choice of substrate materials including bulk thermal conductivity, CTE, electrical conductivity, electromagnetic compatibility, and cost. In addition, demand for lightweight SOP systems makes specific conductivity (defined as conductivity divided by specific gravity) another important parameter. Conventional epoxy-based materials are not very attractive due to their low thermal conductivity. Recently ceramic composites and other advanced materials have been studied and reported as promising candidates for SOP and other traditional microsystem applications [45–49]. Advanced packaging materials can be classified into six broad categories including monolithic carbonaceous materials, metal matrix composites (MMC), carbon-carbon composites (CCCs), ceramic matrix composites (CMCs), polymer matrix composites (PMCs), and advanced metallic alloys [46]. A detailed review of these materials can be found in the aforementioned literature.

Notably some of these materials, such as highly oriented pyrolytic graphite (HOPG), offer very high bulk conductivity about four times higher than copper (1700 W/m · K versus 400 W/m · K for copper), an order of magnitude higher than specific thermal

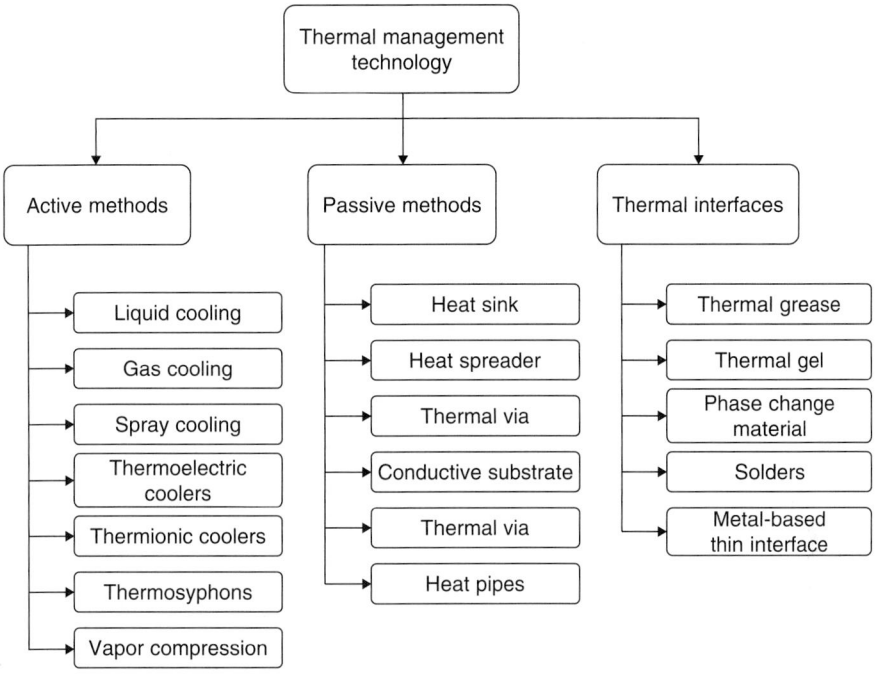

FIGURE 11.22 Thermal management methodologies for system-on-package based devices.

conductivity (740 to 850 W/m · K versus 81 W/m · K for aluminum or 45 for copper) and easy to tailor CTE to the desired value. A detailed description of these advanced materials can be found in the literature [45–49]. Nanomaterials offer the most promise for thermal management of SOP with their ultrahigh thermal conductivity at 6600 W/m · K for carbon-nanotube and nanogranular graphite-reinforced nanocomposite materials [46].

Thermal Vias As stated earlier, increased component density, such as by 3D stacking, and embedded active and passive components are some of the characteristics of SOP systems. The ability to route the generated heat into regions better connected thermally to the heat sink is therefore one of the key thermal design options. As stated previously, this must be included in the design state, just like routing electric power and signal distributions. Thermal vias are high thermal conducting pathways from the IC to the heat sinks [50]. They use thermal conduction and possibly heat spreading mechanisms to carry the heat. Factors that affect their performance are their thermal conductivity, cross-sectional area, and length. Proper attachment to the source and sink are important to reduce interface resistance. Via density needs to be optimized, and vias need to be allocated near the hot regions to increase their efficiency. Several research groups have reported algorithms for thermal via floor planning and allocations [51–54]

Thermal via allocation algorithms can be broadly divided into two classes—one that is based on steady-state thermal analysis and another that is based on transient thermal analysis. It should be noted that the new generation of SOP-based devices use various power management techniques such as dynamic power management (DPM) and need-based power supply (NBP). Such techniques ensure a power supply based on the workload, and thus the power in these devices witnesses both temporal and spatial variations. Steady-state thermal analysis ignores these variations and assumes a maximum power supply throughout operation. Such assumptions may lead to an excessive thermal via density [55]. Transient thermal via allocation method, however, considers these variations. Such schemes can reduce the number of thermal vias by as much as half as compared to a steady-state thermal analysis method under the same temperature bound [55].

A comparative study between various algorithms, such as area and wire-length-driven floor planning (WAF), thermal-driven floor planning (TDF), and integrated thermal via floor planning (IVF), shows that for 3D stacked chips, WAF can achieve a 17 percent decrease in temperature with less than a 3 percent average thermal via density at a cost of 4 percent area expansion and 1 percent wire length [54]. IVF, on the other hand, can decrease temperature by 38 percent with 2.5 percent thermal via density at a cost of 47 percent area and 22 percent wire length increases.

Thermal vias effectively enhance the heat dissipation rate and decrease the temperature of active regions in steady state, as shown in Figure 11.23 [56]. Thermal vias also provide cooling of hot spots such as the emitter-base junction of bipolar transistors. Figure 11.24 illustrates localized heat removal in a MOSFET using thermal vias [57].

Heat Spreader Heat spreading is a process by which the conduction problem is geometrically unconstrained and the heat is allowed to spread in two or three dimensions. This generally results in a shorter thermal path and lower source temperature. Heat spreading is particularly important when confronted with a highly nonuniform thermal generation problem. The overall power dissipation may be readily supported with the primary thermal design solution, but nonuniformities in components can lead to unacceptable hot spots. These can often be mitigated by heat spreading across the thermal path to reduce the impact of the nonuniformity.

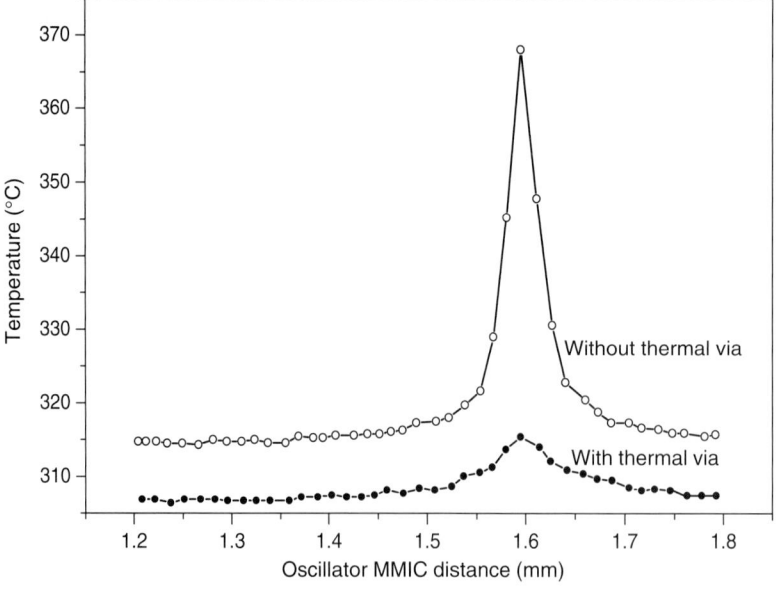

Figure 11.23 Temperature distribution in MMIC with or without thermal via. [56]

FIGURE 11.24 Hot-spot thermal management using thermal via. [57]

There is a potential downside to heat spreading. If the thermal solution mechanism depends on the thermal gradient, the presence of a heat spreader can drastically compromise the overall efficiency. In addition, misapplication of a heat spreader can severely reduce the overall thermal management performance. For example, consider a heated surface applied to a convective heat transfer. Application of a heat spreader increases the overall area and thereby enhances the convective heat flow. However, added thermal resistance due to the TIM and heat spreader can undermine this effect and thus can be useless.

Extended Surfaces Extended surfaces are a logical extension of thermal spreading. Greater surface areas for heat rejection to air, liquid, or even by radiation to space are relatively low cost and straightforward thermal enhancements. The concept of extended surfaces or heat sinks applies in both passive and active cooling. The most common extended surfaces are fins for convective heat transfer enhancement (see Figure 11.25). These fins increase the surface area for the fluid to remove the heat. Fin optimization is an important topic and is covered extensively in the literature [58]. A balance between cost, space, and weight offset against the increased thermal performance must be struck to determine the optimum extended surface for any given system. Extended surfaces are also important for boiling heat transfer mechanism where they can not only increase the heat transfer surfaces, but may also provide nucleation sites for boiling as well.

Heat Pipes Heat pipes are prevalent in laptop computers. As seen in Figure 11.26, they consist of two phase change sections, an evaporator and a condenser, and fluid transfer between the ends. The phase change mechanism can move significant amounts of heat. In this approach, the source, typically a microprocessor, is attached to the evaporator section of the heat pipe. The vapor then progresses to the condenser section where the heat is given off to the heat sink with a high heat transfer rate from the condensation. The condensed liquid then goes back to the evaporator section under capillary and the wicking action. Gravity may also assist if appropriately designed for. Wickless, gravity-driven heat pipes, also called single-chamber thermosyphons have also been reported.

The heat pipe, if not too long, can provide heat transfer at a very low differential temperature such that the evaporator and condenser sections are nearly isothermal. This results in very low thermal resistances. Heat pipe operating temperatures are set

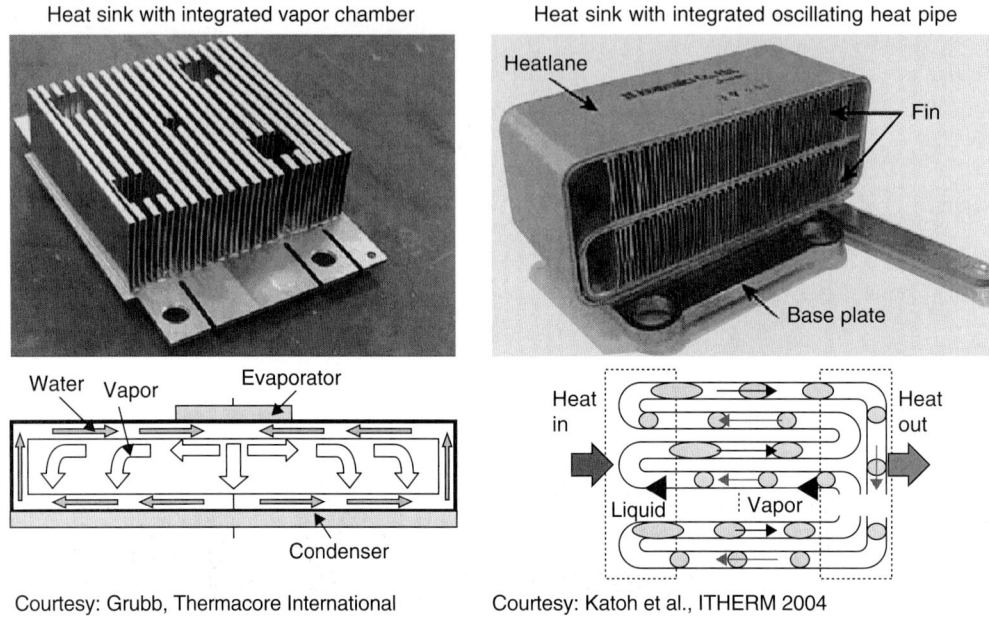

FIGURE 11.25 Advanced heat sinks.

by the saturation temperature of the working fluid. Different fluids may be used; however, the most prevalent is water, with the heat pipe charged at varying internal pressures to set the saturation temperature at the desired value.

The heat pipe is a very simple component from the outside, but its internal physics are quite complex [59–60]. Steady-state performance prediction models balance the available capillary head with the various pressure drops for the liquid and vapor phases, including inertial, viscous, and hydrostatic phases to determine the capillary limit and

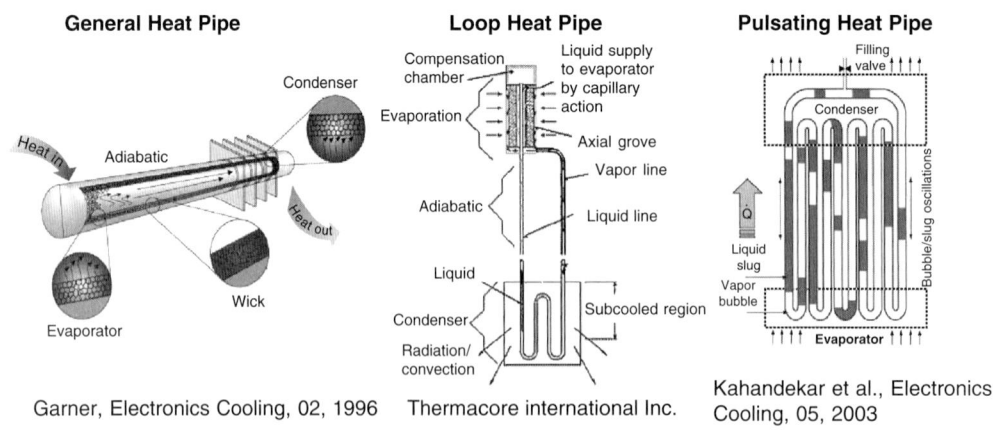

FIGURE 11.26 Schematic of various heat pipes.

the heat transport capability of the heat pipe, beyond which the device will experience a dryout. Other operating limits, including viscous, sonic, boiling, and entrainment, may further limit the heat transportability of the heat pipe, depending upon the operating temperature range. Most manufacturers have simplified models of their designs, but a full experimental characterization should be performed to confirm the performance of any individual design.

Heat pipes are relatively inexpensive with very good heat transfer properties. That is why they are often considered as the yardstick. For example, a heat pipe could replace a copper rod of the same diameter and length but increase heat transfer by more than an order of magnitude. There are no moving parts and failure is rare. Very simple low-power systems may require only heat conduction and heat spreading to an extended surface but as the SOP integration levels rise and power increases, the need for higher capability heat pipes arises.

Active Methods of Thermal Management

Active cooling requires external energy to use the cooling solution. Typically this is a fan, moving air across a heat sink or extended surface, or a pump moving liquid through a heat exchanger. Other examples are refrigeration systems where the compressor is driven to compress and move the refrigerant. Thermoelectric cooling can also be considered as active cooling in that while there are no moving parts, additional power is needed to move the heat from one section to another.

Active cooling typically outperforms passive cooling, the fan driving much more air past the heat sink than the buoyancy-driven flow. However, the costs are higher, power consumption is increased, and noise can be an issue. Another tradeoff to be considered is the volume occupied by the cooling solution. A driven flow such as forced convection needs smaller extended surfaces than buoyant flow by free convection for the same heat transfer rate.

Liquid Loops Liquid-cooled loops, particularly pumped liquid, deserve special consideration. They have very high performance capabilities but are not yet fully commercialized. Challenges remain in reliability of the pumps and the liquid leakage from the system. As these challenges are solved, liquid cooling will provide a significant boost to power density and thus address the SOP thermal challenges. One good example is the microchannel cooling as illustrated in Figure 11.27. Microchannels with single-phase flow can provide heat transfer rates in the range associated with phase change, but with higher stability and ease of control than the boiling and condensation approaches. Microchannels are particularly well suited to SOP due to their ability to be fabricated in silicon or other substrates at very small sizes. Their configurations are limited only by the ever-expanding MEMS capabilities used to build them. MEMS-based micropumps are also used as fluid drivers, but their application may be limited by flow and pressure drop requirements. A description of transport processes in microchannels can be found in [61].

Spray Cooling Spray cooling, as shown in Figure 11.28, is currently applied to system-level cooling in harsh environments. The cooling loop uses a dielectric fluid, sprayed directly on the electronics. The dielectric fluid undergoes a phase change and thus cools the electronics. The fluid is then captured and condensed, rejecting heat from the overall system, either to the ambient air or through a liquid-to-liquid heat exchanger.

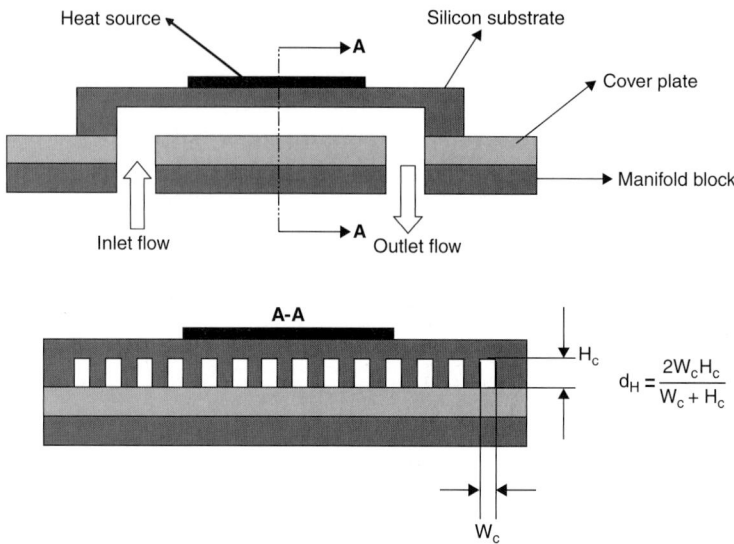

FIGURE 11.27 Schematic of microchannels.

Thermosyphon A thermosyphon is a two-phase flow loop driven by buoyancy forces. It can be implemented as a single chamber device or, as shown in Figure 11.29, as a two-chamber device. The latter segregates the heat input region from the heat rejection region for greater flexibility.

Thermoelectric Cooling Thermoelectric cooling, as shown in Figure 11.30, has great promise for local hot-spot cooling to deal with nonuniformities. Nonuniformity in heat generation of SOP modules could be solved by thermoelectric cooling moving the heat away from the high-density hot spot to a location in the package where it can be more readily transferred to the overall solution. It should be noted that the current generation of commercially available thermoelectric devices has a coefficient of performance significantly

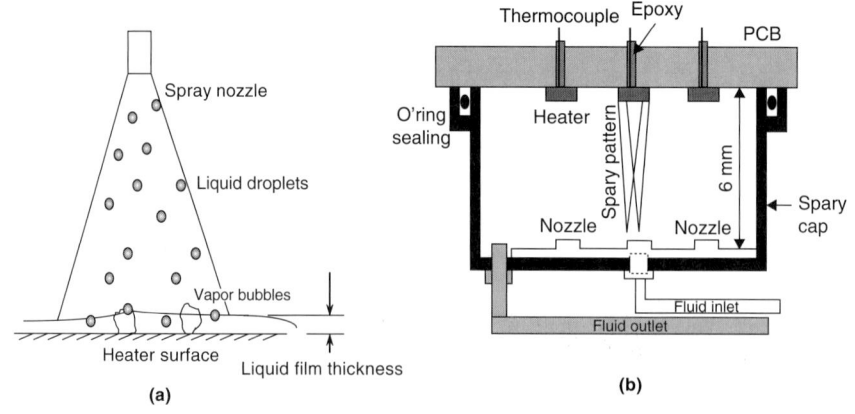

FIGURE 11.28 Schematic of spray cooling device.

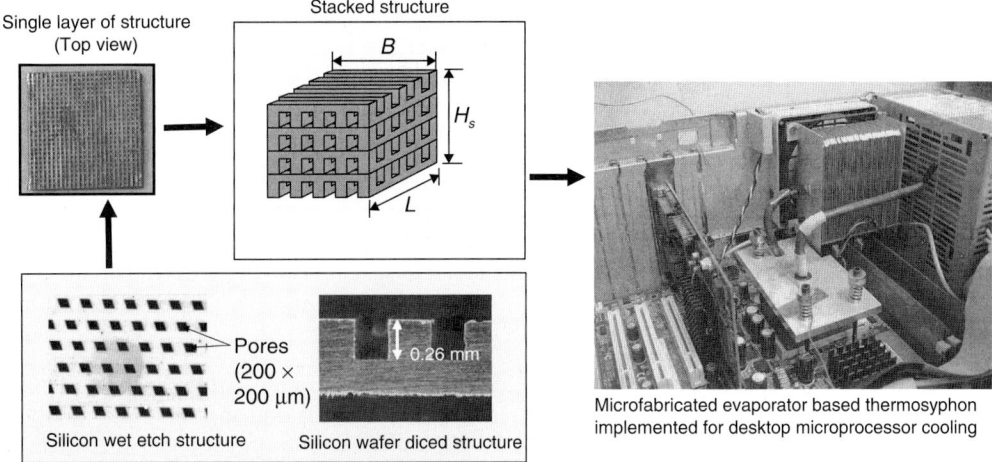

FIGURE 11.29 Schematics of thermosyphon.

below the mechanical refrigeration technologies, such as vapor compression. Even with the most optimistic predictions for thermoelectric cooling efficiencies, it may be problematic to use them as the primary cooling solution, due to the extra power draw. Power limitations on battery life or heat release to a room such as data centers may limit all active cooling solutions, particularly the thermoelectric cooling. Application of thermoelectrics to the cooling of electronics is discussed in [62].

Thermionic Cooling Thermionic cooling depends on quantum-mechanical phenomena where cooling takes place by ejection of electrons due to their high thermal energy. The basic structure of such devices consists of a cathode attached to the hot surface and an

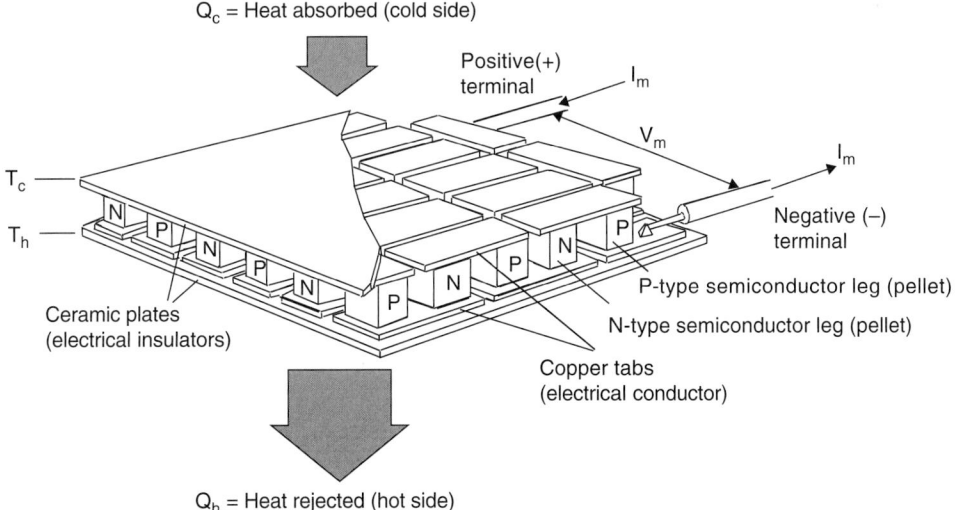

FIGURE 11.30 Schematic of a thermoelectric cooler.

anode that acts as a heat sink. One of the greatest advantages of such devices is that they can be easily integrated with power dissipating devices. Various heterostructures, based on this principle, have been reported [63–64]. These devices are especially useful for thermal management in optoelectronic SOP modules. Such devices can theoretically reduce the temperature by as much as 40°C in a single step of cooling [64–65].

Vapor Compression Refrigeration Vapor compression refrigeration (VCR) systems, as shown in Figure 11.31 [66], are being investigated for their miniaturization and applicability to electronics cooling. While promising greater efficiencies than thermoelectric cooling, they find greatest applications in local hot-spot cooling, or very high power density applications [67] such as cooling of CMOS-based systems that improve computing performance as their temperature is lowered.

Thermal Interface Materials (TIM) The heat transfer path from the source, such as an IC, to the eventual heat sink consists of various interfaces. Microscale surface roughness at these interfaces results in regions of contacts and air gaps at these interfaces, and thus effective heat conduction may be severely reduce. The interfacial resistance can, however, be decreased using compliant and gap filling interface materials, known as thermal interface materials. These can be fabricated to minimize the voids by providing a conformal and planar surface, thereby reducing the interfacial thermal resistance. Thermal resistance of TIM is governed by both the bulk, as well as the contact resistance [68–69], as given by

$$R_{TIM} = \frac{BLT}{A \times K_{TIM}} + R_{C1} + R_{C2} \tag{11.48}$$

where BLT is the bond line thickness, A is the contact area, K_{TIM} is the bulk thermal conductivity of TIM, and R_{C1} and R_{C2} are the contact resistances. From Equation (11.48) it is apparent that the TIM should be as thin as possible, while still filling all the voids between the two higher conducting parts. BLT, however, can't be reduced below a limiting thickness due to reliability concerns. An increase in thermal conductivity and reduction of contact resistance are other ways to reduce the effective thermal resistance. Contact resistance is dependent on the surface roughness, contact pressure, and

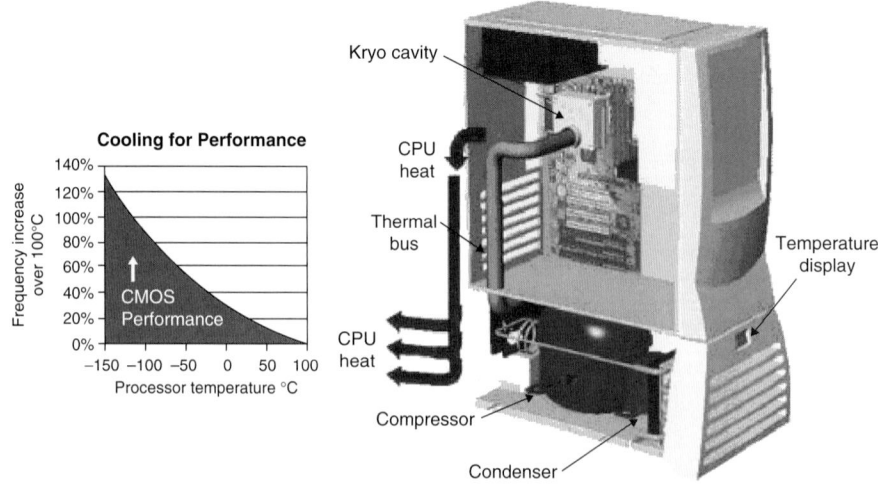

FIGURE 11.31 Schematics of vapor compression refrigeration system. [66]

compressive modulus [69]. Various limiting factors can affect the selection of TIM including thermal conductivity, thermal resistance, operating temperature range, electrical conductivity, phase change temperature, pressure, viscosity, outgassing, surface finish, mechanical stability, reliability, and cost [69].

A variety of thermal interface materials have been studied and reported in the literature and used in the products, including thermal grease, thermal gel, phase change materials (PCM), phase change metallic alloys (PCMA), and solders. Table 11.1 lists various popular

TIM	Characteristic Composition	Advantages	Disadvantages
Thermal grease	Inorganic powders (e.g., AlN, ZnO, Ag) in oil	• High thermal conductivity • Good conformity to rough surfaces • Low cost • Less delamination • High reworkability	• Pump out and phase separation during thermal cycling • Migration of what? • Difficult thickness control • Low reliability over time due to dry out
Thermal gel	Carbon black, high-conductivity metal oxide or metal powders in olefin, silicone oil, etc.	• Easy application • Less susceptibility to pump-out and migration before core. • Reworkability	• Curing needed • Less thermal conductivity • Low adhesion
PCM	Low-melting-point polymers (e.g., polyolefin, epoxies) filled with high-conducting inorganic salts (Al_2O_3, BN, AlN, etc.). Sometimes carbon nanotubes.	• Curing not needed • Good surface conformity • Less susceptibility to pump out • No delamination • No dry-out • Reworkable • Easy handling	• Nonuniform BLT • Lower thermal conductivity than grease • Contact pressure required
PCMA	Low-melting-point metals and alloys (e.g. In, InAg, InSnBi, SnAgCu)	• High thermal conductivity • Easy handling • No cure required	• Formation of intermetallics • Susceptibility of high temperature corrosion
Solder	Low-melting-point metal, eutectic binary or ternary alloys (e.g., same as above)	• High thermal conductivity • Easy to handle • No pumping out	• Reflow needed • Thermomechanical stress, delamination and crack possible • Possibility of void formation • Not reusable

TABLE 11.1 Characteristic Composition, Features, and Shortcomings of Various TIM

648   Chapter Eleven

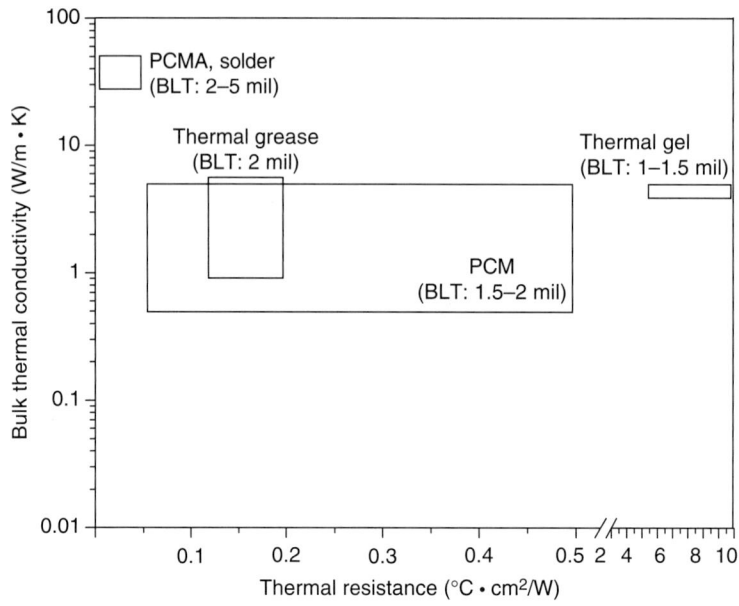

FIGURE 11.32 Trends in thermal interface materials.

TIMs along with their salient features and shortcomings [69–73]. Figure 11.32 illustrates the possible range of BLT, bulk thermal conductivity, and effective thermal impedance of various TIMs.

11.6 Power Minimization Methodologies

The rising functionality and miniaturization such as in SOP-based systems, pose two challenges—the power required to operate these devices and thermal management required to achieve their reliability. Miniaturized batteries such as by fuel cells or nano batteries are being developed to address the former challenge, but later challenge requires not only new and innovative thermal technologies but also power minimization approaches. In the absence of new power sources, current batteries in today's portable systems such as cell phones constitute one of the bulkiest components. In addition, they add to the cost.

Figure 11.33 shows the cost to dissipate power as a function of total power, showing an almost exponential cost increase. Power requirement minimization is the best way to address this cost [74].

The intent of power minimization is to control the power dissipation in a manageable low-cost range. This requires optimization at all levels including design, materials, and more importantly more efficient transistor technologies. There are various ways to minimize the power requirement of the system without significantly affecting the performance of the device such as

- Parallel processing
- Decrease in time sharing
- Bus coding to reduce activity

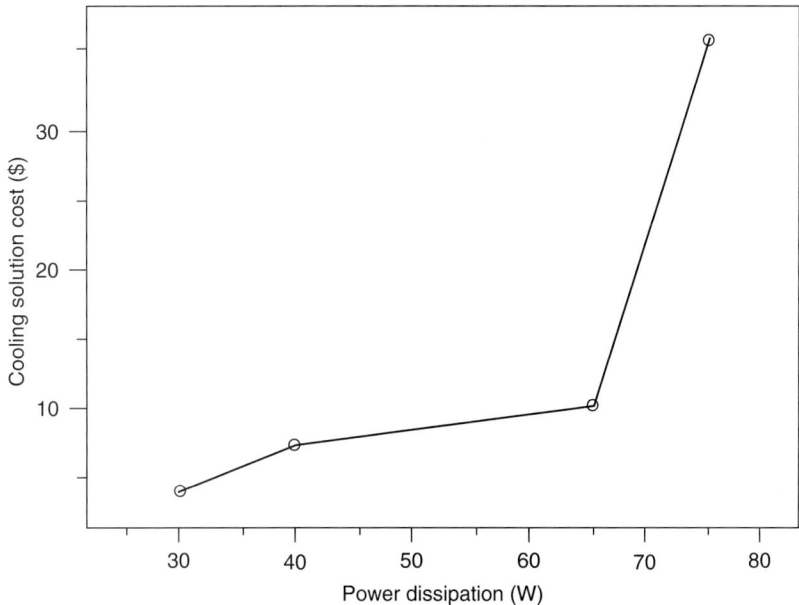

FIGURE 11.33 Cooling cost as a function of power dissipation. [74]

- Reduction of path and glitches
- Dynamic voltage and frequency scaling
- Use of novel materials

11.6.1 Parallel Processing

Parallel processing is one of the most important techniques to increase the energy efficiency and to reduce the system-level power consumption. It gives the processor the capability to simultaneously process more than one instruction. Parallelism can be achieved in a single processor by adopting a software solution, or by using multiple processors operating simultaneously. Such architecture is known as simultaneous multiprocessing (SMP). A few single-core processors have the ability to simultaneous process two or more software threads. Intel's Hyper Threading (HT) technology is one example of such processors [75]. Such a scheme can improve the performance of the processor by 30 percent without significantly increasing the power dissipation. Parallelism can also be implemented by using multicore processors and Explicitly Parallel Instruction Computing (EPIC). Multiple cores enhance the parallelism and performance stalls, and thus enhance the energy efficiency. By optimizing capability at the hardware level as well as at the compiler level, EPIC enhances simultaneous processing capability. Itanium 2 processors have been reported to use such a scheme and enable Intel to maintain the same thermal envelope across three generations of processors [75].

11.6.2 Dynamic Voltage and Frequency Scaling (DVFS)

A typical microprocessor runs at a fixed voltage and frequency, irrespective of the workload. This therefore requires the processor to consume maximum power to ensure

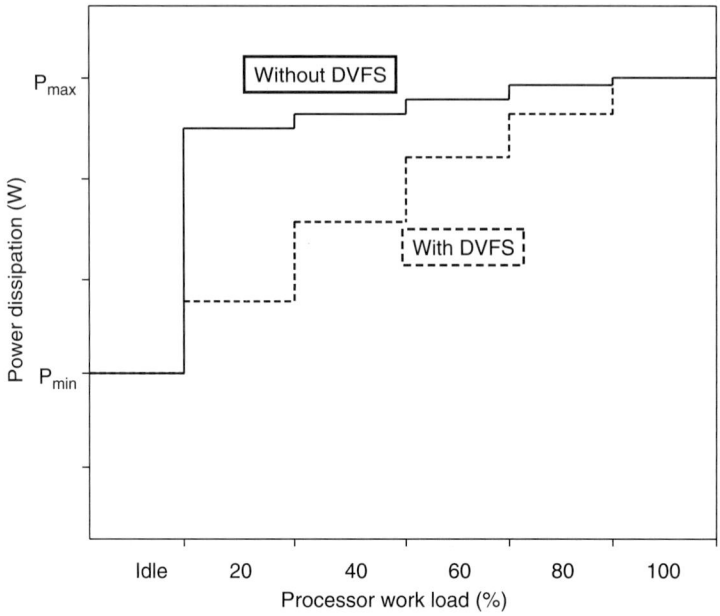

Figure 11.34 Power dissipation in a processor with or without DFVS.

optimum performance. Dynamic voltage and frequency scaling of a microprocessor enables real-time management of voltage and frequency scales of the processor according to the need and workload. Thus the energy requirement of the processor can be reduced significantly. Some of the potential energy reduction solutions are commercially available. For example, Intel's Speed Step processor technology adopts a demand-based switching (DBS) scheme based on the fundamentals of DVFS. It is designed to work on multiple power levels based on the need, as can be seen in Figure 11.34. Such a scheme is reported to reduce the average system power consumption and cooling cost by 25 percent with a minimal performance penalty [76].

11.6.3 Application-Specific Processors (ASP)

Power minimization can also be realized by switching from general-purpose processors to application-specific processors. These ASPs have the capability to execute only the targeted workload, and thus they can help achieve a higher level of performance and efficiency [77–78].

11.6.4 Cache Power Minimization

The increasing demand for high on-chip memory in the microprocessor has made cache memories among the most energy-intensive components. They consumed more than 40 percent of the total chip power in microprocessors, about 27 percent of which is devoted to the instructional cache (I-cache) alone [79]. Leakage power constitutes a large percentage of the total power dissipation in the cache. For example, leakage in the L1 cache constitutes 30 percent of the total L1 cache energy and that in the L2 cache constitutes

80 percent of the total L2 cache energy [11]. Borker estimates a fivefold increase in the total leakage energy dissipation with every new generation of chip [12]. Various approaches are reported in the literature to reduce leakage of LX caches including dual-threshold voltage (dual-V_t) [80], multithreshold CMOS (MT-CMOS) [81], and gated-V_{dd} [9]. According to the reported data, gated V_{dd} can reduce up to 97 percent standby mode energy consumption [80]. Dual-V_t techniques in NC-SRAM can reduce the leakage power by 77 and 55 percent, respectively, for 100- and 70-nm nodes [82].

I-Caches are other important power dissipating elements. All the above techniques to reduce leakage power can be applicable to I-caches also. In addition, various techniques are reported that may reduce dynamic (switching) power dissipation. One of the most studied techniques is code compression technique [83–91]. Some of the examples are dual-instruction-set processors (DISP) such as ARM-thumb [92]; MIPS 16, 32, and 64 [93]; ST 100 [94]; and the framework-based instruction-set processors tuning synthesis (FITS) [91].

11.6.5 Power Harnessing

Power harnessing can not only provide an alternative solution to heat dissipation, but can also prove fruitful as an alternative power source. Some of the innovative power harnessing techniques are especially relevant to thermal SOP technology such as solid-state electricity generation techniques by thermoelectrics, thermionics, and quantum electron thermo-tunneling-based devices. Recently there has been a significant amount of progress in improving the efficiency of these devices [95].

11.7 Summary

This chapter defined thermal SOP and discussed its thermal implications as the system size is miniaturized. Heat generating sources were also identified. The state-of-the-art in thermal characterization needs and methods were reviewed. Up-front thermal design for not only thermal dissipation but also for hot-spot management and, more importantly, for thermomechanical reliability are key to the success of the system. The up-front design should include thermal resistor network analysis for the understanding of the relative importance of various heat transfer paths and for the assessment of the thermal management needs. Subsequent detailed design may require CFD/HT tools. This chapter also reviewed various advanced methodologies for thermal management of SOP systems. These include new material innovations in ultrahigh conducting substrates, advanced thermal vias, and thermal interface materials technologies. Finally power minimization and harnessing technologies at the system operational level are summarized.

Acknowledgments

The authors would like to acknowledge the contributions of all member of Microsystems Packaging Research Center, Georgia Tech for their fruitful contributions in the chapter. Authors are especially thankful to Dr. Tummala for giving the opportunity to be a part of this chapter and also for his constructive comments and suggestions. We are also thankful to Reed Crouch for his help in improving the quality of the artwork. Last but not the least; we are grateful to all other people who helped in writing this chapter in any capacity.

References

1. "Moore's law", *Intel Technology and Research, Intel Corporation*, http://www.intel.com/technology/mooreslaw/index.htm.
2. *International Technology Roadmap for Semiconductors*, 2006 updates.
3. B. S. Glassman, "Spray cooling for land, sea, air and space based applications, a fluid management system for multiple nozzle spray cooling and a guide to high heat flux heater design," *Department of Mechanical Material and Aerospace Engineering*, University of Central Florida: Orlando, FL, Spring 2005 (Thesis, Master of Science).
4. R. R. Tummala (ed.), *Fundamentals of Microsystems Packaging*, New York: McGraw-Hill Professional, 2001.
5. R. R. Tummala and E. J. Rymaszewski (eds.), *Microelectronics Packaging Handbook*, New York: Van Nostrand Reinhold, 1989.
6. M. Pecht, P. Lall, and E. Hakim, *The Influence of Temperature on Integrated Circuit Failure Mechanisms*, vol. 3 in A. Bar-Cohen and A. D. Kraus (eds.), *Advances in Thermal Modeling of Electronic Components and Systems*, ASME Press, 1993, pp. 61–152.
7. K. Aygun et al., "Power delivery for high performance microprocessors," *Intel Technology Journal*, vol. 9, no. 4, 2005, pp. 273–83.
8. S. V. Kosonocky et al., "Low-power circuits and technology for wireless digital systems," *IBM Journal of Research and Development*, vol. 47, nos. 2–3, 2003. pp. 283–98.
9. M. Powell et al., "Gated-V_{dd}: a circuit technique to reduce leakage in deep-submicron cache memories," *Proc. International Symposium on Low Power Electronics and Design (ISLPED)*, Rapallo, Italy: ACM, 2000.
10. S. Manne, A. Klauser, and D. Grunwald, "Pipeline gating: speculation control for energy reduction," *Proc. 25th Annual International Symposium on Computer Architecture*, Barcelona, Spain: IEEE Comput. Soc., 1998.
11. C. H. Kim and K. Roy, "Dynamic V_t SRAM: a leakage tolerant cache memory for low voltage microprocessors," *Proc. International Symposium on Lower Power Electronics and Design (ISLPED'02)*, Monterey, CA: ACM, 2002.
12. S. Borkar, "Design challenges of technology scaling," *IEEE Micro*, vol. 19, no. 4, 1999 pp. 23–9.
13. Y. Bao, W. Peng, and A. Bar-Cohen, "Thermoelectric mini-contact cooler for hot-spot removal in high power devices," *Proc. 56th Electronic Components & Technology Conference*, San Diego, CA: IEEE, 2006.
14. R. Viswanath et al., "Thermal performance challenges from silicon to systems," *Intel Technology Journal*, vol. 4, no. 3, 2000.
15. J. Noonan, *The Design of a High Efficiency RF Power Amplifier for an MCM Process*, M.Engg Thesis in *Computer Science and Engineering*, 2005, Massachusetts Institute of Technology.
16. S. T. Allen et al., "Silicon carbide MESFET's with 2 W/mm and 50% P.A.E. at 1.8 GHz," *IEEE MTT-S International Microwave Symposium Digest*, vol. 2, no. 17–21, 1996, pp. 681–84.
17. J.-L. Lee et al., "68% PAE, GaAs power MESFET operating at 2.3 V drain bias for low distortion power applications," *IEEE Transactions on Electron Devices*, vol. 43, no. 4, 1996, pp. 519–26.
18. M. J. Drinkwine et al., "An Ion-Implanted GaAs MESFET Process for 28V S-band MMIC Applications", proc. *CS Mantech Conference*, Vancouver, British Columbia, Canada, April 24–27, 2006, pp. 187–190.

19. D.-W. Wu et al., "2 W, 65% PAE single-supply enhancement-mode power PHEMT for 3 V PCS applications," *IEEE MTT-S International Microwave Symposium Digest,* vol. 3, 1997, pp. 1319–22.
20. S. H. Chen, E. Y. Chang, and Y. C. Lin, "2.4 V-operated enhancement mode PHEMT with 32 dBm output power and 61% power efficiency," *Asia-Pacific Microwave Conference (APMC 2001),* IEEE Taipei, Taiwan, 2001.
21. E. Suhir, "Modeling of thermal stress in microelectronic and photonic structures: role, attributes, challenges, and brief review," *Transactions of the ASME, Journal of Electronic Packaging,* vol. 125, no. 2, 2003, pp. 261–67.
22. M. Leers et al., "Next generation heat sinks for high-power diode laser bars," *IEEE Semiconductor Thermal Measurement, Modeling and Management (SEMI-THERM) Symposium,* San Jose, CA, March 18–22, 2007.
23. X. Ling et al., "Optimization of thermal management techniques for low cost optoelectronic packages," *Proc. Electronics Packaging Technology Conference (EPTC),* Singapore: IEEE, 2002.
24. Y.-C. Lee et al., "Thermal management of VCSEL-based optoelectronic modules," *IEEE Transactions on Components, Packaging, and Manufacturing Technology, Part B: Advanced Packaging,* vol. 19, no. 3, 1996, pp. 540–47.
25. C. LaBounty et al., "Integrated cooling for optoelectronic devices," *Proc. SPIE—The International Society for Optical Engineering,* San Jose, CA: Society of Photo-Optical Instrumentation Engineers, Bellingham, WA, 2000.
26. L. L. W. Chow et al., "Skin-effect self-heating in air-suspended RF MEMS transmission-line structures," *Journal of Microelectromechanical Systems,* vol. 15, no. 6, 2006, pp. 1622–31.
27. D. L. DeVoe, "Thermal issues in MEMS and microscale systems," *IEEE Transactions on Components and Packaging Technologies,* vol. 25, no. 4, 2002, pp. 576–83.
28. D. P. Kennedy, "Spreading resistance in cylindrical semi-conductor devices," *Journal of Applied Physics,* vol. 31, no. 8, 1960, pp. 1490–97.
29. S. V. Patankar, *Numerical Heat Transfer and Fluid Flow,* New York: Hemisphere, 1980.
30. H. K. Versteed and W. Malasekera, *An Introduction to Computational Fluid Dynamics: The Finite Volume Method,* Longman Scientific & Technical, Harlow, England, 1995.
31. J. H. Ferziger and M. Peric, *Computational Methods for Fluid Dynamics,* 2nd ed., Berlin: Springer, 1999.
32. S. R. Mathur and J. Y. Murthy, "A pressure-based method for unstructured meshes," *Numerical Heat Transfer, Part B (Fundamentals),* vol. 31, no. 2, 1997, pp. 195–215.
33. http://www.fluent.com.
34. http://www.flomerics.com.
35. http://www.software.aeat.com/cfx.
36. http://www.cd-adapco.com.
37. V. H. Adams, Y. Joshi, and D. L. Blackburn, "Three-dimensional study of combined conduction, radiation, and natural convection from discrete heat sources in a horizontal narrow-aspect-ratio enclosure," *Transactions of the ASME, Journal of Heat Transfer,* vol. 12, no. 4, 1999, pp. 992–1001.
38. www.ansys.com.
39. http://www.fluent.com/software/fidap/.
40. http://www.inres.com.

41. R. Steinbrecher et al. *Use of Flow Network Modeling (FNM) for the Design of Air-cooled Servers*, Maui, HI: American Society of Mechanical Engineers, 1999.
42. http://www.crtech.com/sinda.html.
43. http://www.tecplot.com.
44. D. L. Blackburn, "Temperature measurements of semiconductor devices—a review," *Twentieth Annual IEEE Semiconductor Thermal Measurement and Management Symposium*, San Jose, CA: IEEE, 2004, pp. 70–80.
45. C. Zweben, "Advances in high-performance thermal management materials—a review," *Journal of Advanced Materials*, vol. 39, no. 1, 2007, pp. 3–10.
46. C. Zweben, "Ultrahigh-thermal-conductivity packaging materials," *IEEE Semiconductor Thermal Measurement and Management Symposium*, San Jose, CA: IEEE, 2005.
47. C. Zweben, "Advanced electronic packaging materials," *Advanced Materials and Processes*, vol. 163, no. 10, 2005, p. 33–7.
48. C. Zweben, "Advances in materials for optoelectronic, microelectronic and MOEMS/MEMS packaging," Proc. IEEE Semiconductor Thermal Measurement and Management Symposium, San Jose, CA: IEEE, 2002.
49. C. Zweben, "Thermal materials solve power electronics challenges," *Power Electronics Technology*, vol. 32, no. 2, 2006, p. 40–7.
50. E. Wong, J. Minz, and S. K. Lim, "Effective thermal via and decoupling capacitor insertion for 3D system-on-package," *Proc Electronic Components and Technology Conference*, San Diego, CA, 2006.
51. B. Goplen and S. Sapatnekar, "Thermal Via Placement in 3D ICs", *Proc International Symposium on Physical Design*, San Francisco, CA, 2005, pp. 167–174.
52. J. Cong and Y. Zhang, "Thermal Via planning for 3-D ICs". Proc. IEEE/ACM International Conference on Computer-Aided Design, Digest of Technical Papers, ICCAD, San Jose, CA, 2005, pp. 744–751.
53. Z. Li et al., "Efficient thermal via planning approach and its application in 3-D floorplanning," *IEEE Transactions on Computer-Aided Design of Integrated Circuits and Systems*, vol. 26, no. 4, 2007, pp. 645–58.
54. E. Wong and L. Sung Kyu, *3D Floorplanning with Thermal Vias* Munich, Germany: IEEE, 2006.
55. H. Yu et al., "Thermal via allocation for 3D ICs considering temporally and spatially variant thermal power," *Proc. International Symposium on Low Power Electronics and Design*, Tegernsee, Bavaria, Germany: Institute of Electrical and Electronics Engineers Inc., Piscataway, NJ, 2006.
56. J. Ding and D. Linton, "3D modeling and simulation of the electromagnetic and thermal properties of microwave and millimeter wave electronics packages," *Proc. COMSOL Users Conference*, Birmingham, 2006.
57. C. Hill, "Enhance MOSFET cooling with thermal vias," *Power Electronics Technology*, February 2006, pp. 28–33.
58. D. O. Kern and A. D. Kraus, *Extended Surface Heat Transfer*, New York: McGraw-Hill, 1972.
59. A. Faghri, *Heat Pipe Science and Technology*, Taylor and Francis, Washington, DC, 1995.
60. G. P. Peterson, *An Introduction to Heat Pipes: Modeling Testing and Applications*, Wiley series in thermal management of microelectronic & electronic systems, Allan D. Kraus and Avram Bar-Cohen (eds), John Wiley and Sons, NY, 1994.

61. S. V. Garimella and C. B. Sobhan, "Transport in microchannels—a critical review," *Annual Review of Heat Transfer*, 2003.
62. R. E. Simons and R. C. Chu, "Application of thermoelectric cooling to electronic equipment: a review and analysis," *Annual IEEE Semiconductor Thermal Measurement and Management Symposium*, 2000, pp. 1–9.
63. A. Shakouri et al., "Thermionic emission cooling in single barrier heterostructures," *Applied Physics Letters*, vol. 74, no. 1, 1999, pp. 88–89.
64. A. Shakouri and J. E. Bowers, "Heterostructure integrated thermionic coolers," *Applied Physics Letters*, vol. 71, no. 9, 1997, pp. 1234–36.
65. A. Shakouri et al., "Thermoelectric effects in submicron heterostructure barriers," *Microscale Thermophysical Engineering*, vol. 2, no. 1, 1998, pp. 37–47.
66. www.kyotech.com.
67. L. Jaeseon and I. Mudawar, "Implementation of microchannel evaporator for high-heat-flux refrigeration cooling applications," *Transactions of the ASME, Journal of Electronic Packaging*, vol. 128, no. 1, 2006, pp. 30–37.
68. C. Blazej, "Thermal interface materials," *Electronic Cooling*, vol. 9, no. 4, November 2003, pp. 14–20.
69. F. Sarvar, D. C. Whalley, and P. P. Conway, "Thermal interface materials—a review of the state of the art," *Electronics System Integration Technology Conference*, Dresden, Germany: IEEE, 2006.
70. E. C. Samson et al., "Interface material selection and a thermal management technique in second-generation platforms built on Intel® Centrino™ mobile technology," *Intel Technology Journal*, vol. 9, no. 1, 2005, pp. 75–86.
71. "Thermal interface material comparison: thermal pads vs. thermal grease," http://www.amd.com/us-en/assets/content_type/white_papers_and_tech_docs/26951.pdf, 2004.
72. T. Ollila, "Selection criteria for thermal interface materials," http://www.parker.com/chomerics/products/Therm_mgmt_Artcls/TIMarticle.PDF.
73. T. A. Howe, C.-K. Leong, and D. D. L. Chung, "Comparative evaluation of thermal interface materials for improving the thermal contact between an operating computer microprocessor and its heat sink," *Journal of Electronic Materials*, vol. 35, no. 8, 2006, pp. 1628–35.
74. *Thermal Management*. International Electronic Manufacturing Initiative (iNEMI) Technology Roadmap, 2006.
75. *Intel Incorporation*, www.intel.com.
76. "White Paper on Addressing Power and Thermal Challenges in the Datacenter," Intel Inc., http://download.intel.com/design/servers/technologies/thermal.pdf, 2004.
77. L. Wu, C. Weaver, and T. Austin, "CryptoManiac: a fast flexible architecture for secure communication," *Proc. 28th Annual International Symposium on Computer Architecture*, Goteborg, Sweden: IEEE Comput. Soc., 2001.
78. N. Clark, Z. Hongtao, and S. Mahlke, "Processor acceleration through automated instruction set customization," *36th International Symposium on Microarchitecture*, San Diego, CA: IEEE Comput. Soc., 2003.
79. J. Montanaro et al., "160-MHz, 32-b, 0.5-W CMOS RISC microprocessor," *IEEE Journal of Solid-State Circuits*, vol. 31, no. 11, 1996, pp. 1703–14.
80. K. Roy, "Leakage power reduction in low-voltage CMOS designs," *International Conference on Electronics, Circuits and Systems*, Lisboa, Portugal: IEEE, 1998.
81. H. Makino et al., "An auto-backgate-controlled MT-CMOS circuit," *Symposium on VLSI Circuits. Digest of Technical Papers*, Honolulu, HI: IEEE, 1998.

82. P. Elakkumanan, A. Narasimhan, and R. Sridhar, "NC-SRAM—a low-leakage memory circuit for ultra deep submicron designs," *Proc. IEEE International SOC Conference*, Portland, OR: IEEE, 2003.
83. Lekatsas, H., J. Henkel, and W. Wolf. *Code Compression for Low Power Embedded System Design*, Los Angeles: ACM, 2000.
84. H. Benini, F. Menichelli, and M. Olivieri, "A class of code compression schemes for reducing power consumption in embedded microprocessor systems," *IEEE Transactions on Computers*, vol. 53, no. 4, 2004, pp. 467–82.
85. H. Lekatsas, J. Henkel, and W. Wolf, "Arithmetic coding for low power embedded system design," Snowbird, UT: IEEE Comput. Soc., 2000.
86. I. Kadayif and M. T. Kandemir, *Instruction Compression and Encoding for Low-Power Systems*, Rochester, NY: IEEE, 2002.
87. Y. Yoshida et al, *An Object Code Compression Approach to Embedded Processors*, Monterey, CA: ACM, 1997.
88. L. Benini et al, *Selective Instruction Compression for Memory Energy Reduction in Embedded Systems*, San Diego, CA: IEEE, 1999.
89. L. Benini et al., "Minimizing memory access energy in embedded systems by selective instruction compression," *IEEE Transactions on Very Large Scale Integration (VLSI) Systems*, vol. 10, no. 5, 2002 pp. 521–31.
90. G. Chen et al., "Using memory compression for energy reduction in an embedded Java system," *Journal of Circuits, Systems and Computers*, vol. 11, no. 5, 2002, pp. 537–55.
91. A. C. Cheng and G. S. Tyson, "High-quality ISA synthesis for low-power cache designs in embedded microprocessors," *IBM Journal of Research and Development*, vol. 50, no. 2–3, 2006, pp. 299–309.
92. R. Phelan, "Improving ARM Code Density and Performance," http://www.arm.com/pdfs/Thumb-2%20Core%20Technology%20Whitepaper%20-%20Final4.pdf, June 2003.
93. www.mips.com.
94. http://www.st.com/stonline/books/pdf/docs/10071.pdf.
95. N. A. Rider, *Geothermal Power Generator* October 24, 2006, Borealis Technical Limited: United States Patent # 7124583.

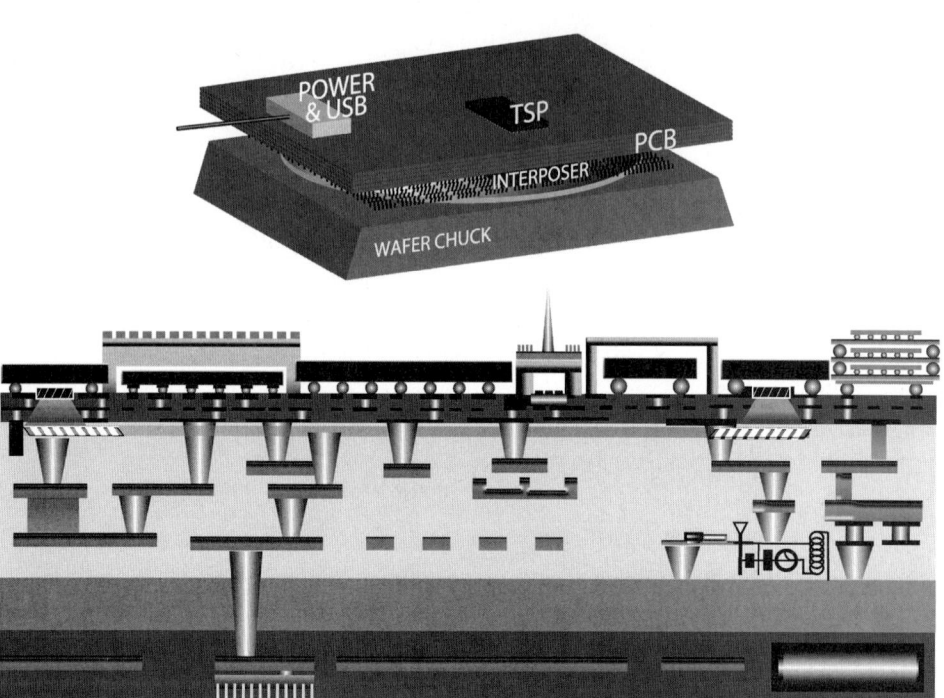

CHAPTER 12
Electrical Test of SOP Modules and Systems

S. S. Akbay, S. Bhattacharya, Prof. D. Keezer, and Prof. A. Chatterjee
Georgia Institute of Technology

12.1	SOP Electrical Test Challenges 660	12.4	KGEM Test of Mixed-Signal and RF Subsystems 685
12.2	Known Good Embedded Substrate Test 664	12.5	Summary 707
12.3	Known Good Embedded Module Test of Digital Subsystems 677		References 707

The goal of testing is to classify manufactured devices or systems in such a way that absolutely no defective parts are shipped to the customer and very few parts are wasted. Different stages of production call for different test paradigms that emphasize verification, coverage, or throughput. The defective SOPs are due to process variations and imperfections. A successful test strategy should not only pass defect-free SOPs but also ensure that their performance meets the specifications. The cost of a production test is of key importance as this is a recurring expense for every part that is manufactured.

Scaling has been the fundamental driver for achieving low cost in semiconductor manufacturing at the device level, enabling more complex devices with sustainable yields. By employing the same principle, SOP drives system integration just like Moore's law has done for IC integration. Testing, like all other components in SOP, needs to be scalable; otherwise, a larger percentage of manufacturing cost will be allocated to testing. However, as more functionality is integrated in SOP, more functionality needs to be tested, increasing the test effort and cost. Therefore, a high-volume manufacturing (HVM) test of SOPs needs to follow alternate test methods rather than classical test techniques, carefully partitioning the test process across levels of integration, avoiding any overlaps during testing.

This chapter introduces the SOP test concepts at the design, characterization, and HVM levels. It provides a methodology for testing of embedded passives, substrate interconnections, and modules including digital, analog, mixed-signal, and "radio-frequency (RF)". Known good embedded substrate and known good embedded module testing methods are discussed in detail with examples and references.

12.1 SOP Electrical Test Challenges

Integration is being defined at a new level, thanks to SOP technologies. These systems bring together not only analog and digital subsystems but also RF components for high-bandwidth communication, optics as the media for multi-gigahertz data transfer, and micro-electro-mechanical systems (MEMS) with moving elements as an interface with the outside world. The design of such systems makes use of many advantages that are not present in the traditional design flow. Ultraminiaturization and uniformity of the medium in these systems result in shorter data paths and smaller parasitics, and hence increase the speed and bandwidth of communication subsystems far beyond classical techniques. Using the SOP approach, the designer is no longer limited with the discrete properties of passives but can make use of embedded passive components tailored specifically to the application. Furthermore, SOP versions of sensor-processor, actuator-driver, and electrical-optical subsystems benefit from increased bandwidth, minimized package parasitics, and lower power consumption when compared to traditional interfaces. Testing of these complex SOPs proceeds in three phases: (1) verification testing and design debugging on prototypes, (2) characterization testing on low-volume samples, and (3) high-volume manufacturing testing.

The first step of verification testing is conducted on a few prototype samples to ensure the correctness of the design and viability of the manufacturing process. The design verification procedure consists of an application of a set of "verification tests" followed by design diagnosis. If necessary, the design can be altered, new prototypes fabricated, and a new set of tests applied until eventually a "perfect" design is obtained [1–2]. Since this is an iterative procedure, verification and debugging need to deal with the problem of partial testing: the process of testing manufactured parts of the SOP as interconnects and embedded passives is fabricated in a sequential or parallel fashion. This process presents a major challenge, because some specifications of such components are only defined after the full integration, or these specifications differ before and after integration. The problem is handled in classical design using the concept of modularity, such that systems are built with well-partitioned components that can be studied and verified individually. Systems that rely on enhanced performances of components using integration technologies are usually avoided. On the other hand, SOP makes use of integration using thin-film technologies, namely properties enhanced under package-level integration. Such SOP modules can become very challenging from a test standpoint.

Following design debugging, characterization tests are performed, which extensively test all critical design parameters of the "perfect" design and make sure that all the performance goals are met. Characterization tests are performed on a sample set of devices and also document the variations in device performance. Both verification and characterization tests make use of benchtop test equipment, which is a collection of test devices specialized for different measurements. These tests emphasize completeness and accuracy rather than testing time. Characterization tests, in addition, include functional tests with known good dies attached. Such tests emulate typical operations of the modules and check for the integrity of the overall system. Functional tests require a significant amount of time, from tens of minutes to hours, because many combinations of possible modes of operation need to be validated.

High-volume manufacturing calls for a different kind of test paradigm, which emphasizes throughput and sufficiency. These tests are performed for every one of the units shipped to customers; hence, functional tests cannot be afforded at this level. Also, HVM tests are performed by automated test equipment (ATE) rather than a collection of benchtop equipment. ATE integrates a subset of the functionality covered by the characterization test, yet it is a generic device that can be programmed to use different test plans for different devices under test. The HVM test is fully automated as shown in Figure 12.1. First, a handler places one or a few manufactured devices onto a load board with sockets for each device. The socket provides a temporary connection to the signal paths on the load board, which connects different ATE ports to the devices under test (DUT). The ATE provides the control signals and test stimuli for a bunch of sequential tests and collects the corresponding responses. Each response is compared to a decision boundary. ATE evaluates the results for all decision boundaries and assigns the DUT to an appropriate "bin." A simple bin structure has "pass" and "fail" classes, while more complicated devices may be sorted to different bins depending on their performance; for example, microprocessors capable of running at higher speeds can be marked and sold with higher profit margins. Once the proper bin is determined, the handler breaks the mechanical connection and moves the DUT into the proper pile. All these events have to be performed in a very short amount of time given that all manufactured devices need to pass through HVM testing. The test time varies from a fraction of a second to a few seconds depending on the complexity of the device; hence, it is one of the most significant factors defining the manufacturing cost of SOP.

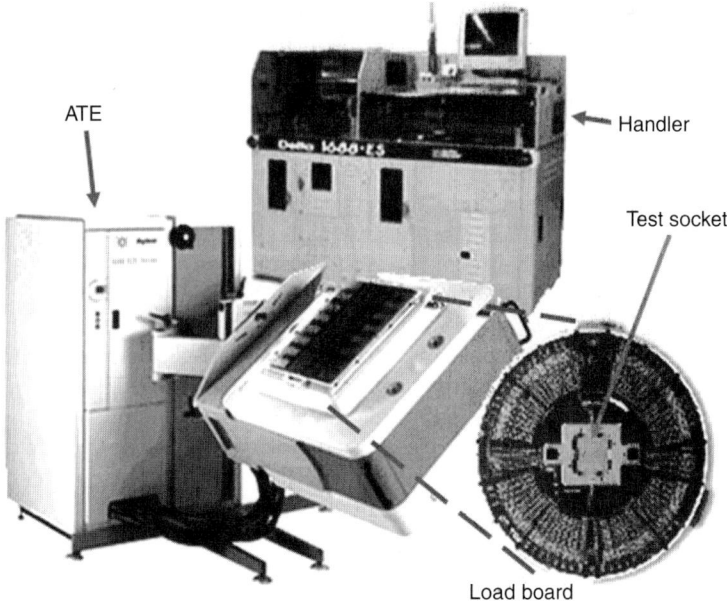

FIGURE 12.1 Components of HVM test flow. (*a*) ATE. (*b*) Test socket. (*c*) Handler. (*d*) Load board.

662 Chapter Twelve

At this point, it is important to state the fundamental principle of HVM SOP tests: namely, scaling. Scaling has been the fundamental driver for semiconductor manufacturing at the device level. Besides design rules, good process control, and design-for-test/yield, scaling has been the dominant enabler for manufacturing more and more complex devices with financially sustainable yields. By employing the same principle, SOP drives system integration just like Moore's law drives IC integration. Testing, like all other components of SOP, needs to be scalable; otherwise, a larger percentage of manufacturing cost will be allocated to testing and eventually testing will dominate the overall cost. On the other hand, as functionality increases, more functionality needs to be tested, increasing the test effort and cost. Therefore, HVM testing of SOPs needs to use alternate test methods rather than classical functional tests. In traditional systems using standard SMD components, conventional test methods work since the components are minimally tested before being assembled. However, in a highly integrated system such as SOP, new test methods are necessary since the components are integrated into the layers of the package and not all the nodes of the device are accessible.

12.1.1 Objectives of the HVM Test Process and Challenges for SOPs

The goal of testing is to identify SOPs that are defective through application of test stimuli during volume production. The SOP defects are due to manufacturing process variations and imperfections. The cost of production testing is very important, as this is a recurring expense for every part that is manufactured and is the primary focus of this chapter. Figure 12.2 shows the stages in the product development cycle and the associated costs. Verification, debugging, characterization, and HVM tests total up to 45 percent of the overall cost [3]. The HVM test gears toward two main types of failure mechanisms:

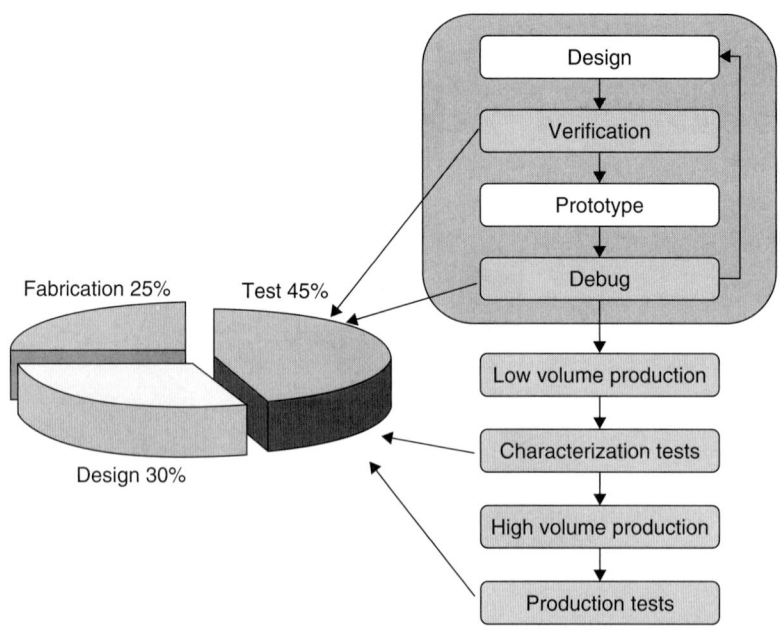

Figure 12.2 Cost of testing: test, 45 percent; design, 30 percent; fabrication, 25 percent.

(1) catastrophic faults, which impair the ability of the device to function, and (2) parametric faults, which make the device fall short of satisfying the specifications related to a function.

SOP comes with its own challenges in terms of HVM testing. A typical SOP encapsulates many of its internal functions, and production testing is performed by application of test signals to the SOP using external ATE. The key problem is that the external ATE does not have direct access to all the internal embedded functions of the SOP. It may be possible to route some of the internal electrical signals out of the package to the external tester; however, these internal signals operate at frequencies that cannot be observed directly by an external tester due to the frequency limitations of the encapsulating package and lower speed of external I/O. A similar speed and integrity concern is applicable to validating the subcomponents of the system. While traditional systems have test nodes to individually verify the operation of their subsystems, a classical test approach to SOP suffers from limited controllability and observability of its subsystems. Furthermore, the system specifications "guaranteed by design" depend on validation of associated subsystem specifications, which may no longer be accessible in a SOP configuration. This proposition is especially important for embedded passives [101] constituting a part of the package. For example, a high-K resistor manufactured with a thin-film component embedded into the substrate may have a direct path to the substrate surface through microvias; however, once the associated ASIC is flip-chip bonded, there will be no direct accessible path for the ATE. Even if a separate design for testability (DfT) path is provided, the ATE will only be able to measure the combined effects of the resistor and the bonding parasitics of the ASIC. This example shows that HVM tests need to be partitioned across several steps of the integration process. Bare specifications of the ASIC and the substrate need to be tested before final assembly. Furthermore, the test set for the final system can be tweaked and reduced depending on those results. This multistep test strategy is discussed in Section 12.1.2.

12.1.2 HVM Test Flow for SOPs

The test flow for SOP can be summarized in three steps: (1) known good die (KGD) test, (2) known good embedded substrate (KGES) test, and (3) known good embedded module (KGEM) test after assembly, as shown in Figure 12.3 [4–5]. The KGD test is

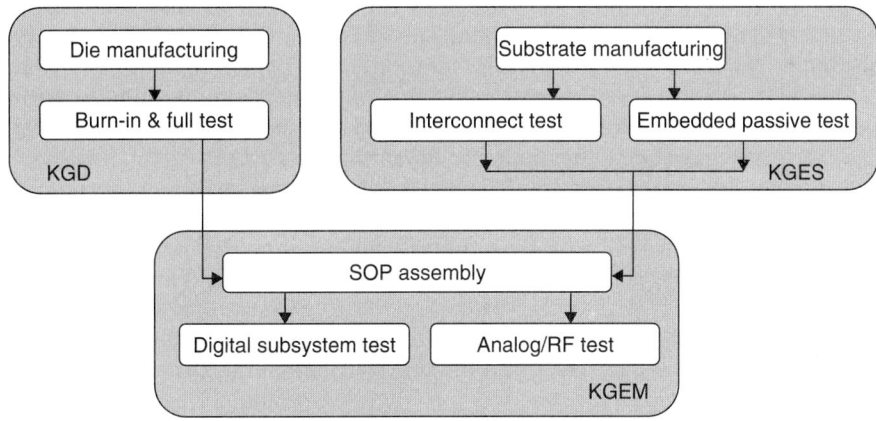

FIGURE 12.3 Test flow for SOP: known good die (KGD), embedded passives and substrate (KGES), and assembled SOP with known good embedded module (KGEM).

performed separately on each bare die before assembly on the package. This test guarantees that the bare die or unpackaged ICs have the same quality and reliability as equivalent packaged devices [6]. More details about KGD tests, which will not be addressed in this chapter, can be found in [7–10]. Before attaching the bare die to the substrate, the substrate needs to be tested. Substrate testing includes verifying the performance of interconnections and the embedded passives fabricated in the layers of the substrate. Finally the SOP is assembled and the final known good embedded module test is performed. In Sections 12.2 and 12.3, we elaborate on different aspects of KGES and KGEM tests.

12.2 Known Good Embedded Substrate Test

Electronic packages provide a means for interconnecting, powering, cooling, and protecting IC chips. In SOP, the system is the package along with the die. The substrate hosts embedded passive and active components as well as provides interconnections between flip-chip bonded active devices, ASICs, sensors, optics, discrete passives, and embedded components. Unless it is possible to completely rework the connections, expensive KGDs will be wasted if assembled on a defective substrate. Hence, a testing scheme is necessary to ensure the integrity of all the substrate interconnection paths and embedded passives prior to assembling the die. The matching of the substrate interconnection attributes with a set of design requirements can be performed based on a set of design criteria such as insulation resistance, conductor resistance, continuity, and net capacitance that guarantees the functionality of the substrate interconnections. Embedded passives can also be verified with high-frequency at-speed tests, which require measurement of network parameters as a function of frequency.

12.2.1 Substrate Interconnect Tests

Package interconnections consist of single or multiple layers of metallization that connect active circuitry and/or embedded passives to form a function. If economically viable, the various interconnection layers can be optically inspected during the processing of the layers for the presence or absence of conductive material along the interconnection length. This allows for the immediate detection and repair of process-related defects during fabrication. In very high density interconnect (HDI) solutions, typical in SOP, the optical inspection is not feasible, so the methodology relies on good process control. Even when every layer is optically inspected, temperature and process stressing of subsequent layers lead to defects on interconnections that need to be diagnosed prior to die attachment. Hence, a test is required after all the layers are fabricated which cannot be done through optical inspection.

In the older paradigm with laminated substrates and plated-through holes, the substrate test can be easily performed through the top and bottom surfaces of the substrate by electrically probing the interconnections. In SOP solutions, HDI with microvias is the norm, and this method mostly creates blind and buried interconnect layers that reach neither of the substrate surfaces. Surface-to-surface as well as ground and power interconnects can still be tested by probing both exposed ends as shown in Figure 12.4a. Some interconnects connect a surface contact to an embedded thin-film passive, which is in turn connected to a power or ground net, as shown in Figure 12.4b. Testing of these interconnects can be performed in a single step with the embedded passive they are connected to. In this section, embedded passive testing is discussed in detail. Other interconnects, as shown in Figure 12.4c, connect an embedded module to

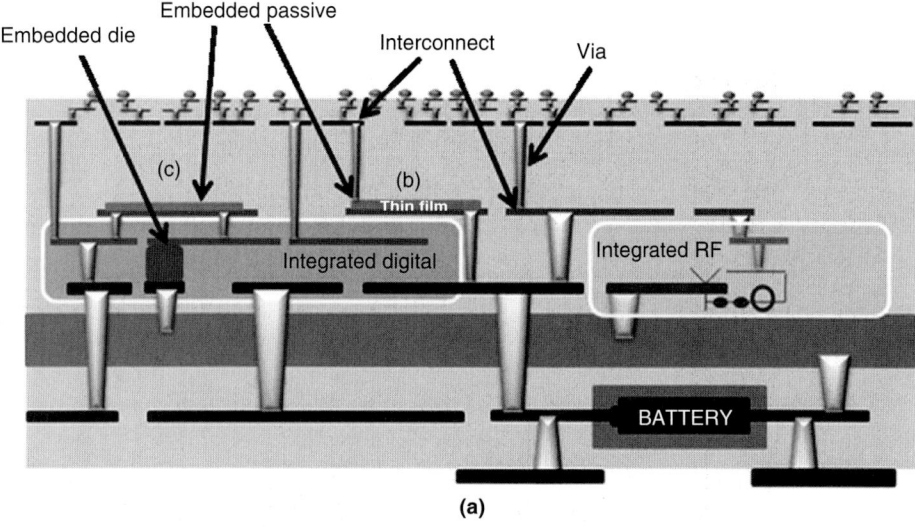

FIGURE 12.4 SOP interconnect test. (*a*) Interconnects exposed on both surfaces. (*b*) Embedded passive with one end connected to power/ground and the other end exposed. (*c*) Embedded passive connected to embedded die.

another embedded structure or to power/ground. These interconnects can be tested in a single step with the embedded modules, as discussed in Sections 12.3 and 12.4.

Both open-circuit defects of interconnections and short-circuit defects between interconnections require detection through substrate testing. Sometimes the substrate test technique requires high resolution to detect latent or near defects. These are physical imperfections that do not render an interconnection functionally open or short, but may degrade to an open or shorted condition at a later date. These classes of defects are of particular concern since they can result in unexpected failures in further processing or during customer use. Examples of latent defects are shown in Figure 12.5 [13]. The use of very narrow metal lines and polymer insulators (as in thin-film substrates) can increase sensitivity to latent open failures under thermal, mechanical, or bias stress during use. Ionic residues or extraneous metal from photolithographic processing can result in current leakage paths or latent shorts. Thus near opens and shorts must be considered as well as the incidence of time-zero defects, during substrate testing.

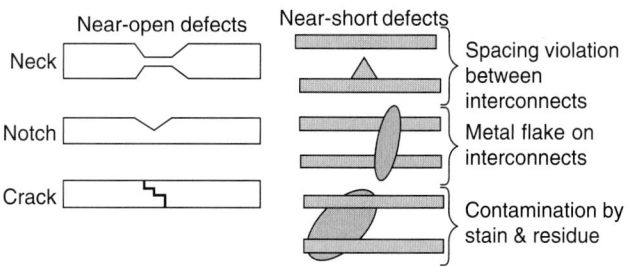

FIGURE 12.5 Classes of latent defects. [13]

Capacitance Testing

Consider two metal traces on the surface of an interconnect substrate as illustrated in Figure 12.6. Each may be viewed as one terminal of a parallel-plate capacitor, where the other terminal is a ground plane or power plane. Since the capacitance (C_i) of a parallel-plate structure is proportional to its area (A_i), we have the following equation:

$$C_i = kA_i \tag{12.1}$$

where k is the constant of proportionality that depends upon the spacing between the trace and the reference plane as well as the dielectric constant of the material between the plates.

For the two-interconnect example, we have

$$C_1 = kA_1 \tag{12.2}$$

$$C_2 = kA_2 \tag{12.3}$$

However, if there is a short between the traces, then the effective area of the faulty net is approximately

$$A_{short} = A_1 + A_2 \tag{12.4}$$

(neglecting the area of the bridging metal). Therefore, the capacitance of the shorted net is

$$C_{short} = k(A_1 + A_2) \tag{12.5}$$

Comparing this with Equations (12.2) and (12.3), we see that

$$C_{short} = C_1 + C_2 \tag{12.6}$$

Using a similar argument, we can deduce that each piece of a broken ("open") net will have a reduced value of capacitance (again depending upon its area):

$$C_{open} = kA_{open} < kA_i \tag{12.7}$$

Therefore, depending upon the type of fault present, the net capacitance will either increase or decrease from the fault-free value.

Figure 12.7 illustrates capacitance measurements. For example, net 1 has a lower than expected capacitance (45 pF as compared with the expected value of 55 pF). It is identified as an open-type fault. If nets 2 and 3 have a short between them, their capacitance measurements will be higher than expected (30 pF in the Figure).

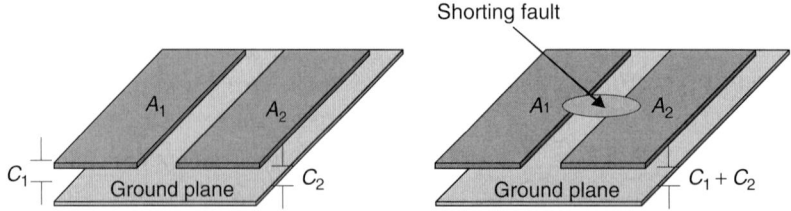

FIGURE 12.6 Capacitance test for a shorting-type fault. (a) Fault-free network. (b) Faulty network.

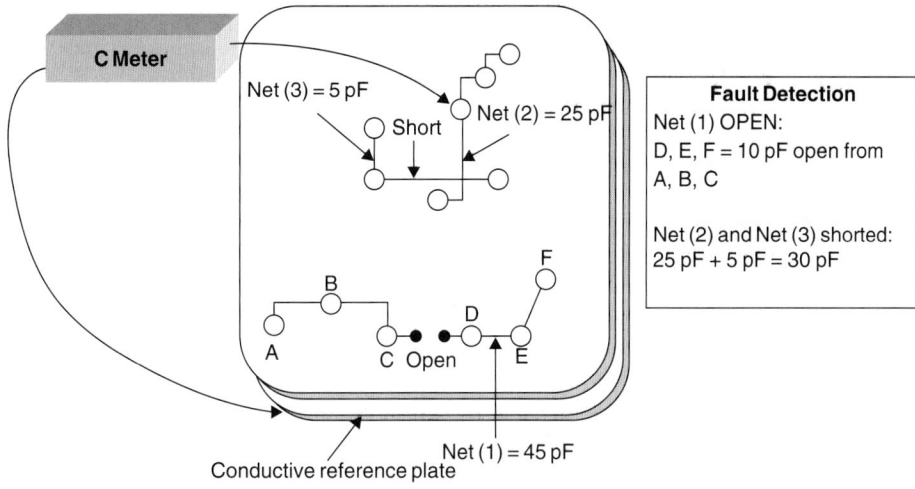

FIGURE 12.7 Interconnect fault detection by capacitance measurements.

Marshall et al. [12], Economikos et al. [13], Hamel et al. [14], and Wedwick [15] provide further details on capacitance testing of substrates.

Resistance Testing

Resistance testing is based upon Ohm's law which shows the relationship between the current (I_{AB}) flowing between two points A and B in a circuit and the voltage difference (V_{AB}) between those points (see Figure 12.8). The ratio of these two quantities is called the resistance, R_{AB}.

$$R_{AB} = V_{AB}/I_{AB} \qquad (12.8)$$

In practice, a small current (I_{AB}) is forced through the interconnection while the voltage (V_{AB}) between the contact points is measured. The net resistance is calculated using Equation (12.8). This value is compared against the expected resistance for the net that can be predicted based upon the trace dimensions and material characteristics. A measured value that is significantly larger than expected indicates the presence of a "resistive" fault.

The advantage of resistance testing is that it measures opens and shorts directly and can detect low-resistance opens and high-resistance shorts. The large number of probe movements for shorts testing heavily favors the use of a matrix, or cluster probe

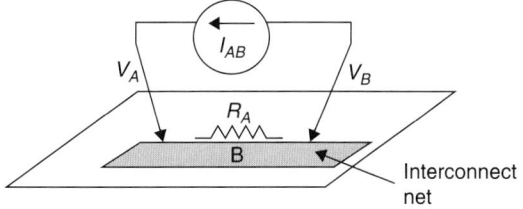

FIGURE 12.8 Resistance measurement of an interconnect net.

(bed of nails) for complex products. This approach employs an array of test probes placed in contact with the substrate and switched, thus greatly minimizing the mechanical movements required of a tester. A cluster probe coupled with mechanical relays or solid-state switching can deliver current and voltage stress to the product in an extremely efficient manner. This method requires a significant outlay for fixturing, but provides the best quality and fastest test available.

Electron Beam Testing

Noncontact testing has shown promise and offers the advantage of eliminating mechanical probing on the substrate. Electron beam test technology is an attractive alternative for high-throughput, layout independent, noncontact, nondestructive testing of unpopulated substrates. Since electrons can be positioned at any location on the substrate using computer control, this mode of testing provides high flexibility with respect to layout changes and hence can be used to test different products. Electron beams have no mechanical contact and thus do not destroy the pad surface or crack fragile substrates. Very fast beam deflection and charge storage allow for high test speed and thus high throughput.

As shown in Figure 12.9, electron beam testing is accomplished by analyzing the energy of secondary electrons that are emitted from the surface of a material as a result of the injection of high-energy "primary" electrons. The secondary electron current is sensitive to the voltage at the point where the primary beam is focused. A higher surface voltage will result in a greater potential barrier for the secondary electrons to overcome. This will reduce the current at the detector. In this way, the surface voltage can be deduced. Since the voltage information is sometimes displayed as a gray scale on a video screen, the effect is called "voltage contrast."

To test for continuity in an interconnection net, the line is first charged using a high-current electron beam. An isolated line will retain the charge for a considerable time, and this will be detected using the voltage contrast effect. By scanning the electron beam across the surface, the charged net can be imaged. A break in the line will be evidenced by the lack of charge on certain segments of the net. If the net is shorted to a power or ground plane, then the charge will not remain on the line.

Electron beam testing can deliver effective high-resistance shorts capture and is limited in the resolution of open defects. For substrate stressing, electron beam testing

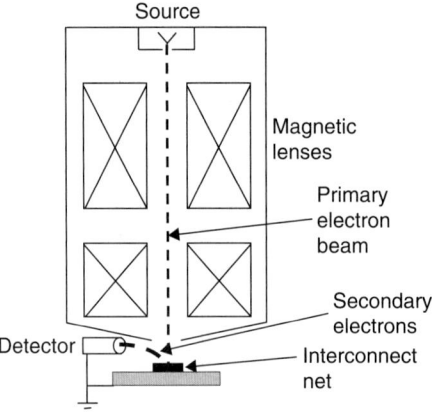

FIGURE 12.9 Electron beam measurement of surface voltage using the voltage contrast effect.

is plagued by severe current carrying limitations. The test system is more expensive for volume production testing than the more commonly used resistance or capacitance test technique. Integrated resistors and capacitors affect defect detection. As an example, a net terminated by an integrated resistor to ground cannot be charged by the electron beam, and hence shorts to other nets cannot be detected as explained by Brunner et al. [16]. However, opens can be detected on these terminated nets since the termination is disconnected due to the defect. Integrated capacitors to ground limit the resolution of defects due to their influence on the charge time. Like other test methods, electron beam testing does not provide details on the precise location or cause of a short or open defect.

Latent Open Testing

As wiring densities increase with a consequent decrease in line width and spacing, defects such as latent opens have a greater chance to occur. Latent opens are near opens that transform into complete opens on the field, producing failures. Detection of latent opens such as cracks, notches, and via-line connections are usually based on optical inspection or stress testing. However, optical testing can be applied only to visible areas. Similarly, stress testing such as thermal cycling or mechanical fatigue test is time consuming. For expensive substrates, the lack of a latent open detection technique makes the substrate vulnerable to failure due to the subsequent assembly process.

Testing for latent opens as a result of burn-in stressing on thin-film substrates has been addressed by IBM through the proprietary latent opens test or electrical module test (EMT). Burn-in stressing involves the application of heat cycles and/or electrical bias to the substrate in order to reduce the failure of latent opens. Properly chosen burn-in conditions can help to weed out unreliable substrates; however, the stress conditions must be developed such that defective circuits can be forced to fail during burn-in without significantly weakening others. The burn-in should also not introduce failure modes inconsistent with use conditions. Burn-in testing is often performed with IC chips mounted on the substrate. However, this could result in expensive chips being lost, or rework being required due to substrate failures. Burn-in, or other steps taken at points well removed from the substrate build, can result in an extended feedback time for latent defect understanding. Once corrective action is implemented, initial parts must reach a measurement point such as burn-in to verify the solution. Thus the cycle time to reach a measurement point is a key parameter in determining reliability improvement rates.

Resonator Band Testing of Substrate Interconnections

The resonator band testing was developed at Georgia Tech [21], which applies a stimulus through a high-Q resonator at only one end of the interconnect using a single-ended probe, as shown in Figure 12.10. The resonator modulates the ac response of the interconnections, producing a change in the frequency response in the presence of defects. By measuring the attenuation of the test stimulus due to pole movement relative to known attenuation measurements, interconnect faults such as near-opens, near-shorts, opens, and shorts can be detected. The total test time is projected to be similar to a capacitance method, and the hardware cost of test equipment is low. The key advantage is the ability to detect latent defects such as near-opens and near-shorts from a lookup table of transfer functions.

Comparison of Test Methods

Table 12.1 provides a qualitative and quantitative comparison between test techniques practiced in the industry. Though time domain network analysis (TDNA) has been used

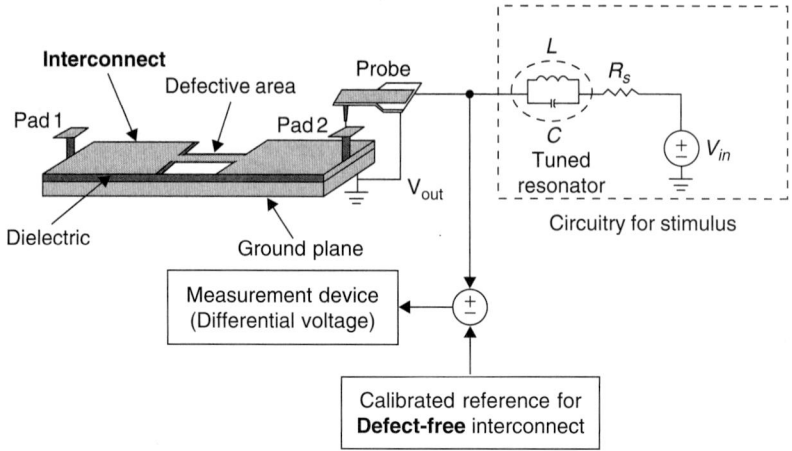

FIGURE 12.10 Resonator band testing using a high-Q tuned resonator at the probe tip.

for high-frequency characterization of interconnects and is not a viable test method, it has been included in the table. The number of probe heads depends on whether one end or both ends of the interconnect require probing for a two-terminal net. This is related to probe movements with the complexity arising due to the necessity for the two probe heads to be in synchronization. The complexity also manifests itself through the requirement for expensive test equipment. The test time required for implementing each test method is based on the number of probings required and assumes that the setup time is similar and small for all the methods. A qualitative comparison, for all methods except resonator band testing, is provided for the total test time based on high

	Capacitance	Resistance	Electron Beam	Latent Open	TDNA	Resonator
Frequency	10 MHz	Direct current	—	1 MHz	30–70 GHz	1 GHz
Probe heads	1	2	—	2	2	1
Probe movement	Simple	Complex	Complex	Complex	Complex	Simple
Total test time	Medium	Large	Small	Large	Large	Medium
Opens resolution	1 MΩ	10 MΩ	10–100 MΩ	3–10 MΩ	Small	Small
Shorts resolution	1 MΩ	300 MΩ	1 Ω–100 MΩ	—	Large	Small
Equipment cost	Small	Small	Large	Large	Large	Small

TABLE 12.1 Comparison of Substrate Test Techniques [17,21]

wiring densities with further details available in Woodward [17]. The opens and shorts resolutions in the table provide a measure of the nature of the defects that are detectable by the various test techniques. A small value for opens and a large value for shorts represent good resolution. For example, capacitance testing is ideal for opens testing but has poor resolution for shorts testing. However, E-beam testing is ideally suited for shorts testing. Among the methods shown in Table 12.1, only the resonator-based approach has the capability to detect both latent opens and shorts.

12.2.2 Testing Embedded Passives

With advancements in SOP manufacturing technology, embedded passive components are playing a key role in the development of these systems [18–19]. The integration of passive components into a system offers inherent benefits, such as a reduction in the size of the system, a reduction of parasitic effects, lower assembly cost, and superior electrical performance. However, testing these embedded integrated passives is more difficult due to the inaccessibility of internal circuit nodes, possibility of many different failure modes due to the analog nature of these components, and the high frequencies of circuit operation especially for RF passives. Three important parameters critical in the design of embedded passive components are [28]: (1) variation of resistance/reactance with frequency, (2) variation of quality factor of the components with frequency, and (3) the resonance behavior of the components.

The main research in testing of embedded passives originates from advancements in multichip modules (MCM) [20,22,24–26]. In this section, we will concentrate on two test techniques: (1) a diagnosis technique for RF passives using single-probe S_{11} measurements [27] and (2) a pole-zero analysis technique using two-probe Y_{11} measurements [23]. These two methods are high-frequency at-speed tests, which require measurement of network parameters as a function of frequency and were developed at Georgia Tech.

Diagnosis of Faults in Embedded RF Passives with Sensitivity Analysis

In chapter 4 of [22], Yoon proposes a fault diagnosis methodology for embedded RF passives based on single-probe S_{11} measurements, as shown in Figure 12.11. This technique can detect as well as diagnose catastrophic and parametric faults in a passive

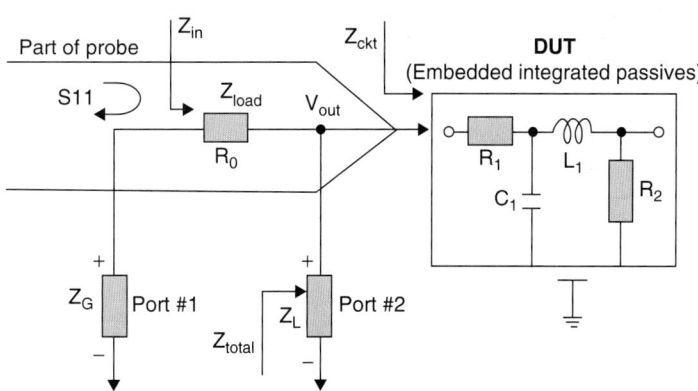

FIGURE 12.11 Single-probe S_{11} measurement setup for fault diagnosis with embedded passive as the device under test (DUT).

network with a single fault assumption. Fault detection is the process of identifying a source that does not meet specifications based on a test stimulus. Diagnosis represents identifying the cause of the defect. The goal is to find the minimum set of frequencies that is sufficient for diagnosis.

The automated test pattern generation (ATPG) algorithm starts with determining the sensitivities of S_{11} for each passive component in the network at specific frequencies. The frequencies are selected such that the magnitude of S_{11} is a maximum at one frequency for a given component. These sensitivities are arranged in a matrix form, and a minimum set of frequencies is selected such that all 2×2 submatrices formed from rows corresponding to the minimum set are nonsingular. Once the minimum set is established, the parametric variation in a single component can be determined by measuring S_{11} of the DUT at the corresponding frequencies. However, for catastrophic faults, two different faults can still lead to the same S_{11} measurement. In order to successfully diagnose these pairs, one can compute the probability that two given faults will yield the same measurement value. This probability is called the fault resolution (FR). If FR is larger than a threshold value for all pairs of faults, then the minimum set of frequencies is adequate to diagnose the DUT. If not, the minimum set of frequencies is extended by the next best choice of candidate frequency.

As an example, consider a filter implemented using embedded passives in Figure 12.12 [29] with 11 resistors, four inductors, and four capacitors yielding a 19×19 sensitivity matrix. The test measurements are available only at node 1 or 4 based on the single-probe measurement in Figure 12.11. For selected frequencies of $f_{C1} = 187.93$ MHz and $f_{L1} = 145.32$ MHz, the corresponding 2×19 matrix is checked against trivial solutions. This can be achieved by checking the nonsingularity condition for all 2×2 submatrices of the 2×19 matrix. In this case, all twenty-eight 2×2 submatrices are singular: two different faults yield the same measurement. By adding a third measurement frequency at $f_{R6} = 59.77$ MHz, full fault resolution is achieved.

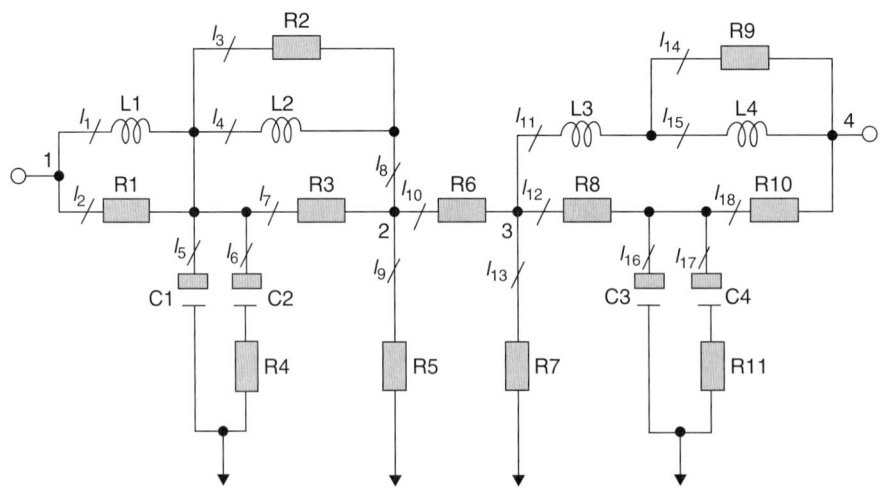

Figure 12.12 Bourns filter implemented with embedded passives.

Diagnosis of Faults in Embedded Integrated Passives with Pole-Zero Analysis

It is also possible to diagnose faults in embedded passives by a lookup operation defined on a prepopulated dictionary. The dictionary can be composed of signature responses with respect to a specified input pattern or composed of possible variations of the network transfer function.

In a passive network, good candidates for representing the transfer function are the poles and zeros of the system, because pole-zero values can be obtained from Y, Z, T, or H parameter measurements using interpolation technique [30]. Component tolerances will result in variations in pole-zero values, and these variation-signature pairs can be precompiled into a fault dictionary by running Monte Carlo simulations on circuit parameters.

One such solution is presented in [23]. First, a concurrent fault simulator [30] is used to generate the results of Monte Carlo simulations with process variations. Then, pole-zero locations for each Monte Carlo instance are extracted, and the resulting signature is either marked as "good" or "faulty" depending on the specifications for that instance. Any pole or zero that has the maximum sensitivity to one or more circuit parameters is determined to be critical. The selected critical poles and zeros are plotted on the real-imaginary plane marking those that correspond to "good" circuits and those that correspond to "faulty" circuits. For each pole or zero, a fault-free region that results in 100 percent yield coverage is computed by applying a grid to the left half plane as in Figure 12.13. In the event that the fault-free regions of two poles in the real-imaginary plane intersect, a combined region for the two is computed as shown in Figure 12.14.

Given a DUT, the Y parameters can be measured on the two ports with the setup in Figure 12.15. Then, the pole-zero positions are extracted using a macromodel [24]. If the critical pole-zero positions correspond to the fault-free region for the given pole-zero value, then the DUT is a fault-free circuit.

If it is a faulty circuit, a region-matching algorithm is used to determine the fault list that the measured poles and zeros of the DUT most closely match. If the positions of critical poles and zeros of the given DUT in the real-imaginary plane match the simulated

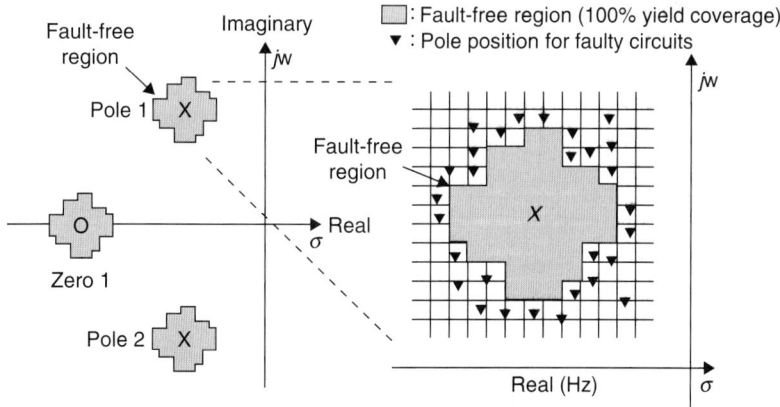

FIGURE 12.13 Diagnosis of faults by pole-zero analysis, applying a fault-free grid to the left half plane.

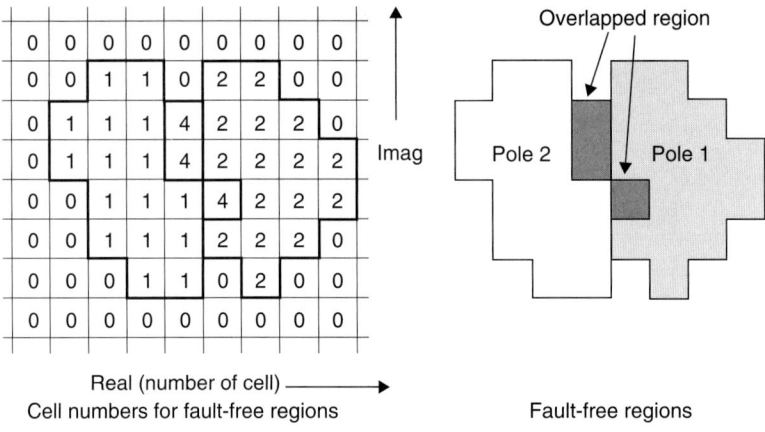

FIGURE 12.14 Combined grid for intersecting fault-free regions of two faults.

locations under specific catastrophic faults, then the specific open or short failure can be isolated. Otherwise, the given DUT has a parametric failure, which can be identified by computing the minimum distance between measured pole-zero values and all possible parametric faults.

As an example, consider the embedded integrated lowpass filter as shown in Figure 12.16 with two complex poles. One-thousand circuit instances are generated by varying every component randomly with $3\sigma = 10$ percent normal distribution. Figure 12.17 shows the pole positions for each instance. Fault-free and faulty circuits are represented with X and V with respect to four circuit specifications: A_v, f_{3dB}, Z_{in}, and Z_{out}. Figure 12.18 shows the fault-free region marked with a lighter shade for a 200 by 200 cell array together with the measured pole locations for a DUT in the darker shade. One can clearly see that the given DUT is faulty.

For catastrophic fault diagnosis, all catastrophic faults (four opens and six shorts) were simulated to populate the pole-zero positions. To diagnose catastrophic faults, 10 different fault arrays were used to compare with the measured pole-zero positions. Since the measured pole positions do not match with any array of catastrophic faults, the given DUT must have a parametric fault. Figure 12.19 shows the pole positions for the parametric faults R_1, C_1, and L_1. In this example, the DUT component of C_1 is obtained to be 1.8 pF as opposed to the nominal value of 2 pF.

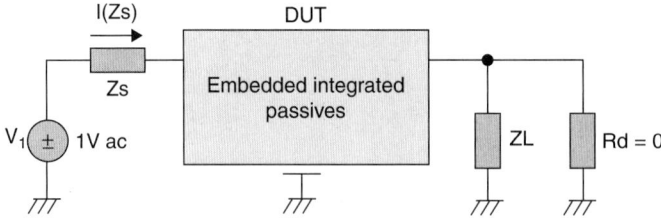

FIGURE 12.15 Measurement setup for two-port Y parameters.

Electrical Test of SOP Modules and Systems 675

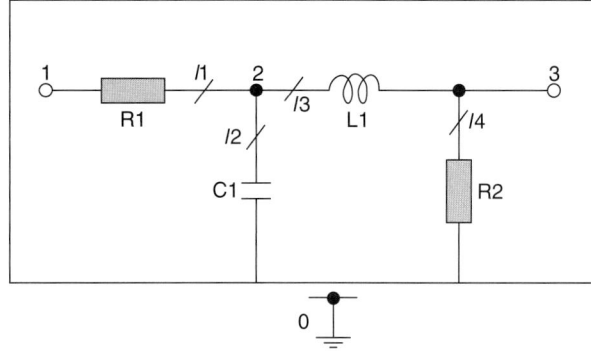

FIGURE 12.16 Embedded integrated lowpass filter with two complex poles.

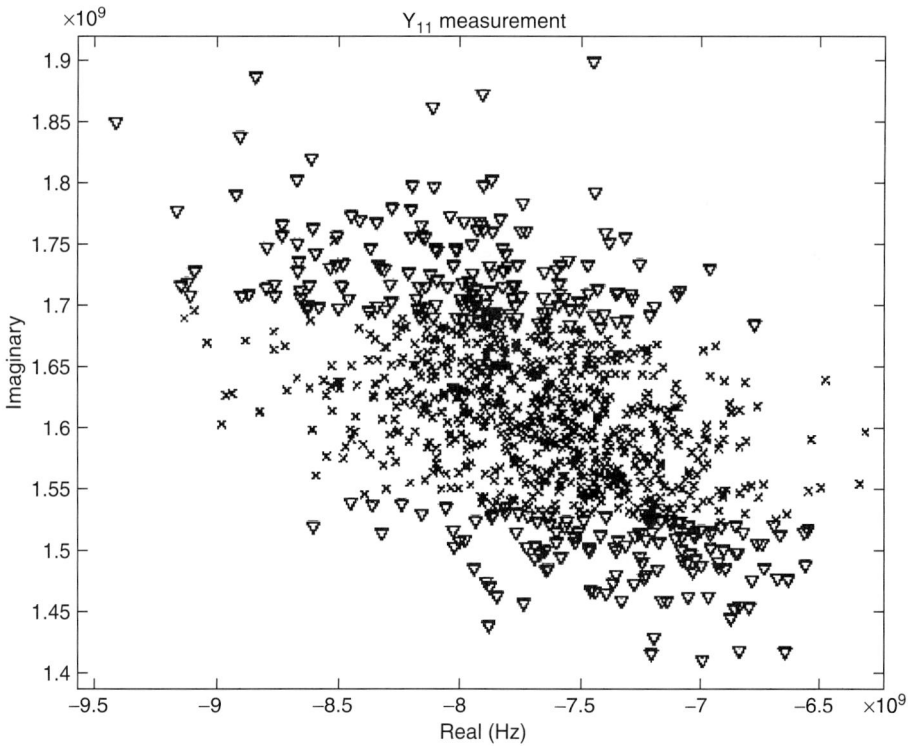

FIGURE 12.17 Pole positions for 1000 instances of the embedded lowpass filter.

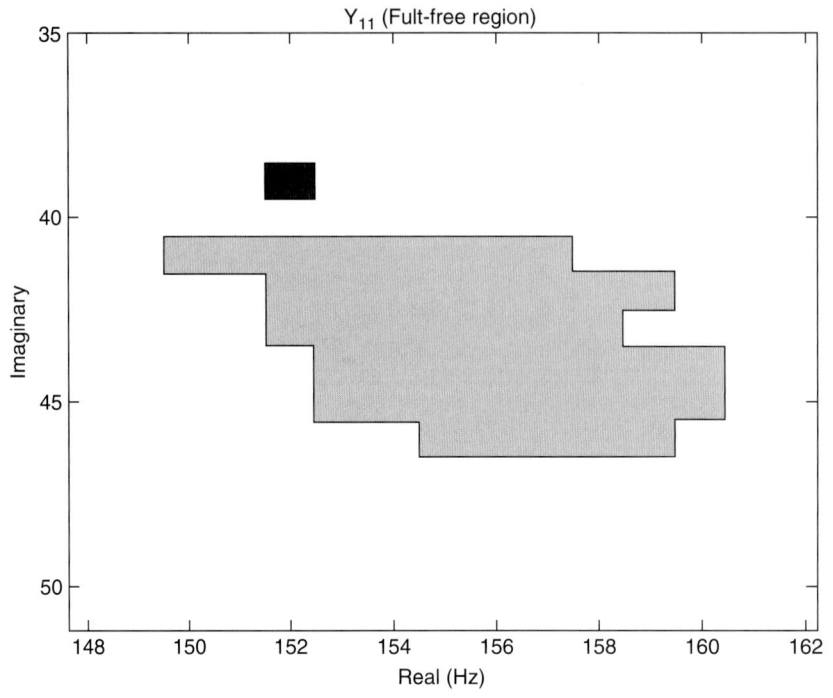

FIGURE 12.18 Fault-free region for the embedded lowpass filter.

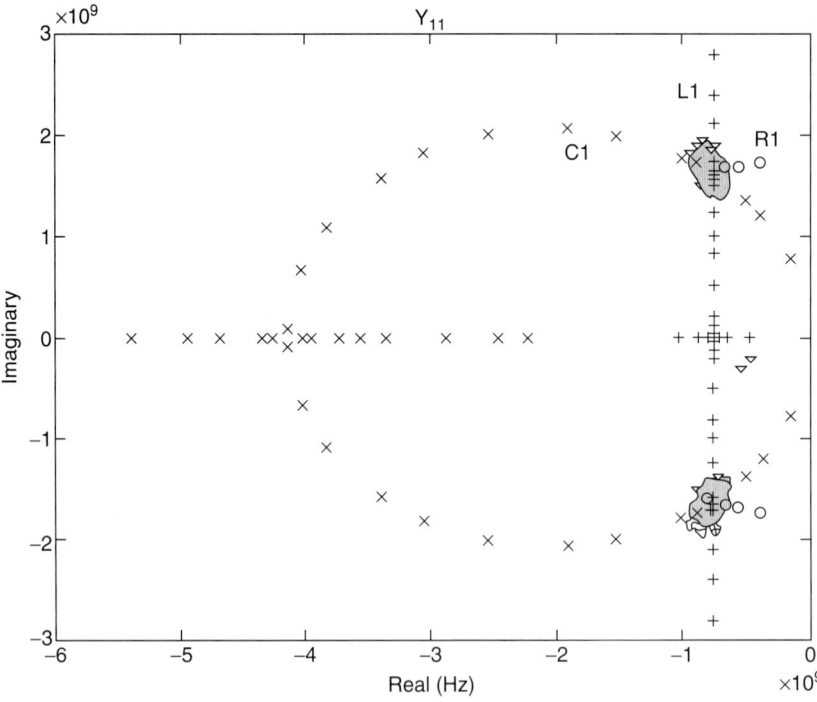

FIGURE 12.19 Pole positions for catastrophic faults in R_1 (marked with circle), L_1 (marked with plus), and C_1 (marked with cross).

12.3 Known Good Embedded Module Test of Digital Subsystems

12.3.1 Boundary Scan—IEEE 1149.1

Boundary scan is a *design* approach and *test interface standard* that allows digital data to be serially "scanned" to and from the I/O cells of ICs even when they are assembled within a larger system (printed circuit board, MCM, SOP, etc). This enables the critical test functions of "controllability" and "observability" in large complex digital systems.

The principal usage of the boundary scan is to provide an efficient means for transmitting *test* patterns. These are used to test both the internal IC logic as well as the interconnections between logic chips. However, innovative designers continue to find other applications for boundary scan logic beyond that of test-enabler.

The strategy is to create chains of storage elements (flip flops) associated with the normal primary I/O cells on the periphery ("boundary") of the IC. These chains are linked together to form serial shift registers (SRs) that are loaded at one port [test data in (TDI)] with test data. The results of tests are shifted out to another port [tests data out (TDO)] for analysis. What happens in between depends on the functional specifics of the system and other aspects of the test strategy. The role of a boundary scan is to enable the test data to be injected into otherwise inaccessible parts of the system and to allow the results to be read out to a convenient port.

Beyond its basic role of facilitating the transmission of test data within a digital system, a boundary scan also provides two other important capabilities: (1) the physical structure of the scan elements at the chip boundaries enable *testing of the interconnections between chips (tests for "opens" and "shorts")*, and (2) enabling and controlling built-in self-test (BIST) logic.

History and Motivation

As far back as the 1970s, developers of digital systems were finding that the cost of developing and applying tests was growing even faster than the complexity of the systems themselves. As small-scale integration (SSI) led to medium-scale integration (MSI), and then to large-scale integration (LSI), very-large-scale integration (VLSI), and ultra-large-scale integration (ULSI), more and more logic was included within a single IC and therefore the tests for each IC became more extensive. At the same time, the density of I/Os and chips within a system was also increasing. Furthermore, the ability to physically access the chip I/O was limited if not prevented by the new surface-mount technologies (notably BGAs) that replaced through-hole mounting as the principal packaging strategy.

Before surface-mount packaging, populated circuit boards were tested using a "bed of nails" fixture that gave electrical access to the chip I/Os. A technique called "in-circuit test" was used to force the chip inputs to a desired state. Since the chip was embedded within a larger system, the terminology "in-circuit" was adopted. Unfortunately, this strategy also required that the system signals that normally drive the IC inputs be "overdriven" by the external test instrumentation. Since common logic families have well-defined current limits, it was possible to design the test instruments to be more powerful. Therefore, the test equipment could force test patterns onto the IC inputs regardless of the state of the rest of the system (providing "controllability"). There was some concern that such overdriving might cause damage to the normal system gate outputs. Eventually this question became academic when surface-mount technology made bed-of-nails testing impractical.

678 Chapter Twelve

Nevertheless, the need for applying test patterns to chips within digital systems continued to grow. On the other hand, the increased level of integration within ICs opened up some new possibilities. Perhaps the very logic that required testing could be used to help solve the problem. Rather than physically probe the chip I/Os, a solution was required that would take better advantage of these abundant logic resources available within the IC.

In the mid-1980s the Joint European Test Access Group (JETAG) was established to address this problem. Shortly thereafter, when American companies joined the group, it was renamed simply JTAG. A formal proposal was published in 1988 that included the boundary scan approach. It was then adopted as an IEEE standard (1149.1) in 1990. Since then the standard was widely adopted and is currently used within most standard and custom VLSI chips. It could be argued that the boundary scan standard represents the most significant development in digital testing technology during the last 15 years of the twentieth century.

Key Elements of the Boundary Scan

The basic structure of the boundary scan is illustrated in Figure 12.20. Here it is assumed that each I/O cell has additional test logic (beyond that needed for normal system functionality). Each cell has a serial input and a serial output, and by connecting the output of one cell to the input of another, the scan "chain" is formed. During test mode, this additional logic allows data to be serially shifted (or "scanned") from one cell to another. At one end of the chain is an unconnected serial input, called test data in (TDI).

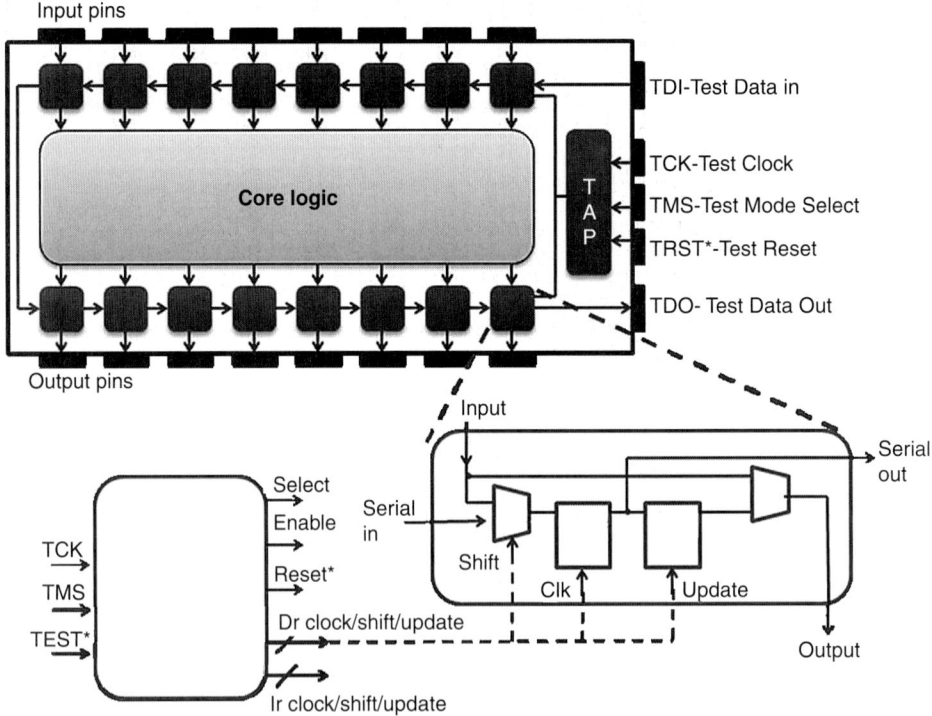

FIGURE 12.20 Boundary scan structure.

At the other end of the chain is the test data out (TDO). In order to clock test data through the chain, a test clock (TCK) is used which is separate from any other system clocks that might be needed for normal functions. A test mode select (TMS) signal is used to control the switching of the scan logic between scan mode and normal system mode. To summarize, the four test signals *required* by the IEEE 1149.1 boundary scan standard are

- Test data in (TDI)
- Test data out (TDO)
- Test clock (TCK)
- Test mode select (TMS)

Beyond these required signals is the optional test reset (TRST*), which is active-low. It is used to reset the test control logic and to deactivate the boundary scan register.

Also shown in Figure 12.20 is the test access port (TAP) controller which is needed within each boundary scan chip. The function (but not the exact structure) of this finite state machine is defined by the boundary scan standard (see Figure 12.21). It provides a standard logical interface for test control and responds to a standardized set of serial instructions that is transmitted through the TDI and puts the chip into various test or

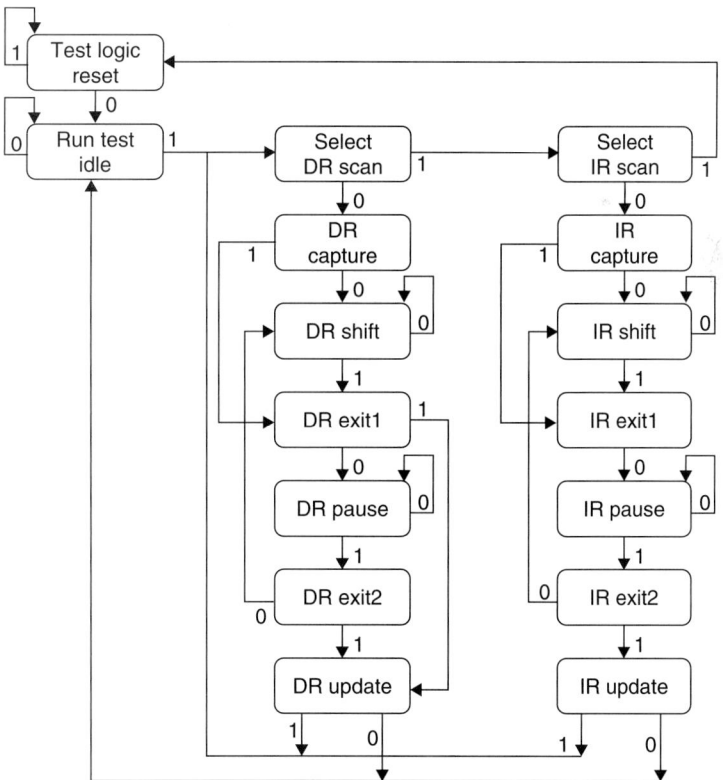

FIGURE 12.21 The finite-state machine defined by boundary scan.

normal function modes. These modes fall into two classes: (1) normal operation mode (during which the boundary scan register is "transparent") and (2) test operation mode (the boundary scan register isolates the internal chip logic from the normal I/O signals).

Boundary Scan for SOP

A scalable structure of the boundary scan has been successfully applied to testing at the system level. The classical paper from Zorian [5] lays out the structured-testability approach well accepted in industry. Figure 12.22 shows this approach: specific dies are tested by on-chip BIST, and the system test is handled by boundary scan, such that all the boundary scan cells in the dies are connected together into a single chain. The TDO of one die is connected to the TDI of another. The initial TDI and the last TDO act as the system boundary scan terminals. System-level TCK and TMS are routed to each die. The system approach has been recently adapted to high-speed tester resources [31–32], I/O [34], and interconnect [35] signal integrity testing, and SIP testing [33] with proposed solutions for scalability.

The main challenge for boundary scan testing of SOPs is the reliability of the solution. As the number of dice increase, the simple scan chain in Figure 12.22 gets longer. Motivating the test hardware design is the need to test the stack electronics reliably in the presence of faults in the boundary-scan circuitry, such as an open TDI line. The work at Georgia Tech [36] evaluates possible scan test methodologies in terms of test reliability, scalability, and testing time overhead introduced by the system-level approach. The best tradeoff is a partitioned approach with separate multichip test controller ICs (MTCs) integrated into the SOP. Each MTC has the ability to control a boundary scan for a group of ICs by acting as a gateway for the TDI, TDO, TCK, TMS,

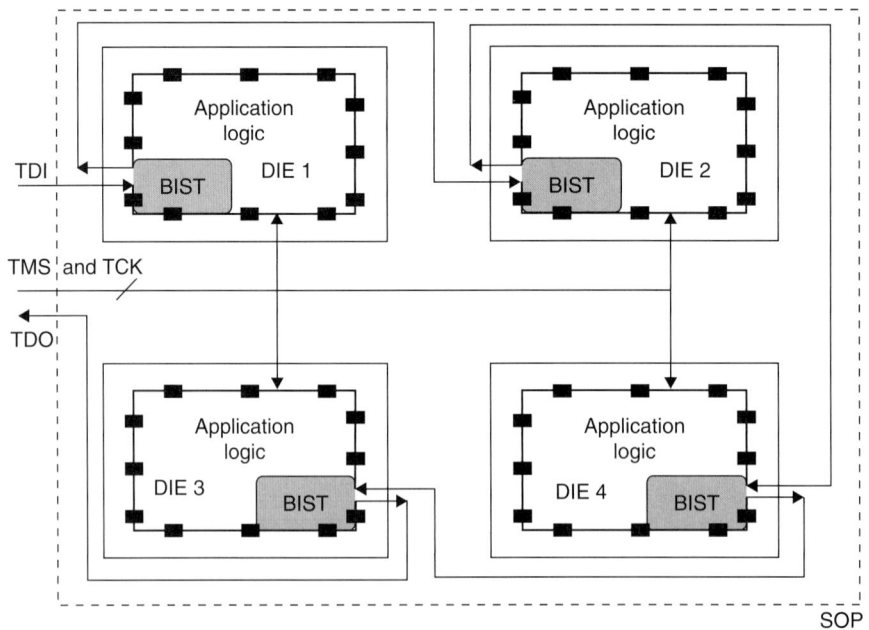

Figure 12.22 System-level boundary scan.

and TRST signals coming from the tester. MTCs are reconfigurable such that the tests can be performed in series (as in Figure 12.22) or in parallel (individual TDI, TDO, TMS, and TCK for each die) depending on the test profile of the group. One MTC is selected as the master test controller for the rest of the MTCs. Redundant test lines enable testing in the event that some lines are faulty. All test lines are implemented as bidirectional and can be configured as input or output lines by the MTCs.

12.3.2 Multi-gigahertz Digital Test: Recent Developments

Figure 12.23 shows the general evolution of high-speed digital ATE from the early 1980s to the present, in terms of performance and size. It is interesting to note that the characteristic size of the ATE roughly scales with the maximum clock period (inverse of frequency). In the early 1980s these systems took up a large room (~10 m) with multiple racks of electronics and ran up to 40 or 50 MHz. By the early 1990s the critical ATE electronics was in the test "head," about the size of a desk (1 to 2 m) and ran up to 500 MHz. The state-of-the-art system in 2007 has ATE running up to 6.4 Gbps, with the critical electronics in modules or chips. In most cases, testing at these rates requires significant on-chip circuitry for built-in self-test (BIST). Eventually, at frequencies above 10 GHz, the ATE size scaling trend suggests that the critical test electronics must be smaller than the chip itself, implying full-BIST.

A snapshot of the recent developments in multi-gigahertz digital testing research can be obtained by examining papers presented at the International Test Conference [37, 56]. The objectives for each of these fall into two broad categories: (1) increased test performance, and (2) reduction in test cost. At any given point in time, these two needs tend to have conflicting requirements, so a tradeoff or balance between them is usually sought. Still, over time, improvements in both are required. Existing ATE is limited to 1 to 1.6 Gbps/s per channel in most cases. Yet devices must be tested in the 2.5- to 3.2-Gbps/s range, and occasionally as high as 10 to 12 Gbps/s. The capital cost of such ATE can exceed several million dollars. That high price is a concern in itself, but it

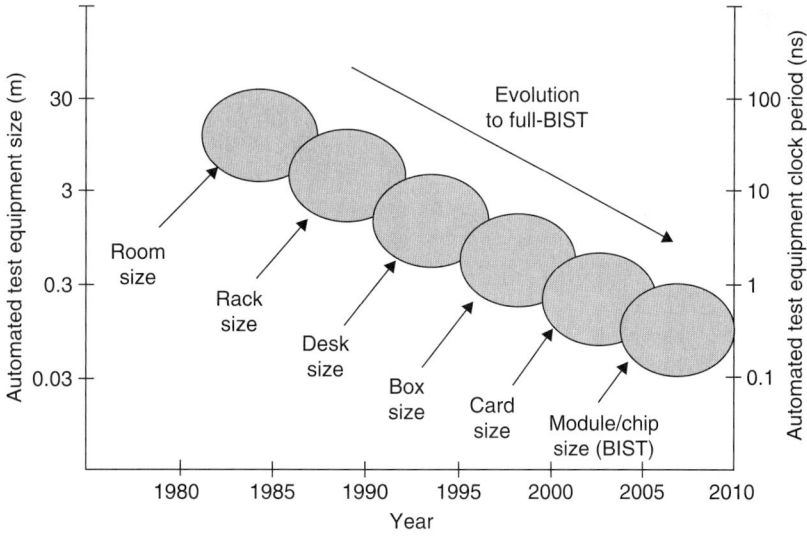

FIGURE 12.23 ATE size scaling and performance evolution.

places a tremendous emphasis on the need to minimize time on the test system (increase test throughput). The multimillion-dollar capital cost translates into significant dollars-per-minute costs on the test floor.

Generally the various tradeoffs between performance and cost issues for multi-gigahertz ATE are discussed in a series of "position papers" [56] on the future directions for ATE development. The authors provide various views as to the specifics of how best to balance these seemingly incompatible requirements.

Increased performance, in this context means: (1) higher speeds (multi-gigabit-per-second data rates), (2) better timing resolution and accuracy (picoseconds), and (3) extended functionality and flexibility (to accommodate new device functions and signaling methods). Obtaining higher data rates alone is usually not the greatest obstacle since the basic technology for achieving higher speeds continues to improve (largely due to reductions in transistor minimum feature sizes). However these same higher speeds greatly impact the need for tighter timing accuracy. So the focus of these recent works tends to be in this area of performance. In some cases new test methods are also required to deal with innovative functions and new signaling conventions such as source-synchronous signaling [48].

The various approaches used to solve the problems of performance and cost for multi-gigahertz digital testing include: built-in self-test (BIST) [39, 44–49], built-out self-test (BOST) [40, 43], ATE instrumentation improvements [37–38, 41, 48, 53–55], and interface improvements [38–40, 43, 44, 46].

The control and/or measurement of jitter at the level of picoseconds is a central issue in many of the recent articles [37–39, 50–53, 56]. Since the ATE is normally designed as a synchronous system, it includes its own clock signals that are distributed throughout the test electronics and eventually determine the timing accuracy of the test. At the high data rates, jitter contributes a significant fraction of the overall uncertainty in edge placement accuracy, so control of jitter within the ATE itself is a challenge. Likewise, the ability to accurately test for jitter tolerance or jitter characteristics of the DUT output signals is difficult because the accuracy required is comparable to the jitter levels of the ATE itself (on the order of a few to several picoseconds for RMS jitter).

In other works [40, 41, 43, 46–56], obtaining higher data rates (usually with an emphasis on maintaining or lowering cost) is a main objective. In these efforts all of the component elements in the ATE are considered for possible improvement in order to economically achieve multi-gigahertz test speeds. These include the ATE "pin electronics" (memory, multiplexers, formatters, timing edge generators, clock distribution, input-output buffer-drivers, etc.), the overall ATE architecture itself, the interface between the ATE and the DUT, circuits and methods for handling new signaling standards, and ways to take advantage of built-in self-test features of the DUT. In [43] for example, additional electronic multiplexers and high-speed samplers are added to the test interface as illustrated in Figure 12.24, forming "Driver" and "Receiver" modules. In this example, the multiplexing modules combine signals from a 1-Gbps ATE to form multi-gigahertz (2 to 3 Gbps) test stimuli. An example is shown in Figure 12.25. In another example [40], based on work done at Georgia Tech, the concept of a "test support processor" is used to create a miniature test module that can provide customized high-speed test signals locally to the DUT during wafer-probe testing as shown in Figure 12.26. This minimizes the transmission line lengths so that high-quality signals can reach the DUT multi-gigabit-per-second rates. An example of the test signals at 5 Gbps is illustrated in Figure 12.27. A potential advantage of this approach

Electrical Test of SOP Modules and Systems 683

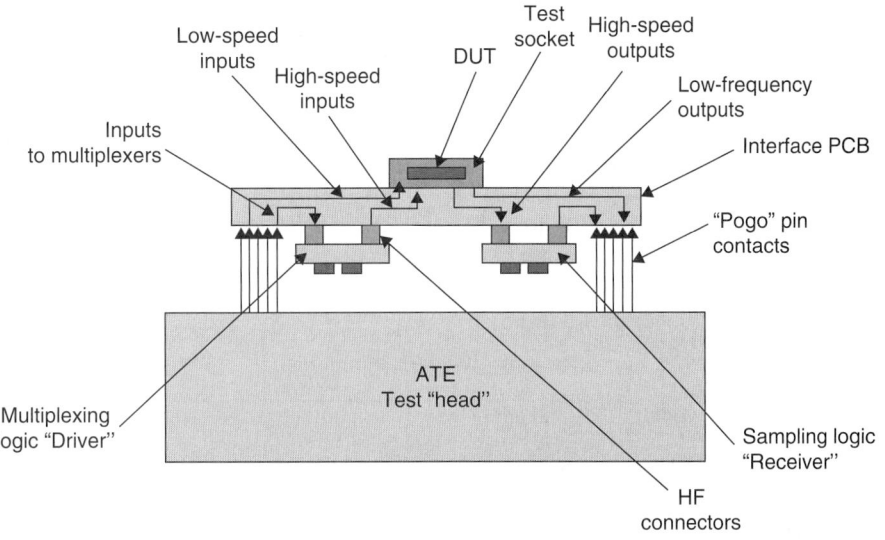

FIGURE 12.24 Multiplexing drivers and high-speed sampling receivers at the DUT-ATE interface [43].

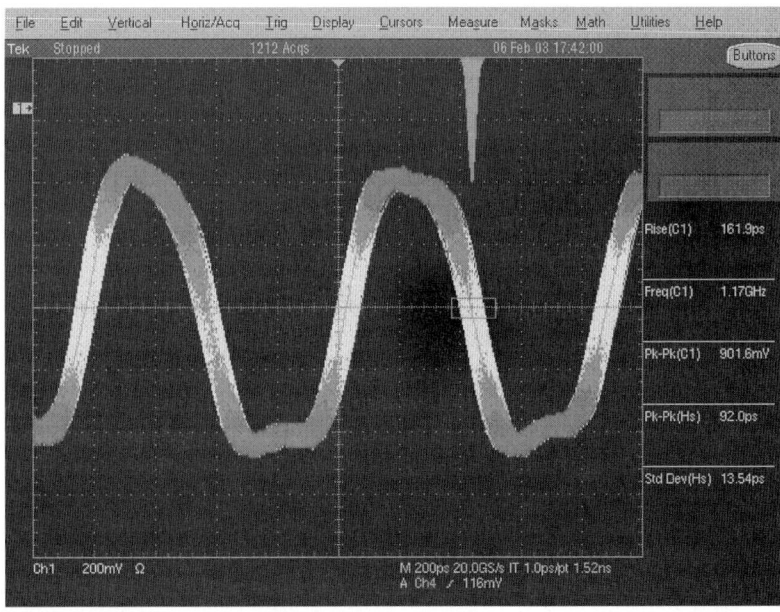

FIGURE 12.25 Example 2.5-Gbps signal from the multiplexing driver module [43].

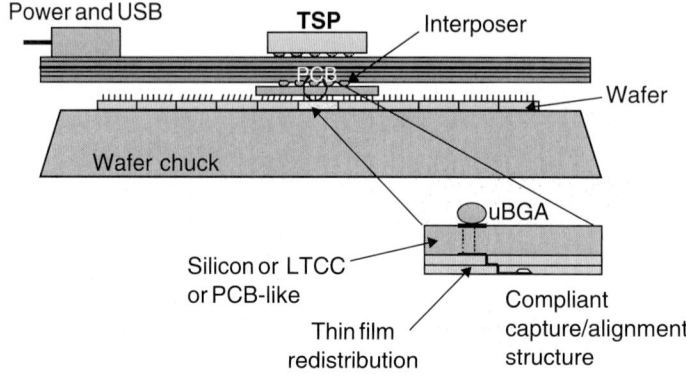

Figure 12.26 Miniature TSP-based tester in a wafer-probe test arrangement [40].

is that it can (in principle) be replicated into an array of miniature testers for parallel multi-gigahertz testing at the wafer level. A comparison of KGEM test methods are summarized in Table 12.2.

In almost all cases, the cost of testing, and specifically the cost of the increased performance required by multi-gigahertz devices, is a major issue, and in some it is the primary objective [41, 42, 49, 52, 54–55]. As pointed out above, the ATE costs are already high for systems that can perform at about 1 Gbps/s. Extrapolating to 10 Gbps/s results in vastly more expensive systems. Therefore innovations are needed for finding ways to obtain the required higher performance without incurring substantial increased test costs.

Extrapolating to 2015, the critical testing of high-performance chips will require complete BIST for validating the internal functionality as is already the case for many

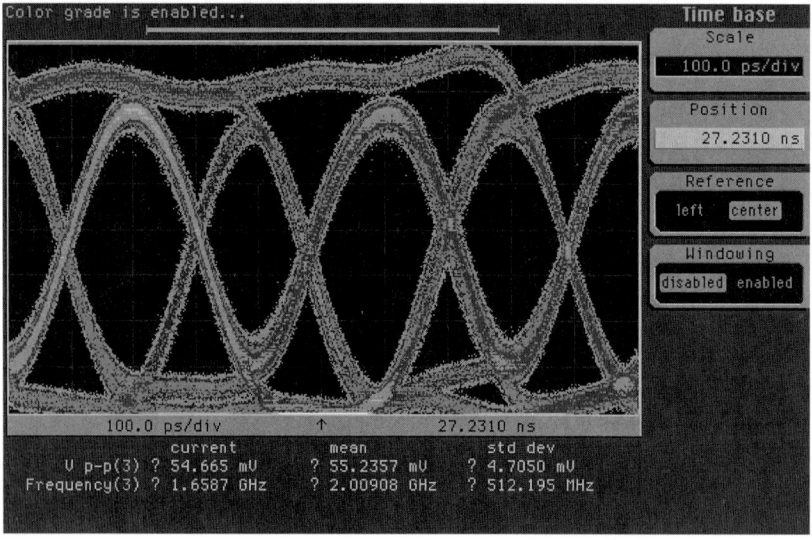

Figure 12.27 Test signals at 5 Gbps from the TSP-based miniature tester.

	Boundary Scan	Dedicated Built-In	Built-Off	Test Support Processor
Frequency	High	Medium	High	High
Total test time	Large	Large	Medium	Medium
Equipment cost	Small	Small	Medium	Large

TABLE 12.2 Comparison of KGEM Test of Digital Subsystems

of today's high-end parts. However, in SOP there will remain a need to validate the chip-to-chip signaling performance. This may be performed using highly specialized test methods that focus entirely on signal integrity and timing. On the other hand there may be innovative "on-line" approaches developed that monitor chip-to-chip signal performance while the SOP is operating in mission mode. As shown in a Figure 12.23, heavy reliance on large external ATE will not be practical as frequencies exceed 10 GHz.

12.4 KGEM Test of Mixed-Signal and RF Subsystems

The call for a testable SOP results in a conflict of interest between the degree of integration afforded by the design process and the level of testability achievable by an external tester. A viable solution is to place the ATE functionalities in close proximity to the SOP module to be tested. This improves the test-access speed, minimizes test signal degradation, and increases controllability and observability of the signals internal to the DUT. One such candidate is the load board itself, where the test functions are migrated from the external tester to the additional circuitry built around the system-under-test. The additional circuitry retains the ability to apply high-speed stimulus to the system-under-test and capture the high-speed test response, which, otherwise, are degraded by the cable parasitics of the low-bandwidth external ATE [2]. The resulting solution, called built-off test (BOT), presents a low-cost alternative to the prohibitive cost of a classical ATE. The other alternative, built-in test (BIT), pushes the external tester functionality into the package and even into the bare dies wherever possible and is consequently a much more aggressive version of built-off test.

Note that without built-off test and built-in test, very high performance SOPs may not be economically testable. This is because the cost of external test equipment for test signal speeds in excess of 1 GHz is very prohibitive. However, multi-gigahertz system designs are now becoming quite routine for high-bandwidth communications. The test economics is greatly improved by having high-speed test functions on the load board (BOT) or the SOP itself (BIT) augmented with low-bandwidth communication with a (low-speed) external tester. This allows very high speed systems to be tested with a low-cost external tester without loss of test quality.

12.4.1 Test Strategies

While migrating external tester functions to the proximity of the device-under-test (DUT) there are two different possibilities: (1) the DUT is considered as an end product without having dedicated test functionality internal to the device and, hence, the test support circuitry is built around the device, or (2) the test support functions are implemented within the device as an integral part of it. The first approach, built-off test, is suitable for applications where the internal design of the DUT cannot be modified for test purposes and the package itself does not constrain the speed of the test signals that can be applied to the DUT. The second approach, built-in test, is more of a DfT methodology. The support functions are implemented within the same package or even in the same chip area. In this approach, the device is modified to incorporate some additional functions within the chip by using dedicated test circuitry [102–105] and by reusing components [107] such as analog-to-digital (ADC) and digital-to-analog (DAC) converters already available at the system level. The introduction of test circuitry into the device may violate original design constraints, for example, device matching and parasitic loading, and, as a result, additional design iterations may be needed during system design. Consequently, built-in test is feasible only when it can be integrated into the system design flow.

Irrespective of whether built-in or built-off testing is used, the load board is a necessary component in a production test environment and typically routes the signal from (to) the test-head of the external ATE to (from) the DUT. Figure 12.28a shows the role of a load board in a high-end conventional ATE environment. In this environment, the load board contains a low-parasitic socket to hold the DUT, power and ground planes, signal traces, and switches and relays to multiplex external tester resources. The external tester generates the entire test stimulus, and the DUT response is directly relayed to it. High-bandwidth data transfer is performed at the operation speed of the DUT.

Alternatively, the test stimulus can be generated on the board and the response can be compressed into a signature by samplers and converters. Figure 12.28b depicts a general built-off test strategy. Built-off test implements complex test signal generation and test signal modulation schemes without employing expensive "feature-enriched" testers at the expense of higher load-board manufacturing cost in production testing. The high-speed test signal processing is all done on the load board itself under external tester control. The tester employed is typically low-cost with low-speed test data transfer (digital and analog) to and from the load board. The high-speed test stimulus and response signature generation is handled at the load board by means of customized signal generators, samplers, converters, modulators, demodulators, multiplexers, and demultiplexers. Modems convert the low-speed stimulus coming from the external tester into high-speed stimulus required by the DUT; similarly the response is down-converted.

Built-in test, Figure 12.28c, pushes the tester functions into the DUT in order to overcome the two main issues in testing: excitation of the DUT and propagation of the response to an external "test" node. As the complexity and integration of SOPs increase, both issues become harder to tackle, and the test paradigm shifts to solutions where DfT [73–75] is employed to improve the controllability and observability of internal nodes [95–96]. The IEEE 1149.1 (JTAG) [98] boundary scan standard boundary scan standard, as explained in Section 12.3.1, provides an effective means for test access to internal modules of the DUT [97] for testing static faults in digital ICs—faults that can be sensitized with a single operation, such as stuck-at or flip; however, its JTAG counterpart in mixed-signal testing, the IEEE 1149.4 standard [100] is limited by its low bandwidth [99]. Hence, built-in testing of analog, RF, and mixed-signal electronics still presents the following major challenges:

- On-chip generation of high-speed test stimulus using low-cost hardware
- High-speed on-chip response acquisition followed by analysis or response compaction

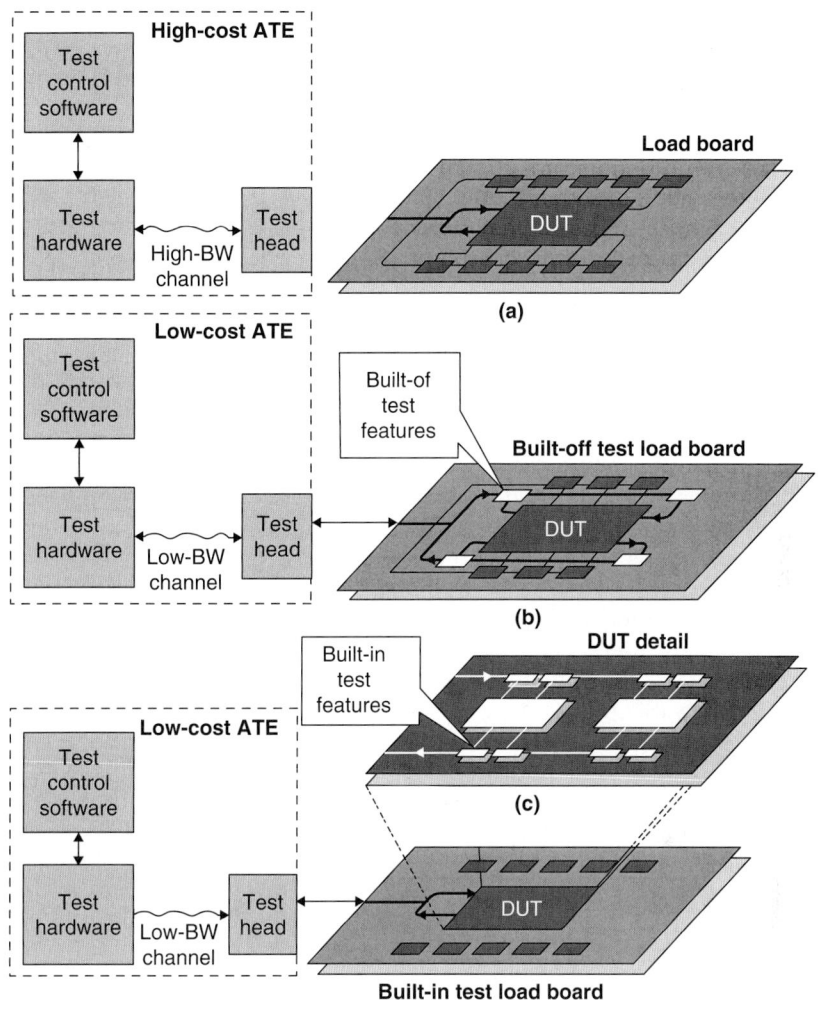

FIGURE 12.28 (*a*) Load board in a high-end conventional ATE environment. (*b*) General built-off test strategy. (*c*) General built-in test strategy.

In built-in testing, low-speed communication takes place between the external tester and the built-in circuitry inside the DUT. This media is used to start or stop a test or run status-checking commands, while the built-in test circuitry performs the rest of the testing in situ. Although this approach addresses the tester cost and test access limitation problems, the large chip-area taken by these circuits, especially in mixed-signal testing, makes it often uneconomical for testing all chip functionalities in situ. With the evolution of highly integrated systems such as SOPs, this area overhead is of less concern thanks to the reuse of already embedded components such as DACs, ADCs, and on-chip digital signal processors (DSP).

The embedded functions in built-off and built-in tests carry different levels of intelligence. They can be implemented in such a way that they create all necessary test vectors and analyze the DUT response on demand, generating a conclusive result about the state of the device. The resulting approaches, built-off self-test (BOST) and built-in self test (BIST), are complete and independent of any external tester help; however, they may require enormous processing power especially when analog and RF components are to be tested. Such components are more likely to benefit from a low-speed, low-pin-count external tester, which analyzes the response signature and generates the test control and low-speed excitation signals. In this kind of "less intelligent" support, the external tester can also be utilized to test the operation of built-off or built-in test components before testing the DUT; this provides flexibility when additional tests are required in the production line. On the other hand, a true "self-test" is not limited to the production line, since it can be applied throughout the lifetime of the device periodically or right before it is turned on. This may be an important criterion for critical systems that are likely to deteriorate over time, such as in space applications. Such schemes are more likely to be implemented as built-in testing since built-off testing requires a load board.

12.4.2 Fault Models and Test Quality

Failures in analog and mixed-signal circuits are broadly classified into two categories: catastrophic, where the circuit fails to operate correctly due to internal manufacturing defects like shorts and opens; and parametric, where one or more specifications of the device deviate from the respective design values due to random variations in the manufacturing process. Defect-oriented tests (DOT) [2] are based on finding a suitable test signal in order to detect the presence of catastrophic failures using different automated fault simulation and test generation techniques [57–61]. The specification-oriented tests (SPOT) [62–66] are concerned with a direct or indirect measurement of the specification on the device data sheet. Under these two categories, test quality metrics are defined so that the effectiveness of a test methodology can be evaluated and that of various test methodologies can be directly compared for a given DUT.

While fault coverage [116] is an accepted test quality metric used for testing digital circuits, its extent and meaning in analog domain is not completely clear in literature. Often the analogy between stuck-at faults in digital circuits and opens and shorts in the analog domain is carried too far into the fault coverage of an analog test and is defined as the percentage of potential shorts and opens the analog test can detect. However, catastrophic failures that result in a significant performance loss of the DUT are detected by simple tests. In reality, the effectiveness of an analog test methodology is largely dictated by its ability to detect parametric failures of the DUT, where performance deviates by a small amount from the nominal. These parametric failures are more likely to occur than catastrophic ones, but are harder to detect than the latter. Furthermore, the meaning of parametric fault is not clearly defined, since any excess variation in a component's value, although considered as a fault, may have little impact on the device specifications. If the test methodologies geared toward increased fault coverage, especially DOT-oriented ones, base their evaluation on parametric faults of individual components, they will eventually end up compromising yield coverage, the probability that a fault-free device passes the test [68].

In built-in test applications, an important test quality metric is the area overhead, the percentage of extra area introduced by tester-related electronics. It is a major concern for practical implementations, since this extra area does not add any value after the device passes the production test barrier. This argument, however, is not valid for some built-in self-test solutions, in which the test can be applied throughout the life span of the product. Often the area overhead is overemphasized when compared to the yield coverage in the qualification of built-in test methodologies. Only the joint figure of these two metrics can define the effective wafer area dedicated to the product.

Any system-level test methodology needs to ensure that all the specifications in the device data sheet are verified in production before the device is shipped to a customer. One possible approach to achieve the above is testing every individual submodule of the device followed by testing of the proper connectivity of the submodules inside the SOP. In effect, this approach breaks down the system testing problem into many smaller module-level testing problems. Although this approach requires physical test access to the individual internal submodules for module-level testing, it is often more effective than end-to-end system-level testing in terms of both the production test feasibility and the test cost. As an example, for wireless transceiver applications, the testing of RF signal blocks, IF signal blocks, and codec blocks can be performed independently and the connectivity of the concerned modules can be verified subsequently to qualify the devices as "good" or "bad" in production tests.

However, in such bottom-up test procedures, algorithms for relating the individual submodule test responses to the (system level) test specifications of the SOP must be devised to aid in the pass/fail decision-making process. *The key is that any circuit, submodule, or system-level failure that causes any of the system-level test specifications of the SOP to be violated is defined to be a "fault."* Any "correct" test methodology must be designed such that it can detect even the smallest of manufacturing defects that can result in such a fault. If no suitable algorithms for determining the system-level test specification values from the SOP submodule test responses can be found, then the only recourse is to directly measure the relevant test specifications at the system level. In general this is more expensive than running submodule-level tests. A typical example is testing of the input referred third-order intercept point (IIP3) specification of circuits exhibiting nonlinear behavior. Measuring end-to-end IIP3 requires high-performance (expensive) measurement instruments. However, if it can be inferred from the results of submodule test, then it can be performed using a simpler setup [112].

12.4.3 Direct Measurement of Specifications Using Dedicated Circuitry

In a traditional production test approach for testing of analog and mixed-signal circuits, the functional specifications are measured using the appropriate tester resources and using the same kind of test stimuli and configuration with respect to which the specification is defined [94], for example, a multitone signal generator for measuring distortion; gain for codec; and a ramp generator for measuring integral nonlinearity (INL) and differential nonlinearity (DNL) of ADCs and DACs. The measurement procedures are in agreement with the general intuition of how the module behaves, and, hence, the results of the measurement are easy to interpret, in contrast to the concept of "alternate test" [70] as described in Section 12.4.4.

In the direct measurement approach using built-in testing, the external ATE functionality is designed inside the DUT for applying appropriate test stimuli and measuring the test response corresponding to the specification. In [117], adjustable delay generators and counters are implemented next to the feedback path of a PLL to measure the RMS jitter. Since the additional circuitry does not modify the operation of the PLL, the same built-in test circuitry can be employed on-line. Reference [117] also discusses different ways to measure properties like loop gain, capture range, and lock-in time by modifying the feedback path to implement dedicated phase delay circuitry. All these built-in test components are automatically synthesized using the digital libraries available in the manufacturing process. This kind of automation provides scalability and easy migration to different technologies. The approach of [118] is similar in the sense that the extra tester circuitry is all-digital and can be easily integrated into an IEEE 1149.1 interface. In this chapter, the built-in test reuses the charge pump and the divide-by-N counter of the PLL in order to generate a defect-oriented test approach, which can structurally verify the PLL. While [117] can also be implemented as a built-off test, [118] is limited to built-in testing since a multiplexer must be inserted into the delay sensitive path between the phase detector and the charge pump. Since both examples employ all-digital test circuitry, their application is limited to a few analog components like PLLs, where digital control is possible.

The works of [107–108,110–111] attempt to implement simple on-chip signal generators and on-chip test response data capture techniques for testing the performance of high-frequency analog circuits. The communication between the built-in test hardware and the external world takes place through a low-frequency digital channel. In particular, [110] measures the spectral content of the test response using direct down-conversion of RF test stimuli and test response waveforms. Although the chip-area taken by additional test circuitry is still a concern, it shows the feasibility of using built-in testing for measuring the performance of high-frequency embedded analog/RF blocks in situ.

With regard to built-off testing, direct measurement techniques for different classes of analog circuits are discussed in [2]. The circuitry for measuring one test specification is reconfigured to measure another using a set of relays and switches on the DUT load board. Typically, the load board test circuitry is designed with the DUT designer's input, unlike the method presented in [115], and takes several weeks to debug.

Although the direct measurement procedures are conceptually simple, this approach has inherent drawbacks as described below:

- Multiple specification measurements require different kind of resources, which are difficult to build either "on-chip" or on the load board due to high area overhead.
- A longer overall test time is required since measurement of multiple specifications cannot be performed simultaneously.

As a result, direct measurement techniques are not suitable for built-in testing as test resource requirements are very high, associated built-in test hardware overhead costs are prohibitive, and the time necessary for testing each specification separately increases the overall manufacturing cost.

12.4.4 Alternate Testing Methods for Mixed-Signal and RF Circuits

Alternate Test Basics
As the cost of conventional testing remains a prohibitive factor for testing of analog and RF circuits, the concept of alternate testing was proposed in [69–71]. The underlying principle of alternate test, also discussed in [67–68] in a different context, is described below. The variation of any process or circuit parameter, such as the width of an FET or value of a resistor, in the process or circuit parameter space P affects the circuit specification S by a corresponding sensitivity factor. If M is the space of measurements (for example, amplitude values of subsystem output spectrum) made on the circuit under test, the variation in the parameters also affects the measurement data in the measurement space M of the circuit by a corresponding sensitivity factor. Figure 12.29

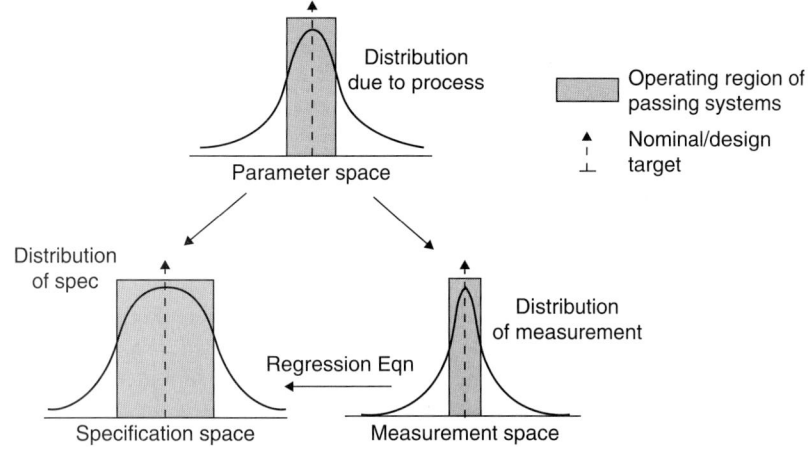

FIGURE 12.29 Variation in process or circuit parameter and its effect on circuit specification and test response measurement.

illustrates the effect of a variation of one such parameter in P on the specification S and the corresponding variation of a particular measurement data in M. Given the parameter space P, for any point in P, a mapping function (nonlinear) onto the specification space S, $f: P \to S$, can be computed. Similarly, for the same point, another mapping function (nonlinear) onto the measurement space in M, $f: P \to M$, can be computed. Therefore, for a region of acceptance in the circuit specification space, there exists a corresponding allowable "acceptable" region of variation of parameters in the parameter space. This in turn defines a region of acceptance of the measurement data in the space M. A circuit can be declared faulty if the measurement data lies outside the acceptance region in M.

Alternatively, as shown in [70–71], a mapping function $f: M \to S$ can be constructed for the circuit specifications S from all the measurements in the measurement space M using nonlinear statistical multivariate regression. Given the existence of the regression model for S, an unknown specification of a DUT can be predicted from the measured data. In the proposed alternate test approaches, multivariate adaptive regression splines (MARS) [114] were used to construct the regression models [92–93] and estimate the test specifications of the subsystem from the frequency spectrum of the test response waveforms. The objective of the alternate test methodology is to find a suitable transient test stimulus and to predict circuit specifications accurately from alternate test responses. Different type of test stimuli, namely, (1) piecewise linear [93], (2) multitone sinusoids [112, 119], and (3) digital pulse trains [106], are used in different cases. The methods are used successfully to test op-amps [93], low-frequency filter circuits [93], and high-frequency RF modules [112]. In particular, using a digital pulse train generated from linear feedback shift registers (LFSR), as in [106] and [109], facilitated built-in test approaches due its low area overhead. In fault diagnosis, which is another key problem in testing SOPs, symbolic formulation and analysis [76–77, 81–83] are usually not possible. However, alternate test generation based diagnosis schemes demonstrated in [78–80] can be applied to solve the fault diagnosis problem.

BIT and BOT Examples

Recent literature addresses applications of alternate testing to RF components. In [113], the load board modulates the baseband test stimulus provided by a low-cost tester and uses the resultant RF signal to stimulate a low-noise amplifier (LNA). The response is down-converted on the load board and lowpass-filtered to generate a signature that can be transferred to the tester through a low-bandwidth channel and analyzed using alternate test principles. The application follows the generic modulator based built-off test scheme in Figure 12.30. An alternative to this scheme is using a simple signal generator that can be implemented on the load board. Reference [119] describes an alternate test generation methodology that seeks an optimal superposition of sinusoids. Simulation results suggest a single sinusoidal, which has two orders of magnitude smaller frequency than the nominal and can be used to excite an RF LNA. The response can be sampled with load board capabilities. A different alternate built-off test approach is reported in [120], which employs the bias control voltage of an RF power amplifier as the test stimulus and measures the bias current to predict critical component specifications like gain, noise figure, and power efficiency. The use of current measurements in response acquisition proposes a noninvasive alternative to voltage measurements in RF applications where tapping into sensitive nodes is prohibitive.

In [109], another version of the built-in test scheme is proposed, which deviates from the "self-test" paradigm in order to minimize additional test-related hardware placed inside the chip and to reuse the existing test hardware already present for testing the digital section of the system IC. Unlike built-in test schemes discussed in [103–105], the DUT response is analyzed externally inside a low-cost ATE. Since the DUT test response waveform is transformed to a digital bit-stream and scanned out through the scan chains of the digital cores, the approach can be integrated with an IEEE 1149.1 based scan structure. Hence, the proposed technique attempts to solve the limited test access problem for embedded analog modules in system ICs to a large extent and can be used for testing the embedded passives at the assembled level as opposed to the substrate level. The test response waveform is reconstructed for analysis in the external tester, and from the reconstructed test response waveform, the DUT's specifications are predicted using the regression analysis discussed in alternate test.

In another kind of built-in test approach, the circuit topology is changed using additional circuit elements to make the circuit behave differently from what it is designed for, and this modified functionality is usually easy to measure in the production test environment. The catastrophic faults that make the original circuit performance fail also causes the reconfigured circuit "performance" to deviate. The latter performance deviation is measured during the production test, and pass/fail decisions for the original circuit are made. Oscillation-based tests (OBT) [87–89] act on the above principle, which reconfigures analog filter circuits into oscillators using additional feedback components. This built-in test technique detects catastrophic faults in DUT by measuring the deviation in oscillation frequency and amplitude. In recent years, the

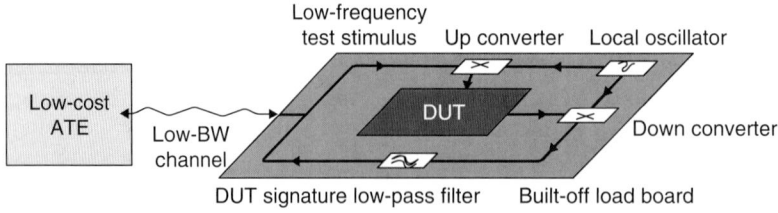

FIGURE 12.30 Modulator-demodulator–based built-off test scheme.

above defect-oriented built-in test technique has been integrated with the regression modeling approach commonly used in alternate test and the modified OBT is used for predicting the specification of DUT under parametric failure conditions. The modified technique relies on the fact that the original circuit and the reconfigured circuit share almost all the circuit components, and, hence, a direct correlation between the original circuit performance and the modified circuit performance (the latter performance is not a design goal) values can be established when the circuit parameters vary. This correlation is computed using circuit simulation under parametric variations using regression analysis previously used in alternate testing. The modified OBT, referred to as predictive oscillation based testing (POBT) [90], predicts the performance of the original circuit by computing the above correlation and measuring the oscillation frequency of the modified circuit during test. One inherent drawback in OBT approaches is that very few circuits other than analog filters can be reconfigured into oscillators.

A built-in response acquisition presents a significant problem in testing RF submodules. In mixed-signal environments with built-in ADCs, the analog response can be fed into the ADC and scanned out to the external tester in digital form after compaction. However, in RF systems, the inherent ADCs are configured to process near--baseband signals, so their performance is not adequate to process high-frequency passband responses. References [121] and [122] tackle this problem by introducing a statistical sampler that compares the analog response with noise. The power spectral density (PSD) of the resultant digital bit stream is a representation of the original PSD with an increased noise floor. Reference [123] extends this methodology by an automatic feature extraction scheme that detects the PSD components above the noise floor introduced by the statistical comparator and uses these components with a nonlinear mapping model to predict device specifications like gain, third-order intercept point, noise figure and power supply rejection ratio. This scheme, given in Figure 12.31, also presents an extension to the alternate test methodology in the sense that the scheme can compensate imperfect tester conditions simulated with a random fluctuation superposed onto the ideal input stimulus.

Direct Measurement versus Alternate Testing

As discussed in Section 12.1 SOP requires nonorthodox test methodologies that can keep up with the test access problems amplified by the inherent integration, and do away with the prohibitive cost of high-end external testers. Built-off and built-in test strategies propose a solution to these problems by placing high-bandwidth test access either next to the package or within the package. The SOP test challenge also calls for automated test solutions, which are generic enough to cut down custom-test support development cost. This requirement ensures that the turnaround time associated with test generation and test hardware development does not have a significant impact on the device manufacturing cost. Although different direct measurement based test approaches reviewed in Section 12.4.3 propose promising results for stand-alone devices, *their application at a system scale is not feasible due to the need for custom test-support hardware for every embedded module to be tested*. Since they do not provide a generic methodology to handle direct measurement of different specifications, each specification to be tested increases the overall turnaround time for product development, as well as increasing the test area overhead and testing time for every device.

On the other hand, alternate test methodologies reviewed in Section 12.4.4 propose generic solutions for embedded analog and RF components, which cover a large range

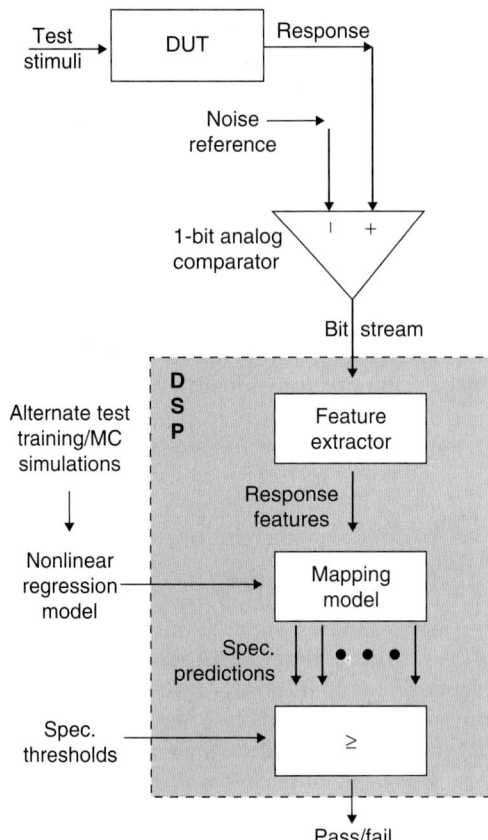

Figure 12.31 Noise-referenced, feature extraction based built-in test scheme.

of system components available in SOP, namely, embedded passives, op-amps, filters, LNAs, mixers, power amplifiers, and others. The ability to predict multiple specifications using a single test reduces the test hardware complicity, area overhead, and testing time. Furthermore, statistical sampler based extensions are compatible with applications utilizing digital scan architecture (IEEE 1149.1), since the resultant bit stream can be relayed to the digital signal processors in the package at no additional cost.

Since alternate built-off test methodologies propose a systematic way to handle a large range of specifications and submodules, their integration into the product flow does not increase the complexity and cost significantly. Although the load board will be populated with extra components to accommodate test-related signal processors, the increase in board design time can be compensated with automation already present in the traditional load board manufacturing flow [115]. The manufacturing cost of traditional load boards is dominated by the quality of the material, the many levels of power planes provided, and the necessity to use only golden boards, boards that very closely follow the specifications of the original board design. In the case of built-off testing, the extra cost of signal processor ICs, their routing and assembly will not be significant when compared to the traditional load board figures. Furthermore, the use of built-off testing will benefit from low-end ATEs, which provide two orders of cost reduction compared to high-end ATEs necessary for traditional tests [124–126]. This reduction is still one order greater than the manufacturing cost of many typical complex load boards. The only practical limit for load board complexity is the fixed area dictated by the interface of the production testing equipment. *It is important that when it comes to testing of very complex systems like SOPs, one of the main problems is feasibility rather than cost [127–128].* As discussed in Section 12.1, even with high-end ATEs, it is not possible to address bandwidth requirements of such systems. Section 12.3.2 further elaborated on feasibility using examples from multi-gigahertz digital testing.

A viable SOP strategy using alternate tests should generate specification-oriented tests considering only the component specifications that progressively develop a violation at the system level. The first step will be analyzing the system specifications to break them down into related component specifications. This process is usually a part of the system design; hence, it will not induce further effort. Then, all related specifications can be tested by a single alternate test per component. Some system-level specifications that cannot be verified by a collection of individual component performances will further be covered by system-level alternate tests. Reference [112] presents an example to this scheme, where system-level specifications of the RF subsystem of a narrowband wireless transceiver (Figure 12.33) are verified by alternate testing generated on high-level models of the system. In this example, multitone sinusoid stimulus is optimized for the test and the test response spectrum is measured for specification prediction. Multiple system specifications, such as gain and IIP3 of the RF subsystem, are simultaneously and accurately predicted using statistical regression. The prediction error in measuring end-to-end specifications is significantly small (within ±1 dB) for these high-frequency complex subsystems. High-level modeling speeds up the simulation intensive features of alternate testing, which are not feasible for SOPs at the netlist level. Furthermore, the inherent complexity of SOPs makes built-in approaches more favorable than built-off test solutions. On the other hand, a joint built-in/built-off approach can add more value to the package area, where module level access and DSP is handled by built-in tester components while more area intensive tester functions like analog signal generators and modulators-demodulators can be migrated onto the load board.

Alternate Test Development Flow
Alternate testing provides a framework for high-volume manufacturing tests of components and systems that are evaluated by analog specification boundaries. This framework takes many diverse forms in implementation depending on test benchmarks and specific requirements of the DUT. Although it is not possible to cover all different implementations in this section, the references provided supply extensive coverage. Different applications share a common flow with greater emphasis on different stages. Figure 12.32 shows alternate test development flow and its application to DUT in HVM. First, the optimization space is defined by the stimulus range and the available measurement equipment. Possible DFT features are also considered at this stage if it is still possible to make an impact on the design. Second, optimization models are created. These models may be netlist level if the IP is available and the simulation time is not a bottleneck. However, for complex systems such as SOP, only high-level models are feasible. These models need to capture enough information to represent expected process variations and facilitate efficient simulation for test optimization. Once these models are created, a sample set with process variations is generated by Monte Carlo analysis or statistical design of experiments (DOE). In the fourth step, the stimulus, signature

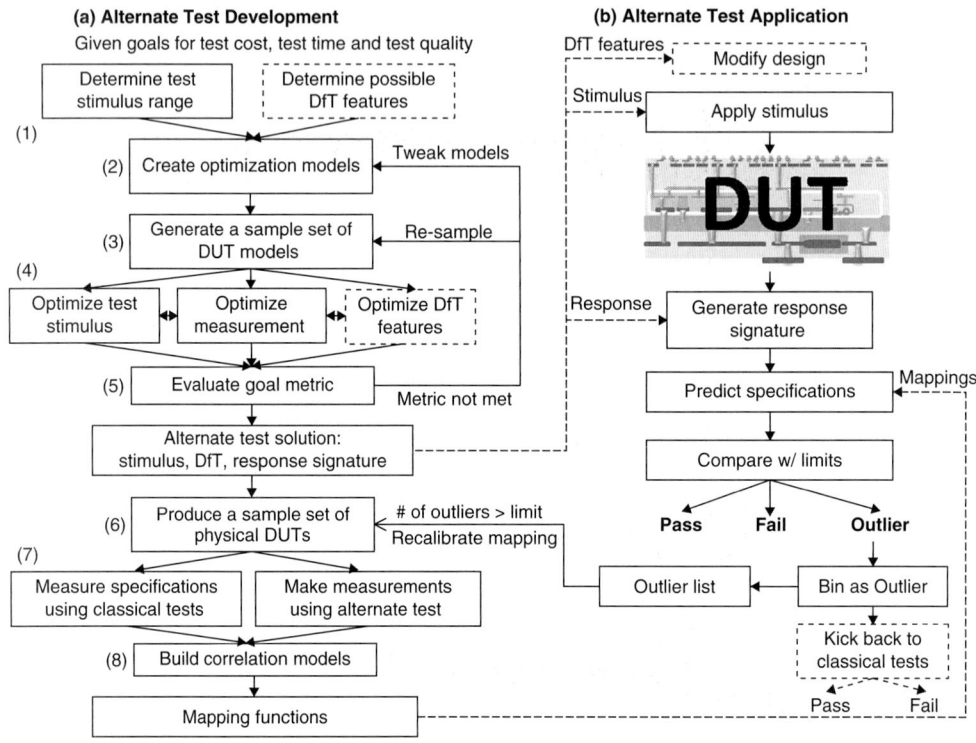

FIGURE 12.32 (a) Alternate test development flow. (b) Alternate test application.

extraction, and, if applicable, DFT features are cooptimized to fit the test envelope defined by test cost, test time, and test quality metrics. The optimization step is usually iterative and makes use of modern techniques such as genetic algorithms or response surface methods that are more suitable for complicated, nonlinear, partially continuous domains typical for test generation. If the optimization does not converge or converges to a suboptimal metric as measured in the fifth step, then the models may be tweaked to better capture the problematic response variables and the sample set can be extended. Steps 2 through 5 are iterated until an alternate solution composed of stimulus, measurement equipment, response signature generation algorithms, and possible DFT features are delivered to satisfy the goal metric. In step 7, both classical specification tests and the optimized alternate tests are performed on a sample set of DUT hardware. Next, correlation models are generated that map response signatures from alternate testing to the actual specification values measured by classical tests. This step usually makes use of supervised learning techniques (the accepted norm is MARS: multivariate adaptive regression splines), but for simple components curve fitting works as well. In HVM (right side of Figure 12.32), alternate stimulus is applied to DUT and the measurements are converted into a response signature. Then the mapping functions from step 8 are used to predict specification values. These values are tested against pass/fail limits unless the response does not fall into the expected envelope defined by the training set in step 6. The result is either a pass/fail decision or the DUT is marked as an outlier. Outliers may be kicked back to classical specification testing for the expense of testing time and tester cost. Once the number of outliers exceeds a certain limit, the mapping functions are recalibrated with an improved sample set including these elements.

In the earlier days of alternate testing, test time reduction was the key focus for stand-alone analog components. The test stimulus optimization cuts down switching time and repeated measurements, trims down the cost associated with testing, and hence creates a competitive advantage. In this context, alternate testing is employed later in the product life cycle and it is possible to use hardware samples for optimization rather than simulation models. The main focus in such applications is defining a cost-effective stimulus domain (step 1) and recalibration (steps 6 to 8).

Later, alternate test focus shifted to systems and end-to-end specifications rather than components. In this case, the bottleneck is step 3; without efficient simulation models either the optimization time is not feasible or the stimulus domain needs to be strictly limited in range.

Recently, alternate testing has been employed as an enabler technology. It makes SOP production possible for acceptable test coverage with the use of embedded sensors and feature extractors. The focus shifted to BIT and DFT features, mainly cooptimization of stimulus and measurement while exploring possible feature extractors. These applications are being extended to postmanufacturing adaptation to account for yield loss from specification failures and real-time power/performance optimizations throughout the product lifetime.

In the following subsections, two examples are selected to emphasize different implementations. The first example is an RF receiver, which demonstrates the use of behavioral models to solve a simulation bottleneck in step 4 of Figure 12.32. It provides significant test cost reduction in terms of ATE requirements. The second example demonstrates test time reduction on an analog-to-digital converter. Mixed-signal applications are especially suitable for high-level models; however, most of the optimization in this example is performed with hardware samples. Later, in the section entitled Sensor-Based Testing of High-Speed Devices, we will briefly discuss DFT features and near dc-level feature extractors for RF systems. Finally the last subsection shows examples for test cost reduction.

Alternate Test of RF Subsystems: Block-Level Receiver Example

In this subsection, a case study is discussed using the receive channel of a transceiver as the DUT, on which a block-level test generation and test validation method have been applied [112]. The circuit component R, L, and C values and sizes of the transistors of different submodules are varied to form the parameter space for the transistor-level description of the system shown in Figure 12.33. In this experiment, the tolerance in the parameter values is assumed to be 10 percent around their respective design values. System gain and system IIP3 are chosen as the specifications of interest. The nominal value of the system gain is 22 dB and for system IIP3 is −12 dB for a −10 dBm input level.

The block-level test generation process makes use of behavioral models instead of transistor-level netlists because of two reasons:

1. Any iterative and deterministic test generation technique (in contrast to a random or pseudorandom test technique) requires repeated simulation of the DUT. Although the use of transistor-level simulations for all the submodules yields high accuracy of simulation, the long simulation time makes the transistor-level system simulation impractical.

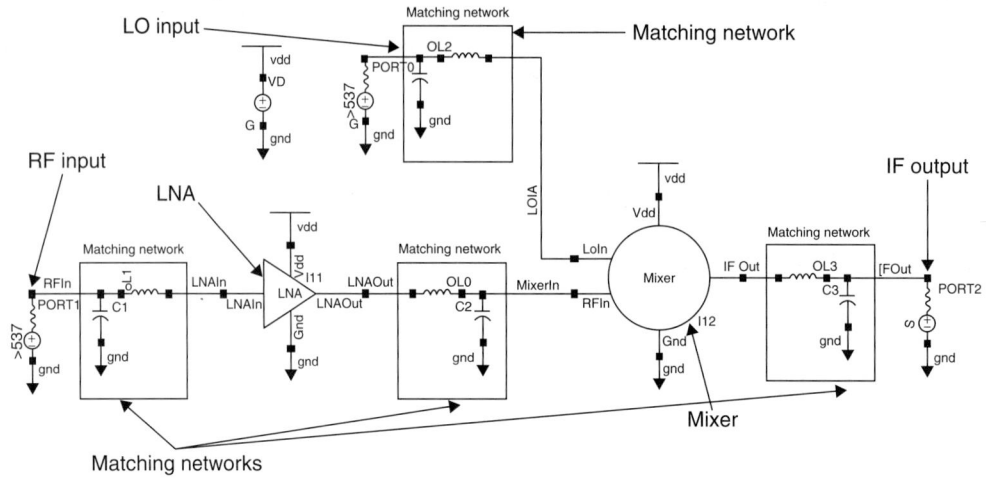

FIGURE 12.33 Block-level diagram of the RF subsystem under test.

2. The primary objective of test generation is to determine the optimal set of test stimuli, which is not a circuit design goal, rather than to verify the functionality of the design. Hence, the loss of accuracy in simulation data while using behavioral models does not hinder the search algorithm as long as the statistical trend in simulation data is maintained under parameter variations.

An RF subsystem in a superheterodyne narrowband wireless RF transceiver architecture contains amplifiers, filters, mixers, and frequency synthesizers operating in a certain range of frequency. The behavioral simulation engine is developed in MATLAB; each submodule in Figure 12.34 is discussed below.

Filter For test generation purposes, the bandpass filters of the RF subsystem are realized as linear transfer functions with different gains at different frequencies. The output of the filter is given by

$$Y(f) = H(f) \cdot \text{diag}(X(f)) \tag{12.9}$$

where f is the frequency of operation.

The different gain values corresponding to different frequencies are characterized by the center frequency, filter-Q, and frequency roll-off. H, X, and Y are complex quantities representing amplitude and phase values together.

Amplifier The amplifiers of the RF subsystem are realized by implementing a nonlinear transfer function of the third order as

$$y(t) = \alpha_0 + \alpha_1 x(t) + \alpha_2 x_2(t) + \alpha_3 x_3(t) \tag{12.10}$$

The coefficients α_0 and α_1 represent dc offset and small signal gain, respectively, while α_2 and α_3 are nonlinearity coefficients such as harmonics and intermodulation terms.

Mixer For test generation purposes, the mixer of the RF subsystem is modeled as a nonlinear transfer function followed by an ideal multiplier. The nonlinear transfer

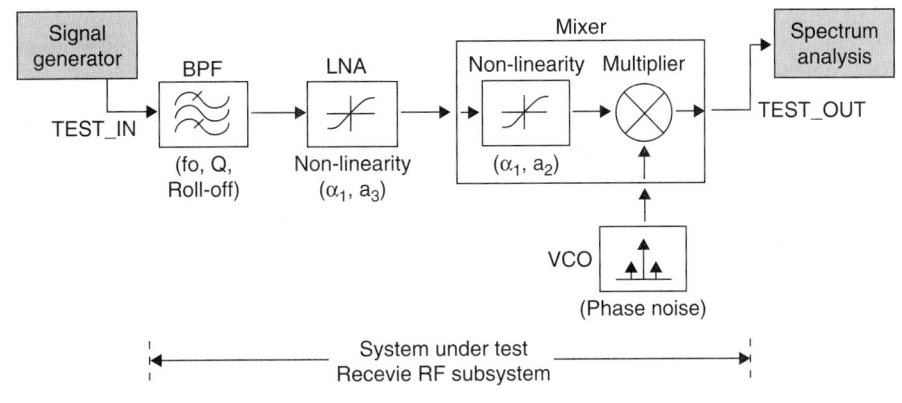

FIGURE 12.34 Behavioral model of the receiver.

function is realized in the same manner as is done for the amplifier. The frequency mixing operation is realized by the multiplication operation.

Oscillator The behavioral model of the frequency synthesizer or oscillator is realized as a set of amplitude values $X(f)$ corresponding to different frequencies. The peak amplitude value corresponds to the local oscillator frequency; the amplitudes adjacent to the frequencies fall off according to the values calculated from the phase-noise of local oscillators.

Once the behavioral models are generated, an iterative greedy algorithm is used to select the parameters of the test stimulus, which is used to perform the specification test of RF subsystems. The goal of the test generation algorithm is to determine the optimal test stimulus waveform and the corresponding test response spectrum from which the specifications of the DUT can be predicted as accurately as possible. Finally, the test stimulus generated by the algorithm is validated against the transistor-level parameter perturbations using a low-level simulator. The resulting test stimulus in this example has two tones, one at 1.5 GHz of 1.2 mV and the other at 1.36 GHz of 1.0 mV. Noise arising from the devices and parasitic elements present in the system was ignored during test generation using behavioral simulations. The test validation process uses transistor-level simulation of the system and hence inherently takes into account the noise present in the system.

In this case study, every iteration during test generation for the RF subsystem using behavioral models for a two-tone test stimulus takes ~1 minute, whereas the generation of the regression models from transistor-level simulations, which is equivalent to one iteration for the test generation algorithm, takes ~10 hours. Hence, the approach presented here shows a significant reduction in test generation time by trading off simulation accuracy in the initial phase of test generation.

Figure 12.35 shows the tracking of the system gain and system IIP3 specifications. For a two-tone test, the proposed approach predicts the system gain and system-IIP3 simultaneously, and it is possible to distinguish between faulty and fault-free circuits if the actual specification values do not lie within the prediction error margins around the upper and lower limits of their respective acceptance region. The results are summarized in Table 12.3.

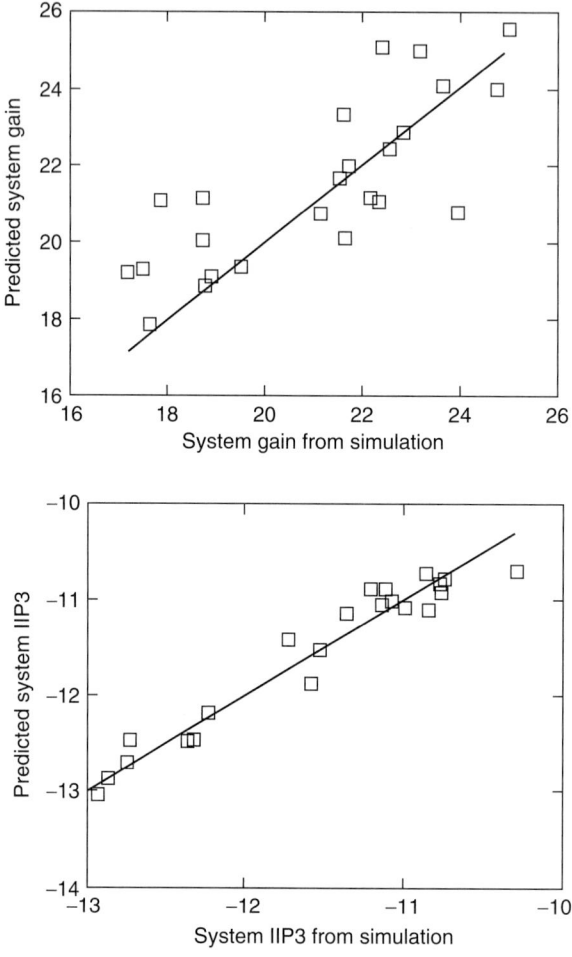

FIGURE 12.35 Receiver gain and IIP3 simulation versus predicted from alternate test.

Alternate Test of Mixed-Signal Subsystems: ADC Example

Recently, there has been a rapid increase in the speed and accuracy of data conversion circuits due to their constant need in direct or high-IF communication systems. Apart from being a design challenge, production testing of high-performance data converters has become a huge challenge for the engineers. High-volume testing of ADCs for dynamic specifications requires faster and more accurate tester circuitry than before. However, using high-speed test equipment for production testing increases the test cost

System Specification	Nominal Value	Max. Error in Spec. Prediction
Gain	22 dB	±3.0% (±0.7 dB)
IIP3	−12 dB	±0.4% (±0.5 dB)

TABLE 12.3 Summary of the Simulation Results for the RF Receiver

significantly. In this example [132], we study an alternate testing based methodology for testing high-speed ADCs using low-speed test resources available on a low-cost tester. The key points of this example are listed here:

1. The method can measure the dynamic specifications of high-speed ADCs using a tester that runs at a comparatively lower speed than the DUT.
2. It can measure the dynamic specifications of the data converters in the presence of nonidealities of test instruments, such as clock jitter and input signal noise.

Dynamic specifications measure the high-frequency nonlinear behavior of ADCs [129–131]. Typical dynamic specifications include signal-to-noise ratio (SNR), spurious-free dynamic range (SFDR), total harmonic distortion (THD), and second and third harmonic power. For testing these specifications, usually a spectrally clean, low phase noise, sinusoidal test stimulus is applied to the ADC. The device is clocked using a low-jitter clock, and the frequency spectrum of the ADC output is computed to measure the specifications of interest (Figure 12.36).

The test stimulus frequency is placed close to the device's maximum rated input frequency. The exact test stimulus frequency is chosen based on the coherent sampling condition in Equation (12.11), where Fin is the input signal frequency, Fsample is the sampling clock frequency, Nwindow is the integer number of cycles within the sampling window, and Nrecord is a power of 2 to enable the use of a radix 2 Fast Fourier Transform (FFT). Nwindow and Nrecord are relatively prime numbers. Coherent sampling is needed to prevent power leakage into the adjacent frequency bins while constructing the frequency spectrum using FFT. This makes it easy to measure the worst-case, nonlinear specifications. However, to do so, the tester must have high-frequency resources.

$$\frac{F_{in}}{F_{sample}} = \frac{N_{window}}{N_{record}} \tag{12.11}$$

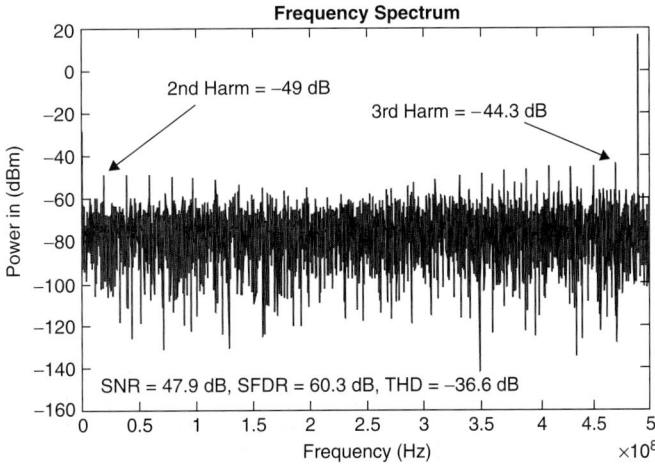

FIGURE 12.36 Frequency spectrum of ADC output.

ADCs are generally characterized by resolution and sampling speed. The resolution of the ADCs ranges from 8 to 24 bits and the sampling speed from 10 samples/s—1 gigasample/s (Gsample/s). Testing issues in the ADCs depend on the resolution and the sampling speed of the device. Sampling clock jitter is a major problem for testing ADCs having medium resolution (10 to 14 bits) and medium sampling speed (5 to 250 Msamples/s). The SNR specification of these converters generally ranges from 60 to 75 dB. To measure the SNR specification using a high-frequency test stimulus, a very low jitter clock is needed. SNR performance of an ADC is given by Equation (12.12), assuming quantization noise is negligible. In the equation, σ_{clk} is the RMS jitter in the sampling clock, σ_{int} is the RMS value of the internal ADC jitter, V_{in} is the RMS noise in the input signal, V_{ts} is the RMS test setup noise, and F_{in} is the input signal frequency.

$$\text{SNR(dB)} = 10\log\left(\frac{A^2}{4\pi^2 A^2 F_{in}^2 (\sigma_{clk}^2 + \sigma_{int}^2) + V_{in}^2 + V_{ts}^2}\right) \quad (12.12)$$

If the sampling clock jitter is assumed to be the only dominant noise source with RMS jitter equal to 2 ps, the SNR measured using Equation (12.12) is equal to 52 dB. Hence, a sampling clock with RMS jitter lesser than 2 ps is required to accurately measure the performance of this ADC. SNR degradation due to the presence of jitter in the sampling clock is shown in Figure 12.37 for a 12-bit ADC, with ideal SNR equal to 74 dB; the measured SNR is 48 dB. The input frequency is 60 MHz, and the RMS jitter in clock is assumed to be 10 ps in this example.

High-speed converters usually have low data resolution, less than 10-bit, and high sampling speed, more than 200 Msamples/s. The test problem for high-speed data converters is usually related to *generation and capture of high-frequency signals*. Test platforms with high-bandwidth resources are needed to test such ADCs. However,

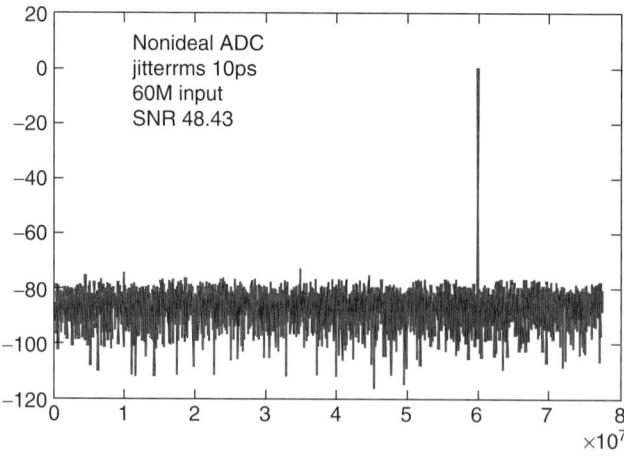

Figure 12.37 SNR measured using a high-jitter sampling clock.

such test systems are expensive. An alternate test methodology can ease the tester resource requirements in production testing of high-speed ADCs.

First, using conventional methodology and high-performance test equipment, dynamic specifications for a set of N devices are measured. A selected test stimulus is then applied to the same set of devices using the low-cost test setup shown in Figure 12.38. The output of these devices is undersampled and stored for building the models. This is called the signature of the device. Based on an alternate test method, a set of prediction models is then generated.

During production test, the devices are tested using this low-cost test setup. The signatures of the devices are measured and the prediction models are used to estimate the dynamic specifications. The low-cost test setup uses an onboard mixer (Figure 12.38). The mixer specifications are given in Table 12.4. A low-frequency test stimulus is sourced from the tester and is up-converted using the onboard mixer. The local oscillator (LO) frequency is generated from the tester to avoid synchronization issues such that the input signal, ADC sampling clock, and output sampling clock have to be synchronized. An external source is used to provide a high-speed clock to the DUT, which is synchronized to the tester. The output of the high-speed ADC is undersampled at a lower frequency $F_{us,}$ as shown in Equation (12.13).

$$F_{us} = \frac{F_s}{n} < F_{max} \qquad n = 2, 4, 8, \ldots, 2^l \qquad (12.13)$$

where F_s is the ADC sampling frequency and F_{max} is the maximum tester sampling frequency.

An ADC is generally tested at the maximum rated input frequency to measure the worst-case specifications. An up-conversion mixer is used to generate the high-frequency input tone from the low-frequency resources available in a low-cost tester. The input to the mixer is a sinusoidal signal generated by the tester. For easier synchronization, the LO is also generated using the low-cost tester. The tones are generated such that the up-converted tone falls at the frequency at which the dynamic specifications of the device need to be measured.

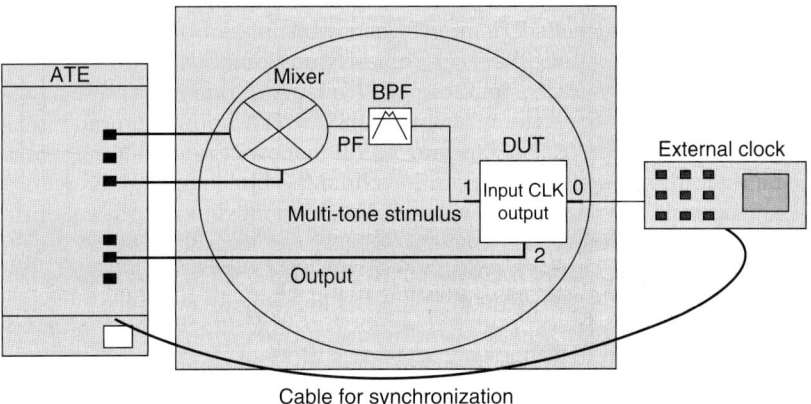

FIGURE 12.38   1Single-mixer low-cost test setup.

Conversion Factor	IIP3	LO Leakage	NF
4.5 dB	24 dBm	−23 dB	10.5 dB

TABLE 12.4 Specifications of the Simulated Mixer

The relation between different frequency tones is given by

$$\omega_{OUT} = \omega_{IF} + \omega_{LO} \tag{12.14}$$

where ω_{IF} is the frequency of the IF tone, ω_{LO} is the frequency of the LO tone, and ω_{OUT} is the frequency at which the dynamic specifications of the device needs to be measured. If an input tone more than two times the maximum tester frequency is needed, a series of mixers can be used to up-convert the output of the first mixer.

A set of 100 instances of the DUT is simulated to validate the proposed approach, of which 60 are used to generate the models and the rest are used to validate the models. The components are modeled using behavioral-level models in Matlab. Code widths were randomly varied from the ideal value of one LSB to insert nonidealities in the device. For an input frequency of 490.11 MHz, the frequencies of the sinusoidal IF and LO signals were 240.11 MHz (ω_{IF}) and 250 MHz (ω_{LO}), respectively; the second and third harmonics present in these signals are less than −60 dBc. The frequency spectrum of the test stimulus at the output of the mixer is shown in Figure 12.39. The sampling clock jitter is assumed to have a gaussian distribution with zero mean and 10 ps standard deviation. The output of the ADC is undersampled at a frequency of 250 Msamples/s and stored for constructing the models.

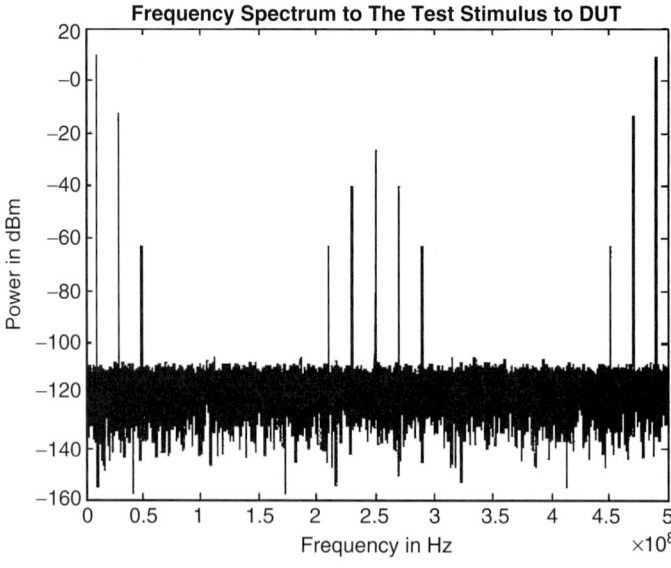

FIGURE 12.39 Frequency spectrum of the test stimulus at the output of the mixer.

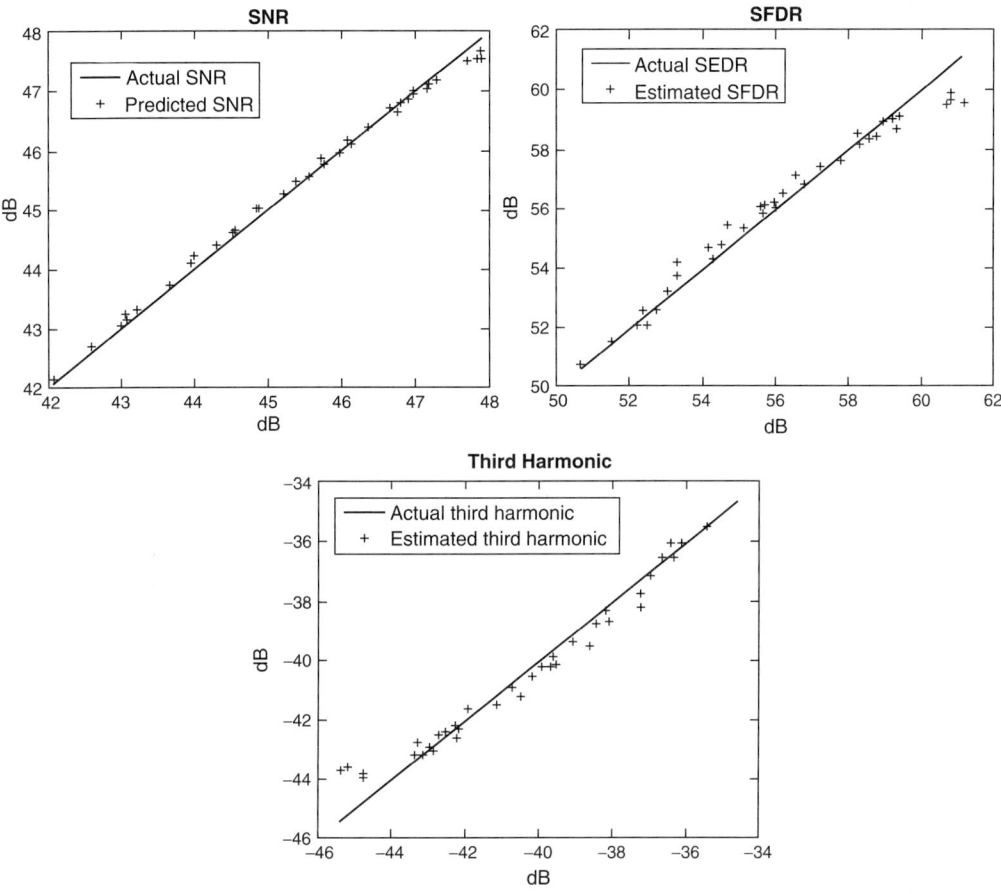

FIGURE 12.40   SNR, SFDR, and THD estimation results.

The predicted dynamic specifications of SNR, SFDR, and THD are shown in Figure 12.40 for the set of validation devices, where a 45° line shows perfect correlation. The maximum and the average error in estimation of the specifications using the proposed approach are shown in Table 12.5.

	Mean Error	Max. Error
SNR	0.37	0.70
SFDR	0.70	1.6
THD	0.72	1.69
Second harmonic	1.08	2.38

TABLE 12.5   Prediction Errors for Single-Mixer Approach

Sensor-Based Testing of High-Speed Devices

Efficient testing of high-speed devices can be performed through embedding low-cost sensors into signal paths for the purpose of built-in test [133–134]. The sensor characteristics are chosen in such a way that the low-frequency or dc signals at sensor outputs are tightly correlated with the target test specification values of the DUT. Hence, instead of testing the devices specifically for complex performance metrics, the outputs of the sensors are used to accurately estimate the target test specification values when the device-under-test is stimulated with sinusoidal stimulus. Instead of sampling the test response, which is difficult to perform on-chip for high-frequency signals, *sensors embedded in the RF signal path are used to "encode" the test response into a low-frequency or dc signal* that can be analyzed using on-chip digital signal processing or off-chip external test equipment and used to predict the target test specifications of interest. This alternate test based approach significantly affects the cost of manufacturing test and allows testing to be performed using low-cost external testers. Using this method, the target test specification values can be estimated with an accuracy of ±5 percent of their actual value.

There are three key steps: (1) for the set of target test specification values to be measured, identify the "best" set of sensors to be used and the "best" set of circuit nodes for sensor insertion, (2) determine the optimal test stimulus to be used for testing, and (3) map the sensor output data to the target test specification values of the DUT.

Because of the use of on-chip sensors that are "designed-in" for test purposes, the problems of high-frequency external tester access and associated signal integrity issues are not present. Low-cost test instrumentation is used for test stimulus capture and analysis, as the sensor output is usually a dc or low-frequency signal. An on-chip digital signal processor can easily also perform this analysis. In addition, since the sensors perform analog processing of high-frequency signals, the test approach is directly scalable to multi-gigahertz electronics. Furthermore, the sensors can be selectively placed at internal test nodes depending on the level of diagnostic granularity desired. That is, a larger number of sensors can be used to provide test data for individual modules as opposed to the system or subsystems. The data obtained from these sensors can be used to directly predict the target test specification values accurately.

There is a key challenge associated with using sensors for BIT: the sensors are affected by the same process variations that affect the performance of the DUT. There are two ways to accommodate these effects:

In general, the degradation of sensor performance is correlated to the degradation in DUT performance due to *systematic* process variations. This can be accounted for to a large degree by the algorithms that determine the target test specification values from the observed sensor output values [133].

A sensor calibration test procedure may be performed before test application. The results of this test along with the results from the rest of the test procedure are then used to determine the target test specification values.

Test Cost Reduction Examples

Several case studies emphasize significant test cost reduction from alternate testing. The test stimulus for a 2-GHz CDMA/TDMA low-noise amplifier was optimized by Ardext in 2004 on an Agilent 8400 tester. The frequency-domain test stimulus reduced test time by half for measuring bias, gain, IIP3, and noise Figure. Another study by Texas Instruments shows that test time is reduced to one-third using an optimized transient alternate test stimulus for a precision op-amp. Both the guardbands and repeatability were better than the original specification test. Another case study for the LM741 amplifier shows that the

	ATE Spec-Based	Dedicated BIT	BOT	Alternate Test
Frequency	High	Medium	High	High
Total test time	Large	Large	Medium	Small
Equipment cost	Large	Small	Medium	Small

TABLE 12.6 Comparison of KGEM Testing of Mixed-Signal and RF Subsystems

original test equipment cost of $200 to $300K can be reduced to less than $10K for a stack-and-rack configuration with a testing time of one-third the original. A joint experiment by Georgia Tech and Intel demonstrates alternate testing of an RF LNA on a mixed-signal tester. The tester cost is reduced to one-tenth with the test time reduction of one-seventh. Table 12.6 shows a comparison of KGEM methodologies.

12.5 Summary

In this chapter, we have discussed known good embedded substrate and known good embedded module testing methods in detail with examples and references. We introduced the concepts of SOP testing at the design, characterization, and high-volume manufacturing stages, providing insight into testing for embedded passives, substrate interconnections, and various modules with digital, analog, and RF functions. The cost of high-volume manufacturing testing for analog or mixed-signal and RF modules is highlighted as the dominant factor in the overall system test, and built-in strategies with alternate testing methods are proposed as solutions. Alternate test methodology makes use of a single-test stimulus and response capture to predict all the performance specifications at once. Therefore, it is a scalable solution with feasible extensions for built-in testing, which is necessary for SOP because of its test challenges in observability and controllability.

Acknowledgments

We would like to acknowledge the following individuals for their contribution to the technology represented in this chapter: Achintya Halder, Shalabh Goyal, Ganesh Srinivasan, Sasi Cherubal, Ram Voorakaranam, Pramodchandran N. Variyam, Heebyung Yoon, Junwei Hou, and Madhavan Swaminathan.

References

1. M. Soma et al., *Analog and Mixed Signal Test*, Prentice Hall, NJ, 1998.
2. M. Burns and G. W. Roberts, *An Introduction to Mixed-Signal IC Test and Measurement*, Oxford University Press, 2001.
3. Rao R. Tummala, *Fundamentals of Microsystems Packaging*, New York: McGraw-Hill, 2001.
4. W. Maly, D. Feltham, A. Gattiker, M. Hobaugh, K. Backus, and M. Thomas, "Smart-substrate multichip-module systems," *IEEE Design & Test of Computers*, vol. 11, 1994, pp. 64–73.

5. Y. Zorian, "A structured testability approach for multi-chip modules based on BIST and boundary-scan," *IEEE Transactions on Advanced Packaging*, vol. 17, 1994, pp. 283–90.
6. L. Gilg, "Known good die: a closer look," *Advanced Packaging*, vol. 14, 2005, pp. 24–27.
7. D. Keezer, "Bare die testing and MCM probing techniques," *Proc. Multi-Chip Module Conference (MCMC-92)*, 1992, pp. 20–23.
8. M. Berry, "How advances in RF and radio SiP affect test strategies," *Advanced Packaging*, 2005.
9. D. Appello, P. Bernardi, M. Grosso, and M. Reorda, "System-in-package testing: problems and solutions," *IEEE Design & Test of Computers*, vol. 23, 2006, pp. 203–11.
10. S. Steps, "Full wafer test: making test more cost effective," *Proc. Known-Good Die Workshop*, 2005.
11. B. Davis, *The Economics of Automatic Testing*, New York: McGraw-Hill, 1982.
12. J. Marshall et al., "CAD-based net capacitance testing of unpopulated MCM substrate," *Advanced Packaging*, vol. 17, no. 1, February 1994.
13. L. Economikos et al., "Electrical test of multichip subtrates," *Advanced Packaging*, vol. 17, no. 1, February 1994.
14. H. Hamel et al., "Capacitance test technique for the MCM of the 90s," *Proc. Int. Electronic Packaging Conf*, September 1993, pp. 855–72.
15. R. W. Wedwick, "Continuity testing of capacitance," *Circuits Manufacturing*, November 1974, pp. 1–61.
16. M. Brunner et al., "Electron-beam MCM testing and probing," *Advanced Packaging*, vol. 17, no. 1, February 1994, pp. 62–68.
17. O. C. Woodard, "High density interconnect verification of unpopulated multichip modules," *Proc. IEEE Int. Electronics Manufacturing Technology Symposium*, 1991, pp. 434–39.
18. K. Lee et al. "Design, fabrication, and reliability assessment of embedded resistors and capacitors on multilayered organic substrates," *Proc. International Symposium on Advanced Packaging Materials: Processes*, 2005, pp. 249–55.
19. R. K. Ulrich and L. W. Schaper (eds.), *Integrated Passive Component Technology*, Wiley-IEEE Press, US, 2003.
20. M. Abadir, A. Parikh, P. Sandborn, K. Drake, and L. Bal, "Analyzing multichip module testing strategies," *IEEE Design & Test of Computers*, vol. 11, 1994, pp. 40–52.
21. B. Kim et al., "A novel test technique for MCM substrates," *IEEE Trans. on Components, Packaging, and Manufacturing Technology, Part B: Advanced Packaging*, vol. 20, 1997, pp. 2–12.
22. H. Yoon, "Fault detection and identification techniques for embedded analog circuits," PhD Thesis, Georgia Institute of Technology, 1998.
23. H. Yoon, J. Hou, S. K. Bhattacharya, A. Chatterjee, and M. Swaminathan, "Fault detection and automated fault diagnosis for embedded integrated electrical passives," *Journal of VLSI Signal Processing Systems for Signal, Image, and Video Technology*, vol. 21, 1999, pp. 265–76.
24. K. L. Choi, "Modeling and simulation of embedded passives using rational functions in multi-layered substrates," PhD Thesis, Georgia Institute of Technology, 1999.
25. K. Kornegay and K. Roy, "Integrated test solutions and test economics for MCMs," *Proc. International Test Conference*, 1995, pp. 193–201.
26. B. Kim and H. Choi, "A new test method for embedded passives in high density package substrates," *Proc. Electronic Components and Technology Conference*, 2001, pp. 1362–66.

27. H. Yoon, A. Chatterjee, M. Swaminathan, and J. L. A. Hughes, "Catastrophic fault diagnosis for embedded MCM RF-passives using single point probing," *Proc. MCM Test Advanced Technology Workshop,* Napa, CA, September 1997.
28. A. Sood, K. Choi, A. Haridass, N. Na, and M. Swaminathan, "Modeling and mixed signal simulation of embedded passive components in high performance packages," *Proc. International Multichip Modules and High Density Packaging Conference,* 1998, pp. 506–11.
29. J. Park, S. Bhattacharya, and M. Allen, "Fully integrated passives modules for filter applications using low temperature processes," *Proc. International Symposium on Microelectronics,* 1997, pp. 592–97.
30. J. Hou and A. Chatterjee, "Concurrent transient fault simulation for analog circuits," *IEEE Transactions on Computer-Aided Design of Integrated Circuits and Systems,* vol. 22, 2003, pp. 1385–98.
31. H. Ehrenberg, "PXI Express based JTAG / Boundary Scan ATE for structural board and system test," *IEEE Systems Readiness Technology Conference,* 2006, pp. 467–73.
32. R. W. Barr, C. Chiang, and E. L. Wallace, "End-to-end testing for boards and systems using boundary scan," *Proc. Int. Test Conference,* 2000, pp. 585–92.
33. F. de Jong, and A. Biewenga, "SiP-TAP: JTAG for SiP," *Proc. Int. Test Conference,* 2006, 10 pages.
34. J. Rearick, S. Patterson, and K. Dorner, "Integrating boundary scan into multi-GHz I/O circuitry," *Proc. Int. Test Conference,* 2004, pp. 560–66.
35. M. H. Tehranipour, N. Ahmed, and M. Nourani, "Testing SoC interconnects for signal integrity using boundary scan," *Proc. VLSI Test Symposium,* 2003, pp. 158–63.
36. S. Koppolu, L. Alkalai, and A. Chatterjee, "Testing NASA's 3D-stack MCM space flight computer," *IEEE Design & Test of Computers,* vol. 15, 1998, pp. 44–55.
37. M. Shimanouchi, "Periodic jitter injection with direct time synthesis by SPP™ ATE for SerDes jitter tolerance test in production," *Proc. International Test Conference,* 2003, pp. 48–57.
38. T. J. Yamaguchi, M. Soma, M. Ishida, M. Kurosawa, and H. Musha, "Effects of deterministic jitter in a cable on jitter tolerance measurements," *Proc. International Test Conference,* 2003, pp. 58–66.
39. H. C. Lin, K. Taylor, A. Chong, E. Chan, M. Soma, H. Haggag, J. Huard, and J. Braatz, "CMOS built-in test architecture for high-speed jitter measurement," *Proc. International Test Conference,* 2003, pp. 67–76.
40. J. S. Davis, D. C. Keezer, O. Liboiron-Ladouceur, and K. Bergman, "Application and demonstration of a digital test core: optoelectronic test bed and wafer-level prober," *Proc. International Test Conference,* 2003, pp. 166–74.
41. A. R. Syed, "RIC/DICMOS—multi-channel CMOS Formatter," *Proc. International Test Conference,* 2003, pp. 175–84.
42. M. Gavardoni, "Data flow within an open architecture tester," *Proc. International Test Conference,* 2003, pp. 185–90.
43. D. C. Keezer, D. Minier, and M. C. Caron, "A production-oriented multiplexing system for testing above 2.5 Gbps," *Proc. International Test Conference,* 2003, pp. 191–200.
44. K. Posse and G. Eide, "Key impediments to DFT-focused test and how to overcome them," *Proc. International Test Conference,* 2003, pp. 503–11.
45. G. Bao, "Challenges in low cost test approach for ARM9™ core based mixed-signal SoC DragonBall™-MX1," *Proc. International Test Conference,* 2003, pp. 512–19.

46. T. P. Warwick, "Mitigating the effects of the DUT interface board and test system parasitics in gigabit-plus measurements," *Proc. International Test Conference,* 2003, pp. 537–44.
47. M. Tripp, T. M. Mak, and A. Meixner, "Elimination of traditional functional testing of interface timings at Intel," *Proc. International Test Conference,* 2003, pp. 1014–22.
48. C. Jia and L. Milor, "A BIST solution for the test of I/O speed," *Proc. International Test Conference,* 2003, pp. 1023–30.
49. T. Newsom, "Future ATE for system on a chip ... some perspectives," *Proc. International Test Conference,* 2003, p. 1301.
50. M. Li, "Production test challenges and possible solutions for multiple GB/s ICs," *Proc. International Test Conference,* 2003, p. 1306.
51. T.J. Yamaguchi, "Open architecture ATE and 250 consecutive UIs," *Proc. International Test Conference,* 2003, p. 1307.
52. J. C. Johnson, "Cost containment for high-volume test of Multi-GB/s Ports," *Proc. International Test Conference,* 2003, p. 1308.
53. M. Li, "Requirements, and solutions for testing multiple GB/s ICs in production," *Proc. International Test Conference,* 2003, p. 1309.
54. U. Schoettmer and B. Laquai, "Managing the multi-Gbit/s test challenges," *Proc. International Test Conference,* 2003, p. 1310.
55. B. G. West, "Multi-GB/s IC test challenges and solutions," *Proc. International Test Conference,* 2003, p. 1311.
56. Y. Cai, "Jitter test in production for high speed serial links," *Proc. International Test Conference,* 2003, p. 1312.
57. A. T. Johnson, Jr., "Efficient fault analysis in linear analog circuits," *IEEE Transactions Circuits Systems,* vol. cs-26, July 1979, pp. 475–84.
58. C. Y. Pan and K. T. Cheng, "Fault macromodeling for analog/mixed-signal circuits," *Proc. International Test Conference,* 1997, pp. 913–22.
59. L. Milor and V. Visvanathan, "Detection of catastrophic faults in analog integrated circuits," *IEEE Transactions Computer-Aided Design,* vol. 8, February 1989, pp. 114–30.
60. R. J. A. Harvey et al., "Analogue fault simulation based on layout dependent fault models," *Proc. International Test Conference,* 1994, pp. 641–49.
61. C. Sebeke, J. P. Teixeira, and M. J. Ohletz, "Automatic fault extraction and simulation of layout realistic faults for integrated analogue circuits," *European Design and Test Conference,* 1995, pp. 464–68.
62. J. A. Starzyk and H. Dai, "Sensitivity based testing of nonlinear circuits," *Proc. ISCAS,* 1990, pp. 1159–62.
63. N. B. Hamida and B. Kaminska, "Multiple fault analog circuit testing by sensitivity analysis," *Journal of Electronic Testing: Theory and Application,* vol. 4, 1993, pp. 331–43.
64. C. Michael and M. Ismail, "Statistical modeling of device mismatch for analog MOS integrated circuits," *IEEE Journal of Solid-State Circuits,* vol. 27, January 1992, pp. 154–65.
65. A. Balivada, H. Zheng, N. Nagi, A. Chatterjee, and J. A. Abraham, "A unified approach for fault simulation of linear mixed-signal circuits," *Journal of Electronic Testing: Theory and Applications,* vol. 9, December 1996, pp. 29–41.
66. N. Nagi, A. Chatterjee, and J. A. Abraham, "Fault simulation of linear analog circuits," *Journal of Electronic Testing: Theory and Applications,* vol. 4, December 1993, pp. 345–60.

67. C. Y. Chao and L. Milor, "Performance modeling of circuits using additive regression splines," *IEEE Transactions Semiconductor Manufacturing,* vol. 8, August 1995, pp. 239–51.
68. W. M. Lindermeir, H. E. Graeb, and K. J. Antreich, "Design based analog testing by characteristic observation inference," *Proc. ICCAD,* 1995, pp. 620–26.
69. P. Variyam, S. Cherubal, and A. Chatterjee, "Prediction of analog performance parameters using fast transient testing," *IEEE Transactions on Computer-Aided Design of Integrated Circuits and Systems,* vol. 21, no. 3, 1992, pp. 349–61.
70. P. Variyam and A. Chatterjee, "Enhancing test effectiveness for analog circuits using synthesized measurements," *Proc. VLSI Test Symposium,* April 1998, pp. 132–37.
71. P. Variyam and A. Chatterjee, "Specification driven test generation for analog circuits," *IEEE Transactions on Computer-Aided Design of Integrated Circuits and Systems,* vol. 19, no. 10, October 2000, pp. 1189–1201.
72. J. L Huertas, A. Rueda, and D. Vazquez, "Testable switched-capacitor filters," *IEEE Journal of Solid-State Circuits,* vol. 28, July 1993, pp. 719–24.
73. K. Singhal and J. F. Pinel, "Statistical design centering and tolerancing using parametric sampling," *IEEE Transactions Circuits and Systems,* vol. CS-28, July 1981, pp. 692–701.
74. G. J. Hemink, B. W. Meijer, and H. G. Kerkhoff, "Testability analysis of analog systems," *IEEE Transactions on Computer-Aided Design,* vol. 9, June 1990, pp. 573–83.
75. K. D. Wagner and T. W. Williams, "Design for testability of analog/digital networks," *IEEE Transactions on Industrial Electronics,* vol. 36, May 1989, pp. 227–30.
76. V. Visvanathan and A. Sangiovanni-Vincentelli, "Diagnosability of nonlinear circuits and systems—Part 1: The DC case," *IEEE Transactions on Circuits and Systems,* vol. CS-28, November 1981, pp. 1093–1102.
77. R. Saeks, A. Sangiovanni-Vincentelli, and V. Visvanathan, "Diagnosability of nonlinear circuits and systems—Part II: Dynamical systems," *IEEE Transactions on Circuits and Systems,* vol. CS-28, November 1981, pp. 1103–08.
78. A. Chatterjee, "Concurrent error detection and fault-tolerance in linear analog circuits using continuous checksums," *IEEE Transactions on VLSI,* vol. 1, no. 2, June 1993, pp. 138–50.
79. S. Cherubal and A. Chatterjee, "Test generation based diagnosis of device parameters for analog circuits," *Proc. Design Automation and Test in Europe,* March 2000, pp. 596–602.
80. S. Cherubal and A. Chatterjee, "Parametric fault diagnosis for analog systems using functional mapping," *Proc. Design, Automation and Test in Europe,* March 1999, pp. 195–200.
81. Z. You, E. Sanchez-Sinencio, and J. Pineda de Gyvez, "Analog system-level fault diagnosis based on a symbolic method in the frequency domain," *IEEE Transactions Instrumentation and Measurement,* vol. 44, February 1995, pp. 28–35.
82. S. Freeman, "Optimum fault isolation by statistical inference," *IEEE Transactions on Circuits and Systems,* vol. CS-26, July 1979, pp. 505–12.
83. A. E. Salama, J. A. Starzyk, and J. W. Bandler, "A unified decomposition approach for fault location in large analog circuits," *IEEE Transactions on Circuit and Systems,* vol. CS-31, July 1984, pp. 609–22.
84. S. D. Huynh, S. Kim, M. Soma, and J Zhang, "Automatic analog test signal generation using multifrequency analysis," *IEEE Transactions on Circuits and System—II: Analog and Digital Signal Processing,* vol. 46, no 5, May 1999, pp. 565–76.

85. N. Sen and R. Saeks, "Fault diagnosis for linear systems via multifrequency measurements," *IEEE Transactions on Circuits and Systems,* vol. CS-26, July 1979, pp. 457–65.
86. G. Iuculano et al., "Multifrequency measurement of testability with application to large linear analog systems," *IEEE Transactions on Circuit and Systems,* vol. CS-33, June 1986, pp. 644–48.
87. G. Huertas, D. Vazquez, E. J. Peralias, A. Rueda, and J. L. Huertas, "Practical oscillation-based test of integrated filters," *IEEE Transactions Design & Test of Computers,* vol. 19, issue 6, November–December 2002, pp. 64–72.
88. K. Arabi and B. Kaminska, "Oscillation-test methodology for low-cost testing of active analog filters," *IEEE Transactions on Instrumentation and Measurement,* vol. 48, issue 4, August 1999, pp. 798–806.
89. K. Arabi and B. Kaminska, "Testing analog and mixed-signal integrated circuits using oscillation-test method," *IEEE Transactions on Computer-Aided Design of Integrated Circuits and Systems,* vol. 16, issue 7, July 1997, pp. 745–53.
90. A. Raghunathan, H. Shin, J. Abraham, and A. Chatterjee, "Prediction of analog performance parameters using oscillation based test," *Proc. VLSI Test Symposium,* April 2004, pp. 377–82.
91. L. Milor and A. L. Sangiovanni-Vincentelli, "Minimizing production test time to detect faults in analog circuits," *IEEE Transactions on Computer-Aided Design,* vol. 13, June 1994, pp. 796–813.
92. P. Variyam and A. Chatterjee, "Test generation for comprehensive testing of linear analog circuits using transient response sampling," *Proc. International Conference on Computer-Aided Design,* November 1997, pp. 382–85.
93. R. Voorakaranam, and A. Chatterjee, "Test generation for accurate prediction of analog specifications," *Proc. VLSI Test Symposium,* April 2000, pp. 137–42.
94. P. Duhamel and J. C. Rault, "Automatic test generation techniques for analog circuits and systems: A review," *IEEE Transactions on Circuits and Systems,* vol. CS-26, July 1979, pp. 411–39.
95. M. Slamani and B. Kaminska, "Fault observability analysis of analog circuits in frequency domain," *IEEE Transactions on Circuits and Systems II,* vol. 43, February 1996, pp. 134–39.
96. G. N. Stenbakken and T. M. Souders, "Test-point selection and testability measures via QR factorization of linear models," *IEEE Transactions on Instrumentation and Measururement,* vol. 36, June 1987, pp. 406–10.
97. P. P. Fasang, "Boundary scan and its application to analog-digital ASIC testing in a board/system environment," *Proc. Custom Integrated Circuits Conference,* 1989, pp. 22.4.1–22.4.4.
98. "IEEE standard test access port and boundary-scan architecture," IEEE Std 1149.1-2001.
99. S. Sunter, "The P1149.4 mixed signal test bus: Costs and benefits," *Proc. International Test Conference,* 1995, pp. 444–50.
100. "IEEE standard for a mixed-signal test bus," IEEE Std 1149.4 -1999.
101. H. Yoon, J. Hou, S. Bhattacharya, A. Chatterjee, and M. Swaminathan, "Fault detection and automated fault diagnosis for embedded integrated electrical passives," *Journal of VLSI Signal Processing Systems,* vol. 21, no. 3, July 1999, pp. 265–76.
102. A. Chatterjee, B. Kim, and N. Nagi, "DC built-in self-test for linear analog circuits," *IEEE Design and Test of Computers,* vol. 13, no. 2, Summer 1996, pp. 26–33.

103. C. L. Wey, "Built-in self-test (BIST) structure for analog circuit fault diagnosis," *IEEE Transactions on Instrumentation and Measurement,* vol. 39, June 1990, pp. 517–21.
104. M. F. Toner and G. W. Roberts, "A BIST SNR, gain tracking and frequency response test of a sigma-delta ADC," *IEEE Transactions on Circuits and Systems II,* vol. 42, January 1995, pp. 1–15.
105. M. J. Ohletz, "Hybrid built-in self test (HBIST) for mixed analogue/digital integrated circuits," *Proc. European Test Conference,* 1991, pp. 307–16.
106. P. Variyam and A. Chatterjee, "Digital-compatible BIST for analog circuits using transient response sampling," *IEEE Design and Test of Computers,* vol. 17, no. 3, July–September 2000, pp. 106–15.
107. B. Dufort and G. W. Roberts, "On-chip analog signal generation for mixed-signal built-in self-test," *IEEE Transactions on Solid State Circuits,* vol. 34, March 1999, pp. 318–30.
108. M. M. Hafed, N. Abaskharoun, and G. W. Roberts, "A 4-GHz effective sample rate integrated test core for analog and mixed-signal circuits," *IEEE Transactions on Solid State Circuits,* vol. 37, April 2002, pp. 499–514.
109. A. Halder and A. Chatterjee, "Specification based digital compatible built-in test of embedded analog circuits," *Proc. Asian Test Symposium,* November 2001, pp. 344–49.
110. M. Mendez-Rivera, J. Silva-Martinez, and E. Sánchez-Sinencio, "On-chip spectrum analyzer for built-in testing analog ICs," *Proc. IEEE International Symposium on Circuits and Systems,* vol. 5, 2002, pp. 61–64.
111. E. M. Hawrysh and G. W. Roberts, "An integrated memory-based analog signal generation into current DFT architectures," *Proc. International Test Conference,* 1996, pp. 528–37.
112. A. Halder, S. Bhattacharya and A. Chatterjee, "Automatic multitone alternate test generation for RF circuits using behavioral models," *Proc. International Test Conference,* 2003, pp. 665–73.
113. R. Voorakaranam, S. Cherubal, and A. Chatterjee, "A signature test framework for rapid production testing of RF circuits," *Proc. Design Automation and Test in Europe,* March 2002.
114. J. H. Friedman, "Multivariate adaptive regression splines," *The Annals of Statistics,* vol. 19, no. 1, 1991, pp. 1–141.
115. W. H. Kao and J. Q. Xia, "Automatic synthesis of DUT board circuits for testing of mixed signal IC's," *Proc. VLSI Test Symposium,* 1993, pp. 230–36.
116. S. Sunter and N. Nagi, "Test metrics for analog parametric faults," *Proc. VLSI Test Symposium,* 1999, pp. 226–34.
117. S. Sunter and A. Roy, "BIST for phase-locked loops in digital applications," *Proc. International Test Conference,* 1999, pp. 532–40.
118. K. Seongwon, and M. Soma, "An all-digital built-in self-test for high-speed phase-locked loops," *IEEE Transactions on Circuits and Systems—II,* vol. 48, issue 2, February 2001, pp. 141–50.
119. S. S. Akbay and A. Chatterjee, "Optimal multisine tests for RF amplifiers," *Wireless Test Workshop,* October 2002.
120. G. Srinivasan, S. Bhattacharya, S. Cherubal, and A. Chatterjee, "Efficient test strategy for TMDA power amplifiers using transient current measurements: uses and benefits," *Proc. Design, Automation, and Test in Europe,* vol. 1, February 2004, pp. 280–85.

121. M. Negreiros, L. Carro, and A. A. Susin, "Statistical sampler for a new on-line analog test method," *Proc. On-Line Testing Workshop,* 2002, pp. 79–83.
122. M. Negreiros, L. Carro, and A. A. Susin, "Ultra low cost analog BIST using spectral analysis," *Proc. VLSI Test Symposium,* 2003, pp. 77–82.
123. S. S. Akbay and A. Chatterjee, "Feature extraction based built-in alternate test of RF components using a noise reference," *Proc. VLSI Test Symposium,* April 2004, pp. 273–78.
124. P. K. Nag, A. Gattiker, W. Sichao, R. D. Blanton, and W. Maly, "Modeling the economics of testing: a DFT perspective," *IEEE Design & Test of Computers,* vol. 19, issue 1, January–February 2002, pp. 29–41.
125. D. Williams and A. P. Ambler, "System manufacturing test cost model," *Proc. International Test Conference,* October 2002, pp. 482–90.
126. J. Turino, "Test economics in the 21st Century," *IEEE Design & Test of Computers,* vol. 14, issue 3, July–September 1997, pp. 41–44.
127. Y. Zorian, "Testing the Monster Chip," *IEEE Spectrum,* vol. 36, issue 7, July 1999, pp. 54–60.
128. International Technology Roadmap for Semiconductors (ITRS), Test and Test Equipment, 2002.
129. Maxim IC application note 728, "Defining and testing dynamic parameters in high-speed ADCs, Part 1," February 2001, http://www.maxim-ic.com/appnotes.cfm/appnote_number/728.
130. D. A. McLeod, "Dynamic testing of analogue to digital converters," *Proc. International Conference on Analogue to Digital and Digital to Analogue Conversion,* September 1991, pp. 29–35.
131. J. A. Mielke, "Frequency domain testing of ADCs," *IEEE Design & Test of Computers,* vol. 13, no. 1, Spring 1996, pp. 64–69.
132. S. Goyal, A. Chatterjee, and M. Purtell, "A low-cost test methodology for dynamic specification testing of high-speed data converters," *Journal of Electronic Testing Theory and Applications,* vol. 23, no. 1, February 2007, pp. 95–106.
133. S. S. Akbay and A. Chatterjee, "Built-in test of RF components using mapped feature extraction sensors," *IEEE VLSI Test Symposium,* Palm Springs, CA, May 2005, pp. 243–48.
134. S. Bhattacharya and A. Chatterjee, "Use of Embedded Sensors for Built-in-Test of RF Circuits," *Proc. International Test Conference,* 2004, pp. 801–09.

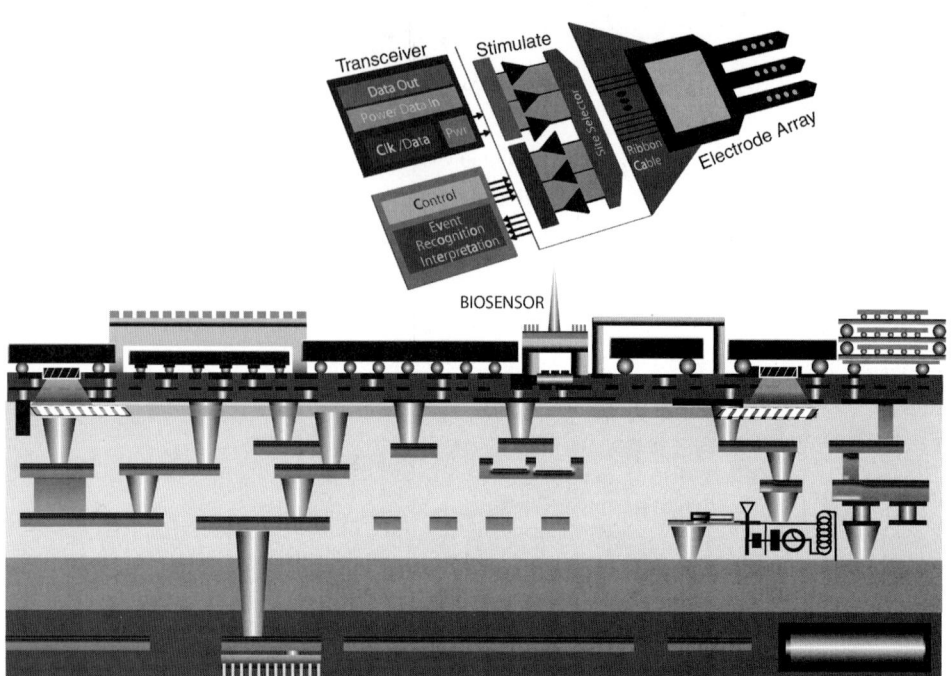

CHAPTER 13
Biosensor SOP

Dasharathan G. Janagama, Jin Liu, and Mahadevan K. Iyer
Georgia Institute of Technology

13.1	Introduction to Biosensor SOP 717		13.4	Signal Detection and Electronic Processing 741
13.2	Biosensing 723			
13.3	Signal Conversion 730		13.5	Summary and Future Trends 745
				References 746

Bioelectronics is an interdisciplinary field encompassing biology, chemistry, physics, material science, and electronics. The SOP described in this book is a highly miniaturized electronics systems technology that is expected to play a key role in all the future bioelectronics systems. As nanotechnology advances across all the above disciplines to enable nanoscale SOP, it promises highly miniaturized SOP-based bioelectronic products for the health care industry, forensic medicine, food and drink industries, environmental protection, personal security, genome analysis of organisms, and communications. The biosensor SOP consists of three building blocks: sensing, signal converging, and signal processing. A key factor in biosensor SOP is to understand the mechanisms and interfaces between the sensing mechanism and the conversion of the detected signal. In this chapter, each of the three building blocks is introduced and explained. The integration of biosensors in SOP technology is reviewed. Finally, the chapter concludes with future trends.

13.1 Introduction to Biosensor SOP

13.1.1 SOP: A Highly Miniaturized Electronic System Technology

Electronic systems rely mostly on integrated circuit (IC) integration (Moore's law) for performance improvement and cost reduction. But ICs make up only 10 to 20 percent of an electronic system size; the remaining 80 to 90 percent is typically made up of bulky power sources, heat sinks, passive components, and interconnects. These are major bottlenecks for reducing size, cost, and power consumption of the system. The system-on-package (SOP) technology paradigm (the second law of electronics) pioneered by Georgia Tech Packaging Research Center since the early 1990s provides system-level miniaturization in a package size that makes today's handheld devices into mega-functional systems, with applications ranging from computing, wireless communications, health care, and personal security. The SOP is a system miniaturization technology that ultimately integrates nanoscale thin-film components for batteries, thermal structures, active and passive components in low-cost organic packaging substrates, leading to micro- to nanoscale modules and systems. True miniaturization of products should take place not only at the IC but also at the system level, the latter made possible by thin-film batteries, thermal structures, and embedded actives and passives in package-size boards. This is the fundamental basis for the SOP concept. The traditional packaging by which components are integrated into systems today presents several sets of barriers in cost, size, performance, and reliability. The SOP concept overcomes these barriers by the best of both the IC and the package integration at the system level, the IC for transistor density, and the system package for component density of RF, optical, digital, and biofunctions. In addition to miniaturization, such a concept leads to lower cost, higher performance, and better reliability of all electronic systems including biosensor systems described in this chapter.

13.1.2 Biosensor SOP for Miniaturized Biomedical Implants and Sensor Systems

Microelectronic devices are becoming increasingly accepted for a variety of biomedical applications. Today, these devices consist of microelectronic components mostly in a single chip form, housed inside a biocompatible protective package. These single-chip or system-on-chip (SOC) solutions are limited due to the fact that they still need to be provided with signal and power distribution, inputs-outputs (I/Os), cooling mechanisms, efficient options for biofluids transport (microfluidic components), and digital, wireless, and optical interfaces. The above constraints are going to be the limiting factors for future biosensors, implants, and bioelectronic systems such as neural prosthetic devices and sensors for unmanned aerial vehicles. SOP-based technologies are able to overcome these constraints and create the biosystems that meet tomorrow's needs by ultraminiaturization technologies.

The bioimplantable systems and biosensing chips are finding their applications in both clinical and nonclinical domains nowadays. These applications include detection of cancer, genetic diseases, AIDS, bacterial and viral diseases [1–3], industrial bioprocesses control, food and drink contamination, and environmental safety [4–8]. These systems usually integrate a number of functional active and passive components, such as sensors, microfluidics, and reagent amplification, into a single chip and package. Lab-on-chip (LOC), which translates the entire laboratory testing processes and functions onto a tiny microelectronic device, is a good example (Figure 13.1). A DNA chip extracts DNA from raw blood or pathogen samples, which are sent through an inlet, and amplifies and detects the targeted DNA. The microfluidics incorporated into the package facilitates the flow of the samples and reagents in the chip. However, for complete biosystem integration, the signal and power distribution, I/Os and suitable interfaces have to be incorporated.

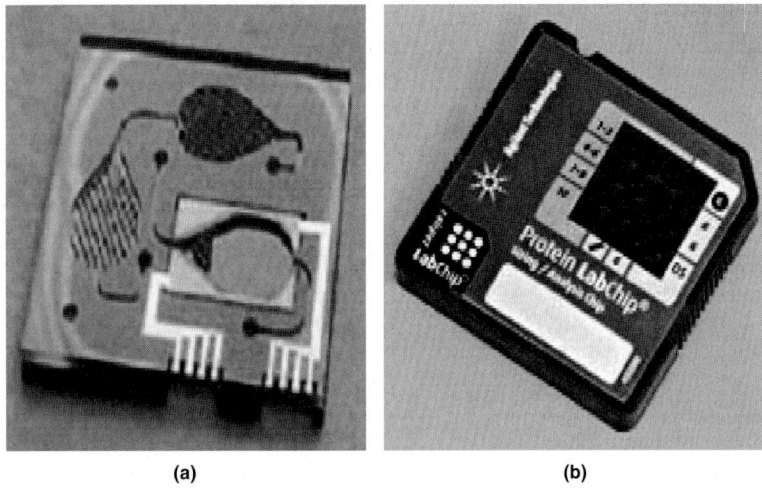

FIGURE 13.1 (a) DNA chip [Courtesy Institute of Microelectronics and SiMEMS (S) PTE Ltd]. (b) Protein chip (Courtesy of Agilent Technologies).

Another example of bio-implantable chip and biosensing chip applications is neural prosthesis. Deep brain stimulation (DBS), which sends electrical impulses to specific parts of the brain, has a microelectronic case (pulse generator) implanted in the chest with a long multiwire cable subcutaneously tunneled to the top of the head and then into the brain (Figure 13.2). The DBS system uses conventional wire bonding for

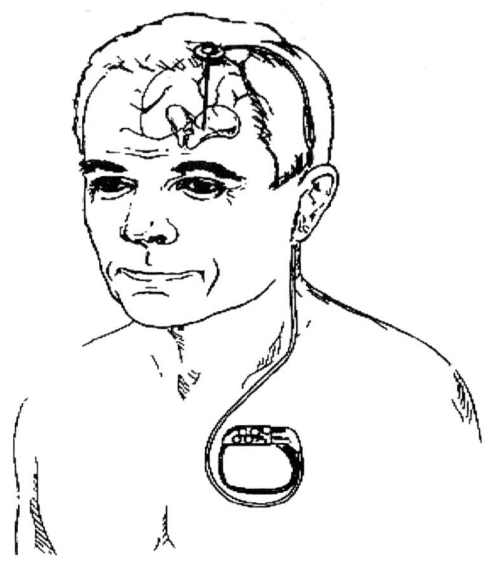

FIGURE 13.2 Electrode connected to a pulse generator. The patient can deliver deep brain stimulation to the targeted area using a handheld magnet to activate the pulse generator, which is implanted subcutaneously in the chest. (Courtesy Medtronic Inc.)

connections to printed circuit boards (PCB). Large passive components like capacitors and diodes are manually added to the system. Inductors, for coupling data and power from an external system to the implants, are hand-wound coils of wire. All these approaches increase the size of the overall system to the point where the physical location of the implant can be distant from the actual tissue. As the next generation of neural prostheses is being developed, the lack of high-density packaging schemes will be a significant limitation. The state-of-the-art in medical-grade implantable microelectronic packaging cannot meet these requirements. The SOP platform, on the other hand, brings embedded ultrathin active and passive components into an ultrathin biocompatible substrate. The SOP technology based system is highly miniaturized such that the implant location could be the stimulation location.

The above discussion clearly establishes a compelling need for high-density systems packaging technology for bioimplant, sensing, and bioelectronics system integration. This forms the basis of the biosensor SOP. This SOP approach has the clear potential to revolutionize biomedical packaging and biosystem integration by integrating multiple components into a single, compact, ultrathin miniaturized package. This technology elevates traditional packaging with advances in thin-film materials, processes and integration of embedded active and passive components, thermal structures, microfluidic components, batteries, microconnectors, and wireless and optical interfaces.

The SOP-based bioelectronic system technology brings together the advantages in miniaturization, reliability, performance, and cost (Figure 13.3). By integrating ultra-miniaturized thin-film components in SOP-based technologies, the size of the traditional bulky electronic systems is expected to be reduced to a few inches in 5 to 10 years and to a few millimeters within a decade after that. In the SOP concept in which the package and system board are merged into the system package, the surface-mounted components are being replaced with embedded thin-film active and passive counterparts. As a result, the traditional soldering joints are eliminated and the length of the interconnection is greatly shortened. These attributes are expected to lead to about three times improvement

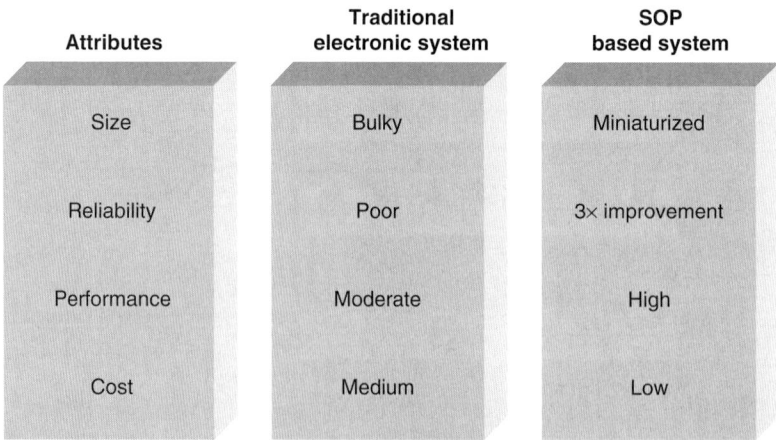

Figure 13.3 Schematic of attributes of SOP-based systems.

in the reliability. Furthermore, SOP substrates are designed with excellent electrical and thermomechanical properties, resulting in enhanced reliability and performance. Finally, SOP manufacturing involving large-area substrates that are more similar to PCB in size and ultrasmall area of system usage reduces the cost, enabled by batch manufacturing capabilities.

Leveraging the above SOP attributes and technologies, the present biosensor systems can be greatly enhanced for their functionalities in both clinical and nonclinical domains.

In the clinical domain, detection of cancer involves the detection of several cancer causing genes. This needs to be a multiple-target detection system with several functionalities to detect all genes involved in the disease at the same time. Similarly, the detection of lung infections needs a multiple-target detection system to detect all pathogens involved in the infection simultaneously. In such cases, detection of targets one at a time has its limitations. Biosensor SOP technologies facilitate the fabrication of multiplex sensor systems. It enables the fabrication of multiplexed sensors with multiple fluidic paths connected to the sensing elements separately, multiple dimensions of mechanical components such as micro-nano switches, valves, pumps, and reservoirs. In the case of disease detection and therapeutic needs, biosensor SOP provides a range of electronic-wireless components for control, feedback, and wireless interfaces for detection and targeted drug delivery. Implantable biosensor SOP-based microsystems with a sensing element and anti-inflammatory drug delivery system could mitigate the problem of inflammatory reactions. Such SOP-based microsystems will allow real-time stimulation and recording of brain activity.

Biosensors in the nonclinical domain, such as food and drink contamination and water safety monitoring, call for unique and application-specific packaging requirements. For example, the aquatic environment requires the water contamination detection system to be moisture resistant. The biosensors in food and drink contamination detection systems not only have to be moisture resistant, but also biocompatible and ultralow-priced. SOP-based technologies address and meet these requirements in their specific situations. By taking advantage of the near-hermetic properties of substrates and packaging materials, the biosensors can provide good moisture resistance. An example in food and drink contamination monitoring is the usage of liquid crystalline polymer (LCP) type substrate, which brings the biocompatibility as well as low cost to the system. Embedded radiofrequency identification (RFID) and other wireless interfaces in biosensor SOP could facilitate the communication as well as the real-time monitoring capabilities.

In addition to the enhancement of the current biosensor technologies, biosensor SOP, with its multifunctional and highly miniaturized platform technologies, finds new applications in the fields such as retinal prosthesis and advanced micro/nano unmanned aerial vehicle (UAV) systems.

In the case of a retinal prostheses biosystem, the optimization for size and performance can be achieved through advanced packaging technologies versus advanced circuit fabrication processes. SOP technology economizes space through embedded ultrathin active and thin-film passive components and by using 3D stacking of chips to reduce the 2D space used. This is particularly important in applications where a strict size constraint comes into play due to biological considerations.

An example of a device that will require the above-mentioned criteria is in retinal prosthesis (Figure 13.4). The device consists of an array of implanted electrodes, which

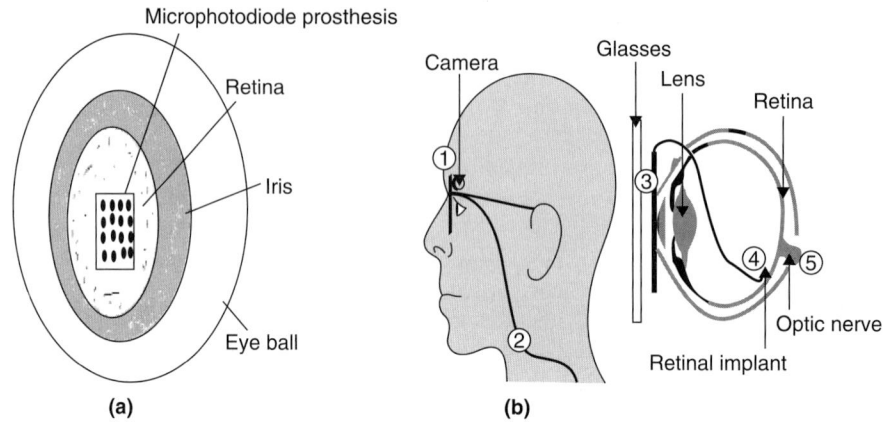

FIGURE 13.4 Retinal implant. (*a*) Schematic of the implant in the eye. (*b*) Retinal implant with associated accessories. (Courtesy Dr. Mark S. Humayun, Dohney Retina Institute.)

is in the backside of the eyeball, and a micro photodiode, which acts as the reception device and locates in the front side of the eyeball. The electrodes stimulate the optic nerve based on the signal transmitted from the micro photodiode. Preliminary results have been obtained on 4 × 4 platinum electrodes. Using this device, subjects who otherwise see only light or dark are able to discriminate between objects in a set and recognize motion. It is suggested that a higher-resolution device may be more effective to enable face recognition and reading [9]. However, the fact that the retinal prosthesis must fit in the eye, which is roughly a sphere of 2.5-cm diameter, highly confines the physical size of any possible device. Current packaging technology does not provide a clear solution for such sophisticated and miniaturized systems. In contrast, biosensor SOP allows not only the integration of high-density electrodes along with advanced active and passive components, which provides enough resolution for the system, but also space utilization through 3D stacking of chips and passive components, thus meeting the size constraints of the system.

Unmanned aerial vehicle (UAV) systems are becoming increasingly important in current military roles, including reconnaissance and attack as they offer the possibility of cheaper, more capable fighting machines that can be used without risk to aircrew. Looking into the future, the advanced micro/nano UAVs may completely overhaul the conventional combat scenario. These systems are expected to be more autonomous and able to perform multiroles. They will have the enhanced capability of rapid navigation, echolocation, and self-flight control. The equipped sensors will detect any explosive, toxic gas, and biohazard species. The embedded camera and microphone allow the crew to monitor the battle field alive without being exposed to it. Packaging technologies have always been one of the major driving forces in the miniaturization of UAV systems. The aforementioned micro/nano UAVs consisting of nano biosensor, micro/nano fluidics, digital, wireless, and/or optical interface, power and thermal solutions could be made possible based on the SOP technologies. The following section describes the building blocks of biosensor SOP, which integrates the sensing elements with signal conversion and processing functions.

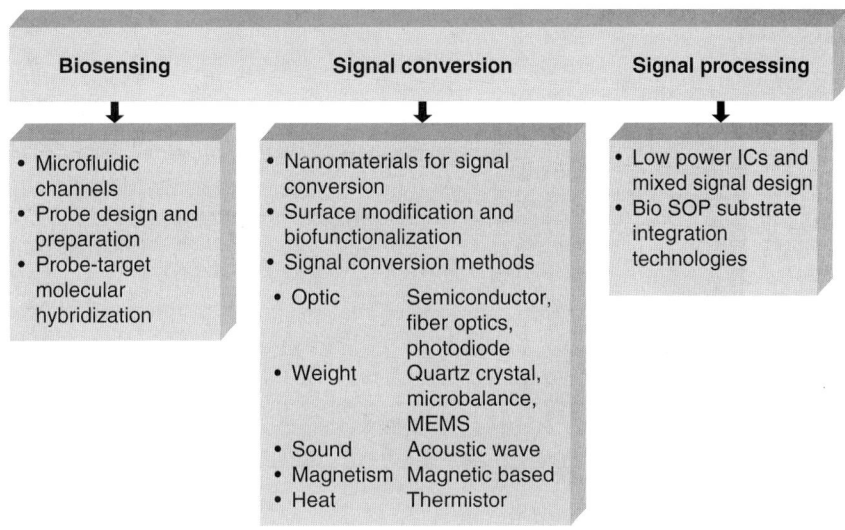

FIGURE 13.5 Building blocks of biosensor SOP.

13.1.3 Building Blocks of Biosensor SOP

Biosensor SOP essentially consists of the following three functional categories, as shown in Figure 13.5:

- Biosensing
- Signal conversion
- Signal processing

13.2 Biosensing

A classical sensor is a device that senses or measures physical contact, motion, heat, or light, and converts the impulses into an analog or digital representation. A biosensor is a device that senses biological molecules by using specific biosensing mechanisms. The biosensing mechanism is centered on two fundamental constituents, namely, bioprobe and hybridization. Bioprobe is defined as a set of specific biological molecules capable of recognizing or probing their complementary set of molecules called the target. Hybridization is defined as a process of probe and target molecular interaction. The bioprobe and target hybridization is a reversible chemical reaction with high specificity. The hybridization reaction takes place in an aquatic environment, which makes the microfluidic system an indispensable component of biosensor SOP. The following sections deal briefly with the introduction of microfluidic channels, preparation of the bioprobe, and hybridization processes.

13.2.1 Microchannels for Biofluid Transport

Microchannels with cross-sectional dimensions on the order of 10 to 100 μm are essential for guiding and helping biofluids carry the target or the sample to be tested onto the

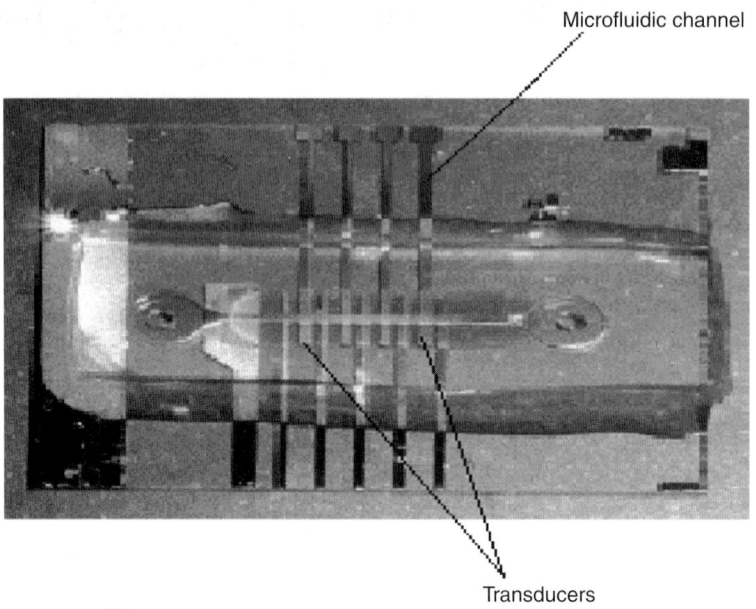

FIGURE 13.6 Integration of transducers into microfluidic channels (Courtesy Yaraliogu GG, Jagarnathan H, and BT Pierre Khwi-Yakub, University of stanford [10]).

sensing element of the biosensor. As all biological molecular reactions take place in a liquid environment, microfluidic channels facilitate controlled flow of microquantities of fluids containing the desired test molecules (Figure 13.6). The common fluids used as the target in microfluidic devices include whole blood samples, bacterial cell suspensions, protein or antibody solutions, and various buffers. Fluidics carries the biological molecules without involving mechanical moving parts that will wear out.

Materials for implantable devices and microfluidic channels need to be biocompatible, which means not producing a toxic, injurious, or host response in living tissue. Biocompatible materials such as polydimethylsiloxane (PDMS) and LCP materials are commonly used for fabrication of microfluidics. Further, laminated plastic microfluidic components made of polyimide, polymethylmethacrylate (PMMA), and polycarbonate materials with a thickness between 25 and 125 µm have also been developed and used. These devices can be potentially manufactured in high volume with a low unit cost because of advancements in the materials and chemical processes such as high-speed full-field excimer laser ablation process (Anvik Incorporated) to generate the fluidic channels. Low-cost PDMS adhesive-based bonding can be used to seal the channels. Fabrication of both the submicron features of the photonic crystal sensor structure and the >10-µm features of a flow channel network in one step at room temperature on a plastic substrate was demonstrated [10].

13.2.2 Biosensing Element (Probe) Design and Preparation

A bioprobe is defined as a set of biological molecules capable of probing or recognizing their complementary (reciprocal) target molecules. Probes are prepared with directed selectivity for the target or analyte of interest. Bioprobes are prepared differently for the detection of different target or test molecules. Protein, DNA, and enzyme catalyst probes are prepared from cells and used in the fabrication of protein, DNA, and enzyme sensors, respectively. Some biological cells are directly used as bioprobes. The synthetic probes are artificial substances that mimic natural biomolecules. The sources of bioprobes are animal cells, bacteria, and viruses. Animal cells range in the size of 1 to 100 μm. Bacteria are generally rod-shaped with dimensions of 1 to 8 μm. Viruses are extremely small, approximately 15 to 25 nm in diameter (Figure 13.7a and b).

Protein Probe

Proteins are the essential structural and functional building blocks of all organisms. A functional protein is formed by a number of amino acids linked through peptide bonds. An amino acid is an organic compound containing an amino group (NH_2), a carboxylic acid group (COOH), and a side chain attached to an alpha carbon atom (H_2N—CR—COOH). The bond between the carboxyl group of one amino acid and the amino group of the adjacent amino acid is known as a peptide bond. Antigens are proteins on the cell surface as well as within the cell. Antibodies are particular proteins produced by cells in response to the antigens, and used as protein probes. Antibodies are produced by injecting antigens into animals. Specific antibodies are raised in the animals in response to a specific antigen. The extracted antibodies are purified and quantified using standard protein extraction protocols, and used as protein probes. Protein probe preparation is an in vivo (in the living body) technique.

DNA-RNA Probe

DNA is a genetic material. It is a unique biological molecule capable of self-replication. DNA is a double-stranded molecule positioned such that the bases of the strands can interact with each other (Figure 13.8). It is a double helix of a complementary strand structure, which contains genetic codes for the operation of all living organisms. DNA is made up of four standard bases in long chains, known as adenosine (A), guanine (G), thymine (T), and cytosine (C). A-T and G-C are the complementary base pairs.

DNA-RNA probes are synthesized in the laboratory using a polymerase chain reaction (PCR) method. It is an in vitro (outside the living body) technique used to exponentially amplify DNA via enzymatic replication. The appropriate primer, Taq DNA polymerase enzyme, deoxyribose nucleotide triphosphates (d-NTPs), and other

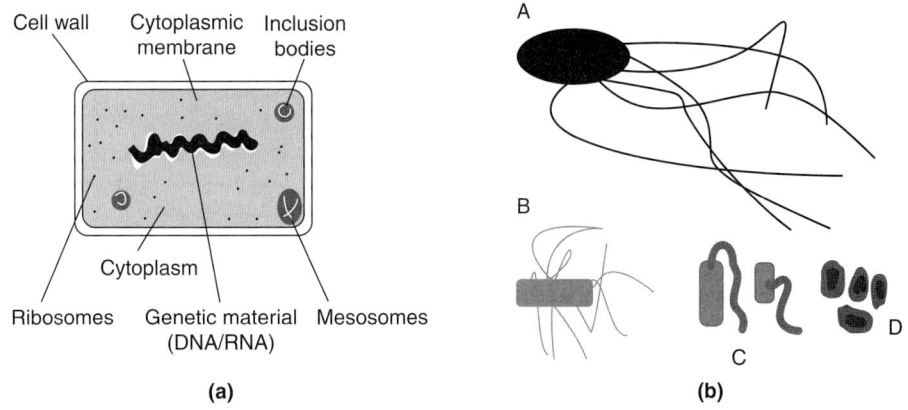

FIGURE 13.7 (a) Schematic of a typical bacterial cell. (b) Schematic of bacteria and viruses: A. *Salmonella*; B. *Bacillus cereus*; C. *Vibrio cholera*; D. *Influenza viruses*.

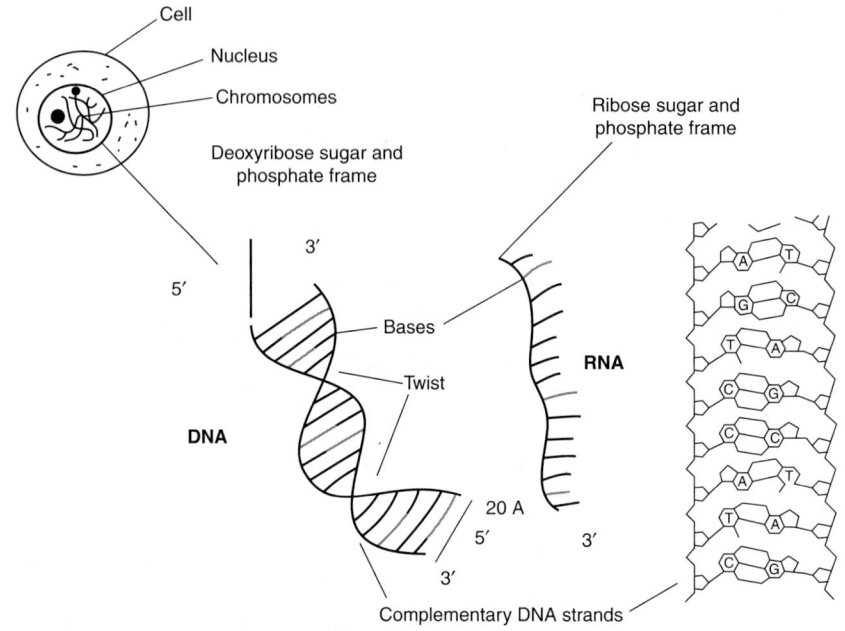

Figure 13.8 Schematic of DNA and RNA structures. DNA double helix interconnected by bases (A-T, G-C). RNA is a single-stranded structure with a uracil (U) base in place of thymine (T). The inset shows the location of DNA and RNA in a cell.

ingredients in the reaction condition can generate DNA-RNA probe synthesis. The primer initiates DNA replication. Taq DNA polymerase facilitates making new DNA at low temperatures. Deoxyribose nucleotide triphosphates (d-NTPs) are the main ingredients in DNA synthesis. The DNA-RNA probes for a particular target are synthesized in a commercially available DNA-RNA synthesizer. DNA probes are different from protein probes mainly in their structural aspects.

Synthetic Probe
A synthetic or artificial probe is designed and fabricated to mimic a bioprobe. For instance, protein engineering techniques are used to construct a specific protein probe that mimics the binding properties of biotin. Biotin is a water soluble B vitamin that efficiently binds with other proteins and DNA, which are used to incorporate bioprobes into a signal conversion component, which will be discussed later.

Cell Probe
In a cell probe, a cell acts as a probe. For instance, a genetically modified bacterium is a kind of cell probe. Cell probes can be directly utilized for the fabrication of cell-based sensors (CBS). Microorganisms such as bacteria and fungi are cultured and used as indicators of toxins in the environment.

Enzyme (Biocatalyst) Probes
In general, a protein that accelerates the rate of chemical reactions is known as an enzyme. Enzymes are used as bioprobes based on their specific binding capabilities and catalytic activity on a particular chemical substance commonly known as a substrate. In

general, two types of enzymes are used as enzyme catalysts, namely oxidases and dehyrogenases. Oxidase is an enzyme that catalyzes a oxidation-reduction reaction involving molecular oxygen (O_2) as the electron acceptor. In this reaction, oxygen reduced ultimately to water (H_2O). For example, glucose oxidase (GOx) enzyme is used in electrochemical biosensors to measure blood glucose. Dehydrogenase is an enzyme that removes hydrogen from a substrate or chemical and transfers the hydrogen to an acceptor. For example, electron mediator-dependent glucose dehydrogenases (GDH) are recognized as ideal constituents of mediator-type enzyme sensors.

13.2.3 Probe-Target Molecular Hybridization

Hybridization is a reversible chemical reaction between bioprobe and target molecules with high specificity. The "target" is an object for detection such as proteins, DNA-RNA, or cells. It is a chemical and molecular basis of recognition, involving dynamic, spatial, and temporal rearrangement of the molecules. These biorecognition elements are specific and sensitive to a particular substance but not others, as in a lock-and-key mechanism. The biomolecules interact as pairs, one of which is called the target (ligand) and the other is called the probe. The existence of complementary structures for proteins (antigens-antibodies), nucleic acid (DNA-DNA), peptides, and synthetic probes (probe-ligand) is a key factor in biosensing. As such, proteins, peptides, DNA, RNA, and whole cells constitute the target molecules, which can be detected by using the similar complementary probe molecules. The molecular hybridization for DNA and RNA, immune complex formation for proteins, peptides, and whole cells is the critical biosensing mechanism of a biosensor.

Protein Hybridization

In the case of protein hybridization, both the protein probe (antibody) and the target (antigen) are proteins. It is a reversible chemical reaction between the protein probe and target with high specificity. This reaction involves the protein probe recognizing the site on the target and then binding to it (Figure 13.9). The main forces that play a role in these reactions are hydrogen bonds, electrostatic forces, van der Waals bonds, and

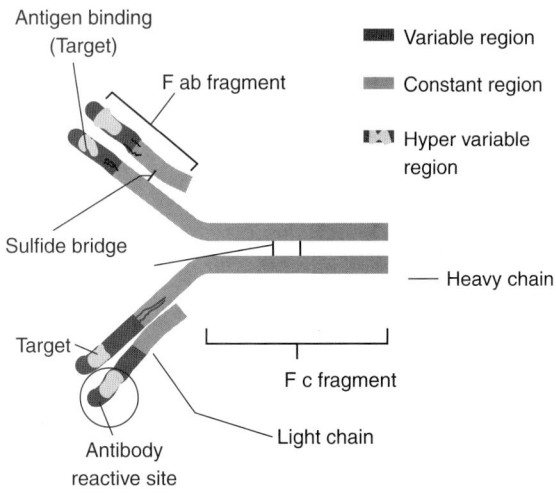

FIGURE 13.9 Schematic of antibody (immunoglobulin) molecular structure. Binding sites (epitopes) of the antigen and antibody reaction is circled.

hydrophobic bonds. The stability of the antigen-antibody binding depends on the valency of antibodies, the so-called number of arms with which the antibody may bind to its antigen. Two models have been proposed to explain the mode of antigen-antibody binding. In the lock-and-key model, the antibody recognizes the sites on the antigen just as in a specific key for a lock (natural fit). In the induced-fit model, the binding site on the antibody is induced to adjust and fit with the site on the antigen and then bind to it.

DNA–RNA Hybridization

In DNA hybridization, both the DNA probe and target are DNA molecules. Hybridization is the process of combining complementary, single-stranded nucleic acids into a double-stranded molecule. It is a reversible chemical reaction between the probe DNA and target DNA with high specificity. DNA hybridization occurs between the DNA probe and target DNA in a base sequence-specific manner. Adenosine and thymine (A-T) and guanine and cytosine (G-C) are the complementary base pairs of the probe and target DNA molecules (Figures 13.8 and 13.10). As in DNA, RNA contains similar bases, except that the base thymine (T) is replaced by uracil (U).

In the case of RNA, hybridization occurs between adenosine and uracil (A-U), and guanine and cytosine (G-C) base pairs of probe and target RNA molecules. Synthetic DNA and RNA probes hybridize with the target molecules essentially in a similar way. As shown in Figure 13.11, the main forces in DNA or RNA hybridization are hydrogen bonds,

Synthetic Probes Hybridization

Synthetic probes are synthetic protein and DNA. They are essentially similar to the natural protein and DNA probes in their basic structures and functions, including hybridization mechanisms. Synthetic probes are synthesized artificially for specific purposes. For instance, the *Staphylococcus aureus* bacterium, which exists in more than one kind of strain or species, is difficult to identify using a conventional single probe. A synthetic enterotoxin B DNA probe with modified features in the DNA structure enables the simultaneous detection of more than one strain or species of *Staphylococcus*.

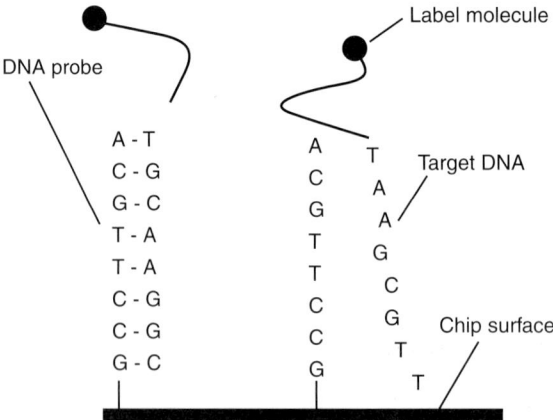

Figure 13.10 Schematic of DNA hybridization. Hybridization occurs between sequence-specific DNA molecules (*left*). Sequence-nonspecific target DNA is not recognized by the DNA probe (*right*).

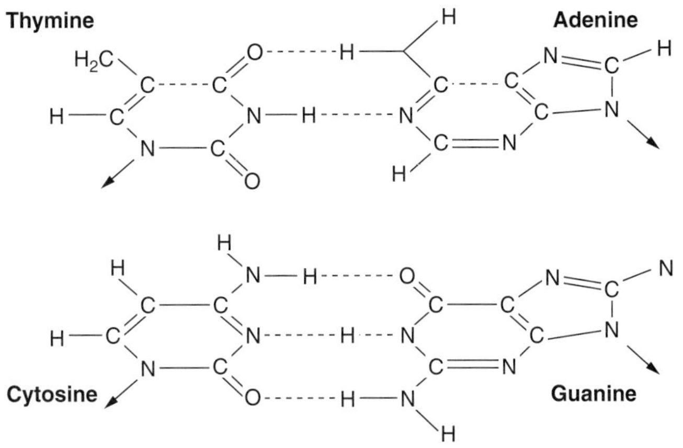

FIGURE 13.11 Schematic of hydrogen bonding between base pair in DNA (Courtesy of Dr. John W Kimball: http://users.rcn.com/jkimball.ma.ultranet/BiologyPages/).

Cell Probe Hybridization

For a cell probe, the cell probe molecules are proteins and other cellular components on the surface of the cells, whereas the target molecules can be environmental pollutant, drugs, or other molecules. For example, some bacteria that are originally white in color turn green upon being stimulated by environmental pollutants. The activated component of the bacterial cell is called a reporter gene, which produces a green protein that imparts a green color to the bacteria. In a clean environment the bacteria remain white. The color change of the bacteria from white to green indicates a pollution status of its environment (Figure 13.12). This color changing property of the bacteria is utilized as a cell-based sensor (CBS) for detection of environmental pollution. Further, the cells grown on biosensor device surfaces act as CBSs, which are used for monitoring pharmacological processes, including drug discovery.

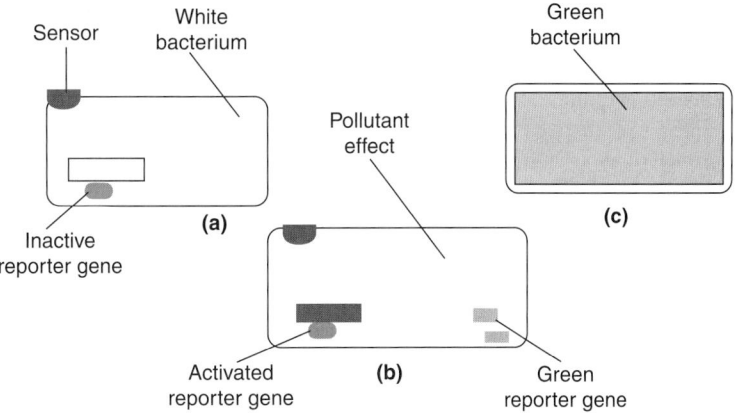

FIGURE 13.12 Schematic of bacteria (*a*) used for detection of environmental pollution. The activation of a reporter gene of white bacteria (*b*) in polluted environment changes into green form (*c*).

Enzyme (Biocatalyst) Reaction

Enzymes are basically proteins with specific biocatalyst activity. They are involved in promoting chemical reactions in electrochemical systems. Unlike protein and DNA probes, enzymes temporarily bind to one or more of the reactants of the reaction they catalyze, and then speed the reaction. In a simple glucose biosensor, the enzymatic oxidation of glucose by glucose oxidase (GOx) in an electrochemical cell produces hydrogen peroxide, which in turn generates electrons from the electrode reaction. The enzyme used in the reaction promotes or catalyzes the reaction. The current density (amperometric) is used as a measure of glucose concentration in the sample. The enzymatic reaction based on glucose oxidase is

$$\text{Glucose} + \text{Oxygen} \xrightarrow{\text{Enzyme}} \text{Gluconolactone} + H_2O_2 \quad (13.1)$$

$$H_2O_2 \longrightarrow O_2 + 2H^+ + e^- \quad (13.2)$$

Thus, the biocatalyst promotes the chemical reaction, which is different from protein, DNA, a synthetic probe, and cell probe hybridizations.

13.3 Signal Conversion

Signal conversion involves converting a biological signal into other forms that can be processed and displayed electronically. In general, a signal conversion component or a transducer in a biosensor is an interface device between the bioprobe and signal processor, which converts biological information into digital or analog signals. Signal conversion is accomplished by employing any one or more of the following techniques: electrochemical, optical, acoustic, electromechanical, magnetic, and thermal.

Signal conversion component development involves the selection of suitable materials and structures, surface modification, and the application of signal conversion techniques.

The following sections describe the fundamentals and advancements in nanomaterials, structures (Section 13.3.1), surface modifications of signal conversion devices (Section 13.3.2), and various signal conversion methods (Section 13.3.3).

13.3.1 Nanomaterials and Nanostructures for Signal Conversion Components

Nanostructured materials are defined as the materials possessing dimensions on the order of a billionth of a meter (nanometer). They exhibit special properties that are different from bulk materials. The novel properties possessed by nanostructured materials open many possibilities toward many applications, including fabricating advanced signal conversion components in nanobiosensors with superior sensitivity and specificity. For instance, magnetic nanoparticles exhibit supermagnetic behavior, and facilitate efficient signal conversion [11]. The semiconducting nanoparticles show resonance tunneling and Coulomb blockade effects [12]. In a conventional detection system, the biosensing and signal converting components are too big compared to the target molecule. Nanomaterial-based signal conversion components such as Si nanowire, carbon nanotubes (CNTs) and quantum dots (QDs) are almost reduced to the size of the target, which confers greater detection functionality. Figure 13.13 shows the size of various objects and molecules. It is obvious from the figure that cells and molecules fall in the range of micro- and nanosize, respectively.

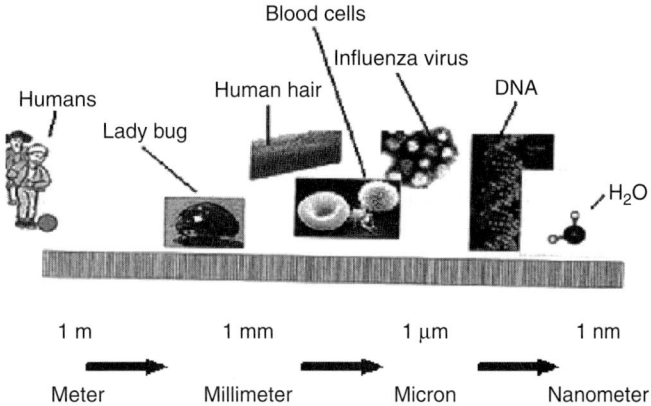

FIGURE 13.13 Schematic of comparison of nanoscale objects. [Source: National Nanotechnology Infrastructure Network (NNIN).]

Nanoparticles

Nanoparticles refer to the particles that are developed at a critical length scale of under 100 nm and possess novel properties. Nanoparticles act as efficient signal conversion components in a nanobiosensor. Different types of nanoparticles are available commercially, including polymeric nanoparticles, metal nanoparticles, liposomes, micelles (aggregate of surfactant molecules), quantum dots, dendrimers (repeatedly branched molecules), and other nanoassemblies (amphiphilic molecular building blocks that produce nanopatterns). The gold nanoparticles and magnetic nanops are widely used in the fabrication of an efficient signal conversion component. The remarkable biorecognition capabilities of the biomolecules such as DNA, RNA, and proteins can be utilized for the fabrication of ultraminiature biological electronic sensing devices, by conjugating them with nanoscale particles. The metal nanoparticles that are incorporated into the interior of DNA enhance thermal stability (Figure 13.14). Similarly, magnetic nanoparticles introduced into the interior of DNA provide unique magnetic properties. Nanoparticle-based signal conversion methods enhance sensitivity, are

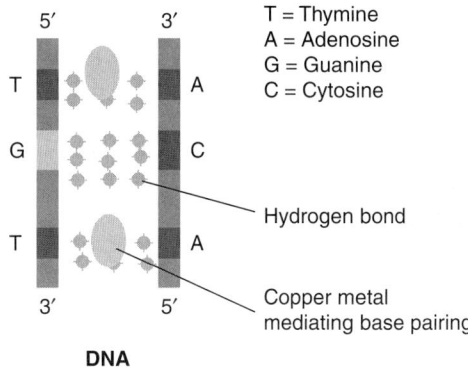

FIGURE 13.14 Schematic of copper metal mediating DNA base pairing.

inexpensive and noninvasive detection systems for measuring biological molecules, and are superior over the existing technologies.

The quantum dots (QDs) are inorganic semiconductor nanocrystals. The structure of quantum dots is such that free electrons are confined in a 3D semiconducting matrix. They are mostly composed of atoms from CdSe, CdTe, InTe, and InAs. A key feature of these materials is that they can be modified with a large number of small molecules and linker groups to optimize the functionality for particular applications. The quantum dots can be covalently linked with biomolecules such as DNA-RNA, antibodies, peptides, and small-molecule inhibitors for use as fluorescent probes [13], which exhibit improved brightness, photo stability, and broader excitation spectra (Figure 13.15b and c). Coulomb charging effect is found in the transport behavior of these quantum dots, and resonant tunneling through the lowest quantized energy level of the QDs [14]. The quantum dots contribute to special optical properties, such as particle-size dependent wavelength of fluorescence, which enables signal conversion in wideband emitter optical sensors. Thus, quantum dots form superior signal conversion components over the existing imaging techniques, which use natural molecules such as organic dyes and proteins (Figure 13.15a).

Transmission electron microscopy (TEM) is used to characterize the nanoparticles and microstructure. TEM enables very high spatial resolution required to access information about morphology, crystal structure, and defects.

Wire-like Nanostructures

Silicon nanowires (Si NWs) and zinc oxide (ZnO) nanobelts are other types of nanostructures used for fabrication of efficient signal conversion components. Additional advantage of ZnO nanowire is being a biocompatible material. Binding of molecules to the nanowire causes a drastic change in the conductance. Therefore, silicon and zinc oxide based signal conversion components are capable of detecting biomolecules in low concentration. Silicon and ZnO based signal conversion components are capable of detecting a single cell or molecule, superior over the conventional detection.

The scanning electron microscope (SEM) is used to characterize the nanostructures. It creates the magnified images by using electrons instead of light waves. As the electron beam hits each spot on the sample, secondary electrons are knocked loose from its surface. A detector counts these electrons and sends the signals to an amplifier. The

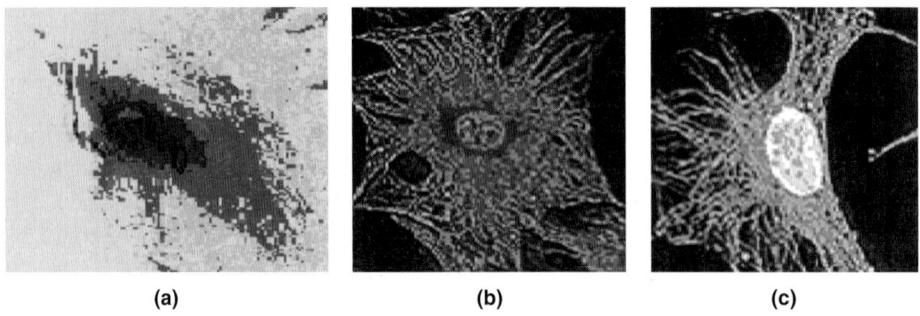

Figure 13.15 (a) Conventional dye. (b) The microtubules of the cell stained with 605-nm fluorescent quantum dot conjugate dye (middle). (c) Nucleus and microtubules labeled with red and green quantum-dot conjugates. (Courtesy Quantum Dot Corp.)

final image is built up from the number of electrons emitted from each spot on the sample. A nonconducting sample is prepared by coating it with a very thin layer of gold using sputter coater during the sample preparation step.

Carbon Nanotubes (CNTs)

Carbon nanotubes are concentric shells of graphite formed by one sheet of conventional graphite rolled up into a cylinder. The lattice of carbon atoms remains continuous around the circumference. A single graphene cylinder is called a single-walled carbon nanotube (SWCNT), whereas a multiconcentric graphene cylinder is called a multiwalled carbon nanotube (MWCNT) (Figure 13.16). SWCNTs and MWCNTs exhibit unique properties such as outstanding charge-transport characteristics, chemical stability, good mechanical properties [15], and high surface area that make them superior over the conventional detection systems.

Carbon nanotubes are carboxylated (addition of —COOH group) for attachment of protein or DNA probe molecules. They are characterized for the presence of —COOH groups using Fourier transform infrared spectroscopy (FT-IR). The Fourier transform infrared spectroscopy technique is used for measuring the infrared (IR) absorption and emission spectra of most materials. The advantages of this technique are in signal-to-noise ratio, resolution, speed, and detection limits. The FT-IR technique is superior over other spectroscopic methods as practically all compounds show characteristic absorption and emission in the IR spectral region, and based on this property they can thus be analyzed both quantitatively and qualitatively.

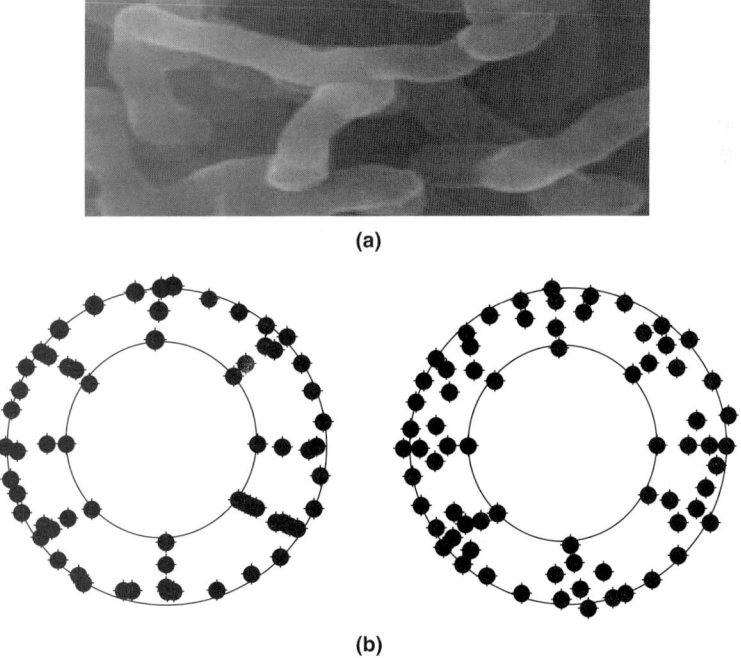

FIGURE 13.16 Schematic of carbon nanotubes. (*a*) Structure of carbon nanotube. (*b*) Schematic of concentric multi-walled carbon nanotubes.

13.3.2 Surface Modification and Biofunctionalization of Signal Conversion Component

Surface modification is a process of treatment by attaching functional inorganic or organic materials molecules on the surface of the material component. It is to achieve improved surface properties for specific applications requirements. For efficient signal conversion, apart from selecting an efficient signal conversion component, proper surface modification is essential.

The attachment or immobilization of probe biomolecules to the surface of the signal conversion component is a critical step in the construction of a nucleic acid and protein-based biosensors. DNA, RNA, and proteins may be passively immobilized through hydrophobic and ionic interactions. However, covalent immobilization is often necessary for proper adsorption, orientation, and conformation to the noncovalent surfaces. The immobilization process should occur rapidly and selectively in the presence of common functional groups, including amines (NH_2), carboxyl (COOH), hydroxyl (OH), thiol (SH), and aldehyde (CHO) groups. The surface density of the molecules should be optimized. Low-density surface coverage yields a correspondingly low frequency of binding sites. High density may result in inaccessible binding sites to the target molecules. The correct orientation of probe molecules on the surface is required to facilitate the probe available for the target to bind. For efficient immo-bilization, the maximum specific interactions of probe and target and minimum nonspecific interactions are required. Alkanethiol of methyl, biotin, COOH, CONHS, and NH_2-terminated are the main functional groups used to establish covalent linkage. The photonic activation of disulfide bridges was shown to orient protein immobilization on biosensor surfaces.

As an example, surface modification and biofunctionalization of carbon nanotubes (CNTs), and detection of biological target molecules are schematically shown below:

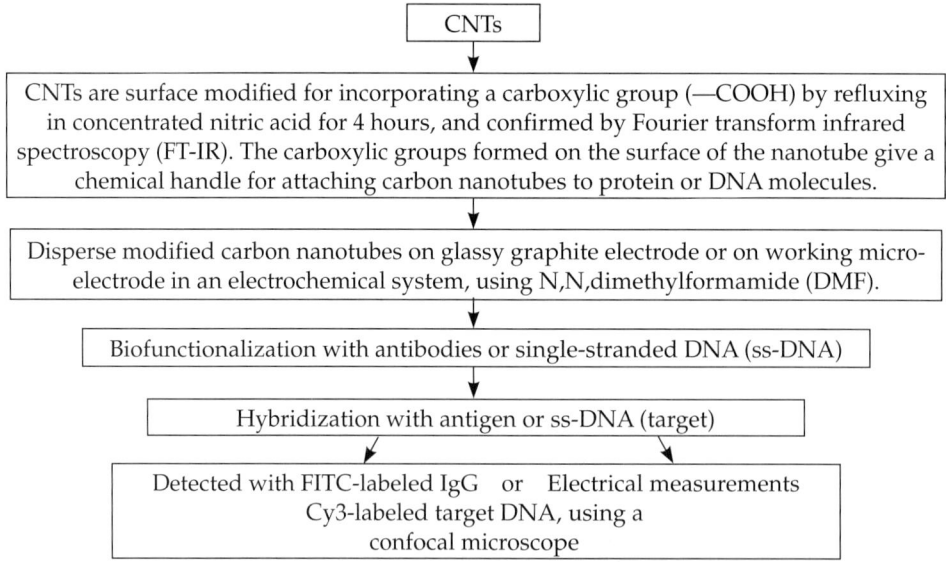

As shown in Figure 13.17, CNTs were functionalized with IgG antibodies and detected with fluoresceir isothiocyanate (FITC) labeled IgG.

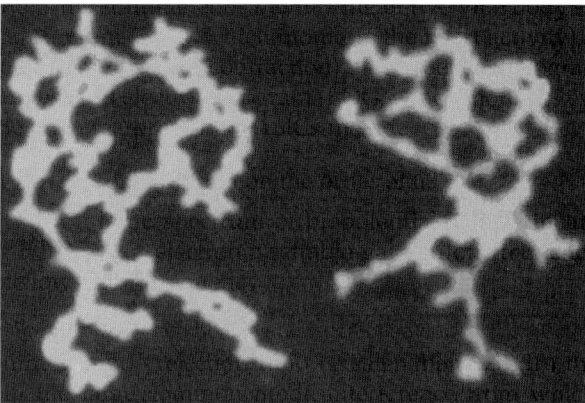

FIGURE 13.17 Carbon nanotubes showing protein hybridization, and visualization using FITC-conjugated antibodies and confocal microscope (Courtesy of Packaging Research Center, Georgia Institute of Technology).

13.3.3 Signal Conversion Methods

The biosignals are converted into electrical, optical, and electromagnetic signals by a variety of signal conversion technologies as described below.

- Electrochemical (conductimetric, amperometric, and potentiometric)
- Optical
- Acoustic wave
- Microelectromechanical (resonant)
- Magnetic
- Thermal

Electrochemical

The chemical reactions brought about by electricity are known as electrochemical reactions. In the electrochemical method, chemical reactions produce or consume ions or electrons which in turn cause some changes in electrical properties (current, impedance, etc.) of the sensor device or its surrounding environment. Electrochemical reactions are measured using a three-electrode electrochemical cell containing a biosensing modified working electrode, an Ag/AgCl reference electrode, and a Pt counter electrode. For example, DNA detection was demonstrated using the electrochemical method [16]. A polydimethylsiloxane (PDMS) fluid cell (5 μl volume) was pressurizedly sealed between a DNA-modified Si sample (8 mm × 4 mm) and the Pt counter electrode (Figure 13.18). Multiwalled carbon nanotubes were dispersed on the working electrode for efficient signal conversion. The probe DNA attached to the working electrode hybridizes with the DNA present in the test sample and causes a change in impedance. Protein also can be detected by amperometric measurements in electrochemical cells.

Microelectrochemical detection is a miniaturized version of a conventional bulky electrochemical biosensor system. It is used to detect glucose and other biological molecules in a small volume of samples and at low concentrations. For example, carbon nanotubes (CNTs) and sol-gel are used for fabrication of efficient signal conversion

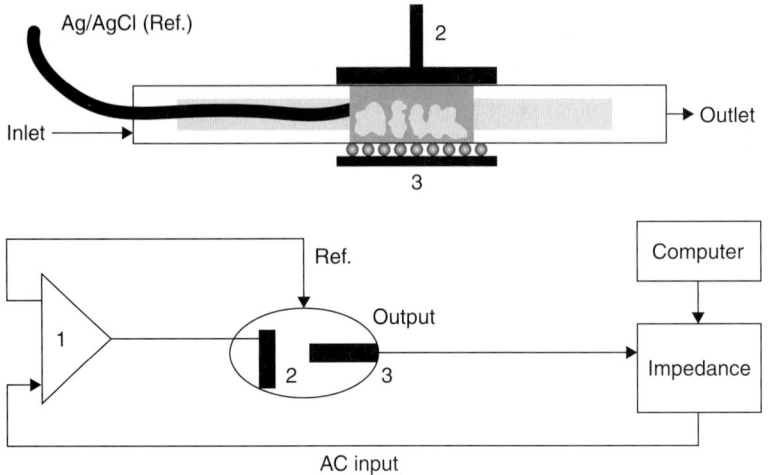

Figure 13.18 Schematic of three-electrode chemical fluidic cell biosensor: 1. Potentiostat. 2. Platinum foil counter electrode. 3. Working electrode as DNA modified silicon sample. (Courtesy Dr. Wei Cai, University of Wisconsin, Madison.)

component in electrochemical system [17]. CNTs possess unique charge-transport characteristics, and zirconia-Nafion sol-gel encapsulation protects the enzyme. The carboxylated CNTs are dispersed on a working electrode, and subsequently zirconia-Nafion sol-gel encapsulated enzyme is cast on CNTs modified working electrode (Figure 13.19a). The enzymatic oxidation of glucose by glucose oxidase (GOx) in the electrochemical cell produces hydrogen peroxide, which in turn generates electrons from the electrode reaction. The enzyme used in the reaction promotes or catalyzes the reaction. The current density (amperometric) is used as a measure of glucose concentration in the sample. The amperometric measurements of glucose concentration are recorded. High sensitivity is demonstrated using a microelectrochemical system (Figure 13.19b).

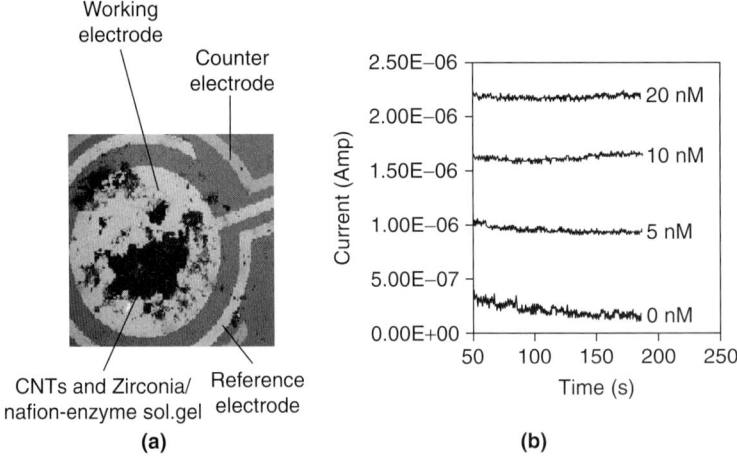

Figure 13.19 (a) Microelectrochemical device for detection of glucose. (b) Data on electrical detection of glucose. (Courtesy Packaging Research Center, Georgia Institute of Technology.)

Optical

In the optical signal conversion method, the bioprobe and target hybridization signals are converted into optical signals. For example, surface plasmon resonance (SPR) is an advanced optical signal conversion technique. The basic SPR configuration consists of a prism or glass slide coated with a thin metal film, usually silver or gold (Figure 13.20). Bioprobe such as DNA or protein is immobilized on the thin gold film. When the light is passed through the prism and onto the gold surface at angles and wavelengths near the surface plasmon resonance condition, the optical reflectivity of the gold changes very sensitively in the presence of biomolecules on the gold surface. The high sensitivity of the optical response is due to very efficient collective excitation of conduction electrons near the gold surface.

The optical components in an optical sensor could be fiber-optics, waveguides, photodiodes, spectroscopy, charge coupled devices (CCDs), and interferometers. The changes in intensity, frequency, phase shift, and polarization are measured. The types of measurements involved are absorbance, fluorescence, refractive index, and light scattering. Surface plasmon resonance is particularly useful for the detection of biological molecules: (1) they can be extremely sensitive (nanomoles or less) and (2) they are nondestructive to the sample. The optical conversion technique offers an inexpensive system for the detection of heavy metals, particularly the high toxic arsenic in the environment [18]. In the optical method, no reference electrode is needed, and there is no electrical interference.

Acoustic Wave

An acoustic wave is a kind of wave used in piezoelectric components, such as quartz crystal, in circuits. The frequency characteristic varies in response to the surrounding environment. In the acoustic wave signal conversion method, an applied radiofrequency produces mechanical stress in the piezoelectric component. As a result, surface acoustic wave (SAW) is induced. The SAW is received by the electrodes and is translated to voltage (Figure 13.21). The probe and target molecular hybridization on the surface of piezoelectric components is converted into the change of voltage. This method is capable of detecting the presence of low mass such as a single molecule or cell.

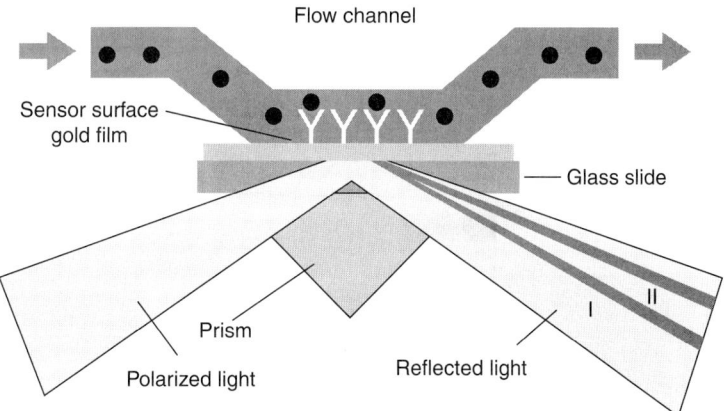

Figure 13.20 Schematic of optical detection of biological samples on gold or silver metal surfaces. (Courtesy Dr. Charles T. Campbell, Univercity of Washington, Seattle).

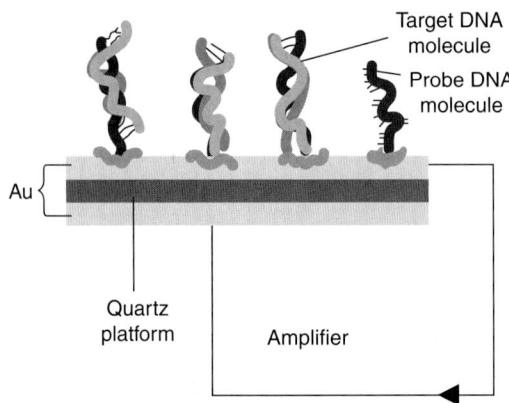

Figure 13.21 Schematic of an acoustic wave biosensor detecting a probe and target molecular binding on the surface of quartz platform causing a difference in the voltage.

Micro-electro-mechanical

Micro-electro-mechanical systems (MEMS) are devices that combine electrical and mechanical functions. MEMS-based devices, such as nanocantilevers and microresonators, can be used as the signal conversion components. They are extremely sensitive and capable of measuring attogram (10^{-18} grams) mass change, aiding in single molecule or cell detection, and superior over other signal conversion methods.

In microcantilevers, when probe and target hybridization occurs on the cantilever beam, molecular binding surface tension causes the cantilever beam to bend and deflect. As shown in Figure 13.22, the deflection of the beam is measured optically.

The microresonators and quartz crystals can be made to vibrate at a specific frequency with the stimulation of an electrical signal at that specific frequency. When the mass increases due to binding of chemicals or biomolecules, the oscillation frequency of the device changes and the resulting change can be measured electrically

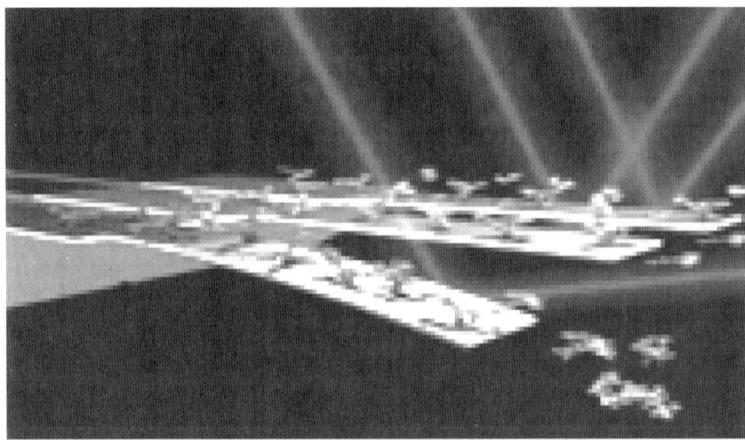

Figure 13.22 Microcantilever device detects prostate-specific antibodies (PSA) in male blood. (Courtesy Dr. Arun Majumdar, University of California, Berkeley.)

and used to determine the additional mass of the device. In resonant-based MEMS, the frequency changes with the change in mass of hybridized probe and target molecules (Figure 13.23c and d).

Semiconducting and piezoelectric nanostructures such as single-crystal zinc oxide (ZnO) nanobelts. (Figure 13.23a and b) are used as sensitive signal conversion components, capable of detecting a single molecule or cell. These nanobelts induce a field effect in the presence of charged biological molecules such as proteins, peptides, DNA, and RNA [19]. ZnO and Si nanowires are used as an efficient signal conversion component [20].

Magnetic

In the magnetism-based signal conversion method, magnetic nanoparticles are used. Metals such as nickel (Ni), cobalt (Co), Fe, Fe_3O_4, and $\gamma\text{-}Fe_2O_3$ and their oxides exhibit easily detectable magnetic properties. They are used to synthesize magnetic beads or particles of length scales ranging from 1 to 100 nm. The magnetic nanoparticles are functionalized with biological molecules such as DNA or protein for signal con-version. The hybridization events of probe and target molecules can be detected by measuring the change in magnetic forces using an atomic force microscope (AFM) [21]. Figure 13.24a and b show antibodies (probe) attached to the nanoparticles and hybridization with antigens (target).

In another example of magnetic-based DNA detection, streptavidin functionalized magnetic beads (Figure 13.25b) hybridize with biotinylated DNA (Figure 13.25a), and the hybridization event is detected by a change in resistance from a giant magnetoresistive (GMR) element [22]. Streptavidin is a protein with a very high affinity for biotin [23]. Because of its high affinity for biotin, it is used to bridge biotinylated probes and biotinylated enzymes. The GMR element shows large resistance changes in a magnetic field for certain materials composed of alternating thin layers of various metallic elements.

The magnetic particles used for force amplified biological sensors (FABSs) must be smooth and spherical so that the entire surface of each particle is available for binding to the flat cantilever surface. Irregular particles effectively have a reduced active area.

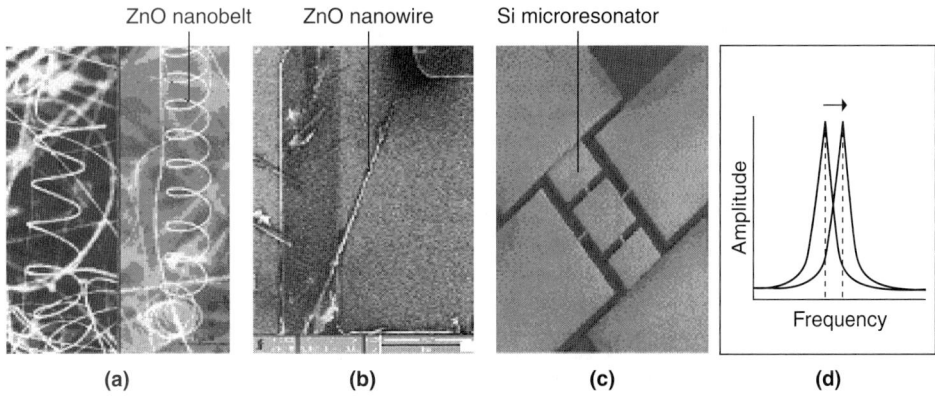

FIGURE 13.23 (a) Single-crystal zinc oxide (ZnO) nanobelts. (b) ZnO nanowire mounted between the electrodes (Courtesy Dr. Z. L. Wang, Georgia Institute of Technology). (c) Si microresonator, capacitive, or piezoelectric MEMS. (d) Frequency shift of MEMS due to mass detection (Courtesy of Dr. Farrokh Ayazi, Georgia Institute of Technology).

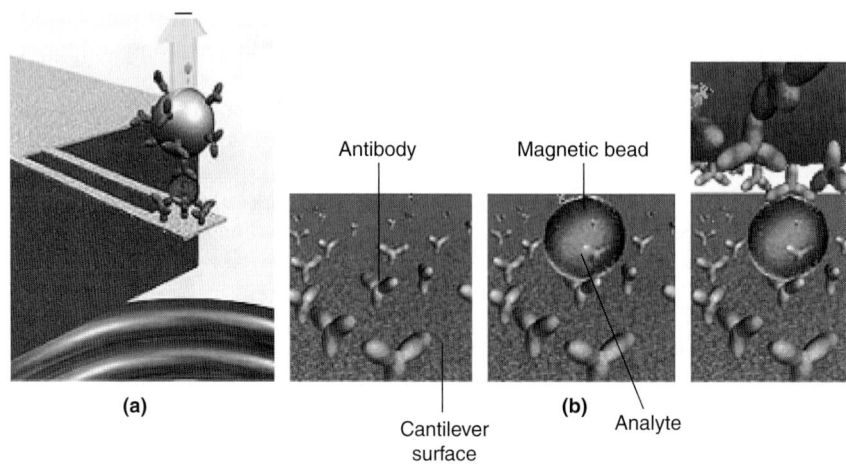

FIGURE 13.24 (a) Magnetic biosensor. (b) Antibodies labeled magnetic particles are attracted in the magnetic field. (Courtesy Dr. Baselt, Institute of Nanoscience, Naval Research Laboratory, Washington DC.)

The particles should exert a uniform force in a magnetic field, and a high magnetic moment for a maximum signal. These particles must be corrosion resistant to physiological solutions, low density, and initially nonmagnetic to avoid coagulation in the solution.

Thermal

Thermal sensing conversion is a measure of fluxes and changes of temperature. In the thermal signal conversion method, the heat generated in enzymatic processes is

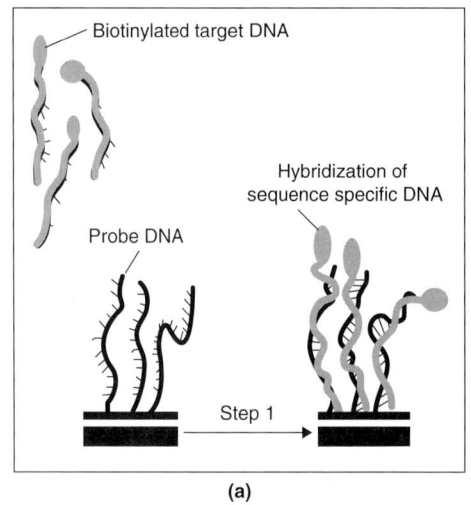
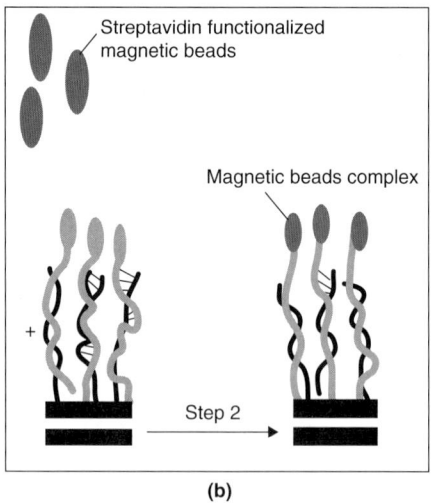

FIGURE 13.25 (a) Schematic of hybridization of target DNA with probe DNA, (b) followed by hybridization of streptavidin functionalized magnetic bead with biotin. (Courtesy Daniel Graham, Hugo Ferreira, Paulo Freitas, INESC-MN, IST, Lisbon, Portugal [22]).

measured. Thermal biosensors are used for monitoring biotechnological processes. The thermal signal conversion is an enzyme thermistor modified for split-flow analysis. For example, a miniaturized thermal binominal conversion involves a flow-injection analysis system for the determination of glucose in whole blood. The concentration of blood glucose is determined by measuring the heat evolved when samples containing glucose are passed through a small column with immobilized glucose oxidase and catalase. A sample of whole blood as small as 1 µL can be measured directly using this thermal biosensor [24].

13.4 Signal Detection and Electronic Processing

Design, fabrication, surface modification, hybridization of sensing elements, and signal conversion methods are dealt with in the previous sections. This section deals with signal processing, detection, and SOP packaging and integration technologies.

Some of the processing and detection components that are essentially needed for bio SOP include low-power application-specific ICs, high-density charge structure components, and high-efficiency power transmission components.

13.4.1 Low-Power Application-Specific Integrated Circuits (ASICs) and Mixed-Signal Design for Biosensor SOP

Application-specific integrated circuit (ASIC) technologies evolved tremendously due to increasing system complexity, increasing cost pressures, continually shrinking available space for electronics, and higher quality that meet specific customer needs. Ultralow-power ICs and new design strategies for mixed-signal functions involving digital, analog, and transducer interface electronics are pertinent for efficient signal processing from biosensors and biomicroelectronic devices [25]. Architectures and circuits are being developed for low-power, low-voltage mixed signals, sensors, and telemetry building blocks integrated in an SOP platform.

Body monitoring wireless sensor nodes that collect and process data from human body sensors and wirelessly transmit the data to a central monitoring and processing system were developed by IMEC, Belgium [26]. A typical two-channel biopotential wireless node targets the simultaneous monitoring of two vital body signs provided by portable electrocardiogram (ECG, which monitors the heart activity), electromyogram (EMG, which monitors muscle contraction), electroencephalogram (EEG, which monitors brain waves) and electrooculogram (EOG, which monitors eye movement). In such a system, the main features of the ASICs include

1. Ultralow power consumption on the order of 60 µW
2. Suitability for a large spectrum of biopotential signals including EEG, ECG, EMG, or EOG by an electronic settable gain cutoff frequency and extremely adjustable low-cutoff frequency
3. Low noise
4. High common mode rejection ratio (greater than 120 dB) that is required to cope with the large common mode interference from which the microvolts range signals suffer
5. Capable of filtering 50-mV dc offset generated by the biopotential electrodes

In the preceding example, the bio SOP system consists of a wireless sensor node that integrates two of the one-channel biopotential EEG, EMG, ECG, and EOG read out ASICs, allowing simultaneous monitoring of two biopotential signals, a commercial microprocessor, and a 2.4-GHz radio link. It also includes a human body–adapted loop antenna offering a range up to 10 m. The system has a power consumption of about 5 to 10 mW, allowing 12 to 24 hours continuous measurements when operated from a prismatic or coin cell rechargeable battery. The SOP technology enables the integration of active devices with ultrathin passive components and integrable antenna on an organic substrate. A typical readout circuitry for MEMS-based biosensor SOP is shown in Figure 13.26.

High-Density Charge Structure Elements: 3D Capacitors

Bio SOP with implantable biosensors and biomicroelectronic devices require a continuous low-power source for functioning of the implants and signal transduction. As an example, 3D ferroelectric capacitors display a large polarization increase compared to planar capacitors due to the additional electrical contribution of the capacitor sidewalls, as shown in Figure 13.27. Capacitance per unit of cell area in silicon MOS IC technologies can be increased by etching convolutions into the semiconductor surface beneath the electrode. Three-dimensional capacitors are efficient storage devices useful in aerospace, military, satellite, biomedical, industrial, and communication electronics.

High-Efficiency Power Transmission: Microelectrode Stimulation and Inductive Telemetry

Another important system processing element of bio SOP is power generation and signal transmission. The miniaturized modules such as biosensors and biomicroelectronic devices are implanted in the body with an objective of providing a better quality of life. An important issue is how to power the modules and transmit the data. A strategy to reduce the power consumption and energy scavenging extracting energy from ambient sources such as vibrations, heat, light, and water needs to be developed. As shown in Figure 13.28a, in a wireless implantable microsystem, sensors and on-chip signal

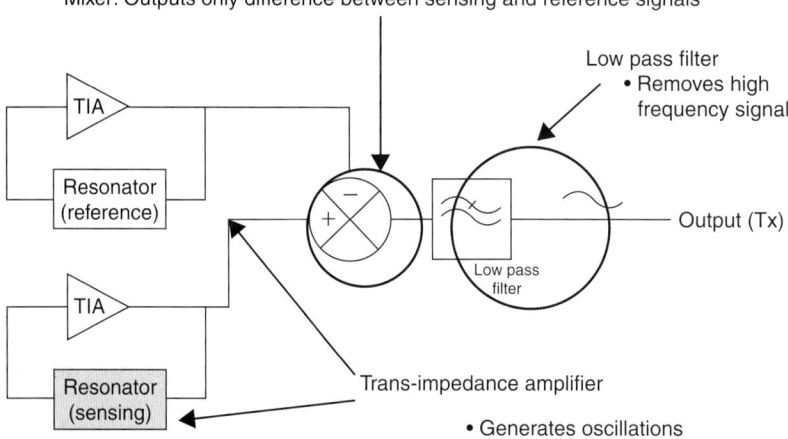

Figure 13.26 Schematic of sensing and readout circuitry for MEMS-based biosensor. (Courtesy Dr. Farrokh Ayazi, Georgia Institute of Technology).

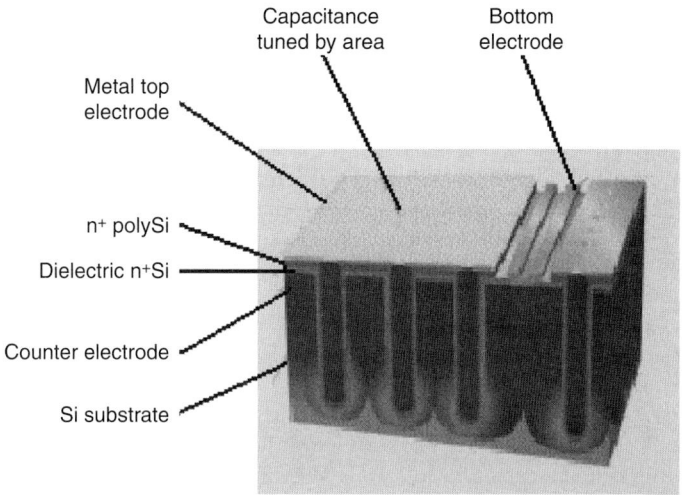

FIGURE 13.27 3-D Capacitors. (Courtesy Aden Dekker et al. NXP Semiconductors and Phillips Applied Technologies, The Netherlands).

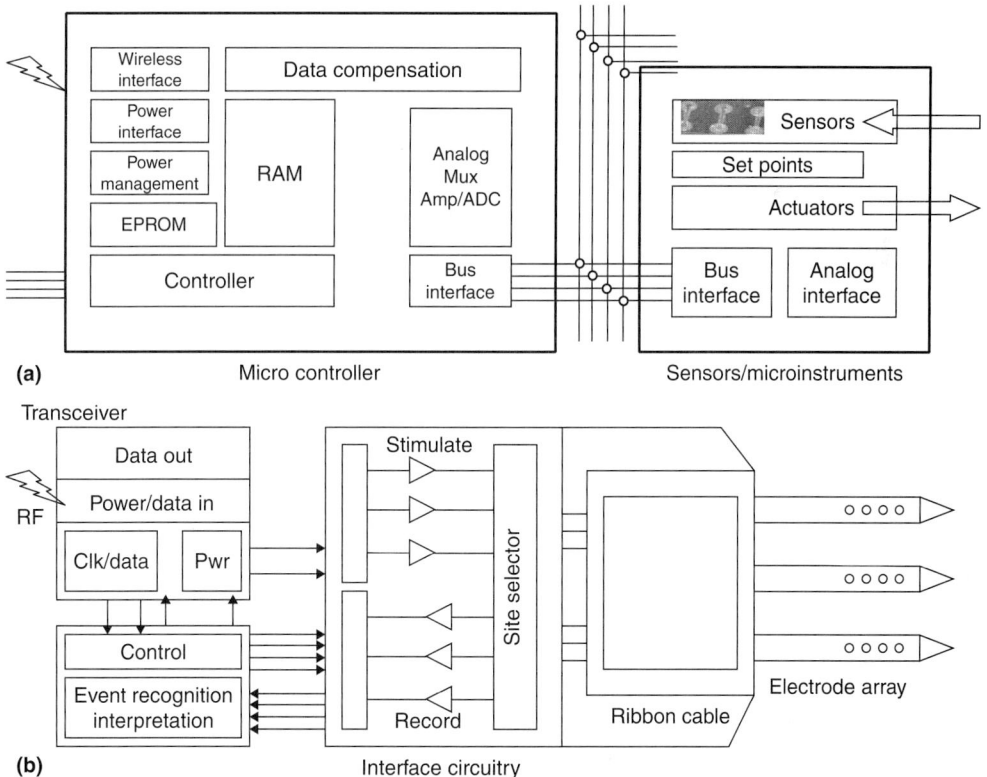

FIGURE 13.28 (a) Schematic of a wireless integrated microsystem. (b) Implantable version of the same microsystem. (Courtesy Dr. Kensall and D. Wise, University of Michigan.)

conditioning circuitry are interfaced with an embedded microprocessor. The energy source for many biomedical microsystems is through an external RF link (Figure 13.28b). The system communicates with the outside world over an inductively coupled bidirectional wireless link [27].

13.4.2 Bio-SOP Substrate Integration Technologies

Bio-compatible Packaging

New generation of biocompatible functional materials is of primary importance for the fabrication of bio SOP with medical implants, biosensors and biochips, and scaffolds for tissue engineering. Biocompatible biosensor and biochip packaging requires specialized material properties and characteristics of biocompatibility, reliability, and manufacturing for in vivo applications.

In vivo medical implants, including biosensors and biochips, should not interact with body tissues and fluids, not promote inflammatory reactions, and not exert toxicity to the implanted host or individual. The implanted biosensors and biochips should remain in the host body inert and stable. In vitro application of biosensors and biochips also should remain inert to corrosion, itching, pH change, and chemical reactivity. The sensing or biorecognition elements such as protein, peptide, enzyme, DNA-RNA, whole cell, or tissue should remain intact structurally and functionally during fabrication and in vitro or in vivo applications.

Integration of Microfluidic and Nanofluidic Channels

Microfluidic channels for guiding biofluids, interconnects, and interfaces connecting mechanical components such as microvalves and micropumps to electronic devices can be realized in biocompatible polydimethylsiloxane (PDMS), LCP-type SOP substrate materials. The supporting polymer network is cross linked and covalently bonded to the substrate. The polymer network should be hydrophilic and nonswellable in water. In the presence of water or body fluids, the hydrophilic polymer adsorbs water molecules to create a watery interface at the surface of the device. The water layer reduces wet friction, protein adsorption, and cell adhesion. The plasma treatment in the parylene-PDMS bilayer coatings offer enhanced protection capability for advanced packaging applications. The hybrid gels prepared using methyltriethoxysilane (MTES) and dimethyldiethoxysilane (DMDES) provide efficient hermetic barriers against humidity for microelectronics, micro-electro-mechanical systems (MEMS), liquid-crystal displays (LCDs), and thin-film transistors (TFTs).

Electromagnetic and Thermal Interference

Implanted biosensors and medical devices are prone to interfere with their functions due to exposure to several electromagnetic sources, including power-line, theft-protection gates, cell phones, home electronic appliances, and security zones. They are also prone to interference with medical equipment such as CT and MRI scanners. The magnetic field effect on the implanted medical devices should be prevented by careful fabrication design and packaging. Medical devices like defibrillators generate significant localized heat in the high-voltage-charge circuit when in use. The temperature rise could result in a dielectric breakdown of the printed circuit board (PCB) and/or substrate, or the failure of incorporated field-effect transistors (FET) and capacitors. Necessary packaging strategies are needed for optimum functioning of the implants.

13.5 Summary and Future Trends

In the past decade, there have been dramatic improvements in the development of biosensor technologies for their miniaturization. These changes are attributed mainly to the need for inexpensive, easy-to-handle, miniaturized biosensor devices for fast, reliable, label-free, and electrical detection of biological molecules such as proteins, peptides, enzymes, DNA, RNA, and whole cells. At present, commercially available biosensors are bulky and expensive and their market therefore is not pervasive. The discovery of nanomaterials with novel properties, their micro- and nanofabrication technology advances, together with advances in molecular biology and bioengineering, combined with ultraminiaturization of electronics enabled by SOP are expected to revolutionize the biosensor systems and products. In addition, genetically engineered biological receptors and genetically transformed cells will form the basis for biorecognition components of the future biosensor devices. The commercial value of biosensors is reported to be immense. The global market for biosensors and other bioelectronics is projected to grow from $8.2 billion in 2009, at an average annual growth rate (AAGR) of 6.3 percent. The U.S. National Science Foundation predicts that the market for nanotechnology-based products, including biosensors, will reach $1 trillion within two decades [28]. The projected market for nano biosensors is $100 million [29].

The integration biosensors with wireless communication systems are expected to provide high-quality information from networked, autonomous analytical stations. One example is Toumaz's Sensium sensor interface platform [30], which includes a reconfigurable sensor interface, a digital clock with low-power microprocessor core, and an RF transceiver block. The system needs a very small battery enabling it to be body-worn with complete freedom of movement. The system can even be attached to the body with a sticking plaster to form what is known as "digital plaster," offering weeks or months of operational time and not requiring any battery charge.

From the foregoing discussion, it is reasonable to assume that the future of biosensors relies on the successful implementation of micro- and nanotechnologies in the form of ultrathin embedded components using ultrathin SOP substrates. The nano-biosensing components, micro-nano fluidic elements, and SOP challenges faced by integration technologies are illustrated in Figure 13.29.

13.5.1 Nano Bio-SOP Integration Challenges

- The research challenges that need to be addressed in integrating biosensors with system-on-package (SOP) technology are as follows:
- Optimized performance of the sensor with associated electronics, fluidics, and integrating with system-on-package (SOP) technology
- The scaling issues involved in miniaturization process without detriment to overall functionality of biosensor
- Development of organic-compatible substrates for the embedded biocomponents
- Development of biocompatible materials for implantable device packaging
- Development of toxic-free implantable sensor devices for in vivo applications
- Proper retention of specificity and sensitivity of complex biological molecules without inherent instability

746 Chapter Thirteen

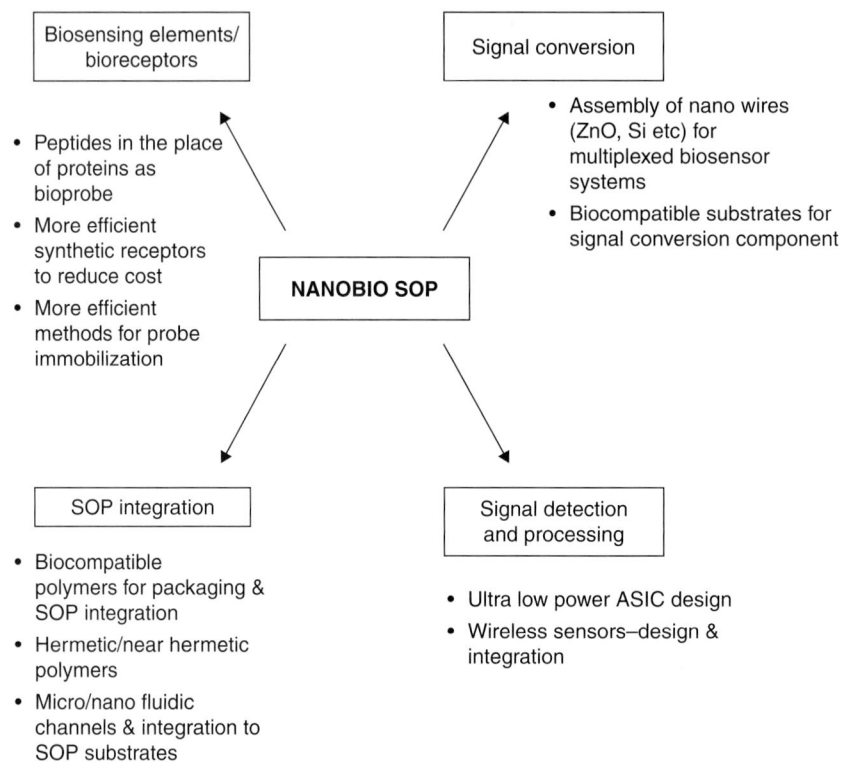

FIGURE 13.29 Schematic of illustration of building blocks of nano bio-SOP and associated challenges.

- Development of toxic-free implantable sensor devices for in vivo applications such as targeted drug delivery, MRI contrast agents, and pathogen cleansing
- Preventing biosensor devices from biofouling, as protein builds up on the biological active interfaces
- Development of new low-cost manufacturing processes for large-scale production

References

1. D. Dell'Atti, S. Tomball, M. Minnie, and M Mascagni, "Detection of clinically relevant point mutations by a novel piezoelectric biosensor," *J. Biosensors and Bioelectronics*, 2006, vol. 27, no. 10, 2006, pp. 1876–79.
2. A. R. Toppozada, J. Wright, A. T. Eldefrawi, M. E. Eldefrawi, E. L. Johnson, S. D. Emche, and C. S. Helling, "Evaluation of a fiber optic immunosensor for quantitating cocaine in coca leaf extracts," *Biosense Bioelectron*, vol. 12, no. 2, 1997, pp. 113–24.
3. F. Nuwaysir, W. Huang, T. J. Albert, S. Jaz, K. Nuwaysir, A. Pitas, T. Richmond, T. Gorski, J. P. Berg, J. Ballin, M. McCormick, J. Norton, T. Pollock, T. Sumwalt, L. Butcher, D. Porter, M. Molla, C. Hall, F. Blattner, M. R. Sussman, R. L. Wallace, F. Cerrina, and R. D. Green, "Gene expression analysis using oligonucleotide arrays

produced by maskless photolithography," *Genome Research*, vol. 12, no. 11, 2002, pp. 1749–55.
4. C. Zhou, P. Pivarnik, A. G. Rand, and S. V. Letcher, "Acoustic standing-wave enhancement of a fiber-optic Salmonella biosensor," *Biosense Bioelectron*, vol 13, no. 5, 1998, pp. 495–500.
5. J. Liu, G. D. Janagama, M. K. Iyer, R. R. Tummala, and Z. L. Wang, "ZnO nanobelts/wire for electronic detection of enzymatic hydrolysis of starch," *International Symposium and Exhibition on Advanced Packaging Materials—Processes, Properties and Interfaces,* Atlanta, USA, March 15–17,2006, pp.104–06.
6. L. C. Shriver-Lake, B. L. Donner, and F. S. Ligler, "On-site detection of TNT with a portable fiber optic biosensor," *Environ. Sci. Technol.*, vol. 31, 1997, pp. 837–41.
7. F. Ragan, M. Meaney, J. G. Vos, B. D. MacCraith, and J. E. Walsh, "Determination of pesticides in water using ATR-FTIR spectroscopy on PVC/chloroparaffin coatings," *Anal. Chim. Acta*, vol. 334, 1996, pp. 85–92.
8. R. A. Potyrailo, S. E. Hobbs, and G. M. Hieftje, "Near-ultraviolet evanescent-wave absorption sensor based on a multimode optical fiber," *Anal. Chem.*, vol. 70, 1998, pp. 1639–45.
9. D. James Weiland et al., "Retinal prosthesis testbed," University of Southern California (USC) website, http://bmes-erc.usc.edu/research/retinal-prosthesis-testbed.htm, 2007.
10. C. J. Choi and B. T. Cunningham, "Single-step fabrication and characterization of photonic crystal biosensors with polymer microfluidic channel," *Lab Chip*, vol. 6, 2006, pp. 1373–80.
11. S. I. Woods, J. R. Kirtley, S. Sun, and R. H. Koch, "Direct investigation of superparamagnetism in Co nanoparticle films," *Phys. Rev. Lett.*, vol. 87, no. 13, 2001, pp. 137205-1–1327205-4.
12. D. K. Kaplan, V. A. Sverdlov, and K. K. Likharev, "Coulomb gap, coulomb blockade, and dynamic activation energy in frustrated single-electron arrays," *Phys. Rev. B.*, 2003, vol. 68, 2003, pp. 045321-1–045321-6.
13. M. Han, X. Gao, J. Su, and Shuming Nie, "Quantum-dot-tagged microbeads for multiplexed optical coding of biomolecules," *Nature Biotechnology*, vol. 19, 2001, pp. 631–35.
14. I. Shorubalko, P. Ramvall, H. Q. Xu, I. Maximov, W. Seifert, P. Omling, and L. Samuelson, "Coulomb blockade and resonant tunneling in etched and regrown Ga0.25 In 0.75As/InP quantum dots," *Semicond. Sci. Technol.*, vol. 16, 2001, pp. 741–44.
15. S. Ijima, "Helical microtubules of graphitic carbon," *Nature*, vol. 222, no. 354, 1991, pp. 56–58.
16. W. Cai, J. R. Peck, D. W. van der Weide, and R. J. Hamers, "Direct electrical detection of hybridization at DNA-modified silicon surfaces," *Biosensors and Bioelectronics,* vol. 19, no. 9, 2004, pp. 1013–19.
17. D. G. Janagama, J. Liu, P.M. Raj, M. K. Iyer, and R. R. Tummala, "Electrochemical Biosensors and microfluidics in organic system-on-package," *Proceedings of 57th Electronic Components and Technology Conference*, Reno, USA, May 2007 pp. 1550–55.
18. C. Durrieu and C Tran-MinhOptical algal biosensor using alkaline phosphatase for determination of heavy metals," *Ecol. Toxi. and Eviron. Safety*, vol. 51, no. 3, 2002, pp. 206–09.

19. J. Liu, D.G. Janagama, P. M. Raj, M. K. Iyer, R. R. Tummala and Z. L. Wang, "Label-free protein detection by ZnO nanowire based biosensors", Proceedings of *57th Electronic Components and Technology Conference*, Reno, USA, May 2007, pp. 1971–76.
20. Yi Cui, W. Qingqiao, P. Hongkun, and C. M. Lieber, "Nanowire nanosensors for highly sensitive and selective detection of biological and chemical species," *Science*, vol. 17, no. 293, 2001, pp. 1289–92.
21. D. R. Baselt, G. U. Lee, M. Natesan, S. W. Metzger, P. E. Sheehan, and R. J. Colten, "A biosensor based on magnetoresistance technology," *Biosensors and Bioelectronics*, vol. 13, 1998, pp. 731–39.
22. D. L. Graham, H. A. Ferreira, and P. P. Freitas, "Magnetoresistive-based biosensors and biochips," TRENDS in Biotechnology, vol. 22, no. 9, 2004, pp.1–8.
23. P. R. Langer, A. A. Waldrop, and D. C. Ward, "Enzymatic synthesis of biotin labeling polynucleotides: novel nucleic acid affinity probe," *Proc. Natl. Acad. Sci USA*, vol. 78, 1981, pp. 6633–37.
24. U. Harborn, B. Xie, R. Venkatesh, and B. Danielsson, "Evaluation of a miniaturized thermal biosensor for the determination of glucose in whole blood," *Clin Chim Acta*, vol. 267, no. 2, 1997, pp. 225–37.
25. G. Wang, L. Wentai, S. Mohansankar, M. Zhou, J. D. Weiland, and M. S. Humayun, "A dual band wireless power and data telemetry for retinal prosthesis," *Proc. of 3rd International IEEE EMBS Conf. on Neural Engineering*, NY, USA, August 2006, pp. 4392–95.
26. M. Quwerker; F. Pasyeer and N. Engin, "SAND: a modular application development platform for miniature wireless sensors," Proceedings of International Workshop on Wearable and implantable body sensor networks; Cambridge, MA, USA, April 2006, pp. 5.
27. K. D. Wise, "Wireless integrated microsystems: coming breakthroughs in health care," *Proc. International Electron Devices Meeting*, San Francisco, USA, Dec. 11–13, 2006, pp. 1.1.1–1.1.8.
28. "The projected market for biosensors and bioelectronics," BCC research market study report, April 2005.
29. "The projected market for nanobio sensors," BCC research report ID: NANOO35A, December 2004.
30. "Advance product information," Toumag Technology Limited. Ultra low power wireless body monitoring, *www.toumaz.com*.

Index

1D (one-dimensional) EBGs, 240–244
1G (first generation) cellular standards, 264
2D (two-dimensional) EBGs, 240–241, 244–246
2G (second generation) cellular standards, 264
3D (three-dimensional) integration, 16
3D capacitors, 742
3D integration. *See* system-in-package technology
3D lightwave circuits, 354–356
3D optoelectronic SOP, 354
3D silicon circuits/wiring, 83
3D WLSOP, 590–591
3G (third-generation) cellular standards, 264
100-im flip-chip assemblies, 471–476
850-nm VCSELs, 334
1310-nm EELs, 334

A

abstraction, 67–68
ACAs (anisotropic conducting adhesives), 485
accelerated thermal cycling (ATC) conditions, 452
acoustic wave signal conversion, 737–738
active amplifiers, 613
active cooling, 643
active methods of thermal management
  liquid loops, 643
    spray cooling, 643–644
      thermionic cooling, 645–646
      thermoelectric cooling, 644–645
      thermosyphones, 644
      TIMs, 646–648
      vapor compression refrigeration, 646
active optoelectronic components, 341
active power dissipation, 228
ADCs (analog-to-digital converters), 74, 613, 686, 703
add-in cards, 337
adenosine and thymine (A-T), 728
adenosine and uracil (A-U), 728
adhesion-related effects, 405
adhesive-substrate interfaces, 480
adiabatic conditions, 621
ADS (Agilent Advanced Design System), 197, 216–217
adsorption theory, 480–481
AFMs (atomic force microscopes), 739
Agilent Advanced Design System (ADS), 197, 216–217
AI-EBGs (alternating impedance-EBGs), 241, 246–250
air-prism-grating configuration, 352
AlGaInAs (aluminum-gallium-indium-arsenic) based transmitter lasers, 334
Alpha 21264 processors, 612
alternating impedance-EBGs (AI-EBGs), 241, 246–250
aluminum-gallium-indium-arsenic (AlGaInAs) based transmitter lasers, 334
ALUs (arithmetic logic units), 50
ambient heat rejection, 625
Amdahl, Gene, 9
amino acid, 725
amorphous transistors, 363

amplifiers
  active, 613
  block-level receivers, 698
  limiting (LA), 340, 358–359
  low noise (LNAs), 25, 181–184, 214, 217
  microwave, 312, 326
  PD array, 332, 358–359
  photodetector, 321, 333, 354
  power (PAs), 25, 613
  transimpedance (TIAs), 340, 358–359
analog signal generators, 695
analog-to-analog coupling, 214–222
analog-to-digital converters (ADCs), 74, 613, 686, 704
anisotropic conducting adhesives (ACAs), 485
anisotropic elastic bimaterials, 478
anodization, 422
antennas
  conformal, 270–274
  integration, 25
  on magneto-dielectric substrates, 274–277
  mixed-signal architecture, 157
  multiband, 270
  overview, 269–271
Anthony, T. R., 123
antibodies, 725
antigen-antibody binding, 728
antigens, 725
anti-inflammatory drug delivery systems, 721
antireflection coating, 352
aperture coupling, 224–227
application specific integrated circuits (ASICs)
  applications, 486
  digital SOP, 611
  low-power, 741–744
  vendors, 64
application-driven platforms, 61
application-specific instruction-set processors (ASIPs), 48, 650
arched solder column interconnections, 108–109
architectural-level partitioning, 57–59

area array interconnection PoP stacking, 114
arithmetic logic units (ALUs), 50
aromatic polyester polymers, 393
ASET (Association of Super-Advanced Electronics Technologies), 125
Ashai Chemical, 292
ASICs. *See* application specific integrated circuits
ASIPs (application-specific instruction-set processors), 48, 650
assembly-induced stress, 86
Association of Super-Advanced Electronics Technologies (ASET), 125
A-T (adenosine and thymine), 728
ATC (accelerated thermal cycling) conditions, 452
atomic force microscopes (AFMs), 739
ATPG (automated test pattern generation) algorithm, 672
attenuators, 289–290
A-U (adenosine and uracil), 728
automated test pattern generation (ATPG) algorithm, 672
automatic feature extraction scheme, 693
axial flow fans, 634
axial stress, 474

B

Back end of line (BEOL), 14, 17, 136, 435, 541–543
back-grinding, 96
balanced bandpass filters, 176
ball grid array interconnects, 113
ball grid arrays, 55, 340–341, 470
ballout patterns, 86
baluns
  cascaded filter, 177–178
  compensated, 171–172
  filter-balun networks, 175–178
  lumped element, 174–175
  Marchand, 171–172, 175–176
  mixed-signal architecture, 157
  narrow-band, 174
  planar, 173–174
  RF SOP technologies, 296–297
  stripline multilayer, 298
  wideband LCP, 173

Index

bandpass filters, 297, 698
bandwidth (BW), antenna, 269
barium strontium titanate (BST), 420
barium titanate, 416
bathtub failure rate distribution, 334
BBUL (bumpless buildup layer), 381
BCB (benzocyclobutene), 132–133, 391–392, 430
BDV (breakdown voltage), 389, 416, 422
bed-of-nails testing, 677
benzocyclobutene (BCB), 132–133, 391–392, 430
BEOL (Back end of line), 14, 17, 136, 435, 541–543
Bi (Biot number), 622
BiCMOS (bipolar CMOS), 76, 125
bimodal/multimodal distribution, 421
bin structure, 661
biochip packaging, 744
biocompatible biosensors, 744
bio-compatible packaging, 744
biofluid transport, 723–724
biofunctionalization, 734
bioimplantable systems, 718
biological molecular reactions, 724
biomedical implants, 718–723
biomicroelectronic devices, 742
biosensor SOP technologies
   building blocks of, 723
   element design and preparation
     cell probes, 726
     DNA-RNA probes, 725–726
     enzyme probes, 726–727
     overview, 724–725
     protein probes, 724
     synthetic probes, 726
   future trends, 745–746
   highly miniaturized electronic system technology, 717–718
   integration challenges, 745–746
   microchannels for biofluid transport, 723–724
   for miniaturized biomedical implants and sensor systems, 718–723
   overview, 717–723
   probe-target molecular hybridization
     cell probe hybridization, 729
     DNA-RNA hybridization, 728
     enzyme reaction, 730
     protein hybridization, 727–728
     synthetic probes hybridization, 728–729
   signal conversion
     methods of, 735–741
     nanomaterials and nanostructures for, 730–733
     surface modification and biofunctionalization, 734–735
   signal detection and processing, 741–742
bio-SOP substrate integration technologies, 742
Biot number (Bi), 622
biotin, 726
bipolar CMOS (BiCMOS), 73, 125
bismaleimide triazine (BT), 162, 378
BIST (built-in self-test) techniques, 63, 677, 681–682, 688
BIT (built-in test), 685, 690, 692
block-level receivers, 698–700
block-level test generation process, 697
board-level technologies, 450
board-to-board optical interconnects, 332–333
board-to-board optical wiring, 335–337
BOF (bump-on-flex) chip stacking, 109
boiling curve, 626
bond pads, 63, 104
Bosch Gmbh, Robert, 124
Bosch process, 124, 127
BOST (built-off self-test), 682, 688
BOT (built-off test), 685, 680
boundary scan
   history of, 677–678
   key elements of, 678–680
   for SOP, 680–681
BPTs (branch prediction tables), 612
Bragg condition, 349–350
branch prediction tables (BPTs), 612
breakdown voltage (BDV), 391, 416, 422
bridge effect, 405
Brunner, M., 669
BST (barium strontium titanate), 420
BT (bismaleimide triazine), 162, 378
buffer layer, 356

buildup wiring and redistribution
   IC-Package pitch gap
      BEOL materials and processes, 541–543
      overview, 540–541
      Si substrate with polymer-copper wiring, 541
      redistribution layers on Si to close pitch gap, 543–544
built-in self-test (BIST) techniques, 63, 677, 681–682, 688
built-in test (BIT), 685, 690, 692
built-off self-test (BOST), 682, 688
built-off test (BOT), 685, 690
bulk diffraction, 349
bump pitch, 63
bumpless buildup layer (BBUL), 381
bumpless interconnects, 112
bump-on-flex (BOF) chip stacking, 109
buoyancy-induced convection, 623
burn-in stressing, 669
bus usage, 331
BW (bandwidth), antenna, 269

C

Cable Modem Termination Systems (CMTSs), 43
cache memory, 612
cache power minimization, 650–651
cantilever beams, 738
capacitance testing, 666–667
capacitors
   biosensor SOP, 720
   decoupling, 332, 415, 460, 461
   edge-probed, 462
   embedded
      attributes of, 415–416
      classification of, 416–420
      decoupling, 230–239
      design of, 166–167
      films, 435
      processes, 420–424
   integrated, composite-type, 286
   parallel plate, 166–167, 282–283
   planar, 229, 231
   RF
      MIM structures, 282–283
      parameters, 281–282
      TCC properties, 283–287
      thick- or thin-film, 229, 232–239
      ultrathin deposition, 548
capture range, 689
carbon composite cores, 386
carbon nanotubes (CNTs), 389, 580–581, 730, 733–734
carbon-carbon composites (CCCs), 638
carbon-silicon carbide (C-SiC) substrate, 472
card-to-card optical links, 337
card-to-card optoelectronic data transceivers, 336
cascade LNA architecture, 181–182
cascaded filter-baluns, 177–178
catastrophic faults, 663, 674
cavity backed patch antennas (CBPAs), 308
cavity resonator method, 202–203
CBPAs (cavity backed patch antennas), 308
CBSs (cell-based sensors), 726, 729
CCCs (carbon-carbon composites), 638
cell probes, 726, 729
cell-based sensors (CBSs), 726, 729
ceramic crystallization, 418
ceramic insulating cores, 386
ceramic matrix composites (CMCs), 638
ceramic packaging, 83
ceramic polymer composites, 418
ceramic substrate, 267–268
CFD (computational fluid dynamics) codes, 90, 629, 634
CFD/HT (computational fluid dynamics/heat transfer), 630
characterization tests, 660–661
Chebychev filters, 187
chemical and mechanical polishing (CMP), 66, 97
chemical resistance, 389
chemical vapor deposition (CVD)
   CNTs, 581
   laser-assisted, 129
   low-pressure (LPCVD), 130
   metalorganic (MOCVD), 281
   plasma-enhanced (PECVD), 126, 136, 581
chip assembly technologies, 81
chip carriers, 92
chip power, 56, 606, 650
chip stacking

embedded IC, 109–110
versus package, 119–120
side termination, 105–109
silicon integration comparison, 83
TAB, 111–112
wafer thinning, 95–98
wire bonded, 98–105
chip-first embedded actives, 430–432
chip-in-polymer (CIP) technology, 381, 432
chip-last in cavity embedded actives, 433–434
chip-middle embedded actives, 432–433
chip-on-board technology, 23
chip-on-chip (COC) stacking, 104–105
chip-package codesign
concurrent oscillator, 184–190
LNAs, 181–184
overview, 180–181
SOC definition phase, 54–57
chip-package interconnects, 157
chip/PWB thickness ratios, 475
chip-scale MEMS packaging, 497–499
chip-scale packaging (CSP), 96, 265, 378
chipset modules, 28
chip-to-chip optical interconnects, 338–340
chip-to-chip signal performance, 685
chip-to-substrate interconnects, 448, 477, 484
CIP (chip-in-polymer) technology, 381, 432
circuit nodes, 706
circuit parameters, 673, 690
clock jitter, 702
clock signal distribution, 363
clock synchronization, 323, 362
cluster probes, 667
CMCs (ceramic matrix composites), 638
CMOSs. *See* complementary metal-oxide semiconductors
CMP (chemical and mechanical polishing), 66, 97
CMTSs (Cable Modem Termination Systems), 43
CNTs (carbon nanotubes), 399, 580–581, 730, 733–734
coarse grinding, 96
COC (chip-on-chip) stacking, 104–105
coefficient of thermal expansion (CTE), 268, 383–384, 471–472, 586, 607
cognitive radios, 308

cointegrated optoelectronic circuit layer, 355
cointegration, 321
collimating lenses, 353
colloidal chemistry concepts, 421
combiners, 297–299
compact thermal models
multiscale modeling, 635–636
overview, 634–635
system-level simulation using lumped models, 635
compensated baluns, 171–172
complaint interconnects, 484
complementary metal-oxide semiconductors (CMOSs)
bipolar (BiCMOS), 74, 125
compatible optical modulators, 362
digital SOP, 611
evolution of, 9
laser drivers, 322
LNAs, 181–182
processors, 333
silicon wafers, 436
single-chip, 25
SOP component, 20
transistors, 354
compliant helical interconnections, 553–555
compliant metal and polymer interconnections, 556–559
component-level thermal modeling, 635–636
composites with conducting fillers, 419
composite-type integrated capacitors, 286
computational fluid dynamics (CFD) codes, 90, 629, 634
computational fluid dynamics/heat transfer (CFD/HT), 630
concurrent engineering, 51
concurrent oscillator design, 184–190
conducting cores, 386–387
conduction
across solid interfaces, 620
multidimensional, 620–622
one-dimensional, 619–620
overview, 618–619
thermal vias, 639
conductive polymers, 108
conformal antennas, 270–274
consumer mobile products, 142, 155–156
contact resistance, 646

control volumes (CVs), 631
controllability test functions, 677
convection
  film condensation, 626
    governing equations and nondimensional parameters, 623–625
  phase change, 625
  pool boiling, 626
  thermal design, 638
conventional combiners, 297
cooling, 607, 718
coplanar waveguides (CPWs), 278, 359
copper (Cu)
  backplanes, 354
  bus drivers, 322
  electroplating systems, 484
  foils, 423
  interconnects, 330, 332, 569–570
  lines, 321
  low-K reliability, 470
  pillar bumps, 484
copper-ceramic–filled vias, 129
copper-filled via process, 411
copper-tin (Cu-Sn) eutectic bonding, 130–131
core substrate materials
  attributes of, 384–385
  classification of, 385, 386
  future trends, 435
  through-via processing, 386–387
coreless multilayer wiring integration process, 414–415
cost
  SOC development, 42, 50
  SOP substrate, 383
Coulomb blockade effects, 730
couplers, 343–353, 614
coupling
  analog-to-analog, 214–222
  digital-to-analog, 222–227
  efficiency, 348
  overview, 214
CPWs (coplanar waveguides), 278, 359
Cray T-90 motherboards, 338
create phase, 50
cryogenic DRIE, 125
crystallization, 420

C-SiC (carbon-silicon carbide) substrate, 472
CSP (chip-scale packaging), 95, 265, 378
CTE (coefficient of thermal expansion), 268, 383–384, 471–472, 586, 607
Cu. *See* copper
cure temperature, 454–455
curie temperature, 415–416
curing process, 395–396
Cu-Sn (copper-tin) eutectic bonding, 130–131
CVD. *See* chemical vapor deposition
CVs (control volumes), 631
cyanate ester, 472

D

DACs (digital-to-analog converters), 46, 74, 613, 686
DARPA (Defense Advanced Research Projects Agency), 299
DBG (dice before grind), 98
DBS (deep brain stimulation), 719
DBS (demand-based switching) scheme, 650
DC plating, 407
DDR (double data rate) bus, 103
decomposition of sacrificial polymers, 509–514
decoupling
  capacitors, 332, 415, 460, 461
  digital applications, 228–229
  embedded capacitors
    characterization of, 235–239
    individual capacitors, 232–235
    overview, 230
    planar capacitors, 231
  overview, 227
  SMD capacitors, 229–230
deep brain stimulation (DBS), 719
deep reactive ion etching (DRIE), 126
deep submicron (DSM) technology, 68–69
defective parts per million (DPPM), 44
defect-oriented tests (DOTs), 688
Defense Advanced Research Projects Agency (DARPA), 299
defibrillators, 744
degree of planarization (DOP), 389
dehydrogenase, 727

demand-based switching (DBS) scheme, 650
dendrimers, 731
deoxyribose nucleotide triphosphates (d-NTPs), 725–726
deposition, 389, 394–395
Design Automation Conference 2004, 60
design closure, 72–74
design diagnosis, 660
design for manufacturing
   overview, 208–210
   parametric yield, 212
   probabilistic diagnosis, 213–214
   statistical analysis, 210–212
design for testability (DfT) path, 63, 663
design of experiments (DOE), 210–211, 267, 695
design optimization models, 445
design planning, 69–71
design tools
   embedded RF circuits, 195–198
   for manufacturing
      overview, 208–210
      parametric yield, 212
      probabilistic diagnosis, 213–214
      statistical analysis, 208–212
   network synthesis, 204–207
   overview, 194–195
   rational functions, 204
   signal and power delivery networks, 198–204
   transient simulation, 207–208
design-for-reliability, 447–449
DFB (distributed feedback) lasers, 358, 618
DfT (design for testability) path, 63, 663
dice before grind (DBG), 98
dicing, thin-wafer, 98
die adhesive, 102
die cracking, 476
die stresses, 474, 476
die thickness/size, 86
dielectric, 382, 387
dielectric constant (K), 284–285, 385, 388, 415–416, 666
dielectric fluid, 643
dielectric layer, 411, 454
dielectric loss, 388, 416
dielectric processes
   curing, 395–397
   deposition, 394–395
   via formation processes, 397–399
dielectrics, 383
die-to-die stacking, 123
differential nonlinearity (DNL), 689
diffraction gratings, 343
diffractive couplers, 343–353
diffusion theory, 480
digital applications, decoupling, 228–229
digital communication analyzer, 358
digital convergence. *See* system technologies
   electronic system trend toward, 5–7
   miniaturization trend, 22–23
digital function reliability and verification
   high-density wiring, 458
   material length scale effects, 458–460
   metal-polymer interface delamination, 460–461
   substrate and interlayer dielectric, 460
digital function target parameters, 468
digital functional blocks, 448
digital functional boards, 442–443
digital integration, 152
digital multimedia processor, 49
digital pulse trains, 691
digital scan architecture, 694
digital signal processors (DSP), 40, 687, 694
digital signaling, 340
digital SOP, 611–613
digital subscriber line (DSL) systems, 40
digital-RF-opto integrated systems, 448
digital-RF-opto mixed-signal system, 445
digital-to-analog converters (DACs), 46, 74, 613, 686
digital-to-analog coupling
   aperture coupling, 224–227
   overview, 222–223
   planes, 224
   split planes, 223
dimethyldiethoxysilane (DMDES), 744
dimethylformamide (DMF), 734
diodes
   biosensor SOP, 720
   laser (LDs), 614, 618
   varactor, 178

diphenylthiourea (DPTU), 404
diplexers, 158
DIPs (dual-in-line packages), 92
direct bonding, 500
direct Cu-Cu bonding, 132
direct foil lamination, 287–288
Direct Memory Access (DMA), 53
discrete capacitor layers, 232–233
discrete components, 265
discretization process, 631–632
dispersion effect, 480
DISPs (dual-instruction-set processors), 651
dissolvable glue, 98
distance from neutral point (DNP) effect, 86
distributed feedback (DFB) lasers, 358, 616
DM642 digital media processor, 54
DMA (Direct Memory Access), 53
DMDES (dimethyldiethoxysilane), 744
DMF (dimethylformamide), 734
DNA-RNA probes, 725–726, 738
DNL (differential nonlinearity), 691
DNP (distance from neutral point) effect, 86
d-NTPs (deoxyribose nucleotide triphosphates), 725–726
DOE (design of experiments), 210–211, 267, 695
domains, 416
DOP (degree of planarization), 389
DOTs (defect-oriented tests), 688
double data rate (DDR) bus, 103
double-layer inductors, 277
doublelayer/interfacial polarization, 419
double-sided tape substrate, 116
downstream plasma, 97
DPM (dynamic power management), 635
DPPM (defective parts per million), 44
DPTU (diphenylthiourea), 404
DRAM (dynamic random-access memory), 611
DRIE (deep reactive ion etching), 126
drilling, 126, 389
dry etching, 97
dry polishing, 97
DSL (digital subscriber line) systems, 40
DSL central offices, 43
DSM (deep submicron) technology, 68–69
DSP (digital signal processors), 40, 687, 694
dual in-line packages, 23
dual polarization structure, 305–307
dual-band filters, 168–171
dual-body configuration, 307
dual-core architectures, 66
dual-frequency oscillators, 185–186, 189–190
dual-in-line packages (DIPs), 92
dual-instruction-set processors (DISPs), 651
DuPont Interra, 292
DVFS (dynamic voltage and frequency scaling), 649–650
dynamic power management (DPM), 639
dynamic random-access memory (DRAM), 611
dynamic voltage and frequency scaling (DVFS), 649–650

E

E-beam testing, 671
EBG structures. *See* electromagnetic bandgap structures
ECGs (electrocardiograms), 741
Economikos, L., 667
EDA (electronic design automation), 52, 195
edge connectors, 332
edge-emitting lasers (EELs), 325, 333–334, 342, 353, 357
edge-probed capacitors, 462
edge-viewing photodetectors (EVPDs), 325, 333, 342, 368
EEGs (electroencephalograms), 741
EELs (edge-emitting lasers), 325, 333–334, 342, 353, 357
effective radiation heat transfer coefficient, 628
elastic modulus, 384, 389
electric field components, 347
electric flux density, 425
electrical contacts, 353
electrical drivers, 614
electrical module test (EMT), 669
electrical parameters, 84

electrical properties
  core substrate materials, 385
  thin-film buildup organic dielectrics, 388–389
electrical resistance, 388
electrical routing, 105
electrical testing
  challenges in, 660–664
  HVM test process, 662–664
  KGEM testing
    boundary scan, 677–681
    of mixed-signal and RF subsystems, 685–707
    multi-gigahertz digital test, 681–685
  KGES testing
    interconnects, 664–671
    passives, 671–676
  overview, 659–660
electrical-optical circuit board (EOCB) concept, 338
electrical-optical devices, 341
electrocardiograms (ECGs), 741
electrochemical cells, 730
electrochemical signal conversion, 735–736
electroencephalograms (EEGs), 741
electroless copper plating, 404
electroless nickel immersion gold (ENIG) metallization, 140
electroless plating, 287, 404
electroless-plated resistors, 426
electrolytic plating process, 407–408
electromagnetic analysis, 455
electromagnetic bandgap (EBG) structures
  analysis and design, 241–245
  digital-to-analog coupling, 223
  mixed-signal SOP design, 152
  overview, 239–242
  planes, 224
  power supply noise suppression, 246–247
  radiation analysis, 248–250
  shielding solutions, 312
electromagnetic boundary conditions, 347
electromagnetic compatibility (EMC), 88
electromagnetic interference (EMI), 25, 44, 88, 152, 159, 248–250, 324, 467
electromechanical MEMS switches, 301–302
electromigration, 74
electromyograms (EMGs), 741
electron beam testing, 658–659
electronic design automation (EDA), 51, 195
electronic polarization, 415
electronic serializer-deserializers, 321–322
electronic switches, 178
electronic system-level (ESL) methods, 57
electronic theory, 480
electrooculograms (EOGs), 741
electroplating, 407–408, 484–485
electrostatic forces, 98
electrostatic switches, 300
EMAP (embedded actives and passives), 29
embedded actives
  chip-first, 430–432
  chip-last in cavity, 433–434
  chip-middle, 432–433
embedded actives and passives (EMAP), 29
embedded capacitors
  attributes of
    BDV, 416
    dielectric constant, 415–416
    dielectric loss, 416
    TCC, 416
  classification of
    composites with conducting fillers, 419
    overview, 416–417
    polymer ceramic composites, 418–419
    polymers, 418
    thin-film ceramics, 419–420
  decoupling
    characterization of, 235–239
    individual capacitors, 232–235
    overview, 230
    planar capacitors, 231
  films, 435
  processes
    polymer-ceramic composite, 420–422
    thin-film ceramic, 422–424
embedded conductors, 399–409
embedded decoupling, 382
embedded IC technology, 109–110, 381
embedded opto functional boards, 456–457

embedded passives
  baluns, 171–175
  capacitors, 166–167
  embedded capacitors
    attributes of, 415–416
    classification of, 416–420
    processes, 420–424
  embedded resistors
    overview, 424–426
    PTF, 426–428
    thin-film, 428–430
  faults in, 672
  filter-balun networks, 175–178
  filters, 167–171
  inductors, 161–165
  overview, 29, 160–161
  testing, 664, 671
  tunable filters, 178–180
embedded programmable processor cores, 47–48
embedded resistors
  overview, 424–426
  PTF, 426–428
  thin-film, 428–430
embedded RF circuits, 195–198
embedded RF functional boards, 453–456
embedded RF passive components, 463
embedded solder interconnection PoP stacking, 114
embedded thin-film components, 265, 605, 720
embedded wafer-level packages (e-WLPs), 381
EMC (electromagnetic compatibility), 88
EMGs (electromyograms), 741
EMI (electromagnetic interference), 25, 44, 88, 152, 159, 248–250, 324, 467
empirical relations, 635
EMT (electrical module test), 669
encapsulant-chip interfaces, 478
end-product system, 12–13
end-to-end system-level testing, 689
energy emission, 626–627
energy losses, 388
enhancement-mode pHEMTs (E-pHEMTs), 614
ENIG (electroless nickel immersion gold) metallization, 140

enzymatic oxidation, 730, 736
enzyme (biocatalyst) probes, 726–727, 730
EOCB (electrical-optical circuit board) concept, 338
EOGs (electrooculograms), 741
E-pHEMTs (enhancement-mode pHEMTs), 614
EPIC (Explicitly Parallel Instruction Computing), 649
epoxies, 390
epoxy dielectrics, 403, 428
epoxy thermosets, 391
epoxy-based laminates, 471
epoxy-copper interfaces, 480–481
ESL (electronic system-level) methods, 57
esoteric technologies, 24
etching sacrificial material, 505–509
EVPDs (edge-viewing photodetectors), 325, 333, 342, 358
e-WLPs (embedded wafer-level packages), 381
experimental characterization, 446, 449
Explicitly Parallel Instruction Computing (EPIC), 649
extended surfaces, 641
extending solder, 483–484
external convection, 623

F

FA (film adhesive), 102
FABSs (force amplified biological sensors), 739
face-to-back bonding, 17
face-to-back die stacking, 17
face-to-face bonding, 17
face-to-face die stacking, 17
failure mechanisms, 446–447, 662
failure mode interaction, 468
failures in time (FIT), 334
far-field pattern, 273
fast cells, 65
fast Fourier transforms (FFTs), 701
fault coverage, 688
fault resolution (FR), 672
fault tolerance, 44
fault-free circuits, 674
FDTD (finite-difference time-domain) method, 266

feature-based parametric models, 458
FEM (finite-element method), 90, 191, 203, 266, 631
FEOL (front end of line), 17, 136
ferrite beads, 222
ferrite composite, 277
ferroelectrics, 415–416, 418–420, 742
FET (field-effect transistors), 744
FFTs (fast Fourier transforms), 701
field programmable gate array (FPGA) technology, 43, 309, 611
field solvers, 216
field-effect transistors (FET), 744
film adhesive (FA), 102
film condensation, 626
filter-baluns
   cascaded, 178–179
   networks, 176–179
filters
   balanced bandpass, 176
   bandpass, 698
   block-level receivers, 698
   Chebychev, 187
   dual-band, 168–171
   embedded, 167–171
   four-pole, 297
   lattice, 176
   layouts, 196–198
   lowpass (LPFs), 613
   mixed-signal architecture, 157
   multimode, 167–168
   RF SOP technologies, 294–296
   tunable, 178–180
fine grinding, 97
finepitch flip-chip routing, 383
fine-pitch metal bumps, 586–587
finite-difference method, 202, 631
finite-difference time-domain (FDTD) method, 266
finite-element method (FEM), 89, 191, 203, 266, 631
finite-volume method (FVM), 631–633
firm IP blocks, 60
first generation (1G) cellular standards, 264
FIT (failures in time), 334
FITS (framework-based instruction-set processors tuning synthesis), 651
flexible optical interconnects, 339
flexible optical straps, 332
flip chip and wire bonding, 14, 16, 103–104
flip chip–on–chip, 14, 16
flip chipping of top/bottom die, 103
flip-chip BGAs, 55
flip-chip interconnection system, 484
flip-chip packages, 54–55, 470
flip-chip solder joints, 460, 478
flip-chip technology, 23
FLOTHERM packages, 634
flow network modeling (FNM), 635
fluorescein isothiocyanate (FTIR), 734
fluorescent probes, 732
fluorinated acrylates, 335
fluoropolymers, 393
FNM (flow network modeling), 635
focusing couplers, 350
foil lamination, 423
folded cases, 272
folded flex substrate, 116–117
folded-stacked chip-scale packages (FSCSPs), 84, 116–118
force amplified biological sensors (FABSs), 739
forced convection, 623, 638
form factor, 43
forward wire bonding, 101
Fourier transform infrared spectroscopy (FT-IR), 733
Fourier's law, 618
four-pole filters on LCP, 297
FPGA (field programmable gate array) technology, 43, 309, 611
FR (fault resolution), 672
framework-based instruction-set processors tuning synthesis (FITS), 651
free convection, 638
free space (FS), 614
front end of line (FEOL), 17, 136
FS (free space), 614
FSCSPs (folded-stacked chip-scale packages), 84, 116–118
FTIR (fluorescein isothiocyanate), 734
FT-IR (Fourier transform infrared spectroscopy), 733
fused junction technology, 105
fused silica prism, 352

future trends
    in biosensor SOP technologies, 745–746
    in core substrate materials, 435
    in MEMS packaging, 527–528
    in mixed-signal reliability
        alternatives to solder and nano interconnects, 484–486
        complaint interconnects, 484
        extending solder, 483–484
        overview, 482
    in optoelectronic SOP technology, 364
    in RF SOP technologies, 311–312
    in SIP technology, 142–143
    in SOP substrates, 435–437
FVM (finite-volume method), 631–633

G

GaAs (gallium arsenide) edge-emitting lasers, 327
gain, antenna, 269
gallium arsenide (GaAs) edge-emitting lasers, 327
gap filling interface materials, 646
gate oxide thickness, 64
gaussian distribution, 704
G-C (guanine and cytosine), 728
GDH (glucose dehydrogenases), 727
genetic algorithms, 267, 696
geometry parameters, 476
Georgia Tech
    filters and, 294
    Interface Optical Coupling (GT IOC), 353–355, 358
    Packaging Research Center (GT-PRC), 433
German Manufacturing Labs (GMTC), 124
getters
    improving reliability through, 520–522
    nonevaporable, 516–517
    thin-film, 517–520
giant magnetoresistive (GMR) elements, 739
glass transition temperature, 389
global clock distribution, 363
global interconnects, 363, 448
global thermal mismatches, 470
glucose biosensors, 730
glucose dehydrogenases (GDH), 727
glucose oxidase (GOx), 730, 736
GMR (giant magnetoresistive) elements, 739
GMTC (German Manufacturing Labs), 124
gold nanoparticles, 731
golden boards, 694
Gould Nichrome, 292–294
GOx (glucose oxidase), 730, 736
GPM (gross profit margin), 50
Grashof number, 625
grating, 345–352
gray diffuse enclosures, 627–628
gross profit margin (GPM), 50
guanine and cytosine (G-C), 728
guided wave Bragg condition (GWBC), 349–350

H

handling, thin-wafer, 97–98
hard IP blocks, 60
hardware description languages (HDLs), 59
hardware-software (HW-SW) codesign
    coverification, 62
    creation phase, 57–59
    definition phase, 54–57
HBTs (heterojunction bipolar transistors), 614
HDI (high density interconnect) solutions, 664
HDI PWBs (high-density interconnect printed wiring boards), 426
HDIs (high-density interconnects), 431, 451, 468
HDLs (hardware description languages), 59
HDTVs (high-definition TVs), 142–143
heat flow across solid interfaces, 620
heat flux, 612, 619, 629, 638
heat pipes, 641–643
heat sinks, 641
heat spreading, 620–622, 640–641
heat transfer modes
    conduction
        across solid interfaces, 620
        multidimensional, 620–622
        one-dimensional, 619–620
        overview, 618–619

convection
    film condensation, 626
    governing equations and nondimensional parameters, 623–625
    phase change, 625
    pool boiling, 626
radiation
    in gray diffuse enclosures, 627–628
    heat transfer coefficient, 628
    overview, 626–627
heat transfer paths, 89, 638
HEDTA (*N*-(2-hydroxyethyl)ethylenediamine triacetic acid trisodium salt hydrate), 404
helical inductors, 280
hermetic seals, 337
heterogeneous integration, 363
heterojunction bipolar transistors (HBTs), 614
heterostructures, 646
HFSSs (high-frequency structure simulators), 266
hierarchical design, 71–72, 194
high density interconnect (HDI) solutions, 664
high-aspect-ratio compliant copper interconnects, 485
high-conductivity package materials, 638–639
high-CTE rigid substrate, 586
high-definition TVs (HDTVs), 142–143
high-density charge structure elements, 742
high-density interconnect printed wiring boards (HDI PWBs), 426
high-density interconnects (HDIs), 431, 451, 468
high-density wiring, 458, 469–470
high-dielectric-constant oxides, 65
high-efficiency power transmission, 742–744
high-frequency photodetectors, 324
high-frequency structure simulators (HFSSs), 266
highly oriented pyrolytic graphite (HOPG), 638
high-performance computing, 322–323
high-speed digital systems, 322–323, 706
high-speed electrical wiring, 330–331
high-surface-area electrodes, 419

high-temperature cofired ceramics (HTCCs), 13, 268, 399
high-temperature direct Cu-to-Cu bonding, 589
high-temperature synthesis methods, 422
high-thermal-conductivity materials, 434, 638
high-throughput, 668
high-volume manufacturing (HVM), 659, 662–664
Hitachi, 125
homogeneous isotropic medium, 618
homogeneous orthotropic medium, 618
homogeneously integrated MSM PDs, 363
HOPG (highly oriented pyrolytic graphite), 638
HT (Hyper Threading) technology, 649
HTCCs (high-temperature cofired ceramics), 13, 268, 399
humidity testing, 462–463
HVM (high-volume manufacturing), 659, 662–664
HW-SW codesign. *See* hardware-software codesign
hybrid chip stackings, 103
hybrid silicon lasers, 26–27
hybridization, 723, 727, 734
hydrogen peroxide, 730
hydrothermal process, 422
Hyper Threading (HT) technology, 649
HyperBGA™ packages, 470
hypophosphite, 404

I

IBM, 125, 336
ICEPAK packages, 634
ICs. *See* integrated circuits
ILD (inter layer dielectric), 98
IME (Institute of Microelectronics), 26, 125
IMEC (Interuniversity Microelectronics Center), 28, 127
impedance matching, 323
implanted electrodes, 721
in vivo medical implants, 744
INC (Intelligent Network Communicator) system, 30, 334–336, 451
in-die variations, 66
inductive telemetry, 742–744

inductors
    embedded, 161–165
    integration, 281
    layouts, 195–196
    RF SOP technologies, 277–281
    single-layer spiral, 162–164
    two-layer spiral, 163–165
infrared (IR) absorption, 733
infrastructure-type applications, 44
INL (integral nonlinearity), 689
inorganic polymer glass (IPG), 343
inorganic reinforcement, 472
inorganic semiconductor nanocrystals, 732
in-plane beam turning single-mode
        waveguides, 343
in-plane lightwave circuit, 358
input-estimate-refine process, 70–71
insertion loss, 293
"inside the box", 323
Institute of Microelectronics (IME), 26, 125
insulating cores, 386–387
integral nonlinearity (INL), 689
integrated capacitors, 286
integrated circuits (ICs)
    evolution of, 19
    integration, 12, 717
    miniaturization trend, 5
    package pitch gap
        BEOL materials and processes,
            541–543
        overview, 540–541
        silicon substrate with polymer-copper
            wiring, 541
integrated passive devices (IPDs), 264
integrated RF modules
    INC system, 309–311
    WLANs, 307–308
integrated thermal via floor planning (IVF),
        640
integrated wireless modules, 154
integration approaches
    in SIP, 15
    SOC development, 59–61
    for system barriers, 8
Intel Corp., 153, 362
intellectual property (IP), 14, 46, 59
Intelligent Network Communicator (INC)
        system, 30, 309–311, 451

inter layer dielectric (ILD), 98
interboard optical links, 335
interconnects
    capacitance testing, 666–667
    electron beam testing, 668–669
    geometry trends, 63
    latent open testing, 669
    MEMS packaging, 522–524
    overview, 664–665
    resistance testing, 667–668
    resonator band testing, 669
    scaling, 76, 363
interface optical coupling, 356–361
interfacial adhesion
    effect of moisture on, 480–482
    overview, 478–480
interfacial delamination, 470
interfacial failure, 478, 482
interfacial polarization, 418
interfacial roughness, 403
interferometric recording configuration, 350
intermediate layer bonding, 500–505
internal convection, 623
internal reflection, 341
International Technology Roadmap for
        Semiconductors (ITRS), 40, 606, 608
intersymbol interference (ISI) phase, 333
Interuniversity Microelectronics Center
        (IMEC), 28, 127
intraboard optical links, 333
IP (intellectual property), 14, 46, 59
IPDs (integrated passive devices), 264
IPG (inorganic polymer glass), 343
IR (infrared) absorption, 733
Irvine Sensors, 82
ISA buses, 336
ISI (intersymbol interference) phase, 333
isotropic elastic bimaterials, 478
ITRS (International Technology Roadmap
        for Semiconductors), 40, 606, 608
IVF (integrated thermal via floor planning),
        640

J

JEDEC JC15 committee, 90
JETAG (Joint European Test Access Group),
        678
jitter, 682

J-leaded chip carriers (JLCC), 94
Joint European Test Access Group (JETAG), 678
junction temperature, 608

K

K (dielectric constant), 284–285, 385, 388, 415–416, 666
K_a-band power divider, 292–293
Kennedy, D.P., 621
KGDs (known good dies), 119, 433, 663
KGEM testing. *See* known good embedded module testing
KGES testing. *See* known good embedded substrate testing
Kinloch, A.J., 480
Kirchhoff's law, 627
known good dies (KGDs), 119, 433, 663
known good embedded module (KGEM) testing
  boundary scan
    history of, 677–678
    key elements of, 678–680
    for SOP, 680–681
  of mixed-signal and RF subsystems, 685–707
    alternate testing methods, 690–707
    direct measurement using dedicated circuitry, 689–690
    fault models and test quality, 688–689
    strategies, 685–688
  multi-gigahertz digital test, 681–685
  overview, 663
known good embedded substrate (KGES) testing
  interconnects
    capacitance testing, 666–667
    comparison of methods, 669–671
    electron beam testing, 668–669
    latent open testing, 669
    overview, 664–665
    resistance testing, 667–668
    resonator band testing, 669
  overview, 663

passives
  with pole-zero analysis, 673–676
  with sensitivity analysis, 671–672
Koyanagi, Mitsumasa, 354

L

L (package bond wire inductance), 74
lab-on-chip (LOC) technology, 718
laminates, 386, 394
land-pads, 112
large-area thin carbon silicon carbide based composite boards, 472
large-scale integration (LSI), 677
LAs (limiting amplifiers), 340, 358–359
laser array driver chips, 332
laser diodes (LDs), 614, 616
laser drivers, 333, 354
laser via process, 398
lasers
  AlGaInAs based, 334
  distributed feedback (DFB), 358, 616
  edge-emitting (EELs), 325, 333–334, 342, 353, 357
  hybrid silicon, 26–27
  silicon Raman, 362
  thin-film, 321
  vertical cavity surface-emitting (VCSELs), 309, 325, 334, 354, 614, 616
latency, 21
latent open testing, 669
lattice filters, 177
lattice phonon interaction, 637
lattices, 240, 242–246
LCDs (liquid crystal displays), 744
LCPs (liquid crystalline polymers), 161–162, 260, 268, 393, 430, 721
LDs (laser diodes), 614, 616
lead-free devices, 44
lead-free solders
  electromigration, 564–567
  at fine pitch, 567–569
  interconnects, 468, 483
  intermetallics in, 562–563
  materials, 562
  mechanical properties of, 563–564
  overview, 561
leakage energy dissipation, 612

Index

leakage power, 64, 612, 650
LEDs (light-emitting diodes), 614, 616
LFSRs (linear feedback shift registers), 691
Lichtenecker's law, 285
light-emitting diodes (LEDs), 614, 616
Lightsmyth, 362
limiting amplifiers (LAs), 340, 358–359
linear feedback shift registers (LFSRs), 691
linear-elastic fracture mechanics, 476
liposomes, 731
liquid cooling, 643
liquid crystal displays (LCDs), 744
liquid crystalline polymers (LCPs), 161–162, 260, 268, 393, 430, 721
liquid crystals, 393, 637
liquid loops, 643
liquid-vapor phase change processes, 625
lithographic mask, 357
lithography, 435
LNAs (low noise amplifiers), 25, 181–184, 214, 217, 613, 692
load boards, 694
LOC (lab-on-chip) technology, 718
local oscillators (LOSCs), 613, 703
logarithmic scale variation, 626
logic ICs, 117
logic switching, 74
loop gain, 689
loop inductors, 276
LOSCs (local oscillators), 613, 703
low noise amplifiers (LNAs), 25, 181–184, 214, 217, 613, 692
low-loss polymer encapsulation techniques, 514–516
low-multimode waveguides, 342
lowpass filters (LPFs), 613
low-power ASICs, 741–744
low-pressure CVD (LPCVD) technique, 130
low-stress electroless copper plating, 405
low-temperature ceramics, 380
low-temperature coefficient of capacitance, 282
low-temperature cofired ceramics (LTCCs), 13, 159–160, 260, 267–268, 282, 380, 399
low-temperature metal-to-metal bonding, 587–589
low-temperature synthesis methods, 422
LPCVD (low-pressure CVD) technique, 130
LPFs (lowpass filters), 613
LSI (large-scale integration), 677
LTCCs (low-temperature cofired ceramics), 13, 159–160, 260, 267–268, 282, 380, 399
lumped capacitance convective transient, 622
lumped element baluns, 174–175
lumped models, 635
Luxtera, 362

M

MacDermid M-Pass, 292
Mach-Zehnder modulators, 309, 362
MACROFLOW packages, 635
magnetic field components, 347
magnetic nanoparticles, 730–731, 739
magnetic signal conversion, 739–740
magneto-dielectric substrate, 274–277
Malaviya National Institute of Technology, 29
mapping functions, 691, 696
Marchand baluns, 171–172, 175–176
Marshall, J., 666–667
MARSs (multivariate adaptive regression splines), 691
massively parallel processing (MPP), 330
material characterization, 449, 460
material length scale effects, 458–460
Matsushita SIMPACT technology, 31
MC (Monte Carlo) method, 211, 673
McASP (multichannel audio serial ports), 46
McDonald et al., 124
MCMs (multichip modules), 9, 11, 13, 121, 264, 380, 671
MCM/TFI (multichip module/thin-film interconnect) package, 457
meander line
    antennas, 273, 275–276
    RFIDs, 305
mechanical interlocking, 480
mechanical stress, 447
mechanism-based strain gradient plasticity (MSG) theory, 459
median life, 334
medium interconnects, 363
medium-scale integration (MSI), 677
megafunctionality, 605
memory stacking, 95

memory subsystems, 50, 52–54
MEMS packaging. *See* micro-electro-mechanical systems packaging
Mertens, Robert, 28
MESFETs (metal semiconductor field-effect transistors), 614
metal bumps with compliant polymers, 559–560
metal cores, 386
metal interconnections with polymer cores, 559
metal matrix composites (MMCs), 386, 638
metal nanoparticles, 731
metal semiconductor field-effect transistors (MESFETs), 614
metal-based stress-relief interconnections, 553
metal-insulator-metal (MIM) structures, 282–283
metallization stacking, 106–108
metallized flexible polymer tapes, 111
metallurgical interconnections, 433
metal-metal bonding, 130–132
metal-on-compliant polymers, 555–556
metalorganic chemical vapor deposition (MOCVD) thin-film RF capacitors, 281
metal-polymer interfaces, 460–461
metal-polymer–based stress relief
   compliant metal and polymer interconnections, 556–559
      metal bumps with compliant polymers, 559–560
      metal interconnections with polymer cores, 559
      metal-on-compliant polymers, 555–556
method of moments (MoM), 267
methyltriethoxysilane (MTES), 744
MFDM (multilayered finite-difference method), 201
MGI (microvia and global interconnect) system, 445
micelles, 731
Micro Channel buses, 337
micro photodiodes, 722
micro solder bump formation method, 105
microchannels for biofluid transport, 723–724

microelectrochemical detection, 735
microelectrode stimulation, 742–744
micro-electro-mechanical systems (MEMS) packaging
   assembly, 524–526
   challenges in, 496–497
   chip-scale versus wafer-scale, 497–499
   electrical test challenges, 660
   future trends, 527–528
   getters
      improving reliability through, 520–522
      nonevaporable, 516–517
      thin-film, 517–520
   interconnections, 522–524
   low-loss polymer encapsulation techniques, 514–516
   microfluidic and nanofluidic channels, 744
   overview, 29, 495–496
   RF switches
      application, 303–304
      challenges, 302–303
      electromechanical versus solid state, 301–302
      history and role of, 299
      operating principle, 300–301
   sacrificial film-based sealing techniques
      decomposition, 508–514
      etching, 505–509
   signal conversion, 738–739
   thermal SOP, 606, 617
   wafer bonding techniques
      direct bonding, 500
      intermediate layer bonding, 500–505
      overview, 499–500
microelectronics technology, 11, 345
microetching, 454
microfluidic components, 718, 744
microlens relays, 342, 356
microlenses, 353
microminiaturization, 605
micromirror relays, 356
micromirrors, 353
microorganisms, 726
micropumps, 617
microresonators, 738

"microscopic" SOC design challenges, 68
microstrip copper lines, 340
microsystems packaging (MSP), 5
microvalves, 617
microvia and global interconnect (MGI) system, 445
microvias, 458–459, 462, 471
microwave amplifiers, 312, 324
MIM (metal-insulator-metal) structures, 282–283
MIMO (multiple-input and multiple-output) architectures, 307
miniaturization, 22–23, 151, 156, 261, 321, 382, 434–435, 464, 718
miniaturized batteries, 648
mission-critical applications, 44
mixed-signal design. *See* system-on-package technology
   integration, 74–76
   KGEM test of
      alternate testing methods, 690–707
      direct measurement using dedicated circuitry, 689–690
      fault models and test quality, 688–689
      strategies, 685–688
   market, 152
   reliability
      future trends, 482–486
      multifunction SOP substrate, 450–468
      overview, 443–445
      substrate-to-IC interconnection, 468–482
      system-level, 445–450
mixers, 698–699
MLCs (multilayer ceramics), 389
MLTS (multilayer thin substrate), 414
MMCs (metal matrix composites), 386, 638
MMI (multimode interference) devices, 343, 359
mobile products, 142, 155–156
MOCVD (metalorganic chemical vapor deposition) thin-film RF capacitors, 281
modal decomposition technique, 199–202

modeling
   multiscale, 635–636
   of RF SOP technologies, 266–267
   system-level failure mechanisms, 468
modulators-demodulators, 625
module-level miniaturization, 5
moisture preconditioning, 480, 482
moisture uptake, 389
molecular binding surface tension, 738
molecular hybridization, 737
MoM (method of moments), 266
Monte Carlo (MC) method, 211, 673
Moore, Gordon E., 606
Moore's law, 40, 141, 153, 263, 606
Motorola, 32
MPP (massively parallel processing), 330
MQW (multiple quantum well) modulators, 363
MRTD (multiresolution time-domain), 266
MSG (mechanism-based strain gradient plasticity) theory, 459
MSI (medium-scale integration), 677
MSP (microsystems packaging), 5
MTCs (multichip test controller ICs), 680
MTES (methyltriethoxysilane), 744
Muller method, 347–348
multiband antennas, 270
multiband architecture, 155, 158, 184–185
multichannel audio serial ports (McASP), 46
multichip modules (MCMs), 9, 11, 13, 121, 264, 380, 671
multichip module/thin-film interconnect (MCM/TFI) package, 457
multichip test controller ICs (MTCs), 680
multidimensional conduction, 620–622
multifunction SOP substrate reliability
   digital function
      high-density wiring reliability, 458
      material length scale effects, 458–460
      metal-polymer interface delamination, 460–461
      substrate and interlayer dielectric reliability, 460

Index

materials and process
  digital functional boards, 452–453
  embedded opto functional boards, 456–457
  embedded RF functional boards, 453–456
  overview, 450–452
  up-front process optimization, 457–458
multifunction system, 467–468
optical function
  overview, 463–465
  reliability testing, 465–466
  stress-induced birefringence, 465
  waveguide refractive index stability, 466
RF function, 461–463
multifunction system reliability, 467–468
multi-gigahertz data transfer, 660
multi-gigahertz digital testing, 681–685, 695
multilayer ceramics (MLCs), 389
multilayer filter structure, 296
multilayer substrate, 56
multilayer thin substrate (MLTS), 414
multilayer wiring integration processes
  coreless, 414–415
  parallel process
    overview, 411–413
    paste via process, 413–414
  sequential process, 409–411
multilayered finite-difference method (MFDM), 201
multimode filters, 167–168
multimode interference (MMI) devices, 343, 359
multimode oscillators, 185
multimode radio architecture, 168
multiple CPU engines, 66
multiple length scale hierarchy, 605
multiple plane pairs, 200–201, 225–227
multiple quantum well (MQW) modulators, 363
multiple-input and multiple-output (MIMO) architectures, 307
multiple-mode heat transfer, 618
multiple-target detection systems, 721
multiplexer-demultiplexers, 342
multiresolution time-domain (MRTD), 266

multiscale modeling, 635–636
multi-threshold voltage libraries, 64
multitone sinusoids, 691
multivariate adaptive regression splines (MARSs), 691
multiwalled carbon nanotubes (MWCNTs), 733
"mushroom-type" EBGs, 240
MWCNTs (multiwalled carbon nanotubes), 733

N

N-(2-hydroxyethyl)ethylenediaminetriacetic acid trisodium salt hydrate (HEDTA), 404
nano bio-SOP integration, 745–746
nano interconnects, alternatives to, 484–486
nanocantilevers, 738
nanocapacitors, 418
nanofluidic channels, 744
nanometer scale, 612
nanoparticle conductive adhesives, 581–584
nanoparticles, 731–732
nanopores, 481
nanoscale interconnections
  alternatives to, 486
  CNTs, 580–581
  nanoparticle conductive adhesives, 581–584
  nanostructured metal interconnections, 577–580
  NCA bonding, 584
nanostructured interconnects, 485, 577–580
narrow-band baluns, 174
natural convection, 89, 623, 638
NBPs (need-based power supplies), 639
NBTI (negative bias temperature instability), 66
NCA (nonconductive adhesive) bonding, 584
NCE (nonconductive epoxy), 102
near-opens, 669
near-shorts, 669
NEC, 362
need-based power supplies (NBPs), 639
negative bias temperature instability (NBTI), 66

negative resistance single-frequency oscillator design, 186
neo-wafers, 107
network access latency, 331
network synthesis, 204–207
neural prosthetic devices, 718
Newton's law, 628
NF (noise factor), 181–184
"N-Gage" gaming device platform, 153
NIM (Nuclear Instrumentation Module) bins, 335
noise coupling, 158–159
noise factor (NF), 181–184
noise suppression, power supply, 246–247
Nokia Corp., 153
nonconductive adhesive (NCA) bonding, 584
nonconductive epoxy (NCE), 102
noncontact testing, 668
nondestructive testing, 668
nonequalized microstrip copper transmission lines, 340
nonevaporable getters, 516–517
nonhomogeneous anisotropic medium, 618
nonlinear transfer function, 698–699
non-return-to-zero (NRZ), 333
non-TSV SIP technology
   chip stacking
      embedded IC stacking, 109–110
      versus package stacking, 119–120
      side termination stacking, 105–109
      TAB stacking, 111–112
      thin-wafer dicing, 98
      thin-wafer handling, 97–98
      wafer thinning, 95–97
      wire bonded stacking, 98–105
   evolution of, 92–95
   overview, 81
   package stacking
      versus chip stacking, 119–120
      FSCSP stacking, 116–118
      PiP stacking, 115–116
      PoP stacking, 112–115
norbornene, 392
NRZ (non-return-to-zero), 333
Nuclear Instrumentation Module (NIM) bins, 335
Nusselt number, 624

O

observability test functions, 677
OBTs (oscillation-based tests), 692
OECB (optical electric circuit board), 341, 354
OF (optical fiber), 335, 337, 614
off-chip external test equipment, 706
Ohmega-Ply, 292
ohmic power dissipation, 614
Ohm's law, 667
OLEDs (organic light-emitting diodes), 321
on-chip digital signal processing, 706
on-chip interconnects, 45, 470
on-chip memory structures, 612
on-chip optical circuits, 362–364
on-chip optical clock distribution, 322
on-chip signal generators, 690
on-chip test response data capture techniques, 690
one-dimensional conduction, 619–620
one-dimensional EBGs, 240–244
one-dimensional unit-cell analysis, 242–244
one-port embedded inductors, 204–206
on-substrate microvias, 448
optical absorption, 326
optical alignment, 325, 357
optical backplanes, 322, 354
optical communication link, 337
optical coupling efficiency, 614
Optical CrossLinks, 337
optical electric circuit board (OECB), 341, 354
optical fiber (OF), 335, 337, 614
optical frequency isolation, 324
optical function reliability and verification
   overview, 463–465
      reliability testing, 465–466
      stress-induced birefringence, 465
      waveguide refractive index stability, 466
optical functional blocks, 448
optical interconnects, 322, 338–340, 343–353, 464
optical interposers, 363
optical leakage, 323
optical links, 323
optical local area networks, 331
optical mode analysis, 465
optical pin, 338

optical polymers, 334–335
optical power splitters, 343
optical signal conversion, 737
optical signal distribution architecture, 325
optical transceivers, 27–28, 332, 335
optical transmitter modules, 335
optical waveguides, 342–343
optical wiring, 333, 335–337
optical-digital signaling, 323
optically suitable polymers, 326
optimal buffer layer design, 342
"opto bumps," 338
optodigital cards, 335
optoelectronic data communication, 335
optoelectronic devices, 353, 614
optoelectronic integration process, 357
optoelectronic SOP technology
  advantages of
    versus high-speed electrical wiring, 330–331
    power dissipation, 333–334
    reliability, 334–335
    wiring density, 331–333
  applications of
    high-performance computing, 322–323
    high-speed digital systems, 322–323
    RF-optical communication systems, 323–324
  evolution of
    board-to-board optical wiring, 335–337
    chip-to-chip optical interconnects, 338–340
  future trends, 364
  interface optical coupling, 356–361
  on-chip optical circuits, 362–364
  overview, 26–28, 320–321, 615–617
  thin-film
    3D lightwave circuits, 354–356
    active components, 353–354
    optical alignment, 325
    overview, 324–325, 340–341
    passive thin-film lightwave circuits, 341–353
    properties of waveguide materials, 325–330

optoelectronics, 22
organic cores, 432
organic dielectric materials, 452
organic light-emitting diodes (OLEDs), 321
organic packaging, 84
organic polymer waveguides, 325, 327
organically modified ceramics (ORMOCERs), 334
oscillation-based tests (OBTs), 692
oscillators
  block-level receivers, 699–700
  concurrent design, 184–190
  dual-frequency, 185–186, 189–190
  local (LOSCs), 613, 703
  multimode, 185
  voltage-controlled (VCOs), 181, 188–189, 309
out-coupling angle, 349
output CPWs, 359
output sampling clocks, 703
output SMA edge connectors, 359
"outside the box," 323
oxidase, 727
oxidation-resistant electrodes, 424
oxide bonding, 130

P

package bond wire inductance (L), 74
package flow impedance, 630
package integration cycle time, 21
package partition, 263
package stacking
  non-TSV SIP
    versus chip stacking, 119–120
    FSCSP stacking, 116–118
    PiP stacking, 115–116
    PoP stacking, 112–115
  SIP by, 18
package substrate interposers, 379
packaged microwave amplifiers, 312
package-enabled integration, 12
package-in-package (PiP) stacking, 18, 84, 115–116, 118
package-level thermal characterization, 90
package-on-package (PoP) stacking, 18, 84, 112–115, 118, 379
packaging misregistration, 464
Packaging Research Center (PRC), 4, 9, 451

packaging-induced stress, 86
PAE (power added efficiency), 614
paraelectric ceramics, 286
parallel multilayer wiring integration process
   overview, 411–413
   paste via process, 413–414
parallel plate capacitors, 166–167, 282–283
parallel processing, 649
parallel waveguides, 323–324
parallel-plate electrostatic actuation, 301
parallel-plate structures, 666
parametric faults, 663, 671, 674, 688
parametric modular component models, 468
parametric yield, 212
parasitic extraction, 194
parasitic inductance, 88, 239
parasitic passbands, 218
partitions, 72
parts synthesis approach (PSA), 29
parylene-PDMS bilayer coatings, 744
PAs (power amplifiers), 25, 613
passivation layer, 107
passive components, 262
passive lightwave circuit, 341, 356
passive methods of thermal management
   extended surfaces, 641
   heat pipes, 641–643
   heat spreader, 640–641
   high-conductivity package materials, 638–639
   thermal vias, 639–640
passive optical fibers, 356
passive thin-film lightwave circuits
   diffractive couplers for optical interconnects, 343–353
   optical power splitters, 343
   optical waveguides, 342–343
   overview, 341–353
passives
   with pole-zero analysis, 673–676
   with sensitivity analysis, 671–672
paste via process, 413–414
patch antennas, 269
PBGA (peripheral BGA) packages, 341, 342
PCBs (printed circuit boards), 56, 159–160, 332, 720, 744
PCI Express Mini card form factor, 191

PCMAs (phase change metallic alloys), 647
PCMCIA (Personal Computer Memory Card International Association), 43
PCR (polymerase chain reaction) method, 725
PD array amplifiers, 332, 358–359
PDMS (polydimethylsiloxane), 724, 735, 744
PDNs (power distribution networks), 227
PECVD (plasma-enhanced chemical vapor deposition), 126, 136, 581
peel stress, 451
performance headroom, 43
periodic variation, 345
peripheral BGA (PBGA) packages, 341–342
permanganate process, 403
Personal Computer Memory Card International Association (PCMCIA), 43
PEV (plasma-etched via) process, 398
phase change, 625
phase change metallic alloys (PCMAs), 647
Phase Locked Loop (PLL) blocks, 74
phase shifters, 303–304
pHEMTs (pseudomorphic high-electron mobility transistors), 614
photo via process, 398
photodetector amplifiers, 322, 333, 354
photodetector arrays, 342
photodetectors, 340, 355, 357, 359
photolithography, 342, 401, 405, 408, 451, 484, 665
photoresist (PR) materials, 108, 454
photoresist patterning, 405
physical adsorption, 480
physics-based parametric models, 458
physics-based reliability models, 445
physics-based thermomechanical models, 463
physics-of-failure approach, 448
pi networks, 214–216
picoseconds, 682
piezoelectric nanostructures, 739
piggyback sockets, 92
pin assignment, 55–56
pin density, 332
PIN diodes, 301
"pin electronics," 682
pin-through hole (PTH) interconnections, 92
PiP (package-in-package) stacking, 18, 84, 115–116, 118

Index

PIP (polymer infiltration and pyrolysis), 472
planar baluns, 173–174
planar capacitors, 229, 231
planar lightwave circuits, 341
planar photodiodes, 325
planarization, 389
planes, 224
plane-strain analysis, 465
plasma etching, 484
plasma-enhanced chemical vapor deposition (PECVD), 126, 136, 581
plasma-etched via (PEV) process, 398
plastic strain, 453
plated thin-film resistors, 428–430
platform-based design, 60–62
PLL (Phase Locked Loop) blocks, 74
plug-in optical cards, 335
PMCs (polymer matrix composites), 638
PMMAs (polymethylmethacrylates), 334, 724
POBT (predictive oscillation based testing), 693
polarization-dependent couplers, 350
polarization-independent couplers, 350
polarizing beam splitters, 350
pole-zero analysis, 671, 673–676
polishing methods, wafer, 97
polydimethylsiloxane (PDMS), 724, 735, 744
polyimide films, 431
polyimide thermosets, 391
polymer bonding, 132–133
polymer ceramic composites, 418–419
polymer infiltration and pyrolysis (PIP), 472
polymer matrix composites (PMCs), 638
polymer matrix nanointerconnects, 485
polymer optical waveguides, 325, 354
polymer thick film (PTF), 287, 426–428, 430
polymer waveguides, 323, 337
polymerase chain reaction (PCR) method, 725
polymer-based composites, 284–287
polymer-ceramic composite processing, 418, 420–422
polymer-dispersed liquid crystals, 345
polymeric nanoparticles, 731
polymethylmethacrylates (PMMAs), 334, 724
polynorbornene, 392

polyphenyl ethers (PPEs), 390
polytetrafluoroethylene (PTFE), 390, 393
pool boiling, 626
PoP (package-on-package) stacking, 18, 84, 112–115, 118, 379
position papers, 682
postprocessing packages, 636
power added efficiency (PAE), 614
power amplifiers (PAs), 25, 613
power calculation, 348
power delivery networks, 198–204
power dissipation, 42–43, 64, 333–334
power distribution, 382, 718
power distribution networks (PDNs), 227
power harnessing, 651
power minimization methodologies
   ASPs, 650
   cache power minimization, 650–651
   DVFS, 649–650
   overview, 648–649
   parallel processing, 649
   power harnessing, 651
power spectral density (PSD), 693
power supply noise suppression, 246–247
power transistors, 614
Poynting vector (PV), 348
PPEs (polyphenyl ethers), 390
PR (photoresist) materials, 108, 454
Prandtl number, 624
PRC (Packaging Research Center), 4, 9, 451
predictive oscillation based testing (POBT), 693
preferential coupling ratio, 345, 348
prepregs, 471
printed antennas, 269
printed circuit boards (PCBs), 56, 159–160, 332, 720, 744
printed wiring assemblies (PWAs), 629
printed wiring boards (PWBs), 29, 94, 159–160, 470, 629
probabilistic diagnosis, 213–214
probe-target molecular hybridization
   cell probe hybridization, 729
   DNA-RNA hybridization, 728
   enzyme reaction, 730
   protein hybridization, 727–728
   synthetic probes hybridization, 728–729

process characterization, 449
process thermal loading, 455
processability, 385, 389
processor performance states, 65
processor temperature, 323
programmability, 43
protein probes, 725, 727–728
prototyping methodology, 69
PSA (parts synthesis approach), 29
PSD (power spectral density), 693
pseudomorphic high-electron mobility transistors (pHEMTs), 614
PTF (polymer thick film), 287, 426–428, 430
PTFE (polytetrafluoroethylene), 390, 393
PTH (pin-through hole) interconnections, 92
pull-down voltage, 301
pulse reverse plating, 408
PV (Poynting vector), 348
PWAs (printed wiring assemblies), 629
PWBs (printed wiring boards), 29, 94, 159–160, 470, 629

Q

Q (quality factor), 23, 158–159, 160, 183–184, 277
QDs (quantum dots), 730–732
quad flat packages (QFPs), 23, 94, 264
quality factor (Q), 23, 158–159, 160, 183–184, 277
quantum dots (QDs), 730–732
quartz crystals, 738
quasi-TEM approaches, 281

R

R&D. *See* research and development
radiation
    analysis in EBG structures, 248–250
    effective radiation heat transfer coefficient, 628
    in gray diffuse enclosures, 627–628
    heat transfer, 630
    overview, 626–627
    pattern plots, 275
"Radio Free Intel" initiative, 153
radiofrequency (RF) design
    embedded circuits, 195–198
    embedded passive design
        baluns, 171–175
        capacitors, 166–167
        filter-balun networks, 175–178
        filters, 166–171
        inductors, 161–165
        overview, 160–161
        tunable filters, 178–180
    function reliability and verification, 461–463
    integration, 152
    KGEM test of
        alternate testing methods, 690–707
        direct measurement, 689–690
        fault models and test quality, 688–689
        strategies, 685–688
    sputtering, 423
radiofrequency identification tags (RFIDs)
    biosensor SOP, 721
    dual polarization structure, 305–307
    dual-body configuration, 307
    meander line, 305
    overview, 304–305
radiofrequency system-on-package (RF SOP) technologies
    antennas
        conformal, 270–274
        on magneto-dielectric substrates, 274–277
        multiband, 270
        overview, 268–270
    baluns, 296–297
    capacitors
        MIM structures, 282–283
        parameters, 281–282
        TCC properties, 283–287
    combiners, 297–299
    concept of, 261–264
    evolution of, 264–265
    filters, 294–296
    future trends, 311–312
    inductors, 277–281
    integrated modules
        INC system, 309–311
        WLANs, 307–308
    MEMS switches
        application, 303–304
        challenges, 302–303

Index

electromechanical versus solid state, 301–302
history and role of, 299
operating principle, 300–301
modeling and optimization, 266–267
overview, 28, 260–261, 613–615
resistors
   applications, 288–292
   TCR properties, 292–294
   technologies, 287–288
RFIDs
   dual polarization structure, 305–307
   dual-body configuration, 307
   meander line, 305
   overview, 304–305
substrate materials, 267–268
radiofrequency-optical communication systems, 323–324
rational functions, 204
RC (Resistance and Capacitance) extraction engines, 66
RDL. *See* redistribution layer
reactive ion etch (RIE), 338
real-time performance, 51
redistribution layer (RDL)
   inductors, 545–547
   overview, 544–545
   ultrathin deposition, 547–548
reflective substrate, 466
refractive index, 464, 466
region-matching algorithm, 673
Register Transfer Level (RTL) encapsulation, 60
regression equations, 211
relative humidity (RH), 462
reliability, 334–335, 383, 445, 448–449, 465
Renesas, 125
research and development (R&D)
   embedded passives SOP, 29
   MEMS SOP, 29
   opto SOP, 26–28
   RF SOP, 28
residual stresses, 470
Resistance and Capacitance (RC) extraction engines, 66
resistance testing, 667–668
"resistive" faults, 667
resistors
   applications
      attenuators, 289–290
      termination resistors, 288–289
      Wilkinson power dividers, 290–292
   electroless-plated, 426
   embedded
      PTF, 426–428
      thin-film, 428–430
   TCR properties, 292–294
   technologies, 287–288
   termination, 288–289, 415
   ultrathin deposition, 547–548
resonator band testing, 669
response surface methods (RSMs), 267, 696
retinal prosthesis, 721–722
reverse wire bonding, 101
Reynolds number, 624
RF design. *See* radiofrequency design
RF SOP technologies. *See* radiofrequency system-on-package technologies
RFIDs. *See* radiofrequency identification tags
RH (relative humidity), 462
RIE (reactive ion etch), 338
rigid interconnections
   alternatives to solder, 569–570
   lead-free solders
      electromigration, 564–567
      at fine pitch, 567–569
      intermetallics in, 562–563
      materials, 562
      mechanical properties of, 563–564
      overview, 561
   nanoscale interconnections
      CNTs, 580–581
      nanoparticle conductive adhesives, 581–584
      nanostructured metal interconnections, 577–580
      NCA bonding, 584
   overview, 560–561
   stud bumping, 570–571
   wafer-level packaging with wafer-applied underfill, 573–574

rolled cases, 272
RSMs (response surface methods), 267, 696
RTL (Register Transfer Level) encapsulation, 60

S

sacrificial film-based sealing techniques
   decomposition, 509–514
   etching, 505–509
Samsung Electronics, 125
saturated pool boiling, 626
saturation concentration, 481
SAWs (surface acoustic waves), 737
scalability, 331, 659, 662
scanning electron microscopes (SEMs), 740
scanning thermal probes, 637
Scarbrough, A. C., 82
Scarbrough, Alfred D., 123
scattering parameters, 297
Schneider, D., 461
second generation (2G) cellular standards, 264
second law of electronics, 18
seed layer plating, 402, 404, 411
selective process bias (SPB), 66
self-resonant frequency (SRF), 281
semi-additive plating, 402
semiconducting cores, 386
semiconducting nanoparticles, 730
semiconducting nanostructures, 739
SEMs (scanning electron microscopes), 732
sensitivity analysis, 671–672
sensitivity functions, 210–212
sensor calibration test procedure, 706
sensor-based testing of high-speed devices, 706
sequential multilayer wiring integration process, 409–411
SER (soft error rate), 44
SFDR (spurious-free dynamic range) specifications, 701
shielding, 332
shift registers (SRs), 677
Shipley Insite, 292
shorting stubs, 305
SHS (soft-hard surfaces), 277
Si. *See* silicon
Si NWs (silicon nanowires), 730, 732
side termination stacking
   with conductive polymer, 108
   metallization stacking, 106–108
   overview, 105–106
   with solder edge interconnect, 108–109
side-metallization chip stacking, 105–107
side-metallization PoP stacking, 114
SiGe (silicon germanium), 14, 125
signal conversion
   methods of
     acoustic wave, 737–738
     electrochemical, 735–736
     magnetic, 739–740
     micro-electro-mechanical, 738–739
     optical, 737
     thermal, 740–741
   nanomaterials and nanostructures for
     CNTs, 733
     nanoparticles, 731–732
     overview, 730–731
     wire-like nanostructures, 732–733
   surface modification and biofunctionalization, 734–735
signal delivery networks, 198–204
signal detection and electronic processing, 741–744
signal integrity, 382
signal integrity – power integrity (SIPI) simulation, 199
signal-to-noise ratio (SNR), 701
silica waveguides, 465
silicon (Si)
   carrier technology, 16–17, 83, 140–142
   cores, 386
   holographic nanostructures, 362
   integration comparison, 83
   photonics, 25, 362
   processing, 281
   substrate with polymer-copper wiring, 541
   trench based capacitors, 425
   wafers, 359
   waveguides, 343
silicon germanium (SiGe), 12, 125
silicon nanowires (Si NWs), 730, 732
silicon Raman lasers, 362

silicon-on-insulator (SOI), 277
silicon-on-silicon technologies, 8
silicon-through-via connections, 8
SIMPACT technology, 30
SIMPLE algorithm, 632
SIMPLEC algorithm, 632
simulation
   electromagnetic interference, 248–250
   transient, 207–208
simultaneous multiprocessing (SMP), 649
SINDA/FLUINT packages, 635
single-chip CMOS, 24
single-chip DSL modem SOC, 40
single-chip package miniaturization, 4
single-crystal zinc oxide (ZnO) nanobelts, 739
single-input–single-output (SISO) components, 158, 169
single-layer inductors, 162–164, 277
single-mode waveguides, 342
single-supported capacitive MEMS switches, 300
single-system packages, 10
single-walled carbon nanotubes (SWCNTs), 733
SIP technology. *See* system-in-package technology
SIPI (signal integrity – power integrity) simulation, 199
SISO (single-input–single-output) components, 158, 169
SJR (solder joint reliability), 86, 87, 470–471, 476–478
SLC (surface laminar circuitry), 381
slow cells, 65
small-scale integration (SSI), 677
SMD (surface-mount discrete) capacitors, 224, 229–231, 233
SMDs (surface-mount devices), 262
SMP (simultaneous multiprocessing), 649
SMP (symmetric multiprocessing), 330
SMT (surface-mount technologies), 94, 264, 338
Snell's law, 352
SNR (signal-to-noise ratio), 701
SOB (system-on-board) technology, 7, 10
SOC technology. *See* system-on-chip technology

soft blocks, 72
soft error rate (SER), 44
soft IP blocks, 60
soft-hard surfaces (SHS), 277
software radios, 311
software simulators, 53
SOI (silicon-on-insulator), 277
solder balls, 115
solder interconnects, 108–109, 484–486
solder joint reliability (SJR), 86, 87, 470–471, 476–478
solder mask, 455
solder reflow, assembly with
   CNT, 587
   with fine-pitch metal bumps, 586–587
   on high-CTE rigid substrates, 586
   overview, 585–586
solderless assembly, 587–589
solders
   alternatives to, 569–570
   comparison with other TIMs, 647
   lead-free
      electromigration, 564–567
      at fine pitch, 567–569
      intermetallics in, 562–563
      materials, 562
      mechanical properties of, 563–564
      overview, 561
sol-gel processes, 423–424
solid state MEMS switches, 301–302
solid-liquid PCMs, 610
solidstate electricity generation techniques, 651
solution convergence, 636
SOM (system-on-module) technology, 30
SOP technology. *See* system-on-package technology
source-to-drain leakage current, 64
spacer technology, 102, 357
SPB (selective process bias), 66
specification-oriented tests (SPOTs), 688
spectral absorption, 326
spectral directional absorptivity, 627
spectral directional emissivity, 627
Spelt, J., 482
spin-coating, 394
spin-on polyimide dielectric, 380

SPIRIT (Structure for Packaging, Integrating, and Re-using IP with Tool-flows), 60
split planes, 223
SPOTs (specification-oriented tests), 688
SPR (surface plasmon resonance), 737
spray cooling, 643–644
spurious-free dynamic range (SFDR) specifications, 701
sputtered thin-film resistors, 430
sputtering plating, 404
SRAM (static random-access memory), 611
SRF (self-resonant frequency), 281
SRs (shift registers), 677
SSI (small-scale integration), 677
stacked die packages
    mechanical challenges, 86
    thermal challenges, 88–89
stacked ICs and packages (SIP). *See* system-in-package technology
stacked TAB, 111–112
standby mode, 43
Staphylococcus aureus bacterium, 728
static random-access memory (SRAM), 611
statistical analysis, 210–212
steady-state thermal analysis, 640
Stefan-Boltzmann constant, 628
Stoney's equation, 384
straight cases, 272
streptavidin, 739
stresses, 476
stress-free temperature, 454–455
stress-induced birefringence, 465
stress-optical effects, 464
stress-relieved interconnections
    compliant helical, 553–555
    metal-based stress-relief, 553
    stretched solder column, 555
stress-strain analysis, 449
stretched solder column interconnections, 555
stripline multilayer baluns, 298
striplines, 201–202
StrongARM processors, 612
Structure for Packaging, Integrating, and Re-using IP with Tool-flows (SPIRIT), 60

Strydom, J.T., 461
stud bumping, 570–573
stud plating process, 411
substrates
    bio-SOP integration technologies, 744
    ceramic substrates, 267–268
    core materials
        attributes of, 384–385
        classification of, 385–386
        future trends, 435
        through-via processing, 386–387
    C-SiC, 480
    dielectric reliability, 460
    double-sided tape, 116
    folded flex, 116–117
    high-CTE rigid, 486
    KGES testing
        interconnects, 664–671
        passives, 671–676
    magneto-dielectric, 274–277
    multifunction SOP reliability
        digital function, 458–461
        materials and process, 450–458
        multifunction system, 467–468
        optical function, 463–466
        RF function, 461–463
    multilayer, 56
    reflective, 466
    RF technologies, 267–268
    SOP
        drivers for, 381–383
        embedded actives, 430–434
        embedded passives, 415–430
        evolution of, 380–381
        future trends, 435–437
        miniaturized thermal materials and structures, 434–435
        overview, 377–380
        ultrathin-film wiring with embedded components, 384–415
    stressing, 668
    testing, 664–665
    unfolded flex, 116
    warpage, 470
substrate-to-IC interconnection reliability
    100-ìm flip-chip assemblies, 471–476
    barriers
        to Cu low-K reliability, 470

Index 779

to high-density wiring, 469–470
to microvia and dielectric reliability, 471
to solder joint reliability, 470–471
against die cracking, 476
interfacial adhesion
effect of moisture on, 480–482
overview, 478–480
overview, 468–469
solder joint
overview, 476–477
vibration effects on solder joint reliability, 477–478
subtractive etching process, 401
supercapacitors, 416
superheterodyne narrowband wireless RF transceiver architecture, 798
surface acoustic waves (SAWs), 737
surface convective resistance, 620
surface laminar circuitry (SLC), 381
surface plasmon resonance (SPR), 737
surface-mount devices (SMDs), 262
surface-mount discrete (SMD) capacitors, 227, 229–231, 233
surface-mount technologies (SMT), 94, 264, 338
SWCNTs (single-walled carbon nanotubes), 733
switch-cycle lifetimes, 303
symmetric multiprocessing (SMP), 330
symmetry plane, 459
synthesis, network, 204–207
synthetic probes, 726, 728–729
system barriers, 7
system drivers, 23
system motherboards, 379
system technologies
building blocks of, 6–7
evolution of, 7–10
overview, 2–4
size comparisons, 25
trend toward digital convergence, 4–6
types of
comparison of, 22–25
MCM technology, 12
SIP technology, 12–17
SOB technology, 10
SOC technology, 10–12
SOP technology, 17–22

system-in-package (SIP) technology. *See* non-TSV SIP technology; through-silicon via SIP technology
applications of, 81
categories of
flip chip and wire bonding, 13, 15
flip chip–on–chip, 13, 15
package stacking, 17
TSV technology, 13, 15–16
wire bonding, 13, 15
defined, 10, 81
design challenges
electrical, 87–88
materials and process, 84–85
mechanical, 86–87
thermal, 88–92
evolution of, 81–84
future trends, 142–143
integration approaches in, 15
non-TSV
chip stacking, 95–112, 119–120
evolution of, 92–95
package stacking, 112–118
overview, 12–13, 80–81
Si carriers, 16–17
TSV
drilling technology, 126
evolution of, 123–125
filling technology, 126–130
overview, 120–123
Si carrier technology, 140–142
via-first scheme, 134–136
via-last scheme, 136–140
wafer bonding technology, 130–133
system-level failure modes, 447
system-level heat removal, 609
system-level reliability
assessment modeling strategy, 467–468
design-for-reliability, 447–449
failure mechanisms, 446–447
overview, 445–446
verification of, 449–450
system-level thermal characterization, 90
system-on-board (SOB) technology, 7, 10

system-on-chip (SOC) technology
  applications of Internet era, 41
  architecture, 44–49
  create phase, 50
  defined, 10
  design phases
    creation phase, 57–76
    definition phase, 50–57
  key requirements, 42–44
  overview, 10–12, 39–41, 159–160
  partitions, 72
system-on-module (SOM) technology, 31
system-on-package (SOP) technology. *See*
    biosensor SOP technologies; coupling;
    decoupling; multifunction SOP
    substrate reliability; optoelectronic SOP
    technologyability; radiofrequency SOP
    technologies; thermal SOP technologies;
    wafer-level system-on-package
    technologies
  defined, 10
  design
    chip-package codesign, 180–190
    design tools, 194–208
    electromagnetic bandgap structures,
      239–250
    embedded passives in RF front end,
      160–180
    fabrication technologies, 159–160
    integration in mobile applications,
      155–156
    manufacturing design, 208–214
    mixed-signal architecture, 153–159
    overview, 151–153
    WLAN front-end module, 191–193
  global R&D
    embedded passives SOP, 28
    MEMS SOP, 28
    opto SOP, 25–27
    RF SOP, 27
  implementations of, 28–32
  integration, 11
  miniaturization trend, 21–22
  overview, 17–21
  substrates
    drivers for, 381–383
    embedded actives, 430–434
    embedded passives, 415–430

evolution of, 380–381
future trends, 435–437
miniaturized thermal materials and
  structures, 434–435
overview, 377–380
ultrathin-film wiring with embedded
  components,
  384–415

T

T (thymine), 728
TAB (tape automated bonding), 81, 95,
  111–112
TAP (test access port) controllers, 678
tape automated bonding (TAB), 81, 95,
  111–112
Taq DNA polymerase enzyme, 725
TCC (thermal coefficient of capacitance), 84,
  284–287, 416
TCE (thermal coefficient of expansion), 85
TCKs (test clocks), 678, 680
TCR (temperature coefficient of resistance),
  84, 292–294
TDF (thermal-driven floor planning), 640
TDMA (tridiagonal matrix algorithm), 632
TDNA (time domain network analysis),
  669
TE (thermoelectric) polarized light, 345
technology scaling, 63–67
technology-driven platforms, 61
TECPLOT packages, 636
TEL (Tokyo Electron Limited) probes,
  592–593
telecom transceivers, 332
TEM (transmission electron microscopy),
  732
temperature coefficient of resistance (TCR),
  84, 292–294
temperature fields, 629
temporal distribution, 629
TEOS (tetra-ethoxysilane)-type oxides, 126
termination resistors, 288–289, 415
termination structures, 290
test access port (TAP) controllers, 678
test clocks (TCKs), 678, 680
test cost reduction, 706–707
test mode select (TMS) signals, 678
test vehicle design, 248

Index

tetra-ethoxysilane (TEOS)-type oxides, 126
TFOS (thin film on silicon), 21, 159–160
TFTs (thin-film transistors), 744
THD (total harmonic distortion), 701
thermal analysis software, 636–637
thermal biosensors, 741
thermal characterization
    experimental methods for, 637
    numerical methods for
        available modeling tools, 634
        combined mode heat transfer modeling, 629–630
        compact thermal models, 634–636
        dedicated packages for electronics thermal management, 634
        general-purpose codes, 634
        limitations of modeling, 633
        methodology, 630–633
        thermal analysis software, 636–637
thermal coefficient of capacitance (TCC), 84, 284–287, 416
thermal coefficient of expansion (TCE), 85
thermal conductivity, 384
thermal cycling tests, 465
thermal gel, 647
thermal grease, 647
thermal interface materials (TIMs), 620, 646–648
thermal interface resistance, 620
thermal management technologies
    active methods of
        liquid loops, 643
        spray cooling, 643–644
        thermionic cooling, 643–644
        thermoelectric cooling, 645–646
        thermosyphones, 644
        TIMs, 646–648
        vapor compression refrigeration, 646
    passive methods of
        extended surfaces, 641
        heat pipes, 641–643
        heat spreader, 640–641
        high-conductivity package materials, 638–639
        thermal vias, 639–640
thermal microactuators, 617
thermal mismatch stresses, 470
thermal parameters, 85
thermal radiation, 626
thermal release tapes, 97
thermal resistance, 621, 630, 638
thermal signal conversion, 740–741
thermal SOP technologies
    heat transfer modes
        conduction, 618–622
        convection, 623–626
        radiation, 626–628
    overview, 605–607
    power minimization methodologies
        ASPs, 650
        cache power minimization, 650–651
        DVFS, 649–650
        overview, 648–649
        parallel processing, 649
        power harnessing, 651
    sources
        digital SOP, 611–613
        MEMS SOP, 617
        optoelectronic SOP, 615–617
        overview, 610–611
        RF SOP, 613–615
    system-level thermal constraints in, 609–610
    thermal characterization
        experimental methods for, 637
        numerical methods for, 629–637
    thermal implications of, 607–609
    thermal management technologies
        active methods, 643–648
        overview, 637–638
        passive methods, 638–643
thermal stability, 389
thermal stresses, 463
thermal switches, 300
thermal via density, 640
thermal vias, 639–640
thermal-driven floor planning (TDF), 640
thermionic cooling, 645–646
thermo mechanical analysis (TMA), 472
thermocouples, 637
thermodynamic arguments, 480
thermoelectric (TE) polarized light, 345
thermoelectric cooling, 337, 643–645
thermographic phosphors, 637
thermomechanical deformation, 455–456, 462–463

thermomechanical environments, 477
thermomechanical fatigue models, 468
thermomechanical properties
    core substrate materials, 384–385
    thin-film buildup organic dielectrics, 389
thermomechanical reliability, 85, 447, 461, 468, 485
thermomechanical stresses, 447, 455, 464, 629
thermomechanical tests, 449
thermoplastics, 393–394
thermosets, 391–393
thermosyphones, 625, 644
Thermount laminates, 471
thickness control, 389
thick-/thin-film capacitors, 229, 232–239
thin film on silicon (TFOS), 22, 159–160
thin-film buildup organic dielectrics
    attributes of, 388–389
    classification of
        overview, 389–391
        thermoplastics, 393–394
        thermosets, 391–393
    dielectric processes
        curing, 395–397
        deposition, 394–395
        via formation processes, 397–399
    overview, 387–388
thin-film ceramics, 419–420, 422–424
thin-film flexible optical transceivers, 320
thin-film getters, 517–520
thin-film interfaces, 460
thin-film lasers, 321
thin-film optoelectronic SOP technology
    3D lightwave circuits, 354–356
    active components, 353–354
    optical alignment, 325
    overview, 324–325, 340–341
    passive thin-film lightwave circuits
        diffractive couplers for optical interconnects, 343–353
        optical power splitters, 343
        optical waveguides, 342–343
        overview, 341–353
    properties of waveguide materials, 325–330
thin-film photodetectors, 321

thin-film resistors, 428–430
thin-film technologies, 20–21
thin-film transistors (TFTs), 744
thinning, wafer, 95–97, 125
thin-wafer dicing, 98
thin-wafer handling, 97–98
thiourea (TU), 404
third-generation (3G) cellular standards, 264
three-dimensional. *See* 3D
three-part-structure (TPS) probes, 593–595
three-tier SOB-based systems, 10
through, reflect, line (TRL) calibration, 289
throughput-driven cost, 21
through-silicon via (TSV) SIP technology
    drilling, 126
    evolution of, 123–125
    filling, 126–130
    overview, 14, 16–17, 123–125
    Si carrier technology, 140–142
    stacking, 379
    via-first scheme
        process 1, 134–136
        process 2, 136
    via-last scheme
        process 1, 136
        process 2, 136–139
        process 3, 139–140
    wafer bonding
        metal-metal bonding, 130–132
        oxide bonding, 130
        polymer bonding, 132–133
through-via processing, 386–387
thymine (T), 728
TIAs (transimpedance amplifiers), 340, 358–359
time domain network analysis (TDNA), 669
time-to-market, 51
time-variant polarization, 388
time-zero defects, 665
TIMs (thermal interface materials), 620, 646–648
TLBs (translational look-aside buffers), 612
TLM (transmission-line matrix) method, 266
TM polarized light, 345
TMA (thermo mechanical analysis), 472
TMM (transmission matrix method), 207, 237

TMS (test mode select) signals, 678
Tokyo Electron Limited (TEL) probes, 592–593
total harmonic distortion (THD), 701
TPS (three-part-structure) probes, 593–595
transceivers, 355
transformer-feedback VCOs (TVCOs), 218–224
transient simulation, 207–208, 209
transient thermal via allocation method, 640
transimpedance amplifiers (TIAs), 340, 358–359
transistors
  amorphous, 363
  CMOS, 354
  power, 614
  thin-film (TFTs), 744
translational look-aside buffers (TLBs), 612
transmission electron microscopy (TEM), 732
transmission matrix method (TMM), 207, 237
transmission-line matrix (TLM) method, 266
triband antennas, 271
tridiagonal matrix algorithm (TDMA), 632
TRL (through, reflect, line) calibration, 289
Tru-Si Technologies, 125
T-splitters, 343
TSV SIP technology. *See* through-silicon via SIP technology
TU (thiourea), 404
tunable filters, 178–180
turbulence modeling, 633, 636
TVCOs (transformer-feedback VCOs), 218–222
two-dimensional. *See* 2D
two-layer spiral inductors, 163–165
two-port frequency domain measurement methodology, 235
two-tier SOP-based systems, 10

U

U (uracil), 728
UAV (unmanned aerial vehicle) systems, 721–722
ULSI (ultra-large-scale integration), 685
ultrafine lines, 402
ultrafine space, 402
ultrafine-pitch flip-chip assembly, 483
ultrahigh wiring densities, 26
ultrahigh-density Cu-polyimide multilayer wiring, 380
ultrahigh-density sop substrate, 377
ultra-large-scale integration (ULSI), 677
ultralow-K dielectric integration, 425
ultrathin conductors, 382
ultrathin deposition, 547–548
ultrathin dielectrics, 382
ultrathin-film wiring with embedded components
  core substrate materials
    attributes of, 384–385
    classification of, 385–386
    through-via processing, 386–387
  embedded conductors, 399–409
  multilayer wiring integration processes
    coreless, 414–415
    parallel process, 411–414
    sequential process, 409–411
  thin-film buildup organic dielectrics
    attributes of, 388–389
    classification of, 389–394
    dielectric processes, 394–399
    overview, 387–388
underfills, 470
unfolded flex substrate, 116
unit-cell analysis, 242–246
University of Arkansas, 29
unmanned aerial vehicle (UAV) systems, 721–722
up-conversion mixers, 711
up-front process optimization, 457–458
uracil (U), 728

V

vapor compression refrigeration, 645–646
varactor diodes, 178
variable frequency microwave (VFM) curing process, 395
VCOs (voltage-controlled oscillators), 181, 188–189, 309
VCSELs (vertical cavity surface-emitting lasers), 309, 325, 334, 354, 614, 616
VelociTI Very Long Instruction Word (VLIW) architecture, 48

verification
  SOC development, 61–63
  of system-level mixed-signal reliability, 449–450
vertical cavity surface-emitting lasers (VCSELs), 309, 325, 334, 354, 614, 616
vertical interconnection process, 108
very-large-scale integration (VLSI), 677
VFM (variable frequency microwave) curing process, 395
VGCs (volume grating couplers), 345
via filling with conductor material, 125
via formation processes, 125, 397–399
via-first scheme, 133–137
via-last scheme, 133–140
vibration effects, 477–478
Virtual Socket Interface Alliance (VSIA), 59
VLIW (VelociTI Very Long Instruction Word) architecture, 48
VLSI (very-large-scale integration), 677
voltage contrast, 668
voltage regulator modules (VRMs), 227, 233
voltage-controlled oscillators (VCOs), 181, 188–189, 309
volume grating couplers (VGCs), 345
volumetric heat generation rate, 629
von Mises strain, 453
VRMs (voltage regulator modules), 227, 233
VSIA (Virtual Socket Interface Alliance), 59

W

WAF (wire-length-driven floor planning), 640
wafer bonding
  direct bonding, 500
  intermediate layer bonding, 500–505
  overview, 499–500
  TSV SIP technology
    metal-metal bonding, 130–132
    oxide bonding, 130
    polymer bonding, 132–133
wafer scale integration (WSI), 9
wafer thinning, 95–97, 125
wafer throughput, 63
wafer-applied underfill, 573–576
wafer-level packaging (WLP), 281, 432, 445, 446
wafer-level packaging and interconnections (WLPI)
  3D WLSOP, 590–591
  assembly
    placement techniques, 589
    with solder reflow, 585–587
    solderless, 587–589
  challenges in
    assembly and thermomechanical reliability, 551–552
    electrical, 549–550
    material, 550
    process, 551
  classes of
    metal-polymer–based stress relief, 555–560
    overview, 552–553
    stress-relieved interconnection, 553–555
  overview, 548–549
  rigid interconnections
    alternatives to solder, 569–570
    lead-free solders, 561–569
    nanoscale interconnections, 577–584
    overview, 560–561
    stud bumping, 570–573
    wafer-applied underfill, 573–576
wafer-level system-on-package (WLSOP) technologies
  buildup wiring and redistribution
    IC-package pitch gap, 540–543
    redistribution layers on Si, 543–544
  defined, 536–537
  historical evolution, 537–540
  overview, 535–536
  probing and burn-in
    overview, 591–592
    TEL probes, 592–593
    TPS probes, 593–595
  thin-film embedded components, 544–548
  WLPI
    3D WLSOP, 590–591
    assembly, 585–589
    challenges in, 549–552
    classes of, 552–560
    overview, 548–549
    rigid interconnections, 560–584

wafer-level thin-film embedded components
  approaches to, 544
  embedded in RDL
    inductors, 545–547
    overview, 544–545
    ultrathin deposition, 547–548
wafer-scale MEMS packaging, 497–499
wafer-to-wafer stacking, 123
warpage control, 470
waveguide core path definition, 333
waveguide core/cladding polymers, 456
waveguide fabrication process, 357, 451
waveguide layer, 348
waveguide optical interconnections, 465
waveguide refractive index stability, 466
waveguides (WGs), 342, 353, 615
wavevectors, 347, 350–351
wax bonding, 97
wearout region, 334
wet etching, 97, 402–403
wet polishing, 97
WGs (waveguides), 342, 353, 614
wide area optical clock distribution, 338
wideband LCP baluns, 173
Wilkinson power dividers, 290–292
wire bonded stacking
  with bottom dies flipped, 104
  for chip stacking, 101–102
  COC stacking, 104–105
  configurations of, 99
  die adhesive, 102
  electrical routing, 103
  flip chip stacking, 103–104
  molding, 102
  overview, 98–100
  spacer technology, 102
wire bonding
  DBS systems, 719–720
  interconnections, 115
  loop height profiles, 101
  SIP by, 13, 15
  for three-dimensional integration, 16
  TSV SIP, 121–122
wirebond BGA, 55
wire-length-driven floor planning (WAF), 640

wireless chipset modules, 28
wireless handset electronics, 45
wireless handsets, 153
wireless LANs (WLANs), 191–193, 260, 307–308
wireless radios, 158
wireless signal speed, 142
wire-like nanostructures, 732–733
wire-load models (WLMs), 71
wire-on-bumps (WOB) chip stacking, 109
wiring density
  optoelectronic SOP technology
    board-to-board optical interconnects, 332–333
    copper-wire interconnects, 331
    overview, 333–332
  ultrahigh, 26
Witwit, A.M.R., 461
WLANs (wireless LANs), 191–193, 260, 307–308
WLMs (wire-load models), 71
WLP (wafer-level packaging), 281, 432, 445, 446
WLPI. *See* wafer-level packaging and interconnections
WLSOP technologies. *See* wafer-level system-on-package technologies
WOB (wire-on-bumps) chip stacking, 109
"world phones", 155
WSI (wafer scale integration), 9
Wylde, J., 482

X

x-ray photoelectron spectroscopy, 481

Y

yield-driven-layout, 67
Yoon, Heebyung, 671
Y-splitters, 343

Z

Z interconnections, 387
zettabyte level, 323
zinc oxide (ZnO) nanobelts, 732, 739
zirconia-Nafion sol-gel encapsulation, 736
ZnO (zinc oxide) nanobelts, 732, 739